Reference Manual on Scientific Evidence

Third Edition

Committee on the Development of the Third Edition of the
Reference Manual on Scientific Evidence

Committee on Science, Technology, and Law
Policy and Global Affairs

FEDERAL JUDICIAL CENTER

NATIONAL RESEARCH COUNCIL
OF THE NATIONAL ACADEMIES

THE NATIONAL ACADEMIES PRESS
Washington, D.C.
www.nap.edu

THE NATIONAL ACADEMIES PRESS 500 Fifth Street, N.W. Washington, DC 20001

The Federal Judicial Center contributed to this publication in furtherance of the Center's statutory mission to develop and conduct educational programs for judicial branch employees. The views expressed are those of the authors and not necessarily those of the Federal Judicial Center.

NOTICE: The project that is the subject of this report was approved by the Governing Board of the National Research Council, whose members are drawn from the councils of the National Academy of Sciences, the National Academy of Engineering, and the Institute of Medicine. The members of the committee responsible for the report were chosen for their special competences and with regard for appropriate balance.

The development of the third edition of the *Reference Manual on Scientific Evidence* was supported by Contract No. B5727.R02 between the National Academy of Sciences and the Carnegie Corporation of New York and a grant from the Starr Foundation. The views expressed in this publication are those of the authors and do not necessarily reflect those of the National Academies or the organizations that provided support for the project.

International Standard Book Number-13: 978-0-309-21421-6
International Standard Book Number-10: 0-309-21421-1

Library of Congress Cataloging-in-Publication Data

Reference manual on scientific evidence. — 3rd ed.
p. cm.
Includes bibliographical references and index.
ISBN-13: 978-0-309-21421-6 (pbk.)
ISBN-10: 0-309-21421-1 (pbk.)
1. Evidence, Expert—United States. I. Federal Judicial Center.
KF8961.R44 2011
347.73′67—dc23

2011031458

Additional copies of this report are available from the National Academies Press, 500 Fifth Street, N.W., Lockbox 285, Washington, DC 20055; (800) 624-6242 or (202) 334-3313 (in the Washington metropolitan area); Internet, http://www.nap.edu.

Printed in the United States of America

THE FEDERAL JUDICIAL CENTER

The Federal Judicial Center is the research and education agency of the federal judicial system. It was established by Congress in 1967 (28 U.S.C. §§ 620–629), on the recommendation of the Judicial Conference of the United States, with the mission to "further the development and adoption of improved judicial administration in the courts of the United States." By statute, the Chief Justice of the United States chairs the Federal Judicial Center's Board, which also includes the director of the Administrative Office of the U.S. Courts and seven judges elected by the Judicial Conference.

The Center undertakes empirical and exploratory research on federal judicial processes, court management, and sentencing and its consequences, often at the request of the Judicial Conference and its committees, the courts themselves, or other groups in the federal system. In addition to orientation and continuing education programs for judges and court staff on law and case management, the Center produces publications, videos, and online resources. The Center provides leadership and management education for judges and court employees, and other training as needed. Center research informs many of its educational efforts. The Center also produces resources and materials on the history of the federal courts, and it develops resources to assist in fostering effective judicial administration in other countries.

Since its founding, the Center has had nine directors. Judge Barbara J. Rothstein became director of the Federal Judicial Center in 2003

www.fjc.gov

THE NATIONAL ACADEMIES

Advisers to the Nation on Science, Engineering, and Medicine

The **National Academy of Sciences** is a private, nonprofit, self-perpetuating society of distinguished scholars engaged in scientific and engineering research, dedicated to the furtherance of science and technology and to their use for the general welfare. Upon the authority of the charter granted to it by the Congress in 1863, the Academy has a mandate that requires it to advise the federal government on scientific and technical matters. Dr. Ralph J. Cicerone is president of the National Academy of Sciences.

The **National Academy of Engineering** was established in 1964, under the charter of the National Academy of Sciences, as a parallel organization of outstanding engineers. It is autonomous in its administration and in the selection of its members, sharing with the National Academy of Sciences the responsibility for advising the federal government. The National Academy of Engineering also sponsors engineering programs aimed at meeting national needs, encourages education and research, and recognizes the superior achievements of engineers. Dr. Charles M. Vest is president of the National Academy of Engineering.

The **Institute of Medicine** was established in 1970 by the National Academy of Sciences to secure the services of eminent members of appropriate professions in the examination of policy matters pertaining to the health of the public. The Institute acts under the responsibility given to the National Academy of Sciences by its congressional charter to be an adviser to the federal government and, upon its own initiative, to identify issues of medical care, research, and education. Dr. Harvey V. Fineberg is president of the Institute of Medicine.

The **National Research Council** was organized by the National Academy of Sciences in 1916 to associate the broad community of science and technology with the Academy's purposes of furthering knowledge and advising the federal government. Functioning in accordance with general policies determined by the Academy, the Council has become the principal operating agency of both the National Academy of Sciences and the National Academy of Engineering in providing services to the government, the public, and the scientific and engineering communities. The Council is administered jointly by both Academies and the Institute of Medicine. Dr. Ralph J. Cicerone and Dr. Charles M. Vest are chair and vice chair, respectively, of the National Research Council.

www.national-academies.org

Committee on the Development of the Third Edition of the Reference Manual on Scientific Evidence

Co-Chairs:

JEROME P. KASSIRER (IOM), Distinguished Professor, Tufts University School of Medicine

GLADYS KESSLER, Judge, U.S. District Court for the District of Columbia

Members:

MING W. CHIN, Associate Justice, The Supreme Court of California

PAULINE NEWMAN, Judge, U.S. Court of Appeals for the Federal Circuit

KATHLEEN MCDONALD O'MALLEY, Judge, U.S. Court of Appeals for the Federal Circuit

JED S. RAKOFF, Judge, U.S. District Court, Southern District of New York

CHANNING R. ROBERTSON, Ruth G. and William K. Bowes Professor, School of Engineering, and Professor, Department of Chemical Engineering, Stanford University

JOSEPH V. RODRICKS, Principal, Environ

ALLEN WILCOX, Senior Investigator, Institute of Environmental Health Sciences

SANDY L. ZABELL, Professor of Statistics and Mathematics, Weinberg College of Arts and Sciences, Northwestern University

Consultant to the Committee:

JOE S. CECIL, Project Director, Program on Scientific and Technical Evidence, Division of Research, Federal Judicial Center

Staff:

ANNE-MARIE MAZZA, Director

STEVEN KENDALL, Associate Program Officer

GURUPRASAD MADHAVAN, Program Officer (until November 2010)

Board of the Federal Judicial Center

Committee on Science, Technology, and Law
National Research Council

DAVID KORN (*Co-Chair*), Professor of Pathology, Harvard Medical School, and formerly, Inaugural Vice Provost for Research, Harvard University

RICHARD A. MESERVE (*Co-Chair*), President, Carnegie Institution for Science, and Senior of Counsel, Covington & Burling LLP

FREDERICK R. ANDERSON, JR., Partner, McKenna, Long & Aldridge LLP

ARTHUR I. BIENENSTOCK, Special Assistant to the President for Federal Research Policy, and Director, Wallenberg Research Link, Stanford University

BARBARA E. BIERER, Professor of Medicine, Harvard Medical School, and Senior Vice President, Research, Brigham and Women's Hospital

ELIZABETH H. BLACKBURN, Morris Herzstein Professor of Biology and Physiology, University of California, San Francisco

JOHN BURRIS, President, Burroughs Wellcome Fund

ARTURO CASADEVALL, Leo and Julia Forchheimer Professor of Microbiology and Immunology; Chair, Department of Biology and Immunology; and Professor of Medicine, Albert Einstein College of Medicine

JOE S. CECIL, Project Director, Program on Scientific and Technical Evidence, Division of Research, Federal Judicial Center

ROCHELLE COOPER DREYFUSS, Pauline Newman Professor of Law and Director, Engelberg Center on Innovation Law and Policy, New York University School of Law

DREW ENDY, Assistant Professor, Bioengineering, Stanford University, and President, The BioBricks Foundation

PAUL G. FALKOWSKI, Board of Governors Professor in Geological and Marine Science, Department of Earth and Planetary Science, Rutgers, The State University of New Jersey

MARCUS FELDMAN, Burnet C. and Mildred Wohlford Professor of Biological Sciences, Stanford University

ALICE P. GAST, President, Lehigh University

JASON GRUMET, President, Bipartisan Policy Center

BENJAMIN W. HEINEMAN, JR., Senior Fellow, Harvard Law School and Harvard Kennedy School of Government

D. BROCK HORNBY, U.S. District Judge for the District of Maine

ALAN B. MORRISON, Lerner Family Associate Dean for Public Interest and Public Service, George Washington University Law School

PRABHU PINGALI, Deputy Director of Agricultural Development, Global Development Program, Bill and Melinda Gates Foundation

HARRIET RABB, Vice President and General Counsel, Rockefeller University
BARBARA JACOBS ROTHSTEIN, Director, The Federal Judicial Center
DAVID S. TATEL, Judge, U.S. Court of Appeals for the District of Columbia Circuit
SOPHIE VANDEBROEK, Chief Technology Officer and President, Xerox Innovation Group, Xerox Corporation

Staff

ANNE-MARIE MAZZA, Director
STEVEN KENDALL, Associate Program Officer

Foreword

In 1993, in the case *Daubert v. Merrell Dow Pharmaceuticals, Inc.*, the Supreme Court instructed trial judges to serve as "gatekeepers" in determining whether the opinion of a proffered expert is based on scientific reasoning and methodology. Since *Daubert,* scientific and technical information has become increasingly important in all types of decisionmaking, including litigation. As a result, the science and legal communities have searched for expanding opportunities for collaboration.

Our two institutions have been at the forefront of trying to improve the use of science by judges and attorneys. In *Daubert,* the Supreme Court cited an *amicus curiae* brief submitted by the National Academy of Sciences and the American Association for the Advancement of Science to support the view of science as "a process for proposing and refining theoretical explanations about the world that are subject to further testing and refinement." Similarly, in *Kumho Tire Co. v. Carmichael* (1999) the Court cited an amicus brief filed by the National Academy of Engineering for its assistance in explaining the process of engineering.

Soon after the *Daubert* decision the Federal Judicial Center published the first edition of the *Reference Manual on Scientific Evidence*, which has become the leading reference source for federal judges for difficult issues involving scientific testimony. The Center also undertook a series of research studies and judicial education programs intended to strengthen the use of science in courts.

More recently the National Research Council through its Committee on Science, Technology, and Law has worked closely with the Federal Judicial Center to organize discussions, workshops, and studies that would bring the two communities together to explore the nature of science and engineering, and the processes by which science and technical information informs legal issues. It is in that spirit that our organizations joined together to develop the third edition of the *Reference Manual on Scientific Evidence.* This third edition, which was supported by grants from the Carnegie Foundation and the Starr Foundation, builds on the foundation of the first two editions, published by the Center. This edition was overseen by a National Research Council committee composed of judges and scientists and engineers who share a common vision that together scientists and engineers and members of the judiciary can play an important role in informing judges about the nature and work of the scientific enterprise.

Our organizations benefit from the contributions of volunteers who give their time and energy to our efforts. During the course of this project, two of the chapter authors passed away: Margaret Berger and David Friedman. Both Margaret and David served on NRC committees and were frequent contributors to Center judicial education seminars. Both were involved in the development of the *Reference Manual* from the beginning, both have aided each of our institutions through their services on committees, and both have made substantial contributions to our understanding of law and science through their individual scholarship.

They will be missed but their work will live on in the thoughtful scholarship they have left behind.

We extend our sincere appreciation to Dr. Jerome Kassirer and Judge Gladys Kessler and all the members of the committee who gave so generously to make this edition possible.

THE HONORABLE BARBARA J. ROTHSTEIN
Director
Federal Judicial Center

RALPH J. CICERONE
President
National Academy of Sciences

Acknowledgments

This report has been reviewed in draft form by individuals chosen for their diverse perspectives and technical expertise, in accordance with procedures approved by the National Academies' Report Review Committee. The purpose of this independent review is to provide candid and critical comments that will assist the institution in making its published report as sound as possible and to ensure that the report meets institutional standards for objectivity, accuracy, and responsiveness to the study charge. The review comments and draft manuscript remain confidential to protect the integrity of the process.

We wish to thank the following individuals for their review of selected chapters of this report: Bert Black, Mansfield, Tanick & Cohen; Richard Bjur, University of Nevada; Michael Brick, Westat; Edward Cheng, Vanderbilt University; Joel Cohen, Rockefeller University; Morton Corn, Morton Corn and Associates; Carl Cranor, University of California, Riverside; Randall Davis, Massachusetts Institute of Technology; John Doull, University of Kansas; Barry Fisher, Los Angeles County Sheriff's Department; Edward Foster, University of Minnesota; David Goldston, Natural Resources Defense Council; James Greiner, Harvard University; Susan Haack, University of Miami; David Hillis, University of Texas; Karen Kafadar, Indiana University; Graham Kalton, Westat; Randy Katz, University of California, Berkeley; Alan Leshner, American Association for the Advancement of Science; Laura Liptai, Biomedical Forensics; Patrick Malone, Patrick Malone & Associates; Geoffrey Mearns, Cleveland State University; John Monahan, The University of Virginia; William Nordhaus, Yale University; Fernando Olguin, U.S. District Court for the Central District of California; Jonathan Samet, University of Southern California; Nora Cate Schaeffer, University of Wisconsin; Shira Scheindlin, U.S. District Court for the Southern District of New York; and Reggie Walton, U.S. District Court for the District of Columbia.

Although the reviewers listed above have provided many constructive comments and suggestions, they were not asked to endorse the report, nor did they see the final draft of the report before its release. The review of this report was overseen by D. Brock Hornby, U.S. District Judge for the District of Maine. Appointed by the National Academies, he was responsible for making certain that an independent examination of this report was carried out in accordance with institutional procedures and that all review comments were carefully considered. Responsibility for the final content of this report rests entirely with the authoring committee and the institution.

Preface

Supreme Court decisions during the last decade of the twentieth century mandated that federal courts examine the scientific basis of expert testimony to ensure that it meets the same rigorous standard employed by scientific researchers and practitioners outside the courtroom. Needless to say, this requirement places a demand on judges not only to comprehend the complexities of modern science but to adjudicate between parties' differing interpretations of scientific evidence. Science, meanwhile, advances. Methods change, new fields are born, new tests are introduced, the lexicon expands, and fresh approaches to the interpretation of causal relations evolve. Familiar terms such as enzymes and molecules are replaced by microarray expression and nanotubes; single-author research studies have now become multi-institutional, multi-author, international collaborative efforts.

No field illustrates the evolution of science better than forensics. The evidence provided by DNA technology was so far superior to other widely accepted methods and called into question so many earlier convictions that the scientific community had to reexamine many of its time-worn forensic science practices. Although flaws of some types of forensic science evidence, such as bite and footprint analysis, lineup identification, and bullet matching were recognized, even the most revered form of forensic science—fingerprint identification—was found to be fallible. Notably, even the "gold standard" of forensic evidence, namely DNA analysis, can lead to an erroneous conviction if the sample is contaminated, if specimens are improperly identified, or if appropriate laboratory protocols and practices are not followed.

Yet despite its advances, science has remained fundamentally the same. In its ideal expression, it examines the nature of nature in a rigorous, disciplined manner in, whenever possible, controlled environments. It still is based on principles of hypothesis generation, scrupulous study design, meticulous data collection, and objective interpretation of experimental results. As in other human endeavors, however, this ideal is not always met. Feverish competition between researchers and their parent institutions, fervent publicity seeking, and the potential for dazzling financial rewards can impair scientific objectivity. In recent years we have experienced serious problems that range from the introduction of subtle bias in the design and interpretation of experiments to overt fraudulent studies. In this welter of modern science, ambitious scientists, self-designated experts, billion-dollar corporate entities, and aggressive claimants, judges must weigh evidence, judge, and decide.

As with previous editions of the *Reference Manual*, this edition is organized according to many of the important scientific and technological disciplines likely to be encountered by federal (or state) judges. We wish to highlight here two critical issues germane to the interpretation of all scientific evidence, namely issues of causation and conflict of interest. Causation is the task of attributing cause and effect, a normal everyday cognitive function that ordinarily takes little or no effort. Fundamentally, the task is an inferential process of weighing evidence and using judgment to conclude whether or not an effect is the result of some stimulus. Judgment is required even when using sophisticated statistical methods. Such methods can provide powerful evidence of associations between variables, but they cannot prove that a causal relationship exists. Theories of causation (evolution, for example) lose their designation as theories only if the scientific community has rejected alternative theories and accepted the causal relationship as fact. Elements that are often considered in helping to establish a causal relationship include predisposing factors, proximity of a stimulus to its putative outcome, the strength of the stimulus, and the strength of the events in a causal chain. Unfortunately, judges may be in a less favorable position than scientists to make causal assessments. Scientists may delay their decision while they or others gather more data. Judges, on the other hand, must rule on causation based on existing information. Concepts of causation familiar to scientists (no matter what stripe) may not resonate with judges who are asked to rule on general causation (i.e., is a particular stimulus known to produce a particular reaction) or specific causation (i.e., did a particular stimulus cause a particular consequence in a specific instance). In the final analysis, a judge does not have the option of suspending judgment until more information is available, but must decide after considering the best available science. Finally, given the enormous amount of evidence to be interpreted, expert scientists from different (or even the same) disciplines may not agree on which data are the most relevant, which are the most reliable, and what conclusions about causation are appropriate to be derived.

Like causation, conflict of interest is an issue that cuts across most, if not all, scientific disciplines and could have been included in each chapter of the *Reference Manual.* Conflict of interest manifests as bias, and given the high stakes and adversarial nature of many courtroom proceedings, bias can have a major influence on evidence, testimony, and decisionmaking. Conflicts of interest take many forms and can be based on religious, social, political, or other personal convictions. The biases that these convictions can induce may range from serious to extreme, but these intrinsic influences and the biases they can induce are difficult to identify. Even individuals with such prejudices may not appreciate that they have them, nor may they realize that their interpretations of scientific issues may be biased by them. Because of these limitations, we consider here only financial conflicts of interest; such conflicts are discoverable. Nonetheless, even though financial conflicts can be identified, having such a conflict, even one involving huge sums of money, does not necessarily mean that a given individual will be biased. Having a financial relationship with a commercial entity produces a conflict of interest, but it does not inevitably evoke bias. In science, financial conflict of interest is often accompanied by disclosure of the relationship, leaving to the public the decision whether the interpretation might be tainted. Needless to say, such an assessment may be difficult. The problem is compounded in scientific publications by obscure ways in which the conflicts are reported and by a lack of disclosure of dollar amounts.

Judges and juries, however, must consider financial conflicts of interest when assessing scientific testimony. The threshold for pursuing the possibility of bias must be low. In some instances, judges have been frustrated in identifying expert witnesses who are free of conflict of interest because entire fields of science seem to be co-opted by payments from industry. Judges must also be aware that the research methods of studies funded specifically for purposes of litigation could favor one of the parties. Though awareness of such financial conflicts in itself is not necessarily predictive of bias, such information should be sought and evaluated as part of the deliberations.

The Reference Manual on Scientific Evidence, here in its third edition, is formulated to provide the tools for judges to manage cases involving complex scientific and technical evidence. It describes basic principles of major scientific fields from which legal evidence is typically derived and provides examples of cases in which such evidence was used. Authors of the chapters were asked to provide an overview of principles and methods of the science and provide relevant citations. We expect that few judges will read the entire manual; most will use the volume in response to a need when a particular case arises involving a technical or scientific issue. To help in this endeavor, the *Reference Manual* contains completely updated chapters as well as new ones on neuroscience, exposure science, mental health, and forensic science. This edition of the manual has also gone through the thorough review process of the National Academy of Sciences.

As in previous editions, we continue to caution judges regarding the proper use of the reference guides. They are not intended to instruct judges concerning what evidence should be admissible or to establish minimum standards for acceptable scientific testimony. Rather, the guides can assist judges in identifying the issues most commonly in dispute in these selected areas and in reaching an informed and reasoned assessment concerning the basis of expert evidence. They are designed to facilitate the process of identifying and narrowing issues concerning scientific evidence by outlining for judges the pivotal issues in the areas of science that are often subject to dispute. Citations in the reference guides identify cases in which specific issues were raised; they are examples of other instances in which judges were faced with similar problems. By identifying scientific areas commonly in dispute, the guides should improve the quality of the dialogue between the judges and the parties concerning the basis of expert evidence.

In our committee discussions, we benefited from the judgment and wisdom of the many distinguished members of our committee, who gave time without compensation. They included Justice Ming Chin of the Supreme Court of California; Judge Pauline Newman of the U.S. Court of Appeals for the Federal Circuit in Washington, D.C.; Judge Kathleen MacDonald O'Malley of the U.S. Court of Appeals for the Federal Circuit; Judge Jed Rakoff of the U.S. District Court for the Southern District of New York; Channing Robertson, Ruth G. and William K. Bowes Professor, School of Enginering, and Professor, Department of Chemical Engineering, Stanford University; Joseph Rodricks,

Principal, Environ, Arlington, Virginia; Allen Wilcox, Senior Investigator, Institute of Environmental Health Sciences, Research Triangle Park, North Carolina; and Sandy Zabell, Professor of Statistics and Mathematics, Weinberg College of Arts and Sciences, Northwestern University.

Special commendation, however, goes to Anne-Marie Mazza, Director of the Committee on Science, Technology, and Law, and Joe Cecil of the Federal Judicial Center. These individuals not only shepherded each chapter and its revisions through the process, but provided critical advice on content and editing. They, not we, are the real editors.

Finally, we would like to express our gratitude for the superb assistance of Steven Kendall and for the diligent work of Guru Madhavan, Sara Maddox, Lillian Maloy, and Julie Phillips.

JEROME P. KASSIRER AND GLADYS KESSLER
Committee Co-Chairs

Summary Table of Contents

A detailed Table of Contents appears at the front of each chapter.

Introduction

STEPHEN BREYER

Stephen Breyer, L.L.B., is Associate Justice of the Supreme Court of the United States.

Portions of this Introduction appear in Stephen Breyer, *The Interdependence of Science and Law*, 280 Science 537 (1998).

In this age of science, science should expect to find a warm welcome, perhaps a permanent home, in our courtrooms. The reason is a simple one. The legal disputes before us increasingly involve the principles and tools of science. Proper resolution of those disputes matters not just to the litigants, but also to the general public—those who live in our technologically complex society and whom the law must serve. Our decisions should reflect a proper scientific and technical understanding so that the law can respond to the needs of the public.

Consider, for example, how often our cases today involve statistics—a tool familiar to social scientists and economists but, until our own generation, not to many judges. In 2007, the U.S. Supreme Court heard *Zuni Public Schools District No. 89 v. Department of Education*,[1] in which we were asked to interpret a statistical formula to be used by the U.S. Secretary of Education when determining whether a state's public school funding program "equalizes expenditures" among local school districts. The formula directed the Secretary to "disregard" school districts with "per-pupil expenditures . . . above the 95th percentile or below the 5th percentile of such expenditures . . . in the State." The question was whether the Secretary, in identifying the school districts to be disregarded, could look to the number of pupils in a district as well as the district's expenditures per pupil. Answering that question in the affirmative required us to draw upon technical definitions of the term "percentile" and to consider five different methods by which one might calculate the percentile cutoffs.

In another recent Term, the Supreme Court heard two cases involving consideration of statistical evidence. In *Hunt v. Cromartie*,[2] we ruled that summary judgment was not appropriate in an action brought against various state officials, challenging a congressional redistricting plan as racially motivated in violation of the Equal Protection Clause. In determining that disputed material facts existed regarding the motive of the state legislature in redrawing the redistricting plan, we placed great weight on a statistical analysis that offered a plausible alternative interpretation that did not involve an improper racial motive. Assessing the plausibility of this alternative explanation required knowledge of the strength of the statistical correlation between race and partisanship, understanding of the consequences of restricting the analysis to a subset of precincts, and understanding of the relationships among alternative measures of partisan support.

In *Department of Commerce v. United States House of Representatives*,[3] residents of a number of states challenged the constitutionality of a plan to use two forms of statistical sampling in the upcoming decennial census to adjust for expected "undercounting" of certain identifiable groups. Before examining the constitutional issue, we had to determine if the residents challenging the plan had standing to sue because of injuries they would be likely to suffer as a result of the sampling

1. 127 S. Ct. 1534 (2007).
2. 119 S. Ct. 1545 (1999).
3. 119 S. Ct. 765 (1999).

plan. In making this assessment, it was necessary to apply the two sampling strategies to population data in order to predict the changes in congressional apportionment that would most likely occur under each proposed strategy. After resolving the standing issue, we had to determine if the statistical estimation techniques were consistent with a federal statute.

In each of these cases, we judges were not asked to become expert statisticians, but we were expected to understand how the statistical analyses worked. Trial judges today are asked routinely to understand statistics at least as well, and probably better.

But science is far more than tools, such as statistics. And that "more" increasingly enters directly into the courtroom. The Supreme Court, for example, has recently decided cases involving basic questions of human liberty, the resolution of which demanded an understanding of scientific matters. Recently we were asked to decide whether a state's method of administering a lethal injection to condemned inmates constituted cruel and unusual punishment in violation of the Eighth Amendment.[4] And in 1997, we were asked to decide whether the Constitution protects a right to physician-assisted suicide.[5] Underlying the legal questions in these cases were medical questions: What effect does a certain combination of drugs, administered in certain doses, have on the human body, and to what extent can medical technology reduce or eliminate the risk of dying in severe pain? The medical questions did not determine the answer to the legal questions, but to do our legal job properly, we needed to develop an informed—although necessarily approximate—understanding of the science.

Nor were the lethal-injection and "right-to-die" cases unique in this respect. A different case concerned a criminal defendant who was found to be mentally competent to stand trial but not mentally competent to represent himself. We held that a state may insist that such a defendant proceed to trial with counsel.[6] Our opinion was grounded in scientific literature suggesting that mental illness can impair functioning in different ways, and consequently that a defendant may be competent to stand trial yet unable to carry out the tasks needed to present his own defense.

The Supreme Court's docket is only illustrative. Scientific issues permeate the law. Criminal courts consider the scientific validity of, say, DNA sampling or voiceprints, or expert predictions of defendants' "future dangerousness," which can lead courts or juries to authorize or withhold the punishment of death. Courts review the reasonableness of administrative agency conclusions about the safety of a drug, the risks attending nuclear waste disposal, the leakage potential of a toxic waste dump, or the risks to wildlife associated with the building of a dam. Patent law cases can turn almost entirely on an understanding of the underlying technical

4. Baze v. Rees, 128 S. Ct. 1520 (2008).

5. Washington v. Glucksberg, 521 U.S. 702 (1997); Vacco v. Quill, 521 U.S. 793 (1997).

6. Indiana v. Edwards, 128 S. Ct. 2379 (2008).

or scientific subject matter. And, of course, tort law often requires difficult determinations about the risk of death or injury associated with exposure to a chemical ingredient of a pesticide or other product.

The importance of scientific accuracy in the decision of such cases reaches well beyond the case itself. A decision wrongly denying compensation in a toxic substance case, for example, can not only deprive the plaintiff of warranted compensation but also discourage other similarly situated individuals from even trying to obtain compensation and encourage the continued use of a dangerous substance. On the other hand, a decision wrongly granting compensation, although of immediate benefit to the plaintiff, can improperly force abandonment of the substance. Thus, if the decision is wrong, it will improperly deprive the public of what can be far more important benefits—those surrounding a drug that cures many while subjecting a few to less serious risk, for example. The upshot is that we must search for law that reflects an understanding of the relevant underlying science, not for law that frees companies to cause serious harm or forces them unnecessarily to abandon the thousands of artificial substances on which modern life depends.

The search is not a search for scientific precision. We cannot hope to investigate all the subtleties that characterize good scientific work. A judge is not a scientist, and a courtroom is not a scientific laboratory. But consider the remark made by the physicist Wolfgang Pauli. After a colleague asked whether a certain scientific paper was wrong, Pauli replied, "That paper isn't even good enough to be wrong!"[7] Our objective is to avoid legal decisions that reflect that paper's so-called science. The law must seek decisions that fall within the boundaries of scientifically sound knowledge.

Even this more modest objective is sometimes difficult to achieve in practice. The most obvious reason is that most judges lack the scientific training that might facilitate the evaluation of scientific claims or the evaluation of expert witnesses who make such claims. Judges typically are generalists, dealing with cases that can vary widely in subject matter. Our primary objective is usually process-related: seeing that a decision is reached fairly and in a timely way. And the decision in a court of law typically (though not always) focuses on a particular event and specific individualized evidence.

Furthermore, science itself may be highly uncertain and controversial with respect to many of the matters that come before the courts. Scientists often express considerable uncertainty about the dangers of a particular substance. And their views may differ about many related questions that courts may have to answer. What, for example, is the relevance to human cancer of studies showing that a substance causes some cancers, perhaps only a few, in test groups of mice or rats? What is the significance of extrapolations from toxicity studies involving high doses to situations where the doses are much smaller? Can lawyers or judges or anyone else expect scientists always to be certain or always to have uniform views

7. Peter W. Huber, Galileo's Revenge: Junk Science in the Courtroom 54 (1991).

with respect to an extrapolation from a large dose to a small one, when the causes of and mechanisms related to cancer are generally not well known? Many difficult legal cases fall within this area of scientific uncertainty.

Finally, a court proceeding, such as a trial, is not simply a search for dispassionate truth. The law must be fair. In our country, it must always seek to protect basic human liberties. One important procedural safeguard, guaranteed by our Constitution's Seventh Amendment, is the right to a trial by jury. A number of innovative techniques have been developed to strengthen the ability of juries to consider difficult evidence.[8] Any effort to bring better science into the courtroom must respect the jury's constitutionally specified role—even if doing so means that, from a scientific perspective, an incorrect result is sometimes produced.

Despite the difficulties, I believe there is an increasingly important need for law to reflect sound science. I remain optimistic about the likelihood that it will do so. It is common to find cooperation between governmental institutions and the scientific community where the need for that cooperation is apparent. Today, as a matter of course, the President works with a science adviser, Congress solicits advice on the potential dangers of food additives from the National Academy of Sciences, and scientific regulatory agencies often work with outside scientists, as well as their own, to develop a product that reflects good science.

The judiciary, too, has begun to look for ways to improve the quality of the science on which scientifically related judicial determinations will rest. The Federal Judicial Center is collaborating with the National Academy of Sciences through the Academy's Committee on Science, Technology, and Law.[9] The Committee brings together on a regular basis knowledgeable scientists, engineers, judges, attorneys, and corporate and government officials to explore areas of interaction and improve communication among the science, engineering, and legal communities. The Committee is intended to provide a neutral, nonadversarial forum for promoting understanding, encouraging imaginative approaches to problem solving, and discussing issues at the intersection of science and law.

In the Supreme Court, as a matter of course, we hear not only from the parties to a case but also from outside groups, which file amicus curiae briefs that help us to become more informed about the relevant science. In the "right-to-die" case, for example, we received about 60 such documents from organizations of doctors, psychologists, nurses, hospice workers, and handicapped persons, among others. Many discussed pain-control technology, thereby helping us to identify areas of technical consensus and disagreement. Such briefs help to educate the justices on potentially relevant technical matters, making us not experts, but moderately educated laypersons, and that education improves the quality of our decisions.

8. *See generally* Jury Trial Innovations (G. Thomas Munsterman et al. eds., 1997).

9. A description of the program can be found at Committee on Science, Technology, and Law, http://www.nationalacademies.org/stl (last visited Aug. 10, 2011).

Moreover, our Court has made clear that the law imposes on trial judges the duty, with respect to scientific evidence, to become evidentiary gatekeepers.[10] The judge, without interfering with the jury's role as trier of fact, must determine whether purported scientific evidence is "reliable" and will "assist the trier of fact," thereby keeping from juries testimony that, in Pauli's sense, isn't even good enough to be wrong. This requirement extends beyond scientific testimony to all forms of expert testimony.[11] The purpose of *Daubert*'s gatekeeping requirement "is to make certain that an expert, whether basing testimony upon professional studies or personal experience, employs in the courtroom the same level of intellectual rigor that characterizes the practice of an expert in the relevant field."[12]

Federal trial judges, looking for ways to perform the gatekeeping function better, increasingly have used case-management techniques such as pretrial conferences to narrow the scientific issues in dispute, pretrial hearings where potential experts are subject to examination by the court, and the appointment of specially trained law clerks or scientific special masters. For example, Judge Richard Stearns of Massachusetts, acting with the consent of the parties in a highly technical genetic engineering patent case,[13] appointed a Harvard Medical School professor to serve "as a sounding board for the court to think through the scientific significance of the evidence" and to "assist the court in determining the validity of any scientific evidence, hypothesis or theory on which the experts base their testimony."[14] Judge Robert E. Jones of Oregon appointed experts from four different fields to help him assess the scientific reliability of expert testimony in silicone gel breast implant litigation.[15] Judge Gladys Kessler of the District of Columbia hired a professor of environmental science at the University of California at Berkeley "to answer the Court's technical questions regarding the meaning of terms, phrases, theories and rationales included in or referred to in the briefs and exhibits" of the parties.[16] Judge A. Wallace Tashima of the Ninth Circuit has described the role of technical advisor as "that of a … tutor who aids the court in understanding the 'jargon and theory' relevant to the technical aspects of the evidence."[17]

Judge Jack B. Weinstein of New York suggests that courts should sometimes "go beyond the experts proffered by the parties" and "appoint indepen-

10. Gen. Elec. Co. v. Joiner, 522 U.S. 136 (1997); Daubert v. Merrell Dow Pharms., Inc., 509 U.S. 579 (1993).

11. Kumho Tire Co. v. Carmichael, 119 S. Ct. 1167 (1999).

12. *Id.* at 1176.

13. Biogen, Inc. v. Amgen, Inc., 973 F. Supp. 39 (D. Mass. 1997).

14. MediaCom Corp. v. Rates Tech., Inc., 4 F. Supp. 2d 17 app. B at 37 (D. Mass. 1998) (quoting the Affidavit of Engagement filed in Biogen, Inc. v. Amgen, Inc., 973 F. Supp. 39 (D. Mass. 1997) (No. 95-10496)).

15. Hall v. Baxter Healthcare Corp., 947 F. Supp. 1387 (D. Or. 1996).

16. Conservation Law Found. v. Evans, 203 F. Supp. 2d 27, 32 (D.D.C. 2002).

17. Ass'n of Mexican-American Educators v. State of California, 231 F.3d 572, 612 (9th Cir. 2000) (en banc) (Tashima, J., dissenting).

dent experts" as the Federal Rules of Evidence allow.[18] Judge Gerald Rosen of Michigan appointed a University of Michigan Medical School professor to testify as an expert witness for the court, helping to determine the relevant facts in a case that challenged a Michigan law prohibiting partial-birth abortions.[19] Chief Judge Robert Pratt of Iowa hired two experts—a professor of insurance and an actuary—to help him review the fairness of a settlement agreement in a complex class-action insurance-fraud case.[20] And Judge Nancy Gertner of Massachusetts appointed a professor from Brandeis University to assist the court in assessing a criminal defendant's challenge to the racial composition of the jury venire in the Eastern Division of the District of Massachusetts.[21]

In what one observer has described as "the most comprehensive attempt to incorporate science, as scientists practice it, into law,"[22] Judge Sam Pointer, Jr., of Alabama appointed a "neutral science panel" of four scientists from different disciplines to prepare a report and testimony on the scientific basis of claims in silicone gel breast implant product liability cases consolidated as part of a multidistrict litigation process.[23] The panel's report was cited in numerous decisions excluding expert testimony that connected silicone gel breast implants with systemic injury.[24] The scientists' testimony was videotaped and made part of the record so that judges and jurors could consider it in cases returned to the district courts from the multidistrict litigation process. The use of such videotape testimony can result in more consistent decisions across courts, as well as great savings of time and expense for individual litigants and courts.

These case-management techniques are neutral, in principle favoring neither plaintiffs nor defendants. When used, they have typically proved successful. Nonetheless, judges have not often invoked their rules-provided authority to appoint their own experts.[25] They may hesitate simply because the process is unfamiliar or because the use of this kind of technique inevitably raises questions. Will use of an independent expert, in effect, substitute that expert's judgment for that of the court? Will it inappropriately deprive the parties of control over the presentation of the case? Will it improperly intrude on the proper function of the jury? Where is one to find a truly neutral expert? After all, different experts, in total honesty, often interpret the same data differently. Will the search for the expert

18. Jack B. Weinstein, Individual Justice in Mass Tort Litigation: The Effect of Class Actions, Consolidations, and Other Multiparty Devices 116 (1995).

19. Evans v. Kelley, 977 F. Supp. 1283 (E.D. Mich. 1997).

20. Grove v. Principal Mutual Life Ins. Co., 200 F.R.D. 434, 443 (S.D. Iowa 2001).

21. United States v. Green, 389 F. Supp. 2d 29, 48 (D. Mass. 2005).

22. Olivia Judson, *Slide-Rule Justice,* Nat'l J., Oct. 9, 1999, at 2882, 2885.

23. *In re* Silicone Gel Breast Implant Prod. Liab. Litig., Order 31 (N.D. Ala. filed May 30, 1996) (MDL No. 926).

24. *See* Laura L. Hooper et al., *Assessing Causation in Breast Implant Litigation: The Role of Science Panels,* 64 Law & Contemp. Probs. 139, 181 n.217 (collecting cases).

25. Joe S. Cecil & Thomas E. Willging, *Accepting Daubert's Invitation: Defining a Role for Court-Appointed Experts in Assessing Scientific Validity,* 43 Emory L.J. 995, 1004 (1994).

create inordinate delay or significantly increase costs? Who will pay the expert? Judge William Acker, Jr., of Alabama writes:

> Unless and until there is a national register of experts on various subjects and a method by which they can be fairly compensated, the federal amateurs wearing black robes will have to overlook their new gatekeeping function lest they assume the intolerable burden of becoming experts themselves in every discipline known to the physical and social sciences, and some as yet unknown but sure to blossom.[26]

A number of scientific and professional organizations have come forward with proposals to aid the courts in finding skilled experts. The National Conference of Lawyers and Scientists, a joint committee of the American Association for the Advancement of Science (AAAS) and the Science and Technology Section of the American Bar Association, has developed a program to assist federal and state judges, administrative law judges, and arbitrators in identifying independent experts in cases that present technical issues, when the adversarial system is unlikely to yield the information necessary for a reasoned and principled resolution of the disputed issues. The program locates experts through professional and scientific organizations and with the help of a Recruitment and Screening Panel of scientists, engineers, and health care professionals.[27]

The Private Adjudication Center at Duke University—which unfortunately no longer exists—established a registry of independent scientific and technical experts who were willing to provide advice to courts or serve as court-appointed experts.[28] Registry services also were available to arbitrators and mediators and to parties and lawyers who together agreed to engage an independent expert at the early stages of a dispute. The registry recruited experts primarily from major academic institutions and conducted targeted searches to find experts with the qualifications required for particular cases. Registrants were required to adhere to a code of conduct designed to ensure confidence in their impartiality and integrity.

Among those judges who have thus far experimented with court-appointed scientific experts, the reaction has been mixed, ranging from enthusiastic to disappointed. The Federal Judicial Center has examined a number of questions arising from the use of court-appointed experts and, based on interviews with participants in Judge Pointer's neutral science panel, has offered lessons to guide courts in future cases. We need to learn how better to identify impartial experts, to screen for possible conflicts of interest, and to instruct experts on the scope of

26. Letter from Judge William Acker, Jr., to the Judicial Conference of the United States et al. (Jan. 2, 1998).

27. Information on the AAAS program can be found at Court Appointed Scientific Experts, http://www.aaas.org/spp/case/case.htm (last visited Aug. 10, 2011).

28. Letter from Corinne A. Houpt, Registry Project Director, Private Adjudication Center, to Judge Rya W. Zobel, Director, Federal Judicial Center (Dec. 29, 1998) (on file with the Research Division of the Federal Judicial Center).

their duties. Also, we need to know how better to protect the interests of the parties and the experts when such extraordinary procedures are used. We also need to know how best to prepare a scientist for the sometimes hostile legal environment that arises during depositions and cross-examination.[29]

It would also undoubtedly be helpful to recommend methods for efficiently educating (i.e., in a few hours) willing scientists in the ways of the courts, just as it would be helpful to develop training that might better equip judges to understand the ways of science and the ethical, as well as practical and legal, aspects of scientific testimony.[30]

In this age of science we must build legal foundations that are sound in science as well as in law. Scientists have offered their help. We in the legal community should accept that offer. We are in the process of doing so. This manual seeks to open legal institutional channels through which science—its learning, tools, and principles—may flow more easily and thereby better inform the law. The manual represents one part of a joint scientific–legal effort that will further the interests of truth and justice alike.

29. Laura L. Hooper et al., Neutral Science Panels: Two Examples of Panels of Court-Appointed Experts in the Breast Implants Product Liability Litigation 93–98 (Federal Judicial Center 2001); Barbara S. Hulka et al., *Experience of a Scientific Panel Formed to Advise the Federal Judiciary on Silicone Breast Implants,* 342 New Eng. J. Med. 812 (2000).

30. Gilbert S. Omenn, *Enhancing the Role of the Scientific Expert Witness,* 102 Envtl. Health Persp. 674 (1994).

The Admissibility of Expert Testimony

MARGARET A. BERGER

Margaret A. Berger, J.D., was the Trustee Professor of Law, Brooklyn Law School, Brooklyn, New York.

[Editor's Note: While revising this chapter Professor Berger became ill and, tragically, passed away. We have published her last revision, with a few edits to respond to suggestions by reviewers.]

CONTENTS

I. Supreme Court Cases

In 1993, the Supreme Court's opinion in *Daubert v. Merrell Dow Pharmaceuticals*[1] ushered in a new era with regard to the admissibility of expert testimony. As expert testimony has become increasingly essential in a wide variety of litigated cases, the *Daubert* opinion has had an enormous impact. If plaintiffs' expert proof is excluded on a crucial issue, plaintiffs cannot win and usually cannot even get their case to a jury. This discussion begins with a brief overview of the Supreme Court's three opinions on expert testimony—often called the *Daubert* trilogy[2]—and their impact. It then examines a fourth Supreme Court case that relates to expert testimony, before turning to a variety of issues that judges are called upon to resolve, particularly when the proffered expert testimony hinges on scientific knowledge.

A. Daubert v. Merrell Dow Pharmaceuticals, Inc.

In the seminal *Daubert* case, the Court granted certiorari to decide whether the so-called *Frye* (or "general acceptance") test,[3] which some federal circuits (and virtually all state courts) used in determining the admissibility of scientific evidence, had been superseded by the enactment of the Federal Rules of Evidence in 1973. The Court held unanimously that the *Frye* test had not survived. Six justices joined Justice Blackmun in setting forth a new test for admissibility after concluding that "Rule 702 . . . clearly contemplates some degree of regulation of the subjects and theories about which an expert may testify."[4] While the two other members of the Court agreed with this conclusion about the role of Rule 702, they thought that the task of enunciating a new rule for the admissibility of expert proof should be left to another day.[5]

The majority opinion in *Daubert* sets forth a number of major themes that run throughout the trilogy. First, it recognized the trial judge as the "gatekeeper" who must screen proffered expert testimony.[6] Second, the objective of the screening is to ensure that expert testimony, in order to be admissible, must be "not only relevant, but reliable."[7] Although there was nothing particularly novel about the Supreme Court finding that a trial judge has the *power* to make an admissibility determination—Federal Rules of Evidence 104(a) and 702 pointed to such a conclusion—and federal trial judges had excluded expert testimony long before

1. 509 U.S. 579 (1993).

2. The other two cases are *Gen. Elec. Co. v. Joiner*, 522 U.S. 136 (1997) and *Kumho Tire Co. v. Carmichael*, 526 U.S. 137 (1999). The disputed issue in all three cases was causation.

3. Frye v. United States, 293 F. 1013 (D.C. Cir. 1923).

4. *Daubert*, 509 U.S. at 589.

5. *Id.* at 601.

6. *Id.* at 589.

7. *Id.*

Daubert, the majority opinion in *Daubert* stated that the trial court has not only the power but the *obligation* to act as gatekeeper.[8]

The Court then considered the meaning of its two-pronged test of relevancy and reliability in the context of scientific evidence. With regard to relevancy, the Court explained that expert testimony cannot assist the trier in resolving a factual dispute, as required by Rule 702, unless the expert's theory is tied sufficiently to the facts of the case. "Rule 702's 'helpfulness' standard requires a valid scientific connection to the pertinent inquiry as a precondition to admissibility."[9] This consideration, the Court remarked, "has been aptly described by Judge Becker as one of 'fit.'"[10]

To determine whether proffered scientific testimony or evidence satisfies the standard of evidentiary reliability,[11] a judge must ascertain whether it is "ground[ed] in the methods and procedures of science."[12] The Court, emphasizing that "[t]he inquiry envisioned by Rule 702 is . . . a flexible one,"[13] then examined the characteristics of scientific methodology and set out a nonexclusive list of four factors that bear on whether a theory or technique has been derived by the scientific method.[14] First and foremost, the Court viewed science as an empirical endeavor: "[W]hether [a theory or technique] can be (and has been) tested" is the "methodology [that] distinguishes science from other fields of human inquiry."[15] The Court also mentioned as indicators of good science whether the technique or theory has been subjected to peer review or publication, whether the existence of known or potential error rates has been determined, and whether standards exist for controlling the technique's operation.[16] In addition, although general acceptance of the methodology within the scientific community is no longer dispositive, it remains a factor to be considered.[17]

The Court did not apply its new test to the eight experts for the plaintiffs who sought to testify on the basis of in vitro, animal, and epidemiological studies

8. *Id.*

9. *Id.* at 591–92.

10. *Id.* at 591. Judge Becker used this term in discussing the admissibility of expert testimony about factors that make eyewitness testimony unreliable. *See* United States v. Downing, 753 F.2d 1224, 1242 (3d Cir. 1985) (on remand court rejected the expert testimony on ground of "fit" because expert discussed factors such as the high likelihood of inaccurate cross-racial identifications that were not present in the case) *and* United States v. Downing, 609 F. Supp. 784, 791–92 (E.D. Pa. 1985), *aff'd*, 780 F.2d 1017 (3d Cir. 1985).

11. Commentators have faulted the Court for using the label "reliability" to refer to the concept that scientists term "validity." The Court's choice of language was deliberate. It acknowledged that scientists typically distinguish between validity and reliability and that "[i]n a case involving scientific evidence, evidentiary reliability will be based upon scientific validity." *Daubert*, 509 U.S. at 590 n.9.

12. *Id.* at 590.

13. *Id.* at 594.

14. *Id.* at 593–94. "[W]e do not presume to set out a definitive checklist or test." *Id.* at 593.

15. *Id.*

16. *Id.* at 593–94.

17. *Id.* at 594.

that the drug Bendectin taken by the plaintiffs' mothers during pregnancy could cause or had caused the plaintiffs' birth defects. Instead, it reversed and remanded the case. Nor did the Court deal with any of the procedural issues raised by the *Daubert* opinion, such as the burden, if any, on the party seeking a ruling excluding expert testimony, or the standard of review on appeal.

The *Daubert* opinion soon led to *Daubert* motions followed by *Daubert* hearings as parties moved *in limine* to have their opponents' experts precluded from testifying at trial for failure to satisfy the new requirements for expert testimony. The motions raised numerous questions that the Court had not dealt with, some of which were dealt with in the next two opinions by the Supreme Court.

B. General Electric v. Joiner

In *General Electric Co. v. Joiner,*[18] the second case in the trilogy, certiorari was granted in order to determine the appropriate standard an appellate court should apply in reviewing a trial court's *Daubert* decision to admit or exclude scientific expert testimony. In *Joiner,* the 37-year-old plaintiff, a longtime smoker with a family history of lung cancer, claimed that exposure to polychlorinated biphenyls (PCBs) and their derivatives had promoted the development of his small-cell lung cancer. The trial court applied the *Daubert* criteria, excluded the opinions of the plaintiff's experts, and granted the defendants' motion for summary judgment.[19] The court of appeals reversed the decision, stating that "[b]ecause the Federal Rules of Evidence governing expert testimony display a preference for admissibility, we apply a particularly stringent standard of review to the trial judge's exclusion of expert testimony."[20]

All the justices joined Chief Justice Rehnquist in holding that abuse of discretion is the correct standard for an appellate court to apply in reviewing a district court's evidentiary ruling, regardless of whether the ruling allowed or excluded expert testimony.[21] The Court unequivocally rejected the suggestion that a more stringent standard is permissible when the ruling, as in *Joiner,* is "outcome determinative" because it resulted in a grant of summary judgment for the defendant because the plaintiff failed to produce evidence of causation.[22] In a concurring opinion, Justice Breyer urged judges to avail themselves of techniques, such as the use of court-appointed experts, that would assist them in making determinations about the admissibility of complex scientific or technical evidence.[23]

18. 522 U.S. 136 (1997).
19. Joiner v. Gen. Elec. Co., 864 F. Supp. 1310 (N.D. Ga. 1994).
20. Joiner v. Gen. Elec. Co., 78 F.3d 524, 529 (11th Cir. 1996).
21. Gen. Elec. Co. v. Joiner, 522 U.S. at 141–43.
22. *Id.* at 142–43.
23. *Id.* at 147–50. This issue is discussed in further detail in Justice Breyer's introduction to this manual.

With the exception of Justice Stevens, who dissented from this part of the opinion, the justices then did what they had not done in *Daubert*—they examined the record, found that the plaintiff's experts had been properly excluded, and reversed the court of appeals decision without a remand to the lower court. The Court concluded that it was within the district court's discretion to find that the statements of the plaintiff's experts with regard to causation were nothing more than speculation. The Court noted that the plaintiff never explained "how and why the experts could have extrapolated their opinions"[24] from animal studies far removed from the circumstances of the plaintiff's exposure.[25] It also observed that the district court could find that the four epidemiological studies the plaintiff relied on were insufficient as a basis for his experts' opinions.[26] Consequently, the court of appeals had erred in reversing the district court's determination that the studies relied on by the plaintiff's experts "were not sufficient, whether individually or in combination, to support their conclusions that Joiner's exposure to PCBs contributed to his cancer."[27]

The plaintiff in *Joiner* had argued that the epidemiological studies showed a link between PCBs and cancer if the results of all the studies were pooled, and that this weight-of-the-evidence methodology was reliable. Therefore, according to the plaintiff, the district court erred when it excluded a conclusion based on a scientifically reliable methodology because it thereby violated the Court's precept in *Daubert* that the "focus, of course, must be solely on principles and methodology, not on the conclusions that they generate."[28] The Supreme Court responded to this argument by stating that

> conclusions and methodology are not entirely distinct from one another. Trained experts commonly extrapolate from existing data. But nothing in either *Daubert* or the Federal Rules of Evidence requires a district court to admit opinion evidence which is connected to existing data only by the *ipse dixit* of the expert. A court may conclude that there is simply too great an analytical gap between the data and the opinion proffered.[29]

24. *Id.* at 144.

25. The studies involved infant mice that had massive doses of PCBs injected directly into their bodies; Joiner was an adult who was exposed to fluids containing far lower concentrations of PCBs. The infant mice developed a different type of cancer than *Joiner* did, and no animal studies showed that adult mice exposed to PCBs developed cancer or that PCBs lead to cancer in other animal species. *Id.*

26. The authors of the first study of workers at an Italian plant found lung cancer rates among ex-employees somewhat higher than might have been expected but refused to conclude that PCBs had caused the excess rate. A second study of workers at a PCB production plant did not find the somewhat higher incidence of lung cancer deaths to be statistically significant. The third study made no mention of exposure to PCBs, and the workers in the fourth study who had a significant increase in lung cancer rates also had been exposed to numerous other potential carcinogens. *Id.* at 145–46.

27. *Id.* at 146–47.

28. *Id.* at 146 (quoting *Daubert*, 509 U.S. at 595).

29. *Id.* at 146.

Justice Stevens, in his partial dissent, assumed that the plaintiff's expert was entitled to rely on such a methodology, which he noted is often used in risk assessment, and that a district court that admits expert testimony based on a weight-of-the-evidence methodology does not abuse its discretion.[30] Justice Stevens would have remanded the case for the court below to determine if the trial court had abused its discretion when it excluded the plaintiff's experts.[31]

C. Kumho Tire Co. v. Carmichael

Less than one year after deciding *Joiner,* the Supreme Court granted certiorari in *Kumho* to decide if the trial judge's gatekeeping obligation under *Daubert* applies only to scientific evidence or if it extends to proffers of "technical, or other specialized knowledge," the other categories of expertise recognized in Federal Rule of Evidence 702. In addition, there was uncertainty about whether disciplines such as economics, psychology, and other "soft" sciences were governed by this standard; about when the four factors endorsed in *Daubert* as indicators of reliability had to be applied; and how experience factors into the gatekeeping process. Although Rule 702 specifies that an expert may be qualified through experience, the Court's emphasis in *Daubert* on "testability" suggested that an expert should not be allowed to base a conclusion solely on experience if the conclusion can easily be tested.

In *Kumho,* the plaintiffs brought suit after a tire blew out on a minivan, causing an accident in which one passenger died and others were seriously injured. The tire, which was manufactured in 1988, had been installed on the minivan sometime before it was purchased as a used car by the plaintiffs in 1993. In their diversity action against the tire's maker and its distributor, the plaintiffs claimed that the tire was defective. To support this allegation, the plaintiffs relied primarily on deposition testimony by an expert in tire-failure analysis, who concluded on the basis of a visual inspection of the tire that the blowout was caused by a defect in the tire's manufacture or design.

When the defendants moved to exclude the plaintiffs' expert, the district court agreed with the defendants that the *Daubert* gatekeeping obligation applied not only to scientific knowledge but also to "technical analyses."[32] The district court excluded the plaintiffs' expert and granted summary judgment. Although the court conceded on a rehearing that it had erred in treating the four factors discussed in *Daubert* as mandatory, it adhered to its original determination because the court simply found the *Daubert* factors appropriate, analyzed them, and discerned no competing criteria sufficiently strong to outweigh them.[33]

30. *Id.* at 153–54.

31. *Id.* at 150–51.

32. Carmichael v. Samyang Tire, Inc., 923 F. Supp. 1514, 1522 (S.D. Ala. 1996), *rev'd,* 131 F.3d 1433 (11th Cir. 1997), *rev'd sub nom.* Kumho Tire Co. v. Carmichael, 526 U.S. 137 (1999).

33. *Id.* at 1522, 1524.

The Eleventh Circuit reversed the district court's decision in *Kumho,* holding, as a matter of law under a de novo standard of review, that *Daubert* applies only to scientific opinions.[34] The court of appeals drew a distinction between expert testimony that relies on the application of scientific theories or principles—which would be subject to a *Daubert* analysis—and testimony that is based on the expert's "skill- or experience-based observation."[35] The court then found that the testimony proffered by plaintiff was "non-scientific" and that "the district court erred as a matter of law by applying *Daubert* in this case."[36] The circuit court agreed that the trial court has a gatekeeping obligation; its quarrel with the district court was with that court's assumption that *Daubert*'s four factors had to be applied.

All of the justices of the Supreme Court, in an opinion by Justice Breyer, held that the trial court's gatekeeping obligation extends to all expert testimony,[37] and unanimously rejected the Eleventh Circuit's dichotomy between the expert who "relies on the application of scientific principles" and the expert who relies on "skill- or experience-based observation."[38] The Court noted that Federal Rule of Evidence 702 "makes no relevant distinction between 'scientific' knowledge and 'technical' or 'other specialized' knowledge," and "applies its reliability standard to all . . . matters within its scope."[39] Furthermore, said the Court, "no clear line" can be drawn between the different kinds of knowledge, and "no one denies that an expert might draw a conclusion from a set of observations based on extensive and specialized experience."[40]

The Court also unanimously found that the court of appeals had erred when it used a de novo standard, instead of the *Joiner* abuse-of-discretion standard, to determine that *Daubert's* criteria were not reasonable measures of the reliability of the expert's testimony.[41] As in *Joiner,* and again over the dissent of Justice Stevens,[42] the Court then examined the record and concluded that the trial court had not abused its discretion when it excluded the testimony of the witness. Accordingly, it reversed the opinion of the Eleventh Circuit.

The opinion adopts a flexible approach that stresses the importance of identifying "the particular circumstances of the particular case at issue."[43] The court must then make sure that the proffered expert will observe the same standard of "intellectual rigor" in testifying as he or she would employ when dealing with similar matters outside the courtroom.[44]

34. Carmichael v. Samyang Tire, Inc., 131 F.3d 1433, 1435 (11th Cir. 1997).
35. *Id.*
36. *Id.* at 1436 (footnotes omitted).
37. Kumho Tire Co. v. Carmichael, 526 U.S. 137 (1999).
38. *Id.* at 151.
39. *Id.* at 148.
40. *Id.* at 156.
41. *Id.* at 152.
42. *Id.* at 158.
43. *Id.* at 150.
44. *Id.* at 152.

How this extremely flexible approach of the Court is to be applied emerges in Part III of the opinion when the Court engages in a remarkably detailed analysis of the record that illustrates its comment in *Joiner* that an expert must account for "how and why" he or she reached the challenged opinion.[45]

The Court illustrated the application of this standard to the facts of the case and its deference to the district court findings as follows:

> After examining the transcript in some detail, and after considering respondents' defense of Carlson's methodology, the District Court determined that Carlson's testimony was not reliable. It fell outside the range where experts might reasonably differ, and where the jury must decide among the conflicting views of different experts, even though the evidence is shaky. In our view, the doubts that triggered the District Court's initial inquiry here were reasonable, as was the court's ultimate conclusion.[46]

Although *Kumho* is the most recent pronouncement by the Supreme Court on how to determine whether proffered testimony by an expert is admissible, and Rule 702 of the Federal Rules of Evidence was amended in 2000 to provide "some general standards that the trial court must use to assess the reliability and helpfulness of proffered expert testimony," it is still *Daubert* that trial courts cite and rely on most frequently when ruling on a motion to preclude expert testimony.[47] Even though *Daubert* interprets a federal rule of evidence, and rules of evidence are designed to operate at trial, *Daubert's* greatest impact has been pretrial: If plaintiff's experts can be excluded from testifying about an issue crucial to plaintiff's case, the litigation may end with summary judgment for the defendant. Furthermore, although summary judgment grants are reviewed de novo by an appellate court, there is nothing to review if plaintiff failed to submit admissible evidence on a material issue. Consequently, only the less stringent abuse-of-discretion standard will apply, and there will be less chance for a reversal on appeal.

D. Weisgram v. Marley

Plaintiff is entitled to only one chance to select an expert who can withstand a *Daubert* motion. In a fourth Supreme Court case, *Weisgram v. Marley*,[48] the district court ruled for plaintiffs on a *Daubert* motion and the plaintiffs won a jury verdict. On appeal, the circuit court found that, despite the abuse-of-discretion standard, plaintiff's experts should have been excluded and granted judgment as a matter of law for the defendants. Plaintiffs argued that they now had the right to a new trial at which they could introduce more expert testimony. The Supreme Court

45. Gen. Elec. Co v. Joiner, 522 U.S. 136, 144 (1997).

46. Kumho Tire Co. v. Carmichael, 526 U.S. at 153.

47. A search of federal cases on Westlaw after *Kumho* was decided indicates that the *Daubert* decision has been cited more than twice as often as the *Kumho* decision.

48. 528 U.S. 440 (2000).

granted certiorari limited to the new trial issue (it did not review the *Daubert* determination) but refused to grant a new trial. Justice Ginsberg explained:

> Since *Daubert,* moreover, parties relying on expert testimony have had notice of the exacting standards of reliability such evidence must meet. . . . It is implausible to suggest, post-*Daubert*, that parties will initially present less than their best expert evidence in the expectation of a second chance should their first trial fail.[49]

Weisgram causes tactical problems for plaintiffs about how much to spend for expert testimony. Should they pay for additional expensive expert testimony even though they think the district court would rule in their favor on a *Daubert* motion, or is the risk of a reversal on *Daubert* grounds and a consequent judgment for the defendant too great despite the abuse-of-discretion standard? *Weisgram* may indeed push plaintiffs to bring the very best expertise into litigation—a stated goal of the trilogy, but it may also make it difficult to litigate legitimate claims because of the cost of expert testimony. Is access to the federal courts less important than regulating the admissibility of expert testimony? Even if plaintiffs successfully withstand a *Daubert* motion, that does not guarantee they will win were the case to be tried. But very few cases now go to trial, and an inability by the defendant to exclude plaintiffs' experts undoubtedly affects the willingness of the defendant to negotiate a settlement.

II. Interpreting *Daubert*

Although almost 20 years have passed since *Daubert* was decided, a number of basic interpretive issues remain.

A. Atomization

When there is a *Daubert* challenge to an expert, should the court look at all the studies on which the expert relies for their collective effect or should the court examine the reliability of each study independently? The issue arises with proof of causation in toxic tort cases when plaintiff's expert relies on studies from different scientific disciplines, or studies within a discipline that present different strengths and weaknesses, in concluding that defendant's product caused plaintiff's adverse health effects. Courts rarely discuss this issue explicitly, but some appear to look at each study separately and give no consideration to those studies that cannot alone prove causation.

Although some use the language in *Joiner* as the basis for this slicing-and-dicing approach,[50] scientific inference typically requires consideration of numerous

49. 528 U.S. at 445 (internal citations omitted).

50. See discussion, *supra* notes 28–31 and related text.

findings, which, when considered alone, may not individually prove the contention.[51] It appears that many of the most well-respected and prestigious scientific bodies (such as the International Agency for Research on Cancer (IARC), the Institute of Medicine, the National Research Council, and the National Institute for Environmental Health Sciences) consider all the relevant available scientific evidence, taken as a whole, to determine which conclusion or hypothesis regarding a causal claim is best supported by the body of evidence. In applying the scientific method, scientists do not review each scientific study individually for whether by itself it reliably supports the causal claim being advocated or opposed. Rather, as the Institute of Medicine and National Research Council noted, "summing, or synthesizing, data addressing different linkages [between kinds of data] forms a more complete causal evidence model and can provide the biological plausibility needed to establish the association" being advocated or opposed.[52] The IARC has concluded that "[t]he final overall evaluation is a matter of scientific judgment reflecting the weight of the evidence derived from studies in humans, studies in experimental animals, and mechanistic and other relevant data."[53]

B. *Conflating Admissibility with Sufficiency*

In *Daubert,* Justice Blackmun's opinion explicitly acknowledges that in some cases admissible evidence may not suffice to support a verdict in favor of plaintiffs. In other words, it seems to recognize that the admissibility determination comes first and is separate from the sufficiency determination. But in *Joiner* the Court pays little attention to this distinction and suggests that plaintiff's expert testimony may be excluded if the evidence on which he seeks to rely is itself deemed insufficient.

But what difference does it make if sufficiency is conflated with admissibility?[54] After all, the case's final outcome will be the same. As *Daubert* recognizes, the trial judge's authority to decide whether the plaintiff has produced sufficient evidence to withstand a dispositive motion under Rule 56 or 50 is indisputable; a one-step process that considers sufficiency when adjudicating a *Daubert* motion is arguably

51. *See e.g.,* Susan Haack, *An Epistemologist in the Bramble-Bush: At the Supreme Court with Mr. Joiner,* 26 J. Health Pol. Pol'y & L. 217–37 (1999) (discussing the individual studies that lead to the compelling inference of a double-helical structure of a DNA molecule, which, when considered separately, fail to compel that inference). *See also* Milward v. Acuity Specialty Products Group, Inc., __ F.3d __, 2011 WL 982385, *10 639 F.3d 11, 26 (1st Cir. 2011) (reversing the district court's exclusion of expert testimony based on an assessment of the direct causal effect of the individual studies, finding that the "weight of the evidence" properly supported the expert's opinion that exposure to benzene can cause acute promyelocytic leukemia).

52. Institute of Medicine and National Research Council, Dietary Supplements: A Framework for Evaluating Safety 262 (2005).

53. Vincent J. Cogliano et al., *The Science and Practice of Carcinogen Identification and Evaluation,* 112 Envtl. Health Persp. 1272 (2004).

54. The distinction between admissibility and sufficiency is also discussed in Michael D. Green et al., Reference Guide on Epidemiology, Section VII, in this manual.

more efficient than a two-step process that requires the district judge to analyze admissibility before it can turn to sufficiency.

There are, however, consequences to conflating admissibility and sufficiency. The de novo standard of review that ordinarily applies to judgments as a matter of law following a determination of insufficient evidence is converted into the lower abuse-of-discretion standard that governs evidentiary rulings on admissibility, and thereby undermines the jury trial mandate of the Seventh Amendment. Science proceeds by cumulating and synthesizing evidence until there is enough for a new paradigm. That does not mean that every study meets the most rigorous scientific standards. Judgment is required in determining which inferences are appropriate, but an approach that encourages looking at studies sequentially rather than holistically has costs that must be considered.

C. Credibility

Daubert and the expense of litigation make it difficult for courts to hew to the line that assigns credibility issues to the jury rather than the court. One troublesome area is conflicts of interest. To what extent should a court permit the plaintiff to inquire into the defense expert's relationship with the defendant? If the expert testified at trial, information that could have skewed the expert's testimony could be brought to the attention of the jury through cross-examination or extrinsic evidence. Impeachment by bias suffers from fewer constraints than other forms of impeachment.[55] But suppose the defendant seeks through a *Daubert* challenge to exclude the plaintiff's expert witness as relying on unreliable evidence to show causation in a toxic tort action. The defendant supports its argument with testimony by an academic from a highly respected institution whose research shows that the defendant's product is safe. Should the court permit the plaintiff to inquire whether the expert was on the payroll of the defendant corporation, or attended conferences paid for by the defendant, or received gifts from the defendant? What about corporate employees ghostwriting reports about their products that are then submitted in someone else's name? Other ties that an expert may have to industry have also been reported: royalties, stock ownership, working in an institution that receives considerable funding from the defendant. These are all practices that have been reported in the media and are practices that the plaintiff would like to question the expert about under oath.[56] A court is unlikely to allow a wide-ranging

55. *See* United States v. Abel, 469 U.S. 45, 50 (1984) (explaining that "proof of bias is almost always relevant because the jury, as finder of fact and weigher of credibility, has historically been entitled to assess all evidence which might bear on the accuracy and truth of a witness' testimony").

56. *See, e.g., In re* Welding Fume Products, 534 F. Supp. 2d 761, 764 (N.D. Ohio 2008) (requiring all parties to the litigation to "disclose the fact of, and the amounts of, payments they made, either directly or indirectly, to any entity (whether an individual or organization) that has authored or published any study, article, treatise, or other text upon which any expert in this MDL litigation relies, or has relied").

fishing expedition if the plaintiff has no proof that the defense expert engaged in such behavior. But even if the plaintiff has extrinsic evidence available that points to conflicts of interest on the part of the expert, how should a court assess this information in ruling on the admissibility of plaintiff's experts? Is this a credibility determination? Should allegations about conflicts be resolved by the judge at an in limine hearing, or should the plaintiff's expert be permitted to testify so that this issue can be explored at trial?

Another troublesome issue about credibility arises when an expert seeks to base an opinion on controverted evidence in the case. May the court exclude the expert's opinion on a *Daubert* motion if it finds that the expert's model did not incorporate the appropriate data that fit the facts of the case, or is this an issue for the jury?[57]

Does the court avoid a credibility determination if it finds that the expert is qualified but the court disagrees with the theory on which the expert is relying? In *Kochert v. Greater Lafayette Health Serv. Inc.*,[58] a complex antitrust case, the court held that the trial court properly excluded the plaintiff's economic experts on the ground that the plaintiff's antitrust theory was based on the wrong legal standard after ruling for the plaintiff on *Daubert* challenges.

III. Applying *Daubert*

Application of *Daubert* raises a number of persistent issues, many of which relate to proof of causation. The three cases in the trilogy and *Weisgram* all turned on questions of causation, and the plaintiffs in each of the cases ultimately lost because they failed to introduce admissible expert testimony on this issue.

Causation questions have been particularly troubling in cases in which plaintiffs allege that the adverse health effects for which they seek damages are a result of exposure to the defendant's product.

A. *Is the Expert Qualified?*

As a threshold matter, the witness must be qualified as an expert to present expert opinion testimony. An expert needs more than proper credentials, whether grounded in "skill, experience, training or education" as set forth in Rule 702 of the Federal Rules of Evidence. A proposed expert must also have "knowledge."

57. *Compare* Consol. Insured Benefits, Inc. v. Conseco Med. Ins. Co., No. 03-cv-3211, 2006 WL 3423891 (D.S.C. 2006) (fraud case; *Daubert* motion to exclude plaintiff's expert economist's testimony on damages; court finds that testimony question of weight, not admissibility) *with* Concord Boat Corp. v. Brunswick Corp., 207 F.3d 1039, 1055-56 (8th Cir. 2000) (excluding expert's testimony as "mere speculation" that ignored inconvenient evidence).

58. 463 F.3d 710 (7th Cir. 2006).

For example, an expert who seeks to testify about the findings of epidemiological studies must be knowledgeable about the results of the studies and must take into account those studies that reach conclusions contrary to the position the expert seeks to advocate.

B. Assessing the Scientific Foundation of Studies from Different Disciplines

Expert opinion is typically based on multiple studies, and those studies may come from different scientific disciplines. Some courts have explicitly stated that certain types of evidence proffered to prove causation have no probative value and therefore cannot be reliable.[59] Opinions based on animal studies have been rejected because of reservations about extrapolating from animals to humans or because the plaintiff's extrapolated dose was lower than the animals'—which is invariably the case because one would have to study unmanageable, gigantic numbers of animals to see results if animals were not given high doses. The field of toxicology, which, unlike epidemiology, is an experimental science, is rapidly evolving, and prior case law regarding such studies may not take into account important new developments.

But even when there are epidemiological studies, a court may conclude that they cannot prove causation because they are not conclusive and therefore unreliable. And if they are unreliable, they cannot be combined with other evidence.[60]

Experts will often rely on multiple studies, each of which has some probative value but, when considered separately, cannot prove general causation.

As noted above, trial judges have great discretion under *Daubert* and a court is free to choose an atomistic approach that evaluates the available studies one by one. Some judges have found this practice contrary to that of scientists who look at knowledge incrementally.[61] But there are no hard-and-fast scientific rules for synthesizing evidence, and most research can be critiqued on a variety of grounds.

59. *See, e.g., In re* Rezulin, 2004 WL 2884327, at *3 (S.D.N.Y. 2004); Cloud v. Pfizer Inc., 198 F. Supp. 2d 1118, 1133 (D. Ariz. 2001) (stating that case reports were merely compilations of occurrences and have been rejected as reliable scientific evidence supporting an expert opinion that *Daubert* requires); Haggerty v. Upjohn Co., 950 F. Supp. 1160, 1164 (S.D. Fla. 1996), *aff'd,* 158 F.3d 588 (11th Cir. 1998) ("scientifically valid cause and effect determinations depend on controlled clinical trials and epidemiological studies"); Wade-Greaux v. Whitehall Labs., Inc., 874 F. Supp. 1441, 1454 (D.V.I. 1994), *aff'd,* 46 F.3d 1120 (3d Cir. 1994) (stating there is a need for consistent epidemiological studies showing statistically significant increased risks).

60. *See* Hollander v. Sandoz Pharm. Corp., 289 F.3d 1193, 1216 n.21 (10th Cir. 2002) ("To suggest that those individual categories of evidence deemed unreliable by the district court may be added to form a reliable theory would be to abandon 'the level of intellectual rigor of the expert in the field.'").

61. *See, e.g., In re* Ephedra, 393 F. Supp. 2d 181, 190 (S.D.N.Y. 2005) (allowing scientific expert testimony regarding "a confluence of suggestive, though non-definitive, scientific studies [that] make[s] it more-probable-than-not that a particular substance . . . contributed to a particular result. . . ."; after a two-week *Daubert* hearing in a case in which there would never be epidemiological evidence, the court concluded that some of plaintiffs' experts could testify on the basis of animal studies, analogous

Few studies are flawless. Epidemiology is vulnerable to attack because of problems with confounders and bias. Furthermore, epidemiological studies are grounded in statistical models. What role should statistical significance play in assessing the value of a study? Epidemiological studies that are not conclusive but show some increased risk do not prove a lack of causation. Some courts find that they therefore have some probative value,[62] at least in proving general causation.[63]

Even, however, if plaintiffs convince the trial judge that their experts relied on reliable and relevant evidence in establishing general causation, that is, in opining that the defendant's product can cause the adverse effects for which plaintiffs seek compensation, plaintiffs must also present admissible expert testimony that the defendant's product caused their specific injuries. For example, in the Zyprexa litigation,[64] the court found that plaintiffs' expert's conclusion that Zyprexa may cause excessive weight gain leading to diabetes was well supported, but the expert's assertion that Zyprexa had a direct adverse effect on cells essential to the production of insulin by the body in cases in which there was no documented weight gain lacked scientific support. The record demonstrates that the expert's opinions relied on a subjective methodology, a fast-and-loose application of his scientific theories to the facts, and conclusion-driven assessments on the issues of causation in the cases on which he proposed to testify. He was not allowed to testify because his opinions were neither "based upon sufficient facts or data," nor were they "the product of reliable principles and methods," and he had not "applied the principles and methods reliably to the facts of the case."[65]

Courts handling *Daubert* motions sometimes sound as though only one possible answer is legitimate. If scientists seeking to testify for opposing sides disagree, some courts conclude that one side must be wrong.[66] The possibility that both sides are offering valid scientific inferences is rarely recognized, even though this happens often in the world of science.

As noted above, district courts have great discretion in deciding how to proceed when faced with evidence from different scientific disciplines and presenting different degrees of scientific rigor. In assessing the proffered testimony of the

human studies, plausible theories of the mechanisms involved, etc.); Milward v. Acuity Specialty Prods. Group, Inc., 639 F.3d 11 (1st Cir. 2011).

62. *See* Cook v. Rockwell Int'l Corp., 580 F. Supp. 2d 1071 (D. Colo. 2006) (discussing why the court excluded expert's testimony, even though his epidemiological study did not produce statistically significant results).

63. *In re* Viagra Prods., 572 F. Supp. 2d 1071 (D. Minn. 2008) (extensive review of all expert evidence proffered in multidistricted product liability case).

64. *See In re* Zyprexa Prods., 2009 WL 1357236 (E.D.N.Y. May 12, 2009) (providing citations to opinions dealing with *Daubert* rulings and summary judgment motions in the Zyprexa litigation).

65. *See* Fed. R. Evid. 702; *cf. Gen. Elec. Co. v. Joiner,* 522 U.S. 136, 146 (opinion that "is connected to existing data only by the *ipse dixit* of the expert" need not be admitted).

66. *See* Soldo v. Sandoz Pharm. Corp., 2003 WL 22005007 (W.D. Pa. 2003) (stating that court appointed three experts to assist it pursuant to Fed. R. Evid. 706 and then rejected opinion expressing minority view).

expert in light of the studies on which the testimony is based, courts may choose to limit the opinion that the expert would be allowed to express if the case went to trial.[67] Given the expense of trials, the paucity of trials, and the uncertainty about how jurors would evaluate such testimony, limiting an expert's opinion may lead to settlements.[68]

The abuse-of-discretion standard may lead to inconsistent results in how courts handle proof of causation. There can be inconsistencies even within circuits when district judges disagree on whether plaintiffs' experts have met their burden of proof.[69]

C. *How Should the Courts Assess Exposure?*

Another difficulty in proving causation in toxic tort cases is that plaintiff must establish that he or she was exposed to defendant's product. Obviously this is not a problem with prescription drugs, but in other types of cases, such as environmental torts, establishing exposure and the extent of the exposure can be difficult.[70] Although exact data on exposure need not be required, an expert should, however, be able to provide reasonable explanations for his or her conclusions about the amount of exposure and that it sufficed to cause plaintiffs' injuries.[71]

67 *See, e.g., In re* Ephedra, 393 F. Supp. 2d 181 (S.D.N.Y. 2005) (stating that qualified experts may testify to a reliable basis for believing that ephedra may contribute to cardiac injury and strokes in persons with high blood pressure, certain serious heart conditions, or a genetic sensitivity to ephedra; experts would have to acknowledge that none of this has been the subject of definitive studies and may yet be disproved).

68. *But cf.* Giles v. Wyeth, 556 F.3d 596 (7th Cir. 2009) (plaintiff won *Daubert* challenge but lost at trial).

69. *Compare* Bonner v. ISP Techs., Inc., 259 F.3d 924 (8th Cir. 2001) (affirming jury verdict that exposure to solvent caused plaintiff's psychological and cognitive impairment and Parkinsonian symptoms; defendant argued that expert's opinion based on case reports, animal studies, structural analysis studies should have been excluded on *Daubert* grounds; the court stated: "The first several victims of a new toxic tort should not be barred from having their day in court simply because the medical literature, which will eventually show the connection between the victims' condition and the toxic substance, has not yet been completed.") *with* Glastetter v. Novartis Pharm. Corp., 107 F. Supp. 2d 1015 (E.D. Mo. 2000), *aff'd per curiam,* 252 F.3d 986 (8th Cir. 2001) (plaintiff claimed that drug she had taken for lactation suppression had caused her stroke; trial court held that *Daubert* precluded experts from finding causation on the basis of case reports, animal studies, human dechallenge/rechallenge data, internal documents from defendant, and Food and Drug Administration's revocation of drug for lactation suppression; appellate court stated: "We do not discount the possibility that stronger evidence of causation exists, or that, in the future, physicians will demonstrate to a degree of medical certainty that Parlodel can cause ICHs. Such evidence has not been presented in this case, however, and we have no basis for concluding that the district court abused its discretion in excluding Glastetter's expert evidence." *Id.* at 992.

70. Issues involving assessment of exposure are discussed in Joseph V. Rodricks, Reference Guide on Exposure Science, in this manual.

71. Anderson v. Hess Corp., 592 F. Supp. 2d 1174, 1178 (D.N.D. 2009) ("[A] plaintiff [in a toxic tort case] is not required to produce a mathematically precise table equating levels of exposure

Suppose, for example, that plaintiff alleges that her unborn child suffered injuries when her room was sprayed with an insecticide. Plaintiff's expert is prepared to testify that she relied on another expert's opinion that the insecticide can cause harm of the sort suffered by the child and that academic studies have found injuries when less than the amount sprayed in this case was used. But the expert who offered this opinion reached this conclusion without considering the size of the house, or the area treated, or how it was applied, or the amount applied to the outside of the house. And no one had measured this substance in the mother. Consequently, the court found that plaintiff had not provided adequate proof of exposure.[72]

A recent case that illustrates the complex problems that arise with exposure issues is *Henricksen v. ConocoPhilips Co.*[73] In *Henricksen*, the plaintiff who drove a gasoline tanker truck for 30 years alleged that his acute myelogenous leukemia (AML) was caused by his occupational exposure to benzene, a component of gasoline. Although some studies show that AML, or at least some forms of AML, may be caused by exposure to benzene, the same is not true with regard to gasoline. The court rejected testimony by plaintiff's experts that sought to link the exposure to the benzene in the gasoline to plaintiff's claim. There were numerous problems: Did plaintiff manifest symptoms typical of AML that was chemically induced and not idiopathic? How could one calculate how much benzene plaintiff would have been exposed to considering how many hours he worked and how the gasoline was delivered? How much benzene exposure is required to support the conclusion that general causation has been established? Each of these issues is discussed in considerable detail, suggesting that the studies that would logically be needed to conclude that the alleged exposure can be linked to causation may simply not have been done. Because the plaintiff bears the burden of proof, this means that plaintiff's experts often will be excluded.

IV. Forensic Science

To date, *Daubert* has rarely been raised in the forensic context, but this may be about to change.[74] We do not know as yet what shifts may occur in response to the National Academies' highly critical report on the forensic sciences.[75] We do know that the report played a role in the Supreme Court's opinion in *Melendez-*

with levels of harm—plaintiff must only produce evidence from which a reasonable person could conclude that the defendant's emissions probably caused the plaintiff's harms.").

72. Junk v. Terminix Int'l. Co., 594 F. Supp. 2d 1062 (S.D. Iowa 2008).

73. 605 F. Supp. 2d 1142 (E.D. Wash. 2009).

74. These issues are discussed at greater length in Paul C. Giannelli et al., Reference Guide on Forensic Identification Expertise, in this manual.

75. National Research Council, Strengthening Forensic Science in the United States: A Path Forward (2009).

Diaz v. Massachusetts[76] concerning the application of the Confrontation Clause to expert forensic testimony. But it will take some time to understand the repercussions this opinion will cause in the criminal justice system.

Even aside from this constitutional development and in the absence of congressional or other institutional action, the extensive coverage of the National Academies' report by the media and academia may bring about change. Furthermore, analysts of the more than 200 DNA exonerations to date claim that in more than 50% of the cases, invalid, or improperly conducted, or misleadingly interpreted forensic science contributed to the wrongful convictions.[77] The seriousness of these mistakes is aggravated because some of the inmates were on death row. These developments may affect judicial approaches to opinions offered by prosecution experts. Also, as judges write more sharply focused opinions in civil cases, the very different approach they use in criminal cases stands out in vivid contrast. Supposedly, the federal rules are trans-substantive, and it is certainly arguable that errors that bear on life and liberty should weigh more heavily than errors in civil cases concerned primarily with money.

To date, however, few prosecution experts have been excluded as witnesses in criminal prosecutions.[78] Usually judges have allowed them to testify or, at most, have curtailed some of the conclusions that prosecution experts sought to offer.[79] However, there are a number of issues in forensic sciences that may become the object of *Daubert* challenges.

A. Validity

As the discussion in Chapter 5 of the National Academies' report recounts, forensic fields vary considerably with regard to the quantity and quality of research done to substantiate that a given technique is capable of making reliable individualized

76. 129 S. Ct. 2527, 2536 (2009).

77. The Innocence Project, *available at* www.innocenceproject.org.

78. *See* Maryland v. Rose, Case No. K06-0545 at 31 (Balt. County Cir. Ct. Oct. 19, 2007) (excluding fingerprint evidence in a death penalty case as a "subjective, untested, unverifiable identification procedure that purports to be infallible").

79. *See, e.g.*, United States v. Green, 405 F. Supp. 2d 104 (D. Mass. 2005) (explaining that an expert would be permitted to describe similarities between shell casings but prohibited from testifying to match; Judge Gertner acknowledged that toolmark identification testimony should be excluded under *Daubert*, but that every single court post-*Daubert* admitted the testimony); United States v. Glynn, 578 F. Supp. 2d 567 (S.D.N.Y. 2008) (explaining that testimony linking bullet and casings to the defendant was inadmissible under *Daubert*, but testimony that the evidence was "more likely than not" from the firearm was admissible under Federal Rule of Evidence 401); United States v. Rutherford, 104 F. Supp. 2d 1190, 1193 (D. Neb. 2000) (handwriting experts permitted to testify to similarities between sample from defendant and document in question but not permitted to conclude that defendant was the author). *See* United States v. Rutherford, 104 F. Supp. 2d 1190, 1193 (D. Neb. 2000); United States v. Hines, 55 F. Supp. 2d 530 (D. Md. 2002).

identifications. Non-DNA forensic techniques often turn on subjective analyses.[80] But making *Daubert* objections in these fields requires defense counsel to understand in detail how the particular technique works, as well as to be knowledgeable about the scientific method and statistical issues.[81]

B. *Proficiency*

Non-DNA forensic techniques often rely on subjective judgments, and the proficiency of the expert to make such judgments may become the focus of a *Daubert* challenge. In theory, proficiency tests could determine whether well-trained experts in those fields can reach results with low error rates. In practice, however, there are numerous obstacles to such tests. Sophisticated proficiency tests are difficult and expensive to design. If the tests are too easy, the results will not assess the ability of examiners to draw correct conclusions when forensic evidence presents a difficult challenge in identifying a specific individual or source.[82] Furthermore, in many jurisdictions, forensic examiners are not independent of law enforcement agencies and/or prosecutors' offices and can often obtain information about a proficiency testing program through those sources.

C. *Malfunctioning Laboratories*

Numerous problems have been identified in crime laboratories ranging from uncertified laboratory professionals and unaccredited laboratories performing incompetent work to acts of deliberate fraud, such as providing falsified results from tests that were never done.[83] Although outright fraud may be rare, unintended inaccurate results that stem from inadequate supervision, training, and record keeping, failure to prevent contamination, and failure to follow proper statistical procedures can have devastating effects. Evidence that a laboratory has engaged in such practices should certainly lead to *Daubert* challenges for lack of reliability, but this requires that such investigations be undertaken and the defense have access to the results. Whether courts can be persuaded to almost automatically reject laboratory results in the absence of proper accreditation of laboratories and certification

80. *See* National Research Council, *supra* note 75, at 133.

81. Specific forensic science techniques are discussed in Paul C. Giannelli et al., Reference Guide on Forensic Identification Expertise, Sections V–X, in this manual.

82. United States v. Llera Plaza, 188 F. Supp. 2d 549 (E.D. Pa. 2002) (court acknowledged that defense raised real questions about the adequacy of proficiency tests taken by FBI fingerprint examiners but concluded that fingerprint testimony satisfied *Daubert* in part because no examples were shown of erroneous identifications by FBI examiners). An erroneous FBI identification was made in the Brandon Mayfield case discussed in the introduction to *Strengthening Forensic Science in the United States*, *supra* note 75, at 45–46.

83. *See* National Research Council, *supra* note 75, at 183–215 and Paul C. Giannelli et al., Reference Guide on Forensic Identification Expertise, Section IV, in this manual.

of forensic practitioners remains to be seen. Laboratory techniques, such as drug analyses, that do not suffer from the same uncertainties regarding validity as the forensic identification techniques can, of course, also produce erroneous results if the laboratory is failing to follow proper procedures.

D. Interpretation

Forensic techniques that rest on subjective judgments are susceptible to cognitive biases.[84] We have seen instances of contextual bias, but as yet there has been little research on contextual or other types of cognitive bias. We do not yet know whether courts will consider this type of evidence when expertise is challenged.

E. Testimony

Defense counsel may of course object to testimony that a prosecution expert seeks to give. When the prosecution relies on a subjective identification technique, lawyers for the defense should attempt to clarify what "match" means if the expert uses this terminology and to explain to the jury that studies to date do not permit conclusions about individualization. To do this, the defense may have to call its own experts and ask for jury instructions. Defense counsel must also remain alert and object to prosecution testimony in which the witness claims to know probabilities—that have not been established in a particular field—on the basis of extensive personal experience. Objections also should be raised to testimony about zero error rates. The defense must also remember that the *Daubert* opinion itself recognized that testimony can be excluded under Federal Rule of Evidence 403 if its prejudicial effect substantially outweighs its probative value.

F. Assistance for the Defense and Judges

Perhaps the most troubling aspect of trying to apply *Daubert* to forensic evidence is that very few defense counsel are equipped to take on this challenge. Such counsel lack the training and resources to educate judges on these complex issues. Judges in the state criminal justice system that handle the great majority of criminal cases often have overloaded dockets and little or no assistance. Whether a defendant in a particular case is constitutionally entitled to expert assistance is a complicated issue that defense counsel needs to explore.[85] Possibly the best chance for the defense to get meaningful help that also would assist the court is to get pro bono assistance

84. National Research Council, *supra* note 75, at 184–185.

85. *See* Ake v. Oklahoma, 470 U.S. 68 (1985) (recognizing indigent's right to psychiatric expert assistance in a capital case in which defendant raised insanity defense). Jurisdictions differ widely in how they interpret *Ake*.

from other counsel who are knowledgeable about *Daubert* and have a sophisticated understanding of statistical reasoning. Lawyers who have handled complex issues about causation may be able to transfer their expertise to other difficult issues relating to expert testimony.[86] Judges might also consider asking for amicus briefs from appropriate organizations or governmental units.

G. Confrontation Clause

The majority in *Melendez-Diaz v. Massachusetts*, in an opinion by Justice Scalia over a strong dissent by Justice Kennedy, held that the defendant has a constitutional right to demand that a forensic analyst whose conclusions the prosecution wishes to introduce into evidence must be produced in court for cross-examination. In a drug case, for example, the prosecution may not simply introduce a report or an affidavit from the analyst if the defendant demands production of the analyst for cross-examination. When the analyst is produced, this will gave the defense the opportunity through cross-examination to raise questions about fraud, incompetence, and carelessness and to ask questions about laboratory procedures and other issues discussed in the National Research Council report. Effective cross-examination will demand of defense counsel the same type of expertise needed to succeed on *Daubert* challenges. Numerous unanswered questions about the operation of *Melendez-Diaz* will have to be litigated. It remains to be seen how often, if at all, defense counsel will take advantage of the Confrontation Clause or whether they will waive the defendant's right to confront expert witnesses.[87]

V. Procedural Context

Apart from their effect on admissibility of expert testimony, *Daubert* and its subsequent interpretations have also affected the broader context in which such cases are litigated and have altered the role of testifying experts in the pretrial stages of litigation.

A. Class Certification Proceedings

One question that arises with increasing frequency is whether and how *Daubert* is to be applied at class certification proceedings. The problem arises because of the commonality and predominance requirements in Rule 23(a) of the Federal Rules

86. *Cf.* Kitzmiller v. Dover Area School Dist., 400 F. Supp. 2d 707 (M.D. Pa. 2005) (attorneys who specialized in defense product liability litigation and had expertise about the nature of science participated in case objecting to teaching intelligent design in public schools).

87. Both defendants and prosecutors face concerns about the resources required to fully implement such protections. *See* National Research Council, *supra* note 75, at 187.

of Civil Procedure and has emerged with regard to a wide variety of substantive claims that plaintiffs seek to bring as a class action. For example, in *Sanneman v. Chrysler Corp.*,[88] plaintiff sought class certification of a common-law fraud action and a breach-of-warranty action, the gist of which was "that Chrysler had fraudulently concealed a paint defect in many of the vehicles it manufactured beginning on or about 1990."[89] Plaintiff's expert testified at the class certification hearing that the paint problem is always caused by ultraviolet rays, but acknowledged "that other causes may contribute to or exacerbate the problem."[90] After oral argument, the court concluded that plaintiff's expert's testimony satisfied *Daubert,* but because ultraviolet rays are not always the only cause of problems with paint, proof of damages would probably have to be made vehicle by vehicle. The motion for class certification was therefore denied. *Daubert* challenges have been raised to class certification in numerous other cases.[91]

As of this writing, there is a decided trend toward rejecting class certification on the ground that plaintiff's proffered expert testimony does not satisfy the Rule 23(a) requirements, although the circuits are not unanimous in how rigorous the examination of expert proof needs to be. Must the expert testimony be subjected to the same rigorous scrutiny to determine whether it is relevant and reliable as when the issue is admissibility at trial, or is a less searching analysis appropriate at the certification stage? In other words, should the trial judge conduct a *Daubert* hearing and analysis identical to that undertaken when a defendant seeks to preclude a plaintiff's witness from testifying at trial? Not only "should" the trial judge conduct a *Daubert* hearing, but, as the Seventh Circuit has ruled in *American Honda*, the trial judge "must" do so. If a full *Daubert* hearing is required in every class certification case, what has happened to the broad and case-familiar discretion that a trial judge is supposed to exercise?"

The trial judge in *Rhodes v. E.I. du Pont de Nemours & Co.*[92] concluded that the expert opinions offered in support of class certification should be subjected to a full-scale *Daubert* analysis, including a *Daubert* hearing. The judge explained

88. 191 F.R.D. 441 (E.D. Pa. 2000).

89. *Id.* at 443.

90. *Id.* at 451.

91. *See, e.g.,* Blades v. Monsanto Co., 400 F.3d 562 (8th Cir. 2005) (antitrust price-fixing conspiracy); Rhodes v. E.I. du Pont de Nemours & Co., 2008 WL 2400944 (S.D. W. Va. June 11, 2008) (medical monitoring claim in toxic tort action); Gutierrez v. Johnson & Johnson, 2006 WL 3246605 (D.N.J. Nov. 6, 2006) (employment discrimination); Nichols v. SmithKline Beecham Corp., 2003 WL 302352 (E.D. Pa. Jan. 29, 2003) (violation of Sherman Antitrust Act); *In re* St. Jude Med., Inc., 2003 WL 1589527 (D. Minn. Mar. 27, 2003) (product liability action); Bacon v. Honda of Am. Mfg Inc., 205 F.R.D. 466 (S.D. Ohio 2001) (same); Midwestern Mach v. Northwest Airlines, Inc., 211 F.R.D. 562 (D. Minn. 2001) (violation of Clayton Act); *In re* Polypropylene Carpet, 996 F. Supp. 18 (N.D. Ga. 1997) (same); *In re* Monosodium Glutamate, 205 F.R.D. 229 (D. Minn. 2001).

92. 2008 WL 2400944 (S.D. W. Va. June 11, 2008). *See also* American Honda Motor Co. v. Allen, 600 F.3d 813, 816 (7th Cir. 2010) (district court must perform a full *Daubert* analysis before certifying a class action where the expert's report or testimony is critical to class certification).

that decisions that see a more limited role for *Daubert* in class certification hearings stem in part from misinterpreting the Supreme Court's opinion in *Eisen v. Carlisle & Jacquelin.*[93] In *Eisen*, which predated *Daubert* by 19 years, the Court instructed district courts to refrain from conducting "a preliminary inquiry into the merits of a proposed class action" when they consider certification.[94] At this time, only the Ninth Circuit forbids the lower courts from examining evidence that relates to the merits and from requiring a rigorous examination of the expert testimony and Rule 23(a) requirements.[95] The *Rhodes* case deplored this approach because the overwhelming majority of class actions settle and therefore allowing the action to proceed as a class action "might invite plaintiffs to seek class status for settlement purposes." On the other hand, knocking out the possibility of class certification early in the proceedings affects the possibility of settling cases in which liability is debatable. A possible compromise is partial certification that would allow a common issue to be established at a class trial, leaving individual issues for separate proceedings.

B. Discovery

1. Amended discovery rules

Rule 26 of the Federal Rules of Civil Procedure—the core rule on civil discovery—was amended in 1993 more or less contemporaneously with *Daubert* to allow judges to exert greater control of expert testimony. Those amendments required experts retained or specially employed to provide expert testimony, or whose duties as the party's employee regularly involve giving expert testimony, to furnish an extensive report prior to his or her deposition.[96] These reports were required to indicate

93. 417 U.S. 156 (1974).

94. *Id.* at 177–78.

95. *See* Dukes v. Wal-Mart, Inc., 474 F.3d 1214 (9th Cir. 2007). The Supreme Court declined an opportunity to address the role of *Daubert* in class certification when it granted certiorari in *Dukes*, even though the issue was raised in some of the petitions. The Court subsequently granted a petition for certiorari in *Erica P. John Fund Inc. v. Halliburton Co.* (U.S. Jan. 7, 2011) (No. 09-1403), which raises related questions regarding the extent to which the district court may consider the merits of the underlying litigation and require that loss causation be demonstrated by a preponderance of admissible evidence at the class certification stage under Federal Rule of Civil Procedure 23. Other courts accord *Daubert* a limited role, such as requiring the trial judge to determine only that the expert testimony is "not fatally flawed." *See* Fogarazzo v. Lehman Bros., Inc., 2005 WL 361205 (S.D.N.Y. Feb. 16, 2005).

96. Fed R. Civ. P. 26(a)(2)(B), as amended December 1, 2010, made substantial changes to the 1993 amendments. The 1993 amendments also recognized a second category of testifying experts who were not retained or specially employed in anticipation of litigation, such as treating physicians, who were not required to provide reports. *But see* 3M v. Signtech USA, 177 F.R.D. 459 (D. Minn. 1998) (requiring report from employee experts who do not regularly provide expert testimony because it eliminates surprise and is consistent with the spirit of Rule 26(a)(2)(B)). Under the 2010 amendments the attorney must submit a report indicating the subject matter and the facts and opinions to which an unretained testifying expert is expected to testify. Fed. R. Civ. P. 26(a)(2)(C) (amended Dec. 1, 2010).

"the data *or other information* considered by the expert witness in forming the opinions" (emphasis added). Many, although not all, courts construed this language as opening the door to discovery of anything conveyed by counsel to the expert.[97] Courts taking this approach found that all communications between counsel and experts were discoverable even if the communication was opinion work product. In other words, these courts found that the protection for opinion work product in Rule 26(b)(3) was trumped by the disclosure provisions in Rule 26(a)(2)(B). These courts also required disclosure of all the expert's draft reports and notes.

Trigon Ins. Co. v. United States,[98] went a step further. It held that drafts prepared with the assistance of consultants who would not testify, as well as all communications between the consultants and the experts, including e-mails, were discoverable. In *Trigon,* many of these materials had been destroyed. The court ordered the defendant to hire an outside technology consultant to retrieve as much of these data as possible, allowed adverse inferences to be drawn against the defendant, and awarded more than $179,000 in fees and costs to plaintiff.[99]

Those who favor the free discovery of communications between counsel and experts and draft reports justified these results as shedding light on whether the expert's opinions are his or her own or those of counsel. Critics of this approach found it costly and time-consuming and point out that lawyers have developed strategies to overcome transparency, such as retaining two sets of experts—one to consult and the other to testify—which makes discovery even more expensive.

After a series of public hearings the Advisory Committee on Civil Rules determined that the disclosure rules increased the cost of litigation with no offsetting advantage to the conduct of litigation. The report of the Advisory Committee noted that such an extensive inquiry into expert communications with attorneys did not lead to better testing of expert opinions "because attorneys and expert witnesses go to great lengths to forestall discovery."[100]

Under amended rules that became effective in December 2010, disclosure is limited to "the facts or data" considered by the expert, and does not extend to "other information." Draft reports are no longer discoverable, and communications between counsel and an expert are protected from discovery unless the communications: (1) relate to compensation for the expert's study or testimony;

97. *See* Karn v. Ingersoll Rand, 168 F.R.D. 633 (N.D. Ind. 1996) (requiring disclosure of all documents reviewed by experts in forming their opinions); Reg'l Airport Auth. v. LFG, LLC, 460 F.3d 697, 716 (6th Cir. 2006) ("other information" interpreted to include all communications by counsel to expert).

98. 204 F.R.D. 277 (E.D. Va. 2002).

99. *Id. See also* Semtech Corp. v. Royal Ins. Co., 2007 WL 5462339 (C.D. Cal. Oct. 24, 2007) (explaining that preclusion of expert from testifying for failure to disclose drafts and failing to disclose input of counsel at hearing made it impossible to discern the basis for his opinion).

100. Report of the Civil Rules Advisory Committee, from Honorable Mark R. Kravitz, Chair, Advisory Committee on Federal Rules of Civil Procedure, to Honorable Lee H. Rosenthal, Chair, Standing Committee on Rules of Practice and Procedure (May 8, 2008), *available at* http://www.uscourts.gov/uscourts/RulesAndPolicies/rules/Reports/CV05-2009.pdf.

(2) identify facts or data provided by counsel and considered by the expert; or (3) identify assumptions furnished by counsel that the expert relied upon in forming opinions. Testifying experts who were not required to provide a report under the previous rules—such as treating physicians—are now required to provide a summary of the facts or opinions to which the witness expects to testify. While this requirement relating to experts not required to file a report would provide more disclosure than under the 1993 amendments, the main thrust of the 2010 amendments is to narrow expert discovery with an eye toward minimizing expense and focusing attention on the expert's opinion.

Nothing in the amendments precludes asking an expert at a deposition to explain the bases or foundations for his or her opinions or asking whether the expert considered other possible approaches, but inquiries into counsel's input would be severely curtailed. Aside from communications with counsel relating to compensation, or inquiring into "facts or data" provided by counsel that the expert considered, the expert may also be asked if counsel furnished him or her with assumptions on which he or she relied. Now that the amended rules have become effective, it remains to be seen how broadly courts and magistrates will interpret the "assumptions" provision. Are there instances in which it will be inferred that counsel was seeking to have the expert make an assumption although this was never explicitly stated? Those who think more transparency is desirable in dealing with expert testimony will certainly push to expand this category. Whether these amendments if adopted can constrain the gamesmanship that surrounds expert testimony remains to be seen.

2. *E-discovery*

Also uncertain is whether experts will be needed to determine the proper scope of e-discovery. Rule 26(b)(2)(B) provides the following:

> A party need not provide discovery of electronically stored information from sources that the party identifies as not reasonably accessible because of undue burden or cost.

The burden is on the party from whom discovery is sought to show this undue burden or cost, but the court may nevertheless order discovery if the requesting party can show good cause.

May the requesting party making a motion to compel proffer expert testimony to show that the requested information would have been readily accessible if the party with the information had used a different search methodology? Recent opinions by a magistrate judge so suggest.[101] Magistrate Judge John Facciola notes that "[w]hether search terms or 'keywords' will yield the information sought is a complicated question involving the interplay, at least, of the sciences of computer

101. *See e.g.* United States v. O'Keefe, 537 F. Supp. 2d 14 (D.D.C. 2008); Equity Analytics, LLC v. Lunden, 248 F.R.D. 331 (D.D.C. 2008).

technology, statistics and linguistics. . . . This topic is clearly beyond the ken of a layman and requires that any conclusion be based on evidence that, for example, meets the criteria of Rule 702 of the Federal Rules of Evidence."[102]

Superimposing *Daubert* hearings on top of e-discovery proceedings will make an already costly procedure even more costly, one of the consequences that Rule 26(b)(2)(B) seeks to avoid. On the other hand, a search that would not lead to the information sought defeats the objectives of discovery. A helpful opinion on how these factors should be balanced that examines the issues a court must consider can be found in *Victor Shirley, Inc. v. Creative Pipe, Inc.*,[103] which also contains a very brief overview of the various techniques for conducting searches of electronically stored information. A court may well require technical assistance in dealing with these issues. In some instances, a court-appointed expert or a special master appointed pursuant to Rule 53 of the Federal Rules of Civil Procedure might be more desirable than a full-fledged *Daubert* battle among experts, particularly if one of the parties has far fewer resources than its opponent.

C. Daubert *Hearings*

When a *Daubert* issue arises, the trial court has discretion about how to proceed.[104] It need not grant an evidentiary hearing and has leeway to decide when and how issues about the admissibility of expert testimony should be determined. The burden is on the parties to persuade the court that a particular procedure is needed.[105]

The generally unfettered power of the trial judge to make choices emerges clearly if we look at *United States v. Nacchio*,[106] a criminal case. The defendant claimed that the trial judge erred in granting the government's *Daubert* motion to exclude his expert in the middle of the trial without an evidentiary hearing, leading to his conviction. On appeal, a divided panel of the Tenth Circuit reversed on the ground that the expert testimony had been improperly excluded and remanded for a new trial. After a rehearing, the conviction was reinstated in a 5-4 opinion. The majority rejected the defense's central argument that the court had to take into account that this was a criminal case; the majority saw this purely as a *Daubert* issue and found that the burden of satisfying *Daubert* and convincing the trial judge to hold a hearing rested solely on the defendant. Although there may be some cases in which a reviewing court would find that the trial court abused its discretion in the procedures it used in handling a *Daubert* motion,[107] this has

102. *See Equity Analytics*, 248 F.R.D. at 333.

103. 250 F.R.D. 251 (D. Md. 2008).

104. Kumho Tire Co. v. Carmichael, 526 U.S. at 137, 150 (1999).

105. For example, in the government's RICO tobacco case, all *Daubert* issues were decided on the papers without any testimony being presented. United States v. Phillip Morris Inc., 2002 WL 34233441, at *1 (D.D.C. Sept. 30, 2002).

106. 555 F.3d 1234 (10th Cir. 2009).

107. *See* Padillas v. Stork-Gamco, Inc., 186 F.3d 412 (3d Cir. 1999).

become more and more unlikely in civil cases as *Daubert* rulings have accumulated and courts increasingly expect litigators to understand their obligations.

VI. Conclusion

The *Daubert* trilogy has dramatically changed the legal landscape with regard to expert witness testimony. The Supreme Court attempted in *Daubert* to articulate basic principles to guide trial judges in making decisions about the admissibility of complex scientific and technological expert testimony. Unfortunately, the *Daubert* trilogy has, in actuality, spawned a huge, and expensive, new subject of litigation and have left many procedural and substantive questions unanswered. Moreover, there are serious concerns about whether the guidelines enunciated by the Court have been interpreted by lower courts to limit, rather than respect, the discretion of trial judges to manage their complex cases, whether the guidelines conflict with the preference for admissibility contained in both the Federal Rules of Evidence and *Daubert* itself, and whether the guidelines have resulted in trial judges encroaching on the province of the jury to decide highly contested factual issues and to judge the overall credibility of expert witnesses and their scientific theories. Perhaps most disturbingly, there are serious concerns on the part of many scientists as to whether the courts are, as *Daubert* prescribed, making admissibility decisions—decisions that may well determine the ultimate outcome of a case—which are in fact "ground[ed] in the methods and procedures of science."[108]

108. Daubert v. Merrill Dow Pharms., 509 U.S. at 579, 590 (1993).

How Science Works

DAVID GOODSTEIN

David Goodstein, Ph.D., is Professor of Physics and Applied Physics, and the Frank J. Gilloon Distinguished Teaching and Service Professor, Emeritus, California Institute of Technology, Pasadena, California, where he also served for 20 years as Vice Provost.

CONTENTS

I. Introduction

Recent Supreme Court decisions have put judges in the position of having to decide what is scientific and what is not.[1] Some judges may not be entirely comfortable making such decisions, despite the guidance supplied by the Court and illuminated by learned commentators.[2] The purpose of this chapter is not to resolve the practical difficulties that judges will encounter in reaching those decisions; it is to demystify somewhat the business of science and to help judges understand the *Daubert* decision, at least as it appears to a scientist. In the hope of accomplishing these tasks, I take a mildly irreverent look at some formidable subjects. I hope the reader will accept this chapter in that spirit.

II. A Bit of History

Modern science can reasonably be said to have come into being during the time of Queen Elizabeth I of England and William Shakespeare. Almost immediately, it came into conflict with the law.

While Shakespeare was composing his sonnets and penning his plays in England, Galileo Galilei in Italy was inventing the idea that careful experiments in a laboratory could reveal universal truths about the way objects move through space. A bit later, after hearing about the newly invented telescope, he made one for himself, and with it he made discoveries in the heavens that astonished and thrilled all of Europe. Nonetheless, in 1633, Galileo was put on trial for his scientific teachings. The trial of Galileo is usually portrayed as a conflict between science and the Roman Catholic Church, but it was, after all, a trial, with judges and lawyers, and all the other trappings of a formal legal procedure.

Another great scientist of the day, William Harvey, who discovered the circulation of blood, worked not only at the same time as Galileo, but also at the same place—the University of Padua, not far from Venice. If you visit the University of Padua today and tour the old campus at the heart of the city, you will be shown Galileo's *cattedra,* the wooden pulpit from which he lectured (and curiously, one of his vertebrae in a display case just outside the rector's office—maybe the rector needs to be reminded to have a little spine). You will also be shown the lecture

1. These Supreme Court decisions are discussed in Margaret A. Berger, The Admissibility of Expert Testimony, Sections II–III, IV.A, in this manual. For a discussion of the difficulty in distinguishing between science and engineering, see Channing R. Robertson et al., Reference Guide on Engineering, in this manual.

2. Since publication of the first edition of this manual, a number of works have been developed to assist judges and attorneys in understanding a wide range of scientific evidence. *See, e.g.,* 1 & 2 Modern Scientific Evidence: The Law and Science of Expert Testimony (David L. Faigman et al. eds., 1997); Expert Evidence: A Practitioner's Guide to Law, Science, and the FJC Manual (Bert Black & Patrick W. Lee eds., 1997).

theater in which Harvey dissected cadavers while eager students peered downward from tiers of overhanging balconies. Because dissecting cadavers was illegal in Harvey's time, the floor of the theater was equipped with a mechanism that whisked the body out of sight when a lookout gave the word that the authorities were coming. Obviously, both science and the law have changed a great deal since the seventeenth century.

Another important player who lived in the same era was not a scientist at all, but a lawyer who rose to be Lord Chancellor of England in the reign of Elizabeth's successor, James I. His name was Sir Francis Bacon, and in his magnum opus, which he called *Novum Organum,* he put forth the first theory of the scientific method. In Bacon's view, the scientist should be an impartial observer of nature, collecting observations with a mind cleansed of harmful preconceptions that might cause error to creep into the scientific record. Once enough such observations were gathered, patterns would emerge, giving rise to truths about nature.

Bacon's theory has been remarkably influential down through the centuries, even though in his own time there were those who knew better. "That's exactly how a Lord Chancellor *would* do science," William Harvey is said to have grumbled.

III. Theories of Science

Today, in contrast to the seventeenth century, few would deny the central importance of science to our lives, but not many would be able to give a good account of what science is. To most, the word probably brings to mind not science itself, but the fruits of science, the pervasive complex of technology and discoveries that has transformed all of our lives. However, science also might equally be thought to include the vast body of knowledge we have accumulated about the natural world. There are still mysteries, and there always will be mysteries, but the fact is that, by and large, we understand how nature works.

A. Francis Bacon's Scientific Method

But science is even more than that. Ask a scientist what science is, and the answer will almost surely be that it is a process—a way of examining the natural world and discovering important truths about it. In short, the essence of science is the scientific method.[3]

3. The Supreme Court, in *Daubert v. Merrell Dow Pharmaceuticals, Inc.*, acknowledged the importance of defining science in terms of its methods as follows: "'Science is not an encyclopedic body of knowledge about the universe. Instead, it represents a *process* for proposing and refining theoretical explanations about the world that are subject to further testing and refinement'" (emphasis in original). 509 U.S. 579, 590 (1993) (quoting Brief for the American Association for the Advancement of Science and the National Academy of Sciences as Amici Curiae at 7–8).

This stirring description suffers from an important shortcoming. We do not really know what the scientific method is.[4] There have been many attempts at formulating a general theory of how science works, or at least how it should work, starting, as we have seen, with the theory of Sir Francis Bacon. But Bacon's idea, that science proceeds through the collection of observations without prejudice, has been rejected by all serious thinkers. Everything about the way we do science—the language we use, the instruments we use, the methods we use—depends on clear presuppositions about how the world works. Modern science is full of things that cannot be observed at all, such as force fields and complex molecules. At the most fundamental level, it is impossible to observe nature without having some reason to choose what is and is not worth observing. Once that elementary choice is choice is made, Bacon has been left behind.

B. Karl Popper's Falsification Theory

Over the past century, the ideas of the Vienna-born philosopher Sir Karl Popper have had a profound effect on theories of the scientific method.[5] In contrast to Bacon, Popper believed that all science begins with a prejudice, or perhaps more politely, a theory or hypothesis. Nobody can say where the theory comes from. Formulating the theory is the creative part of science, and it cannot be analyzed within the realm of philosophy. However, once the theory is in hand, Popper tells us, it is the duty of the scientist to extract from it logical but unexpected predictions that, if they are shown by experiment not to be correct, will serve to render the theory invalid.

Popper was deeply influenced by the fact that a theory can never be proved right by agreement with observation, but it can be proved wrong by disagreement with observation. Because of this asymmetry, science uniquely makes progress by proving that good ideas are wrong so that they can be replaced by even better ideas. Thus, Bacon's impartial observer of nature is replaced by Popper's skeptical theorist. The good Popperian scientist somehow comes up with a hypothesis that fits all or most of the known facts, then proceeds to attack that hypothesis at its weakest point by extracting from it predictions that can be shown to be false. This process is known as falsification.[6]

4. For a general discussion of theories of the scientific method, see Alan F. Chalmers, What Is This Thing Called Science? (1982). For a discussion of the ethical implications of the various theories, see James Woodward & David Goodstein, *Conduct, Misconduct and the Structure of Science*, 84 Am. Scientist 479 (1996).

5. *See, e.g.*, Karl R. Popper, The Logic of Scientific Discovery (Karl R. Popper trans., 1959).

6. The Supreme Court in *Daubert* recognized Popper's conceptualization of scientific knowledge by noting that "[o]rdinarily, a key question to be answered in determining whether a theory or technique is scientific knowledge that will assist the trier of fact will be whether it can be (and has been) tested." 509 U.S. at 593. In support of this point, the Court cited as parenthetical passages from both Carl Gustav Hempel, Philosophy of Natural Science 49 (1966) ("'[T]he statements constituting

Popper's ideas have been fruitful in weaning the philosophy of science away from the Baconian view and some other earlier theories, but they fall short in a number of ways in describing correctly how science works. The first of these is the observation that, although it may be impossible to prove a theory is true by observation or experiment, it is as almost equally impossible to prove one is false by these same methods. Almost without exception, in order to extract a falsifiable prediction from a theory, it is necessary to make additional assumptions beyond the theory itself. Then, when the prediction turns out to be false, it may well be one of the other assumptions, rather than the theory itself, that is false. To take a simple example, early in the twentieth century it was found that the orbits of the outermost planets did not quite obey the predictions of Newton's laws of gravity and mechanics. Rather than take this to be a falsification of Newton's laws, astronomers concluded that the orbits were being perturbed by an additional unseen body out there. They were right. That is precisely how Pluto was discovered.

The apparent asymmetry between falsification and verification that lies at the heart of Popper's theory thus vanishes. But the difficulties with Popper's view go even beyond that problem. It takes a great deal of hard work to come up with a new theory that is consistent with nearly everything that is known in any area of science. Popper's notion that the scientist's duty is then to attack that theory at its most vulnerable point is fundamentally inconsistent with human nature. It would be impossible to invest the enormous amount of time and energy necessary to develop a new theory in any part of modern science if the primary purpose of all that work was to show that the theory was wrong.

This point is underlined by the fact that the behavior of the scientific community is not consistent with Popper's notion of how it should be. Credit in science is most often given for offering correct theories, not wrong ones, or for demonstrating the correctness of unexpected predictions, not for falsifying them. I know of no example of a Nobel Prize awarded to a scientist for falsifying his or her own theory.

C. Thomas Kuhn's Paradigm Shifts

Another towering figure in the twentieth century theory of science is Thomas Kuhn.[7] Kuhn was not a philosopher but a historian (more accurately, a physicist who retrained himself as a historian). It is Kuhn who popularized the word *paradigm*, which has today come to seem so inescapable.

A paradigm, for Kuhn, is a kind of consensual worldview within which scientists work. It comprises an agreed-upon set of assumptions, methods, language, and

a scientific explanation must be capable of empirical test'") and Karl R. Popper, Conjectures and Refutations: The Growth of Scientific Knowledge 37 (5th ed. 1989) ("'[T]he criterion of the scientific status of a theory is its falsifiability, or refutability, or testability'").

7. Thomas S. Kuhn, The Structure of Scientific Revolutions (1962).

everything else needed to do science. Within a given paradigm, scientists make steady, incremental progress, doing what Kuhn calls "normal science."

As time goes on, difficulties and contradictions arise that cannot be resolved, but the tendency among scientists is to resist acknowledging them. One way or another they are swept under the rug, rather than being allowed to threaten the central paradigm. However, at a certain point, enough of these difficulties accumulate to make the situation intolerable. At that point, a scientific revolution occurs, shattering the paradigm and replacing it with an entirely new one.

This new paradigm, says Kuhn, is so radically different from the old that normal discourse between the practitioners of the two paradigms becomes impossible. They view the world in different ways and speak different languages. It is not even possible to tell which of the two paradigms is superior, because they address different sets of problems. They are incommensurate. Thus, science does not progress incrementally, as the science textbooks would have it, except during periods of normal science. Every once in a while, a scientific revolution brings about a paradigm shift, and science heads off in an entirely new direction.

Kuhn's view was formed largely on the basis of two important historical revolutions. One was the original scientific revolution that started with Nicolaus Copernicus and culminated with the new mechanics of Isaac Newton. The very word *revolution*, whether it refers to the scientific kind, the political kind, or any other kind, refers metaphorically to the revolutions in the heavens that Copernicus described in a book, *De Revolutionibus Orbium Caelestium*, published as he lay dying in 1543.[8] Before Copernicus, the dominant paradigm was the worldview of ancient Greek philosophy, frozen in the fourth century B.C.E. ideas of Plato and Aristotle. After Newton, whose masterwork, *Philosophiæ Naturalis Principia Mathematica*, was published in 1687, every scientist was a Newtonian, and Aristotelianism was banished forever from the world stage. It is even possible that Sir Francis Bacon's disinterested observer was a reaction to Aristotelian authority. Look to nature, not to the ancient texts, Bacon may have been saying.

The second revolution that served as an example for Kuhn occurred early in the twentieth century. In a headlong series of events that lasted a mere 25 years, the Newtonian paradigm was overturned and replaced with the new physics, in the form of quantum mechanics and Einstein's theories of special and general relativity. This second revolution, although it happened much faster, was no less profound than the first.

The idea that science proceeds by periods of normal activity punctuated by shattering breakthroughs that make scientists rethink the whole problem is an appealing one, especially to the scientists themselves, who believe from personal experience that it really happens that way. Kuhn's contribution is important. It offers us a useful context (a paradigm, one might say) for organizing the entire history of science.

8. I. Bernard Cohen, Revolution in Science (1985).

Nonetheless, Kuhn's theory does suffer from a number of shortcomings. One of them is that it contains no measure of how big the change must be in order to qualify as a revolution or paradigm shift. Most scientists will say that there is a paradigm shift in their laboratory every 6 months or so (or at least every time it becomes necessary to write another proposal for research support). That is not exactly what Kuhn had in mind.

Another difficulty is that even when a paradigm shift is truly profound, the paradigms it separates are not necessarily incommensurate. The new sciences of quantum mechanics and relativity, for example, did indeed show that Newton's laws of mechanics were not the most fundamental laws of nature. However, they did not show that they were wrong. Quite the contrary, they showed why Newton's laws were right: Newton's laws arose out of newly discovered laws that were even deeper and that covered a wider range of circumstances unimagined by Newton and his followers—that is, phenomena as small as atoms, or nearly as fast as the speed of light, or as dense as black holes. In our more familiar realms of experience, Newton's laws go on working just as well as they always did. Thus, there is no quarrel and no ambiguity at all about which paradigm is "better." The new laws of quantum mechanics and relativity subsume and enhance the older Newtonian worldview.

D. An Evolved Theory of Science

If neither Bacon nor Popper nor Kuhn gives us a perfect description of what science is or how it works, all three of them help us to gain a much deeper understanding of it.

Scientists are not Baconian observers of nature, but all scientists become Baconians when it comes to describing their observations. With very few exceptions, scientists are rigorously, even passionately, honest about reporting scientific results and how they were obtained. Scientific data are the coin of the realm in science, and they are always treated with reverence. Those rare instances in which scientists are found to have fabricated or altered their data in some way are always traumatic scandals of the first order.[9]

Scientists are also not Popperian falsifiers of their own theories, but they do not have to be. They do not work in isolation. If a scientist has a rival with a different theory of the same phenomenon, the rival will be more than happy to perform the Popperian duty of attacking the scientist's theory at its weakest point.

9. Such instances are discussed in David Goodstein, *Scientific Fraud*, 60 Am. Scholar 505 (1991). For a summary of recent investigations into scientific fraud and lesser instances of scientific misconduct, see Office of Research Integrity, Department of Health and Human Services, Scientific Misconduct Investigations: 1993–1997, http://ori.dhhs.gov/PDF/scientific.pdf (last visited Nov. 21, 1999) (summarizing 150 scientific misconduct investigations closed by the Office of Research Integrity).

Moreover, if falsification is no more definitive than verification, and scientists prefer in any case to be right rather than wrong, they nonetheless know how to hold verification to a very high standard. If a theory makes novel and unexpected predictions, and those predictions are verified by experiments that reveal new and useful or interesting phenomena, then the chances that the theory is correct are greatly enhanced. And, even if it is not correct, it has been fruitful in the sense that it has led to the discovery of previously unknown phenomena that might prove useful in themselves and that will have to be explained by the next theory that comes along.

Finally, science does not, as Kuhn seemed to think, periodically self-destruct and need to start over again. It does, however, undergo startling changes of perspective that lead to new and, invariably, better ways of understanding the world. Thus, although science does not proceed smoothly and incrementally, it is one of the few areas of human endeavor that is genuinely progressive. There is no doubt at all that the quality of twentieth century science is better than nineteenth century science, and we can be absolutely confident that the quality of science in the twenty-first century will be better still. One cannot say the same about, say, art or literature.[10]

To all of this, a few things must be added. The first is that science is, above all, an adversarial process. It is an arena in which ideas do battle, with observations and data the tools of combat. The scientific debate is very different from what happens in a court of law, but just as in the law, it is crucial that every idea receive the most vigorous possible advocacy, just in case it might be right. Thus, the Popperian ideal of holding one's hypothesis in a skeptical and tentative way is not merely inconsistent with reality; it would be harmful to science if it were pursued. As will be discussed shortly, not only ideas, but the scientists themselves, engage in endless competition according to rules that, although they are not written down, are nevertheless complex and binding.

In the competition among ideas, the institution of peer review plays a central role. Scientific articles submitted for publication and proposals for funding often are sent to anonymous experts in the field, in other words, to peers of the author, for review. Peer review works superbly to separate valid science from nonsense,

10. The law, too, can claim to be progressive. The development of legal constructs, such as due process, equal protection, and individual privacy, reflects notable progress in the betterment of mankind. *See* Laura Kalman, The Strange Career of Legal Liberalism 2–4 (1996) (recognizing the "faith" of legal liberalists in the use of law as an engine for progressive social change in favor of society's disadvantaged). Such progress is measured by a less precise form of social judgment than the consensus that develops regarding scientific progress. *See* Steven Goldberg, *The Reluctant Embrace: Law and Science in America,* 75 Geo. L.J. 1341, 1346 (1987) ("Social judgments, however imprecise, can sometimes be reached on legal outcomes. If a court's decision appears to lead to a sudden surge in the crime rate, it may be judged wrong. If it appears to lead to new opportunities for millions of citizens, it may be judged right. The law does gradually change to reflect this kind of social testing. But the process is slow, uncertain, and controversial; there is nothing in the legal community like the consensus in the scientific community on whether a particular result constitutes progress.").

or, in Kuhnian terms, to ensure that the current paradigm has been respected.[11] It works less well as a means of choosing between competing valid ideas, in part because the peer doing the reviewing is often a competitor for the same resources (space in prestigious journals, funds from government agencies or private foundations) being sought by the authors. It works very poorly in catching cheating or fraud, because all scientists are socialized to believe that even their toughest competitor is rigorously honest in the reporting of scientific results, which makes it easy for a purposefully dishonest scientist to fool a referee. Despite all of this, peer review is one of the venerated pillars of the scientific edifice.

IV. Becoming a Professional Scientist

Science as a profession or career has become highly organized and structured.[12] It is not, relatively speaking, a very remunerative profession—that would be inconsistent with the Baconian ideal—but it is intensely competitive, and material well-being does tend to follow in the wake of success (successful scientists, one might say, do get to bring home the Bacon).

A. The Institutions

These are the institutions of science: Research is done in the Ph.D.-granting universities and, to a lesser extent, in colleges that do not grant Ph.D.s. It is also done in national laboratories and in industrial laboratories. Before World War II, basic science was financed mostly by private foundations (Rockefeller, Carnegie), but since the war, the funding of science (except in industrial laboratories) has largely been taken over by agencies of the federal government, notably the National Science Foundation (an independent agency), the National Institutes of

11. The Supreme Court received differing views regarding the proper role of peer review. *Compare* Brief for Amici Curiae Daryl E. Chubin et al. at 10, Daubert v. Merrell Dow Pharms., Inc., 509 U.S. 579 (1993) (No. 92-102) ("peer review referees and editors limit their assessment of submitted articles to such matters as style, plausibility, and defensibility; they do not duplicate experiments from scratch or plow through reams of computer-generated data in order to guarantee accuracy or veracity or certainty"), *with* Brief for Amici Curiae New England Journal of Medicine, Journal of the American Medical Association, and Annals of Internal Medicine in Support of Respondent, Daubert v. Merrell Dow Pharm., Inc., 509 U.S. 579 (1993) (No. 92-102) (proposing that publication in a peer-reviewed journal be the primary criterion for admitting scientific evidence in the courtroom). *See generally* Daryl E. Chubin & Edward J. Hackett, Peerless Science: Peer Review and U.S. Science Policy (1990); Arnold S. Relman & Marcia Angell, *How Good Is Peer Review?* 321 New Eng. J. Med. 827–29 (1989). As a practicing scientist and frequent peer reviewer, I can testify that Chubin's view is correct.

12. The analysis that follows is based on David Goodstein & James Woodward, *Inside Science*, 68 Am. Scholar 83 (1999).

Health (part of the Department of Health and Human Services), and parts of the Department of Energy and the Department of Defense.

Scientists who work at all these organizations—universities, colleges, national and industrial laboratories, and funding agencies—belong to scientific societies that are organized mostly by discipline. There are large societies, such as the American Physical Society and the American Chemical Society; societies for subdisciplines, such as optics and spectroscopy; and even organizations of societies, such as FASEB, the Federation of American Societies for Experimental Biology.

Scientific societies are private organizations that elect their own officers, hold scientific meetings, publish journals, and finance their operations from the collection of dues and from the proceeds of their publishing and educational activities. The American Association for the Advancement of Science also holds meetings and publishes *Science*, a famous journal, but it is not restricted to any one discipline. The National Academy of Sciences holds meetings and publishes the *Proceedings of the National Academy of Sciences*, and, along with the National Academy of Engineering, Institute of Medicine, and its operational arm, the National Research Council, advises various government agencies on matters pertaining to science, engineering, and health. In addition to the advisory activities, one of its most important activities is to elect its own members.

These are the basic institutions of American science. It should not come as news that the universities and colleges engage in a fierce but curious competition, in which no one knows who is keeping score, but everyone knows roughly what the score is. (In recent years, some national and international media outlets have found it worthwhile to appoint themselves scorekeepers in this competition. Academic officials dismiss these journalistic judgments, except when their own institutions come out on top.) Departments in each discipline compete with one another, as do national and industrial laboratories and even funding agencies. Competition in science is at its most refined, however, at the level of individual careers.

B. The Reward System and Authority Structure

To regulate competition among scientists, there is a reward system and an authority structure. The fruits of the reward system are fame, glory, and immortality. The purposes of the authority structure are power and influence. The reward system and the authority structure are closely related to one another, but scientists distinguish sharply between them. When they speak of a colleague who has become president of a famous university, they will say sadly, "It's a pity—he was still capable of good work," sounding like warriors lamenting the loss of a fallen comrade. The university president is a kingpin of the authority structure, but, with rare exceptions, he is a dropout from the reward system. Similar kinds of behavior can be observed in industrial and government laboratories, but a description of what goes on in universities will be enough to illustrate how the system works.

A career in academic science begins at the first step on the reward system ladder, earning a Ph.D., followed in many areas by one or two stints as a post-doctoral fellow. The Ph.D. and postdoctoral positions had best be at universities (or at least departments) that are high up in that fierce but invisible competition, because all subsequent steps are more likely than not to take the individual sideways or downward on the list. The next step is a crucial one: appointment to a tenure-track junior faculty position. About two-thirds of all postdoctoral fellows in biology in American universities believe that they are going to make this step, but in fact, only about a quarter of them succeed. This step and all subsequent steps require growing renown as a scientist beyond the individual's own circle of acquaintances. Thus, it is essential by this time that the individual has accomplished something. The remaining steps up the reward system ladder are promotion to an academic tenured position and full professorship; various prizes, medals, and awards given out by the scientific societies; an endowed chair (the virtual equivalent of Galileo's wooden *cattedra*); election to the National Academy of Sciences; particularly prestigious awards up to and including the Nobel Prize; and, finally, a reputation equivalent to immortality.

Positions in the authority structure are generally rewards for having achieved a certain level in the reward system. For example, starting from the Ph.D. or junior faculty level, it is possible to step sideways temporarily or even permanently into a position as contract officer in a funding agency. Because contract officers influence the distribution of research funds, they have a role in deciding who will succeed in the climb up the reward system ladder. At successively higher levels one can become a journal editor; department chair; dean, provost, director of national research laboratory or president of a university; and even the head of a funding agency, a key player in determining national policy as it relates to science and technology. People in these positions have stepped out of the traditional reward system, but they have something to say about who succeeds within it.

V. Some Myths and Facts About Science

"In matters of science," Galileo wrote, "the authority of thousands is not worth the humble reasoning of one single person."[13] Doing battle with the Aristotelian professors of his day, Galileo believed that kowtowing to authority was the enemy of reason. But, contrary to Galileo's famous remark, the fact is that within the scientific community itself, authority is of fundamental importance. If a paper's

13. I found this statement framed on the office wall of a colleague in Italy in the form, "*In questioni di scienza L'autorità di mille non vale l'umile ragionare di un singolo*." However, I have not been able to find the famous remark in this form in Galileo's writings. An equivalent statement in different words can be found in Galileo's Il Saggiatore (1623). *See* Andrea Frova & Mariapiera Marenzona, Parola di Galileo 473 (1998).

author is a famous scientist, the paper is probably worth reading. The triumph of reason over authority is just one of the many myths about science. Following is a brief list of some others:

Myth: Scientists must have open minds, being ready to discard old ideas in favor of new ones.

Fact: Because science is an adversarial process through which each idea deserves the most vigorous possible defense, it is useful for the successful progress of science that scientists tenaciously cling to their own ideas, even in the face of contrary evidence.

Myth: The institution of peer review assures that all published papers are sound and dependable.

Fact: Peer review generally will catch something that is completely out of step with majority thinking at the time, but it is practically useless for catching outright fraud, and it is not very good at dealing with truly novel ideas. Peer review mostly assures that all papers follow the current paradigm (see comments on Kuhn, above). It certainly does not ensure that the work has been fully vetted in terms of the data analysis and the proper application of research methods.

Myth: Science must be an open book. For example, every new experiment must be described so completely that any other scientist can reproduce it.

Fact: There is a very large component of skill in making cutting-edge experiments work. Often, the only way to import a new technique into a laboratory is to hire someone (usually a postdoctoral fellow) who has already made it work elsewhere. Nonetheless, scientists have a solemn responsibility to describe the methods they use as fully and accurately as possible. And, eventually, the skill will be acquired by enough people to make the new technique commonplace.

Myth: When a new theory comes along, the scientist's duty is to falsify it.

Fact: When a new theory comes along, the scientist's instinct is to verify it. When a theory is new, the effect of a decisive experiment that shows it to be wrong is that both the theory and the experiment are in most cases quickly forgotten. This result leads to no progress for anybody in the reward system. Only when a theory is well established and widely accepted does it pay off to prove that it is wrong.

Myth: University-based research is pure and free of conflicts of interest.

Fact: The Bayh-Dole Act of the early 1980s permits universities to patent the results of research supported by the federal government. Many universities have become adept at obtaining such patents. In many cases this raises conflict-of-interest problems when the universities' interest in pursuing knowledge comes into conflict with its need for revenue. This is an area that has generated considerable scrutiny. For instance, the recent Institute of Medicine report *Conflict of Interest in Medical Research, Education, and Practice* sheds light on the changing dimensions of conflicts of interest associated with growing interdisciplinary collaborations between individuals, universities, and industry especially in life sciences and biomedical research.[14]

Myth: Real science is easily distinguished from pseudoscience.

Fact: This is what philosophers call the problem of demarcation: One of Popper's principal motives in proposing his standard of falsifiability was precisely to provide a means of demarcation between real science and impostors. For example, Einstein's general theory of relativity (with which Popper was deeply impressed) made clear predictions that could certainly be falsified if they were not correct. In contrast, Freud's theories of psychoanalysis (with which Popper was far less impressed) could never be proven wrong. Thus, to Popper, relativity was science but psychoanalysis was not.

Real scientists do not behave as Popper says they should, and there is another problem with Popper's criterion (or indeed any other criterion) for demarcation: Would-be scientists read books too. If it becomes widely accepted (and to some extent it has) that falsifiable predictions are the signature of real science, then pretenders to the throne of science will make falsifiable predictions too.[15] There is no simple, mechanical criterion for distinguishing real science from something that is not real science. That certainly does not mean, however, that the job cannot be done. As I discuss below, the Supreme Court, in the *Daubert* decision, has made a respectable stab at showing how to do it.[16]

14. Institute of Medicine, Conflict of Interest in Medical Research, Education, and Practice (Bernard Lo & Marilyn Field eds., 2009).

15. For a list of such pretenders, see Larry Laudan, Beyond Positivism and Relativism 219 (1996).

16. The Supreme Court in *Daubert* identified four nondefinitive factors that were thought to be illustrative of characteristics of scientific knowledge: testability or falsifiability, peer review, a known or potential error rate, and general acceptance within the scientific community. 509 U.S. at 590. Subsequent cases have expanded on these factors. *See, e.g., In re* TMI Litig. Cases Consol. II, 911 F. Supp. 775, 787 (M.D. Pa. 1995) (which considered the following additional factors: the relationship of the technique to methods that have been established to be reliable, the qualifications of the expert witness testifying based on the methodology, the nonjudicial uses of the method, logical or internal consistency of the hypothesis, the consistency of the hypothesis with accepted theories, and the precision of the hypothesis or theory). *See generally* Bert Black et al., *Science and the Law in the Wake of* Daubert*: A New Search for Scientific Knowledge,* 72 Tex. L. Rev. 715, 783–84 (1994) (discussion of expanded list of factors).

Myth: Scientific theories are just that: theories. All scientific theories are eventually proved wrong and are replaced by other theories.

Fact: The things that science has taught us about how the world works are the most secure elements in all of human knowledge. Here I must distinguish between science at the frontiers of knowledge (where by definition we do not yet understand everything and where theories are indeed vulnerable) and textbook science that is known with great confidence. Matter is made of atoms, DNA transmits the blueprints of organisms from generation to generation, light is an electromagnetic wave—these things are not likely to be proved wrong. The theory of relativity and the theory of evolution are in the same class and are still called "theories" for historic reasons only.[17] The GPS device in my car routinely uses the general theory of relativity to make calculations accurate enough to tell me exactly where I am and to take me to my destination with unerring precision. The phenomenon of natural selection has been observed under numerous field conditions as well as in controlled laboratory experiments.

In recent times, the courts have had much to say about the teaching of the theory of evolution in public schools.[18] In one instance the school district decided that students should be taught the "gaps/problems" in Darwin's theory and given "Intelligent Design" as an alternative explanation. The courts (Judge Jones of the United States District Court for the Middle District of Pennsylvania) came down hard on the side of Darwin, ruling that "Intelligent Design" was thinly disguised religion that had no place in the science classroom.

It should be said here that the incorrect notion that all theories must eventually be wrong is fundamental to the work of both Popper and Kuhn, and these theorists have been crucial in helping us understand how science works. Thus, their theories, like good scientific theories at the frontiers of knowledge, can be both useful and wrong.

Myth: Scientists are people of uncompromising honesty and integrity.

Fact: They would have to be if Bacon were right about how science works, but he was not. Most scientists are rigorously honest where honesty matters most to them: in the reporting of scientific procedures and data in peer-reviewed publications. In all else, they are ordinary mortals.

17. According to the National Academy of Sciences and Institute of Medicine's 2008 report Science, Evolution, and Creationism, "the strength of a theory rests in part on providing scientists with the basis to explain observed phenomena and to predict what they are likely to find when exploring new phenomena and observations." The report also helps differentiate a theory from a hypothesis, the latter being testable natural explanations that may offer tentative scientific insights.

18. Kitzmiller v. Dover Area School District, 400 F. Supp. 2d 707 (M.D. Pa. 2005).

VI. Comparing Science and the Law

Science and the law differ both in the language they use and the objectives they seek to accomplish.

A. Language

Oscar Wilde (and G.B. Shaw too) once remarked that the United States and England are two nations divided by a common language. Something similar can be said, with perhaps more truth (if less wit), of science and the law. There are any number of words commonly used in both disciplines, but with different meanings.

For example, the word *force*, as it is used by lawyers, has connotations of violence and the domination of one person's will over another, when used in phrases such as "excessive use of force" and "forced entry." In science, *force* is something that when applied to a body, causes its speed and direction of motion to change. Also, all forces arise from a few fundamental forces, most notably gravity and the electric force. The word carries no other baggage.

In contrast, the word *evidence* is used much more loosely in science than in law. The law has precise rules of evidence that govern what is admissible and what is not. In science, the word merely seems to mean something less than "proof." A certain number of the papers in any issue of a scientific journal will have titles that begin with "Evidence for (or against) . . ." What that means is, the authors were not able to prove their point, but are presenting their results anyway.

The word *theory* is a particularly interesting example of a word that has different meanings in each discipline. A legal theory is a proposal that fits the known facts and legal precedents and that favors the attorney's client. What's required of a theory in science is that it make new predictions that can be tested by new experiments or observations and falsified or verified (as discussed above).

Even the word *law* has different meanings in the two disciplines. To a legal practitioner, a law is something that has been promulgated by some human authority, such as a legislature or parliament. In science, a law is a law of nature, something that humans can hope to discover and describe accurately, but that can never be changed by any human authority or intervention.

My final example is, to me, the most interesting of all. It is the word *error*. In the law, and in common usage, *error* and *mistake* are more or less synonymous. A legal decision can be overturned if it is found to be contaminated by judicial error. In science, however, *error* and *mistake* have different meanings. Anyone can make a mistake, and scientists have no obligation to report theirs in the scientific literature. They just clean up the mess and go on to the next attempt. Error, on the other hand, is intrinsic to any measurement, and far from ignoring it or covering it up or even attempting to eliminate it, authors of every paper about a scientific experiment will include a careful analysis of the errors to put limits on the uncertainty in the measured result. To make mistakes is human, one might say, but error is intrinsic to our interaction with nature, and is therefore part of science.

B. *Objectives*

Beyond the meanings of certain key words, science and the law differ fundamentally in their objectives. The objective of the law is justice; that of science is truth.[19] These are among the highest goals to which humans can aspire, but they are not the same thing. Justice, of course, also seeks truth, but it requires that clear decisions be made in a reasonable and limited period of time. In the scientific search for truth there are no time limits and no point at which a final decision must be made.

And yet, despite all these differences, science and the law share, at the deepest possible level, the same aspirations and many of the same methods. Both disciplines seek, in structured debate and using empirical evidence, to arrive at rational conclusions that transcend the prejudices and self-interest of individuals.

VII. A Scientist's View of *Daubert*

In the 1993 *Daubert* decision, the U.S. Supreme Court took it upon itself to resolve, once and for all, the knotty problem of the demarcation between science and pseudoscience. Better yet, it undertook to enable every federal judge to solve that problem in deciding the admissibility of each scientific expert witness in every case that arises. In light of all the uncertainties discussed in this chapter, it must be considered an ambitious thing to do.[20]

The presentation of scientific evidence in a court of law is a kind of shotgun marriage between the two disciplines. Both are obliged to some extent to yield

19. This point was made eloquently by D. Allen Bromley in Science and the Law, Address at the 1998 Annual Meeting of the American Bar Association (Aug. 2, 1998).

20. Chief Justice Rehnquist, responding to the majority opinion in *Daubert*, was the first to express his uneasiness with the task assigned to federal judges, as follows: "I defer to no one in my confidence in federal judges; but I am at a loss to know what is meant when it is said that the scientific status of a theory depends on its 'falsifiability,' and I suspect some of them will be, too." 509 U.S. at 579 (Rehnquist, C.J., concurring in part and dissenting in part). His concern was then echoed by Judge Alex Kozinski when the case was reconsidered by the U.S. Court of Appeals for the Ninth Circuit following remand by the Supreme Court. 43 F.3d 1311, 1316 (9th Cir. 1995) ("Our responsibility, then, unless we badly misread the Supreme Court's opinion, is to resolve disputes among respected, well-credentialed scientists about matters squarely within their expertise, in areas where there is no scientific consensus as to what is and what is not 'good science,' and occasionally to reject such expert testimony because it was not 'derived by the scientific method.' Mindful of our position in the hierarchy of the federal judiciary, we take a deep breath and proceed with this heady task.").

to the central imperatives of the other's way of doing business, and it is likely that neither will be shown in its best light. The *Daubert* decision is an attempt (not the first, of course) to regulate that encounter. Judges are asked to decide the "evidential reliability" of the intended testimony, based not on the conclusions to be offered, but on the methods used to reach those conclusions.

In particular, *Daubert* says, the methods should be judged by the following four criteria:

1. The theoretical underpinnings of the methods must yield testable predictions by means of which the theory could be falsified.
2. The methods should preferably be published in a peer-reviewed journal.
3. There should be a known rate of error that can be used in evaluating the results.
4. The methods should be generally accepted within the relevant scientific community.

In reading these four illustrative criteria mentioned by the Court, one is struck immediately by the specter of Karl Popper looming above the robed justices. (It is no mere illusion. The dependence on Popper is explicit in the written decision.) Popper alone is not enough, however, and the doctrine of falsification is supplemented by a bow to the institution of peer review, an acknowledgment of the scientific meaning of error, and a paradigm check (really, an inclusion of the earlier *Frye* standard).[21]

The *Daubert* case and two others (*General Electric v. Joiner*,[22] and *Kumho Tires v. Carmichael*[23]) have led to increasing attention on the part of judges to scientific and technical issues and have led to the increased exclusion of expert testimony, but the *Daubert* criteria seem too general to resolve many of the difficult decisions the courts face when considering scientific evidence. Nonetheless, despite some inconsistency in rulings by various judges, the *Daubert* decision has given the courts new flexibility, and so far, it has stood the test of time.

All in all, I would give the decision pretty high marks.[24] The justices ventured into the treacherous crosscurrents of the philosophy of science—where even most scientists fear to tread—and emerged with at least their dignity intact. Falsifiability may not be a good way of doing science, but it is not the worst a posteriori way to judge science, and that is all that's required here. At least they managed to avoid the Popperian trap of demanding that the scientists be skeptical of their own ideas.

21. In *Frye v. United States*, 293 F. 1013, 1014 (D.C. Cir. 1923), the court stated that expert opinion based on a scientific technique is inadmissible unless the technique is "generally accepted" as reliable in the relevant scientific community.

22. 522 U.S. 136 (1997).

23. 526 U.S. 137 (1999).

24. For a contrary view, see Gary Edmond & David Mercer, *Recognizing Daubert: What Judges Should Know About Falsification*, 5 Expert Evid. 29–42 (1996).

The other considerations help lend substance and flexibility.[25] The jury is still out (so to speak) on how well this decision will work in practice, but it is certainly an impressive attempt to serve justice, if not truth. Applying it in practice will never be easy, but then that is what this manual is about.[26]

25. *See supra* note 16.

26. For further reading, see John Ziman, PublicKnowledge: An Essay Concerning the Social Dimension of Science (Cambridge University Press 1968).

Reference Guide on Forensic Identification Expertise

PAUL C. GIANNELLI, EDWARD J. IMWINKELRIED, AND JOSEPH L. PETERSON

Paul C. Giannelli, L.L.M, is Albert J. Weatherhead III and Richard W. Weatherhead Professor of Law, and Distinguished University Professor, Case Western Reserve University.

Edward J. Imwinkelried, J.D., is Edward L. Barrett, Jr. Professor of Law and Director of Trial Advocacy, University of California, Davis.

Joseph L. Peterson, D.Crim., is Professor of Criminal Justice and Criminalistics, California State University, Los Angeles.

CONTENTS

I. Introduction

Forensic identification expertise encompasses fingerprint, handwriting, and firearms ("ballistics"), and toolmark comparisons, all of which are used by crime laboratories to associate or dissociate a suspect with a crime. Shoe and tire prints also fall within this large pattern evidence domain. These examinations consist of comparing a known exemplar with evidence collected at a crime scene or from a suspect. Bite mark analysis can be added to this category, although it developed within the field of forensic dentistry as an adjunct of dental identification and is not conducted by crime laboratories. In a broad sense, the category includes trace evidence such as the analysis of hairs, fibers, soil, glass, and wood. Some forensic disciplines attempt to individuate and thus attribute physical evidence to a particular source—a person, object, or location.[1] Other techniques are useful because they narrow possible sources to a discrete category based upon what are known as "class characteristics" (as opposed to "individual characteristics"). Moreover, some techniques are valuable because they eliminate possible sources.

Following this introduction, Part II of this guide sketches a brief history of the development of forensic expertise and crime laboratories. Part III discusses the impact of the advent of DNA analysis and the Supreme Court's 1993 *Daubert* decision,[2] developments that prompted a reappraisal of the trustworthiness of testimony by forensic identification experts. Part IV focuses on the 2009 National Research Council (NRC) report on forensic science.[3] Parts V through X examine specific identification techniques: (1) fingerprint analysis, (2) questioned document examination, (3) firearms and toolmark identification, (4) bite mark comparison, and (5) microscopic hair analysis. Part XI considers recurrent problems, including the clarity of expert testimony, limitations on its scope, and restrictions on closing arguments. Part XII addresses procedural issues—pretrial discovery and access to defense experts.

1. Some forensic scientists believe the word *individualization* is more accurate than *identification*. Paul L. Kirk, *The Ontogeny of Criminalistics*, 54 J. Crim. L., Criminology & Police Sci. 235, 236 (1963). The identification of a substance as heroin, for example, does not individuate, whereas a fingerprint identification does.

2. Daubert v. Merrell Dow Pharms., Inc., 509 U.S. 579 (1993). *Daubert* is discussed in Margaret A. Berger, The Admissibility of Expert Testimony, in this manual.

3. National Research Council, Strengthening Forensic Science in the United States: A Path Forward (2009) [hereinafter NRC Forensic Science Report], *available at* http://www.nap.edu/catalog.php?record_id=12589.

II. Development of Forensic Identification Techniques

An understanding of the current issues requires some appreciation of the past. The first reported fingerprint case was decided in 1911.[4] This case preceded the establishment of the first American crime laboratory, which was created in Los Angeles in 1923.[5] The Federal Bureau of Investigation (FBI) laboratory came online in 1932. At its inception, the FBI laboratory staff included only firearms identification and fingerprint examination.[6] Handwriting comparisons, trace evidence examinations, and serological testing of blood and semen were added later. When initially established, crime laboratories handled a modest number of cases. For example, in its first full year of operation, the FBI laboratory processed fewer than 1000 cases.[7]

Several sensational cases in these formative years highlighted the value of forensic identification evidence. The Sacco and Vanzetti trial in 1921 was one of the earliest cases to rely on firearms identification evidence.[8] In 1935, the extensive use of handwriting comparison testimony[9] and wood evidence[10] at the Lindbergh kidnapping trial raised the public consciousness of identification expertise and solidified its role in the criminal justice system. Crime laboratories soon sprang up in other large cities such as Chicago and New York.[11] The num-

4. People v. Jennings, 96 N.E. 1077 (Ill. 1911).

5. *See* John I. Thornton, *Criminalistics: Past, Present and Future*, 11 Lex et Scientia 1, 23 (1975) ("In 1923, Vollmer served as Chief of Police of the City of Los Angeles for a period of one year. During that time, a crime laboratory was established at his direction.").

6. *See* Federal Bureau of Investigation, U.S. Department of Justice, FBI Laboratory 3 (1981), *available at* http://www.ncjrs.gov/App/publications/Abstract.aspx?id=78689.

7. *See Anniversary Report, 40 Years of Distinguished Scientific Assistance to Law Enforcement*, FBI Law Enforcement Bull., Nov. 1972, at 4 ("During its first month of service, the FBI Laboratory examiners handled 20 cases. In its first full year of operation, the volume increased to a total of 963 examinations. By the next year that figure more than doubled.").

8. *See* G. Louis Joughin & Edmund M. Morgan, The Legacy of Sacco & Vanzetti 15 (1948); *see also* James E. Starrs, *Once More Unto the Breech: The Firearms Evidence in the Sacco and Vanzetti Case Revisited*, Parts I & II, 31 J. Forensic Sci. 630, 1050 (1986).

9. *See* D. Michael Risinger et al., *Exorcism of Ignorance as a Proxy for Rational Knowledge: The Lessons of Handwriting Identification "Expertise,"* 137 U. Pa. L. Rev. 731, 738 (1989).

10. *See* Shirley A. Graham, *Anatomy of the Lindbergh Kidnapping*, 42 J. Forensic Sci. 368 (1997). The kidnapper had used a wooden ladder to reach the second-story window of the child's bedroom. Arthur Koehler, a wood technologist and identification expert for the Forest Products Laboratory of the U.S. Forest Service, traced part of the ladder's wood from its mill source to a lumberyard near the home of the accused. Relying on plant anatomical comparisons, he also testified that a piece of the ladder came from a floorboard in the accused's attic.

11. *See* Joseph L. Peterson, *The Crime Lab*, *in* Thinking About Police 184, 185 (Carl Klockars ed., 1983) ("[T]he Chicago Crime Laboratory has the distinction of being one of the oldest in the country. Soon after, however, many other jurisdictions also built police laboratories in an attempt to cope with the crimes of violence associated with the 1930s gangster era.").

ber of laboratories gradually grew and then skyrocketed. The national campaign against drug abuse led most crime laboratories to create forensic chemistry units, and today the analysis of suspected contraband drugs constitutes more than 50% of the caseload of many laboratories.[12] By 2005, the nation's crime laboratories were handling approximately 2.7 million cases every year.[13] According to a 2005 census, there are now 389 publicly funded crime laboratories in the United States: 210 state or regional laboratories, 84 county laboratories, 62 municipal laboratories, and 33 federal laboratories.[14] Currently, these laboratories employ more than 11,900 full-time staff members.[15]

The establishment of crime laboratories represented a significant reform in the types of evidence used in criminal trials. Previously, prosecutors had relied primarily on eyewitness testimony and confessions. The reliability of physical evidence is often superior to that of other types of proof.[16] However, the seeds of the current controversies over forensic identification expertise were sown during this period. Even though the various techniques became the stock and trade of crime laboratories, many received their judicial imprimatur without a critical evaluation of the supporting scientific research.[17]

This initial lack of scrutiny resulted, in part, from the deference that previous standards of admissibility accorded the community of specialists in the various fields of expert testimony. In 1923, the D.C. Circuit adopted the "general accep-

12. J. Peterson & M. Hickman, Bureau of Just. Stat. Bull. (Feb. 2005), NCJ 207205. In most cases, the forensic chemist simply identifies the unknown as a particular drug. However, in some cases the chemist attempts to individuate and establish that several drug samples originated from the same production batch at a particular illegal drug laboratory. *See* Fabrice Besacier et al., *Isotopic Analysis of 13C as a Tool for Comparison and Origin Assignment of Seized Heroin Samples,* 42 J. Forensic Sci. 429 (1997); C. Sten et al., *Computer Assisted Retrieval of Common-Batch Members in Leukart Amphetamine Profiling,* 38 J. Forensic Sci. 1472 (1993).

13. Matthew R. Durose, *Crime Labs Received an Estimated 2.7 Million Cases in 2005*, Bureau of Just. State. Bull. (July 2008) NCJ 222181, *available at* http://pjs.ojp.usdoj.gov/index.cfm?ty=pbdetail&lid=490 (summarizing statistics compiled by the Justice Department's Bureau of Justice Statistics).

14. NRC Forensic Science Report, *supra* note 3, at 58.

15. *Id.* at 59.

16. For example, in 1927, Justice Frankfurter, then a law professor, sharply critiqued the eyewitness identifications in the Sacco and Vanzetti case. *See* Felix Frankfurter, The Case of Sacco and Vanzetti 30 (1927) ("What is the worth of identification testimony even when uncontradicted? The identification of strangers is proverbially untrustworthy."). In 1936, the Supreme Court expressed grave reservations about the trustworthiness of confessions wrung from a suspect by abusive interrogation techniques. *See* Brown v. Mississippi, 297 U.S. 278 (1936) (due process violated by beating a confession out of a suspect).

17. "[F]ingerprints were accepted as an evidentiary tool without a great deal of scrutiny or skepticism" of their underlying assumptions. Jennifer L. Mnookin, *Fingerprint Evidence in an Age of DNA Profiling*, 67 Brook. L. Rev. 13, 17 (2001); *see also* Risinger et al., *supra* note 9, at 738 ("Our literature search for empirical evaluation of handwriting identification turned up one primitive and flawed validity study from nearly 50 years ago, one 1973 paper that raises the issue of consistency among examiners but presents only uncontrolled impressionistic and anecdotal information not qualifying as data in any rigorous sense, and a summary of one study in a 1978 government report. Beyond this, nothing.").

tance" test for determining the admissibility of scientific evidence. The case, *Frye v. United States*,[18] involved a precursor of the modern polygraph. Although the general acceptance test was limited to mostly polygraph cases for several decades, it eventually became the majority pre-*Daubert* standard.[19] However, under that test, scientific testimony is admissible if the underlying theory or technique is generally accepted by the specialists within the expert's field. The *Frye* test did not require foundational proof of the empirical validity of the technique's scientific premises.

III. Reappraisal of Forensic Identification Expertise

The advent of DNA profiling in the late 1980s, quickly followed by the Supreme Court's 1993 *Daubert* decision (rejecting *Frye*), prompted a reassessment of identification expertise.[20]

A. DNA Profiling and Empirical Testing

In many ways, DNA profiling revolutionized the use of expert testimony in criminal cases.[21] Population geneticists, often affiliated with universities, used statistical techniques to define the extent to which a match of DNA markers individuated the accused as the possible source of the crime scene sample.[22] Typically, the experts testified to a random-match probability, supporting their opinions by pointing to extensive empirical testing.

The fallout from the introduction of DNA analysis in criminal trials was significant in three ways. First, DNA profiling became the gold standard, regarded as the most reliable of all forensic techniques.[23] NRC issued two reports on the

18. 293 F. 1013 (D.C. Cir. 1923).

19. *Frye* was cited only five times in published opinions before World War II, mostly in polygraph cases. After World War II, it was cited 6 times before 1950, 20 times in the 1950s, and 21 times in the 1960s. Bert Black et al., *Science and the Law in the Wake of* Daubert: *A New Search for Scientific Knowledge*, 72 Tex. L. Rev. 715, 722 n.30 (1994).

20. *See* Michael J. Saks & Jonathan J. Koehler, *The Coming Paradigm Shift in Forensic Identification Science*, 309 Science 892 (2005).

21. *See* People v. Wesley, 533 N.Y.S.2d 643, 644 (County Ct. 1988) (calling DNA evidence the "single greatest advance in the 'search for truth' . . . since the advent of cross-examination").

22. DNA Profiling is examined in detail in David H. Kaye & George Sensabaugh, Reference Guide on DNA Identification Evidence, in this manual.

23. *See* Michael Lynch, *God's Signature: DNA Profiling, The New Gold Standard in Forensic Science*, 27 Endeavour 2, 93 (2003); Joseph L. Peterson & Anna S. Leggett, *The Evolution of Forensic Science: Progress Amid the Pitfalls*, 36 Stetson L. Rev. 621, 654 (2007) ("The scientific integrity and reliability of DNA testing have helped DNA replace fingerprinting and made DNA evidence the new 'gold standard' of forensic evidence"); *see also* NRC Forensic Science Report, *supra* note 3, at 40–41 (the ascendancy of DNA).

subject, emphasizing the importance of certain practices: "No laboratory should let its results with a new DNA typing method be used in court, unless it has undergone . . . proficiency testing via blind trials."[24] Commentators soon pointed out the broader implications of this development:

> The increased use of DNA analysis, which has undergone extensive validation, has thrown into relief the less firmly credentialed status of other forensic science identification techniques (fingerprints, fiber analysis, hair analysis, ballistics, bite marks, and tool marks). These have not undergone the type of extensive testing and verification that is the hallmark of science elsewhere.[25]

Second, the DNA admissibility battles highlighted the absence of mandatory regulation of crime laboratories.[26] This situation began to change with the passage of the DNA Identification Act of 1994,[27] the first federal statute regulating a crime laboratory procedure. The Act authorized the creation of a national database for the DNA profiles of convicted offenders as well as a database for unidentified profiles from crime scenes: the Combined DNA Index System (CODIS). Bringing CODIS online was a major undertaking, and its successful operation required an effective quality assurance program. As one government report noted, "the integrity of the data contained in CODIS is extremely important since the DNA matches provided by CODIS are frequently a key piece of evidence linking a suspect to a crime."[28] The statute also established a DNA Advisory Board (DAB) to assist in promulgating quality assurance standards[29] and required proficiency

24. National Research Council, DNA Technology in Forensic Science 55 (1992) [hereinafter NRC I], *available at* http://www.nap.edu/catalog.php?record _id=1866. A second report followed. *See* National Research Council, The Evaluation of Forensic DNA Evidence (1996), *available at* http://www.nap.edu/catalog.php/record_id=5141. The second report also recommended proficiency testing. *Id.* at 88 (Recommendation 3.2: "Laboratories should participate regularly in proficiency tests, and the results should be available for court proceedings.").

25. Donald Kennedy & Richard A. Merrill, *Assessing Forensic Science*, 20 Issues in Sci. & Tech. 33, 34 (2003); *see also* Michael J. Saks & Jonathan J. Koehler, *What DNA "Fingerprinting" Can Teach the Law About the Rest of Forensic Science*, 13 Cardozo L. Rev. 361, 372 (1991) ("[F]orensic scientists, like scientists in all other fields, should subject their claims to methodologically rigorous empirical tests. The results of these tests should be published and debated."); Sandy L. Zabell, *Fingerprint Evidence*, 13 J.L. & Pol'y 143, 143 (2005) ("DNA identification has not only transformed and revolutionized forensic science, it has also created a new set of standards that have raised expectations for forensic science in general.").

26. In 1989, Eric Lander, a prominent molecular biologist who became enmeshed in the early DNA admissibility disputes, wrote: "At present, forensic science is virtually unregulated—with the paradoxical result that clinical laboratories must meet higher standards to be allowed to diagnose strep throat than forensic labs must meet to put a defendant on death row." Eric S. Lander, *DNA Fingerprinting on Trial*, 339 Nature 501, 505 (1989).

27. 42 U.S.C. § 14131 (2004).

28. Office of Inspector General, U.S. Department of Justice, Audit Report, The Combined DNA Index System, ii (2001), *available at* http://www.justice.gov/oig/reports/FBI/a0126/final.pdf.

29. 42 U.S.C. § 14131(b). The legislation contained a "sunset" provision; DAB would expire after 5 years unless extended by the Director of the FBI. The board was extended for several months and then ceased to exist. The FBI had established the Technical Working Group on DNA Identifica-

testing for FBI analysts as well as those in laboratories participating in the national database or receiving federal funding.[30]

Third, the use of DNA evidence to exonerate innocent convicts led to a reexamination of the evidence admitted to secure their original convictions.[31] Some studies indicated that, after eyewitness testimony, forensic identification evidence was one of the most common types of testimony that jurors relied on at the earlier trials in returning erroneous verdicts.[32] These studies suggested that flawed forensic analyses may have contributed to the convictions.[33]

B. Daubert *and Empirical Testing*

The second major development prompting a reappraisal of forensic identification evidence was the *Daubert* decision.[34] Although there was some uncertainty about the effect of the decision at the time *Daubert* was decided, the Court's subsequent cases, *General Electric Co. v. Joiner*[35] and *Kumho Tire Co. v. Carmichael,*[36] signaled

tion Methods (TWGDAM) in 1988 to develop standards. TWGDAM functioned under DAB. It was renamed the Scientific Working Group on DNA Analysis Methods (SWGDAM) in 1999 and replaced DAB when the latter expired.

30. 42 U.S.C. § 14132(b)(2) (2004) (external proficiency testing for CODIS participation); *id.* § 14133(a)(1)(A) (2004) (FBI examiners). DAB Standard 13 implements this requirement. The Justice for All Act, enacted in 2004, amended the statute, requiring all DNA labs to be accredited within 2 years "by a nonprofit professional association of persons actively involved in forensic science that is nationally recognized within the forensic science community" and to "undergo external audits, not less than once every 2 years, that demonstrate compliance with standards established by the Director of the Federal Bureau of Investigation." 42 U.S.C. § 14132(b)(2).

31. *See* Samuel R. Gross et al., *Exonerations in the United States 1989 Through 2003*, 95 J. Crim. L. & Criminology 523, 543 (2005).

32. A study of 200 DNA exonerations found that expert testimony (55%) was the second leading type of evidence (after eyewitness identifications, 79%) used in the wrongful conviction cases. Pre-DNA serology of blood and semen evidence was the most commonly used technique (79 cases). Next came hair evidence (43 cases), soil comparison (5 cases), DNA tests (3 cases), bite mark evidence (3 cases), fingerprint evidence (2 cases), dog scent (2 cases), spectrographic voice evidence (1 case), shoe prints (1 case), and fibers (1 case). Brandon L. Garrett, *Judging Innocence*, 108 Colum. L. Rev. 55, 81 (2008). These data do not necessarily mean that the forensic evidence was improperly used. For example, serological testing at the time of many of these convictions was simply not as discriminating as DNA profiling. Consequently, a person could be included using these serological tests but be excluded by DNA analysis. Yet, some evidence was clearly misused. *See also* Paul C. Giannelli, *Wrongful Convictions and Forensic Science: The Need to Regulate Crime Labs*, 86 N.C. L. Rev. 163, 165–70, 172–207 (2007).

33. *See* Melendez-Diaz v. Massachusetts, 129 S. Ct. 2527, 2537 (2009) (citing Brandon L. Garrett & Peter J. Neufeld, *Invalid Forensic Science Testimony and Wrongful Convictions*, 95 Va. L. Rev. 1, 34–84 (2009)). *See also* Brandon L. Garrett, Convicting the Innocent: Where Criminal Prosecutions Go Wrong, ch. 4 (2011).

34. *Daubert* is discussed in detail in Margaret A. Berger, The Admissibility of Expert Testimony, in this manual.

35. 522 U.S. 136 (1997).

36. 526 U.S. 137 (1999).

that the *Daubert* standard may often be more demanding than the traditional *Frye* standard.[37] *Kumho* extended the reliability requirement to all types of expert testimony, and in 2000, the Court characterized *Daubert* as imposing an "exacting" standard for the admissibility of expert testimony.[38]

Daubert's impact in civil cases is well documented.[39] Although *Daubert's* effect on criminal litigation has been less pronounced,[40] it nonetheless has partially changed the legal landscape. Defense attorneys invoked *Daubert* as the basis for mounting attacks on forensic identification evidence, and a number of courts view the *Daubert* trilogy as "inviting a reexamination even of 'generally accepted' venerable, technical fields."[41] Several courts have held that a forensic technique is not exempt from Rule 702 scrutiny simply because it previously qualified for admission under *Frye*'s general acceptance standard.[42]

In addition to enunciating a new reliability test, *Daubert* listed several factors that trial judges may consider in assessing reliability. The first and most important *Daubert* factor is testability. Citing scientific authorities, the *Daubert* Court noted that a hallmark of science is empirical testing. The Court quoted Hempel:

37. *See* United States v. Horn, 185 F. Supp. 2d 530, 553 (D. Md. 2002) ("Under *Daubert*, . . . it was expected that it would be easier to admit evidence that was the product of new science or technology. In practice, however, it often seems as though the opposite has occurred—application of *Daubert/Kumho Tire* analysis results in the exclusion of evidence that might otherwise have been admitted under *Frye*.").

38. Weisgram v. Marley Co., 528 U.S. 440, 455 (2000).

39. *See* Lloyd Dixon & Brian Gill, Changes in the Standards of Admitting Expert Evidence in Federal Civil Cases Since the *Daubert* Decision 25 (2002) ("[S]ince *Daubert*, judges have examined the reliability of expert evidence more closely and have found more evidence unreliable as a result."); Margaret A. Berger, *Upsetting the Balance Between Adverse Interests: The Impact of the Supreme Court's Trilogy on Expert Testimony in Toxic Tort Litigation*, 64 Law & Contemp. Probs. 289, 290 (2001) ("The Federal Judicial Center conducted surveys in 1991 and 1998 asking federal judges and attorneys about expert testimony. In the 1991 survey, seventy-five percent of the judges reported admitting all proffered expert testimony. By 1998, only fifty-nine percent indicated that they admitted all proffered expert testimony without limitation. Furthermore, sixty-five percent of plaintiff and defendant counsel stated that judges are less likely to admit some types of expert testimony since *Daubert*.").

40. *See* Jennifer L. Groscup et al., *The Effects of* Daubert *on the Admissibility of Expert Testimony in State and Federal Criminal Cases*, 8 Psychol. Pub. Pol'y & L. 339, 364 (2002) ("[T]he *Daubert* decision did not impact on the admission rates of expert testimony at either the trial or the appellate court levels."); D. Michael Risinger, *Navigating Expert Reliability: Are Criminal Standards of Certainty Being Left on the Dock?* 64 Alb. L. Rev. 99, 149 (2000) ("[T]he heightened standards of dependability imposed on expertise proffered in civil cases has continued to expand, but . . . expertise proffered by the prosecution in criminal cases has been largely insulated from any change in pre-*Daubert* standards or approach.").

41. United States v. Hines, 55 F. Supp. 2d 62, 67 (D. Mass. 1999) (handwriting comparison); *see also* United States v. Hidalgo, 229 F. Supp. 2d 961, 966 (D. Ariz. 2002) ("Courts are now confronting challenges to testimony, as here, whose admissibility had long been settled"; discussing handwriting comparison).

42. *See, e.g.*, United States v. Williams, 506 F.3d 151, 162 (2d Cir. 2007) ("Nor did [*Daubert*] 'grandfather' or protect from *Daubert* scrutiny evidence that had previously been admitted under *Frye*."); United States v. Starzecpyzel, 880 F. Supp. 1027, 1040 n.14 (S.D.N.Y. 1995).

"[T]he statements constituting a scientific explanation must be capable of empirical test,"[43] and then Popper: "[T]he criterion of the scientific status of a theory is its falsifiability, or refutability, or testability."[44] The other factors listed by the Court are generally complementary. For example, the second factor, peer review and publication, is a means to verify the results of the testing mentioned in the first factor; and in turn, verification can lead to general acceptance of the technique within the broader scientific community.[45] These factors serve as circumstantial evidence that other experts have examined the underlying research and found it to be sound. Similarly, another factor, an error rate, is derived from testing.

IV. National Research Council Report on Forensic Science

In 2005, the Science, State, Justice, Commerce, and Related Agencies Appropriations Act became law.[46] The accompanying Senate report commented that, "[w]hile a great deal of analysis exists of the requirements of the discipline of DNA, there exists little or no analysis of the . . . needs of the [forensic] community outside of the area of DNA."[47] In the Act, Congress authorized the National Academy of Sciences (NAS) to conduct a comprehensive study of the current state of forensic science to develop recommendations. In fall 2006, the Academy established the Committee on Identifying the Needs of the Forensic Science Community within NRC to fulfill the task appointed by Congress. In February 2009, NRC released the report *Strengthening Forensic Science in the United States: A Path Forward*.[48]

43. Carl G. Hempel, Philosophy of Natural Science 49 (1966).

44. Karl R. Popper, Conjectures and Refutations: The Growth of Scientific Knowledge 37 (5th ed. 1989).

45. In their amici brief in *Daubert*, the *New England Journal of Medicine* and other medical journals observed:

> "Good science" is a commonly accepted term used to describe the scientific community's system of quality control which protects the community and those who rely upon it from unsubstantiated scientific analysis. It mandates that each proposition undergo a rigorous trilogy of publication, replication and verification before it is relied upon.

Brief for the New England Journal of Medicine, Journal of the American Medical Association, and Annals of Internal Medicine as Amici Curiae supporting Respondent at *2, Daubert v. Merrell Dow Pharms., Inc., 509 U.S. 579 (1993) (No. 92-102), 1993 WL 13006387. Peer review's "role is to promote the publication of well-conceived articles so that the most important review, the consideration of the reported results by the scientific community, may occur after publication." *Id.* at *3.

46. Pub. L. No. 109-108, 119 Stat. 2290 (2005).

47. S. Rep. No. 109-88, at 46 (2005).

48. NRC Forensic Science Report, *supra* note 3. The Supreme Court cited the report 3 months later. Melendez-Diaz v. Massachusetts, 129 S. Ct. 2527 (2009).

In keeping with its congressional charge, the NRC committee did not address admissibility issues. The NRC report stated: "No judgment is made about past convictions and no view is expressed as to whether courts should reassess cases that already have been tried."[49] When the report was released, the co-chair of the NRC committee stated:

> I want to make it clear that the committee's report does not mean to offer any judgments on any cases in the judicial system. The report does not assess past criminal convictions, nor does it speculate about pending or future cases. And the report offers no proposals for law reform. That was beyond our charge. Each case in the criminal justice system must be decided on the record before the court pursuant to the applicable law, controlling precedent, and governing rules of evidence. The question whether forensic evidence in a particular case is admissible under applicable law is not coterminous with the question whether there are studies confirming the scientific validity and reliability of a forensic science discipline.[50]

Yet, in one passage, the report remarked: "Much forensic evidence—including, for example, bite marks and firearm and toolmark identifications—is introduced in criminal trials without any meaningful scientific validation, determination of error rates, or reliability testing to explain the limits of the discipline."[51] Moreover, the report did discuss a number of forensic techniques and, where relevant, passages from the report are cited throughout this chapter.

As the NRC report explained, its primary focus is forward-looking—to outline an "agenda for progress."[52] The report's recommendations are wide-ranging, covering diverse topics such as medical examiner systems,[53] interoperability of the automated fingerprint systems,[54] education and training in the forensic sciences,[55] codes of ethics,[56] and homeland security issues.[57] Some recommendations are

49. *Id.* at 85. The report goes on to state:

> The report finds that the existing legal regime—including the rules governing the admissibility of forensic evidence, the applicable standards governing appellate review of trial court decisions, the limitations of the adversary process, and judges and lawyers who often lack the scientific expertise necessary to comprehend and evaluate forensic evidence—is inadequate to the task of curing the documented ills of the forensic science disciplines.

Id.

50. Harry T. Edwards, Co-Chair, Forensic Science Committee, Opening Statement of Press Conference (Feb. 18, 2009), transcript *available at* http://www.nationalacademies.org/includes/OSEdwards.pdf.

51. NRC Forensic Science Report, *supra* note 3, at 107–08.

52. *Id.* at xix.

53. Recommendation 10 (urging the replacement of the coroner with medical examiner system in medicolegal death investigation).

54. Recommendation 11.

55. Recommendation 2.

56. Recommendation 9.

57. Recommendation 12.

structural—that is, the creation of an independent federal entity (to be named the National Institute of Forensic Sciences) to oversee the field[58] and the removal of crime laboratories from the "administrative" control of law enforcement agencies.[59] The National Institute of Forensic Sciences would be responsible for (1) establishing and enforcing best practices for forensic science professionals and laboratories; (2) setting standards for the mandatory accreditation of crime laboratories and the mandatory certification of forensic scientists; (3) promoting scholarly, competitive, peer-reviewed research and technical development in the forensic sciences; and (4) developing a strategy to improve forensic science research. Congressional action would be needed to establish the institute. Several other recommendations are discussed below.

A. Research

The NRC report urged funding for additional research "to address issues of accuracy, reliability, and validity in the forensic science disciplines."[60] In the report's words, "[a]mong existing forensic methods, only nuclear DNA analysis has been rigorously shown to have the capacity to consistently, and with a high degree of certainty, demonstrate a connection between an evidentiary sample and a specific individual or source."[61] In another passage, the report discussed the need for further research into the premises underlying forensic disciplines other than DNA:

> A body of research is required to establish the limits and measures of performance and to address the impact of sources of variability and potential bias. Such research is sorely needed, but it seems to be lacking in most of the forensic disciplines that rely on subjective assessments of matching characteristics. These disciplines need to develop rigorous protocols to guide these subjective interpretations and pursue equally rigorous research and evaluation programs.[62]

58. Recommendation 1.

59. Recommendation 4.

60. *Id.* at 22 (Recommendation 3).

61. *Id.* at 100; *see also id.* at 7 & 87.

62. *Id.* at 8; *see also id.* at 15 ("Of the various facets of underresourcing, the committee is most concerned about the knowledge base. Adding more dollars and people to the enterprise might reduce case backlogs, but it will not address fundamental limitations in the capabilities of forensic science disciplines to discern valid information from crime scene evidence."); *id.* at 22 ("[S]ome forensic science disciplines are supported by little rigorous systematic research to validate the discipline's basic premises and techniques. There is no evident reason why such research cannot be conducted.").

B. Observer Effects

Another recommendation focuses on research to investigate observer bias and other sources of human error in forensic examinations.[63] According to psychological theory of observer effects, external information provided to persons conducting analyses may taint their conclusions—a serious problem in techniques with a subjective component.[64] A growing body of modern research, noted in the report,[65] demonstrates that exposure to such information can affect forensic science experts. For example, a handwriting examiner who is informed that an exemplar belongs to the prime suspect in a case may be subconsciously influenced by this information.[66]

One of the first studies to document the biasing effect was a research project involving hair analysts.[67] Some recent studies involving fingerprints have found biasing.[68] Another study concluded that external information had an effect but not toward making errors. Instead, these researchers found fewer definitive and

63. Recommendation 8:

> Such programs might include studies to determine the effects of contextual bias in forensic practice (e.g., studies to determine whether and to what extent the results of forensic analyses are influenced by knowledge regarding the background of the suspect and the investigator's theory of the case). In addition, research on sources of human error should be closely linked with research conducted to quantify and characterize the amount of error.

64. *See generally* D. Michael Risinger et al., *The* Daubert/Kumho *Implications of Observer Effects in Forensic Science: Hidden Problems of Expectation and Suggestion*, 90 Cal. L. Rev. 1 (2002).

65. NRC Forensic Science Report, *supra* note 3, at 139 n.23 & 185 n.2.

66. *See* L.S. Miller, *Bias Among Forensic Document Examiners: A Need for Procedural Change*, 12 J. Police Sci. & Admin. 407, 410 (1984) ("The conclusions and opinions reported by the examiners supported the bias hypothesis."). Confirmation bias is another illustration. The FBI noted the problem in its internal investigation of the Mayfield case. A review by another examiner was not conducted blind—that is, the reviewer knew that a positive identification had already been made—and thus was subject to the influence of confirmation bias. Robert B. Stacey, *A Report on the Erroneous Fingerprint Individualization in the Madrid Train Bombing Case*, 54 J. Forensic Identification 707 (2004).

67. *See* Larry S. Miller, *Procedural Bias in Forensic Science Examinations of Human Hair*, 11 Law & Hum. Behav. 157 (1987). In the conventional method, the examiner is given hair samples from a known suspect along with a report including other facts and information relating to the guilt of the suspect. "The findings of the present study raise some concern regarding the amount of unintentional bias among human hair identification examiners. . . . A preconceived conclusion that a questioned hair sample and a known hair sample originated from the same individual may influence the examiner's opinion when the samples are similar." *Id.* at 161.

68. *See* Itiel Dror & Robert Rosenthal, *Meta-analytically Quantifying the Reliability and Biasability of Forensic Experts*, 53 J. Forensic Sci. 900 (2008); Itiel E. Dror et al., *Contextual Information Renders Experts Vulnerable to Making Erroneous Identifications*, 156 Forensic Sci. Int'l 74 (2006); Itiel Dror et al., *When Emotions Get the Better of Us: The Effect of Contextual Tap-Down Processing on Matching Fingerprints*, 19 App. Cognit. Psychol. 799 (2005).

erroneous judgments.[69] In any event, forensic examinations should, to the extent feasible, be conducted "blind."[70]

C. Accreditation and Certification

The NRC report called for the mandatory accreditation of crime labs and the certification of examiners.[71] Accreditation and certification standards should be based on recognized international standards, such as those published by the International Organization for Standardization (ISO). According to the report, no person (public or private) ought to practice or testify as a forensic expert without certification.[72] In addition, laboratories should establish "quality assurance and quality control procedures to ensure the accuracy of forensic analyses and the work of forensic practitioners."[73]

The American Society of Crime Lab Directors/Laboratory Accreditation Board (ASCLD/LAB) is the principal accrediting organization in the United States. Accreditation requirements generally include ensuring the integrity of evidence, adhering to valid and generally accepted procedures, employing qualified examiners, and operating quality assurance programs—that is, proficiency testing, technical reviews, audits, and corrective action procedures.[74] Currently, accreditation is mostly voluntary. Only a few states require accreditation of crime

69. Glenn Langenburg et al., *Testing for Potential Contextual Bias Effects During the Verification Stage of the ACE-V Methodology When Conducting Fingerprint Comparisons*, 54 J. Forensic Sci. 571 (2009). As the researchers acknowledge, the examiners knew that they were being tested.

70. *See* Mike Redmayne, Expert Evidence and Criminal Justice 16 (2001) ("To the extent that we are aware of our vulnerability to bias, we may be able to control it. In fact, a feature of good scientific practice is the institution of processes—such as blind testing, the use of precise measurements, standardized procedures, statistical analysis—that control for bias.").

71. Recommendation 3; *see also* NRC Forensic Science Report, *supra* note 3, at 23 ("In short, oversight and enforcement of operating standards, certification, accreditation, and ethics are lacking in most local and state jurisdictions.").

72. *Id.*, Recommendation 7. The recommendation goes on to state:

> Certification requirements should include, at a minimum, written examinations, supervised practice, proficiency testing, continuing education, recertification procedures, adherence to a code of ethics, and effective disciplinary procedures. All laboratories (public or private) should be accredited and all forensic science professionals should be certified, when eligible, within a time period estbalished by NIFS.

73. *Id.*, Recommendation 8. The recommendation further comments: "Quality control procedures should be designed to: identify mistakes, fraud, and bias; confirm the continued validity and reliability of standard operating procedures and protocols; ensure that best practices are being followed; and correct procedures and protocols that are found to need improvement."

74. *See* Jan S. Bashinski & Joseph L. Peterson, *Forensic Sciences, in* Local Government: Police Management 559, 578 (William Geller & Darrel Stephens eds., 4th ed. 2004).

laboratories.[75] New York mandated accreditation in 1994.[76] Texas[77] and Oklahoma[78] followed after major crime laboratory failures.

D. Proficiency Testing

Several of the report's recommendations referred to proficiency testing,[79] of which there are several types: internal or external, and blind or nonblind (declared).[80] The results of the first Laboratory Proficiency Testing Program, sponsored by the Law Enforcement Assistance Administration (LEAA), were reported in 1978.[81] Voluntary proficiency testing continued after this study.[82] The DNA Identification Act of 1994 mandated proficiency testing for examiners at the FBI as well as for

75. The same is true for certification. NRC Forensic Science Report, *supra* note 3, at 6 ("[M]ost jurisdictions do not require forensic practitioners to be certified, and most forensic science disciplines have no mandatory certification program.").

76. N.Y. Exec. Law § 995-b (McKinney 2003) (requiring accreditation by the state Forensic Science Commission); *see also* Cal. Penal Code § 297 (West 2004) (requiring accreditation of DNA units by ASCLD/LAB or any certifying body approved by ASCLD/LAB); Minn. Stat. Ann. § 299C.156(2)(4) (West Supp. 2006) (specifying that the Forensic Science Advisory Board should encourage accreditation by ASCLD/LAB or other accrediting body).

77. Tex. Code Crim. Proc. Ann. art. 38.35 (Vernon 2004) (requiring accreditation by the Department of Public Safety). Texas also created a Forensic Science Commission. *Id.* art. 38.01 (2007).

78. Okla. Stat. Ann. tit. 74, § 150.37(D) (West 2004) (requiring accreditation by ASCLD/LAB or the American Board of Forensic Toxicology).

79. Recommendations 6 & 7.

80. Proficiency testing does not automatically correlate with a technique's "error rate." There is a question whether error rate should be based on the results of declared and/or blind proficiency tests of simulated evidence administered to crime laboratories, or if this rate should be based on the retesting of actual case evidence drawn randomly (1) from the files of crime laboratories or (2) from evidence presented to courts in prosecuted and/or contested cases.

81. Joseph L. Peterson et al., Crime Laboratory Proficiency Testing Research Program (1978) [hereinafter Laboratory Proficiency Test]. The report concluded: "A wide range of proficiency levels among the nation's laboratories exists, with several evidence types posing serious difficulties for the laboratories. . . ." *Id.* at 3. Although the proficiency tests identified few problems in certain forensic disciplines such as glass analysis, tests of other disciplines such as hair analysis produced very high rates of "unacceptable proficiency." According to the report, unacceptable proficiency was most often caused by (1) misinterpretation of test results due to carelessness or inexperience, (2) failure to employ adequate or appropriate methodology, (3) mislabeling or contamination of primary standards, and (4) inadequate databases or standard spectra. *Id.* at 258.

82. *See* Joseph L. Peterson & Penelope N. Markham, *Crime Laboratory Proficiency Testing Results, 1978–1991, Part I: Identification and Classification of Physical Evidence*, 40 J. Forensic Sci. 994 (1995); Joseph L. Peterson & Penelope N. Markham, *Crime Laboratory Proficiency Testing Results, 1978–1991, Part II: Resolving Questions of Common Origin*, 40 J. Forensic Sci. 1009 (1995). After collaborating with the Forensic Sciences Foundation in the initial LEAA-funded crime laboratory proficiency testing research program, Collaborative Testing Services, Inc. (CTS) began in 1978 to offer a fee-based testing program. Today, CTS offers samples in many scientific evidence testing areas to more than 500 forensic science laboratories worldwide. See test results at www.collaborativetesting.com/.

analysts in laboratories that participate in the national DNA database or receive federal funding.[83]

E. Standard Terminology

The NRC report voiced concern about the use of terms such as "match," "consistent with," "identical," "similar in all respects tested," and "cannot be excluded as the source of." These terms can have "a profound effect on how the trier of fact in a criminal or civil matter perceives and evaluates scientific evidence."[84] Such terms need to be defined and standardized, according to the report.

F. Laboratory Reports

A related recommendation concerns laboratory reports and the need for model formats.[85] The NRC report commented:

> As a general matter, laboratory reports generated as the result of a scientific analysis should be complete and thorough. They should contain, at minimum, "methods and materials," "procedures," "results," "conclusions," and, as appropriate, sources and magnitudes of uncertainty in the procedures and conclusions (e.g., levels of confidence). Some forensic science laboratory reports meet this standard of reporting, but many do not. Some reports contain only identifying and agency information, a brief description of the evidence being submitted, a brief description of the types of analysis requested, and a short statement of the results (e.g., "the greenish, brown plant material in item #1 was identified as marijuana"), and they include no mention of methods or any discussion of measurement uncertainties.[86]

In addition, reports "must include clear characterizations of the limitations of the analyses, including measures of uncertainty in reported results and associated estimated probabilities where possible."[87]

83. 42 U.S.C. § 14131(c) (2005). The DNA Act authorized a study of the feasibility of blind proficiency testing; that study raised questions about the cost and practicability of this type of examination, as well as its effectiveness when compared with other methods of quality assurance such as accreditation and more stringent external case audits. Joseph L. Peterson et al., *The Feasibility of External Blind DNA Proficiency Testing. 1. Background and Findings*, 48 J. Forensic Sci. 21, 30 (2003) ("In the extreme, blind proficiency testing is possible, but fraught with problems (including costs), and it is recommended that a blind proficiency testing program be deferred for now until it is more clear how well implementation of the first two recommendations [accreditation and external case audits] are serving the same purposes as blind proficiency testing.").

84. NRC Forensic Science Report, *supra* note 3, at 21.

85. *Id.* at 22, Recommendation 2.

86. *Id.* at 21.

87. *Id.* at 21–22.

V. Specific Techniques

The broad field of forensic science includes disparate disciplines such as forensic pathology, forensic anthropology, arson investigation, and gunshot residue testing.[88] The NRC report explained:

> Some of the forensic science disciplines are laboratory based (e.g., nuclear and mitochondrial DNA analysis, toxicology and drug analysis); others are based on expert interpretation of observed patterns (e.g., fingerprints, writing samples, toolmarks, bite marks, and specimens such as hair). . . . There are also sharp distinctions between forensic practitioners who have been trained in chemistry, biochemistry, biology, and medicine (and who bring these disciplines to bear in their work) and technicians who lend support to forensic science enterprises.[89]

The report devoted special attention to forensic disciplines in which the expert's final decision is subjective in nature: "In terms of scientific basis, the analytically based disciplines generally hold a notable edge over disciplines based on expert interpretation."[90] Moreover, many of the subjective techniques attempt to render the most specific conclusions—that is, opinions concerning "individualization."[91] Following the report's example, the remainder of this chapter focuses on "pattern recognition" disciplines, each of which contains a subjective component. These disciplines exemplify most of the issues that a trial judge may encounter in ruling on the admissibility of forensic testimony. Each part describes the technique, the available empirical research, and contemporary case law.

A. Terminology

Although courts often use the terms "validity" and "reliability" interchangeably, the terms have distinct meanings in scientific disciplines. "Validity" refers to the ability of a test to measure what it is supposed to measure—its accuracy. "Reliability" refers to whether the same results are obtained in each instance in which the test is performed—its consistency. Validity includes reliability, but the converse is not necessarily true. Thus, a reliable, invalid technique will consistently

88. Other examples include drug analysis, blood spatter examinations, fiber comparisons, toxicology, entomology, voice spectrometry, and explosives and bomb residue analysis. As the Supreme Court noted in Melendez-Diaz v. Massachusetts, 129 S. Ct. 2527, 2537–38 (2009), errors can be made when instrumental techniques, such as gas chromatography/mass spectrometry analysis, are used.

89. NRC Forensic Science Report, *supra* note 3, at 7.

90. *Id.*

91. "Often in criminal prosecutions and civil litigation, forensic evidence is offered to support conclusions about 'individualization' (sometimes referred to as 'matching' a specimen to a particular individual or other source) or about classification of the source of the specimen into one of several categories. With the exception of nuclear DNA analysis, however, no forensic method has been rigorously shown to have the capacity to consistently, and with a high degree of certainty, demonstrate a connection between evidence and a specific individual or source." *Id.*

yield inaccurate results. The Supreme Court acknowledged this distinction in *Daubert*, but the Court indicated that it was using the term "reliability" in a different sense. The Court wrote that its concern was "evidentiary reliability—that is, trustworthiness. . . . In a case involving scientific evidence, *evidentiary reliability* will be based upon *scientific validity*."[92]

In forensic science, class and individual characteristics are distinguished. Class characteristics are shared by a group of persons or objects (e.g., ABO blood types).[93] Individual characteristics are unique to an object or person. The term "match" is ambiguous because it is sometimes used to indicate the "matching" of individual characteristics, but on other occasions it is used to refer to "matching" class characteristics (e.g., blood type A at a crime scene "matches" suspect's type A blood). Expert opinions involving "individual" and "class" characteristics raise different issues. In the former, the question is whether an individuation determination rests on a firm scientific foundation.[94] For the latter, the question is determining the size of the class.[95]

VI. Fingerprint Evidence

Sir William Herschel, an Englishman serving in the Indian civil service, and Henry Faulds, a Scottish physician serving as a missionary in Japan, were among the first to suggest the use of fingerprints as a means of personal identification. Since 1858, Herschel had been collecting the handprints of natives for that purpose. In 1880, Faulds published an article entitled "On the Skin—Furrows

92. 509 U.S. at 590 n.9 ("We note that scientists typically distinguish between 'validity' (does the principle support what it purports to show?) and 'reliability' (does application of the principle produce consistent results?). . . .").

93. *See* Bashinski & Peterson, *supra* note 74, at 566 ("The forensic scientist first investigates whether items possess similar 'class' characteristics—that is, whether they possess features shared by all objects or materials in a single class or category. (For firearms evidence, bullets of the same caliber, bearing rifling marks of the same number, width, and direction of twist, share class characteristics. They are consistent with being fired from the same *type* of weapon.) The forensic scientist then attempts to determine an item's 'individuality'—the features that make one thing different from all others similar to it, including those with similar class characteristics.").

94. *See* Michael Saks & Jonathan Koehler, *The Individualization Fallacy in Forensic Science Evidence*, 61 Vand. L. Rev. 199 (2008).

95. *See* Margaret A. Berger, *Procedural Paradigms for Applying the* Daubert *Test*, 78 Minn. L. Rev. 1345, 1356–57 (1994) ("We allow eyewitnesses to testify that the person fleeing the scene wore a yellow jacket and permit proof that a defendant owned a yellow jacket without establishing the background rate of yellow jackets in the community. Jurors understand, however, that others than the accused own yellow jackets. When experts testify about samples matching in every respect, the jurors may be oblivious to the probability concerns if no background rate is offered, or may be unduly prejudiced or confused if the probability of a match is confused with the probability of guilt, or if a background rate is offered that does not have an adequate scientific foundation.").

of the Hand" in *Nature*.[96] Sir Francis Galton authored the first textbook on the subject.[97] Individual ridge characteristics came to be known as "Galton details."[98] Subsequently, Edward Henry, the Inspector General of Police in Bengal, realized the potential of fingerprinting for law enforcement and helped establish the Fingerprint Branch at Scotland Yard when he was recalled to England in 1901.[99]

English and American courts have accepted fingerprint identification testimony for just over a century. "The first English appellate endorsement of fingerprint identification testimony was the 1906 opinion in *Rex v. Castleton*. . . . In 1906 and 1908, Sergeant Joseph Faurot, a New York City detective who had in 1904 been posted to Scotland Yard to learn about fingerprinting, used his new training to break open two celebrated cases: in each instance fingerprint identification led the suspect to confess. . . ."[100] A 1911 Illinois Supreme Court decision, *People v. Jennings*,[101] is the first published American appellate opinion sustaining the admission of fingerprint testimony.

Over the years, fingerprint analysis became the gold standard of forensic identification expertise. In fact, proponents of new, emerging techniques in forensics would sometimes attempt to invoke onto the new techniques the prestige of fingerprint analysis. Thus, advocates of sound spectrography referred to it as "voiceprint" analysis.[102] Likewise, some early proponents of DNA typing alluded to it as "DNA fingerprinting."[103] However, as previously noted, DNA analysis has replaced fingerprint analysis as the gold standard.

A. The Technique

Even a cursory study of fingerprints establishes that there is "intense variability . . . in even small areas of prints."[104] Given that variability, it is generally assumed that an identification is possible if the comparison involves two sets of clear images of all 10 fingerprints. These are known as "record" prints and are typically rolled onto a fingerprint card or digitized and scanned into an electronic file. Two complete fingerprint sets are available for comparison in some settings such as

96. Henry Faulds, *On the Skin—Furrows of the Hand,* 22 Nature 605 (1881). *See generally* Simon Cole, Suspect Identities: A History of Fingerprint and Criminal Identification (2001).

97. Francis Galton, Fingerprints (1892).

98. *See* Andre A. Moenssens, Scientific Evidence in Civil and Criminal Cases § 10.02, at 621 (5th ed. 2007).

99. United States v. Llera Plaza, 188 F. Supp. 2d 549, 554 (E.D. Pa. 2002).

100. *Id.* at 572.

101. 96 N.E. 1077 (Ill. 1911).

102. Kenneth Thomas, *Voiceprint—Myth or Miracle, in* Scientific and Expert Evidence 1015 (2d ed. 1981).

103. Colin Norman, *Maine Case Deals Blow to DNA Fingerprinting,* 246 Science 1556 (Dec. 22, 1989).

104. David A. Stoney, *Scientific Status, in* 4 David L. Faigman et al., Modern Scientific Evidence: The Law and Science of Expert Testimony § 32:45, at 361 (2007–2008 ed.).

immigration matters. However, in the law enforcement setting, the task is more challenging because only a partial impression (latent print) of a single finger may be left by a criminal.

Fingerprint evidence is based on three assumptions: (1) the uniqueness of each person's friction ridges, (2) the permanence of those ridges throughout a person's life, and (3) the transferability of an impression of that uniqueness to another surface. The last point raises the most significant issue of reliability because a crime scene (latent) impression is often only a fifth of the size of the record print. Furthermore, variations in pressure and skin elasticity almost inevitably distort the impression.[105] Consequently, fingerprint impressions from the same person typically differ in some respects each time the impression is left on an object.[106]

Although fingerprint analysis is based on physical characteristics, the final step in the analysis—the formation of an opinion regarding individuation—is subjective.[107] Examiners lack population frequency data to quantify how rare or common a particular type of fingerprint characteristic is.[108] Rather, in making that judgment, the examiner relies on personal experience and discussions with colleagues. Although examiners in some countries must find a certain minimum number of points of similarities between the latent and the known before declaring a match,[109] neither the FBI nor New Scotland Yard requires any set number.[110] A single inexplicable difference between the two impressions precludes finding a match. Because there are frequently "dissimilarities" between the crime scene and record prints, the examiner must decide whether there is a *true* dis-

105. *See* United States v. Mitchell, 365 F.3d 215, 220–21 (3d Cir. 2004) ("Criminals generally do not leave behind full fingerprints on clean, flat surfaces. Rather, they leave fragments that are often distorted or marred by artifacts. . . . Testimony at the *Daubert* hearing suggested that the typical latent print is a fraction—perhaps 1/5th—of the size of a full fingerprint."). "In the jargon, artifacts are generally small amounts of dirt or grease that masquerade as parts of the ridge impressions seen in a fingerprint, while distortions are produced by smudging or too much pressure in making the print, which tends to flatten the ridges on the finger and obscure their detail." *Id.* at 221 n.1.

106. NRC Forensic Science Report, *supra* note 3, at 144 ("The impression left by a given finger will differ every time, because of inevitable variations in pressure, which change the degree of contact between each part of the ridge structure and the impression medium.").

107. *See* Commonwealth v. Patterson, 840 N.E.2d 12, 15, 16–17 (Mass. 2005) ("These latent print impressions are almost always partial and may be distorted due to less than full, static contact with the object and to debris covering or altering the latent impression"; "In the evaluation stage, . . . the examiner relies on his subjective judgment to determine whether the quality and quantity of those similarities are sufficient to make an identification, an exclusion, or neither"); Zabell, *supra* note 25, at 158 ("In contrast to the scientifically-based statistical calculations performed by a forensic scientist in analyzing DNA profile frequencies, each fingerprint examiner renders an opinion as to the similarity of friction ridge detail based on his subjective judgment.").

108. NRC Forensic Science Report, *supra* note 3, at 139–40 & 144.

109. Stoney, *supra* note 104, § 32:34, at 354–55.

110. United States v. Llera Plaza, 188 F. Supp. 2d 549, 566–71 (E.D. Pa. 2002).

similarity, or whether the apparent dissimilarity can be discounted as an artifact or resulting from distortion.[111]

Three levels of details may be scrutinized: Level 1 details are general flow ridge patterns such as whorls, loops, and arches.[112] Level 2 details are fine ridges or minutiae such as bifurcations, dots, islands, and ridge endings.[113] These minutiae are essentially ridge discontinuities.[114] Level 3 details are "microscopic ridge attributes such as the width of a ridge, the shape of its edge, or the presence of a sweat pore near a particular ridge."[115] Within the fingerprint community there is disagreement about the usefulness and reliability of Level 3 details.[116]

FBI examiners generally follow a procedure known as analysis, comparison, evaluation, and verification (ACE-V). In the *analysis* stage, the examiner studies the latent print to determine whether the quantity and quality of details in the print are sufficient to permit further evaluation.[117] The latent print may be so fragmentary or smudged that analysis is impossible. In the *evaluation* stage, the examiner considers at least the Level 2 details, including "the type of minutiae (forks or ridge endings), their direction (loss or production of a ridge) and their relative position (how many intervening ridges there are between minutiae and how far along the ridges it is from one minutiae to the next)."[118] Again, if the examiner finds a single, inexplicable difference between the two prints, the examiner concludes that there is no match.[119] Alternatively, if the examiner concludes that there is a match, the examiner seeks *verification* by a second examiner. "[T]he friction ridge community actively discourages its members from testifying in terms of the probability of a match; when a latent print examiner testifies that two impressions

111. *Patterson*, 840 N.E.2d at 17 ("There is a rule of examination, the 'one-discrepancy' rule, that provides that a nonidentification finding should be made if a single discrepancy exists. However, the examiner has the discretion to ignore a possible discrepancy if he concludes, based on his experience and the application of various factors, that the discrepancy might have been caused by distortions of the fingerprint at the time it was made or at the time it was collected.").

112. *See id.* at 16 ("Level one detail involves the general ridge flow of a fingerprint, that is, the pattern of loops, arches, and whorls visible to the naked eye. The examiner compares this information to the exemplar print in an attempt to exclude a print that has very clear dissimilarities.").

113. *See id.* ("Level two details include ridge characteristics (or Galton Points) like islands, dots, and forks, formed as the ridges begin, end, join or bifurcate."). *See generally* FBI, The Science of Fingerprints (1977).

114. Stoney, *supra* note 104, § 32:31, at 350.

115. *See Patterson*, 840 N.E.2d at 16.

116. *See* Office of the Inspector General, U.S. Dep't of Justice, A Review of the FBI's Handling of the Brandon Mayfield Case, Unclassified Executive Summary 8 (Jan. 2006) *available at* www.justice.gov/oig/special/s0601/PDF list.htm. ("Because Level 3 details are so small, the appearance of such details in fingerprints is highly variable, even between different fingerprints made by the same finger. As a result, the reliability of Level 3 details is the subject of some controversy within the latent fingerprint community.").

117. NRC Forensic Science Report, *supra* note 3, at 137–38.

118. Stoney, *supra* note 104, § 32:31, at 350–51.

119. NRC Forensic Science Report, *supra* note 3, at 140.

'match,' they are communicating the notion that the prints could not possibly have come from two different individuals."[120] The typical fingerprint analyst will give one of only three opinions: (1) the prints are unsuitable for analysis, (2) the suspect is definitely excluded, or (3) the latent print is definitely that of the suspect.

B. The Empirical Record

At several points, the 2009 NRC report noted that there is room for human error in fingerprint analysis. For example, the report stated that because "the ACE-V method does not specify particular measurements or a standard test protocol, . . . examiners must make subjective assessments throughout."[121] The report further commented that the ACE-V method is too "broadly stated" to "qualify as a validated method for this type of analysis."[122] The report added that "[t]he latent print community in the United States has eschewed numerical scores and corresponding thresholds" and consequently relies "on primarily subjective criteria" in making the ultimate attribution decision.[123] In making the decision, the examiner must draw on his or her personal experience to evaluate such factors as "inevitable variations" in pressure, but to date these factors have not been "characterized, quantified, or compared."[124] At the conclusion of the section devoted to fingerprint analysis, the report outlined an agenda for the research it considered necessary "[t]o properly underpin the process of friction ridge identification."[125] The report noted that some of these research projects have already begun.[126]

Fingerprint analysis raises a number of scientific issues. For example, do the salient features of fingerprints remain constant throughout a person's life?[127] Few of the underlying scientific premises have been subjected to rigorous empirical investigation,[128] although some experiments have been conducted, and proficiency test results are available.

Two experimental studies were discussed at the 2000 trial in *United States v. Mitchell*[129]:

> One of the studies conducted by the government for the *Daubert* hearing [in *Mitchell*] employed the two actual latent and the known prints that were at issue in the case. These prints were submitted to 53 state law enforcement agency

120. *Id.* at 140–41.

121. *Id.* at 139.

122. *Id.* at 142.

123. *Id.* at 141.

124. *Id.* at 144.

125. *Id.*

126. *Id.*

127. Stoney, *supra* note 104, § 32:21, at 342.

128. *See* Zabell, *supra* note 25, at 164 ("Although there is a substantial literature on the uniqueness of fingerprints, it is surprising how little true scientific support for the proposition exists.").

129. 365 F.3d 215 (3d Cir. 2004).

> crime laboratories around the country for their evaluation. Though, of the 35 that responded, most concluded that the latent and known prints matched, eight said that no match could be made to one of the prints and six said that no match could be made to the other print.[130]

Although there were no false positives, a significant percentage of the participating laboratories reported at best inconclusive findings.

Lockheed-Martin conducted the second test, the FBI-sponsored 50K study. This was an empirical study of 50,000 fingerprint images taken from the FBI's Automated Fingerprint System, a computer database. The study

> was an effort to obtain an estimate of the probability that one person's fingerprints would be mistaken for those of another person, at least to a computer system designed to match fingerprints. The FBI asked Lockheed-Martin, the manufacturer of its . . . automated fingerprint identification system, . . . to help it run a comparison of the images of 50,000 single fingerprints against the same 50,000 images, and produce a similarity score for each comparison. The point of this exercise was to show that the similarity score for an image matched against itself was far higher than the scores obtained when it was compared to the others.[131]

The comparisons between the two identical images yielded "extremely high scores."[132] Nonetheless, some commentators disputed whether the Lockheed-Martin study demonstrated the validity of fingerprint analysis.[133] The study compared a computerized image of a fingerprint impression against other computerized images in the database. The study did not address the problem examiners encounter in the real world; it did not attempt to match a partial fingerprint impression against images in the database. As noted earlier, crime scene prints are typically distorted from pressure and sometimes only one-fifth the size of record prints.[134] Even the same finger will not leave the exact impression each time: "The impression left by a given finger will differ every time, because of inevitable variations in pressure, which change the degree of contact between each part of the ridge structure and the impression medium."[135] Thus, one scholar asserted that the "study addresses the irrelevant question of whether one image of a fingerprint is immensely more similar to itself than to other images—including those of the same finger."[136] Citing

130. Stoney, *supra* note 104, § 32:3, at 287.

131. *Id.* § 32:3, at 288.

132. *Id.* (quoting James L. Wayman, Director, U.S. National Biometric Test Center at the College of Engineering, San Jose State University).

133. *E.g.*, David H. Kaye, *Questioning a Courtroom Proof of the Uniqueness of Fingerprints*, 71 Int'l Statistical Rev. 521 (2003); S. Pankanti et al., *On the Individuality of Fingerprints*, 24 IEEE Trans. Pattern Analysis Mach. Intelligence 1010 (2002).

134. *See supra* note 105 & accompanying text.

135. NRC Forensic Science Report, *supra* note 3, at 144.

136. Kaye, *supra* note 133, at 527–28. In another passage, he wrote: "[T]he study merely demonstrates the trivial fact that the same two-dimensional representation of the surface of a finger is far

this assertion, the 2009 NRC report stated that the Lockheed-Martin study "has several major design and analysis flaws."[137]

1. Proficiency testing

In *United States v. Llera Plaza*,[138] the district court described internal and external proficiency tests of FBI fingerprint analysts and their supervisors. Between 1995 and 2001, the supervisors participated in 16 external tests created by CTS.[139] One false-positive result was reported among the 16 tests.[140] During the same period, there was a total of 431 internal tests of FBI fingerprint personnel. These personnel committed no false-positive errors, but there were three false eliminations.[141] Hence, the overall error rate was approximately 0.8%.[142]

Although these proficiency tests yielded impressive accuracy rates, the quality of the tests became an issue. First, the examinees participating in the tests knew that they were being tested and, for that reason, may have been more meticulous than in regular practice. Second, the rigor of proficiency testing was questioned. The *Llera Plaza* court concluded that the FBI's internal proficiency tests were "less demanding than they should be."[143] In the judge's words, "the FBI examiners got very high proficiency grades, but the tests they took did not."[144]

more similar to itself than to such representation of the source of finger from any other person in the data set." *Id.* at 527.

137. NRC Forensic Science Report, *supra* note 3, at 144 n.35.

138. 188 F. Supp. 2d 549 (E.D. Pa. 2002).

139. *Id.* at 556.

140. However, a later inquiry led Stephen Meagher, Unit Chief of Latent Print Unit 3 of the Forensic Analysis Section of the FBI Laboratory "to conclude that the error was not one of faulty evaluation but of faulty recording of the evaluation—i.e., a clerical error rather than a technical error." *Id.*

141. *Id.*

142. Sharon Begley, *Fingerprint Matches Come Under More Fire as Potentially Fallible,* Wall St. J., Oct. 7, 2005, at B1.

143. *Llera Plaza*, 188 F. Supp. 2d at 565. A fingerprint examiner from New Scotland Yard with 25 years' experience testified that the FBI tests were deficient:

> Mr. Bayle had reviewed copies of the internal FBI proficiency tests. . . . He found the latent prints utilized in those tests to be, on the whole, markedly unrepresentative of the latent prints that would be lifted at a crime scene. In general, Mr. Bayle found the test latent prints to be far clearer than the prints an examiner would routinely deal with. The prints were too clear—they were, according to Mr. Bayle, lacking in the "background noise" and "distortion" one would expect in latent prints lifted at a crime scene. Further, Mr. Bayle testified, the test materials were deficient in that there were too few latent prints that were not identifiable; according to Mr. Bayle, at a typical crime scene only about ten percent of the lifted latent prints will turn out to be matched. In Mr. Bayle's view the paucity of non-identifiable latent prints "makes the test too easy. It's not testing their ability. . . . [I]f I gave my experts these tests, they'd fall about laughing."

Id. at 557–58.

144. *Id.* at 565; *see also* United States v. Crisp, 324 F.3d 261, 274 (4th Cir. 2003) (Michael, J., dissenting) ("Proficiency testing is typically based on a study of prints that are far superior to those usually retrieved from a crime scene.").

In an earlier proficiency study (1995), the examiners did not do as well,[145] although many of the subjects were not certified FBI examiners. Of the 156 examiners who participated, only 44% reached the correct conclusion on all the identification tasks. Eighty-eight examiners or 56% provided divergent (wrong, incorrect, erroneous) answers. Six examiners failed to identify any of the latent prints. Forty eight of the 156 examiners made erroneous identifications—representing 22% of the total identifications made by the examiners.

A 2006 study resurrected some of the questions raised by the 1995 test. In that study, examiners were presented with sets of prints that they had previously reviewed.[146] The researchers found that "experienced examiners do not necessarily agree with even their own past conclusions when the examination is presented in a different context some time later."[147]

These studies call into question the soundness of testimonial claims that fingerprint analysis is infallible[148] or has a zero error rate.[149] In 2008, Haber and Haber reviewed the literature describing the ACE-V technique and the supporting research.[150] Although many practitioners professed using the technique, Haber and Haber found that the practitioners' "descriptions [of their technique] differ, no single protocol has been officially accepted by the profession and the standards upon which the method's conclusion rest[s] have not been specified quantitatively."[151] After considering the Haber study, NRC concluded that the ACE-V "framework is not specific enough to qualify as a validated method for this type of analysis."[152]

2. *The Mayfield case*

Like the empirical data, several reports of fingerprint misidentifications raised questions about the reliability of fingerprint analysis. The FBI misidentified Brandon Mayfield as the source of the crime scene prints in the terrorist train bombing in Madrid, Spain, on March 11, 2004.[153] The mistake was attributed in part to several types of cognitive bias. According to an FBI review, the "power" of the automated

145. *See* David L. Grieve, *Possession of Truth,* 46 J. Forensic Identification 521, 524–25 (1996); James Starrs, *Forensic Science on the Ropes: An Upper Cut to Fingerprinting,* 20 Sci. Sleuthing Rev. 1 (1996).

146. Itiel E. Dror et al., *Contextual Information Renders Experts Vulnerable to Making Erroneous Identifications,* 156 Forensic Sci. Int'l 74, 76 (2006) (Four of five examiners changed their opinions; three directly contradicted their prior identifications, and the fourth concluded that data were insufficient to reach a definite conclusion); *see also* I. E. Dror & D. Charlton, *Why Experts Make Errors,* 56 J. Forensic Identification 600 (2006).

147. NRC Forensic Science Report, *supra* note 3, at 139.

148. *Id.* at 104.

149. *Id.* at 143–44.

150. Lyn Haber & Ralph Norman Haber, *Scientific Validation of Fingerprint Evidence Under* Daubert, 7 Law, Probability & Risk 87 (2008).

151. NRC Forensic Science Report, *supra* note 3, at 143.

152. *Id.* at 142.

153. *Id.* at 46 & 105.

fingerprint correlation "was thought to have influenced the examiner's initial judgment and subsequent examination."[154] Thus, he was subject to confirmation bias. Moreover, a second review by another examiner was not conducted blind—that is, the reviewer knew that a positive identification had already been made and was thus subject to expectation (context) bias. Indeed, a third expert from outside the FBI, one appointed by the court, also erroneously confirmed the identification.[155] In addition to the Bureau's review, the Inspector General of the Department of Justice investigated the case.[156] The Mayfield case is not an isolated incident.[157]

The Mayfield case led to a more extensive FBI review of the scientific basis of fingerprints.[158] In January 2006, the FBI created a three-person review committee to evaluate the fundamental basis of fingerprint analysis. The committee identified two possible approaches. One approach would be to "develop a quantifiable minimum threshold based on objective criteria"—if possible.[159] "Any minimum threshold must consider both the clarity (quality) and the quantity of features and include all levels of detail, not simply points or minutiae."[160] Apparently, some FBI examiners use an unofficial seven-point cutoff, but this standard has never been tested.[161] As the FBI Review cautioned: "It is compelling to focus on a quantifiable threshold; however, quality/clarity, that is, distortion and degradation of prints, is the fundamental issue that needs to be addressed."[162]

154. Stacey, *supra* note 66, at 713.

155. In addition, the culture at the laboratory was poorly suited to detect mistakes: "To disagree was not an expected response." *Id.*

156. *See* Office of the Inspector General, U.S. Dep't of Justice, A Review of the FBI's Handling of the Brandon Mayfield Case, Unclassified Executive Summary 9 (Jan. 2006). The I.G. made several recommendations that went beyond the FBI's internal report:

> These include recommendations that the Laboratory [1] develop criteria for the use of Level 3 details to support identifications, [2] clarify the "one discrepancy rule" to assure that it is applied in a manner consistent with the level of certainty claimed for latent fingerprint identifications, [3] require documentation of features observed in the latent fingerprint before the comparison phase to help prevent circular reasoning, [4] adopt alternate procedures for blind verifications, [5] review prior cases in which the identification of a criminal suspect was made on the basis of only one latent fingerprint searched through IAFIS, and [6] require more meaningful and independent documentation of the causes of errors as part of the Laboratory's corrective action procedures.

157. In 2005, Professor Cole released an article identifying 23 cases of documented fingerprint misidentifications. *See* Simon A. Cole, *More Than Zero: Accounting for Error in Latent Fingerprint Identification*, 95 J. Crim. L. & Criminology 985 (2005). The misidentification cases include some that involved (1) verification by one or more other examiners, (2) examiners certified by the International Association of Identification, (3) procedures using a 16-point standard, and (4) defense experts who corroborated misidentifications made by prosecution experts.

158. *See* Bruce Budowle et al., *Review of the Scientific Basis for Friction Ridge Comparisons as a Means of Identification: Committee Findings and Recommendations*, 8 Forensic Sci. Comm. (Jan. 2006) [hereinafter FBI Review].

159. *Id.* at 5.

160. *Id.*

161. There is also a 12-point cutoff, under which a supervisor's approval is required.

162. *Id.*

The second approach would treat the examiner as a "black box." This methodology would be necessary if minimum criteria for rendering an identification cannot be devised—in other words, there is simply too much subjectivity in the process to formulate meaningful, quantitative guidelines. Under this approach, it becomes critical to determine just how good a "black box" each examiner is: "The examiner(s) can be tested with various inputs of a range of defined categories of prints. This approach would demonstrate whether or not it is possible to obtain a degree of accuracy (that is, assess the performance of the black-box examiner for rendering an identification)."[163] The review committee noted that this approach would provide the greatest assurance of reliability if it incorporated blind technical review. According to the review committee's report, "[t]o be truly blind, the second examiner should have no knowledge of the interpretation by the first examiner (to include not seeing notes or reports)."[164]

Although the FBI Review concluded that reliable identifications could be made, it conceded that "there are scientific areas where improvements in the practice can be made particularly regarding validation, more objective criteria for certain aspects of the ACE-V process, and data collection."[165] Efforts to improve fingerprint analysis appear to be under way. In 2008, a symposium on validity testing of fingerprint examinations was published.[166] In late 2008, the National Institute of Standards and Technology formed the Expert Group on Human Factors in Latent Print Analysis tasked to identify the major sources of human error in fingerprint examination and to develop strategies to minimize such errors.

C. Case Law Development

As noted earlier, the seminal American decision is the Illinois Supreme Court's 1911 opinion in *Jennings*.[167] Fingerprint testimony was routinely admitted in later

163. *Id.* at 4.

164. *Id.*

165. *Id.* at 10.

166. The lead article is Lyn Haber & Ralph Norman Haber, *supra* note 150. Other contributors are Christopher Champod, *Fingerprint Examination: Towards More Transparency*, 7 Law, Probability & Risk 111 (2008); Simon A. Cole, *Comment on "Scientific Validation of Fingerprint Evidence Under* Daubert," 7 Law, Probability & Risk 119 (2008); Jennifer Mnookin, *The Validity of Latent Fingerprint Identification: Confessions of a Fingerprinting Moderate*, 7 Law, Probability & Risk 127 (2008).

167. People v. Jennings, 96 N.E. 1077 (Ill. 1911); *see* Donald Campbell, *Fingerprints: A Review*, [1985] Crim. L. Rev. 195, 196 ("Galton gave evidence to the effect that the chance of agreement would be in the region of 1 in 64,000,000,000."). As Professor Mnookin has noted, however, "fingerprints were accepted as an evidentiary tool without a great deal of scrutiny or skepticism." Mnookin, *supra* note 17, at 17. She elaborated:

> Even if no two people had identical *sets* of fingerprints, this did not establish that no two people could have a *single* identical print, much less an identical *part* of a print. These are necessarily matters of prob-

years. Some courts stated that fingerprint evidence was the strongest proof of a person's identity.[168]

With the exception of one federal district court decision that was later withdrawn,[169] the post-*Daubert* federal cases have continued to accept fingerprint testimony about individuation at least as sufficiently reliable nonscientific expertise.[170]

Two subsequent state court decisions also deserve mention. In one, a Maryland trial judge excluded fingerprint evidence under the *Frye* test, which still controls in that state.[171] In the other case, *Commonwealth v. Patterson*,[172] the Supreme Judicial

> ability, but neither the court in *Jennings* nor subsequent judges ever required that fingerprint identification be placed on a secure statistical foundation.

Id. at 19.

168. People v. Adamson, 165 P.2d 3, 12 (Cal. 1946), *aff'd*, 332 U.S. 46 (1947).

169. United States v. Llera Plaza, 179 F. Supp. 2d 492 (E.D. Pa.), *vacated, mot. granted on recons.*, 188 F. Supp. 2d 549 (E.D. Pa. 2002). The ruling was limited to excluding expert testimony that two sets of prints "matched"—that is, a positive identification to the exclusion of all other persons:

> Accordingly, this court will permit the government to present testimony by fingerprint examiners who, suitabl[y] qualified as "expert" examiners by virtue of training and experience, may (1) describe how the rolled and latent fingerprints at issue in this case were obtained, (2) identify and place before the jury the fingerprints and such magnifications thereof as may be required to show minute details, and (3) point out observed similarities (and differences) between any latent print and any rolled print the government contends are attributable to the same person. What such expert witnesses will not be permitted to do is to present "evaluation" testimony as to their "opinion" (Rule 702) that a particular latent print is in fact the print of a particular person.

Id. at 516. On rehearing, however, the court reversed itself. A spate of legal articles followed. *See, e.g.*, Simon A. Cole, *Grandfathering Evidence: Fingerprint Admissibility Rulings from* Jennings *to* Llera Plaza *and Back Again*, 41 Am. Crim. L. Rev. 1189 (2004); Robert Epstein, *Fingerprints Meet* Daubert: *The Myth of Fingerprint "Science" Is Revealed*, 75 S. Cal. L. Rev. 605 (2002); Kristin Romandetti, *Recognizing and Responding to a Problem with the Admissibility of Fingerprint Evidence Under* Daubert, 45 Jurimetrics J. 41 (2004).

170. *See, e.g.*, United States v. Baines, 573 F.3d 979, 990 (10th Cir. 2009) ("[U]nquestionably the technique has been subject to testing, albeit less rigorous than a scientific ideal, in the world of criminal investigation, court proceedings, and other practical applications, such as identification of victims of disasters. Thus, while we must agree with defendant that this record does not show that the technique has been subject to testing that would meet all of the standards of science, it would be unrealistic in the extreme for us to ignore the countervailing evidence. Fingerprint identification has been used extensively by law enforcement agencies all over the world for almost a century."); United States v. Abreu, 406 F.3d 1304, 1307 (11th Cir. 2005) ("We agree with the decisions of our sister circuits and hold that the fingerprint evidence admitted in this case satisfied *Daubert*."); United States v. Janis, 387 F.3d 682, 690 (8th Cir. 2004) (finding fingerprint evidence to be reliable); United States v. Mitchell, 365 F.3d 215, 234–52 (3d Cir. 2004); United States v. Crisp, 324 F.3d 261, 268–71 (4th Cir. 2003); United States v. Collins, 340 F.3d 672, 682 (8th Cir. 2002) ("Fingerprint evidence and analysis is generally accepted."); United States v. Hernandez, 299 F.3d 984, 991 (8th Cir. 2002); United States v. Sullivan, 246 F. Supp. 2d 700, 704 (E.D. Ky. 2003); United States v. Martinez-Cintron, 136 F. Supp. 2d 17, 20 (D.P.R. 2001).

171. State v. Rose, No. K06-0545, 2007 WL 5877145 (Cir. Ct. Baltimore, Md., Oct. 19, 2007). *See* NRC Forensic Science Report, *supra* note 3, at 43 & 105. However, in a parallel federal case, the evidence was admitted. United States v. Rose, 672 F. Supp. 2d 723 (D. Md. 2009).

172. 840 N.E.2d 12 (Mass. 2005).

Court of Massachusetts considered the reliability of applying the ACE-V methodology to simultaneous impressions. Simultaneous impressions "are two or more friction ridge impressions from the fingers and/or palm on one hand that are determined to have been deposited at the same time."[173] The key is deciding whether the impressions were left at the same time and therefore came from the same person, rather than having been left by two different people at different times.[174] Although the court found that the ACE-V method is generally accepted by the relevant scientific community, the record did not demonstrate similar acceptance of that methodology as applied to simultaneous impressions. The court consequently remanded the case to the trial court.[175]

VII. Handwriting Evidence

The Lindbergh kidnapping trial showcased testimony by questioned document examiners. Later, in the litigation over Howard Hughes' alleged will, both sides relied on handwriting comparison experts.[176] Thanks in part to such cases, questioned document examination expertise has enjoyed widespread use and judicial acceptance.

A. The Technique

Questioned document examiners are called on to perform a variety of tasks such as determining the sequence of strokes on a page and whether a particular ink formulation existed on the purported date of a writing.[177] However, the most common task performed is signature authentication—that is, deciding whether to attribute the handwriting on a document to a particular person. Here, the examiner compares known samples of the person's writing to the questioned

173. FBI Review, *supra* note 158, at 7.

174. *Patterson*, 840 N.E.2d at 18 ("[T]he examiner apparently may take into account the distance separating the latent impressions, the orientation of the impressions, the pressure used to make the impression, and any other facts the examiner deems relevant. The record does not, however, indicate that there is any approved standardized method for making the determination that two or more print impressions have been made simultaneously.").

175. The FBI review addressed this subject: "[I]f an item could only be held in a certain manner, then the only way of explaining the evidence is that the multiple prints are from the single person. In some cases, identifying simultaneous prints may infer, for example, the manner in which a knife was held." FBI Review, *supra* note 158, at 8. However, the review found that there was not agreement on what constitutes a "simultaneous impression," and therefore, more explicit guidelines were needed.

176. Irby Todd, *Do Experts Frequently Disagree?* 18 J. Forensic Sci. 455, 457–59 (1973).

177. Questioned document examinations cover a wide range of analyses: handwriting, hand printing, typewriting, mechanical impressions, altered documents, obliterated writing, indented writing, and charred documents. *See* 2 Paul C. Giannelli & Edward J. Imwinkelried, Scientific Evidence ch. 21 (4th ed. 2007).

document. In performing this comparison, examiners consider (1) class and (2) individual characteristics. Of class characteristics, two types are weighed: system[178] and group. People exhibiting system characteristics would include, for example, those who learned the Palmer method of cursive writing, taught in many schools. Such people should manifest some of the characteristics of that writing style. An example of people exhibiting group characteristics would include persons of certain nationalities who tend to have some writing mannerisms in common.[179] The writing of arthritic or blind persons also tends to exhibit some common general characteristics.[180]

Individual characteristics take several forms: (1) the manner in which the author begins or ends the word, (2) the height of the letters, (3) the slant of the letters, (4) the shading of the letters, and (5) the distance between the words. An identification rarely rests on a single characteristic. More commonly, a combination of characteristics is the basis for an identification. As in fingerprint analysis, there is no universally accepted number of points of similarity required for an individuation opinion. As with fingerprints, the examiner's ultimate judgment is subjective.

There is one major difference, though, between the approaches taken by fingerprint analysts and questioned document examiners. As previously stated, the typical fingerprint analyst will give one of only three opinions: (1) the prints are unsuitable for analysis, (2) the suspect is definitely excluded, or (3) the latent print is definitely that of the suspect. In contrast, questioned document examiners recognize a wider range of permissible opinions: (1) definite identification, (2) strong probability of identification, (3) probable identification, (4) indication of identification, (5) no conclusion, (6) indication of nonauthorship, (7) probability of nonauthorship, (8) strong probability of nonauthorship, and (9) elimination.[181] In short, in many cases, a questioned document examiner explicitly acknowledges the uncertainty of his or her opinion.[182] Whether such a nine-level scale is justified is another matter.[183]

178. *See* James A. Kelly, *Questioned Document Examination, in* Scientific and Expert Evidence 695, 698 (2d ed. 1981).

179. *See* Nellie Chang et al., *Investigation of Class Characteristics in English Handwriting of the Three Main Racial Groups: Chinese, Malay, and Indian in Singapore,* 50 J. Forensic Sci. 177 (2005); Robert J. Muehlberger, *Class Characteristics of Hispanic Writing in the Southeastern United States,* 34 J. Forensic Sci. 371 (1989); Sandra L. Ramsey, *The Cherokee Syllabary,* 39 J. Forensic Sci. 1039 (1994) (one of the landmark questioned document cases, *Hickory v. United States,* 151 U.S. 303 (1894), involved Cherokee writing); Marvin L. Simner et al., *A Comparison of the Arabic Numerals One Through Nine, Written by Adults from Native English-Speaking vs. Non-Native English-Speaking Countries,* 15 J. Forensic Doc. Examination (2003).

180. *See* Larry S. Miller, *Forensic Examination of Arthritic Impaired Writings,* 15 J. Police Sci. & Admin. 51 (1987).

181. NRC Forensic Science Report, *supra* note 3, at 166.

182. *See id.* at 47.

183. *See* United States v. Starzecpyzel, 880 F. Supp. 1027, 1048 (S.D.N.Y. 1995) ("No showing has been made, however, that FDEs can combine their first stage observations into such accurate conclusions as would justify a nine level scale.").

B. The Empirical Record

The 2009 NRC report included a section discussing questioned document examination. The report acknowledged that some tasks performed by examiners are similar in nature "to other forensic chemistry work."[184] For example, some ink and paper analyses use the same hardware and rely on criteria as objective as many tests in forensic chemistry. In contrast, other analyses depend heavily on the examiner's subjective judgment and do not have as "firm [a] scientific foundation" as the analysis of inks and paper.[185] In particular, the report focused on the typical task of deciding common authorship. With respect to that task, the report stated:

> The scientific basis for handwriting comparisons needs to be strengthened. Recent studies have increased our understanding of the individuality and consistency of handwriting . . . and suggest that there may be a scientific basis for handwriting comparison, at least in the absence of intentional obfuscation or forgery. Although there has been only limited research to quantify the reliability and replicability of the practices used by trained document examiners, the committee agrees that there may be some value in handwriting analysis.[186]

Until recently, the empirical record for signature authentication was sparse. Even today there are no population frequency studies establishing, for example, the incidence of persons who conclude their "w" with a certain lift. As a 1989 article commented,

> our literature search for empirical evaluation of handwriting identification turned up one primitive and flawed validity study from nearly 50 years ago, one 1973 paper that raises the issue of consistency among examiners but presents only uncontrolled impressionistic and anecdotal information not qualifying as data in any rigorous sense, and a summary of one study in a 1978 government report. Beyond this, nothing.[187]

This 1989 article then surveyed five proficiency tests administered by CTS in 1975, 1984, 1985, 1986, and 1987. The article set out the results from each of the tests[188] and then aggregated the data by computing the means for the various categories of answers: "A rather generous reading of the data would be that in 45% of the reports forensic document examiners reached the correct finding, in 36% they erred partially or completely, and in 19% they were unable to draw a conclusion."[189]

The above studies were conducted prior to *Daubert*, which was decided in 1993. After the first post-*Daubert* admissibility challenge to handwriting evidence

184. NRC Forensic Science Report, *supra* note 3, at 164.
185. *Id.* at 167.
186. *Id.* at 166–67.
187. Risinger et al., *supra* note 9, at 747.
188. *Id.* at 744 (1975 test), at 745 (1984 and 1985 tests), at 746 (1986 test), and at 747 (1987 test).
189. *Id.* at 747.

in 1995,[190] a number of research projects investigated two questions: (1) are experienced document examiners better at signature authentication than laypersons and (2) do experienced document examiners reach correct signature authentication decisions at a rate substantially above chance?

1. Comparison of experts and laypersons

Two Australian studies support the claim that experienced examiners are more competent at signature authentication tasks than laypersons. The first study was reported in 1999.[191] In this study, document examiners chose the "inconclusive" option far more frequently than did the laypersons. However, in the cases in which a conclusion was reached, the overall error rate for lay subjects was 28%, compared with 2% for experts. More specifically, the lay error rate for false authentication was 7% while it was 0% for the experts. The second Australian study was released in 2002.[192] Excluding "inconclusive" findings, the error rate for forensic document examiners was 5.8%; for laypersons, it was 23.5%.

In the United States, Dr. Moshe Kam, a computer scientist at Drexel University, has been the leading researcher in signature authentication. Dr. Kam and his colleagues have published five articles reporting experiments comparing the signature authentication expertise of document examiners and laypersons. Although the last study involved printing,[193] the initial four were related to cursive writing. In the first, excluding inconclusive findings, document examiners were correct 92.41% of the time and committed false elimination errors in 7.59% of their decisions.[194] Lay subjects were correct 72.84% of the time and made false elimination errors in 27.16% of their decisions. In the second through fourth studies, the researchers provided the laypersons with incentives, usually monetary, for correct decisions. In the fourth study, forgeries were called genuine only 0.5% of the time by experts but 6.5% of the time by laypersons.[195] Laypersons were 13 times more likely to err in concluding that a simulated document was genuine.

Some critics of Dr. Kam's research have asserted that the tasks performed in the tests do not approximate the signature authentication challenges faced by

190. *See* United States v. Starzecpyzel, 880 F. Supp. 1027 (S.D.N.Y. 1995).

191. Bryan Found et al., *The Development of a Program for Characterizing Forensic Handwriting Examiners' Expertise: Signature Examination Pilot Study,* 12 J. Forensic Doc. Examination 69, 72–76 (1999).

192. Jodi Sita et al., *Forensic Handwriting Examiners' Expertise for Signature Comparison,* 47 J. Forensic Sci. 1117 (2002).

193. Moshe Kam et al., *Writer Identification Using Hand-Printed and Non-Hand-Printed Questioned Documents,* 48 J. Forensic Sci. 1 (2003).

194. Moshe Kam et al., *Proficiency of Professional Document Examiners in Writer Identification,* 39 J. Forensic Sci. 5 (1994).

195. Moshe Kam et al., *Signature Authentication by Forensic Document Examiners,* 46 J. Forensic Sci. 884 (2001); Moshe Kam et al., *The Effects of Monetary Incentives on Performance of Nonprofessionals in Document Examiners Proficiency Tests,* 43 J. Forensic Sci. 1000 (1998); Moshe Kam et al., *Writer Identification by Professional Document Examiners,* 42 J. Forensic Sci. 778 (1997).

examiners in real life.[196] In addition, critics have claimed that even the monetary incentives for the laypersons do not come close to equaling the powerful incentives that experts have to be careful in these tests.[197] Yet by now the empirical research record includes a substantial number of studies. With the exception of a 1975 German study,[198] the studies uniformly conclude that professional examiners are much more adept at signature authentication than laypersons.[199]

2. Proficiency studies comparing experts' performance to chance

Numerous proficiency studies have been conducted in the United States[200] and Australia.[201] Some of the American tests reported significant error rates. For example, on a 2001 test, excluding inconclusive findings, the false authentication rate was 22%, while the false elimination rate was 0%. Moreover, as previously stated, on the five CTS proficiency tests mentioned in the 1989 article, 36% of the participating examiners erred partially or completely.[202] Further, critics have claimed that some of the proficiency tests were far easier than the tasks encountered in actual practice,[203] and that consequently, the studies tend to overstate examiners' proficiency.

196. D. Michael Risinger, *Cases Involving the Reliability of Handwriting Identification Expertise Since the Decision in* Daubert, 43 Tulsa L. Rev. 477, 490 (2007).

197. *Id.*

198. The German study included 25 experienced examiners, laypersons with no handwriting background, and some university students who had taken courses in handwriting psychology and comparison. On the one hand, the professional examiners outperformed the regular laypersons. The experts had a 14.7% error rate compared with the 34.4% rate for laypersons without any training. On the other hand, the university students had a lower aggregate error rate than the professional questioned document examiners. Wolfgang Conrad, *Empirische Untersuchungen uber die Urteilsgute vershiedener Gruppen von Laien und Sachvertstandigen bei der Unterscheidung authentischer und gefalschter Unterschriften* [Empirical Studies Regarding the Quality of Assessments of Various Groups of Lay Persons and Experts in Differentiating Between Authentic and Forged Signatures], 156 Archiv für Kriminologie 169–83 (1975).

199. *See* Roger Park, *Signature Identification in the Light of Science and Experience,* 59 Hastings L.J. 1101, 1135–36 (2008).

200. *E.g.*, Collaborative Testing Service (CTS), Questioned Document Examination, Report No. 92-6 (1992); CTS, Questioned Document Examination, Report No. 9406 (1994), CTS, Questioned Document Examination, Report No. 9606 (1996); CTS, Forensic Testing Program, Handwriting Examination, Report No. 9714 (1997); CTS, Forensic Testing Program, Handwriting Examination, Report No. 9814 (1998); CTS, Forensic Testing Program, Handwriting Examination, Test No. 99-524 (1999); CTS, Forensic Testing Program, Handwriting Examination, Test No. 00-524 (2000); CTS, Forensic Testing Program, Handwriting Examination, Test No. 01-524 (2001); CTS, Forensic Testing Program, Handwriting Examination, Test No. 02-524 (2003); *available at* http://www.ctsforensics.com/reports/main.aspx.

201. Bryan Found & Doug Rogers, *The Probative Character of Forensic Handwriting Examiners' Identification and Elimination Opinions on Questioned Signatures*, 178 Forensic Sci. Int'l 54 (2008).

202. Risinger et al., *supra* note 9, at 747–48.

203. Risinger, *supra* note 196, at 485.

The CTS proficiency test results for the 1978–2005 period addressed the comparison of known and questioned signatures and other writings to determine authorship. In other exercises participants were asked to examine a variety of mechanical impressions on paper and the use of photocopying and inks.

- Between 1978 and 1999,[204] fewer than 5% of the mechanical impression comparisons were in error, but 10% of the replies were inconclusive where the examiner should have excluded the impressions as having a common source. With regard to handwriting comparisons, the examiners did very well on the straightforward comparisons, with almost 100% of the comparisons correct. However, in more challenging tests, such as those involving multiple authors, as high as 25% of the replies were inconclusive and nearly 10% of the author associations were incorrect.
- In the 2000–2005 time period, the participants generally performed very well (some approaching 99% correct responses) in determining the genuineness of documents where text in a document had been manipulated or where documents had been altered with various pens and inks. The handwriting exercises were not as successful; in those exercises, comparisons of questioned and known writings were correct about 92% of the time, inconclusive 7% of the time, and incorrect 1% of the time. Nearly all incorrect responses occurred where participants reported handwriting to be of common origin when it was not.

During these tests, some examiners characterized the tests as too easy, while others described them as realistic and very challenging.

Thus, the results of the most recent proficiency studies are encouraging. Moreover, the data in the five proficiency tests discussed in the 1989 article[205] can be subject to differing interpretation. The critics of questioned document examination sometimes suggest that the results of the 1985 test in particular prove that signature authentication has "a high error rate."[206] However,

> [t]hese results can be characterized in different ways. [Another] way of viewing the result would be to disaggregate the specific decisions made by the experts. . . . [S]uppose that a teacher gives a multiple-choice test containing fifty questions. There are different ways that the results could be reported. One could calculate the percentage of students who got any of the fifty questions wrong, and report that as the error rate. A more customary approach would be to treat

204. John I. Thornton & Joseph L. Peterson, *The General Assumptions and Rationale of Forensic Identification, in* 4 Modern Scientific Evidence: The Law and Science of Expert Testimony, *supra* note 104, § 29:40, at 54.

205. Risinger et al., *supra* note 9.

206. Park, *supra* note 199, at 1113.

each question as a separate task, and report the error rate as the mean percentage of questions answered incorrectly.[207]

If the specific decisions made by the examiners were disaggregated, each examiner had to make 66 decisions regarding whether certain pairs of signatures were written by the same person.[208] Under this approach, the false authentication error rate was 3.8%, and the false elimination error rate was 4.5%.[209] In that light, even the 1985 study supports the contention that examiners perform signature authentication tasks at a validity rate considerably exceeding chance.

C. Case Law Development

Although the nineteenth-century cases were skeptical of handwriting expertise,[210] in the twentieth century the testimony in leading cases, such as the Lindbergh prosecution, helped the discipline gain judicial acceptance. There was little dispute that handwriting comparison testimony was admissible at the time the Federal Rules of Evidence were enacted in 1975. Rule 901(b)(3) recognized that a document could be authenticated by an expert, and the drafters explicitly mentioned handwriting comparison "testimony of expert witnesses."[211]

The first significant admissibility challenge under *Daubert* was mounted in *United States v. Starzecpyzel*.[212] In that case, the district court concluded that "forensic document examination, despite the existence of a certification program, professional journals and other trappings of science, cannot, after *Daubert*, be regarded as 'scientific . . . knowledge.'"[213] Nonetheless, the court did not exclude handwriting comparison testimony. Instead, the court admitted the individuation testimony as nonscientific "technical" evidence.[214] *Starzecpyzel* prompted more attacks that questioned the lack of empirical validation in the field.[215]

207. *Id.* at 1114.

208. *Id.* at 1115.

209. *Id.* at 1116.

210. *See* Strother v. Lucas, 31 U.S. 763, 767 (1832); Phoenix Fire Ins. Co. v. Philip, 13 Wend. 81, 82–84 (N.Y. Sup. Ct. 1834).

211. Fed. R. Evid. 901(b)(3) advisory committee's note.

212. 880 F. Supp. 1027 (S.D.N.Y. 1995).

213. *Id.* at 1038.

214. *Kumho Tire* later called this aspect of the *Starzecpyzel* opinion into question because *Kumho* held that the reliability requirement applies to all types of expertise—"scientific," "technical," or "specialized." Moreover, the Supreme Court indicated that the *Daubert* factors, including empirical testing, may be applicable to technical expertise. Some aspects of handwriting can and have been tested.

215. *See, e.g.*, United States v. Hidalgo, 229 F. Supp. 2d 961, 967 (D. Ariz. 2002) ("Because the principle of uniqueness is without empirical support, we conclude that a document examiner will not be permitted to testify that the maker of a known document is the maker of the questioned document. Nor will a document examiner be able to testify as to identity in terms of probabilities.").

As of the date of this publication, there is a three-way split of authority. The majority of courts permit examiners to express individuation opinions.[216] As one court noted, "all six circuits that have addressed the admissibility of handwriting expert [testimony] . . . [have] determined that it can satisfy the reliability threshold" for nonscientific expertise.[217] In contrast, several courts have excluded expert testimony,[218] although one involved handprinting[219] and another Japanese handprinting.[220] Many district courts have endorsed a third view. These courts limit the reach of the examiner's opinion, permitting expert testimony about similarities and dissimilarities between exemplars but not an ultimate conclusion that the defendant was the author ("common authorship" opinion) of the questioned document.[221] The expert is allowed to testify about "the specific similarities and idiosyncrasies between the known writings and the questioned writings, as well as testimony regarding, for example, how frequently or infrequently in his experience, [the expert] has seen a particular idiosyncrasy."[222] As the justification for this limitation, these courts often state that the examiners' claimed ability to individuate lacks "empirical support."[223]

216. *See, e.g.*, United States v. Prime, 363 F.3d 1028, 1033 (9th Cir. 2004); United States v. Crisp, 324 F.3d 261, 265–71 (4th Cir. 2003); United States v. Jolivet, 224 F.3d 902, 906 (8th Cir. 2000) (affirming the introduction of expert testimony that it was likely that the accused wrote the questioned documents); United States v. Velasquez, 64 F.3d 844, 848–52 (3d Cir. 1995); United States v. Ruth, 42 M.J. 730, 732 (A. Ct. Crim. App. 1995), *aff'd on other grounds*, 46 M.J. 1 (C.A.A.F. 1997); United States v. Morris, No. 06-87-DCR, 2006 U.S. Dist. LEXIS 53983, *5 (E.D. Ky. July 20, 2006); Orix Fin. Servs. v. Thunder Ridge Energy, Inc., No. 01Civ. 4788 (RJH) (HBP). 2005 U.S. Dist. LEXIS 41889 (S.D.N.Y. Dec. 29, 2005).

217. *Prime*, 363 F.3d at 1034.

218. United States v. Lewis, 220 F. Supp. 2d 548 (S.D. W. Va. 2002).

219. United States v. Saelee, 162 F. Supp. 2d 1097 (D. Alaska 2001).

220. United States v. Fujii, 152 F. Supp. 2d 939, 940 (N.D. Ill. 2000) (holding expert testimony concerning Japanese handprinting inadmissible: "Handwriting analysis does not stand up well under the *Daubert* standards. Despite its long history of use and acceptance, validation studies supporting its reliability are few, and the few that exist have been criticized for methodological flaws.").

221. *See, e.g.*, United States v. Oskowitz, 294 F. Supp. 2d 379, 384 (E.D.N.Y. 2003) ("Many other district courts have similarly permitted a handwriting expert to analyze a writing sample for the jury without permitting the expert to offer an opinion on the ultimate question of authorship."); United States v. Rutherford, 104 F. Supp. 2d 1190, 1194 (D. Neb. 2000) ("[T]he Court concludes that FDE Rauscher's testimony meets the requirements of Rule 702 to the extent that he limits his testimony to identifying and explaining the similarities and dissimilarities between the known exemplars and the questioned documents. FDE Rauscher is precluded from rendering any ultimate conclusions on authorship of the questioned documents and is similarly precluded from testifying to the degree of confidence or certainty on which his opinions are based."); United States v. Hines, 55 F. Supp. 2d 62, 69 (D. Mass. 1999) (expert testimony concerning the general similarities and differences between a defendant's handwriting exemplar and a stick-up note was admissible while the specific conclusion that the defendant was the author was not).

222. United States v. Van Wyk, 83 F. Supp. 2d 515, 524 (D.N.J. 2000).

223. United States v. Hidalgo, 229 F. Supp. 2d 961, 967 (D. Ariz. 2002).

VIII. Firearms Identification Evidence

It is widely considered that the first written reference to firearms identification (popularly known as "ballistics") in the United States appeared in 1900.[224] In the 1920s, the technique gained considerable attention because of the work of Calvin Goddard[225] and played a controversial role in the Sacco and Vanzetti case during the same decade.[226] Goddard also analyzed the bullet evidence in the St. Valentine's Day Massacre in 1929, in which five gangsters and two acquaintances were gunned down in Chicago.[227] In 1923, the Illinois Supreme Court wrote that positive identification of a bullet was not only impossible but "preposterous."[228] Seven years later, however, that court did an about-face and became one of the first courts in this country to admit firearms identification evidence.[229] The technique subsequently gained widespread judicial acceptance and was not seriously challenged until recently.

A. The Technique

1. Firearms

Typically, three types of firearms—rifles, handguns, and shotguns—are encountered in criminal investigations.[230] The barrels of modern rifles and handguns are *rifled*; that is, parallel spiral grooves are cut into the inner surface (bore) of the barrel. The surfaces between the grooves are called *lands*. The lands and grooves twist in a direction: right twist or left twist. For each type of firearm produced, the manufacturer specifies the number of lands and grooves, the direction of twist, the angle of twist (pitch), the depth of the grooves, and the width of the lands and grooves. As a bullet passes through the bore, the lands and grooves force the

224. *See* Albert Llewellyn Hall, *The Missile and the Weapon*, 39 Buff. Med. J. 727 (1900).

225. Calvin Goddard, often credited as the "father" of firearms identification, was responsible for much of the early work on the subject. *E.g.*, Calvin Goddard, *Scientific Identification of Firearms and Bullets*, 17 J. Crim. L., Criminology & Police Sci. 254 (1926).

226. *See* Joughin & Morgan, *supra* note 8, at 15 (The firearms identification testimony was "carelessly assembled, incompletely and confusedly presented, and . . . beyond the comprehension" of the jury); Starrs, *supra* note 8, at 630 (Part I), 1050 (Part II).

227. *See* Calvin Goddard, *The Valentine Day Massacre: A Study in Ammunition-Tracing*, 1 Am. J. Police Sci. 60, 76 (1930) ("Since two of the members of the execution squad had worn police uniforms, and since it had been subsequently intimated by various persons that the wearers of the uniforms might really have been policeman rather than disguised gangsters, it became a matter of no little importance to ascertain, if possible, whether these rumors had any foundation in fact."); Jim Ritter, *St. Valentine's Hit Spurred Creation of Nation's First Lab*, Chicago Sun-Times, Feb. 9, 1997, at 40 ("Sixty-eight years ago this Friday, Al Capone's hit men, dressed as cops, gunned down seven men in the Clark Street headquarters of rival mobster Bugs Moran.").

228. People v. Berkman, 139 N.E. 91, 94 (Ill. 1923).

229. People v. Fisher, 172 N.E. 743, 754 (Ill. 1930).

230. Other types of firearms, such as machine guns, tear gas guns, zip guns, and flare guns, may also be examined.

bullet to rotate, giving it stability in flight and thus increased accuracy. Shotguns are smooth-bore firearms; they do not have lands and grooves.

Rifles and handguns are classified according to their caliber. The caliber is the diameter of the bore of the firearm; the caliber is expressed in either hundredths or thousandths of an inch (e.g., .22, .45, .357 caliber) or millimeters (e.g., 7.62 mm).[231] The two major types of handguns are revolvers[232] and semiautomatic pistols. A major difference between the two is that when a semiautomatic pistol is fired, the cartridge case is automatically ejected and, if recovered at the crime scene, could help link the case to the firearm from which it was fired. In contrast, when a revolver is discharged the case is not ejected.

2. Ammunition

Rifle and handgun cartridges consist of the projectile (bullet),[233] case,[234] propellant (powder), and primer. The primer contains a small amount of an explosive mixture, which detonates when struck by the firing pin. When the firing pin detonates the primer, an explosion occurs that ignites the propellant. The most common modern propellant is smokeless powder.

3. Class characteristics

Firearms identifications may be based on either bullet or cartridge case examinations. Identifying features include class, subclass, and individual characteristics.

The class characteristics of a firearm result from design factors and are determined prior to manufacture. They include the following caliber and rifling specifications: (1) the land and groove diameters, (2) the direction of rifling (left or right twist), (3) the number of lands and grooves, (4) the width of the lands and grooves, and (5) the degree of the rifling twist.[235] Generally, a .38-caliber bullet with six land and groove impressions and with a right twist could have been fired only from a firearm with these same characteristics. Such a bullet could not have been fired from a .32-caliber firearm, or from a .38-caliber firearm with a different number of lands and grooves or a left twist. In sum, if the class characteristics do not match, the firearm could not have fired the bullet and is excluded.

231. The caliber is measured from land to land in a rifled weapon. Typically, the designated caliber is more an approximation than an accurate measurement. *See* 1 J. Howard Mathews, Firearms Identification 17 (1962) ("'nominal caliber' would be a more proper term").

232. Revolvers have a cylindrical magazine that rotates behind the barrel. The cylinder typically holds five to nine cartridges, each within a separate chamber. When a revolver is fired, the cylinder rotates and the next chamber is aligned with the barrel. A single-action revolver requires the manual cocking of the hammer; in a double-action revolver the trigger cocks the hammer.

233. Bullets are generally composed of lead and small amounts of other elements (hardeners). They may be completely covered (jacketed) with another metal or partially covered (semijacketed).

234. Cartridge cases are generally made of brass.

235. 1 Mathews, *supra* note 231, at 17.

4. Subclass characteristics

Subclass characteristics are produced at the time of manufacture and are shared by a discrete subset of weapons in a production run or "batch." According to the Association of Firearm and Tool Mark Examiners (AFTE),[236] subclass characteristics are discernible surface features that are more restrictive than class characteristics in that they are (1) "produced incidental to manufacture," (2) "relate to a smaller group source (a subset to which they belong)," and (3) can arise from a source that changes over time.[237] The AFTE states that "[c]aution should be exercised in distinguishing subclass characteristics from class characteristics."[238]

5. Individual characteristics

Bullet identification involves a comparison of the evidence bullet and a test bullet fired from the firearm.[239] The two bullets are examined by means of a comparison microscope, which permits a split-screen view of the two bullets and manipulation in order to attempt to align the striations (marks) on the two bullets.

Barrels are machined during the manufacturing process, and imperfections in the tools used in the machining process are imprinted on the bore.[240] The subsequent use of the firearm adds further individual imperfections. For example, mechanical action (erosion) caused by the friction of bullets passing through the bore of the firearm produces accidental imperfections. Similarly, chemical action (corrosion) caused by moisture (rust), as well as primer and propellant chemicals, produce other imperfections.

When a bullet is fired, microscopic striations are imprinted on the bullet surface as it passes through the bore of the firearm. These bullet markings are produced by the imperfections in the bore. Because these imperfections are randomly produced, examiners assume that they are unique to each firearm.[241] Although the assumption is plausible, there is no statistical basis for this assumption.[242]

236. AFTE is the leading professional organization in the field. There is also the Scientific Working Group for Firearms and Toolmarks (SWGGUN), which promulgates guidelines for examiners.

237. *Theory of Identification, Association of Firearm and Toolmark Examiners*, 30 AFTE J. 86, 88 (1998) [hereinafter *AFTE Theory*].

238. *Id.*

239. Test bullets are obtained by firing a firearm into a recovery box or bullet trap, which is usually filled with cotton, or into a recovery tank, which is filled with water.

240. "No two barrels are microscopically identical, as the surfaces of their bores all possess individual and characteristic markings." Gerald Burrard, The Identification of Firearms and Forensic Ballistics 138 (1962).

241. 1 Mathews, *supra* note 231, at 3 ("Experience has shown that no two firearms, even those of the same make and model and made consecutively by the same tools, will produce the same markings on a bullet or a cartridge.").

242. Alfred A. Biasotti, *The Principles of Evidence Evaluation as Applied to Firearms and Tool Mark Identification*, 9 J. Forensic Sci. 428, 432 (1964) ("[W]e lack the fundamental statistical data needed to

Although an identification is based on objective data (the striations on the bullet surface), the AFTE explains that the examiner's individuation is essentially a subjective judgment. The AFTE describes the traditional pattern recognition methodology as "subjective in nature, founded on scientific principles and based on the examiner's training and experience."[243] There are no objective criteria governing this determination: "Ultimately, unless other issues are involved, it remains for the examiner to determine for himself the modicum of proof necessary to arrive at a definitive opinion."[244]

The condition of a firearm or evidence bullet may preclude an identification. For example, there may be insufficient marks on the bullet or, because of mutilation, an insufficient amount of the bullet may have been recovered. Likewise, if the bore of the firearm has changed significantly as a result of erosion or corrosion, an identification may be impossible. (Unlike fingerprints, firearms change over time.) In these situations, the examiner may render a "no conclusion" determination. Such a conclusion, however, may have some evidentiary value even if the examiner cannot form an individuation opinion; that is, the firearm could have fired the bullet if the class characteristics match.

6. *Consecutive matching striae*

In an attempt to make firearms identification more objective, some commentators advocate a technique known as consecutive matching striae (CMS). As the name implies, this method is based on finding a specified number of consecutive matching striae on two bullets. Other commentators have questioned this approach,[245] and it remains a minority position.[246]

7. *Cartridge identification*

Cartridge case identification is based on the same theory of random markings as bullet identification.[247] As with barrels, defects produced in the manufacturing

develop verifiable criteria."); *see also* Alfred A. Biasotti, *A Statistical Study of the Individual Characteristics of Fired Bullets*, 4 J. Forensic Sci. 34 (1959).

243. *AFTE Theory*, *supra* note 237, at 86.

244. Laboratory Proficiency Test, *supra* note 81, at 207; *see also* Alfred A. Biasotti, *The Principles of Evidence Evaluation as Applied to Firearms and Tool Mark Identification*, *supra* note 242, at 429 ("In general, the texts on firearms identification take the position that each practitioner must develop his own intuitive criteria of identity gained through practical experience.").

245. *See* Stephen G. Bunch, *Consecutive Matching Striation Criteria: A General Critique,* 45 J. Forensic Sci. 955, 955 (2000) (finding the traditional methodology superior: "[P]resent-day firearm identification, in the final analysis is subjective.").

246. Roger C. Nichols, *Firearm and Toolmark Identification Criteria: A Review of the Literature, Part II*, 48 J. Forensic Sci. 318, 326 (2003) (CMS "has not been promoted as an alternative [to traditional pattern recognition], but as a numerical threshold.").

247. Burrard, *supra* note 240, at 107. However, bullet and cartridge case identifications differ in several respects. Because the bullet is traveling through the barrel at the time it is imprinted with the

process leave distinctive characteristics on the breech face, firing pin, chamber, extractor, and ejector. Subsequent use of the firearm produces additional defects. When the trigger is pulled, the firing pin strikes the primer of the cartridge, causing the primer to detonate. This detonation ignites the propellant (powder). In the process of combustion, the powder is converted rapidly into gases. The pressure produced by this process propels the bullet from the weapon and also forces the base of the cartridge case backward against the breech face, imprinting breech face marks on the base of the cartridge case. Similarly, the firing pin, ejector, and extractor may leave characteristic marks on a cartridge case.[248]

Cartridge case identification involves a comparison of the cartridge case recovered at the crime scene and a test cartridge case obtained from the firearm after it has been fired. Shotgun shell casings may be identified in this way, as well. As in bullet identification, the comparison microscope is used in the examination. According to AFTE, "interpretation of toolmark individualization and identification is still considered to be subjective in nature, based on one's training and experience."[249]

8. Automated identification systems

"These ballistic imaging systems use the powerful searching capabilities of the computer to match the images of recovered crime scene evidence against digitized images stored in a computer database."[250] The current system is the Integrated Ballistics Information System (IBIS).[251] Automated systems "give[ ] firearms examiners the ability to screen virtually unlimited numbers of bullets and cartridge casings for possible matches."[252] These systems identify a number of candidate matches. They do not replace the examiner, who still must make the final comparison: "'High Confidence' candidates (likely hits) are referred to a firearms examiner for examination on a comparison microscope."[253] The examiner need

bore imperfections, these marks are "sliding" imprints, called striated marks. In contrast, the cartridge case receives "static" imprints, called impressed marks. *Id.* at 145.

248. Ejector and extractor marks by themselves may indicate only that the cartridge case had been *loaded in,* not fired from, a particular firearm.

249. Eliot Springer, *Toolmark Examinations—A Review of Its Development in the Literature*, 40 J. Forensic Sci. 964, 966–67 (1995).

250. *Benchmark Evaluation Studies of the Bulletproof and Drugfire Ballistic Imaging Systems*, 22 Crime Lab. Digest 51 (1995); *see also* Jan De Kinder & Monica Bonfanti, *Automated Comparisons of Bullet Striations Based on 3D Topography*, 101 Forensic Sci. Int'l 85, 86 (1999) ("[A]n automatic system will cut the time demanding and tedious manual searches for one specific item in large open case files.").

251. *See* Jan De Kinder et al., *Reference Ballistic Imaging Database Performance*, 140 Forensic Sci. Int'l 207 (2004); Ruprecht Nennstiel & Joachim Rahm, *An Experience Report Regarding the Performance of the IBIS™ Correlator*, 51 J. Forensic Sci. 24 (2006).

252. Richard E. Tontarski & Robert M. Thompson, *Automated Firearms Evidence Comparison: A Forensic Tool for Firearms Identification—An Update*, 43 J. Forensic Sci. 641, 641 (1998).

253. *Id.*

not accept the highest ranked candidate identified by the system. For that matter, the examiner may reject all the candidates.

9. Toolmarks

Toolmark identifications rest on essentially the same theory as firearms identifications.[254] Tools have both (1) class characteristics and (2) individual characteristics; the latter are accidental imperfections produced by the machining process and subsequent use. When the tool is used, these characteristics are sometimes imparted onto the surface of another object struck by the tool. Toolmarks may be impressions (compression marks), striations (friction or scrape marks), or a combination of both.[255] Fracture matches constitute another type of examination.

The marks may be left on a variety of different materials, such as wood or metal. In some cases, only class characteristics can be matched. For example, it may be possible to identify a mark (impression) left on a piece of wood as having been produced by a hammer, punch, or screwdriver. A comparison of the mark and the evidence tool may establish the size of the tool (another class characteristic). Unusual features of the tool, such as a chip, may permit a positive identification. Striations caused by scraping with a tool can also produce distinguishing marks in much the same way that striations are imprinted on a bullet when a firearm is discharged. This type of examination has the same limitations as firearms identification: "[T]he characteristics of a tool will change with use."[256]

Firearms identification could be considered a subspecialty of toolmark identification; the firearm (tool) imprints its individual characteristics on the bullet. However, the markings on a bullet or cartridge case are imprinted in roughly the same way every time a firearm is fired. In contrast, toolmark analysis can be more complicated because a tool can be employed in a variety of different ways, each producing a different mark: "[I]n toolmark work the angle at which the tool was used must be duplicated in the test standard, pressures must be dealt with, and the degree of hardness of metals and other materials must be taken into account."[257]

The comparison microscope is also used in this examination. As with firearms identification testimony, toolmark identification testimony is based on the subjective judgment of the examiner, who determines whether sufficient marks of

254. *See* Biasotti, *The Principles of Evidence Evaluation as Applied to Firearms and Tool Mark Identification*, *supra* note 242; *see also* Springer, *supra* note 249, at 964 ("The identification is based . . . on a series of scratches, depressions, and other marks which the tool leaves on the object it comes into contact with. The combination of these various marks ha[s] been termed toolmarks and the claim is that every instrument can impart a mark individual to itself.").

255. David Q. Burd & Roger S. Greene, *Tool Mark Examination Techniques*, 2 J. Forensic Sci. 297, 298 (1957).

256. Emmett M. Flynn, *Toolmark Identification*, 2 J. Forensic Sci. 95, 102 (1957).

257. *Id.* at 105.

similarity are present to permit an identification.[258] There are no objective criteria governing the determination of whether there is a match.[259]

B. The Empirical Record

In its 2009 report, NRC summarized the state of the research as follows:

> Because not enough is known about the variabilities among individual tools and guns, we are not able to specify how many points of similarity are necessary for a given level of confidence in the result. Sufficient studies have not been done to understand the reliability and repeatability of the methods. The committee agrees that class characteristics are helpful in narrowing the pool of tools that may have left a distinctive mark. Individual patterns from manufacture or from wear might, in some cases, be distinctive enough to suggest one particular source, but additional studies should be performed to make the process of individualization more precise and repeatable.[260]

The 1978 Crime Laboratory Proficiency Testing Program reported mixed results on firearms identification tests. In one test, 5.3% of the participating laboratories misidentified firearms evidence, and in another test 13.6% erred. These tests involved bullet and cartridge case comparisons. The Project Advisory Committee considered these errors "particularly grave in nature" and concluded that they probably resulted from carelessness, inexperience, or inadequate supervision.[261] A third test required the examination of two bullets and two cartridge cases to identify the "most probable weapon" from which each was fired. The error rate was 28.2%.

In later tests,

> [e]xaminers generally did very well in making the comparisons. For all fifteen tests combined, examiners made a total of 2106 [bullet and cartridge case] comparisons and provided responses which agreed with the manufacturer responses 88% of the time, disagreed in only 1.4% of responses, and reported inconclusive results in 10% of cases.[262]

258. *See* Springer, *supra* note 249, at 966–67 ("According to the Association of Firearms and Toolmarks Examiners' Criteria for Identification Committee, interpretation of toolmark individualization and identification is still considered to be subjective in nature, based on one's training and experience.").

259. As one commentator has noted: "[I]t is not possible at present to categorically state the number and percentage of the [striation] lines which must correspond." Burd & Greene, *supra* note 255, at 310.

260. *Id.*

261. Laboratory Proficiency Test, *supra* note 81, at 207–08.

262. Peterson & Markham, *supra* note 82, at 1018. The authors also stated:

> The performance of laboratories in the firearms tests was comparable to that of the earlier LEAA study, although the rate of successful identifications actually was slightly over—88% vs. 91%. Laboratories cut the rate of errant identifications by half (3% to 1.4%) but the rate of inconclusive responses doubled, from 5% to 10%.

Id. at 1019.

Proficiency testing on toolmark examinations has also been reported.[263]

For the period 1978–1999, firearms examiners performed well on their CTS proficiency tests, with only 2% to 3% of their comparisons incorrect, but with 10% to 13% of their responses inconclusive.[264] The scenarios that accompanied the test materials asked examiners to compare test-fired bullets and/or cartridge cases with evidence projectiles found at a crime scene. Between 2000 and 2005, participants, again, performed very well, averaging less than 1% incorrect responses, but with inconclusive results about 10% of the time. Most of the inconclusive results in these tests occurred where bullets and/or cartridge cases were actually fired from different weapons. Examiners frequently stated they were unable to reach the proper conclusion because they did not have the actual weapon with which they could perform their own test fires of ammunition.

In CTS toolmark proficiency comparisons, laboratories were asked to compare marks made with such tools as screwdrivers, bolt cutters, hammers, and handstamps. In some cases, tools were supplied to participants, but in most cases they were given only test marks. Over the entire 1978–2005 period, fewer than 5% of responses were in error, but individual test results varied substantially. In some cases, 30% to 40% of replies were inconclusive, because laboratories were unsure if the blade of the tool in question might have been altered between the time(s) different markings had been made. During the final 6-year period reviewed (2000–2005), laboratories averaged a 1% incorrect comparison rate for toolmarks. Inconclusive responses remained high (30% and greater) and, together with firearms testing, constitute the evidence category where evidence comparisons have the highest rates of inconclusive responses.

Questions have arisen concerning the significance of these tests. First, such testing is not required of all firearms examiners, only those working in laboratories voluntarily seeking accreditation by the ASCLD. In short, "the sample is self-selecting and may not be representative of the complete universe of firearms examiners."[265] Second, the examinations are not blind—that is, examiners know when they are being tested. Thus, the examiner may be more meticulous and careful than in ordinary case work. Third, the results of an evaluation can vary, depending on whether an "inconclusive" answer is counted. Fourth, the rigor of the examinations has been questioned. According to one witness, in a 2005 test involving cartridge case comparisons, *none* of the 255 test-takers nationwide answered incorrectly. The court observed: "One could read these results to mean that the technique is foolproof, but the results might instead indicate that the test was somewhat elementary."[266]

263. *Id.* at 1025 ("Overall, laboratories performed not as well on the toolmark tests as they did on the firearms tests.").

264. Thornton & Peterson, *supra* note 204, § 29:47, at 66.

265. United States v. Monteiro, 407 F. Supp. 2d 351, 367 (D. Mass. 2006).

266. *Id.*

In 2008, NAS published a report on computer imaging of bullets.[267] Although firearms identification was not the primary focus of the investigation, a section of the report commented on this subject.[268] After surveying the literature on the uniqueness, reproducibility, and permanence of individual characteristics, the committee noted that "[m]ost of these studies are limited in scale and have been conducted by firearms examiners (and examiners in training) in state and local law enforcement laboratories as adjuncts to their regular casework."[269] The report concluded: "The validity of the fundamental assumptions of uniqueness and reproducibility of firearms-related toolmarks has not yet been fully demonstrated."[270] This statement, however, was qualified:

> There is one baseline level of credibility . . . that must be demonstrated lest any discussion of ballistic imaging be rendered moot—namely, that there is at least some "signal" that may be detected. In other words, the creation of toolmarks must not be so random and volatile that there is no reason to believe that any similar and matchable marks exist on two exhibits fired from the same gun. The existing research, and the field's general acceptance in legal proceedings for several decades, is more than adequate testimony to that baseline level. Beyond that level, we neither endorse nor oppose the fundamental assumptions. Our review in this chapter is not—and is not meant to be—a full weighing of evidence for or against the assumptions, but it is ample enough to suggest that they are not fully settled, mechanically or empirically.
>
> Another point follows directly: *Additional general research on the uniqueness and reproducibility of firearms-related toolmarks would have to be done if the basic premises of firearms identification are to be put on a more solid scientific footing.*[271]

The 2008 report cautioned:

> *Conclusions drawn in firearms identification should not be made to imply the presence of a firm statistical basis when none has been demonstrated.* Specifically, . . . examiners tend to cast their assessments in bold absolutes, commonly asserting that a match can be made "to the exclusion of all other firearms in the world." Such comments cloak an inherently subjective assessment of a match with an extreme probability statement that has no firm grounding and unrealistically implies an error rate of zero.[272]

267. National Research Council, Ballistic Imaging (2008), *available at* http://www.nap.edu/catalog.php?record_id=12162.

268. The committee was asked to assess the feasibility, accuracy, reliability, and technical capability of developing and using a national ballistic database as an aid to criminal investigations. It concluded: (1) "A national reference ballistic image database of all new and imported guns is not advisable at this time." (2) "NIBIN can and should be made more effective through operational and technological improvements." *Id.* at 5.

269. *Id.* at 70.

270. *Id.* at 81.

271. *Id.* at 81–82.

272. *Id.* at 82.

The issue of the adequacy of the empirical basis of firearms identification expertise remains in dispute,[273] and research is ongoing. A recent study reported testing concerning 10 consecutively rifled Ruger pistol barrels. In 463 tests during the study, no false positives were reported; 8 inconclusive results were reported.[274] "But the capsule summaries [in this study] suggest a heavy reliance on the subjective findings of examiners rather than on the rigorous quantification and analysis of sources of variability."[275]

C. Case Law Development

Firearms identification developed in the early part of the last century, and by 1930, courts were admitting evidence based on this technique.[276] Subsequent cases followed these precedents, admitting evidence of bullet,[277] cartridge case,[278] and shot shell[279] identifications. A number of courts have also permitted an expert to

273. *Compare* Roger G. Nichols, *Defending the Scientific Foundations of the Firearms and Tool Mark Identification Discipline: Responding to Recent Challenges*, 52 J. Forensic Sci. 586 (2007), *with* Adina Schwartz, *Commentary on "Nichols, R.G., Defending the scientific foundations of the firearms and tool mark identification discipline: Responding to recent challenges, J. Forensic Sci. 52(3):586-94 (2007),"* 52 J. Forensic Sci. 1414 (2007) (responding to Nichols). Moreover, AFTE disputed the Academy's conclusions. *See The Response of the Association of Firearm and Tool Mark Examiners to the National Academy of Sciences 2008 Report Assessing the Feasibility, Accuracy, and Capability of a National Ballistic Database August 20, 2008*, 40 AFTE J. 234 (2008) (concluding that underlying assumptions of uniqueness and reproducibility have been demonstrated, and the implication that there is no statistical basis is unwarranted); *see also* Adina Schwartz, *A Systemic Challenge to the Reliability and Admissibility of Firearms and Toolmark Identification,* 6 Colum. Sci. & Tech. L. Rev. 2 (2005).

274. James E. Hamby et al., *The Identification of Bullets Fired from 10 Consecutively Rifled 9mm Ruger Pistol Barrels—A Research Project Involving 468 Participants from 19 Countries*, 41 AFTE J. 99 (Spring 2009).

275. NRC Forensic Science Report, *supra* note 3, at 155.

276. *E.g.*, People v. Fisher, 172 N.E. 743 (Ill. 1930); Evans v. Commonwealth, 19 S.W.2d 1091 (Ky. 1929); Burchett v. State, 172 N.E. 555 (Ohio Ct. App. 1930).

277. *E.g.*, United States v. Wolff, 5 M.J. 923, 926 (N.C.M.R. 1978); State v. Mack, 653 N.E.2d 329, 337 (Ohio 1995) (The examiner "compared the test shot with the morgue bullet recovered from the victim, . . . and the spent shell casings recovered from the crime scene, concluding that all had been discharged from appellant's gun.").

278. *E.g.*, Bentley v. Scully, 41 F.3d 818, 825 (2d Cir. 1994) ("[A] ballistic expert found that the spent nine millimeter bullet casing recovered from the scene of the shooting was fired from the pistol found on the rooftop."); State v. Samonte, 928 P.2d 1, 6 (Haw. 1996) ("Upon examining the striation patterns on the casings, [the examiner] concluded that the casing she had fired matched six casings that police had recovered from the house.").

279. *E.g.*, Williams v. State, 384 So. 2d 1205, 1210–11 (Ala. Crim. App. 1980); Burge v. State, 282 So. 2d 223, 229 (Miss. 1973); Commonwealth v. Whitacre, 878 A.2d 96, 101 (Pa. Super. Ct. 2005) ("no abuse of discretion in the trial court's decision to permit admission of the evidence regarding comparison of the two shell casings with the shotgun owned by Appellant").

testify that a bullet *could have been fired* from a particular firearm;[280] that is, the class characteristics of the bullet and the firearm are consistent.[281]

The early post-*Daubert* challenges to the admissibility of firearms identification evidence failed.[282] This changed in 2005 in *United States v. Green*,[283] where the court ruled that the expert could describe only the ways in which the casings were similar but not that the casings came from a specific weapon "to the exclusion of every other firearm in the world."[284] In *United States v. Monteiro*[285] the expert had not made any sketches or taken photographs and thus adequate documentation was lacking: "Until the basis for the identification is described in such a way that the procedure performed by [the examiner] is reproducible and verifiable, it is inadmissible under Rule 702."[286]

In 2007 in *United States v. Diaz*,[287] the court found that the record did not support the conclusion that identifications could be made to the exclusion of all other firearms in the world. Thus, "the examiners who testify in this case may only testify that a match has been made to a 'reasonable degree of certainty in the ballistics field.'"[288] In 2008, *United States v. Glynn*[289] ruled that the expert could

280. *E.g.*, People v. Horning, 102 P.3d 228, 236 (Cal. 2004) (expert "opined that both bullets and the casing could have been fired from the same gun . . . ; because of their condition he could not say for sure"); Luttrell v. Commonwealth, 952 S.W.2d 216, 218 (Ky. 1997) (expert "testified only that the bullets which killed the victim could have been fired from Luttrell's gun"); State v. Reynolds, 297 S.E.2d 532, 539–40 (N.C. 1982); Commonwealth v. Moore, 340 A.2d 447, 451 (Pa. 1975).

281. This type of evidence has some probative value and satisfies the minimal evidentiary test for logical relevancy. *See* Fed. R. Evid. 401. As one court commented, the expert's "testimony, which established that the bullet which killed [the victim] could have been fired from the same caliber and make of gun found in the possession of [the defendant], significantly advanced the inquiry." Commonwealth v. Hoss, 283 A.2d 58, 68 (Pa. 1971).

282. *See* United States v. Hicks, 389 F.3d 514, 526 (5th Cir. 2004) (ruling that "the matching of spent shell casings to the weapon that fired them has been a recognized method of ballistics testing in this circuit for decades"); United States v. Foster, 300 F. Supp. 2d 375, 377 n.1 (D. Md. 2004) ("Ballistics evidence has been accepted in criminal cases for many years. . . . In the years since *Daubert,* numerous cases have confirmed the reliability of ballistics identification."); United States v. Santiago, 199 F. Supp. 2d 101, 111 (S.D.N.Y. 2002) ("The Court has not found a single case in this Circuit that would suggest that the entire field of ballistics identification is unreliable."); State v. Anderson, 624 S.E.2d 393, 397–98 (N.C. Ct. App. 2006) (no abuse of discretion in admitting bullet identification evidence); *Whitacre*, 878 A.2d at 101 ("no abuse of discretion in the trial court's decision to permit admission of the evidence regarding comparison of the two shell casings with the shotgun owned by Appellant").

283. 405 F. Supp. 2d 104 (D. Mass. 2005).

284. *Id.* at 107. The court had followed the same approach in a handwriting case. *See* United States v. Hines, 55 F. Supp. 2d 62, 67 (D. Mass. 1999) (expert testimony concerning the general similarities and differences between a defendant's handwriting exemplar and a stick-up note was admissible but not the specific conclusion that the defendant was the author).

285. 407 F. Supp. 2d 351 (D. Mass. 2006).

286. *Id.* at 374.

287. No. CR 05-00167 WHA, 2007 WL 485967 (N.D. Cal. Feb. 12, 2007).

288. *Id.* at *1.

289. 578 F. Supp. 2d 567 (S.D.N.Y. 2008).

not use the term "reasonable scientific certainty" in testifying. Rather, the expert would be permitted to testify only that it was "more likely than not" that recovered bullets and cartridge cases came from a particular weapon.

Yet other courts continued to uphold admission.[290] By way of example, in *United States v. Williams*,[291] the Second Circuit upheld the admissibility of firearms identification evidence—bullets and cartridge casings. The opinion, however, contained some cautionary language: "We do not wish this opinion to be taken as saying that any proffered ballistic expert should be routinely admitted."[292] Several cases limited testimony after the 2009 NAS Report was published.[293] In the past, courts often have admitted toolmark identification evidence,[294] includ-

290. *See* United States v. Natson, 469 F. Supp. 2d 1253, 1261 (M.D. Ga. 2007) ("According to his testimony, these toolmarks were sufficiently similar to allow him to identify Defendant's gun as the gun that fired the cartridge found at the crime scene. He opined that he held this opinion to a 100% degree of certainty. . . . The Court also finds [the examiner's] opinions reliable and based upon a scientifically valid methodology. Evidence was presented at the hearing that the toolmark testing methodology he employed has been tested, has been subjected to peer review, has an ascertainable error rate, and is generally accepted in the scientific community."); Commonwealth v. Meeks, Nos. 2002-10961, 2003-10575, 2006 WL 2819423, at * 50 (Mass. Super. Ct. Sept. 28, 2006) ("The theory and process of firearms identification are generally accepted and reliable, and the process has been reliably applied in these cases. Accordingly, the firearms identification evidence, including opinions as to matches, may be presented to the juries for their consideration, but only if that evidence includes a detailed statement of the reasons for those opinions together with appropriate documentation.").

291. 506 F.3d 151, 161–62 (2d Cir. 2007) ("*Daubert* did make plain that Rule 702 embodies a more liberal standard of admissibility for expert opinions than did *Frye*. . . . But this shift to a more permissive approach to expert testimony did not abrogate the district court's gatekeeping function. Nor did it 'grandfather' or protect from *Daubert* scrutiny evidence that had previously been admitted under *Frye*.") (citations omitted).

292. *Id.* at 161.

293. *See* United States v. Willock, 696 F. Supp. 2d 536, 546, 549 (D. Md. 2010) (holding, based on a comprehensive magistrate's report, that "Sgt. Ensor shall not opine that it is a 'practical impossibility' for a firearm to have fired the cartridges other than the common 'unknown firearm' to which Sgt. Ensor attributes the cartridges." Thus, "Sgt. Ensor shall state his opinions and conclusions without any characterization as to the degree of certainty with which he holds them."); United States v. Taylor, 663 F. Supp. 2d 1170, 1180 (D.N.M. 2009) ("[B]ecause of the limitations on the reliability of firearms identification evidence discussed above, Mr. Nichols will not be permitted to testify that his methodology allows him to reach this conclusion as a matter of scientific certainty. Mr. Nichols also will not be allowed to testify that he can conclude that there is a match to the exclusion, either practical or absolute, of all other guns. He may only testify that, in his opinion, the bullet came from the suspect rifle to within a reasonable degree of certainty in the firearms examination field.").

294. In 1975, the Ninth Circuit noted that toolmark identification "rests upon a scientific basis and is a reliable and generally accepted procedure." United States v. Bowers, 534 F.2d 186, 193 (9th Cir. 1976).

ing screwdrivers,[295] crowbars,[296] punches,[297] knives,[298] as well as other objects.[299] An expert's opinion is admissible even if the expert cannot testify to a positive identification.[300]

IX. Bite Mark Evidence

Bite mark analysis has been used for more than 50 years to establish a connection between a defendant and a crime.[301] The specialty developed within the field of forensic dentistry as an adjunct of dental identification, rather than originating in

295. *E.g.,* State v. Dillon, 161 N.W.2d 738, 741 (Iowa 1968) (screwdriver and nail bar fit marks on door frame); State v. Wessling, 150 N.W.2d 301 (Iowa 1967) (screwdriver); State v. Hazelwood, 498 P.2d 607, 612 (Kan. 1972) (screwdriver and imprint on window molding); State v. Wade, 465 S.W.2d 498, 499–500 (Mo. 1971) (screwdriver and pry marks on door jamb); State v. Brown, 291 S.W.2d 615, 618–19 (Mo. 1956) (crowbar and screwdriver marks on window sash and door); State v. Eickmeier, 191 N.W.2d 815, 816 (Neb. 1971) (screwdriver and marks on door).

296. *E.g., Brown,* 291 S.W.2d at 618–19 (Mo. 1956) (crowbar and screwdriver marks on window sash and door); State v. Raines, 224 S.E.2d 232, 234 (N.C. Ct. App. 1976).

297. *E.g.,* State v. Montgomery, 261 P.2d 1009, 1011–12 (Kan. 1953) (punch marks on safe).

298. *E.g.,* State v. Baldwin, 12 P. 318, 324–25 (Kan. 1886) (experienced carpenters could testify that wood panel could have been cut by accused's knife); Graves v. State, 563 P.2d 646, 650 (Okla. Crim. App. 1977) (blade and knife handle matched); State v. Clark, 287 P. 18, 20 (Wash. 1930) (knife and cuts on tree branches); State v. Bernson, 700 P.2d 758, 764 (Wash. Ct. App. 1985) (knife tip comparison).

299. *E.g.,* United States v. Taylor, 334 F. Supp. 1050, 1056–57 (E.D. Pa. 1971) (impressions on stolen vehicle and impressions made by dies found in defendant's possession), *aff'd,* 469 F.2d 284 (3d Cir. 1972); State v. McClelland, 162 N.W.2d 457, 462 (Iowa 1968) (pry bar and marks on "jimmied" door); Adcock v. State, 444 P.2d 242, 243–44 (Okla. Crim. App. 1968) (tool matched pry marks on door molding); State v. Olsen, 317 P.2d 938, 940 (Or. 1957) (hammer marks on the spindle of a safe).

300. For example, in *United States v. Murphy,* 996 F.2d 94 (5th Cir. 1993), an FBI expert gave limited testimony "that the tools such as the screwdriver associated with Murphy 'could' have made the marks on the ignitions but that he could not positively attribute the marks to the tools identified with Murphy." *Id.* at 99; *see also* State v. Genrich, 928 P.2d 799, 802 (Colo. App. 1996) (upholding expert testimony that three different sets of pliers recovered from the accused's house were used to cut wire and fasten a cap found in the debris from pipe bombs: "The expert's premise, that no two tools make exactly the same mark, is not challenged by any evidence in this record. Hence, the lack of a database and points of comparison does not render the opinion inadmissible.").

Although most courts have been receptive to toolmark evidence, a notable exception was *Ramirez v. State,* 810 So. 2d 836, 849–51(Fla. 2001). In *Ramirez,* the Florida Supreme Court rejected the testimony of five experts who claimed general acceptance for a process of matching a knife with a cartilage wound in a murder victim—a type of "toolmark" comparison. Although the court applied *Frye,* it emphasized the lack of testing, the paucity of "meaningful peer review," the absence of a quantified error rate, and the lack of developed objective standards. In *Sexton v. State,* 93 S.W.3d 96 (Tex. Crim. App. 2002), an expert testified that cartridge cases from *unfired* bullets found in the appellant's apartment had distinct marks that matched fired cartridge cases found at the scene of the offense. The court ruled the testimony inadmissible: "This record qualifies Crumley as a firearms identification expert, but does not support his capacity to identify cartridge cases on the basis of magazine marks only." *Id.* at 101.

301. *See* E.H. Dinkel, *The Use of Bite Mark Evidence as an Investigative Aid,* 19 J. Forensic Sci. 535 (1973).

crime laboratories. Courts have admitted bite mark comparison evidence in homicide, rape, and child abuse cases. In virtually all the cases, the evidence was first offered by the prosecution. The typical bite mark case has involved the identification of the defendant by matching his dentition with a mark left on the victim. In several cases, however, the victim's teeth have been compared with marks on the defendant's body. One bite mark case involved dentures[302] and another braces.[303] A few cases have entailed bite impressions on foodstuff found at a crime scene: apple,[304] piece of cheese,[305] and sandwich.[306] Still other cases involved dog bites.[307]

Bite marks occur primarily in sex-related crimes, child abuse cases, and offenses involving physical altercations, such as homicide. A survey of 101 cases reported these findings: "More than one bitemark was present in 48% of all the bite cases studied. Bitemarks were found on adults in 81.3% of the cases and on children under 18 years-of-age in 16.7% of cases. Bitemarks were associated with the following types of crimes: murder, including attempted murder (53.9%), rape (20.8%), sexual assault (9.7%), child abuse (9.7%), burglary (3.3%), and kidnapping (12.6%)."[308]

A. The Technique

Bite mark identification is an offshoot of the dental identification of deceased persons, which is often used in mass disasters. Dental identification is based on the assumption that every person's dentition is unique. The human adult dentition consists of 32 teeth, each with 5 anatomic surfaces. Thus, there are 160 dental surfaces that can contain identifying characteristics. Restorations, with varying shapes, sizes, and restorative materials, may offer numerous additional points of individuality. Moreover, the number of teeth, prostheses, decay, malposition,

302. *See* Rogers v. State, 344 S.E.2d 644, 647 (Ga. 1986) ("Bite marks on one of Rogers' arms were consistent with the dentures worn by the elderly victim.").

303. *See* People v. Shaw, 664 N.E.2d 97, 101, 103 (Ill. App. Ct. 1996) (In a murder and aggravated sexual assault prosecution, the forensic odontologist opined that the mark on the defendant was caused by the orthodontic braces on the victim's teeth; "Dr. Kenney admitted that he was not a certified toolmark examiner"; no abuse of discretion to admit evidence).

304. *See* State v. Ortiz, 502 A.2d 400, 401 (Conn. 1985).

305. *See* Doyle v. State, 263 S.W.2d 779, 779 (Tex. Crim. App. 1954); Seivewright v. State, 7 P.3d 24, 26 (Wyo. 2000) ("On the basis of his comparison of the impressions from the cheese with Seivewright's dentition, Dr. Huber concluded that Seivewright was the person who bit the cheese.").

306. *See* Banks v. State, 725 So. 2d 711, 714–16 (Miss. 1997) (finding a due process violation when prosecution expert threw away sandwich after finding the accused's teeth consistent with the sandwich bite).

307. *See* Davasher v. State, 823 S.W.2d 863, 870 (Ark. 1992) (expert testified that victim's dog could be eliminated as the source of mark found on defendant); State v. Powell, 446 S.E.2d 26, 27–28 (N.C. 1994) ("A forensic odontologist testified that dental impressions taken from Bruno and Woody [accused's dogs] were compatible with some of the lacerations in the wounds pictured in scale photographs of Prevette's body.").

308. Iain A. Pretty & David J. Sweet, *Anatomical Location of Bitemarks and Associated Findings in 101 Cases from the United States,* 45 J. Forensic Sci. 812, 812 (2000).

malrotation, peculiar shapes, root canal therapy, bone patterns, bite relationship, and oral pathology may also provide identifying characteristics.[309] The courts have accepted dental identification as a means of establishing the identity of a homicide victim,[310] with some cases dating back to the nineteenth century.[311] According to one court, "it cannot be seriously disputed that a dental structure may constitute a means of identifying a deceased person . . . where there is some dental record of that person with which the structure may be compared."[312]

1. *Theory of uniqueness*

Identification of a suspect by matching his or her dentition with a bite mark found on the victim of a crime rests on the theory that each person's dentition is unique. However, there are significant differences between the use of forensic dental techniques to identify a decedent and the use of bite mark analysis to identify a perpetrator.[313] In 1969, when bite mark comparisons were first studied, one authority raised the following problems:

> [Bite]marks can never be taken to reproduce accurately the dental features of the originator. This is due partially to the fact that bite marks generally include only a limited number of teeth. Furthermore, the material (whether food stuff or human skin) in which the mark has been left is usually found to be a very unsatisfactory impression material with shrinkage and distortion characteristics that are unknown. Finally, these marks represent only the remaining and fixed picture of an action, the mechanism of which may vary from case to case. For instance, there is as yet no precise knowledge of the possible differences between biting off a morsel of food and using one's teeth for purposes of attack or defense.[314]

309. The identification is made by comparing the decedent's teeth with antemortem dental records, such as charts and, more importantly, radiographs.

310. *E.g.*, Wooley v. People, 367 P.2d 903, 905 (Colo. 1961) (dentist compared his patient's record with dentition of a corpse); Martin v. State, 636 N.E.2d 1268, 1272 (Ind. Ct. App. 1994) (dentist qualified to compare X rays of one of his patients with skeletal remains of murder victim and make a positive identification); Fields v. State, 322 P.2d 431, 446 (Okla. Crim. App. 1958) (murder case in which victim was burned beyond recognition).

311. *See* Commonwealth v. Webster, 59 Mass. (5 Cush.) 295, 299–300 (1850) (remains of the incinerated victim, including charred teeth and parts of a denture, were identified by the victim's dentist); Lindsay v. People, 63 N.Y. 143, 145–46 (1875).

312. People v. Mattox, 237 N.E.2d 845, 846 (Ill. App. Ct. 1968).

313. *See* Iain A. Pretty & David J. Sweet, *The Scientific Basis for Human Bitemark Analyses—A Critical Review,* 41 Sci. & Just. 85, 88 (2001) ("A distinction must be drawn from the ability of a forensic dentist to identify an individual from their dentition by using radiographs and dental records and the science of bitemark analysis.").

314. S. Keiser-Nielson, *Forensic Odontology*, 1 U. Tol. L. Rev. 633, 636 (1969); *see also* NRC Forensic Science Report, *supra* note 3, at 174 ("[B]ite marks on the skin will change over time and can be distorted by the elasticity of the skin, the unevenness of the surface bite, and swelling and healing. These features may severely limit the validity of forensic odontology. Also, some practical difficulties, such as distortions in photographs and changes over time in the dentition of suspects, may limit the accuracy of the results.").

Dental identifications of decedents do not pose any of these problems; the expert can often compare all 32 teeth with X rays depicting all those teeth. However, in the typical bite mark case, all 32 teeth cannot be compared; often only 4 to 8 are biting teeth that can be compared. Similarly, all five anatomic surfaces are not engaged in biting; only the edges of the front teeth come into play. In sum, bite mark identification depends not only on the uniqueness of each person's dentition but also on "whether there is a [sufficient] representation of that uniqueness in the mark found on the skin or other inanimate object."[315]

2. *Methods of comparison*

Several methods of bite mark analysis have been reported. All involve three steps: (1) registration of both the bite mark and the suspect's dentition, (2) comparison of the dentition and bite mark, and (3) evaluation of the points of similarity or dissimilarity. The reproductions of the bite mark and the suspect's dentition are analyzed through a variety of methods.[316] The comparison may be either direct or indirect. A model of the suspect's teeth is used in direct comparisons; the model is compared to life-size photographs of the bite mark. Transparent overlays made from the model are used in indirect comparisons.

Although the expert's conclusions are based on objective data, the ultimate opinion regarding individuation is essentially a subjective one.[317] There is no accepted minimum number of points of identity required for a positive identification.[318] The experts who have appeared in published bite mark cases have testified to a wide range of points of similarity, from a low of eight points to a

315. Raymond D. Rawson et al., *Statistical Evidence for the Individuality of the Human Dentition*, 29 J. Forensic Sci. 252 (1984).

316. *See* David J. Sweet, *Human Bitemarks: Examination, Recovery, and Analysis*, in Manual of Forensic Odontology 162 (American Society of Forensic Odontology, 3d ed. 1997) [hereinafter ASFO Manual] ("The analytical protocol for bitemark comparison is made up of two broad categories. Firstly, the measurement of specific traits and features called a *metric analysis*, and secondly, the physical matching or comparison of the configuration and pattern of the injury called a *pattern association*."); *see also* David J. Sweet & C. Michael Bowers, *Accuracy of Bite Mark Overlays: A Comparison of Five Common Methods to Produce Exemplars from a Suspect's Dentition*, 43 J. Forensic Sci. 362, 362 (1998) ("A review of the forensic odontology literature reveals multiple techniques for overlay production. There is an absence of reliability testing or comparison of these methods to known or reference standards.").

317. *See* Roland F. Kouble & Geoffrey T. Craig, *A Comparison Between Direct and Indirect Methods Available for Human Bite Mark Analysis*, 49 J. Forensic Sci. 111, 111 (2004) ("It is important to remember that computer-generated overlays still retain an element of subjectivity, as the selection of the biting edge profiles is reliant on the operator placing the 'magic wand' onto the areas to be highlighted within the digitized image.").

318. *See* Keiser-Nielson, *supra* note 314, at 637–38; *see also* Stubbs v. State, 845 So. 2d 656, 669 (Miss. 2003) ("There is little consensus in the scientific community on the number of points which must match before any positive identification can be announced.").

high of 52 points.[319] Moreover, disagreements among experts in court appear commonplace: "Although bite mark evidence has demonstrated a high degree of acceptance, it continues to be hotly contested in 'battles of the experts.' Review of trial transcripts reveals that distortion and the interpretation of distortion is a factor in most cases."[320] Because of the subjectivity, some odontologists have argued that "bitemark evidence should only be used to exclude a suspect. This [argument] is supported by research which shows that the exclusion of non-biters within a population of suspects is extremely accurate; far more so than the positive identification of biters."[321]

3. ABFO Guidelines

In an attempt to develop an objective method, in 1984 the American Board of Forensic Odontology (ABFO) promulgated guidelines for bite mark analysis, including a uniform scoring system.[322] According to the drafting committee, "[t]he scoring system . . . has demonstrated a method of evaluation that produced a high degree of reliability among observers."[323] Moreover, the committee characterized "[t]he scoring guide . . . [as] the beginning of a truly scientific approach to bite mark analysis."[324] In a subsequent letter, however, the drafting committee wrote:

> While the Board's published guidelines suggest use of the scoring system, the authors' present recommendation is that all odontologists await the results of further research before relying on precise point counts in evidentiary proceedings. . . . [T]he authors believe that further research is needed regarding the quantification of bite mark evidence before precise point counts can be relied upon in court proceedings.[325]

319. *E.g.*, State v. Garrison, 585 P.2d 563, 566 (Ariz. 1978) (10 points); People v. Slone, 143 Cal. Rptr. 61, 67 (Cal. Ct. App. 1978) (10 points); People v. Milone, 356 N.E.2d 1350, 1356 (Ill. App. Ct. 1976) (29 points); State v. Sager, 600 S.W.2d 541, 564 (Mo. Ct. App. 1980) (52 points); State v. Green, 290 S.E.2d 625, 630 (N.C. 1982) (14 points); State v. Temple, 273 S.E.2d 273, 279 (N.C. 1981) (8 points); Kennedy v. State, 640 P.2d 971, 976 (Okla. Crim. App. 1982) (40 points); State v. Jones, 259 S.E.2d 120, 125 (S.C. 1979) (37 points).

320. Raymond D. Rawson et al., *Analysis of Photographic Distortion in Bite Marks: A Report of the Bite Mark Guidelines Committee*, 31 J. Forensic Sci. 1261, 1261–62 (1986). The committee noted: "[P]hotographic distortion can be very difficult to understand and interpret when viewing prints of bite marks that have been photographed from unknown angles." *Id.* at 1267.

321. Iain A. Pretty, *A Web-Based Survey of Odontologist's Opinions Concerning Bitemark Analyses*, 48 J. Forensic Sci. 1117, 1120 (2003) [hereinafter *Web-Based Survey*].

322. ABFO, *Guidelines for Bite Mark Analysis*, 112 J. Am. Dental Ass'n 383 (1986).

323. Raymond D. Rawson et al., *Reliability of the Scoring System of the American Board of Forensic Odontology for Human Bite Marks*, 31 J. Forensic Sci. 1235, 1259 (1986).

324. *Id.*

325. Letter, *Discussion of "Reliability of the Scoring System of the American Board of Forensic Odontology for Human Bite Marks*," 33 J. Forensic Sci. 20 (1988).

B. The Empirical Record

The 2009 NRC report concluded:

> More research is needed to confirm the fundamental basis for the science of bite mark comparison. Although forensic odontologists understand the anatomy of teeth and the mechanics of biting and can retrieve sufficient information from bite marks on skin to assist in criminal investigations and provide testimony at criminal trials, the scientific basis is insufficient to conclude that bite mark comparisons can result in a conclusive match.[326]

Moreover, "[t]here is no science on the reproducibility of the different methods of analysis that lead to conclusions about the probability of a match."[327] Another passage provides: "Despite the inherent weaknesses involved in bite mark comparison, it is reasonable to assume that the process can sometimes reliably exclude suspects."[328]

Although bitemark identifications are accepted by forensic dentists, only a few empirical studies have been conducted[329] and only a small number of forensic dentists have addressed the empirical issue. In the words of one expert,

> The research suggests that bitemark evidence, at least that which is used to identify biters, is a potentially valid and reliable methodology. It is generally accepted within the scientific [dental] community, although the basis of this acceptance within the peer-reviewed literature is thin. Only three studies have examined the ability of odontologists to utilise bitemarks for the identification of biters, and only two studies have been performed in what could be considered a contemporary framework of attitudes and techniques.[330]

326. NRC Forensic Science Report, *supra* note 3, at 175. *See also id.* at 176. ("Although the majority of forensic odontologists are satisfied that bite marks can demonstrate sufficient detail for positive identification, no scientific studies support this assessment, and no large population studies have been conducted."),

327. *Id.* at 174.

328. *Id.* at 176.

329. *See* C. Michael Bowers, Forensic Dental Evidence: An Investigator's Handbook 189 (2004) ("As a number of legal commentators have observed, bite mark analysis has never passed through the rigorous scientific examination that is common to most sciences. The literature does not go far in disputing that claim."); Iain A. Pretty, *Unresolved Issues in Bitemark Analysis*, *in* Bitemark Evidence 547, 547 (Robert B.J. Dorion ed., 2005) ("As a general rule, case reports add little to the scientific knowledge base, and therefore, if these, along with noncritical reviews, are discarded, very little new empirical evidence has been developed in the past five years."); *id.* at 561 ("[T]he final question in the recent survey asked, 'Should an appropriately trained individual positively identify a suspect from a bitemark on skin'—70% of the respondents stated yes. However, it is the judicial system that must assess validity, reliability, and a sound scientific base for expert forensic testimony. A great deal of further research is required if odontology hopes to continue to be a generally accepted science.").

330. Iain A. Pretty, *Reliability of Bitemark Evidence*, *in* Bitemark Evidence at 543 (Robert B.J. Dorion ed., 2005).

Commentators have highlighted the following areas of controversy: "a) accuracy of the bitemark itself, b) uniqueness of the human dentition, and c) analytical techniques."[331]

One part of a 1975 study involved identification of bites made on pigskin: "Incorrect identification of the bites made on pigskin ranged from 24% incorrect identifications under ideal laboratory conditions to as high as 91% incorrect identifications when the bites were photographed 24 hours after the bites made."[332] A 1999 ABFO Workshop, "where ABFO diplomats attempted to match four bitemarks to seven dental models, resulted in 63.5% false positives."[333] A 2001 study of bites on pigskin "found false positive identifications of 11.9–22.0% for various groups of forensic odontologists (15.9% false positives for ABFO diplomats), with some ABFO diplomats faring far worse."[334] Other commentators take a more favorable view of these studies.[335]

1. DNA exonerations

In several cases, subsequent DNA testing has demonstrated the error in a prior bite mark identification. In *State v. Krone*,[336] two experienced experts concluded that the defendant had made the bite mark found on a murder victim. The defendant, however, was later exonerated through DNA testing.[337] In *Otero v. Warnick*,[338] a forensic dentist testified that the "plaintiff was the only person in the world who

331. Pretty & Sweet, *supra* note 313, at 87. Commentators had questioned the lack of research in the field as long ago as 1985. Two commentators wrote:

> There is effectively no valid documented scientific data to support the hypothesis that bite marks are demonstrably unique. Additionally, there is no documented scientific data to support the hypothesis that a latent bite mark, like a latent fingerprint, is a true and accurate reflection of this uniqueness. To the contrary, what little scientific evidence that does exist clearly supports the conclusion that crime-related bite marks are grossly distorted, inaccurate, and therefore unreliable as a method of identification.

Allen P. Wilkinson & Ronald M. Gerughty, *Bite Mark Evidence: Its Admissibility Is Hard to Swallow*, 12 W. St. U. L. Rev. 519, 560 (1985).

332. C. Michael Bowers, *Problem-Based Analysis of Bitemark Misidentifications: The Role of DNA*, 159S Forensic Sci. Int'l S104, S106 (2006) (citing D.K. Whittaker, *Some Laboratory Studies on the Accuracy of Bite Mark Comparison*, 25 Int'l Dent. J. 166 (1975)) [hereinafter *Problem-Based Analysis*].

333. Bowers, *Problem-Based Analysis*, *supra* note 332, at S106. *But see* Kristopher L. Arheart & Iain A. Pretty, *Results of the 4th ABFO Bitemark Workshop 1999*, 124 Forensic Sci. Int'l 104 (2001).

334. Bowers, *Problem-Based Analysis*, *supra* note 332, at S106 (citing Iain A. Pretty & David J. Sweet, *Digital Bitemark Overlays—An Analysis of Effectiveness*, 46 J. Forensic Sci. 1385, 1390 (2001) ("While the overall effectiveness of overlays has been established, the variation in individual performance of odontologists is of concern.")).

335. *See* Pretty, *Reliability of Bitemark Evidence*, *in* Bitemark Evidence, *supra* note 330, at 538–42.

336. 897 P.2d 621, 622, 623 (Ariz. 1995) ("The bite marks were crucial to the State's case because there was very little other evidence to suggest Krone's guilt."; "Another State dental expert, Dr. John Piakis, also said that Krone made the bite marks. . . . Dr. Rawson himself said that Krone made the bite marks. . . .").

337. *See* Mark Hansen, *The Uncertain Science of Evidence*, A.B.A. J. 49 (2005) (discussing *Krone*).

338. 614 N.W.2d 177 (Mich. Ct. App. 2000).

could have inflicted the bite marks on [the murder victim's] body. On January 30, 1995, the Detroit Police Crime Laboratory released a supplemental report that concluded that plaintiff was excluded as a possible source of DNA obtained from vaginal and rectal swabs taken from [the victim's] body."[339] In *Burke v. Town of Walpole*,[340] the expert concluded that "Burke's teeth matched the bite mark on the victim's left breast to a 'reasonable degree of scientific certainty.' That same morning . . . DNA analysis showed that Burke was excluded as the source of male DNA found in the bite mark on the victim's left breast."[341] In the future, the availability of nuclear DNA testing may reduce the need to rely on bite mark identifications.[342]

C. Case Law Development

People v. Marx (1975)[343] emerged as the leading bite mark case. After *Marx*, bite mark evidence became widely accepted.[344] By 1992, it had been introduced or noted in 193 reported cases and accepted as admissible in 35 states.[345] Some courts described bite mark comparison as a "science,"[346] and several cases took judicial notice of its validity.[347]

339. *Id.* at 178.

340. 405 F.3d 66, 73 (1st Cir. 2005).

341. *See also* Bowers, *Problem-Based Analysis, supra* note 332, at S104 (citing several cases involving bitemarks and DNA exonerations: *Gates, Bourne, Morris, Krone, Otero, Young,* and *Brewer*); Mark Hansen, *Out of the Blue*, A.B.A. 50, 51 (1996) (DNA analysis of skin taken from fingernail scrapings of the victim conclusively excluded Bourne).

342. *See* Pretty, *Web-Based Survey, supra* note 321, at 1119 ("The use of DNA in the assessment of bitemarks has been established for some time, although previous studies have suggested that the uptake of this technique has been slow. It is encouraging to note that nearly half of the respondents in this case have employed biological evidence in a bitemark case.").

343. 126 Cal. Rptr. 350 (Cal. Ct. App. 1975). The court in *Marx* avoided applying the *Frye* test, which requires acceptance of a novel technique by the scientific community as a prerequisite to admissibility. According to the court, the *Frye* test "finds its rational basis in the degree to which the trier of fact must accept, on faith, scientific hypotheses not capable of proof or disproof in court and not even generally accepted outside the courtroom." *Id.* at 355–56.

344. Two Australian cases, however, excluded bite mark evidence. *See* Lewis v. The Queen (1987) 29 A. Crim. R. 267 (odontological evidence was improperly relied on, in that this method has not been scientifically accepted); R v. Carroll (1985) 19 A. Crim. R. 410 ("[T]he evidence given by the three odontologist is such that it would be unsafe or dangerous to allow a verdict based upon it to stand.").

345. Steven Weigler, *Bite Mark Evidence: Forensic Odontology and the Law*, 2 Health Matrix: J.L.-Med. 303 (1992).

346. *See* People v. Marsh, 441 N.W.2d 33, 35 (Mich. Ct. App. 1989) ("the science of bite mark analysis has been extensively reviewed in other jurisdictions"); State v. Sager, 600 S.W.2d 541, 569 (Mo. Ct. App. 1980) ("an exact science").

347. *See* State v. Richards, 804 P.2d 109, 112 (Ariz. Ct. App. 1990) ("[B]ite mark evidence is admissible without a preliminary determination of reliability. . . ."); People v. Middleton, 429 N.E.2d 100, 101 (N.Y. 1981) ("The reliability of bite mark evidence as a means of identification is sufficiently

1. *Specificity of opinion*

In some cases, experts testified only that a bite mark was "consistent with" the defendant's teeth.[348] In other cases, they went further and opined that it is "highly probable" or "very highly probable" that the defendant made the mark.[349] In still other cases, experts made positive identifications (to the exclusion of all other persons).[350] It is not unusual to find experts disagreeing in individual cases—often over the threshold question of whether a wound was even a bite mark.[351]

established in the scientific community to make such evidence admissible in a criminal case, without separately establishing scientific reliability in each case. . . ."); State v. Armstrong, 369 S.E.2d 870, 877 (W. Va. 1988) (judicially noticing the reliability of bite mark evidence).

348. *E.g.*, Rogers v. State, 344 S.E.2d 644, 647 (Ga. 1986) ("Bite marks on one of Rogers' arms were consistent with the dentures worn by the elderly victim."); People v. Williams, 470 N.E.2d 1140, 1150 (Ill. App. Ct. 1984) ("could have"); State v. Hodgson, 512 N.W.2d 95, 98 (Minn. 1994) (en banc) (Board-certified forensic odontologist testified that "there were several similarities between the bite mark and the pattern of [the victim's] teeth, as revealed by known molds of his mouth."); State v. Routh, 568 P.2d 704, 705 (Or. Ct. App. 1977) ("similarity"); Williams v. State, 838 S.W.2d 952, 954 (Tex. Ct. App. 1992) ("One expert, a forensic odontologist, testified that Williams's dentition was consistent with the injury (bite mark) on the deceased."); State v. Warness, 893 P.2d 665, 669 (Wash. Ct. App. 1995) ("[T]he expert testified that his opinion was not conclusive, but the evidence was consistent with the alleged victim's assertion that she had bitten Warness. . . . Its probative value was therefore limited, but its relevance was not extinguished.").

349. *E.g.*, People v. Slone, 143 Cal. Rptr. 61, 67 (Cal. Ct. App. 1978); People v. Johnson, 289 N.E.2d 722, 726 (Ill. App. Ct. 1972).

350. *E.g.*, Morgan v. State, 639 So. 2d 6, 9 (Fla. 1994) ("[T]he testimony of a dental expert at trial positively matched the bite marks on the victim with Morgan's teeth."); Duboise v. State, 520 So. 2d 260, 262 (Fla. 1988) (Expert "testified at trial that within a reasonable degree of dental certainty Duboise had bitten the victim."); Brewer v. State, 725 So. 2d 106, 116 (Miss. 1998) ("Dr. West opined that Brewer's teeth inflicted the five bite mark patterns found on the body of Christine Jackson."); State v. Schaefer, 855 S.W.2d 504, 506 (Mo. Ct. App. 1993) ("[A] forensic dentist testified that the bite marks on Schaefer's shoulder matched victim's dental impression, and concluded that victim caused the marks."); State v. Lyons, 924 P.2d 802, 804 (Or. 1996) (forensic odontologist "had no doubt that the wax models were made from the same person whose teeth marks appeared on the victim's body"); State v. Cazes, 875 S.W.2d 253, 258 (Tenn. 1994) (A forensic odontologist "concluded to a reasonable degree of dental certainty that Cazes' teeth had made the bite marks on the victim's body at or about the time of her death.").

351. *E.g.*, Ege v. Yukins, 380 F. Supp. 2d 852, 878 (E.D. Mich. 2005) ("[T]he defense attempted to rebut Dr. Warnick's testimony with the testimony of other experts who opined that the mark on the victim's cheek was the result of *livor mortis* and was not a bite mark at all."); Czapleski v. Woodward, No. C-90-0847 MHP, 1991 U.S. Dist. LEXIS 12567, at *3–4 (N.D. Cal. Aug. 30, 1991) (dentist's initial report concluded that "bite" marks found on child were consistent with dental impressions of mother; several experts later established that the marks on child's body were postmortem abrasion marks and not bite marks); Kinney v. State, 868 S.W.2d 463, 464–65 (Ark. 1994) (disagreement that marks were human bite marks); People v. Noguera, 842 P.2d 1160, 1165 n.1 (Cal. 1992) ("At trial, extensive testimony by forensic ondontologists [sic] was presented by both sides, pro and con, as to whether the wounds were human bite marks and, if so, when they were inflicted."); State v. Duncan, 802 So. 2d 533, 553 (La. 2001) ("Both defense experts testified that these marks on the victim's body were not bite marks."); Stubbs v. State, 845 So. 2d 656, 668 (Miss. 2003) ("Dr. Galvez denied the impressions found on Williams were the results of bite marks.").

2. *Post-*Daubert *cases*

Although some commentators questioned the underlying basis for the technique after *Daubert*,[352] courts have continued to admit the evidence.[353]

X. Microscopic Hair Evidence

The first reported use of forensic hair analysis occurred more than 150 years ago in 1861 in Germany.[354] The first published American opinion was an 1882 Wisconsin decision, *Knoll v. State*.[355] Based on a microscopic comparison, the expert testified that the hair samples shared a common source. Hair and the closely related fiber analysis played a prominent role in two of the most famous twentieth-century American prosecutions: Ted Bundy in Florida and Wayne Williams, the alleged Atlanta child killer.[356] Although hair comparison evidence has been judicially accepted for decades, it is another forensic identification discipline that is being reappraised today.

A. The Technique

Generally, after assessing whether a sample is a hair and not a fiber, an analyst may be able to determine: (1) whether the hair is of human or animal origin, (2) the part of the body that the hair came from, (3) whether the hair has been dyed, (4) whether the hair was pulled or fell out as a result of natural causes or disease,[357] and (5) whether the hair was cut or crushed.[358]

352. *See* Pretty & Sweet, *supra* note 313, at 86 ("Despite the continued acceptance of bitemark evidence in European, Oceanic and North American Courts the fundamental scientific basis for bitemark analysis has never been established.").

353. *See* State v. Timmendequas, 737 A.2d 55, 114 (N.J. 1999) ("Judicial opinion from other jurisdictions establish that bite-mark analysis has gained general acceptance and therefore is reliable. Over thirty states considering such evidence have found it admissible and no state has rejected bitemark evidence as unreliable.") (citations omitted); *Stubbs*, 845 So. 2d at 670; Howard v. State, 853 So. 2d 781, 795–96 (Miss. 2003); Seivewright v. State, 7 P.3d 24, 30 (Wyo. 2000) ("Given the wide acceptance of bite mark identification testimony and Seivewright's failure to present evidence challenging the methodology, we find no abuse of discretion in the district court's refusal to hold an evidentiary hearing to analyze Dr. Huber's testimony.").

354. E. James Crocker, *Trace Evidence*, *in* Forensic Evidence in Canada 259, 265 (1991) (the analyst was Rudolf Virchow, a Berliner).

355. 12 N.W. 369 (Wis. 1882).

356. Edward J. Imwinkelried, *Forensic Hair Analysis: The Case Against the Underemployment of Scientific Evidence*, 39 Wash. & Lee L. Rev. 41, 43 (1982).

357. *See* Delaware v. Fensterer, 474 U.S. 15. 16–17 (1985) (FBI analyst testified hair found at a murder scene had been forcibly removed.).

358. *See* 2 Giannelli & Imwinkelried, *supra* note 177, § 24-2.

The most common subject for hair testimony involves an attempt to individuate the hair sample, at least to some degree. If the unknown is head hair, the expert might gather approximately 50 hair strands from five different areas of the scalp (the top, front, back, and both sides) from the known source.[359] Before the microscopic analysis, the expert examines the hair macroscopically to identify obvious features visible to the naked eye such as the color of the hair and its form, that is, whether it is straight, wavy, or curved.[360] The expert next mounts the unknown hair and the known samples on microscope slides for a more detailed examination of characteristics such as scale patterns, size, color, pigment distribution, maximum diameter, shaft length, and scale count. Some of these comparative judgments are subjective in nature: "Human hair characteristics (e.g., scale patterns, pigmentation, size) vary within a single individual. . . . Although the examination procedure involves objective methods of analysis, the subjective weights associated with the characteristics rest with the examiner."[361]

Often the examiner determines only whether the hair samples from the crime scene and the accused are "microscopically indistinguishable." Although this finding is consistent with the hypothesis that the samples had the same source, its probative value would, of course, vary if only a hundred people had microscopically indistinguishable hair as opposed to several million. As discussed below, experts have often gone beyond this "consistent with" testimony.

B. The Empirical Record

The 2009 NRC report contained an assessment of hair analysis. The report began the assessment by observing that there are neither "scientifically accepted [population] frequency" statistics for various hair characteristics nor "uniform standards on the number of features on which hairs must agree before an examiner may declare a 'match.'"[362] The report concluded,

> [T]estimony linking microscopic hair analysis with particular defendants is highly unreliable. In cases where there seems to be a morphological match (based on microscopic examination), it must be confirmed using mtDNA analysis; microscopic studies are of limited probative value. The committee found no scientific support for the use of hair comparisons for individualization in the absence of nuclear DNA. Microscopy and mtDNA analysis can be used in tandem and add to one another's value for classifying a common source, but no studies have been performed specifically to quantify the reliability of their joint use.[363]

359. NRC Forensic Science Report, *supra* note 3, at 157.
360. *Id.*
361. Miller, *supra* note 67, at 157–58.
362. NRC Forensic Science Report, *supra* note 3, at 160.
363. *Id.* at 8.

There is a general consensus that hair examination can yield reliable information about class characteristics of hair strands.[364] Indeed, experts can identify major as well as secondary characteristics. Major characteristics include such features as color, shaft form, and hair diameter.[365] Secondary characteristics are such features as pigment size and shaft diameter.[366] These characteristics can help narrow the class of possible sources for the unknown hair sample.

There have been several major efforts to provide an empirical basis for individuation opinions in hair analysis. In the 1940s, Gamble and Kirk investigated whether hair samples from different persons could be distinguished on the basis of scale counts.[367] However, they used a small database of only thirty-nine hair samples, and a subsequent attempt to replicate the original experiment yielded contradictory results.[368]

In the 1960s, neutron activation analysis was used in an effort to individuate hair samples. The research focused on determining the occurrence of various trace element concentrations in human hair.[369] Again, subsequent research tended to show that there are significant hair-to-hair variations in trace element concentration among the hairs of a single person.[370]

In the 1970s, two Canadian researchers, Gaudette and Keeping, attempted to develop a "ballpark" estimate of the probability of a false match in hair analysis. They published articles describing three studies: (1) a 1974 study involving scalp hair,[371] (2) a

364. *Id.* at 157.

365. *Id.* at 5–23.

366. *Id.*

367. Their initial research indicated that: (1) the scale count of even a single hair strand is nearly always representative of all scalp hairs; and (2) while the average or mean scale count is constant for the individual, the count differs significantly from person to person. Lucy L. Gamble & Paul L. Kirk, *Human Hair Studies II. Scale Counts,* 31 J. Crim. L. & Criminology 627, 629 (1941); Paul L. Kirk & Lucy L. Gamble, *Further Investigation of the Scale Count of Human Hair,* 33 J. Crim. L. & Criminology 276, 280 (1942).

368. Joseph Beeman, *The Scale Count of Human Hair,* 32 J. Crim. L. & Criminology 572, 574 (1942).

369. Rita Cornelis, *Is It Possible to Identify Individuals by Neutron Activation Analysis of Hair?* 12 Med. Sci. & L. 188 (1972); Lima et al., *Activation Analysis Applied to Forensic Investigation: Some Observations on the Problem of Human Hair Individualization,* 1 Radio Chem. Methods of Analysis 119 (Int'l Atomic Energy Agency 1965); A.K. Perkins, *Individualization of Human Head Hair, in* Proceedings of the First Int'l Conf. on Forensic Activation Analysis 221 (V. Guin ed., 1967).

370. Rita Cornelis, *Truth Has Many Facets: The Neutron Activation Analysis Story,* 20 J. Forensic Sci. 93, 95 (1980) ("I am convinced that irrefutable hair identification from its trace element composition still belongs to the realm of wishful thinking. . . . The state of the art can be said to be that nearly all interest for trace elements present in hair, as a practical identification tool, has faded."); Dennis S. Karjala, *Evidentiary Uses of Neutron Activation Analysis,* 59 Cal. L. Rev. 977, 1039 (1971).

371. B.D. Gaudette & E.S. Keeping, *An Attempt at Determining Probabilities in Human Scalp Hair Comparison,* 19 J. Forensic Sci. 599 (1974).

1976 study using pubic hair,[372] and (3) a 1978 followup.[373] In the two primary studies (1974 and 1976), hair samples were analyzed to determine whether hairs from different persons were microscopically indistinguishable. The analysts used 23 different characteristics such as color, pigment distribution, maximum diameter, shaft length, and scale count.[374] Based on those data, they estimated the probability of a false match in scalp hair to be 1 in 4500 and the probability of a false match in pubic hair to be 1 in 800.

In the view of one commentator, Gaudette and Keeping's probability estimates "are easily challenged."[375] One limitation was the relatively small database in the study.[376] Moreover, the studies involved samples from different individuals and sought the probability that the samples from different persons would nonetheless appear microscopically indistinguishable. In a criminal trial, the question is quite different: Assuming the samples appear microscopically indistinguishable, what is the probability that they came from the same person?[377]

Early in the twenty-first century, the Verma research team revisited the individualization issue and attempted to develop an objective, automated method for identifying matches.[378] The authors claimed that their "system accurately judged whether two populations of hairs came from the same person or from different persons 83% of the time." However, a close inspection of the authors' tabular data indicates that (1) relying on this method, researchers characterized "9 of 73 different pairs as 'same' for a false positive rate of 9/73 = 12%"; and (2) the

372. B.D. Gaudette, *Probabilities and Human Pubic Hair Comparisons,* 21 J. Forensic Sci. 514, 514 (1976).

373. B.D. Gaudette, *Some Further Thoughts on Probabilities in Human Hair Comparisons,* 23 J. Forensic Sci. 758 (1978); *see also* Ray A. Wickenhaiser & David G. Hepworth, *Further Evaluation of Probabilities in Human Scalp Hair Comparisons,* 35 J. Forensic Sci. 1323 (1990).

374. They prescribed that with respect to each characteristic, the analysts had to classify into one of a number of specified subcategories. For example, the length characteristic was subdivided into five groups, depending on the strand's length in inches. They computed both the total number of comparisons made by the analysts and recorded the number of instances in which the analysts reported finding samples indistinguishable under the specified criteria.

375. D. Kaye, Science in Evidence 28 (1997); *see also* NRC Forensic Science Report, *supra* note 3, at 158 ([T]he "assignment of probabilities [by Gaudette and Keeping] has since been shown to be unreliable."); P.D. Barnett & R.R. Ogle, *Probabilities and Human Hair Comparisons,* 27 J. Forensic Sci. 272, 273–74 (1982); Dalva Moellenberg, *Splitting Hairs in Criminal Trials: Admissibility of Hair Comparison Probability Estimates,* 1984 Ariz. St. L.J. 521. *See generally* Nicholas Petrarco et al., *The Morphology and Evidential Significance of Human Hair Roots,* 33 J. Forensic Sci. 68, 68 (1988) ("Although many instrumental techniques to the individualization of human hair have been tried in recent years, these have not proved to be useful or reliable.").

376. For example, the pubic hair study involved a total of 60 individuals. In addition, the experiments involved primarily Caucasians. While the scalp hair study included 92 Caucasians, there were only 6 Asians and 2 African Americans in the study.

377. A Tawshunsky, *Admissibility of Mathematical Evidence in Criminal Trials,* 21 Am. Crim. L. Rev. 55, 57–66 (1983).

378. M.S. Verma et al., *Hair-MAP: A Prototype Automated System for Forensic Hair Comparison and Analysis,* 129 Forensic Sci. Int'l 168 (2002).

researchers characterized "4 sets of hairs from the same person as 'different' for a false negative rate of 4/9 = 44%."[379]

The above studies do not provide the only data relevant to the validity of hair analysis. There are also comparative studies of microscopic analysis and mtDNA, proficiency tests, and DNA exoneration cases involving microscopic analysis.

1. *Mitochondrial DNA*[380]

An FBI study compared microscopic ("consistent with" testimony) and mtDNA analysis of hair: "Of the 80 hairs that were microscopically associated, nine comparisons were excluded by mtDNA analysis."[381]

2. *Proficiency testing*

Early proficiency tests indicated a high rate of laboratory error in microscopic comparisons of hair samples. In the 1970s the LEAA conducted its Laboratory Proficiency Testing Program.[382] The crime laboratories' performance on hair analysis was the weakest. Fifty-four percent misanalyzed hair sample C and 67% submitted unacceptable responses on hair sample D.[383] Followup studies between 1980 and 1991 yielded similar results.[384] Summarizing the results of this series of tests, two commentators concluded: "Animal and human (body area) hair identifications are clearly the most troublesome of all categories tested."[385]

In another series of hair tests, the examiners were asked to "include" or "exclude" in comparing known and unknown samples: "Laboratories reported inclusions and exclusions which agreed with the manufacturer in approximately 74% of their comparisons. About 18% of the responses were inconclusive, and 8% in disagreement with the manufacturers' information."[386]

379. NRC Forensic Science Report, *supra* note 3, at 159.

380. For a detailed discussion of mitochondrial DNA, see David H. Kaye & George Sensabaugh, Reference Guide on DNA Identification Evidence, Section V.A, in this manual

381. Max M. Houck & Bruce Budowle, *Correlation of Microscopic and Mitochondrial DNA Hair Comparisons*, 47 J. Forensic Sci. 964, 966 (2002).

382. Laboratory Proficiency Test, *supra* note 81.

383. *Id.* at 251. By way of comparison, 20% of the laboratories failed a paint analysis (test #5); 30% failed glass analysis (test #9).

384. Peterson & Markham, *supra* note 82, at 1007 ("In sum, laboratories were no more successful in identifying the correct species of origin of animal hair . . . than they were in the earlier LEAA study.").

385. *Id.*

386. *Id.* at 1023; *see also id.* at 1022 ("Examiners warned that they needed to employ particular caution in interpreting the hair results given the virtual impossibility of achieving complete sample homogeneity.").

3. DNA exonerations

The publication of the Department of Justice study of the first 28 DNA exonerations spotlighted the significant role that hair analysis played in several of these miscarriages of justice.[387] For example, in the trial of Edward Honeker, an expert testified that the crime scene hair sample "was unlikely to match anyone" else[388]—a clear overstatement. Moreover, an exoneration in Canada triggered a judicial inquiry, which recommended that "[t]rial judges should undertake a more critical analysis of the admissibility of hair comparison evidence as circumstantial evidence of guilt."[389] One study of 200 DNA exoneration cases reported that hair testimony had been presented at 43 of the original trials.[390] A subsequent examination of 137 trial transcripts in exoneration cases concluded: "Sixty-five of the trials examined involved microscopic hair comparison analysis. Of those, 25—or 38%—had invalid hair comparison testimony. Most (18) of these cases involved invalid individualizing claims."[391] The other cases contained flawed probability testimony.

C. Case Law Development

Prior to *Daubert*, an overwhelming majority of courts accepted expert testimony that hair samples are microscopically indistinguishable.[392] Experts often conceded that microscopic analysis did not permit a positive identification of the source.[393]

387. Edward Connors et al., Convicted by Juries, Exonerated by Science: Case Studies in the Use of DNA Evidence to Establish Innocence After Trial (1996). *See id.* at 73 (discussing David Vasquez case); *id.* at 64–65 (discussing Steven Linscott case).

388. Barry Scheck et al., Actual Innocence: Five Days to Execution and Other Dispatches from the Wrongly Convicted 146 (2000).

389. Hon. Fred Kaufman, The Commission on Proceedings Involving Guy Paul Morin (Ontario Ministry of the Attorney General 1998) (Recommendation 2). Morin was erroneously convicted based, in part, on hair evidence.

390. Brandon L. Garrett, *Judging Innocence*, 108 Colum. L. Rev. 55, 81 (2008).

391. Garrett & Neufeld, *supra* note 33, at 47.

392. *See, e.g.*, United States v. Hickey, 596 F.2d 1082, 1089 (1st Cir. 1979); United States v. Brady, 595 F.2d 359, 362–63 (6th Cir. 1979); United States v. Cyphers, 553 F.2d 1064, 1071–73 (7th Cir. 1977); Jent v. State, 408 So. 2d 1024, 1028–29 (Fla.1981), Commonwealth v. Tarver, 345 N.E.2d 671, 676–77 (Mass. 1975); State v. White, 621 S.W.2d 287, 292–93 (Mo. 1981); State v. Smith, 637 S.W.2d 232, 236 (Mo. Ct. App. 1982); People v. Allweiss, 396 N.E.2d 735 (N.Y. 1979); State v. Green, 290 S.E.2d 625, 629–30 (N.C. 1982); State v. Watley, 788 P.2d 375, 381 (N.M. 1989).

393. Moore v. Gibson, 195 F.3d 1152, 1167 (10th Cir. 1999); Butler v. State, 108 S.W.3d 18, 21 (Mo. Ct. App. 2003); *see also* Thompson v. State, 539 A.2d 1052, 1057 (Del. 1988) ("it is now universally recognized that although fingerprint comparisons can result in the positive identification of an individual, hair comparisons are not this precise"). *But see* People v. Kosters, 467 N.W.2d 311, 313 (Mich. 1991) (Cavanaugh, C.J., dissenting) (the "minuscule probative value" of such opinions is "clearly . . . outweighed by the unfair prejudicial effect"); State v. Wheeler, 1981 WL 139588, at *4 (Wis. Ct. App. Feb. 8, 1981) (in an unpublished opinion, the appellate court held that the trial judge did not err in finding that the expert's opinion that the accused "could have been the source" of the hair lacked probative value, because it "only include[d] defendant in a broad class of possible assailants").

Nonetheless, the courts varied in how far they permitted the expert to go. In some cases, analysts testified only that the samples matched[394] or were similar[395] and thus consistent with the hypothesis that the samples had the same source.[396] Other courts permitted experts to directly opine that the accused was the source of the crime scene sample.[397] However, a 1990 decision held it error to admit testimony that "it would be improbable that these hairs would have originated from another individual."[398] In the court's view, this testimony amounted "effectively, [to] a positive identification of defendant. . . ."[399]

On the basis of Gaudette and Keeping research, several courts admitted opinions in statistical terms (e.g., 1 in 4500 chance of a false match).[400] In contrast, other courts, including a federal court of appeals, reached a contrary conclusion.[401]

The most significant post-*Daubert* challenge to microscopic hair analysis came in *Williamson v. Reynolds*,[402] a habeas case decided in 1995. There, an expert testified that, after considering approximately 25 characteristics, he concluded that the hair samples were "consistent microscopically." He then elaborated: "In other words, hairs are not an absolute identification, but they either came from this individual or there is—could be another individual *somewhere in the world* that would have the same characteristics to their hair."[403] The district court was "unsuccessful in its attempts to locate *any* indication that expert hair comparison testimony

394. Garland v. Maggio, 717 F.2d 199, 207 n.9 (5th Cir. 1983).

395. United States v. Brady, 595 F.2d 359, 362–63 (6th Cir. 1979).

396. People v. Allen, 115 Cal. Rptr. 839, 842 (Cal. Ct. App. 1974).

397. In the 1986 Mississippi prosecution of Randy Bevill for murder, the expert testified that "there was a transfer of hair from the Defendant to the body of" the victim. Clive A. Stafford Smith & Patrick D. Goodman, *Forensic Hair Comparison Analysis: Nineteenth Century Science or Twentieth Century Snake Oil?* 27 Colum. Hum. Rts. L. Rev. 227, 273 (1996).

398. State v. Faircloth, 394 S.E.2d 198, 202–03 (N.C. Ct. App. 1990).

399. *Id.* at 202.

400. United States v. Jefferson, 17 M.J. 728, 731 (N.M.C.M.R. 1983); People v. DiGiacomo, 388 N.E.2d 1281, 1283 (Ill. App. Ct. 1979); *see also* United States *ex rel.* DiGiacomo v. Franzen, 680 F.2d 515, 516 (7th Cir. 1982) (During its deliberations, the jury submitted the following question to the judge: "Has it been established by sampling of hair specimens that the defendant was positively proven to have been in the automobile?").

401. United States v. Massey, 594 F.2d 676, 679–80 (8th Cir. 1979) (the expert testified that he "had microscopically examined 2,000 cases and in only one or two cases was he ever unable to make identification"; the expert cited a study for the proposition that there was a 1 in 4500 chance of a random match; the expert added that "there was only 'one chance in a 1,000' that hair comparisons could be in error"); State v. Carlson, 267 N.W.2d 170, 176 (Minn. 1978).

402. 904 F. Supp. 1529, 1554 (E.D. Okla. 1995), *rev'd on this issue sub nom.* Williamson v. Ward, 110 F.3d 1508, 1523 (10th Cir. 1997). The district court noted that the "expert did not explain which of the 'approximately' 25 characteristics were consistent, any standards for determining whether the samples were consistent, how many persons could be expected to share this same combination of characteristics, or how he arrived at his conclusions." *Id.* at 1554.

403. *Id.* (emphasis added).

meets *any* of the requirements of *Daubert*."[404] Finally, the prosecutor in closing argument declared, "There's a match."[405] Even the state court had misinterpreted the evidence, writing that the "hair evidence placed [petitioner] at the decedent's apartment."[406] Although the Tenth Circuit did not fault the district judge's reading of the empirical record relating to hair analysis and ultimately upheld habeas relief, that court reversed the district judge on this issue. The Tenth Circuit ruled that the district had committed legal error because the due process (fundamental fairness), not the more stringent *Daubert* (reliability), standard controls evidentiary issues in habeas corpus proceedings.[407] Before retrial, the defendant was exonerated by exculpatory DNA evidence.[408]

Post-*Daubert,* many cases have continued to admit testimony about microscopic hair analysis.[409] In 1999, one state court judicially noticed the reliability of hair evidence,[410] implicitly finding this evidence to be not only admissible but also based on a technique of indisputable validity.[411] In contrast, a Missouri court reasoned that, without the benefit of population frequency data, an expert overreached in opining to "a reasonable degree of certainty that the unidentified hairs were in fact from" the defendant.[412] The NRC report commented that there appears to be growing judicial support for the view that "testimony linking microscopic hair analysis with particular defendants is highly unreliable."[413]

404. *Id.* at 1558. The court also observed: "Although the hair expert may have followed procedures accepted in the community of hair experts, the human hair comparison results in this case were, nonetheless, scientifically unreliable." *Id.*

405. *Id.* at 1557.

406. *Id.* (quoting Williamson v. State, 812 P.2d 384, 387 (Okla. Crim. 1991)).

407. Williamson v. Ward, 110 F.3d 1508, 1523 (10th Cir. 1997).

408. Scheck et al., *supra* note 388, at 146 (hair evidence was shown to be "patently unreliable."); *see also* John Grisham, The Innocent Man (2006) (examining the Williamson case).

409. *E.g.,* State v. Fukusaku, 946 P.2d 32, 44 (Haw. 1997) ("Because the scientific principles and procedures underlying hair and fiber evidence are well-established and of proven reliability, the evidence in the present case can be treated as 'technical knowledge.' Thus, an independent reliability determination was unnecessary."); McGrew v. State, 682 N.E.2d 1289, 1292 (Ind. 1997) (concluding that hair comparison is "more a 'matter of observation by persons with specialized knowledge' than 'a matter of scientific principles'"); *see also* NRC Forensic Science Report, *supra* note 3, at 161 n.88 (citing State v. West, 877 A.2d 787 (Conn. 2005), and Bookins v. State, 922 A.2d 389 (Del. Super. Ct. 2007)).

410. *See* Johnson v. Commonwealth, 12 S.W.3d 258, 262 (Ky. 1999).

411. *See* Fed. R. Evid. 201(b); *Daubert*, 509 U.S. at 593 n.11 ("[T]heories that are so firmly established as to have attained the status of scientific law, such as the laws of thermodynamics, properly are subject to judicial notice under Federal Rule [of] Evidence 201.").

412. Butler v. State, 108 S.W.3d 18, 21–22 (Mo. Ct. App. 2003).

413. NRC Forensic Science Report, *supra* note 3, at 161.

XI. Recurrent Problems

The discussions of specific techniques in this chapter, as well as the 2009 NRC report, reveal several recurrent problems in the presentation of testimony about forensic expertise.

A. Clarity of Testimony

As noted earlier, the report voiced concern about the use of terms such as "match," "consistent with," "identical," "similar in all respects tested," and "cannot be excluded as the source of." These terms can have "a profound effect on how the trier of fact in a criminal or civil matter perceives and evaluates scientific evidence."[414]

The comparative bullet lead cases are illustrative of this point.[415] The technique was used when conventional firearms identification was not possible because the recovered bullet was so deformed that the striations were destroyed. In the bullet lead cases, the phrasing of the experts' opinions varied widely. In some, experts testified only to the limited opinion that two exhibits were "analytically indistinguishable."[416] In other cases, examiners concluded that samples *could have* come from the same "source" or "batch."[417] In still others, they stated that the samples *came* from the same source.[418] In several cases, the experts went even further and identified a particular "box" of ammunition (usually 50 loaded cartridges, sometimes 20) as the source of the bullet recovered at the crime scene. For example, experts opined that two specimens:

- Could have come from the same box.[419]
- Could have come from the same box or a box manufactured on the same day.[420]

414. *Id.* at 21.

415. The technique compared trace chemicals found in bullets at crime scenes with ammunition found in the possession of a suspect. It was used when firearms ("ballistics") identification could not be employed. FBI experts used various analytical techniques (first, neutron activation analysis, and then inductively coupled plasma-atomic emission spectrometry) to determine the concentrations of seven elements—arsenic, antimony, tin, copper, bismuth, silver, and cadmium—in the bullet lead alloy of both the crime-scene and suspect's bullets. Statistical tests were then used to compare the elements in each bullet and determine whether the fragments and suspect's bullets were "analytically indistinguishable" for each of the elemental concentration means.

416. *See* Wilkerson v. State, 776 A.2d 685, 689 (Md. Ct. Spec. App. 2001).

417. *See* State v. Krummacher, 523 P.2d 1009, 1012–13 (Or. 1974) (en banc).

418. *See* United States v. Davis, 103 F.3d 660, 673–74 (8th Cir. 1996); People v. Lane, 628 N.E.2d 682, 689–90 (Ill. App. Ct. 1993).

419. *See* State v. Jones, 425 N.E.2d 128, 131 (Ind. 1981); State v. Strain, 885 P.2d 810, 817 (Utah Ct. App. 1994).

420. *See* State v. Grube, 883 P.2d 1069, 1078 (Idaho 1994); People v. Johnson, 499 N.E.2d 1355, 1366 (Ill. 1986); Earhart v. State, 823 S.W.2d 607, 614 (Tex. Crim. App. 1991) (en banc) ("He

- Were consistent with their having come from the same box of ammunition.[421]
- Probably came from the same box.[422]
- Must have come from the same box or from another box that would have been made by the same company on the same day.[423]

Moreover, these inconsistent statements were not supported by empirical research. According to a 2004 NRC report, the number of bullets that can be produced from an "analytically indistinguishable" melt "can range from the equivalent of as few as 12,000 to as many as 35 million 40 grain, .22 caliber long-rifle bullets."[424] Consequently, according to the 2004 NRC report, the "available data do not support any statement that a crime bullet came from a particular box of ammunition. [R]eferences to 'boxes' of ammunition in any form should be excluded as misleading under Federal Rule of Evidence 403."[425]

B. Limitations on Testimony

Some courts have limited the scope of the testimony, permitting expert testimony about the similarities and dissimilarities between exemplars but not the specific conclusion that the defendant was the author ("common authorship" opinion).[426] Although the courts have used this approach most frequently in questioned docu-

later modified that statement to acknowledge that analytically indistinguishable bullets which do not come from the same box most likely would have been manufactured at the same place on or about the same day; that is, in the same batch."), *vacated*, 509 U.S. 917 (1993).

421. *See* State v. Reynolds, 297 S.E.2d 532, 534 (N.C. 1982).

422. *See* Bryan v. State, 935 P.2d 338, 360 (Okla. Crim. App. 1997).

423. *See Davis*, 103 F.3d at 666–67 ("An expert testified that such a finding is rare and that the bullets must have come from the same box or from another box that would have been made by the same company on the same day."); Commonwealth v. Daye, 587 N.E.2d 194, 207 (Mass. 1992); State v. King, 546 S.E.2d 575, 584 (N.C. 2001) (Kathleen Lundy "opined that, based on her lead analysis, the bullets she examined either came from the same box of cartridges or came from different boxes of the same caliber, manufactured at the same time.").

424. National Research Council, Forensic Analysis: Weighing Bullet Lead Evidence 6 (2004), [hereinafter NRC Bullet Lead Evidence], *available at* http://www.nap.edu/catalog.php?record_id=10924.

425. *Id.*

426. *See* United States v. Oskowitz, 294 F. Supp. 2d 379, 384 (E.D.N.Y. 2003) ("Many other district courts have similarly permitted a handwriting expert to analyze a writing sample for the jury without permitting the expert to offer an opinion on the ultimate question of authorship."); United States v. Rutherford, 104 F. Supp. 2d 1190, 1194 (D. Neb. 2000) ("[T]he Court concludes that FDE Rauscher's testimony meets the requirements of Rule 702 to the extent that he limits his testimony to identifying and explaining the similarities and dissimilarities between the known exemplars and the questioned documents. FDE Rauscher is precluded from rendering any ultimate conclusions on authorship of the questioned documents and is similarly precluded from testifying to the degree of confidence or certainty on which his opinions are based."); United States v. Hines, 55 F. Supp. 2d 62, 69 (D. Mass. 1999) (expert testimony concerning the general similarities and differences between a defendant's handwriting exemplar and a stick-up note was admissible but not the specific conclusion that the defendant was the author).

ment cases, they have sometimes applied the same approach to other types of forensic expertise such as firearms examination as well.[427]

The NRC report criticized "exaggerated"[428] testimony such as claims of perfect accuracy,[429] infallibility,[430] or a zero error rate.[431] Several courts have barred excessive expert claims for lack of empirical support. For example, in *United States v. Mitchell*,[432] the court commented: "Testimony at the *Daubert* hearing indicated that some latent fingerprint examiners insist that there is no error rate associated with their activities. . . . This would be out-of-place under Rule 702."[433] Similarly, in a firearms identification case, one court noted that

> during the testimony at the hearing, the examiners testified to the effect that they could be 100 percent sure of a match. Because an examiner's bottom line opinion as to an identification is largely a subjective one, there is no reliable statistical or scientific methodology which will currently permit the expert to testify that it is a 'match' to an absolute certainty, or to an arbitrary degree of statistical certainty.[434]

Other courts have excluded the use of terms such as "science" or "scientific," because of the risk that jurors may bestow the aura of the infallibility of science on the testimony.[435]

In particular, some courts are troubled by the use of the expression "reasonable scientific certainty" by some forensic experts. The term "reasonable scientific certainty" is problematic. Although it is used frequently in cases, its legal meaning is ambiguous.[436] Sometimes it is used in lieu of a confidence statement (i.e., "high degree of certainty"), in which case the expert could altogether avoid the term and directly testify how confident he or she is in the opinion.

In other cases, courts have interpreted reasonable scientific certainty to mean that the expert must testify that a sample *probably* came from the defendant and not

427. United States v. Green, 405 F. Supp. 2d 104, 124 (D. Mass. 2005).

428. NRC Forensic Science Report, *supra* note 3, at 4.

429. *Id*. at 47.

430. *Id*. at 104.

431. *Id*. at 142–43.

432. 365 F.3d 215 (3d Cir. 2004).

433. *Id*. at 246.

434. United States v. Monteiro, 407 F. Supp. 2d 351, 372 (D. Mass. 2006).

435. United States v. Starzecpyzel, 880 F. Supp. 1027, 1038 (S.D.N.Y. 1995).

436. James E. Hullverson, *Reasonable Degree of Medical Certainty: A Tort et a Travers*, 31 St. Louis U. L.J. 577, 582 (1987) ("[T]here is nevertheless an undercurrent that the expert in federal court express some basis for both the confidence with which his conclusion is formed, and the probability that his conclusion is accurate."); Edward J. Imwinkelried & Robert G. Scofield, *The Recognition of an Accused's Constitutional Right to Introduce Expert Testimony Attacking the Weight of Prosecution Science Evidence: The Antidote for the Supreme Court's Mistaken Assumption in* California v. Trombetta, 33 Ariz. L. Rev. 59, 69 (1991) ("Many courts continue to exclude opinions which fall short of expressing a probability or certainty. . . . These opinions have been excluded in jurisdictions which have adopted the Federal Rules of Evidence.").

that it *possibly* came from the defendant.[437] However, experts frequently testify that two samples "could have come from the same source." Such testimony meets the relevancy standard of Federal Rule 401, and there is no requirement in Article VII of the Federal Rules that an expert's opinion be expressed in terms of "probabilities." Thus, in *United States v. Cyphers*[438] the expert testified that hair samples found on items used in a robbery "could have come" from the defendants.[439] The defendants argued that the testimony was inadmissible because the expert did not express his opinion in terms of reasonable scientific certainty. The court wrote: "There is no such requirement."[440]

In *Burke v. Town of Walpole*,[441] a bite mark identification case, the court of appeals had to interpret the term as used in an arrest warrant:

> [W]e must assume that the magistrate who issued the arrest warrant assigned no more than the commonly accepted meaning among lawyers and judges to the term "reasonable degree of scientific certainty"—"a standard requiring a showing that the injury was more likely than not caused by a particular stimulus, based on the general consensus of recognized [scientific] thought." Black's Law Dictionary 1294 (8th ed. 2004) (defining "reasonable medical probability," or "reasonable medical certainty," as used in tort actions). That standard, of course, is fully consistent with the probable cause standard.[442]

The case involved the guidelines adopted by ABFO that recognized several levels of certainty ("reasonable medical certainty," "high degree of certainty," and "virtual certainty"). The guidelines described "reasonable medical certainty" as "convey[ing] the connotation of virtual certainty or beyond reasonable doubt."[443] This is not the way that some courts use the term.

437. State v. Holt, 246 N.E.2d 365, 368 (Ohio 1969). The expert testified, based on neutron activation analysis, that two hair samples were "similar *and* . . . *likely* to be from the same source" (emphasis in original).

438. 553 F.2d 1064 (7th Cir. 1977).

439. *Id.* at 1072; *see also* United States v. Davis, 44 M.J. 13, 16 (C.A.A.F. 1996) ("Evidence was also admitted that appellant owned sneakers which 'could have' made these prints.").

440. *Cyphers*, 553 F.2d at 1072; *see also* United States v. Oaxaca, 569 F.2d 518, 526 (9th Cir. 1978) (expert's opinion regarding hair comparison admissible even though expert was less than certain); United States v. Spencer, 439 F.2d 1047, 1049 (2d Cir. 1971) (expert's opinion regarding handwriting comparison admissible even though expert did not make a positive identification); United States v. Longfellow, 406 F.2d 415, 416 (4th Cir. 1969) (expert's opinion regarding paint comparison admissible, even though expert did not make a positive identification); State v. Boyer, 406 So. 2d 143, 148 (La. 1981) (reasonable scientific certainty not required where expert testifies concerning the presence of gunshot residue based on neutron activation analysis).

441. 405 F.3d 66 (1st Cir. 2005).

442. *Id.* at 91.

443. *Id.* at 91 n.30 (emphasis omitted).

Moreover, the term may be problematic for a different reason—misleading the jury. One court ruled that the term "reasonable scientific certainty" could not be used because of the subjective nature of the opinion.[444]

C. Restriction of Final Argument

In a number of cases, in summation counsel has overstated the content of the expert testimony. In *People v. Linscott*,[445] for example, "the prosecutor argued that hairs found in the victim's apartment and on the victim's body were in fact defendant's hairs."[446] Reversing, the Illinois Supreme Court wrote: "With these statements, the prosecutor improperly argued that the hairs removed from the victim's apartment were conclusively identified as coming from defendant's head and pubic region. There simply was no testimony at trial to support these statements. In fact, [the prosecution experts] and the defense hair expert . . . testified that no such identification was possible."[447] DNA testing exculpated Linscott.[448] Trial judges can police the attorneys' descriptions of the testimony during closing argument as well as the content of expert testimony presented.

XII. Procedural Issues

The *Daubert* standard operates in a procedural setting, not a vacuum. In *Daubert*, the Supreme Court noted that "[v]igorous cross-examination, presentation of contrary evidence, and careful instruction on the burden of proof are the traditional and appropriate means of attacking shaky but admissible evidence."[449] Adversarial testing presupposes advance notice of the content of the expert's testimony and access to comparable expertise to evaluate that testimony. This section discusses some of the procedural mechanisms that trial judges may use to assure that jurors properly evaluate any expert testimony by forensic identification experts.

444. United States v. Glynn, 578 F. Supp. 2d 567, 568–75 (S.D.N.Y. 2008) (firearms identification case).

445. 566 N.E.2d 1355 (Ill. 1991).

446. *Id.* at 1358.

447. *Id.* at 1359.

448. *See* Connors et al., *supra* note 387, at 65 ("The State's expert on the hair examination testified that only 1 in 4,500 persons would have consistent hairs when tested for 40 different characteristics. He only tested between 8 and 12 characteristics, however, and could not remember which ones. The appellate court ruled on July 29, 1987, that his testimony, coupled with the prosecution's use of it at closing arguments, constituted denial of a fair trial.") (citation omitted).

449. 509 U.S. at 596 (citing Rock v. Arkansas, 483 U.S. 44, 61 (1987)).

A. Pretrial Discovery

Judges can monitor discovery in scientific evidence cases to ensure that disclosure is sufficiently comprehensive.[450] Federal Rule 16 requires discovery of laboratory reports[451] and a summary of the expert's opinion.[452] The efficacy of these provisions depends on the content of the reports and the summary. The *Journal of Forensic Sciences*, the official publication of the American Academy of Forensic Sciences, published a symposium on the ethical responsibilities of forensic scientists in 1989. One symposium article described a number of unacceptable laboratory reporting practices, including (1) "preparation of reports containing minimal information in order not to give the 'other side' ammunition for cross-examination," (2) "reporting of findings without an interpretation on the assumption that if an interpretation is required it can be provided from the witness box," and (3) "[o]mitting some significant point from a report to trap an unsuspecting cross-examiner."[453]

NRC has recommended extensive discovery in DNA cases: "All data and laboratory records generated by analysis of DNA samples should be made freely available to all parties. Such access is essential for evaluating the analysis."[454] The NRC report on bullet lead contained similar comments about the need for a thorough report in bullet lead cases:

> The conclusions in laboratory reports should be expanded to include the limitations of compositional analysis of bullet lead evidence. In particular, a further

450. *See* Fed. R. Crim. P. 16 (1975) advisory committee's note ("[I]t is difficult to test expert testimony at trial without advance notice and preparation."), *reprinted in* 62 F.R.D. 271, 312 (1974); Paul C. Giannelli, *Criminal Discovery, Scientific Evidence, and DNA*, 44 Vand. L. Rev. 791 (1991). "Early disclosure can have the following benefits: [1] Avoiding surprise and unnecessary delay. [2] Identifying the need for defense expert services. [3] Facilitating exoneration of the innocent and encouraging plea negotiations if DNA evidence confirms guilt." National Institute of Justice, President's DNA Initiative: Principles of Forensic DNA for Officers of the Court (2005), *available at* http://www.dna.gov/training/otc.

451. Fed. R. Crim. P. 16(a)(1)(F).

452. *Id.* 16(a)(1)(G).

453. Douglas M. Lucas, *The Ethical Responsibilities of the Forensic Scientist: Exploring the Limits*, 34 J. Forensic Sci. 719, 724 (1989). Lucas was the Director of The Centre of Forensic Sciences, Ministry of the Solicitor General, Toronto, Ontario.

454. National Research Council, DNA Technology in Forensic Science 146 (1992) ("The prosecutor has a strong responsibility to reveal fully to defense counsel and experts retained by the defendant all material that might be necessary in evaluating the evidence."); *see also id.* at 105 ("Case records—such as notes, worksheets, autoradiographs, and population databanks—and other data or records that support examiners' conclusions are prepared, retained by the laboratory, and made available for inspection on court order after review of the reasonableness of a request."); National Research Council, The Evaluation of Forensic DNA Evidence 167–69 (1996) ("Certainly, there are no strictly scientific justifications for withholding information in the discovery process, and in Chapter 3 we discussed the importance of full, written documentation of all aspects of DNA laboratory operations. Such documentation would facilitate technical review of laboratory work, both within the laboratory and by outside experts. . . . Our recommendations that all aspects of DNA testing be fully documented is most valuable when this documentation is discoverable in advance of trial.").

> explanatory comment should accompany the laboratory conclusions to portray the limitations of the evidence. Moreover, a section of the laboratory report translating the technical conclusions into language that a jury could understand would greatly facilitate the proper use of this evidence in the criminal justice system. Finally, measurement data (means and standard deviations) for all of the crime scene bullets and those deemed to match should be included.[455]

As noted earlier, the recent NRC report made similar comments:

> Some reports contain only identifying and agency information, a brief description of the evidence being submitted, a brief description of the types of analysis requested, and a short statement of the results (e.g., "the greenish, brown plant material in item #1 was identified as marijuana"), and they include no mention of methods or any discussion of measurement uncertainties.[456]

Melendez-Diaz v. Massachusetts[457] illustrates the problem. The laboratory report in that case "contained only the bare-bones statement that '[t]he substance was found to contain: Cocaine.' At the time of trial, petitioner did not know what tests the analysts performed, whether those tests were routine, and whether interpreting their results required the exercise of judgment or the use of skills that the analysts may not have possessed."[458]

1. Testifying beyond the report

Experts should generally not be allowed to testify beyond the scope of the report without issuing a supplemental report. *Troedel v. Wainwright*,[459] a capital murder case, illustrates the problem. In that case, a report of a gunshot residue test based on neutron activation analysis stated the opinion that swabs "from the hands of Troedel and Hawkins contained antimony and barium in amounts typically found on the hands of a person who has discharged a firearm or has had his hands in close proximity to a discharging firearm."[460] An expert testified consistently with this report at Hawkins' trial but embellished his testimony at Troedel's trial by adding the more inculpatory opinion that "Troedel had fired the murder weapon."[461] In contrast, at a deposition during federal habeas proceedings, the *same* expert testified that "he could not, from the results of his tests, determine or say to a scientific certainty who had fired the murder weapon" and the "amount of barium and antimony on the hands of Troedel and Hawkins were basically insignificant."[462] The district court found the trial testimony, "at the very least," misleading and

455. *See* NRC Bullet Lead Evidence, *supra* note 424, at 110–11.
456. NRC Forensic Science Report, *supra* note 3, at 21.
457. 129 S. Ct. 2527 (2009).
458. *Id.* at 2537.
459. 667 F. Supp. 1456 (S.D. Fla. 1986), *aff'd*, 828 F.2d 670 (11th Cir. 1987).
460. *Id.* at 1458.
461. *Id.*
462. *Id.* at 1459.

granted relief.[463] The expert claimed that the prosecutor had "pushed" him to embellish his testimony, a claim the prosecutor substantiated.[464]

B. *Defense Experts*

In appropriate cases, trial judges can provide the opposition with access to expert resources. Defense experts are often important in cases involving forensic identification expertise. Counsel will frequently need expert guidance to determine whether a research study is methodologically sound and, if so, whether the data adequately support the specific opinion proffered, and the role, if any, that subjective judgment played in forming the opinion.

The NAS 1992 DNA report stressed that experts are necessary for an adequate defense in many cases: "Defense counsel must have access to adequate expert assistance, even when the admissibility of the results of analytical techniques is not in question because there is still a need to review the quality of the laboratory work and the interpretation of results."[465] According to the *President's DNA Initiative*, "[e]ven if DNA evidence is admitted, there still may be disagreement about its interpretation—what do the DNA results mean in a particular case?"[466]

The need for defense experts is not limited to cases involving DNA evidence. In *Ake v. Oklahoma*,[467] the Supreme Court recognized a due process right to a defense expert under certain circumstances.[468] In federal trials, the Criminal Justice Act of 1964[469] provides for expert assistance for indigent defendants.

463. "[T]he Court concludes that the opinion Troedel had fired the weapon was known by the prosecution not to be based on the results of the neutron activation analysis tests, or on any scientific certainty or even probability. Thus, the subject testimony was not only misleading, but also was used by the State knowing it to be misleading." *Id.* at 1459–60.

464. *Id.* at 1459 ("[A]s Mr. Riley candidly admitted in his deposition, he was 'pushed' further in his analysis at Troedel's trial than at Hawkins' trial. . . . [At the] evidentiary hearing held before this Court, one of the prosecutors testified that, at Troedel's trial, after Mr. Riley had rendered his opinion which was contained in his written report, the prosecutor *pushed* to 'see if more could have been gotten out of this witness.'").

465. NRC I, *supra* note 24, at 149 ("Because of the potential power of DNA evidence, authorities must make funds available to pay for expert witnesses. . . .").

466. President's DNA Initiative, *supra* note 450.

467. 470 U.S. 68 (1985); *see* Paul C. Giannelli, Ake v. Oklahoma*: The Right to Expert Assistance in a Post-*Daubert, *Post-DNA World*, 89 Cornell L. Rev. 1305 (2004).

468. *Ake*, 470 U.S. at 74.

469. 18 U.S.C. § 3006(A).

Reference Guide on DNA Identification Evidence

DAVID H. KAYE AND GEORGE SENSABAUGH

David H. Kaye, M.A., J.D., is Distinguished Professor of Law, Weiss Family Scholar, and Graduate Faculty Member, Forensic Science Program, The Pennsylvania State University, University Park, and Regents' Professor Emeritus, Arizona State University Sandra Day O'Connor College of Law and School of Life Sciences, Tempe.

George Sensabaugh, D.Crim., is Professor of Biomedical and Forensic Sciences, School of Public Health, University of California, Berkeley.

CONTENTS

I. Introduction

Deoxyribonucleic acid, or DNA, is a molecule that encodes the genetic information in all living organisms. Its chemical structure was elucidated in 1954. More than 30 years later, samples of human DNA began to be used in the criminal justice system, primarily in cases of rape or murder. The evidence has been the subject of extensive scrutiny by lawyers, judges, and the scientific community. It is now admissible in all jurisdictions, but there are many types of forensic DNA analysis, and still more are being developed. Questions of admissibility arise as advancing methods of analysis and novel applications of established methods are introduced.[1]

This reference guide addresses technical issues that are important when considering the admissibility of and weight to be accorded analyses of DNA, and it identifies legal issues whose resolution requires scientific information. The goal is to present the essential background information and to provide a framework for resolving the possible disagreements among scientists or technicians who testify about the results and import of forensic DNA comparisons.

A. Summary of Contents

Section I provides a short history of DNA evidence and outlines the types of scientific expertise that go into the analysis of DNA samples.

Section II provides an overview of the scientific principles behind DNA typing. It describes the structure of DNA and how this molecule differs from person to person. These are basic facts of molecular biology. The section also defines the more important scientific terms and explains at a general level how DNA differences are detected. These are matters of analytical chemistry and laboratory procedure. Finally, the section indicates how it is shown that these differences permit individuals to be identified. This is accomplished with the methods of probability and statistics.

Section III considers issues of sample quantity and quality as well as laboratory performance. It outlines the types of information that a laboratory should produce to establish that it can analyze DNA reliably and that it has adhered to established laboratory protocols.

Section IV examines issues in the interpretation of laboratory results. To assist the courts in understanding the extent to which the results incriminate the defendant, it enumerates the hypotheses that need to be considered before concluding that the defendant is the source of the crime scene samples, and it explores the

1. For a discussion of other forensic identification techniques, see Paul C. Giannelli et al., Reference Guide on Forensic Identification Expertise, in this manual. *See also* David H. Kaye et al., The New Wigmore, A Treatise on Evidence: Expert Evidence (2d ed. 2011).

issues that arise in judging the strength of the evidence. It focuses on questions of statistics, probability, and population genetics.[2]

Section V describes special issues in human DNA testing for identification. These include the detection and interpretation of mixtures, Y-STR testing, mitochondrial DNA testing, and the evidentiary implications of DNA database searches of various kinds.

Finally, Section VI discusses the forensic analysis of nonhuman DNA. It identifies questions that can be useful in judging whether a new method or application of DNA science has the scientific merit and power claimed by the proponent of the evidence.

A glossary defines selected terms and acronyms encountered in genetics, molecular biology, and forensic DNA work.

B. A Brief History of DNA Evidence

"DNA evidence" refers to the results of chemical or physical tests that directly reveal differences in the structure of the DNA molecules found in organisms as diverse as bacteria, plants, and animals.[3] The technology for establishing the identity of individuals became available to law enforcement agencies in the mid to late 1980s.[4] The judicial reception of DNA evidence can be divided into at least five phases.[5] The first phase was one of rapid acceptance. Initial praise for RFLP (restriction fragment length polymorphism) testing in homicide, rape, paternity, and other cases was effusive. Indeed, one judge proclaimed "DNA fingerprinting" to be "the single greatest advance in the 'search for truth' . . . since the advent of cross-examination."[6] In this first wave of cases, expert testimony for the prosecution rarely was countered, and courts readily admitted DNA evidence.

In a second wave of cases, however, defendants pointed to problems at two levels—controlling the experimental conditions of the analysis and interpreting the results. Some scientists questioned certain features of the procedures for extracting and analyzing DNA employed in forensic laboratories, and it became apparent

2. For a broader discussion of statistics, see David H. Kaye & David A. Freedman, Reference Guide on Statistics, in this manual.

3. Differences in DNA also can be revealed by differences in the proteins that are made according to the "instructions" in a DNA molecule. Blood group factors, serum enzymes and proteins, and tissue types all reveal information about the DNA that codes for these chemical structures. Such immunogenetic testing predates the "direct" DNA testing that is the subject of this chapter. On the nature and admissibility of the "indirect" DNA testing, see, for example, David H. Kaye, The Double Helix and the Law of Evidence 5–19 (2010); 1 McCormick on Evidence § 205(B) (Kenneth Broun ed., 6th ed. 2006).

4. The first reported appellate opinion is *Andrews v. State,* 533 So. 2d 841 (Fla. Dist. Ct. App. 1988).

5. The description that follows is adapted from 1 McCormick on Evidence, *supra* note 3, § 205(B).

6. People v. Wesley, 533 N.Y.S.2d 643, 644 (Alb. County. Ct. 1988).

that declaring matches or nonmatches in the DNA variations being compared was not always trivial. Despite these concerns, most cases continued to find the DNA analyses to be generally accepted, and a number of states provided for admissibility of DNA tests by legislation. Concerted attacks by defense experts of impressive credentials, however, produced a few cases rejecting specific proffers on the ground that the testing was not sufficiently rigorous.[7]

A different attack on DNA profiling begun in cases during this period proved far more successful and led to a third wave of cases in which many courts held that estimates of the probability of a coincidentally matching DNA profile were inadmissible. These estimates relied on a simple population genetics model for the frequencies of DNA profiles, and some prominent scientists claimed that the applicability of the mathematical model had not been adequately verified. A heated debate on this point spilled over from courthouses to scientific journals and convinced the supreme courts of several states that general acceptance was lacking. A 1992 report of the National Academy of Sciences proposed a more "conservative" computational method as a compromise,[8] and this seemed to undermine the claim of scientific acceptance of the less conservative procedure that was in general use.

In response to the population genetics criticism and the 1992 report came an outpouring of critiques of the report and new studies of the distribution of the DNA variations in many populations. Relying on the burgeoning literature, a second National Academy panel concluded in 1996 that the usual method of estimating frequencies in broad racial groups generally was sound, and it proposed improvements and additional procedures for estimating frequencies in subgroups within the major population groups.[9] In the corresponding fourth phase of judicial scrutiny of DNA evidence, the courts almost invariably returned to the earlier view that the statistics associated with DNA profiling are generally accepted and scientifically valid.

In the fifth phase of the judicial evaluation of DNA evidence, results obtained with the newer "PCR-based methods" entered the courtroom. Once again, courts considered whether the methods rested on a solid scientific foundation and were generally accepted in the scientific community. The opinions are practically unanimous in holding that the PCR-based procedures satisfy these standards. Before long, forensic scientists settled on the use of one type of DNA variation (known as short tandem repeats, or STRs) to include or exclude individuals as the source of crime scene DNA.

7. Moreover, a minority of courts, perhaps concerned that DNA evidence might be conclusive in the minds of jurors, added a "third prong" to the general-acceptance standard of *Frye v. United States*, 293 F. 1013 (D.C. Cir. 1923). This augmented *Frye* test requires not only proof of the general acceptance of the ability of science to produce the type of results offered in court, but also of the proper application of an approved method on the particular occasion. For criticism of this approach, see David H. Kaye et al., *supra* note 1, § 6.3.3(a)(2).

8. National Research Council, DNA Technology in Forensic Science (1992) [hereinafter NRC I].

9. National Research Council, The Evaluation of Forensic DNA Evidence (1996) [hereinafter NRC II].

Throughout these phases, DNA tests also exonerated an increasing number of men who had been convicted of capital and other crimes, posing a challenge to traditional postconviction remedies and raising difficult questions of postconviction access to DNA samples.[10] The value of DNA evidence in solving older crimes also prompted extensions of some statutes of limitations.[11]

In sum, in little more than a decade, forensic DNA typing made the transition from a novel set of methods for identification to a relatively mature and well-studied forensic technology. However, one should not lump all forms of DNA identification together. New techniques and applications continue to emerge, ranging from the use of new genetic systems and new analytical procedures to the typing of DNA from plants and animals. Before admitting such evidence, courts normally inquire into the biological principles and knowledge that would justify inferences from these new technologies or applications. As a result, this guide describes not only the predominant STR technology, but also newer analytical techniques that can be used for forensic DNA identification.

C. Relevant Expertise

Human DNA identification can involve testimony about laboratory findings, about the statistical interpretation of those findings, and about the underlying principles of molecular biology. Consequently, expertise in several fields might be required to establish the admissibility of the evidence or to explain it adequately to the jury. The expert who is qualified to testify about laboratory techniques might not be qualified to testify about molecular biology, to make estimates of population frequencies, or to establish that an estimation procedure is valid.[12]

10. *See, e.g.*, Osborne v. District Attorney's Office for Third Judicial District, 129 S. Ct. 2308 (2009) (narrowly rejecting a convicted offender's claim of a due process right to DNA testing at his expense, enforceable under 42 U.S.C. § 1983, to establish that he is probably innocent of the crime for which he was convicted after a fair trial, when (1) the convicted offender did not seek extensive DNA testing before trial even though it was available, (2) he had other opportunities to prove his innocence after a final conviction based on substantial evidence against him, (3) he had no new evidence of innocence (only the hope that more extensive DNA testing than that done before the trial would exonerate him), and (4) even a finding that he was not source of the DNA would not conclusively demonstrate his innocence); Skinner v. Switzer, 131 S. Ct. 1289 (2011); Brandon L. Garrett, *Judging Innocence*, 108 Colum. L. Rev. 55 (2008); Brandon L. Garrett, *Claiming Innocence*, 92 Minn. L. Rev. 1629 (2008).

11. *See, e.g.*, Veronica Valdivieso, *DNA Warrants: A Panacea for Old, Cold Rape Cases?* 90 Geo. L.J. 1009 (2002).

12. Nonetheless, if previous cases establish that the testing and estimation procedures are legally acceptable, and if the computations are essentially mechanical, then highly specialized statistical expertise might not be essential. Reasonable estimates of DNA characteristics in major population groups can be obtained from standard references, and many quantitatively literate experts could use the appropriate formulae to compute the relevant profile frequencies or probabilities. NRC II, *supra* note 9, at 170. Limitations in the knowledge of a technician who applies a generally accepted statistical procedure can be explored on cross-examination. *See* Kaye et al., *supra* note 1, § 2.2. *Accord* Roberson v. State, 16 S.W.3d 156, 168 (Tex. Crim. App. 2000).

Trial judges ordinarily are accorded great discretion in evaluating the qualifications of a proposed expert witness, and the decisions depend on the background of each witness. Courts have noted the lack of familiarity of academic experts—who have done respected work in other fields—with the scientific literature on forensic DNA typing and on the extent to which their research or teaching lies in other areas.[13] Although such concerns may affect the persuasiveness of particular testimony, they rarely result in exclusion on the grounds that the witness simply is not qualified as an expert.

The scientific and legal literature on the objections to DNA evidence is extensive. By studying the scientific publications, or perhaps by appointing a special master or expert adviser to assimilate this material, a court can ascertain where a party's expert falls within the spectrum of scientific opinion. Furthermore, an expert appointed by the court under Federal Rule of Evidence 706 could testify about the scientific literature generally or even about the strengths or weaknesses of the particular arguments advanced by the parties.

Given the great diversity of forensic questions to which DNA testing might be applied, it is not feasible to list the specific scientific expertise appropriate to all applications. Assessing the value of DNA analyses of a novel application involving unfamiliar species can be especially challenging. If the technology is novel, expertise in molecular genetics or biotechnology might be necessary. If testing has been conducted on a particular organism or category of organisms, expertise in that area of biology may be called for. If a random-match probability has been presented, one might seek expertise in statistics as well as the population biology or population genetics that goes with the organism tested. Given the penetration of molecular technology into all areas of biological inquiry, it is likely that individuals can be found who know both the technology and the population biology of the organism in question. Finally, when samples come from crime scenes, the expertise and experience of forensic scientists can be crucial. Just as highly focused specialists may be unaware of aspects of an application outside their field of expertise, so too scientists who have not previously dealt with forensic samples can be unaware of case-specific factors that can confound the interpretation of test results.

II. Variation in Human DNA and Its Detection

DNA is a complex molecule that contains the "genetic code" of organisms as diverse as bacteria and humans. Although the DNA molecules in human cells are

13. *E.g.*, State v. Copeland, 922 P.2d 1304, 1318 n.5 (Wash. 1996) (noting that defendant's statistical expert "was also unfamiliar with publications in the area," including studies by "a leading expert in the field" whom he thought was "a 'guy in a lab somewhere'").

largely identical from one individual to another, there are detectable variations—except for identical twins, every two human beings have some differences in the detailed structure of their DNA. This section describes the basic features of DNA and some ways in which it can be analyzed to detect these differences.

A. What Are DNA, Chromosomes, and Genes?

The DNA molecule is made of subunits that include four chemical structures known as nucleotide bases. The names of these bases (adenine, thymine, guanine, and cytosine) usually are abbreviated as A, T, G, and C. The physical structure of DNA is often described as a double helix because the molecule has two spiraling strands connected to each other by weak bonds between the nucleotide bases. As shown in Figure 1, A pairs only with T and G only with C. Thus, the order of the single bases on either strand reveals the order of the pairs from one end of the molecule to the other, and the DNA molecule could be said to be like a long sequence of As, Ts, Gs, and Cs.

Figure 1. Sketch of a small part of a double-stranded DNA molecule. Nucleotide bases are held together by weak bonds. A pairs with T; C pairs with G.

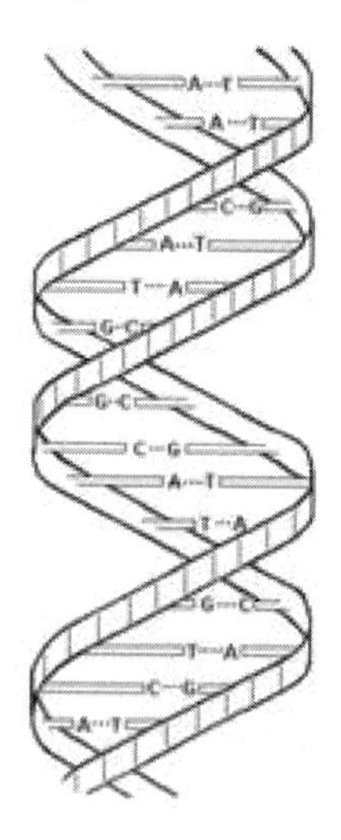

Most human DNA is tightly packed into structures known as chromosomes, which come in different sizes and are located in the nuclei of cells. The chromosomes are numbered (in descending order of size) 1 through 22, with the remaining chromosome being an X or a much smaller Y. If the bases are like letters, then each chromosome is like a book written in this four-letter alphabet, and the nucleus is like a bookshelf in the interior of the cell. All the cells in one individual contain identical copies of the same collection of books. The sequence of the As, Ts, Gs, and Cs that constitutes the "text" of these books is referred to as the individual's nuclear genome.

All told, the genome comprises more than three billion "letters" (As, Ts, Gs, and Cs). If these letters were printed in books, the resulting pile would be as high as the Washington Monument. About 99.9% of the genome is identical between any two individuals. This similarity is not really surprising—it accounts for the common features that make humans an identifiable species (and for features that we share with many other species as well). The remaining 0.1% is particular to an individual. This variation makes each person (other than identical twins) genetically unique. This small percentage may not sound like a lot, but it adds up to some three million sites for variation among individuals.

The process that gives rise to this variation among people starts with the production of special sex cells—sperm cells in males and egg cells in females. All the nucleated cells in the body other than sperm and egg cells contain two versions of each of the 23 chromosomes—two copies of chromosome 1, two copies of chromosome 2, and so on, for a total of 46 chromosomes. The X and Y chromosomes are the sex-determining chromosomes. Cells in females contain two X chromosomes, and cells in males contain one X and one Y chromosome. An egg cell, however, contains only 23 chromosomes—one chromosome 1, one chromosome 2, . . . , and one X chromosome—each selected at random from the woman's full complement of 23 chromosome pairs. Thus, each egg carries half the genetic information present in the mother's 23 chromosome pairs, and because the assortment of the chromosomes is random, each egg carries a different complement of genetic information. The same situation exists with sperm cells. Each sperm cell contains a single copy of each of the 23 chromosomes selected at random from a man's 23 pairs, and each sperm differs in the assortment of the 23 chromosomes it carries. Fertilization of an egg by a sperm therefore restores the full number of 46 chromosomes, with the 46 chromosomes in the fertilized egg being a new combination of those in the mother and father. The process resembles taking two decks of cards (a male and a female deck) and shuffling a random half from the male deck into a random half from the female deck, to produce a new deck.

During pregnancy, the fertilized cell divides to form two cells, each of which has an identical copy of the 46 chromosomes. The two then divide to form four, the four form eight, and so on. As gestation proceeds, various cells specialize ("differentiate") to form different tissues and organs. Although cell differentiation yields many different kinds of cells, the process of cell division results in each progeny cell having the same genomic complement as the cell that divided. Thus, each of the approximately 100 trillion cells in the adult human body has the same DNA text as was present in the original 23 pairs of chromosomes from the fertilized egg, one member of each pair having come from the mother and one from the father.

A second mechanism operating during the chromosome reduction process in sperm and egg cells further shuffles the genetic information inherited from mother and father. In the first stage of the reduction process, each chromosome of a chromosome pair aligns with its partner. The maternally inherited chromosome 1 aligns with the paternally inherited chromosome 1, and so on through the 22 pairs; X chromosomes align with each other as well, but X and Y chromosomes do not. While the chromosome pairs are aligned, they exchange pieces to create new combinations. The recombined chromosomes are passed on in the sperm and eggs. As a consequence, the chromosomes we inherit from our parents are not exact copies of their chromosomes, but rather are mosaics of these parental chromosomes.

The swapping of material between chromosome pairs (as they align in the emerging sex cells) and the random selection (of half of each parent's 46 chromosomes) in making sex cells is called recombination. Recombination is the principal source of diversity in individual human genomes.

The diverse variations occur both within the genes and in the regions of DNA sequences between the genes. A gene can be defined as a segment of DNA, usually from 1000 to 10,000 base pairs long, that "codes" for a protein. The cell produces specific proteins that correspond to the order of the base pairs (the "letters") in the coding part of the gene.[14] Human genes also contain noncoding sequences that regulate the cell type in which a protein will be synthesized and how much protein will be produced.[15] Many genes contain interspersed noncoding, nonregulatory sequences that no longer participate in protein synthesis. These sequences, which have no apparent function, constitute about 23% of the base pairs within human genes.[16] In terms of the metaphor of DNA as text, the gene is like an important paragraph in the book, often with some gibberish in it.

Proteins perform all sorts of functions in the body and thus produce observable characteristics. For example, a tiny part of the sequence that directs the production of the human group-specific complement protein (a protein that binds to vitamin D and transports it to certain tissues) is

G C A A A A T T G C C T G A T G C C A C A C C C A A G G A A C T G G C A.

14. The sequence in which the building blocks (amino acids) of a protein are arranged corresponds to the sequence of base pairs within a gene. (A sequence of three base pairs specifies a particular 1 of the 20 possible amino acids in the protein. The mapping of a set of three nucleotide bases to a particular amino acid is the genetic code. The cell makes the protein through intermediate steps involving coding RNA transcripts.) About 1.5% of the human genome codes for the amino acid sequences.

15. These noncoding but functional sequences include promoters, enhancers, and repressors.

16. This gene-related DNA consists of introns (which interrupt the coding sequences, called exons, in genes and which are edited out of the RNA transcript for the protein), pseudogenes (evolutionary remnants of once-functional genes), and gene fragments. The idea of a gene as a block of DNA (some of which is coding, some of which is regulatory, and some of which is functionless) is an oversimplification, but it is useful enough here. *See, e.g.*, Mark B. Gerstein et al., *What Is a Gene, Post-ENCODE? History and Updated Definition*, 17 Genome Res. 669 (2007).

This gene always is located at the same position, or locus, on chromosome 4. As we have seen, most individuals have two copies of each gene at a given locus—one from the father and one from the mother.

A locus where almost all humans have the same DNA sequence is called monomorphic ("of one form"). A locus where the DNA sequence varies among significant numbers of individuals (more than 1% or so of the population possesses the variant) is called polymorphic ("of many forms"), and the alternative forms are called alleles. For example, the GC protein gene sequence has three common alleles that result from substitutions in a base at a given point. Where an A appears in one allele, there is a C in another. The third allele has the A, but at another point a G is swapped for a T. These changes are called single nucleotide polymorphisms (SNPs, pronounced "snips").

If a gene is like a paragraph in a book, a SNP is a change in a letter somewhere within that paragraph (a substitution, a deletion, or an insertion), and the two versions of the gene that result from this slight change are the alleles. An individual who inherits the same allele from both parents is called a homozygote. An individual with distinct alleles is a heterozygote.

DNA sequences used for forensic analysis usually are not genes. They lie in the vast regions between genes (about 75% of the genome is extragenic) or in the apparently nonfunctional regions within genes. These extra- and intragenic regions of DNA have been found to contain considerable sequence variation, which makes them particularly useful in distinguishing individuals. Although the terms "locus," "allele," "homozygous," and "heterozygous" were developed to describe genes, the nomenclature has been carried over to describe all DNA variation—coding and noncoding alike. Both types are inherited from mother and father in the same fashion.

B. *What Are DNA Polymorphisms and How Are They Detected?*

By determining which alleles are present at strategically chosen loci, the forensic scientist ascertains the genetic profile, or genotype, of an individual (at those loci). Although the differences among the alleles arise from alterations in the order of the ATGC letters, genotyping does not necessarily require "reading" the full DNA sequence. Here we outline the major types of polymorphisms that are (or could be) used in identity testing and the methods for detecting them.

1. *Sequencing*

Researchers are investigating radically new and efficient technologies to sequence entire genomes, one base pair at a time, but the direct sequencing methods now in existence are technically demanding, expensive, and time-consuming for whole-genome sequencing. Therefore, most genetic typing focuses on identifying only those variations that define the alleles and does not attempt to "read out" each and every base pair as it appears. The exception is mitochondrial DNA, described in Section V. As next-generation sequencing technologies are perfected, however (*see infra* Section II.F), this situation could change.

2. Sequence-specific probes and SNP chips

Simple sequence variation, such as that for the GC locus, is conveniently detected using sequence-specific oligonucleotide (SSO) probes. A probe is a short, single strand of DNA. With GC typing, for example, probes for the three common alleles are attached to designated locations on a membrane. Copies of the variable sequence region of the GC gene in the crime scene sample are made with the polymerase chain reaction (PCR), which is discussed in the next section. These copies (in the form of single strands) are poured onto the membrane. Whichever allele is present in a single-stranded DNA fragment will cause the fragment to stick to the corresponding, immobilized probe strands. To permit the fragments of this type to be seen, a chemical "label" that catalyzes a color change at the spot where the DNA binds to its probe can be attached when the copies are made. A colored spot showing that the allele is present thus should appear on the membrane at the location of the probe that corresponds to this particular allele. If only one allele is present in the crime scene DNA (because of homozygosity), there will be no change at the spots where the other probes are located. If two alleles are present (heterozygosity), the corresponding two spots will change color.

This approach can be miniaturized and automated by embedding probes for many loci on a silicon chip. Commercially available "SNP chips" for disease research incorporate enough different probes to detect on the order of a million different known SNPs throughout the human genome. These chips have become a basic tool in searches for genetic changes associated with human diseases. They are described further in Section II.F.

3. VNTRs and RFLP testing

Another category of DNA variations comes from the insertion of a variable number of tandem repeats (VNTR) at a locus. These were the first polymorphisms to find widespread use in identity testing and hence were the subject of most of the court opinions on the admissibility of DNA in the late 1980s and early 1990s. The core unit of a VNTR is a particular short DNA sequence that is repeated many times end-to-end. The first VNTRs to be used in genetic and forensic testing had core repeat sequences of 15–35 base pairs. In this testing, bacterial enzymes (known as "restriction enzymes") were used to cut the DNA molecule both before and after the VNTR sequence. A small number of repeats in the VNTR region gives rise to a small "restriction fragment," and a large number of repeats yields a large fragment. A substantial quantity of DNA from a crime scene sample is required to give a detectable number of VNTR fragments with this procedure.

The detection is accomplished by applying a probe that binds when it encounters the repeated core sequence. A radioactive or fluorescent molecule attached to the probe provides a way to mark the VNTR fragment. The probe ignores DNA fragments that do not include the VNTR core sequence. (There are many of these unwanted fragments, because the restriction enzymes chop up the DNA throughout the genome—not just at the VNTR loci.) The restriction fragments are sorted by a process known as electrophoresis, which separates DNA fragments based on size. Many early court opinions refer to this process as RFLP testing.[17]

4. STRs

Although RFLP-VNTR profiling is highly discriminating,[18] it has several drawbacks. Not only does it require a substantial sample size, but it also is time-consuming and does not measure the fragment lengths to the nearest number of repeats. The measurement error inherent in the form of electrophoresis used (known as "gel electrophoresis") is not a fundamental obstacle, but it complicates the determination of which profiles match and how often other profiles in the population would be declared to match.[19] Consequently, forensic scientists have moved from VNTRs to another form of repetitive DNA known as short tandem repeats (STRs) or microsatellites. STRs have very short core repeats, two to seven base pairs in length, and they typically extend for only some 50 to 350 base pairs.[20] Like the larger VNTRs, which extend for thousands of base pairs, STR sequences do not code for proteins, and the ones used in identity testing convey little or no information about an individual's propensity for disease.[21] Because STR alleles

17. It would be clearer to call it RFLP-VNTR testing, because the fragments being measured contain the VNTRs rather than some simpler polymorphisms that were used in genetic research and disease testing. A more detailed exposition of the steps in RFLP-VNTR profiling (including gel electrophoresis, Southern blotting, and autoradiography) can be found in the previous edition of this guide and in many judicial opinions circa 1990.

18. Alleles at VNTR loci generally are too long to be measured precisely by electrophoretic methods—alleles differing in size by only a few repeat units may not be distinguished. Although this makes for complications in deciding whether two length measurements that are close together result from the same allele, these loci are quite powerful for the genetic differentiation of individuals, because they tend to have many alleles that occur relatively rarely in the population. At a locus with only 20 such alleles (and most loci typically have many more), there are 210 possible genotypes. With five such loci, the number of possible genotypes is 210^5, which is more than 400 billion.

19. For a case reversing a conviction as a result of an expert's confusion on this score, see *People v. Venegas,* 954 P.2d 525 (Cal. 1998). More suitable procedures for match windows and probabilities are described in NRC II, *supra* note 9.

20. The numbers, and the distinction between "minisatellites" (VNTRs) and microsatellites (STRs), are not precise, but the mechanisms that give rise to the shorter tandem repeats differ from those that produce the longer ones. *See* Benjamin Lewin, Genes IX 124–25 (9th ed. 2008).

21. *See* David H. Kaye, *Please, Let's Bury the Junk: The CODIS Loci and the Revelation of Private Information*, 102 Nw. U. L. Rev. Colloquy 70 (2007), *available at* http://www.law.northwestern.edu/lawreview/colloquy/2007/25/.

are much smaller than VNTR alleles, however, they can be amplified with PCR designed to copy only the locus of interest. This obviates the need for restriction enzymes, and it allows laboratories to analyze STR loci much more quickly. Because the amplified fragments are shorter, electrophoretic detection permits the exact number of base pairs in an STR to be determined, allowing alleles to be defined as discrete entities. Figure 2 illustrates the nature of allelic variation at an STR locus found on chromosome 16.

Figure 2. Three alleles of the D16S539 STR. The core sequence is GATA. The first allele listed has 9 tandem repeats, the second has 10, and the third has 11. The locus has other alleles (different numbers of repeats), shown in Figure 4.

Nine-repeat allele:
GATAGATAGATAGATAGATAGATAGATAGATAGATA
Ten-repeat allele:
GATAGATAGATAGATAGATAGATAGATAGATAGATAGATA
Eleven-repeat allele:
GATAGATAGATAGATAGATAGATAGATAGATAGATAGATAGATA

Although there are fewer alleles per locus for STRs than for VNTRs, there are many STRs, and they can be analyzed simultaneously. Such "multiplex" systems now permit the simultaneous analysis of 16 loci. A subset of 13 is standard in the United States (*see infra* Section II.D), and these are capable of distinguishing among almost everyone in the population.[22]

5. Summary

DNA contains the genetic information of an organism. In humans, most of the DNA is found in the cell nucleus, where it is organized into separate chromosomes. Each chromosome is like a book, and each cell has the same library (genome) of books of various sizes and shapes. There are two copies of each book of a particular size and shape, one that came from the father, the other from the mother. Thus, there are two copies of the book entitled "Chromosome One," two copies of "Chromosome Two," and so on. Genes are the most meaningful paragraphs in the books. Other parts of the text appear to have no coherent message. Two individuals sometimes have different versions (alleles) of the same paragraph. Some alleles result from the substitution of one letter for another. These are SNPs. Others come about from the insertion or deletion of single letters, and still

22. Usually, there are between 7 and 15 STR alleles per locus. Thirteen loci that have 10 STR alleles each can give rise to 55^{13}, or 42 billion trillion, possible genotypes.

others represent a kind of stuttering repetition of a string of extra letters. These are the VNTRs and STRs.[23] The locations within a chromosome where these interpersonal variations occur are called loci.

C. How Is DNA Extracted and Amplified?

DNA usually can be found in biological materials such as blood, bone, saliva, hair, semen, and urine. A combination of routine chemical and physical methods permits DNA to be extracted from cell nuclei and isolated from the other chemicals in a sample. PCR then is used to make exponentially large numbers of copies of targeted regions of the extracted DNA. PCR might be applied to the double-stranded DNA segments extracted and purified from a forensic sample as follows: First, the purified DNA is separated into two strands by heating it to near the boiling point of water. This "denaturing" takes about a minute. Second, the single strands are cooled, and "primers" attach themselves to the points at which the copying will start and stop. (Primers are small, manmade pieces of DNA, usually between 15 and 30 nucleotides long, of known sequences. If a locus of interest starts near the sequence ATCGAATCGGTAGCCATATG on one strand, a suitable primer would have the complementary sequence TAGCTTAGCCATCGGTATAC.) "Annealing" these primers takes about 45 seconds. Finally, the soup containing the annealed DNA strands, the enzyme DNA polymerase, and lots of the four nucleotide building blocks (A, C, G, and T) is warmed to a comfortable working temperature for the polymerase to insert the complementary base pairs one at a time, building a matching second strand bound to the original "template" and thus replicating part of the DNA strand that was separated from its partner in the first step. The same replication occurs with the separated partner as the template. This "extension" step for both templates takes about 2 minutes. The result is two identical double-stranded DNA segments, one made from each strand of the original DNA. The three-step cycle is repeated, usually 20 to 35 times in automated machines known as thermocyclers. Ideally, the first cycle results in two double-stranded DNA segments. The second cycle produces four, the third eight, and so on, until the number of copies of the original DNA is enormous. In practice, there is some inefficiency in the doubling process, but the yield from a 30-cycle amplification is generally about 1 million to 10 million copies of the targeted sequence.[24] In this way, PCR magnifies short sequences of interest in a small number of DNA fragments into millions of exact copies. Machines that automate the PCR process are commercially available.

For PCR amplification to work properly and yield copies of only the desired sequence, however, care must be taken to achieve the appropriate chemical con-

23. In addition to the 23 pairs of books in the cell nucleus, other scraps of text reside in each of the mitochondria, the power plants of the cell. *See infra* Section V.

24. NRC II, *supra* note 9, at 69–70.

ditions and to avoid excessive contamination of the sample. A laboratory should be able to demonstrate that it can amplify targeted sequences faithfully with the equipment and reagents that it uses and that it has taken suitable precautions to avoid or detect contamination from foreign DNA. With small samples, it is possible that some alleles will be amplified and others missed (preferential amplification, discussed *infra* Section III.A.1), and mutations in the region of a primer can prevent the amplification of the allele downstream of the primer (null alleles).[25]

D. How Is STR Profiling Done with Capillary Electrophoresis?

In the most commonly used analytical method for detecting STRs, the STR fragments in the sample are amplified using primers with fluorescent tags. Each new STR fragment made in a PCR cycle bears a fluorescent dye. When struck by a source light, each dye glows with a particular color. The fragments are separated according to their length by electrophoresis in automated "genetic analyzer" machinery—a byproduct of the technology developed for the Human Genome Project that first sequenced most of the entire genome. In these machines, a long, narrow tube (a "capillary") is filled with an entangled polymer or comparable sieving medium, and an electric field is applied to pull DNA fragments placed at one end of the tube through the medium. Shorter fragments slip through the medium more quickly than larger, bulkier ones. A laser beam is sent through a small glass window in the tube. The laser light excites the dye, causing it to fluoresce at a characteristic wavelength as the tagged fragments pass under the light. The intensity of the light emitted by the dye is recorded by a kind of electronic camera and transformed into a graph (an electropherogram), which shows a peak as an STR flashes by. A shorter allele will pass by the window and fluoresce first; a longer fragment will come by later, giving rise to another peak on the graph. Figure 3 provides a sketch of how the alleles with five and eight repeats of the GATA sequence at the D16S539 STR locus might appear in an electropherogram.

Medical and human geneticists were interested in STRs as markers in family studies to locate the genes that are associated with inherited diseases, and papers on their potential for identity testing appeared in the early 1990s. Developmental research to pick suitable loci moved into high gear in England, Europe, and Canada. Britain's Forensic Science Service applied a four-locus testing system in 1994. Then it introduced the "second generation multiplex" (SGM)—for simultaneously typing six loci in 1996. These soon would be used to build England's National DNA Database. The database system allows a computer to check the STR types of millions of known or suspected criminals against thousands of crime

25. A null allele will not lead to a false exclusion if the two DNA samples from the same individual are amplified with the same primer system, but it could lead to an exclusion at one locus when searching a database of STR profiles if the database profile was determined with a different PCR kit than the one used to analyze the crime scene DNA.

Figure 3. Sketch of an electropherogram for two D16S539 alleles. One allele has five repeats of the sequence GATA; the other has eight. Each GATA repeat is depicted as a small rectangle. Although only one copy of each allele (with a fluorescent molecule, or "tag" attached) is shown here, PCR generates a great many copies from the DNA sample with these alleles at the D16S539 locus. These copies are drawn through the capillary tube, and the tags glow as the STR fragments move through the laser beam. An electronic camera measures the colored light from the tags. Finally, a computer processes the signal from the camera to produce the electropherogram. Source: David H. Kaye, The Double Helix and the Law of Evidence 189, fig. 9.1 (2010).

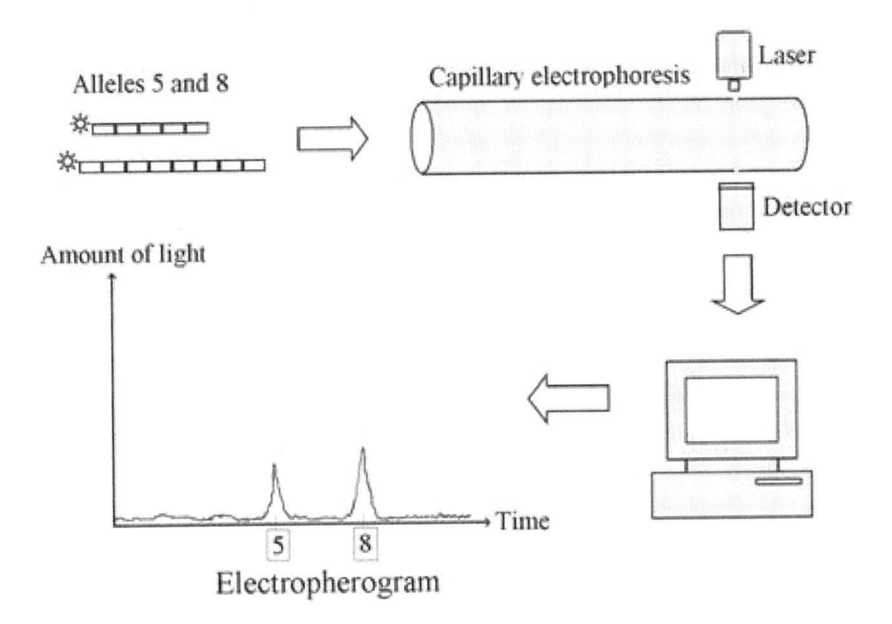

scene samples. A six-locus STR profile can be represented as a string of 12 digits; each digit indicates the number of repeat units in the alleles at each locus. These discrete, numerical DNA profiles are far easier to compare mechanically than the complex patterns of fingerprints. In the United States, the FBI settled on 13 "core loci" to use in the U.S. national DNA database system. These are often called the "CODIS core loci," and an additional 7 STR loci are under consideration.[26]

Modern genetic analyzers produce electropherograms for many loci at once. This "multiplexing" is accomplished by using dyes that fluoresce at distinct colors

26. Douglas R. Hares, *Expanding the CODIS Core Loci in the United States*, Forensic Sci. Int'l: Genetics (forthcoming 2011). CODIS stands for "convicted offender DNA index system."

to label the alleles from different groups of loci. A separate set of fragments of known sizes that comigrate through the capillary function as a kind of ruler (an "internal-lane size standard") to determine the lengths of the allelic fragments. Software processes the raw data to generate an electropherogram of the separate allele peaks of each color. By comparing the positions of the allele peaks to the size standard, the program determines the number of repeats in each allele. The plotted heights of the peaks (measured in relative fluorescent units, or RFUs) are proportional to the amount of the PCR product.

Figure 4 is an electropherogram of all 203 major alleles at 15 STR loci that can be typed in a single "multiplex" PCR reaction. (In addition, it shows the two alleles of the gene used to determine the sex of the contributor of a DNA

Figure 4. Alleles of 15 STR loci and the amelogenin sex-typing test from the AmpFlSTR Identifiler kit. The bottom panel is a "sizing standard"—a set of peaks from DNA sequences of known lengths (in base pairs). The numbers in the vertical axis in each panel are relative fluorescence units (RFUs) that indicate the amount of light emitted after the laser beam strikes the fluorescent tag on an STR fragment.

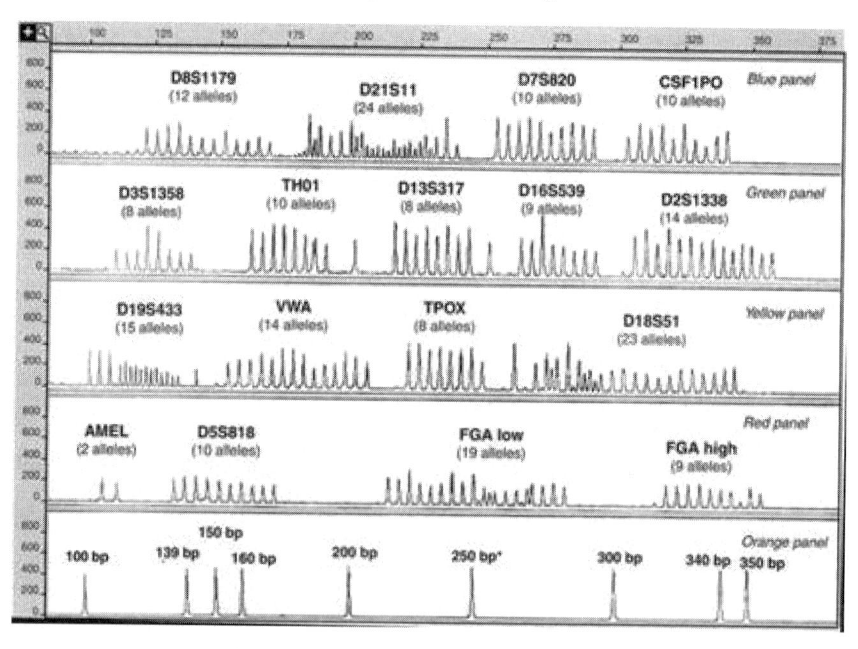

Note: Applied Biosystems makes the kit that produced these allelic ladders.

Source: John M. Butler, Forensic DNA Typing: Biology, Technology, and Genetics of STR Markers 128 (2d ed. 2005), Copyright Elsevier 2005, with the permission of Elsevier Academic Press. John Butler supplied the illustration.

sample.[27]) An electropherogram from an individual's DNA would have only one or two peaks at each of these 15 STR loci (depending on whether the person is homozygous or heterozygous). These "allelic ladders" aid in deciding which allele a peak from an unknown sample represents.

Figure 5 is an electropherogram from the vaginal epithelial cells of the body of a girl who had been sexually assaulted and killed in *People v. Pizarro*.[28] It was produced for the retrial in 2008 of the defendant who was linked to the victim by VNTR typing at his first trial in 1990.

Figure 5. Electropherogram for nine STR loci of the victim's DNA in *People v. Pizzaro*. (The amelogenin locus and a sizing standard at the bottom also are included.) Some STR loci have small peaks, indicating that there was not much PCR product for those loci, likely because of DNA degradation. All of the STR loci have two peaks, as would be expected when the source is heterozygous at those loci.

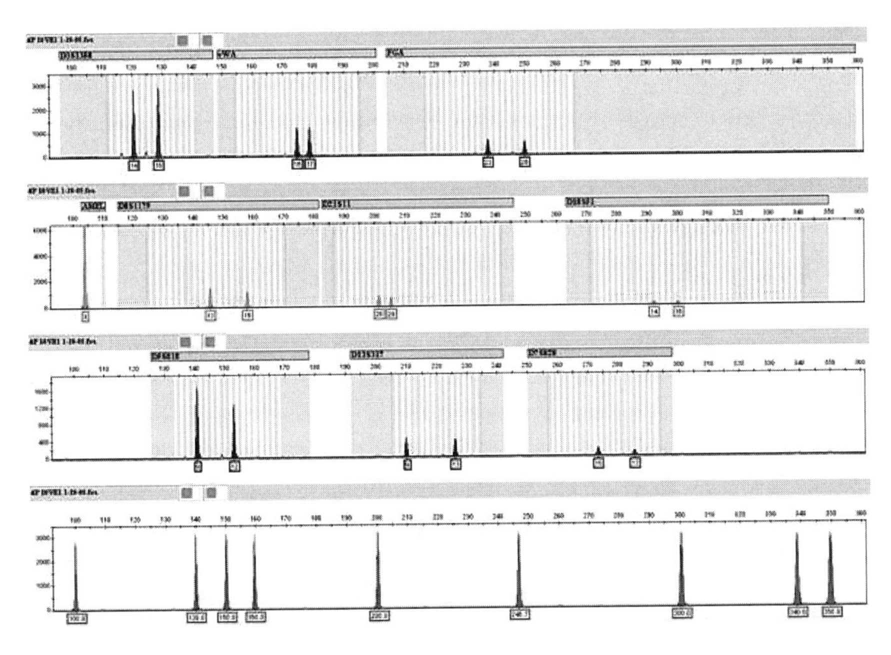

Source: Steven Myers and Jeanette Wallin, California Department of Justice, provided the image.

27. The amelogenin gene, which is found on the X and the Y chromosomes, codes for a protein that is a major component of tooth enamel matrix. The copy on the X chromosome is 112 bp long. The copy on the Y chromosome has a string of six base pairs deleted, making it slightly shorter (106 bp). A female (XX) will have one peak at 112 bp. A male (XY) will have two peaks (at 106 and 112 bp).

28. 12 Cal. Rptr. 2d 436 (Ct. App. 1992), *after remand*, 3 Cal. Rptr. 3d 21 (Ct. App. 2003), *review denied* (Oct 15, 2003).

E. What Can Be Done to Validate a Genetic System for Identification?

Regardless of the kind of genetic system used for typing—STRs, SNPs, or still other polymorphisms—some general principles and questions can be applied to each system that is offered for courtroom use. First, the nature of the polymorphism should be well characterized. Is it a simple sequence polymorphism or a fragment length polymorphism? This information should be in the published literature or in archival genome databanks.

Second, the published scientific literature can be consulted to verify claims that a particular method of analysis can produce accurate profiles under various conditions. Although such validation studies have been conducted for all the systems ordinarily used in forensic work, determining the point at which the empirical validation of a particular system is sufficiently convincing to pass scientific muster may well require expert assistance.

Finally, the population genetics of the system should be characterized. As new systems are discovered, researchers typically analyze convenient collections of DNA samples from various human populations and publish studies of the relative frequencies of each allele in these population samples. These studies measure the extent of genetic variability at the polymorphic locus in the various populations, and thus of the potential probative power of the marker for distinguishing among individuals.

At this point, the capability of PCR-based procedures to ascertain DNA genotypes accurately cannot be doubted. Of course, the fact that scientists have shown that it is possible to extract DNA, to amplify it, and to analyze it in ways that bear on the issue of identity does not mean that a particular laboratory has adopted a suitable protocol and is proficient in following it. These case-specific issues are considered in Sections III and IV.

F. What New Technologies Might Emerge?

1. Miniaturized "lab-on-a-chip" devices

Miniaturized capillary electrophoresis (CE) devices have been developed for rapid detection of STRs (described in Section II.D) and other genetic analyses. The mini-CE systems consist of microchannels roughly the diameter of a hair etched on glass wafers ("chips") using technology borrowed from the computer industry. The principles of electrophoretic separation are the same as with conventional CE systems. With microfluidic technologies, it is possible to integrate DNA extraction and PCR amplification processes with the CE separation in a single device, a so-called lab on a chip. Once a sample is added to the device, all the analytical steps are performed on the chip without further human contact. These integrated devices combine the benefits of simplified sample handling with rapid analysis and are under active development for point-of-care medical diagnostics.[29] Efforts are under way to develop an integrated microdevice for STR analysis that would improve the speed and efficiency of forensic DNA profiling. A portable device for rapid and secure analysis of samples in the field is a distinct possibility.[30]

2. *High-throughput sequencing*

The initial success of the Human Genome Project and the promise of "personalized medicine" is driving research to develop technologies for DNA analysis that are faster, cheaper, and less labor intensive. In 2004, the National Human Genome Research Institute announced funding for research leading to the "$1000 genome," an achievement that would permit sequencing an individual's genome for medical diagnosis and improved drug therapies. Advances in the years since 2004 suggest that this goal will be achieved before the target date of 2014,[31] and the successful innovations could provide major advances in forensic DNA testing. However, it is too soon to identify which of the nascent sequencing technologies might emerge from the pack.

As of 2009, three different next-generation sequencing technologies were commercially available, and more instruments are in the pipeline.[32] These new technologies generate massive amounts of DNA sequence data (100 million to 1 billion base pairs per run) at very low cost (under $50 per megabase). They do so by simultaneously sequencing millions of short fragments, then applying bioinformatics software to assemble the sequences in the correct order. These high-throughput sequencing technologies have demonstrated their usefulness in research applications. Two of these applications, the analysis of highly degraded DNA[33] and the identification of microbial bioterrorism agents, are of forensic relevance.[34] As the speed and cost of sequencing diminish and the necessary bioinformatics software becomes more accessible and effective, full-genome sequence

29. P. Yager et al., *Microfluidic Diagnostic Technologies for Global Public Health*, 442 Nature 412 (2006).

30. K.M. Horsman et al., *Forensic DNA Analysis on Microfluidic Devices: A Review*, 52 J. Forensic Sci. 784 (2007). As indicated in this review, there remain challenges to overcome before the forensic lab on a chip comes to fruition. However, given the progress being made on multiple research fronts in chip fabrication design and in microfluidic technology, these challenges seem surmountable.

31. R.F. Service, *The Race for the $1000 Genome*, 311 Science 1544 (2006).

32. Michael L. Metzker, *Sequencing Technologies—The Next Generation*, 11 Nature Rev. Genetics 31 (2010).

33. The next-generation technologies have been used to sequence highly degraded DNA from Neanderthal bones and from the hair of the extinct woolly mammoth. R.E. Green et al., *Analysis of One Million Base Pairs of Neanderthal DNA*, 444 Nature 330 (2006); W. Miller, *Sequencing the Nuclear Genome of the Extinct Woolly Mammoth*, 456 Nature 387 (2008). The approaches used in these studies are readily translatable to SNP typing of highly degraded DNA such as found in cases involving victims of mass disasters.

34. By sequencing entire bacterial genomes, researchers can rapidly differentiate organisms that have been genetically modified for biological warfare or terrorism from routine clinical and envi-

analysis or something approaching it could become a practical tool for human identification.

3. Microarrays

Hybridization microarrays are the third technological innovation with readily foreseeable forensic application. A microarray consists of a two-dimensional grid of many thousands of microscopic spots on a glass or plastic surface, each containing many copies of a short piece of single-stranded DNA tethered to the surface at one end; each spot can be thought of as a dense cluster of tiny, single-stranded DNA "whiskers" with their own particular sequence. A solution containing single-stranded target DNA is washed over the microarray surface. The whiskers on the array serve as probes to detect DNA (or RNA) with the corresponding complementary sequence. The spots that capture target DNA are identified, indicating the presence of that sequence in the target sample. (The hybridization can be detected in several different ways.) Microarrays are commercially available for the detection of SNPs in the human genome and for sequencing human mitochondrial DNA.[35]

4. What questions do the new technologies raise?

As these or other emerging technologies are introduced in court, certain basic questions will need to be answered. What is the principle of the new technology? Is it simply an extension of existing technologies, or does it invoke entirely new concepts? Is the new technology used in research or clinical applications independent of forensic science? Does the new technology have limitations that might affect its application in the forensic sphere? Finally, what testing has been done and with what outcomes to establish that the new technology is reliable when used on forensic samples? For next-generation sequencing technologies and microarray technologies, the questions may be directed as well to the bioinformatics methods used to analyze and interpret the raw data. Obtaining answers to these questions would likely require input both from experts involved in technology development and application and from knowledgeable forensic experts.

ronmental strains. B. La Scola et al., *Rapid Comparative Genomic Analysis for Clinical Microbiology: The Francisella Tularensis Paradigm*, 18 Genome Res. 742 (2008).

35. One study of 3000 Europeans used a commercial microarray with over half a million SNPs "to infer [the individuals'] geographic origin with surprising accuracy—often to within a few hundred kilometers." John Novembre et al., *Genes Mirror Geography Within Europe*, 456 Nature 98, 98 (2008). Microarrays also are used in studies of variation in the number of copies of certain genes in different people's genomes (copy number variation). Microarrays to detect pathogens and other targets also have been developed.

III. Sample Collection and Laboratory Performance

A. Sample Collection, Preservation, and Contamination

The primary determinants of whether DNA typing can be done on any particular sample are (1) the quantity of DNA present in the sample and (2) the extent to which it is degraded. Generally speaking, if a sufficient quantity of reasonable quality DNA can be extracted from a crime scene sample, no matter what the nature of the sample, DNA typing can be done without problem. Thus, DNA typing has been performed on old blood stains, semen stains, vaginal swabs, hair, bone, bite marks, cigarette butts, urine, and fecal material. This section discusses what constitutes sufficient quantity and reasonable quality in the context of STR typing. Complications from contaminants and inhibitors also are discussed. The special technique of mitotyping and the treatment of samples that contain DNA from two or more contributors are discussed in Section V.

1. Did the sample contain enough DNA?

Amounts of DNA present in some typical kinds of samples vary from a trillionth or so of a gram for a hair shaft to several millionths of a gram for a postcoital vaginal swab. Most PCR test protocols recommend samples on the order of 1 billionth to 5 billionths of a gram for optimum yields. Normally, the number of amplification cycles for nuclear DNA is limited to 28 or so to ensure that there is no detectable product for samples containing less than about 20 cell equivalents of DNA.[36]

Procedures for typing still smaller samples—down to a single cell's worth of nuclear DNA—have been studied. These have been shown to work, to some extent, with trace or contact DNA left on the surface of an object such as the steering wheel of a car. The most obvious strategy is to increase the number of amplification cycles. The danger is that chance effects might result in one allele being amplified much more than another. Alleles then could drop out, small peaks from unusual alleles at other loci might "drop in," and a bit of extraneous DNA could contribute to the profile. Other protocols have been developed for typing such "low copy number" (LCN) or "low template" (LT) DNA.[37] LT-STR

36. This is about 100 to 200 trillionths of a gram. A lower limit of about 10 to 15 cells' worth of DNA has been determined to give balanced amplification.

37 *See, e.g.*, John Buckleton & Peter Gill, *Low Copy Number, in* Forensic DNA Evidence Interpretation 275 (John S. Buckleton et al. eds., 2005); Pamela J. Smith & Jack Ballantyne, *Simplified Low-Copy-Number DNA Analysis by Post-PCR Purification*, 52 J. Forensic Sci. 820 (2007).

profiles have been admitted in courts in a few countries,[38] and they are beginning to appear in prosecutions in the United States.[39]

Although there are tests to estimate the quantity of DNA in a sample, whether a particular sample contains enough human DNA to allow typing cannot always be predicted accurately. The best strategy is to try. If a result is obtained, and if the controls (samples of known DNA and blank samples) have behaved properly, then the sample had enough DNA. The appearance of the same peaks in repeated runs helps assure that these alleles are present.[40]

2. *Was the sample of sufficient quality?*

The primary determinant of DNA quality for forensic analysis is the extent to which the long DNA molecules are intact. Within the cell nucleus, each molecule of DNA extends for millions of base pairs. Outside the cell, DNA spontaneously degrades into smaller fragments at a rate that depends on temperature, exposure to oxygen, and, most importantly, the presence of water.[41] In dry biological samples, protected from air, and not exposed to temperature extremes, DNA degrades very slowly. STR testing has proved effective with old and badly degraded material such as the remains of the Tsar Nicholas family (buried in 1918 and recovered in 1991).[42]

38. *E.g.*, R. v. Reed [2009] (CA Crim. Div.) EWCA Crim. 2698, ¶ 74 (reviewing expert submissions and concluding that "Low Template DNA can be used to obtain profiles capable of reliable interpretation if the quantity of DNA that can be analysed is above the stochastic threshold [of] between 100 and 200 picograms").

39. People v. Megnath, 898 N.Y.S.2d 408 (N.Y. Sup. Ct. 2010) (reasoning that "LCN DNA analysis" uses the same steps as STR analysis of larger samples and that the modifications in the procedure used by the laboratory in the case were generally accepted); *cf.* United States v. Davis, 602 F. Supp. 2d 658 (D. Md. 2009) (avoiding "making a finding with regard to the dueling definitions of LCN testing advocated by the parties" by finding that "the amount of DNA present in the evidentiary samples tested in this case" was in the normal range). These cases and the admissibility of low-template DNA analysis are discussed in Kaye et al., *supra* note 1, § 9.2.3(c).

40. John M. Butler & Cathy R. Hill, *Scientific Issues with Analysis of Low Amounts of DNA*, LCN Panel on Scientific Issues with Low Amounts of DNA, Promega Int'l Symposium on Human Identification, Oct. 15, 2009, *available at* http://www.cstl.nist.gov/strbase/pub_pres/Butler_Promega2009-LCNpanel-for-STRBase.pdf.

41. Other forms of chemical alteration to DNA are well studied, both for their intrinsic interest and because chemical changes in DNA are a contributing factor in the development of cancers in living cells. Some forms of DNA modification, such as that produced by exposure to ultraviolet radiation, inhibit the amplification step in PCR-based tests, whereas other chemical modifications appear to have no effect. C.L. Holt et al., *TWGDAM Validation of AmpFlSTR PCR Amplification Kits for Forensic DNA Casework*, 47 J. Forensic Sci. 66 (2002); George F. Sensabaugh & Cecilia von Beroldingen, *The Polymerase Chain Reaction: Application to the Analysis of Biological Evidence*, *in* Forensic DNA Technology 63 (Mark A. Farley & James J. Harrington eds., 1991).

42. Peter Gill et al., *Identification of the Remains of the Romanov Family by DNA Analysis*, 6 Nature Genetics 130 (1994).

The extent to which degradation affects a PCR-based test depends on the size of the DNA segment to be amplified. For example, in a sample in which the bulk of the DNA has been degraded to fragments well under 1000 base pairs in length, it may be possible to amplify a 100-base-pair sequence, but not a 1000-base-pair target. Consequently, the shorter alleles may be detected in a highly degraded sample, but the larger ones may be missed. Fortunately, the size differences among STR alleles at a locus are quite small (typically no more than 50 base pairs). Therefore, if there is a degradation effect on STR typing, it is usually "locus dropout"—in cases involving severe degradation, loci yielding larger products (greater than 200 base pairs) may not be detected.[43]

DNA can be exposed to a great variety of environmental insults without any effect on its capacity to be typed correctly. Exposure studies have shown that contact with a variety of surfaces, both clean and dirty, and with gasoline, motor oil, acids, and alkalis either have no effect on DNA typing or, at worst, render the DNA untypable.[44]

Although contamination with microbes generally does little more than degrade the human DNA, other problems sometimes can occur. Therefore, the validation of DNA typing systems should include tests for interference with a variety of microbes to see if artifacts occur. If artifacts are observed, then control tests should be applied to distinguish between the artifactual and the true results.

B. *Laboratory Performance*

1. *What forms of quality control and assurance should be followed?*

DNA profiling is valid and reliable, but confidence in a particular result depends on the quality control and quality assurance procedures in the laboratory. Quality control refers to measures to help ensure that a DNA-typing result (and its interpretation) meets a specified standard of quality. Quality assurance refers to monitoring, verifying, and documenting laboratory performance. A quality assurance program helps demonstrate that a laboratory is meeting its quality control objectives and thus justifies confidence in the quality of its product.[45]

43. Holt et al., *supra* note 41. Special primers and very short STRs give better results with extremely degraded samples. *See* Michael D. Coble & John M. Butler, *Characterization of New MiniSTR Loci to Aid Analysis of Degraded DNA*, 50 J. Forensic Sci. 43 (2005).

44. Holt et al., *supra* note 41. Most of the effects of environmental insult readily can be accounted for in terms of basic DNA chemistry. For example, some agents produce degradation or damaging chemical modifications. Other environmental contaminants inhibit restriction enzymes or PCR. (This effect sometimes can be reversed by cleaning the DNA extract to remove the inhibitor.) But environmental insult does not result in the selective loss of an allele at a locus or in the creation of a new allele at that locus.

45. For a review of the history of quality assurance in forensic DNA testing, *see* J.L. Peterson et al., *The Feasibility of External Blind DNA Proficiency Testing. I. Background and Findings*, 48 J. Forensic Sci. 21, 22 (2003).

Professional bodies within forensic science have described procedures for quality assurance. Guidelines for DNA analysis have been prepared by FBI-appointed groups (the current incarnation is known as SWGDAM);[46] a number of states require forensic DNA laboratories to be accredited;[47] and federal law requires accreditation or other safeguards of laboratories that receive certain federal funds[48] or participate in the national DNA database system.[49]

a. Documentation

Quality assurance guidelines normally call for laboratories to document laboratory organization and management, personnel qualifications and training, facilities, evidence control procedures, validation of methods and procedures, analytical procedures, equipment calibration and maintenance, standards for case documentation and report writing, procedures for reviewing case files and testimony, proficiency testing, corrective actions, audits, safety programs, and review of subcontractors.

46. The FBI established the Technical Working Group on DNA Analysis Methods (TWGDAM) in 1988 to develop standards. The DNA Identification Act of 1994, 42 U.S.C. § 14131(a) & (c) (2006), created a DNA Advisory Board (DAB) to assist in promulgating quality assurance standards, but the legislation allowed the DAB to expire after 5 years (unless extended by the Director of the FBI). 42 U.S.C. § 14131(b) (2008). TWGDAM functioned under DAB, 42 U.S.C. § 14131(a) (2006), and was renamed the Scientific Working Group on DNA Analysis Methods (SWGDAM) in 1999. When the FBI allowed DAB to expire, SWGDAM replaced DAB. *See* Norah Rudin & Keith Inman, An Introduction to Forensic DNA Analysis 180 (2d ed. 2002); Paul C. Giannelli, *Regulating Crime Laboratories: The Impact of DNA Evidence*, 15 J.L. & Pol'y 59, 82–83 (2007).

47. New York was the first state to impose this requirement. N.Y. Exec. Law § 995-b (McKinney 2006) (requiring accreditation by the state Forensic Science Commission).

48. The Justice for All Act, enacted in 2004, required DNA labs to be accredited within 2 years "by a nonprofit professional association of persons actively involved in forensic science that is nationally recognized within the forensic science community" and to "undergo external audits, not less than once every 2 years, that demonstrate compliance with standards established by the Director of the Federal Bureau of Investigation." 42 U.S.C. § 14132(b)(2) (2006). Established in 1981, the American Society of Crime Laboratory Directors–Laboratory Accreditation Board (ASCLD-LAB) accredits forensic laboratories. Giannelli, *supra* note 46, at 75. The 2004 Act also requires applicants for federal funds for forensic laboratories to certify that the laboratories use "generally accepted laboratory practices and procedures, established by accrediting organizations or appropriate certifying bodies," 42 U.S.C. § 3797k(2) (2004), and that "a government entity exists and an appropriate process is in place to conduct independent external investigations into allegations of serious negligence or misconduct substantially affecting the integrity of the forensic results committed by employees or contractors of any forensic laboratory system, medical examiner's office, coroner's office, law enforcement storage facility, or medical facility in the State that will receive a portion of the grant amount." *Id.* § 3797k(4). There have been problems in implementing the § 3797k(4) certification requirement. *See* Office of the Inspector General, U.S. Dep't of Justice, Review of the Office of Justice Programs' Paul Coverdell Forensic Science Improvement Grants Program, Evaluation and Inspections Report I-2008-001 (2008), *available at* http://www.usdoj.gov/oig/reports/OJP/e0801/index.htm.

49. *See* 42 U.S.C § 14132 (b)(2) (2006) (requiring as of late 2006, that records in the database come from laboratories that "have been accredited by a nonprofit professional association . . . and . . . undergo external audits, not less than once every 2 years [and] that demonstrate compliance with standards established by the Director of the Federal Bureau of Investigation. . . .").

Of course, maintaining documentation and records alone does not guarantee the correctness of results obtained in any particular case. Errors in analysis or interpretation might occur as a result of a deviation from an established procedure, analyst misjudgment, or an accident. Although case review procedures within a laboratory should be designed to detect errors before a report is issued, it is always possible that some incorrect result will slip through. Accordingly, determination that a laboratory maintains a strong quality assurance program does not eliminate the need for case-by-case review.

b. Validation

The validation of procedures is central to quality assurance. Developmental validation is undertaken to determine the applicability of a new test to crime scene samples; it defines conditions that give reliable results and identifies the limitations of the procedure. For example, a new genetic marker being considered for use in forensic analysis will be tested to determine if it can be typed reliably in both fresh samples and in samples typical of those found at crime scenes. The validation would include testing samples originating from different tissues—blood, semen, hair, bone, samples containing degraded DNA, samples contaminated with microbes, samples containing DNA mixtures, and so on. Developmental validation of a new set of loci also includes the generation of population databases and the testing of alleles for statistical independence. Developmental validation normally results in publication in the scientific literature, but a new procedure can be validated in multiple laboratories well ahead of publication.

Internal validation, on the other hand, involves the capacity of a specific laboratory to analyze the new loci. The laboratory should verify that it can reliably perform an established procedure that already has undergone developmental validation. In particular, before adopting a new procedure, the laboratory should verify its ability to use the system in a proficiency trial.[50]

c. Proficiency testing

Proficiency testing in forensic genetic testing is designed to ascertain whether an analyst can correctly determine genetic types in a sample whose origin is unknown to the analyst but is known to a tester. Proficiency is demonstrated by making correct genetic typing determinations in repeated trials. The laboratory also can be tested to verify that it correctly computes random-match probabilities or similar statistics.

An internal proficiency trial is conducted within a laboratory. One person in the laboratory prepares the sample and administers the test to another person in the labo-

50. Both forms of validation build on the accumulated body of knowledge and experience. Thus, some aspects of validation testing need be repeated only to the extent required to verify that previously established principles apply.

ratory. In an external trial, the test sample originates from outside the laboratory—from another laboratory, a commercial vendor, or a regulatory agency. In a declared (or open) proficiency trial, the analyst knows the sample is a proficiency sample. The DNA Identification Act of 1994 requires proficiency testing for analysts in the FBI as well as those in laboratories participating in the national database or receiving federal funding,[51] and the standards of accrediting bodies typically call for periodic open, external proficiency testing.[52]

In a blind (or, more properly, "full blind") trial, the sample is submitted so that the analyst does not recognize it as a proficiency sample. A full-blind trial provides a better indication of proficiency because it ensures that the analyst will not give the trial sample any special attention, and it tests more steps in the laboratory's processing of samples. However, full-blind proficiency trials entail considerably more organizational effort and expense than open proficiency trials. Obviously, the "evidence" samples prepared for the trial have to be sufficiently realistic that the laboratory does not suspect the legitimacy of the submission. A police agency and prosecutor's office have to submit the "evidence" and respond to laboratory inquiries with information about the "case." Finally, the genetic profile from a proficiency test must not be entered into regional and national databases. Consequently, although some forensic DNA laboratories participate in full-blind testing, they are not required to do so.[53]

2. *How should samples be handled?*

Sample mishandling, mislabeling, or contamination, whether in the field or in the laboratory, is more likely to compromise a DNA analysis than is an error in genetic typing. For example, a sample mixup due to mislabeling reference blood samples taken at the hospital could lead to incorrect association of crime scene samples to a reference individual or to incorrect exclusions. Similarly, packaging two items with wet bloodstains into the same bag could result in a transfer of stains between the items, rendering it difficult or impossible to determine whose blood was originally on each item. Contamination in the laboratory may result in artifactual

51. 42 U.S.C. § 14132(b)(2) (requiring external proficiency testing of laboratories for participation in the national database); *id.* § 14133(a)(1)(A) (2006) (same for FBI examiners).

52. *See* Peterson et al., *supra* note 45, at 24 (describing the ASCL-LAB standards). Certification by the American Board of Criminalistics as a specialist in forensic biology DNA analysis requires one proficiency trial per year. Accredited laboratories must maintain records documenting compliance with required proficiency test standards.

53. The DNA Identification Act of 1994 required the director of the National Institute of Justice to report to Congress on the feasibility of establishing an external blind proficiency testing program for DNA laboratories. 42 U.S.C. § 14131(c) (2006). A National Forensic DNA Review Panel advised the Director that "blind proficiency testing is possible, but fraught with problems" of the kind listed above). Peterson et al., *supra* note 46, at 30. It "recommended that a blind proficiency testing program be deferred for now until it is more clear how well implementation of the first two recommendations [the promulgation of guidelines for accreditation, quality assurance, and external audits of casework] are serving the same purposes as blind proficiency testing." *Id.*

typing results or in the incorrect attribution of a DNA profile to an individual or to an item of evidence. Procedures should be prescribed and implemented to guard against such error.

Mislabeling or mishandling can occur when biological material is collected in the field, when it is transferred to the laboratory, when it is in the analysis stream in the laboratory, when the analytical results are recorded, or when the recorded results are transcribed into a report. Mislabeling and mishandling can happen with any kind of physical evidence and are of great concern in all fields of forensic science. Checkpoints should be established to detect mislabeling and mishandling along the line of evidence flow. Investigative agencies should have guidelines for evidence collection and labeling so that a chain of custody is maintained. Similarly, there should be guidelines, produced with input from the laboratory, for handling biological evidence in the field.

Professional guidelines and recommendations require documented procedures to ensure sample integrity and to avoid sample mixups, labeling errors, recording errors, and the like.[54] They also mandate case review to identify inadvertent errors before a final report is released. Finally, laboratories must retain, when feasible, portions of the crime scene samples and extracts to allow reanalysis.[55] However, retention is not always possible. For example, retention of original items is not to be expected when the items are large or immobile (e.g., a wall or sidewalk). In such situations, a swabbing or scraping of the stain from the item would typically be collected and retained. There also are situations where the sample is so small that it will be consumed in the analysis.

Assuming that appropriate chain-of-custody and evidence-handling protocols are in place, the critical question is whether there are deviations in the particular case. This may require a review of the total case documentation as well as the laboratory findings. In addition, the opportunity to retest original evidence items or the material extracted from them is an important safeguard against error because of mislabeling and mishandling. Should mislabeling or mishandling have occurred, reanalysis of the original sample and the intermediate extracts should detect not only the fact of the error but also the point at which it occurred.[56]

54. SWGDAM guidelines are published as FBI, Standards for Forensic DNA Testing Labs, *available at* http://www.fbi.gov/hq/lab/codis/forensic.htm (last visited Feb. 16, 2010).

55. Forensic laboratories have a professional responsibility to preserve retained evidence so as to minimize degradation. *See id.*, standard 7.2.1. Furthermore, failure to preserve potentially exculpatory evidence has been treated as a denial of due process and grounds for suppression. People v. Nation, 604 P.2d 1051, 1054–55 (Cal. 1980). In *Arizona v. Youngblood,* 488 U.S. 51 (1988), however, the Supreme Court held that a police agency's failure to preserve evidence not known to be exculpatory does not constitute a denial of due process unless "bad faith" can be shown. Ironically, DNA testing that was not available at Youngblood's trial established that he had been falsely convicted. Maurice Possley, *DNA Exonerates Inmate Who Lost Key Test Case: Prosecutors Ruined Evidence in Original Trial,* Chi. Trib., Aug. 10, 2000, at 6.

56. Of course, retesting cannot correct all errors that result from mishandling of samples, but it is even possible in some cases to detect mislabeling at the point of sample collection if the genetic

Contamination describes any situation in which foreign material is mixed with a sample of DNA. As noted in Section III.A.2, contamination by nonbiological materials, such as gasoline or grit, can cause test failures, but they are not a source of genetic typing errors. Similarly, contamination with nonhuman biological materials, such as bacteria, fungi, or plant materials, is generally not a problem. These contaminants may accelerate DNA degradation, but they do not generate spurious human genetic types.

The contamination of greatest concern is that resulting from the addition of human DNA. This sort of contamination can occur three ways. First, the crime scene samples by their nature may contain a mixture of fluids or tissues from different individuals. Examples include vaginal swabs collected as sexual assault evidence and bloodstain evidence from scenes where several individuals shed blood. Mixtures are the subject of Section V.C.

Second, the crime scene samples may be inadvertently contaminated in the course of sample handling in the field or in the laboratory. Inadvertent contamination of crime scene DNA with DNA from a reference sample could lead to a false inclusion.

Third, carryover contamination in PCR-based typing can occur if the amplification products of one typing reaction are carried over into the reaction mix for a subsequent PCR reaction. If the carryover products are present in sufficient quantity, they could be preferentially amplified over the target DNA. The primary strategy used in most forensic laboratories to protect against carryover contamination is to keep PCR products away from sample materials and test reagents by having separate work areas for pre-PCR and post-PCR sample handling, by preparing samples in controlled-air-flow biological safety hoods, by using dedicated equipment (such as pipetters) for each of the various stages of sample analysis, by decontaminating work areas after use (usually by wiping down or by irradiating with ultraviolet light), and by having a one-way flow of sample from the pre-PCR to post-PCR work areas. Additional protocols are used to detect any carryover contamination.[57]

In the end, whether a laboratory has conducted proper tests and whether it conducted them properly depends both on the general standard of practice and

typing results on a particular sample are inconsistent with an otherwise consistent reconstruction of events. For example, a mislabeling of husband and wife samples in a paternity case might result in an apparent maternal exclusion, a very unlikely event. The possibility of mislabeling could be confirmed by testing the samples for gender and ultimately verified by taking new samples from each party under better controlled conditions.

57. Standard protocols include the amplification of blank control samples—those to which no DNA has been added. If carryover contaminants have found their way into the reagents or sample tubes, these will be detected as amplification products. Outbreaks of carryover contamination can also be recognized by monitoring test results. Detection of an unexpected and persistent genetic profile in different samples indicates a contamination problem. When contamination outbreaks are detected, appropriate corrective actions should be taken, and both the outbreak and the corrective action should be documented.

on the questions posed in the particular case. There is no universal checklist, but the selection of tests and the adherence to the correct test procedures can be reviewed by experts and by reference to professional standards such as the SWGDAM guidelines.

IV. Inference, Statistics, and Population Genetics in Human Nuclear DNA Testing

The results of DNA testing can be presented in various ways. With discrete allele systems, such as STRs, it is natural to speak of "matching" and "nonmatching" profiles. If the genetic profile obtained from the biological sample taken from the crime scene or the victim (the "trace evidence sample") matches that of a particular individual, that individual is included as a possible source of the sample. But other individuals also might possess a matching DNA profile. Accordingly, the expert should be asked to provide some indication of how significant the match is. If, on the other hand, the genetic profiles are different, then the individual is excluded as the source of the trace evidence. Typically, proof tending to show that the defendant is the source incriminates the defendant, whereas proof that someone else is the source exculpates the defendant.[58] This section elaborates on these ideas, indicating issues that can arise in connection with an expert's testimony interpreting the results of a DNA test.

A. What Constitutes a Match or an Exclusion?

When the DNA from the trace evidence clearly does not match the DNA sample from the suspect, the DNA analysis demonstrates that the suspect's DNA is not in the forensic sample. Indeed, if the samples have been collected, handled, and analyzed properly, then the suspect is excluded as a possible source of the DNA in the forensic sample. As a practical matter, such exclusionary results normally would keep charges from being filed against the excluded suspect.

At the other extreme, the genotypes at a large number of loci can be clearly identical. In these cases, the DNA evidence is quite incriminating, and the challenge for the legal system lies in explaining just how probative it is. Naturally, as with exclusions, inclusions are most powerful when the samples have been

58. Whether being the source of the forensic sample is incriminating and whether someone else being the source is exculpatory depends on the circumstances. For example, a suspect who might have committed the offense without leaving the trace evidence sample still could be guilty. In a rape case with several rapists, a semen stain could fail to incriminate one assailant because insufficient semen from that individual is present in the sample.

collected, handled, and analyzed properly. But there is one logical difference between exclusions and inclusions. If it is accepted that the samples have different genotypes, then the conclusion that the DNA in them came from different individuals is essentially inescapable.[59] In contrast, even if two samples have the same genotype, there is a chance that the forensic sample came not from the defendant, but from another individual who has the same genotype. This complication has produced extensive arguments over the statistical procedures for assessing this chance or related quantities. This problem of describing the significance of an unequivocal match is the subject of the remaining parts of this section.

Some cases lie between the poles of a clear inclusion or a definite exclusion. For example, when the trace evidence sample is small and extremely degraded, STR profiling can be afflicted with allelic "drop-in" and "drop-out," requiring judgments as to whether true peaks are missing and whether spurious peaks are present. Experts then might disagree about whether a suspect is included or excluded—or whether any conclusion can be drawn.[60]

B. What Hypotheses Can Be Formulated About the Source?

If the defendant is the source of DNA of sufficient quantity and quality found at a crime scene, then a DNA sample from the defendant and the crime scene sample should have the same profile. The inference required in assessing the evidence, however, runs in the opposite direction. The forensic scientist reports that the sample of DNA from the crime scene and a sample from the defendant have the same genotype. The prosecution's hypothesis is that the defendant is the source of the crime scene sample.[61]

Conceivably, other hypotheses could account for the matching profiles. One possibility is laboratory error—the genotypes are not actually the same even though the laboratory thinks that they are. This situation could arise from mistakes

59. The legal implications of this fact are discussed in Kaye et al., *supra* note 1, § 13.3.2.

60. *See, e.g.*, State v. Murray, 174 P.3d 407, 417–18 (Kan. 2008) (inconclusive Y-STR results were presented as consistent with the defendant's blood). Since the early days of DNA testing, concerns have been expressed about subjective aspects of specific procedures that leave room for "observer effects" in interpreting data. *See* William C. Thompson & Simon Ford, *The Meaning of a Match: Sources of Ambiguity in the Interpretation of DNA Prints*, *in* Forensic DNA Technology (M. Farley & J. Harrington eds., 1990); *see generally* D. Michael Risinger et al., *The* Daubert/Kumho *Implications of Observer Effects in Forensic Science: Hidden Problems of Expectation and Suggestion*, 90 Calif. L. Rev. 1 (2002). A number of commentators have proposed that the analyst determine the profile of a trace evidence sample before knowing the profile of any suspects. Dan E. Krane et al., *Sequential Unmasking: A Means of Minimizing Observer Effects in Forensic DNA Interpretation*, 53 J. Forensic Sci. 1006 (2008).

61. That the defendant is the source does not necessarily mean that the defendant is guilty of the offense charged. Aside from issues of intent or knowledge that have nothing to do with DNA, there remains, for example, the possibility that the two samples match because someone framed the defendant by putting a sample of defendant's DNA at the crime scene or in the container of DNA thought to have come from the crime scene.

in labeling or handling samples or from cross-contamination of the samples. As the 1992 NRC report cautioned, "[e]rrors happen, even in the best laboratories, and even when the analyst is certain that every precaution against error was taken."[62]

Another possibility is that the laboratory analysis is correct—the genotypes are truly identical—but the forensic sample came from another individual. In general, the true source might be a close relative of the defendant[63] or an unrelated person who, as luck would have it, just happens to have the same profile as the defendant. The former hypothesis we shall refer to as kinship, and the latter as coincidence. To infer that the defendant is the source of the crime scene DNA, one must reject these alternative hypotheses of laboratory error, kinship, and coincidence. Table 1 summarizes the logical possibilities.

Table 1. Hypotheses That Might Explain a Match Between Defendant's DNA and DNA at a Crime Scene[a]

IDENTITY:	Same genotype, defendant's DNA at crime scene
NONIDENTITY:	
Lab error	Different genotypes mistakenly found to be the same
Kinship	Same genotype, relative's DNA at crime scene
Coincidence	Same genotype, unrelated individual's DNA

[a] *Cf.* N.E. Morton, *The Forensic DNA Endgame*, 37 Jurimetrics J. 477, 480 tbl. 1 (1997).

Some scientists have urged that probabilities associated with false-positive error, kinship, or coincidence be presented to juries. Although it is not clear that this goal is feasible, scientific knowledge and more conventional evidence can help in assessing the plausibility of these alternative hypotheses. If laboratory error, kinship, and coincidence are rejected as implausible, then only the hypothesis of identity remains. We turn, then, to the considerations that affect the chances of a match when the defendant is not the source of the trace evidence.

C. Can the Match Be Attributed to Laboratory Error?

Although many experts would concede that even with rigorous protocols, the chance of a laboratory error exceeds that of a coincidental match, quantifying the former probability is a formidable task. Some commentary proposes using the proportion of false positives that the particular laboratory has experienced in blind

62. NRC I, *supra* note 8, at 89.

63. A close relative, for these purposes, would be a brother, uncle, nephew, etc. For relationships more distant than second cousins, the probability of a chance match is nearly as small as for persons of the same ethnic subgroup. Bernard Devlin & Kathryn Roeder, *DNA Profiling: Statistics and Population Genetics*, *in* 1 Modern Scientific Evidence: The Law and Science of Expert Testimony § 18-3.1.3, at 724 (David L. Faigman et al. eds., 1997).

proficiency tests or the rate of false positives on proficiency tests averaged across all laboratories.[64] Indeed, the 1992 NRC Report remarks that "proficiency tests provide a measure of the false-positive and false-negative rates of a laboratory."[65] Yet the same report recognizes that "errors on proficiency tests do not necessarily reflect permanent probabilities of false-positive or false-negative results,"[66] and the 1996 NRC report suggests that a probability of a false-positive error that would apply to a specific case cannot be estimated objectively.[67] If the false-positive probability were, say, 0.001, it would take tens of thousands of proficiency tests to estimate that probability accurately, and the application of an historical industry-wide error rate to a particular laboratory at a later time would be debatable.[68]

Most commentators who urge the use of proficiency tests to estimate the probability that a laboratory has erred in a particular case agree that blind proficiency testing cannot be done in sufficient numbers to yield an accurate estimate of a small error rate. However, they maintain that proficiency tests, blind or otherwise, should be used to provide a conservative estimate of the false-positive error probability.[69] For example, if there were no errors in 100 tests, a 95% confidence interval would include the possibility that the error rate could be almost as high as 3%.[70]

Whether or not a case-specific probability of laboratory error can be estimated with proficiency tests, traditional legal and scientific procedures can help to assess the possibilities of errors in handling or analyzing the samples. Scrutinizing the chain of custody, examining the laboratory's protocol, verifying that it adhered to that protocol, and conducting confirmatory tests (including testing by the defense) can help show that the profiles really do match.

D. Could a Close Relative Be the Source?

With enough loci to test, all individuals except identical twins should be distinguishable. With existing technology and small sample sizes of DNA recovered from crime scenes, however, this ideal is not always attainable. A thorough inves-

64. *E.g.*, Jonathan J. Koehler, *Error and Exaggeration in the Presentation of DNA Evidence at Trial*, 34 Jurimetrics J. 21, 37–38 (1993).

65. NRC I, *supra* note 8, at 94.

66. *Id.* at 89.

67. NRC II, *supra* note 9, at 85–87.

68. *Id.* at 85–86; Devlin & Roeder, *supra* note 63, § 18-5.3, at 744–45. Such arguments have not persuaded the proponents of estimating the probability of error from industry-wide proficiency testing. *E.g.*, Jonathan J. Koehler, *Why DNA Likelihood Ratios Should Account for Error (Even When a National Research Council Report Says They Should Not)*, 37 Jurimetrics J. 425 (1997).

69. *E.g.*, Jonathan J. Koehler, *DNA Matches and Statistics: Important Questions, Surprising Answers*, 76 Judicature 222, 228 (1993); Richard Lempert, *After the DNA Wars: Skirmishing with NRC II*, 37 Jurimetrics J. 439, 447–48, 453 (1997).

70. *See* NRC II, *supra* note 9, at 86 n.1. For an explanation of confidence intervals, see David H. Kaye & David A. Freedman, Reference Guide on Statistics, in this manual.

tigation might extend to all known relatives, but this is not feasible in every case, and there is always the chance that some unknown relatives are in the suspect population. Formulas are available for computing the probability that any person with a specified degree of kinship to the defendant also possesses the incriminating genotype. For example, the probability that an untested brother (or sister) would match at four loci (with alleles that each occur in 10% of the population) is about 1/380;[71] the probability that an aunt (or uncle) would match is about 1/100,000.[72]

E. Could an Unrelated Person Be the Source?

Another rival hypothesis is coincidence: The defendant is not the source of the crime scene DNA but happens to have the same genotype as an unrelated individual who is the true source. Various procedures for assessing the plausibility of this hypothesis are available. In principle, one could test all conceivable suspects. If everyone except the defendant has a nonmatching profile, then the defendant must be the source. But exhaustive, error-free testing of the population of conceivable suspects is almost never feasible. The suspect population normally defies any enumeration, and in the typical crime where DNA evidence is found, the population of possible perpetrators is so huge that even if all of its members could be listed, they could not all be tested.[73]

An alternative procedure would be to take a sample of people from the suspect population, find the relative frequency of the profile in this sample, and use that statistic to estimate the frequency in the entire suspect population. The smaller the frequency, the less likely it is that the defendant's DNA would match if the defendant were not the source of trace evidence. Again, however, the suspect population is difficult to define, so some surrogate must be used. The procedure commonly followed is to estimate the relative frequency of the incriminating

71. For a case with conflicting calculations of the probability of an untested brother having a matching genotype, see *McDaniel v. Brown*, 130 S. Ct. 665 (2010) (per curiam). The correct computation is given in David H. Kaye, *"False, but Highly Persuasive": How Wrong Were the Probability Estimates in* McDaniel v. Brown*?* 108 Mich. L. Rev. First Impressions 1 (2009), *available at* http://www.michiganlawreview.org/assets/fi/108/kaye.pdf.

72. These figures follow from the equations in NRC II, *supra* note 9, at 113. The large discrepancy between two siblings on the one hand, and an uncle and nephew on the other, reflects the fact that the siblings have far more shared ancestry. All their genes are inherited through the same two parents. In contrast, a nephew and an uncle inherit from two unrelated mothers, and so will have few maternal alleles in common. As for paternal alleles, the nephew inherits not from his uncle, but from his uncle's brother, who shares by descent only about one-half of his alleles with the uncle.

73. As the cost of DNA profiling drops, it will become technically and economically feasible to have a comprehensive, population-wide DNA database that could be used to produce a list of nearly everyone whose DNA profile is consistent with the trace evidence DNA. Whether such a system would be constitutionally and politically acceptable is another question. *See* David H. Kaye & Michael S. Smith, *DNA Identification Databases: Legality, Legitimacy, and the Case for Population-Wide Coverage*, 2003 Wis. L. Rev. 413.

genotype in a large population. But even this cannot be done directly because each possible multilocus profile is so rare that it is not likely to show up in any sample of a reasonable size. However, the frequencies of most alleles can be determined accurately by sampling the population to construct databases that reveal how often each allele occurs. Principles of population genetics then can be applied to combine the estimated allele frequencies into an estimate of the probability that a person born in the population will have the multilocus genotype. This probability often is referred to as the random-match probability. This section describes how the allele frequencies are estimated from samples and how the random-match probability is computed from allele frequencies.

1. Estimating allele frequencies from samples

As we saw in Section II.B, the loci currently used in forensic testing have been chosen partly because their alleles tend to be different in different people. For example, 2% of the population might have the alleles with 7 and 10 repeats at a particular STR locus; 1% might have the combination of 5 and 6; and so on. If we take a DNA molecule's view of the population, human beings are containers for DNA and machines for copying and propagating them to the next generation of human beings. The different DNA molecules are swimming, so to speak, in a huge pool of humanity. All the possible alleles (the fives, sixes, sevens, and so on) form a large population, or pool, of alleles. Each allele constitutes a certain proportion of allele pool. Suppose, then, that a five-repeat allele represents 12% of all of the allele pool, a six-repeat allele contributes 20%, and so on, for all the alleles at a locus.

The first step in computing a random-match probability is to estimate these allele frequencies. Ideally, a probability sample from the human population of interest would be taken.[74] We would start with a list of everyone who might have left the trace evidence, take a random sample of these people, and count the numbers of alleles of each length that are present in the sample. Unfortunately, a list of the people who comprise the entire population of possible suspects is almost never available; consequently, probability sampling from the directly relevant population is impossible. Probability sampling from a comparable population (with regard to the individuals' DNA) is possible, but it is not the norm in studies of the distributions of genes in populations. Typically, convenience samples (from blood banks or paternity cases) are used.[75] Rela-

74. Probability sampling is described in Kaye & Freedman, *supra* note 2, and Shari Seidman Diamond, Reference Guide on Survey Research, in this manual.

75. A few experts have testified that no meaningful conclusions can be drawn in the absence of random sampling. *E.g.*, People v. Soto, 88 Cal. Rptr. 2d 34 (1999); State v. Anderson, 881 P.2d 29, 39 (N.M. 1994). The 1996 NRC report suggests that for the purpose of estimating allele frequencies, convenience sampling should give results comparable to random sampling, and it discusses procedures for estimating the random sampling error. NRC II, *supra* note 9, at 126–27, 146–48, 186. The courts

tively small samples can produce fairly accurate estimates of individual allele frequencies.[76]

Once the allele frequencies have been estimated, the next step in arriving at a random-match probability is to combine them. This requires some knowledge of how DNA is copied and recombined in the course of sexual reproduction and how human beings choose their mates.

2. *The product rule for a randomly mating population*

All scientists use simplified models of a complex reality. Physicists solve equations of motion in the absence of friction. Economists model exchanges among rational agents who bargain freely with no transaction costs. Population geneticists compute genotype frequencies in an infinite population of individuals who choose their mates independently of their alleles at the loci in question. Although geneticists describe this situation as random mating, geneticists know that people do not choose their mates by a lottery. "Random mating" simply indicates that the choices are uncorrelated with the specific alleles that make up the genotypes in question.

In a randomly mating population, the expected frequency of a pair of alleles at any single locus depends on whether the two alleles are distinct. If the offspring happens to inherit the same allele from each parent, the expected single-locus genotype frequency is the square of the allele frequency (p^2). If a different allele is inherited from each parent, the expected single-locus genotype frequency is twice the product of the two individual allele frequencies (often written as $2p_1p_2$).[77] These proportions are known as Hardy-Weinberg proportions. Even if two populations with distinct allele frequencies are thrown together, within the limits of chance variation, random mating produces Hardy-Weinberg equilibrium in a single generation.

generally have rejected the argument that random samples are essential to valid or generally accepted random-match probabilities. *See* D.H. Kaye, *Bible Reading: DNA Evidence in Arizona*, 28 Ariz. St. L.J. 1035 (1996).

76. In the formative years of forensic DNA testing, defendants frequently contended that forensic databases were too small to give accurate estimates, but this argument generally proved unpersuasive. *E.g.*, United States v. Shea, 957 F. Supp. 331, 341–43 (D.N.H. 1997); State v. Dishon, 687 A.2d 1074, 1090 (N.J. Super. Ct. App. Div. 1997); State v. Copeland, 922 P.2d 1304, 1321 (Wash. 1996). To the extent that the databases are comparable to random samples, confidence intervals are a standard method for indicating the uncertainty resulting from sample size. Unfortunately, the meaning of a confidence interval is subtle, and the estimate commonly is misconstrued. *See* Kaye & Freedman, *supra* note 2.

77. Suppose that 10% of the sperm in the gene pool of the population carry allele 1 (A1), and 50% carry allele 2 (A2). Similarly, 10% of the eggs carry A1, and 50% carry A2. (Other sperm and eggs carry other types.) With random mating, we expect 10% × 10% = 1% of all the fertilized eggs to be A1A1, and another 50% × 50% = 25% to be A2A2. These constitute two distinct homozygote profiles. Likewise, we expect 10% × 50% = 5% of the fertilized eggs to be A1A2 and another 50% × 10% = 5% to be A2A1. These two configurations produce indistinguishable profiles—a peak, band, or dot for A1 and another mark for A2. So the expected proportion of heterozygotes A1A2 is 5% + 5% = 10%.

Once the proportion of the population that has each of the single-locus genotypes for the forensic profile has been estimated, the proportion of the population that is expected to share the combination of them—the multilocus profile frequency—is given by multiplying all the single-locus proportions. This multiplication is exactly correct when the single-locus genotypes are statistically independent. In that case, the population is said to be in linkage equilibrium.

Early estimates of DNA genotype frequencies assumed that alleles were inherited independently within and across loci (Hardy-Weinberg and linkage equilibrium, respectively). Because the frequencies of the VNTR loci then in use were shown to vary across census groups (whites, blacks, Hispanics, Asians, and Native Americans), it became common to present the estimated genotype frequencies within each of these groups (in cases in which the "race" of the source of the trace evidence was unknown) or only in a particular census group (if the "race" of the source was known).[78]

3. *The product rule for a structured population*

Population geneticists understood that the equilibrium frequencies were only approximations and that the major racial populations are composed of ethnic subpopulations whose members tend to mate among themselves. Within each ethnic subpopulation, mating still can be random, but if, say, Italian Americans have allele frequencies that are markedly different than the average for all whites, and if Italian Americans only mate among themselves, then using the average frequencies for all whites in the basic product formula could understate—or overstate—a multilocus profile frequency for the subpopulation of Italian Americans. Similarly, using the population frequencies could understate—or overstate—the profile frequencies in the white population itself.

Consequently, if we want to know the frequency of an incriminating profile among Italian Americans, the basic product rule applied to the allele frequencies for whites in general could be in error; and there is even some chance that the rule will understate the profile frequency in the white population as a whole. Experts have disagreed, however, as to whether the major population groups are so severely structured that the departures from equilibrium would be substantial. Courts applying the *Daubert* and *Frye* rules for scientific evidence issued conflicting opinions as to the admissibility of basic product-rule estimates.[79] A 1992 report from a committee of the National Academy of Sciences did not resolve the question, but a second committee concluded in 1996 that the basic product rule provided reasonable estimates in most cases, and it described a modified version of the product rule

78. The use of a range of estimates conditioned on race is defended, and several alternatives are discussed in Kaye, *supra* note 3, at 192–97; David H. Kaye, *The Role of Race in DNA Evidence: What Experts Say, What California Courts Allow*, 37 Sw. U. L. Rev. 303 (2008).

79. These legal and scientific developments are chronicled in detail in Kaye, *supra* note 3.

to account for population structure.[80] By the mid-1990s, the population-structure objection to admitting random-match probabilities had lost its power.[81]

F. Probabilities, Probative Value, and Prejudice

Up to this point, we have described the random-match probabilities that commonly are presented in conjunction with the finding that the trace evidence sample contains DNA of the same type as the defendant's. We have concentrated on the methods used to compute the probabilities. Assuming that these methods meet *Daubert*'s demand for scientific validity and reliability (or, in many states, *Frye*'s requirement of general acceptance in the scientific community) and thus satisfy Federal Rule of Evidence 702, a further issue can arise under Rule 403: To what extent will the presentation assist the jury in understanding the meaning of a match so that the jury can give the evidence the weight that it deserves? This question involves psychology and law, and we summarize the arguments about probative value and prejudice that have been made in litigation and in the legal and scientific literature. We take no position on how the legal issue of the admissibility of any particular statistic generally should be resolved under the balancing standard of Rule 403. The answer may turn not only on the general features of the evidence described here, but on the context and circumstances of particular cases.

1. Frequencies and match probabilities

a. Argument: Frequencies or probabilities are prejudicial because they are so small

The most common form of expert testimony about matching DNA involves an explanation of how the laboratory ascertained that the defendant's DNA has the profile of the forensic sample plus an estimate of the profile frequency or random-match probability. It has been suggested, however, that jurors do not understand probabilities in general, and that infinitesimal match probabilities will so bedazzle jurors that they will not appreciate the other evidence in the case or any innocent explanations for the match.[82] Empirical research into this hypothesis has been limited,[83] and commentators have noted that remedies short of exclusion

80. The 1996 committee's recommendations for computing random-match probabilities with broad populations and particular subpopulations are summarized in the previous edition of this guide. The 1992 committee had proposed a more conservative (and less elegant) method of dealing with variations across subpopulations (the "ceiling principle"), also described in the previous edition.

81. *See, e.g.*, Kaye, *supra* note 3.

82. *Cf.* Gov't of the Virgin Islands v. Byers, 941 F. Supp. 513, 527 (D.V.I. 1996) ("Vanishingly small probabilities of a random match may tend to establish guilt in the minds of jurors and are particularly suspect.").

83. This research is tabulated in David H. Kaye et al., *Statistics in the Jury Box: Do Jurors Understand Mitochondrial DNA Match Probabilities?* 4 J. Empirical Legal Stud. 797 (2007). The findings do

are available.[84] Thus, although there once was a line of cases that excluded probability testimony in criminal matters, by the mid-1990s, no jurisdiction excluded DNA match probabilities on this basis.[85] The opposite argument—that relatively large random-match probabilities are prejudicial—also has been advanced without success.[86]

b. Argument: Frequencies or probabilities are prejudicial because they might be transposed

A related concern is that the jury will misconstrue the random-match probability as the probability that the evidence DNA came from a random individual.[87] The words are almost identical, but the probabilities can be quite different. The random-match probability is the probability that the suspect has the DNA genotype of the crime scene sample *if he is not the true source of that sample* (and is unrelated to the true source). The tendency to invert or transpose the probability—to go from a one-in-a-million chance *if the suspect is not the source* to a million-to-one chance that *the suspect is the source* is known as the fallacy of the transposed conditional.[88] To appreciate that the transposition is fallacious, consider the probability

not clearly support the argument that jurors will overweight the probability, but the details of how the probability is presented and countered may be important.

84. According to the 1996 NRC committee, suitable cross-examination, defense experts, and jury instructions might reduce the risk that small estimates of the match probability will produce an unwarranted sense of certainty and lead a jury to disregard other evidence. NRC II, *supra* note 9, at 197.

85. *E.g.*, United States v. Chischilly, 30 F.3d 1144 (9th Cir. 1994) (citing cases); State v. Weeks, 891 P.2d 477, 489 (Mont. 1995) (rejecting the argument that "the exaggerated opinion of the accuracy of DNA testing is prejudicial, as juries would give undue weight and deference to the statistical evidence" and "that the probability aspect of the DNA analysis invades the province of the jury to decide the guilt or innocence of the defendant").

86. *See* United States v. Morrow, 374 F. Supp. 2d 51, 65 (D.D.C. 2005) (rejecting the argument because "the DNA evidence remains probative, and helps to corroborate other evidence and support the Government's case as to the identity of the relevant perpetrators. Indeed, the low statistical significance actually benefits Defendants, as Defendants can argue that having random match probabilities running between 1:12 and 1:1 means that hundreds, if not thousands, of others in the Washington, D.C. area cannot be excluded as possible contributors as well.").

87. Numerous opinions or experts present the random-match probability in this manner. *E.g.*, State v. Davolt, 84 P.3d 456, 475 (Ariz. 2004) (stating that "the chance the saliva found on cigarette remains in the house did not belong to [the defendant] was one in 280 quadrillion for the Caucasian population"); Kaye et al., *supra* note 1, § 14.1.2(a) (collecting opinions reflecting this fallacy).

88. The transposition fallacy also is called the "prosecutor's fallacy" in the legal literature—despite the fact that it hardly is limited to prosecutors. Our description of the fallacy is imprecise. In this context, the random-match probability is the chance that (A) the suspect has the crime scene genotype given that (B) he is not the true source. The probability that the match is random is the probability that (B) the individual tested has been selected at random given that (A) the individual has the requisite genotype. In general, for two events A and B, the probability of A given B, which we can write as *P*(A given B), does not equal *P*(B given A). *See* Kaye & Freedman, *supra* note 2. The claim that the probabilities are necessarily equal is the transposition fallacy. *Id.* (also noting instances of the fallacy in other types of litigation).

that a lawyer picked at random from all lawyers in the United States is an appellate judge. This "random-judge probability" is practically zero. But the probability that a person randomly selected from the current appellate judiciary is a lawyer is one. The random-judge probability, *P*(judge given lawyer), does not equal the transposed probability *P*(lawyer given judge). Likewise, the random-match probability *P*(genotype given unrelated source) does not necessarily equal *P*(unrelated source given genotype).[89]

No federal court has excluded a random-match probability (or, for that matter, an estimate of the small frequency of a DNA profile in the general population) as unfairly prejudicial simply because the jury might misinterpret it as a probability that the defendant is the source of the forensic DNA.[90] Courts, however, have noted the need to have the concept "properly explained,"[91] and prosecutorial or expert misrepresentations of the random-match probabilities for

89. To avoid this fallacious reasoning by jurors, some scientific and legal commentators have urged the exclusion of random-match probabilities. In response, the 1996 NRC committee suggested that "if the initial presentation of the probability figure, cross-examination, and opposing testimony all fail to clarify the point, the judge can counter [the fallacy] by appropriate instructions to the jurors that minimize the possibility of cognitive errors." NRC II, *supra* note 9, at 198 (footnote omitted). The committee suggested the following instruction to define the random-match probability:

> In evaluating the expert testimony on the DNA evidence, you were presented with a number indicating the probability that another individual drawn at random from the [specify] population would coincidentally have the same DNA profile as the [bloodstain, semen stain, etc.]. That number, which assumes that no sample mishandling or laboratory error occurred, indicates how distinctive the DNA profile is. It does not by itself tell you the probability that the defendant is innocent.

Id. at 198 n.93. An alternative adopted in England is to confine the prosecution to stating a frequency rather than a probability. *See* Kaye et al., *supra* note 1, § 14.1.2(b); *cf.* D.H. Kaye, *The Admissibility of "Probability Evidence" in Criminal Trials—Part II*, 27 Jurimetrics J. 160, 168 (1987) (similar proposal).

The NRC committee also noted the opposing "defendant's fallacy" of dismissing or undervaluing the matches with high likelihood ratios because other matches are to be expected in unrealistically large populations of potential suspects. For example, defense counsel might argue that (1) with a random-match probability of one in a million, we would expect to find three or four unrelated people with the requisite genotypes in a major metropolitan area with a population of 3.6 million; (2) the defendant just happens to be one of these three or four, which means that the chances are at least 2 out of 3 that someone unrelated to the defendant is the source; so (3) the DNA evidence does nothing to incriminate the defendant. The problem with this argument is that in a case involving both DNA and non-DNA evidence against the defendant, it is unrealistic to assume that there are 3.6 million equally likely suspects. When juries are confronted with both fallacies, the defendant's fallacy seems to dominate. NRC II, *supra* note 9, at 198; *cf.* Jonathan J. Koehler, *The Psychology of Numbers in the Courtroom: How to Make DNA-Match Statistics Seem Impressive or Insufficient*, 74 S. Cal. L. Rev. 1275 (2001) (discussing ways of framing the evidence that make it more or less persuasive).

90. *See, e.g.,* United States v. Morrow, 374 F. Supp. 2d 51, 66 (D.D.C. 2005) ("careful oversight by the district court and proper explanation can easily thwart this issue").

91. United States v. Shea, 957 F. Supp. 331, 345 (D.N.H. 1997); *see also* United States v. Chischilly, 30 F.3d 1144, 1158 (9th Cir. 1994) (stating that the government must be "careful to frame the DNA profiling statistics presented at trial as the probability of a random match, not the probability of the defendant's innocence that is the crux of the prosecutor's fallacy").

DNA and other trace evidence have produced reversals or contributed to the setting aside of verdicts.[92]

c. Argument: Random-match probabilities that are smaller than false-positive error probabilities are irrelevant or prejudicial

Some scientists and lawyers have maintained that match probabilities are logically irrelevant when they are far smaller than the probability of a frameup, a blunder in labeling samples, cross-contamination, or other events that would yield a false positive.[93] The argument is that the jury should concern itself only with the chance that the forensic sample is reported to match the defendant's profile even though the defendant is not the source. Match probabilities do not express this chance unless the probability of a false-positive report (because of fraud or an error in the collection, handling, or analysis of the DNA samples) is essentially zero. The mathematical observation has led to the argument that because these other possible explanations for a match are more probable than the very small random-match probabilities for most STR profiles, the latter probabilities are irrelevant. Commentators have crafted theoretical, doctrinal, and practical rejoinders to this claim.[94] The essence of the counterargument is that it is logical to give jurors information about kinship or random-match probabilities because, even if these numbers do not give the whole picture, they address pertinent hypotheses about the true source of the trace evidence.

It also has been argued that even if very small match probabilities are logically relevant, they are unfairly prejudicial in that they will cause jurors to neglect the probability of a match arising due to a false-positive laboratory error.[95] A court

92. *E.g.*, United States v. Massey, 594 F.2d 676, 681 (8th Cir. 1979) (explaining that in closing argument about hair evidence, "the prosecutor 'confuse[d] the probability of concurrence of the identifying marks with the probability of mistaken identification'") (alteration in original). The Supreme Court noted the transposition fallacy in the prosecution's presentation of DNA evidence as a basis for a federal writ of habeas corpus in *McDaniel v. Brown*, 130 S. Ct. 665 (2010) (per curiam). The Court unanimously held that the prisoner had not properly raised the issue of whether this error amounted to a violation of due process. For comments on that issue, *see* Kaye, *supra* note 71.

93. *E.g.*, Jonathan J. Koehler et al., *The Random Match Probability in DNA Evidence: Irrelevant and Prejudicial?* 35 Jurimetrics J. 201 (1995); Richard C. Lewontin & Daniel L. Hartl, *Population Genetics in Forensic DNA Typing*, 254 Science 1745, 1749 (1991) ("[p]robability estimates like 1 in 738,000,000,000,000 . . . are terribly misleading because the rate of laboratory error is not taken into account").

94. *See* Kaye et al., *supra* note 1, § 14.1.1 (discussing the issue).

95. Some commentators believe that this prejudice is so likely and so serious that "jurors ordinarily should receive *only* the laboratory's false positive rate. . . ." Richard Lempert, *Some Caveats Concerning DNA as Criminal Identification Evidence: With Thanks to the Reverend Bayes*, 13 Cardozo L. Rev. 303, 325 (1991) (emphasis added). The 1996 NRC committee was skeptical of this view, especially when the defendant has had a meaningful opportunity to retest the DNA at a laboratory of his or her choice, and it suggested that judicial instructions can be crafted to avoid this form of prejudice. NRC II, *supra* note 9, at 199. Pertinent psychological research includes Dale A. Nance & Scott B. Morris, *Juror Understanding of DNA Evidence: An Empirical Assessment of Presentation Formats for Trace*

that shares this concern might require the expert who presents a random-match probability also to report a probability that the laboratory is mistaken about the profiles. Of course, for reasons given in Section III.B.2, some experts would deny that they can provide a meaningful statistic for the case at hand, but it has been pointed out that they could report the results of proficiency tests and leave it to the jury to use this figure as best it can in considering whether a false-positive error has occurred.[96] In any event, the courts have been unreceptive to efforts to replace random-match probabilities with a blended figure that incorporates the risk of a false-positive error[97] or to exclude random-match probabilities that are not accompanied by a separate false-positive error probability.[98]

Evidence with a Relatively Small Random Match Probability, 34 J. Legal Stud. 395 (2005); Dale A. Nance & Scott B. Morris, *An Empirical Assessment of Presentation Formats for Trace Evidence with a Relatively Large and Quantifiable Random Match Probability*, 42 Jurimetrics J. 1 (2002); Jason Schklar & Shari Seidman Diamond, *Juror Reactions to DNA Evidence: Errors and Expectancies*, 23 Law & Hum. Behav. 159, 179 (1999) (concluding that separate figures for laboratory error and a random match to a correctly ascertained profile are desirable in that "[j]urors . . . may need to know the disaggregated elements that influence the aggregated estimate as well as how they were combined in order to evaluate the DNA test results in the context of their background beliefs and the other evidence introduced at trial").

96. *Cf.* Williams v. State, 679 A.2d 1106, 1120 (Md. 1996) (reversing because the trial court restricted cross-examination about the results of proficiency tests involving other DNA analysts at the same laboratory). *But see* United States v. Shea, 957 F. Supp. 331, 344 n.42 (D.N.H. 1997) ("The parties assume that error rate information is admissible at trial. This assumption may well be incorrect. Even though a laboratory or industry error rate may be logically relevant, a strong argument can be made that such evidence is barred by Fed. R. Evid. 404 because it is inadmissible propensity evidence.").

97. United States v. Ewell, 252 F. Supp. 2d 104, 113–14 (D.N.J. 2003) (stating that exclusion of the random-match probability is not justified when "the defendant's argument is not based on evidence of actual errors by the laboratory, but instead has simply challenged the Government's failure to quantify the rate of laboratory error," while "the Government has demonstrated the scientific method has a virtually zero rate of error, and that it employs sufficient procedures and controls to limit laboratory error," and the defendant had an expert who could testify to the probability of error); United States v. Shea, 957 F. Supp. 331, 334–45 (D.N.H. 1997) (holding that separate figures for match and error probabilities are not prejudicial); People v. Reeves, 109 Cal. Rptr. 2d 728, 753 (Ct. App. 2001) (holding that probability of laboratory error need not be combined with random-match probability); Armstead v. State, 673 A.2d 221, 245 (Md. 1996) (finding that the failure to combine a random-match probability with an error rate on proficiency tests that was many orders of magnitude greater (and that was placed before the jury) did not deprive the defendant of due process); State v. Tester, 968 A.2d 895 (Vt. 2009).

98. United States v. Trala, 162 F. Supp. 2d 336, 350–51 (D. Del. 2001) (stating that presenting a nonzero laboratory error rate is not a condition of admissibility, and *Daubert* does not require separate figures for match and error probabilities to be combined); United States v. Lowe, 954 F. Supp. 401, 415–16 (D. Mass. 1997), *aff'd*, 145 F.3d 45 (1st Cir. 1998) (finding that a "theoretical" error rate need not be presented when quality assurance standards have been followed and defendant had the opportunity to retest the sample); Roberts v. United States, 916 A.2d 922, 930–31 (D.C. 2007) (finding that presenting a laboratory error rate is not a condition of admissibility); Roberson v. State, 16 S.W.3d 156, 168 (Tex. Crim. App. 2000) (finding that error rate not needed when laboratory was accredited and underwent blind proficiency testing); *Tester*, 968 A.2d 895 (stating that when the laboratory chemist stated that "[t]here is no error rate to report" because the number of proficiency

2. Likelihood ratios

Sufficiently small probabilities of a match for close relatives and unrelated members of the suspect population undermine the hypotheses of kinship and coincidence. Adequate safeguards and checks for possible laboratory error make that explanation of the finding of matching genotypes implausible. The inference that the defendant is the source of the crime scene DNA is then secure. But this mode of reasoning by elimination is not the only way to analyze DNA evidence. This subsection and the next describe alternatives—likelihoods and posterior probabilities—that some statisticians prefer and that have been used in a growing number of court cases.

To choose between two competing hypotheses, one can compare how probable the evidence is under each hypothesis. Suppose that the probability of a match in a well-run laboratory is close to 1 when the samples both contain only the defendant's DNA, while both the probability of a coincidental match and the probability of a match to a close relative are close to 0. In these circumstances, the DNA profiling strongly supports the claim that the defendant is the source, because the observed outcome—the match—is many times more probable when the defendant is the source than when someone else is. How many times more probable? Suppose that there is a 1% chance that the laboratory would miss a true match, so that the probability of its finding a match when the defendant is the source is 0.99. Suppose further that $p = 0.00001$ is the random-match probability. Then the match is 0.99/0.00001, or 99,000 times more likely to be seen if the defendant is the source than if an unrelated individual is. Such a ratio is called a likelihood ratio, and a likelihood ratio of 99,000 means that the DNA profiling supports the claim of identity 99,000 times more strongly than it supports the hypothesis of coincidence.[99]

Likelihood ratios have been presented in court in many cases. They are routinely introduced under the name "paternity index" in civil and criminal cases that involve DNA testing for paternity.[100] Experts also have used them in cases in which the issue is whether two samples originated from the same individual. For example, in one California case, an expert stated that "for the Caucasian population, the evidence DNA profile was approximately 1.9 trillion times more likely to match appellant's DNA profile if he was the contributor of that DNA rather than some unknown, unrelated individual; for the Hispanic population, it was 2.6 trillion times more likely; and for the African-American population, it was about 9.1 trillion times more likely."[101] And, as explained below (Section V.C), likeli-

trials was insufficient, the random-match probability was admissible and preferable to presenting the finding of a match with no accompanying statistic).

99. Another likelihood ratio would give the relative likelihood of the hypotheses of identity and a falsely declared match arising from an error in the laboratory. *See supra* Section IV.F.1.

100. *See* Kaye, *supra* note 3; 1 McCormick on Evidence, *supra* note 3, § 211.

101. People v. Prince, 36 Cal. Rptr. 3d 300, 310 (Ct. App. 2005), *review denied*, 142 P.3d 1184 (Cal. 2006).

hood ratios are especially useful for samples that are mixtures of DNA from several people.

The major objection to likelihoods is not statistical, but psychological.[102] As with random-match probabilities, they are easily transposed.[103] With random-match probabilities, we saw that courts have reasoned that the possibility of transposition does not justify a blanket rule of exclusion. The same issue has not been addressed directly for likelihood ratios.

3. Posterior probabilities

The likelihood ratio expresses the relative strength of two hypotheses, but the judge or jury ultimately must assess a different type of quantity—the probability of the hypotheses themselves. An elementary rule of probability theory known as Bayes' theorem yields this probability. The theorem states that the odds in light of the data (here, the observed profiles) are the odds as they were known prior to receiving the data times the likelihood ratio. More succinctly, posterior odds = likelihood ratio × prior odds.[104] For example, if the relevant match probability[105] were 1/100,000, and if the chance that the laboratory would report a match between samples from the same source were 0.99, then the likelihood ratio would be 99,000, and the jury could be told how the DNA evidence raises various prior probabilities that the defendant's DNA is in the evidence sample.[106]

102. For legal commentary and additional cases upholding the admission of likelihood ratios over objections based on *Frye* and *Daubert*, see Kaye et al., *supra* note 1, § 14.2.2.

103. United States v. Thomas, 43 M.J. 626 (A.F. Ct. Crim. App. 1995), provides an example. In this murder case, a military court described testimony from a population geneticist that "conservatively, it was 76.5 times more likely that the samples . . . came from the victim than from someone else in the Filipino population." *Id.* at 635. Yet, this is not what the DNA testing showed. A more defensible statement is that "the match between the bloodstains was 76.5 times more probable if the stains came from the victim than from an unrelated Filipino" or "the match supports the hypothesis that the stains came from the victim 76.5 times more than it supports the hypothesis that they came from an unrelated Filipino woman." Kaye et al., *supra* note 7, § 14.2.2.

104. Odds and probabilities are two ways to express chances quantitatively. If the probability of an event is P, the odds are $P/(1 - P)$. If the odds are O, the probability is $O/(O + 1)$. For instance, if the probability of rain is 2/3, the odds of rain are 2 to 1 because $(2/3)/(1 - 2/3) = (2/3)/(1/3) = 2$. If the odds of rain are 2 to 1, then the probability is $2/(2 + 1) = 2/3$.

105. By "relevant match probability," we mean the probability of a match given a specified type of kinship or the probability of a random match in the relevant suspect population. For relatives more distantly related than second cousins, the probability of a chance match is nearly as small as for persons of the same subpopulation. Devlin & Roeder, *supra* note 63, § 18-3.1.3, at 724.

106. If this procedure is followed, the analyst could explain that these calculations rest on many premises, including the premise that the genotypes have been correctly determined. *See, e.g.*, Richard Lempert, *The Honest Scientist's Guide to DNA Evidence*, 96 Genetica 119 (1995). If the jury accepted these premises and also decided to accept the hypothesis of identity over those of kinship and coincidence, it still would be open to the defendant to offer explanations of how the forensic samples came to include his or her DNA even though he or she is innocent.

One difficulty with this use of Bayes' theorem is that the computations consider only one alternative to the claim of identity at a time. As indicated earlier, however, several rival hypotheses might apply in a given case. If the DNA in the crime scene sample is not the defendant's, is it from his father, his brother, his uncle, or another relative? Is the true source a member of the same subpopulation? Or is the source a member of a different subpopulation in the same general population? In principle, the likelihood ratio can be generalized to a likelihood function that takes on suitable values for every person in the world, and the prior probability for each person can be cranked into a general version of Bayes' rule to yield the posterior probability that the defendant is the source. In this vein, some commentators suggest that Bayes' rule be used to combine the various likelihood ratios for all possible degrees of kinship and subpopulations.[107]

As with likelihood ratios, Bayes' rule is routine in cases involving parentage testing. Some courts have held that the "probability of paternity" derived from the formula is inadmissible in criminal cases, but most have reached the opposite conclusion, at least when the prior odds used in the calculation are disclosed to the jury.[108] An extended literature has grown up on the subject of how posterior probabilities might be useful in criminal cases.[109]

G. Verbal Expressions of Probative Value

Having surveyed the issues related to the value and dangers of probabilities and statistics for DNA evidence, we turn to a related issue that can arise under Rules 702 and 403: Should an expert be permitted to offer a nonnumerical judgment about the DNA profiles? Many courts have held that a DNA match is inadmissible unless the expert attaches a scientifically valid number to the match. Indeed, some opinions state that this requirement flows from the nature of science itself. However, this view has been challenged,[110] and not all courts agree that an expert must explain the power of a DNA match in purely numerical terms.

107. David J. Balding, Weight-of-Evidence for Forensic DNA Profiles (2005); David J. Balding & Peter Donnelly, *Inference in Forensic Identification*, 158 J. Royal Stat. Soc'y Ser. A 21 (1995); *cf.* Lempert, *supra* note 69, at 458 (describing a similar procedure).

108. Kaye et al., *supra* note 1, § 14.3.2.

109. *See id.*; 1 McCormick on Evidence, *supra* note 3, § 211; David H. Kaye, *Rounding Up the Usual Suspects: A Legal and Logical Analysis of DNA Database Trawls*, 87 N.C. L. Rev. 425 (2009) (defending a Bayesian presentation by a defendant identified by a "cold hit" in a DNA database).

110. *See, e.g.*, Commonwealth v. Crews, 640 A.2d 395, 402 (Pa. 1994) (explaining that "[t]he factual evidence of the physical testing of the DNA samples and the matching alleles, even without statistical conclusions, tended to make appellant's presence more likely than it would have been without the evidence, and was therefore relevant."). The 1996 NRC committee wrote that science only demands "underlying data that permit some reasonable estimate of how rare the matching characteristics actually are," and "[o]nce science has established that a methodology has some individualizing power, the legal system must determine whether and how best to import that technology into the trial process." NRC II, *supra* note 9, at 192.

1. "Rarity" or "strength" testimony

Instead of presenting numerical frequencies or match probabilities, a scientist could characterize a 13-locus STR profile as "rare," "extremely rare," or the like. Instead of quoting a numerical likelihood ratio, the analyst could refer to the match as "powerful," "very strong evidence," and so on. At least one state supreme court has endorsed this qualitative approach as a substitute to the presentation of quantitative estimates.[111]

2. Source or uniqueness testimony

The most extreme case of a purely verbal description of the infrequency of a profile occurs when that profile can be said to be unique. Of course, the uniqueness of any object, from a snowflake to a fingerprint, in a population that cannot be enumerated never can be proved directly. As with all sample evidence, one must generalize from the sample to the entire population. There is always some probability that a census would prove the generalization to be false. Over a decade ago, the second NRC committee therefore wrote that "[t]here is no 'bright-line' standard in law or science that can pick out exactly how small the probability of the existence of a given profile in more than one member of a population must be before assertions of uniqueness are justified. . . . There might already be cases in which it is defensible for an expert to assert that, assuming that there has been no sample mishandling or laboratory error, the profile's probable uniqueness means that the two DNA samples come from the same person."[112] Before concluding that a DNA profile is unique in a given population, however, a careful expert also should consider not only the random-match probability (which pertains to unrelated individuals) but also the chance of a match to a close relative. Indeed, the possible existence of an unknown, identical twin also means that a scientist never can be absolutely certain that crime scene evidence could have come from only the defendant.

Courts have accepted or approved of expert assertions of uniqueness or of individual source identification.[113] For these assertions to be justified, a large

111. State v. Bloom, 516 N.W.2d 159, 166–67 (Minn. 1994) ("Since it may be pointless to expect ever to reach a consensus on how to estimate, with any degree of precision, the probability of a random match and that, given the great difficulty in educating the jury as to precisely what that figure means and does not mean, it might make sense to simply try to arrive at a fair way of explaining the significance of the match in a verbal, qualitative, nonquantitative, nonstatistical way."). A related question is whether an expert should be allowed to declare a match without adding any information on how common or rare the profile is. For discussion of such pure "defendant-not-excluded" testimony, see United States v. Morrow, 374 F. Supp. 2d 51 (D.D.C. 2005); Kaye et al., *supra* note 1, § 15.4.

112. NRC II, *supra* note 9, at 194.

113. *E.g.*, United States v. Davis, 602 F. Supp. 2d 658 (D. Md. 2009) ("the random match probability figures . . . are sufficiently low so that the profile can be considered unique"); People v. Baylor, 118 Cal. Rptr. 2d 518, 522 (Ct. App. 2002) (testimony that "defendant had a unique DNA

number of sufficiently polymorphic loci must have been tested, making the probabilities of matches to both relatives and unrelated individuals so tiny that the probability of finding another person who could be the source within the relevant population is negligible.[114]

V. Special Issues in Human DNA Testing

A. Mitochondrial DNA

Mitochondria are small structures, with their own membranes, found inside the cell but outside its nucleus. Inside these organelles, molecules are broken down to supply energy. Mitochondria have a small genome—a circle of 16,569 nucleotide base pairs within the mitochondrion—that bears no relation to the comparatively monstrous chromosomal genome in the cell nucleus.[115]

Mitochondrial DNA (mtDNA) has four features that make it useful for forensic DNA testing. First, the typical cell, which has but one nucleus, contains hundreds or thousands of nearly identical mitochondria. Hence, for every copy of

profile that 'probably does not exist in anyone else in the world.'"); State v. Hauge, 79 P.3d 131 (Haw. 2003) (uniqueness); Young v. State, 879 A.2d 44, 46 (Md. 2005) (holding that "when a DNA method analyzes genetic markers at sufficient locations to arrive at an infinitesimal random match probability, expert opinion testimony of a match and of the source of the DNA evidence is admissible"; hence, it was permissible to introduce a report providing no statistics but stating that "(in the absence of an identical twin), Anthony Young (K1) is the source of the DNA obtained from the sperm fraction of the Anal Swab (R1)."); State v. Buckner, 941 P.2d 667, 668 (Wash. 1997) (finding that in light of 1996 NRC Report, "we now conclude there should be no bar to an expert giving his or her expert opinion that, based upon an exceedingly small probability of a defendant's DNA profile matching that of another in a random human population, the profile is unique.").

114. We apologize for the length of this sentence, but there are three distinct probabilities that arise in speaking of the uniqueness of DNA profiles. First, there is the probability of a match to a single, randomly selected individual in the population. This is the random-match probability. Second, there is the probability that the particular profile is unique. This probability involves pairing the profile with every member of the population. Third, there is the probability that all pairs of all profiles are unique. The first probability is larger than the second, which is many times larger than the third. Uniqueness or source testimony need only establish that *the one* DNA profile in the trace evidence is unique—and not that *all* DNA profiles are unique. Thus, it is the second probability, properly computed, that must be quite small to warrant the conclusion that no one but the defendant (and any identical twins) could be the source of the crime scene DNA. *See* David H. Kaye, *Identification, Individuality, and Uniqueness: What's the Difference?* 8 Law, Probability & Risk 85 (2009).

Formulas for estimating all these probabilities are given in NRC II, *supra* note 9, but DNA analysts and judges sometimes infer uniqueness on the basis of incorrect intuitions about the size of the random-match probability. *See* Balding, *supra* note 107, at 148 (2005) (describing "the uniqueness fallacy"); *cf.* State v. Lee, 976 So. 2d 109, 117 (La. 2008) (incorrect but harmless miscalculation).

115. Mitochondria probably started out as bacteria that were engulfed by cells eons ago. Some of their genes have migrated to the chromosomes, but STR and other DNA sequences in the nucleus are not physically or statistically associated with the sequences of the DNA in the mitochondria.

chromosomal DNA, there are hundreds or thousands of copies of mitochondrial DNA. This means that it is possible to detect mtDNA in samples, such as bone and hair shafts, that contain too little nuclear DNA for conventional typing.

Second, two "hypervariable" regions that tend to be different in different individuals lie within the "control region" or "D-loop" (displacement loop) of the mitochondrial genome.[116] These regions extend for a bit more than 300 base pairs each—short enough to be typable even in highly degraded samples such as very old human remains.

Third, mtDNA comes solely from the egg cell.[117] For this reason, mtDNA is inherited maternally, with no fatherly contribution:[118] Siblings, maternal half-siblings, and others related through maternal lineage normally possess the same mtDNA sequence. This feature makes mtDNA particularly useful for associating persons related through their maternal lineage. It has been exploited to identify the remains of the last Russian tsar and other members of the royal family, of soldiers missing in action, and of victims of mass disasters.

Finally, point mutations accumulate in the noncoding D-loop without altering how the mitochondrion functions. Hence, a single individual can develop distinct internal populations of mitochondria.[119] As discussed below, this phenomenon, known as heteroplasmy, complicates the interpretation of mtDNA sequences. Yet, it is mutations that make mtDNA polymorphic and hence useful in identifying individuals. Over time, mutations in egg cells can propagate to later generations, producing more heterogeneity in mitochondrial genomes in the human population.[120] This polymorphism allows scientists to compare mtDNA from crime scenes to mtDNA from given individuals to ascertain whether the tested individuals are within the maternal line (or another coincidentally matching maternal line) of people who could have been the source of the trace evidence.

The small mitochondrial genome can be analyzed with a PCR-based method that gives the order of all the base pairs.[121] The sequences of two samples—say, DNA extracted from a hair shaft found at a crime scene and hairs plucked from a suspect—then can be compared. Most analysts describe the results in terms on

116. A third, somewhat less polymorphic, region in the D-loop can be used for additional discrimination. The remainder of the control region, although noncoding, consists of DNA sequences that are involved in the transcription of the mitochondrial genes. These control sequences are essentially the same in everyone (monomorphic).

117. The relatively few mitochondria in the spermatozoan that fertilizes the egg cell soon degrade and are not replicated in the multiplying cells of the pre-embryo.

118. The possibility of paternal contributions to mtDNA in humans is discussed in, *e.g.*, John Buckleton et al., *Nonautosomal Forensic Markers*, *in* Forensic DNA Evidence Interpretation 299, 302 (John Buckleton et al. eds., 2005).

119. A single tissue has only one mitotype; another tissue from the same individual might have another mitotype; a third might have both mitotypes.

120. Evolutionary studies suggest an average mutation rate for the mtDNA control region of as little as one nucleotide difference every 300 generations, or one difference every 6000 years.

121. Other methods to ascertain the base-pair sequences also are available.

inclusions and exclusions, although, in principle, a likelihood ratio is better suited to cases in which there are slight sequence differences.[122] In the simplest case, the two sequences show a good number of differences (a clear exclusion), or they are identical (an inclusion). In such cases, mitotyping can exclude individuals as the source of stray hairs even when the hairs are microscopically indistinguishable.[123]

As with nuclear DNA, to indicate the significance of the match, analysts usually estimate the frequency of the sequence in some population. The estimation procedure is actually much simpler with mtDNA. It is not necessary to combine any allele frequencies because the entire mtDNA sequence, whatever its internal structure may be, is inherited as a single unit (a "haplotype"). In other words, the sequence itself is like a single allele, and one can simply see how often it occurs in a sample of unrelated people.[124]

Laboratories therefore refer to databases of mtDNA sequences to see how often the type in question has been seen before. Often, the mtDNA sequence from the crime scene is not represented in the database, indicating that it is a relatively rare sequence. For example, in *State v. Pappas*,[125] the reference database consisted of 1219 mtDNA sequences from whites, and it did not include the sequence that was present in the hairs near the crime scene and in the defendant. Thus, this particular sequence was observed once (at the crime scene) out of 1220 times (adding the new sequence to the 1219 different sequences on file). This would correspond to a population frequency of 0.082%. However, to account for sampling error (the inevitable differences between random samples and the population from which they are drawn), a laboratory might use a slightly different estimate. In general, laboratories count the occurrences in the database and take the upper end of a 95% confidence interval around the corresponding proportion.[126] Applying this logic, an FBI analyst in *Pappas* testified to "the *maximum* match probability . . . of three in 1000. . . . [O]ut of 1000 randomly selected persons, it could be expected that three persons would share the same mtDNA type as the defendant."[127] The basic idea is that even if 3/1000 people in the white population have the sequence, there still is a 5% chance that it would not show up in a specific (randomly drawn) database of size 1219; hence, 3/1000 is a reasonable

122. The likelihood-ratio approach is developed in Buckleton et al., *supra* note 118.

123. The implications of this fact for the admissibility of microscopic hair analysis is discussed in Kaye, *supra* note 3.

124. In this context, "unrelated people" means individuals with a different maternal lineage.

125. 776 A.2d 1091 (Conn. 2001).

126. The Reference Manual on Statistics discusses the meaning of a confidence interval. It has been argued that instead of using x/N in forming the confidence interval, one should use the proportion $(x + 1)/(N + 1)$, where x is the number of matching sequences in the database and N is the size of the database. After all, following the testing, $x + 1$ is the number of times that the sequence has been seen in $N + 1$ individuals. This is the reasoning that produced the point estimate of 1/1220 rather than 0/1219. For large databases, this alteration will make little difference in the confidence interval.

127. 776 A.2d at 1111 (emphasis added).

upper estimate for the population frequency.[128] If the population frequency of the sequence in unrelated whites were much larger, the chance that the sequence would have been missed in the database sampling would be even less than 5%.

Computations that rely on databases of major population groups (such as whites) assume that the reference database is a representative sample of a population of unrelated individuals who might have committed the alleged crime. This assumption is justified if there has been sufficient random mating within the racial population. In principle, the adjustment that accounts for population structure (*see supra* Section IV.E.3) could be used, but how large the adjustment should be is not clear.[129] Statistics derived from many databases from different locations also have been proposed.[130] An alternative is to develop local databases that would reflect the proportion of all the people in the vicinity of the crime possessing each possible mitotype.[131] Until these databases exist, an expert might give rather restricted quantitative testimony. In *Pappas*, for example, the expert could have said that the hairs and the defendants have the same mitotype and that this mitotype did not appear in a group of 1219 other people in a national sample, and the expert could have refrained from offering any estimate of the frequency in all whites. This restricted presentation suggests that the match has some probative value, but a court might need to consider whether it is sufficient to leave it to the jury to decide how to weigh the fact of the match and the absence of the same sequence in a convenience sample that might—or might not—be representative of the local white population.

Another issue is heteroplasmy. The simple inclusion-exclusion approach must be modified to account for the fact that the same individual can have detectably different mitotypes in different tissues or even in different cells in the same tissue. To understand the implications of heteroplasmy, we need to understand how it comes into existence.[132] Heteroplasmy can occur because of mtDNA mutations during the division of adult cells, such as those at the roots of hair shafts. These new mitotypes are confined to the individual. They will not be passed on to future generations. Heteroplasmy also can result from a mutation contained in the egg cell that grew into an individual. Such mutations can make their way into succeeding generations, establishing new mitotypes in the population. But this is

128. In general, if the sequence does not exist in the database of size *N*, the upper 95% confidence limit is approximately 3/*N*. *E.g.*, J.A. Hanley & A. Lipp-Hand, *If Nothing Goes Wrong, Is Everything All Right? Interpreting Zero Numerators*, 249 JAMA 1743 (1983). In *Pappas*, 3/*N* is 3/1219 = 0.25%, which rounds off to the 3 per 1000 figure quoted by the FBI analyst.

129. *See* Buckleton et al., *supra* note 118.

130. T. Egeland & A. Salas, *Statistical Evaluation of Haploid Genetic Evidence*, 1 Open Forensic Sci. J. 4 (2008).

131. *Id.*; *see also* F.A. Kaestle et al., *Database Limitations on the Evidentiary Value of Forensic Mitochondrial DNA Evidence*, 43 Am. Crim. L. Rev. 53 (2006).

132. An entertaining discussion can be found in Brian Sykes, The Seven Daughters of Eve: The Science That Reveals Our Genetic Ancestry 55–57, 62, 77–78 (2001).

an uncertain process. Eggs cells contain many mitochondria, and the mature egg cell will not contain just the mutation—it will house a mixed population of the old-style mitochondria and a number of the mutated ones (with DNA that usually differs from the original at a single base pair). Figuratively speaking, the original mtDNA sequence and the mutated version fight it out for several generations until one of them becomes "fixed" in the population. In the interim, the progeny of the mutated egg cell will harbor both strains of mitochondria.

When mtDNA from a crime scene sample is compared to a suspect's sample, there are three possibilities: (1) neither sample is detectably heteroplasmic; (2) one sample displays heteroplasmy, but the other does not; (3) both samples display heteroplasmy. In each scenario, the comparison can produce an exclusion or an inclusion:

1. *Neither sample heteroplasmic.* In the first situation, if the sequence in the crime scene sample is markedly different from the sequence in the suspect's sample, then the suspect is excluded. But heteroplasmy could be the reason for a difference of only a single base or so. For example, the sequence in a hair shaft coming from the suspect could be a slight mutation of the dominant sequence in the suspect. Therefore, the FBI treats a difference at a single base pair as inconclusive.[133] When the one mtDNA sequence characteristic of each sample is identical, the issue becomes how to use the reference database of mtDNA sequences, as discussed above.
2. *Suspect's sample heteroplasmic, crime scene sample not.* One version of the second scenario arises when heteroplasmy is seen in the suspect's tissues but not in the crime scene sample. If the crime scene sequence is not close to either of the suspect's sequences, then the suspect is excluded. If it is identical to one of the suspect's sequences, then the suspect is included, and a suitable reference database should indicate how infrequent such an inclusion would be. If crime scene DNA is one base pair removed from either of the suspect's sequences, then the result is inconclusive.

133. Scientific Working Group on DNA Analysis Methods (SWGDAM), *Guidelines for Mitochondrial DNA (mtDNA) Nucleotide Sequence Interpretation*, Forensic Sci. Comm., Apr. 2003, *available at* http://www.fbi.gov/hq/lab/fsc/backissu/april2003/swgdammitodna.htm. *But see* Vaughn v. State, 646 S.E.2d 212, 215 (Ga. 2007) (apparently transforming the statement that a suspect "cannot be excluded" when "there is a single base pair difference" into "a match"). These inconclusive sequences contribute to the number of people who would not be excluded. Therefore, in *Pappas*, it is misleading to conclude "that approximately 99.75% of the Caucasian population could be excluded as the source of the mtDNA in the sample." 776 A.2d 1091, 1104 (Conn. 2001) (footnote omitted). This percentage neglects the individuals whose mtDNA sequences are off by one base pair. Along with the 0.25% who are included because their mtDNA matches completely, these one-off people would not be excluded. An analyst who speaks of the fraction of people who would not be excluded should report a nonexclusion rate that accounts for these inconclusive cases. Of course, the difference may be fairly small. In *Pappas*, a defense expert reported that the actual nonexclusion rate was still "99.3 percent of the Caucasian population." *Id.* at 1105 (footnote omitted). *See* Kaye et al., *supra* note 83.

3. *Both samples heteroplasmic.* In this third scenario, multiple sequences are seen in each sample. To keep track of things, we can call the sequences in the crime scene sample C1 and C2, and those in the suspect's sample S1 and S2. If either C1 or C2 is very different from both S1 and S2, the suspect is excluded. If C1 and C2 are the same as S1 and S2, the suspect is included. Because detectable heteroplasmy is not very common, this inclusion is stronger evidence of identity than the simple match in the first scenario. Finally, in the middle range, where C1 is very close to S1 or S2, or C2 is very close to S1 or S2, the result is inconclusive.

A number of courts have rejected objections that the methods for mtDNA sequencing do not comport with *Frye*[134] or *Daubert*[135] and that the phenomenon of heteroplasmy or the limitations in the statistical analysis preclude the forensic use of this technology under either Rule 702 or Rule 403.[136]

B. Y Chromosomes

Y chromosomes contain genes that result in development as a male rather than a female. Therefore, men are type XY and women are XX. A male child receives an X chromosome from his mother and a Y from his father; females receive two different X chromosomes, one from each parent. Like all chromosomes, the Y chromosome contains STRs and SNPs.

Because there is limited recombination between Y and X chromosomes, Y-STRs and Y-SNPs are inherited as a single block—a haplotype—from father to son. This means that the issues of population genetics and statistics are similar to those for mtDNA. No matter how many Y-STRs are in the haplotype, all the men in the same paternal line (up to the last mutation giving rise to a new line in the family tree) would match the crime scene sample.

134. *E.g.*, Magaletti v. State, 847 So. 2d 523, 528 (Fla. Dist. Ct. App. 2003) ("[T]he mtDNA analysis conducted [on hair] determined an exclusionary rate of 99.93 percent. In other words, the results indicate that 99.93 percent of people randomly selected would not match the unknown hair sample found in the victim's bindings."); People v. Sutherland. 860 N.E.2d 178, 271–72 (Ill. 2006); People v. Holtzer, 660 N.W.2d 405, 411 (Mich. Ct. App. 2003); Wagner v. State, 864 A.2d 1037, 1043–49 (Md. Ct. Spec. App. 2005) (mtDNA sequencing admissible despite contamination and heteroplasmy).

135. *E.g.*, United States v. Beverly, 369 F.3d 516, 531 (6th Cir. 2004) ("The scientific basis for the use of such DNA is well established."); United States v. Coleman, 202 F. Supp. 2d 962, 967 (E.D. Mo. 2002) ("'[a]t the most,' seven out of 10,000 people would be expected to have that exact sequence of As, Ts, Cs, and Gs."), *aff'd*, 349 F.3d 1077 (8th Cir. 2003); *Pappas*, 776 A.2d at 1095; State v. Underwood, 518 S.E.2d 231, 240 (N.C. Ct. App. 1999); State v. Council, 515 S.E.2d 508, 518 (S.C. 1999).

136. *E.g.*, *Beverly*, 369 F.3d at 531 ("[T]he mathematical basis for the evidentiary power of the mtDNA evidence was carefully explained, and was not more prejudicial than probative."); *Pappas*, 776 A.2d 1091.

Consequently, multiplication of allele frequencies is inappropriate, and an estimate of how many men might share the haplotype must be based on the frequency of that one haplotype in a relevant population. Population structure is a concern, and obtaining a suitable sample to estimate the frequency in a local population could be a challenge. If such a database is not available, DNA analysts might consider limiting their testimony on direct examination to the size of the available database, the population sampled, and the number of individuals in the database who share the crime scene haplotype. This presentation is less ambitious than a random-match probability, and courts must decide whether it gives the jury sufficient information to fairly assess the probative value of the match, which could be substantial.

When a standard DNA profile (involving a reasonable number of STRs or other polymorphisms of the other chromosomes) is available, there is little reason to add a Y-STR test. The profiles already are extremely rare. In some cases, however, standard STR typing will fail. Consider, for example, what happens when a PCR primer that targets an STR locus on, say, chromosome 16 is applied to a sample that contains a small number of sperm (from, say, a vasectomized man) and a huge number of cells from a woman who is a victim of sexual assault. Almost never will the primer lock onto the man's chromosome 16. Therefore, his alleles on this chromosome will not produce a detectable peak in an electropherogram. But a primer for a Y-STR will not bind to the victim's chromosomes—her chromosomes swamp the sample, but they are essentially invisible to the Y-STR primer. Because this primer binds only to the Y chromosomes from the man, only his STRs will be amplified. This is one example of how Y-STR profiling can be valuable in dealing with a mixture of DNA from several individuals. The next section provides other examples and describes other ways in which analysis of the Y chromosome can be valuable in mixture cases.

Although the statistics and population genetics of Y-STRs are different from the other STRs, the underlying technology for obtaining the profile is the same. On this basis, some courts have upheld the admission of these markers.[137]

C. *Mixtures*

Samples of biological trace evidence recovered from crime scenes often contain a mixture of fluids or tissues from different individuals. Examples include vaginal swabs collected as sexual assault evidence and bloodstain evidence from scenes where several individuals shed blood. However, not all mixed samples produce mixed STR profiles.[138] Consider a sample in which 99% of the DNA comes from

137. *E.g.*, Shabazz v. State, 592 S.E.2d 876, 879 (Ga. Ct. App. 2004); Curtis v. State, 205 S.W.3d 656, 660–61 (Tex. Ct. App. 2006).

138. The discussion in this section is limited to electropherograms of STR alleles. A recent paper reports a statistical technique that compares the known SNP genotypes (involving hundreds of

the defendant and 1% comes from a different individual. Even if some of the molecules from the minor contributor come in contact with the polymerase and an STR is amplified, the resulting signal might be too small to be detected—the peak in an electropherogram will blend into the background. Because the vast bulk of the amplified STRs will come from the defendant's DNA, the electropherogram should show only one STR profile. In these situations, the interpretation of the single DNA profile is the same as when 100% of the DNA molecules in the sample are the defendant's.

When the mixtures are more evenly balanced among contributors, however, the STRs from multiple contributors can appear as "extra" peaks. As a rule, because DNA from a single individual can have no more than two alleles at each locus,[139] the presence of three or more peaks at several loci indicates that a mixture of DNA is in the sample.[140] Figure 6 shows another electropherogram from DNA recovered in *People v. Pizarro*.[141] The fact that there are as many as four alleles at some loci and that many of the peaks match the victim's) suggests that the sample is a mixture of the victim's and another person's DNA. Furthermore, a peak at the amelogenein locus shows that male DNA is part of the mixture. Because all the peaks that do not match the victim are part of the defendant's STR profile, the mixture is consistent with the state's theory that the defendant raped the victim.

Five approaches are available to cope with detectable mixtures. First, if a laboratory has other samples that do not show evidence of mixing, it can avoid the problem of deciphering the convoluted set of profiles. Even across a single stain, the proportions of a mixture can vary, and it might be possible to extract a DNA sample that does not produce a mixed signal.

Second, a chemical procedure exists to separate the DNA from sperm from a rape victim's vaginal epithelial cell DNA.[142] When this procedure works, the

thousands of SNPs) of a set of individuals to the SNPs detected in complex mixtures. The report states that the technique is able to discern "whether an individual is within a series of complex mixtures (2 to 200 individuals) when the individual contributes trace levels (at and below 1%) of the total genomic DNA." Nils Homer et al., *Resolving Individuals Contributing Trace Amounts of DNA to Highly Complex Mixtures Using High-Density SNP Genotyping Microarrays*, 4 PLoS Genetics No. 8 (2008), *available at* http://www.plosgenetics.org/article/info%3Adoi%2F10.1371%2Fjournal.pgen.1000167.

139. This follows from the fact that individuals inherit chromosomes in pairs, one from each parent. An individual who inherits the same allele from each parent (a homozygote) can contribute only that one allele to a sample, and an individual who inherits a different allele from each parent (a heterozygote) will contribute those two alleles. Finding three or more alleles at several loci therefore indicates a mixture.

140. On rare occasions, an individual exhibits a phenotype with three alleles at a locus. This can be the result of a chromosome anomaly (such as a duplicated gene on one chromosome or a mutation). A sample from such an individual is usually easily distinguished from a mixed sample. The three-allele variant is seen at only the affected locus, whereas with mixtures, more than two alleles typically are evident at several loci.

141. *See supra* Figure 5.

142. The nucleus of a sperm cell lies behind a protective structure that does not break down as easily as the membrane in an epithelial cell. This makes it possible to disrupt the epithelial cells first and extract their DNA, and then to use a harsher treatment to disrupt the sperm cells.

Figure 6. Electropherogram in *People v. Pizarro* that can be interpreted as a mixture of DNA from the victim and the defendant.

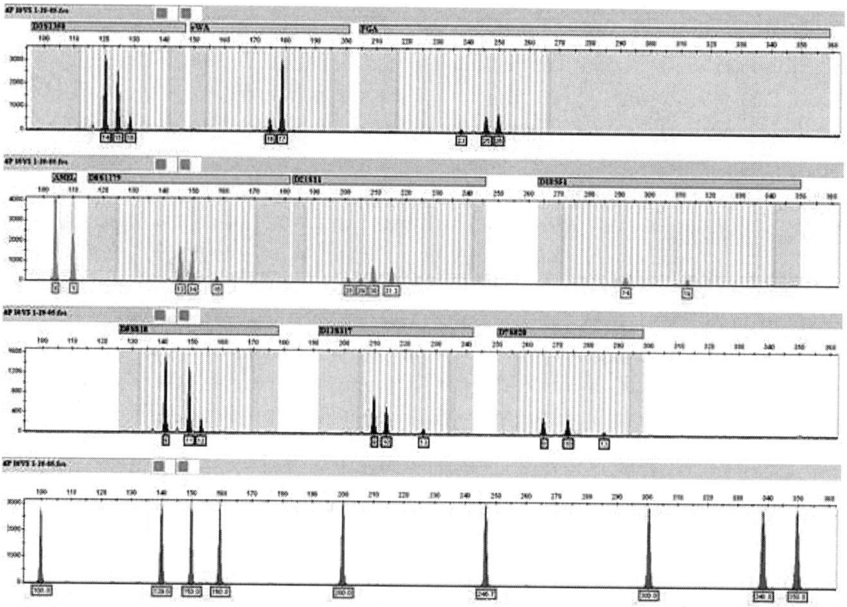

Source: Steven Myers and Jeanette Wallin, California Department of Justice, provided the image.

laboratory can assign the DNA profiles to the different individuals because it has created, in effect, two samples that are not mixtures.

Third, in sexual assault cases, Y chromosome testing can reveal the number of men (from different paternal lines) whose DNA is being detected and whether the defendant's Y chromosome is consistent with his being in one of these paternal lines.[143] Because only males have Y chromosomes, the female DNA in a mixture has no effect.

Fourth, a laboratory simply can report that a defendant's profile is consistent with the mixed profile, and it can provide an estimate of the proportion of the relevant population that also cannot be excluded (or would be included).[144] When

143 *E.g.*, State v. Polizzi, 924 So. 2d 303, 308–09 (La. Ct. App. 2006) (testing for Y-STRs on "the genital swab with the DNA profile from the Defendant's buccal swab, . . . the Defendant or any of his paternal relatives could not be excluded as having been a donor to the sample from the victim," while "99.7 percent of the Caucasian population, 99.8 percent of the African American population, and 99.3 percent of the Hispanic population could be excluded as donors of the DNA in the sample.").

144. *E.g.*, State v. Roman Nose, 667 N.W.2d 386, 394 n.5 (Minn. 2003). If the laboratory can explain why one or more of the defendant's alleles do not appear in the mixed profile from the

an individual's DNA—for example, the victim's—is known to be in a two-person crime scene sample, the profile of the unknown person is readily deduced. In those situations, the analysis of a remaining single-person profile can proceed in the ordinary fashion.

Finally, a laboratory can try to determine (or make assumptions about) how many contributors are present and then deduce which set of alleles is likely to be from each contributor. To accomplish this, DNA analysts look to such clues as the number of peaks in an expected allele-size range and the imbalance in the heights of the peaks.[145] A good deal of judgment can go into the determination of which peaks are real, which are artifacts, which are "masked," and which are absent for some other reason.[146] Courts generally have rejected arguments that mixture analysis is so unreliable or so open to manipulation that the results are inadmissible.[147] In addition, expert computer systems have been devised for facilitating the analysis and for automatically "deconvoluting" mixtures.[148] Once they are validated, these systems can make the process more standardized.

The five approaches listed here are not mutually exclusive (and not all apply to every case). When the number of contributors to a mixture is in doubt, for example, a laboratory is not limited to giving the overall probability of excluding (or including) an individual as a possible contributor (the statistic mentioned as part of the fourth method). The 1996 NRC report observed that "when the contributors to a mixture are not known or cannot otherwise be distinguished, a likelihood-ratio approach offers a clear advantage [over the simplistic exclusion-inclusion statistic] and is particularly suitable."[149] Despite the arguments of some

crime scene, it might be willing to declare a match not withstanding this discrepancy. Of course, as the number of alleles that must be present for there to be a match declines, the proportion of the population that would be included goes up.

145. *See, e.g.*, Roberts v. United States, 916 A.2d 922, 932–35 (D.C. 2007) (holding such inferences to be admissible).

146. The proportion of the population included in a mixture and the likelihood ratios conditioned on a particular genotype do not take into account the other possible genotypes that the expert eliminated in a subjective analysis. William C. Thompson, *Painting the Target Around the Matching Profile: The Texas Sharpshooter Fallacy in Forensic DNA Interpretation*, 8 Law, Probability & Risk 257 (2009). Adhering to preestablished standards and protocols for interpreting mixtures reduces the range of judgment in settling on the most likely set of genotypes to consider. Recent recommendations appear in Bruce Budowle et al., *Mixture Interpretation: Defining the Relevant Features for Guidelines for the Assessment of Mixed DNA Profiles in Forensic Casework*, 54 J. Forensic Sci. 810 (2009) (with commentary at 55 J. Forensic Sci. 265 (2010)); Peter Gill et al., *National Recommendations of the Technical UK DNA Working Group on Mixture Interpretation for the NDNAD and for Court Going Purposes*, 2 Forensic Sci. Int'l Genetics 76 (2008).

147. *Roberts*, 916 A.2d at 932 n.9 (citing cases).

148. *See, e.g.*, Tim Clayton & John Buckleton, *Mixtures*, *in* Forensic DNA Evidence Interpretation 217 (John Buckleton et al. eds., 2005); Mark W. Perlin et al., *Validating TrueAllele® DNA Mixture Interpretation*, 56 J. Forensic Sci. (forthcoming 2011).

149. NRC II, *supra* note 9, at 129.

legal commentators that likelihood ratios are inherently prejudicial,[150] and despite objections based on *Frye* or *Daubert*, almost all courts have found likelihood ratios admissible in mixture cases.[151]

D. Offender and Suspect Database Searches

1. Which statistics express the probative value of a match to a defendant located by searching a DNA database?

States and the federal government are amassing huge databases consisting of the DNA profiles of suspected or convicted offenders.[152] If the DNA profile from a crime scene stain matches one of those on file, the person identified by this "cold hit" will become the target of the investigation. Prosecution may follow.

These database-trawl cases can be contrasted with traditional "confirmation cases" in which the defendant already was a suspect and the DNA testing provided additional evidence against him. In confirmation cases, statistics such as the estimated frequency of the matching DNA profile in various populations, the equivalent random-match probabilities, or the corresponding likelihood ratios can be used to indicate the probative value of the DNA match.[153]

In trawl cases, however, an additional question arises—does the fact that the defendant was selected for prosecution by trawling require some adjustment to the usual statistics? The legal issues are twofold. First, is a particular quantity—be it the unadjusted random-match probability or some adjusted probability—scientifically valid (or generally accepted) in the case of a database search? If not, it must be excluded under the *Daubert* (or *Frye*) standards. Second, is the statistic irrelevant or unduly misleading? If so, it must be excluded under the rules that

150. *E.g.*, William C. Thompson, *DNA Evidence in the O.J. Simpson Trial*, 67 U. Colo. L. Rev. 827, 855–56 (1996); *see also* R.C. Lewontin, *Population Genetic Issues in the Forensic Use of DNA*, *in* 1 Modern Scientific Evidence: The Law and Science of Expert Testimony § 17-5.0, at 703–05 (Faigman et al. eds, 1st ed. 1998).

151. *E.g.*, State v. Garcia, 3 P.3d 999 (Ariz. Ct. App. 1999) (likelihood ratios admissible under *Frye* to explain mixed sample); Commonwealth v. Gaynor, 820 N.E.2d 233, 252 (Mass. 2005) ("Likelihood ratio analysis is appropriate for test results of mixed samples when the primary and secondary contributors cannot be distinguished. . . . It need not be applied when a primary contributor can be identified.") (citation omitted); People v. Coy, 669 N.W.2d 831, 835–39 (Mich. Ct. App. 2003) (incorrectly treating mixed-sample likelihood ratios as a part of the statistics on single-source DNA matches that had already been held to be generally accepted); State v. Ayers, 68 P.3d 768, 775 (Mont. 2003) (affirming trial court's admission of expert testimony where expert used likelihood ratios to explain DNA results from a sample known to contain a mixture of DNA); *cf.* Coy v. Renico, 414 F. Supp. 2d 744, 762–63 (E.D. Mich. 2006) (stating that the use of likelihood ratio and other statistics for a mixed stain in *People v. Coy*, *supra*, was sufficiently accepted in the scientific community to be consistent with due process).

152. *See supra* Section II.E.

153. On the computation and admissibility of such statistics, see *supra* Section IV.

require all evidence to be relevant and not unfairly prejudicial. To clarify, we summarize the statistical literature on this point. Then, we describe the emerging case law.

a. The statistical analyses of adjustment

All statisticians agree that, in principle, the search strategy affects the probative value of a DNA match. One group describes and emphasizes the impact of the database match on the hypothesis that the database does not contain the source of the crime scene DNA. This is a "frequentist" view. It asks how frequently searches of innocent databases—those for which the true source is someone outside the database—will generate cold hits. From this perspective, trawling is a form of "data mining" that produces a "selection effect" or "ascertainment bias." If we pick a lottery ticket at random, the probability p that we have the winning ticket is negligible. But if we search through all the tickets, sooner or later we will find the winning one. And even if we search through some smaller number N of tickets, the probability of picking a winning ticket is no longer p, but Np.[154] Likewise, if DNA from N innocent people is examined to determine if any of them match the crime scene DNA, then the probability of a match in this group is not p, but some quantity that could be as large as Np. This type of reasoning led the 1996 NRC committee to recommend that "[w]hen the suspect is found by a search of DNA databases, the random-match probability should be multiplied by N, the number of persons in the database."[155] The 1992 committee[156] and the FBI's former DNA Advisory Board[157] took a similar position.

154. The analysis of the DNA database search is more complicated than the lottery example suggests. In the simple lottery, there was exactly one winner. The trawl case is closer to a lottery in which we hold a ticket with a winning number, but it might be counterfeit, and we are not sure how many counterfeit copies of the winning ticket were in circulation when we bought our N tickets.

155. NRC II, *supra* note 9, at 161 (Recommendation 5.1).

156. Initially, the board explained that

> Two questions arise when a match is derived from a database search: (1) What is the rarity of the DNA profile? and (2) What is the probability of finding such a DNA profile in the database searched? These two questions address different issues. That the different questions produce different answers should be obvious. The former question addresses the random match probability, which is often of particular interest to the fact finder. Here we address the latter question, which is especially important when a profile found in a database search matches the DNA profile of an evidence sample.

DNA Advisory Board, *Statistical and Population Genetics Issues Affecting the Evaluation of the Frequency of Occurrence of DNA Profiles Calculated from Pertinent Population Database(s)*, 2 Forensic Sci. Comm., July 2000, *available at* http://www.fbi.gov/hq/lab/fsc/backissu/july2000/dnastat.htm. After a discussion of the literature as of 2000, the Board wrote that "we continue to endorse the recommendation of the NRC II Report for the evaluation of DNA evidence from a database search."

157. The first NRC committee wrote that "[t]he distinction between finding a match between an evidence sample and a suspect sample and finding a match between an evidence sample and one of many entries in a DNA profile databank is important." It used the same Np formula in a numerical example to show that "[t]he chance of finding a match in the second case is considerably higher,

No one questions the mathematics that show that when the database size N is very small compared with the size of the population, Np is an upper bound on the expected frequency with which searches of databases will incriminate innocent individuals when the true source of the crime scene DNA is not represented in the databases. The "Bayesian" school of thought, however, suggests that the frequency with which innocent databases will be falsely accused of harboring the source of the crime scene DNA is basically irrelevant. The question of interest to the legal system is whether the one individual whose database DNA matches the trace-evidence DNA is the source of that trace. As the size of a database approaches that of the entire population, finding one and only one matching individual should be more, not less, convincing evidence against that person. Thus, instead of looking at how surprising it would be to find a match in a large group of innocent suspects, this school of thought asks how much the result of the database search enhances the probability that the individual so identified is the source. The database search is actually more probative than the confirmation search because the DNA evidence in the trawl case is much more extensive. Trawling through large databases excludes millions of people, thereby reducing the number of people who might have left the trace evidence if the suspect did not. This additional information increases the likelihood that the defendant is the source, although the effect is indirect and generally small.[158]

Of course, when the cold hit is the only evidence against the defendant, the total package of evidence in the trawl case is less than in the confirmation case. Nonetheless, the Bayesian treatment shows that the DNA part of the total evidence is more powerful in a cold-hit case because this part of the evidence is more complete than when the search for matching DNA is limited to a single suspect. This reasoning suggests that the random-match probability (or, equivalently, the frequency p in the population) understates the probative value of the unique DNA match in the trawl case. And if this is so, then the unadjusted random-match probability or frequency p can be used as a conservative indication of the probative value of the finding that, of the many people in the database, only the defendant matches.[159]

because one . . . fishes through the databank, trying out many hypotheses." NRC I, *supra* note 8, at 124. Rather than proposing a statistical adjustment to the match probability, however, that committee recommended using only a few loci in the databank search, then confirming the match with additional loci, and presenting only "the statistical frequency associated with the additional loci. . . ." *Id.* at 124 tbl. 1.1.

158. When the size of the database approaches the size of the entire population, the effect is large. Finding that only one individual in a large database has a particular profile also raises the probability that this profile is very rare, further enhancing the probative value of the DNA evidence.

159. This analysis was developed by David Balding and Peter Donnelly. For informal expositions, see, for example, Peter Donnelly & Richard D. Friedman, *DNA Database Searches and the Legal Consumption of Scientific Evidence*, 97 Mich. L. Rev. 931 (1999); Kaye, *supra* note 109; Simon Walsh & John Buckleton, *DNA Intelligence Databases*, *in* Forensic DNA Evidence Interpretation 439 (John Buckleton et al. eds., 2005). For a related analysis directed at the average probability that an individual

b. The judicial opinions on adjustment

The need for an adjustment has been vigorously debated in the statistical, and to a lesser extent, the legal literature.[160] The dominant view in the journal articles is that the random-match probability or frequency need *not* be inflated to protect the defendant. The major opinions to confront this issue agree that p is admissible against the defendant in a trawl case.[161] They reason that all the statistics are admissible under *Frye* and *Daubert* because there is no controversy over how they are computed. They then assume that both p and Np are logically relevant and not prejudicial in a trawl case.

But commentators have pointed out that if the frequentist position that trawling degrades the probative value of the match is correct, then it is hard to see what p offers the jury. Conversely, if the Bayesian position that trawling enhances the probative value of the match is correct, then it is hard to see what Np offers the jury.[162] Thus, it has been argued that, to decide whether p should be admissible when offered by the prosecution and whether Np (or some variant of it) should be admissible when offered by the defense, the law needs to directly confront the schism between the frequentist and Bayesian perspectives on characterizing the probative value of a cold hit.[163]

2. *Near-miss (familial) searching*

Normally, police trawl a DNA database to see if any recorded STR profiles match a crime scene profile. It is not generally necessary to inform the jury that the defendant was located in this manner. Indeed, the rules of evidence sometimes prohibit this proof over the objection of the defendant.[164] Another search pro-

identified through a database trawl is the source of a crime scene DNA sample, see Yun S. Song et al., *Average Probability That a "Cold Hit" in a DNA Database Search Results in an Erroneous Attribution*, 54 J. Forensic Sci. 22, 23–24 (2009).

160. For citations to this literature, see Kaye, *supra* note 109; Walsh & Buckleton, *supra* note 159, at 464.

161. People v. Nelson, 185 P.3d 49 (Cal. 2008); United States v. Jenkins, 887 A.2d 1013 (D.C. 2005). The cases are analyzed in Kaye, *supra* note 109.

162. Furthermore, even if an adjustment is logically required, Np might be too extreme because the offender databases include the profiles of individuals—those in prison at the time of the offense, for instance—who could not have been the source of the crime scene sample. To that extent, Np overstates the expected frequency of matches to innocent individuals in the database. Kaye, *supra* note 109; David H. Kaye, People v. Nelson: *A Tale of Two Statistics*, 7 Law, Probability & Risk 249 (2008).

163. Kaye, *supra* note 109; Kaye, *supra* note 162.

164. The common law of evidence and Federal Rule of Evidence 404 prevent the government from proving that a defendant has committed other crimes when the only purpose of the revelation is to suggest a general propensity toward criminality. *See, e.g.*, 1 McCormick on Evidence, *supra* note 3, § 190. Proof that the defendant was identified through a database search is likely to suggest the existence of a criminal record, because it is widely known that existing law enforcement DNA databases are largely filled with the profiles of convicted offenders. Nonetheless, where the bona fide and important purpose of the disclosure is "to complete the story" (*id.*) or to help the jury to understand an Np

cedure can lead to charges against a defendant who is not even in the database. The clearest illustration is the case of identical twins, one of whom is a convicted offender whose DNA type is on file and the other who has never been typed. If the convicted twin was in prison when the crime under investigation was committed, and if the police realized that he had an identical twin, suspicion should fall on the identical twin. Presumably, the police would seek a sample from this twin, and at trial it would not be necessary for the prosecution to explain the roundabout process through which he was identified.

In this example, the defendant was found because a relative's DNA led the police to him. More generally, the fact that close relatives share more alleles than other members of the same subpopulation can be exploited as an investigative tool. Rather than search for a match at all 13 loci in an STR profile, police could search for a near miss—a partial match that is much more probable when the partly matching profile in the database comes from a close relative than when it comes from an unrelated person. (Analysis of Y-STRs or mtDNA then could determine whether the offender who provided the partially matching DNA in the database probably is in the same paternal or maternal lineage as the unknown individual who left DNA at the scene of the crime.)

Such "familial searching" raises technical and policy questions. The technical and practical challenge is to devise a search strategy that keeps the number of false leads to a tolerable level. The policy question is whether exposing relatives to the possibility of being investigated on the basis of genetic leads from their kin is appropriate.[165]

In receiving the DNA evidence, courts might consider having the prosecution describe the match without revealing that the defendant's close relative is a known or suspected criminal. In addition, if database trawls degrade the probative value of a perfect match in the database—a theory discussed in the previous subsection—then the usual random-match probability or estimated frequency exaggerates the value of the match derived from a database search. From the frequentist perspective, one must ask how often trawling databases for leads to individuals (both within and outside the database) will produce false accusations. From the Bayesian perspective, however, the usual match probabilities and likeli-

statistic, Rule 404 itself arguably does not prevent the prosecution (and certainly not the defense) from revealing that the defendant was found through a DNA database trawl. In the absence of a categorical rule of exclusion (like the one in Rule 404), a case-by-case balancing of the value of the information for its legitimate purposes as against its potential prejudice to the defendant is required. *See id.*

165. *See, e.g.*, Bruce Budowle et al., *Clarification of Statistical Issues Related to the Operation of CODIS*, *in* Genetic Identity Conference Proceedings: 18th Int'l Symposium on Human Identification (2006), *available at* http://www.promega.com/GENETICIDPROC/ussymp17proc/oralpresentations/budowle.pdf; Henry T. Greely et al., *Family Ties: The Use of DNA Offender Databases to Catch Offenders' Kin*, 34 J.L. Med. & Ethics 248 (2006); Erica Haimes, *Social and Ethical Issues in the Use of Familiar Searching in Forensic Investigations: Insights from Family and Kinship Studies*, 34 J.L. Med. & Ethics 262 (2006).

hoods can be used because, if anything, they understate the probative value of the DNA information.[166]

3. All-pairs matching within a database to verify estimated random-match probabilities

A third and final use of police intelligence databases has evidentiary implications. Section IV.E explained how population genetics models and reference samples for determining allele frequencies are used to estimate DNA genotype frequencies. Large databases can be used to check these theoretical computations. In New Zealand, for example, researchers compared every six-locus STR profile in the national database with every other profile.[167] At the time, there were 10,907 profiles. This means that there were about 59 million distinct pairs.[168] Because the theoretical random-match probability was about 1 in 50 million, if all the individuals represented in the database were unrelated, one would expect that an exhaustive comparison of the profiles for these 59 million pairs would produce only about one match. In fact, the 59 million comparisons revealed 10 matches. The excess number of matches is evidence that not all the individuals in the database were unrelated, that the true match probability was smaller than the theoretical calculation, or both. In fact, eight of the pairs were twins or brothers. The ninth was a duplicate (because one person gave a sample as himself and then again pretending to be someone else). The tenth was apparently a match between two unrelated people. This exercise thus confirmed the theoretical computation of the random-match probability. On average, the theoretical match probability was about 1/50,000,000, and the rate of matches in the unrelated pairs within the database was 1/59,000,000.

In the United States, defendants have sought discovery of the criminal-offender databases to determine whether the number of matching and partially matching pairs exceeds the predictions made with the population genetics model.[169] An early report about partial matches in a state database in Arizona was said to show extraordinarily large numbers of partial matches (without accounting for the combinatorial explosion in the number of comparisons in an all-pairs data-

166. *See* Kaye, *supra* note 3; *supra* Section V.D.1.

167. Walsh & Buckleton, *supra* note 159, at 463.

168. Altogether, nearly 11,000 people were represented in the New Zealand database. Hence, about 10,907 × 10,907 pairs such as (1,1), (1,2), (1,3), … , (1,10907), (2,1), (2,2), (2,3), . . . , (2,10907), . . . (10907,1), (10907,2), (10907,3), (10907,10907) can be formed. This amounts to almost 119 million possible pairs. Of course, there is no point in checking the pairs (1,1), (2,2), . . . (10907,10907). Thus, the number of ordered pairs with different individuals is 119 million minus a mere 10,907. The subtraction hardly changes anything. Finally, ordered pairs such as (1,5) and (5,1) involve the same two people. Therefore, the number of distinct pairs of people is about half of 119 million—the 59 million figure in the text.

169. Jason Felch & Maura Dolan, *How Reliable Is DNA in Identifying Suspects?* L.A. Times, July 20, 2008.

base search).[170] However, some scientists question the utility of the investigative databases for population genetics research.[171] They observe that these databases contain an unknown number of relatives, that they might contain duplicates, and that the population in the offender databases is highly structured. These complicating factors would need to be considered in testing for an excess of matches or partial matches. Studies of offender databases in Australia and New Zealand that make adjustments for population structure and close relatives have shown substantial agreement between the expected and observed numbers of partial matches, at least up to the nine STR loci used in those databases.[172]

The existence of large databases also provides a means of estimating a random-match probability without making any modeling assumptions. For the New Zealand study, even ignoring the possibility of relatives and duplicates, there were only 10 matches out of 59 million comparisons. The empirical estimate of the random-match probability is therefore about 1 in 5.9 million. This is about 10 times larger than the theoretical estimate, but still quite small. As this example indicates, crude but simple empirical estimates from all-pairs comparisons in large databases may well produce random-match probabilities that are larger than the theoretical estimates (as expected when full siblings or other close relatives are in the databases), but the estimated probabilities are likely to remain impressively small.

170. *Id.*; Kaye, *supra* note 3. As illustrated *supra* note 168, an all-pairs search in a large database of size N will involve $N(N-1)/2$, or about $N^2/2$ comparisons. For example, a database of 6 million samples gives rise to some 18,000,000,000,000 comparisons. Even with no population structure, relatives, and duplicates, and with random-match probabilities in the trillionths, one would expect to find a large number of matches or near-matches. An analogy can be made to the famous "birthday problem" mentioned in the 1996 NRC Report, *supra* note 9, at 165. In its simplest form, the birthday problem assumes that equal numbers of people are born every day of the year. The problem is to determine the minimum number of people in a room such that the odds favor there being at least two of them who were born on the same day of the same month. Focusing solely on the random-match probability of 1/365 for a specified birthday makes it appear that a huge number of people must be in the room for a match to be likely. After all, the chance of a match between two individuals having a given birthday (say, January 1) is (ignoring leap years) a miniscule $1/365 \times 1/365 = 1/133{,}225$. But because the matching birthday can be any one of the 365 days in the year and because there are $N(N-1)/2$ ways to have a match, it takes only $N = 23$ people before it is more likely than not that at least two people share a birthday. The birthday problem thus shows that surprising coincidences commonly occur even in relatively small databases. *See, e.g.*, Persi Diaconis & Frederick Mosteller, *Methods for Studying Coincidences*, 84 J. Am. Statistical Ass'n 853 (1989).

171. Bruce Budowle et al., *Partial Matches in Heterogeneous Offender Databases Do Not Call into Question the Validity of Random Match Probability Calculations*, 123 Int'l J. Legal Med. 59 (2009).

172. James M. Curran et al., *Empirical Support for the Reliability of DNA Evidence Interpretation in Australia and New Zealand*, 40 Australian J. Forensic Sci. 99, 102–06 (2008); Bruce S. Weir, *The Rarity of DNA Profiles*, 1 Annals Applied Stat. 358 (2007); B.S. Weir, *Matching and Partially-Matching DNA Profiles*, 49 J. Forensic Sci. 1009, 1013 (2004); *cf.* Laurence D. Mueller, *Can Simple Population Genetic Models Reconcile Partial Match Frequencies Observed in Large Forensic Databases?* 87 J. Genetics (India) 101 (2008) (maintaining that excess partial matches in an Arizona offender database are not easily reconciled with theoretical expectations). This literature is reviewed in David H. Kaye, *Trawling DNA Databases for Partial Matches: What Is the FBI Afraid of?* 19 Cornell J.L. & Pub. Pol'y 145 (2009).

VI. Nonhuman DNA Testing

Most routine applications of DNA technology in the forensic setting involve the identification of human beings—suspects in criminal cases, missing persons, or victims of mass disasters. However, inasmuch as DNA analysis might be informative in any kind of case involving biological material, DNA analysis has found application in such diverse situations as identification of individual plants and animals that link suspects to crime scenes, enforcement of endangered species and other wildlife regulations, investigation of patent issues involving specific animal breeds and plant cultivars, identification of fraudulently labeled foodstuffs, identification of sources of bacterial and viral epidemic outbreaks, and identification of agents of bioterrorism.[173] These applications are directed either at identifying the species origin of an item or at distinguishing among individuals (or subgroups) within a species. In deciding whether the evidence is scientifically sound, it can be important to consider the novelty of the application, the validity of the underlying scientific theory, the validity of any statistical interpretations, and the relevant scientific community to consult in assessing the application. This section considers these factors in the context of nonhuman DNA testing.

A. Species and Subspecies

Evolution is a branching process. Over time, populations may split into distinct species. Ancestral species and some or all of their branches become extinct. Phylogenetics uses DNA sequences to elucidate these evolutionary "trees." This information can help determine the species of the organism from which material has been obtained. For example, the most desirable Russian black caviar originates from three species of wild sturgeon inhabiting the Volga River–Caspian Sea basin. But caviar from other sturgeon species is sometimes falsely labeled as originating from these three species—in violation of food labeling laws. Moreover, the three sturgeon species are listed as endangered, and trade in their caviar is restricted. A test of caviar species based on DNA sequence variation in a mitochondrial gene found that 23% of caviar products in the New York City area were mislabeled,[174] and in *United States v. Yazback*, caviar species testing was used to convict an

173. *See, e.g.*, R.G. Breeze et al., Microbial Forensics (2005); Laurel A. Neme, Animal Investigators: How the World's First Wildlife Forensics Lab Is Solving Crimes and Saving Endangered Species (2009). In still other situations, DNA testing has been used to establish the identity of a missing or stolen animal. *E.g.*, Augillard v. Madura, 257 S.W.3d 494 (Tex. App. 2008) (action for conversion to recover dog lost during Hurricane Katrina); Guillermo Giovambattista et al., *DNA Typing in a Cattle Stealing Case*, 46 J. Forensic Sci. 1484 (2001).

174. Rob DeSalle & Vadim J. Birstein, *PCR Identification of Black Caviar*, 381 Nature 197 (1996); Vadim J. Birstein et al., *Population Aggregation Analysis of Three Caviar-Producing Species of Sturgeons and Implications for the Species Identification of Black Caviar*, 12 Conservation Biology 766 (1998).

importer of gourmet foods for falsely labeling fish eggs from the environmentally protected American paddlefish as the more prized Russian sevruga caviar.[175]

Phylogenetic analysis also is used to study changes in populations of organisms of the same species. In *State v. Schmidt*,[176] a physician was convicted of attempting to murder his former lover by injecting her with the HIV virus obtained from an infected patient. The virus evolves rapidly—its sequence can change by as much as 1% per year over the course of infection in a single individual. In time, an infected individual will harbor new strains, but these will be more closely related to the particular strain (or strains) that originally infected the individual than to the diverse strains of the virus in the geographic area. The victim in *Schmidt* had fewer strains of HIV than the patient—indicating a later infection—and all the victim's strains were closely related to a subset of the patient's strains—indicating that the victim's strains originated from that subset then in the patient. This technique of examining the genetic similarities and differences in two populations of viruses has been used in other cases across the world.[177]

The FBI employed similar reasoning to conclude that the anthrax spores in letters sent through the mail in 2001 came from the descendants of bacteria first cultured from an infected cow in Texas in 1981. This "Ames strain" was disseminated to various research laboratories over the years, and the FBI also attempted to associate the letter spores with particular collections of anthrax bacteria (all derived from the one Ames strain) now housed in different laboratories.[178]

Both the caviar and the HIV cases exemplify the translation of established scientific methods into a forensic application. The mitochondrial gene used for species identification in *Yazback* was the cytochrome b gene. Having accumulated mutations over time, this gene commonly is used for assessing species relationships among vertebrates, and the database of cytochrome b sequences is extensive. In particular, this gene sequence previously had been used to determine the evolutionary placement of sturgeons among other species of fish.[179] Likewise, the use of phylogenetic analysis for assessing relationships among HIV strains has provided critical insights into the biology of this deadly virus.

175. Dep't of Justice, Caviar Company and President Convicted in Smuggling Conspiracy, *available at* http://www.usdoj.gov/opa/pr/2002/January/02_enrd_052.htm. An earlier case is described in Andrew Cohen, *Sturgeon Poaching and Black Market Caviar: A Case Study*, 48 J. Env'l Biology Fishes 423 (1997).

176. 699 So. 2d 448 (La. Ct. App. 1997) (holding that the evidence satisfied *Daubert*).

177. Edwin J Bernard et al., *The Use of Phylogenetic Analysis as Evidence in Criminal Investigation of HIV Transmission*, Feb. 2007, *available at* http://www.nat.org.uk/Media%20library/Files/PDF%20 Documents/HIV-Forensics.pdf.

178. *See* National Research Council, Committee on the Review of the Scientific Approaches Used During the FBI's Investigation of the 2001 *Bacillus Anthracis* Mailings, Review of the Scientific Approaches Used During the FBI's Investigation of the 2001 Anthrax Letters (2011).

179. Sturgeon Biodiversity and Conservation (Vadim J. Birstein et al. eds., 1997).

That said, how the phylogenetic analysis is implemented and the genes used for the analysis may prompt questions in some cases.[180] Both the computer algorithm used to align DNA sequences prior to the construction of the phylogenetic tree and the computer algorithms used to build the tree contain assumptions that can influence the outcomes. Consequently, alignments generated by different software can result in different trees, and different tree-building algorithms can yield different trees from the same alignments. Thus, phylogenetic analysis should not be looked upon as a simple mechanical process. In *Schmidt*, the investigators anticipated and addressed potential problem areas in their choice of sequence data to collect and by using different algorithms for phylogenetic analysis. The results from the multiple analyses were the same, supporting the overall conclusion.[181]

B. Individual Organisms

DNA analysis to determine that trace evidence originated from a particular individual within a species requires both a valid analytical procedure for forensic samples and at least a rough assessment of how rare the DNA types are in the population. In human DNA testing, suitable reference databases permit reasonable estimates of allele frequencies among groups of human beings (*see supra* Section IV), but adequate databases will not always be available for other organisms. Nonetheless, a match between the DNA at a crime scene and the organism that could be the source of that trace evidence still may be informative. In these cases, a court may consider admitting testimony about the matching features along with circumscribed, qualitative explanations of the significance of the similarities.[182]

Such cases began appearing in the 1990s. In *State v. Bogan*,[183] for example, a woman's body was found in the desert, near several Palo Verde trees. A detective noticed two Palo Verde seed pods in the bed of a truck that the suspect was driving before the murder. However, genetic variation in Palo Verde tree DNA had not been widely studied, and no one knew how much variation actually

180. One cannot assume that cytochrome b gene testing, for example, is automatically appropriate for all species identification. Mitochondria are maternally inherited, and one can ask whether cross-breeding between different species of sturgeon could make a sturgeon of one species appear to be another species because it carries mitochondrial DNA originating from the other species. Mitochondrial introgression has been detected in several vertebrate species. Coyote mtDNA in wolves and cattle mtDNA in bison are notable examples. Introgression in sturgeon has been reported—some individual sturgeon appearing to be of one of the prized Volga region caviar species were found to carry cytochrome b genes from a lesser regarded non-Volga species. These examples indicate the need for specialized knowledge of the basic biology and ecology of the species in question.

181. On the need for caution in the interpretation of HIV sequence similarities, see Bernard et al., *supra* note 177.

182. *See generally* Kaye et al., *supra* note 1 (discussing various ways to explain "matches" in forensic identification tests).

183. 905 P.2d 515 (Ariz. Ct. App. 1995) (holding that the admission of the DNA match was proper under *Frye*).

existed within the species. Accordingly, a method for genetic analysis had to be developed, assessed for what it revealed about genetic variability, and evaluated for reliability. A university biologist chose random amplified polymorphic DNA (RAPD) analysis, a PCR-based method then commonly used for the detection of variation within species for which no genomic sequence information exists. This approach employs a single short piece of DNA with a random, arbitrary sequence as the PCR primer; the amplification products are DNA fragments of unknown sequence and variable length that can be separated by electrophoresis into a barcode-like "fingerprint" pattern. In a blind trial, the biologist was able to show that the DNA from nearly 30 Palo Verde trees yielded distinct RAPD patterns.[184] He testified that the two pods "were identical" and "matched completely with" a particular tree and "didn't match any of the [other] trees." In fact, he went so as to say that he felt "quite confident in concluding that" the tree's DNA would be distinguishable from that of "any tree that might be furnished" to him. Numerical estimates of the random-match probability were not introduced.[185]

The first example of an animal identification using STR typing involved linking evidence cat hairs to a particular cat. In *R. v. Beamish*, a woman disappeared from her home on Prince Edward Island, on Canada's eastern seaboard. Weeks later a man's brown leather jacket stained with blood was discovered in a plastic bag in the woods. In the jacket's lining were white cat hairs. After the missing woman's body was found in a shallow grave, her estranged common-law husband was arrested and charged with murder. He lived with his parents and a white cat named Snowball. A laboratory already engaged in the study of genetic diversity in cats showed that the DNA profile of the evidence cat hairs matched Snowball at 10 STR loci. Based on a survey of genetic variation in domestic cats generated for this case, the probability of a chance match was offered as evidence in support of the hypothesis that the hairs originated from Snowball.[186]

184. He analyzed samples from the nine trees near the body and another 19 trees from across the county. He "was not informed, until after his tests were completed and his report written, which samples came from" which trees. *Id.* at 521. Furthermore, unbeknownst to the experimenter, two apparently distinct samples were prepared from the tree at the crime scene that appeared to have been abraded by the defendant's truck. The biologist correctly identified the two samples from the one tree as matching, and he "distinguished the DNA from the seed pods in the truck bed from the DNA of all twenty-eight trees except" that one. *Id.*

185. RAPD analysis does not provide systematic information about sequence variation at defined loci. As a result, it is not possible to make a reliable estimate of allele or genotype frequencies at a locus, nor can one make the assumption of genetic independence required to legitimately multiply frequencies across multiple loci, as one can with STR markers. Furthermore, RAPD profile results are not generally portable between laboratories. Often, profiles generated by different laboratories will differ in their details. Therefore, RAPD profile data are not amenable to the generation of large databases. Nonetheless, the state's expert estimated a random match probability of 1 in 1,000,000, and the defense expert countered with 1 in 136,000. The trial court excluded both estimates because of the then-existing controversy (*see* Section IV) over analogous estimates for human RFLP genotypes.

186. *See* Marilyn A. Menott-Haymond et al., *Pet Cat Hair Implicates Murder Suspect*, 386 Nature 774 (1997).

In *Beamish*, there was no preexisting population database characterizing STR polymorphism in domestic cats, but the premise that cats exhibit substantial genetic variation at STR loci was in accord with knowledge of STR variation in other mammals. Moreover, testing done on two small cat populations provided evidence that the STR loci chosen for analysis were polymorphic and behaved as independent genetic characteristics, allowing allele frequency estimates to be used for the calculation of random-match probabilities as is done with human STR data. On this basis, the random-match probability for Snowball's STR profile was estimated to be one in many millions, and the trial court admitted this statistic.[187]

An animal-DNA random-match probability prompted a reversal, however, in a Washington case. In *State v. Leuluaialii*,[188] the prosecution offered testimony of an STR match with a dog's blood that linked the defendants to the victims' bodies. The defendants objected, seeking a *Frye* hearing, but the trial court denied this motion and admitted testimony that included the report that "the probability of finding another dog with Chief's DNA profile was 1 in 18 billion [or] 1 in 3 trillion."[189] The state court of appeals remanded the case for a hearing on general acceptance, cautioning that "[b]ecause PE Zoogen has not yet published sufficient data to show that its DNA markers and associated probability estimates are reliable, we would suggest that other courts tread lightly in these waters and closely examine canine DNA results before accepting them at trial."[190]

The scientific literature shows continued use of STR profiling[191] (as well as the use of SNP typing)[192] to characterize individuals in plant and animal populations. STR databases have been established for domestic and agriculturally significant animals such as dogs, cats, cattle, and horses as well as for a number of plant species.[193] Critical to the use of these databases is an understanding of the

187. David N. Leff, *Killer Convicted by a Hair: Unprecedented Forensic Evidence from Cat's DNA Convinced Canadian Jury*, Bioworld Today, Apr. 24, 1997, *available in* 1997 WL 7473675 ("the frequency of the match came out to be on the order of about one in 45 million," quoting Steven O'Brien); *All Things Considered: Cat DNA* (NPR broadcast, Apr. 23, 1997), *available in* 1997 WL 12832754 ("it was less than one in two hundred million," quoting Steven O'Brien).

188. 77 P.3d 1192 (Wash. Ct. App. 2003).

189. *Id.* at 1196.

190. *Id.* at 1201.

191. *E.g.*, Kathleen J. Craft et al., *Application of Plant DNA Markers in Forensic Botany: Genetic Comparison of* Quercus *Evidence Leaves to Crime Scene Trees Using Microsatellites*, 165 Forensic Sci. Int'l 64 (2007) (differentiation of oak tree leaves); Christine Kubik et al., *Genetic Diversity in Seven Perennial Ryegrass (*Lolium perenne *L.) Cultivars Based on SSR Markers*, 41 Crop Sci. 1565 (2001) (210 ryegrass samples correctly assigned to seven cultivars).

192. *E.g.*, Bridgett M. vonHoldt et al., *Genome-wide SNP and Haplotype Analyses Reveal a Rich History Underlying Dog Domestication*, 464 Nature 898 (2010) (48,000 SNPs in 912 dogs and 225 wolves).

193. Joy Halverson & Christopher J. Basten, *A PCR Multiplex and Database for Forensic DNA Identification of Dogs*, 50 J. Forensic Sci. 352 (2005); Marilyn A. Menotti-Raymond et al., *An STR Forensic Typing System for Genetic Individualization of Domestic Cat (*Felis catus*) Samples*, 50 J. Forensic Sci. 1061 (2005); L.H.P. van de Goor et al., *Population Studies of 16 Bovine STR Loci for Forensic*

basic reproductive patterns of the species in question. The simple product rule (Section IV) assumes that the sexually reproducing species mates at random with regard to the STR loci used for typing. When that is the case, the usual STR alleles and loci can be regarded as independent. But if mating is nonrandom, as occurs when individuals within a species are selectively bred to obtain some property such as coat color, body type, or behavioral repertoire, or as occurs when a species exists in geographically distinct subpopulations, the inheritance of loci may no longer be independent. Because it cannot be assumed a priori that a crime scene sample originates from a mixed-breed animal, inbreeding normally must be accounted for.[194]

A different approach is called for if the species is not sexually reproducing. For example, many plants, some simple animals, and bacteria reproduce asexually. With asexual reproduction, most offspring are genetically identical to the parent. All the individuals that originate from a common parent constitute, collectively, a clone. The major source of genetic variation in asexually reproducing species is mutation. When a mutation occurs, a new clonal lineage is created. Individuals in the original clonal lineage continue to propagate, and two clonal lineages now exist where before there was one. Thus, in species that reproduce asexually, genetic testing distinguishes clones, not individuals; hence, the product rule cannot be applied to estimate genotype frequencies for individuals. Rather, the frequency of a particular clone in a population of clones must be determined by direct observation. For example, if a rose thorn found on a suspect's clothing were to be identified as originating from a particular cultivar of rose, the relevant question becomes how common that variety of rose bush is and where it is located in the community.

In short, the approach for estimating a genotype frequency depends on the reproductive pattern and population genetics of the species. In cases involving unusual organisms, a court will need to rely on experts with sufficient knowledge of the species to verify that the method for estimating genotype frequencies is appropriate.

Purposes, 125 Int'l J.L. Med. 111 (2009). *But see* People v. Sutherland, 860 N.E.2d 178 (Ill. 2006) (conflicting expert testimony on the representativeness of dog databases); Barbara van Asch & Filipe Pereira, *State-of-the-Art and Future Prospects of Canine STR-Based Genotyping*, 3 Open Forensic Sci. J. 45 (2010). (recommending collaborative efforts for standardization and additional development of population databases)..

194. This can be done either by using the affinal model for a structured population or by using the probability of a match to a littermate or other closely related animal in lieu of the general random-match probability. *See Sutherland*, 860 N.E.2d 178 (describing such testimony).

Glossary of Terms

adenine (A). One of the four bases, or nucleotides, that make up the DNA molecule. Adenine binds only to thymine. See nucleotide.

affinal method. A method for computing the single-locus profile probabilities for a theoretical subpopulation by adjusting the single-locus profile probability, calculated with the product rule from the mixed population database, by the amount of heterogeneity across subpopulations. The model is appropriate even if there is no database available for a particular subpopulation, and the formula always gives more conservative probabilities than the product rule applied to the same database.

allele. In classical genetics, an allele is one of several alternative forms of a gene. A biallelic gene has two variants; others have more. Alleles are inherited separately from each parent, and for a given gene, an individual may have two different alleles (heterozygosity) or the same allele (homozygosity). In DNA analysis, the term is applied to any DNA region (even if it is not a gene) used for analysis.

allelic ladder. A mixture of all the common alleles at a given locus. Periodically producing electropherograms of the allelic ladder aids in designating the alleles detected in an unknown sample. The positions of the peaks for the unknown can be compared to the positions in a ladder electropherogram produced near the time when the unknown was analyzed. Peaks that do not match up with the ladder require further analysis.

Alu sequences. A family of short interspersed elements (SINEs) distributed throughout the genomes of primates.

amplification. Increasing the number of copies of a DNA region, usually by PCR.

amplified fragment length polymorphism (AMP-FLP). A DNA identification technique that uses PCR-amplified DNA fragments of varying lengths. The DS180 locus is a VNTR whose alleles can be detected with this technique.

antibody. A protein (immunoglobulin) molecule, produced by the immune system, that recognizes a particular foreign antigen and binds to it; if the antigen is on the surface of a cell, this binding leads to cell aggregation and subsequent destruction.

antigen. A molecule (typically found in the surface of a cell) whose shape triggers the production of antibodies that will bind to the antigen.

autoradiograph (autoradiogram, autorad). In RFLP analysis, the X-ray film (or print) showing the positions of radioactively marked fragments (bands) of DNA, indicating how far these fragments have migrated, and hence their molecular weights.

autosome. A chromosome other than the X and Y sex chromosomes.

band. See autoradiograph.

band shift. Movement of DNA fragments in one lane of a gel at a different rate than fragments of an identical length in another lane, resulting in the same pattern "shifted" up or down relative to the comparison lane. Band shift does not necessarily occur at the same rate in all portions of the gel.

base pair (bp). Two complementary nucleotides bonded together at the matching bases (A and T or C and G) along the double helix "backbone" of the DNA molecule. The length of a DNA fragment often is measured in numbers of base pairs (1 kilobase (kb) = 1000 bp); base-pair numbers also are used to describe the location of an allele on the DNA strand.

Bayes' theorem. A formula that relates certain conditional probabilities. It can be used to describe the impact of new data on the probability that a hypothesis is true. See the chapter on statistics in this manual.

bin, fixed. In VNTR profiling, a bin is a range of base pairs (DNA fragment lengths). When a database is divided into fixed bins, the proportion of bands within each bin is determined and the relevant proportions are used in estimating the profile frequency.

binning. Grouping VNTR alleles into sets of similar sizes because the alleles' lengths are too similar to differentiate.

bins, floating. In VNTR profiling, a bin is a range of base pairs (DNA fragment lengths). In a floating bin method of estimating a profile frequency, the bin is centered on the base-pair length of the allele in question, and the width of the bin can be defined by the laboratory's matching rule (e.g., ±5% of band size).

blind proficiency test. See proficiency test.

capillary electrophoresis. A method for separating DNA fragments (including STRs) according to their lengths. A long, narrow tube is filled with an entangled polymer or comparable sieving medium, and an electric field is applied to pull DNA fragments placed at one end of the tube through the medium. The procedure is faster and uses smaller samples than gel electrophoresis, and it can be automated.

ceiling principle. A procedure for setting a minimum DNA profile frequency proposed in 1992 by a committee of the National Academy of Sciences. One hundred persons from each of 15 to 20 genetically homogeneous populations spanning the range of racial groups in the United States are sampled. For each allele, the higher frequency among the groups sampled (or 5%, whichever is larger) is used in calculating the profile frequency. Compare interim ceiling principle.

chip. A miniaturized system for genetic analysis. One such chip mimics capillary electrophoresis and related manipulations. DNA fragments, pulled by small voltages, move through tiny channels etched into a small block of glass, silicon, quartz, or plastic. This system should be useful in analyzing STRs. Another technique mimics reverse dot blots by placing a large array of oligonucleotide probes on a solid surface. Such hybridization arrays are useful in identifying SNPs and in sequencing mitochondrial DNA.

chromosome. A rodlike structure composed of DNA, RNA, and proteins. Most normal human cells contain 46 chromosomes, 22 autosomes and a sex chromosome (X) inherited from the mother, and another 22 autosomes and one sex chromosome (either X or Y) inherited from the father. The genes are located along the chromosomes. See also homologous chromosomes.

coding and noncoding DNA. The sequence in which the building blocks (amino acids) of a protein are arranged corresponds to the sequence of base pairs within a gene. (A sequence of three base pairs specifies a particular one of the 20 possible amino acids in the protein. The mapping of a set of three nucleotide bases to a particular amino acid is the genetic code. The cell makes the protein through intermediate steps involving coding RNA transcripts.) About 1.5% of the human genome codes for the amino acid sequences. Another 23.5% of the genome is classified as genetic sequence but does not encode proteins. This portion of the noncoding DNA is involved in regulating the activity of genes. It includes promoters, enhancers, and repressors. Other gene-related DNA consists of introns (that interrupt the coding sequences, called exons, in genes and that are edited out of the RNA transcript for the protein), pseudogenes (evolutionary remnants of once-functional genes), and gene fragments. The remaining, extragenic DNA (about 75% of the genome) also is noncoding.

CODIS (combined DNA index system). A collection of databases on STR and other loci of convicted felons, maintained by the FBI.

complementary sequence. The sequence of nucleotides on one strand of DNA that corresponds to the sequence on the other strand. For example, if one sequence is CTGAA, the complementary bases are GACTT.

control region. See D-loop.

cytoplasm. A jelly-like material (80% water) that fills the cell.

cytosine (C). One of the four bases, or nucleotides, that make up the DNA double helix. Cytosine binds only to guanine. See nucleotide.

database. A collection of DNA profiles.

degradation. The breaking down of DNA by chemical or physical means.

denature, denaturation. The process of splitting, as by heating, two complementary strands of the DNA double helix into single strands in preparation for hybridization with biological probes.

deoxyribonucleic acid (DNA). The molecule that contains genetic information. DNA is composed of nucleotide building blocks, each containing a base (A, C, G, or T), a phosphate, and a sugar. These nucleotides are linked together in a double helix—two strands of DNA molecules paired up at complementary bases (A with T, C with G). See adenine, cytosine, guanine, thymine.

diploid number. See haploid number.

D-loop. A portion of the mitochrondrial genome known as the "control region" or "displacement loop" instrumental in the regulation and initiation of mtDNA gene products. Two short "hypervariable" regions within the D-loop do not appear to be functional and are the sequences used in identity or kinship testing.

DNA polymerase. The enzyme that catalyzes the synthesis of double-stranded DNA.

DNA probe. See probe.

DNA profile. The alleles at each locus. For example, a VNTR profile is the pattern of band lengths on an autorad. A multilocus profile represents the combined results of multiple probes. *See* genotype.

DNA sequence. The ordered list of base pairs in a duplex DNA molecule or of bases in a single strand.

DQ. The antigen that is the product of the DQA gene. See DQA, human leukocyte antigen.

DQA. The gene that codes for a particular class of human leukocyte antigen (HLA). This gene has been sequenced completely and can be used for forensic typing. See human leukocyte antigen.

EDTA. A preservative added to blood samples.

electropherogram. The PCR products separated by capillary electrophoresis can be labeled with a dye that glows at a given wavelength in response to light shined on it. As the tagged fragments pass the light source, an electronic camera records the intensity of the fluorescence. Plotting the intensity as a function of time produces a series of peaks, with the shorter fragments producing peaks sooner. The intensity is measured in relative fluorescent units and is proportional to the number of glowing fragments passing by the detector. The graph of the intensity over time is an electropherogram.

electrophoresis. See capillary electrophoresis, gel electrophoresis.

endonuclease. An enzyme that cleaves the phosphodiester bond within a nucleotide chain.

environmental insult. Exposure of DNA to external agents such as heat, moisture, and ultraviolet radiation, or chemical or bacterial agents. Such exposure can interfere with the enzymes used in the testing process or otherwise make DNA difficult to analyze.

enzyme. A protein that catalyzes (speeds up or slows down) a reaction.

epigenetic. Heritable changes in phenotype (appearance) or gene expression caused by mechanisms other than changes in the underlying DNA sequence. Epigenetic marks are molecules attached to DNA that can determine whether genes are active and used by the cell.

ethidium bromide. A molecule that can intercalate into DNA double helices when the helix is under torsional stress. Used to identify the presence of DNA in a sample by its fluorescence under ultraviolet light.

exon. See coding and noncoding DNA.

fallacy of the transposed conditional. See transposition fallacy.

false match. Two samples of DNA that have different profiles could be declared to match if, instead of measuring the distinct DNA in each sample, there is an error in handling or preparing samples such that the DNA from a single sample is analyzed twice. The resulting match, which does not reflect the true profiles of the DNA from each sample, is a false match. Some people use "false match" more broadly, to include cases in which the true profiles of each sample are the same, but the samples come from different individuals. Compare true match. See also match, random match.

gel, agarose. A semisolid medium used to separate molecules by electrophoresis.

gel electrophoresis. In RFLP analysis, the process of sorting DNA fragments by size by applying an electric current to a gel. The different-size fragments move at different rates through the gel.

gene. A set of nucleotide base pairs on a chromosome that contains the "instructions" for controlling some cellular function such as making an enzyme. The gene is the fundamental unit of heredity; each simple gene "codes" for a specific biological characteristic.

gene frequency. The relative frequency (proportion) of an allele in a population.

genetic drift. Random fluctuation in a population's allele frequencies from generation to generation.

genetics. The study of the patterns, processes, and mechanisms of inheritance of biological characteristics.

genome. The complete genetic makeup of an organism, including roughly 23,000 genes and many other DNA sequences in humans. Over three billion nucleotide base pairs comprise the haploid human genome.

genotype. The particular forms (alleles) of a set of genes possessed by an organism (as distinguished from phenotype, which refers to how the genotype expresses itself, as in physical appearance). In DNA analysis, the term is applied to the variations within all DNA regions (whether or not they constitute genes) that are analyzed.

genotype, multilocus. The alleles that an organism possesses at several sites in its genome.

genotype, single locus. The alleles that an organism possesses at a particular site in its genome.

guanine (G). One of the four bases, or nucleotides, that make up the DNA double helix. Guanine binds only to cytosine. See nucleotide.

haploid number. Human sex cells (egg and sperm) contain 23 chromosomes each. This is the haploid number. When a sperm cell fertilizes an egg cell, the number of chromosomes doubles to 46. This is the diploid number.

haplotype. A specific combination of linked alleles at several loci.

Hardy-Weinberg equilibrium. A condition in which the allele frequencies within a large, random, intrabreeding population are unrelated to patterns of mating. In this condition, the occurrence of alleles from each parent will be independent and have a joint frequency estimated by the product rule. See independence, linkage disequilibrium.

heteroplasmy, heteroplasty. The condition in which some copies of mitochondrial DNA in the same individual have different base pairs at certain points.

heterozygous. Having a different allele at a given locus on each of a pair of homologous chromosomes. See allele. Compare homozygous.

homologous chromosomes. The 44 autosomes (nonsex chromosomes) in the normal human genome are in homologous pairs (one from each parent) that share an identical set of genes, but may have different alleles at the same loci.

homozygous. Having the same allele at a given locus on each of a pair of homologous chromosomes. See allele. Compare heterozygous.

human leukocyte antigen (HLA). Antigen (foreign body that stimulates an immune system response) located on the surface of most cells (excluding red blood cells and sperm cells). HLAs differ among individuals and are associated closely with transplant rejection. See DQA.

hybridization. Pairing up of complementary strands of DNA from different sources at the matching base-pair sites. For example, a primer with the sequence AGGTCT would bond with the complementary sequence TCCAGA on a DNA fragment.

independence. Two events are said to be independent if one is neither more nor less likely to occur when the other does.

interim ceiling principle. A procedure proposed in 1992 by a committee of the National Academy of Sciences for setting a minimum DNA profile frequency. For each allele, the highest frequency (adjusted upward for sampling error) found in any major racial group (or 10%, whichever is higher), is used in product-rule calculations. Compare ceiling principle.

intron. See coding and noncoding DNA.

kilobase (kb). A measure of DNA length (1000 bases).

likelihood ratio. A measure of the support that an observation provides for one hypothesis as opposed to an alternative hypothesis. The likelihood ratio is computed by dividing the conditional probability of the observation given that one hypothesis is true by the conditional probability of the observation given the alternative hypothesis. For example, the likelihood ratio for the hypothesis that two DNA samples with the same STR profile originated from the same individual (as opposed to originating from two unrelated individuals) is the reciprocal of the random-match probability. Legal scholars have introduced the likelihood ratio as a measure of the probative value of evidence. Evidence that is 100 times more probable to be observed when one hypothesis is true as opposed to another has more probative value than evidence that is only twice as probable.

linkage. The inheritance together of two or more genes on the same chromosome.

linkage equilibrium. A condition in which the occurrence of alleles at different loci is independent.

locus. A location in the genome, that is, a position on a chromosome where a gene or other structure begins.

mass spectroscopy. The separation of elements or molecules according to their molecular weight. In the version being developed for DNA analysis, small quantities of PCR-amplified fragments are irradiated with a laser to form gaseous ions that traverse a fixed distance. Heavier ions have longer times of flight, and the process is known as matrix-assisted laser desorption-ionization time-of-flight mass spectroscopy. MALDI-TOF-MS, as it is abbreviated, may be useful in analyzing STRs.

match. The presence of the same allele or alleles in two samples. Two DNA profiles are declared to match when they are indistinguishable in genetic type. For loci with discrete alleles, two samples match when they display the same set of alleles. For RFLP testing of VNTRs, two samples match when the pattern of the bands is similar and the positions of the corresponding bands at each locus fall within a preset distance. See match window, false match, true match.

match window. If two RFLP bands lie within a preset distance, called the match window, that reflects normal measurement error, they can be declared to match.

microsatellite. Another term for an STR.

minisatellite. Another term for a VNTR.

mitochondria. A structure (organelle) within nucleated (eukaryotic) cells that is the site of the energy-producing reactions within the cell. Mitochondria contain their own DNA (often abbreviated as mtDNA), which is inherited only from mother to child.

molecular weight. The weight in grams of 1 mole (approximately 6.02×10^{23} molecules) of a pure, molecular substance.

monomorphic. A gene or DNA characteristic that is almost always found in only one form in a population. Compare polymorphism.

multilocus probe. A probe that marks multiple sites (loci). RFLP analysis using a multilocus probe will yield an autorad showing a striped pattern of 30 or more bands. Such probes are no longer used in forensic applications.

multilocus profile. See profile.

multiplexing. Typing several loci simultaneously.

mutation. The process that produces a gene or chromosome set differing from the type already in the population; the gene or chromosome set that results from such a process.

nanogram (ng). A billionth of a gram.

nucleic acid. RNA or DNA.

nucleotide. A unit of DNA consisting of a base (A, C, G, or T) and attached to a phosphate and a sugar group; the basic building block of nucleic acids. See deoxyribonucleic acid.

nucleus. The membrane-covered portion of a eukaryotic cell containing most of the DNA and found within the cytoplasm.

oligonucleotide. A synthetic polymer made up of fewer than 100 nucleotides; used as a primer or a probe in PCR. See primer.

paternity index. A number (technically, a likelihood ratio) that indicates the support that the paternity test results lend to the hypothesis that the alleged father is the biological father as opposed to the hypothesis that another man selected at random is the biological father. Assuming that the observed phenotypes correctly represent the phenotypes of the mother, child, and alleged father tested, the number can be computed as the ratio of the probability of the phenotypes under the first hypothesis to the probability under the second hypothesis. Large values indicate substantial support for the hypothesis of paternity; values near zero indicate substantial support for the hypothesis that someone other than the alleged father is the biological father; and values near unity indicate that the results do not help in determining which hypothesis is correct.

pH. A measure of the acidity of a solution.

phenotype. A trait, such as eye color or blood group, resulting from a genotype.

point mutation. See SNP.

polymarker. A commercially marketed set of PCR-based tests for protein polymorphisms.

polymerase chain reaction (PCR). A process that mimics DNA's own replication processes to make up to millions of copies of short strands of genetic material in a few hours.

polymorphism. The presence of several forms of a gene or DNA characteristic in a population.

population genetics. The study of the genetic composition of groups of individuals.

population structure. When a population is divided into subgroups that do not mix freely, that population is said to have structure. Significant structure can lead to allele frequencies being different in the subpopulations.

primer. An oligonucleotide that attaches to one end of a DNA fragment and provides a point for more complementary nucleotides to attach and replicate the DNA strand. See oligonucleotide.

probe. In forensics, a short segment of DNA used to detect certain alleles. The probe hybridizes, or matches up, to a specific complementary sequence. Probes allow visualization of the hybridized DNA, either by a radioactive tag (usually used for RFLP analysis) or a biochemical tag (usually used for PCR-based analyses).

product rule. When alleles occur independently at each locus (Hardy-Weinberg equilibrium) and across loci (linkage equilibrium), the proportion of the population with a given genotype is the product of the proportion of each allele at each locus, times factors of two for heterozygous loci.

proficiency test. A test administered at a laboratory to evaluate its performance. In a blind proficiency study, the laboratory personnel do not know that they are being tested.

prosecutor's fallacy. See transposition fallacy.

protein. A class of biologically important molecules made up of a linear string of building blocks called amino acids. The order in which these components are arranged is encoded in the DNA sequence of the gene that expresses the protein. See coding DNA.

pseudogenes. Genes that have been so disabled by mutations that they can no longer produce proteins. Some pseudogenes can still produce noncoding RNA.

quality assurance. A program conducted by a laboratory to ensure accuracy and reliability.

quality audit. A systematic and independent examination and evaluation of a laboratory's operations.

quality control. Activities used to monitor the ability of DNA typing to meet specified criteria.

random match. A match in the DNA profiles of two samples of DNA, where one is drawn at random from the population. See also random-match probability.

random-match probability. The chance of a random match. As it is usually used in court, the random-match probability refers to the probability of a true match when the DNA being compared to the evidence DNA comes from a person drawn at random from the population. This random true match probability reveals the probability of a true match when the samples of DNA come from different, unrelated people.

random mating. The members of a population are said to mate randomly with respect to particular genes of DNA characteristics when the choice of mates is independent of the alleles.

recombination. In general, any process in a diploid or partially diploid cell that generates new gene or chromosomal combinations not found in that cell or in its progenitors.

reference population. The population to which the perpetrator of a crime is thought to belong.

relative fluorescent unit (RFU). See electropherogram.

replication. The synthesis of new DNA from existing DNA. See polymerase chain reaction.

restriction enzyme. Protein that cuts double-stranded DNA at specific base-pair sequences (different enzymes recognize different sequences). See restriction site.

restriction fragment length polymorphism (RFLP). Variation among people in the length of a segment of DNA cut at two restriction sites.

restriction fragment length polymorphism (RFLP) analysis. Analysis of individual variations in the lengths of DNA fragments produced by digesting sample DNA with a restriction enzyme.

restriction site. A sequence marking the location at which a restriction enzyme cuts DNA into fragments. See restriction enzyme.

reverse dot blot. A detection method used to identify SNPs in which DNA probes are affixed to a membrane, and amplified DNA is passed over the probes to see if it contains the complementary sequence.

ribonucleic acid (RNA). A single-stranded molecule "transcribed" from DNA. "Coding" RNA acts as a template for building proteins according the sequences in the coding DNA from which it is transcribed. Other RNA transcripts can be a sensor for detecting signals that affect gene expression, a switch for turning genes off or on, or they may be functionless.

sequence-specific oligonucleotide (SSO) probe. Also, allele-specific oligonucleotide (ASO) probe. Oligonucleotide probes used in a PCR-associated detection technique to identify the presence or absence of certain base-pair sequences identifying different alleles. The probes are visualized by an array of dots rather than by the electrophoretograms associated with STR analysis.

sequencing. Determining the order of base pairs in a segment of DNA.

short tandem repeat (STR). See variable number tandem repeat.

single-locus probe. A probe that only marks a specific site (locus). RFLP analysis using a single-locus probe will yield an autorad showing one band if the individual is homozygous, two bands if heterozygous. Likewise, the probe will produce one or two peaks in an STR electrophoretogram.

SNP (single nucleotide polymorphism). A substitution, insertion, or deletion of a single base pair at a given point in the genome.

SNP chip. See chip.

Southern blotting. Named for its inventor, a technique by which processed DNA fragments, separated by gel electrophoresis, are transferred onto a nylon membrane in preparation for the application of biological probes.

thymine (T). One of the four bases, or nucleotides, that make up the DNA double helix. Thymine binds only to adenine. See nucleotide.

transposition fallacy. Also called the prosecutor's fallacy, the transposition fallacy confuses the conditional probability of A given B [$P(A|B)$] with that of B given A [$P(B|A)$]. Few people think that the probability that a person speaks Spanish (A) given that he or she is a citizen of Chile (B) equals the probability that a person is a citizen of Chile (B) given that he or she speaks Spanish (A). Yet, many court opinions, newspaper articles, and even some expert witnesses speak of the probability of a matching DNA genotype (A) given that someone other than the defendant is the source of the crime scene DNA (B) as if it were the probability of someone else being the source (B) given the matching profile (A). Transposing conditional probabilities correctly requires Bayes' theorem.

true match. Two samples of DNA that have the same profile should match when tested. If there is no error in the labeling, handling, and analysis of the samples and in the reporting of the results, a match is a true match. A true match establishes that the two samples of DNA have the same profile. Unless the profile is unique, however, a true match does not conclusively prove that the two samples came from the same source. Some people use "true match" more narrowly, to mean only those matches among samples from the same source. Compare false match. See also match, random match.

variable number tandem repeat (VNTR). A class of RFLPs resulting from multiple copies of virtually identical base-pair sequences, arranged in succession at a specific locus on a chromosome. The number of repeats varies from individual to individual, thus providing a basis for individual recognition. VNTRs are longer than STRs.

window. See match window.

X chromosome. See chromosome.

Y chromosome. See chromosome.

References on DNA

Forensic DNA Interpretation (John Buckleton et al. eds., 2005).

John M. Butler, Fundamentals of Forensic DNA Typing (2010).

Ian W. Evett & Bruce S. Weir, Interpreting DNA Evidence: Statistical Genetics for Forensic Scientists (1998).

William Goodwin et al., An Introduction to Forensic Genetics (2d ed. 2011).

David H. Kaye, The Double Helix and the Law of Evidence (2010).

National Research Council Committee on DNA Forensic Science: An Update, The Evaluation of Forensic DNA Evidence (1996).

National Research Council Committee on DNA Technology in Forensic Science, DNA Technology in Forensic Science (1992).

The President's DNA Initiative, Forensic DNA Resources for Specific Audiences, *available at* www.dna.gov/audiences/.

Reference Guide on Statistics

DAVID H. KAYE AND DAVID A. FREEDMAN

David H. Kaye, M.A., J.D., is Distinguished Professor of Law and Weiss Family Scholar, The Pennsylvania State University, University Park, and Regents' Professor Emeritus, Arizona State University Sandra Day O'Connor College of Law and School of Life Sciences, Tempe.

David A. Freedman, Ph.D., was Professor of Statistics, University of California, Berkeley.

[Editor's Note: Sadly, Professor Freedman passed away during the production of this manual.]

CONTENTS

I. Introduction

Statistical assessments are prominent in many kinds of legal cases, including antitrust, employment discrimination, toxic torts, and voting rights cases.[1] This reference guide describes the elements of statistical reasoning. We hope the explanations will help judges and lawyers to understand statistical terminology, to see the strengths and weaknesses of statistical arguments, and to apply relevant legal doctrine. The guide is organized as follows:

- Section I provides an overview of the field, discusses the admissibility of statistical studies, and offers some suggestions about procedures that encourage the best use of statistical evidence.
- Section II addresses data collection and explains why the design of a study is the most important determinant of its quality. This section compares experiments with observational studies and surveys with censuses, indicating when the various kinds of study are likely to provide useful results.
- Section III discusses the art of summarizing data. This section considers the mean, median, and standard deviation. These are basic descriptive statistics, and most statistical analyses use them as building blocks. This section also discusses patterns in data that are brought out by graphs, percentages, and tables.
- Section IV describes the logic of statistical inference, emphasizing foundations and disclosing limitations. This section covers estimation, standard errors and confidence intervals, *p*-values, and hypothesis tests.
- Section V shows how associations can be described by scatter diagrams, correlation coefficients, and regression lines. Regression is often used to infer causation from association. This section explains the technique, indicating the circumstances under which it and other statistical models are likely to succeed—or fail.
- An appendix provides some technical details.
- The glossary defines statistical terms that may be encountered in litigation.

1. *See generally* Statistical Science in the Courtroom (Joseph L. Gastwirth ed., 2000); Statistics and the Law (Morris H. DeGroot et al. eds., 1986); National Research Council, The Evolving Role of Statistical Assessments as Evidence in the Courts (Stephen E. Fienberg ed., 1989) [hereinafter The Evolving Role of Statistical Assessments as Evidence in the Courts]; Michael O. Finkelstein & Bruce Levin, Statistics for Lawyers (2d ed. 2001); 1 & 2 Joseph L. Gastwirth, Statistical Reasoning in Law and Public Policy (1988); Hans Zeisel & David Kaye, Prove It with Figures: Empirical Methods in Law and Litigation (1997).

A. Admissibility and Weight of Statistical Studies

Statistical studies suitably designed to address a material issue generally will be admissible under the Federal Rules of Evidence. The hearsay rule rarely is a serious barrier to the presentation of statistical studies, because such studies may be offered to explain the basis for an expert's opinion or may be admissible under the learned treatise exception to the hearsay rule.[2] Because most statistical methods relied on in court are described in textbooks or journal articles and are capable of producing useful results when properly applied, these methods generally satisfy important aspects of the "scientific knowledge" requirement in *Daubert v. Merrell Dow Pharmaceuticals, Inc.*[3] Of course, a particular study may use a method that is entirely appropriate but that is so poorly executed that it should be inadmissible under Federal Rules of Evidence 403 and 702.[4] Or, the method may be inappropriate for the problem at hand and thus lack the "fit" spoken of in *Daubert.*[5] Or the study might rest on data of the type not reasonably relied on by statisticians or substantive experts and hence run afoul of Federal Rule of Evidence 703. Often, however, the battle over statistical evidence concerns weight or sufficiency rather than admissibility.

B. Varieties and Limits of Statistical Expertise

For convenience, the field of statistics may be divided into three subfields: probability theory, theoretical statistics, and applied statistics. Probability theory is the mathematical study of outcomes that are governed, at least in part, by chance. Theoretical statistics is about the properties of statistical procedures, including error rates; probability theory plays a key role in this endeavor. Applied statistics draws on both of these fields to develop techniques for collecting or analyzing particular types of data.

2. *See generally* 2 McCormick on Evidence §§ 321, 324.3 (Kenneth S. Broun ed., 6th ed. 2006). Studies published by government agencies also may be admissible as public records. *Id.* § 296.

3. 509 U.S. 579, 589–90 (1993).

4. *See* Kumho Tire Co. v. Carmichael, 526 U.S. 137, 152 (1999) (suggesting that the trial court should "make certain that an expert, whether basing testimony upon professional studies or personal experience, employs in the courtroom the same level of intellectual rigor that characterizes the practice of an expert in the relevant field."); Malletier v. Dooney & Bourke, Inc., 525 F. Supp. 2d 558, 562–63 (S.D.N.Y. 2007) ("While errors in a survey's methodology usually go to the weight accorded to the conclusions rather than its admissibility, . . . 'there will be occasions when the proffered survey is so flawed as to be completely unhelpful to the trier of fact.'") (quoting AHP Subsidiary Holding Co. v. Stuart Hale Co., 1 F.3d 611, 618 (7th Cir.1993)).

5. *Daubert*, 509 U.S. at 591; Anderson v. Westinghouse Savannah River Co., 406 F.3d 248 (4th Cir. 2005) (motion to exclude statistical analysis that compared black and white employees without adequately taking into account differences in their job titles or positions was properly granted under *Daubert*); *Malletier*, 525 F. Supp. 2d at 569 (excluding a consumer survey for "a lack of fit between the survey's questions and the law of dilution" and errors in the execution of the survey).

Statistical expertise is not confined to those with degrees in statistics. Because statistical reasoning underlies many kinds of empirical research, scholars in a variety of fields—including biology, economics, epidemiology, political science, and psychology—are exposed to statistical ideas, with an emphasis on the methods most important to the discipline.

Experts who specialize in using statistical methods, and whose professional careers demonstrate this orientation, are most likely to use appropriate procedures and correctly interpret the results. By contrast, forensic scientists often lack basic information about the studies underlying their testimony. *State v. Garrison*[6] illustrates the problem. In this murder prosecution involving bite mark evidence, a dentist was allowed to testify that "the probability factor of two sets of teeth being identical in a case similar to this is, approximately, eight in one million," even though "he was unaware of the formula utilized to arrive at that figure other than that it was 'computerized.'"[7]

At the same time, the choice of which data to examine, or how best to model a particular process, could require subject matter expertise that a statistician lacks. As a result, cases involving statistical evidence frequently are (or should be) "two expert" cases of interlocking testimony. A labor economist, for example, may supply a definition of the relevant labor market from which an employer draws its employees; the statistical expert may then compare the race of new hires to the racial composition of the labor market. Naturally, the value of the statistical analysis depends on the substantive knowledge that informs it.[8]

C. Procedures That Enhance Statistical Testimony

1. Maintaining professional autonomy

Ideally, experts who conduct research in the context of litigation should proceed with the same objectivity that would be required in other contexts. Thus, experts who testify (or who supply results used in testimony) should conduct the analysis required to address in a professionally responsible fashion the issues posed by the litigation.[9] Questions about the freedom of inquiry accorded to testifying experts,

6. 585 P.2d 563 (Ariz. 1978).

7. *Id.* at 566, 568. For other examples, see David H. Kaye et al., The New Wigmore: A Treatise on Evidence: Expert Evidence § 12.2 (2d ed. 2011).

8. In *Vuyanich v. Republic National Bank*, 505 F. Supp. 224, 319 (N.D. Tex. 1980), *vacated*, 723 F.2d 1195 (5th Cir. 1984), defendant's statistical expert criticized the plaintiffs' statistical model for an implicit, but restrictive, assumption about male and female salaries. The district court trying the case accepted the model because the plaintiffs' expert had a "very strong guess" about the assumption, and her expertise included labor economics as well as statistics. *Id.* It is doubtful, however, that economic knowledge sheds much light on the assumption, and it would have been simple to perform a less restrictive analysis.

9. *See* The Evolving Role of Statistical Assessments as Evidence in the Courts, *supra* note 1, at 164 (recommending that the expert be free to consult with colleagues who have not been retained

as well as the scope and depth of their investigations, may reveal some of the limitations to the testimony.

2. *Disclosing other analyses*

Statisticians analyze data using a variety of methods. There is much to be said for looking at the data in several ways. To permit a fair evaluation of the analysis that is eventually settled on, however, the testifying expert can be asked to explain how that approach was developed. According to some commentators, counsel who know of analyses that do not support the client's position should reveal them, rather than presenting only favorable results.[10]

3. *Disclosing data and analytical methods before trial*

The collection of data often is expensive and subject to errors and omissions. Moreover, careful exploration of the data can be time-consuming. To minimize debates at trial over the accuracy of data and the choice of analytical techniques, pretrial discovery procedures should be used, particularly with respect to the quality of the data and the method of analysis.[11]

II. How Have the Data Been Collected?

The interpretation of data often depends on understanding "study design"—the plan for a statistical study and its implementation.[12] Different designs are suited to answering different questions. Also, flaws in the data can undermine any statistical analysis, and data quality is often determined by study design.

In many cases, statistical studies are used to show causation. Do food additives cause cancer? Does capital punishment deter crime? Would additional disclosures

by any party to the litigation and that the expert receive a letter of engagement providing for these and other safeguards).

10. *Id.* at 167; *cf.* William W. Schwarzer, *In Defense of "Automatic Disclosure in Discovery,"* 27 Ga. L. Rev. 655, 658–59 (1993) ("[T]he lawyer owes a duty to the court to make disclosure of core information."). The National Research Council also recommends that "if a party gives statistical data to different experts for competing analyses, that fact be disclosed to the testifying expert, if any." The Evolving Role of Statistical Assessments as Evidence in the Courts, *supra* note 1, at 167.

11. *See* The Special Comm. on Empirical Data in Legal Decision Making, Recommendations on Pretrial Proceedings in Cases with Voluminous Data, *reprinted in* The Evolving Role of Statistical Assessments as Evidence in the Courts, *supra* note 1, app. F; *see also* David H. Kaye, *Improving Legal Statistics*, 24 Law & Soc'y Rev. 1255 (1990).

12. For introductory treatments of data collection, see, for example, David Freedman et al., Statistics (4th ed. 2007); Darrell Huff, How to Lie with Statistics (1993); David S. Moore & William I. Notz, Statistics: Concepts and Controversies (6th ed. 2005); Hans Zeisel, Say It with Figures (6th ed. 1985); Zeisel & Kaye, *supra* note 1.

in a securities prospectus cause investors to behave differently? The design of studies to investigate causation is the first topic of this section.[13]

Sample data can be used to describe a population. The population is the whole class of units that are of interest; the sample is the set of units chosen for detailed study. Inferences from the part to the whole are justified when the sample is representative. Sampling is the second topic of this section.

Finally, the accuracy of the data will be considered. Because making and recording measurements is an error-prone activity, error rates should be assessed and the likely impact of errors considered. Data quality is the third topic of this section.

A. Is the Study Designed to Investigate Causation?

1. Types of studies

When causation is the issue, anecdotal evidence can be brought to bear. So can observational studies or controlled experiments. Anecdotal reports may be of value, but they are ordinarily more helpful in generating lines of inquiry than in proving causation.[14] Observational studies can establish that one factor is associ-

13. *See also* Michael D. Green et al., Reference Guide on Epidemiology, Section V, in this manual; Joseph Rodricks, Reference Guide on Exposure Science, Section E, in this manual.

14. In medicine, evidence from clinical practice can be the starting point for discovery of cause-and-effect relationships. For examples, see David A. Freedman, *On Types of Scientific Enquiry, in* The Oxford Handbook of Political Methodology 300 (Janet M. Box-Steffensmeier et al. eds., 2008). Anecdotal evidence is rarely definitive, and some courts have suggested that attempts to infer causation from anecdotal reports are inadmissible as unsound methodology under *Daubert v. Merrell Dow Pharmaceuticals, Inc.*, 509 U.S. 579 (1993). *See, e.g.*, McClain v. Metabolife Int'l, Inc., 401 F.3d 1233, 1244 (11th Cir. 2005) ("simply because a person takes drugs and then suffers an injury does not show causation. Drawing such a conclusion from temporal relationships leads to the blunder of the *post hoc ergo propter hoc* fallacy."); *In re* Baycol Prods. Litig., 532 F. Supp. 2d 1029, 1039–40 (D. Minn. 2007) (excluding a meta-analysis based on reports to the Food and Drug Administration of adverse events); Leblanc v. Chevron USA Inc., 513 F. Supp. 2d 641, 650 (E.D. La. 2007) (excluding plaintiffs' experts' opinions that benzene causes myelofibrosis because the causal hypothesis "that has been generated by case reports . . . has not been confirmed by the vast majority of epidemiologic studies of workers being exposed to benzene and more generally, petroleum products."), *vacated*, 275 Fed. App'x. 319 (5th Cir. 2008) (remanding for consideration of newer government report on health effects of benzene); *cf.* Matrixx Initiatives, Inc. v. Siracusano, 131 S. Ct. 1309, 1321 (2011) (concluding that adverse event reports combined with other information could be of concern to a reasonable investor and therefore subject to a requirement of disclosure under SEC Rule 10b-5, but stating that "the mere existence of reports of adverse events . . . says nothing in and of itself about whether the drug is causing the adverse events"). Other courts are more open to "differential diagnoses" based primarily on timing. *E.g.*, Best v. Lowe's Home Ctrs., Inc., 563 F.3d 171 (6th Cir. 2009) (reversing the exclusion of a physician's opinion that exposure to propenyl chloride caused a man to lose his sense of smell because of the timing in this one case and the physician's inability to attribute the change to anything else); Kaye et al., *supra* note 7, §§ 8.7.2 & 12.5.1. *See also Matrixx Initiatives*, *supra*, at 1322 (listing "a temporal relationship" in a single patient as one indication of "a reliable causal link").

ated with another, but work is needed to bridge the gap between association and causation. Randomized controlled experiments are ideally suited for demonstrating causation.

Anecdotal evidence usually amounts to reports that events of one kind are followed by events of another kind. Typically, the reports are not even sufficient to show association, because there is no comparison group. For example, some children who live near power lines develop leukemia. Does exposure to electrical and magnetic fields cause this disease? The anecdotal evidence is not compelling because leukemia also occurs among children without exposure.[15] It is necessary to compare disease rates among those who are exposed and those who are not. If exposure causes the disease, the rate should be higher among the exposed and lower among the unexposed. That would be association.

The next issue is crucial: Exposed and unexposed people may differ in ways other than the exposure they have experienced. For example, children who live near power lines could come from poorer families and be more at risk from other environmental hazards. Such differences can create the appearance of a cause-and-effect relationship. Other differences can mask a real relationship. Cause-and-effect relationships often are quite subtle, and carefully designed studies are needed to draw valid conclusions.

An epidemiological classic makes the point. At one time, it was thought that lung cancer was caused by fumes from tarring the roads, because many lung cancer patients lived near roads that recently had been tarred. This is anecdotal evidence. But the argument is incomplete. For one thing, most people—whether exposed to asphalt fumes or unexposed—did not develop lung cancer. A comparison of rates was needed. The epidemiologists found that exposed persons and unexposed persons suffered from lung cancer at similar rates: Tar was probably not the causal agent. Exposure to cigarette smoke, however, turned out to be strongly associated with lung cancer. This study, in combination with later ones, made a compelling case that smoking cigarettes is the main cause of lung cancer.[16]

A good study design compares outcomes for subjects who are exposed to some factor (the treatment group) with outcomes for other subjects who are

15. *See* National Research Council, Committee on the Possible Effects of Electromagnetic Fields on Biologic Systems (1997); Zeisel & Kaye, *supra* note 1, at 66–67. There are problems in measuring exposure to electromagnetic fields, and results are inconsistent from one study to another. For such reasons, the epidemiological evidence for an effect on health is inconclusive. National Research Council, *supra*; Zeisel & Kaye, *supra*; Edward W. Campion, *Power Lines, Cancer, and Fear*, 337 New Eng. J. Med. 44 (1997) (editorial); Martha S. Linet et al., *Residential Exposure to Magnetic Fields and Acute Lymphoblastic Leukemia in Children*, 337 New Eng. J. Med. 1 (1997); Gary Taubes, *Magnetic Field-Cancer Link: Will It Rest in Peace?*, 277 Science 29 (1997) (quoting various epidemiologists).

16. Richard Doll & A. Bradford Hill, *A Study of the Aetiology of Carcinoma of the Lung*, 2 Brit. Med. J. 1271 (1952). This was a matched case-control study. Cohort studies soon followed. *See* Green et al., *supra* note 13. For a review of the evidence on causation, see 38 International Agency for Research on Cancer (IARC), World Health Org., IARC Monographs on the Evaluation of the Carcinogenic Risk of Chemicals to Humans: Tobacco Smoking (1986).

not exposed (the control group). Now there is another important distinction to be made—that between controlled experiments and observational studies. In a controlled experiment, the investigators decide which subjects will be exposed and which subjects will go into the control group. In observational studies, by contrast, the subjects themselves choose their exposures. Because of self-selection, the treatment and control groups are likely to differ with respect to influential factors other than the one of primary interest. (These other factors are called lurking variables or confounding variables.)[17] With the health effects of power lines, family background is a possible confounder; so is exposure to other hazards. Many confounders have been proposed to explain the association between smoking and lung cancer, but careful epidemiological studies have ruled them out, one after the other.

Confounding remains a problem to reckon with, even for the best observational research. For example, women with herpes are more likely to develop cervical cancer than other women. Some investigators concluded that herpes caused cancer: In other words, they thought the association was causal. Later research showed that the primary cause of cervical cancer was human papilloma virus (HPV). Herpes was a marker of sexual activity. Women who had multiple sexual partners were more likely to be exposed not only to herpes but also to HPV. The association between herpes and cervical cancer was due to other variables.[18]

What are "variables?" In statistics, a variable is a characteristic of units in a study. With a study of people, the unit of analysis is the person. Typical variables include income (dollars per year) and educational level (years of schooling completed): These variables describe people. With a study of school districts, the unit of analysis is the district. Typical variables include average family income of district residents and average test scores of students in the district: These variables describe school districts.

When investigating a cause-and-effect relationship, the variable that represents the effect is called the dependent variable, because it depends on the causes. The variables that represent the causes are called independent variables. With a study of smoking and lung cancer, the independent variable would be smoking (e.g., number of cigarettes per day), and the dependent variable would mark the presence or absence of lung cancer. Dependent variables also are called outcome variables or response variables. Synonyms for independent variables are risk factors, predictors, and explanatory variables.

17. For example, a confounding variable may be correlated with the independent variable and act causally on the dependent variable. If the units being studied differ on the independent variable, they are also likely to differ on the confounder. The confounder—not the independent variable—could therefore be responsible for differences seen on the dependent variable.

18. For additional examples and further discussion, see Freedman et al., *supra* note 12, at 12–28, 150–52; David A. Freedman, *From Association to Causation: Some Remarks on the History of Statistics,* 14 Stat. Sci. 243 (1999). Some studies find that herpes is a "cofactor," which increases risk among women who are also exposed to HPV. Only certain strains of HPV are carcinogenic.

2. Randomized controlled experiments

In randomized controlled experiments, investigators assign subjects to treatment or control groups at random. The groups are therefore likely to be comparable, except for the treatment. This minimizes the role of confounding. Minor imbalances will remain, due to the play of random chance; the likely effect on study results can be assessed by statistical techniques.[19] The bottom line is that causal inferences based on well-executed randomized experiments are generally more secure than inferences based on well-executed observational studies.

The following example should help bring the discussion together. Today, we know that taking aspirin helps prevent heart attacks. But initially, there was some controversy. People who take aspirin rarely have heart attacks. This is anecdotal evidence for a protective effect, but it proves almost nothing. After all, few people have frequent heart attacks, whether or not they take aspirin regularly. A good study compares heart attack rates for two groups: people who take aspirin (the treatment group) and people who do not (the controls). An observational study would be easy to do, but in such a study the aspirin-takers are likely to be different from the controls. Indeed, they are likely to be sicker—that is why they are taking aspirin. The study would be biased against finding a protective effect. Randomized experiments are harder to do, but they provide better evidence. It is the experiments that demonstrate a protective effect.[20]

In summary, data from a treatment group without a control group generally reveal very little and can be misleading. Comparisons are essential. If subjects are assigned to treatment and control groups at random, a difference in the outcomes between the two groups can usually be accepted, within the limits of statistical error (*infra* Section IV), as a good measure of the treatment effect. However, if the groups are created in any other way, differences that existed before treatment may contribute to differences in the outcomes or mask differences that otherwise would become manifest. Observational studies succeed to the extent that the treatment and control groups are comparable—apart from the treatment.

3. Observational studies

The bulk of the statistical studies seen in court are observational, not experimental. Take the question of whether capital punishment deters murder. To conduct a randomized controlled experiment, people would need to be assigned randomly to a treatment group or a control group. People in the treatment group would know they were subject to the death penalty for murder; the

19. Randomization of subjects to treatment or control groups puts statistical tests of significance on a secure footing. Freedman et al., *supra* note 12, at 503–22, 545–63; *see infra* Section IV.

20. In other instances, experiments have banished strongly held beliefs. *E.g.*, Scott M. Lippman et al., Effect of Selenium and Vitamin E on Risk of Prostate Cancer and Other Cancers: The Selenium and Vitamin E Cancer Prevention Trial (SELECT), 301 JAMA 39 (2009).

controls would know that they were exempt. Conducting such an experiment is not possible.

Many studies of the deterrent effect of the death penalty have been conducted, all observational, and some have attracted judicial attention. Researchers have catalogued differences in the incidence of murder in states with and without the death penalty and have analyzed changes in homicide rates and execution rates over the years. When reporting on such observational studies, investigators may speak of "control groups" (e.g., the states without capital punishment) or claim they are "controlling for" confounding variables by statistical methods.[21] However, association is not causation. The causal inferences that can be drawn from analysis of observational data—no matter how complex the statistical technique—usually rest on a foundation that is less secure than that provided by randomized controlled experiments.

That said, observational studies can be very useful. For example, there is strong observational evidence that smoking causes lung cancer (*supra* Section II.A.1). Generally, observational studies provide good evidence in the following circumstances:

- The association is seen in studies with different designs, on different kinds of subjects, and done by different research groups.[22] That reduces the chance that the association is due to a defect in one type of study, a peculiarity in one group of subjects, or the idiosyncrasies of one research group.
- The association holds when effects of confounding variables are taken into account by appropriate methods, for example, comparing smaller groups that are relatively homogeneous with respect to the confounders.[23]
- There is a plausible explanation for the effect of the independent variable; alternative explanations in terms of confounding should be less plausible than the proposed causal link.[24]

21. A procedure often used to control for confounding in observational studies is regression analysis. The underlying logic is described *infra* Section V.D and in Daniel L. Rubinfeld, Reference Guide on Multiple Regression, Section II, in this manual. *But see* Richard A. Berk, Regression Analysis: A Constructive Critique (2004); Rethinking Social Inquiry: Diverse Tools, Shared Standards (Henry E. Brady & David Collier eds., 2004); David A. Freedman, Statistical Models: Theory and Practice (2005); David A. Freedman, *Oasis or Mirage,* Chance, Spring 2008, at 59.

22. For example, case-control studies are designed one way and cohort studies another, with many variations. *See, e.g.,* Leon Gordis, Epidemiology (4th ed. 2008); *supra* note 16.

23. The idea is to control for the influence of a confounder by stratification—making comparisons separately within groups for which the confounding variable is nearly constant and therefore has little influence over the variables of primary interest. For example, smokers are more likely to get lung cancer than nonsmokers. Age, gender, social class, and region of residence are all confounders, but controlling for such variables does not materially change the relationship between smoking and cancer rates. Furthermore, many different studies—of different types and on different populations—confirm the causal link. That is why most experts believe that smoking causes lung cancer and many other diseases. For a review of the literature, see International Agency for Research on Cancer, *supra* note 16.

24. A. Bradford Hill, *The Environment and Disease: Association or Causation?*, 58 Proc. Royal Soc'y Med. 295 (1965); Alfred S. Evans, Causation and Disease: A Chronological Journey 187 (1993). Plausibility, however, is a function of time and circumstances.

Thus, evidence for the causal link does not depend on observed associations alone.

Observational studies can produce legitimate disagreement among experts, and there is no mechanical procedure for resolving such differences of opinion. In the end, deciding whether associations are causal typically is not a matter of statistics alone, but also rests on scientific judgment. There are, however, some basic questions to ask when appraising causal inferences based on empirical studies:

- Was there a control group? Unless comparisons can be made, the study has little to say about causation.
- If there was a control group, how were subjects assigned to treatment or control: through a process under the control of the investigator (a controlled experiment) or through a process outside the control of the investigator (an observational study)?
- If the study was a controlled experiment, was the assignment made using a chance mechanism (randomization), or did it depend on the judgment of the investigator?

If the data came from an observational study or a nonrandomized controlled experiment,

- How did the subjects come to be in treatment or in control groups?
- Are the treatment and control groups comparable?
- If not, what adjustments were made to address confounding?
- Were the adjustments sensible and sufficient?[25]

4. *Can the results be generalized?*

Internal validity is about the specifics of a particular study: Threats to internal validity include confounding and chance differences between treatment and control groups. *External validity* is about using a particular study or set of studies to reach more general conclusions. A careful randomized controlled experiment on a large but unrepresentative group of subjects will have high internal validity but low external validity.

Any study must be conducted on certain subjects, at certain times and places, and using certain treatments. To extrapolate from the conditions of a study to more general conditions raises questions of external validity. For example, studies suggest that definitions of insanity given to jurors influence decisions in cases of incest. Would the definitions have a similar effect in cases of murder? Other studies indicate that recidivism rates for ex-convicts are not affected by provid-

25. Many courts have noted the importance of confounding variables. *E.g.*, People Who Care v. Rockford Bd. of Educ., 111 F.3d 528, 537–38 (7th Cir. 1997) (educational achievement); Hollander v. Sandoz Pharms. Corp., 289 F.3d 1193, 1213 (10th Cir. 2002) (stroke); *In re* Proportionality Review Project (II), 757 A.2d 168 (N.J. 2000) (capital sentences).

ing them with temporary financial support after release. Would similar results be obtained if conditions in the labor market were different?

Confidence in the appropriateness of an extrapolation cannot come from the experiment itself. It comes from knowledge about outside factors that would or would not affect the outcome.[26] Sometimes, several studies, each having different limitations, all point in the same direction. This is the case, for example, with studies indicating that jurors who approve of the death penalty are more likely to convict in a capital case.[27] Convergent results support the validity of generalizations.

B. Descriptive Surveys and Censuses

We now turn to a second topic—choosing units for study. A census tries to measure some characteristic of every unit in a population. This is often impractical. Then investigators use sample surveys, which measure characteristics for only part of a population. The accuracy of the information collected in a census or survey depends on how the units are selected for study and how the measurements are made.[28]

1. What method is used to select the units?

By definition, a census seeks to measure some characteristic of every unit in a whole population. It may fall short of this goal, in which case one must ask

26. Such judgments are easiest in the physical and life sciences, but even here, there are problems. For example, it may be difficult to infer human responses to substances that affect animals. First, there are often inconsistencies across test species: A chemical may be carcinogenic in mice but not in rats. Extrapolation from rodents to humans is even more problematic. Second, to get measurable effects in animal experiments, chemicals are administered at very high doses. Results are extrapolated—using mathematical models—to the very low doses of concern in humans. However, there are many dose–response models to use and few grounds for choosing among them. Generally, different models produce radically different estimates of the "virtually safe dose" in humans. David A. Freedman & Hans Zeisel, *From Mouse to Man: The Quantitative Assessment of Cancer Risks,* 3 Stat. Sci. 3 (1988). For these reasons, many experts—and some courts in toxic tort cases—have concluded that evidence from animal experiments is generally insufficient by itself to establish causation. *See, e.g.,* Bruce N. Ames et al., *The Causes and Prevention of Cancer,* 92 Proc. Nat'l Acad. Sci. USA 5258 (1995); National Research Council, Science and Judgment in Risk Assessment 59 (1994) ("There are reasons based on both biologic principles and empirical observations to support the hypothesis that many forms of biologic responses, including toxic responses, can be extrapolated across mammalian species, including *Homo sapiens*, but the scientific basis of such extrapolation is not established with sufficient rigor to allow broad and definitive generalizations to be made.").

27. Phoebe C. Ellsworth, *Some Steps Between Attitudes and Verdicts, in* Inside the Juror 42, 46 (Reid Hastie ed., 1993). Nonetheless, in *Lockhart v. McCree*, 476 U.S. 162 (1986), the Supreme Court held that the exclusion of opponents of the death penalty in the guilt phase of a capital trial does not violate the constitutional requirement of an impartial jury.

28. *See* Shari Seidman Diamond, Reference Guide on Survey Research, Sections III, IV, in this manual.

whether the missing data are likely to differ in some systematic way from the data that are collected.[29] The methodological framework of a scientific survey is different. With probability methods, a sampling frame (i.e., an explicit list of units in the population) must be created. Individual units then are selected by an objective, well-defined chance procedure, and measurements are made on the sampled units.

To illustrate the idea of a sampling frame, suppose that a defendant in a criminal case seeks a change of venue: According to him, popular opinion is so adverse that it would be difficult to impanel an unbiased jury. To prove the state of popular opinion, the defendant commissions a survey. The relevant population consists of all persons in the jurisdiction who might be called for jury duty. The sampling frame is the list of all potential jurors, which is maintained by court officials and is made available to the defendant. In this hypothetical case, the fit between the sampling frame and the population would be excellent.

In other situations, the sampling frame is more problematic. In an obscenity case, for example, the defendant can offer a survey of community standards.[30] The population comprises all adults in the legally relevant district, but obtaining a full list of such people may not be possible. Suppose the survey is done by telephone, but cell phones are excluded from the sampling frame. (This is usual practice.) Suppose too that cell phone users, as a group, hold different opinions from landline users. In this second hypothetical, the poll is unlikely to reflect the opinions of the cell phone users, no matter how many individuals are sampled and no matter how carefully the interviewing is done.

Many surveys do not use probability methods. In commercial disputes involving trademarks or advertising, the population of all potential purchasers of a product is hard to identify. Pollsters may resort to an easily accessible subgroup of the population, for example, shoppers in a mall.[31] Such convenience samples may be biased by the interviewer's discretion in deciding whom to approach—a form of

29. The U.S. Decennial Census generally does not count everyone that it should, and it counts some people who should not be counted. There is evidence that net undercount is greater in some demographic groups than others. Supplemental studies may enable statisticians to adjust for errors and omissions, but the adjustments rest on uncertain assumptions. *See* Lawrence D. Brown et al., *Statistical Controversies in Census 2000*, 39 Jurimetrics J. 347 (2007); David A. Freedman & Kenneth W. Wachter, *Methods for Census 2000 and Statistical Adjustments, in* Social Science Methodology 232 (Steven Turner & William Outhwaite eds., 2007) (reviewing technical issues and litigation surrounding census adjustment in 1990 and 2000); 9 Stat. Sci. 458 (1994) (symposium presenting arguments for and against adjusting the 1990 census).

30. On the admissibility of such polls, see *State v. Midwest Pride IV, Inc.*, 721 N.E.2d 458 (Ohio Ct. App. 1998) (holding one such poll to have been properly excluded and collecting cases from other jurisdictions).

31. *E.g.*, Smith v. Wal-Mart Stores, Inc., 537 F. Supp. 2d 1302, 1333 (N.D. Ga. 2008) (treating a small mall-intercept survey as entitled to much less weight than a survey based on a probability sample); R.J. Reynolds Tobacco Co. v. Loew's Theatres, Inc., 511 F. Supp. 867, 876 (S.D.N.Y. 1980) (questioning the propriety of basing a "nationally projectable statistical percentage" on a suburban mall intercept study).

selection bias—and the refusal of some of those approached to participate—nonresponse bias (*infra* Section II.B.2). Selection bias is acute when constituents write their representatives, listeners call into radio talk shows, interest groups collect information from their members, or attorneys choose cases for trial.[32]

There are procedures that attempt to correct for selection bias. In quota sampling, for example, the interviewer is instructed to interview so many women, so many older people, so many ethnic minorities, and the like. But quotas still leave discretion to the interviewers in selecting members of each demographic group and therefore do not solve the problem of selection bias.[33]

Probability methods are designed to avoid selection bias. Once the population is reduced to a sampling frame, the units to be measured are selected by a lottery that gives each unit in the sampling frame a known, nonzero probability of being chosen. Random numbers leave no room for selection bias.[34] Such procedures are used to select individuals for jury duty. They also have been used to choose "bellwether" cases for representative trials to resolve issues in a large group of similar cases.[35]

32. *E.g.*, Pittsburgh Press Club v. United States, 579 F.2d 751, 759 (3d Cir. 1978) (tax-exempt club's mail survey of its members to show little sponsorship of income-producing uses of facilities was held to be inadmissible hearsay because it "was neither objective, scientific, nor impartial"), *rev'd on other grounds*, 615 F.2d 600 (3d Cir. 1980). *Cf. In re* Chevron U.S.A., Inc., 109 F.3d 1016 (5th Cir. 1997). In that case, the district court decided to try 30 cases to resolve common issues or to ascertain damages in 3000 claims arising from Chevron's allegedly improper disposal of hazardous substances. The court asked the opposing parties to select 15 cases each. Selecting 30 extreme cases, however, is quite different from drawing a random sample of 30 cases. Thus, the court of appeals wrote that although random sampling would have been acceptable, the trial court could not use the results in the 30 extreme cases to resolve issues of fact or ascertain damages in the untried cases. *Id.* at 1020. Those cases, it warned, were "not cases calculated to represent the group of 3000 claimants." *Id. See infra* note 35.

A well-known example of selection bias is the 1936 *Literary Digest* poll. After successfully predicting the winner of every U.S. presidential election since 1916, the *Digest* used the replies from 2.4 million respondents to predict that Alf Landon would win the popular vote, 57% to 43%. In fact, Franklin Roosevelt won by a landslide vote of 62% to 38%. *See* Freedman et al., *supra* note 12, at 334–35. The *Digest* was so far off, in part, because it chose names from telephone books, rosters of clubs and associations, city directories, lists of registered voters, and mail order listings. *Id.* at 335, A-20 n.6. In 1936, when only one household in four had a telephone, the people whose names appeared on such lists tended to be more affluent. Lists that overrepresented the affluent had worked well in earlier elections, when rich and poor voted along similar lines, but the bias in the sampling frame proved fatal when the Great Depression made economics a salient consideration for voters.

33. *See* Freedman et al., *supra* note 12, at 337–39.

34. In simple random sampling, units are drawn at random without replacement. In particular, each unit has the same probability of being chosen for the sample. *Id.* at 339–41. More complicated methods, such as stratified sampling and cluster sampling, have advantages in certain applications. In systematic sampling, every fifth, tenth, or hundredth (in mathematical jargon, every *n*th) unit in the sampling frame is selected. If the units are not in any special order, then systematic sampling is often comparable to simple random sampling.

35. *E.g.*, *In re* Simon II Litig., 211 F.R.D. 86 (E.D.N.Y. 2002), *vacated*, 407 F.3d 125 (2d Cir. 2005), *dismissed*, 233 F.R.D. 123 (E.D.N.Y. 2006); *In re* Estate of Marcus Human Rights Litig., 910

2. *Of the units selected, which are measured?*

Probability sampling ensures that within the limits of chance (*infra* Section IV), the sample will be representative of the sampling frame. The question remains regarding which units actually get measured. When documents are sampled for audit, all the selected ones can be examined, at least in principle. Human beings are less easily managed, and some will refuse to cooperate. Surveys should therefore report nonresponse rates. A large nonresponse rate warns of bias, although supplemental studies may establish that nonrespondents are similar to respondents with respect to characteristics of interest.[36]

In short, a good survey defines an appropriate population, uses a probability method for selecting the sample, has a high response rate, and gathers accurate information on the sample units. When these goals are met, the sample tends to be representative of the population. Data from the sample can be extrapolated

F. Supp. 1460 (D. Haw. 1995), *aff'd sub nom.* Hilao v. Estate of Marcos, 103 F.3d 767 (9th Cir. 1996); Cimino v. Raymark Indus., Inc., 751 F. Supp. 649 (E.D. Tex. 1990), *rev'd*, 151 F.3d 297 (5th Cir. 1998); *cf. In re* Chevron U.S.A., Inc., 109 F.3d 1016 (5th Cir. 1997) (discussed *supra* note 32). Although trials in a suitable random sample of cases can produce reasonable estimates of average damages, the propriety of precluding individual trials raises questions of due process and the right to trial by jury. *See* Thomas E. Willging, Mass Torts Problems and Proposals: A Report to the Mass Torts Working Group (Fed. Judicial Ctr. 1999); cf. Wal-Mart Stores, Inc. v. Dukes, 131 S. Ct. 2541, 2560–61 (2011). The cases and the views of commentators are described more fully in David H. Kaye & David A. Freedman, *Statistical Proof, in* 1 Modern Scientific Evidence: The Law and Science of Expert Testimony § 6:16 (David L. Faigman et al. eds., 2009–2010).

36. For discussions of nonresponse rates and admissibility of surveys conducted for litigation, see *Johnson v. Big Lots Stores, Inc.*, 561 F. Supp. 2d 567 (E.D. La. 2008) (fair labor standards); *United States v. Dentsply Int'l, Inc.*, 277 F. Supp. 2d 387, 437 (D. Del. 2003), *rev'd on other grounds*, 399 F.3d 181 (3d Cir. 2005) (antitrust).

The 1936 *Literary Digest* election poll (*supra* note 32) illustrates the dangers in nonresponse. Only 24% of the 10 million people who received questionnaires returned them. Most of the respondents probably had strong views on the candidates and objected to President Roosevelt's economic program. This self-selection is likely to have biased the poll. Maurice C. Bryson, *The* Literary Digest *Poll: Making of a Statistical Myth*, 30 Am. Statistician 184 (1976); Freedman et al., *supra* note 12, at 335–36. Even when demographic characteristics of the sample match those of the population, caution is indicated. *See* David Streitfeld, *Shere Hite and the Trouble with Numbers*, 1 Chance 26 (1988); Chamont Wang, Sense and Nonsense of Statistical Inference: Controversy, Misuse, and Subtlety 174–76 (1993).

In *United States v. Gometz*, 730 F.2d 475, 478 (7th Cir. 1984) (en banc), the Seventh Circuit recognized that "a low rate of response to juror questionnaires could lead to the underrepresentation of a group that is entitled to be represented on the qualified jury wheel." Nonetheless, the court held that under the Jury Selection and Service Act of 1968, 28 U.S.C. §§ 1861–1878 (1988), the clerk did not abuse his discretion by failing to take steps to increase a response rate of 30%. According to the court, "Congress wanted to make it possible for all qualified persons to serve on juries, which is different from forcing all qualified persons to be available for jury service." *Gometz*, 730 F.2d at 480. Although it might "be a good thing to follow up on persons who do not respond to a jury questionnaire," the court concluded that Congress "was not concerned with anything so esoteric as nonresponse bias." *Id.* at 479, 482; *cf. In re* United States, 426 F.3d 1 (1st Cir. 2005) (reaching the same result with respect to underrepresentation of African Americans resulting in part from nonresponse bias).

to describe the characteristics of the population. Of course, surveys may be useful even if they fail to meet these criteria. But then, additional arguments are needed to justify the inferences.

C. Individual Measurements

1. Is the measurement process reliable?

Reliability and validity are two aspects of accuracy in measurement. In statistics, reliability refers to reproducibility of results.[37] A reliable measuring instrument returns consistent measurements. A scale, for example, is perfectly reliable if it reports the same weight for the same object time and again. It may not be accurate—it may always report a weight that is too high or one that is too low—but the perfectly reliable scale always reports the same weight for the same object. Its errors, if any, are systematic: They always point in the same direction.

Reliability can be ascertained by measuring the same quantity several times; the measurements must be made independently to avoid bias. Given independence, the correlation coefficient (*infra* Section V.B) between repeated measurements can be used as a measure of reliability. This is sometimes called a test-retest correlation or a reliability coefficient.

A courtroom example is DNA identification. An early method of identification required laboratories to determine the lengths of fragments of DNA. By making independent replicate measurements of the fragments, laboratories determined the likelihood that two measurements differed by specified amounts.[38] Such results were needed to decide whether a discrepancy between a crime sample and a suspect sample was sufficient to exclude the suspect.[39]

Coding provides another example. In many studies, descriptive information is obtained on the subjects. For statistical purposes, the information usually has to be reduced to numbers. The process of reducing information to numbers is called "coding," and the reliability of the process should be evaluated. For example, in a study of death sentencing in Georgia, legally trained evaluators examined short summaries of cases and ranked them according to the defendant's culpability.[40]

37. Courts often use "reliable" to mean "that which can be relied on" for some purpose, such as establishing probable cause or crediting a hearsay statement when the declarant is not produced for confrontation. *Daubert v. Merrell Dow Pharms., Inc.*, 509 U.S. 579, 590 n.9 (1993), for example, distinguishes "evidentiary reliability" from reliability in the technical sense of giving consistent results. We use "reliability" to denote the latter.

38. *See* National Research Council, The Evaluation of Forensic DNA Evidence 139–41 (1996).

39. *Id.*; National Research Council, DNA Technology in Forensic Science 61–62 (1992). Current methods are discussed in David H. Kaye & George Sensabaugh, Reference Guide on DNA Identification Evidence, Section II, in this manual.

40. David C. Baldus et al., Equal Justice and the Death Penalty: A Legal and Empirical Analysis 49–50 (1990).

Two different aspects of reliability should be considered. First, the "within-observer variability" of judgments should be small—the same evaluator should rate essentially identical cases in similar ways. Second, the "between-observer variability" should be small—different evaluators should rate the same cases in essentially the same way.

2. *Is the measurement process valid?*

Reliability is necessary but not sufficient to ensure accuracy. In addition to reliability, validity is needed. A valid measuring instrument measures what it is supposed to. Thus, a polygraph measures certain physiological responses to stimuli, for example, in pulse rate or blood pressure. The measurements may be reliable. Nonetheless, the polygraph is not valid as a lie detector unless the measurements it makes are well correlated with lying.[41]

When there is an established way of measuring a variable, a new measurement process can be validated by comparison with the established one. Breathalyzer readings can be validated against alcohol levels found in blood samples. LSAT scores used for law school admissions can be validated against grades earned in law school. A common measure of validity is the correlation coefficient between the predictor and the criterion (e.g., test scores and later performance).[42]

Employment discrimination cases illustrate some of the difficulties. Thus, plaintiffs suing under Title VII of the Civil Rights Act may challenge an employment test that has a disparate impact on a protected group, and defendants may try to justify the use of a test as valid, reliable, and a business necessity.[43] For validation, the most appropriate criterion variable is clear enough: job performance. However, plaintiffs may then turn around and challenge the validity of performance ratings. For reliability, administering the test twice to the same group of people may be impractical. Even if repeated testing is practical, it may be statistically inadvisable, because subjects may learn something from the first round of testing that affects their scores on the second round. Such "practice effects" are likely to compromise the independence of the two measurements, and independence is needed to estimate reliability. Statisticians therefore use internal evidence

41. *See* United States v. Henderson, 409 F.3d 1293, 1303 (11th Cir. 2005) ("while the physical responses recorded by a polygraph machine may be tested, 'there is no available data to prove that those specific responses are attributable to lying.'"); National Research Council, The Polygraph and Lie Detection (2003) (reviewing the scientific literature).

42. As the discussion of the correlation coefficient indicates, *infra* Section V.B, the closer the coefficient is to 1, the greater the validity. For a review of data on test reliability and validity, see Paul R. Sackett et al., *High-Stakes Testing in Higher Education and Employment: Appraising the Evidence for Validity and Fairness*, 63 Am. Psychologist 215 (2008).

43. *See, e.g.*, Washington v. Davis, 426 U.S. 229, 252 (1976); Albemarle Paper Co. v. Moody, 422 U.S. 405, 430–32 (1975); Griggs v. Duke Power Co., 401 U.S. 424 (1971); Lanning v. S.E. Penn. Transp. Auth., 308 F.3d 286 (3d Cir. 2002).

from the test itself. For example, if scores on the first half of the test correlate well with scores from the second half, then that is evidence of reliability.

A further problem is that test-takers are likely to be a select group. The ones who get the jobs are even more highly selected. Generally, selection attenuates (weakens) the correlations. There are methods for using internal measures of reliability to estimate test-retest correlations; there are other methods that correct for attenuation. However, such methods depend on assumptions about the nature of the test and the procedures used to select the test-takers and are therefore open to challenge.[44]

3. Are the measurements recorded correctly?

Judging the adequacy of data collection involves an examination of the process by which measurements are taken. Are responses to interviews coded correctly? Do mistakes distort the results? How much data are missing? What was done to compensate for gaps in the data? These days, data are stored in computer files. Cross-checking the files against the original sources (e.g., paper records), at least on a sample basis, can be informative.

Data quality is a pervasive issue in litigation and in applied statistics more generally. A programmer moves a file from one computer to another, and half the data disappear. The definitions of crucial variables are lost in the sands of time. Values get corrupted: Social security numbers come to have eight digits instead of nine, and vehicle identification numbers fail the most elementary consistency checks. Everybody in the company, from the CEO to the rawest mailroom trainee, turns out to have been hired on the same day. Many of the residential customers have last names that indicate commercial activity ("Happy Valley Farriers"). These problems seem humdrum by comparison with those of reliability and validity, but—unless caught in time—they can be fatal to statistical arguments.[45]

44. *See* Thad Dunning & David A. Freedman, *Modeling Selection Effects, in* Social Science Methodology 225 (Steven Turner & William Outhwaite eds., 2007); Howard Wainer & David Thissen, *True Score Theory: The Traditional Method, in* Test Scoring 23 (David Thissen & Howard Wainer eds., 2001).

45. *See, e.g.,* Malletier v. Dooney & Bourke, Inc., 525 F. Supp. 2d 558, 630 (S.D.N.Y. 2007) (coding errors contributed "to the cumulative effect of the methodological errors" that warranted exclusion of a consumer confusion survey); EEOC v. Sears, Roebuck & Co., 628 F. Supp. 1264, 1304, 1305 (N.D. Ill. 1986) ("[E]rrors in EEOC's mechanical coding of information from applications in its hired and nonhired samples also make EEOC's statistical analysis based on this data less reliable." The EEOC "consistently coded prior experience in such a way that less experienced women are considered to have the same experience as more experienced men" and "has made so many general coding errors that its data base does not fairly reflect the characteristics of applicants for commission sales positions at Sears."), *aff'd*, 839 F.2d 302 (7th Cir. 1988). *But see* Dalley v. Mich. Blue Cross-Blue Shield, Inc., 612 F. Supp. 1444, 1456 (E.D. Mich. 1985) ("although plaintiffs show that there were some mistakes in coding, plaintiffs still fail to demonstrate that these errors were so generalized and so pervasive that the entire study is invalid.").

D. What Is Random?

In the law, a selection process sometimes is called "random," provided that it does not exclude identifiable segments of the population. Statisticians use the term in a far more technical sense. For example, if we were to choose one person at random from a population, in the strict statistical sense, we would have to ensure that everybody in the population is chosen with exactly the same probability. With a randomized controlled experiment, subjects are assigned to treatment or control at random in the strict sense—by tossing coins, throwing dice, looking at tables of random numbers, or more commonly these days, by using a random number generator on a computer. The same rigorous definition applies to random sampling. It is randomness in the technical sense that provides assurance of unbiased estimates from a randomized controlled experiment or a probability sample. Randomness in the technical sense also justifies calculations of standard errors, confidence intervals, and *p*-values (*infra* Sections IV–V). Looser definitions of randomness are inadequate for statistical purposes.

III. How Have the Data Been Presented?

After data have been collected, they should be presented in a way that makes them intelligible. Data can be summarized with a few numbers or with graphical displays. However, the wrong summary can mislead.[46] Section III.A discusses rates or percentages and provides some cautionary examples of misleading summaries, indicating the kinds of questions that might be considered when summaries are presented in court. Percentages are often used to demonstrate statistical association, which is the topic of Section III.B. Section III.C considers graphical summaries of data, while Sections III.D and III.E discuss some of the basic descriptive statistics that are likely to be encountered in litigation, including the mean, median, and standard deviation.

A. Are Rates or Percentages Properly Interpreted?

1. Have appropriate benchmarks been provided?

The selective presentation of numerical information is like quoting someone out of context. Is a fact that "over the past three years," a particular index fund of large-cap stocks "gained a paltry 1.9% a year" indicative of poor management? Considering that "the average large-cap value fund has returned just 1.3% a year,"

46. *See generally* Freedman et al., *supra* note 12; Huff, *supra* note 12; Moore & Notz, *supra* note 12; Zeisel, *supra* note 12.

a growth rate of 1.9% is hardly an indictment.[47] In this example and many others, it is helpful to find a benchmark that puts the figures into perspective.

2. Have the data collection procedures changed?

Changes in the process of collecting data can create problems of interpretation. Statistics on crime provide many examples. The number of petty larcenies reported in Chicago more than doubled one year—not because of an abrupt crime wave, but because a new police commissioner introduced an improved reporting system.[48] For a time, police officials in Washington, D.C., "demonstrated" the success of a law-and-order campaign by valuing stolen goods at $49, just below the $50 threshold then used for inclusion in the Federal Bureau of Investigation's Uniform Crime Reports.[49] Allegations of manipulation in the reporting of crime from one time period to another are legion.[50]

Changes in data collection procedures are by no means limited to crime statistics. Indeed, almost all series of numbers that cover many years are affected by changes in definitions and collection methods. When a study includes such time-series data, it is useful to inquire about changes and to look for any sudden jumps, which may signal such changes.

3. Are the categories appropriate?

Misleading summaries also can be produced by the choice of categories to be used for comparison. In *Philip Morris, Inc. v. Loew's Theatres, Inc.*,[51] and *R.J. Reynolds Tobacco Co. v. Loew's Theatres, Inc.*,[52] Philip Morris and R.J. Reynolds sought an injunction to stop the maker of Triumph low-tar cigarettes from running advertisements claiming that participants in a national taste test preferred Triumph to other brands. Plaintiffs alleged that claims that Triumph was a "national taste test winner" or Triumph "beats" other brands were false and misleading. An exhibit introduced by the defendant contained the data shown in Table 1.[53] Only 14% + 22% = 36% of the sample preferred Triumph to Merit, whereas

47. Paul J. Lim, *In a Downturn, Buy and Hold or Quit and Fold?*, N.Y. Times, July 27, 2008.

48. James P. Levine et al., Criminal Justice in America: Law in Action 99 (1986) (referring to a change from 1959 to 1960).

49. D. Seidman & M. Couzens, *Getting the Crime Rate Down: Political Pressure and Crime Reporting*, 8 Law & Soc'y Rev. 457 (1974).

50. Michael D. Maltz, *Missing UCR Data and Divergence of the NCVS and UCR Trends*, *in* Understanding Crime Statistics: Revisiting the Divergence of the NCVS and UCR 269, 280 (James P. Lynch & Lynn A. Addington eds., 2007) (citing newspaper reports in Boca Raton, Atlanta, New York, Philadelphia, Broward County (Florida), and Saint Louis); Michael Vasquez, *Miami Police: FBI: Crime Stats Accurate*, Miami Herald, May 1, 2008.

51. 511 F. Supp. 855 (S.D.N.Y. 1980).

52. 511 F. Supp. 867 (S.D.N.Y. 1980).

53. *Philip Morris*, 511 F. Supp. at 866.

29% + 11% = 40% preferred Merit to Triumph. By selectively combining categories, however, the defendant attempted to create a different impression. Because 24% found the brands to be about the same, and 36% preferred Triumph, the defendant claimed that a clear majority (36% + 24% = 60%) found Triumph "as good [as] or better than Merit."[54] The court resisted this chicanery, finding that defendant's test results did not support the advertising claims.[55]

Table 1. Data Used by a Defendant to Refute Plaintiffs' False Advertising Claim

	Triumph Much Better Than Merit	Triumph Somewhat Better Than Merit	Triumph About the Same as Merit	Triumph Somewhat Worse Than Merit	Triumph Much Worse Than Merit
Number	45	73	77	93	36
Percentage	14	22	24	29	11

There was a similar distortion in claims for the accuracy of a home pregnancy test. The manufacturer advertised the test as 99.5% accurate under laboratory conditions. The data underlying this claim are summarized in Table 2.

Table 2. Home Pregnancy Test Results

	Actually Pregnant	Actually not Pregnant
Test says pregnant	197	0
Test says not pregnant	1	2
Total	198	2

Table 2 does indicate that only one error occurred in 200 assessments, or 99.5% overall accuracy, but the table also shows that the test can make two types of errors: It can tell a pregnant woman that she is not pregnant (a false negative), and it can tell a woman who is not pregnant that she is (a false positive). The reported 99.5% accuracy rate conceals a crucial fact—the company had virtually no data with which to measure the rate of false positives.[56]

54. *Id.*

55. *Id.* at 856–57.

56. Only two women in the sample were not pregnant; the test gave correct results for both of them. Although a false-positive rate of 0 is ideal, an estimate based on a sample of only two women is not. These data are reported in Arnold Barnett, *How Numbers Can Trick You*, Tech. Rev., Oct. 1994, at 38, 44–45.

4. *How big is the base of a percentage?*

Rates and percentages often provide effective summaries of data, but these statistics can be misinterpreted. A percentage makes a comparison between two numbers: One number is the base, and the other number is compared to that base. Putting them on the same base (100) makes it easy to compare them.

When the base is small, however, a small change in absolute terms can generate a large percentage gain or loss. This could lead to newspaper headlines such as "Increase in Thefts Alarming," even when the total number of thefts is small.[57] Conversely, a large base will make for small percentage increases. In these situations, actual numbers may be more revealing than percentages.

5. *What comparisons are made?*

Finally, there is the issue of which numbers to compare. Researchers sometimes choose among alternative comparisons. It may be worthwhile to ask why they chose the one they did. Would another comparison give a different view? A government agency, for example, may want to compare the amount of service now being given with that of earlier years—but what earlier year should be the baseline? If the first year of operation is used, a large percentage increase should be expected because of startup problems. If last year is used as the base, was it also part of the trend, or was it an unusually poor year? If the base year is not representative of other years, the percentage may not portray the trend fairly. No single question can be formulated to detect such distortions, but it may help to ask for the numbers from which the percentages were obtained; asking about the base can also be helpful.[58]

B. *Is an Appropriate Measure of Association Used?*

Many cases involve statistical association. Does a test for employee promotion have an exclusionary effect that depends on race or gender? Does the incidence of murder vary with the rate of executions for convicted murderers? Do consumer purchases of a product depend on the presence or absence of a product warning? This section discusses tables and percentage-based statistics that are frequently presented to answer such questions.[59]

Percentages often are used to describe the association between two variables. Suppose that a university alleged to discriminate against women in admitting

57. Lyda Longa, *Increase in Thefts Alarming*, Daytona News-J. June 8, 2008 (reporting a 35% increase in armed robberies in Daytona Beach, Florida, in a 5-month period, but not indicating whether the number had gone up by 6 (from 17 to 23), by 300 (from 850 to 1150), or by some other amount).

58. For assistance in coping with percentages, see Zeisel, *supra* note 12, at 1–24.

59. Correlation and regression are discussed *infra* Section V.

students consists of only two colleges—engineering and business. The university admits 350 out of 800 male applicants; by comparison, it admits only 200 out of 600 female applicants. Such data commonly are displayed as in Table 3.[60]

As Table 3 indicates, 350/800 = 44% of the males are admitted, compared with only 200/600 = 33% of the females. One way to express the disparity is to subtract the two percentages: 44% – 33% = 11 percentage points. Although such subtraction is commonly seen in jury discrimination cases,[61] the difference is inevitably small when the two percentages are both close to zero. If the selection rate for males is 5% and that for females is 1%, the difference is only 4 percentage points. Yet, females have only one-fifth the chance of males of being admitted, and that may be of real concern.

Table 3. Admissions by Gender

Decision	Male	Female	Total
Admit	350	200	550
Deny	450	400	850
Total	800	600	1400

For Table 3, the selection ratio (used by the Equal Employment Opportunity Commission in its "80% rule") is 33/44 = 75%, meaning that, on average, women have 75% the chance of admission that men have.[62] However, the selection ratio has its own problems. In the last example, if the selection rates are 5% and 1%, then the exclusion rates are 95% and 99%. The ratio is 99/95 = 104%, meaning that females have, on average, 104% the risk of males of being rejected. The underlying facts are the same, of course, but this formulation sounds much less disturbing.

60. A table of this sort is called a "cross-tab" or a "contingency table." Table 3 is "two-by-two" because it has two rows and two columns, not counting rows or columns containing totals.

61. *See, e.g.*, State v. Gibbs, 758 A.2d 327, 337 (Conn. 2000); Primeaux v. Dooley, 747 N.W.2d 137, 141 (S.D. 2008); D.H. Kaye, *Statistical Evidence of Discrimination in Jury Selection, in* Statistical Methods in Discrimination Litigation 13 (David H. Kaye & Mikel Aickin eds., 1986).

62. A procedure that selects candidates from the least successful group at a rate less than 80% of the rate for the most successful group "will generally be regarded by the Federal enforcement agencies as evidence of adverse impact." EEOC Uniform Guidelines on Employee Selection Procedures, 29 C.F.R. § 1607.4(D) (2008). The rule is designed to help spot instances of substantially discriminatory practices, and the commission usually asks employers to justify any procedures that produce selection ratios of 80% or less.

The analogous statistic used in epidemiology is called the relative risk. *See* Green et al., *supra* note 13, Section III.A. Relative risks are usually quoted as decimals; for example, a selection ratio of 75% corresponds to a relative risk of 0.75.

The odds ratio is more symmetric. If 5% of male applicants are admitted, the odds on a man being admitted are 5/95 = 1/19; the odds on a woman being admitted are 1/99. The odds ratio is (1/99)/(1/19) = 19/99. The odds ratio for rejection instead of acceptance is the same, except that the order is reversed.[63] Although the odds ratio has desirable mathematical properties, its meaning may be less clear than that of the selection ratio or the simple difference.

Data showing disparate impact are generally obtained by aggregating—putting together—statistics from a variety of sources. Unless the source material is fairly homogeneous, aggregation can distort patterns in the data. We illustrate the problem with the hypothetical admission data in Table 3. Applicants can be classified not only by gender and admission but also by the college to which they applied, as in Table 4.

Table 4. Admissions by Gender and College

	Engineering		Business	
Decision	Male	Female	Male	Female
Admit	300	100	50	100
Deny	300	100	150	300

The entries in Table 4 add up to the entries in Table 3. Expressed in a more technical manner, Table 3 is obtained by aggregating the data in Table 4. Yet there is no association between gender and admission in either college; men and women are admitted at identical rates. Combining two colleges with no association produces a university in which gender is associated strongly with admission. The explanation for this paradox is that the business college, to which most of the women applied, admits relatively few applicants. It is easier to be accepted at the engineering college, the college to which most of the men applied. This example illustrates a common issue: Association can result from combining heterogeneous statistical material.[64]

63. For women, the odds on rejection are 99 to 1; for men, 19 to 1. The ratio of these odds is 99/19. Likewise, the odds ratio for an admitted applicant being a man as opposed to a denied applicant being a man is also 99/19.

64. Tables 3 and 4 are hypothetical, but closely patterned on a real example. *See* P.J. Bickel et al., *Sex Bias in Graduate Admissions: Data from Berkeley*, 187 Science 398 (1975). The tables are an instance of Simpson's Paradox.

C. Does a Graph Portray Data Fairly?

Graphs are useful for revealing key characteristics of a batch of numbers, trends over time, and the relationships among variables.

1. How are trends displayed?

Graphs that plot values over time are useful for seeing trends. However, the scales on the axes matter. In Figure 1, the rate of all crimes of domestic violence in Florida (per 100,000 people) appears to decline rapidly over the 10 years from 1998 through 2007; in Figure 2, the same rate appears to drop slowly.[65] The moral is simple: Pay attention to the markings on the axes to determine whether the scale is appropriate.

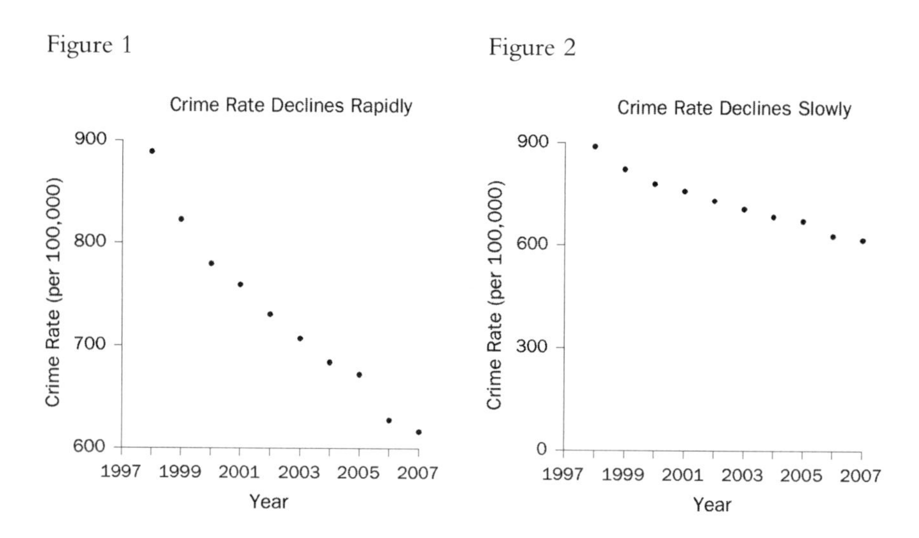

2. How are distributions displayed?

A graph commonly used to display the distribution of data is the histogram. One axis denotes the numbers, and the other indicates how often those fall within

65. Florida Statistical Analysis Center, Florida Department of Law Enforcement, Florida's Crime Rate at a Glance, *available at* http://www.fdle.state.fl.us/FSAC/Crime_Trends/domestic_violence/index.asp. The data are from the Florida Uniform Crime Report statistics on crimes ranging from simple stalking and forcible fondling to murder and arson. The Web page with the numbers graphed in Figures 1 and 2 is no longer posted, but similar data for all violent crime is available at http://www.fdle.state.fl.us/FSAC/Crime_Trends/Violent-Crime.aspx.

specified intervals (called "bins" or "class intervals"). For example, we flipped a quarter 10 times in a row and counted the number of heads in this "batch" of 10 tosses. With 50 batches, we obtained the following counts:[66]

7 7 5 6 8   4 2 3 6 5   4 3 4 7 4   6 8 4 7 4   7 4 5 4 3
4 4 2 5 3   5 4 2 4 4   5 7 2 3 5   4 6 4 9 10   5 5 6 6 4

The histogram is shown in Figure 3.[67] A histogram shows how the data are distributed over the range of possible values. The spread can be made to appear larger or smaller, however, by changing the scale of the horizontal axis. Likewise, the shape can be altered somewhat by changing the size of the bins.[68] It may be worth inquiring how the analyst chose the bin widths.

Figure 3. Histogram showing how frequently various numbers of heads appeared in 50 batches of 10 tosses of a quarter.

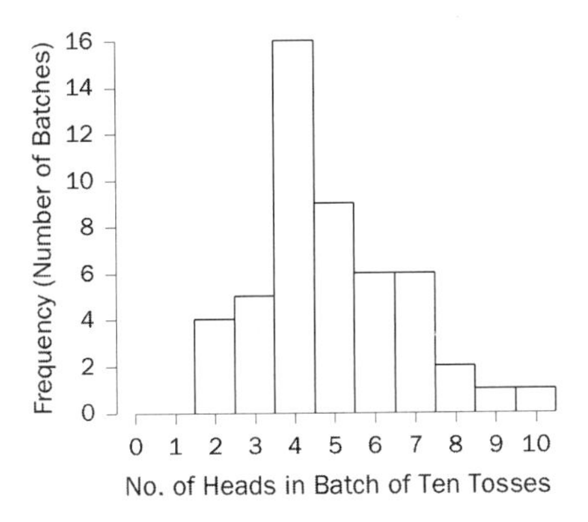

66. The coin landed heads 7 times in the first 10 tosses; by coincidence, there were also 7 heads in the next 10 tosses; there were 5 heads in the third batch of 10 tosses; and so forth.

67. In Figure 3, the bin width is 1. There were no 0s or 1s in the data, so the bars over 0 and 1 disappear. There is a bin from 1.5 to 2.5; the four 2s in the data fall into this bin, so the bar over the interval from 1.5 to 2.5 has height 4. There is another bin from 2.5 to 3.5, which catches five 3s; the height of the corresponding bar is 5. And so forth.

All the bins in Figure 3 have the same width, so this histogram is just like a bar graph. However, data are often published in tables with unequal intervals. The resulting histograms will have unequal bin widths; bar heights should be calculated so that the areas (height × width) are proportional to the frequencies. In general, a histogram differs from a bar graph in that it represents frequencies by area, not height. *See* Freedman et al., *supra* note 12, at 31–41.

68. As the width of the bins decreases, the graph becomes more detailed, but the appearance becomes more ragged until finally the graph is effectively a plot of each datum. The optimal bin width depends on the subject matter and the goal of the analysis.

D. Is an Appropriate Measure Used for the Center of a Distribution?

Perhaps the most familiar descriptive statistic is the mean (or "arithmetic mean"). The mean can be found by adding all the numbers and dividing the total by how many numbers were added. By comparison, the median cuts the numbers into halves: half the numbers are larger than the median and half are smaller.[69] Yet a third statistic is the mode, which is the most common number in the dataset. These statistics are different, although they are not always clearly distinguished.[70] The mean takes account of all the data—it involves the total of all the numbers; however, particularly with small datasets, a few unusually large or small observations may have too much influence on the mean. The median is resistant to such outliers.

Thus, studies of damage awards in tort cases find that the mean is larger than the median.[71] This is because the mean takes into account (indeed, is heavily influenced by) the magnitudes of the relatively few very large awards, whereas the median merely counts their number. If one is seeking a single, representative number for the awards, the median may be more useful than the mean.[72] Still, if the issue is whether insurers were experiencing more costs from jury verdicts, the mean is the more appropriate statistic: The total of the awards is directly related to the mean, not to the median.[73]

69. Technically, at least half the numbers are at the median or larger; at least half are at the median or smaller. When the distribution is symmetric, the mean equals the median. The values diverge, however, when the distribution is asymmetric, or skewed.

70. In ordinary language, the arithmetic mean, the median, and the mode seem to be referred to interchangeably as "the average." In statistical parlance, however, the average is the arithmetic mean. The mode is rarely used by statisticians, because it is unstable: Small changes to the data often result in large changes to the mode.

71. In a study using a probability sample of cases, the median compensatory award in wrongful death cases was $961,000, whereas the mean award was around $3.75 million for the 162 cases in which the plaintiff prevailed. Thomas H. Cohen & Steven K. Smith, U.S. Dep't of Justice, Bureau of Justice Statistics Bulletin NCJ 202803, Civil Trial Cases and Verdicts in Large Counties 2001, 10 (2004). In *TXO Production Corp. v. Alliance Resources Corp.*, 509 U.S. 443 (1993), briefs portraying the punitive damage system as out of control pointed to mean punitive awards. These were some 10 times larger than the median awards described in briefs defending the system of punitive damages. Michael Rustad & Thomas Koenig, *The Supreme Court and Junk Social Science: Selective Distortion in Amicus Briefs*, 72 N.C. L. Rev. 91, 145–47 (1993).

72. In passing on proposed settlements in class-action lawsuits, courts have been advised to look to the magnitude of the settlements negotiated by the parties. But the mean settlement will be large if a higher number of meritorious, high-cost cases are resolved early in the life cycle of the litigation. This possibility led the court in *In re Educational Testing Service Praxis Principles of Learning and Teaching, Grades 7-12 Litig.*, 447 F. Supp. 2d 612, 625 (E.D. La. 2006), to regard the smaller median settlement as "more representative of the value of a typical claim than the mean value" and to use this median in extrapolating to the entire class of pending claims.

73. To get the total award, just multiply the mean by the number of awards; by contrast, the total cannot be computed from the median. (The more pertinent figure for the insurance industry is

Research also has shown that there is considerable stability in the ratio of punitive to compensatory damage awards, and the Supreme Court has placed great weight on this ratio in deciding whether punitive damages are excessive in a particular case. In *Exxon Shipping Co. v. Baker*,[74] Exxon contended that an award of $2.5 billion in punitive damages for a catastrophic oil spill in Alaska was unreasonable under federal maritime law. The Court looked to a "comprehensive study of punitive damages awarded by juries in state civil trials [that] found a median ratio of punitive to compensatory awards of just 0.62:1, but a mean ratio of 2.90:1."[75] The higher mean could reflect a relatively small but disturbing proportion of unjustifiably large punitive awards.[76] Looking to the median ratio as "the line near which cases like this one largely should be grouped," the majority concluded that "a 1:1 ratio, which is above the median award, is a fair upper limit in such maritime cases [of reckless conduct]."[77]

E. Is an Appropriate Measure of Variability Used?

The location of the center of a batch of numbers reveals nothing about the variations exhibited by these numbers.[78] Statistical measures of variability include the range, the interquartile range, and the standard deviation. The range is the difference between the largest number in the batch and the smallest. The range seems natural, and it indicates the maximum spread in the numbers, but the range is unstable because it depends entirely on the most extreme values.[79] The interquartile range is the difference between the 25th and 75th percentiles.[80] The interquartile range contains 50% of the numbers and is resistant to changes in extreme values. The standard deviation is a sort of mean deviation from the mean.[81]

not the total of jury awards, but actual claims experience including settlements; of course, even the risk of large punitive damage awards may have considerable impact.)

74. 128 S. Ct. 2605 (2008).

75. *Id.* at 2625.

76. According to the Court, "the outlier cases subject defendants to punitive damages that dwarf the corresponding compensatories," and the "stark unpredictability" of these rare awards is the "real problem." *Id.* This perceived unpredictability has been the subject of various statistical studies and much debate. *See* Anthony J. Sebok, *Punitive Damages: From Myth to Theory*, 92 Iowa L. Rev. 957 (2007).

77. 128 S. Ct. at 2633.

78. The numbers 1, 2, 5, 8, 9 have 5 as their mean and median. So do the numbers 5, 5, 5, 5, 5. In the first batch, the numbers vary considerably about their mean; in the second, the numbers do not vary at all.

79. Moreover, the range typically depends on the number of units in the sample.

80. By definition, 25% of the data fall below the 25th percentile, 90% fall below the 90th percentile, and so on. The median is the 50th percentile.

81. When the distribution follows the normal curve, about 68% of the data will be within 1 standard deviation of the mean, and about 95% will be within 2 standard deviations of the mean. For other distributions, the proportions will be different.

There are no hard and fast rules about which statistic is the best. In general, the bigger the measures of spread are, the more the numbers are dispersed.[82] Particularly in small datasets, the standard deviation can be influenced heavily by a few outlying values. To assess the extent of this influence, the mean and the standard deviation can be recomputed with the outliers discarded. Beyond this, any of the statistics can (and often should) be supplemented with a figure that displays much of the data.

IV. What Inferences Can Be Drawn from the Data?

The inferences that may be drawn from a study depend on the design of the study and the quality of the data (*supra* Section II). The data might not address the issue of interest, might be systematically in error, or might be difficult to interpret because of confounding. Statisticians would group these concerns together under the rubric of "bias." In this context, bias means systematic error, with no connotation of prejudice. We turn now to another concern, namely, the impact of random chance on study results ("random error").[83]

If a pattern in the data is the result of chance, it is likely to wash out when more data are collected. By applying the laws of probability, a statistician can assess the likelihood that random error will create spurious patterns of certain kinds. Such assessments are often viewed as essential when making inferences from data.

Technically, the standard deviation is the square root of the variance; the variance is the mean square deviation from the mean. For example, if the mean is 100, then 120 deviates from the mean by 20, and the square of 20 is $20^2 = 400$. If the variance (i.e., the mean of the squared deviations) is 900, then the standard deviation is the square root of 900, that is, $\sqrt{900} = 30$. Taking the square root gets back to the original scale of the measurements. For example, if the measurements are of length in inches, the variance is in square inches; taking the square root changes back to inches.

82. In *Exxon Shipping Co. v. Baker*, 554 U.S. 471 (2008), along with the mean and median ratios of punitive to compensatory awards of 0.62 and 2.90, the Court referred to a standard deviation of 13.81. *Id.* at 498. These numbers led the Court to remark that "[e]ven to those of us unsophisticated in statistics, the thrust of these figures is clear: the spread is great, and the outlier cases subject defendants to punitive damages that dwarf the corresponding compensatories." *Id.* at 499-500. The size of the standard deviation compared to the mean supports the observation that ratios in the cases of jury award studies are dispersed. A graph of each pair of punitive and compensatory damages offers more insight into how scattered these figures are. *See* Theodore Eisenberg et al., *The Predictability of Punitive Damages*, 26 J. Legal Stud. 623 (1997); *infra* Section V.A (explaining scatter diagrams).

83. Random error is also called sampling error, chance error, or statistical error. Econometricians use the parallel concept of random disturbance terms. *See* Rubinfeld, *supra* note 21. Randomness and cognate terms have precise technical meanings; it is randomness in the technical sense that justifies the probability calculations behind standard errors, confidence intervals, and *p*-values (*supra* Section II.D, *infra* Sections IV.A–B). For a discussion of samples and populations, see *supra* Section II.B.

Thus, statistical inference typically involves tasks such as the following, which will be discussed in the rest of this guide.

- *Estimation*. A statistician draws a sample from a population (*supra* Section II.B) and estimates a parameter—that is, a numerical characteristic of the population. (The average value of a large group of claims is a parameter of perennial interest.) Random error will throw the estimate off the mark. The question is, by how much? The precision of an estimate is usually reported in terms of the standard error and a confidence interval.
- *Significance testing*. A "null hypothesis" is formulated—for example, that a parameter takes a particular value. Because of random error, an estimated value for the parameter is likely to differ from the value specified by the null—even if the null is right. ("Null hypothesis" is often shortened to "null.") How likely is it to get a difference as large as, or larger than, the one observed in the data? This chance is known as a *p*-value. Small *p*-values argue against the null hypothesis. Statistical significance is determined by reference to the *p*-value; significance testing (also called hypothesis testing) is the technique for computing *p*-values and determining statistical significance.
- *Developing a statistical model*. Statistical inferences often depend on the validity of statistical models for the data. If the data are collected on the basis of a probability sample or a randomized experiment, there will be statistical models that suit the occasion, and inferences based on these models will be secure. Otherwise, calculations are generally based on analogy: This group of people is like a random sample; that observational study is like a randomized experiment. The fit between the statistical model and the data collection process may then require examination—how good is the analogy? If the model breaks down, that will bias the analysis.
- *Computing posterior probabilities*. Given the sample data, what is the probability of the null hypothesis? The question might be of direct interest to the courts, especially when translated into English; for example, the null hypothesis might be the innocence of the defendant in a criminal case. Posterior probabilities can be computed using a formula called Bayes' rule. However, the computation often depends on prior beliefs about the statistical model and its parameters; such prior beliefs almost necessarily require subjective judgment. According to the frequentist theory of statistics,[84]

84. The frequentist theory is also called objectivist, by contrast with the subjectivist version of Bayesian theory. In brief, frequentist methods treat probabilities as objective properties of the system being studied. Subjectivist Bayesians view probabilities as measuring subjective degrees of belief. *See infra* Section IV.D and Appendix, Section A, for discussion of the two positions. The Bayesian position is named after the Reverend Thomas Bayes (England, c. 1701–1761). His essay on the subject was published after his death: *An Essay Toward Solving a Problem in the Doctrine of Chances*, 53 Phil. Trans. Royal Soc'y London 370 (1763–1764). For discussion of the foundations and varieties of Bayesian and

prior probabilities rarely have meaning and neither do posterior probabilities.[85]

Key ideas of estimation and testing will be illustrated by courtroom examples, with some complications omitted for ease of presentation and some details postponed (*see infra* Section V.D on statistical models, and the Appendix on the calculations).

The first example, on estimation, concerns the Nixon papers. Under the Presidential Recordings and Materials Preservation Act of 1974, Congress impounded Nixon's presidential papers after he resigned. Nixon sued, seeking compensation on the theory that the materials belonged to him personally. Courts ruled in his favor: Nixon was entitled to the fair market value of the papers, with the amount to be proved at trial.[86]

The Nixon papers were stored in 20,000 boxes at the National Archives in Alexandria, Virginia. It was plainly impossible to value this entire population of material. Appraisers for the plaintiff therefore took a random sample of 500 boxes. (From this point on, details are simplified; thus, the example becomes somewhat hypothetical.) The appraisers determined the fair market value of each sample box. The average of the 500 sample values turned out to be $2000. The standard deviation (*supra* Section III.E) of the 500 sample values was $2200. Many boxes had low appraised values whereas some boxes were considered to be extremely valuable; this spread explains the large standard deviation.

A. Estimation

1. What estimator should be used?

With the Nixon papers, it is natural to use the average value of the 500 sample boxes to estimate the average value of all 20,000 boxes comprising the population.

other forms of statistical inference, see, e.g., Richard M. Royall, Statistical Inference: A Likelihood Paradigm (1997); James Berger, *The Case for Objective Bayesian Analysis*, 1 Bayesian Analysis 385 (2006), *available at* http://ba.stat.cmu.edu/journal/2006/vol01/issue03/berger.pdf; Stephen E. Fienberg, *Does It Make Sense to be an "Objective Bayesian"? (Comment on Articles by Berger and by Goldstein)*, 1 Bayesian Analysis 429 (2006); David Freedman, *Some Issues in the Foundation of Statistics*, 1 Found. Sci. 19 (1995), *reprinted in* Topics in the Foundation of Statistics 19 (Bas C. van Fraasen ed., 1997); see also D.H. Kaye, *What Is Bayesianism? in* Probability and Inference in the Law of Evidence: The Uses and Limits of Bayesianism (Peter Tillers & Eric Green eds., 1988), *reprinted in* 28 Jurimetrics J. 161 (1988) (distinguishing between "Bayesian probability," "Bayesian statistical inference," "Bayesian inference writ large," and "Bayesian decision theory").

85. Prior probabilities of repeatable events (but not hypotheses) can be defined within the frequentist framework. See *infra* note 122. When this happens, prior and posterior probabilities for these events are meaningful according to both schools of thought.

86. Nixon v. United States, 978 F.2d 1269 (D.C. Cir. 1992); Griffin v. United States, 935 F. Supp. 1 (D.D.C. 1995).

With the average value for each box having been estimated as $2000, the plaintiff demanded compensation in the amount of

$$20{,}000 \times \$2{,}000 = \$40{,}000{,}000.$$

In more complex problems, statisticians may have to choose among several estimators. Generally, estimators that tend to make smaller errors are preferred; however, "error" might be quantified in more than one way. Moreover, the advantage of one estimator over another may depend on features of the population that are largely unknown, at least before the data are collected and analyzed. For complicated problems, professional skill and judgment may therefore be required when choosing a sample design and an estimator. In such cases, the choices and the rationale for them should be documented.

2. *What is the standard error? The confidence interval?*

An estimate based on a sample is likely to be off the mark, at least by a small amount, because of random error. The standard error gives the likely magnitude of this random error, with smaller standard errors indicating better estimates.[87] In our example of the Nixon papers, the standard error for the sample average can be computed from (1) the size of the sample—500 boxes—and (2) the standard deviation of the sample values; *see infra* Appendix. Bigger samples give estimates that are more precise. Accordingly, the standard error should go down as the sample size grows, although the rate of improvement slows as the sample gets bigger. ("Sample size" and "the size of the sample" just mean the number of items in the sample; the "sample average" is the average value of the items in the sample.) The standard deviation of the sample comes into play by measuring heterogeneity. The less heterogeneity in the values, the smaller the standard error. For example, if all the values were about the same, a tiny sample would give an accurate estimate. Conversely, if the values are quite different from one another, a larger sample would be needed.

With a random sample of 500 boxes and a standard deviation of $2200, the standard error for the sample average is about $100. The plaintiff's total demand was figured as the number of boxes (20,000) times the sample average ($2000). Therefore, the standard error for the total demand can be computed as 20,000 times the standard error for the sample average[88]:

87. We distinguish between (1) the standard deviation of the sample, which measures the spread in the sample data and (2) the standard error of the sample average, which measures the likely size of the random error in the sample average. The standard error is often called the standard deviation, and courts generally use the latter term. *See, e.g.*, Castaneda v. Partida, 430 U.S. 482 (1977).

88. We are assuming a simple random sample. Generally, the formula for the standard error must take into account the method used to draw the sample and the nature of the estimator. In fact, the Nixon appraisers used more elaborate statistical procedures. Moreover, they valued the material as of

$$20{,}000 \times \$100 = \$2{,}000{,}000.$$

How is the standard error to be interpreted? Just by the luck of the draw, a few too many high-value boxes may have come into the sample, in which case the estimate of $40,000,000 is too high. Or, a few too many low-value boxes may have been drawn, in which case the estimate is too low. This is random error. The net effect of random error is unknown, because data are available only on the sample, not on the full population. However, the net effect is likely to be something close to the standard error of $2,000,000. Random error throws the estimate off, one way or the other, by something close to the standard error. The role of the standard error is to gauge the likely size of the random error.

The plaintiff's argument may be open to a variety of objections, particularly regarding appraisal methods. However, the sampling plan is sound, as is the extrapolation from the sample to the population. And there is no need for a larger sample: The standard error is quite small relative to the total claim.

Random errors larger in magnitude than the standard error are commonplace. Random errors larger in magnitude than two or three times the standard error are unusual. Confidence intervals make these ideas more precise. Usually, a confidence interval for the population average is centered at the sample average; the desired confidence level is obtained by adding and subtracting a suitable multiple of the standard error. Statisticians who say that the population average falls within 1 standard error of the sample average will be correct about 68% of the time. Those who say "within 2 standard errors" will be correct about 95% of the time, and those who say "within 3 standard errors" will be correct about 99.7% of the time, and so forth. (We are assuming a large sample; the confidence levels correspond to areas under the normal curve and are approximations; the "population average" just means the average value of all the items in the population.[89]) In summary,

- To get a 68% confidence interval, start at the sample average, then add and subtract 1 standard error.
- To get a 95% confidence interval, start at the sample average, then add and subtract twice the standard error.

1995, extrapolated backward to the time of taking (1974), and then added interest. The text ignores these complications.

89. *See infra* Appendix. The area under the normal curve between –1 and +1 is close to 68.3%. Likewise, the area between –2 and +2 is close to 95.4%. Many academic statisticians would use ±1.96 SE for a 95% confidence interval. However, the normal curve only gives an approximation to the relevant chances, and the error in that approximation will often be larger than a few tenths of a percent. For simplicity, we use ±1 SE for the 68% confidence level, and ±2 SE for 95% confidence. The normal curve gives good approximations when the sample size is reasonably large; for small samples, other techniques should be used. *See infra* notes 106–07.

- To get a 99.7% confidence interval, start at the sample average, then add and subtract three times the standard error.

With the Nixon papers, the 68% confidence interval for plaintiff's total demand runs

$$\begin{aligned} &\text{from } \$40{,}000{,}000 - \$2{,}000{,}000 = \$38{,}000{,}000 \\ &\text{to } \$40{,}000{,}000 + \$2{,}000{,}000 = \$42{,}000{,}000. \end{aligned}$$

The 95% confidence interval runs

$$\begin{aligned} &\text{from } \$40{,}000{,}000 - (2 \times \$2{,}000{,}000) = \$36{,}000{,}000 \\ &\text{to } \$40{,}000{,}000 + (2 \times \$2{,}000{,}000) = \$44{,}000{,}000. \end{aligned}$$

The 99.7% confidence interval runs

$$\begin{aligned} &\text{from } \$40{,}000{,}000 - (3 \times \$2{,}000{,}000) = \$34{,}000{,}000 \\ &\text{to } \$40{,}000{,}000 + (3 \times \$2{,}000{,}000) = \$46{,}000{,}000. \end{aligned}$$

To write this more compactly, we abbreviate standard error as SE. Thus, 1 SE is one standard error, 2 SE is twice the standard error, and so forth. With a large sample and an estimate like the sample average, a 68% confidence interval is the range

$$\text{estimate} - 1 \text{ SE to estimate} + 1 \text{ SE}.$$

A 95% confidence interval is the range

$$\text{estimate} - 2 \text{ SE to estimate} + 2 \text{ SE}.$$

The 99.7% confidence interval is the range

$$\text{estimate} - 3 \text{ SE to estimate} + 3 \text{ SE}.$$

For a given sample size, increased confidence can be attained only by widening the interval. The 95% confidence level is the most popular, but some authors use 99%, and 90% is seen on occasion. (The corresponding multipliers on the SE are about 2, 2.6, and 1.6, respectively; *see infra* Appendix.) The phrase "margin of error" generally means twice the standard error. In medical journals, "confidence interval" is often abbreviated as "CI."

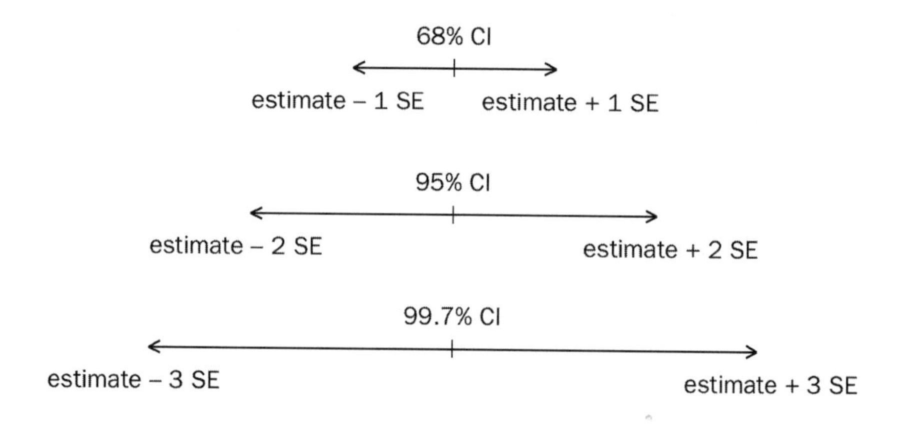

The main point is that an estimate based on a sample will differ from the exact population value, because of random error. The standard error gives the likely size of the random error. If the standard error is small, random error probably has little effect. If the standard error is large, the estimate may be seriously wrong. Confidence intervals are a technical refinement, and bias is a separate issue to consider (*infra* Section IV.A.4).

3. *How big should the sample be?*

There is no easy answer to this sensible question. Much depends on the level of error that is tolerable and the nature of the material being sampled. Generally, increasing the size of the sample will reduce the level of random error ("sampling error"). Bias ("nonsampling error") cannot be reduced that way. Indeed, beyond some point, large samples are harder to manage and more vulnerable to nonsampling error. To reduce bias, the researcher must improve the design of the study or use a statistical model more tightly linked to the data collection process.

If the material being sampled is heterogeneous, random error will be large; a larger sample will be needed to offset the heterogeneity (*supra* Section IV.A.1). A pilot sample may be useful to estimate heterogeneity and determine the final sample size. Probability samples require some effort in the design phase, and it will rarely be sensible to draw a sample with fewer than, say, two or three dozen items. Moreover, with such small samples, methods based on the normal curve (*supra* Section IV.A.2) will not apply.

Population size (i.e., the number of items in the population) usually has little bearing on the precision of estimates for the population average. This is surprising. On the other hand, population size has a direct bearing on estimated totals. Both points are illustrated by the Nixon papers (*see supra* Section IV.A.2 and *infra* Appendix). To be sure, drawing a probability sample from a large population may involve a lot of work. Samples presented in the courtroom have ranged from 5 (tiny) to 1.7 million (huge).[90]

4. *What are the technical difficulties?*

To begin with, "confidence" is a term of art. The confidence level indicates the percentage of the time that intervals from repeated samples would cover the true value. The confidence level does not express the chance that repeated estimates would fall into the confidence interval.[91] With the Nixon papers, the 95% confidence interval should not be interpreted as saying that 95% of all random samples will produce estimates in the range from $36 million to $44 million. Moreover, the confidence level does not give the probability that the unknown parameter lies within the confidence interval.[92] For example, the 95% confidence level should not be translated to a 95% probability that the total value of the papers is in the range from $36 million to $44 million. According to the frequentist theory of statistics, probability statements cannot be made about population characteristics: Probability statements apply to the behavior of samples. That is why the different term "confidence" is used.

The next point to make is that for a given confidence level, a narrower interval indicates a more precise estimate, whereas a broader interval indicates less

90. *See* Lebrilla v. Farmers Group, Inc., No. 00-CC-017185 (Cal. Super. Ct., Orange County, Dec. 5, 2006) (preliminary approval of settlement), a class action lawsuit on behalf of plaintiffs who were insured by Farmers and had automobile accidents. Plaintiffs alleged that replacement parts recommended by Farmers did not meet specifications: Small samples were used to evaluate these allegations. At the other extreme, it was proposed to adjust Census 2000 for undercount and overcount by reviewing a sample of 1.7 million persons. *See* Brown et al., *supra* note 29, at 353.

91. Opinions reflecting this misinterpretation include *In re* Silicone Gel Breast Implants Prods. Liab. Litig, 318 F. Supp. 2d 879, 897 (C.D. Cal. 2004) ("a margin of error between 0.5 and 8.0 at the 95% confidence level . . . means that 95 times out of 100 a study of that type would yield a relative risk value somewhere between 0.5 and 8.0."); United States *ex rel.* Free v. Peters, 806 F. Supp. 705, 713 n.6 (N.D. Ill. 1992) ("A 99% confidence interval, for instance, is an indication that if we repeated our measurement 100 times under identical conditions, 99 times out of 100 the point estimate derived from the repeated experimentation will fall within the initial interval estimate. . . ."), *rev'd in part*, 12 F.3d 700 (7th Cir. 1993). The more technically correct statement in the *Silicone Gel* case, for example, would be that "the confidence interval of 0.5 to 8.0 means that the relative risk in the population could fall within this wide range and that in roughly 95 times out of 100, random samples from the same population, the confidence intervals (however wide they might be) would include the population value (whatever it is)."

92. *See, e.g.*, Freedman et al., *supra* note 12, at 383–86; *infra* Section IV.B.1. Consequently, it is misleading to suggest that "[a] 95% confidence interval means that there is a 95% probability that the 'true' relative risk falls within the interval" or that "the probability that the true value was . . . within two standard deviations of the mean . . . would be 95 percent." DeLuca v. Merrell Dow Pharms., Inc., 791 F. Supp. 1042, 1046 (D.N.J. 1992), *aff'd*, 6 F.3d 778 (3d Cir. 1993); SmithKline Beecham Corp. v. Apotex Corp., 247 F. Supp. 2d 1011, 1037 (N.D. Ill. 2003), *aff'd on other grounds*, 403 F.3d 1331 (Fed. Cir. 2005).

precision.[93] A high confidence level with a broad interval means very little, but a high confidence level for a small interval is impressive, indicating that the random error in the sample estimate is low. For example, take a 95% confidence interval for a damage claim. An interval that runs from $34 million to $44 million is one thing, but –$10 million to $90 million is something else entirely. Statements about confidence without mention of an interval are practically meaningless.[94]

Standard errors and confidence intervals are often derived from statistical models for the process that generated the data. The model usually has parameters—numerical constants describing the population from which samples were drawn. When the values of the parameters are not known, the statistician must work backward, using the sample data to make estimates. That was the case here.[95] Generally, the chances needed for statistical inference are computed from a model and estimated parameter values.

If the data come from a probability sample or a randomized controlled experiment (*supra* Sections II.A–B), the statistical model may be connected tightly to the actual data collection process. In other situations, using the model may be tantamount to assuming that a sample of convenience is like a random sample, or that an observational study is like a randomized experiment. With the Nixon papers, the appraisers drew a random sample, and that justified the statistical

93. In *Cimino v. Raymark Industries, Inc.*, 751 F. Supp. 649 (E.D. Tex. 1990), *rev'd*, 151 F.3d 297 (5th Cir. 1998), the district court drew certain random samples from more than 6000 pending asbestos cases, tried these cases, and used the results to estimate the total award to be given to all plaintiffs in the pending cases. The court then held a hearing to determine whether the samples were large enough to provide accurate estimates. The court's expert, an educational psychologist, testified that the estimates were accurate because the samples matched the population on such characteristics as race and the percentage of plaintiffs still alive. *Id.* at 664. However, the matches occurred only in the sense that population characteristics fell within 99% confidence intervals computed from the samples. The court thought that matches within the 99% confidence intervals proved more than matches within 95% intervals. *Id.* This is backward. To be correct in a few instances with a 99% confidence interval is not very impressive—by definition, such intervals are broad enough to ensure coverage 99% of the time.

94. In *Hilao v. Estate of Marcos*, 103 F.3d 767 (9th Cir. 1996), for example, "an expert on statistics . . . testified that . . . a random sample of 137 claims would achieve 'a 95% statistical probability that the same percentage determined to be valid among the examined claims would be applicable to the totality of [9541 facially valid] claims filed.'" *Id.* at 782. There is no 95% "statistical probability" that a percentage computed from a sample will be "applicable" to a population. One can compute a confidence interval from a random sample and be 95% confident that the interval covers some parameter. The computation can be done for a sample of virtually any size, with larger samples giving smaller intervals. What is missing from the opinion is a discussion of the widths of the relevant intervals. For the same reason, it is meaningless to testify, as an expert did in *Ayyad v. Sprint Spectrum, L.P.*, No. RG03-121510 (Cal. Super. Ct., Alameda County) (transcript, May 28, 2008, at 730), that a simple regression equation is trustworthy because the coefficient of the explanatory variable has "an extremely high indication of reliability to more than 99% confidence level."

95. With the Nixon papers, one parameter is the average value of all 20,000 boxes, and another parameter is the standard deviation of the 20,000 values. These parameters can be used to approximate the distribution of the sample average. *See infra* Appendix. Regression models and their parameters are discussed *infra* Section V and in Rubinfeld, *supra* note 21.

calculations—if not the appraised values themselves. In many contexts, the choice of an appropriate statistical model is less than obvious. When a model does not fit the data collection process, estimates and standard errors will not be probative.

Standard errors and confidence intervals generally ignore systematic errors such as selection bias or nonresponse bias (*supra* Sections II.B.1–2). For example, after reviewing studies to see whether a particular drug caused birth defects, a court observed that mothers of children with birth defects may be more likely to remember taking a drug during pregnancy than mothers with normal children. This selective recall would bias comparisons between samples from the two groups of women. The standard error for the estimated difference in drug usage between the groups would ignore this bias, as would the confidence interval.[96]

B. Significance Levels and Hypothesis Tests

1. What Is the p-value?

In 1969, Dr. Benjamin Spock came to trial in the U.S. District Court for Massachusetts. The charge was conspiracy to violate the Military Service Act. The jury was drawn from a panel of 350 persons selected by the clerk of the court. The panel included only 102 women—substantially less than 50%—although a majority of the eligible jurors in the community were female. The shortfall in women was especially poignant in this case: "Of all defendants, Dr. Spock, who had given wise and welcome advice on child-rearing to millions of mothers, would have liked women on his jury."[97]

Can the shortfall in women be explained by the mere play of random chance? To approach the problem, a statistician would formulate and test a null hypothesis. Here, the null hypothesis says that the panel is like 350 persons drawn at random from a large population that is 50% female. The expected number of women drawn would then be 50% of 350, which is 175. The observed number of women is 102. The shortfall is 175 − 102 = 73. How likely is it to find a disparity this large or larger, between observed and expected values? The probability is called p, or the p-value.

96. Brock v. Merrell Dow Pharms., Inc., 874 F.2d 307, 311–12 (5th Cir.), *modified*, 884 F.2d 166 (5th Cir. 1989). In *Brock*, the court stated that the confidence interval took account of bias (in the form of selective recall) as well as random error. 874 F.2d at 311–12. This is wrong. Even if the sampling error were nonexistent—which would be the case if one could interview every woman who had a child during the period that the drug was available—selective recall would produce a difference in the percentages of reported drug exposure between mothers of children with birth defects and those with normal children. In this hypothetical situation, the standard error would vanish. Therefore, the standard error could disclose nothing about the impact of selective recall.

97. Hans Zeisel, *Dr. Spock and the Case of the Vanishing Women Jurors*, 37 U. Chi. L. Rev. 1 (1969). Zeisel's reasoning was different from that presented in this text. The conviction was reversed on appeal without reaching the issue of jury selection. United States v. Spock, 416 F.2d 165 (1st Cir. 1965).

The *p*-value is the probability of getting data as extreme as, or more extreme than, the actual data—given that the null hypothesis is true. In the example, *p* turns out to be essentially zero. The discrepancy between the observed and the expected is far too large to explain by random chance. Indeed, even if the panel had included 155 women, the *p*-value would only be around 0.02, or 2%.[98] (If the population is more than 50% female, *p* will be even smaller.) In short, the jury panel was nothing like a random sample from the community.

Large *p*-values indicate that a disparity can easily be explained by the play of chance: The data fall within the range likely to be produced by chance variation. On the other hand, if *p* is very small, something other than chance must be involved: The data are far away from the values expected under the null hypothesis. Significance testing often seems to involve multiple negatives. This is because a statistical test is an argument by contradiction.

With the Dr. Spock example, the null hypothesis asserts that the jury panel is like a random sample from a population that is 50% female. The data contradict this null hypothesis because the disparity between what is observed and what is expected (according to the null) is too large to be explained as the product of random chance. In a typical jury discrimination case, small *p*-values help a defendant appealing a conviction by showing that the jury panel is not like a random sample from the relevant population; large *p*-values hurt. In the usual employment context, small *p*-values help plaintiffs who complain of discrimination—for example, by showing that a disparity in promotion rates is too large to be explained by chance; conversely, large *p*-values would be consistent with the defense argument that the disparity is just due to chance.

Because *p* is calculated by assuming that the null hypothesis is correct, *p* does not give the chance that the null is true. The *p*-value merely gives the chance of getting evidence against the null hypothesis as strong as or stronger than the evidence at hand. Chance affects the data, not the hypothesis. According to the frequency theory of statistics, there is no meaningful way to assign a numerical probability to the null hypothesis. The correct interpretation of the *p*-value can therefore be summarized in two lines:

> *p* is the probability of extreme data given the null hypothesis.
> *p* is not the probability of the null hypothesis given extreme data.[99]

98. With 102 women out of 350, the *p*-value is about $2/10^{15}$, where 10^{15} is 1 followed by 15 zeros, that is, a quadrillion. See *infra* Appendix for the calculations.

99. Some opinions present a contrary view. *E.g.*, Vasquez v. Hillery, 474 U.S. 254, 259 n.3 (1986) ("the District Court . . . ultimately accepted . . . a probability of 2 in 1000 that the phenomenon was attributable to chance"); Nat'l Abortion Fed. v. Ashcroft, 330 F. Supp. 2d 436 (S.D.N.Y. 2004), *aff'd in part*, 437 F.3d 278 (2d Cir. 2006), *vacated*, 224 Fed. App'x. 88 (2d Cir. 2007) ("According to Dr. Howell, . . . a 'P value' of 0.30 . . . indicates that there is a thirty percent probability that the results of the . . . [s]tudy were merely due to chance alone."). Such statements confuse the probability of the

To recapitulate the logic of significance testing: If p is small, the observed data are far from what is expected under the null hypothesis—too far to be readily explained by the operations of chance. That discredits the null hypothesis.

Computing p-values requires statistical expertise. Many methods are available, but only some will fit the occasion. Sometimes standard errors will be part of the analysis; other times they will not be. Sometimes a difference of two standard errors will imply a p-value of about 5%; other times it will not. In general, the p-value depends on the model, the size of the sample, and the sample statistics.

2. Is a difference statistically significant?

If an observed difference is in the middle of the distribution that would be expected under the null hypothesis, there is no surprise. The sample data are of the type that often would be seen when the null hypothesis is true. The difference is not significant, as statisticians say, and the null hypothesis cannot be rejected. On the other hand, if the sample difference is far from the expected value—according to the null hypothesis—then the sample is unusual. The difference is significant, and the null hypothesis is rejected. Statistical significance is determined by comparing p to a preset value, called the significance level.[100] The null hypothesis is rejected when p falls below this level.

In practice, statistical analysts typically use levels of 5% and 1%.[101] The 5% level is the most common in social science, and an analyst who speaks of significant results without specifying the threshold probably is using this figure. An unexplained reference to highly significant results probably means that p is less

kind of outcome observed, which is computed under some model of chance, with the probability that chance is the explanation for the outcome—the "transposition fallacy."

Instances of the transposition fallacy in criminal cases are collected in David H. Kaye et al., The New Wigmore: A Treatise on Evidence: Expert Evidence §§ 12.8.2(b) & 14.1.2 (2d ed. 2011). In *McDaniel v. Brown*, 130 S. Ct. 665 (2010), for example, a DNA analyst suggested that a random match probability of 1/3,000,000 implied a .000033 probability that the DNA was not the source of the DNA found on the victim's clothing. *See* David H. Kaye, *"False But Highly Persuasive": How Wrong Were the Probability Estimates in* McDaniel v. Brown? 108 Mich. L. Rev. First Impressions 1 (2009).

100. Statisticians use the Greek letter alpha (α) to denote the significance level; α gives the chance of getting a significant result, assuming that the null hypothesis is true. Thus, α represents the chance of a false rejection of the null hypothesis (also called a false positive, a false alarm, or a Type I error). For example, suppose $\alpha = 5\%$. If investigators do many studies, and the null hypothesis happens to be true in each case, then about 5% of the time they would obtain significant results—and falsely reject the null hypothesis.

101. The Supreme Court implicitly referred to this practice in *Castaneda v. Partida*, 430 U.S. 482, 496 n.17 (1977), and *Hazelwood School District v. United States*, 433 U.S. 299, 311 n.17 (1977). In these footnotes, the Court described the null hypothesis as "suspect to a social scientist" when a statistic from "large samples" falls more than "two or three standard deviations" from its expected value under the null hypothesis. Although the Court did not say so, these differences produce p-values of about 5% and 0.3% when the statistic is normally distributed. The Court's standard deviation is our standard error.

than 1%. These levels of 5% and 1% have become icons of science and the legal process. In truth, however, such levels are at best useful conventions.

Because the term "significant" is merely a label for a certain kind of *p*-value, significance is subject to the same limitations as the underlying *p*-value. Thus, significant differences may be evidence that something besides random error is at work. They are not evidence that this something is legally or practically important. Statisticians distinguish between statistical and practical significance to make the point. When practical significance is lacking—when the size of a disparity is negligible—there is no reason to worry about statistical significance.[102]

It is easy to mistake the *p*-value for the probability of the null hypothesis given the data (*supra* Section IV.B.1). Likewise, if results are significant at the 5% level, it is tempting to conclude that the null hypothesis has only a 5% chance of being correct.[103] This temptation should be resisted. From the frequentist perspective, statistical hypotheses are either true or false. Probabilities govern the samples, not the models and hypotheses. The significance level tells us what is likely to happen when the null hypothesis is correct; it does not tell us the probability that the hypothesis is true. Significance comes no closer to expressing the probability that the null hypothesis is true than does the underlying *p*-value.

3. *Tests or interval estimates?*

How can a highly significant difference be practically insignificant? The reason is simple: *p* depends not only on the magnitude of the effect, but also on the sample size (among other things). With a huge sample, even a tiny effect will be

102. *E.g.*, Waisome v. Port Auth., 948 F.2d 1370, 1376 (2d Cir. 1991) ("though the disparity was found to be statistically significant, it was of limited magnitude."); United States v. Henderson, 409 F.3d 1293, 1306 (11th Cir. 2005) (regardless of statistical significance, excluding law enforcement officers from jury service does not have a large enough impact on the composition of grand juries to violate the Jury Selection and Service Act); *cf.* Thornburg v. Gingles, 478 U.S. 30, 53–54 (1986) (repeating the district court's explanation of why "the correlation between the race of the voter and the voter's choice of certain candidates was [not only] statistically significant," but also "so marked as to be substantively significant, in the sense that the results of the individual election would have been different depending upon whether it had been held among only the white voters or only the black voters.").

103. *E.g.*, *Waisome*, 948 F.2d at 1376 ("Social scientists consider a finding of two standard deviations significant, meaning there is about one chance in 20 that the explanation for a deviation could be random . . . ."); Adams v. Ameritech Serv., Inc., 231 F.3d 414, 424 (7th Cir. 2000) ("Two standard deviations is normally enough to show that it is extremely unlikely (. . . less than a 5% probability) that the disparity is due to chance"); Magistrini v. One Hour Martinizing Dry Cleaning, 180 F. Supp. 2d 584, 605 n.26 (D.N.J. 2002) (a "statistically significant . . . study shows that there is only 5% probability that an observed association is due to chance."); *cf.* Giles v. Wyeth, Inc., 500 F. Supp. 2d 1048, 1056 (S.D. Ill. 2007) ("While [plaintiff] admits that a *p*-value of .15 is three times higher than what scientists generally consider statistically significant—that is, a *p*-value of .05 or lower—she maintains that this "represents 85% certainty, which meets any conceivable concept of preponderance of the evidence.").

highly significant.[104] For example, suppose that a company hires 52% of male job applicants and 49% of female applicants. With a large enough sample, a statistician could compute an impressively small *p*-value. This *p*-value would confirm that the difference does not result from chance, but it would not convert a trivial difference (52% versus 49%) into a substantial one.[105] In short, the *p*-value does not measure the strength or importance of an association.

A "significant" effect can be small. Conversely, an effect that is "not significant" can be large. By inquiring into the magnitude of an effect, courts can avoid being misled by *p*-values. To focus attention on more substantive concerns—the size of the effect and the precision of the statistical analysis—interval estimates (e.g., confidence intervals) may be more valuable than tests. Seeing a plausible range of values for the quantity of interest helps describe the statistical uncertainty in the estimate.

4. Is the sample statistically significant?

Many a sample has been praised for its statistical significance or blamed for its lack thereof. Technically, this makes little sense. Statistical significance is about the difference between observations and expectations. Significance therefore applies to statistics computed from the sample, but not to the sample itself, and certainly not to the size of the sample. Findings can be statistically significant. Differences can be statistically significant (*supra* Section IV.B.2). Estimates can be statistically significant (*infra* Section V.D.2). By contrast, samples can be representative or unrepresentative. They can be chosen well or badly (*supra* Section II.B.1). They can be large enough to give reliable results or too small to bother with (*supra* Section IV.A.3). But samples cannot be "statistically significant," if this technical phrase is to be used as statisticians use it.

C. Evaluating Hypothesis Tests

1. What is the power of the test?

When a *p*-value is high, findings are not significant, and the null hypothesis is not rejected. This could happen for at least two reasons:

104. *See supra* Section IV.B.2. Although some opinions seem to equate small *p*-values with "gross" or "substantial" disparities, most courts recognize the need to decide whether the underlying sample statistics reveal that a disparity is large. *E.g.*, Washington v. People, 186 P.3d 594 (Colo. 2008) (jury selection).

105. *Cf.* Frazier v. Garrison Indep. Sch. Dist., 980 F.2d 1514, 1526 (5th Cir. 1993) (rejecting claims of intentional discrimination in the use of a teacher competency examination that resulted in retention rates exceeding 95% for all groups); *Washington*, 186 P.2d 594 (although a jury selection practice that reduced the representation of "African-Americans [from] 7.7 percent of the population [to] 7.4 percent of the county's jury panels produced a highly statistically significant disparity, the small degree of exclusion was not constitutionally significant.").

1. The null hypothesis is true.
2. The null is false—but, by chance, the data happened to be of the kind expected under the null.

If the power of a statistical study is low, the second explanation may be plausible. Power is the chance that a statistical test will declare an effect when there is an effect to be declared.[106] This chance depends on the size of the effect and the size of the sample. Discerning subtle differences requires large samples; small samples may fail to detect substantial differences.

When a study with low power fails to show a significant effect, the results may therefore be more fairly described as inconclusive than negative. The proof is weak because power is low. On the other hand, when studies have a good chance of detecting a meaningful association, failure to obtain significance can be persuasive evidence that there is nothing much to be found.[107]

2. *What about small samples?*

For simplicity, the examples of statistical inference discussed here (*supra* Sections IV.A–B) were based on large samples. Small samples also can provide useful

106. More precisely, power is the probability of rejecting the null hypothesis when the alternative hypothesis (*infra* Section IV.C.5) is right. Typically, this probability will depend on the values of unknown parameters, as well as the preset significance level α. The power can be computed for any value of α and any choice of parameters satisfying the alternative hypothesis. *See infra* Appendix for an example. Frequentist hypothesis testing keeps the risk of a false positive to a specified level (such as $\alpha = 5\%$) and then tries to maximize power.

Statisticians usually denote power by the Greek letter beta (β). However, some authors use β to denote the probability of *accepting* the null hypothesis when the alternative hypothesis is true; this usage is fairly standard in epidemiology. Accepting the null hypothesis when the alternative holds true is a false negative (also called a Type II error, a missed signal, or a false acceptance of the null hypothesis).

The chance of a false negative may be computed from the power. Some commentators have claimed that the cutoff for significance should be chosen to equalize the chance of a false positive and a false negative, on the ground that this criterion corresponds to the more-probable-than-not burden of proof. The argument is fallacious, because α and β do not give the probabilities of the null and alternative hypotheses; *see supra* Sections IV.B.1–2; *supra* note 34. *See also* D.H. Kaye, *Hypothesis Testing in the Courtroom, in* Contributions to the Theory and Application of Statistics: A Volume in Honor of Herbert Solomon 331, 341–43 (Alan E. Gelfand ed., 1987).

107. Some formal procedures (meta-analysis) are available to aggregate results across studies. *See, e.g., In re* Bextra and Celebrex Marketing Sales Practices and Prod. Liab. Litig., 524 F. Supp. 2d 1166, 1174, 1184 (N.D. Cal. 2007) (holding that "[a] meta-analysis of all available published and unpublished randomized clinical trials" of certain pain-relief medicine was admissible). In principle, the power of the collective results will be greater than the power of each study. However, these procedures have their own weakness. *See, e.g.*, Richard A. Berk & David A. Freedman, *Statistical Assumptions as Empirical Commitments, in* Punishment and Social Control: Essays in Honor of Sheldon Messinger 235, 244–48 (T.G. Blomberg & S. Cohen eds., 2d ed. 2003); Michael Oakes, Statistical Inference: A Commentary for the Social and Behavioral Sciences (1986); Diana B. Petitti, Meta-Analysis, Decision Analysis, and Cost-Effectiveness Analysis Methods for Quantitative Synthesis in Medicine (2d ed. 2000).

information. Indeed, when confidence intervals and *p*-values can be computed, the interpretation is the same with small samples as with large ones.[108] The concern with small samples is not that they are beyond the ken of statistical theory, but that

1 The underlying assumptions are hard to validate.
2. Because approximations based on the normal curve generally cannot be used, confidence intervals may be difficult to compute for parameters of interest. Likewise, *p*-values may be difficult to compute for hypotheses of interest.[109]
3. Small samples may be unreliable, with large standard errors, broad confidence intervals, and tests having low power.

3. *One tail or two?*

In many cases, a statistical test can be done either one-tailed or two-tailed; the second method often produces a *p*-value twice as big as the first method. The methods are easily explained with a hypothetical example. Suppose we toss a coin 1000 times and get 532 heads. The null hypothesis to be tested asserts that the coin is fair. If the null is correct, the chance of getting 532 or more heads is 2.3%. That is a one-tailed test, whose *p*-value is 2.3%. To make a two-tailed test, the statistician computes the chance of getting 532 or more heads—or 500 – 32 = 468 heads or fewer. This is 4.6%. In other words, the two-tailed *p*-value is 4.6%. Because small *p*-values are evidence against the null hypothesis, the one-tailed test seems to produce stronger evidence than its two-tailed counterpart. However, the advantage is largely illusory, as the example suggests. (The two-tailed test may seem artificial, but it offers some protection against possible artifacts resulting from multiple testing—the topic of the next section.)

Some courts and commentators have argued for one or the other type of test, but a rigid rule is not required if significance levels are used as guidelines rather than as mechanical rules for statistical proof.[110] One-tailed tests often make it

108. Advocates sometimes contend that samples are "too small to allow for meaningful statistical analysis," United States v. New York City Bd. of Educ., 487 F. Supp. 2d 220, 229 (E.D.N.Y. 2007), and courts often look to the size of samples from earlier cases to determine whether the sample data before them are admissible or convincing. *Id.* at 230; Timmerman v. U.S. Bank, 483 F.3d 1106, 1116 n.4 (10th Cir. 2007). However, a meaningful statistical analysis yielding a significant result can be based on a small sample, and reliability does not depend on sample size alone (*see supra* Section IV.A.3, *infra* Section V.C.1). Well-known small-sample techniques include the sign test and Fisher's exact test. *E.g.*, Michael O. Finkelstein & Bruce Levin, Statistics for Lawyers 154–56, 339–41 (2d ed. 2001); *see generally* E.L. Lehmann & H.J.M. d'Abrera, Nonparametrics (2d ed. 2006).

109. With large samples, approximate inferences (e.g., based on the central limit theorem, *see infra* Appendix) may be quite adequate. These approximations will not be satisfactory for small samples.

110. *See, e.g.*, United States v. State of Delaware, 93 Fair Empl. Prac. Cas. (BNA) 1248, 2004 WL 609331, *10 n.4 (D. Del. 2004). According to formal statistical theory, the choice between one

easier to reach a threshold such as 5%, at least in terms of appearance. However, if we recognize that 5% is not a magic line, then the choice between one tail and two is less important—as long as the choice and its effect on the *p*-value are made explicit.

4. *How many tests have been done?*

Repeated testing complicates the interpretation of significance levels. If enough comparisons are made, random error almost guarantees that some will yield "significant" findings, even when there is no real effect. To illustrate the point, consider the problem of deciding whether a coin is biased. The probability that a fair coin will produce 10 heads when tossed 10 times is $(1/2)^{10} = 1/1024$. Observing 10 heads in the first 10 tosses, therefore, would be strong evidence that the coin is biased. Nonetheless, if a fair coin is tossed a few thousand times, it is likely that at least one string of ten consecutive heads will appear. Ten heads in the first ten tosses means one thing; a run of ten heads somewhere along the way to a few thousand tosses of a coin means quite another. A test—looking for a run of ten heads—can be repeated too often.

Artifacts from multiple testing are commonplace. Because research that fails to uncover significance often is not published, reviews of the literature may produce an unduly large number of studies finding statistical significance.[111] Even a single researcher may examine so many different relationships that a few will achieve statistical significance by mere happenstance. Almost any large dataset—even pages from a table of random digits—will contain some unusual pattern that can be uncovered by diligent search. Having detected the pattern, the analyst can perform a statistical test for it, blandly ignoring the search effort. Statistical significance is bound to follow.

There are statistical methods for dealing with multiple looks at the data, which permit the calculation of meaningful *p*-values in certain cases.[112] However, no general solution is available, and the existing methods would be of little help in the typical case where analysts have tested and rejected a variety of models before arriving at the one considered the most satisfactory (*see infra* Section V on regression models). In these situations, courts should not be overly impressed with

tail or two can sometimes be made by considering the exact form of the alternative hypothesis (*infra* Section IV.C.5). *But see* Freedman et al., *supra* note 12, at 547–50. One-tailed tests at the 5% level are viewed as weak evidence—no weaker standard is commonly used in the technical literature. One-tailed tests are also called one-sided (with no pejorative intent); two-tailed tests are two-sided.

111. *E.g.*, Philippa J. Easterbrook et al., *Publication Bias in Clinical Research*, 337 Lancet 867 (1991); John P.A. Ioannidis, *Effect of the Statistical Significance of Results on the Time to Completion and Publication of Randomized Efficacy Trials*, 279 JAMA 281 (1998); Stuart J. Pocock et al., *Statistical Problems in the Reporting of Clinical Trials: A Survey of Three Medical Journals*, 317 New Eng. J. Med. 426 (1987).

112. *See, e.g.*, Sandrine Dudoit & Mark J. van der Laan, Multiple Testing Procedures with Applications to Genomics (2008).

claims that estimates are significant. Instead, they should be asking how analysts developed their models.[113]

5. *What are the rival hypotheses?*

The *p*-value of a statistical test is computed on the basis of a model for the data: the null hypothesis. Usually, the test is made in order to argue for the alternative hypothesis: another model. However, on closer examination, both models may prove to be unreasonable. A small *p*-value means something is going on besides random error. The alternative hypothesis should be viewed as one possible explanation, out of many, for the data.

In *Mapes Casino, Inc. v. Maryland Casualty Co.*,[114] the court recognized the importance of explanations that the proponent of the statistical evidence had failed to consider. In this action to collect on an insurance policy, Mapes sought to quantify its loss from theft. It argued that employees were using an intermediary to cash in chips at other casinos. The casino established that over an 18-month period, the win percentage at its craps tables was 6%, compared to an expected value of 20%. The statistics proved that *something* was wrong at the craps tables—the discrepancy was too big to explain as the product of random chance. But the court was not convinced by plaintiff's alternative hypothesis. The court pointed to other possible explanations (Runyonesque activities such as skimming, scamming, and crossroading) that might have accounted for the discrepancy without implicating the suspect employees.[115] In short, rejection of the null hypothesis does not leave the proffered alternative hypothesis as the only viable explanation for the data.[116]

113. Intuition may suggest that the more variables included in the model, the better. However, this idea often turns out to be wrong. Complex models may reflect only accidental features of the data. Standard statistical tests offer little protection against this possibility when the analyst has tried a variety of models before settling on the final specification. *See* authorities cited, *supra* note 21.

114. 290 F. Supp. 186 (D. Nev. 1968).

115. *Id.* at 193. Skimming consists of "taking off the top before counting the drop," scamming is "cheating by collusion between dealer and player," and crossroading involves "professional cheaters among the players." *Id.* In plainer language, the court seems to have ruled that the casino itself might be cheating, or there could have been cheaters other than the particular employees identified in the case. At the least, plaintiff's statistical evidence did not rule out such possibilities. *Compare* EEOC v. Sears, Roebuck & Co., 839 F.2d 302, 312 & n.9, 313 (7th Cir. 1988) (EEOC's regression studies showing significant differences did not establish liability because surveys and testimony supported the rival hypothesis that women generally had less interest in commission sales positions), *with* EEOC v. General Tel. Co., 885 F.2d 575 (9th Cir. 1989) (unsubstantiated rival hypothesis of "lack of interest" in "nontraditional" jobs insufficient to rebut prima facie case of gender discrimination); *cf. supra* Section II.A (problem of confounding).

116. *E.g.*, Coleman v. Quaker Oats Co., 232 F.3d 1271, 1283 (9th Cir. 2000) (a disparity with a *p*-value of "3 in 100 billion" did not demonstrate age discrimination because "Quaker never contends that the disparity occurred by chance, just that it did not occur for discriminatory reasons. When other pertinent variables were factored in, the statistical disparity diminished and finally disappeared.").

D. Posterior Probabilities

Standard errors, p-values, and significance tests are common techniques for assessing random error. These procedures rely on sample data and are justified in terms of the operating characteristics of statistical procedures.[117] However, frequentist statisticians generally will not compute the probability that a particular hypothesis is correct, given the data.[118] For example, a frequentist may postulate that a coin is fair: There is a 50-50 chance of landing heads, and successive tosses are independent. This is viewed as an empirical statement—potentially falsifiable—about the coin. It is easy to calculate the chance that a fair coin will turn up heads in the next 10 tosses: The answer (*see supra* Section IV.C.4) is 1/1024. Therefore, observing 10 heads in a row brings into serious doubt the initial hypothesis of fairness.

But what of the converse probability: If the coin does land heads 10 times, what is the chance that it is fair?[119] To compute such converse probabilities, it is necessary to postulate initial probabilities that the coin is fair, as well as probabilities of unfairness to various degrees. In the frequentist theory of inference, such postulates are untenable: Probabilities are objective features of the situation that specify the chances of events or effects, not hypotheses or causes.

By contrast, in the Bayesian approach, probabilities represent subjective degrees of belief about hypotheses or causes rather than objective facts about observations. The observer must quantify beliefs about the chance that the coin is unfair to various degrees—in advance of seeing the data.[120] These subjective probabilities, like the probabilities governing the tosses of the coin, are set up to obey the axioms of probability theory. The probabilities for the various hypotheses about the coin, specified before data collection, are called prior probabilities.

117. Operating characteristics include the expected value and standard error of estimators, probabilities of error for statistical tests, and the like.

118. In speaking of "frequentist statisticians" or "Bayesian statisticians," we do not mean to suggest that all statisticians fall on one side of the philosophical divide or the other. These are archetypes. Many practicing statisticians are pragmatists, using whatever procedure they think is appropriate for the occasion, and not concerning themselves greatly with what the numbers they obtain really mean.

119. We call this a converse probability because it is of the form $P(H_0 | \text{data})$ rather than $P(\text{data} | H_0)$; an equivalent phrase, "inverse probability," also is used. Treating $P(\text{data} | H_0)$ as if it were the converse probability $P(H_0 | \text{data})$ is the transposition fallacy. For example, most U.S. senators are men, but few men are senators. Consequently, there is a high probability that an individual who is a senator is a man, but the probability that an individual who is a man is a senator is practically zero. For examples of the transposition fallacy in court opinions, see cases cited *supra* notes 98, 102. The frequentist p-value, $P(\text{data} | H_0)$, is generally not a good approximation to the Bayesian $P(H_0 | \text{data})$; the latter includes considerations of power and base rates.

120. For example, let p be the unknown probability that the coin lands heads. What is the chance that p exceeds 0.1? 0.6? The Bayesian statistician must be prepared to answer such questions. Bayesian procedures are sometimes defended on the ground that the beliefs of any rational observer must conform to the Bayesian rules. However, the definition of "rational" is purely formal. *See* Peter C. Fishburn, *The Axioms of Subjective Probability*, 1 Stat. Sci. 335 (1986); Freedman, *supra* note 84; David Kaye, *The Laws of Probability and the Law of the Land*, 47 U. Chi. L. Rev. 34 (1979).

Prior probabilities can be updated, using Bayes' rule, given data on how the coin actually falls. (The Appendix explains the rule.) In short, a Bayesian statistician can compute posterior probabilities for various hypotheses about the coin, given the data. These posterior probabilities quantify the statistician's confidence in the hypothesis that a coin is fair.[121] Although such posterior probabilities relate directly to hypotheses of legal interest, they are necessarily subjective, for they reflect not just the data but also the subjective prior probabilities—that is, degrees of belief about hypotheses formulated prior to obtaining data.

Such analyses have rarely been used in court, and the question of their forensic value has been aired primarily in the academic literature. Some statisticians favor Bayesian methods, and some commentators have proposed using these methods in some kinds of cases.[122] The frequentist view of statistics is more conventional; subjective Bayesians are a well-established minority.[123]

121. Here, confidence has the meaning ordinarily ascribed to it, rather than the technical interpretation applicable to a frequentist confidence interval. Consequently, it can be related to the burden of persuasion. *See* D.H. Kaye, *Apples and Oranges: Confidence Coefficients and the Burden of Persuasion,* 73 Cornell L. Rev. 54 (1987).

122. *See* David H. Kaye et al., The New Wigmore: A Treatise on Evidence: Expert Evidence §§ 12.8.5, 14.3.2 (2d ed. 2010); David H. Kaye, *Rounding Up the Usual Suspects: A Legal and Logical Analysis of DNA Database Trawls,* 87 N.C. L. Rev. 425 (2009). In addition, as indicated in the Appendix, Bayes' rule is crucial in solving certain problems involving conditional probabilities of related events. For example, if the proportion of women with breast cancer in a region is known, along with the probability that a mammogram of an affected woman will be positive for cancer and that the mammogram of an unaffected woman will be negative, then one can compute the numbers of false-positive and false-negative mammography results that would be expected to arise in a population-wide screening program. Using Bayes' rule to diagnose a specific patient, however, is more problematic, because the prior probability that the patient has breast cancer may not equal the population proportion. Nevertheless, to overcome the tendency to focus on a test result without considering the "base rate" at which a condition occurs, a diagnostician can apply Bayes' rule to plausible base rates before making a diagnosis. Finally, Bayes' rule also is valuable as a device to explicate the meaning of concepts such as error rates, probative value, and transposition. *See, e.g.,* David H. Kaye, The Double Helix and the Law of Evidence (2010); Wigmore, *supra,* § 7.3.2; David H. Kaye & Jonathan J. Koehler, *The Misquantification of Probative Value,* 27 Law & Hum. Behav. 645 (2003).

123. "Objective Bayesians" use Bayes' rule without eliciting prior probabilities from subjective beliefs. One strategy is to use preliminary data to estimate the prior probabilities and then apply Bayes' rule to that empirical distribution. This "empirical Bayes" procedure avoids the charge of subjectivism at the cost of departing from a fully Bayesian framework. With ample data, however, it can be effective and the estimates or inferences can be understood in frequentist terms. Another "objective" approach is to use "noninformative" priors that are supposed to be independent of all data and prior beliefs. However, the choice of such priors can be questioned, and the approach has been attacked by frequentists and subjective Bayesians. *E.g.,* Joseph B. Kadane, *Is "Objective Bayesian Analysis" Objective, Bayesian, or Wise?,* 1 Bayesian Analysis 433 (2006), *available at* http://ba.stat.cmu.edu/journal/2006/vol01/issue03/kadane.pdf; Jon Williamson, *Philosophies of Probability, in* Philosophy of Mathematics 493 (Andrew Irvine ed., 2009) (discussing the challenges to objective Bayesianism).

V. Correlation and Regression

Regression models are used by many social scientists to infer causation from association. Such models have been offered in court to prove disparate impact in discrimination cases, to estimate damages in antitrust actions, and for many other purposes. Sections V.A, V.B, and V.C cover some preliminary material, showing how scatter diagrams, correlation coefficients, and regression lines can be used to summarize relationships between variables.[124] Section V.D explains the ideas and some of the pitfalls.

A. Scatter Diagrams

The relationship between two variables can be graphed in a scatter diagram (also called a scatterplot or scattergram). We begin with data on income and education for a sample of 178 men, ages 25 to 34, residing in Kansas.[125] Each person in the sample corresponds to one dot in the diagram. As indicated in Figure 5, the horizontal axis shows education, and the vertical axis shows income. Person A completed 12 years of schooling (high school) and had an income of $20,000. Person B completed 16 years of schooling (college) and had an income of $40,000.

Figure 5. Plotting a scatter diagram. The horizontal axis shows educational level and the vertical axis shows income.

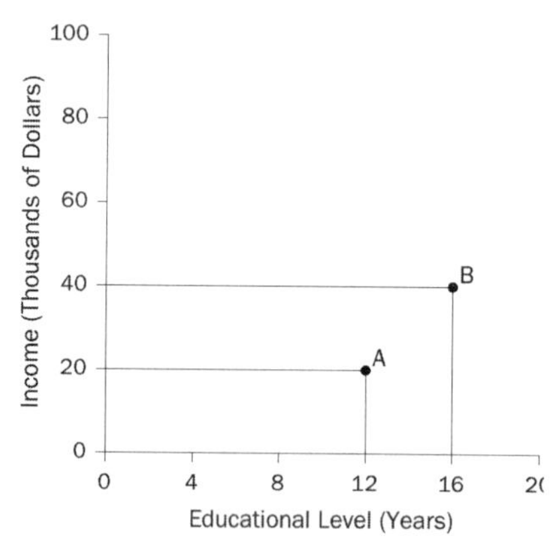

124. The focus is on simple linear regression. *See also* Rubinfeld, *supra* note 21, and the Appendix, *infra*, and Section II, *supra*, for further discussion of these ideas with an emphasis on econometrics.

125. These data are from a public-use CD, Bureau of the Census, U.S. Department of Commerce, for the March 2005 Current Population Survey. Income and education are self-reported. Income is censored at $100,000. For additional details, see Freedman et al., *supra* note 12, at A-11. Both variables in a scatter diagram have to be quantitative (with numerical values) rather than qualitative (nonnumerical).

Figure 6 is the scatter diagram for the Kansas data. The diagram confirms an obvious point. There is a positive association between income and education. In general, persons with a higher educational level have higher incomes. However, there are many exceptions to this rule, and the association is not as strong as one might expect.

Figure 6. Scatter diagram for income and education: men ages 25 to 34 in Kansas.

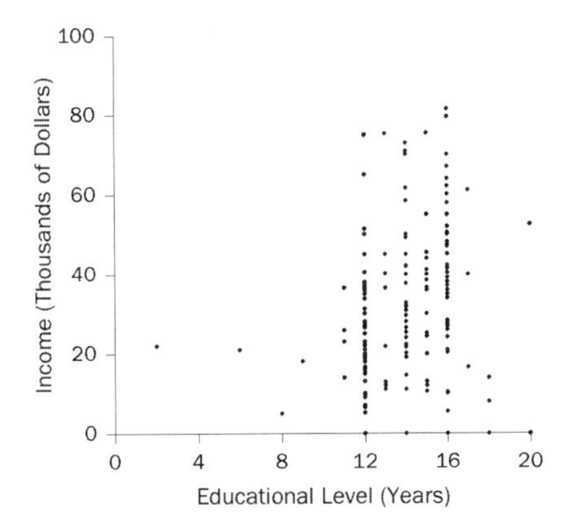

B. Correlation Coefficients

Two variables are positively correlated when their values tend to go up or down together, such as income and education in Figure 5. The correlation coefficient (usually denoted by the letter *r*) is a single number that reflects the sign of an association and its strength. Figure 7 shows *r* for three scatter diagrams: In the first, there is no association; in the second, the association is positive and moderate; in the third, the association is positive and strong.

A correlation coefficient of 0 indicates no linear association between the variables. The maximum value for the coefficient is +1, indicating a perfect linear relationship: The dots in the scatter diagram fall on a straight line that slopes up. Sometimes, there is a negative association between two variables: Large values of one tend to go with small values of the other. The age of a car and its fuel economy in miles per gallon illustrate the idea. Negative association is indicated by negative values for *r*. The extreme case is an *r* of –1, indicating that all the points in the scatter diagram lie on a straight line that slopes down.

Weak associations are the rule in the social sciences. In Figure 5, the correlation between income and education is about 0.4. The correlation between college grades and first-year law school grades is under 0.3 at most law schools, while the

Figure 7. The correlation coefficient measures the sign of a linear association and its strength.

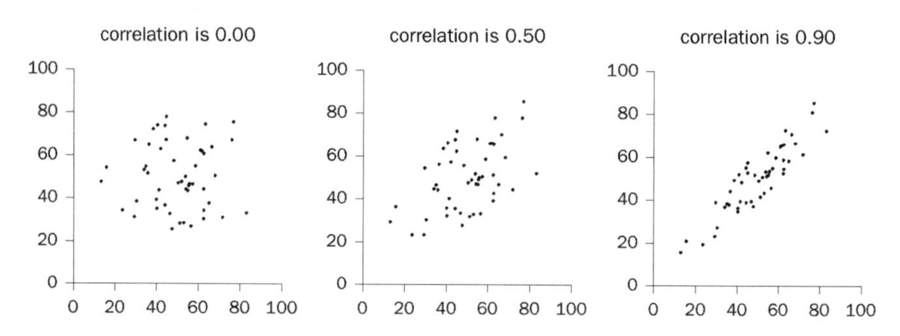

correlation between LSAT scores and first-year grades is generally about 0.4.[126] The correlation between heights of fraternal twins is about 0.5. By contrast, the correlation between heights of identical twins is about 0.95.

1. Is the association linear?

The correlation coefficient has a number of limitations, to be considered in turn. The correlation coefficient is designed to measure linear association. Figure 8 shows a strong nonlinear pattern with a correlation close to zero. The correlation coefficient is of limited use with nonlinear data.

2. Do outliers influence the correlation coefficient?

The correlation coefficient can be distorted by outliers—a few points that are far removed from the bulk of the data. The left-hand panel in Figure 9 shows that one outlier (lower right-hand corner) can reduce a perfect correlation to nearly nothing. Conversely, the right-hand panel shows that one outlier (upper right-hand corner) can raise a correlation of zero to nearly one. If there are extreme outliers in the data, the correlation coefficient is unlikely to be meaningful.

3. Does a confounding variable influence the coefficient?

The correlation coefficient measures the association between two variables. Researchers—and the courts—are usually more interested in causation. Causation is not the same as association. The association between two variables may be driven by a lurking variable that has been omitted from the analysis (*supra*

126. Lisa Anthony Stilwell et al., Predictive Validity of the LSAT: A National Summary of the 2001–2002 Correlation Studies 5, 8 (2003).

Figure 8. The scatter diagram shows a strong nonlinear association with a correlation coefficient close to zero. The correlation coefficient only measures the degree of linear association.

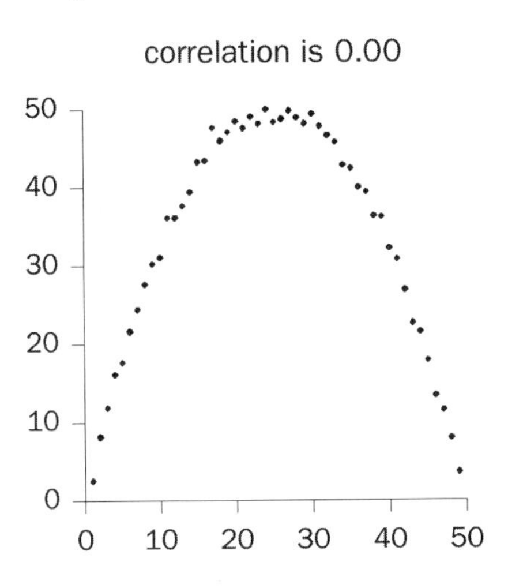

Figure 9. The correlation coefficient can be distorted by outliers.

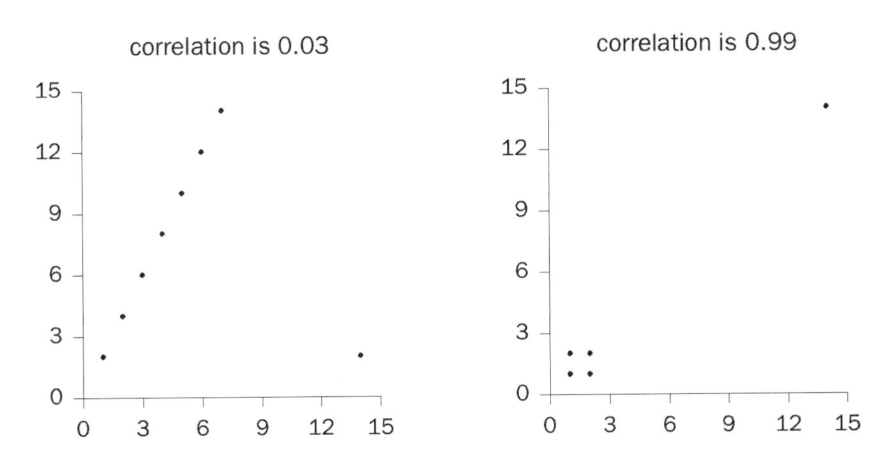

Section II.A). For an easy example, there is an association between shoe size and vocabulary among schoolchildren. However, learning more words does not cause the feet to get bigger, and swollen feet do not make children more articulate. In this case, the lurking variable is easy to spot—age. In more realistic examples, the lurking variable is harder to identify.[127]

127. Green et al., *supra* note 13, Section IV.C, provides one such example.

In statistics, lurking variables are called confounders or confounding variables. Association often does reflect causation, but a large correlation coefficient is not enough to warrant causal inference. A large value of r only means that the dependent variable marches in step with the independent one: Possible reasons include causation, confounding, and coincidence. Multiple regression is one method that attempts to deal with confounders (*infra* Section V.D).[128]

C. Regression Lines

The regression line can be used to describe a linear trend in the data. The regression line for income on education in the Kansas sample is shown in Figure 10. The height of the line estimates the average income for a given educational level. For example, the average income for people with 8 years of education is estimated at $21,100, indicated by the height of the line at 8 years. The average income for people with 16 years of education is estimated at $34,700.

Figure 10. The regression line for income on education and its estimates.

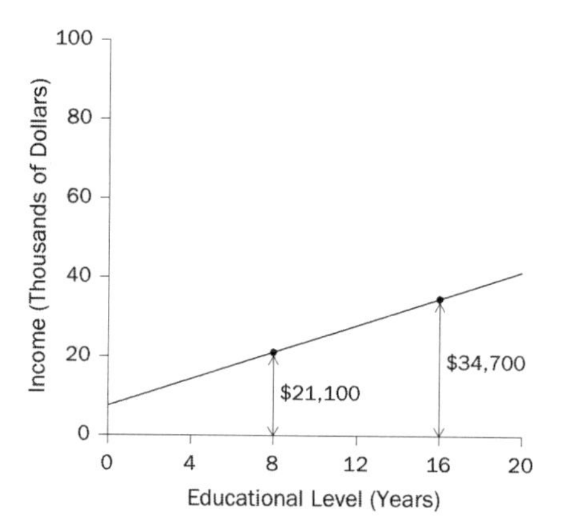

Figure 11 combines the data in Figures 5 and 10: it shows the scatter diagram for income and education, with the regression line superimposed. The line shows the average trend of income as education increases. Thus, the regression line indicates the extent to which a change in one variable (income) is associated with a change in another variable (education).

128. *See also* Rubinfeld, *supra* note 21. The difference between experiments and observational studies is discussed *supra* Section II.B.

Figure 11. Scatter diagram for income and education, with the regression line indicating the trend.

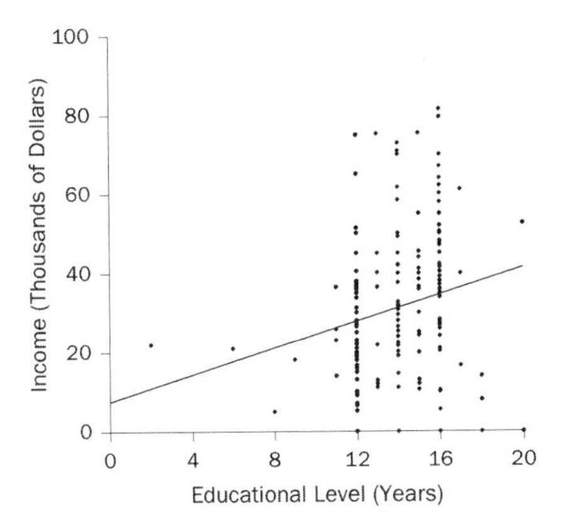

1. *What are the slope and intercept?*

The regression line can be described in terms of its intercept and slope. Often, the slope is the more interesting statistic. In Figure 11, the slope is $1700 per year. On average, each additional year of education is associated with an additional $1700 of income. Next, the intercept is $7500. This is an estimate of the average income for (hypothetical) persons with zero years of education.[129] Figure 10 suggests this estimate may not be especially good. In general, estimates based on the regression line become less trustworthy as we move away from the bulk of the data.

The slope of the regression line has the same limitations as the correlation coefficient: (1) The slope may be misleading if the relationship is strongly nonlinear and (2) the slope may be affected by confounders. With respect to (1), the slope of $1700 per year in Figure 10 presents each additional year of education as having the same value, but some years of schooling surely are worth more and

129. The regression line, like any straight line, has an equation of the form $y = a + bx$. Here, a is the intercept (the value of y when $x = 0$), and b is the slope (the change in y per unit change in x). In Figure 9, the intercept of the regression line is $7500 and the slope is $1700 per year. The line estimates an average income of $34,700 for people with 16 years of education. This may be computed from the intercept and slope as follows:

$$\$7500 + (\$1700 \text{ per year}) \times 16 \text{ years} = \$7500 + \$22,200 = \$34,700.$$

The slope b is the same anywhere along the line. Mathematically, that is what distinguishes straight lines from other curves. If the association is negative, the slope will be negative too. The slope is like the grade of a road, and it is negative if the road goes downhill. The intercept is like the starting elevation of a road, and it is computed from the data so that the line goes through the center of the scatter diagram, rather than being generally too high or too low.

others less. With respect to (2), the association between education and income is no doubt causal, but there are other factors to consider, including family background. Compared to individuals who did not graduate from high school, people with college degrees usually come from richer and better educated families. Thus, college graduates have advantages besides education. As statisticians might say, the effects of family background are confounded with the effects of education. Statisticians often use the guarded phrases "on average" and "associated with" when talking about the slope of the regression line. This is because the slope has limited utility when it comes to making causal inferences.

2. *What is the unit of analysis?*

If association between characteristics of individuals is of interest, these characteristics should be measured on individuals. Sometimes individual-level data are not to be had, but rates or averages for groups are available. "Ecological" correlations are computed from such rates or averages. These correlations generally overstate the strength of an association. For example, average income and average education can be determined for men living in each state and in Washington, D.C. The correlation coefficient for these 51 pairs of averages turns out to be 0.70. However, states do not go to school and do not earn incomes. People do. The correlation for income and education for men in the United States is only 0.42. The correlation for state averages overstates the correlation for individuals—a common tendency for ecological correlations.[130]

Ecological analysis is often seen in cases claiming dilution in voting strength of minorities. In this type of voting rights case, plaintiffs must prove three things: (1) the minority group constitutes a majority in at least one district of a proposed plan; (2) the minority group is politically cohesive, that is, votes fairly solidly for its preferred candidate; and (3) the majority group votes sufficiently as a bloc to defeat the minority-preferred candidate.[131] The first requirement is compactness; the second and third define polarized voting.

130. Correlations are computed from the March 2005 Current Population Survey for men ages 25–64. Freedman et al., *supra* note 12, at 149. The ecological correlation uses only the average figures, but within each state there is a lot of spread about the average. The ecological correlation smoothes away this individual variation. *Cf.* Green et al., *supra* note 13, Section II.B.4 (suggesting that ecological studies of exposure and disease are "far from conclusive" because of the lack of data on confounding variables (a much more general problem) as well as the possible aggregation bias described here); David A. Freedman, *Ecological Inference and the Ecological Fallacy, in* 6 Int'l Encyclopedia of the Social and Behavioral Sciences 4027 (Neil J. Smelser & Paul B. Baltes eds., 2001).

131. *See* Thornburg v. Gingles, 478 U.S. 30, 50–51 (1986) ("First, the minority group must be able to demonstrate that it is sufficiently large and geographically compact to constitute a majority in a single-member district. . . . Second, the minority group must be able to show that it is politically cohesive. . . . Third, the minority must be able to demonstrate that the white majority votes sufficiently as a bloc to enable it . . . usually to defeat the minority's preferred candidate."). In subsequent cases, the Court has emphasized that these factors are not sufficient to make out a violation of section 2 of

The secrecy of the ballot box means that polarized voting cannot be directly observed. Instead, plaintiffs in voting rights cases rely on ecological regression, with scatter diagrams, correlations, and regression lines to estimate voting behavior by groups and demonstrate polarization. The unit of analysis typically is the precinct. For each precinct, public records can be used to determine the percentage of registrants in each demographic group of interest, as well as the percentage of the total vote for each candidate—by voters from all demographic groups combined. Plaintiffs' burden is to determine the vote by each demographic group separately.

Figure 12 shows how the argument unfolds. Each point in the scatter diagram represents data for one precinct in the 1982 Democratic primary election for auditor in Lee County, South Carolina. The horizontal axis shows the percentage of registrants who are white. The vertical axis shows the turnout rate for the white candidate. The regression line is plotted too. The slope would be interpreted as the difference between the white turnout rate and the black turnout rate for the white candidate. Furthermore, the intercept would be interpreted as the black turnout rate for the white candidate.[132] The validity of such estimates is contested in the statistical literature.[133]

the Voting Rights Act. *E.g.*, Johnson v. De Grandy, 512 U.S. 997, 1011 (1994) ("*Gingles* . . . clearly declined to hold [these factors] sufficient in combination, either in the sense that a court's examination of relevant circumstances was complete once the three factors were found to exist, or in the sense that the three in combination necessarily and in all circumstances demonstrated dilution.").

132. By definition, the turnout rate equals the number of votes for the candidate, divided by the number of registrants; the rate is computed separately for each precinct. The intercept of the line in Figure 11 is 4%, and the slope is 0.52. Plaintiffs would conclude that only 4% of the black registrants voted for the white candidate, while 4% + 52% = 56% of the white registrants voted for the white candidate, which demonstrates polarization.

133. For further discussion of ecological regression in this context, see D. James Greiner, *Ecological Inference in Voting Rights Act Disputes: Where Are We Now, and Where Do We Want to Be?*, 47 Jurimetrics J. 115 (2007); Bernard Grofman & Chandler Davidson, Controversies in Minority Voting: The Voting Rights Act in Perspective (1992); Stephen P. Klein & David A. Freedman, *Ecological Regression in Voting Rights Cases*, 6 Chance 38 (Summer 1993). The use of ecological regression increased considerably after the Supreme Court noted in *Thornburg v. Gingles*, 478 U.S. 30, 53 n.20 (1986), that "[t]he District Court found both methods [extreme case analysis and bivariate ecological regression analysis] standard in the literature for the analysis of racially polarized voting." *See, e.g.*, Cottier v. City of Martin, 445 F.3d 1113, 1118 (8th Cir. 2006) (ecological regression is one of the "proven approaches to evaluating elections"); Bruce M. Clarke & Robert Timothy Reagan, Fed. Judicial Ctr., Redistricting Litigation: An Overview of Legal, Statistical, and Case-Management Issues (2002); Greiner, *supra*, at 117, 121. Nevertheless, courts have cautioned against "overreliance on bivariate ecological regression" in light of the inherent limitations of the technique. Lewis v. Alamance County, 99 F.3d 600, 604 n.3 (4th Cir. 1996); Johnson v. Hamrick, 296 F.3d 1065, 1080 n.4 (11th Cir. 2002) ("as a general rule, homogenous precinct analysis may be more reliable than ecological regression."). However, there are problems with both methods. *See, e.g.*, Greiner, *supra*, at 123–39 (arguing that homogeneous precinct analysis is fundamentally flawed and that courts need to be more discerning in dealing with ecological regression).

Redistricting plans based predominantly on racial considerations are unconstitutional unless narrowly tailored to meet a compelling state interest. Shaw v. Reno, 509 U.S. 630 (1993). Whether compliance with the Voting Rights Act can be considered a compelling interest is an open ques-

Figure 12. Turnout rate for the white candidate plotted against the percentage of registrants who are white. Precinct-level data, 1982 Democratic Primary for Auditor, Lee County, South Carolina.

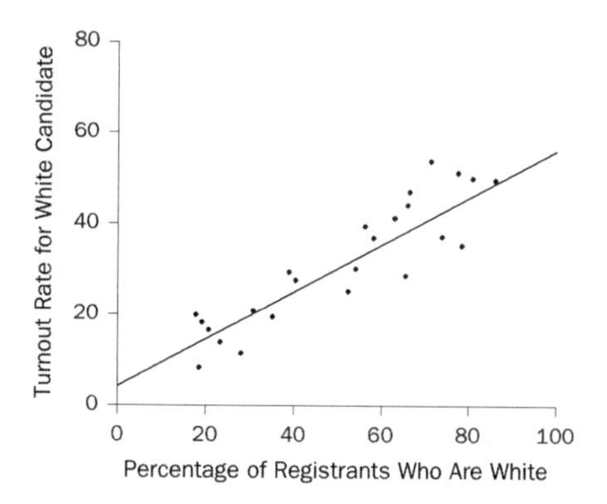

Source: Data from James W. Loewen & Bernard Grofman, *Recent Developments in Methods Used in Vote Dilution Litigation*, 21 Urb. Law. 589, 591 tbl.1 (1989).

D. Statistical Models

Statistical models are widely used in the social sciences and in litigation. For example, the census suffers an undercount, more severe in certain places than others. If some statistical models are to be believed, the undercount can be corrected—moving seats in Congress and millions of dollars a year in tax funds.[134] Other models purport to lift the veil of secrecy from the ballot box, enabling the experts to determine how minority groups have voted—a crucial step in voting rights litigation (*supra* Section V.C). This section discusses the statistical logic of regression models.

A regression model attempts to combine the values of certain variables (the independent variables) to get expected values for another variable (the dependent variable). The model can be expressed in the form of a regression equation. A simple regression equation has only one independent variable; a multiple regression equation has several independent variables. Coefficients in the equation will be interpreted as showing the effects of changing the corresponding variables. This is justified in some situations, as the next example demonstrates.

tion, but efforts to sustain racially motivated redistricting on this basis have not fared well before the Supreme Court. *See* Abrams v. Johnson, 521 U.S. 74 (1997); Shaw v. Hunt, 517 U.S. 899 (1996); Bush v. Vera, 517 U.S. 952 (1996).

134. *See* Brown et al., *supra* note 29; *supra* note 89.

Hooke's law (named after Robert Hooke, England, 1653–1703) describes how a spring stretches in response to a load: Strain is proportional to stress. To verify Hooke's law experimentally, a physicist will make a number of observations on a spring. For each observation, the physicist hangs a weight on the spring and measures its length. A statistician could develop a regression model for these data:

$$\text{length} = a + b \times \text{weight} + \varepsilon. \qquad (1)$$

The error term, denoted by the Greek letter epsilon ε, is needed because measured length will not be exactly equal to $a + b \times$ weight. If nothing else, measurement error must be reckoned with. The model takes ε as "random error"—behaving like draws made at random with replacement from a box of tickets. Each ticket shows a potential error, which will be realized if that ticket is drawn. The average of the potential errors in the box is assumed to be zero.

Equation (1) has two parameters, a and b. These constants of nature characterize the behavior of the spring: a is length under no load, and b is elasticity (the increase in length per unit increase in weight). By way of numerical illustration, suppose a is 400 and b is 0.05. If the weight is 1, the length of the spring is expected to be

$$400 + 0.05 = 400.05.$$

If the weight is 3, the expected length is

$$400 + 3 \times 0.05 = 400 + 0.15 = 400.15.$$

In either case, the actual length will differ from expected, by a random error ε.

In standard statistical terminology, the ε's for different observations on the spring are assumed to be independent and identically distributed, with a mean of zero. Take the ε's for the first two observations. Independence means that the chances for the second ε do not depend on outcomes for the first. If the errors are like draws made at random with replacement from a box of tickets, as we assumed earlier, that box will not change from one draw to the next—independence. "Identically distributed" means that the chance behavior of the two ε's is the same: They are drawn at random from the same box. (*See infra* Appendix for additional discussion.)

The parameters a and b in equation (1) are not directly observable, but they can be estimated by the method of least squares.[135] Statisticians often denote esti-

135. It might seem that a is observable; after all, we can measure the length of the spring with no load. However, the measurement is subject to error, so we observe not a, but $a + \varepsilon$. *See* equation (1). The parameters a and b can be estimated, even estimated very well, but they cannot be observed directly. The least squares estimates of a and b are the intercept and slope of the regression

mates by hats. Thus, $\hat{a}$ is the estimate for a, and $\hat{b}$ is the estimate for b. The values of $\hat{a}$ and $\hat{b}$ are chosen to minimize the sum of the squared prediction errors. These errors are also called residuals. They measure the difference between the actual length of the spring and the predicted length, the latter being $\hat{a} + \hat{b} \times$ weight:

$$\text{actual length} = \hat{a} + \hat{b} \times \text{weight} + \text{residual}. \quad (2)$$

Of course, no one really imagines there to be a box of tickets hidden in the spring. However, the variability of physical measurements (under many but by no means all circumstances) does seem to be remarkably like the variability in draws from a box.[136] In short, the statistical model corresponds rather closely to the empirical phenomenon.

Equation (1) is a statistical model for the data, with unknown parameters a and b. The error term ε is not observable. The model is a theory—and a good one—about how the data are generated. By contrast, equation (2) is a regression equation that is fitted to the data: The intercept $\hat{a}$, the slope $\hat{b}$, and the residual can all be computed from the data. The results are useful because $\hat{a}$ is a good estimate for a, and $\hat{b}$ is a good estimate for b. (Similarly, the residual is a good approximation to ε.) Without the theory, these estimates would be less useful. Is there a theoretical model behind the data processing? Is the model justifiable? These questions can be critical when it comes to making statistical inferences from the data.

In social science applications, statistical models often are invoked without an independent theoretical basis. We give an example involving salary discrimination in the Appendix.[137] The main ideas of such regression modeling can be captured in a hypothetical exchange between a plaintiff seeking to prove salary discrimination and a company denying the allegation. Such a dialog might proceed as follows:

1. Plaintiff argues that the defendant company pays male employees more than females, which establishes a prima facie case of discrimination.
2. The company responds that the men are paid more because they are better educated and have more experience.
3. Plaintiff refutes the company's theory by fitting a regression equation that includes a particular, presupposed relationship between salary (the dependent variable) and some measures of education and experience. Plaintiff's expert reports that even after adjusting for differences in education and

line. *See supra* Section V.C.1; Freedman et al., *supra* note 12, at 208–10. The method of least squares was developed by Adrien-Marie Legendre (France, 1752–1833) and Carl Friedrich Gauss (Germany, 1777–1855) to fit astronomical orbits.

136. This is the Gauss model for measurement error. *See* Freedman et al., *supra* note 12, at 450–52.

137. The Reference Guide to Multiple Regression in this manual describes a comparable example.

experience in this specific manner, men earn more than women. This remaining difference in pay shows discrimination.

4. The company argues that the difference could be the result of chance, not discrimination.
5. Plaintiff replies that because the coefficient for gender in the model is statistically significant, chance is not a good explanation for the data.[138]

In step 3, the three explanatory variables are education (years of schooling completed), experience (years with the firm), and a dummy variable for gender (1 for men and 0 for women). These are supposed to predict salaries (dollars per year). The equation is a formal analog of Hooke's law (equation 1). According to the model, an employee's salary is determined as if by computing

$$a + (b \times \text{education}) + (c \times \text{experience}) + (d \times \text{gender}), \qquad (3)$$

and then adding an error ε drawn at random from a box of tickets.[139] The parameters *a*, *b*, *c*, and *d*, are estimated from the data by the method of least squares.

In step 5, the estimated coefficient *d* for the dummy variable turns out to be positive and statistically significant and is offered as evidence of disparate impact. Men earn more than women, even after adjusting for differences in background factors that might affect productivity. This showing depends on many assumptions built into the model.[140] Hooke's law—equation (1)—is relatively easy to test experimentally. For the salary discrimination model, validation would be difficult. When expert testimony relies on statistical models, the court may well inquire, what are the assumptions behind the model, and why do they apply to the case at hand? It might then be important to distinguish between two situations:

- The nature of the relationship between the variables is known and regression is being used to make quantitative estimates of parameters in that relationship, or
- The nature of the relationship is largely unknown and regression is being used to determine the nature of the relationship—or indeed whether any relationship exists at all.

138. In some cases, the *p*-value has been interpreted as the probability that defendants are innocent of discrimination. However, as noted earlier, such an interpretation is wrong: *p* merely represents the probability of getting a large test statistic, given that the model is correct and the true coefficient for gender is zero (*see supra* Section IV.B, *infra* Appendix, Section D.2). Therefore, even if we grant the model, a *p*-value less than 50% does not demonstrate a preponderance of the evidence against the null hypothesis.

139. Expression (3) is the expected value for salary, given the explanatory variables (education, experience, gender). The error term is needed to account for deviations from expected: Salaries are not going to be predicted very well by linear combinations of variables such as education and experience.

140. *See infra* Appendix.

Regression was developed to handle situations of the first type, with Hooke's law being an example. The basis for the second type of application is analogical, and the tightness of the analogy is an issue worth exploration.

In employment discrimination cases, and other contexts too, a wide variety of models can be used. This is only to be expected, because the science does not dictate specific equations. In a strongly contested case, each side will have its own model, presented by its own expert. The experts will reach opposite conclusions about discrimination. The dialog might continue with an exchange about which model is better. Although statistical assumptions are challenged in court from time to time, arguments more commonly revolve around the choice of variables. One model may be questioned because it omits variables that should be included—for example, skill levels or prior evaluations.[141] Another model may be challenged because it includes tainted variables reflecting past discriminatory behavior by the firm.[142] The court must decide which model—if either—fits the occasion.[143]

The frequency with which regression models are used is no guarantee that they are the best choice for any particular problem. Indeed, from one perspective, a regression or other statistical model may seem to be a marvel of mathematical rigor. From another perspective, the model is a set of assumptions, supported only by the say-so of the testifying expert. Intermediate judgments are also possible.[144]

141. *E.g.*, Bazemore v. Friday, 478 U.S. 385 (1986); *In re* Linerboard Antitrust Litig., 497 F. Supp. 2d 666 (E.D. Pa. 2007).

142. *E.g.*, McLaurin v. Nat'l R.R. Passenger Corp., 311 F. Supp. 2d 61, 65–66 (D.D.C. 2004) (holding that the inclusion of two allegedly tainted variables was reasonable in light of an earlier consent decree).

143. *E.g.*, Chang v. Univ. of R.I., 606 F. Supp. 1161, 1207 (D.R.I. 1985) ("it is plain to the court that [defendant's] model comprises a better, more useful, more reliable tool than [plaintiff's] counterpart."); Presseisen v. Swarthmore College, 442 F. Supp. 593, 619 (E.D. Pa. 1977) ("[E]ach side has done a superior job in challenging the other's regression analysis, but only a mediocre job in supporting their own . . . and the Court is . . . left with nothing."), *aff'd*, 582 F.2d 1275 (3d Cir. 1978).

144. *See, e.g.*, David W. Peterson, *Reference Guide on Multiple Regression*, 36 Jurimetrics J. 213, 214–15 (1996) (review essay); *see supra* note 21 for references to a range of academic opinion. More recently, some investigators have turned to graphical models. However, these models have serious weaknesses of their own. *See, e.g.*, David A. Freedman, *On Specifying Graphical Models for Causation, and the Identification Problem*, 26 Evaluation Rev. 267 (2004).

Appendix

A. Frequentists and Bayesians

The mathematical theory of probability consists of theorems derived from axioms and definitions. Mathematical reasoning is seldom controversial, but there may be disagreement as to how the theory should be applied. For example, statisticians may differ on the interpretation of data in specific applications. Moreover, there are two main schools of thought about the foundations of statistics: frequentist and Bayesian (also called objectivist and subjectivist).[145]

Frequentists see probabilities as empirical facts. When a fair coin is tossed, the probability of heads is 1/2; if the experiment is repeated a large number of times, the coin will land heads about one-half the time. If a fair die is rolled, the probability of getting an ace (one spot) is 1/6. If the die is rolled many times, an ace will turn up about one-sixth of the time.[146] Generally, if a chance experiment can be repeated, the relative frequency of an event approaches (in the long run) its probability. By contrast, a Bayesian considers probabilities as representing not facts but degrees of belief: In whole or in part, probabilities are subjective.

Statisticians of both schools use conditional probability—that is, the probability of one event given that another has occurred. For example, suppose a coin is tossed twice. One event is that the coin will land HH. Another event is that at least one H will be seen. Before the coin is tossed, there are four possible, equally likely, outcomes: HH, HT, TH, TT. So the probability of HH is 1/4. However, if we know that at least one head has been obtained, then we can rule out two tails TT. In other words, given that at least one H has been obtained, the conditional probability of TT is 0, and the first three outcomes have conditional probability 1/3 each. In particular, the conditional probability of HH is 1/3. This is usually written as $P(HH \mid \text{at least one H}) = 1/3$. More generally, the probability of an event C is denoted P(C); the conditional probability of D given C is written as $P(D \mid C)$.

Two events C and D are independent if the conditional probability of D given that C occurs is equal to the conditional probability of D given that C does not occur. Statisticians use "~C" to denote the event that C does not occur. Thus C and D are independent if $P(D \mid C) = P(D \mid {\sim}C)$. If C and D are independent, then the probability that both occur is equal to the product of the probabilities:

$$P(C \text{ and } D) = P(C) \times P(D). \tag{A1}$$

145. But see *supra* note 123 (on "objective Bayesianism").

146. Probabilities may be estimated from relative frequencies, but probability itself is a subtler idea. For example, suppose a computer prints out a sequence of 10 letters H and T (for heads and tails), which alternate between the two possibilities H and T as follows: H T H T H T H T H T. The relative frequency of heads is 5/10 or 50%, but it is not at all obvious that the chance of an H at the next position is 50%. There are difficulties in both the subjectivist and objectivist positions. *See* Freedman, *supra* note 84.

This is the multiplication rule (or product rule) for independent events. If events are dependent, then conditional probabilities must be used:

$$P(C \text{ and } D) = P(C) \times P(D \mid C). \tag{A2}$$

This is the multiplication rule for dependent events.

Bayesian statisticians assign probabilities to hypotheses as well as to events; indeed, for them, the distinction between hypotheses and events may not be a sharp one. We turn now to Bayes' rule. If H_0 and H_1 are two hypotheses[147] that govern the probability of an event A, a Bayesian can use the multiplication rule (A2) to find that

$$P(A \text{ and } H_0) = P(A \mid H_0)P(H_0) \tag{A3}$$

and

$$P(A \text{ and } H_1) = P(A \mid H_1)P(H_1). \tag{A4}$$

Moreover,

$$P(A) = P(A \text{ and } H_0) + P(A \text{ and } H_1). \tag{A5}$$

The multiplication rule (A2) also shows that

$$P(H_1 \mid A) = \frac{P(A \text{ and } H_1)}{P(A)}. \tag{A6}$$

We use (A4) to evaluate $P(A \text{ and } H_1)$ in the numerator of (A6), and (A3), (A4), and (A5) to evaluate P(A) in the denominator:

$$P(H_1 \mid A) = \frac{P(A \mid H_1)P(H_1)}{P(A \mid H_0)P(H_0) + P(A \mid H_1)P(H_1)}. \tag{A7}$$

This is a special case of Bayes' rule. It yields the conditional probability of hypothesis H_0 given that event A has occurred.

For a stylized example in a criminal case, H_0 is the hypothesis that blood found at the scene of a crime came from a person other than the defendant; H_1 is the hypothesis that the blood came from the defendant; A is the event that blood from the crime scene and blood from the defendant are both type A. Then $P(H_0)$ is the prior probability of H_0, based on subjective judgment, while $P(H_0 \mid A)$ is the posterior probability—updated from the prior using the data.

147. H_0 is read "H-sub-zero," while H_1 is "H-sub-one."

Type A blood occurs in 42% of the population. So $P(A \mid H_0) = 0.42$.[148] Because the defendant has type A blood, $P(A \mid H_1) = 1$. Suppose the prior probabilities are $P(H_0) = P(H_1) = 0.5$. According to (A7), the posterior probability that the blood is from the defendant is

$$P(H_1 \mid A) = \frac{1 \times 0.5}{0.42 \times 0.5 + 1 \times 0.5} = 0.70. \quad \text{(A8)}$$

Thus, the data increase the likelihood that the blood is the defendant's. The probability went up from the prior value of $P(H_1) = 0.50$ to the posterior value of $P(H_1 \mid A) = 0.70$.

More generally, H_0 and H_1 refer to parameters in a statistical model. For a stylized example in an employment discrimination case, H_0 asserts equal selection rates in a population of male and female applicants; H_1 asserts that the selection rates are not equal; A is the event that a test statistic exceeds 2 in absolute value. In such situations, the Bayesian proceeds much as before. However, the frequentist computes $P(A \mid H_0)$, and rejects H_0 if this probability falls below 5%. Frequentists have to stop there, because they view $P(H_0 \mid A)$ as poorly defined at best. In their setup, $P(H_0)$ and $P(H_1)$ rarely make sense, and these prior probabilities are needed to compute $P(H_1 \mid A)$: *See supra* equation (A7).

Assessing probabilities, conditional probabilities, and independence is not entirely straightforward, either for frequentists or Bayesians. Inquiry into the basis for expert judgment may be useful, and casual assumptions about independence should be questioned.[149]

B. The Spock Jury: Technical Details

The rest of this Appendix provides some technical backup for the examples in Sections IV and V, *supra*. We begin with the *Spock* jury case. On the null hypothesis, a sample of 350 people was drawn at random from a large population that was 50% male and 50% female. The number of women in the sample follows the binomial distribution. For example, the chance of getting exactly 102 women in the sample is given by the binomial formula[150]

$$\frac{n!}{j! \times (n-j)!} f^j (1-f)^{n-j}. \quad \text{(A9)}$$

148. Not all statisticians would accept the identification of a population frequency with $P(A \mid H_0)$. Indeed, H_0 has been translated into a hypothesis that the true donor has been selected from the population at random (i.e., in a manner that is uncorrelated with blood type). This step needs justification. See *supra* note 123.

149. For problematic assumptions of independence in litigation, see, e.g., *Wilson v. State*, 803 A.2d 1034 (Md. 2002) (error to admit multiplied probabilities in a case involving two deaths of infants in same family); 1 McCormick, *supra* note 2, § 210; *see also supra* note 29 (on census litigation).

150. The binomial formula is discussed in, e.g., Freedman et al., *supra* note 12, at 255–61.

In the formula, n stands for the sample size, and so $n = 350$; and $j = 102$. The f is the fraction of women in the population; thus, $f = 0.50$. The exclamation point denotes factorials: $1! = 1$, $2! = 2 \times 1 = 2$, $3! = 3 \times 2 \times 1 = 6$, and so forth. The chance of 102 women works out to 10^{-15}. In the same way, we can compute the chance of getting 101 women, or 100, or any other particular number. The chance of getting 102 women or fewer is then computed by addition. The chance is $p = 2 \times 10^{-15}$, as reported *supra* note 98. This is very bad news for the null hypothesis.

With the binomial distribution given by (9), the expected the number of women in the sample is

$$nf = 350 \times 0.5 = 175. \tag{A10}$$

The standard error is

$$\sqrt{n} \times \sqrt{f \times (1-f)} = \sqrt{350} \times \sqrt{0.5 \times 0.5} = 9.35. \tag{A11}$$

The observed value of 102 is nearly 8 SEs below the expected value, which is a lot of SEs.

Figure 13 shows the probability histogram for the number of women in the sample.[151] The graph is drawn so that the area between two values is proportional to the chance that the number of women will fall in that range. For example, take the rectangle over 175; its base covers the interval from 174.5 to 175.5. The area of this rectangle is 4.26% of the total area. So the chance of getting exactly 175 women is 4.26%. Next, take the range from 165 to 185 (inclusive): 73.84% of the area falls into this range. This means there is a 73.84% chance that the number of women in the sample will be in the range from 165 to 185 (inclusive).

According to a fundamental theorem in statistics (the central limit theorem), the histogram follows the normal curve.[152] Figure 13 shows the curve for comparison: The normal curve is almost indistinguishable from the top of the histogram. For a numerical example, suppose the jury panel had included 155 women. On the null hypothesis, there is about a 1.85% chance of getting 155 women or fewer. The normal curve gives 1.86%. The error is nil. Ordinarily, we would just report $p = 2\%$, as in the text (*supra* Section IV.B.1).

Finally, we consider power. Suppose we reject the null hypothesis when the number of women in the sample is 155 or less. Let us assume a particular alternative hypothesis that quantifies the degree of discrimination against women: The jury panel is selected at random from a population that is 40% female, rather than 50%. Figure 14 shows the probability histogram for the number of women, but now the histogram is computed according to the alternative hypothesis. Again,

151. Probability histograms are discussed in, e.g., *id.* at 310–13.

152. The central limit theorem is discussed in, e.g., *id.* at 315–27.

Figure 13. Probability histogram for the number of women in a random sample of 350 people drawn from a large population that is 50% female and 50% male. The normal curve is shown for comparison. About 2% of the area under the histogram is to the left of 155 (marked by a heavy vertical line).

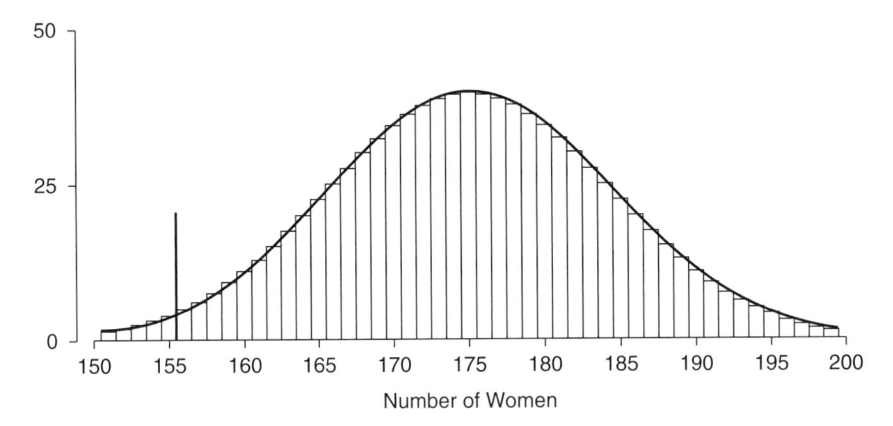

Note: The vertical line is placed at 155.5, and so the area to the left of it includes the rectangles over 155, 154, . . . ; the area represents the chance of getting 155 women or fewer. *Cf.* Freedman et al., *supra* note 12, at 317. The units on the vertical axis are "percent per standard unit"; *cf. id.* at 80, 315.

Figure 14. Probability histogram for the number of women in a random sample of 350 people drawn from a large population that is 40% female and 60% male. The normal curve is shown for comparison. The area to the left of 155 (marked by a heavy vertical line) is about 95%.

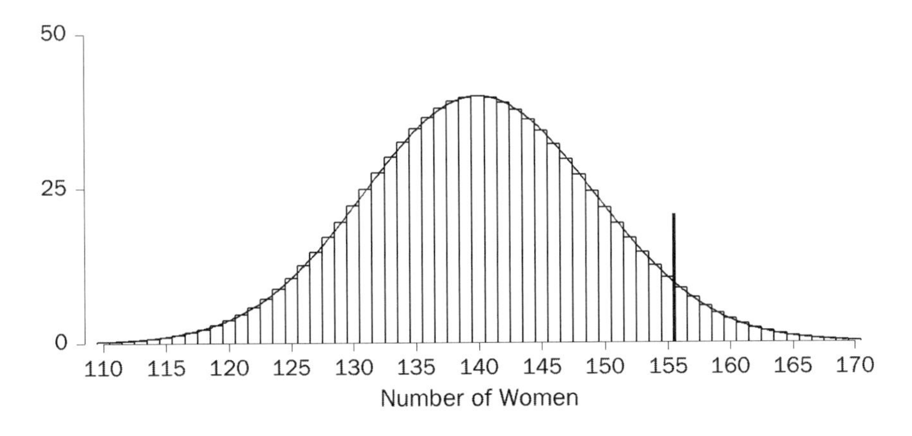

the histogram follows the normal curve. About 95% of the area is to the left of 155, and so power is about 95%. The area can be computed exactly by using the binomial distribution, or to an excellent approximation using the normal curve.

Figures 13 and 14 have the same shape: The central limit theorem is at work. However, the histograms are centered differently. Figure 13 is centered at 175, according to requirements of the null hypothesis. Figure 14 is centered at 140, because the alternative hypothesis is used to determine the center, not the null hypothesis. Thus, 155 is well to the left of center in Figure 13, and well to the right in Figure 14: The figures have different centers. The main point of Figures 13 and 14 is that chances can often be approximated by areas under the normal curve, justifying the large-sample theory presented *supra* Sections IV.A–B.

C. The Nixon Papers: Technical Details

With the Nixon papers, the population consists of 20,000 boxes. A random sample of 500 boxes is drawn and each sample box is appraised. Statistical theory enables us to make some precise statements about the behavior of the sample average.

- The expected value of the sample average equals the population average. Even more tersely, the sample average is an unbiased estimate of the population average.
- The standard error for the sample average equals

$$\sqrt{\frac{N-n}{N-1}} \times \frac{\sigma}{\sqrt{n}}. \qquad \text{(A12)}$$

In (A12), the N stands for the size of the population, which is 20,000; and n stands for the size of the sample, which is 500. The first factor in (A12), with the square root, is the finite sample correction factor. Here, as in many other such examples, the correction factor is so close to 1 that it can safely be ignored. (This is why the size of population usually has no bearing on the precision of the sample average as an estimator for the population average.) Next, σ is the population standard deviation. This is unknown, but it can be estimated by the sample standard deviation, which is \$2200. The SE for the sample mean is therefore estimated from the data as $\$2200/\sqrt{500}$, which is nearly \$100. Plaintiff's total claim is 20,000 times the sample average. The SE for the total claim is therefore 20,000 × \$100 = \$2,000,000. (Here, the size of the population comes into the formula.)

With a large sample, the probability histogram for the sample average follows the normal curve quite closely. That is a consequence of the central limit theorem. The center of the histogram is the population average. The SE is given by (A12), and is about \$100.

- What is the chance that the sample average differs from the population average by 1 SE or less? This chance is equal to the area under the probability histogram within 1 SE of average, which by the central limit theorem is almost equal to the area under the standard normal curve between –1 and 1; that normal area is about 68%.
- What is the chance that the sample average differs from the population average by 2 SE or less? By the same reasoning, this chance is about equal to the area under the standard normal curve between –2 and 2, which is about 95%.
- What is the chance that the sample average differs from the population average by 3 SE or less? This chance is about equal to the area under the standard normal curve between –3 and 3, which is about 99.7%.

To sum up, the probability histogram for the sample average is centered at the population average. The spread is given by the standard error. The histogram follows the normal curve. That is why confidence levels can be based on the standard error, with confidence levels read off the normal curve—for estimators that are essentially unbiased, and obey the central limit theorem (*supra* Section IV.A.2, Appendix Section B).[153] These large-sample methods generally work for sums, averages, and rates, although much depends on the design of the sample.

More technically, the normal curve is the density of a normal distribution. The standard normal density has mean equal to 0 and standard error equal to 1. Its equation is

$$y = e^{-x^2/2} / \sqrt{2\pi}$$

where $e = 2.71828\ldots$ and $\pi = 3.14159\ldots$. This density can be rescaled to have any desired mean and standard error. The resulting densities are the famous "normal curves" or "bell-shaped curves" of statistical theory. In Figure 12, the density is scaled to match the probability histogram in terms of the mean and standard error; likewise in Figure 13.

D. A Social Science Example of Regression: Gender Discrimination in Salaries

1. The regression model

To illustrate social science applications of the kind that might be seen in litigation, Section V referred to a stylized example on salary discrimination. A particular

153. *See, e.g., id.* at 409–24. On the standard deviation, see *supra* Section III.E; Freedman et al., *supra* note 12, at 67–72. The finite sample correction factor is discussed in *id.* at 367–70.

regression model was used to predict salaries (dollars per year) of employees in a firm. It had three explanatory variables: education (years of schooling completed), experience (years with the firm), and a dummy variable for gender (1 for men and 0 for women). The regression equation is

$$\text{salary} = a + b \times \text{education} + c \times \text{experience} + d \times \text{gender} + \varepsilon. \qquad \text{(A13)}$$

Equation (A13) is a statistical model for the data, with unknown parameters a, b, c, and d. Here, a is the intercept and the other parameters are regression coefficients. The ε at the end of the equation is an unobservable error term. In the right-hand side of (A3) and similar expressions, by convention, the multiplications are done before the additions.

As noted in Section V, the equation is a formal analog of Hooke's law (1). According to the model, an employee's salary is determined as if by computing

$$a + b \times \text{education} + c \times \text{experience} + d \times \text{gender} \qquad \text{(A14)}$$

and then adding an error ε drawn at random from a box of tickets. Expression (A14) is the expected value for salary, given the explanatory variables (education, experience, gender). The error term is needed to account for deviations from expected: Salaries are not going to be predicted very well by linear combinations of variables such as education and experience.

The parameters are estimated from the data using least squares. If the estimated coefficient for the dummy variable turns out to be positive and statistically significant, that would be evidence of disparate impact. Men earn more than women, even after adjusting for differences in background factors that might affect productivity. Suppose the estimated equation turns out as follows:

$$\text{predicted salary} = \$7100 + \$1300 \times \text{education} + \$2200 \times \text{experience} + \$700 \times \text{gender}. \qquad \text{(A15)}$$

According to (A15), the estimated value for the intercept a in (A14) is \$7100; the estimated value for the coefficient b is \$1300, and so forth. According to equation (A15), every extra year of education is worth \$1300. Similarly, every extra year of experience is worth \$2200. And, most important, the company gives men a salary premium of \$700 over women with the same education and experience.

A male employee with 12 years of education (high school) and 10 years of experience, for example, would have a predicted salary of

$$\$7100 + \$1300 \times 12 + \$2200 \times 10 + \$700 \times 1 = \$7100 + \$15{,}600 + \$22{,}000 + \$700 = \$45{,}400. \qquad \text{(A16)}$$

A similarly situated female employee has a predicted salary of only

$$\$7100 + \$1300 \times 12 + \$2200 \times 10 + \$700 \times 0$$
$$= \$7100 + \$15{,}600 + \$22{,}000 + \$0 = \$44{,}700. \qquad \text{(A17)}$$

Notice the impact of the gender variable in the model: $700 is added to equation (A16), but not to equation (A17).

A major step in proving discrimination is showing that the estimated coefficient of the gender variable—$700 in the numerical illustration—is statistically significant. This showing depends on the assumptions built into the model. Thus, each extra year of education is assumed to be worth the same across all levels of experience. Similarly, each extra year of experience is worth the same across all levels of education. Furthermore, the premium paid to men does not depend systematically on education or experience. Omitted variables such as ability, quality of education, or quality of experience do not make any systematic difference to the predictions of the model.[154] These are all assumptions made going into the analysis, rather than conclusions coming out of the data.

Assumptions are also made about the error term—the mysterious ε at the end of (A13). The errors are assumed to be independent and identically distributed from person to person in the dataset. Such assumptions are critical when computing *p*-values and demonstrating statistical significance. Regression modeling that does not produce statistically significant coefficients will not be good evidence of discrimination, and statistical significance cannot be established unless stylized assumptions are made about unobservable error terms.

The typical regression model, like the one sketched above, therefore involves a host of assumptions. As noted in Section V, Hooke's law—equation (1)—is relatively easy to test experimentally. For the salary discrimination model—equation (A13)—validation would be difficult. That is why we suggested that when expert testimony relies on statistical models, the court may well inquire about the assumptions behind the model and why they apply to the case at hand.

2. Standard errors, t*-statistics, and statistical significance*

Statistical proof of discrimination depends on the significance of the estimated coefficient for the gender variable. Significance is determined by the *t*-test, using the standard error. The standard error measures the likely difference between the estimated value for the coefficient and its true value. The estimated value is $700—the coefficient of the gender variable in equation (A5); the true value *d* in (A13), remains unknown. According to the model, the difference between the estimated value and the true value is due to the action of the error term ε in (A3). Without ε, observed values would line up perfectly with expected values,

154. Technically, these omitted variables are assumed to be independent of the error term in the equation.

and estimated values for parameters would be exactly equal to true values. This does not happen.

The *t*-statistic is the estimated value divided by its standard error. For example, in (A15), the estimate for *d* is \$700. If the standard error is \$325, then *t* is \$700/\$325 = 2.15. This is significant—that is, hard to explain as the product of random error. Under the null hypothesis that *d* is zero, there is only about a 5% chance that the absolute value of *t* is greater than 2. (We are assuming the sample is large.) Thus, statistical significance is achieved (*supra* Section IV.B.2). Significance would be taken as evidence that *d*—the true parameter in the model (A13)—does not vanish. According to a viewpoint often presented in the social science journals and the courtroom, here is statistical proof that gender matters in determining salaries. On the other hand, if the standard error is \$1400, then *t* is \$700/\$1400 = 0.5. The difference between the estimated value of *d* and zero could easily result from chance. So the true value of *d* could well be zero, in which case gender does not affect salaries.

Of course, the parameter *d* is only a construct in a model. If the model is wrong, the standard error, *t*-statistic, and significance level are rather difficult to interpret. Even if the model is granted, there is a further issue. The 5% is the chance that the absolute value of *t* exceeds 2, given the model and given the null hypothesis that *d* is zero. However, the 5% is often taken to be the chance of the null hypothesis given the data. This misinterpretation is commonplace in the social science literature, and it appears in some opinions describing expert testimony.[155] For a frequentist statistician, the chance that *d* is zero given the data makes no sense: Parameters do not exhibit chance variation. For a Bayesian statistician, the chance that *d* is zero given the data makes good sense, but the computation via the *t*-test could be seriously in error, because the prior probability that *d* is zero has not been taken into account.[156]

The mathematical terminology in the previous paragraph may need to be deciphered: The "absolute value" of *t* is the magnitude, ignoring sign. Thus, the absolute value of both +3 and −3 is 3.

155. *See supra* Section IV.B & notes 102 & 116.

156. *See supra* Section IV & *supra* Appendix.

Glossary of Terms

The following definitions are adapted from a variety of sources, including Michael O. Finkelstein & Bruce Levin, Statistics for Lawyers (2d ed. 2001), and David A. Freedman et al., Statistics (4th ed. 2007).

absolute value. Size, neglecting sign. The absolute value of +2.7 is 2.7; so is the absolute value of −2.7.

adjust for. See control for.

alpha (α). A symbol often used to denote the probability of a Type I error. See Type I error; size. Compare beta.

alternative hypothesis. A statistical hypothesis that is contrasted with the null hypothesis in a significance test. See statistical hypothesis; significance test.

area sample. A probability sample in which the sampling frame is a list of geographical areas. That is, the researchers make a list of areas, choose some at random, and interview people in the selected areas. This is a cost-effective way to draw a sample of people. See probability sample; sampling frame.

arithmetic mean. See mean.

average. See mean.

Bayes' rule. In its simplest form, an equation involving conditional probabilities that relates a "prior probability" known or estimated before collecting certain data to a "posterior probability" that reflects the impact of the data on the prior probability. In Bayesian statistical inference, "the prior" expresses degrees of belief about various hypotheses. Data are collected according to some statistical model; at least, the model represents the investigator's beliefs. Bayes' rule combines the prior with the data to yield the posterior probability, which expresses the investigator's beliefs about the parameters, given the data. See Appendix A. Compare frequentist.

beta (β). A symbol sometimes used to denote power, and sometimes to denote the probability of a Type II error. See Type II error; power. Compare alpha.

between-observer variability. Differences that occur when two observers measure the same thing. Compare within-observer variability.

bias. Also called systematic error. A systematic tendency for an estimate to be too high or too low. An estimate is unbiased if the bias is zero. (Bias does not mean prejudice, partiality, or discriminatory intent.) See nonsampling error. Compare sampling error.

bin. A class interval in a histogram. See class interval; histogram.

binary variable. A variable that has only two possible values (e.g., gender). Called a dummy variable when the two possible values are 0 and 1.

binomial distribution. A distribution for the number of occurrences in repeated, independent "trials" where the probabilities are fixed. For example, the number of heads in 100 tosses of a coin follows a binomial distribution. When the probability is not too close to 0 or 1 and the number of trials is large, the binomial distribution has about the same shape as the normal distribution. See normal distribution; Poisson distribution.

blind. See double-blind experiment.

bootstrap. Also called resampling; Monte Carlo method. A procedure for estimating sampling error by constructing a simulated population on the basis of the sample, then repeatedly drawing samples from the simulated population.

categorical data; categorical variable. See qualitative variable. Compare quantitative variable.

central limit theorem. Shows that under suitable conditions, the probability histogram for a sum (or average or rate) will follow the normal curve. See histogram; normal curve.

chance error. See random error; sampling error.

chi-squared (χ^2). The chi-squared statistic measures the distance between the data and expected values computed from a statistical model. If the chi-squared statistic is too large to explain by chance, the data contradict the model. The definition of "large" depends on the context. See statistical hypothesis; significance test.

class interval. Also, bin. The base of a rectangle in a histogram; the area of the rectangle shows the percentage of observations in the class interval. See histogram.

cluster sample. A type of random sample. For example, investigators might take households at random, then interview all people in the selected households. This is a cluster sample of people: A cluster consists of all the people in a selected household. Generally, clustering reduces the cost of interviewing. See multistage cluster sample.

coefficient of determination. A statistic (more commonly known as *R*-squared) that describes how well a regression equation fits the data. See *R*-squared.

coefficient of variation. A statistic that measures spread relative to the mean: SD/mean, or SE/expected value. See expected value; mean; standard deviation; standard error.

collinearity. See multicollinearity.

conditional probability. The probability that one event will occur given that another has occurred.

confidence coefficient. See confidence interval.

confidence interval. An estimate, expressed as a range, for a parameter. For estimates such as averages or rates computed from large samples, a 95% confidence interval is the range from about two standard errors below to two standard errors above the estimate. Intervals obtained this way cover the true value about 95% of the time, and 95% is the confidence level or the confidence coefficient. See central limit theorem; standard error.

confidence level. See confidence interval.

confounding variable; confounder. A confounder is correlated with the independent variable and the dependent variable. An association between the dependent and independent variables in an observational study may not be causal, but may instead be due to confounding. See controlled experiment; observational study.

consistent estimator. An estimator that tends to become more and more accurate as the sample size grows. Inconsistent estimators, which do not become more accurate as the sample gets larger, are frowned upon by statisticians.

content validity. The extent to which a skills test is appropriate to its intended purpose, as evidenced by a set of questions that adequately reflect the domain being tested. See validity. Compare reliability.

continuous variable. A variable that has arbitrarily fine gradations, such as a person's height. Compare discrete variable.

control for. Statisticians may control for the effects of confounding variables in nonexperimental data by making comparisons for smaller and more homogeneous groups of subjects, or by entering the confounders as explanatory variables in a regression model. To "adjust for" is perhaps a better phrase in the regression context, because in an observational study the confounding factors are not under experimental control; statistical adjustments are an imperfect substitute. See regression model.

control group. See controlled experiment.

controlled experiment. An experiment in which the investigators determine which subjects are put into the treatment group and which are put into the control group. Subjects in the treatment group are exposed by the investigators to some influence—the treatment; those in the control group are not so exposed. For example, in an experiment to evaluate a new drug, subjects in the treatment group are given the drug, and subjects in the control group are given some other therapy; the outcomes in the two groups are compared to see whether the new drug works.

Randomization—that is, randomly assigning subjects to each group—is usually the best way to ensure that any observed difference between the two groups comes from the treatment rather than from preexisting differences. Of course, in many situations, a randomized controlled experiment is impractical, and investigators must then rely on observational studies. Compare observational study.

convenience sample. A nonrandom sample of units, also called a grab sample. Such samples are easy to take but may suffer from serious bias. Typically, mall samples are convenience samples.

correlation coefficient. A number between –1 and 1 that indicates the extent of the linear association between two variables. Often, the correlation coefficient is abbreviated as *r*.

covariance. A quantity that describes the statistical interrelationship of two variables. Compare correlation coefficient; standard error; variance.

covariate. A variable that is related to other variables of primary interest in a study; a measured confounder; a statistical control in a regression equation.

criterion. The variable against which an examination or other selection procedure is validated. See validity.

data. Observations or measurements, usually of units in a sample taken from a larger population.

degrees of freedom. See *t*-test.

dependence. Two events are dependent when the probability of one is affected by the occurrence or non-occurrence of the other. Compare independence; dependent variable.

dependent variable. Also called outcome variable. Compare independent variable.

descriptive statistics. Like the mean or standard deviation, used to summarize data.

differential validity. Differences in validity across different groups of subjects. See validity.

discrete variable. A variable that has only a small number of possible values, such as the number of automobiles owned by a household. Compare continuous variable.

distribution. See frequency distribution; probability distribution; sampling distribution.

disturbance term. A synonym for error term.

double-blind experiment. An experiment with human subjects in which neither the diagnosticians nor the subjects know who is in the treatment group or the control group. This is accomplished by giving a placebo treatment to patients in the control group. In a single-blind experiment, the patients do not know whether they are in treatment or control; the diagnosticians have this information.

dummy variable. Generally, a dummy variable takes only the values 0 or 1, and distinguishes one group of interest from another. See binary variable; regression model.

econometrics. Statistical study of economic issues.

epidemiology. Statistical study of disease or injury in human populations.

error term. The part of a statistical model that describes random error, i.e., the impact of chance factors unrelated to variables in the model. In econometrics, the error term is called a disturbance term.

estimator. A sample statistic used to estimate the value of a population parameter. For example, the sample average commonly is used to estimate the population average. The term "estimator" connotes a statistical procedure, whereas an "estimate" connotes a particular numerical result.

expected value. See random variable.

experiment. See controlled experiment; randomized controlled experiment. Compare observational study.

explanatory variable. See independent variable; regression model.

external validity. See validity.

factors. See independent variable.

Fisher's exact test. A statistical test for comparing two sample proportions. For example, take the proportions of white and black employees getting a promotion. An investigator may wish to test the null hypothesis that promotion does not depend on race. Fisher's exact test is one way to arrive at a *p*-value. The calculation is based on the hypergeometric distribution. For details, see Michael O. Finkelstein and Bruce Levin, Statistics for Lawyers 154–56 (2d ed. 2001). See hypergeometric distribution; *p*-value; significance test; statistical hypothesis.

fitted value. See residual.

fixed significance level. Also alpha; size. A preset level, such as 5% or 1%; if the *p*-value of a test falls below this level, the result is deemed statistically significant. See significance test. Compare observed significance level; *p*-value.

frequency; relative frequency. Frequency is the number of times that something occurs; relative frequency is the number of occurrences, relative to a total. For example, if a coin is tossed 1000 times and lands heads 517 times, the frequency of heads is 517; the relative frequency is 0.517, or 51.7%.

frequency distribution. Shows how often specified values occur in a dataset.

frequentist. Also called objectivist. Describes statisticians who view probabilities as objective properties of a system that can be measured or estimated. Compare Bayesian. *See* Appendix.

Gaussian distribution. A synonym for the normal distribution. See normal distribution.

general linear model. Expresses the dependent variable as a linear combination of the independent variables plus an error term whose components may be dependent and have differing variances. See error term; linear combination; variance. Compare regression model.

grab sample. See convenience sample.

heteroscedastic. See scatter diagram.

highly significant. See *p*-value; practical significance; significance test.

histogram. A plot showing how observed values fall within specified intervals, called bins or class intervals. Generally, matters are arranged so that the area under the histogram, but over a class interval, gives the frequency or relative frequency of data in that interval. With a probability histogram, the area gives the chance of observing a value that falls in the corresponding interval.

homoscedastic. See scatter diagram.

hypergeometric distribution. Suppose a sample is drawn at random, without replacement, from a finite population. How many times will items of a certain type come into the sample? The hypergeometric distribution gives the probabilities. For more details, see 1 William Feller, An Introduction to Probability Theory and Its Applications 41–42 (2d ed. 1957). Compare Fisher's exact test.

hypothesis. See alternative hypothesis; null hypothesis; one-sided hypothesis; significance test; statistical hypothesis; two-sided hypothesis.

hypothesis test. See significance test.

identically distributed. Random variables are identically distributed when they have the same probability distribution. For example, consider a box of numbered tickets. Draw tickets at random with replacement from the box. The draws will be independent and identically distributed.

independence. Also, statistical independence. Events are independent when the probability of one is unaffected by the occurrence or non-occurrence of the other. Compare conditional probability; dependence; independent variable; dependent variable.

independent variable. Independent variables (also called explanatory variables, predictors, or risk factors) represent the causes and potential confounders in a statistical study of causation; the dependent variable represents the effect. In an observational study, independent variables may be used to divide the population up into smaller and more homogenous groups ("stratification"). In a regression model, the independent variables are used to predict the dependent variable. For example, the unemployment rate has been used as the independent variable in a model for predicting the crime rate; the unemployment rate is the independent variable in this model, and the crime rate is the dependent variable. The distinction between independent and dependent variables is unrelated to statistical independence. See regression model. Compare dependent variable; dependence; independence.

indicator variable. See dummy variable.

internal validity. See validity.

interquartile range. Difference between 25th and 75th percentile. See percentile.

interval estimate. A confidence interval, or an estimate coupled with a standard error. See confidence interval; standard error. Compare point estimate.

least squares. See least squares estimator; regression model.

least squares estimator. An estimator that is computed by minimizing the sum of the squared residuals. See residual.

level. The level of a significance test is denoted alpha (α). See alpha; fixed significance level; observed significance level; *p*-value; significance test.

linear combination. To obtain a linear combination of two variables, multiply the first variable by some constant, multiply the second variable by another constant, and add the two products. For example, $2u + 3v$ is a linear combination of u and v.

list sample. See systematic sample.

loss function. Statisticians may evaluate estimators according to a mathematical formula involving the errors—that is, differences between actual values and estimated values. The "loss" may be the total of the squared errors, or the total of the absolute errors, etc. Loss functions seldom quantify real losses, but may be useful summary statistics and may prompt the construction of useful statistical procedures. Compare risk.

lurking variable. See confounding variable.

mean. Also, the average; the expected value of a random variable. The mean gives a way to find the center of a batch of numbers: Add the numbers and divide by how many there are. Weights may be employed, as in "weighted mean" or "weighted average." See random variable. Compare median; mode.

measurement validity. See validity. Compare reliability.

median. The median, like the mean, is a way to find the center of a batch of numbers. The median is the 50th percentile. Half the numbers are larger, and half are smaller. (To be very precise: at least half the numbers are greater than or equal to the median; At least half the numbers are less than or equal to the median; for small datasets, the median may not be uniquely defined.) Compare mean; mode; percentile.

meta-analysis. Attempts to combine information from all studies on a certain topic. For example, in the epidemiological context, a meta-analysis may attempt to provide a summary odds ratio and confidence interval for the effect of a certain exposure on a certain disease.

mode. The most common value. Compare mean; median.

model. See probability model; regression model; statistical model.

multicollinearity. Also, collinearity. The existence of correlations among the independent variables in a regression model. See independent variable; regression model.

multiple comparison. Making several statistical tests on the same dataset. Multiple comparisons complicate the interpretation of a p-value. For example, if 20 divisions of a company are examined, and one division is found to have a disparity significant at the 5% level, the result is not surprising; indeed, it would be expected under the null hypothesis. Compare p-value; significance test; statistical hypothesis.

multiple correlation coefficient. A number that indicates the extent to which one variable can be predicted as a linear combination of other variables. Its magnitude is the square root of R-squared. See linear combination; R-squared; regression model. Compare correlation coefficient.

multiple regression. A regression equation that includes two or more independent variables. See regression model. Compare simple regression.

multistage cluster sample. A probability sample drawn in stages, usually after stratification; the last stage will involve drawing a cluster. See cluster sample; probability sample; stratified random sample.

multivariate methods. Methods for fitting models with multiple variables; in statistics, multiple response variables; in other fields, multiple explanatory variables. See regression model.

natural experiment. An observational study in which treatment and control groups have been formed by some natural development; the assignment of subjects to groups is akin to randomization. See observational study. Compare controlled experiment.

nonresponse bias. Systematic error created by differences between respondents and nonrespondents. If the nonresponse rate is high, this bias may be severe.

nonsampling error. A catch-all term for sources of error in a survey, other than sampling error. Nonsampling errors cause bias. One example is selection bias: The sample is drawn in a way that tends to exclude certain subgroups in the population. A second example is nonresponse bias: People who do not respond to a survey are usually different from respondents. A final example: Response bias arises, for example, if the interviewer uses a loaded question.

normal distribution. Also, Gaussian distribution. When the normal distribution has mean equal to 0 and standard error equal to 1, it is said to be "standard normal." The equation for the density is then

$$y = e^{-x^2/2}/\sqrt{2\pi}$$

where $e = 2.71828\ldots$ and $\pi = 3.14159\ldots$. The density can be rescaled to have any desired mean and standard error, resulting in the famous "bell-shaped curves" of statistical theory. Terminology notwithstanding, there need be nothing wrong with a distribution that differs from normal.

null hypothesis. For example, a hypothesis that there is no difference between two groups from which samples are drawn. See significance test; statistical hypothesis. Compare alternative hypothesis.

objectivist. See frequentist.

observational study. A study in which subjects select themselves into groups; investigators then compare the outcomes for the different groups. For example, studies of smoking are generally observational. Subjects decide whether or not to smoke; the investigators compare the death rate for smokers to the death rate for nonsmokers. In an observational study, the groups may differ in important ways that the investigators do not notice; controlled experiments minimize this problem. The critical distinction is that in a controlled experiment, the investigators intervene to manipulate the circumstances of the subjects; in an observational study, the investigators are passive observers. (Of course, running a good observational study is hard work, and may be quite useful.) Compare confounding variable; controlled experiment.

observed significance level. A synonym for *p*-value. See significance test. Compare fixed significance level.

odds. The probability that an event will occur divided by the probability that it will not. For example, if the chance of rain tomorrow is 2/3, then the odds on rain are (2/3)/(1/3) = 2/1, or 2 to 1; the odds against rain are 1 to 2.

odds ratio. A measure of association, often used in epidemiology. For example, if 10% of all people exposed to a chemical develop a disease, compared with 5% of people who are not exposed, then the odds of the disease in the exposed group are 10/90 = 1/9, compared with 5/95 = 1/19 in the unexposed group. The odds ratio is (1/9)/(1/19) = 19/9 = 2.1. An odds ratio of 1 indicates no association. Compare relative risk.

one-sided hypothesis; one-tailed hypothesis. Excludes the possibility that a parameter could be, for example, less than the value asserted in the null hypothesis. A one-sided hypothesis leads to a one-sided (or one-tailed) test. See significance test; statistical hypothesis; compare two-sided hypothesis.

one-sided test; one-tailed test. See one-sided hypothesis.

outcome variable. See dependent variable.

outlier. An observation that is far removed from the bulk of the data. Outliers may indicate faulty measurements and they may exert undue influence on summary statistics, such as the mean or the correlation coefficient.

***p*-value.** Result from a statistical test. The probability of getting, just by chance, a test statistic as large as or larger than the observed value. Large *p*-values are consistent with the null hypothesis; small *p*-values undermine the null hypothesis. However, *p* does not give the probability that the null hypothesis is true. If *p* is smaller than 5%, the result is statistically significant. If *p* is smaller than 1%, the result is highly significant. The *p*-value is also called the observed significance level. See significance test; statistical hypothesis.

parameter. A numerical characteristic of a population or a model. See probability model.

percentile. To get the percentiles of a dataset, array the data from the smallest value to the largest. Take the 90th percentile by way of example: 90% of the values fall below the 90th percentile, and 10% are above. (To be very precise: At least 90% of the data are at the 90th percentile or below; at least 10% of the data are at the 90th percentile or above.) The 50th percentile is the median: 50% of the values fall below the median, and 50% are above. On the LSAT, a score of 152 places a test taker at the 50th percentile; a score of 164 is at the 90th percentile; a score of 172 is at the 99th percentile. Compare mean; median; quartile.

placebo. See double-blind experiment.

point estimate. An estimate of the value of a quantity expressed as a single number. See estimator. Compare confidence interval; interval estimate.

Poisson distribution. A limiting case of the binomial distribution, when the number of trials is large and the common probability is small. The parameter of the approximating Poisson distribution is the number of trials times the common probability, which is the expected number of events. When this number is large, the Poisson distribution may be approximated by a normal distribution.

population. Also, universe. All the units of interest to the researcher. Compare sample; sampling frame.

population size. Also, size of population. Number of units in the population.

posterior probability. See Bayes' rule.

power. The probability that a statistical test will reject the null hypothesis. To compute power, one has to fix the size of the test and specify parameter values outside the range given by the null hypothesis. A powerful test has a good chance of detecting an effect when there is an effect to be detected. See beta; significance test. Compare alpha; size; *p*-value.

practical significance. Substantive importance. Statistical significance does not necessarily establish practical significance. With large samples, small differences can be statistically significant. See significance test.

practice effects. Changes in test scores that result from taking the same test twice in succession, or taking two similar tests one after the other.

predicted value. See residual.

predictive validity. A skills test has predictive validity to the extent that test scores are well correlated with later performance, or more generally with outcomes that the test is intended to predict. See validity. Compare reliability.

predictor. See independent variable.

prior probability. See Bayes' rule.

probability. Chance, on a scale from 0 to 1. Impossibility is represented by 0, certainty by 1. Equivalently, chances may be quoted in percent; 100% corresponds to 1, 5% corresponds to .05, and so forth.

probability density. Describes the probability distribution of a random variable. The chance that the random variable falls in an interval equals the area below the density and above the interval. (However, not all random variables have densities.) See probability distribution; random variable.

probability distribution. Gives probabilities for possible values or ranges of values of a random variable. Often, the distribution is described in terms of a density. See probability density.

probability histogram. See histogram.

probability model. Relates probabilities of outcomes to parameters; also, statistical model. The latter connotes unknown parameters.

probability sample. A sample drawn from a sampling frame by some objective chance mechanism; each unit has a known probability of being sampled. Such samples minimize selection bias, but can be expensive to draw.

psychometrics. The study of psychological measurement and testing.

qualitative variable; quantitative variable. Describes qualitative features of subjects in a study (e.g., marital status—never-married, married, widowed, divorced, separated). A quantitative variable describes numerical features of the subjects (e.g., height, weight, income). This is not a hard-and-fast distinction, because qualitative features may be given numerical codes, as with a dummy variable. Quantitative variables may be classified as discrete or continuous. Concepts such as the mean and the standard deviation apply only to quantitative variables. Compare continuous variable; discrete variable; dummy variable. See variable.

quartile. The 25th or 75th percentile. See percentile. Compare median.

***R*-squared (R^2).** Measures how well a regression equation fits the data. *R*-squared varies between 0 (no fit) and 1 (perfect fit). *R*-squared does not measure the extent to which underlying assumptions are justified. See regression model. Compare multiple correlation coefficient; standard error of regression.

random error. Sources of error that are random in their effect, like draws made at random from a box. These are reflected in the error term of a statistical model. Some authors refer to random error as chance error or sampling error. See regression model.

random variable. A variable whose possible values occur according to some probability mechanism. For example, if a pair of dice are thrown, the total number of spots is a random variable. The chance of two spots is 1/36, the chance of three spots is 2/36, and so forth; the most likely number is 7, with chance 6/36.

The expected value of a random variable is the weighted average of the possible values; the weights are the probabilities. In our example, the expected value is

$$\frac{1}{36}\times 2+\frac{2}{36}\times 3+\frac{3}{36}\times 4+\frac{5}{36}\times 6+\frac{6}{36}\times 7$$
$$+\frac{5}{36}\times 8+\frac{4}{36}\times 9+\frac{3}{36}\times 10+\frac{2}{36}\times 11+\frac{1}{36}\times 12$$

In many problems, the weighted average is computed with respect to the density; then sums must be replaced by integrals. The expected value need not be a possible value for the random variable.

Generally, a random variable will be somewhere around its expected value, but will be off (in either direction) by something like a standard error (SE) or so. If the random variable has a more or less normal distribution, there is about a 68% chance for it to fall in the range expected value – SE to expected value + SE. See normal curve; standard error.

randomization. See controlled experiment; randomized controlled experiment.

randomized controlled experiment. A controlled experiment in which subjects are placed into the treatment and control groups at random—as if by a lottery. See controlled experiment. Compare observational study.

range. The difference between the biggest and the smallest values in a batch of numbers.

rate. In an epidemiological study, the number of events, divided by the size of the population; often cross-classified by age and gender. For example, the death rate from heart disease among American men ages 55–64 in 2004 was about three per thousand. Among men ages 65–74, the rate was about seven per thousand. Among women, the rate was about half that for men. Rates adjust for differences in sizes of populations or subpopulations. Often, rates are computed per unit of time, e.g., per thousand persons per year. Data source: Statistical Abstract of the United States tbl. 115 (2008).

regression coefficient. The coefficient of a variable in a regression equation. See regression model.

regression diagnostics. Procedures intended to check whether the assumptions of a regression model are appropriate.

regression equation. See regression model.

regression line. The graph of a (simple) regression equation.

regression model. A regression model attempts to combine the values of certain variables (the independent or explanatory variables) in order to get expected values for another variable (the dependent variable). Sometimes, the phrase "regression model" refers to a probability model for the data; if no qualifications are made, the model will generally be linear, and errors will be assumed independent across observations, with common variance, The coefficients in the linear combination are called regression coefficients; these are parameters. At times, "regression model" refers to an equation ("the regression equation") estimated from data, typically by least squares.

For example, in a regression study of salary differences between men and women in a firm, the analyst may include a dummy variable for gender, as well as statistical controls such as education and experience to adjust for productivity differences between men and women. The dummy variable would be defined as 1 for the men and 0 for the women. Salary would be the dependent variable; education, experience, and the dummy would be the independent variables. See least squares; multiple regression; random error; variance. Compare general linear model.

relative frequency. See frequency.

relative risk. A measure of association used in epidemiology. For example, if 10% of all people exposed to a chemical develop a disease, compared to 5% of people who are not exposed, then the disease occurs twice as frequently among the exposed people: The relative risk is 10%/5% = 2. A relative risk of 1 indicates no association. For more details, see Leon Gordis, Epidemiology (4th ed. 2008). Compare odds ratio.

reliability. The extent to which a measurement process gives the same results on repeated measurement of the same thing. Compare validity.

representative sample. Not a well-defined technical term. A sample judged to fairly represent the population, or a sample drawn by a process likely to give samples that fairly represent the population, for example, a large probability sample.

resampling. See bootstrap.

residual. The difference between an actual and a predicted value. The predicted value comes typically from a regression equation, and is better called the fitted value, because there is no real prediction going on. See regression model; independent variable.

response variable. See independent variable.

risk. Expected loss. "Expected" means on average, over the various datasets that could be generated by the statistical model under examination. Usually, risk cannot be computed exactly but has to be estimated, because the parameters in the statistical model are unknown and must be estimated. See loss function; random variable.

risk factor. See independent variable.

robust. A statistic or procedure that does not change much when data or assumptions are modified slightly.

sample. A set of units collected for study. Compare population.

sample size. Also, size of sample. The number of units in a sample.

sample weights. See stratified random sample.

sampling distribution. The distribution of the values of a statistic, over all possible samples from a population. For example, suppose a random sample is drawn. Some values of the sample mean are more likely; others are less likely. The sampling distribution specifies the chance that the sample mean will fall in one interval rather than another.

sampling error. A sample is part of a population. When a sample is used to estimate a numerical characteristic of the population, the estimate is likely to differ from the population value because the sample is not a perfect microcosm of the whole. If the estimate is unbiased, the difference between the estimate and the exact value is sampling error. More generally,

$$\text{estimate} = \text{true value} + \text{bias} + \text{sampling error}$$

Sampling error is also called chance error or random error. See standard error. Compare bias; nonsampling error.

sampling frame. A list of units designed to represent the entire population as completely as possible. The sample is drawn from the frame.

sampling interval. See systematic sample.

scatter diagram. Also, scatterplot; scattergram. A graph showing the relationship between two variables in a study. Each dot represents one subject. One variable is plotted along the horizontal axis, the other variable is plotted along the vertical axis. A scatter diagram is homoscedastic when the spread is more or less the same inside any vertical strip. If the spread changes from one strip to another, the diagram is heteroscedastic.

selection bias. Systematic error due to nonrandom selection of subjects for study.

sensitivity. In clinical medicine, the probability that a test for a disease will give a positive result given that the patient has the disease. Sensitivity is analogous to the power of a statistical test. Compare specificity.

sensitivity analysis. Analyzing data in different ways to see how results depend on methods or assumptions.

sign test. A statistical test based on counting and the binomial distribution. For example, a Finnish study of twins found 22 monozygotic twin pairs where 1 twin smoked, 1 did not, and at least 1 of the twins had died. That sets up a race to death. In 17 cases, the smoker died first; in 5 cases, the nonsmoker died first. The null hypothesis is that smoking does not affect time to death, so the chances are 50-50 for the smoker to die first. On the null hypothesis, the chance that the smoker will win the race 17 or more times out of 22 is

8/1000. That is the *p*-value. The *p*-value can be computed from the binomial distribution. For additional detail, see Michael O. Finkelstein & Bruce Levin, Statistics for Lawyers 339–41 (2d ed. 2001); David A. Freedman et al., Statistics 262–63 (4th ed. 2007).

significance level. See fixed significance level; *p*-value.

significance test. Also, statistical test; hypothesis test; test of significance. A significance test involves formulating a statistical hypothesis and a test statistic, computing a *p*-value, and comparing *p* to some preestablished value (α) to decide if the test statistic is significant. The idea is to see whether the data conform to the predictions of the null hypothesis. Generally, a large test statistic goes with a small *p*-value; and small *p*-values would undermine the null hypothesis.

For example, suppose that a random sample of male and female employees were given a skills test and the mean scores of the men and women were different—in the sample. To judge whether the difference is due to sampling error, a statistician might consider the implications of competing hypotheses about the difference in the population. The null hypothesis would say that on average, in the population, men and women have the same scores: The difference observed in the data is then just due to sampling error. A one-sided alternative hypothesis would be that on average, in the population, men score higher than women. The one-sided test would reject the null hypothesis if the sample men score substantially higher than the women—so much so that the difference is hard to explain on the basis of sampling error.

In contrast, the null hypothesis could be tested against the two-sided alternative that on average, in the population, men score differently than women—higher or lower. The corresponding two-sided test would reject the null hypothesis if the sample men score substantially higher or substantially lower than the women.

The one-sided and two-sided tests would both be based on the same data, and use the same *t*-statistic. However, if the men in the sample score higher than the women, the one-sided test would give a *p*-value only half as large as the two-sided test; that is, the one-sided test would appear to give stronger evidence against the null hypothesis. ("One-sided" and "one-tailed" are synonymous; so are "two-sided and "two-tailed.") See *p*-value; statistical hypothesis; *t*-statistic.

significant. See *p*-value; practical significance; significance test.

simple random sample. A random sample in which each unit in the sampling frame has the same chance of being sampled. The investigators take a unit at random (as if by lottery), set it aside, take another at random from what is left, and so forth.

simple regression. A regression equation that includes only one independent variable. Compare multiple regression.

size. A synonym for alpha (α).

skip factor. See systematic sample.

specificity. In clinical medicine, the probability that a test for a disease will give a negative result given that the patient does not have the disease. Specificity is analogous to $1 - \alpha$, where α is the significance level of a statistical test. Compare sensitivity.

spurious correlation. When two variables are correlated, one is not necessarily the cause of the other. The vocabulary and shoe size of children in elementary school, for example, are correlated—but learning more words will not make the feet grow. Such noncausal correlations are said to be spurious. (Originally, the term seems to have been applied to the correlation between two rates with the same denominator: Even if the numerators are unrelated, the common denominator will create some association.) Compare confounding variable.

standard deviation (SD). Indicates how far a typical element deviates from the average. For example, in round numbers, the average height of women age 18 and over in the United States is 5 feet 4 inches. However, few women are exactly average; most will deviate from average, at least by a little. The SD is sort of an average deviation from average. For the height distribution, the SD is 3 inches. The height of a typical woman is around 5 feet 4 inches, but is off that average value by something like 3 inches.

For distributions that follow the normal curve, about 68% of the elements are in the range from 1 SD below the average to 1 SD above the average. Thus, about 68% of women have heights in the range 5 feet 1 inch to 5 feet 7 inches. Deviations from the average that exceed 3 or 4 SDs are extremely unusual. Many authors use standard deviation to also mean standard error. See standard error.

standard error (SE). Indicates the likely size of the sampling error in an estimate. Many authors use the term standard deviation instead of standard error. Compare expected value; standard deviation.

standard error of regression. Indicates how actual values differ (in some average sense) from the fitted values in a regression model. See regression model; residual. Compare *R*-squared.

standard normal. See normal distribution.

standardization. See standardized variable.

standardized variable. Transformed to have mean zero and variance one. This involves two steps: (1) subtract the mean; (2) divide by the standard deviation.

statistic. A number that summarizes data. A statistic refers to a sample; a parameter or a true value refers to a population or a probability model.

statistical controls. Procedures that try to filter out the effects of confounding variables on non-experimental data, for example, by adjusting through statistical procedures such as multiple regression. Variables in a multiple regression equation. See multiple regression; confounding variable; observational study. Compare controlled experiment.

statistical dependence. See dependence.

statistical hypothesis. Generally, a statement about parameters in a probability model for the data. The null hypothesis may assert that certain parameters have specified values or fall in specified ranges; the alternative hypothesis would specify other values or ranges. The null hypothesis is tested against the data with a test statistic; the null hypothesis may be rejected if there is a statistically significant difference between the data and the predictions of the null hypothesis.

Typically, the investigator seeks to demonstrate the alternative hypothesis; the null hypothesis would explain the findings as a result of mere chance, and the investigator uses a significance test to rule out that possibility. See significance test.

statistical independence. See independence.

statistical model. See probability model.

statistical test. See significance test.

statistical significance. See *p*-value.

stratified random sample. A type of probability sample. The researcher divides the population into relatively homogeneous groups called "strata," and draws a random sample separately from each stratum. Dividing the population into strata is called "stratification." Often the sampling fraction will vary from stratum to stratum. Then sampling weights should be used to extrapolate from the sample to the population. For example, if 1 unit in 10 is sampled from stratum A while 1 unit in 100 is sampled from stratum B, then each unit drawn from A counts as 10, and each unit drawn from B counts as 100. The first kind of unit has weight 10; the second has weight 100. See Freedman et al., Statistics 401 (4th ed. 2007).

stratification. See independent variable; stratified random sample.

study validity. See validity.

subjectivist. See Bayesian.

systematic error. See bias.

systematic sample. Also, list sample. The elements of the population are numbered consecutively as 1, 2, 3, ... The investigators choose a starting point and a "sampling interval" or "skip factor" k. Then, every kth element is selected into the sample. If the starting point is 1 and $k = 10$, for example, the sample would consist of items 1, 11, 21, ... Sometimes the starting point is chosen at random from 1 to k: this is a random-start systematic sample.

***t*-statistic.** A test statistic, used to make the *t*-test. The *t*-statistic indicates how far away an estimate is from its expected value, relative to the standard error. The expected value is computed using the null hypothesis that is being tested.

Some authors refer to the *t*-statistic, others to the *z*-statistic, especially when the sample is large. With a large sample, a *t*-statistic larger than 2 or 3 in absolute value makes the null hypothesis rather implausible—the estimate is too many standard errors away from its expected value. See statistical hypothesis; significance test; *t*-test.

***t*-test.** A statistical test based on the *t*-statistic. Large *t*-statistics are beyond the usual range of sampling error. For example, if t is bigger than 2, or smaller than –2, then the estimate is statistically significant at the 5% level; such values of t are hard to explain on the basis of sampling error. The scale for *t*-statistics is tied to areas under the normal curve. For example, a *t*-statistic of 1.5 is not very striking, because 13% = 13/100 of the area under the normal curve is outside the range from –1.5 to 1.5. On the other hand, $t = 3$ is remarkable: Only 3/1000 of the area lies outside the range from –3 to 3. This discussion is predicated on having a reasonably large sample; in that context, many authors refer to the *z*-test rather than the *t*-test.

Consider testing the null hypothesis that the average of a population equals a given value; the population is known to be normal. For small samples, the *t*-statistic follows Student's *t*-distribution (when the null hypothesis holds) rather than the normal curve; larger values of t are required to achieve significance. The relevant *t*-distribution depends on the number of degrees of freedom, which in this context equals the sample size minus one. A *t*-test is not appropriate for small samples drawn from a population that is not normal. See *p*-value; significance test; statistical hypothesis.

test statistic. A statistic used to judge whether data conform to the null hypothesis. The parameters of a probability model determine expected values for the data; differences between expected values and observed values are measured by a test statistic. Such test statistics include the chi-squared statistic (χ^2) and the *t*-statistic. Generally, small values of the test statistic are consistent with the null hypothesis; large values lead to rejection. See *p*-value; statistical hypothesis; *t*-statistic.

time series. A series of data collected over time, for example, the Gross National Product of the United States from 1945 to 2005.

treatment group. See controlled experiment.

two-sided hypothesis; two-tailed hypothesis. An alternative hypothesis asserting that the values of a parameter are different from—either greater than or less than—the value asserted in the null hypothesis. A two-sided alternative hypothesis suggests a two-sided (or two-tailed) test. See significance test; statistical hypothesis. Compare one-sided hypothesis.

two-sided test; two-tailed test. See two-sided hypothesis.

Type I error. A statistical test makes a Type I error when (1) the null hypothesis is true and (2) the test rejects the null hypothesis, i.e., there is a false positive. For example, a study of two groups may show some difference between samples from each group, even when there is no difference in the population. When a statistical test deems the difference to be significant in this situation, it makes a Type I error. See significance test; statistical hypothesis. Compare alpha; Type II error.

Type II error. A statistical test makes a Type II error when (1) the null hypothesis is false and (2) the test fails to reject the null hypothesis, i.e., there is a false negative. For example, there may not be a significant difference between samples from two groups when, in fact, the groups are different. See significance test; statistical hypothesis. Compare beta; Type I error.

unbiased estimator. An estimator that is correct on average, over the possible datasets. The estimates have no systematic tendency to be high or low. Compare bias.

uniform distribution. For example, a whole number picked at random from 1 to 100 has the uniform distribution: All values are equally likely. Similarly, a uniform distribution is obtained by picking a real number at random between 0.75 and 3.25: The chance of landing in an interval is proportional to the length of the interval.

validity. Measurement validity is the extent to which an instrument measures what it is supposed to, rather than something else. The validity of a standardized test is often indicated by the correlation coefficient between the test scores and some outcome measure (the criterion variable). See content validity; differential validity; predictive validity. Compare reliability.

Study validity is the extent to which results from a study can be relied upon. Study validity has two aspects, internal and external. A study has high internal validity when its conclusions hold under the particular circumstances of the study. A study has high external validity when its results are generalizable. For example, a well-executed randomized controlled double-blind experiment performed on an unusual study population will have high internal validity because the design is good; but its external validity will be debatable because the study population is unusual.

Validity is used also in its ordinary sense: assumptions are valid when they hold true for the situation at hand.

variable. A property of units in a study, which varies from one unit to another, for example, in a study of households, household income; in a study of people, employment status (employed, unemployed, not in labor force).

variance. The square of the standard deviation. Compare standard error; covariance.

weights. See stratified random sample.

within-observer variability. Differences that occur when an observer measures the same thing twice, or measures two things that are virtually the same. Compare between-observer variability.

z-statistic. See t-statistic.

z-test. See t-test.

References on Statistics

General Surveys

David Freedman et al., Statistics (4th ed. 2007).

Darrell Huff, How to Lie with Statistics (1993).

Gregory A. Kimble, How to Use (and Misuse) Statistics (1978).

David S. Moore & William I. Notz, Statistics: Concepts and Controversies (2005).

Michael Oakes, Statistical Inference: A Commentary for the Social and Behavioral Sciences (1986).

Statistics: A Guide to the Unknown (Roxy Peck et al. eds., 4th ed. 2005).

Hans Zeisel, Say It with Figures (6th ed. 1985).

Reference Works for Lawyers and Judges

David C. Baldus & James W.L. Cole, Statistical Proof of Discrimination (1980 & Supp. 1987) (continued as Ramona L. Paetzold & Steven L. Willborn, The Statistics of Discrimination: Using Statistical Evidence in Discrimination Cases (1994) (updated annually).

David W. Barnes & John M. Conley, Statistical Evidence in Litigation: Methodology, Procedure, and Practice (1986 & Supp. 1989).

James Brooks, A Lawyer's Guide to Probability and Statistics (1990).

Michael O. Finkelstein & Bruce Levin, Statistics for Lawyers (2d ed. 2001).

Modern Scientific Evidence: The Law and Science of Expert Testimony (David L. Faigman et al. eds., Volumes 1 and 2, 2d ed. 2002) (updated annually).

David H. Kaye et al., The New Wigmore: A Treatise on Evidence: Expert Evidence § 12 (2d ed. 2011) (updated annually).

National Research Council, The Evolving Role of Statistical Assessments as Evidence in the Courts (Stephen E. Fienberg ed., 1989).

Statistical Methods in Discrimination Litigation (David H. Kaye & Mikel Aickin eds., 1986).

Hans Zeisel & David Kaye, Prove It with Figures: Empirical Methods in Law and Litigation (1997).

General Reference

Encyclopedia of Statistical Sciences (Samuel Kotz et al. eds., 2d ed. 2005).

Reference Guide on Multiple Regression

DANIEL L. RUBINFELD

Daniel L. Rubinfeld, Ph.D., is Robert L. Bridges Professor of Law and Professor of Economics Emeritus, University of California, Berkeley, and Visiting Professor of Law at New York University Law School.

CONTENTS

I. Introduction and Overview

Multiple regression analysis is a statistical tool used to understand the relationship between or among two or more variables.[1] Multiple regression involves a variable to be explained—called the dependent variable—and additional explanatory variables that are thought to produce or be associated with changes in the dependent variable.[2] For example, a multiple regression analysis might estimate the effect of the number of years of work on salary. Salary would be the dependent variable to be explained; the years of experience would be the explanatory variable.

Multiple regression analysis is sometimes well suited to the analysis of data about competing theories for which there are several possible explanations for the relationships among a number of explanatory variables.[3] Multiple regression typically uses a single dependent variable and several explanatory variables to assess the statistical data pertinent to these theories. In a case alleging sex discrimination in salaries, for example, a multiple regression analysis would examine not only sex, but also other explanatory variables of interest, such as education and experience.[4] The employer-defendant might use multiple regression to argue that salary is a function of the employee's education and experience, and the employee-plaintiff might argue that salary is also a function of the individual's sex. Alternatively, in an antitrust cartel damages case, the plaintiff's expert might utilize multiple regression to evaluate the extent to which the price of a product increased during the period in which the cartel was effective, after accounting for costs and other variables unrelated to the cartel. The defendant's expert might use multiple

1. A variable is anything that can take on two or more values (e.g., the daily temperature in Chicago or the salaries of workers at a factory).

2. Explanatory variables in the context of a statistical study are sometimes called independent variables. *See* David H. Kaye & David A. Freedman, Reference Guide on Statistics, Section II.A.1, in this manual. The guide also offers a brief discussion of multiple regression analysis. *Id.*, Section V.

3. Multiple regression is one type of statistical analysis involving several variables. Other types include matching analysis, stratification, analysis of variance, probit analysis, logit analysis, discriminant analysis, and factor analysis.

4. Thus, in *Ottaviani v. State University of New York,* 875 F.2d 365, 367 (2d Cir. 1989) (citations omitted), *cert. denied*, 493 U.S. 1021 (1990), the court stated:

> In disparate treatment cases involving claims of gender discrimination, plaintiffs typically use multiple regression analysis to isolate the influence of gender on employment decisions relating to a particular job or job benefit, such as salary.

The first step in such a regression analysis is to specify all of the possible "legitimate" (i.e., nondiscriminatory) factors that are likely to significantly affect the dependent variable and which could account for disparities in the treatment of male and female employees. By identifying those legitimate criteria that affect the decisionmaking process, individual plaintiffs can make predictions about what job or job benefits similarly situated employees should ideally receive, and then can measure the difference between the predicted treatment and the actual treatment of those employees. If there is a disparity between the predicted and actual outcomes for female employees, plaintiffs in a disparate treatment case can argue that the net "residual" difference represents the unlawful effect of discriminatory animus on the allocation of jobs or job benefits.

regression to suggest that the plaintiff's expert had omitted a number of price-determining variables.

More generally, multiple regression may be useful (1) in determining whether a particular effect is present; (2) in measuring the magnitude of a particular effect; and (3) in forecasting what a particular effect would be, but for an intervening event. In a patent infringement case, for example, a multiple regression analysis could be used to determine (1) whether the behavior of the alleged infringer affected the price of the patented product, (2) the size of the effect, and (3) what the price of the product would have been had the alleged infringement not occurred.

Over the past several decades, the use of multiple regression analysis in court has grown widely. Regression analysis has been used most frequently in cases of sex and race discrimination[5] antitrust violations,[6] and cases involving class cer-

5. Discrimination cases using multiple regression analysis are legion. *See, e.g.*, Bazemore v. Friday, 478 U.S. 385 (1986), *on remand*, 848 F.2d 476 (4th Cir. 1988); Csicseri v. Bowsher, 862 F. Supp. 547 (D.D.C. 1994) (age discrimination), *aff'd*, 67 F.3d 972 (D.C. Cir. 1995); EEOC v. General Tel. Co., 885 F.2d 575 (9th Cir. 1989), *cert. denied*, 498 U.S. 950 (1990); Bridgeport Guardians, Inc. v. City of Bridgeport, 735 F. Supp. 1126 (D. Conn. 1990), *aff'd*, 933 F.2d 1140 (2d Cir.), *cert. denied*, 502 U.S. 924 (1991); Bickerstaff v. Vassar College, 196 F.3d 435, 448–49 (2d Cir. 1999) (sex discrimination); McReynolds v. Sodexho Marriott, 349 F. Supp. 2d 1 (D.C. Cir. 2004) (race discrimination); Hnot v. Willis Group Holdings Ltd., 228 F.R.D. 476 (S.D.N.Y. 2005) (gender discrimination); Carpenter v. Boeing Co., 456 F.3d 1183 (10th Cir. 2006) (sex discrimination); Coward v. ADT Security Systems, Inc., 140 F.3d 271, 274–75 (D.C. Cir. 1998); Smith v. Virginia Commonwealth Univ., 84 F.3d 672 (4th Cir. 1996) (en banc); Hemmings v. Tidyman's Inc., 285 F.3d 1174, 1184–86 (9th Cir. 2000); Mehus v. Emporia State University, 222 F.R.D. 455 (D. Kan. 2004) (sex discrimination); Guiterrez v. Johnson & Johnson, 2006 WL 3246605 (D.N.J. Nov. 6, 2006 (race discrimination); Morgan v. United Parcel Service, 380 F.3d 459 (8th Cir. 2004) (racial discrimination). *See also* Keith N. Hylton & Vincent D. Rougeau, *Lending Discrimination: Economic Theory, Econometric Evidence, and the Community Reinvestment Act*, 85 Geo. L.J. 237, 238 (1996) ("regression analysis is probably the best empirical tool for uncovering discrimination").

6. *E.g.*, United States v. Brown Univ., 805 F. Supp. 288 (E.D. Pa. 1992) (price fixing of college scholarships), *rev'd*, 5 F.3d 658 (3d Cir. 1993); Petruzzi's IGA Supermarkets, Inc. v. Darling-Delaware Co., 998 F.2d 1224 (3d Cir.), *cert. denied*, 510 U.S. 994 (1993); Ohio v. Louis Trauth Dairy, Inc., 925 F. Supp. 1247 (S.D. Ohio 1996); *In re* Chicken Antitrust Litig., 560 F. Supp. 963, 993 (N.D. Ga. 1980); New York v. Kraft Gen. Foods, Inc., 926 F. Supp. 321 (S.D.N.Y. 1995); Freeland v. AT&T, 238 F.R.D. 130 (S.D.N.Y. 2006); *In re* Pressure Sensitive Labelstock Antitrust Litig., 2007 U.S. Dist. LEXIS 85466 (M.D. Pa. Nov. 19, 2007); *In re* Linerboard Antitrust Litig., 497 F. Supp. 2d 666 (E.D. Pa. 2007) (price fixing by manufacturers of corrugated boards and boxes); *In re* Polypropylene Carpet Antitrust Litig., 93 F. Supp. 2d 1348 (N.D. Ga. 2000); *In re* OSB Antitrust Litig., 2007 WL 2253418 (E.D. Pa. Aug. 3, 2007) (price fixing of Oriented Strand Board, also known as "waferboard"); *In re* TFT-LCD (Flat Panel) Antitrust Litig., 267 F.R.D. 583 (N.D. Cal. 2010).

For a broad overview of the use of regression methods in antitrust, see ABA Antitrust Section, Econometrics: Legal, Practical and Technical Issues (John Harkrider & Daniel Rubinfeld, eds. 2005). *See also* Jerry Hausman et al., *Competitive Analysis with Differenciated Products*, 34 Annales D'Économie et de Statistique 159 (1994); Gregory J. Werden, *Simulating the Effects of Differentiated Products Mergers: A Practical Alternative to Structural Merger Policy*, 5 Geo. Mason L. Rev. 363 (1997).

tification (under Rule 23).[7] However, there are a range of other applications, including census undercounts,[8] voting rights,[9] the study of the deterrent effect of the death penalty,[10] rate regulation,[11] and intellectual property.[12]

7. In antitrust, the circuits are currently split as to the extent to which plaintiffs must prove that common elements predominate over individual elements. *E.g., compare In Re* Hydrogen Peroxide Litig., 522 F.2d 305 (3d Cir. 2008) *with In Re* Cardizem CD Antitrust Litig., 391 F.3d 812 (6th Cir. 2004). For a discussion of use of multiple regression in evaluating class certification, see Bret M. Dickey & Daniel L. Rubinfeld, *Antitrust Class Certification: Towards an Economic Framework,* 66 N.Y.U. Ann. Surv. Am. L. 459 (2010) and John H. Johnson & Gregory K. Leonard, *Economics and the Rigorous Analysis of Class Certification in Antitrust Cases,* 3 J. Competition L. & Econ. 341 (2007).

8. *See, e.g.,* City of New York v. U.S. Dep't of Commerce, 822 F. Supp. 906 (E.D.N.Y. 1993) (decision of Secretary of Commerce not to adjust the 1990 census was not arbitrary and capricious), *vacated,* 34 F.3d 1114 (2d Cir. 1994) (applying heightened scrutiny), *rev'd sub nom.* Wisconsin v. City of New York, 517 U.S. 565 (1996); Carey v. Klutznick, 508 F. Supp. 420, 432–33 (S.D.N.Y. 1980) (use of reasonable and scientifically valid statistical survey or sampling procedures to adjust census figures for the differential undercount is constitutionally permissible), *stay granted,* 449 U.S. 1068 (1980), *rev'd on other grounds,* 653 F.2d 732 (2d Cir. 1981), *cert. denied,* 455 U.S. 999 (1982); Young v. Klutznick, 497 F. Supp. 1318, 1331 (E.D. Mich. 1980), *rev'd on other grounds,* 652 F.2d 617 (6th Cir. 1981), *cert. denied,* 455 U.S. 939 (1982).

9. Multiple regression analysis was used in suits charging that at-large areawide voting was instituted to neutralize black voting strength, in violation of section 2 of the Voting Rights Act, 42 U.S.C. § 1973 (1988). Multiple regression demonstrated that the race of the candidates and that of the electorate were determinants of voting. *See* Williams v. Brown, 446 U.S. 236 (1980); Rodriguez v. Pataki, 308 F. Supp. 2d 346, 414 (S.D.N.Y. 2004); United States v. Vill. of Port Chester, 2008 U.S. Dist. LEXIS 4914 (S.D.N.Y. Jan. 17, 2008); Meza v. Galvin, 322 F. Supp. 2d 52 (D. Mass. 2004) (violation of VRA with regard to Hispanic voters in Boston); Bone Shirt v. Hazeltine, 336 F. Supp. 2d 976 (D.S.D. 2004) (violations of VRA with regard to Native American voters in South Dakota); Georgia v. Ashcroft, 195 F. Supp. 2d 25 (D.D.C. 2002) (redistricting of Georgia's state and federal legislative districts); Benavidez v. City of Irving, 638 F. Supp. 2d 709 (N.D. Tex. 2009) (challenge of city's at-large voting scheme). For commentary on statistical issues in voting rights cases, see, e.g., *Statistical and Demographic Issues Underlying Voting Rights Cases,* 15 Evaluation Rev. 659 (1991); Stephen P. Klein et al., *Ecological Regression Versus the Secret Ballot,* 31 Jurimetrics J. 393 (1991); James W. Loewen & Bernard Grofman, *Recent Developments in Methods Used in Vote Dilution Litigation,* 21 Urb. Law. 589 (1989); Arthur Lupia & Kenneth McCue, *Why the 1980s Measures of Racially Polarized Voting Are Inadequate for the 1990s,* 12 Law & Pol'y 353 (1990).

10. *See, e.g.,* Gregg v. Georgia, 428 U.S. 153, 184–86 (1976). For critiques of the validity of the deterrence analysis, see National Research Council, Deterrence and Incapacitation: Estimating the Effects of Criminal Sanctions on Crime Rates (Alfred Blumstein et al. eds., 1978); Richard O. Lempert, *Desert and Deterrence: An Assessment of the Moral Bases of the Case for Capital Punishment,* 79 Mich. L. Rev. 1177 (1981); Hans Zeisel, *The Deterrent Effect of the Death Penalty: Facts v. Faith,* 1976 Sup. Ct. Rev. 317; and John Donohue & Justin Wolfers, *Uses and Abuses of Statistical Evidence in the Death Penalty Debate,* 58 Stan. L. Rev. 787 (2005).

11. *See, e.g.,* Time Warner Entertainment Co. v. FCC, 56 F.3d 151 (D.C. Cir. 1995) (challenge to FCC's application of multiple regression analysis to set cable rates), *cert. denied,* 516 U.S. 1112 (1996); Appalachian Power Co. v. EPA, 135 F.3d 791 (D.C. Cir. 1998) (challenging the EPA's application of regression analysis to set nitrous oxide emission limits); Consumers Util. Rate Advocacy Div. v. Ark. PSC, 99 Ark. App. 228 (Ark. Ct. App. 2007) (challenging an increase in nongas rates).

12. *See* Polaroid Corp. v. Eastman Kodak Co., No. 76-1634-MA, 1990 WL 324105, at *29, *62–63 (D. Mass. Oct. 12, 1990) (damages awarded because of patent infringement), *amended by* No.

Multiple regression analysis can be a source of valuable scientific testimony in litigation. However, when inappropriately used, regression analysis can confuse important issues while having little, if any, probative value. In *EEOC v. Sears, Roebuck & Co.*,[13] in which Sears was charged with discrimination against women in hiring practices, the Seventh Circuit acknowledged that "[m]ultiple regression analyses, designed to determine the effect of several independent variables on a dependent variable, which in this case is hiring, are an accepted and common method of proving disparate treatment claims."[14] However, the court affirmed the district court's findings that the "E.E.O.C.'s regression analyses did not 'accurately reflect Sears' complex, nondiscriminatory decision-making processes'" and that the "'E.E.O.C.'s statistical analyses [were] so flawed that they lack[ed] any persuasive value.'"[15] Serious questions also have been raised about the use of multiple regression analysis in census undercount cases and in death penalty cases.[16]

The Supreme Court's rulings in *Daubert* and *Kumho Tire* have encouraged parties to raise questions about the admissibility of multiple regression analyses.[17] Because multiple regression is a well-accepted scientific methodology, courts have frequently admitted testimony based on multiple regression studies, in some cases over the strong objection of one of the parties.[18] However, on some occasions courts have excluded expert testimony because of a failure to utilize a multiple regression methodology.[19] On other occasions, courts have rejected regression

76-1634-MA, 1991 WL 4087 (D. Mass. Jan. 11, 1991); Estate of Vane v. The Fair, Inc., 849 F.2d 186, 188 (5th Cir. 1988) (lost profits were the result of copyright infringement), *cert. denied*, 488 U.S. 1008 (1989); Louis Vuitton Malletier v. Dooney & Bourke, Inc., 525 F. Supp. 2d 576, 664 (S.D.N.Y. 2007) (trademark infringement and unfair competition suit). The use of multiple regression analysis to estimate damages has been contemplated in a wide variety of contexts. *See, e.g.*, David Baldus et al., *Improving Judicial Oversight of Jury Damages Assessments: A Proposal for the Comparative Additur/Remittitur Review of Awards for Nonpecuniary Harms and Punitive Damages*, 80 Iowa L. Rev. 1109 (1995); Talcott J. Franklin, *Calculating Damages for Loss of Parental Nurture Through Multiple Regression Analysis*, 52 Wash. & Lee L. Rev. 271 (1997); Roger D. Blair & Amanda Kay Esquibel, *Yardstick Damages in Lost Profit Cases: An Econometric Approach*, 72 Denv. U. L. Rev. 113 (1994). Daniel Rubinfeld, *Quantitative Methods in Antitrust*, *in* 1 Issues in Competition Law and Policy 723 (2008).

13. 839 F.2d 302 (7th Cir. 1988).

14. *Id.* at 324 n.22.

15. *Id.* at 348, 351 (quoting EEOC v. Sears, Roebuck & Co., 628 F. Supp. 1264, 1342, 1352 (N.D. Ill. 1986)). The district court commented specifically on the "severe limits of regression analysis in evaluating complex decision-making processes." 628 F. Supp. at 1350.

16. *See* David H. Kaye & David A. Freedman, Reference Guide on Statistics, Sections II.A.3, B.1, in this manual.

17. Daubert v. Merrill Dow Pharms., Inc. 509 U.S. 579 (1993); Kumho Tire Co. v. Carmichael, 526 U.S. 137, 147 (1999) (expanding the *Daubert's* application to nonscientific expert testimony).

18. *See* Newport Ltd. v. Sears, Roebuck & Co., 1995 U.S. Dist. LEXIS 7652 (E.D. La. May 26, 1995). *See also* Petruzzi's IGA Supermarkets, *supra* note 6, 998 F.2d at 1240, 1247 (finding that the district court abused its discretion in excluding multiple regression-based testimony and reversing the grant of summary judgment to two defendants).

19. *See, e.g.*, *In re* Executive Telecard Ltd. Sec. Litig., 979 F. Supp. 1021 (S.D.N.Y. 1997).

studies that did not have an adequate foundation or research design with respect to the issues at hand.[20]

In interpreting the results of a multiple regression analysis, it is important to distinguish between correlation and causality. Two variables are correlated—that is, associated with each other—when the events associated with the variables occur more frequently together than one would expect by chance. For example, if higher salaries are associated with a greater number of years of work experience, and lower salaries are associated with fewer years of experience, there is a positive correlation between salary and number of years of work experience. However, if higher salaries are associated with less experience, and lower salaries are associated with more experience, there is a negative correlation between the two variables.

A correlation between two variables does not imply that one event causes the second. Therefore, in making causal inferences, it is important to avoid *spurious correlation*.[21] Spurious correlation arises when two variables are closely related but bear no causal relationship because they are both caused by a third, unexamined variable. For example, there might be a negative correlation between the age of certain skilled employees of a computer company and their salaries. One should not conclude from this correlation that the employer has necessarily discriminated against the employees on the basis of their age. A third, unexamined variable, such as the level of the employees' technological skills, could explain differences in productivity and, consequently, differences in salary.[22] Or, consider a patent infringement case in which increased sales of an allegedly infringing product are associated with a lower price of the patented product.[23] This correlation would be spurious if the two products have their own noncompetitive market niches and the lower price is the result of a decline in the production costs of the patented product.

Pointing to the possibility of a spurious correlation will typically not be enough to dispose of a statistical argument. It may be appropriate to give little weight to such an argument absent a showing that the correlation is relevant. For example, a statistical showing of a relationship between technological skills

20. *See* City of Tuscaloosa v. Harcros Chemicals, Inc., 158 F.2d 548 (11th Cir. 1998), in which the court ruled plaintiffs' regression-based expert testimony inadmissible and granted summary judgment to the defendants. *See also* American Booksellers Ass'n v. Barnes & Noble, Inc., 135 F. Supp. 2d 1031, 1041 (N.D. Cal. 2001), in which a model was said to contain "too many assumptions and simplifications that are not supported by real-world evidence," *and* Obrey v. Johnson, 400 F.3d 691 (9th Cir. 2005).

21. *See* David H. Kaye & David A. Freedman, Reference Guide on Statistics, Section V.B.3, in this manual.

22. *See, e.g.*, Sheehan v. Daily Racing Form Inc., 104 F.3d 940, 942 (7th Cir.) (rejecting plaintiff's age discrimination claim because statistical study showing correlation between age and retention ignored the "more than remote possibility that age was correlated with a legitimate job-related qualification"), *cert. denied*, 521 U.S. 1104 (1997).

23. In some particular cases, there are statistical tests that allow one to reject claims of causality. For a brief description of these tests, which were developed by Jerry Hausman, see Robert S. Pindyck & Daniel L. Rubinfeld, Econometric Models and Economic Forecasts § 7.5 (4th ed. 1997).

and worker productivity might be required in the age discrimination example, above.[24]

Causality cannot be inferred by data analysis alone; rather, one must infer that a causal relationship exists on the basis of an underlying causal theory that explains the relationship between the two variables. Even when an appropriate theory has been identified, causality can never be inferred directly. One must also look for empirical evidence that there is a causal relationship. Conversely, the fact that two variables are correlated does not guarantee the existence of a relationship; it could be that the model—a characterization of the underlying causal theory—does not reflect the correct interplay among the explanatory variables. In fact, the absence of correlation does not guarantee that a causal relationship does not exist. Lack of correlation could occur if (1) there are insufficient data, (2) the data are measured inaccurately, (3) the data do not allow multiple causal relationships to be sorted out, or (4) the model is specified wrongly because of the omission of a variable or variables that are related to the variable of interest.

There is a tension between any attempt to reach conclusions with near certainty and the inherently uncertain nature of multiple regression analysis. In general, the statistical analysis associated with multiple regression allows for the expression of uncertainty in terms of probabilities. The reality that statistical analysis generates probabilities concerning relationships rather than certainty should not be seen in itself as an argument against the use of statistical evidence, or worse, as a reason to not admit that there is uncertainty at all. The only alternative might be to use less reliable anecdotal evidence.

This reference guide addresses a number of procedural and methodological issues that are relevant in considering the admissibility of, and weight to be accorded to, the findings of multiple regression analyses. It also suggests some standards of reporting and analysis that an expert presenting multiple regression analyses might be expected to meet. Section II discusses research design—how the multiple regression framework can be used to sort out alternative theories about a case. The guide discusses the importance of choosing the appropriate specification of the multiple regression model and raises the issue of whether multiple regression is appropriate for the case at issue. Section III accepts the regression framework and concentrates on the interpretation of the multiple regression results from both a statistical and a practical point of view. It emphasizes the distinction between regression results that are statistically significant and results that are meaningful to the trier of fact. It also points to the importance of evaluating the robustness

24. *See, e.g.*, Allen v. Seidman, 881 F.2d 375 (7th Cir. 1989) (judicial skepticism was raised when the defendant did not submit a logistic regression incorporating an omitted variable—the possession of a higher degree or special education; defendant's attack on statistical comparisons must also include an analysis that demonstrates that comparisons are flawed). The appropriate requirements for the defendant's showing of spurious correlation could, in general, depend on the discovery process. *See, e.g.*, Boykin v. Georgia Pac. Co., 706 F.2d 1384 (1983) (criticism of a plaintiff's analysis for not including omitted factors, when plaintiff considered all information on an application form, was inadequate).

of regression analyses, i.e., seeing the extent to which the results are sensitive to changes in the underlying assumptions of the regression model. Section IV briefly discusses the qualifications of experts and suggests a potentially useful role for court-appointed neutral experts. Section V emphasizes procedural aspects associated with use of the data underlying regression analyses. It encourages greater pretrial efforts by the parties to attempt to resolve disputes over statistical studies.

Throughout the main body of this guide, hypothetical examples are used as illustrations. Moreover, the basic "mathematics" of multiple regression has been kept to a bare minimum. To achieve that goal, the more formal description of the multiple regression framework has been placed in the Appendix. The Appendix is self-contained and can be read before or after the text. The Appendix also includes further details with respect to the examples used in the body of this guide.

II. Research Design: Model Specification

Multiple regression allows the testifying economist or other expert to choose among alternative theories or hypotheses and assists the expert in distinguishing correlations between variables that are plainly spurious from those that may reflect valid relationships.

A. What Is the Specific Question That Is Under Investigation by the Expert?

Research begins with a clear formulation of a research question. The data to be collected and analyzed must relate directly to this question; otherwise, appropriate inferences cannot be drawn from the statistical analysis. For example, if the question at issue in a patent infringement case is what price the plaintiff's product would have been but for the sale of the defendant's infringing product, sufficient data must be available to allow the expert to account statistically for the important factors that determine the price of the product.

B. What Model Should Be Used to Evaluate the Question at Issue?

Model specification involves several steps, each of which is fundamental to the success of the research effort. Ideally, a multiple regression analysis builds on a theory that describes the variables to be included in the study. A typical regression model will include one or more dependent variables, each of which is believed to be causally related to a series of explanatory variables. Because we cannot be certain that the explanatory variables are themselves unaffected or independent of the influence of the dependent variable (at least at the point of initial study), the explanatory variables are often termed *covariates*. Covariates are known to have an association with the dependent or outcome variable, but causality remains an open question.

For example, the theory of labor markets might lead one to expect salaries in an industry to be related to workers' experience and the productivity of workers' jobs. A belief that there is job discrimination would lead one to create a model in which the dependent variable was a measure of workers' salaries and the list of covariates included a variable reflecting discrimination in addition to measures of job training and experience.

In a perfect world, the analysis of the job discrimination (or any other) issue might be accomplished through a controlled "natural experiment," in which employees would be randomly assigned to a variety of employers in an industry under study and asked to fill positions requiring identical experience and skills. In this observational study, where the only difference in salaries could be a result of discrimination, it would be possible to draw clear and direct inferences from an analysis of salary data. Unfortunately, the opportunity to conduct observational studies of this kind is rarely available to experts in the context of legal proceedings. In the real world, experts must do their best to interpret the results of real-world "*quasi-experiments*," in which it is impossible to control all factors that might affect worker salaries or other variables of interest.[25]

Models are often characterized in terms of parameters—numerical characteristics of the model. In the labor market discrimination example, one parameter might reflect the increase in salary associated with each additional year of prior job experience. Another parameter might reflect the reduction in salary associated with a lack of current on-the-job experience. Multiple regression uses a sample, or a selection of data, from the population (all the units of interest) to obtain estimates of the values of the parameters of the model. An estimate associated with a particular explanatory variable is an estimated regression coefficient.

Failure to develop the proper theory, failure to choose the appropriate variables, or failure to choose the correct form of the model can substantially bias the statistical results—that is, create a systematic tendency for an estimate of a model parameter to be too high or too low.

1. Choosing the dependent variable

The variable to be explained, the dependent variable, should be the appropriate variable for analyzing the question at issue.[26] Suppose, for example, that pay dis-

25. In the literature on natural and quasi-experiments, the explanatory variables are characterized as "treatments" and the dependent variable as the "outcome." For a review of natural experiments in the criminal justice arena, see David P. Farrington, *A Short History of Randomized Experiments in Criminology*, 27 Evaluation Rev. 218–27 (2003).

26. In multiple regression analysis, the dependent variable is usually a continuous variable that takes on a range of numerical values. When the dependent variable is categorical, taking on only two or three values, modified forms of multiple regression, such as probit analysis or logit analysis, are

crimination among hourly workers is a concern. One choice for the dependent variable is the hourly wage rate of the employees; another choice is the annual salary. The distinction is important, because annual salary differences may in part result from differences in hours worked. If the number of hours worked is the product of worker preferences and not discrimination, the hourly wage is a good choice. If the number of hours worked is related to the alleged discrimination, annual salary is the more appropriate dependent variable to choose.[27]

2. Choosing the explanatory variable that is relevant to the question at issue

The explanatory variable that allows the evaluation of alternative hypotheses must be chosen appropriately. Thus, in a discrimination case, the variable of interest may be the race or sex of the individual. In an antitrust case, it may be a variable that takes on the value 1 to reflect the presence of the alleged anticompetitive behavior and the value 0 otherwise.[28]

3. Choosing the additional explanatory variables

An attempt should be made to identify additional known or hypothesized explanatory variables, some of which are measurable and may support alternative substantive hypotheses that can be accounted for by the regression analysis. Thus, in a discrimination case, a measure of the skills of the workers may provide an alternative explanation—lower salaries may have been the result of inadequate skills.[29]

appropriate. For an example of the use of the latter, see EEOC v. Sears, Roebuck & Co., 839 F.2d 302, 325 (7th Cir. 1988) (EEOC used logit analysis to measure the impact of variables, such as age, education, job-type experience, and product-line experience, on the female percentage of commission hires).

27. In job systems in which annual salaries are tied to grade or step levels, the annual salary corresponding to the job position could be more appropriate.

28. Explanatory variables may vary by type, which will affect the interpretation of the regression results. Thus, some variables may be continuous and others may be categorical.

29. In *James v. Stockham Valves*, 559 F. 2d 310 (5th Cir. 1977), the Court of Appeals rejected the employer's claim that skill level rather than race determined assignment and wage levels, noting the circularity of defendant's argument. In *Ottaviani v. State University of New York*, 679 F. Supp. 288, 306–08 (S.D.N.Y. 1988), *aff'd*, 875 F.2d 365 (2d Cir. 1989), *cert. denied*, 493 U.S. 1021 (1990), the court ruled (in the liability phase of the trial) that the university showed that there was no discrimination in either placement into initial rank or promotions between ranks, and so rank was a proper variable in multiple regression analysis to determine whether women faculty members were treated differently than men.

However, in *Trout v. Garrett*, 780 F. Supp. 1396, 1414 (D.D.C. 1991), the court ruled (in the damage phase of the trial) that the extent of civilian employees' prehire work experience was not an appropriate variable in a regression analysis to compute back pay in employment discrimination. According to the court, including the prehire level would have resulted in a finding of no sex discrimination, despite a contrary conclusion in the liability phase of the action. *Id. See also* Stuart v. Roache, 951 F.2d 446 (1st Cir. 1991) (allowing only 3 years of seniority to be considered as the result of prior

Not all possible variables that might influence the dependent variable can be included if the analysis is to be successful; some cannot be measured, and others may make little difference.[30] If a preliminary analysis shows the unexplained portion of the multiple regression to be unacceptably high, the expert may seek to discover whether some previously undetected variable is missing from the analysis.[31]

Failure to include a major explanatory variable that is correlated with the variable of interest in a regression model may cause an included variable to be credited with an effect that actually is caused by the excluded variable.[32] In general, omitted variables that are correlated with the dependent variable reduce the probative value of the regression analysis. The importance of omitting a relevant variable depends on the strength of the relationship between the omitted variable and the dependent variable and the strength of the correlation between the omitted variable and the explanatory variables of interest. Other things being equal, the greater the correlation between the omitted variable and the variable of interest, the greater the bias caused by the omission. As a result, the omission of an important variable may lead to inferences made from regression analyses that do not assist the trier of fact.[33]

discrimination), *cert. denied*, 504 U.S. 913 (1992). Whether a particular variable reflects "legitimate" considerations or itself reflects or incorporates illegitimate biases is a recurring theme in discrimination cases. *See, e.g.*, Smith v. Virginia Commonwealth Univ., 84 F.3d 672, 677 (4th Cir. 1996) (en banc) (suggesting that whether "performance factors" should have been included in a regression analysis was a question of material fact); *id.* at 681–82 (Luttig, J., concurring in part) (suggesting that the failure of the regression analysis to include "performance factors" rendered it so incomplete as to be inadmissible); *id.* at 690–91 (Michael, J., dissenting) (suggesting that the regression analysis properly excluded "performance factors"); *see also* Diehl v. Xerox Corp., 933 F. Supp. 1157, 1168 (W.D.N.Y. 1996).

30. The summary effect of the excluded variables shows up as a random error term in the regression model, as does any modeling error. *See* Appendix, *infra,* for details. *But see* David W. Peterson, *Reference Guide on Multiple Regression*, 36 Jurimetrics J. 213, 214 n.2 (1996) (review essay) (asserting that "the presumption that the combined effect of the explanatory variables omitted from the model are uncorrelated with the included explanatory variables" is "a knife-edge condition . . . not likely to occur").

31. A very low *R*-squared (R^2) is one indication of an unexplained portion of the multiple regression model that is unacceptably high. However, the inference that one makes from a particular value of R^2 will depend, of necessity, on the context of the particular issues under study and the particular dataset that is being analyzed. For reasons discussed in the Appendix, a low R^2 does not necessarily imply a poor model (and vice versa).

32. Technically, the omission of explanatory variables that are correlated with the variable of interest can cause biased estimates of regression parameters.

33. *See* Bazemore v. Friday, 751 F.2d 662, 671–72 (4th Cir. 1984) (upholding the district court's refusal to accept a multiple regression analysis as proof of discrimination by a preponderance of the evidence, the court of appeals stated that, although the regression used four variable factors (race, education, tenure, and job title), the failure to use other factors, including pay increases that varied by county, precluded their introduction into evidence), *aff'd in part, vacated in part*, 478 U.S. 385 (1986).

Note, however, that in *Sobel v. Yeshiva University*, 839 F.2d 18, 33, 34 (2d Cir. 1988), *cert. denied*, 490 U.S. 1105 (1989), the court made clear that "a [Title VII] defendant challenging the validity of

Omitting variables that are not correlated with the variable of interest is, in general, less of a concern, because the parameter that measures the effect of the variable of interest on the dependent variable is estimated without bias. Suppose, for example, that the effect of a policy introduced by the courts to encourage husbands to pay child support has been tested by randomly choosing some cases to be handled according to current court policies and other cases to be handled according to a new, more stringent policy. The effect of the new policy might be measured by a multiple regression using payment success as the dependent variable and a 0 or 1 explanatory variable (1 if the new program was applied; 0 if it was not). Failure to include an explanatory variable that reflected the age of the husbands involved in the program would not affect the court's evaluation of the new policy, because men of any given age are as likely to be affected by the old policy as they are the new policy. Randomly choosing the court's policy to be applied to each case has ensured that the omitted age variable is not correlated with the policy variable.

Bias caused by the omission of an important variable that is related to the included variables of interest can be a serious problem.[34] Nonetheless, it is possible for the expert to account for bias qualitatively if the expert has knowledge (even if not quantifiable) about the relationship between the omitted variable and the explanatory variable. Suppose, for example, that the plaintiff's expert in a sex discrimination pay case is unable to obtain quantifiable data that reflect the skills necessary for a job, and that, on average, women are more skillful than men. Suppose also that a regression analysis of the wage rate of employees (the dependent variable) on years of experience and a variable reflecting the sex of each employee (the explanatory variable) suggests that men are paid substantially more than women with the same experience. Because differences in skill levels have not been taken into account, the expert may conclude reasonably that the

a multiple regression analysis [has] to make a showing that the factors it contends ought to have been included would weaken the showing of salary disparity made by the analysis," by making a specific attack and "a showing of relevance for each particular variable it contends . . . ought to [be] includ[ed]" in the analysis, rather than by simply attacking the results of the plaintiffs' proof as inadequate for lack of a given variable. *See also* Smith v. Virginia Commonwealth Univ., 84 F.3d 672 (4th Cir. 1996) (en banc) (finding that whether certain variables should have been included in a regression analysis is a question of fact that precludes summary judgment); Freeland v. AT&T, 238 F.R.D. 130, 145 (S.D.N.Y. 2006) ("Ordinarily, the failure to include a variable in a regression analysis will affect the probative value of the analysis and not its admissibility").

Also, in *Bazemore v. Friday,* the Court, declaring that the Fourth Circuit's view of the evidentiary value of the regression analyses was plainly incorrect, stated that "[n]ormally, failure to include variables will affect the analysis' probativeness, not its admissibility. Importantly, it is clear that a regression analysis that includes less than 'all measurable variables' may serve to prove a plaintiff's case." 478 U.S. 385, 400 (1986) (footnote omitted).

34. *See also* David H. Kaye & David A. Freedman, Reference Guide on Statistics, Section V.B.3, in this manual.

wage difference measured by the regression is a conservative estimate of the true discriminatory wage difference.

The precision of the measure of the effect of a variable of interest on the dependent variable is also important.[35] In general, the more complete the explained relationship between the included explanatory variables and the dependent variable, the more precise the results. Note, however, that the inclusion of explanatory variables that are irrelevant (i.e., not correlated with the dependent variable) reduces the precision of the regression results. This can be a source of concern when the sample size is small, but it is not likely to be of great consequence when the sample size is large.

4. Choosing the functional form of the multiple regression model

Choosing the proper set of variables to be included in the multiple regression model does not complete the modeling exercise. The expert must also choose the proper form of the regression model. The most frequently selected form is the linear regression model (described in the Appendix). In this model, the magnitude of the change in the dependent variable associated with the change in any of the explanatory variables is the same no matter what the level of the explanatory variables. For example, one additional year of experience might add $5000 to salary, regardless of the previous experience of the employee.

In some instances, however, there may be reason to believe that changes in explanatory variables will have differential effects on the dependent variable as the values of the explanatory variables change. In these instances, the expert should consider the use of a nonlinear model. Failure to account for nonlinearities can lead to either overstatement or understatement of the effect of a change in the value of an explanatory variable on the dependent variable.

One particular type of nonlinearity involves the interaction among several variables. An interaction variable is the product of two other variables that are included in the multiple regression model. The interaction variable allows the expert to take into account the possibility that the effect of a change in one variable on the dependent variable may change as the level of another explanatory variable changes. For example, in a salary discrimination case, the inclusion of a term that interacts a variable measuring experience with a variable representing the sex of the employee (1 if a female employee; 0 if a male employee) allows the expert to test whether the sex differential varies with the level of experience. A significant negative estimate of the parameter associated with the sex variable suggests that inexperienced women are discriminated against, whereas a significant

35. A more precise estimate of a parameter is an estimate with a smaller standard error. *See* Appendix, *infra,* for details.

negative estimate of the interaction parameter suggests that the extent of discrimination increases with experience.[36]

Note that insignificant coefficients in a model with interactions may suggest a lack of discrimination, whereas a model without interactions may suggest the contrary. It is especially important to account for interaction terms that could affect the determination of discrimination; failure to do so may lead to false conclusions concerning discrimination.

5. Choosing multiple regression as a method of analysis

There are many multivariate statistical techniques other than multiple regression that are useful in legal proceedings. Some statistical methods are appropriate when nonlinearities are important;[37] others apply to models in which the dependent variable is discrete, rather than continuous.[38] Still others have been applied predominantly to respond to methodological concerns arising in the context of discrimination litigation.[39]

It is essential that a valid statistical method be applied to assist with the analysis in each legal proceeding. Therefore, the expert should be prepared to explain why any chosen method, including multiple regression, was more suitable than the alternatives.

36. For further details concerning interactions, see the Appendix, *infra*. Note that in *Ottaviani v. State University of New York*, 875 F.2d 365, 367 (2d Cir. 1989), *cert. denied*, 493 U.S. 1021 (1990), the defendant relied on a regression model in which a dummy variable reflecting gender appeared as an explanatory variable. The female plaintiff, however, used an alternative approach in which a regression model was developed for men only (the alleged protected group). The salaries of women predicted by this equation were then compared with the actual salaries; a positive difference would, according to the plaintiff, provide evidence of discrimination. For an evaluation of the methodological advantages and disadvantages of this approach, see Joseph L. Gastwirth, *A Clarification of Some Statistical Issues in* Watson v. Fort Worth Bank and Trust, 29 Jurimetrics J. 267 (1989).

37. These techniques include, but are not limited to, piecewise linear regression, polynomial regression, maximum likelihood estimation of models with nonlinear functional relationships, and autoregressive and moving-average time-series models. *See, e.g.*, Pindyck & Rubinfeld, *supra* note 23, at 117–21, 136–37, 273–84, 463–601.

38. For a discussion of probit analysis and logit analysis, techniques that are useful in the analysis of qualitative choice, see *id.* at 248–81.

39. The correct model for use in salary discrimination suits is a subject of debate among labor economists. As a result, some have begun to evaluate alternative approaches, including urn models (Bruce Levin & Herbert Robbins, *Urn Models for Regression Analysis, with Applications to Employment Discrimination Studies*, Law & Contemp. Probs., Autumn 1983, at 247) and, as a means of correcting for measurement errors, reverse regression (Delores A. Conway & Harry V. Roberts, *Reverse Regression, Fairness, and Employment Discrimination*, 1 J. Bus. & Econ. Stat. 75 (1983)). *But see* Arthur S. Goldberger, *Redirecting Reverse Regressions*, 2 J. Bus. & Econ. Stat. 114 (1984); Arlene S. Ash, *The Perverse Logic of Reverse Regression*, *in* Statistical Methods in Discrimination Litigation 85 (D.H. Kaye & Mikel Aickin eds., 1986).

III. Interpreting Multiple Regression Results

Multiple regression results can be interpreted in purely statistical terms, through the use of significance tests, or they can be interpreted in a more practical, nonstatistical manner. Although an evaluation of the practical significance of regression results is almost always relevant in the courtroom, tests of statistical significance are appropriate only in particular circumstances.

A. What Is the Practical, as Opposed to the Statistical, Significance of Regression Results?

Practical significance means that the magnitude of the effect being studied is not de minimis—it is sufficiently important substantively for the court to be concerned. For example, if the average wage rate is $10.00 per hour, a wage differential between men and women of $0.10 per hour is likely to be deemed practically insignificant because the differential represents only 1% ($0.10/$10.00) of the average wage rate.[40] That same difference could be statistically significant, however, if a sufficiently large sample of men and women was studied.[41] The reason is that statistical significance is determined, in part, by the number of observations in the dataset.

As a general rule, the statistical significance of the magnitude of a regression coefficient increases as the sample size increases. Thus, a $1.00 per hour wage differential between men and women that was determined to be insignificantly different from zero with a sample of 20 men and women could be highly significant if the sample size were increased to 200.

Often, results that are practically significant are also statistically significant.[42] However, it is possible with a large dataset to find statistically significant coeffi-

40. There is no specific percentage threshold above which a result is practically significant. Practical significance must be evaluated in the context of a particular legal issue. *See also* David H. Kaye & David A. Freedman, Reference Guide on Statistics, Section IV.B.2, in this manual.

41. Practical significance also can apply to the overall credibility of the regression results. Thus, in *McCleskey v. Kemp,* 481 U.S. 279 (1987), coefficients on race variables were statistically significant, but the Court declined to find them legally or constitutionally significant.

42. In *Melani v. Board of Higher Education,* 561 F. Supp. 769, 774 (S.D.N.Y. 1983), a Title VII suit was brought against the City University of New York (CUNY) for allegedly discriminating against female instructional staff in the payment of salaries. One approach of the plaintiff's expert was to use multiple regression analysis. The coefficient on the variable that reflected the sex of the employee was approximately $1800 when all years of data were included. Practically (in terms of average wages at the time) and statistically (in terms of a 5% significance test), this result was significant. Thus, the court stated that "[p]laintiffs have produced statistically *significant* evidence that women hired as CUNY instructional staff since 1972 received *substantially* lower salaries than similarly qualified men." *Id.* at 781 (emphasis added). For a related analysis involving multiple comparison, see Csicseri v. Bowsher,

cients that are practically insignificant. Similarly, it is also possible (especially when the sample size is small) to obtain results that are practically significant but fail to achieve statistical significance. Suppose, for example, that an expert undertakes a damages study in a patent infringement case and predicts "but-for sales"—what sales would have been had the infringement not occurred—using data that predate the period of alleged infringement. If data limitations are such that only 3 or 4 years of preinfringement sales are known, the difference between but-for sales and actual sales during the period of alleged infringement could be practically significant but statistically insignificant. Alternatively, with only 3 or 4 data points, the expert would be unable to detect an effect, even if one existed.

1. When should statistical tests be used?

A test of a specific contention, a hypothesis test, often assists the court in determining whether a violation of the law has occurred in areas in which direct evidence is inaccessible or inconclusive. For example, an expert might use hypothesis tests in race and sex discrimination cases to determine the presence of a discriminatory effect.

Statistical evidence alone never can prove with absolute certainty the worth of any substantive theory. However, by providing evidence contrary to the view that a particular form of discrimination has not occurred, for example, the multiple regression approach can aid the trier of fact in assessing the likelihood that discrimination has occurred.[43]

Tests of hypotheses are appropriate in a cross-sectional analysis, in which the data underlying the regression study have been chosen as a sample of a population at a particular point in time, and in a time-series analysis, in which the data being evaluated cover a number of time periods. In either analysis, the expert may want to evaluate a specific hypothesis, usually relating to a question of liability or to the determination of whether there is measurable impact of an alleged violation. Thus, in a sex discrimination case, an expert may want to evaluate a null hypothesis of no discrimination against the alternative hypothesis that discrimination takes a par-

862 F. Supp. 547, 572 (D.D.C. 1994) (noting that plaintiff's expert found "statistically significant instances of discrimination" in 2 of 37 statistical comparisons, but suggesting that "2 of 37 amounts to roughly 5% and is hardly indicative of a pattern of discrimination"), *aff'd*, 67 F.3d 972 (D.C. Cir. 1995).

43. *See* International Brotherhood. of Teamsters v. United States, 431 U.S. 324 (1977) (the Court inferred discrimination from overwhelming statistical evidence by a preponderance of the evidence); Ryther v. KARE 11, 108 F.3d 832, 844 (8th Cir. 1997) ("The plaintiff produced overwhelming evidence as to the elements of a prima facie case, and strong evidence of pretext, which, when considered with indications of age-based animus in [plaintiff's] work environment, clearly provide sufficient evidence as a matter of law to allow the trier of fact to find intentional discrimination."); Paige v. California, 291 F.3d 1141 (9th Cir. 2002) (allowing plaintiffs to rely on aggregated data to show employment discrimination).

ticular form.[44] Alternatively, in an antitrust damages proceeding, the expert may want to test a null hypothesis of no legal impact against the alternative hypothesis that there was an impact. In either type of case, it is important to realize that rejection of the null hypothesis does not in itself prove legal liability. It is possible to reject the null hypothesis and believe that an alternative explanation other than one involving legal liability accounts for the results.[45]

Often, the null hypothesis is stated in terms of a particular regression coefficient being equal to 0. For example, in a wage discrimination case, the null hypothesis would be that there is no wage difference between sexes. If a negative difference is observed (meaning that women are found to earn less than men, after the expert has controlled statistically for legitimate alternative explanations), the difference is evaluated as to its statistical significance using the *t*-test.[46] The *t*-test uses the *t*-statistic to evaluate the hypothesis that a model parameter takes on a particular value, usually 0.

2. *What is the appropriate level of statistical significance?*

In most scientific work, the level of statistical significance required to reject the null hypothesis (i.e., to obtain a statistically significant result) is set conventionally at 0.05, or 5%.[47] The significance level measures the probability that the null hypothesis will be rejected incorrectly. In general, the lower the percentage required for statistical significance, the more difficult it is to reject the null hypothesis; therefore, the lower the probability that one will err in doing so. Although the 5% criterion is typical, reporting of more stringent 1% significance tests or less stringent 10% tests can also provide useful information.

In doing a statistical test, it is useful to compute an observed significance level, or *p*-value. The *p*-value associated with the null hypothesis that a regression coefficient is 0 is the probability that a coefficient of this magnitude or larger could have occurred by chance if the null hypothesis were true. If the *p*-value were less than or equal to 5%, the expert would reject the null hypothesis in favor of the

44. Tests are also appropriate when comparing the outcomes of a set of employer decisions with those that would have been obtained had the employer chosen differently from among the available options.

45. *See* David H. Kaye & David A. Freedman, Reference Guide on Statistics, Section IV.C.5, in this manual.

46. The *t*-test is strictly valid only if a number of important assumptions hold. However, for many regression models, the test is approximately valid if the sample size is sufficiently large. See Appendix, *infra,* for a more complete discussion of the assumptions underlying multiple regression..

47. *See, e.g.*, Palmer v. Shultz, 815 F.2d 84, 92 (D.C. Cir. 1987) ("'the .05 level of significance . . . [is] certainly sufficient to support an inference of discrimination'" (quoting Segar v. Smith, 738 F.2d 1249, 1283 (D.C. Cir. 1984), *cert. denied,* 471 U.S. 1115 (1985))); United States v. Delaware, 2004 U.S. Dist. LEXIS 4560 (D. Del. Mar. 22, 2004) (stating that .05 is the normal standard chosen).

alternative hypothesis; if the *p*-value were greater than 5%, the expert would fail to reject the null hypothesis.[48]

3. Should statistical tests be one-tailed or two-tailed?

When the expert evaluates the null hypothesis that a variable of interest has no linear association with a dependent variable against the alternative hypothesis that there is an association, a two-tailed test, which allows for the effect to be either positive or negative, is usually appropriate. A one-tailed test would usually be applied when the expert believes, perhaps on the basis of other direct evidence presented at trial, that the alternative hypothesis is either positive or negative, but not both. For example, an expert might use a one-tailed test in a patent infringement case if he or she strongly believes that the effect of the alleged infringement on the price of the infringed product was either zero or negative. (The sales of the infringing product competed with the sales of the infringed product, thereby lowering the price.) By using a one-tailed test, the expert is in effect stating that prior to looking at the data it would be very surprising if the data pointed in the direct opposite to the one posited by the expert.

Because using a one-tailed test produces *p*-values that are one-half the size of *p*-values using a two-tailed test, the choice of a one-tailed test makes it easier for the expert to reject a null hypothesis. Correspondingly, the choice of a two-tailed test makes null hypothesis rejection less likely. Because there is some arbitrariness involved in the choice of an alternative hypothesis, courts should avoid relying solely on sharply defined statistical tests.[49] Reporting the *p*-value or a confidence interval should be encouraged because it conveys useful information to the court, whether or not a null hypothesis is rejected.

48. The use of 1%, 5%, and, sometimes, 10% levels for determining statistical significance remains a subject of debate. One might argue, for example, that when regression analysis is used in a price-fixing antitrust case to test a relatively specific alternative to the null hypothesis (e.g., price fixing), a somewhat lower level of confidence (a higher level of significance, such as 10%) might be appropriate. Otherwise, when the alternative to the null hypothesis is less specific, such as the rather vague alternative of "effect" (e.g., the price increase is caused by the increased cost of production, increased demand, a sharp increase in advertising, or price fixing), a high level of confidence (associated with a low significance level, such as 1%) may be appropriate. *See, e.g.*, Vuyanich v. Republic Nat'l Bank, 505 F. Supp. 224, 272 (N.D. Tex. 1980) (noting the "arbitrary nature of the adoption of the 5% level of [statistical] significance" to be required in a legal context); Cook v. Rockwell Int'l Corp., 2006 U.S. Dist. LEXIS 89121 (D. Colo. Dec. 7, 2006).

49. Courts have shown a preference for two-tailed tests. *See, e.g.*, Palmer v. Shultz, 815 F.2d 84, 95–96 (D.C. Cir. 1987) (rejecting the use of one-tailed tests, the court found that because some appellants were claiming overselection for certain jobs, a two-tailed test was more appropriate in Title VII cases); Moore v. Summers, 113 F. Supp. 2d 5, 20 (D.D.C. 2000) (reiterating the preference for a two-tailed test). *See also* David H. Kaye & David A. Freedman, Reference Guide on Statistics, Section IV.C.2, in this manual; Csicseri v. Bowsher, 862 F. Supp. 547, 565 (D.D.C. 1994) (finding that although a one-tailed test is "not without merit," a two-tailed test is preferable).

B. Are the Regression Results Robust?

The issue of robustness—whether regression results are sensitive to slight modifications in assumptions (e.g., that the data are measured accurately)—is of vital importance. If the assumptions of the regression model are valid, standard statistical tests can be applied. However, when the assumptions of the model are violated, standard tests can overstate or understate the significance of the results.

The violation of an assumption does not necessarily invalidate a regression analysis, however. In some instances in which the assumptions of multiple regression analysis fail, there are other statistical methods that are appropriate. Consequently, experts should be encouraged to provide additional information that relates to the issue of whether regression assumptions are valid, and if they are not valid, the extent to which the regression results are robust. The following questions highlight some of the more important assumptions of regression analysis.

1. What evidence exists that the explanatory variable causes changes in the dependent variable?

In the multiple regression framework, the expert often assumes that changes in explanatory variables affect the dependent variable, but changes in the dependent variable do not affect the explanatory variables—that is, there is no feedback.[50] In making this assumption, the expert draws the conclusion that a correlation between a covariate and the dependent outcome variable results from the effect of the former on the latter and not vice versa. Were it the case that the causality was reversed so that the outcome variable affected the covariate, and not vice versa, spurious correlation is likely to cause the expert and the trier of fact to reach the wrong conclusion. Finally, it is possible in some cases that both the outcome variable and the covariate each affect the other; if the expert does not take this more complex relationship into account, the regression coefficient on the variable of interest could be either too high or too low.[51]

Figure 1 illustrates this point. In Figure 1(a), the dependent variable, price, is explained through a multiple regression framework by three covariate explanatory variables—demand, cost, and advertising—with no feedback. Each of the three covariates is assumed to affect price causally, while price is assumed to have no effect on the three covariates. However, in Figure 1(b), there is feedback, because price affects demand, and demand, cost, and advertising affect price. Cost and advertising, however, are not affected by price. In this case both price and demand are jointly determined; each has a causal effect on the other.

50. The assumption of no feedback is especially important in litigation, because it is possible for the defendant (if responsible, for example, for price fixing or discrimination) to affect the values of the explanatory variables and thus to bias the usual statistical tests that are used in multiple regression.

51. When both effects occur at the same time, this is described as "simultaneity."

Figure 1. Feedback.

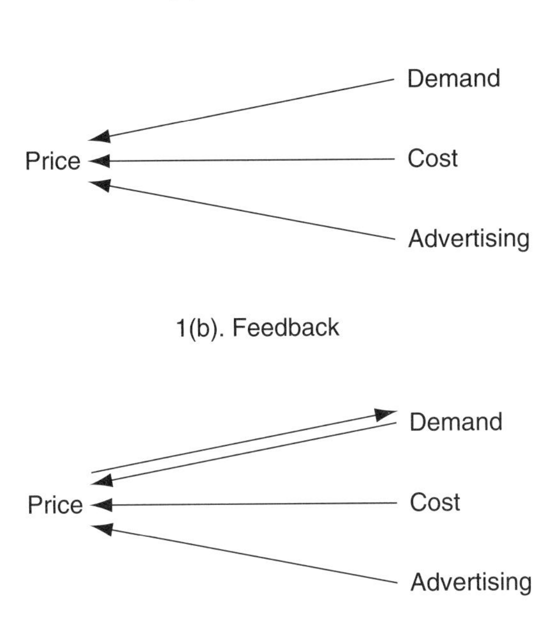

As a general rule, there are no basic direct statistical tests for determining the direction of causality; rather, the expert, when asked, should be prepared to defend his or her assumption based on an understanding of the underlying behavior evidence relating to the businesses or individuals involved.[52]

Although there is no single approach that is entirely suitable for estimating models when the dependent variable affects one or more explanatory variables, one possibility is for the expert to drop the questionable variable from the regression to determine whether the variable's exclusion makes a difference. If it does not, the issue becomes moot. Another approach is for the expert to expand the multiple regression model by adding one or more equations that explain the relationship between the explanatory variable in question and the dependent variable.

Suppose, for example, that in a salary-based sex discrimination suit the defendant's expert considers employer-evaluated test scores to be an appropriate explanatory variable for the dependent variable, salary. If the plaintiff were to provide information that the employer adjusted the test scores in a manner that penalized women, the assumption that salaries were determined by test scores and not that test scores were affected by salaries might be invalid. If it is clearly inappropriate,

52. There are statistical time-series tests for particular formulations of causality; see Pindyck & Rubinfeld, *supra* note 23, § 9.2.

the test-score variable should be removed from consideration. Alternatively, the information about the employer's use of the test scores could be translated into a second equation in which a new dependent variable—test score—is related to workers' salary and sex. A test of the hypothesis that salary and sex affect test scores would provide a suitable test of the absence of feedback.

2. To what extent are the explanatory variables correlated with each other?

It is essential in multiple regression analysis that the explanatory variable of interest not be correlated perfectly with one or more of the other explanatory variables. If there were perfect correlation between two variables, the expert could not separate out the effect of the variable of interest on the dependent variable from the effect of the other variable. In essence, there are two explanations for the same pattern in the data. Suppose, for example, that in a sex discrimination suit, a particular form of job experience is determined to be a valid source of high wages. If all men had the requisite job experience and all women did not, it would be impossible to tell whether wage differentials between men and women were the result of sex discrimination or differences in experience.

When two or more explanatory variables are correlated perfectly—that is, when there is *perfect collinearity*—one cannot estimate the regression parameters. The existing dataset does not allow one to distinguish between alternative competing explanations of the movement in the dependent variable. However, when two or more variables are highly, but not perfectly, correlated—that is, when there is *multicollinearity*—the regression can be estimated, but some concerns remain. The greater the multicollinearity between two variables, the less precise are the estimates of individual regression parameters, and an expert is less able to distinguish among competing explanations for the movement in the outcome variable (even though there is no problem in estimating the joint influence of the two variables and all other regression parameters).[53]

Fortunately, the reported regression statistics take into account any multicollinearity that might be present.[54] It is important to note as a corollary, however, that a failure to find a strong relationship between a variable of interest and

53. *See* Griggs v. Duke Power Co., 401 U.S. 424 (1971) (The court argued that an education requirement was one rationalization of the data, but racial discrimination was another. If you had put both race and education in the regression, it would have been asking too much of the data to tell which variable was doing the real work, because education and race were so highly correlated in the market at that time.).

54. *See* Denny v. Westfield State College, 669 F. Supp. 1146, 1149 (D. Mass. 1987) (The court accepted the testimony of one expert that "the presence of multicollinearity would merely tend to *overestimate* the amount of error associated with the estimate. . . . In other words, *p*-values will be artificially higher than they would be if there were no multicollinearity present.") (emphasis added); *In re* High Fructose Corn Syrup Antitrust Litig., 295 F.3d 651, 659 (7th Cir. Ill. 2002) (refusing to second-guess district court's admission of regression analyses that addressed multicollinearity in different ways).

a dependent variable need not imply that there is no relationship.[55] A relatively small sample, or even a large sample with substantial multicollinearity, may not provide sufficient information for the expert to determine whether there is a relationship.

3. To what extent are individual errors in the regression model independent?

If the expert calculated the parameters of a multiple regression model using as data the entire population, the estimates might still measure the model's population parameters with error. Errors can arise for a number of reasons, including (1) the failure of the model to include the appropriate explanatory variables, (2) the failure of the model to reflect any nonlinearities that might be present, and (3) the inclusion of inappropriate variables in the model. (Of course, further sources of error will arise if a sample, or subset, of the population is used to estimate the regression parameters.)

It is useful to view the cumulative effect of all of these sources of modeling error as being represented by an additional variable, the error term, in the multiple regression model. An important assumption in multiple regression analysis is that the error term and each of the explanatory variables are independent of each other. (If the error term and an explanatory variable are independent, they are not correlated with each other.) To the extent this is true, the expert can estimate the parameters of the model without bias; the magnitude of the error term will affect the precision with which a model parameter is estimated, but will not cause that estimate to be consistently too high or too low.

The assumption of independence may be inappropriate in a number of circumstances. In some instances, failure of the assumption makes multiple regression analysis an unsuitable statistical technique; in other instances, modifications or adjustments within the regression framework can be made to accommodate the failure.

The independence assumption may fail, for example, in a study of individual behavior over time, in which an unusually high error value in one time period is likely to lead to an unusually high value in the next time period. For example, if an economic forecaster underpredicted this year's Gross Domestic Product, he or she is likely to underpredict next year's as well; the factor that caused the prediction error (e.g., an incorrect assumption about Federal Reserve policy) is likely to be a source of error in the future.

55. If an explanatory variable of concern and another explanatory variable are highly correlated, dropping the second variable from the regression can be instructive. If the coefficient on the explanatory variable of concern becomes significant, a relationship between the dependent variable and the explanatory variable of concern is suggested.

Alternatively, the assumption of independence may fail in a study of a group of firms at a particular point in time, in which error terms for large firms are systematically higher than error terms for small firms. For example, an analysis of the profitability of firms may not accurately account for the importance of advertising as a source of increased sales and profits. To the extent that large firms advertise more than small firms, the regression errors would be large for the large firms and small for the small firms. A third possibility is that the dependent variable varies at the individual level, but the explanatory variable of interest varies only at the level of a group. For example, an expert might be viewing the price of a product in an antitrust case as a function of a variable or variables that measure the marketing channel through which the product is sold (e.g., wholesale or retail). In this case, errors within each of the marketing groups are likely not to be independent. Failure to account for this could cause the expert to overstate the statistical significance of the regression parameters.

In some instances, there are statistical tests that are appropriate for evaluating the independence assumption.[56] If the assumption has failed, the expert should ask first whether the source of the lack of independence is the omission of an important explanatory variable from the regression. If so, that variable should be included when possible, or the potential effect of its omission should be estimated when inclusion is not possible. If there is no important missing explanatory variable, the expert should apply one or more procedures that modify the standard multiple regression technique to allow for more accurate estimates of the regression parameters.[57]

4. *To what extent are the regression results sensitive to individual data points?*

Estimated regression coefficients can be highly sensitive to particular data points. Suppose, for example, that one data point deviates greatly from its expected value, as indicated by the regression equation, while the remaining data points show

56. In a time-series analysis, the correlation of error values over time, the "serial correlation," can be tested (in most instances) using a number of tests, including the Durbin-Watson test. The possibility that some error terms are consistently high in magnitude and others are systematically low, heteroscedasticity can also be tested in a number of ways. *See, e.g.*, Pindyck & Rubinfeld, *supra* note 23, at 146–59. When serial correlation and/or heteroscedasticity are present, the standard errors associated with the estimated coefficients must be modified. For a discussion of the use of such "robust" standard errors, see Jeffrey M. Wooldridge, Introductory Econometrics: A Modern Approach, ch. 8 (4th ed. 2009).

57. When serial correlation is present, a number of closely related statistical methods are appropriate, including generalized differencing (a type of generalized least squares) and maximum likelihood estimation. When heteroscedasticity is the problem, weighted least squares and maximum likelihood estimation are appropriate. *See, e.g.*, *id.* All these techniques are readily available in a number of statistical computer packages. They also allow one to perform the appropriate statistical tests of the significance of the regression coefficients.

little deviation. It would not be unusual in this situation for the coefficients in a multiple regression to change substantially if the data point in question were removed from the sample.

Evaluating the robustness of multiple regression results is a complex endeavor. Consequently, there is no agreed set of tests for robustness that analysts should apply. In general, it is important to explore the reasons for unusual data points. If the source is an error in recording data, the appropriate corrections can be made. If all the unusual data points have certain characteristics in common (e.g., they all are associated with a supervisor who consistently gives high ratings in an equal pay case), the regression model should be modified appropriately.

One generally useful diagnostic technique is to determine to what extent the estimated parameter changes as each data point in the regression analysis is dropped from the sample. An *influential* data point—a point that causes the estimated parameter to change substantially—should be studied further to determine whether mistakes were made in the use of the data or whether important explanatory variables were omitted.[58]

5. To what extent are the data subject to measurement error?

In multiple regression analysis it is assumed that variables are measured accurately.[59] If there are measurement errors in the dependent variable, estimates of regression parameters will be less accurate, although they will not necessarily be biased. However, if one or more independent variables are measured with error, the corresponding parameter estimates are likely to be biased, typically toward zero (and other coefficient estimates are likely to be biased as well).

To understand why, suppose that the dependent variable, salary, is measured without error, and the explanatory variable, experience, is subject to measurement error. (Seniority or years of experience should be accurate, but the type of experience is subject to error, because applicants may overstate previous job responsibilities.) As the measurement error increases, the estimated parameter associated with the experience variable will tend toward zero, that is, eventually, there will be no relationship between salary and experience.

It is important for any source of measurement error to be carefully evaluated. In some circumstances, little can be done to correct the measurement-error prob-

58. A more complete and formal treatment of the robustness issue appears in David A. Belsley et al., Regression Diagnostics: Identifying Influential Data and Sources of Collinearity 229–44 (1980). For a useful discussion of the detection of outliers and the evaluation of influential data points, see R.D. Cook & S. Weisberg, Residuals and Influence in Regression (Monographs on Statistics and Applied Probability No. 18, 1982). For a broad discussion of robust regression methods, see Peer J. Rouseeuw & Annick M. Leroy, Robust Regression and Outlier Detection (2004).

59. Inaccuracy can occur not only in the precision with which a particular variable is measured, but also in the precision with which the variable to be measured corresponds to the appropriate theoretical construct specified by the regression model.

lem; the regression results must be interpreted in that light. In other circumstances, however, the expert can correct measurement error by finding a new, more reliable data source. Finally, alternative estimation techniques (using related variables that are measured without error) can be applied to remedy the measurement-error problem in some situations.[60]

IV. The Expert

Multiple regression analysis is taught to students in extremely diverse fields, including statistics, economics, political science, sociology, psychology, anthropology, public health, and history. Nonetheless, the methodology is difficult to master, necessitating a combination of technical skills (the science) and experience (the art). This naturally raises two questions:

1. Who should be qualified as an expert?
2. When and how should the court appoint an expert to assist in the evaluation of statistical issues, including those relating to multiple regression?

A. Who Should Be Qualified as an Expert?

Any individual with substantial training in and experience with multiple regression and other statistical methods may be qualified as an expert.[61] A doctoral degree in a discipline that teaches theoretical or applied statistics, such as economics, history, and psychology, usually signifies to other scientists that the proposed expert meets this preliminary test of the qualification process.

The decision to qualify an expert in regression analysis rests with the court. Clearly, the proposed expert should be able to demonstrate an understanding of the discipline. Publications relating to regression analysis in peer-reviewed journals, active memberships in related professional organizations, courses taught on regression methods, and practical experience with regression analysis can indicate a professional's expertise. However, the expert's background and experience with the specific issues and tools that are applicable to a particular case should also be considered during the qualification process. Thus, if the regression methods are being utilized to evaluate damages in an antitrust case, the qualified expert should have sufficient qualifications in economic analysis as well as statistics. An individual whose expertise lies solely with statistics will be limited in his or her ability to evaluate the usefulness of alternative economic models. Similarly, if a case involves

60. *See, e.g.*, Pindyck & Rubinfeld, *supra* note 23, at 178–98 (discussion of instrumental variables estimation).

61. A proposed expert whose only statistical tool is regression analysis may not be able to judge when a statistical analysis should be based on an approach other than regression analysis.

eyewitness identification, a background in psychology as well as statistics may provide essential qualifying elements.

B. Should the Court Appoint a Neutral Expert?

There are conflicting views on the issue of whether court-appointed experts should be used. In complex cases in which two experts are presenting conflicting statistical evidence, the use of a "neutral" court-appointed expert can be advantageous. There are those who believe, however, that there is no such thing as a truly "neutral" expert. In any event, if an expert is chosen, that individual should have substantial expertise and experience—ideally, someone who is respected by both plaintiffs and defendants.[62]

The appointment of such an expert is likely to influence the presentation of the statistical evidence by the experts for the parties in the litigation. The neutral expert will have an incentive to present a balanced position that relies on broad principles for which there is consensus in the community of experts. As a result, the parties' experts can be expected to present testimony that confronts core issues that are likely to be of concern to the court and that is sufficiently balanced to be persuasive to the court-appointed expert.[63]

Rule 706 of the Federal Rules of Evidence governs the selection and instruction of court-appointed experts. In particular:

1. The expert should be notified of his or her duties through a written court order or at a conference with the parties.
2. The expert should inform the parties of his or her findings orally or in writing.
3. If deemed appropriate by the court, the expert should be available to testify and may be deposed or cross-examined by any party.
4. The court must determine the expert's compensation.[64]
5. The parties should be free to utilize their own experts.

Although not required by Rule 706, it will usually be advantageous for the court to opt for the appointment of a neutral expert as early in the litigation process as possible. It will also be advantageous to minimize any ex parte contact with

62. Judge Posner notes in *In re High Fructose Corn Syrup Antitrust Litig.*, 295 F.2d 651, 665 (7th Cir., 2002), "the judge and jury can repose a degree of confidence in his testimony that it could not repose in that of a party's witness. The judge and the jury may not understand the neutral expert perfectly but at least they will know that he has no axe to grind, and so, to a degree anyway, they will be able to take his testimony on faith."

63. For a discussion of the presentation of expert evidence generally, including the use of court-appointed experts, see Samuel R. Gross, *Expert Evidence*, 1991 Wis. L. Rev. 1113 (1991).

64. Although Rule 706 states that the compensation must come from public funds, complex litigation may be sufficiently costly as to require that the parties share the costs of the neutral expert.

the neutral expert; this will diminish the possibility that one or both parties will come to the view that the court's ultimate opinion was unreasonably influenced by the neutral expert.

Rule 706 does not offer specifics as to the process of appointment of a court-appointed expert. One possibility is to have the parties offer a short list of possible appointees. If there was no common choice, the court could select from the combined list, perhaps after allowing each party to exercise one or more peremptory challenges. Another possibility is to obtain a list of recommended experts from a selection of individuals known to be experts in the field.

V. Presentation of Statistical Evidence

The costs of evaluating statistical evidence can be reduced and the precision of that evidence increased if the discovery process is used effectively. In evaluating the admissibility of statistical evidence, courts should consider the following issues:

1. Has the expert provided sufficient information to replicate the multiple regression analysis?
2. Are the expert's methodological choices reasonable, or are they arbitrary and unjustified?

A. What Disagreements Exist Regarding Data on Which the Analysis Is Based?

In general, a clear and comprehensive statement of the underlying research methodology is a requisite part of the discovery process. The expert should be encouraged to reveal both the nature of the experimentation carried out and the sensitivity of the results to the data and to the methodology.

The following suggestions are useful requirements that can substantially improve the discovery process:

1. To the extent possible, the parties should be encouraged to agree to use a common database. Even if disagreement about the significance of the data remains, early agreement on a common database can help focus the discovery process on the important issues in the case.
2. A party that offers data to be used in statistical work, including multiple regression analysis, should be encouraged to provide the following to the other parties: (a) a hard copy of the data when available and manageable in size, along with the underlying sources; (b) computer disks or tapes on which the data are recorded; (c) complete documentation of the disks or tapes; (d) computer programs that were used to generate the data (in hard

copy if necessary, but preferably on a computer disk or tape, or both); and (e) documentation of such computer programs. The documentation should be sufficiently complete and clear so that the opposing expert can reproduce all of the statistical work.

3. A party offering data should make available the personnel involved in the compilation of such data to answer the other parties' technical questions concerning the data and the methods of collection or compilation.
4. A party proposing to offer an expert's regression analysis at trial should ask the expert to fully disclose (a) the database and its sources,[65] (b) the method of collecting the data, and (c) the methods of analysis. When possible, this disclosure should be made sufficiently in advance of trial so that the opposing party can consult its experts and prepare cross-examination. The court must decide on a case-by-case basis where to draw the disclosure line.
5. An opposing party should be given the opportunity to object to a database or to a proposed method of analysis of the database to be offered at trial. Objections may be to simple clerical errors or to more complex issues relating to the selection of data, the construction of variables, and, on occasion, the particular form of statistical analysis to be used. Whenever possible, these objections should be resolved before trial.
6. The parties should be encouraged to resolve differences as to the appropriateness and precision of the data to the extent possible by informal conference. The court should make an effort to resolve differences before trial.

These suggestions are motivated by the objective of improving the discovery process to make it more informative. The fact that these questions may raise some doubts or concerns about a particular regression model should not be taken to mean that the model does not provide useful information. It does, however, take considerable skill for an expert to determine the extent to which information is useful when the model being utilized has some shortcomings.

B. Which Database Information and Analytical Procedures Will Aid in Resolving Disputes over Statistical Studies?[66]

To help resolve disputes over statistical studies, experts should follow the guidelines below when presenting database information and analytical procedures:

65. These sources would include all variables used in the statistical analyses conducted by the expert, not simply those variables used in a final analysis on which the expert expects to rely.

66. For a more complete discussion of these requirements, see *The Evolving Role of Statistical Assessments as Evidence in the Courts,* app. F at 256 (Stephen E. Fienberg ed., 1989) (Recommended

1. The expert should state clearly the objectives of the study, as well as the time frame to which it applies and the statistical population to which the results are being projected.
2. The expert should report the units of observation (e.g., consumers, businesses, or employees).
3. The expert should clearly define each variable.
4. The expert should clearly identify the sample for which data are being studied,[67] as well as the method by which the sample was obtained.
5. The expert should reveal if there are missing data, whether caused by a lack of availability (e.g., in business data) or nonresponse (e.g., in survey data), and the method used to handle the missing data (e.g., deletion of observations).
6. The expert should report investigations into errors associated with the choice of variables and assumptions underlying the regression model.
7. If samples were chosen randomly from a population (i.e., probability sampling procedures were used),[68] the expert should make a good-faith effort to provide an estimate of a sampling error, the measure of the difference between the sample estimate of a parameter (such as the mean of a dependent variable under study), and the (unknown) population parameter (the population mean of the variable).[69]
8. If probability sampling procedures were not used, the expert should report the set of procedures that was used to minimize sampling errors.

Standards on Disclosure of Procedures Used for Statistical Studies to Collect Data Submitted in Evidence in Legal Cases).

67. The sample information is important because it allows the expert to make inferences about the underlying population.

68. In probability sampling, each representative of the population has a known probability of being in the sample. Probability sampling is ideal because it is highly structured, and in principle, it can be replicated by others. Nonprobability sampling is less desirable because it is often subjective, relying to a large extent on the judgment of the expert.

69. Sampling error is often reported in terms of standard errors or confidence intervals. *See* Appendix, *infra,* for details.

Appendix: The Basics of Multiple Regression

A. Introduction

This appendix illustrates, through examples, the basics of multiple regression analysis in legal proceedings. Often, visual displays are used to describe the relationship between variables that are used in multiple regression analysis. Figure 2 is a scatterplot that relates scores on a job aptitude test (shown on the *x*-axis) and job performance ratings (shown on the *y*-axis). Each point on the scatterplot shows where a particular individual scored on the job aptitude test and how his or her job performance was rated. For example, the individual represented by Point A in Figure 2 scored 49 on the job aptitude test and had a job performance rating of 62.

Figure 2. Scatterplot of scores on a job aptitude test relative to job performance rating.

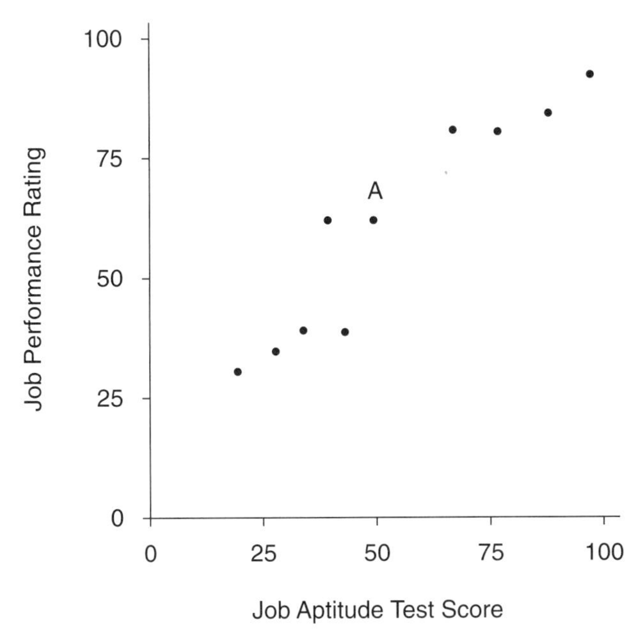

The relationship between two variables can be summarized by a correlation coefficient, which ranges in value from –1 (a perfect negative relationship) to +1 (a perfect positive relationship). Figure 3 depicts three possible relationships between the job aptitude variable and the job performance variable. In Figure 3(a), there is a positive correlation: In general, higher job performance ratings are associated with higher aptitude test scores, and lower job performance ratings are associated with lower aptitude test scores. In Figure 3(b), the correlation is negative: Higher job performance ratings are associated with lower aptitude test scores, and lower job performance ratings are associated with higher aptitude test scores. Positive and negative correlations can be relatively strong or relatively weak. If the relationship is sufficiently weak, there is effectively no correlation, as is illustrated in Figure 3(c).

Figure 3. Correlation between the job aptitude variable and the job performance variable: (a) positive correlation, (b) negative correlation, (c) weak relationship with no correlation.

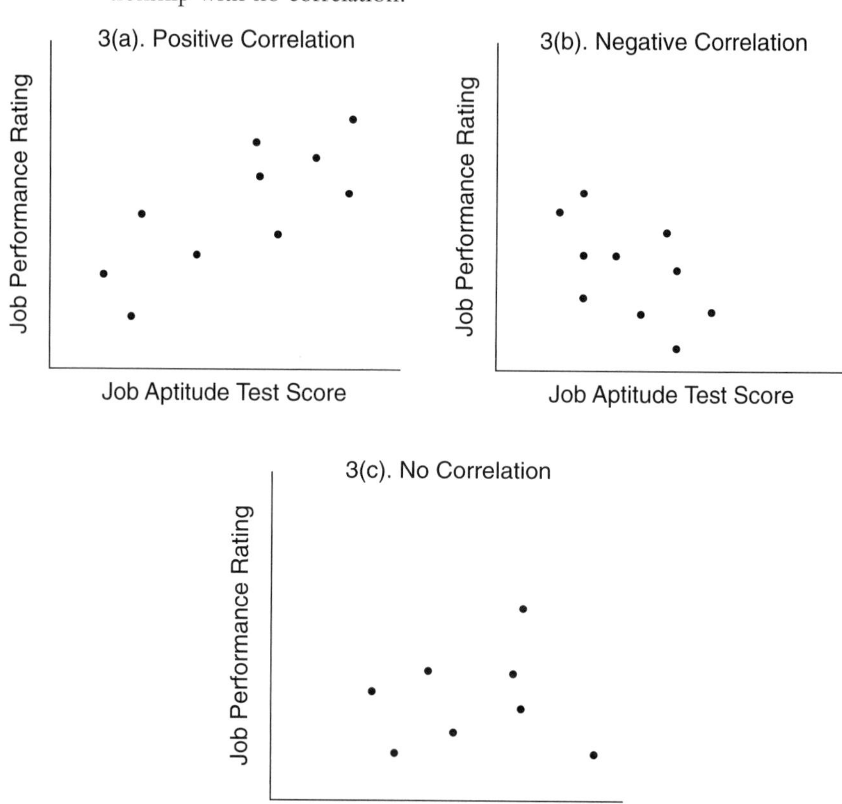

Multiple regression analysis goes beyond the calculation of correlations; it is a method in which a regression line is used to relate the average of one variable—the dependent variable—to the values of other explanatory variables. As a result, regression analysis can be used to predict the values of one variable using the values of others. For example, if average job performance ratings depend on aptitude test scores, regression analysis can use information about test scores to predict job performance.

A regression line is the best-fitting straight line through a set of points in a scatterplot. If there is only one explanatory variable, the straight line is defined by the equation

$$Y = a + bX. \tag{1}$$

In equation (1), a is the intercept of the line with the y-axis when X equals 0, and b is the slope—the change in the dependent variable associated with a 1-unit change in the explanatory variable. In Figure 4, for example, when the aptitude test score is 0, the predicted (average) value of the job performance rating is the intercept, 18.4. Also, for each additional point on the test score, the job performance rating increases .73 units, which is given by the slope .73. Thus, the estimated regression line is

$$Y = 18.4 + .73X. \tag{2}$$

The regression line typically is estimated using the standard method of least squares, where the values of a and b are calculated so that the sum of the squared deviations of the points from the line are minimized. In this way, positive deviations and negative deviations of equal size are counted equally, and large deviations are counted more than small deviations. In Figure 4 the deviation lines are verti-

Figure 4. Regression line.

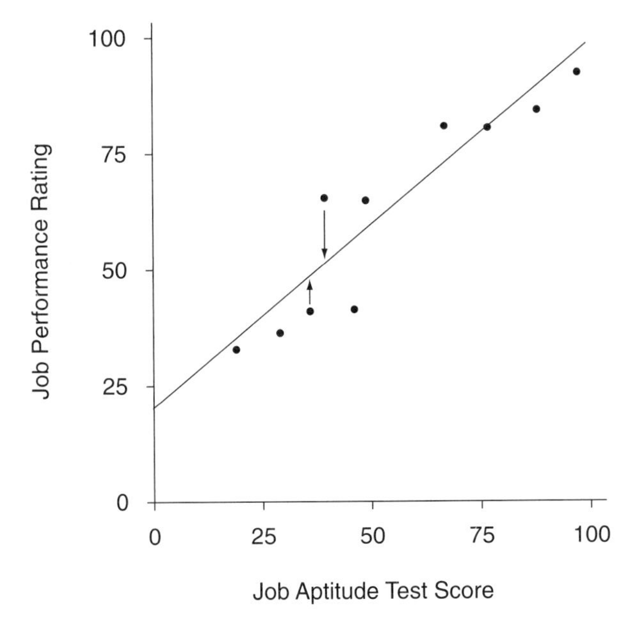

cal because the equation is predicting job performance ratings from aptitude test scores, not aptitude test scores from job performance ratings.

The important variables that systematically might influence the dependent variable, and for which data can be obtained, typically should be included explicitly in a statistical model. All remaining influences, which should be small individually, but can be substantial in the aggregate, are included in an additional random error term.[70] Multiple regression is a procedure that separates the systematic effects (associated with the explanatory variables) from the random effects (associated with the error term) and also offers a method of assessing the success of the process.

B. Linear Regression Model

When there are an arbitrary number of explanatory variables, the linear regression model takes the following form:

$$Y = \beta_0 + \beta_1 X_1 + \beta_2 X_2 + \ldots + \beta_k X_k + \varepsilon \qquad (3)$$

where Y represents the dependent variable, such as the salary of an employee, and $X_1 \ldots X_k$ represent the explanatory variables (e.g., the experience of each employee and his or her sex, coded as a 1 or 0, respectively). The error term, ε, represents the collective unobservable influence of any omitted variables. In a linear regression, each of the terms being added involves unknown parameters, $\beta_0, \beta_1, \ldots \beta_k$,[71] which are estimated by "fitting" the equation to the data using least squares.

Each estimated coefficient β_k measures how the dependent variable Y responds, on average, to a change in the corresponding covariate X_k, after "controlling for" all the other covariates. The informal phrase "controlling for" has a specific statistical meaning. Consider the following three-step procedure. First, we calculate the residuals from a regression of Y on all covariates other than X_k. Second, we calculate the residuals of a regression of X_k on all the other covariates. Third, and finally, we regress the first residual variable on the second residual variable. The resulting coefficient will be identically equal to β_k. Thus, the coeffi-

70. It is clearly advantageous for the random component of the regression relationship to be small relative to the variation in the dependent variable.

71. The variables themselves can appear in many different forms. For example, Y might represent the logarithm of an employee's salary, and X_1 might represent the logarithm of the employee's years of experience. The logarithmic representation is appropriate when Y increases exponentially as X increases—for each unit increase in X, the corresponding increase in Y becomes larger and larger. For example, if an expert were to graph the growth of the U.S. population (Y) over time (t), the following equation might be appropriate:

$$\log(Y) = \beta_0 + \beta_1 \log(t).$$

cient in a multiple regression represents the slope of the line "Y, adjusted for all covariates other than X_k versus X_k adjusted for all the other covariates."[72]

Most statisticians use the least squares regression technique because of its simplicity and its desirable statistical properties. As a result, it also is used frequently in legal proceedings.

1. *Specifying the regression model*

Suppose an expert wants to analyze the salaries of women and men at a large publishing house to discover whether a difference in salaries between employees with similar years of work experience provides evidence of discrimination.[73] To begin with the simplest case, Y, the salary in dollars per year, represents the dependent variable to be explained, and X_1 represents the explanatory variable—the number of years of experience of the employee. The regression model would be written

$$Y = \beta_0 + \beta_1 X_1 + \varepsilon. \tag{4}$$

In equation (4), β_0 and β_1 are the parameters to be estimated from the data, and ε is the random error term. The parameter β_0 is the average salary of all employees with no experience. The parameter β_1 measures the average effect of an additional year of experience on the average salary of employees.

2. *Regression line*

Once the parameters in a regression equation, such as equation (3), have been estimated, the fitted values for the dependent variable can be calculated. If we denote the estimated regression parameters, or regression coefficients, for the model in equation (3) by $\beta_{0,}\ \beta_1, \ldots \beta_k$, the fitted values for Y, denoted $\hat{Y}$, are given by

$$\hat{Y} = \beta_0 + \beta_1 X_1 + \beta_2 X_2 + \ldots \beta_k X_k. \tag{5}$$

Figure 5 illustrates this for the example involving a single explanatory variable. The data are shown as a scatter of points; salary is on the vertical axis, and years of experience is on the horizontal axis. The estimated regression line is drawn through the data points. It is given by

$$\hat{Y} = \$15{,}000 + \$2000 X_1. \tag{6}$$

72. In econometrics, this is known as the Frisch–Waugh–Lovell theorem.

73. The regression results used in this example are based on data for 1715 men and women, which were used by the defense in a sex discrimination case against the *New York Times* that was settled in 1978. Professor Orley Ashenfelter, Department of Economics, Princeton University, provided the data.

Figure 5. Goodness of fit.

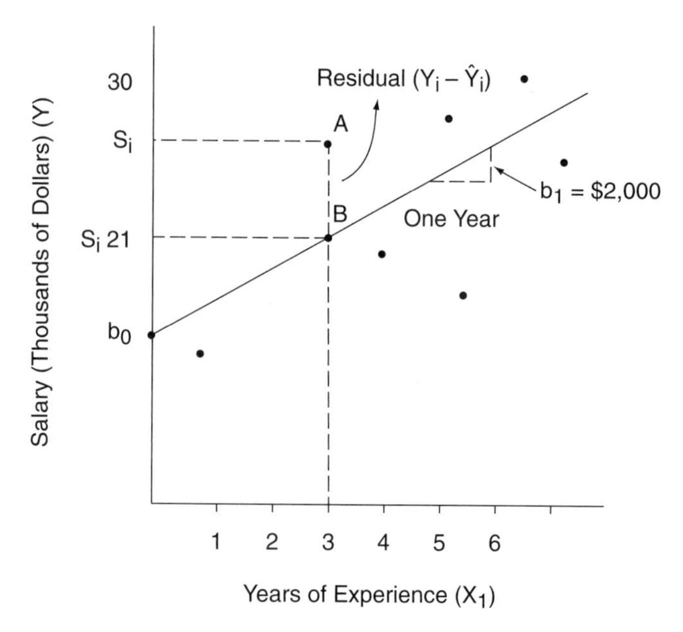

Thus, the fitted value for the salary associated with an individual's years of experience X_{1i} is given by

$$\hat{Y}_i = \beta_0 + \beta_1 X_{1i} \text{ (at Point B).} \qquad (7)$$

The intercept of the straight line is the average value of the dependent variable when the explanatory variable or variables are equal to 0; the intercept β_0 is shown on the vertical axis in Figure 5. Similarly, the slope of the line measures the (average) change in the dependent variable associated with a unit increase in an explanatory variable; the slope β_1 also is shown. In equation (6), the intercept $15,000 indicates that employees with no experience earn $15,000 per year. The slope parameter implies that each year of experience adds $2000 to an "average" employee's salary.

Now, suppose that the salary variable is related simply to the sex of the employee. The relevant indicator variable, often called a dummy variable, is X_2, which is equal to 1 if the employee is male, and 0 if the employee is female. Suppose the regression of salary Y on X_2 yields the following result: $Y = \$30,449 + \$10,979X_2$. The coefficient $10,979 measures the difference between the average salary of men and the average salary of women.[74]

74. To understand why, note that when X_2 equals 0, the average salary for women is $30,449 + $10,979*0 = $30,449. Correspondingly, when X_2 = 1, the average salary for men is $30,449 + $10,979*1 = $41,428. The difference, $41,428 – $30,449, is $10,979.

a. Regression residuals

For each data point, the regression residual is the difference between the actual values and fitted values of the dependent variable. Suppose, for example, that we are studying an individual with 3 years of experience and a salary of $27,000. According to the regression line in Figure 5, the average salary of an individual with 3 years of experience is $21,000. Because the individual's salary is $6000 higher than the average salary, the residual (the individual's salary minus the average salary) is $6000. In general, the residual *e* associated with a data point, such as Point A in Figure 5, is given by $e_i = Y_i - \hat{Y}_i$. Each data point in the figure has a residual, which is the error made by the least squares regression method for that individual.

b. Nonlinearities

Nonlinear models account for the possibility that the effect of an explanatory variable on the dependent variable may vary in magnitude as the level of the explanatory variable changes. One useful nonlinear model uses interactions among variables to produce this effect. For example, suppose that

$$S = \beta_1 + \beta_2 \text{SEX} + \beta_3 \text{EXP} + \beta_4 (\text{EXP})(\text{SEX}) + \varepsilon \qquad (8)$$

where S is annual salary, SEX is equal to 1 for women and 0 for men, EXP represents years of job experience, and ε is a random error term. The coefficient β_2 measures the difference in average salary (across all experience levels) between men and women for employees with no experience. The coefficient β_3 measures the effect of experience on salary for men (when SEX = 0), and the coefficient β_4 measures the difference in the effect of experience on salary between men and women. It follows, for example, that the effect of 1 year of experience on salary for men is β_3, whereas the comparable effect for women is $\beta_3 + \beta_4$.[75]

C. Interpreting Regression Results

To explain how regression results are interpreted, we can expand the earlier example associated with Figure 5 to consider the possibility of an additional explanatory variable—the square of the number of years of experience, X_3. The X_3 variable is designed to capture the fact that for most individuals, salaries increase with experience, but eventually salaries tend to level off. The estimated regression line using the third additional explanatory variable, as well as the first explanatory variable for years of experience (X_1) and the dummy variable for sex (X_2), is

75. Estimating a regression in which there are interaction terms for all explanatory variables, as in equation (8), is essentially the same as estimating two separate regressions, one for men and one for women.

$$\hat{Y} = \$14{,}085 + \$2323X_1 + \$1675X_2 - \$36X_3. \quad (9)$$

The importance of including relevant explanatory variables in a regression model is illustrated by the change in the regression results after the X_3 and X_1 variables are added. The coefficient on the variable X_2 measures the difference in the salaries of men and women while controlling for the effect of experience. The differential of $1675 is substantially lower than the previously measured differential of $10,979. Clearly, failure to control for job experience in this example leads to an overstatement of the difference in salaries between men and women.

Now consider the interpretation of the explanatory variables for experience, X_1 and X_3. The positive sign on the X_1 coefficient shows that salary increases with experience. The negative sign on the X_3 coefficient indicates that the rate of salary increase decreases with experience. To determine the combined effect of the variables X_1 and X_3, some simple calculations can be made. For example, consider how the average salary of women ($X_2 = 0$) changes with the level of experience. As experience increases from 0 to 1 year, the average salary increases by $2251, from $14,085 to $16,336. However, women with 2 years of experience earn only $2179 more than women with 1 year of experience, and women with 1 year of experience earn only $2127 more than women with 2 years. Furthermore, women with 7 years of experience earn $28,582 per year, which is only $1855 more than the $26,727 earned by women with 6 years of experience.[76] Figure 6 illustrates the results: The regression line shown is for women's salaries; the corresponding line for men's salaries would be parallel and $1675 higher.

D. Determining the Precision of the Regression Results

Least squares regression provides not only parameter estimates that indicate the direction and magnitude of the effect of a change in the explanatory variable on the dependent variable, but also an estimate of the reliability of the parameter estimates and a measure of the overall goodness of fit of the regression model. Each of these factors is considered in turn.

1. Standard errors of the coefficients and t-statistics

Estimates of the true but unknown parameters of a regression model are numbers that depend on the particular sample of observations under study. If a different sample were used, a different estimate would be calculated.[77] If the expert continued to collect more and more samples and generated additional estimates, as might happen when new data became available over time, the estimates of each

76. These numbers can be calculated by substituting different values of X_1 and X_3 in equation (9).

77. The least squares formula that generates the estimates is called the least squares estimator, and its values vary from sample to sample.

Figure 6. Regression slope for women's salaries and men's salaries.

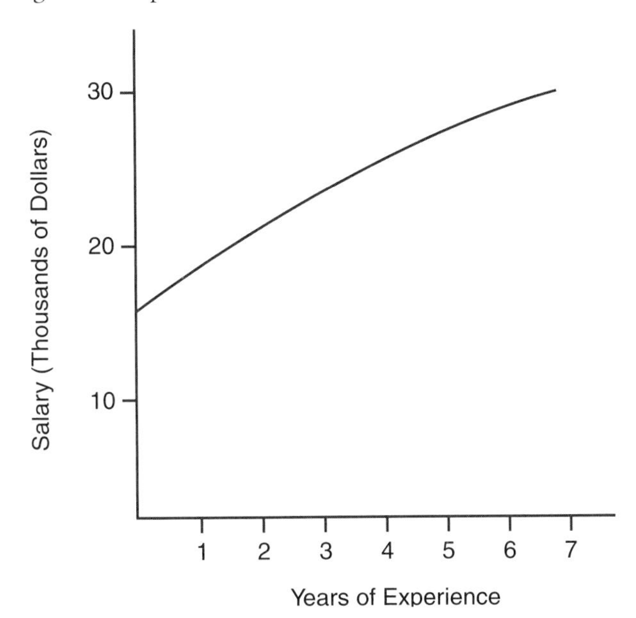

parameter would follow a probability distribution (i.e., the expert could determine the percentage or frequency of the time that each estimate occurs). This probability distribution can be summarized by a mean and a measure of dispersion around the mean, a standard deviation, which usually is referred to as the standard error of the coefficient, or the standard error (SE).[78]

Suppose, for example, that an expert is interested in estimating the average price paid for a gallon of unleaded gasoline by consumers in a particular geographic area of the United States at a particular point in time. The mean price for a sample of 10 gas stations might be $1.25, while the mean for another sample might be $1.29, and the mean for a third, $1.21. On this basis, the expert also could calculate the overall mean price of gasoline to be $1.25 and the standard deviation to be $0.04.

Least squares regression generalizes this result, by calculating means whose values depend on one or more explanatory variables. The standard error of a regression coefficient tells the expert how much parameter estimates are likely to vary from sample to sample. The greater the variation in parameter estimates from sample to sample, the larger the standard error and consequently the less reliable the regression results. Small standard errors imply results that are likely to

78. *See* David H. Kaye & David A. Freedman, Reference Guide on Statistics, Section IV.A, in this manual.

be similar from sample to sample, whereas results with large standard errors show more variability.

Under appropriate assumptions, the least squares estimators provide "best" determinations of the true underlying parameters.[79] In fact, least squares has several desirable properties. First, least squares estimators are unbiased. Intuitively, this means that if the regression were calculated repeatedly with different samples, the average of the many estimates obtained for each coefficient would be the true parameter. Second, least squares estimators are consistent; if the sample were very large, the estimates obtained would come close to the true parameters. Third, least squares is efficient, in that its estimators have the smallest variance among all (linear) unbiased estimators.

If the further assumption is made that the probability distribution of each of the error terms is known, statistical statements can be made about the precision of the coefficient estimates. For relatively large samples (often, thirty or more data points will be sufficient for regressions with a small number of explanatory variables), the probability that the estimate of a parameter lies within an interval of 2 standard errors around the true parameter is approximately .95, or 95%. A frequent, although not always appropriate, assumption in statistical work is that the error term follows a normal distribution, from which it follows that the estimated parameters are normally distributed. The normal distribution has the property that the area within 1.96 standard errors of the mean is equal to 95% of the total area. Note that the normality assumption is not necessary for least squares to be used, because most of the properties of least squares apply regardless of normality.

In general, for any parameter estimate b, the expert can construct an interval around b such that there is a 95% probability that the interval covers the true parameter. This 95% confidence interval[80] is given by[81]

$$b \pm 1.96 \text{ (SE of } b). \tag{10}$$

The expert can test the hypothesis that a parameter is actually equal to 0 (often stated as testing the null hypothesis) by looking at its t-statistic, which is defined as

$$t = \frac{b}{\text{SE}(b)}. \tag{11}$$

79. The necessary assumptions of the regression model include (a) the model is specified correctly, (b) errors associated with each observation are drawn randomly from the same probability distribution and are independent of each other, (c) errors associated with each observation are independent of the corresponding observations for each of the explanatory variables in the model, and (d) no explanatory variable is correlated perfectly with a combination of other variables.

80. Confidence intervals are used commonly in statistical analyses because the expert can never be certain that a parameter estimate is equal to the true population parameter.

81. If the number of data points in the sample is small, the standard error must be multiplied by a number larger than 1.96.

If the *t*-statistic is less than 1.96 in magnitude, the 95% confidence interval around *b* must include 0.[82] Because this means that the expert cannot reject the hypothesis that β equals 0, the estimate, whatever it may be, is said to be not statistically significant. Conversely, if the *t*-statistic is greater than 1.96 in absolute value, the expert concludes that the true value of β is unlikely to be 0 (intuitively, *b* is "too far" from 0 to be consistent with the true value of β being 0). In this case, the expert rejects the hypothesis that β equals 0 and calls the estimate statistically significant. If the null hypothesis β equals 0 is true, using a 95% confidence level will cause the expert to falsely reject the null hypothesis 5% of the time. Consequently, results often are said to be significant at the 5% level.[83]

As an example, consider a more complete set of regression results associated with the salary regression described in equation (9):

$$\begin{array}{lllll} \hat{Y} = & \$14{,}085 + & \$2323X_1 + & \$1675X_2 - & \$36X_3 \\ & (1577) & (140) & (1435) & (3.4) \\ t = & 8.9 & 16.5 & 1.2 & -10.8. \end{array} \quad (12)$$

The standard error of each estimated parameter is given in parentheses directly below the parameter, and the corresponding *t*-statistics appear below the standard error values.

Consider the coefficient on the dummy variable X_2. It indicates that \$1675 is the best estimate of the mean salary difference between men and women. However, the standard error of \$1435 is large in relation to its coefficient \$1675. Because the standard error is relatively large, the range of possible values for measuring the true salary difference, the true parameter, is great. In fact, a 95% confidence interval is given by

$$\$1675 \pm \$1435 \cdot 1.96 = \$1675 \pm \$2813. \quad (13)$$

In other words, the expert can have 95% confidence that the true value of the coefficient lies between –\$1138 and \$4488. Because this range includes 0, the effect of sex on salary is said to be insignificantly different from 0 at the 5% level. The *t* value of 1.2 is equal to \$1675 divided by \$1435. Because this *t*-statistic is less than 1.96 in magnitude (a condition equivalent to the inclusion of a 0 in the above confidence interval), the sex variable again is said to be an insignificant determinant of salary at the 5% level of significance.

82. The *t*-statistic applies to any sample size. As the sample gets large, the underlying distribution, which is the source of the *t*-statistic (Student's *t*-distribution), approximates the normal distribution.

83. A *t*-statistic of 2.57 in magnitude or greater is associated with a 99% confidence level, or a 1% level of significance, that includes a band of 2.57 standard deviations on either side of the estimated coefficient.

Note also that experience is a highly significant determinant of salary, because both the X_1 and the X_3 variables have *t*-statistics substantially greater than 1.96 in magnitude. More experience has a significant positive effect on salary, but the size of this effect diminishes significantly with experience.

2. *Goodness of fit*

Reported regression results usually contain not only the point estimates of the parameters and their standard errors or *t*-statistics, but also other information that tells how closely the regression line fits the data. One statistic, the standard error of the regression (SER), is an estimate of the overall size of the regression residuals.[84] An SER of 0 would occur only when all data points lie exactly on the regression line—an extremely unlikely possibility. Other things being equal, the larger the SER, the poorer the fit of the data to the model.

For a normally distributed error term, the expert would expect approximately 95% of the data points to lie within 2 SERs of the estimated regression line, as shown in Figure 7 (in Figure 7, the SER is approximately $5000).

Figure 7. Standard error of the regression.

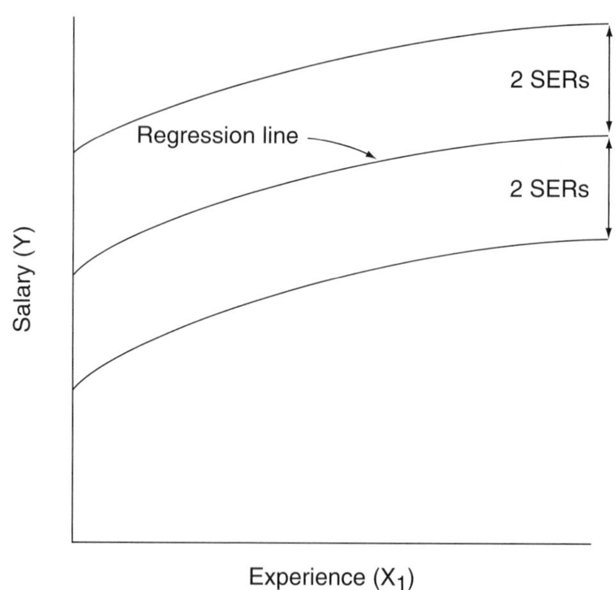

84. More specifically, it is a measure of the standard deviation of the regression error ε. It sometimes is called the root mean squared error of the regression line.

R-squared (R^2) is a statistic that measures the percentage of variation in the dependent variable that is accounted for by all the explanatory variables.[85] Thus, R^2 provides a measure of the overall goodness of fit of the multiple regression equation. Its value ranges from 0 to 1. An R^2 of 0 means that the explanatory variables explain none of the variation of the dependent variable; an R^2 of 1 means that the explanatory variables explain all of the variation. The R^2 associated with equation (12) is .56. This implies that the three explanatory variables explain 56% of the variation in salaries.

What level of R^2, if any, should lead to a conclusion that the model is satisfactory? Unfortunately, there is no clear-cut answer to this question, because the magnitude of R^2 depends on the characteristics of the data being studied and, in particular, whether the data vary over time or over individuals. Typically, an R^2 is low in cross-sectional studies in which differences in individual behavior are explained. It is likely that these individual differences are caused by many factors that cannot be measured. As a result, the expert cannot hope to explain most of the variation. In time-series studies, in contrast, the expert is explaining the movement of aggregates over time. Because most aggregate time series have substantial growth, or trend, in common, it will not be difficult to "explain" one time series using another time series, simply because both are moving together. It follows as a corollary that a high R^2 does not by itself mean that the variables included in the model are the appropriate ones.

As a general rule, courts should be reluctant to rely solely on a statistic such as R^2 to choose one model over another. Alternative procedures and tests are available.[86]

3. *Sensitivity of least squares regression results*

The least squares regression line can be sensitive to extreme data points. This sensitivity can be seen most easily in Figure 8. Assume initially that there are only three data points, A, B, and C, relating information about X_1 to the variable Y. The least squares line describing the best-fitting relationship between Points A, B, and C is represented by Line 1. Point D is called an *outlier* because it lies far from the regression line that fits the remaining points. When a new, best-fitting least squares line is reestimated to include Point D, Line 2 is obtained. Figure 8 shows that the outlier Point D is an *influential* data point, because it has a dominant effect on the slope and intercept of the least squares line. Because least squares attempts to minimize the sum of squared deviations, the sensitivity of the line to individual points sometimes can be substantial.[87]

85. The variation is the square of the difference between each *Y* value and the average *Y* value, summed over all the *Y* values.

86. These include *F*-tests and specification error tests. *See* Pindyck & Rubinfeld, *supra* note 23, at 88–95, 128–36, 194–98.

87. This sensitivity is not always undesirable. In some instances it may be much more important to predict Point D when a big change occurs than to measure the effects of small changes accurately.

Figure 8. Least squares regression.

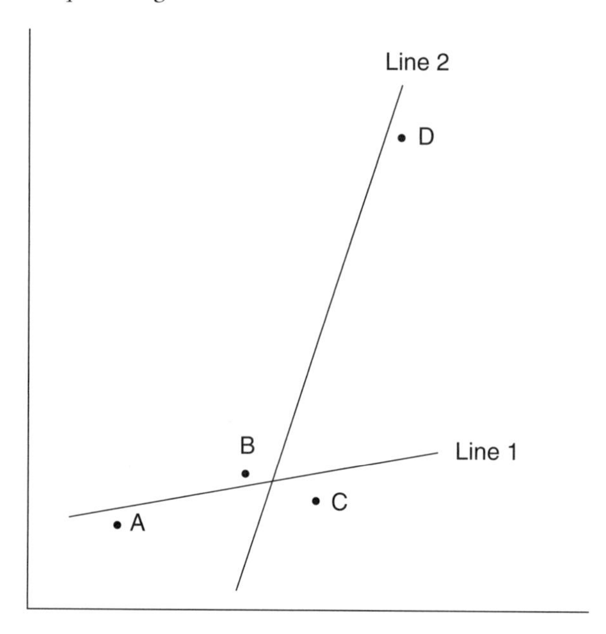

What makes the influential data problem even more difficult is that the effect of an outlier may not be seen readily if deviations are measured from the final regression line. The reason is that the influence of Point D on Line 2 is so substantial that its deviation from the regression line is not necessarily larger than the deviation of any of the remaining points from the regression line.[88] Although they are not as popular as least squares, alternative estimation techniques that are less sensitive to outliers, such as robust estimation, are available.

E. Reading Multiple Regression Computer Output

Statistical computer packages that report multiple regression analyses vary to some extent in the information they provide and the form that the information takes. Table 1 contains a sample of the basic computer output that is associated with equation (9).

88. The importance of an outlier also depends on its location in the dataset. Outliers associated with relatively extreme values of explanatory variables are likely to be especially influential. *See, e.g.*, Fisher v. Vassar College, 70 F.3d 1420, 1436 (2d Cir. 1995) (court required to include assessment of "service in academic community," because concept was too amorphous and not a significant factor in tenure review), *rev'd on other grounds*, 114 F.3d 1332 (2d Cir. 1997) (en banc).

Table 1. Regression Output

Dependent variable: Y		SSE	62346266124	F-test	174.71
		DFE	561	Prob > F	0.0001
		MSE	111134164	R^2	0.556

Variable	DF	Parameter Estimate	Standard Error	t-Statistic	Prob > $\|t\|$
Intercept	1	14,084.89	1577.484	8.9287	.0001
X_1	1	2323.17	140.70	16.5115	.0001
X_2	1	1675.11	1435.422	1.1670	.2437
X_3	1	−36.71	3.41	−10.7573	.0001

Note: SSE = sum of squared errors; DFE = degrees of freedom associated with the error term; MSE = mean squared error; DF = degrees of freedom; Prob = probability.

In the lower portion of Table 1, note that the parameter estimates, the standard errors, and the t-statistics match the values given in equation (12).[89] The variable "Intercept" refers to the constant term b_0 in the regression. The column "DF" represents degrees of freedom. The "1" signifies that when the computer calculates the parameter estimates, each variable that is added to the linear regression adds an additional constraint that must be satisfied. The column labeled "Prob > $|t|$" lists the two-tailed p-values associated with each estimated parameter; the p-value measures the observed significance level—the probability of getting a test statistic as extreme or more extreme than the observed number if the model parameter is in fact 0. The very low p-values on the variables X_1 and X_3 imply that each variable is statistically significant at less than the 1% level—both highly significant results. In contrast, the X_2 coefficient is only significant at the 24% level, implying that it is insignificant at the traditional 5% level. Thus, the expert cannot reject with confidence the null hypothesis that salaries do not differ by sex after the expert has accounted for the effect of experience.

The top portion of Table 1 provides data that relate to the goodness of fit of the regression equation. The sum of squared errors (SSE) measures the sum of the squares of the regression residuals—the sum that is minimized by the least squares procedure. The degrees of freedom associated with the error term (DFE) are given by the number of observations minus the number of parameters that were estimated. The mean squared error (MSE) measures the variance of the error term (the square of the standard error of the regression). MSE is equal to SSE divided by DFE.

89. Computer programs give results to more decimal places than are meaningful. This added detail should not be seen as evidence that the regression results are exact.

The R^2 of 0.556 indicates that 55.6% of the variation in salaries is explained by the regression variables, X_1, X_2, and X_3. Finally, the F-test is a test of the null hypothesis that all regression coefficients (except the intercept) are jointly equal to 0—that there is no linear association between the dependent variable and any of the explanatory variables. This is equivalent to the null hypothesis that R^2 is equal to 0. In this case, the F-ratio of 174.71 is sufficiently high that the expert can reject the null hypothesis with a very high degree of confidence (i.e., with a 1% level of significance).

F. Forecasting

In general, a forecast is a prediction made about the values of the dependent variable using information about the explanatory variables. Often, ex ante forecasts are performed; in this situation, values of the dependent variable are predicted beyond the sample (e.g., beyond the time period in which the model has been estimated). However, ex post forecasts are frequently used in damage analyses.[90] An ex post forecast has a forecast period such that all values of the dependent and explanatory variables are known; ex post forecasts can be checked against existing data and provide a direct means of evaluation.

For example, to calculate the forecast for the salary regression discussed above, the expert uses the estimated salary equation

$$\hat{Y} = \$14{,}085 + \$2323X_1 + \$1675X_2 - \$36X_3. \tag{14}$$

To predict the salary of a man with 2 years' experience, the expert calculates

$$\hat{Y}(2) = \$14{,}085 + (\$2323 \cdot 2) + \$1675 - (\$36 \cdot 2) = \$20{,}262. \tag{15}$$

The degree of accuracy of both ex ante and ex post forecasts can be calculated provided that the model specification is correct and the errors are normally distributed and independent. The statistic is known as the standard error of forecast (SEF). The SEF measures the standard deviation of the forecast error that is made within a sample in which the explanatory variables are known with certainty.[91] The

90. Frequently, in cases involving damages, the question arises, what the world would have been like had a certain event not taken place. For example, in a price-fixing antitrust case, the expert can ask what the price of a product would have been had a certain event associated with the price-fixing agreement not occurred. If prices would have been lower, the evidence suggests impact. If the expert can predict how much lower they would have been, the data can help the expert develop a numerical estimate of the amount of damages.

91. There are actually two sources of error implicit in the SEF. The first source arises because the estimated parameters of the regression model may not be exactly equal to the true regression parameters. The second source is the error term itself; when forecasting, the expert typically sets the error equal to 0 when a turn of events not taken into account in the regression model may make it appropriate to make the error positive or negative.

SEF can be used to determine how accurate a given forecast is. In equation (15), the SEF associated with the forecast of $20,262 is approximately $5000. If a large sample size is used, the probability is roughly 95% that the predicted salary will be within 1.96 standard errors of the forecasted value. In this case, the appropriate 95% interval for the prediction is $10,822 to $30,422. Because the estimated model does not explain salaries effectively, the SEF is large, as is the 95% interval. A more complete model with additional explanatory variables would result in a lower SEF and a smaller 95% interval for the prediction.

A danger exists when using the SEF, which applies to the standard errors of the estimated coefficients as well. The SEF is calculated on the assumption that the model includes the correct set of explanatory variables and the correct functional form. If the choice of variables or the functional form is wrong, the estimated forecast error may be misleading. In some instances, it may be smaller, perhaps substantially smaller, than the true SEF; in other instances, it may be larger, for example, if the wrong variables happen to capture the effects of the correct variables.

The difference between the SEF and the SER is shown in Figure 9. The SER measures deviations within the sample. The SEF is more general, because it calculates deviations within or without the sample period. In general, the difference between the SEF and the SER increases as the values of the explanatory variables increase in distance from the mean values. Figure 9 shows the 95% prediction interval created by the measurement of two SEFs about the regression line.

Figure 9. Standard error of forecast.

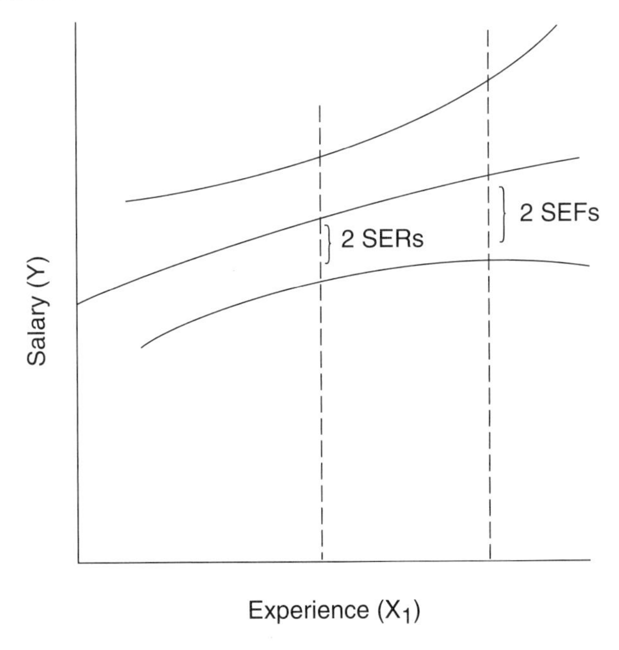

G. A Hypothetical Example

Jane Thompson filed suit in federal court alleging that officials in the police department discriminated against her and a class of other female police officers in violation of Title VII of the Civil Rights Act of 1964, as amended. On behalf of the class, Ms. Thompson alleged that she was paid less than male police officers with equivalent skills and experience. Both plaintiff and defendant used expert economists with econometric expertise to present statistical evidence to the court in support of their positions.

Plaintiff's expert pointed out that the mean salary of the 40 female officers was $30,604, whereas the mean salary of the 60 male officers was $43,077. To show that this difference was statistically significant, the expert put forward a regression of salary (SALARY) on a constant term and a dummy indicator variable (FEM) equal to 1 for each female and 0 for each male. The results were as follows:

	SALARY =	$43,077	−$12,373*FEM
Standard Error		($1528)	($2416)
p-value		<.01	<.01
$R^2 = .22$			

The −$12,373 coefficient on the FEM variable measures the mean difference between male and female salaries. Because the standard error is approximately one-fifth of the value of the coefficient, this difference is statistically significant at the 5% (and indeed at the 1%) level. If this is an appropriate regression model (in terms of its implicit characterization of salary determination), one can conclude that it is highly unlikely that the difference in salaries between men and women is due to chance.

The defendant's expert testified that the regression model put forward was the wrong model because it failed to account for the fact that males (on average) had substantially more experience than females. The relatively low R^2 was an indication that there was substantial unexplained variation in the salaries of male and female officers. An examination of data relating to years spent on the job showed that the average male experience was 8.2 years, whereas the average for females was only 3.5 years. The defense expert then presented a regression analysis that added an additional explanatory variable (i.e., a covariate), the years of experience of each police officer (EXP). The new regression results were as follows:

	SALARY =	$28,049	– $3860*FEM	+ $1833*EXP
Standard Error		(2513)	($2347)	($265)
p-value		<.01	<.11	<.01
$R^2 = .47$				

Experience is itself a statistically significant explanatory variable, with a *p*-value of less than .01. Moreover, the difference between male and female salaries, holding experience constant, is only $3860, and this difference is not statistically significant at the 5% level. The defense expert was able to testify on this basis that the court could not rule out alternative explanations for the difference in salaries other than the plaintiff's claim of discrimination.

The debate did not end here. On rebuttal, the plaintiff's expert made three distinct points. First, whether $3860 was statistically significant or not, it was practically significant, representing a salary difference of more than 10% of the mean female officers' salaries. Second, although the result was not statistically significant at the 5% level, it was significant at the 11% level. If the regression model were valid, there would be approximately an 11% probability that one would err by concluding that the mean salary difference between men and women was a result of chance.

Third, and most importantly, the expert testified that the regression model was not correctly specified. Further analysis by the expert showed that the value of an additional year of experience was $2333 for males on average, but only $1521 for females. Based on supporting testimonial experience, the expert testified that one could not rule out the possibility that the mechanism by which the police department discriminated against females was by rewarding males more for their experience than females. The expert made this point clear by running an additional regression in which a further covariate was added to the model. The new variable was an interaction variable, INT, measured as the product of the FEM and EXP variables. The regression results were as follows:

SALARY =	$35,122	− $5250★FEM	+ $2333★EXP	− $812★FEM★EXP
St. Error	($2825)	($347)	($265)	($185)
p-value	<.01	<.11	<.01	<.04
$R^2 = .65$				

The plaintiff's expert noted that for all males in the sample, FEM = 0, in which case the regression results are given by the equation

SALARY = $35,122 + $2333★EXP

However, for females, FEM = 1, in which the corresponding equation is

SALARY = $29,872 + $1521★EXP

It appears, therefore, that females are discriminated against not only when hired (i.e., when EXP = 0), but also in the reward they get as they accumulate more and more experience.

The debate between the experts continued, focusing less on the statistical interpretation of any one particular regression model, but more on the model choice itself, and not simply on statistical significance, but also with regard to practical significance.

Glossary

The following terms and definitions are adapted from a variety of sources, including A Dictionary of Epidemiology (John M. Last et al., eds., 4th ed. 2000) and Robert S. Pindyck & Daniel L. Rubinfeld, Econometric Models and Economic Forecasts (4th ed. 1998).

alternative hypothesis. See hypothesis test.

association. The degree of statistical dependence between two or more events or variables. Events are said to be associated when they occur more frequently together than one would expect by chance.

bias. Any effect at any stage of investigation or inference tending to produce results that depart systematically from the true values (i.e., the results are either too high or too low). A biased estimator of a parameter differs on average from the true parameter.

coefficient. An estimated regression parameter.

confidence interval. An interval that contains a true regression parameter with a given degree of confidence.

consistent estimator. An estimator that tends to become more and more accurate as the sample size grows.

correlation. A statistical means of measuring the linear association between variables. Two variables are correlated positively if, on average, they move in the same direction; two variables are correlated negatively if, on average, they move in opposite directions.

covariate. A variable that is possibly predictive of an outcome under study; an explanatory variable.

cross-sectional analysis. A type of multiple regression analysis in which each data point is associated with a different unit of observation (e.g., an individual or a firm) measured at a particular point in time.

degrees of freedom (DF). The number of observations in a sample minus the number of estimated parameters in a regression model. A useful statistic in hypothesis testing.

dependent variable. The variable to be explained or predicted in a multiple regression model.

dummy variable. A variable that takes on only two values, usually 0 and 1, with one value indicating the presence of a characteristic, attribute, or effect (1), and the other value indicating its absence (0).

efficient estimator. An estimator of a parameter that produces the greatest precision possible.

error term. A variable in a multiple regression model that represents the cumulative effect of a number of sources of modeling error.

estimate. The calculated value of a parameter based on the use of a particular sample.

estimator. The sample statistic that estimates the value of a population parameter (e.g., a regression parameter); its values vary from sample to sample.

ex ante forecast. A prediction about the values of the dependent variable that go beyond the sample; consequently, the forecast must be based on predictions for the values of the explanatory variables in the regression model.

explanatory variable. A variable that is associated with changes in a dependent variable.

ex post forecast. A prediction about the values of the dependent variable made during a period in which all values of the explanatory and dependent variables are known. Ex post forecasts provide a useful means of evaluating the fit of a regression model.

***F*-test.** A statistical test (based on an *F*-ratio) of the null hypothesis that a group of explanatory variables are jointly equal to 0. When applied to all the explanatory variables in a multiple regression model, the *F*-test becomes a test of the null hypothesis that R^2 equals 0.

feedback. When changes in an explanatory variable affect the values of the dependent variable, and changes in the dependent variable also affect the explanatory variable. When both effects occur at the same time, the two variables are described as being determined simultaneously.

fitted value. The estimated value for the dependent variable; in a linear regression, this value is calculated as the intercept plus a weighted average of the values of the explanatory variables, with the estimated parameters used as weights.

heteroscedasticity. When the error associated with a multiple regression model has a nonconstant variance; that is, the error values associated with some observations are typically high, while the values associated with other observations are typically low.

hypothesis test. A statement about the parameters in a multiple regression model. The null hypothesis may assert that certain parameters have specified values or ranges; the alternative hypothesis would specify other values or ranges.

independence. When two variables are not correlated with each other (in the population).

independent variable. An explanatory variable that affects the dependent variable but that is not affected by the dependent variable.

influential data point. A data point whose deletion from a regression sample causes one or more estimated regression parameters to change substantially.

interaction variable. The product of two explanatory variables in a regression model. Used in a particular form of nonlinear model.

intercept. The value of the dependent variable when each of the explanatory variables takes on the value of 0 in a regression equation.

least squares. A common method for estimating regression parameters. Least squares minimizes the sum of the squared differences between the actual values of the dependent variable and the values predicted by the regression equation.

linear regression model. A regression model in which the effect of a change in each of the explanatory variables on the dependent variable is the same, no matter what the values of those explanatory variables.

mean (sample). An average of the outcomes associated with a probability distribution, where the outcomes are weighted by the probability that each will occur.

mean squared error (MSE). The estimated variance of the regression error, calculated as the average of the sum of the squares of the regression residuals.

model. A representation of an actual situation.

multicollinearity. When two or more variables are highly correlated in a multiple regression analysis. Substantial multicollinearity can cause regression parameters to be estimated imprecisely, as reflected in relatively high standard errors.

multiple regression analysis. A statistical tool for understanding the relationship between two or more variables.

nonlinear regression model. A model having the property that changes in explanatory variables will have differential effects on the dependent variable as the values of the explanatory variables change.

normal distribution. A bell-shaped probability distribution having the property that about 95% of the distribution lies within 2 standard deviations of the mean.

null hypothesis. In regression analysis the null hypothesis states that the results observed in a study with respect to a particular variable are no different from what might have occurred by chance, independent of the effect of that variable. See *hypothesis test*.

one-tailed test. A hypothesis test in which the alternative to the null hypothesis that a parameter is equal to 0 is for the parameter to be either positive or negative, but not both.

outlier. A data point that is more than some appropriate distance from a regression line that is estimated using all the other data points in the sample.

***p*-value.** The significance level in a statistical test; the probability of getting a test statistic as extreme or more extreme than the observed value. The larger the *p*-value, the more likely that the null hypothesis is valid.

parameter. A numerical characteristic of a population or a model.

perfect collinearity. When two or more explanatory variables are correlated perfectly.

population. All the units of interest to the researcher; also, universe.

practical significance. Substantive importance. Statistical significance does not ensure practical significance, because, with large samples, small differences can be statistically significant.

probability distribution. The process that generates the values of a random variable. A probability distribution lists all possible outcomes and the probability that each will occur.

probability sampling. A process by which a sample of a population is chosen so that each unit of observation has a known probability of being selected.

quasi-experiment (or natural experiment). A naturally occurring instance of observable phenomena that yield data that approximate a controlled experiment.

***R*-squared (R^2).** A statistic that measures the percentage of the variation in the dependent variable that is accounted for by all of the explanatory variables in a regression model. *R*-squared is the most commonly used measure of goodness of fit of a regression model.

random error term. A term in a regression model that reflects random error (sampling error) that is the result of chance. As a consequence, the result obtained in the sample differs from the result that would be obtained if the entire population were studied.

regression coefficient. Also, regression parameter. The estimate of a population parameter obtained from a regression equation that is based on a particular sample.

regression residual. The difference between the actual value of a dependent variable and the value predicted by the regression equation.

robust estimation. An alternative to least squares estimation that is less sensitive to outliers.

robustness. A statistic or procedure that does not change much when data or assumptions are slightly modified is robust.

sample. A selection of data chosen for a study; a subset of a population.

sampling error. A measure of the difference between the sample estimate of a parameter and the population parameter.

scatterplot. A graph showing the relationship between two variables in a study; each dot represents one subject. One variable is plotted along the horizontal axis; the other variable is plotted along the vertical axis.

serial correlation. The correlation of the values of regression errors over time.

slope. The change in the dependent variable associated with a one-unit change in an explanatory variable.

spurious correlation. When two variables are correlated, but one is not the cause of the other.

standard deviation. The square root of the variance of a random variable. The variance is a measure of the spread of a probability distribution about its mean; it is calculated as a weighted average of the squares of the deviations of the outcomes of a random variable from its mean.

standard error of forecast (SEF). An estimate of the standard deviation of the forecast error; it is based on forecasts made within a sample in which the values of the explanatory variables are known with certainty.

standard error of the coefficient; standard error (SE). A measure of the variation of a parameter estimate or coefficient about the true parameter. The standard error is a standard deviation that is calculated from the probability distribution of estimated parameters.

standard error of the regression (SER). An estimate of the standard deviation of the regression error; it is calculated as the square root of the average of the squares of the residuals associated with a particular multiple regression analysis.

statistical significance. A test used to evaluate the degree of association between a dependent variable and one or more explanatory variables. If the calculated p-value is smaller than 5%, the result is said to be statistically significant (at the 5% level). If p is greater than 5%, the result is statistically insignificant (at the 5% level).

***t*-statistic.** A test statistic that describes how far an estimate of a parameter is from its hypothesized value (i.e., given a null hypothesis). If a t-statistic is sufficiently large (in absolute magnitude), an expert can reject the null hypothesis.

***t*-test.** A test of the null hypothesis that a regression parameter takes on a particular value, usually 0. The test is based on the t-statistic.

time-series analysis. A type of multiple regression analysis in which each data point is associated with a particular unit of observation (e.g., an individual or a firm) measured at different points in time.

two-tailed test. A hypothesis test in which the alternative to the null hypothesis that a parameter is equal to 0 is for the parameter to be either positive or negative, or both.

variable. Any attribute, phenomenon, condition, or event that can have two or more values.

variable of interest. The explanatory variable that is the focal point of a particular study or legal issue.

References on Multiple Regression

Jonathan A. Baker & Daniel L. Rubinfeld, *Empirical Methods in Antitrust: Review and Critique,* 1 Am. L. & Econ. Rev. 386 (1999).

Gerald V. Barrett & Donna M. Sansonetti, *Issues Concerning the Use of Regression Analysis in Salary Discrimination Cases,* 41 Personnel Psychol. 503 (2006).

Thomas J. Campbell, *Regression Analysis in Title VII Cases: Minimum Standards, Comparable Worth, and Other Issues Where Law and Statistics Meet,* 36 Stan. L. Rev. 1299 (1984).

Catherine Connolly, *The Use of Multiple Regression Analysis in Employment Discrimination Cases,* 10 Population Res. & Pol'y Rev. 117 (1991).

Arthur P. Dempster, *Employment Discrimination and Statistical Science,* 3 Stat. Sci. 149 (1988).

Michael O. Finkelstein, *The Judicial Reception of Multiple Regression Studies in Race and Sex Discrimination Cases,* 80 Colum. L. Rev. 737 (1980).

Michael O. Finkelstein & Hans Levenbach, *Regression Estimates of Damages in Price-Fixing Cases,* Law & Contemp. Probs., Autumn 1983, at 145.

Franklin M. Fisher, *Multiple Regression in Legal Proceedings,* 80 Colum. L. Rev. 702 (1980).

Franklin M. Fisher, *Statisticians, Econometricians, and Adversary Proceedings,* 81 J. Am. Stat. Ass'n 277 (1986).

Joseph L. Gastwirth, *Methods for Assessing the Sensitivity of Statistical Comparisons Used in Title VII Cases to Omitted Variables,* 33 Jurimetrics J. 19 (1992).

Note, *Beyond the Prima Facie Case in Employment Discrimination Law: Statistical Proof and Rebuttal,* 89 Harv. L. Rev. 387 (1975).

Daniel L. Rubinfeld, *Econometrics in the Courtroom,* 85 Colum. L. Rev. 1048 (1985).

Daniel L. Rubinfeld & Peter O. Steiner, *Quantitative Methods in Antitrust Litigation,* Law & Contemp. Probs., Autumn 1983, at 69.

Daniel L. Rubinfeld, *Statistical and Demographic Issues Underlying Voting Rights Cases,* 15 Evaluation Rev. 659 (1991).

The Evolving Role of Statistical Assessments as Evidence in the Courts (Stephen E. Fienberg ed., 1989).

Reference Guide on Survey Research

SHARI SEIDMAN DIAMOND

Shari Seidman Diamond, J.D., Ph.D., is the Howard J. Trienens Professor of Law and Professor of Psychology, Northwestern University, and a Research Professor, American Bar Foundation, Chicago, Illinois.

CONTENTS

I. Introduction

Sample surveys are used to describe or enumerate the beliefs, attitudes, or behavior of persons or other social units.[1] Surveys typically are offered in legal proceedings to establish or refute claims about the characteristics of those individuals or social units (e.g., whether consumers are likely to be misled by the claims contained in an allegedly deceptive advertisement;[2] which qualities purchasers focus on in making decisions about buying new computer systems).[3] In a broader sense, a *survey* can describe or enumerate the attributes of any units, including animals and objects.[4] We focus here primarily on sample surveys, which must deal not only with issues of population definition, sampling, and measurement common to all surveys, but also with the specialized issues that arise in obtaining information from human respondents.

In principle, surveys may count or measure every member of the relevant population (e.g., all plaintiffs eligible to join in a suit, all employees currently working for a corporation, all trees in a forest). In practice, surveys typically count or measure only a portion of the individuals or other units that the survey is intended to describe (e.g., a sample of jury-eligible citizens, a sample of potential job applicants). In either case, the goal is to provide information on the relevant population from which the sample was drawn. Sample surveys can be carried out using probability or nonprobability sampling techniques. Although probability sampling offers important advantages over nonprobability sampling,[5] experts in some fields (e.g., marketing) regularly rely on various forms of nonprobability sampling when conducting surveys. Consistent with Federal Rule of Evidence 703, courts generally have accepted such evidence.[6] Thus, in this reference guide, both the probability sample and the nonprobability sample are discussed. The strengths of probability sampling and the weaknesses of various types of nonprobability sampling are described.

1. Sample surveys conducted by social scientists "consist of (relatively) systematic, (mostly) standardized approaches to collecting information on individuals, households, organizations, or larger organized entities through questioning systematically identified samples." James D. Wright & Peter V. Marsden, *Survey Research and Social Science: History, Current Practice, and Future Prospects*, *in* Handbook of Survey Research 1, 3 (James D. Wright & Peter V. Marsden eds., 2d ed. 2010).

2. *See* Sanderson Farms v. Tyson Foods, 547 F. Supp. 2d 491 (D. Md. 2008).

3. *See* SMS Sys. Maint. Servs. v. Digital Equip. Corp., 118 F.3d 11, 30 (1st Cir. 1999). For other examples, see notes 19–32 and accompanying text.

4. In *J.H. Miles & Co. v. Brown*, 910 F. Supp. 1138 (E.D. Va. 1995), clam processors and fishing vessel owners sued the Secretary of Commerce for failing to use the unexpectedly high results from 1994 survey data on the size of the clam population to determine clam fishing quotas for 1995. The estimate of clam abundance is obtained from surveys of the amount of fishing time the research survey vessels require to collect a specified yield of clams in major fishing areas over a period of several weeks. *Id.* at 1144–45.

5. *See infra* Section III.C.

6. Fed. R. Evid. 703 recognizes facts or data "of a type reasonably relied upon by experts in the particular field. . . ."

As a method of data collection, surveys have several crucial potential advantages over less systematic approaches.[7] When properly designed, executed, and described, surveys (1) economically present the characteristics of a large group of respondents or other units and (2) permit an assessment of the extent to which the measured respondents or other units are likely to adequately represent a relevant group of individuals or other units.[8] All questions asked of respondents and all other measuring devices used (e.g., criteria for selecting eligible respondents) can be examined by the court and the opposing party for objectivity, clarity, and relevance, and all answers or other measures obtained can be analyzed for completeness and consistency. The survey questions should not be the only focus of attention. To make it possible for the court and the opposing party to closely scrutinize the survey so that its relevance, objectivity, and representativeness can be evaluated, the party proposing to offer the survey as evidence should also describe in detail the design, execution, and analysis of the survey. This should include (1) a description of the population from which the sample was selected, demonstrating that it was the relevant population for the question at hand; (2) a description of how the sample was drawn and an explanation for why that sample design was appropriate; (3) a report on response rate and the ability of the sample to represent the target population; and (4) an evaluation of any sources of potential bias in respondents' answers.

The questions listed in this reference guide are intended to assist judges in identifying, narrowing, and addressing issues bearing on the adequacy of surveys either offered as evidence or proposed as a method for developing information.[9] These questions can be (1) raised from the bench during a pretrial proceeding to determine the admissibility of the survey evidence; (2) presented to the contending experts before trial for their joint identification of disputed and undisputed issues; (3) presented to counsel with the expectation that the issues will be addressed during the examination of the experts at trial; or (4) raised in bench trials when a motion for a preliminary injunction is made to help the judge evaluate

7. This does not mean that surveys can be relied on to address all questions. For example, if survey respondents had been asked in the days before the attacks of 9/11 to predict whether they would volunteer for military service if Washington, D.C., were to be bombed, their answers may not have provided accurate predictions. Although respondents might have willingly answered the question, their assessment of what they would actually do in response to an attack simply may have been inaccurate. Even the option of a "do not know" choice would not have prevented an error in prediction if they believed they could accurately predict what they would do. Thus, although such a survey would have been suitable for assessing the *predictions* of respondents, it might have provided a very inaccurate estimate of what an actual response to the attack would be.

8. The ability to quantitatively assess the limits of the likely margin of error is unique to probability sample surveys, but an expert testifying about any survey should provide enough information to allow the judge to evaluate how potential error, including coverage, measurement, nonresponse, and sampling error, may have affected the obtained pattern of responses.

9. *See infra* text accompanying note 31.

what weight, if any, the survey should be given.[10] These questions are intended to improve the utility of cross-examination by counsel, where appropriate, not to replace it.

All sample surveys, whether they measure individuals or other units, should address the issues concerning purpose and design (Section II), population definition and sampling (Section III), accuracy of data entry (Section VI), and disclosure and reporting (Section VII). Questionnaire and interview surveys, whether conducted in-person, on the telephone, or online, raise methodological issues involving survey questions and structure (Section IV) and confidentiality (Section VII.C). Interview surveys introduce additional issues (e.g., interviewer training and qualifications) (Section V), and online surveys raise some new issues and questions that are currently under study (Section VI). The sections of this reference guide are labeled to direct the reader to those topics that are relevant to the type of survey being considered. The scope of this reference guide is necessarily limited, and additional issues might arise in particular cases.

A. Use of Surveys in Court

Fifty years ago the question of whether surveys constituted acceptable evidence still was unsettled.[11] Early doubts about the admissibility of surveys centered on their use of sampling[12] and their status as hearsay evidence.[13] Federal Rule of Evidence

10. Lanham Act cases involving trademark infringement or deceptive advertising frequently require expedited hearings that request injunctive relief, so judges may need to be more familiar with survey methodology when considering the weight to accord a survey in these cases than when presiding over cases being submitted to a jury. Even in a case being decided by a jury, however, the court must be prepared to evaluate the methodology of the survey evidence in order to rule on admissibility. *See* Daubert v. Merrell Dow Pharms., Inc., 509 U.S. 579, 589 (1993).

11. Hans Zeisel, *The Uniqueness of Survey Evidence*, 45 Cornell L.Q. 322, 345 (1960).

12. In an early use of sampling, Sears, Roebuck & Co. claimed a tax refund based on sales made to individuals living outside city limits. Sears randomly sampled 33 of the 826 working days in the relevant working period, computed the proportion of sales to out-of-city individuals during those days, and projected the sample result to the entire period. The court refused to accept the estimate based on the sample. When a complete audit was made, the result was almost identical to that obtained from the sample. *Sears, Roebuck & Co. v. City of Inglewood*, tried in Los Angeles Superior Court in 1955, is described in R. Clay Sprowls, *The Admissibility of Sample Data into a Court of Law: A Case History*, 4 UCLA L. Rev. 222, 226–29 (1956–1957).

13. Judge Wilfred Feinberg's thoughtful analysis in *Zippo Manufacturing Co. v. Rogers Imports, Inc.*, 216 F. Supp. 670, 682–83 (S.D.N.Y. 1963), provides two alternative grounds for admitting opinion surveys: (1) Surveys are not hearsay because they are not offered in evidence to prove the truth of the matter asserted; and (2) even if they are hearsay, they fall under one of the exceptions as a "present sense impression." In *Schering Corp. v. Pfizer Inc.*, 189 F.3d 218 (2d Cir. 1999), the Second Circuit distinguished between perception surveys designed to reflect the present sense impressions of respondents and "memory" surveys designed to collect information about a past occurrence based on the recollections of the survey respondents. The court in *Schering* suggested that if a survey is offered to prove the existence of a specific idea in the public mind, then the survey does constitute hearsay

703 settled both matters for surveys by redirecting attention to the "validity of the techniques employed."[14] The inquiry under Rule 703 focuses on whether facts or data are "of a type reasonably relied upon by experts in the particular field in forming opinions or inferences upon the subject."[15] For a survey, the question becomes, "Was the poll or survey conducted in accordance with generally accepted survey principles, and were the results used in a statistically correct way?"[16] This focus on the adequacy of the methodology used in conducting and analyzing results from a survey is also consistent with the Supreme Court's discussion of admissible scientific evidence in *Daubert v. Merrell Dow Pharmaceuticals, Inc.*[17]

Because the survey method provides an economical and systematic way to gather information and draw inferences about a large number of individuals or other units, surveys are used widely in business, government, and, increasingly,

evidence. As the court observed, Federal Rule of Evidence 803(3), creating "an exception to the hearsay rule for such statements [i.e., state-of-mind expressions] rather than excluding the statements from the definition of hearsay, makes sense only in this light." *Id.* at 230 n.3. *See also* Playtex Prods. v. Procter & Gamble Co., 2003 U.S. Dist. LEXIS 8913 (S.D.N.Y. May 28, 2003), *aff'd*, 126 Fed. Appx. 32 (2d Cir. 2005). Note, however, that when survey respondents are shown a stimulus (e.g., a commercial) and then respond to a series of questions about their impressions of what they viewed, those impressions reflect both respondents' initial perceptions and their memory for what they saw and heard. Concerns about the impact of memory on the trustworthiness of survey responses appropriately depend on the passage of time between exposure and testing and on the likelihood that distorting events occurred during that interval.

Two additional exceptions to the hearsay exclusion can be applied to surveys. First, surveys may constitute a hearsay exception if the survey data were collected in the normal course of a regularly conducted business activity, unless "the source of information or the method or circumstances of preparation indicate lack of trustworthiness." Fed. R. Evid. 803(6); *see also* Ortho Pharm. Corp. v. Cosprophar, Inc., 828 F. Supp. 1114, 1119–20 (S.D.N.Y. 1993) (marketing surveys prepared in the course of business were properly excluded because they lacked foundation from a person who saw the original data or knew what steps were taken in preparing the report), *aff'd*, 32 F.3d 690 (2d Cir. 1994). In addition, if a survey shows guarantees of trustworthiness equivalent to those in other hearsay exceptions, it can be admitted if the court determines that the statement is offered as evidence of a material fact, it is more probative on the point for which it is offered than any other evidence that the proponent can procure through reasonable efforts, and admissibility serves the interests of justice. Fed. R. Evid. 807; *e.g.*, *Schering*, 189 F.3d at 232. Admissibility as an exception to the hearsay exclusion thus depends on the trustworthiness of the survey. New Colt Holding v. RJG Holdings of Fla., 312 F. Supp. 2d 195, 223 (D. Conn. 2004).

14. Fed. R. Evid. 703 Advisory Committee Note.

15. Fed. R. Evid. 703.

16. Manual for Complex Litigation § 2.712 (1982). Survey research also is addressed in the Manual for Complex Litigation, Second § 21.484 (1985) [hereinafter MCL 2d]; the Manual for Complex Litigation, Third § 21.493 (1995) [hereinafter MCL 3d]; and the Manual for Complex Litigation, Fourth §11.493 (2004) [hereinafter MCL 4th]. Note, however, that experts who collect survey data, along with the professions that rely on those surveys, may differ in some of their methodological standards and principles. An assessment of the precision of sample estimates and an evaluation of the sources and magnitude of likely bias are required to distinguish methods that are acceptable from methods that are not.

17. 509 U.S. 579 (1993); *see also* General Elec. Co. v. Joiner, 522 U.S. 136, 147 (1997).

administrative settings and judicial proceedings.[18] Both federal and state courts have accepted survey evidence on a variety of issues. In a case involving allegations of discrimination in jury panel composition, the defense team surveyed prospective jurors to obtain their age, race, education, ethnicity, and income distribution.[19] Surveys of employees or prospective employees are used to support or refute claims of employment discrimination.[20] Surveys provide information on the nature and similarity of claims to support motions for or against class certification.[21] In ruling on the admissibility of scientific claims, courts have examined surveys of scientific experts to assess the extent to which the theory or technique has received widespread acceptance.[22] Some courts have admitted surveys in obscenity cases to provide evidence about community standards.[23] Requests for a change of venue on grounds of jury pool bias often are backed by evidence from a survey of jury-eligible respondents in the area of the original venue.[24] The plaintiff in an antitrust suit conducted a survey to assess what characteristics, including price, affected consumers' preferences. The survey was offered as one way to estimate damages.[25] In a Title IX suit based on allegedly discriminatory scheduling of girls'

18. Some sample surveys are so well accepted that they even may not be recognized as surveys. For example, some U.S. Census Bureau data are based on sample surveys. Similarly, the Standard Table of Mortality, which is accepted as proof of the average life expectancy of an individual of a particular age and gender, is based on survey data.

19. United States v. Green, 389 F. Supp. 2d 29 (D. Mass. 2005), *rev'd on other grounds*, 426 F.3d 1 (1st Cir. 2005) (evaluating minority underrepresentation in the jury pool by comparing racial composition of the voting-age population in the district with the racial breakdown indicated in juror questionnaires returned to court); *see also* People v. Harris, 36 Cal. 3d 36, 679 P.2d 433 (Cal. 1984).

20. John Johnson v. Big Lots Stores, Inc., No. 04-321, 2008 U.S. Dist. LEXIS 35316, at *20 (E.D. La. Apr. 29, 2008); Stender v. Lucky Stores, Inc., 803 F. Supp. 259, 326 (N.D. Cal. 1992); EEOC v. Sears, Roebuck & Co., 628 F. Supp. 1264, 1308 (N.D. Ill. 1986), *aff'd*, 839 F.2d 302 (7th Cir. 1988).

21. John Johnson v. Big Lots Stores, Inc., 561 F. Supp. 2d 567 (E.D. La. 2008); Marlo v. United Parcel Service, Inc., 251 F.R.D. 476 (C.D. Cal. 2008).

22. United States v. Scheffer, 523 U.S. 303, 309 (1998); United States v. Bishop, 64 F. Supp. 2d 1149 (D. Utah 1999); United States v. Varoudakis, No. 97-10158, 1998 WL 151238 (D. Mass. Mar. 27, 1998); State v. Shively, 268 Kan. 573 (2000), *aff'd*, 268 Kan. 589 (2000) (all cases in which courts determined, based on the inconsistent reactions revealed in several surveys, that the polygraph test has failed to achieve general acceptance in the scientific community). *Contra, see* Lee v. Martinez, 136 N.M. 166, 179–81, 96 P.3d 291, 304–06 (N.M. 2004). People v. Williams, 830 N.Y.S.2d 452 (2006) (expert permitted to testify regarding scientific studies of factors affecting the perceptual ability and memory of eyewitnesses to make identifications based in part on general acceptance demonstrated in survey of experts who study eyewitness identification).

23. *E.g.*, People v. Page Books, Inc., 601 N.E.2d 273, 279–80 (Ill. App. Ct. 1992); State v. Williams, 598 N.E.2d 1250, 1256–58 (Ohio Ct. App. 1991).

24. *E.g.*, United States v. Eagle, 586 F.2d 1193, 1195 (8th Cir. 1978); United States v. Tokars, 839 F. Supp. 1578, 1583 (D. Ga. 1993), *aff'd*, 95 F.3d 1520 (11th Cir. 1996); State v. Baumruk, 85 S.W.3d 644 (Mo. 2002); People v. Boss, 701 N.Y.S.2d 342 (App. Div. 1999).

25. Dolphin Tours, Inc. v. Pacifico Creative Servs., Inc., 773 F.2d 1506, 1508 (9th Cir. 1985). *See also* SMS Sys. Maint. Servs., Inc. v. Digital Equip. Corp., 188 F.3d 11 (1st Cir. 1999); Benjamin F. King, *Statistics in Antitrust Litigation*, *in* Statistics and the Law 49 (Morris H. DeGroot et al. eds.,

sports, a survey was offered for the purpose of establishing how girls felt about the scheduling of girls' and boys' sports.[26] A routine use of surveys in federal courts occurs in Lanham Act[27] cases, when the plaintiff alleges trademark infringement[28] or claims that false advertising[29] has confused or deceived consumers. The pivotal legal question in such cases virtually demands survey research because it centers on consumer perception and memory (i.e., is the consumer likely to be confused about the source of a product, or does the advertisement imply a false or misleading message?).[30] In addition, survey methodology has been used creatively to assist federal courts in managing mass torts litigation. Faced with the prospect of conducting discovery concerning 10,000 plaintiffs, the plaintiffs and defendants in *Wilhoite v. Olin Corp.*[31] jointly drafted a discovery survey that was administered

1986). Surveys have long been used in antitrust litigation to help define relevant markets. In *United States v. E.I. du Pont de Nemours & Co.*, 118 F. Supp. 41, 60 (D. Del. 1953), *aff'd*, 351 U.S. 377 (1956), a survey was used to develop the "market setting" for the sale of cellophane. In *Mukand, Ltd. v. United States*, 937 F. Supp. 910 (Ct. Int'l Trade 1996), a survey of purchasers of stainless steel wire rods was conducted to support a determination of competition and fungibility between domestic and Indian wire rod.

26. Alston v. Virginia High Sch. League, Inc., 144 F. Supp. 2d 526, 539–40 (W.D. Va. 1999).

27. Lanham Act § 43(a), 15 U.S.C. § 1125(a) (1946) (amended 2006).

28. *E.g.*, Herman Miller v. Palazzetti Imports & Exports, 270 F.3d 298, 312 (6th Cir. 2001) ("Because the determination of whether a mark has acquired secondary meaning is primarily an empirical inquiry, survey evidence is the most direct and persuasive evidence."); Simon Property Group v. MySimon, 104 F. Supp. 2d 1033, 1038 (S.D. Ind. 2000) ("Consumer surveys are generally accepted by courts as one means of showing the likelihood of consumer confusion."). *See also* Qualitex Co. v. Jacobson Prods. Co., No. CIV-90-1183HLH, 1991 U.S. Dist. LEXIS 21172 (C.D. Cal. Sept. 3, 1991), *aff'd in part & rev'd in part on other grounds*, 13 F.3d 1297 (9th Cir. 1994), *rev'd on other grounds*, 514 U.S. 159 (1995); Union Carbide Corp. v. Ever-Ready, Inc., 531 F.2d 366 (7th Cir.), *cert. denied*, 429 U.S. 830 (1976). According to Neal Miller, *Facts, Expert Facts, and Statistics: Descriptive and Experimental Research Methods in Litigation*, 40 Rutgers L. Rev. 101, 137 (1987), trademark law has relied on the institutionalized use of statistical evidence more than any other area of the law.

29. *E.g.*, Southland Sod Farms v. Stover Seed Co., 108 F.3d 1134, 1142–43 (9th Cir. 1997); American Home Prods. Corp. v. Johnson & Johnson, 577 F.2d 160 (2d Cir. 1978); Rexall Sundown, Inc. v. Perrigo Co., 651 F. Supp. 2d 9 (E.D.N.Y. 2009); Mutual Pharm. Co. v. Ivax Pharms. Inc., 459 F. Supp. 2d 925 (C.D. Cal. 2006); Novartis Consumer Health v. Johnson & Johnson-Merck Consumer Pharms., 129 F. Supp. 2d 351 (D.N.J. 2000).

30. Courts have observed that "the court's reaction is at best not determinative and at worst irrelevant. The question in such cases is, what does the person to whom the advertisement is addressed find to be the message?" American Brands, Inc. v. R.J. Reynolds Tobacco Co., 413 F. Supp. 1352, 1357 (S.D.N.Y. 1976). The wide use of surveys in recent years was foreshadowed in *Triangle Publications, Inc. v. Rohrlich*, 167 F.2d 969, 974 (2d Cir. 1948) (Frank, J., dissenting). Called on to determine whether a manufacturer of girdles labeled "Miss Seventeen" infringed the trademark of the magazine *Seventeen*, Judge Frank suggested that, in the absence of a test of the reactions of "numerous girls and women," the trial court judge's finding as to what was likely to confuse was "nothing but a surmise, a conjecture, a guess," noting that "neither the trial judge nor any member of this court is (or resembles) a teen-age girl or the mother or sister of such a girl." *Id.* at 976–77.

31. No. CV-83-C-5021-NE (N.D. Ala. filed Jan. 11, 1983). The case ultimately settled before trial. *See* Francis E. McGovern & E. Allan Lind, *The Discovery Survey*, Law & Contemp. Probs., Autumn 1988, at 41.

in person by neutral third parties, thus replacing interrogatories and depositions. It resulted in substantial savings in both time and cost.

B. *Surveys Used to Help Assess Expert Acceptance in the Wake of* Daubert

Scientists who offer expert testimony at trial typically present their own opinions. These opinions may or may not be representative of the opinions of the scientific community at large. In deciding whether to admit such testimony, courts applying the *Frye* test must determine whether the science being offered is generally accepted by the relevant scientific community. Under *Daubert* as well, a relevant factor used to decide admissibility is the extent to which the theory or technique has received widespread acceptance. Properly conducted surveys can provide a useful way to gauge acceptance, and courts recently have been offered assistance from surveys that allegedly gauge relevant scientific opinion. As with any scientific research, the usefulness of the information obtained from a survey depends on the quality of research design. Several critical factors have emerged that have limited the value of some of these surveys: problems in defining the relevant target population and identifying an appropriate sampling frame, response rates that raise questions about the representativeness of the results, and a failure to ask questions that assess opinions on the relevant issue.

Courts deciding on the admissibility of polygraph tests have considered results from several surveys of purported experts. Surveys offered as providing evidence of relevant scientific opinion have tested respondents from several populations: (1) professional polygraph examiners,[32] (2) psychophysiologists (members of the Society for Psychophysiological Research),[33] and (3) distinguished psychologists (Fellows of the Division of General Psychology of the American Psychological Association).[34] Respondents in the first group expressed substantial confidence in the scientific accuracy of polygraph testing, and those in the third group expressed substantial doubts about it. Respondents in the second group were asked the same question across three surveys that differed in other aspects of their methodology (e.g., when testing occurred and what the response rate was). Although over 60% of those questioned in two of the three surveys characterized the polygraph as a useful diagnostic tool, one of the surveys was conducted in 1982 and the more recent survey, published in 1984, achieved only a 30% response rate. The third

32. See plaintiff's survey described in Meyers v. Arcudi, 947 F. Supp. 581, 588 (D. Conn. 1996).

33. Susan L. Amato & Charles R. Honts, *What Do Psychophysiologists Think About Polygraph Tests? A Survey of the Membership of SPR*, 31 Psychophysiology S22 [abstract]; Gallup Organization, *Survey of Members of the Society for Psychological Research Concerning Their Opinions of Polygraph Test Interpretation*, 13 Polygraph 153 (1984); William G. Iacono & David T. Lykken, *The Validity of the Lie Detector: Two Surveys of Scientific Opinion*, 82 J. Applied Psychol. 426 (1997).

34. Iacono & Lykken, *supra* note 33.

survey, also conducted in 1984, achieved a response rate of 90% and found that only 44% of respondents viewed the polygraph as a useful diagnostic tool. On the basis of these inconsistent reactions from the several surveys, courts have determined that the polygraph has failed to achieve general acceptance in the scientific community.[35] In addition, however, courts have criticized the relevance of the population surveyed by proponents of the polygraph. For example, in *Meyers v. Arcudi* the court noted that the survey offered by proponents of the polygraph was a survey of "practitioners who estimated the accuracy of the control question technique [of polygraph testing] to be between 86% and 100%."[36] The court rejected the conclusions from this survey on the basis of a determination that the population surveyed was not the relevant scientific community, noting that "many of them . . . do not even possess advanced degrees and are not trained in the scientific method."[37]

The link between specialized expertise and self-interest poses a dilemma in defining the relevant scientific population. As the court in *United States v. Orians* recognized, "The acceptance in the scientific community depends in large part on how the relevant scientific community is defined."[38] In rejecting the defendants' urging that the court consider as relevant only psychophysiologists whose work is dedicated in large part to polygraph research, the court noted that *Daubert* "does not require the court to limit its inquiry to those individuals that base their livelihood on the acceptance of the relevant scientific theory. These individuals are often too close to the science and have a stake in its acceptance; i.e., their livelihood depends in part on the acceptance of the method."[39]

To be relevant to a *Frye* or *Daubert* inquiry on general acceptance, the questions asked in a survey of experts should assess opinions on the quality of the scientific theory and methodology, rather than asking whether or not the instrument should be used in a legal setting. Thus, a survey in which 60% of respondents agreed that the polygraph is "a useful diagnostic tool when considered with other available information," 1% viewed it as sufficiently reliable to be the sole determinant, and the remainder thought it entitled to little or no weight, failed to assess the relevant issue. As the court in *United States v. Cordoba* noted, because "useful" and "other available information" could have many meanings, "there is little wonder why [the response chosen by the majority of respondents] was most frequently selected."[40]

35. United States v. Scheffer, 523 U.S. 303, 309 (1998); United States v. Bishop, 64 F. Supp. 2d 1149 (D. Utah 1999); Meyers v. Arcudi, 947 F. Supp. 581, 588 (D. Conn. 1996); United States v. Varoudakis, 48 Fed. R. Evid. Serv. 1187 (D. Mass. 1998).

36. *Meyers v. Arcudi,* 947 F. Supp. at 588.

37. *Id.*

38. 9 F. Supp. 2d 1168, 1173 (D. Ariz. 1998).

39. *Id.*

40. 991 F. Supp. 1199 (C.D. Cal. 1998), *aff'd*, 194 F.3d 1053 (9th Cir. 1999).

A similar flaw occurred in a survey conducted by experts opposed to the use of the polygraph in trial proceedings. Survey respondents were asked whether they would advocate that courts admit into evidence the outcome of a polygraph test.[41] That question calls for more than an assessment of the accuracy of the polygraph, and thus does not appropriately limit expert opinion to issues within the expert's competence, that is, to the accuracy of the information provided by the test results. The survey also asked whether respondents agreed that the control question technique, the most common form of polygraph test, is accurate at least 85% of the time in real-life applications for guilty and innocent subjects.[42] Although polygraph proponents frequently claim an accuracy level of 85%, it is up to the courts to decide what accuracy level would be required to justify admissibility. A better approach would be to ask survey respondents to estimate the level of accuracy they believe the test is likely to produce.[43]

Surveys of experts are no substitute for an evaluation of whether the testimony an expert witness is offering will assist the trier of fact. Nonetheless, courts can use an assessment of opinion in the relevant scientific community to aid in determining whether a particular expert is proposing to use methods that would be rejected by a representative group of experts to arrive at the opinion the expert will offer. Properly conducted surveys can provide an economical way to collect and present information on scientific consensus and dissensus.

C. *Surveys Used to Help Assess Community Standards:* Atkins v. Virginia

In *Atkins v. Virginia*,[44] the U.S. Supreme Court determined that the Eighth Amendment's prohibition of "cruel and unusual punishment" forbids the execution of mentally retarded persons.[45] Following the interpretation advanced in *Trop v. Dulles*[46] that "The Amendment must draw its meaning from the evolving standards of decency that mark the progress of a maturing society,"[47] the Court examined a variety of sources, including legislative judgments and public opinion polls, to find that a national consensus had developed barring such executions.[48]

41. *See* Iacono & Lykken, *supra* note 33, at 430, tbl. 2 (1997).

42. *Id.*

43. At least two assessments should be made: an estimate of the accuracy for guilty subjects and an estimate of the accuracy for innocent subjects.

44. 536 U.S. 304, 322 (2002).

45. Although some groups have recently moved away from the term "mental retardation" in response to concerns that the term may have pejorative connotations, mental retardation was the name used for the condition at issue in *Atkins* and it continues to be employed in federal laws, in cases determining eligibility for the death penalty, and as a diagnosis by the medical profession.

46. 356 U.S. 86 (1958).

47. *Id.* at 101.

48. *Atkins,* 536 U.S. at 313–16.

In a vigorous dissent, Chief Justice Rehnquist objected to the use of the polls, arguing that legislative judgments and jury decisions should be the sole indicators of national opinion. He also objected to the particular polls cited in the majority opinion, identifying what he viewed as serious methodological weaknesses.

The Court has struggled since *Furman v. Georgia*[49] to develop an adequate way to measure public standards regarding the application of the death penalty to specific categories of cases. In relying primarily on surveys of state legislative actions, the Court has ignored the forces that influence whether an issue emerges on a legislative agenda, and the strong influence of powerful minorities on legislative actions.[50] Moreover, the various members of the Court have disagreed about whether states without any death penalty should be included in the count of states that bar the execution of a particular category of defendant.

The Court has sometimes considered jury verdicts in assessing public standards. In *Coker v. Georgia*,[51] the Court forbade the imposition of the death penalty for rape. Citing *Gregg v. Georgia*[52] for the proposition that "[t]he jury . . . is a significant and reliable objective index of contemporary values because it is so directly involved," the Court noted that "in the vast majority of cases [of rape in Georgia], at least 9 out of 10, juries have not imposed the death sentence."[53] In *Atkins*, Chief Justice Rehnquist complained about the absence of jury verdict data.[54] Had such data been available, however, they would have been irrelevant because a "survey" of the jurors who have served in such cases would constitute a biased sample of the public. A potential juror unwilling to impose the death penalty on a mentally retarded person would have been ineligible to serve in a capital case involving a mentally retarded defendant because the juror would not have been able to promise during voir dire that he or she would be willing to listen to the evidence and impose the death penalty if the evidence warranted it. Thus, the death-qualified jury in such a case would be composed only of representatives from that subset of citizens willing to execute a mentally retarded defendant, an unrepresentative and systematically biased sample.

Public opinion surveys can provide an important supplementary source of information about contemporary values.[55] The Court in *Atkins* was presented with data from 27 different polls and surveys,[56] 8 of them national and 19 statewide.

49. 408 U.S. 238 (1972).

50. *See* Stanford v. Kentucky, 492 U.S. 361 (1989), *abrogated by* Roper v. Simmons, 543 U.S. 551 (2005).

51. 433 U.S. 584, 596 (1977).

52. 428 U.S. 153, 181 (1976).

53. *Coker v. Georgia*, 433 U.S. at 596.

54. *See Atkins,* 536 U.S. at 323 (Rehnquist, C.J., dissenting).

55. *See id.* at 316 n.21 ("[T]heir consistency with the legislative evidence lends further support to our conclusion that there is a consensus").

56. The quality of any poll or survey depends on the methodology used, which should be fully visible to the court and the opposing party. *See* Section VII, *infra*.

The information on the polling data appeared in an amicus brief filed by the American Association on Mental Retardation.[57] Respondents were asked in various ways how they felt about imposing the death penalty on a mentally retarded defendant. In each poll, a majority of respondents expressed opposition to executing the mentally retarded. Chief Justice Rehnquist noted two weaknesses reflected in the data presented to the Court. First, almost no information was provided about the target populations from which the samples were drawn or the methodology of sample selection and data collection. Although further information was available on at least some of the surveys (e.g., the nationwide telephone survey of 1000 voters conducted in 1993 by the Tarrance Group used a sample based on voter turnout in the last three presidential elections), that information apparently was not part of the court record. This omission violates accepted reporting standards in survey research, and the information is needed if the decisionmaker is to intelligently evaluate the quality of the survey. Its absence in this instance occurred because the survey information was obtained from secondary sources.

A second objection raised by Chief Justice Rehnquist was that the wording of some of the questions required respondents to say merely whether they favored or were opposed to the use of the death penalty when the defendant is mentally retarded. It is unclear how a respondent who favors execution of a mentally retarded defendant only in a rare case would respond to that question. Some of the questions, however, did ask whether the respondent felt that it was never appropriate to execute the mentally retarded or whether it was appropriate in some circumstances.[58] In responses to these questions as well, a majority of respondents said that they found the execution of mentally retarded persons unacceptable under any circumstances. The critical point is that despite variations in wording of questions, the year in which the poll was conducted, who conducted it, where it was conducted, and how it was carried out, a majority of respondents (between 56% and 83%) expressed opposition to executing mentally retarded defendants. The Court thus was presented with a consistent set of findings, providing striking reinforcement for the *Atkins* majority's legislative analysis. Opinion poll data and legislative decisions have different strengths and weaknesses as indicators of contemporary values. The value of a multiple-measure approach is that it avoids a potentially misleading reliance on a single source or measure.

57. The data appear as an appendix to the Opinion of Chief Justice Rehnquist in *Atkins*.

58. Appendix to the Opinion of Chief Justice Rehnquist in *Atkins*. "Some people feel that there is nothing wrong with imposing the death penalty on persons who are mentally retarded, depending on the circumstances. Others feel that the death penalty should never be imposed on persons who are mentally retarded under any circumstances. Which of these views comes closest to your own?" The Tarrance Group, Death Penalty Poll, Q. 9 (Mar. 1993), citing Samuel R. Gross, *Update: American Public Opinion on the Death Penalty—It's Getting Personal*, 83 Cornell L. Rev. 1448, 1467 (1998).

D. A Comparison of Survey Evidence and Individual Testimony

To illustrate the value of a survey, it is useful to compare the information that can be obtained from a competently done survey with the information obtained by other means. A survey is presented by a survey expert who testifies about the responses of a substantial number of individuals who have been selected according to an explicit sampling plan and asked the same set of questions by interviewers who were not told who sponsored the survey or what answers were predicted or preferred. Although parties presumably are not obliged to present a survey conducted in anticipation of litigation by a nontestifying expert if it produced unfavorable results,[59] the court can and should scrutinize the method of respondent selection for any survey that is presented.

A party using a nonsurvey method generally identifies several witnesses who testify about their own characteristics, experiences, or impressions. Although the party has no obligation to select these witnesses in any particular way or to report on how they were chosen, the party is not likely to select witnesses whose attributes conflict with the party's interests. The witnesses who testify are aware of the parties involved in the case and have discussed the case before testifying.

Although surveys are not the only means of demonstrating particular facts, presenting the results of a well-done survey through the testimony of an expert is an efficient way to inform the trier of fact about a large and representative group of potential witnesses. In some cases, courts have described surveys as the most direct form of evidence that can be offered.[60] Indeed, several courts have drawn negative inferences from the absence of a survey, taking the position that failure to undertake a survey may strongly suggest that a properly done survey would not support the plaintiff's position.[61]

59. *In re* FedEx Ground Package System, 2007 U.S. Dist. LEXIS 27086 (N.D. Ind. April 10, 2007); Loctite Corp. v. National Starch & Chem. Corp., 516 F. Supp. 190, 205 (S.D.N.Y. 1981) (distinguishing between surveys conducted in anticipation of litigation and surveys conducted for non-litigation purposes which cannot be reproduced because of the passage of time, concluding that parties should not be compelled to introduce the former at trial, but may be required to provide the latter).

60. *See, e.g.*, Morrison Entm't Group v. Nintendo of Am., 56 Fed. App'x. 782, 785 (9th Cir. Cal. 2003).

61. Ortho Pharm. Corp. v. Cosprophar, Inc., 32 F.3d 690, 695 (2d Cir. 1994); Henri's Food Prods. Co. v. Kraft, Inc., 717 F.2d 352, 357 (7th Cir. 1983); Medici Classics Productions LLC v. Medici Group LLC, 590 F. Supp. 2d 548, 556 (S.D.N.Y. 2008); Citigroup v. City Holding Co., 2003 U.S. Dist. LEXIS 1845 (S.D.N.Y. Feb. 10, 2003); Chum Ltd. v. Lisowski, 198 F. Supp. 2d 530 (S.D.N.Y. 2002).

II. Purpose and Design of the Survey

A. Was the Survey Designed to Address Relevant Questions?

The report describing the results of a survey should include a statement describing the purpose or purposes of the survey. One indication that a survey offers probative evidence is that it was designed to collect information relevant to the legal controversy (e.g., to estimate damages in an antitrust suit or to assess consumer confusion in a trademark case). Surveys not conducted specifically in preparation for, or in response to, litigation may provide important information,[62] but they frequently ask irrelevant questions[63] or select inappropriate samples of respondents for study.[64] Nonetheless, surveys do not always achieve their stated goals. Thus, the content and execution of a survey must be scrutinized whether or not the survey was designed to provide relevant data on the issue before the court.[65] Moreover, if a survey was not designed for purposes of litigation, one source of bias is less likely: The party presenting the survey is less likely to have designed and constructed the survey to provide evidence supporting its side of the issue in controversy.

62. *See, e.g.,* Wright v. Jeep Corp., 547 F. Supp. 871, 874 (E.D. Mich. 1982). Indeed, as courts increasingly have been faced with scientific issues, parties have requested in a number of recent cases that the courts compel production of research data and testimony by unretained experts. The circumstances under which an unretained expert can be compelled to testify or to disclose research data and opinions, as well as the extent of disclosure that can be required when the research conducted by the expert has a bearing on the issues in the case, are the subject of considerable current debate. *See, e.g.,* Joe S. Cecil, *Judicially Compelled Disclosure of Research Data,* 1 Cts. Health Sci. & L. 434 (1991); Richard L. Marcus, *Discovery Along the Litigation/Science Interface,* 57 Brook. L. Rev. 381, 393–428 (1991); *see also Court-Ordered Disclosure of Academic Research: A Clash of Values of Science and Law,* Law & Contemp. Probs., Summer 1996, at 1.

63. *See* Loctite Corp. v. National Starch & Chem. Corp., 516 F. Supp. 190, 206 (S.D.N.Y. 1981) (marketing surveys conducted before litigation were designed to test for brand awareness, while the "single issue at hand . . . [was] whether consumers understood the term 'Super Glue' to designate glue from a single source").

64. In *Craig v. Boren,* 429 U.S. 190 (1976), the state unsuccessfully attempted to use its annual roadside survey of the blood alcohol level, drinking habits, and preferences of drivers to justify prohibiting the sale of 3.2% beer to males under the age of 21 and to females under the age of 18. The data were biased because it was likely that the male would be driving if both the male and female occupants of the car had been drinking. As pointed out in 2 Joseph L. Gastwirth, Statistical Reasoning in Law and Public Policy: Tort Law, Evidence, and Health 527 (1988), the roadside survey would have provided more relevant data if all occupants of the cars had been included in the survey (and if the type and amount of alcohol most recently consumed had been requested so that the consumption of 3.2% beer could have been isolated).

65. *See* Merisant Co. v. McNeil Nutritionals, LLC, 242 F.R.D. 315 (E.D. Pa. 2007).

B. *Was Participation in the Design, Administration, and Interpretation of the Survey Appropriately Controlled to Ensure the Objectivity of the Survey?*

An early handbook for judges recommended that survey interviews be "conducted independently of the attorneys in the case."[66] Some courts interpreted this to mean that any evidence of attorney participation is objectionable.[67] A better interpretation is that the attorney should have no part in carrying out the survey.[68] However, some attorney involvement in the survey design is necessary to ensure that relevant questions are directed to a relevant population.[69] The 2009 amendments to Federal Rule of Civil Procedure 26(a)(2)[70] no longer allow an inquiry into the nature of communications between attorneys and experts, and so the role of attorneys in constructing surveys may become less apparent. The key issues for the trier of fact concerning the design of the survey are the objectivity and relevance of the questions on the survey and the appropriateness of the definition of the population used to guide sample selection. These aspects of the survey are visible to the trier of fact and can be judged on their quality, irrespective of who suggested them. In contrast, the interviews themselves are not directly visible, and any potential bias is minimized by having interviewers and respondents blind to the purpose and sponsorship of the survey and by excluding attorneys from any part in conducting interviews and tabulating results.[71]

66. Judicial Conference of the United States, Handbook of Recommended Procedures for the Trial of Protracted Cases 75 (1960).

67. *See, e.g.*, Boehringer Ingelheim G.m.b.H. v. Pharmadyne Lab., 532 F. Supp. 1040, 1058 (D.N.J. 1980).

68. Upjohn Co. v. American Home Prods. Corp., No. 1-95-CV-237, 1996 U.S. Dist. LEXIS 8049, at *42 (W.D. Mich. Apr. 5, 1996) (objection that "counsel reviewed the design of the survey carries little force with this Court because [opposing party] has not identified any flaw in the survey that might be attributed to counsel's assistance"). For cases in which attorney participation was linked to significant flaws in the survey design, see Johnson v. Big Lots Stores, Inc., No. 04-321, 2008 U.S. Dist. LEXIS 35316, at *20 (E.D. La. April 29, 2008); United States v. Southern Indiana Gas & Elec. Co., 258 F. Supp. 2d 884, 894 (S.D. Ind. 2003); Gibson v. County of Riverside, 181 F. Supp. 2d 1057, 1069 (C.D. Cal. 2002).

69. *See* 6 J. Thomas McCarthy, McCarthy on Trademarks and Unfair Competition § 32:166 (4th ed. 2003).

70. www.uscourts.gov/News/TheThirdBranch/10-11-01/Rules_Recommendations_Take_Effect_December_1_2010.aspx.

71. *Gibson*, 181 F. Supp. 2d at 1068.

C. *Are the Experts Who Designed, Conducted, or Analyzed the Survey Appropriately Skilled and Experienced?*

Experts prepared to design, conduct, and analyze a survey generally should have graduate training in psychology (especially social, cognitive, or consumer psychology), sociology, political science, marketing, communication sciences, statistics, or a related discipline; that training should include courses in survey research methods, sampling, measurement, interviewing, and statistics. In some cases, professional experience in teaching or conducting and publishing survey research may provide the requisite background. In all cases, the expert must demonstrate an understanding of foundational, current, and best practices in survey methodology, including sampling,[72] instrument design (questionnaire and interview construction), and statistical analysis.[73] Publication in peer-reviewed journals, authored books, fellowship status in professional organizations, faculty appointments, consulting experience, research grants, and membership on scientific advisory panels for government agencies or private foundations are indications of a professional's area and level of expertise. In addition, some surveys involving highly technical subject matter (e.g., the particular preferences of electrical engineers for various pieces of electrical equipment and the bases for those preferences) or special populations (e.g., developmentally disabled adults with limited cognitive skills) may require experts to have some further specialized knowledge. Under these conditions, the survey expert also should be able to demonstrate sufficient familiarity with the topic or population (or assistance from an individual on the research team with suitable expertise) to design a survey instrument that will communicate clearly with relevant respondents.

D. *Are the Experts Who Will Testify About Surveys Conducted by Others Appropriately Skilled and Experienced?*

Parties often call on an expert to testify about a survey conducted by someone else. The secondary expert's role is to offer support for a survey commissioned by the party who calls the expert, to critique a survey presented by the opposing party, or to introduce findings or conclusions from a survey not conducted in preparation for litigation or by any of the parties to the litigation. The trial court should take into account the exact issue that the expert seeks to testify about and the nature of the expert's field of expertise.[74] The secondary expert who gives an opinion

72. The one exception is that sampling expertise would be unnecessary if the survey were administered to all members of the relevant population. *See, e.g.*, McGovern & Lind, *supra* note 31.

73. If survey expertise is being provided by several experts, a single expert may have general familiarity but not special expertise in all these areas.

74. *See* Margaret A. Berger, The Admissibility of Expert Testimony, Section III.A, in this manual.

about the adequacy and interpretation of a survey not only should have general skills and experience with surveys and be familiar with all of the issues addressed in this reference guide, but also should demonstrate familiarity with the following properties of the survey being discussed:

1. Purpose of the survey;
2. Survey methodology,[75] including
   a. the target population,
   b. the sampling design used in conducting the survey,
   c. the survey instrument (questionnaire or interview schedule), and
   d. (for interview surveys) interviewer training and instruction;
3. Results, including rates and patterns of missing data; and
4. Statistical analyses used to interpret the results.

III. Population Definition and Sampling

A. *Was an Appropriate Universe or Population Identified?*

One of the first steps in designing a survey or in deciding whether an existing survey is relevant is to identify the target population (or universe).[76] The target population consists of all elements (i.e., individuals or other units) whose characteristics or perceptions the survey is intended to represent. Thus, in trademark litigation, the relevant population in some disputes may include all prospective and past purchasers of the plaintiff's goods or services and all prospective and past purchasers of the defendant's goods or services. Similarly, the population for a discovery survey may include all potential plaintiffs or all employees who worked for Company A between two specific dates. In a community survey designed to provide evidence for a motion for a change of venue, the relevant population consists of all jury-eligible citizens in the community in which the trial is to take place.[77]

75. *See* A & M Records, Inc. v. Napster, Inc., 2000 U.S. Dist. LEXIS 20668 (N.D. Cal. Aug. 10, 2000) (holding that expert could not attest credibly that the surveys upon which he relied conformed to accepted survey principles because of his minimal role in overseeing the administration of the survey and limited expert report).

76. Identification of the proper target population or universe is recognized uniformly as a key element in the development of a survey. *See, e.g.*, Judicial Conference of the U.S., *supra* note 66; MCL 4th, *supra* note 16, § 11.493; *see also* 3 McCarthy, *supra* note 69, § 32:166; Council of Am. Survey Res. Orgs., Code of Standards and Ethics for Survey Research § III.A.3 (2010).

77. A second relevant population may consist of jury-eligible citizens in the community where the party would like to see the trial moved. By questioning citizens in both communities, the survey can test whether moving the trial is likely to reduce the level of animosity toward the party requesting the change of venue. *See* United States v. Haldeman, 559 F.2d 31, 140, 151, app. A at 176–79 (D.C. Cir. 1976) (court denied change of venue over the strong objection of Judge MacKinnon, who cited survey evidence that Washington, D.C., residents were substantially more likely to conclude, before

The definition of the relevant population is crucial because there may be systematic differences in the responses of members of the population and nonmembers. For example, consumers who are prospective purchasers may know more about the product category than consumers who are not considering making a purchase.

The universe must be defined carefully. For example, a commercial for a toy or breakfast cereal may be aimed at children, who in turn influence their parents' purchases. If a survey assessing the commercial's tendency to mislead were conducted based on a sample from the target population of prospective and actual adult purchasers, it would exclude a crucial relevant population. The appropriate population in this instance would include children as well as parents.[78]

B. Did the Sampling Frame Approximate the Population?

The target population consists of all the individuals or units that the researcher would like to study. The sampling frame is the source (or sources) from which the sample actually is drawn. The surveyor's job generally is easier if a complete list of every eligible member of the population is available (e.g., all plaintiffs in a discovery survey), so that the sampling frame lists the identity of all members of the target population. Frequently, however, the target population includes members who are inaccessible or who cannot be identified in advance. As a result, reasonable compromises are sometimes required in developing the sampling frame. The survey report should contain (1) a description of the target population, (2) a description of the sampling frame from which the sample is to be drawn, (3) a discussion of the difference between the target population and the sampling frame, and, importantly, (4) an evaluation of the likely consequences of that difference.

A survey that provides information about a wholly irrelevant population is itself irrelevant.[79] Courts are likely to exclude the survey or accord it little

trial, that the defendants were guilty); *see also* People v. Venegas, 31 Cal. Rptr. 2d 114, 117 (Cal. Ct. App. 1994) (change of venue denied because defendant failed to show that the defendant would face a less hostile jury in a different court).

78. *See, e.g.*, Warner Bros., Inc. v. Gay Toys, Inc., 658 F.2d 76 (2d Cir. 1981) (surveying children users of the product rather than parent purchasers). Children and some other populations create special challenges for researchers. For example, very young children should not be asked about sponsorship or licensing, concepts that are foreign to them. Concepts, as well as wording, should be age appropriate.

79. A survey aimed at assessing how persons in the trade respond to an advertisement should be conducted on a sample of persons in the trade and not on a sample of consumers. *See* Home Box Office v. Showtime/The Movie Channel, 665 F. Supp. 1079, 1083 (S.D.N.Y.), *aff'd in part and vacated in part*, 832 F.2d 1311 (2d Cir. 1987); J & J Snack Food Corp. v. Earthgrains Co., 220 F. Supp. 2d 358, 371–72 (N.J. 2002). *But see* Lon Tai Shing Co. v. Koch + Lowy, No. 90-C4464, 1990 U.S. Dist. LEXIS 19123, at *50 (S.D.N.Y. Dec. 14, 1990), in which the judge was willing to find likelihood of consumer confusion from a survey of lighting store salespersons questioned by a survey researcher posing as a customer. The court was persuaded that the salespersons who were misstating the source

weight.[80] Thus, when the plaintiff submitted the results of a survey to prove that the green color of its fishing rod had acquired a secondary meaning, the court gave the survey little weight in part because the survey solicited the views of fishing rod dealers rather than consumers.[81] More commonly, however, the sampling frame and the target population have some overlap, but the overlap is imperfect: The sampling frame excludes part of the target population, that is, it is underinclusive, or the sampling frame includes individuals who are not members of the target population, that is, it is overinclusive relative to the target population. Coverage error is the term used to describe inconsistencies between a sampling frame and a target population. If the coverage is underinclusive, the survey's value depends on the proportion of the target population that has been excluded from the sampling frame and the extent to which the excluded population is likely to respond differently from the included population. Thus, a survey of spectators and participants at running events would be sampling a sophisticated subset of those likely to purchase running shoes. Because this subset probably would consist of the consumers most knowledgeable about the trade dress used by companies that sell running shoes, a survey based on this sampling frame would be likely to substantially overrepresent the strength of a particular design as a trademark, and the extent of that overrepresentation would be unknown and not susceptible to any reasonable estimation.[82]

Similarly, in a survey designed to project demand for cellular phones, the assumption that businesses would be the primary users of cellular service led surveyors to exclude potential nonbusiness users from the survey. The Federal Communications Commission (FCC) found the assumption unwarranted and concluded that the research was flawed, in part because of this underinclusive coverage.[83] With the growth in individual cell phone use over time, noncoverage error would be an even greater problem for this survey today.

of the lamp, whether consciously or not, must have believed reasonably that the consuming public would be likely to rely on the salespersons' inaccurate statements about the name of the company that manufactured the lamp they were selling.

80. *See* Wells Fargo & Co. v. WhenU.com, Inc., 293 F. Supp. 2d 734 (E.D. Mich. 2003).

81. *See* R.L. Winston Rod Co. v. Sage Mfg. Co., 838 F. Supp. 1396, 1401–02 (D. Mont. 1993).

82. *See* Brooks Shoe Mfg. Co. v. Suave Shoe Corp., 533 F. Supp. 75, 80 (S.D. Fla. 1981), *aff'd*, 716 F.2d 854 (11th Cir. 1983); *see also* Hodgdon Power Co. v. Alliant Techsystems, Inc., 512 F. Supp. 2d 1178 (D. Kan. 2007) (excluding survey on gunpowder brands distributed at plaintiff's promotional booth at a shooting tournament); Winning Ways, Inc. v. Holloway Sportswear, Inc., 913 F. Supp. 1454, 1467 (D. Kan. 1996) (survey flawed in failing to include sporting goods customers who constituted a major portion of customers). *But see* Thomas & Betts Corp. v. Panduit Corp., 138 F.3d 277, 294–95 (7th Cir. 1998) (survey of store personnel admissible because relevant market included both distributors and ultimate purchasers).

83. *See* Gencom, Inc., 56 Rad. Reg. 2d (P&F) 1597, 1604 (1984). This position was affirmed on appeal. *See* Gencom, Inc. v. FCC, 832 F.2d 171, 186 (D.C. Cir. 1987); *see also* Beacon Mut. Ins. Co. v. Onebeacon Ins. Corp, 376 F. Supp. 2d 251, 261 (D.R.I. 2005) (sample included only defendant's insurance agents and lack of confusion among those agents was "nonstartling").

In some cases, it is difficult to determine whether a sampling frame that omits some members of the population distorts the results of the survey and, if so, the extent and likely direction of the bias. For example, a trademark survey was designed to test the likelihood of confusing an analgesic currently on the market with a new product that was similar in appearance.[84] The plaintiff's survey included only respondents who had used the plaintiff's analgesic, and the court found that the target population should have included users of other analgesics, "so that the full range of potential customers for whom plaintiff and defendants would compete could be studied."[85] In this instance, it is unclear whether users of the plaintiff's product would be more or less likely to be confused than users of the defendants' product or users of a third analgesic.[86]

An overinclusive sampling frame generally presents less of a problem for interpretation than does an underinclusive sampling frame.[87] If the survey expert can demonstrate that a sufficiently large (and representative) subset of respondents in the survey was drawn from the appropriate sampling frame, the responses obtained from that subset can be examined, and inferences about the relevant population can be drawn based on that subset.[88] If the relevant subset cannot be identified, however, an overbroad sampling frame will reduce the value of the survey.[89] If the sampling frame does not include important groups in the target population, there is generally no way to know how the unrepresented members of the target population would have responded.[90]

84. *See* American Home Prods. Corp. v. Barr Lab., Inc., 656 F. Supp. 1058 (D.N.J.), *aff'd*, 834 F.2d 368 (3d Cir. 1987).

85. *Id.* at 1070.

86. *See also* Craig v. Boren, 429 U.S. 190 (1976).

87. *See* Schwab v. Philip Morris USA, Inc. 449 F. Supp. 2d 992, 1134–35 (E.D.N.Y. 2006) ("Studies evaluating broadly the beliefs of low tar smokers generally are relevant to the beliefs of "light" smokers more specifically.").

88. *See* National Football League Props. Inc. v. Wichita Falls Sportswear, Inc. 532 F. Supp. 651, 657–58 (W.D. Wash. 1982).

89. *See* Leelanau Wine Cellars, Ltd. v. Black & Red, Inc., 502 F.3d 504, 518 (6th Cir. 2007) (lower court was correct in giving little weight to survey with overbroad universe); Big Dog Motorcycles, L.L.C. v. Big Dog Holdings, Inc., 402 F. Supp. 2d 1312, 1334 (D. Kan. 2005) (universe composed of prospective purchasers of all t-shirts and caps overinclusive for evaluating reactions of buyers likely to purchase merchandise at motorcycle dealerships). *See also* Schieffelin & Co. v. Jack Co. of Boca, 850 F. Supp. 232, 246 (S.D.N.Y. 1994).

90. *See, e.g.*, Amstar Corp. v. Domino's Pizza, Inc., 615 F.2d 252, 263–64 (5th Cir. 1980) (court found both plaintiff's and defendant's surveys substantially defective for a systematic failure to include parts of the relevant population); Scott Fetzer Co. v. House of Vacuums, Inc., 381 F.3d 477 (5th Cir. 2004) (universe drawn from plaintiff's customer list underinclusive and likely to differ in their familiarity with plaintiff's marketing and distribution techniques).

C. *Does the Sample Approximate the Relevant Characteristics of the Population?*

Identification of a survey population must be followed by selection of a sample that accurately represents that population.[91] The use of probability sampling techniques maximizes both the representativeness of the survey results and the ability to assess the accuracy of estimates obtained from the survey.

Probability samples range from simple random samples to complex multistage sampling designs that use stratification, clustering of population elements into various groupings, or both. In all forms of probability sampling, each element in the relevant population has a known, nonzero probability of being included in the sample.[92] In simple random sampling, the most basic type of probability sampling, every element in the population has a known, equal probability of being included in the sample, and all possible samples of a given size are equally likely to be selected.[93] Other probability sampling techniques include (1) stratified random sampling, in which the researcher subdivides the population into mutually exclusive and exhaustive subpopulations, or strata, and then randomly selects samples from within these strata; and (2) cluster sampling, in which elements are sampled in groups or clusters, rather than on an individual basis.[94] Note that selection probabilities do not need to be the same for all population elements; however, if the probabilities are unequal, compensatory adjustments should be made in the analysis.

Probability sampling offers two important advantages over other types of sampling. First, the sample can provide an unbiased estimate that summarizes the responses of all persons in the population from which the sample was drawn; that is, the expected value of the sample estimate is the population value being estimated. Second, the researcher can calculate a confidence interval that describes explicitly how reliable the sample estimate of the population is likely to be. If the sample is unbiased, the difference between the estimate and the exact value is called the sampling error.[95] Thus, suppose a survey collected responses from a simple random sample of 400 dentists selected from the population of all dentists

91. MCL 4th, *supra* note 16, § 11.493. *See also* David H. Kaye & David A. Freedman, Reference Guide on Statistics, Section II.B, in this manual.

92. The exception is that population elements omitted from the sampling frame have a zero probability of being sampled.

93. Systematic sampling, in which every *n*th unit in the population is sampled and the starting point is selected randomly, fulfills the first of these conditions. It does not fulfill the second, because no systematic sample can include elements adjacent to one another on the list of population members from which the sample is drawn. Except in unusual situations when periodicities occur, systematic samples and simple random samples generally produce the same results. Thomas Plazza, *Fundamentals of Applied Sampling*, *in* Handbook of Survey Research, *supra* note 1, at 139, 145.

94. *Id.* at 139, 150–63.

95. *See* David H. Kaye & David A. Freedman, *supra* note 91, Glossary, for a definition of sampling error.

licensed to practice in the United States and found that 80, or 20%, of them mistakenly believed that a new toothpaste, Goldgate, was manufactured by the makers of Colgate. A survey expert could properly compute a confidence interval around the 20% estimate obtained from this sample. If the survey were repeated a large number of times, and a 95% confidence interval was computed each time, 95% of the confidence intervals would include the actual percentage of dentists in the entire population who would believe that Goldgate was manufactured by the makers of Colgate.[96] In this example, the margin of error is ±4%, and so the confidence interval is the range between 16% and 24%, that is, the estimate (20%) plus or minus 4%.

All sample surveys produce estimates of population values, not exact measures of those values. Strictly speaking, the margin of error associated with the sample estimate assumes probability sampling. Assuming a probability sample, a confidence interval describes how stable the mean response in the sample is likely to be. The width of the confidence interval depends on three primary characteristics:

1. Size of the sample (the larger the sample, the narrower the interval);
2. Variability of the response being measured; and
3. Confidence level the researcher wants to have.[97]

Traditionally, scientists adopt the 95% level of confidence, which means that if 100 samples of the same size were drawn, the confidence interval expected for at least 95 of the samples would be expected to include the true population value.[98]

Stratified probability sampling can be used to obtain more precise response estimates by using what is known about characteristics of the population that are likely to be associated with the response being measured. Suppose, for example, we anticipated that more-experienced and less-experienced dentists might respond differently to Goldgate toothpaste, and we had information on the year in which each dentist in the population began practicing. By dividing the population of dentists into more- and less-experienced strata (e.g., in practice 15 years or more versus in practice less than 15 years) and then randomly sampling within experience stratum, we would be able to ensure that the sample contained precisely

96. Actually, because survey interviewers would be unable to locate some dentists and some dentists would be unwilling to participate in the survey, technically the population to which this sample would be projectable would be all dentists with current addresses who would be willing to participate in the survey if they were asked. The expert should be prepared to discuss possible sources of bias due to, for example, an address list that is not current.

97. When the sample design does not use a simple random sample, the confidence interval will be affected.

98. To increase the likelihood that the confidence interval contains the actual population value (e.g., from 95% to 99%) without increasing the sample size, the width of the confidence interval can be expanded. An increase in the confidence interval brings an increase in the confidence level. For further discussion of confidence intervals, see David H. Kaye & David A. Freedman, Reference Guide on Statistics, Section IV.A, in this manual.

proportionate representation from each stratum, in this case, more- and less-experienced dentists. That is, if 60% of dentists were in practice 15 years or more, we could select 60% of the sample from the more-experienced stratum and 40% from the less-experienced stratum and be sure that the sample would have proportionate representation from each stratum, reducing the likely sampling error.[99]

In proportionate stratified probability sampling, as in simple random sampling, each individual member of the population has an equal chance of being selected. Stratified probability sampling can also disproportionately sample from different strata, a procedure that will produce more precise estimates if some strata are more heterogeneous than others on the measure of interest.[100] Disproportionate sampling may also used to enable the survey to provide separate estimates for particular subgroups. With disproportionate sampling, sampling weights must be used in the analysis to accurately describe the characteristics of the population as a whole.

Although probability sample surveys often are conducted in organizational settings and are the recommended sampling approach in academic and government publications on surveys, probability sample surveys can be expensive when in-person interviews are required, the target population is dispersed widely, or members of the target population are rare. A majority of the consumer surveys conducted for Lanham Act litigation present results from nonprobability convenience samples.[101] They are admitted into evidence based on the argument that nonprobability sampling is used widely in marketing research and that "results of these studies are used by major American companies in making decisions of considerable consequence."[102] Nonetheless, when respondents are not selected randomly from the relevant population, the expert should be prepared to justify the method used to select respondents. Special precautions are required to reduce the likelihood of biased samples.[103] In addition, quantitative values computed from such samples (e.g., percentage of respondents indicating confusion) should be viewed as rough

99. *See* Pharmacia Corp. v. Alcon Lab., 201 F. Supp. 2d 335, 365 (D.N.J. 2002).

100. Robert M. Groves et al., Survey Methodology, Stratification and Stratified Sampling, 106–18 (2004).

101. Jacob Jacoby & Amy H. Handlin, *Non-Probability Sampling Designs for Litigation Surveys*, 81 Trademark Rep. 169, 173 (1991). For probability surveys conducted in trademark cases, see James Burrough, Ltd. v. Sign of Beefeater, Inc., 540 F.2d 266 (7th Cir. 1976); Nightlight Systems, Inc., v. Nite Lights Franchise Sys., 2007 U.S. Dist. LEXIS 95565 (N.C. Ga. July 17, 2007); National Football League Props., Inc. v. Wichita Falls Sportswear, Inc., 532 F. Supp. 651 (W.D. Wash. 1982).

102. National Football League Props., Inc. v. New Jersey Giants, Inc., 637 F. Supp. 507, 515 (D.N.J. 1986). A survey of members of the Council of American Survey Research Organizations, the national trade association for commercial survey research firms in the United States, revealed that 95% of the in-person independent contacts in studies done in 1985 took place in malls or shopping centers. Jacoby & Handlin, *supra* note 101, at 172–73, 176. More recently, surveys conducted over the Internet have been administered to samples of respondents drawn from panels of volunteers; *see infra* Section IV.G.4 for a discussion of online surveys. Although panel members may be randomly selected from the panel population to complete the survey, the panel population itself is not usually the product of a random selection process.

103. *See infra* Sections III.D–E.

indicators rather than as precise quantitative estimates.[104] Confidence intervals technically should not be computed, although if the calculation shows a wide interval, that may be a useful indication of the limited value of the estimate.

D. *What Is the Evidence That Nonresponse Did Not Bias the Results of the Survey?*

Even when a sample is drawn randomly from a complete list of elements in the target population, responses or measures may be obtained on only part of the selected sample. If this lack of response is distributed randomly, valid inferences about the population can be drawn with assurance using the measures obtained from the available elements in the sample. The difficulty is that nonresponse often is not random, so that, for example, persons who are single typically have three times the "not at home" rate in U.S. Census Bureau surveys as do family members.[105] Efforts to increase response rates include making several attempts to contact potential respondents, sending advance letters,[106] and providing financial or nonmonetary incentives for participating in the survey.[107]

The key to evaluating the effect of nonresponse in a survey is to determine as much as possible the extent to which nonrespondents differ from the respondents in the nature of the responses they would provide if they were present in the sample. That is, the difficult question to address is the extent to which nonresponse has biased the pattern of responses by undermining the representativeness of the sample and, if it has, the direction of that bias. It is incumbent on the expert presenting the survey results to analyze the level and sources of nonresponse, and to assess how that nonresponse is likely to have affected the results. On some occasions, it may be possible to anticipate systematic patterns of nonresponse. For example, a survey that targets a population of professionals may encounter difficulty in obtaining the same level of participation from individuals with high-volume practices that can be obtained from those with lower-volume practices. To enable the researcher to assess whether response rate varies with the volume of practice, it may be possible to identify in advance potential respondents

104. The court in Kinetic Concept, Inc. v. Bluesky Medical Corp., 2006 U.S. Dist. LEXIS 60187, *14 (W.D. Tex. Aug. 11, 2006), found the plaintiff's survey using a nonprobability sample to be admissible and permitted the plaintiff's expert to present results from a survey using a convenience sample. The court then assisted the jury by providing an instruction on the differences between probability and convenience samples and the estimates obtained from each.

105. 2 Gastwirth, *supra* note 64, at 501. This volume contains a useful discussion of sampling, along with a set of examples. *Id.* at 467.

106. Edith De Leeuw et al., *The Influence of Advance Letters on Response in Telephone Surveys: A Meta-analysis*, 71 Pub. Op. Q. 413 (2007) (advance letters effective in increasing response rates in telephone as well as mail and face-to-face surveys).

107. Erica Ryu et al., *Survey Incentives: Cash vs. In-kind; Face-to-Face vs. Mail; Response Rate vs. Nonresponse Error*, 18 Int'l J. Pub. Op. Res. 89 (2005).

with varying years of experience. Even if it is not possible to know in advance the level of experience of each potential member in the target population and to design a sampling plan that will produce representative samples at each level of experience, the survey itself can include questions about volume of practice that will permit the expert to assess how experience level may have affected the pattern of results.[108]

Although high response rates (i.e., 80% or higher)[109] are desirable because they generally eliminate the need to address the issue of potential bias from nonresponse,[110] such high response rates are increasingly difficult to achieve. Survey nonresponse rates have risen substantially in recent years, along with the costs of obtaining responses, and so the issue of nonresponse has attracted substantial attention from survey researchers.[111] Researchers have developed a variety of approaches to adjust for nonresponse, including weighting obtained responses in proportion to known demographic characteristics of the target population, comparing the pattern of responses from early and late responders to mail surveys, or the pattern of responses from easy-to-reach and hard-to-reach responders in telephone surveys, and imputing estimated responses to nonrespondents based on known characteristics of those who have responded. All of these techniques can only approximate the response patterns that would have been obtained if nonrespondents had responded. Nonetheless, they are useful for testing the robustness of the findings based on estimates obtained from the simple aggregation of answers to questions given by responders.

To assess the general impact of the lower response rates, researchers have conducted comparison studies evaluating the results obtained from surveys with

108. In *People v. Williams*, *supra* note 22, a published survey of experts in eyewitness research was used to show general acceptance of various eyewitness phenomena. *See* Saul Kassin et al., *On the "General Acceptance" of Eyewitness Testimony Research: A New Survey of the Experts*, 56 Am. Psychologist 405 (2001). The survey included questions on the publication activity of respondents and compared the responses of those with high and low research productivity. Productivity levels in the respondent sample suggested that respondents constituted a blue ribbon group of leading researchers. *Williams*, 830 N.Y.S.2d at 457 n.16. *See also* Pharmacia Corp. v. Alcon Lab., Inc., 201 F. Supp. 2d 335 (D.N.J. 2002).

109. Note that methods of computing response rates vary. For example, although response rate can be generally defined as the number of complete interviews with reporting units divided by the number of eligible reporting units in the sample, decisions on how to treat partial completions and how to estimate the eligibility of nonrespondents can produce differences in measures of response rate. *E.g.*, American Association of Public Opinion Research, Standard Definitions: Final Dispositions of Case Codes and Outcome Rates for Surveys (rev. 2008), *available at* www. Aapor.org/uploads/ Standard_Definitions_07-08_Final.pdf.

110. Office of Management and Budget, Standards and Guidelines for Statistical Surveys (Sept. 2006), Guideline 1.3.4: Plan for a nonresponse bias analysis if the expected unit response rate is below 80%. *See* Albert v. Zabin, 2009 Mass. App. Unpub. LEXIS 572 (July 14, 2009) reversing summary judgment that had excluded surveys with response rates of 27% and 31% based on a thoughtful analysis of measures taken to assess potential nonresponse bias.

111. *E.g.*, Richard Curtin et al., *Changes in Telephone Survey Nonresponse Over the Past Quarter Century*, 69 Pub. Op. Q. 87 (2005); Survey Nonresponse (Robert M. Groves et al. eds., 2002).

varying response rates.[112] Contrary to earlier assumptions, surprisingly comparable results have been obtained in many surveys with varying response rates, suggesting that surveys may achieve reasonable estimates even with relatively low response rates. The key is whether nonresponse is associated with systematic differences in response that cannot be adequately modeled or assessed.

Determining whether the level of nonresponse in a survey seriously impairs inferences drawn from the results of a survey generally requires an analysis of the determinants of nonresponse. For example, even a survey with a high response rate may seriously underrepresent some portions of the population, such as the unemployed or the poor. If a general population sample is used to chart changes in the proportion of the population that knows someone with HIV, the survey would underestimate the population value if some groups more likely to know someone with HIV (e.g., intravenous drug users) are underrepresented in the sample. The survey expert should be prepared to provide evidence on the potential impact of nonresponse on the survey results.

In surveys that include sensitive or difficult questions, particularly surveys that are self-administered, some respondents may refuse to provide answers or may provide incomplete answers (i.e., item rather than unit nonresponse).[113] To assess the impact of nonresponse to a particular question, the survey expert should analyze the differences between those who answered and those who did not answer. Procedures to address the problem of missing data include recontacting respondents to obtain the missing answers and using the respondent's other answers to predict the missing response (i.e., imputation).[114]

E. What Procedures Were Used to Reduce the Likelihood of a Biased Sample?

If it is impractical for a survey researcher to sample randomly from the entire target population, the researcher still can apply probability sampling to some aspects of respondent selection to reduce the likelihood of biased selection. For example, in many studies the target population consists of all consumers or purchasers of a product. Because it is impractical to randomly sample from that population, research is often conducted in shopping malls where some members of the target population may not shop. Mall locations, however, can be sampled randomly from a list of possible mall sites. By administering the survey at several different

112. *E.g.*, Daniel M. Merkle & Murray Edelman, *Nonresponse in Exit Polls: A Comprehensive Analysis*, *in* Survey Nonresponse, *supra* note 111, at 243–57 (finding minimal nonresponse error associated with refusals to participate in in-person exit polls); *see also* Jon A. Krosnick, *Survey Research*, 50 Ann. Rev. Psychol. 537 (1999).

113. *See* Roger Tourangeau et al., The Psychology of Survey Response (2000).

114. *See* Paul D. Allison, *Missing Data*, *in* Handbook of Survey Research, *supra* note 1, at 630; *see also* Survey Nonresponse, *supra* note 111.

malls, the expert can test for and report on any differences observed across sites. To the extent that similar results are obtained in different locations using different onsite interview operations, it is less likely that idiosyncrasies of sample selection or administration can account for the results.[115] Similarly, because the characteristics of persons visiting a shopping center vary by day of the week and time of day, bias in sampling can be reduced if the survey design calls for sampling time segments as well as mall locations.[116]

In mall intercept surveys, the organization that manages the onsite interview facility generally employs recruiters who approach potential survey respondents in the mall and ascertain if they are qualified and willing to participate in the survey. If a potential respondent agrees to answer the questions and meets the specified criteria, he or she is escorted to the facility where the survey interview takes place. If recruiters are free to approach potential respondents without controls on how an individual is to be selected for screening, shoppers who spend more time in the mall are more likely to be approached than shoppers who visit the mall only briefly. Moreover, recruiters naturally prefer to approach friendly looking potential respondents, so that it is more likely that certain types of individuals will be selected. These potential biases in selection can be reduced by providing appropriate selection instructions and training recruiters effectively. Training that reduces the interviewer's discretion in selecting a potential respondent is likely to reduce bias in selection, as are instructions to approach every *n*th person entering the facility through a particular door.[117]

F. What Precautions Were Taken to Ensure That Only Qualified Respondents Were Included in the Survey?

In a carefully executed survey, each potential respondent is questioned or measured on the attributes that determine his or her eligibility to participate in the survey. Thus, the initial questions screen potential respondents to determine if they are members of the target population of the survey (e.g., Is she at least 14 years old? Does she own a dog? Does she live within 10 miles?). The screening questions must be drafted so that they do not appeal to or deter specific groups within the target population, or convey information that will influence the respondent's

115. Note, however, that differences in results across sites may arise from genuine differences in respondents across geographic locations or from a failure to administer the survey consistently across sites.

116. Seymour Sudman, *Improving the Quality of Shopping Center Sampling*, 17 J. Marketing Res. 423 (1980).

117. In the end, even if malls are randomly sampled and shoppers are randomly selected within malls, results from mall surveys technically can be used to generalize only to the population of mall shoppers. The ability of the mall sample to describe the likely response pattern of the broader relevant population will depend on the extent to which a substantial segment of the relevant population (1) is not found in malls and (2) would respond differently to the interview.

answers on the main survey. For example, if respondents must be prospective and recent purchasers of Sunshine orange juice in a trademark survey designed to assess consumer confusion with Sun Time orange juice, potential respondents might be asked to name the brands of orange juice they have purchased recently or expect to purchase in the next 6 months. They should not be asked specifically if they recently have purchased, or expect to purchase, Sunshine orange juice, because this may affect their responses on the survey either by implying who is conducting the survey or by supplying them with a brand name that otherwise would not occur to them.

The content of a screening questionnaire (or screener) can also set the context for the questions that follow. In *Pfizer, Inc. v. Astra Pharmaceutical Products, Inc.*,[118] physicians were asked a screening question to determine whether they prescribed particular drugs. The survey question that followed the screener asked "Thinking of the practice of cardiovascular medicine, what first comes to mind when you hear the letters XL?" The court found that the screener conditioned the physicians to respond with the name of a drug rather than a condition (long-acting).[119]

The criteria for determining whether to include a potential respondent in the survey should be objective and clearly conveyed, preferably using written instructions addressed to those who administer the screening questions. These instructions and the completed screening questionnaire should be made available to the court and the opposing party along with the interview form for each respondent.

IV. Survey Questions and Structure

A. Were Questions on the Survey Framed to Be Clear, Precise, and Unbiased?

Although it seems obvious that questions on a survey should be clear and precise, phrasing questions to reach that goal is often difficult. Even questions that appear clear can convey unexpected meanings and ambiguities to potential respondents. For example, the question "What is the average number of days each week you have butter?" appears to be straightforward. Yet some respondents wondered whether margarine counted as butter, and when the question was revised to include the introductory phrase "not including margarine," the reported frequency of butter use dropped dramatically.[120]

118. 858 F. Supp. 1305, 1321 & n.13 (S.D.N.Y. 1994).

119. *Id.* at 1321.

120. Floyd J. Fowler, Jr., *How Unclear Terms Affect Survey Data*, 56 Pub. Op. Q. 218, 225–26 (1992).

When unclear questions are included in a survey, they may threaten the validity of the survey by systematically distorting responses if respondents are misled in a particular direction, or by inflating random error if respondents guess because they do not understand the question.[121] If the crucial question is sufficiently ambiguous or unclear, it may be the basis for rejecting the survey. For example, a survey was designed to assess community sentiment that would warrant a change of venue in trying a case for damages sustained when a hotel skywalk collapsed.[122] The court found that the question "Based on what you have heard, read or seen, do you believe that in the current compensatory damage trials, the defendants, such as the contractors, designers, owners, and operators of the Hyatt Hotel, should be punished?" could neither be correctly understood nor easily answered.[123] The court noted that the phrase "compensatory damages," although well-defined for attorneys, was unlikely to be meaningful for laypersons.[124]

A variety of pretest activities may be used to improve the clarity of communication with respondents. Focus groups can be used to find out how the survey population thinks about an issue, facilitating the construction of clear and understandable questions. Cognitive interviewing, which includes a combination of think-aloud and verbal probing techniques, may be used for questionnaire evaluation.[125] Pilot studies involving a dress rehearsal for the main survey can also detect potential problems.

Texts on survey research generally recommend pretests as a way to increase the likelihood that questions are clear and unambiguous,[126] and some courts have recognized the value of pretests.[127] In many pretests or pilot tests,[128] the proposed survey is administered to a small sample (usually between 25 and 75)[129] of the

121. *See id.* at 219.

122. Firestone v. Crown Ctr. Redevelopment Corp., 693 S.W.2d 99 (Mo. 1985) (en banc).

123. *See id.* at 102, 103.

124. *See id.* at 103. When there is any question about whether some respondents will understand a particular term or phrase, the term or phrase should be defined explicitly.

125. Gordon B. Willis et al., *Is the Bandwagon Headed to the Methodological Promised Land? Evaluating the Validity of Cognitive Interviewing Techniques, in* Cognitive and Survey Research 136 (Monroe G. Sirken et al. eds., 1999). *See also* Tourangeau et al., *supra* note 113, at 326–27.

126. *See* Jon A. Krosnick & Stanley Presser, *Questions and Questionnaire Design, in* Handbook of Survey Research, *supra* note 1, at 294 ("No matter how closely a questionnaire follows recommendations based on best practices, it is likely to benefit from pretesting. . ."). *See also* Jean M. Converse & Stanley Presser, Survey Questions: Handcrafting the Standardized Questionnaire 51 (1986); Fred W. Morgan, *Judicial Standards for Survey Research: An Update and Guidelines*, 54 J. Marketing 59, 64 (1990).

127. *See e.g.*, Zippo Mfg. Co. v. Rogers Imports, Inc., 216 F. Supp. 670 (S.D.N.Y. 1963); Scott v. City of New York, 591 F. Supp. 2d 554, 560 (S.D.N.Y. 2008) ("[s]urvey went through multiple pretests in order to insure its usefulness and statistical validity.").

128. The terms *pretest* and *pilot test* are sometimes used interchangeably to describe pilot work done in the planning stages of research. When they are distinguished, the difference is that a pretest tests the questionnaire, whereas a pilot test generally tests proposed collection procedures as well.

129. Converse & Presser, *supra* note 126, at 69. Converse and Presser suggest that a pretest with 25 respondents is appropriate when the survey uses professional interviewers.

same type of respondents who would be eligible to participate in the full-scale survey. The interviewers observe the respondents for any difficulties they may have with the questions and probe for the source of any such difficulties so that the questions can be rephrased if confusion or other difficulties arise.[130] Attorneys who commission surveys for litigation sometimes are reluctant to approve pilot work or to reveal that pilot work has taken place because they are concerned that if a pretest leads to revised wording of the questions, the trier of fact may believe that the survey has been manipulated and is biased or unfair. A more appropriate reaction is to recognize that pilot work is a standard and valuable way to improve the quality of a survey[131] and to anticipate that it often results in word changes that increase clarity and correct misunderstandings. Thus, changes may indicate informed survey construction rather than flawed survey design.[132]

B. Were Some Respondents Likely to Have No Opinion? If So, What Steps Were Taken to Reduce Guessing?

Some survey respondents may have no opinion on an issue under investigation, either because they have never thought about it before or because the question mistakenly assumes a familiarity with the issue. For example, survey respondents may not have noticed that the commercial they are being questioned about guaranteed the quality of the product being advertised and thus they may have no opinion on the kind of guarantee it indicated. Likewise, in an employee survey, respondents may not be familiar with the parental leave policy at their company and thus may have no opinion on whether they would consider taking advantage of the parental leave policy if they became parents. The following three alternative question structures will affect how those respondents answer and how their responses are counted.

First, the survey can ask all respondents to answer the question (e.g., "Did you understand the guarantee offered by Clover to be a 1-year guarantee, a 60-day guarantee, or a 30-day guarantee?"). Faced with a direct question, particularly one that provides response alternatives, the respondent obligingly may supply an

130. Methods for testing respondent understanding include concurrent and retrospective think-alouds, in which respondents describe their thinking as they arrive at, or after they have arrived at, an answer, and paraphrasing (asking respondents to restate the question in their own words). Tourangeau et al., *supra* note 113, at 326–27; *see also* Methods for Testing and Evaluating Survey Questionnaires (Stanley Presser et al. eds., 2004).

131. *See* OMB Standards and Guidelines for Statistical Survey, *supra* note 110, Standard 1.4, Pretesting Survey Systems (specifying that to ensure that all components of a survey function as intended, pretests of survey components should be conducted unless those components have previously been successfully fielded); American Association for Public Opinion Research, Best Practices (2011) ("Because it is rarely possible to foresee all the potential misunderstandings or biasing effects of different questions or procedures, it is vital for a well-designed survey operation to include provision for a pretest.").

132. *See infra* Section VII.B for a discussion of obligations to disclose pilot work.

answer even if (in this example) the respondent did not notice the guarantee (or is unfamiliar with the parental leave policy). Such answers will reflect only what the respondent can glean from the question, or they may reflect pure guessing. The imprecision introduced by this approach will increase with the proportion of respondents who are unfamiliar with the topic at issue.

Second, the survey can use a quasi-filter question to reduce guessing by providing "don't know" or "no opinion" options as part of the question (e.g., "Did you understand the guarantee offered by Clover to be for more than a year, a year, or less than a year, or don't you have an opinion?").[133] By signaling to the respondent that it is appropriate not to have an opinion, the question reduces the demand for an answer and, as a result, the inclination to hazard a guess just to comply. Respondents are more likely to choose a "no opinion" option if it is mentioned explicitly by the interviewer than if it is merely accepted when the respondent spontaneously offers it as a response. The consequence of this change in format is substantial. Studies indicate that, although the relative distribution of the respondents selecting the *listed* choices is unlikely to change dramatically, presentation of an explicit "don't know" or "no opinion" alternative commonly leads to a 20% to 25% increase in the proportion of respondents selecting that response.[134]

Finally, the survey can include full-filter questions, that is, questions that lay the groundwork for the substantive question by first asking the respondent if he or she has an opinion about the issue or happened to notice the feature that the interviewer is preparing to ask about (e.g., "Based on the commercial you just saw, do you have an opinion about how long Clover stated or implied that its guarantee lasts?").[135] The interviewer then asks the substantive question only of those respondents who have indicated that they have an opinion on the issue.

Which of these three approaches is used and the way it is used can affect the rate of "no opinion" responses that the substantive question will evoke.[136] Respondents are more likely to say that they do not have an opinion on an issue if a full filter is used than if a quasi-filter is used.[137] However, in maximizing respondent expressions of "no opinion," full filters may produce an underreporting of opinions. There is some evidence that full-filter questions discourage respondents who actually have opinions from offering them by conveying the implicit suggestion that respondents can avoid difficult followup questions by saying that they have no opinion.[138]

133. Norbert Schwarz & Hans-Jürgen Hippler, *Response Alternatives: The Impact of Their Choice and Presentation Order, in* Measurement Errors in Surveys 41, 45–46 (Paul P. Biemer et al. eds., 1991).

134. Howard Schuman & Stanley Presser, Questions and Answers in Attitude Surveys: Experiments on Question Form, Wording and Context 113–46 (1981).

135. *See, e.g.*, Johnson & Johnson–Merck Consumer Pharmas. Co. v. SmithKline Beecham Corp., 960 F.2d 294, 299 (2d Cir. 1992).

136. Considerable research has been conducted on the effects of filters. For a review, see George F. Bishop et al., *Effects of Filter Questions in Public Opinion Surveys*, 47 Pub. Op. Q. 528 (1983).

137. Schwarz & Hippler, *supra* note 133, at 45–46.

138. *Id.* at 46.

In general, then, a survey that uses full filters provides a conservative estimate of the number of respondents holding an opinion, while a survey that uses neither full filters nor quasi-filters may overestimate the number of respondents with opinions, if some respondents offering opinions are guessing. The strategy of including a "no opinion" or "don't know" response as a quasi-filter avoids both of these extremes. Thus, rather than asking, "Based on the commercial, do you believe that the two products are made in the same way, or are they made differently?"[139] or prefacing the question with a preliminary, "Do you have an opinion, based on the commercial, concerning the way that the two products are made?" the question could be phrased, "Based on the commercial, do you believe that the two products are made in the same way, or that they are made differently, or don't you have an opinion about the way they are made?"

Recent research on the effects of including a "don't know" option shows that quasi-filters as well as full filters may discourage a respondent who would be able to provide a meaningful answer from expressing it.[140] The "don't know" option provides a cue that it is acceptable to avoid the work of trying to provide a more substantive response. Respondents are particularly likely to be attracted to a "don't know" option when the question is difficult to understand or the respondent is not strongly motivated to carefully report an opinion.[141] One solution that some survey researchers use is to provide respondents with a general instruction not to guess at the beginning of an interview, rather than supplying a "don't know" or "no opinion" option as part of the options attached to each question.[142] Another approach is to eliminate the "don't know" option and to add followup questions that measure the strength of the respondent's opinion.[143]

C. *Did the Survey Use Open-Ended or Closed-Ended Questions? How Was the Choice in Each Instance Justified?*

The questions that make up a survey instrument may be open-ended, closed-ended, or a combination of both. Open-ended questions require the respondent to formulate and express an answer in his or her own words (e.g., "What was the main point of the commercial?" "Where did you catch the fish you caught

139. The question in the example without the "no opinion" alternative was based on a question rejected by the court in Coors Brewing Co. v. Anheuser-Busch Cos., 802 F. Supp. 965, 972–73 (S.D.N.Y. 1992). *See also* Procter & Gamble Pharms., Inc. v. Hoffmann-La Roche, Inc., 2006 U.S. Dist. LEXIS 64363 (S.D.N.Y. Sept. 6, 2006).

140. Jon A. Krosnick et al., *The Impact of "No Opinion" Response Options on Data Quality: Non-Attitude Reduction or Invitation to Satisfice?* 66 Pub. Op. Q. 371 (2002).

141. Krosnick & Presser, *supra* note 126, at 284.

142. Anheuser-Busch, Inc. v. VIP Prods, LLC, No. 4:08cv0358, 2008 U.S. Dist. LEXIS 82258, at *6 (E.D. Mo. Oct. 16, 2008).

143. Krosnick & Presser, *supra* note 126, at 285.

in these waters?"[144]). Closed-ended questions provide the respondent with an explicit set of responses from which to choose; the choices may be as simple as *yes* or *no* (e.g., "Is Colby College coeducational?"[145]) or as complex as a range of alternatives (e.g., "The two pain relievers have (1) the same likelihood of causing gastric ulcers; (2) about the same likelihood of causing gastric ulcers; (3) a somewhat different likelihood of causing gastric ulcers; (4) a very different likelihood of causing gastric ulcers; or (5) none of the above."[146]). When a survey involves in-person interviews, the interviewer may show the respondent these choices on a showcard that lists them.

Open-ended and closed-ended questions may elicit very different responses.[147] Most responses are less likely to be volunteered by respondents who are asked an open-ended question than they are to be chosen by respondents who are presented with a closed-ended question. The response alternatives in a closed-ended question may remind respondents of options that they would not otherwise consider or which simply do not come to mind as easily.[148]

The advantage of open-ended questions is that they give the respondent fewer hints about expected or preferred answers. Precoded responses on a closed-ended question, in addition to reminding respondents of options that they might not otherwise consider,[149] may direct the respondent away from or toward a particular response. For example, a commercial reported that in shampoo tests with more than 900 women, the sponsor's product received higher ratings than

144. A relevant example from *Wilhoite v. Olin Corp.* is described in McGovern & Lind, *supra* note 31, at 76.

145. Presidents & Trustees of Colby College v. Colby College–N.H., 508 F.2d 804, 809 (1st Cir. 1975).

146. This question is based on one asked in American Home Products Corp. v. Johnson & Johnson, 654 F. Supp. 568, 581 (S.D.N.Y. 1987), that was found to be a leading question by the court, primarily because the choices suggested that the respondent had learned about aspirin's and ibuprofen's relative likelihood of causing gastric ulcers. In contrast, in McNeilab, Inc. v. American Home Products Corp., 501 F. Supp. 517, 525 (S.D.N.Y. 1980), the court accepted as nonleading the question, "Based only on what the commercial said, would Maximum Strength Anacin contain more pain reliever, the same amount of pain reliever, or less pain reliever than the brand you, yourself, currently use most often?"

147. Howard Schuman & Stanley Presser, *Question Wording as an Independent Variable in Survey Analysis*, 6 Soc. Methods & Res. 151 (1977); Schuman & Presser, *supra* note 134, at 79–112; Converse & Presser, *supra* note 126, at 33.

148. For example, when respondents in one survey were asked, "What is the most important thing for children to learn to prepare them for life?", 62% picked "to think for themselves" from a list of five options, but only 5% spontaneously offered that answer when the question was open-ended. Schuman & Presser, *supra* note 134, at 104–07. An open-ended question presents the respondent with a free-recall task, whereas a closed-ended question is a recognition task. Recognition tasks in general reveal higher performance levels than recall tasks. Mary M. Smyth et al., Cognition in Action 25 (1987). In addition, there is evidence that respondents answering open-ended questions may be less likely to report some information that they would reveal in response to a closed-ended question when that information seems self-evident or irrelevant.

149. Schwarz & Hippler, *supra* note 133, at 43.

other brands.[150] According to a competitor, the commercial deceptively implied that each woman in the test rated more than one shampoo, when in fact each woman rated only one. To test consumer impressions, a survey might have shown the commercial and asked an open-ended question: "How many different brands mentioned in the commercial did each of the 900 women try?"[151] Instead, the survey asked a closed-ended question; respondents were given the choice of "one," "two," "three," "four," or "five or more." The fact that four of the five choices in the closed-ended question provided a response that was greater than one implied that the correct answer was probably more than one.[152] Note, however, that the open-ended question also may suggest that the answer is more than one.

By asking "how many different brands," the question suggests (1) that the viewer should have received some message from the commercial about the number of brands each woman tried and (2) that different brands were tried. Similarly, an open-ended question that asks, "[W]hich company or store do you think puts out this shirt?" indicates to the respondent that the appropriate answer is the name of a company or store. The question would be leading if the respondent would have considered other possibilities (e.g., an individual or Webstore) if the question had not provided the frame of a company or store.[153] Thus, the wording of a question, open-ended or closed-ended, can be leading or non-leading, and the degree of suggestiveness of each question must be considered in evaluating the objectivity of a survey.

Closed-ended questions have some additional potential weaknesses that arise if the choices are not constructed properly. If the respondent is asked to choose one response from among several choices, the response chosen will be meaningful only if the list of choices is exhaustive—that is, if the choices cover all possible answers a respondent might give to the question. If the list of possible choices is incomplete, a respondent may be forced to choose one that does not express his or her opinion.[154] Moreover, if respondents are told explicitly that they are

150. *See* Vidal Sassoon, Inc. v. Bristol-Myers Co., 661 F.2d 272, 273 (2d Cir. 1981).

151. This was the wording of the closed-ended question in the survey discussed in *Vidal Sassoon*, 661 F.2d at 275–76, without the closed-ended options that were supplied in that survey.

152. Ninety-five percent of the respondents who answered the closed-ended question in the plaintiff's survey said that each woman had tried two or more brands. The open-ended question was never asked. *Vidal Sassoon*, 661 F.2d at 276. Norbert Schwarz, *Assessing Frequency Reports of Mundane Behaviors: Contributions of Cognitive Psychology to Questionnaire Construction*, *in* Research Methods in Personality and Social Psychology 98 (Clyde Hendrick & Margaret S. Clark eds., 1990), suggests that respondents often rely on the range of response alternatives as a frame of reference when they are asked for frequency judgments. *See, e.g.*, Roger Tourangeau & Tom W. Smith, *Asking Sensitive Questions: The Impact of Data Collection Mode, Question Format, and Question Context*, 60 Pub. Op. Q. 275, 292 (1996).

153. Smith v. Wal-Mart Stores, Inc, 537 F. Supp. 2d 1302, 1331–32 (N.D. Ga. 2008).

154. *See, e.g.*, American Home Prods. Corp. v. Johnson & Johnson, 654 F. Supp. 568, 581 (S.D.N.Y. 1987).

not limited to the choices presented, most respondents nevertheless will select an answer from among the listed ones.[155]

One form of closed-ended question format that typically produces some distortion is the popular agree/disagree, true/false, or yes/no question. Although this format is appealing because it is easy to write and score these questions and their responses, the format is also seriously problematic. With its simplicity comes acquiescence, "[T]he tendency to endorse any assertion made in a question, regardless of its content," is a systematic source of bias that has produced an inflation effect of 10% across a number of studies.[156] Only when control groups or control questions are added to the survey design can this question format provide reasonable response estimates.[157]

Although many courts prefer open-ended questions on the ground that they tend to be less leading, the value of any open-ended or closed-ended question depends on the information it conveys in the question and, in the case of closed-ended questions, in the choices provided. Open-ended questions are more appropriate when the survey is attempting to gauge what comes first to a respondent's mind, but closed-ended questions are more suitable for assessing choices between well-identified options or obtaining ratings on a clear set of alternatives.

D. *If Probes Were Used to Clarify Ambiguous or Incomplete Answers, What Steps Were Taken to Ensure That the Probes Were Not Leading and Were Administered in a Consistent Fashion?*

When questions allow respondents to express their opinions in their own words, some of the respondents may give ambiguous or incomplete answers, or may ask for clarification. In such instances, interviewers may be instructed to record any answer that the respondent gives and move on to the next question, or they may be instructed to probe to obtain a more complete response or clarify the meaning of the ambiguous response. They may also be instructed what clarification they can provide. In all of these situations, interviewers should record verbatim both what the respondent says and what the interviewer says in the attempt to get or provide clarification. Failure to record every part of the exchange in the order in which it occurs raises questions about the reliability of the survey, because neither the court nor the opposing party can evaluate whether the probe affected the views expressed by the respondent.

155. *See* Howard Schuman, *Ordinary Questions, Survey Questions, and Policy Questions*, 50 Pub. Opinion Q. 432, 435–36 (1986).

156. Jon A. Krosnick, *Survey Research*, 50 Ann. Rev. Psychol. 537, 552 (1999).

157. *See infra* Section IV.F.

If the survey is designed to allow for probes, interviewers must be given explicit instructions on when they should probe and what they should say in probing.[158] Standard probes used to draw out all that the respondent has to say (e.g., "Any further thoughts?" "Anything else?" "Can you explain that a little more?" Or "Could you say that another way?") are relatively innocuous and noncontroversial in content, but persistent continued requests for further responses to the same or nearly identical questions may convey the idea to the respondent that he or she has not yet produced the "right" answer.[159] Interviewers should be trained in delivering probes to maintain a professional and neutral relationship with the respondent (as they should during the rest of the interview), which minimizes any sense of passing judgment on the content of the answers offered. Moreover, interviewers should be given explicit instructions on when to probe, so that probes are administered consistently.

A more difficult type of probe to construct and deliver reliably is one that requires a substantive question tailored to the answer given by the respondent. The survey designer must provide sufficient instruction to interviewers so that they avoid giving directive probes that suggest one answer over another. Those instructions, along with all other aspects of interviewer training, should be made available for evaluation by the court and the opposing party.

E. What Approach Was Used to Avoid or Measure Potential Order or Context Effects?

The order in which questions are asked on a survey and the order in which response alternatives are provided in a closed-ended question can influence the answers.[160] For example, although asking a general question before a more specific question on the same topic is unlikely to affect the response to the specific question, reversing the order of the questions may influence responses to the general question. As a rule, then, surveys are less likely to be subject to order effects if the questions move from the general (e.g., "What do you recall being discussed

158. Floyd J. Fowler, Jr. & Thomas W. Mangione, Standardized Survey Interviewing: Minimizing Interviewer-Related Error 41–42 (1990).

159. *See, e.g.*, Johnson & Johnson–Merck Consumer Pharms. Co. v. Rhone-Poulenc Rorer Pharms., Inc., 19 F.3d 125, 135 (3d Cir. 1994); American Home Prods. Corp. v. Procter & Gamble Co., 871 F. Supp. 739, 748 (D.N.J. 1994).

160. *See* Schuman & Presser, *supra* note 134, at 23, 56–74. Krosnick & Presser, *supra* note 126, at 278–81. In *R.J. Reynolds Tobacco Co. v. Loew's Theatres, Inc.*, 511 F. Supp. 867, 875 (S.D.N.Y. 1980), the court recognized the biased structure of a survey that disclosed the tar content of the cigarettes being compared before questioning respondents about their cigarette preferences. Not surprisingly, respondents expressed a preference for the lower tar product. *See also* E. & J. Gallo Winery v. Pasatiempos Gallo, S.A., 905 F. Supp. 1403, 1409–10 (E.D. Cal. 1994) (court recognized that earlier questions referring to playing cards, board or table games, or party supplies, such as confetti, increased the likelihood that respondents would include these items in answers to the questions that followed).

in the advertisement?") to the specific (e.g., "Based on your reading of the advertisement, what companies do you think the ad is referring to when it talks about rental trucks that average five miles per gallon?").[161]

The mode of questioning can influence the form that an order effect takes. When respondents are shown response alternatives visually, as in mail surveys and other self-administered questionnaires or in face-to-face interviews when respondents are shown a card containing response alternatives, they are more likely to select the first choice offered (a primacy effect).[162] In contrast, when response alternatives are presented orally, as in telephone surveys, respondents are more likely to choose the last choice offered (a recency effect).[163] Although these effects are typically small, no general formula is available that can adjust values to correct for order effects, because the size and even the direction of the order effects may depend on the nature of the question being asked and the choices being offered. Moreover, it may be unclear which order is most appropriate. For example, if the respondent is asked to choose between two different products, and there is a tendency for respondents to choose the first product mentioned,[164] which order of presentation will produce the more accurate response?[165] To control for order effects, the order of the questions and the order of the response choices in a survey should be rotated,[166] so that, for example, one-third of the respondents have Product A listed first, one-third of the respondents have Product B listed first, and one-third of the respondents have Product C listed first. If the three different orders[167] are distributed randomly among respondents, no response alternative will have an inflated chance of being selected because of its position, and the average of the three will provide a reasonable estimate of response level.[168]

161. This question was accepted by the court in *U-Haul Int'l, Inc. v. Jartran, Inc.*, 522 F. Supp. 1238, 1249 (D. Ariz. 1981), *aff'd*, 681 F.2d 1159 (9th Cir. 1982).

162. Krosnick & Presser, *supra* note 126, at 280.

163. *Id.*

164. Similarly, candidates in the first position on the ballot tend to attract extra votes. J.M. Miller & Jon A. Krosnick, *The Impact of Candidate Name Order on Election Outcomes*, 62 Pub. Op. Q. 291 (1998).

165. *See* Rust Env't & Infrastructure, Inc. v. Teunissen, 131 F.3d 1210, 1218 (7th Cir. 1997) (survey did not pass muster in part because of failure to incorporate random rotation of corporate names that were the subject of a trademark dispute).

166. *See, e.g.* Winning Ways, Inc. v. Holloway Sportswear, Inc., 913 F. Supp. 1454, 1465–67 (D. Kan. 1996) (failure to rotate the order in which the jackets were shown to the consumers led to reduced weight for the survey); Procter & Gamble Pharms., Inc. v. Hoffmann-La Roche, Inc., 2006 U.S. Dist. LEXIS 64363, 2006-2 Trade Cas. (CCH) P75465 (S.D.N.Y. Sept. 6, 2006).

167. Actually, there are six possible orders of the three alternatives: ABC, ACB, BAC, BCA, CAB, and CBA. Thus, the optimal survey design would allocate equal numbers of respondents to each of the six possible orders.

168. Although rotation is desirable, many surveys are conducted with no attention to this potential bias. Because it is impossible to know in the abstract whether a particular question suffers much, little, or not at all from an order bias, lack of rotation should not preclude reliance on the answer to the question, but it should reduce the weight given to that answer.

F. *If the Survey Was Designed to Test a Causal Proposition, Did the Survey Include an Appropriate Control Group or Question?*

Many surveys are designed not simply to describe attitudes or beliefs or reported behaviors, but to determine the source of those attitudes or beliefs or behaviors. That is, the purpose of the survey is to test a causal proposition. For example, how does a trademark or the content of a commercial affect respondents' perceptions or understanding of a product or commercial? Thus, the question is not merely whether consumers hold inaccurate beliefs about Product A, but whether exposure to the commercial misleads the consumer into thinking that Product A is a superior pain reliever. Yet if consumers already believe, before viewing the commercial, that Product A is a superior pain reliever, a survey that simply records consumers' impressions after they view the commercial may reflect those preexisting beliefs rather than impressions produced by the commercial.

Surveys that merely record consumer impressions have a limited ability to answer questions about the origins of those impressions. The difficulty is that the consumer's response to any question on the survey may be the result of information or misinformation from sources other than the trademark the respondent is being shown or the commercial he or she has just watched.[169] In a trademark survey attempting to show secondary meaning, for example, respondents were shown a picture of the stripes used on Mennen stick deodorant and asked, "[W]hich [brand] would you say uses these stripes on their package?"[170] The court recognized that the high percentage of respondents selecting "Mennen" from an array of brand names may have represented "merely a playback of brand share";[171] that is, respondents asked to give a brand name may guess the one that is most familiar, generally the brand with the largest market share.[172]

Some surveys attempt to reduce the impact of preexisting impressions on respondents' answers by instructing respondents to focus solely on the stimulus as a basis for their answers. Thus, the survey includes a preface (e.g., "based on the commercial you just saw") or directs the respondent's attention to the mark at issue (e.g., "these stripes on the package"). Such efforts are likely to be only partially successful. It is often difficult for respondents to identify accurately the

169. *See, e.g.*, Procter & Gamble Co. v. Ultreo, Inc., 574 F. Supp. 2d. 339, 351–52 (S.D.N.Y. 2008) (survey was unreliable because it failed to control for the effect of preexisting beliefs).

170. Mennen Co. v. Gillette Co., 565 F. Supp. 648, 652 (S.D.N.Y. 1983), *aff'd*, 742 F.2d 1437 (2d Cir. 1984). To demonstrate secondary meaning, "the [c]ourt must determine whether the mark has been so associated in the mind of consumers with the entity that it identifies that the goods sold by that entity are distinguished by the mark or symbol from goods sold by others." *Id.*

171. *Id.*

172. *See also* Upjohn Co. v. American Home Prods. Corp., No. 1-95-CV-237, 1996 U.S. Dist. LEXIS 8049, at *42–44 (W.D. Mich. Apr. 5, 1996).

source of their impressions.[173] The more routine the idea being examined in the survey (e.g., that the advertised pain reliever is more effective than others on the market; that the mark belongs to the brand with the largest market share), the more likely it is that the respondent's answer is influenced by (1) preexisting impressions; (2) general expectations about what commercials typically say (e.g., the product being advertised is better than its competitors); or (3) guessing, rather than by the actual content of the commercial message or trademark being evaluated.

It is possible to adjust many survey designs so that causal inferences about the effect of a trademark or an allegedly deceptive commercial become clear and unambiguous. By adding one or more appropriate control groups, the survey expert can test directly the influence of the stimulus.[174] In the simplest version of such a survey experiment, respondents are assigned randomly to one of two conditions.[175] For example, respondents assigned to the experimental condition view an allegedly deceptive commercial, and respondents assigned to the control condition either view a commercial that does not contain the allegedly deceptive material or do not view any commercial.[176] Respondents in both the experimental and control groups answer the same set of questions about the allegedly deceptive message. The effect of the commercial's allegedly deceptive message is evaluated by comparing the responses made by the experimental group members with those of the control group members. If 40% of the respondents in the experimental group responded indicating that they received the deceptive message (e.g., the advertised product has fewer calories than its competitor), whereas only 8% of the respondents in the control group gave that response, the difference between 40% and 8% (within the limits of sampling error[177]) can be attributed only to the allegedly deceptive message. Without the control group, it is not possible to determine how much of the 40% is attributable to respondents' preexisting beliefs

173. *See* Richard E. Nisbett & Timothy D. Wilson, *Telling More Than We Can Know: Verbal Reports on Mental Processes*, 84 Psychol. Rev. 231 (1977).

174. *See* Shari S. Diamond, *Using Psychology to Control Law: From Deceptive Advertising to Criminal Sentencing*, 13 Law & Hum. Behav. 239, 244–46 (1989); Jacob Jacoby & Constance Small, *Applied Marketing: The FDA Approach to Defining Misleading Advertising*, 39 J. Marketing 65, 68 (1975). *See also* David H. Kaye & David A. Freedman, Reference Guide on Statistics, Section II.A, in this manual.

175. Random assignment should not be confused with random selection. When respondents are assigned randomly to different treatment groups (e.g., respondents in each group watch a different commercial), the procedure ensures that within the limits of sampling error the two groups of respondents will be equivalent except for the different treatments they receive. Respondents selected for a mall intercept study, and not from a probability sample, may be assigned randomly to different treatment groups. Random selection, in contrast, describes the method of selecting a sample of respondents in a probability sample. *See supra* Section III.C.

176. This alternative commercial could be a "tombstone" advertisement that includes only the name of the product or a more elaborate commercial that does not include the claim at issue.

177. For a discussion of sampling error, see David H. Kaye & David A. Freedman, Reference Guide on Statistics, Section IV.A, in this manual.

or other background noise (e.g., respondents who misunderstand the question or misstate their responses). Both preexisting beliefs and other background noise should have produced similar response levels in the experimental and control groups. In addition, if respondents who viewed the allegedly deceptive commercial respond differently than respondents who viewed the control commercial, the difference cannot be merely the result of a leading question, because both groups answered the same question. The ability to evaluate the effect of the wording of a particular question makes the control group design particularly useful in assessing responses to closed-ended questions,[178] which may encourage guessing or particular responses. Thus, the focus on the response level in a control group design is not on the absolute response level, but on the difference between the response level of the experimental group and that of the control group.[179]

In designing a survey-experiment, the expert should select a stimulus for the control group that shares as many characteristics with the experimental stimulus as possible, with the key exception of the characteristic whose influence is being assessed.[180] Although a survey with an imperfect control group may provide better information than a survey with no control group at all, the choice of an appropriate control group requires some care and should influence the weight that the survey receives. For example, a control stimulus should not be less attractive than the experimental stimulus if the survey is designed to measure how familiar the experimental stimulus is to respondents, because attractiveness may affect perceived familiarity.[181] Nor should the control stimulus share with the experimental stimulus the feature whose impact is being assessed. If, for example, the control stimulus in a case of alleged trademark infringement is itself a likely source of consumer confusion, reactions to the experimental and control stimuli may not

178. The Federal Trade Commission has long recognized the need for some kind of control for closed-ended questions, although it has not specified the type of control that is necessary. *See* Stouffer Foods Corp., 118 F.T.C. 746, No. 9250, 1994 FTC LEXIS 196, at *31 (Sept. 26, 1994).

179. *See, e.g.*, Cytosport, Inc. v. Vital Pharms., Inc., 617 F. Supp. 2d 1051, 1075–76 (E.D. Cal. 2009) (net confusion level of 25.4% obtained by subtracting 26.5% in the control group from 51.9% in the test group).

180. *See, e.g.*, Skechers USA, Inc. v. Vans, Inc., No. CV-07-01703, 2007 WL 4181677, at *8–9 (C.D. Cal. Nov. 20, 2007) (in trade dress infringement case, control stimulus should have retained design elements not at issue); Procter & Gamble Pharms., Inc. v. Hoffman-LaRoche, Inc., No. 06-Civ-0034, 2006 U.S. Dist. LEXIS 64363, at *87 (S.D.N.Y. Sept. 6, 2006) (in false advertising action, disclaimer was inadequate substitute for appropriate control group).

181. *See, e.g.*, Indianapolis Colts, Inc. v. Metropolitan Baltimore Football Club L.P., 34 F.3d 410, 415–16 (7th Cir. 1994) (court recognized that the name "Baltimore Horses" was less attractive for a sports team than the name "Baltimore Colts."); *see also* Reed-Union Corp. v. Turtle Wax, Inc., 77 F.3d 909, 912 (7th Cir. 1996) (court noted that one expert's choice of a control brand with a well-known corporate source was less appropriate than the opposing expert's choice of a control brand whose name did not indicate a specific corporate source); Louis Vuitton Malletier v. Dooney & Bourke, Inc., 525 F. Supp. 2d 576, 595 (S.D.N.Y. 2007) (underreporting of background "noise" likely occurred because handbag used as control was quite dissimilar in shape and pattern to both plaintiff and defendant's bags).

differ because both cause respondents to express the same level of confusion.[182] In an extreme case, an inappropriate control may do nothing more than control for the effect of the nature or wording of the survey questions (e.g., acquiescence).[183] That may not be enough to rule out other explanations for different or similar responses to the experimental and control stimuli. Finally, it may sometimes be appropriate to have more than one control group to assess precisely what is causing the response to the experimental stimulus (e.g., in the case of an allegedly deceptive ad, whether it is a misleading graph or a misleading claim by the announcer; or in the case of allegedly infringing trade dress, whether it is the style of the font used or the coloring of the packaging).

Explicit attention to the value of control groups in trademark and deceptive-advertising litigation is a relatively recent phenomenon, but courts have increasingly come to recognize the central role the control group can play in evaluating claims.[184] A LEXIS search using *Lanham Act* and *control group* revealed only 4 federal district court cases before 1991 in which surveys with control groups were discussed, 16 in the 9 years from 1991 to 1999, and 46 in the 9 years between 2000 and 2008, a rate of growth that far exceeds the growth in Lanham Act litigation. In addition, courts in other cases have described or considered surveys using control group designs without labeling the comparison group a control group.[185] Indeed, one reason why cases involving surveys with control groups may be underrepresented in reported cases is that a survey with a control group produces

182. *See, e.g.*, Western Publ'g Co. v. Publications Int'l, Ltd., No. 94-C-6803, 1995 U.S. Dist. LEXIS 5917, at *45 (N.D. Ill. May 2, 1995) (court noted that the control product was "arguably more infringing than" the defendant's product) (emphasis omitted). *See also* Classic Foods Int'l Corp. v. Kettle Foods, Inc., 2006 U.S. Dist. LEXIS 97200 (C.D. Cal. Mar. 2, 2006); McNeil-PPC, Inc. v. Merisant Co., 2004 U.S. Dist. LEXIS 27733 (D.P.R. July 29, 2004).

183. *See* text accompanying note 156, *supra*.

184. *See, e.g.*, SmithKline Beecham Consumer Healthcare, L.P. v. Johnson & Johnson-Merck, 2001 U.S. Dist. LEXIS 7061, at *37 (S.D.N.Y. June 1, 2001) (survey to assess implied falsity of a commercial not probative in the absence of a control group); Consumer American Home Prods. Corp. v. Procter & Gamble Co., 871 F. Supp. 739, 749 (D.N.J. 1994) (discounting survey results based on failure to control for participants' preconceived notions); ConAgra, Inc. v. Geo. A. Hormel & Co., 784 F. Supp. 700, 728 (D. Neb. 1992) ("Since no control was used, the . . . study, standing alone, must be significantly discounted."), *aff'd*, 990 F.2d 368 (8th Cir. 1993).

185. Indianapolis Colts, Inc. v. Metropolitan Baltimore Football Club L.P., No. 94727-C, 1994 U.S. Dist. LEXIS 19277, at *10–11 (S.D. Ind. June 27, 1994), *aff'd*, 34 F.3d 410 (7th Cir. 1994). In *Indianapolis Colts*, the district court described a survey conducted by the plaintiff's expert in which half of the interviewees were shown a shirt with the name "Baltimore CFL Colts" on it and half were shown a shirt on which the word "Horses" had been substituted for the word "Colts." *Id.* The court noted that the comparison of reactions to the horse and colt versions of the shirt made it possible "to determine the impact from the use of the word 'Colts.'" *Id.* at *11. *See also* Quality Inns Int'l, Inc. v. McDonald's Corp., 695 F. Supp. 198, 218 (D. Md. 1988) (survey revealed confusion between McDonald's and McSleep, but control survey revealed no confusion between McDonald's and McTavish). *See also* Simon Prop. Group L.P. v. MySimon, Inc., 104 F. Supp. 2d 1033 (S.D. Ind. 2000) (court criticized the survey design based on the absence of a control that could show that results were produced by legally relevant confusion).

less ambiguous findings, which may lead to a resolution before a preliminary injunction hearing or trial occurs.

A less common use of control methodology is a control question. Rather than administering a control stimulus to a separate group of respondents, the survey asks all respondents one or more control questions along with the question about the product or service at issue. In a trademark dispute, for example, a survey indicated that 7.2% of respondents believed that "The Mart" and "K-Mart" were owned by the same individuals. The court found no likelihood of confusion based on survey evidence that 5.7% of the respondents also thought that "The Mart" and "King's Department Store" were owned by the same source.[186]

Similarly, a standard technique used to evaluate whether a brand name is generic is to present survey respondents with a series of product or service names and ask them to indicate in each instance whether they believe the name is a brand name or a common name. By showing that 68% of respondents considered Teflon a brand name (a proportion similar to the 75% of respondents who recognized the acknowledged trademark Jell-O as a brand name, and markedly different from the 13% who thought aspirin was a brand name), the makers of Teflon retained their trademark.[187]

Every measure of opinion or belief in a survey reflects some degree of error. Control groups and, as a second choice, control questions are the most reliable means for assessing response levels against the baseline level of error associated with a particular question.

G. *What Limitations Are Associated with the Mode of Data Collection Used in the Survey?*

Three primary methods have traditionally been used to collect survey data: (1) in-person interviews, (2) telephone interviews, and (3) mail questionnaires.[188] Recently, in the wake of increasing use of the Internet, researchers have added Web-based surveys to their arsenal of tools. Surveys using in-person and telephone interviews, too, now regularly rely on computerized data collection.[189]

186. S.S. Kresge Co. v. United Factory Outlet, Inc., 598 F.2d 694, 697 (1st Cir. 1979). Note that the aggregate percentages reported here do not reveal how many of the same respondents were confused by both names, an issue that may be relevant in some situations. *See* Joseph L. Gastwirth, *Reference Guide on Survey Research*, 36 Jurimetrics J. 181, 187–88 (1996) (review essay).

187. E.I. du Pont de Nemours & Co. v. Yoshida Int'l, Inc., 393 F. Supp. 502, 526–27 & n.54 (E.D.N.Y. 1975); *see also* Donchez v. Coors Brewing Co., 392 F.3d 1211, 1218 (10th Cir. 2004) (respondents evaluated eight brand and generic names in addition to the disputed name). A similar approach is used in assessing secondary meaning.

188. Methods also may be combined, as when the telephone is used to "screen" for eligible respondents, who then are invited to participate in an in-person interview.

189. Wright & Marsden, *supra* note 1, at 13–14.

The interviewer conducting a computer-assisted interview (CAI), whether by telephone (CATI) or face-to-face (CAPI), follows the computer-generated script for the interview and enters the respondent's answers as the interview proceeds. A primary advantage of CATI and other CAI procedures is that skip patterns can be built into the program. If, for example, the respondent answers *yes* when asked whether she has ever been the victim of a burglary, the computer will generate further questions about the burglary; if she answers *no*, the program will automatically skip the followup burglary questions. Interviewer errors in following the skip patterns are therefore avoided, making CAI procedures particularly valuable when the survey involves complex branching and skip patterns.[190] CAI procedures also can be used to control for order effects by having the program rotate the order in which the questions or choices are presented.[191]

Recent innovations in CAI procedures include audio computer-assisted self-interviewing (ACASI) in which the respondent listens to recorded questions over the telephone or reads questions from a computer screen while listening to recorded versions of them through headphones. The respondent then answers verbally or on a keypad. ACASI procedures are particularly useful for collecting sensitive information (e.g., illegal drug use and other HIV risk behavior).[192]

All CAI procedures require additional planning to take advantage of the potential for improvements in data quality. When a CAI protocol is used in a survey presented in litigation, the party offering the survey should supply for inspection the computer program that was used to generate the interviews. Moreover, CAI procedures do not eliminate the need for close monitoring of interviews to ensure that interviewers are accurately reading the questions in the interview protocol and accurately entering the respondent's answers.

The choice of any data collection method for a survey should be justified by its strengths and weaknesses.

1. *In-person interviews*

Although costly, in-person interviews generally are the preferred method of data collection, especially when visual materials must be shown to the respondent under controlled conditions.[193] When the questions are complex and the interviewers are skilled, in-person interviewing provides the maximum opportunity to

190. Willem E. Saris, Computer-Assisted Interviewing 20, 27 (1991).

191. *See, e.g.*, Intel Corp. v. Advanced Micro Devices, Inc., 756 F. Supp. 1292, 1296–97 (N.D. Cal. 1991) (survey designed to test whether the term *386* as applied to a microprocessor was generic used a CATI protocol that tested reactions to five terms presented in rotated order).

192. *See, e.g.*, N. Galai et al., *ACASI Versus Interviewer-Administered Questionnaires for Sensitive Risk Behaviors: Results of a Cross-Over Randomized Trial Among Injection Drug Users* (abstract, 2004), *available at* http://gateway.nlm.nih.gov/MeetingAbstracts/ma?f=102280272.html.

193. A mail survey also can include limited visual materials but cannot exercise control over when and how the respondent views them.

clarify or probe. Unlike a mail survey, both in-person and telephone interviews have the capability to implement complex skip sequences (in which the respondent's answer determines which question will be asked next) and the power to control the order in which the respondent answers the questions. Interviewers also can directly verify who is completing the survey, a check that is unavailable in mail and Web-based surveys. As described *infra* Section V.A, appropriate interviewer training, as well as monitoring of the implementation of interviewing, is necessary if these potential benefits are to be realized. Objections to the use of in-person interviews arise primarily from their high cost or, on occasion, from evidence of inept or biased interviewers. In-person interview quality in recent years has been assisted by technology. Using computer-assisted personal interviewing (CAPI), the interviewer reads the questions off the screen of a laptop computer and then enters responses directly.[194] This support makes it easier to follow complex skip patterns and to promptly submit results via the Internet to the survey center.

2. Telephone interviews

Telephone surveys offer a comparatively fast and lower-cost alternative to in-person surveys and are particularly useful when the population is large and geographically dispersed. Telephone interviews (unless supplemented with mailed or e-mailed materials) can be used only when it is unnecessary to show the respondent any visual materials. Thus, an attorney may present the results of a telephone survey of jury-eligible citizens in a motion for a change of venue in order to provide evidence that community prejudice raises a reasonable suspicion of potential jury bias.[195] Similarly, potential confusion between a restaurant called McBagel's and the McDonald's fast-food chain was established in a telephone survey. Over objections from defendant McBagel's that the survey did not show respondents the defendant's print advertisements, the court found likelihood of confusion based on the survey, noting that "by soliciting audio responses[, the telephone survey] was closely related to the radio advertising involved in the case."[196] In contrast, when words are not sufficient because, for example, the survey is assessing reactions to the trade

194. Wright & Marsden, *supra* note 1, at 13.

195. *See, e.g.*, State v. Baumruk, 85 S.W.3d 644 (Mo. 2002). (overturning the trial court's decision to ignore a survey that found about 70% of county residents remembered the shooting that led to the trial and that of those who had heard about the shooting, 98% believed that the defendant was either definitely guilty or probably guilty); State v. Erickstad, 620 N.W.2d 136, 140 (N.D. 2000) (denying change of venue motion based on media coverage, concluding that "defendants [need to] submit qualified public opinion surveys, other opinion testimony, or any other evidence demonstrating community bias caused by the media coverage"). For a discussion of surveys used in motions for change of venue, see Neal Miller, *Facts, Expert Facts, and Statistics: Descriptive and Experimental Research Methods in Litigation, Part II*, 40 Rutgers L. Rev. 467, 470–74 (1988); National Jury Project, Jurywork: Systematic Techniques (2d ed. 2008).

196. McDonald's Corp. v. McBagel's, Inc., 649 F. Supp. 1268, 1278 (S.D.N.Y. 1986).

dress or packaging of a product that is alleged to promote confusion, a telephone survey alone does not offer a suitable vehicle for questioning respondents.[197]

In evaluating the sampling used in a telephone survey, the trier of fact should consider:

1. Whether (when prospective respondents are not business personnel) some form of random-digit dialing[198] was used instead of or to supplement telephone numbers obtained from telephone directories, because a high percentage of all residential telephone numbers in some areas may be unlisted;[199]
2. Whether any attempt was made to include cell phone users, particularly the growing subpopulation of individuals who rely solely on cell phones for telephone services;[200]
3. Whether the sampling procedures required the interviewer to sample within the household or business, instead of allowing the interviewer to administer the survey to any qualified individual who answered the telephone;[201] and
4. Whether interviewers were required to call back multiple times at several different times of the day and on different days to increase the likelihood of contacting individuals or businesses with different schedules.[202]

197. *See* Thompson Med. Co. v. Pfizer Inc., 753 F.2d 208 (2d Cir. 1985); Incorporated Publ'g Corp. v. Manhattan Magazine, Inc., 616 F. Supp. 370 (S.D.N.Y. 1985), *aff'd without op.*, 788 F.2d 3 (2d Cir. 1986).

198. Random-digit dialing provides coverage of households with both listed and unlisted telephone numbers by generating numbers at random from the sampling frame of all possible telephone numbers. James M. Lepkowski, *Telephone Sampling Methods in the United States*, *in* Telephone Survey Methodology 81–91 (Robert M. Groves et al. eds., 1988).

199. Studies comparing listed and unlisted household characteristics show some important differences. *Id.* at 76.

200. According to a 2009 study, an estimated 26.5% of households cannot be reached by landline surveys, because 2.0% have no phone service and 24.5% have only a cell phone. Stephen J. Blumberg & Julian V. Luke, Wireless Substitution: Early Release of Estimates Based on the National Health Interview Survey, July–December 2009 (2010), *available at* http://www.cdc.gov/nchs/data/nhis/earlyrelease/wireless201005.pdf. People who can be reached only by cell phone tend to be younger and are more likely to be African American or Hispanic and less likely to be married or to own their home than individuals reachable on a landline. Although at this point, the effect on estimates from landline-only telephone surveys appears to be minimal on most topics, on some issues (e.g., voter registration) and within the population of young adults, the gap may warrant consideration. Scott Keeter et al., *What's Missing from National RDD Surveys? The Impact of the Growing Cell-Only Population*, Paper presented at the 2007 Conference of AAPOR, May 2007.

201. This is a consideration only if the survey is sampling individuals. If the survey is seeking information on the household, more than one individual may be able to answer questions on behalf of the household.

202. This applied equally to in-person interviews.

Telephone surveys that do not include these procedures may not provide precise measures of the characteristics of a representative sample of respondents, but may be adequate for providing rough approximations. The vulnerability of the survey depends on the information being gathered. More elaborate procedures are advisable for achieving a representative sample of respondents if the survey instrument requests information that is likely to differ for individuals with listed telephone numbers versus individuals with unlisted telephone numbers, individuals rarely at home versus those usually at home, or groups who are more versus less likely to rely exclusively on cell phones.

The report submitted by a survey expert who conducts a telephone survey should specify:

1. The procedures that were used to identify potential respondents, including both the procedures used to select the telephone numbers that were called and the procedures used to identify the qualified individual to question),
2. The number of telephone numbers for which no contact was made; and
3. The number of contacted potential respondents who refused to participate in the survey.[203]

Like CAPI interviewing,[204] computer-assisted telephone interviewing (CATI) facilitates the administration and data entry of large-scale surveys.[205] A computer protocol may be used to generate and dial telephone numbers as well as to guide the interviewer.

3. Mail questionnaires

In general, mail surveys tend to be substantially less costly than both in-person and telephone surveys.[206] Response rates tend to be lower for self-administered mail surveys than for telephone or face-to-face surveys, but higher than for their Web-based equivalents.[207] Procedures that raise response rates include multiple mailings, highly personalized communications, prepaid return envelopes, incentives or gratuities, assurances of confidentiality, first-class outgoing postage, and followup reminders.[208]

203. Additional disclosure and reporting features applicable to surveys in general are described in Section VII.B, *infra*.

204. See text accompanying note 194, *supra*.

205. *See* Roger Tourangeau et al., The Psychology of Survey Response 289 (2000); Saris, *supra* note 190.

206. *See* Chase H. Harrison, *Mail Surveys and Paper Questionnaires*, *in* Handbook of Survey Research, *supra* note 1, at 498, 499.

207. *See* Mick Couper et al., *A Comparison of Mail and E-Mail for a Survey of Employees in Federal Statistical Agencies*, 15 J. Official Stat. 39 (1999); Mick Couper, Web Surveys: A Review of Issues and Approaches 464, 473 (2001).

208. *See, e.g.*, Richard J. Fox et al., *Mail Survey Response Rate: A Meta-Analysis of Selected Techniques for Inducing Response*, 52 Pub. Op. Q. 467, 482 (1988); Kenneth D. Hopkins & Arlen R.

A mail survey will not produce a high rate of return unless it begins with an accurate and up-to-date list of names and addresses for the target population. Even if the sampling frame is adequate, the sample may be unrepresentative if some individuals are more likely to respond than others. For example, if a survey targets a population that includes individuals with literacy problems, these individuals will tend to be underrepresented. Open-ended questions are generally of limited value on a mail survey because they depend entirely on the respondent to answer fully and do not provide the opportunity to probe or clarify unclear answers. Similarly, if eligibility to answer some questions depends on the respondent's answers to previous questions, such skip sequences may be difficult for some respondents to follow. Finally, because respondents complete mail surveys without supervision, survey personnel are unable to prevent respondents from discussing the questions and answers with others before completing the survey and to control the order in which respondents answer the questions. Although skilled design of questionnaire format, question order, and the appearance of the individual pages of a survey can minimize these problems,[209] if it is crucial to have respondents answer questions in a particular order, a mail survey cannot be depended on to provide adequate data.

4. *Internet surveys*

A more recent innovation in survey technology is the Internet survey in which potential respondents are contacted and their responses are collected over the Internet. Internet surveys in principle can reduce substantially the cost of reaching potential respondents. Moreover, they offer some of the advantages of in-person interviews by enabling the respondent to view pictures, videos, and lists of response choices on the computer screen during the survey. A further advantage is that whenever a respondent answers questions presented on a computer screen, whether over the Internet or in a dedicated facility, the survey can build in a variety of controls. In contrast to a mail survey in which the respondent can examine and/or answer questions out of order and may mistakenly skip questions, a computer-administered survey can control the order in which the questions are displayed so that the respondent does not see a later question before answering an earlier one and so that the respondent cannot go back to change an answer previously given to an earlier question in light of the questions that follow it. The order of the questions or response options can be rotated easily to control for order effects. In addition, the structure permits the survey to remind, or even require, the respondent to answer a question before the next question is presented. One advantage of computer-administered surveys over interviewer-administered

Gullickson, *Response Rates in Survey Research: A Meta-Analysis of the Effects of Monetary Gratuities*, 61 J. Experimental Educ. 52, 54–57, 59 (1992); Eleanor Singer et al., *Confidentiality Assurances and Response: A Quantitative Review of the Experimental Literature*, 59 Pub. Op. Q. 66, 71 (1995); *see generally* Don A. Dillman, Internet Mail and Mixed-Mode Surveys: The Tailored Design Method (3d ed. 2009).

209. Dilman, *supra* note 208, at 151–94.

surveys is that they eliminate interviewer error because the computer presents the questions and the respondent records her own answers.

Internet surveys do have limitations, and many questions remain about the extent to which those limitations impair the quality of the data they provide. A key potential limitation is that respondents accessible over the Internet may not fairly represent the relevant population whose responses the survey was designed to measure. Although Internet access has not approached the 95% penetration achieved by the telephone, the proportion of individuals with Internet access has grown at a remarkable rate, as has the proportion of individuals who regularly use a computer. For example, according to one estimate, use of the Internet among adults jumped from 22% in 1997 to 60% in 2003.[210] Despite this rapid expansion, a digital divide still exists, so that the "have-nots" are less likely to be represented in surveys that depend on Internet access. The effect of this divide on survey results will depend on the population the survey is attempting to capture. For example, if the target population consists of computer users, any bias from systematic underrepresentation is likely to be minimal. In contrast, if the target population consists of owners of television sets, a proportion of whom may not have Internet access, significant bias is more likely. The trend toward greater access to the Internet is likely to continue, and the issue of underrepresentation may disappear in time. At this point, a party presenting the results of a Web-based survey should be prepared to provide evidence on how coverage limitations may have affected the pattern of survey results.

Even if noncoverage error is not a significant concern, courts evaluating a Web-based survey must still determine whether the sampling approach is adequate. That evaluation will depend on the type of Internet survey involved, because Web-based surveys vary in fundamental ways.

At one extreme is the list-based Web survey. This Web survey is sent to a closed set of potential respondents drawn from a list that consists of the e-mail addresses of the target individuals (e.g., all students at a university or employees at a company where each student or employee has a known e-mail address).

At the other extreme is the self-selected Web survey in which Web users in general, or those who happen to visit a particular Web site, are invited to express their views on a topic and they participate simply by volunteering. Whereas the list-based survey enables the researcher to evaluate response rates and often to assess the representativeness of respondents on a variety of characteristics, the self-selected Web survey provides no information on who actually participates or how representative the participants are. Thus, it is impossible to evaluate nonresponse error or even participation rates. Moreover, participants are very likely to self-select on the basis of the nature of the topic. These self-selected pseudosurveys resemble reader polls published in magazines and do not meet standard criteria for legitimate surveys

210. Jennifer C. Day et al., Computer and Internet Use in the United States: 2003, 8–9 (U.S. Census Bureau 2005).

admissible in court.[211] Occasionally, proponents of such polls tout the large number of respondents as evidence of the weight the results should be given, but the size of the sample cannot cure the likely participation bias in such voluntary polls.[212]

Between these two extremes is a large category of Web-based survey approaches that researchers have developed to address concerns about sampling bias and nonresponse error. For example, some approaches create a large database of potential participants by soliciting volunteers through appeals on well-traveled sites.[213] Based on the demographic data collected from those who respond to the appeals, a sample of these panel members are asked to participate in a particular survey by invitation only. Responses are weighted to reduce selection bias.[214] An expert presenting the results from such a survey should be prepared to explain why the particular weighting approach can be relied upon to achieve that purpose.[215]

Another approach that is more costly uses probability sampling from the initial contact with a potential respondent. Potential participants are initially contacted by telephone using random-digit dialing procedures. Those who lack Internet access are provided with the technology to participate. Members from the panel are then invited to participate in a particular survey, and the researchers know the characteristics of participants and nonparticipants from the initial telephone contact.[216] For all surveys that rely on preselected panels, whether nonrandomly or randomly selected, questions have been raised about panel conditioning (i.e., the effect of having participants in earlier surveys respond to later surveys) and the relatively low rate of response to survey invitations. An expert presenting results from a Web-based survey should be prepared to address these issues and to discuss how they may have affected the results.

Finally, the recent proliferation of Internet surveys has stimulated a growing body of research on the influence of formatting choices in Web surveys. Evidence from this research indicates that formatting decisions can significantly affect the quality of survey responses.[217]

211. *See, e.g.*, Merisant Co. v. McNeil Nutritionals, LLC, 242 F.R.D. 315 (E.D. Pa. 2007) (report on results from AOL "instant poll" excluded).

212. *See, e.g.*, Couper (2001), *supra* note 207, at 480–81 (a self-selected Web survey conducted by the National Geographic Society through its Web site attracted 50,000 responses; a comparison of the Canadian respondents with data from the Canadian General Social Survey telephone survey conducted using random-digit dialing showed marked differences on a variety of response measures).

213. *See, e.g.*, Ecce Panis, Inc. v. Maple Leaf Bakery, Inc. 2007 U.S. Dist. LEXIS 85780 (D. Ariz. Nov. 7, 2007).

214. *See, e.g.*, Philip Morris USA, Inc. v. Otamedia Limited, 2005 U.S. Dist. LEXIS 1259 (S.D.N.Y. Jan. 28, 2005).

215. *See, e.g.*, A&M Records, Inc. v. Napster, Inc. 2000 WL 1170106 (N.D. Cal. Aug. 10, 2000) (court refused to rely on results from Internet panel survey when expert presenting the results showed lack of familiarity with panel construction and weighting methods).

216. *See, e.g.*, Price v. Philip Morris, Inc., 219 Ill. 2d 182, 848 N.E.2d 1 (2005).

217. *See, e.g.*, Mick P. Couper et al., *What They See Is What We Get: Response Options for Web Surveys*, 22 Soc. Sci. Computer Rev. 111 (2004) (comparing order effects with radio button and

A final approach to data collection does not depend on a single mode, but instead involves a mixed-mode approach. By combining modes, the survey design may increase the likelihood that all sampling members of the target population will be contacted. For example, a person without a landline may be reached by mail or e-mail. Similarly, response rates may be increased if members of the target population are more likely to respond to one mode of contact versus another. For example, a person unwilling to be interviewed by phone may respond to a written or e-mail contact. If a mixed-mode approach is used, the questions and structure of the questionnaires are likely to differ across modes, and the expert should be prepared to address the potential impact of mode on the answers obtained.[218]

V. Surveys Involving Interviewers

A. Were the Interviewers Appropriately Selected and Trained?

A properly defined population or universe, a representative sample, and clear and precise questions can be depended on to produce trustworthy survey results only if "sound interview procedures were followed by competent interviewers."[219] Properly trained interviewers receive detailed written instructions on everything they are to say to respondents, any stimulus materials they are to use in the survey, and how they are to complete the interview form. These instructions should be made available to the opposing party and to the trier of fact. Thus, interviewers should be told, and the interview form on which answers are recorded should indicate, which responses, if any, are to be read to the respondent. Moreover, interviewers should be instructed to record verbatim the respondent's answers, to indicate explicitly whenever they repeat a question to the respondent, and to record any statements they make to or supplementary questions they ask the respondent.

Interviewers require training to ensure that they are able to follow directions in administering the survey questions. Some training in general interviewing techniques is required for most interviews (e.g., practice in pausing to give the respondent enough time to answer and practice in resisting invitations to express the interviewer's beliefs or opinions). Although procedures vary, there is evidence that interviewer performance suffers with less than a day of training in general interviewing skills and techniques for new interviewers.[220]

drop-box formats); Andy Peytchev et al., *Web Survey Design: Paging Versus Scrolling*, 70 Pub. Op. Q. 212 (2006) (comparing the effects of presenting survey questions in a multitude of short pages or in long scrollable pages).

218. Don A. Dillman & Benjamin L. Messer, *Mixed-Mode Surveys, in* Wright & Marsden, *supra* note 1, at 550, 553.

219. Toys "R" Us, Inc. v. Canarsie Kiddie Shop, Inc., 559 F. Supp. 1189, 1205 (E.D.N.Y. 1983).

220. Fowler & Mangione, *supra* note 158, at 117; Nora Cate Schaeffer et al., *Interviewers and Interviewing, in* Handbook of Survey Research, *supra* note 1, at 437, 460.

The more complicated the survey instrument is, the more training and experience the interviewers require. Thus, if the interview includes a skip pattern (where, e.g., Questions 4–6 are asked only if the respondent says *yes* to Question 3, and Questions 8–10 are asked only if the respondent says *no* to Question 3), interviewers must be trained to follow the pattern. Note, however, that in surveys conducted using CAPI or CATI procedures, the interviewer will be guided by the computer used to administer the questionnaire.

If the questions require specific probes to clarify ambiguous responses, interviewers must receive instruction on when to use the probes and what to say. In some surveys, the interviewer is responsible for last-stage sampling (i.e., selecting the particular respondents to be interviewed), and training is especially crucial to avoid interviewer bias in selecting respondents who are easiest to approach or easiest to find.

Training and instruction of interviewers should include directions on the circumstances under which interviews are to take place (e.g., question only one respondent at a time outside the hearing of any other respondent). The trustworthiness of a survey is questionable if there is evidence that some interviews were conducted in a setting in which respondents were likely to have been distracted or in which others could overhear. Such evidence of careless administration of the survey was one ground used by a court to reject as inadmissible a survey that purported to demonstrate consumer confusion.[221]

Some compromises may be accepted when surveys must be conducted swiftly. In trademark and deceptive advertising cases, the plaintiff's usual request is for a preliminary injunction, because a delay means irreparable harm. Nonetheless, careful instruction and training of interviewers who administer the survey, as well as monitoring and validation to ensure quality control,[222] and complete disclosure of the methods used for all of the procedures followed are crucial elements that, if compromised, seriously undermine the trustworthiness of any survey.

B. *What Did the Interviewers Know About the Survey and Its Sponsorship?*

One way to protect the objectivity of survey administration is to avoid telling interviewers who is sponsoring the survey. Interviewers who know the identity of the survey's sponsor may affect results inadvertently by communicating to respondents their expectations or what they believe are the preferred responses of the survey's sponsor. To ensure objectivity in the administration of the survey, it is standard interview practice in surveys conducted for litigation to do double-blind

221. *Toys "R" Us*, 559 F. Supp. at 1204 (some interviews apparently were conducted in a bowling alley; some interviewees waiting to be interviewed overheard the substance of the interview while they were waiting).

222. *See* Section V.C, *infra*.

research whenever possible: Both the interviewer and the respondent are blind to the sponsor of the survey and its purpose. Thus, the survey instrument should provide no explicit or implicit clues about the sponsorship of the survey or the expected responses. Explicit clues could include a sponsor's letterhead appearing on the survey; implicit clues could include reversing the usual order of the *yes* and *no* response boxes on the interviewer's form next to a crucial question, thereby potentially increasing the likelihood that *no* will be checked.[223]

Nonetheless, in some surveys (e.g., some government surveys), disclosure of the survey's sponsor to respondents (and thus to interviewers) is required. Such surveys call for an evaluation of the likely biases introduced by interviewer or respondent awareness of the survey's sponsorship. In evaluating the consequences of sponsorship awareness, it is important to consider (1) whether the sponsor has views and expectations that are apparent and (2) whether awareness is confined to the interviewers or involves the respondents. For example, if a survey concerning attitudes toward gun control is sponsored by the National Rifle Association, it is clear that responses opposing gun control are likely to be preferred. In contrast, if the survey on gun control attitudes is sponsored by the Department of Justice, the identity of the sponsor may not suggest the kinds of responses the sponsor expects or would find acceptable.[224] When interviewers are well trained, their awareness of sponsorship may be a less serious threat than respondents' awareness. The empirical evidence for the effects of interviewers' prior expectations on respondents' answers generally reveals modest effects when the interviewers are well trained.[225]

C. What Procedures Were Used to Ensure and Determine That the Survey Was Administered to Minimize Error and Bias?

Three methods are used to ensure that the survey instrument was implemented in an unbiased fashion and according to instructions. The first, monitoring the interviews as they occur, is done most easily when telephone surveys are used. A supervisor listens to a sample of interviews for each interviewer. Field settings make monitoring more difficult, but evidence that monitoring has occurred provides an additional indication that the survey has been reliably implemented. Some

223. *See* Centaur Communications, Ltd. v. A/S/M Communications, Inc., 652 F. Supp. 1105, 1111 n.3 (S.D.N.Y. 1987) (pointing out that reversing the usual order of response choices, yes or no, to no or yes may confuse interviewers as well as introduce bias), *aff'd*, 830 F.2d 1217 (2d Cir. 1987).

224. *See, e.g.*, Stanley Presser et al., *Survey Sponsorship, Response Rates, and Response Effects*, 73 Soc. Sci. Q. 699, 701 (1992) (different responses to a university-sponsored telephone survey and a newspaper-sponsored survey for questions concerning attitudes toward the mayoral primary, an issue on which the newspaper had taken a position).

225. *See, e.g.*, Seymour Sudman et al., *Modest Expectations: The Effects of Interviewers' Prior Expectations on Responses*, 6 Soc. Methods & Res. 171, 181 (1977).

monitoring systems, both telephone and field, now use recordings, procedures that may require permission from respondents.

Second, validation of interviews occurs when respondents in a sample are recontacted to ask whether the initial interviews took place and to determine whether the respondents were qualified to participate in the survey. Validation callbacks may also collect data on a few key variables to confirm that the correct respondent has been interviewed. The standard procedure for validation of in-person interviews is to telephone a random sample of about 10% to 15% of the respondents.[226] Some attempts to reach the respondent will be unsuccessful, and occasionally a respondent will deny that the interview took place even though it did. Because the information checked is typically limited to whether the interview took place and whether the respondent was qualified, this validation procedure does not determine whether the initial interview as a whole was conducted properly. Nonetheless, this standard validation technique warns interviewers that their work is being checked and can detect gross failures in the administration of the survey. In computer-assisted interviews, further validation information can be obtained from the timings that can be automatically recorded when an interview occurs.

A third way to verify that the interviews were conducted properly is to examine the work done by each individual interviewer. By reviewing the interviews and individual responses recorded by each interviewer and comparing patterns of response across interviewers, researchers can identify any response patterns or inconsistencies that warrant further investigation.

When a survey is conducted at the request of a party for litigation rather than in the normal course of business, a heightened standard for validation checks may be appropriate. Thus, independent validation of a random sample of interviews by a third party rather than by the field service that conducted the interviews increases the trustworthiness of the survey results.[227]

VI. Data Entry and Grouping of Responses

A. What Was Done to Ensure That the Data Were Recorded Accurately?

Analyzing the results of a survey requires that the data obtained on each sampled element be recorded, edited, and often coded before the results can be tabulated

226. *See, e.g.*, Davis v. Southern Bell Tel. & Tel. Co., No. 89-2839, 1994 U.S. Dist. LEXIS 13257, at *16 (S.D. Fla. Feb. 1, 1994); National Football League Properties, Inc. v. New Jersey Giants, Inc., 637 F. Supp. 507, 515 (D.N.J. 1986).

227. In *Rust Environment & Infrastructure, Inc. v. Teunissen*, 131 F.3d 1210, 1218 (7th Cir. 1997), the court criticized a survey in part because it "did not comport with accepted practice for independent validation of the results."

and processed. Procedures for data entry should include checks for completeness, checks for reliability and accuracy, and rules for resolving inconsistencies. Accurate data entry is maximized when responses are verified by duplicate entry and comparison, and when data-entry personnel are unaware of the purposes of the survey.

B. What Was Done to Ensure That the Grouped Data Were Classified Consistently and Accurately?

Coding of answers to open-ended questions requires a detailed set of instructions so that decision standards are clear and responses can be scored consistently and accurately. Two trained coders should independently score the same responses to check for the level of consistency in classifying responses. When the criteria used to categorize verbatim responses are controversial or allegedly inappropriate, those criteria should be sufficiently clear to reveal the source of disagreements. In all cases, the verbatim responses should be available so that they can be recoded using alternative criteria.[228]

VII. Disclosure and Reporting

A. When Was Information About the Survey Methodology and Results Disclosed?

Objections to the definition of the relevant population, the method of selecting the sample, and the wording of questions generally are raised for the first time when the results of the survey are presented. By that time it is often too late to correct methodological deficiencies that could have been addressed in the planning stages of the survey. The plaintiff in a trademark case[229] submitted a set of proposed survey questions to the trial judge, who ruled that the survey results

228. *See, e.g.*, Revlon Consumer Prods. Corp. v. Jennifer Leather Broadway, Inc., 858 F. Supp. 1268, 1276 (S.D.N.Y. 1994) (inconsistent scoring and subjective coding led court to find survey so unreliable that it was entitled to no weight), *aff'd*, 57 F.3d 1062 (2d Cir. 1995); Rock v. Zimmerman, 959 F.2d 1237, 1253 n.9 (3d Cir. 1992) (court found that responses on a change-of-venue survey incorrectly categorized respondents who believed the defendant was insane as believing he was guilty); Coca-Cola Co. v. Tropicana Prods., Inc., 538 F. Supp. 1091, 1094–96 (S.D.N.Y.) (plaintiff's expert stated that respondents' answers to the open-ended questions revealed that 43% of respondents thought Tropicana was portrayed as fresh squeezed; the court's own tabulation found no more than 15% believed this was true), *rev'd on other grounds*, 690 F.2d 312 (2d Cir. 1982); *see also* Cumberland Packing Corp. v. Monsanto Co., 140 F. Supp. 2d 241 (E.D.N.Y. 2001) (court examined verbatim responses that respondents gave to arrive at a confusion level substantially lower than the level reported by the survey expert).

229. Union Carbide Corp. v. Ever-Ready, Inc., 392 F. Supp. 280 (N.D. Ill. 1975), *rev'd*, 531 F.2d 366 (7th Cir. 1976).

would be admissible at trial while reserving the question of the weight the evidence would be given.[230] The Seventh Circuit called this approach a commendable procedure and suggested that it would have been even more desirable if the parties had "attempt[ed] in good faith to agree upon the questions to be in such a survey."[231]

The *Manual for Complex Litigation, Second,* recommended that parties be required, "before conducting any poll, to provide other parties with an outline of the proposed form and methodology, including the particular questions that will be asked, the introductory statements or instructions that will be given, and other controls to be used in the interrogation process."[232] The parties then were encouraged to attempt to resolve any methodological disagreements before the survey was conducted.[233] Although this passage in the second edition of the Manual has been cited with apparent approval,[234] the prior agreement that the Manual recommends has occurred rarely, and the *Manual for Complex Litigation, Fourth,* recommends, but does not advocate requiring, prior disclosure and discussion of survey plans.[235] As the Manual suggests, however, early disclosure can enable the parties to raise prompt objections that may permit corrective measures to be taken before a survey is completed.[236]

Rule 26 of the Federal Rules of Civil Procedure requires extensive disclosure of the basis of opinions offered by testifying experts. However, Rule 26 does not produce disclosure of all survey materials, because parties are not obligated to disclose information about nontestifying experts. Parties considering whether to commission or use a survey for litigation are not obligated to present a survey that produces unfavorable results. Prior disclosure of a proposed survey instrument places the party that ultimately would prefer not to present the survey in the position of presenting damaging results or leaving the impression that the results are not being presented because they were unfavorable. Anticipating such a situation,

230. Before trial, the presiding judge was appointed to the court of appeals, and so the case was tried by another district court judge

231. *Union Carbide*, 531 F.2d at 386. More recently, the Seventh Circuit recommended filing a motion in limine, asking the district court to determine the admissibility of a survey based on an examination of the survey questions and the results of a preliminary survey before the party undertakes the expense of conducting the actual survey. Piper Aircraft Corp. v. Wag-Aero, Inc., 741 F.2d 925, 929 (7th Cir. 1984). On one recent occasion, the parties jointly developed a survey administered by a neutral third-party survey firm. Scott v. City of New York, 591 F. Supp. 2d 554, 560 (S.D.N.Y. 2008) (survey design, including multiple pretests, negotiated with the help of the magistrate judge).

232. MCL 2d, *supra* note 16, § 21.484.

233. *See id.*

234. *See, e.g.*, National Football League Props., Inc. v. New Jersey Giants, Inc., 637 F. Supp. 507, 514 n.3 (D.N.J. 1986).

235. MCL 4th, *supra* note 16, § 11.493 ("including the specific questions that will be asked, the introductory statements or instructions that will be given, and other controls to be used in the interrogation process.").

236. *See id.*

parties do not decide whether an expert will testify until after the results of the survey are available.

Nonetheless, courts are in a position to encourage early disclosure and discussion even if they do not lead to agreement between the parties. In *McNeilab, Inc. v. American Home Products Corp.*,[237] Judge William C. Conner encouraged the parties to submit their survey plans for court approval to ensure their evidentiary value; the plaintiff did so and altered its research plan based on Judge Conner's recommendations. Parties can anticipate that changes consistent with a judicial suggestion are likely to increase the weight given to, or at least the prospects of admissibility of, the survey.[238]

B. Does the Survey Report Include Complete and Detailed Information on All Relevant Characteristics?

The completeness of the survey report is one indicator of the trustworthiness of the survey and the professionalism of the expert who is presenting the results of the survey. A survey report generally should provide in detail:

1. The purpose of the survey;
2. A definition of the target population and a description of the sampling frame;
3. A description of the sample design, including the method of selecting respondents, the method of interview, the number of callbacks, respondent eligibility or screening criteria and method, and other pertinent information;
4. A description of the results of sample implementation, including the number of
    a. potential respondents contacted,
    b. potential respondents not reached,
    c. noneligibles,
    d. refusals,
    e. incomplete interviews or terminations, and
    f. completed interviews;
5. The exact wording of the questions used, including a copy of each version of the actual questionnaire, interviewer instructions, and visual exhibits;[239]

237. 848 F.2d 34, 36 (2d Cir. 1988) (discussing with approval the actions of the district court). *See also* Hubbard v. Midland Credit Mgmt, 2009 U.S. Dist. LEXIS 13938 (S.D. Ind. Feb. 23, 2009) (court responded to plaintiff's motions to approve survey methodology with a critique of the proposed methodology).

238. Larry C. Jones, *Developing and Using Survey Evidence in Trademark Litigation*, 19 Memphis St. U. L. Rev. 471, 481 (1989).

239. The questionnaire itself can often reveal important sources of bias. *See* Marria v. Broaddus, 200 F. Supp. 2d 280, 289 (S.D.N.Y. 2002) (court excluded survey sent to prison administrators based

6. A description of any special scoring (e.g., grouping of verbatim responses into broader categories);
7. A description of any weighting or estimating procedures used;
8. Estimates of the sampling error, where appropriate (i.e., in probability samples);
9. Statistical tables clearly labeled and identified regarding the source of the data, including the number of raw cases forming the base for each table, row, or column; and
10. Copies of interviewer instructions, validation results, and code books.[240]

Additional information to include in the survey report may depend on the nature of sampling design. For example, reported response rates along with the time each interview occurred may assist in evaluating the likelihood that nonresponse biased the results. In a survey designed to assess the duration of employee preshift activities, workers were approached as they entered the workplace; records were not kept on refusal rates or the timing of participation in the study. Thus, it was impossible to rule out the plausible hypothesis that individuals who arrived early for their shift with more time to spend on preshift activities were more likely to participate in the study.[241]

Survey professionals generally do not describe pilot testing in their survey reports. They would be more likely to do so if courts recognized that surveys are improved by pilot work that maximizes the likelihood that respondents understand the questions they are being asked. Moreover, the Federal Rules of Civil Procedure may require that a testifying expert disclose pilot work that serves as a basis for the expert's opinion. The situation is more complicated when a nontestifying expert conducts the pilot work and the testifying expert learns about the pilot testing only indirectly through the attorney's advice about the relevant issues

on questionnaire that began, "We need your help. We are helping to defend the NYS Department of Correctional Service in a case that involves their policy on intercepting Five-Percenter literature. Your answers to the following questions will be helpful in preparing a defense.").

240. These criteria were adapted from the Council of American Survey Research Organizations, *supra* note 76, § III.B. Failure to supply this information substantially impairs a court's ability to evaluate a survey. *In re* Prudential Ins. Co. of Am. Sales Practices Litig., 962 F. Supp. 450, 532 (D.N.J. 1997) (citing the first edition of this manual). *But see* Florida Bar v. Went for It, Inc., 515 U.S. 618, 626–28 (1995), in which a majority of the Supreme Court relied on a summary of results prepared by the Florida Bar from a consumer survey purporting to show consumer objections to attorney solicitation by mail. In a strong dissent, Justice Kennedy, joined by three other Justices, found the survey inadequate based on the document available to the court, pointing out that the summary included "no actual surveys, few indications of sample size or selection procedures, no explanations of methodology, and no discussion of excluded results . . . no description of the statistical universe or scientific framework that permits any productive use of the information the so-called Summary of Record contains." *Id.* at 640.

241. *See* Chavez v. IBP, Inc., 2004 U.S. Dist. LEXIS 28838 (E.D. Wash. Aug. 18, 2004).

in the case. Some commentators suggest that attorneys are obligated to disclose such pilot work.[242]

C. In Surveys of Individuals, What Measures Were Taken to Protect the Identities of Individual Respondents?

The respondents questioned in a survey generally do not testify in legal proceedings and are unavailable for cross-examination. Indeed, one of the advantages of a survey is that it avoids a repetitious and unrepresentative parade of witnesses. To verify that interviews occurred with qualified respondents, standard survey practice includes validation procedures,[243] the results of which should be included in the survey report.

Conflicts may arise when an opposing party asks for survey respondents' names and addresses so that they can re-interview some respondents. The party introducing the survey or the survey organization that conducted the research generally resists supplying such information.[244] Professional surveyors as a rule promise confidentiality in an effort to increase participation rates and to encourage candid responses, although to the extent that identifying information is collected, such promises may not effectively prevent a lawful inquiry. Because failure to extend confidentiality may bias both the willingness of potential respondents to participate in a survey and their responses, the professional standards for survey researchers generally prohibit disclosure of respondents' identities. "The use of survey results in a legal proceeding does not relieve the Survey Research Organization of its ethical obligation to maintain in confidence all Respondent-identifiable information or lessen the importance of Respondent anonymity."[245] Although no surveyor–respondent privilege currently is recognized, the need for surveys and the availability of other means to examine and ensure their trustworthiness argue for deference to legitimate claims for confidentiality in order to avoid seriously compromising the ability of surveys to produce accurate information.[246]

242. *See* Yvonne C. Schroeder, *Pretesting Survey Questions*, 11 Am. J. Trial Advoc. 195, 197–201 (1987).

243. *See supra* Section V.C.

244. *See, e.g.*, Alpo Petfoods, Inc. v. Ralston Purina Co., 720 F. Supp. 194 (D.D.C. 1989), *aff'd in part and vacated in part*, 913 F.2d 958 (D.C. Cir. 1990).

245. Council of Am. Survey Res. Orgs., *supra* note 76, § I.A.3.f. Similar provisions are contained in the By-Laws of the American Association for Public Opinion Research.

246. United States v . Dentsply Int'l, Inc., 2000 U.S. Dist. LEXIS 6994, at *23 (D. Del. May 10, 2000) (Fed. R. Civ. P. 26(a)(1) does not require party to produce the identities of individual survey respondents); Litton Indus., Inc., No. 9123, 1979 FTC LEXIS 311, at *13 & n.12 (June 19, 1979) (Order Concerning the Identification of Individual Survey-Respondents with Their Questionnaires) (citing Frederick H. Boness & John F. Cordes, *The Researcher–Subject Relationship: The Need for Protection and a Model Statute*, 62 Geo. L.J. 243, 253 (1973)); *see also* Applera Corp. v. MJ Research, Inc., 389 F. Supp. 2d 344, 350 (D. Conn. 2005) (denying access to names of survey respondents); Lampshire

Copies of all questionnaires should be made available upon request so that the opposing party has an opportunity to evaluate the raw data. All identifying information, such as the respondent's name, address, and telephone number, should be removed to ensure respondent confidentiality.

VIII. Acknowledgment

Thanks are due to Jon Krosnick for his research on surveys and his always sage advice.

v. Procter & Gamble Co., 94 F.R.D. 58, 60 (N.D. Ga. 1982) (defendant denied access to personal identifying information about women involved in studies by the Centers for Disease Control based on Fed. R. Civ. P. 26(c) giving court the authority to enter "any order which justice requires to protect a party or persons from annoyance, embarrassment, oppression, or undue burden or expense.") (citation omitted).

Glossary of Terms

The following terms and definitions were adapted from a variety of sources, including Handbook of Survey Research (Peter H. Rossi et al. eds., 1st ed. 1983; Peter V. Marsden & James D. Wright eds., 2d ed. 2010); Measurement Errors in Surveys (Paul P. Biemer et al. eds., 1991); Willem E. Saris, Computer-Assisted Interviewing (1991); Seymour Sudman, Applied Sampling (1976).

branching. A questionnaire structure that uses the answers to earlier questions to determine which set of additional questions should be asked (e.g., citizens who report having served as jurors on a criminal case are asked different questions about their experiences than citizens who report having served as jurors on a civil case).

CAI (computer-assisted interviewing). A method of conducting interviews in which an interviewer asks questions and records the respondent's answers by following a computer-generated protocol.

CAPI (computer-assisted personal interviewing). A method of conducting face-to-face interviews in which an interviewer asks questions and records the respondent's answers by following a computer-generated protocol.

CATI (computer-assisted telephone interviewing). A method of conducting telephone interviews in which an interviewer asks questions and records the respondent's answers by following a computer-generated protocol.

closed-ended question. A question that provides the respondent with a list of choices and asks the respondent to choose from among them.

cluster sampling. A sampling technique allowing for the selection of sample elements in groups or clusters, rather than on an individual basis; it may significantly reduce field costs and may increase sampling error if elements in the same cluster are more similar to one another than are elements in different clusters.

confidence interval. An indication of the probable range of error associated with a sample value obtained from a probability sample.

context effect. A previous question influences the way the respondent perceives and answers a later question.

convenience sample. A sample of elements selected because they were readily available.

coverage error. Any inconsistencies between the sampling frame and the target population.

double-blind research. Research in which the respondent and the interviewer are not given information that will alert them to the anticipated or preferred pattern of response.

error score. The degree of measurement error in an observed score (see true score).

full-filter question. A question asked of respondents to screen out those who do not have an opinion on the issue under investigation before asking them the question proper.

mall intercept survey. A survey conducted in a mall or shopping center in which potential respondents are approached by a recruiter (intercepted) and invited to participate in the survey.

multistage sampling design. A sampling design in which sampling takes place in several stages, beginning with larger units (e.g., cities) and then proceeding with smaller units (e.g., households or individuals within these units).

noncoverage error. The omission of eligible population units from the sampling frame.

nonprobability sample. Any sample that does not qualify as a probability sample.

open-ended question. A question that requires the respondent to formulate his or her own response.

order effect. A tendency of respondents to choose an item based in part on the order of response alternatives on the questionnaire (see primacy effect and recency effect).

parameter. A summary measure of a characteristic of a population (e.g., average age, proportion of households in an area owning a computer). Statistics are estimates of parameters.

pilot test. A small field test replicating the field procedures planned for the full-scale survey; although the terms *pilot test* and *pretest* are sometimes used interchangeably, a pretest tests the questionnaire, whereas a pilot test generally tests proposed collection procedures as well.

population. The totality of elements (individuals or other units) that have some common property of interest; the target population is the collection of elements that the researcher would like to study. Also, universe.

population value, population parameter. The actual value of some characteristic in the population (e.g., the average age); the population value is estimated by taking a random sample from the population and computing the corresponding sample value.

pretest. A small preliminary test of a survey questionnaire. See pilot test.

primacy effect. A tendency of respondents to choose early items from a list of choices; the opposite of a recency effect.

probability sample. A type of sample selected so that every element in the population has a known nonzero probability of being included in the sample; a simple random sample is a probability sample.

probe. A followup question that an interviewer asks to obtain a more complete answer from a respondent (e.g., "Anything else?" "What kind of medical problem do you mean?").

quasi-filter question. A question that offers a "don't know" or "no opinion" option to respondents as part of a set of response alternatives; used to screen out respondents who may not have an opinion on the issue under investigation.

random sample. See probability sample.

recency effect. A tendency of respondents to choose later items from a list of choices; the opposite of a primacy effect.

sample. A subset of a population or universe selected so as to yield information about the population as a whole.

sampling error. The estimated size of the difference between the result obtained from a sample study and the result that would be obtained by attempting a complete study of all units in the sampling frame from which the sample was selected in the same manner and with the same care.

sampling frame. The source or sources from which the individuals or other units in a sample are drawn.

secondary meaning. A descriptive term that becomes protectable as a trademark if it signifies to the purchasing public that the product comes from a single producer or source.

simple random sample. The most basic type of probability sample; each unit in the population has an equal probability of being in the sample, and all possible samples of a given size are equally likely to be selected.

skip pattern, skip sequence. A sequence of questions in which some should not be asked (should be skipped) based on the respondent's answer to a previous question (e.g., if the respondent indicates that he does not own a car, he should not be asked what brand of car he owns).

stratified sampling. A sampling technique in which the researcher subdivides the population into mutually exclusive and exhaustive subpopulations, or strata; within these strata, separate samples are selected. Results can be combined to form overall population estimates or used to report separate within-stratum estimates.

survey-experiment. A survey with one or more control groups, enabling the researcher to test a causal proposition.

survey population. See population.

systematic sampling. A sampling technique that consists of a random starting point and the selection of every *n*th member of the population; it is generally analyzed as if it were a simple random sample and generally produces the same results..

target population. See population.

trade dress. A distinctive and nonfunctional design of a package or product protected under state unfair competition law and the federal Lanham Act § 43(a), 15 U.S.C. § 1125(a) (1946) (amended 1992).

true score. The underlying true value, which is unobservable because there is always some error in measurement; the observed score = true score + error score.

universe. See population.

References on Survey Research

Paul P. Biemer, Robert M. Groves, Lars E. Lyberg, Nancy A. Mathiowetz, & Seymour Sudman (eds.), Measurement Errors in Surveys (2004).

Jean M. Converse & Stanley Presser, Survey Questions: Handcrafting the Standardized Questionnaire (1986).

Mick P. Couper, Designing Effective Web Surveys (2008).

Don A. Dillman, Jolene Smyth, & Leah M. Christian, Internet, Mail and Mixed-Mode Surveys: The Tailored Design Method (3d ed. 2009).

Robert M. Groves, Floyd J. Fowler, Jr., Mick P. Couper, James M. Lepkowski, Eleanor Singer, & Roger Tourangeau, Survey Methodology (2004).

Sharon Lohr, Sampling: Design and Analysis (2d ed. 2010).

Questions About Questions: Inquiries into the Cognitive Bases of Surveys (Judith M. Tanur ed., 1992).

Howard Schuman & Stanley Presser, Questions and Answers in Attitude Surveys: Experiments on Question Form, Wording and Context (1981).

Monroe G. Sirken, Douglas J. Herrmann, Susan Schechter, Norbert Schwarz, Judith M. Tanur, & Roger Tourangeau, Cognition and Survey Research (1999).

Seymour Sudman, Applied Sampling (1976).

Survey Nonresponse (Robert M. Groves, Don A. Dillman, John L. Eltinge, & Roderick J. A. Little eds., 2002).

Telephone Survey Methodology (Robert M. Groves, Paul P. Biemer, Lars E. Lyberg, James T. Massey, & William L. Nicholls eds., 1988).

Roger Tourangeau, Lance J. Rips, & Kenneth Rasinski, The Psychology of Survey Response (2000).

Reference Guide on Estimation of Economic Damages

MARK A. ALLEN, ROBERT E. HALL, AND VICTORIA A. LAZEAR

Mark Allen, J.D., is Senior Consultant at Cornerstone Research, Menlo Park, California.

Robert Hall, Ph.D., is Robert and Carole McNeil Hoover Senior Fellow and Professor of Economics, Stanford University, Stanford, California.

Victoria Lazear, M.S., is Vice President at Cornerstone Research, Menlo Park, California.

CONTENTS

I. Introduction

This reference guide identifies areas of dispute that arise when economic losses are at issue in a legal proceeding. Our focus is on explaining the issues in these disputes rather than taking positions on their proper resolutions. We discuss the application of economic analysis within established legal frameworks for damages. We cover topics in economics that arise in measuring damages and provide citations to cases to illustrate the principles and techniques discussed in the text.

We begin by discussing the qualifications required of experts who quantify damages. We then set forth the standard general approach to damages quantification, with particular focus on defining the harmful event and the alternative, often called the but-for scenario. In principle, the difference between the plaintiff's economic value in the but-for scenario and in actuality measures the loss caused by the harmful act of the defendant. We then consider damages estimation for two cases: (1) a discrete loss of market value and (2) the loss of a flow of income over time, where damages are the discounted value of the lost cash flow. Other topics include the role of inflation, issues relating to income taxes and stock options, adjustments for the time value of money, legal limitations on damages, damages for a new business, disaggregation of damages when there are multiple challenged acts, the role of random events occurring between the harmful act and trial, data for damages measurement, standards for disclosing data to opposing parties, special masters and neutral experts, liquidated damages, damages in class actions, and lost earnings.[1]

Our discussion follows the structure of the standard damages study, as shown in Figure 1. Damages quantification operates on the premise that the defendant is liable for damages from the defendant's harmful act. The plaintiff is entitled to recover monetary damages for losses occurring before and possibly after the time of the trial. The top line of Figure 1 measures the losses before trial; the bottom line measures the losses after trial.[2]

The goal of damages measurement is to find the plaintiff's loss of economic value from the defendant's harmful act. The loss of value may have a one-time character, such as the diminished market value of a business or property, or it may take the form of a reduced stream of profit or earnings. The losses are net of any costs avoided because of the harmful act.

The essential elements of a study of losses are the quantification of the reduction in economic value, the calculation of interest on past losses, and the appli-

1. For a discussion of specific issues relating to estimating damages in antitrust, intellectual property, and securities litigation, see Mark A. Allen et al., *Estimation of Economic Damages in Antitrust, Intellectual Property, and Securities Litigation* (June 2011), *available at* http://www.stanford.edu/~rehall/DamagesEstimation.pdf.

2. Our scope here is limited to losses of actual dollar income. However, economists sometimes have a role in the measurement of nondollar damages, including pain and suffering and the hedonic value of life. *See generally* W. Kip Viscusi, Reforming Products Liability (1991).

Figure 1. Standard format for a damages study.

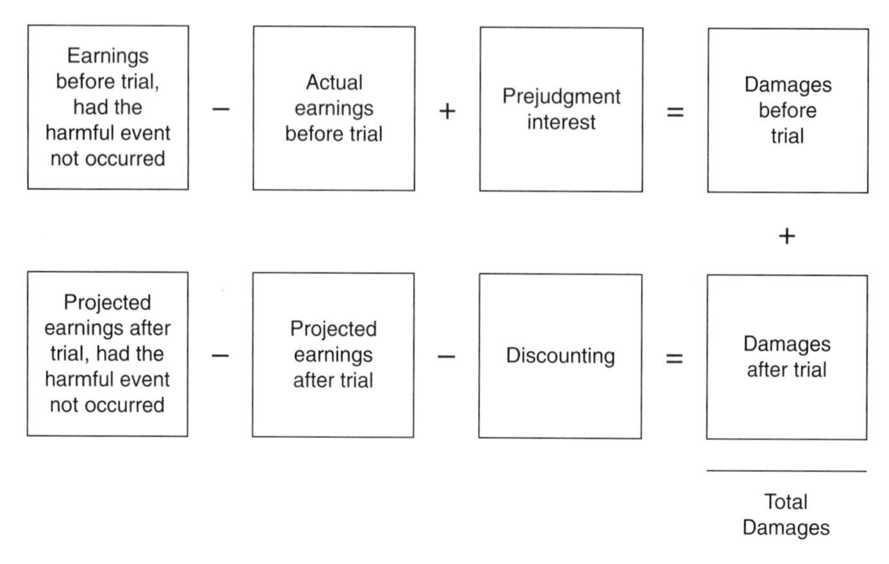

cation of financial discounting to future losses. The losses are the difference between the value the plaintiff would have received if the harmful event had not occurred and the value the plaintiff has or will receive, given the harmful event. The plaintiff may be entitled to interest for losses occurring before trial. Losses occurring after trial are usually discounted to the time of trial. The plaintiff may be due interest on the judgment from the time of trial to the time the defendant actually pays. The majority of damages studies fit this format; thus, we have used such a format as the basic model for this reference guide.

We use numerous brief examples to explain the disputes that can arise. These examples are not full case descriptions; they are deliberately stylized. They attempt to capture the types of disagreements about damages that arise in practical experience, although they are purely hypothetical. In many examples, the dispute involves factual as well as legal issues. We do not try to resolve the disputes in these examples and hope that the examples will help clarify the legal and factual disputes that need to be resolved before or at trial. We introduce many areas of potential dispute with a question, because asking the parties these questions can identify and clarify the majority of disputes over economic damages.

The reader with limited experience in the economic analysis of damages may find it most helpful to begin with Sections II and III and then read Section XII.A, which provides a straightforward application of the principles. Sections IV, V, and VI may be particularly helpful for readers knowledgeable in accounting and valuation. The other sections discuss specific issues relating to damages, and some readers may find it useful to review only those specific to their needs. Section XII.B discusses an application of some of these more specific issues in the context of a damages analysis for a business.

II. Damages Experts' Qualifications

Experts who quantify damages come from a variety of backgrounds. The expert should be trained and experienced in quantitative analysis. For economists, the common qualification is the Ph.D. Damages experts with business or accounting backgrounds often have M.B.A. degrees or other advanced degrees, or C.P.A. credentials. Both the method used and the substance of the damages claim dictate the specific areas of specialization the expert needs. In some cases, participation in original research and authorship of professional publications may add to the qualifications of an expert. However, relevant research and publications are not likely to be on the topic of damages measurement per se but rather on topics and methods encountered in damages analysis. For example, a damages expert may need to restate prices and quantities for a but-for market with more sellers than are present in the actual market. For an expert undertaking this task, direct participation in research on the relation between market structure and performance would be helpful.

Many damages studies use statistical regression analysis.[3] Specific training is required to apply regression analysis. Damages studies sometimes use field surveys.[4] In this case, the damages expert should be trained in survey methods or should work in collaboration with a qualified survey statistician. Because damages estimation often makes use of accounting records, most damages experts need to be able to interpret materials prepared by professional accountants. Some damages issues may require assistance from a professional accountant.

Experts also benefit from professional training and experience in areas relevant to the substance of the damages claim. For example, in antitrust, a background in industrial organization may be helpful; in securities damages, a background in finance may assist the expert; and in the case of lost earnings, an expert may benefit from training in labor economics.

An analysis by even the most qualified expert may face a challenge under the criteria associated with the *Daubert* and *Kumho* cases.[5] These criteria are intended to exclude testimony based on untested and unreliable theories. Relatively few economists serving as damages experts succumb to *Daubert* challenges, because

3. For a discussion of regression analysis, *see generally* Daniel L. Rubinfeld, Reference Guide on Multiple Regression, in this manual.

4. For a discussion of survey methods, *see generally* Shari Seidman Diamond, Reference Guide on Survey Research, in this manual.

5. Daubert v. Merrell Dow Pharms., Inc., 509 U.S. 579 (1993); Kumho Tire Co. v. Carmichael, 526 U.S. 137 (1999). For a discussion of emerging standards of scientific evidence, see Margaret A. Berger, The Admissibility of Expert Testimony, Section IV, in this manual.

most damages analyses operate in the familiar territory of measuring economic values using a combination of professional judgment and standard tools. But the circumstances of each damages analysis are unique, and a party may raise a *Daubert* challenge based on the proposition that the tools have never before been applied to these circumstances. Even if a *Daubert* challenge fails, it can be an effective way for the opposing party to probe the damages analysis prior to trial.

III. The Standard General Approach to Quantification of Damages

In this section, we review the elements of the standard loss measurement in the format of Figure 1. For each element, there are several areas of potential dispute. The sequence of issues discussed here should identify most of the areas of disagreement between the damages analyses of opposing parties.

A. Isolating the Effect of the Harmful Act

The first step in a damages study is the translation of the legal theory of the harmful event into an analysis of the economic impact of that event. In most cases, the analysis considers the difference between the plaintiff's economic position if the harmful event had not occurred and the plaintiff's actual economic position.

In almost all cases, the damages expert proceeds on the hypothesis that the defendant committed the harmful act and that the act was unlawful. Accordingly, throughout this discussion, we assume that the plaintiff is entitled to compensation for losses sustained from a harmful act of the defendant. The characterization of the harmful event begins with a clear statement of what occurred. The characterization also will include a description of the defendant's proper actions in place of its unlawful actions and a statement about the economic situation absent the wrongdoing, with the defendant's proper actions replacing the unlawful ones (the but-for scenario). Damages measurement then determines the plaintiff's hypothetical value in the but-for scenario. Economic damages are the difference between that value and the actual value that the plaintiff achieved.

Because the but-for scenario differs from what actually happened only with respect to the harmful act, damages measured in this way isolate the loss of value caused by the harmful act and exclude any change in the plaintiff's value arising from other sources. Thus, a proper construction of the but-for scenario and measurement of the hypothetical but-for plaintiff's value by definition includes in damages only the loss *caused* by the harmful act. The damages expert using the but-for approach does not usually testify separately about the causal relation between damages and the harmful act, although variations may occur where there are issues about the directness of the causal link.

B. The Damages Quantum Prescribed by Law

In most cases, the law prescribes a damages measure that falls into one of the following five categories:

- *Expectation:* Plaintiff restored to the same financial position as if the defendant had performed as promised.
- *Reliance:* Plaintiff restored to the same position as if the relationship with the defendant or the defendant's misrepresentation (and resulting harm) had not existed in the first place.
- *Restitution:* Plaintiff compensated by the amount of the defendant's gain from the unlawful conduct, also called compensation for unjust enrichment, disgorgement of ill-gotten gains, or compensation for unbargained-for benefits.[6]
- *Statutory:* Plaintiff's compensation is a set amount per occurrence of wrongdoing. This occurs in cases involving violations of state labor codes and in copyright infringement.
- *Punitive:* Compensation rewards the plaintiff for detecting and prosecuting wrongdoing to deter similar future wrongdoing.

Expectation damages[7] often apply to breach of contract claims, where the wrongdoing is the failure to perform as promised, and the but-for scenario hypothesizes the absence of that wrongdoing, that is, proper performance by the defendant. Expectation damages are an amount sufficient to give the plaintiff the same economic value the plaintiff would have received if the defendant had fulfilled the promise or bargain.[8]

6. Courts and commentators often subsume unjust enrichment in defining restitution. Professor Farnsworth, for example, states: "[T]he object of restitution is not the enforcement of a promise, but rather the prevention of unjust enrichment. . . . The party in breach is required to disgorge what he has received in money or services. . . ." *See, e.g.,* E. Allen Farnsworth, Contracts § 12.1, at 814 (1982). However, others have argued that restitution and unjust enrichment are different concepts. *See, e.g.,* James J. Edelman, *Unjust Enrichment, Restitution, and Wrongs,* 79 Tex. L. Rev. 1869 (2001); Peter Birks, *Unjust Enrichment and Wrongful Enrichment,* 79 Tex. L. Rev. 1767 (2001); and Emily Sherwin, *Restitution and Equity: An Analysis of the Principle of Unjust Enrichment,* 79 Tex. L. Rev. 2083 (2001). Judge Posner discusses restitution (defined as returning the breaching party's profits from the breach) in relation to contract damages and unjust enrichment (defined as compensation for unbargained-for benefits) in connection with implied contracts. *See* Richard A. Posner, *Economic Analysis of Law* 130, 151 (1998). *See also* Restatement (Third) of Restitution and Unjust Enrichment (2011).

7. *See* John R. Trentacosta, *Damages in Breach of Contract Cases,* 76 Mich. Bus. J. 1068, 1068 (1997) (describing expectation damages as damages that place the injured party in the same position as if the breaching party completely performed the contract); Bausch & Lomb, Inc. v. Bressler, 977 F.2d 720, 728–29 (2d Cir. 1992) (defining expectation damages as damages that put the injured party in the same economic position the party would have enjoyed if the contract had been performed).

8. *See* Restatement (Second) of Contracts § 344 cmt. a (1981). Expectation has been called "a queer kind of 'compensation,'" because it gives the promisee something it never had, i.e., the benefit

Reliance damages generally apply to torts and to some contract breaches. Such damages restore the plaintiff to the same financial position it would have enjoyed absent the defendant's conduct as well as, in the case of torts, compensation for nonpecuniary losses such as pain and suffering.[9] Reliance most often includes out-of-pocket costs, but may also include compensation for lost opportunities, when appropriate. In such cases, reliance damages may approach expectation damages. For a tort, reliance damages place the plaintiff in a position economically equivalent to the position absent the harmful act.[10] For a breach of contract, measuring damages as the amount of compensation needed to place the plaintiff in the same position as if the contract had not been made in the first place will result in refunding the part of the plaintiff's reliance investment that cannot be recovered in other ways.[11] Thus, reliance damages may be appropriate when the plaintiff made an investment relying on the defendant's performance.

of its bargain. L.L. Fuller & William R. Perdue, Jr., *The Reliance Interest in Contract Damages: 1*, 46 Yale L.J. 52, 53 (1936). The policy underlying expectation damages is that they promote and facilitate reliance on business agreements. *Id.* at 61–62.

9. Generally, the objective of reliance damages is to put the *promisee* or nonbreaching party back to the position in which it would have been had the promise not been made. *See* E. Allan Farnsworth, *Legal Remedies for Breach of Contract*, 70 Colum. L. Rev. 1145, 1148 (1979). *See also* Restatement (Second) of Contracts § 344(b). Reliance damages include expenditures made in preparation for performance and performance itself. Restatement (Second) of Contracts § 349.

10. *See, e.g.*, East River Steamship Corp. v. Transamerica Delaval Inc., 476 U.S. 858, 873 n.9 (1986) ("tort damages generally compensate the plaintiff for loss and return him to the position he occupied before the injury"). The compensatory goal of tort damages is to make the plaintiff whole as nearly as possible through an award of money damages. *See* Randall R. Bovbjerg et al., *Valuing Life and Limb in Tort: Scheduling "Pain and Suffering,"* 83 Nw. U. L. Rev. 908, 910 (1989); John C.P. Goldberg, *Two Conceptions of Tort Damages: Fair v. Full Compensation*, 5 DePaul L. Rev. 435 (2006). Often, the damages expert is not asked to provide guidance relating to estimating damages for nonpecuniary losses such as pain and suffering. However, hedonic analysis may sometimes be used.

11. Economists and legal scholars have debated contract damages and the concepts of expectation and reliance for decades. Fuller and Perdue's definition of reliance included the plaintiff's foregone lost opportunities in addition to his expenditures. But courts that award reliance damages typically award only out-of-pocket expenditures. *See, e.g.*, Michael B. Kelly, *The Phantom Reliance Interest in Contract Damages*, 1992 Wis. L. Rev. 1755, 1771 (1992). Farnsworth has suggested that this is most likely explained by difficulties in damages proof rather than any rule excluding lost opportunities from reliance damages—that is, that the reason for barring the expectation measure (most often lack of proof of damages with reasonable certainty) will apply equally to bar lost opportunities. E Allan Farnsworth, *Precontractual Liability and Preliminary Agreements: Fair Dealing and Failed Negotiations*, 87 Colum. L. Rev. 217, 225 (1987). Reliance damages including lost opportunities may be awarded in cases where the expectation is unavailable because the agreement is illusory or too indefinite to be enforceable. *See, e.g.*, Grouse v. Group Health Plan, Inc., 306 N.W.2d 114 (Minn. 1981), where the plaintiff employee resigned one job and turned down the offer of another in reliance on defendant's promise of employment, but the promised employment would have been at will. The court stated that the proper measure of damages was not what the plaintiff would have earned in his employment with the defendant, but what he lost in quitting his job and turning down an additional offer of employment. *Id.* at 116. Finally, we note that in a competitive market, reliance damages including lost opportunities

Example: Agent contracts with Owner for Agent to sell Owner's farm. The asking price is $1,000,000, and the agreed fee is 6%. Agent incurs costs of $1,000 in listing the property. A potential buyer offers the asking price, but Owner withdraws the listing. Agent calculates damages as $60,000, the agreed fee for selling the property. Owner calculates damages as $1,000, the amount that Agent spent to advertise the property.

Comment: Under the expectation remedy, Agent is entitled to $60,000, the fee for selling the property. However, the Agent has only partly performed under the contract, and thus it may be appropriate to limit damages to $1,000. Some states limit recovery in this situation by law to the $1,000, the reliance measure of damages, unless the property is actually sold.[12]

Restitution damages[13] are often the same, from the perspective of quantification, as reliance damages. If the only loss to the plaintiff from the defendant's harmful act arises from an expenditure that the plaintiff made that cannot otherwise be recovered, the plaintiff receives compensation equal to the amount of that expenditure.[14]

Interesting and often difficult issues arise in cases that involve elements of both contract and tort. Consider a contract for a product that turns out to be defective. Generally, under what has become known as the economic loss rule, if the defective product only causes economic or commercial loss, the dispute is a private matter between the parties, and the contract will likely control their dispute. But if the product causes personal injury or property damage (other than to the product itself), then tort law and tort damages will likely control.[15]

are generally equivalent to expectation damages. *See, e.g.*, Robert Cooter & Melvin Aron Eisenberg, *Damages for Breach of Contract*, 73 Cal. L. Rev. 1432, 1445 (1985).

12. *Compare* Hollinger v. McMichael, 177 Mont. 144, 580 P.2d 927, 929 (1978) (broker earned his commission when he "procured a purchaser able, ready and willing to purchase the seller's property") *with* Ellsworth Dobbs, Inc. v. Johnson, 50 N.J. 528, 236 A.2d 843, 855 (1967) (broker earns commission only when the transaction is completed by closing the title in accordance with the provisions of the contract). *See generally* Steven K. Mulliken, *When Does the Seller Owe the Broker a Commission? A Discussion of the Law and What It Teaches About Listing Agreements*, 132 Mil. L. Rev. 265 (1991).

13. The objective of restitution damages is to put the *promisor* or breaching party back in the position in which it would have been had the promise not been made. Note the traditional legal distinction between restitution and reliance damages: Reliance damages seek to put the *promisee* or nonbreaching party back in the position in which it would have been if the promise had not been made. *See* E. Allan Farnsworth, *Legal Remedies for Breach of Contract*, 70 Colum. L. Rev. 1145, 1148 (1979). Both measures seek to restore the status quo ante. *See also* Restatement (Third) of Restitution and Unjust Enrichment (2011).

14. *See* Restatement (Second) of Contracts § 344(c).

15. Judge Posner has advocated using the term "commercial" rather than "economic" loss because, since personal injuries and property losses destroy values that can be monetized, they are economic losses also. *See* Miller v. United States Steel Corp., 902 F.2d 573, 574 (7th Cir. 1990). *See generally* Dan B. Dobbs, *An Introduction to Non-Statutory Economic Loss Claims*, 48 Ariz. L. Rev. 713

Fraud actions can present particularly difficult problems. For example, if the claim is that the defendant fraudulently induced the plaintiff to enter into an agreement that caused purely commercial losses, the economic loss rule may apply to limit the plaintiff's recovery to only commercial losses for breach of contract, and thus not allow recovery of additional damages recoverable under fraud, such as punitive damages. Generally, courts have taken three approaches to this problem. Some courts have found that the economic loss rule applies to bar the tort claim completely, so that the plaintiff can proceed only under a breach of contract theory. Other courts have found that fraud is an exception to the economic loss doctrine, allowing fraud actions to proceed. A third approach allows a separate fraud action, but only if the fraud is "independent of" or "extraneous to" the contract promises.[16]

A plaintiff asserting fraud can generally recover either out-of-pocket costs or expectation damages,[17] but courts today more commonly award expectation damages to place the plaintiff in the position it would have occupied had the fraudulent statement been true.[18] In cases where the court interprets the fraudulent statement as an actual warranty, then the appropriate remedy is expectation damages. Courts, though, have awarded expectation damages even when the fraudulent statement is not interpreted as an actual warranty. Some of these cases may be situations where a contract exists but is legally unenforceable for technical reasons.

As an alternative, the but-for analysis may consider the value the plaintiff would have received in the absence of the economically detrimental relationship created by the fraud. In this case, the but-for analysis for fraud may adopt the premise that the plaintiff would have entered into a valuable relationship with an entity other than the defendant. For example, if the defendant's misrepresentations have caused the plaintiff to purchase property unsuited to the plaintiff's planned use, then the but-for analysis might consider the value that the plaintiff would have received by purchasing a suitable property from another seller.[19]

(2006); Richard A. Posner, *Common-Law Economic Torts: An Economic and Legal Analysis*, 48 Ariz. L. Rev. 735 (2006).

16. *See, e.g.*, Dan B. Dobbs, *An Introduction to Non-Statutory Economic Loss Claims*, 48 Ariz. L. Rev. 713, 728–30 (2006); Ralph C. Anzivino, *The Fraud in the Inducement Exception to the Economic Loss Doctrine*, 90 Marq. L. Rev. 921, 931–36 (2007); Richard A. Posner, *Common-Law Economic Torts: An Economic and Legal Analysis*, 48 Ariz. L. Rev. 735 (2006); R. Joseph Barton, *Note: Drowning in a Sea of Contract: Application of the Economic Loss Rule to Fraud and Negligent Misrepresentation Claims*, 41 Wm. & Mary L. Rev. 1789 (2000). *See also* Marvin Lumber and Cedar Co. v. PPG Industries, 34 F. Supp. 2d 738 (D.C. Minn. 1999) *aff'd*, 223 F.3d 873 (7th Cir. 2000) (economic loss doctrine barred fraud claim of merchant against manufacturer where facts supporting such claim were not independent of those supporting its UCC contract claims).

17. *See* Restatement (Second) of Torts § 549 (1974). Under the Restatement, expectation damages are available only to "the recipient of a fraudulent misrepresentation in a business transaction," and only for intentional, not negligent, misrepresentation. *Id.* §§ 549(2), 552.

18. *See, e.g.*, Richard Craswell, *Against Fuller and Perdue*, 67 U. Chi. L. Rev. 99, 148 (2000).

19. This measure is equivalent to the reliance interest with recovery for lost opportunities, which can approach expectation damages. *See supra* note 11.

Plaintiffs cannot normally seek punitive damages in a claim for breach of contract,[20] but may seek them in addition to compensatory damages in connection with a tort claim. Although punitive damages are rarely the subject of expert testimony, economists have advanced the concept that punitive damages compensate a plaintiff who brings a case for a wrongdoing that is hard to detect or hard to prosecute. Thus under this concept, punitive damages should be calculated so that the expected recovery for a randomly chosen victim is equal to the victim's loss. To do this, actual damages are multiplied by a factor that is equal to the reciprocal of the probability of both detecting the harmful act and prosecuting the wrongdoer. This adjustment to the damages estimate ensures that the expected recovery from a randomly chosen victim is equal to the victim's loss.[21]

In some situations, the plaintiff may have a choice of remedies under different legal theories. For example, in determining damages for fraud in connection with a contract, damages may be awarded under tort law for deceit or under contract law for breach.[22]

Example: Buyer purchases a condominium from Owner for $900,000. However, the condominium is known by the Owner to be worth only $800,000 at the time of sale because of defects. Buyer chooses to compute damages under the expectation measure of damages as $100,000 and to retain the condominium. Owner computes damages under the reliance measure owed to Buyer as $900,000 and also seeks the return of the condominium to Owner, despite the fact that the condominium is now worth $1,200,000.

Comment: Owner's application of the reliance remedy is incomplete. Absent the fraud, Buyer would have purchased another condominium and enjoyed the general appreciation in the market. Thus, correctly applied, the two measures are likely to be similar.

20. Posner explains that most breaches are either involuntary, where performance is impossible at a reasonable cost, or voluntary but efficient. The policy of contract law is not to compel adherence to contracts, but only to require each party either to perform under the contract or compensate the other party for any resulting injuries. *See* Richard A. Posner, *Economic Analysis of Law, supra* note 6, at 131. For an argument in favor of punitive damages in contracts, see William S. Dodge, *The Case for Punitive Damages in Contracts*, 48 Duke L. J. 629 (1999).

21. *See* A. Mitchell Polinsky & Steven Shavell, *Punitive Damages: An Economic Analysis*, 111 Harv. L. Rev. 879 (1998).

22. This assumes that the economic loss rule does not apply. Generally, plaintiffs will prefer tort remedies to contract remedies because such remedies are broader, affording the possibility of recovery for nonpecuniary losses and punitive damages. For fraud actions, most jurisdictions do not allow recovery for nonpecuniary loses such as emotional distress, although some do if the distress is severe. *See, e.g.*, Nelson v. Progressive Corp., 976 P.2d 859, 868 (Alaska 1999). The Restatement advocates restricting fraud recovery to pecuniary losses. *See* Restatement (Second) of Torts § 549.

A plaintiff may argue that a harmful act has caused significant losses for many years. The defendant may reply that most of the losses that occurred from the injury are the result of causes other than the harmful act. Thus, the defendant may argue that the injury was caused by multiple factors only one of which was the result of the harmful act, or the defendant may argue that the observed injury over time was caused by subsequent events.

Example: Worker is the victim of a disease caused either by exposure to xerxium or by smoking. Worker makes leather jackets tanned with xerxium. Worker sues the producer of the xerxium, Xerxium Mine, and calculates damages as all lost wages. Defendant Xerxium Mine, in contrast, attributes most of the losses to smoking and calculates damages as only a fraction of lost wages.

Comment: The resolution of this dispute will turn on the legal question of comparative or contributory fault. If the law permits the division of damages into parts attributable to exposure to xerxium and smoking, then medical evidence on the likelihood of cause may be needed to make that division. We discuss this topic further in Section VIII.B. on disaggregation of damages.

Example: Real Estate Agent is wrongfully denied affiliation with Broker. Agent's damages study projects past earnings into the future at the rate of growth of the previous 3 years. Broker's study projects that earnings would have declined even without the breach because the real estate market has turned downward.

Comment: The difference between a damages study based on extrapolation from the past, here used by Agent, and a study based on actual data after the harmful act, here used by Broker, is one of the most common sources of disagreement in damages. This is a factual dispute that hinges on the broker demonstrating that there is a relationship between real estate market conditions and the earnings of agents. The example also illustrates how subsequent unexplained events can affect damages calculations, discussed in Section VIII.E.

Frequently, the defendant will calculate damages on the premise that the harmful act had no causal relationship to the plaintiff's losses—that is, that the plaintiff's losses would have occurred without the harmful act. The defendant's but-for scenario will thus describe a situation in which the losses happen anyway. This is equivalent to arguing that the harmful act occurred but the plaintiff suffered no losses.

Example: Contractors conspired to rig bids in a construction deal. State seeks damages for subsequent higher prices. Contractors' damages estimate is zero because they assert that the only effect of the bid rigging was to determine the winner of the contract and that prices were not affected.

Comment: This is a factual dispute about how much effect bid rigging has on the ultimate price. The analysis should go beyond the mechanics of the bid-rigging system to consider how the bids would be different had there been no collaboration among the bidders.

The defendant may also argue that the plaintiff has overstated the scope of the harmful act. Here, the legal character of the harmful act may be critical; the law may limit the scope to proximate effects if the harmful act was negligence, but may require a broader scope if the harmful act was intentional.[23]

Example: Plaintiff Drugstore Network experiences losses because defendant Superstore priced its products predatorily. Drugstore Network reduced prices in all its stores because it has a policy of uniform national pricing. Drugstore Network's damages study considers the entire effect of national price cuts on profits. Defendant Superstore argues that Network should have lowered prices only on the West Coast and its price reductions elsewhere should not be included in damages.

Comment: Whether adherence to a policy of national pricing is the reasonable response to predatory pricing in only part of the market is a question of fact.

C. Is There Disagreement About What Legitimate Conduct of the Defendant Should Be Hypothesized in Projecting the Plaintiff's Earnings but for the Harmful Event?

One party's damages analysis may hypothesize the absence of any act of the defendant that influenced the plaintiff, whereas the other's damages analysis may hypothesize an alternative, legal act. This type of disagreement is particularly common in antitrust and intellectual property disputes. Although disagreement over the alternative scenario in a damages study is generally a legal question, opposing experts may have been given different legal guidance and therefore made different economic assumptions, resulting in major differences in their damages estimates.

Example: Defendant Copier Service's long-term contracts with customers are found to be unlawful because they create a barrier to entry that maintains Copier Service's monopoly power. Rival's damages study hypothesizes

23. *See generally* Prosser and Keeton on the Law of Torts § 65, at 462 (Prosser et al. 5th ed., 1984). Dean Prosser states that simple negligence and intentional wrongdoing differ "not merely in degree but in the kind of fault . . . and in the social condemnation attached to it." *Id.*

no contracts between Copier Service and its customers, so Rival would face no contractual barrier to bidding those customers away from Copier Service. Copier Service's damages study hypothesizes medium-term contracts with its customers and argues that these would not have been found to be unlawful. Under Copier Service's assumption, Rival would have been much less successful in bidding away Copier Service's customers, and damages are correspondingly lower.

Comment: Assessment of damages will depend greatly on the substantive law governing the injury. The proper characterization of Copier Service's permissible conduct usually is an economic issue. However, the expert must also have legal guidance as to the proper legal framework for damages. Counsel for plaintiff may instruct plaintiff's damages expert to use a different legal framework from that of counsel for the defendant.

D. *Does the Damages Analysis Consider All the Differences in the Plaintiff's Situation in the But-For Scenario, or Does It Assume That Many Aspects Would Be the Same as in Actuality?*

The analysis of some types of harmful events requires consideration of effects, such as price erosion,[24] that involve changes in the economic environment caused by the harmful event. For a business, the main elements of the economic environment that may be affected by the harmful event are the prices charged by rivals, the demand facing the seller, and the prices of inputs. For example, misappropriation of intellectual property can cause lower prices because products produced with the misappropriated intellectual property compete with products sold by the owner of the intellectual property. In contrast, some harmful events do not change the plaintiff's economic environment. The theft of some of the plaintiff's products would not change the market price of those products, nor would an injury to a worker change the general level of wages in the labor market. A damages study need not analyze changes in broader markets when the harmful act plainly has minuscule effects in those markets. The plaintiff may assert that, absent the defendant's wrongdoing, a higher price could have been charged and therefore that the defendant's harmful act has eroded the market price. The defendant may reply that the higher price would lower the quantity sold. The parties may then dispute how much the quantity would fall as a result of higher prices.

24. *See, e.g.*, General Am. Transp. Corp. v. Cryo-Trans, Inc., 897 F. Supp. 1121, 1123–24 (N.D. Ill. 1995), *modified*, 93 F.3d 766 (Fed. Cir. 1996); Rawlplug Co., Inc. v. Illinois Tool Works Inc., No. 91 Civ. 1781, 1994 WL 202600, at *2 (S.D.N.Y. May 23, 1994); Micro Motion, Inc. v. Exac Corp., 761 F. Supp. 1420, 1430–31 (N.D. Cal. 1991) (holding in all three cases that the patentee is entitled to recover lost profits due to past price erosion caused by the wrongdoer's infringement).

Example: Valve Maker infringes patent of Rival. Rival calculates lost profits as the profits Rival would have made plus a price-erosion effect. The amount of price erosion is the difference between the higher price that Rival would have been able to charge absent Valve Maker's presence in the market and the actual price. The price-erosion effect is that price difference multiplied by the combined sales volume of Valve Maker and Rival. Defendant Valve Maker counters that the volume would have been lower had the price been higher and measures damages using the lower volume.

Comment: Wrongful competition is likely to cause some price erosion[25] and, correspondingly, some enlargement of the total market because of the lower price. The more elastic the demand, the lower the volume would have been with a higher price. The actual magnitude of the price-erosion effect could be determined by economic analysis.

Price erosion is a common issue in quantifying intellectual property damages. However, price erosion may be an issue in many other commercial disputes. For example, a plaintiff may argue that the disparagement of its product due to false advertising has eroded the product's price.[26]

In more complicated situations, the damages analysis may need to focus on how an entire industry would be affected by the defendant's wrongdoing. For example, one federal appeals court held that a damages analysis for exclusionary conduct must consider that other firms beside the plaintiff would have enjoyed the benefits of the absence of that conduct. Thus, prices would have been lower, and the plaintiff's profits correspondingly less than those posited in the plaintiff's damages analysis.[27]

Example: Computer Printer Maker has used unlawful means to exclude rival suppliers of ink cartridges. Rival calculates damages on the assumption that it would have been the only additional seller in the market absent the exclusionary conduct, and that Rival would have been able to sell its cartridges at the same price actually charged by Printer Maker. Printer Maker counters that other sellers would have entered the market and driven the price down, and so Rival has overstated its damages.

25. *See, e.g., Micro Motion,* 761 F. Supp. at 1430 (citing Yale Lock Mfg. Co. v. Sargent, 117 U.S. 536, 553 (1886), in which the *Micro Motion* court stated that "[i]n most price erosion cases, a patent owner has reduced the actual price of its patented product in response to an infringer's competition").

26. *See, e.g.,* BASF Corp. v. Old World Trading Co., Inc., 41 F.3d 1081 (7th Cir. 1994) (finding that the plaintiff's damages only consisted of lost profits before consideration of price erosion, prejudgment interest, and costs due to the presence of other competitors who would keep prices low).

27. *See* Dolphin Tours, Inc. v. Pacifico Creative Servs., Inc., 773 F.2d 1506, 1512 (9th Cir. 1985).

Comment: Increased competition lowers price in all but the most unusual situations. Again, determination of the number of entrants attracted by the elimination of exclusionary conduct and their effect on the price probably requires a full economic analysis.

A comparison of the parties' statements about the harmful event and the likely impact of its absence will likely reveal differences in legal theories that can result in large differences in damages claims.

Example: Client is the victim of unsuitable investment advice by Broker (all of Client's investments made by Broker are the result of Broker's negligence). Client's damages study measures the sum of the losses of the investments made by Broker, including only the investments that incurred losses. Broker's damages study measures the net loss by including an offset for those investments that achieved gains.

Comment: Client is considering the harmful event to be the recommendation of investments that resulted in losses, whereas Broker is considering the harmful event to be the entire body of investment advice. Under Client's theory, Client would not have made the unsuccessful investments but would have made the successful ones, absent the unsuitable advice. Under Broker's theory, Client would not have made any investments based on Broker's advice.

A clear statement about the plaintiff's situation but for the harmful event is also helpful in avoiding double counting that can arise if a damages study confuses or combines reliance[28] and expectation damages.

Example: Marketer is the victim of defective products made by Manufacturer; Marketer's business fails as a result. Marketer's damages study adds together the out-of-pocket costs of creating the business in the first place and the projected profits of the business had there been no defects. Manufacturer's damages study measures the difference between the profit margin Marketer would have made absent the defects and the profit margin Marketer actually made.

28. *See* Section III.B. Reliance damages are distinguished from expectation damages. Reliance damages are defined as damages that do not place the injured party in as good a position as if the contract had been fully performed (expectation damages) but in the same position as if promises were never made. Reliance damages reimburse the injured party for expenses incurred in reliance of promises made. *See, e.g.*, Satellite Broad. Cable, Inc. v. Telefonica de Espana, S.A., 807 F. Supp. 218 (D.P.R. 1992) (holding that under Puerto Rican law an injured party is entitled to reliance but not expectation damages due to the wrongdoer's willful and malicious termination or withdrawal from precontractual negotiations).

Comment: Marketer has mistakenly added together damages from the reliance principle and the expectation principle.[29] Under the reliance principle, Marketer is entitled to be put back to where it would have been had it not started the business in the first place. Damages are total outlays less the revenue actually received. Under the expectation principle, as applied in Manufacturer's damages study, Marketer is entitled to the profit on the extra sales it would have received had there been no product defects. Out-of-pocket expenses of starting the business would have no effect on expectation damages because they would be present in both the actual and the but-for cases and would offset each other in the comparison of actual and but-for value.

IV. Valuation and Damages

Most damages measurements deal, one way or another, with the question of the economic value of streams of profit or income. In this section, we introduce some of the basic concepts of valuation. In the following two sections, we first address market approaches that use current data on prices and values to estimate value directly. Second, we address income approaches that start by estimating future flows and then discounting them back to a reference date (often referred to as discounting cash flows). The income approaches apply to losses of personal earnings as well as to business losses from lost streams of profits or income, where damages are calculated as the present value of a lost stream of earnings. Although commonly called income approaches, the methods include discounting any form of cash flow, such as revenues and costs as well as income, to arrive at an estimate of damages.

The choice between the two types of approaches is a matter of expert judgment. In some cases, an expert will use both types of approaches. Much of our discussion is stated in terms of business valuation, but the discussion also applies to real estate and other assets.

Some of the ways experts implement a market approach, based on market prices or values, to determine damages include

- Relying on comparables such as a similar business or property,
- Using balance sheet information such as assets and liabilities,

29. The injured party cannot recover both reliance and expectation damages if such recovery would result in double counting. *See, e.g.*, West Haven Sound Development Corp. v. City of West Haven, 514 A.2d 734, 746–47 (Conn. 1986) (plaintiff could seek recovery of reliance expenditures instead of lost profits, but not in addition to lost profits, because reliance expenditures were part of the value of the business as a going concern). *See also* George M. Cohen, *The Fault Lines in Contract Damages,* 80 Va. L. Rev. 1225, 1262 (1994).

- Using known ratios from valuing comparables to measure losses, and
- Multiplying existing valuations by changes in market values from publicly available information.

Different methods that experts use to implement an income approach, based on discounting cash flows, to determine damages include

- Projecting revenues and costs with and without the alleged bad act,
- Adjusting profit streams to present value using measures of inflation and the real rate of interest, and
- Projecting profit streams to present value implicitly using capitalization rates.

Each approach presents challenges. The expert must identify the most appropriate method and implement it properly.

Although these methods may seem different and may rely on different information about the firm, each should generate similar numbers. If not, then there is usually an underlying difference in assumptions. Section V discusses the issues and pitfalls frequently encountered when damages are computed from prices or values, while Section VI discusses the issues and pitfalls frequently encountered when damages are computed relying on discounted cash flows.

V. Quantifying Damages Using a Market Approach Based on Prices or Values

An expert can sometimes measure damages as of the time of the wrongdoing directly from market prices or values. For example, if the defendant's negligence causes the total destruction of the plaintiff's cargo of wheat, worth $17 million at the current market price, damages are simply that amount. The only task for the expert is to restate the damages as the economic equivalent at time of trial, through the calculation of prejudgment interest, a topic we consider in Section VI.G.

In many cases, the expert does not take a market price and apply it directly. The price of the product or object at issue may not itself be known from a market, but the expert can approximate the market value from the prices of similar products or objects. Appraisers are experts whose task is to estimate the fair market value of real estate, equipment, and works of art. Experts who assess the value of businesses—some of whom specialize as business valuation experts—often perform similar functions based on the known market values of comparable businesses.

A. Is One of the Parties Using an Appraisal Approach to the Measurement of Damages?

Damages analyses based on appraisals usually have two parts. The first is an appraisal of the property, and the second is an application of that appraisal to quantify the loss from the harmful act. The starting point for an appraisal is the choice of comparable properties or businesses. For real estate, the comparables are nearby similar properties. For businesses, the comparables are businesses similar in as many ways as possible to the business at issue, based on characteristics such as type of business, type of customers, size, type of location, and so forth. Only in the case of publicly traded companies is there a known market value at virtually all times. For real estate and private businesses, the comparables must have traded hands at a known transaction price fairly recently. Numerous firms sell databases of transaction prices and other data for the use of business valuation experts.

The second step in an appraisal is the adjustment of the comparables to account for differences between each comparable business or property and the one at issue. Business values are often restated as valuation ratios, such as the ratio of price to revenue or to earnings. Real estate is restated as value per square foot of land or interior space. Such ratios usually need to be specific to the type of business or real estate. In particular, rapidly growing businesses and real estate in growing areas have higher valuation ratios than those with zero or negative growth outlooks.

Example: Oil Company deprives Gas Station Operator of the benefits of Operator's business. Oil Company's damages study starts by calculating the ratio of sales value to gasoline sales for five nearby gas station businesses that have sold recently. The ratio is \$0.26 per gallon of sales per year. The Operator sells 1.6 million gallons per year, so the business was worth \$0.26 × 1,600,000 = \$416,000, according to the Oil Company's expert. The Gas Station Expert argues that the sales used by the Oil Company occurred before a major business relocated nearby. Thus, the sales value to gasoline sales should be increased to \$0.30 to reflect the new growth rate as a result of the expected increase in business. He calculates his business to be worth \$0.30 × 1,600,000 = \$480,000.

B. Are the Parties Disputing an Adjustment of an Appraisal for Partial Loss?

In most cases where the appraisal approach is appropriate, the plaintiff has not suffered a total loss of property or business, but rather some impairment of its value. In that case, the damages expert will adapt the appraisal to measure the loss from the impairment. Here, again, the use of valuation ratios is common.

Example: Oil Company breaches an earlier agreement with Gas Station Operator and opens another station near Operator's station. Operator's gasoline sales fall by 700,000 gallons per year. Oil Company's damages study applies the ratio of $0.26 per gallon of sales per year to the loss: $0.26 × 700,000 = $182,000. Operator's damages study uses a regression analysis of the valuation of recently sold businesses and finds that each gallon of *added* sales raises value by $0.47 and so calculates damages as $0.47 × 700,000 = $329,000.

Comment: Because of fixed costs, the average valuation of gasoline sales will be less than the marginal valuation, and the latter is the conceptually correct approach.

C. *Is One of the Parties Using the Assets and Liabilities Approach?*

The assets and liabilities approach starts with the accounting balance sheet of a company and adjusts assets and liabilities to approximate current market values. It then nets the assets and the liabilities to compute the net asset value of the firm. The asset values include the value of intangibles. Because these values are hard to determine, the assets and liabilities method is not generally suited to the valuation of businesses with substantial intangible assets.

D. *Are the Parties Disputing an Adjustment for Market Frictions?*

Purely competitive markets have what economists term a "frictionless" market structure. These markets have (1) a large number of buyers and sellers of a single, homogeneous product; (2) fully informed participants; and (3) the feature that sellers can easily enter or exit from the market. A "friction" is anything that prevents the market from being purely competitive. The markets for businesses and properties have frictions that may make transaction values depart from the usual concept of the price negotiated by a willing seller and a willing buyer. In the case of a forced sale and thus a less willing seller, the transaction price may understate the value. Adverse selection, which occurs when one party knows more about a property or business than the other, may cause severe understatements in some markets.[30] Because equipment with hidden defects is more likely to be offered for sale than equipment in unusually good condition, and sales prices are lower as a result, owners of the good equipment tend not to offer it on the market.

30. *See, e.g.*, George A. Akerlof, *The Market for "Lemons": Quality Uncertainty and the Market Mechanism*, 84 Q. J. Econ. 488 (1970).

Example: Negligence of Tire Maker causes the total loss of a 747 aircraft. Tire Maker's damages analysis uses the prices of 747s of similar age in the used airplane market to set a value of $23 million on the ruined airplane. However, Airline offers the testimony of an economist expert who explains that only a small fraction of 747s are ever put up for sale in the used airplane market. Rather, airlines choose to sell only defective planes because they continue to fly nondefective 747s. He then adjusts for the adverse selection of inferior airplanes in the used market and places a value of $42 million on the plane.

Comment: Although merited in principle, the airline's adjustment is challenging to carry out and is likely to be the subject of expert disagreement.

A major source of friction in property and business markets is the capital gains tax. Because capital gains are taxed only at realization, after-tax sales prices will generally understate the value of a business or property to the existing owners if they have no plans to sell except in the more distant future. The forced sale implicit in any act that harms a business or property imposes a loss on the owners in excess of their after-tax loss. We discuss this topic later in Section VI.E on taxes.

E. Is One of the Parties Relying on Hypothetical Property in Its Damages Analysis?

Plaintiffs may argue that undeveloped land or a business opportunity yet to be pursued was taken from them by the defendant's harmful act and that the value lost should include the value of the still hypothetical improvements. We consider this topic in more detail in Section VIII.A in its most important form, damages for harm to a startup business.

Example: Property Owner sues County for the value of undeveloped property condemned for a rapid transit extension. Owner's damages claim is $18 million, the appraisal value of a hypothetical condominium development on the property less the anticipated cost of building the development. The County's expert, an appraiser, argues that the market value of the property is $2 million, based on comparable undeveloped land nearby.

Comment: In principle, the current market value of undeveloped land and the market value of the same land with proper development, less the cost of that development, should be the same, because buyers would bid based on the value of the undeveloped land. Property Owner probably understated the development costs. But the value of the nearby property may understate the value of the condemned property—it is for sale because it lacks certain features that make it less desirable to develop, such as a view. On the other hand, the Property Owner's valuation does not reflect the probability that the Property Owner may not succeed in building the condominium.

F. What Complications Arise When Anticipation of Damages Affects Market Values?

For publicly traded companies, the harmful act may depress the market value of the company itself. For example, suppose that a manufacturer of wood windows treats its windows with a preservative that is defective, causing the windows to rot. The window manufacturer sues the manufacturer of the preservative for damages from lost sales in addition to the cost of replacing the defective windows. The window manufacturer's expert may be tempted to use the decline following the harm as a measure of damages. In cases when the news of the harm reaches the public discretely, say in a single day, the technique of an event study, commonly used in securities fraud cases, can be used to isolate the special component of the decline in market value.

The problem with using the plaintiff's market value is that the market will anticipate recovery in the form of damages, and this will offset at least some of the decline in market value. In the extreme, if stock traders expect that the plaintiff will receive exactly full compensation, the plaintiff's market value will not change at all when knowledge of the wrongdoing—including the fact that a damages award will be made—hits the stock market. Thus, the use of the observed decline in the value of the plaintiff company at the time of the injury understates the actual amount of harm by an unknown amount, so the expert should consider using other valuation techniques. Note that this understatement arises when the publicly traded company itself stands to recover damages.

Changes in market values have a different role in situations, such as fraud on the market, where the public company is the defendant. If the release of previously fraudulently concealed adverse information causes both a reduction in the value of the company because of the information and a further reduction because the market anticipates that the company will pay damages to investors who overpaid for their shares during the period when the information was concealed, damages may be overstated by an unknown amount.

VI. Quantifying Damages as the Sum of Discounted Lost Cash Flows

The fundamental principle of economics governing the second approach to valuation is that the value of a business is the present value of its expected future cash flows.[31] In forming a present value, the expert multiplies each future year's cash flows by the value today for a dollar received in that future year. This price is the discount factor. Thus, the discount factor reflects the decreased value for a dollar received in the future compared to the value of a dollar received today.

In broad summary, the damages expert using the discounted cash flow approach projects historical and future but-for revenue and cost, actual historical revenue and cost, and projected actual revenue and cost. The difference between revenue and cost is cash flow, and the difference between but-for and actual cash flow is the loss of cash flow attributable to the harmful act. The expert then applies discount rates to each year's lost cash flow to determine damages.

A. Is There Disagreement About But-For Revenues in the Past?

A common source of disagreement about the likely profitability of a business is the absence of a track record of earlier profitability. Whenever the plaintiff is a startup business, the issue will arise of reconstructing the value of a business with no historical benchmark.

Example: Plaintiff Xterm is a failed startup. Defendant VenFund has breached a venture capital financing agreement. Xterm's damages study projects the profits it would have made under its business plan. VenFund's damages estimate, which is much lower, is based on the value of the startup as revealed by sales of Xterm equity made just before the breach.

Comment: Both sides confront factual issues to validate their damages estimates. Xterm needs to show that its business plan was still a reasonable forecast as of the time of the breach. VenFund needs to show that the sale of equity places a reasonable value on the firm, that is, that the equity sale was at arm's length and was not subject to discounts. This dispute can also be characterized as whether the plaintiff is entitled to expectation damages or reliance damages. The jurisdiction may limit damages for firms with no track record.

B. Is There Disagreement About the Costs That the Plaintiff Would Have Incurred but for the Harmful Event?

Where the injury takes the form of lost sales volume, the plaintiff usually has avoided the cost of production for the lost sales. Calculation of these avoided costs is a common area of disagreement about damages. Conceptually, avoided cost is the difference between the cost that would have been incurred at the higher volume of

31. This discussion follows that in Shannon Pratt, Business Valuation Body of Knowledge 85–95 (2d ed. 2003).

sales but for the harmful event and the cost actually incurred at the lower volume of sales achieved. In the format of Figure 1, the avoided-cost calculation is done each year. The following are some of the issues that arise in calculating avoided cost:

- For a firm operating at capacity, expansion of sales is cheaper in the long run than in the short run, whereas, if there is unused capacity, expansion may be cheaper in the short run.
- The costs that can be avoided if sales fall abruptly are smaller in the short run than in the long run.
- Avoided costs may include marketing, selling, and administrative costs as well as the cost of manufacturing.
- Some costs are fixed, at least in the short run, and are not avoided as a result of the reduced volume of sales caused by the harmful act.

Sometimes putting costs into just two categories is useful: those that vary with sales (variable costs) and those that do not vary with sales (fixed costs). This breakdown is approximate, however, and does not do justice to important aspects of avoided costs. In particular, costs that are fixed in the short run may be variable in the longer run. Disputes frequently arise over whether particular costs are fixed or variable. One side may argue that most costs are fixed and were not avoided by losing sales volume, whereas the other side may argue that many costs are variable.

Certain accounting concepts relate to the calculation of avoided cost. Profit-and-loss statements frequently report the "cost of goods sold."[32] Costs in this category are frequently, but not uniformly, avoided when sales volume is lower. But costs in other categories, called "operating costs" or "overhead costs," also may be avoided, especially in the long run. One approach to the measurement of avoided cost is based on an examination of all of a firm's cost categories. The expert determines how much of each category of cost was avoided.

An alternative approach uses regression analysis or some other statistical method to determine how costs vary with sales as a general matter within the firm or across similar firms. The results of such an analysis can be used to measure the costs avoided by the decline in sales volume caused by the harmful act.

C. *Is There Disagreement About the Plaintiff's Actual Revenue After the Harmful Event?*

When the plaintiff has mitigated the adverse effects of the harmful act by making an investment that has not yet paid off at the time of trial, disagreement may arise about the value that the plaintiff has actually achieved.

32. *See, e.g.*, United States v. Arnous, 122 F.3d 321, 323 (6th Cir. 1997) (finding that the district court erred when it relied on government's theory of loss because the theory ignored the cost of goods sold).

Example: Manufacturer breaches agreement with Distributor. Distributor starts a new business that shows no accounting profit as of the time of trial. Distributor's damages study makes no deduction for actual earnings during the period from breach to trial. Manufacturer's damages study places a value on the new business as of the time of trial and deducts that value from damages.

Comment: Some offset for economic value created by Distributor's mitigation efforts may be appropriate. Note that if Distributor made a good-faith effort to create a new business, but was unsuccessful because of adverse events outside its control, the issue of the treatment of unexpected subsequent events will arise.[33]

D. What Is the Role of Inflation?

1. Do the parties use constant dollars for future losses, or are such losses stated in future dollars whose values will be diminished by inflation?

Persistent inflation in the U.S. economy complicates projections of future losses. Although inflation rates in the United States since 1987 have been only in the range of 1% to 3% per year, the cumulative effect of inflation has a pronounced effect on future dollar quantities. At 3% annual inflation, a dollar today buys what $4.38 will buy 50 years from now. Under inflation, the unit of measurement of economic values becomes smaller each year, and this shrinkage must be considered if future losses are measured in the smaller dollars of the future. Calculations of this type are often termed "escalation." Dollar losses grow in the future because of the use of the shrinking unit of measurement. For example, an expert might project that revenues for a firm will rise at approximately 5% per year for the next 10 years—3% because of general inflation and 2% more because of the growth of the firm.[34]

Alternatively, the expert may project future losses in constant dollars without explicitly accounting for escalation for future inflation.[35] The use of constant dollars avoids the problems of dealing with a shrinking unit of measurement. In the example just given, the expert might project that revenues will rise at 2% per year in constant dollars. Constant dollars must be stated with respect to a base year. Thus, a calculation in constant 2009 dollars means that the unit for future measurement is the purchasing power of the dollar in 2009.

33. *See* Section VIII.F.

34. *See* Section VI.D.2.

35. *See, e.g.*, Willamette Indus., Inc. v. Commissioner, 64 T.C.M. (CCH) 202 (1992) (holding expert witness erred in failing to take inflation escalation into account).

2. Are the parties using a discount rate properly matched to the projection?

For future losses, a damages study calculates the amount of compensation needed at the time of trial to replace expected future lost income. The result is discounted future losses;[36] it is also sometimes referred to as the present value of future losses.[37] Discounting is conceptually separate from the adjustment for inflation considered in the preceding section. Discounting is typically carried out in the format shown in Table 1.

Table 1. Calculation of Discounted Loss at 5% Interest

Years in Future	Loss	Discount Factor	Discounted Loss[a]
0	$100	1.000	$100
1	125	0.952	119
2	130	0.907	118
Total			$337

[a]Discounted Loss = Loss × Discount Factor.

"Loss" is the estimated future loss, in either escalated or constant-dollar form. "Discount factor" is a factor that calculates the number of dollars needed at the time of trial to compensate for a lost dollar in the future year. The discount factor is the ratio of the value at a future date of a cash flow received today to its value today. It is calculated from the discount rate, which is the interest rate that values a cash flow at a future date. If the current 1-year interest rate is 5%, then the discount rate is 1.05—the value of $1 will be $1.05 a year from now. The discount factor will therefore be $1/$1.05. The 2-year discount rate is the square of 1.05, and the discount factor will be 1/(1.05 × 1.05) Thus, the discount factor is computed by compounding the discount rate forward from the base year to the future year and then taking the reciprocal.

For example, in Table 1, the interest rate is 5%. As discussed, the discount factor for the next year is calculated as the reciprocal of 1.05, and the discount factor for 2 years in the future is calculated as the reciprocal of 1.05 squared. Future discounts would be obtained by multiplying by 1.05 a suitably larger number of times and then taking the reciprocal. The discounted loss is the loss multiplied by the discount factor for that year. The number of dollars at time of trial that compensates for the loss is the sum of the discounted losses, $337 in this example.

36. *See generally* Michael A. Rosenhouse, Annotation, *Effect of Anticipated Inflation on Damages for Future Losses—Modern Cases*, 21 A.L.R. 4th 21 (1981) (discussing discounted future losses extensively).

37. *See generally* George A. Schieren, *Is There an Advantage in Using Time-Series to Forecast Lost Earnings?* 4 J. Legal Econ. 43 (1994) (discussing effects of different forecasting methods on present discounted value of future losses). *See, e.g.*, Wingad v. John Deere & Co., 523 N.W.2d 274, 277–79 (Wis. Ct. App. 1994) (calculating present discounted value of future losses).

To discount a future loss projected in escalated terms, one should use an ordinary interest rate. For example, in Table 1, if the losses of $125 and $130 are in dollars of those years, and not in constant dollars of the initial year, then the use of a 5% discount rate is appropriate if 5% represents an accurate measure of the current interest rate, also known as the time value of money. The ordinary interest rate is often called the nominal interest rate to distinguish it from the real interest rate.

To discount a future loss projected in constant dollars, one should use a real interest rate as the discount rate. A real interest rate is an ordinary interest rate less an assumed rate of future inflation.[38] In Table 1, the use of a 5% discount rate for discounting constant-dollar losses would be appropriate if the ordinary interest rate was 8% and the rate of inflation was 3%.[39] Then the real interest rate would be 8% minus 3%, or 5%. The deduction of the inflation rate from the discount rate is the counterpart of the omission of escalation for inflation from the projection of future losses.

3. *Is one of the parties assuming that discounting and earnings growth offset each other?*

An expert might make the assumption that future growth of losses will occur at the same rate as the discount rate. Table 2 illustrates the standard format for this method of calculating discounted loss.

Table 2. Calculation of Discounted Loss When Growth and Discounting Offset Each Other

Years in Future	Loss	Discount Factor	Discounted Loss[a]
0	$100	1.000	$100
1	105	0.952	100
2	110	0.907	100
Total			$300

[a]Discounted Loss = Loss × Discount Factor.

When growth and discounting exactly offset each other, the present discounted value is the number of years of lost future earnings multiplied by the

38. Some experts rely on the real interest rate inferred from the price of TIPS (Treasury Inflation Protected Securities).

39. Technically, the formula is: (1 + real rate of interest) = (1 + ordinary rate of interest)/(1 + inflation). However, the difference is diminimus unless the ordinary rate of interest is high. Thus, using this formula, the real interest rate is 4.85%.

current amount of lost earnings.[40] In Table 2, the loss of $300 is exactly three times the base year's loss of $100. Thus the discounted value of future losses can be calculated by a shortcut in this special case. The explicit projection of future losses and the discounting back to the time of trial are unnecessary. However, the parties may dispute whether the assumption that growth and discounting are exactly offsetting is realistic in view of projected rates of growth of losses and market interest rates at the time of trial.

In *Jones & Laughlin Steel Corp. v. Pfeifer*,[41] the Supreme Court considered the issue of escalated dollars with nominal discounting against constant dollars with real discounting. It found both acceptable, although the Court seemed to express a preference for the second format.

E. Are Losses Measured Before or After the Plaintiff's Income Taxes?

A damages award compensates the plaintiff for lost economic value. In principle, the calculation of compensation should measure the plaintiff's loss after taxes and then calculate the magnitude of the pretax award needed to compensate the plaintiff fully, once taxation of the award is considered. In practice, the tax rates applied to the original loss and to the compensation are frequently the same. When the rates are the same, the two tax adjustments are a wash. In that case, the appropriate pretax compensation is simply the pretax loss, and the damages calculation may be simplified by the omission of tax considerations.[42]

In some damages analyses, explicit consideration of taxes is essential, and disagreements between the parties may arise about these tax issues. If the plaintiff's lost income would have been taxed as a capital gain (at a preferential rate), but the damages award will be taxed as ordinary income, the plaintiff can be expected to include an explicit calculation of the extra compensation needed to make up for the loss of the tax advantage. Sometimes tax considerations are paramount in damages calculations.[43]

40. Certain state courts have, in the past, required that the offset rule be used so as to avoid speculation about future earnings growth. In *Beaulieu v. Elliott*, 434 P.2d 665, 671–72 (Alaska 1967), the court ruled that discounting was exactly offset by wage growth. In *Kaczkowki v. Bolubasz*, 421 A.2d 1027, 1036–38 (Pa. 1980), the Pennsylvania Supreme Court ruled that no evidence on price inflation was to be introduced and deemed that inflation was exactly offset by discounting.

41. 462 U.S. 523 (1983).

42. There is a separate issue about the effect of taxes on the interest rate for prejudgment interest and discounting. *See* discussion *infra* Sections VI.G, VI.H.

43 *See generally* John H. Derrick, Annotation, *Damages for Breach of Contract as Affected by Income Tax Considerations*, 50 A.L.R. 4th 452 (1987) (discussing a variety of state and federal cases in which courts ruled on the propriety of tax considerations in damage calculations; courts have often been reluctant to award difference in taxes as damages because it is calling for too much speculation).

Example: Trustee wrongfully sells Beneficiary's property at full market value. Beneficiary would have owned the property until death and deferred the capital gains tax.

Comment: Damages are the difference between the actual capital gains tax and the present value of the future capital gains tax that would have been paid but for the wrongful sale, even though the property sold at its full value.

In some cases, the law requires different tax treatment of loss and compensatory awards. Again, the tax adjustments do not offset each other, and consideration of taxes may be a source of dispute.

Example: Driver injures Victim in a truck accident. A state law provides that awards for personal injury are not taxable, even though the income lost as a result of the injury would have been taxable. Victim calculates damages as lost pretax earnings, but Driver calculates damages as lost earnings after tax.[44] Driver argues that the nontaxable award would exceed actual economic loss if it were not adjusted for the taxation of the lost income.

Comment: Under the principle that damages are to restore the plaintiff to the economic equivalent of the plaintiff's position absent the harmful act, it may be recognized that the income to be replaced by the award would have been taxed. However, the law in a particular jurisdiction may not allow a jury instruction on the taxability of an award.[45]

Example: Worker is wrongfully deprived of tax-free fringe benefits by Employer. Under applicable law, the award is taxable. Worker's damages estimate includes a factor so that the amount of the award, after tax, is sufficient to replace the lost tax-free value.

Comment: Again, to achieve the goal of restoring plaintiff to a position economically equivalent absent the harmful act, an adjustment of this type is

44. *See generally* Brian C. Brush & Charles H. Breedon, *A Taxonomy for the Treatment of Taxes in Cases Involving Lost Earnings*, 6 J. Legal Econ. 1 (1996) (discussing four general approaches for treating tax consequences in cases involving lost future earnings or earning capacity based on the economic objective and the tax treatment of the lump sum award). *See, e.g.*, Myers v. Griffin-Alexander Drilling Co., 910 F.2d 1252 (5th Cir. 1990) (holding loss of past earnings between the time of the accident and the trial could not be based on pretax earnings).

45. *See generally* John E. Theuman, Annotation, *Propriety of Taking Income Tax into Consideration in Fixing Damages in Personal Injury or Death Action*, 16 A.L.R. 4th 589 (1981) (discussing a variety of state and federal cases in which the propriety of jury instructions regarding tax consequences is at issue). *See, e.g.*, Bussell v. DeWalt Prods. Corp., 519 A.2d 1379 (N.J. 1987) (holding that trial court hearing a personal injury case must instruct jury, upon request, that personal injury damages are not subject to state and federal income taxes); Gorham v. Farmington Motor Inn, Inc., 271 A.2d 94 (Conn. 1970) (holding court did not err in refusing to instruct jury that personal injury damages were tax-free).

appropriate. The adjustment is often called "grossing up" damages.[46] To accomplish grossing up, divide the lost tax-free value by one minus the tax rate. For example, if the loss is $100,000 of tax-free income, and the income tax rate is 25%, the award should be $100,000 divided by 0.75, or $133,333.

F. Is There a Dispute About the Costs of Stock Options?

In some firms, employee stock options are a significant part of total compensation. Stock options are often used by startup businesses because the options do not require the business to pay out any cash. However, at a future date, the options may be exercised and the option holder will pay only the price per share at the time the options are received as opposed to the price per share at the time the options are exercised. In this way, the firm transfers part of the compensation costs incurred today to the firm's shareholders.

The parties may dispute whether the value of options should be included in the costs avoided by the plaintiff as a result of lost sales volume. The defendant might argue that stock options should be included, because their issuance is costly to the shareholders. The defendant might place a value on newly issued options and amortize this value over the period from issuance to vesting. The plaintiff, in contrast, might exclude options costs because the options cost the firm no cash payout, even though they impose costs on the firm's shareholders.

Example: Firm A pays its sales manager $2000 for every machine sold at $100,000 as well as options to purchase 1600 shares in a year at the existing price of $10 per share. As a result of B's disparagement of A, A asserts that it lost $10,000,000 in sales (100 machines). In its damages analysis, A states that the lost sales represent lost profits of $5,800,000: $10,000,000 less $4,000,000 in avoided production costs and $200,000 in avoided sales commissions. Defendant B calculates that each stock option is worth $5 today based on an analysis using accepted financial models to value the options. Thus, B asserts that damages are $5,000,000: $10,000,000 less $4,000,000 in avoided production costs and $1,000,000 in sales commissions ($200,000 plus $5 × 100 × 1600).

Comment: The costs of the options will never show up on the profit-and-loss statements for Firm A, even if exercised. However, Firm A will receive a lower value for each share it sells either to an investor or through an IPO to reflect the potential future dilution in its shares outstanding.

46. *See* Cecil D. Quillen, Jr., *Income, Cash, and Lost Profits Damages Awards in Patent Infringement Cases*, 2 Fed. Circuit B.J. 201, 207 (1992) (discussing the importance of taking tax consequences and cash flows into account when estimating damages).

G. Is There a Dispute About Prejudgment Interest?[47]

The law may specify how to calculate interest for losses prior to a verdict on liability, generally termed "prejudgment interest." The law may exclude prejudgment interest, specify prejudgment interest to be a statutory rate, or exclude compounded interest. Table 3 illustrates these alternatives. With simple uncompounded interest, losses from 5 years before trial earn five times the specified interest, and so compensation for a $100 loss from 5 years ago is $135 at 7% interest. With compound interest, the plaintiff earns interest on past interest. Compensation at 7% interest compounded is about $140 for a loss of $100 five years before trial. The difference between simple and compound interest becomes much larger if the time from loss to trial is greater or if the interest rate is higher. Because interest receipts in practice do earn further interest, economic analysis generally supports the use of compound interest.

Table 3. Calculation of Prejudgment Interest (in Dollars)

Years Before Trial	Loss Without Interest	Loss with Compound Interest at 7%	Loss with Simple Uncompounded Interest at 7%
10	100	197	170
9	100	184	163
8	100	172	156
7	100	161	149
6	100	150	142
5	100	140	135
4	100	131	128
3	100	123	121
2	100	114	114
1	100	107	107
0	100	100	100
Total	1100	1579	1485

47. *See generally* Michael S. Knoll, *A Primer on Prejudgment Interest,* 75 Tex. L. Rev. 293 (1996) (discussing prejudgment interest extensively). *See, e.g.,* Ford v. Rigidply Rafters, Inc., 984 F. Supp. 386, 391–92 (D. Md. 1997) (specifying a method of calculating prejudgment interest in an employment discrimination case to ensure plaintiff is fairly compensated rather than given a windfall); Acron/Pacific Ltd. v. Coit, No. C-81-4264-VRW, 1997 WL 578673, at *2 (N.D. Cal. Sept. 8, 1997) (reviewing supplemental interest calculations and applying California state law to determine the appropriate amount of prejudgment interest to be awarded); Prestige Cas. Co. v. Michigan Mut. Ins. Co., 969 F. Supp. 1029 (E.D. Mich. 1997) (analyzing Michigan state law to determine the appropriate prejudgment interest award).

Where the law does not prescribe the form of interest for past losses, the experts will normally apply a reasonable interest rate to bring those losses forward. The parties may disagree on whether the interest rate should be measured before or after tax. The before-tax interest rate is the normally quoted rate. To calculate the corresponding after-tax rate, one subtracts the amount of income tax the recipient would have to pay on the interest. Thus, the after-tax rate depends on the tax situation of the plaintiff. The format for calculation of the after-tax interest rate is shown in the following example:

1. Interest rate before tax: 9%
2. Tax rate: 30%
3. Tax on interest (line 1 times line 2): 2.7%
4. After-tax interest rate (line 1 less line 3): 6.3%

Even where damages are calculated on a pretax basis, economic considerations suggest that the prejudgment interest rate should be on an after-tax basis: Had a taxpaying plaintiff actually received the lost earnings in the past and invested the earnings at the assumed rate, income tax would have been due on the interest. The plaintiff's accumulated value would be the amount calculated by compounding past losses at the after-tax interest rate.

Where there is economic disparity between the parties, there may be a disagreement about whose interest rate should be used—the borrowing rate of the defendant or the lending rate of the plaintiff, or some other rate. There may also be disagreements about adjustment for risk.[48]

Example: Crop Insurance Company disputes payment of insurance to Farmer. Farmer calculates damages as the payment due plus the large amount of interest charged by a personal finance company; no bank was willing to lend to her, given her precarious financial condition. Crop Insurer calculates damages as a lower payment plus the interest on the late payment at the normal bank loan rate.

Comment: The law may limit claims for prejudgment interest to a specified interest rate, and a court may hold that this situation falls within the limit. Economic analysis does support the idea that delays in payments are more costly to people with higher borrowing rates and that the actual rate incurred may be considered damages.

48. *See generally* James M. Patell et al., *Accumulating Damages in Litigation: The Roles of Uncertainty and Interest Rates*, 11 J. Legal Stud. 341 (1982) (extensive discussion of interest rates in damages calculations).

H. Is There Disagreement About the Interest Rate Used to Discount Future Lost Value?

Discount calculations should use a reasonable interest rate drawn from current data at the time of trial for losses projected to occur after trial. The interest rate might be obtained from the rates that could be earned in the bond market from a bond of maturity comparable to the lost stream of receipts. As in the case of prejudgment interest, there is an issue as to whether the interest rate should be on a before- or after-tax basis. The parties may also disagree about adjusting the interest rate for risk. A common approach for determining the interest on lost business profit is to use the Capital Asset Pricing Model (CAPM)[49] to calculate the risk-adjusted discount rate. The CAPM is the standard method in financial economics to analyze the relation between risk and discounting. In the CAPM method, the expert first measures the firm's "beta"—the ratio of the percent variation in one firm's value to the percent variation in the value of all businesses. That is, if the index of value for a representative set of firms[50] increases by 10% over a year and the firm has a beta of 1.5, then its value is expected to increase by 15% over a year. Then the risk-adjusted discount rate is the risk-free rate from a U.S. Treasury security plus the beta multiplied by the historical average risk premium for the stock market.[51] The calculation may be presented in the following format:

1. Risk-free interest rate: 4.0%
2. Beta for this firm: 1.2
3. Market equity premium: 6.0%
4. Equity premium for this firm (line 2 times line 3): 7.2%
5. Discount rate for this firm (line 1 plus line 4): 11.2%

I. Is One of the Parties Using a Capitalization Factor?

Another approach to discounting a stream of losses uses a market capitalization factor. A capitalization factor is the ratio of the value of a future stream of income to the current amount of the stream; for example, if a firm is worth $1 million and its current earnings are $100,000, its capitalization factor is ten.

The capitalization factor generally is obtained from the market values of comparable assets or businesses. For example, the expert might locate a comparable business traded in the stock market and compute the capitalization factor as

49. *See, e.g.*, Cede & Co. v. Technicolor, Inc., No. Civ.A.7129, 1990 WL 161084 (Del. Ch. Oct. 19, 1990) (Mem.) (assessing the propriety of using CAPM to determine the discount rate); Gilbert v. MPM Enters., Inc., No. 14416, 1997 WL 633298, at *8 (Del. Ch. Oct. 9, 1997) (finding that petitioner's expert witnesses' use of CAPM is appropriate).

50. For example, the S&P 500.

51. Richard A. Brealey et al., Principles of Corporate Finance 213–22 (9th ed. 2008).

the ratio of stock market value to operating income. In addition to capitalization factors derived from markets, experts sometimes use rule-of-thumb capitalization factors. For example, the value of a dental practice might be taken as 2 year's gross revenue (the capitalization factor for revenue is 2). Often the parties dispute whether there is reliable evidence that the capitalization factor accurately measures value for the specific asset or business.

Once the capitalization factor is determined, the calculation of the discounted value of the loss is straightforward: It is the current annual loss in operating profit multiplied by the capitalization factor. A capitalization factor approach to valuing future losses may be formatted in the following way:

1. Ratio of market value to current annual earnings in comparable publicly traded firms: 13
2. Plaintiff's lost earnings over past year: $200,000
3. Value of future lost earnings (line 1 times line 2): $2,600,000

The capitalization factor approach might also be applied to revenue, cash flow, accounting profit, or other measures. The expert might adjust market values for any differences between the valuation principles relevant for damages and those that the market applies. For example, the value in the stock market may be considered the value placed on a business for a minority interest, whereas the plaintiff's loss relates to a controlling interest. In this case, the expert would adjust the capitalization factor upward to account for the value of the control rights. The parties may dispute almost every element of the capitalization calculation.

Example: Lender is responsible for failure of Auto Dealer. Plaintiff Auto Dealer's damages study projects rapid growth of future profits based on the current year's profit but for Lender's misconduct. The study uses a discount rate calculated as the after-tax interest rate on Treasury bills. As a result, the application of the discount rate to the future stream of earnings implies a capitalization rate of 12 times the current pretax profit. The resulting estimate of lost value is $10 million. Defendant Lender's damages study uses data on the actual sale prices of similar dealerships in various parts of the country. The data show that the typical sales price of a dealership is six times its 5-year average annual pretax profit. Lender's damages study multiplies the capitalization factor of six by the 5-year average annual pretax profit of Auto Dealer of $500,000 to estimate lost value as $3 million.

Comment: Part of the difference between the two damages studies comes from the higher implied capitalization factor used by Auto Dealer. Another reason for the differences may be that the 5-year average pretax profit is less than the current-year profit.

VII. Limitations on Damages

The law imposes four important limitations on a plaintiff's ability to recover losses as damages: (1) a plaintiff must prove its damages with reasonable certainty, (2) a plaintiff may not recover damages that are too remote, (3) a plaintiff has a duty to mitigate its damages, and (4) a liquidated damages clause may limit the amount of damages by prior agreement.

A. *Is the Defendant Arguing That Plaintiff's Damages Estimate Is Too Uncertain and Speculative?*

In general, damages law holds that a plaintiff may not recover damages beyond an amount proven with reasonable certainty.[52] This rule permits damages estimates that are not mathematically certain but excludes those that are speculative.[53] Failure to prove damages to a reasonable certainty is a common defense. The determination of what constitutes speculation is increasingly a matter of law to be determined prior to trial in a *Daubert* proceeding.[54]

Courts and commentators have long recognized the difficulties in defining what constitutes reasonable certainty or speculation in a damages analysis. The exclusion of damages on grounds of excessive uncertainty regarding the amount of damages may result in an award of zero damages when it is likely that the plaintiff suffered significant damages, even though the actual amount is quite uncertain.

There are three contexts in which reasonable certainty or speculation can arise: (1) where the outcome is uncertain, (2) where it is argued that the expert has not used the best method or data, or (3) where the damages suffered by a specific plaintiff are uncertain.

Traditionally, damages are calculated without reference to uncertainty about outcomes in the but-for scenario. Outcomes are taken as actually occurring if they are the expected outcome and as not occurring if they are not expected to occur. This approach may overcompensate some plaintiffs and undercompensate others. For example, suppose that a drug company was deprived of the opportunity to bring to market a drug that had a 90% chance of receiving Food and Drug

52. *See, e.g.*, Restatement (Second) of Contracts § 352 ("Damages are not recoverable for loss beyond an amount that the evidence permits to be established with reasonable certainty").

53. Comment a to Restatement (Second) of Contracts § 352 states, in pertinent part: "Damages need not be calculable with mathematical accuracy and are often at best approximate."

54. *See, e.g.*, Cole v. Homier Distributing Co., Inc., 599 F.3d 856, 866 (8th Cir. 2010) (expert testimony on lost profits excluded under *Daubert* standard because it "failed to rise above the level of speculation"). *See also* Webb v. Braswell, 930 So. 2d 387 (Miss. 2006). In *Webb*, the plaintiff's expert sought to testify as to future damages resulting from unplanted crops, without establishing that the crops would have been profitable. The court excluded the testimony based on Mississippi's adoption of the *Daubert* standard, stating that "damages for breach of contract must be proven with reasonable certainty and not based merely on speculation and conjecture." *Id.* at 398.

Administration (FDA) approval, at a profit of $2 billion, and a 10% chance of not receiving FDA approval, with losses of $1 billion. The court may treat 90% as near enough to certainty and ignore the 10% risk of failure and award damages of $2 billion.

By contrast, economists quantify losses of uncertain outcomes in terms of expected values, where the value in each outcome is weighted by its probability. Under that approach, economic losses in our example should be calculated as the $2 billion economic loss assuming FDA approval times 90% plus the $1 billion economic loss times 10%, or (0.9) × ($2 billion) + (0.1) × (−$1 billion) = $1.7 billion. The plaintiff would be overcompensated by $300 million under the approach that ignored the small probability of failure.

Now suppose the drug has only a 40% chance of FDA approval with the same economic payoffs. The plaintiff may recover no damages on grounds of uncertainty and speculation even though the economic loss is (0.4) × ($2 billion) + (0.6) × (−$1 billion) = $200 million. This issue also arises with respect to new businesses and is discussed further in Section VIII.A.

The second context where speculation arises when a damages expert fails to conduct his analysis in accordance with the principles discussed in *Daubert*.[55]

In general, the expert should provide all available information about the degree of uncertainty in an estimate of damages particularly when the claim is that inadequate data are available.

Example: A fire destroyed Broker's business including its business records. Defendant Smoke Detector Manufacturer argues that determining the profitability of the Broker's business is impossible without the business records. Therefore, damages are speculative and damages should not be awarded. Broker argues that the information would have been available absent the failure of Smoke Detector Manufacturer's product, and so Broker should be permitted wide latitude to measure damages from fragmentary records.

This issue also arises in labor cases where the defendant has failed to maintain the records as required by law.

Example: A class of workers was denied lunch breaks as required by state law. The class estimates damages assuming that no lunch breaks were ever taken. Defendant Can Maker argues that lunch breaks were often taken and provides testimony by a few employees as proof. The class argues that it is entitled to damages on the hypothesis that no lunch breaks were ever taken because Can Maker failed to keep proper records.

55. *See* Margaret A. Berger, The Admissibility of Expert Testimony, in this manual.

Disputes about what constitutes a reasonable damages analysis can range from the plaintiff's assertion that lack of records entitles it to damages under the worst-case scenario to defendant's assertion that damages are zero because any calculation is speculative. Furthermore, the latitude afforded the plaintiffs sometimes appears to depend on the egregiousness of the defendants' improper actions. The difficulties in finding a middle ground are greater when the defendant fails to make an affirmative estimate of damages but only attacks the plaintiff's quantification as speculative. Defendants frequently avoid offering a jury an affirmative damages analysis for fear that the jury will take the affirmative analysis as a concession of fault.

The question of speculative damages also arises in a third context when the certainty of damages for a specific plaintiff is not knowable at the time of trial.

Example: Vaccine Maker's duck flu vaccine given to children has been proven to harm one-quarter of the children who receive it, but determining which children will be affected is impossible. The harm is the onset of dementia at age 50, with economic losses of $1 million per person. Trial occurs well before any of the vaccinated children has reached this age. The expert for the class measures damages as $250,000 per recipient of the vaccine. The expert for Vaccine Maker argues that damages are zero because it is more likely than not that any given child was not harmed.

Comment: The class might not recover damages even though the average class member's economic loss is the expected value of $250,000. The case might be resolved at an early stage by denial of class certification because it is not possible to define a class in which all members were proven to be harmed.[56] Note that a possible solution would be to create a trust with $250,000 per class member, let it earn market returns, and pay out that amount plus the returns to each class member who develops dementia.

This difficulty in determining the probability of damages may be part of a challenge to class certification and is discussed further in Section XI.

B. Are the Parties Disputing the Remoteness of Damages?

A second legal limitation on damages is that a plaintiff may not recover damages that are too remote. In tort cases, this restriction is expressed in terms of proximate cause,[57] which often is equivalent to reasonable foreseeability. In contract

56. *See, e.g., In Re* New Motor Vehicles Canadian Export Antitrust Litig., 522 F.3d 6 (1st Cir. 2008).

57. *See* William L. Prosser, *Palsgraf Revisited,* 52 U. Mich. L. Rev. 1 (1953); Osborne M. Reynolds, Jr., *Limits on Negligence Liability: Palsgraf at 50,* 32 Okla. L. Rev. 63 (1979).

cases, the limitation is similarly embodied in the idea of foreseeability—a party may not recover damages that were not reasonably foreseeable by the parties at the time of the agreement.[58] The foreseeability rule has two parts. First, a party is liable for what are known as direct or general damages—those damages that arise naturally from the breach itself. Second, a defendant may also be liable for consequential or special damages—damages apart from those arising naturally from the breach—if such damages were reasonably foreseeable at the time of the agreement.[59] Although sometimes there are differences between proximate cause in torts and foreseeability in contracts, the general concept is the same: The law imposes a limit on damages that are too remote.[60]

The rule is often at issue in cases in which the injured party's loss greatly exceeds the benefit the breaching party received in return.

Example: Manufacturer hires Repairman to replace a part in a machine in its plant. Repairman negligently performs the service, causing Manufacturer's plant to cease production for two weeks. Manufacturer's damages demand includes a claim for two weeks of lost profits. Repairman counters that, although he may be liable for the cost of proper repairs, the foreseeability rule bars a claim for lost profits because such damages were not a probable consequence reasonably foreseeable at the time of the agreement.

Similar examples involve cases in which a package delivery firm or courier service is sued for remote consequential damages resulting from its failure to deliver a package.[61]

These limitations on damages are closely related to mitigation and the proper protection from losses resulting from the failure of agents or counterparties. A responsible company would not risk large losses from the failings of a repairman or delivery service. Rather, the company would use redundancy or other standard measures to limit the chances that such a failure would cause huge losses.

C. Are the Parties Disputing the Plaintiff's Efforts to Mitigate Its Losses?

A third limitation on damages is that a party may not recover for losses it could have avoided, and is often expressed by stating that the injured party has a duty to mitigate, or lessen, its damages. The economic justification for the mitigation rule

58. *See* E. Allan Farnsworth, *Legal Remedies for Breach of Contract,* 70 Colum. L. Rev. 1145, 1199–1210.

59. *Id.*

60. *See, e.g.* Richard A. Posner, Economic Analysis of Law 203–04 (1998).

61. *See* Hampton by Hampton v. Fed. Express Corp., 917 F. 2d 1124 (8th Cir. 1990).

is that the injured party should not cause economic waste by needlessly increasing its losses.[62]

In a dispute about mitigation, the law places the burden of proof on the defendant to show that the plaintiff failed to take reasonable steps to mitigate.[63] The defendant will propose that the proper offset is the earnings the plaintiff should have achieved, under proper mitigation, rather than actual earnings. In some cases, the defendant may presume the ability of the plaintiff to mitigate in certain ways unless the defendant has specific knowledge to the contrary at the time of a breach. For example, the defendant might presume that the plaintiff could mitigate by locating another source of supply in the event of a breach of a supply agreement. Damages are limited to the difference between the contract price and the current market price in that situation.

For personal injuries, the issue of mitigation often arises because the defendant believes that the plaintiff's failure to work after the injury is a withdrawal from the labor force or retirement rather than the result of the injury.[64] For commercial torts, mitigation issues can be more subtle. Where the plaintiff believes that the harmful act destroyed a company, the defendant may argue that the company could have been put back together and earned profit, possibly in a different line of business.[65] The defendant will then treat the hypothetical profits as an offset to damages.[66]

Alternatively, where the plaintiff continues to operate the business after the harmful act and includes subsequent losses in damages, the defendant may argue that the proper mitigation was to shut down after the harmful act.[67]

Example: Franchisee Soil Tester starts up a business based on Franchiser's proprietary technology, which Franchiser represents as meeting government standards. During the startup phase, Franchiser notifies Soil Tester that the technology has failed. Soil Tester continues to develop the business but sues Franchiser for profits it would have made from successful technology. Franchiser calculates much lower damages on the theory that Soil Tester should have mitigated by terminating the startup.

62. *See* E. Allan Farnsworth, *Legal Remedies for Breach of Contract,* 70 Colum. L. Rev. 1145, 1183–84.

63. *See, e.g.,* Broadnax v. City of New Haven, 415 F.3d 265, 268 (2d Cir. 2005) (defendant employer seeking to avoid a claim of lost wages bears the burden of proving that the plaintiff failed to mitigate his damages by, among other things, taking reasonable steps to obtain alternate employment).

64. *See* William T. Paulk, Commentary, *Mitigation Through Employment in Personal Injury Cases: The Application of the "Reasonable" Standard and the Wealth Effects of Remedies*, 58 Ala. L. Rev. 647–64 (2007).

65. *See* Seahorse Marine Supplies v. Puerto Rico Sun Oil, 295 F.3d 68, 84–85 (1st Cir. 2002).

66. *Id.* at 84.

67. *Id.* at 85. *Also see In re* First New England Dental Ctrs., Inc., 291 B.R. 240 (D. Mass. 2003).

Comment: This is primarily a factual dispute about mitigation. If the failure of the technology was unambiguous, it would appear that Soil Tester was deliberately trying to increase damages by continuing its business. On the other hand, Soil Tester might argue that the notification overstated the defects of the technology and was an attempt by Franchiser to avoid its obligations under the contract.

Disagreements about mitigation may be hidden within the frameworks of the plaintiff's and the defendant's damages studies.

Example: Defendant Board Maker has breached an agreement to supply circuit boards. Plaintiff Computer Maker's damages study is based on the loss of profits on the computers to be made from the circuit boards. Board Maker's damages study is based on the difference between the contract price for the boards and the market price at the time of the breach.

Comment: There is an implicit disagreement about Computer Maker's duty to mitigate by locating alternative sources for the boards not supplied by the defendant. The Uniform Commercial Code spells out the principles for resolving these legal issues under the contracts it governs.[68]

D. *Are the Parties Disputing Damages That May Exceed the Cost of Avoidance?*

An important consideration in capping damages may be the costs of steps that the plaintiff could have taken that would have eliminated damages. This argument is closely related to mitigation, but has an important difference: The defendant may argue that the plaintiff's failure to undertake a costly step that would have avoided losses was reasonable, but that the failure to take that step shows that the plaintiff knew that damages were much smaller than its later damages claim.

Example: Insurance Company suffered a business interruption because a fire made its offices unusable for a period of time. Insurance Company's damages claim for $10 million includes not only the lost business until the offices were usable but also damages for permanent loss of business from customers who found other sources during the period the

68. *See, e.g.*, Aircraft Guaranty Corp. v. Strato-Lift, Inc., 991 F. Supp. 735, 738–39 (E.D. Pa. 1998) (Mem.) (finding that according to the Uniform Commercial Code, plaintiff-buyer had a duty to mitigate if the duty was reasonable in light of all the facts and circumstances, but that failure to mitigate does not preclude recovery); S.J. Groves & Sons Co. v. Warner Co., 576 F.2d 524 (3d Cir. 1978) (holding that the duty to mitigate is a tool to lessen plaintiff's recovery and is a question of fact); Thomas Creek Lumber & Log Co. v. United States, 36 Fed. Cl. 220 (1996) (finding that under federal common law the U.S. government had a duty to mitigate in breach-of-contract cases).

offices were unusable. Defendant argues that the plaintiff's failure to relocate to temporary quarters shows that their losses were less than the $350,000 cost of that relocation.

Comment: Defendant's argument has the unstated premise that Insurance Company could have carried on its business and avoided any of its later losses by relocating. Insurance Company will likely argue that a decision not to relocate was commercially appropriate because relocation would not have avoided much of the lost business.

E. Are the Parties Disputing a Liquidated Damages Clause?

In addition to legally imposed limitations on damages, the parties themselves may have agreed to impose limits on damages should a dispute arise. Such clauses are common in many types of agreements. Once litigation has begun, the parties may dispute whether these provisions are legally enforceable. The law may limit enforcement of liquidated damages provisions to those that bear a reasonable relation to the actual damages. In particular, the defendant may attack the amount of liquidated damages as an unenforceable penalty. The parties may disagree on whether the harmful event falls within the class intended by the contract provision.

Changes in economic conditions may be an important source of disagreement about the reasonableness of a liquidated damages provision. One party may seek to overturn a liquidated damages provision on the grounds that new conditions make it unreasonable.

Example: Scrap Iron Supplier breaches supply agreement and pays only the specified liquidated damages. Buyer seeks to set aside the liquidated damages provision because the price of scrap iron has risen, and the liquidated damages are a small fraction of actual damages under the expectation principle.

Comment: There may be conflict between the date for judging the reasonableness of a liquidated damages provision and the date for measuring expectation damages, as in this example. Generally, the date for evaluating the reasonableness of liquidated damages is the date the contract is made. In contrast, the date for measuring expectation damages is the date of the breach. The conflict may be resolved by the substantive law of the jurisdiction. Enforcement of the liquidated damages provision in this example will induce inefficient breach.

VIII. Other Issues Arising in General in Damages Measurement

A. Damages for a Startup Business

Failure rates for startups are high even without any actionable harm. More than two-thirds of venture-funded startups return nothing to their founding entrepreneurs, although the expected value of venture outcomes is several million dollars per entrepreneur.[69] Thus, a damages calculation for harm to a startup puts particular stress on the treatment of uncertainty in damages, as we discussed earlier in Section VII.A. At one time, legal principles barred recovery because damages were too speculative, but today most courts will allow a new business to recover damages for lost profits if such damages can be proven with reasonable certainty.[70] Whether a court will award damages for an injured startup if the plaintiff's damages expert testifies that the likelihood was less than 50% that the company would have become profitable is still unresolved.

1. Is the defendant challenging the fact of economic loss?

Expert testimony on damages does not usually include separate consideration of the fact of damages, because an opinion that damages are positive amounts to an opinion about the fact, and a zero-damages opinion amounts to an opinion against the fact. Damages for startups may be an exception. Analysis by the plaintiff's expert may conclude that there is a significant probability that the startup would not have been profitable and, in that contingency, damages would have turned out to be zero (or even negative, in the sense that the defendant's action prevented the plaintiff from incurring a loss). Thus, a defendant may argue that the plaintiff has not proven the fact of damages. In most cases involving an existing business, the fact of economic loss is often self-evident, but in a case involving a new business, the fact of economic loss may be at issue.

2. Is the defendant challenging the use of the expected value approach?

The expected value approach to uncertain damages weights each outcome by its probability of occurring. The expected value can be positive, indicating damages, even if the odds favor a company making a loss. Application of the expected value approach involves studying the various outcomes of the new business in relation to risk factors. Risks can be categorized as idiosyncratic (i.e., risks specific to the

69. *See* Robert E. Hall & Susan E. Woodward, *The Burden of the Nondiversifiable Risk of Entrepreneurship*, 100 Am. Econ. Rev. 1163 (2010).

70. *See* Mark A. Allen & Victoria A. Lazear, *Valuing Losses in New Businesses, in* Litigation Services Handbook: The Role of the Financial Expert §§ 11.1–.26 (Roman L. Weil et al. eds., 4th ed. 2007).

venture) or systematic (i.e., risks that affect the venture in the same way as other businesses). Idiosyncratic risks include whether the venture will succeed, the firm's ability to obtain financing, whether a competitor will develop a similar product, and risks related to the pricing of the product or competitive products, such as the price of inputs. Examples of systematic risks are financial crisis, inflation, collapse of the stock market, and recession.

The expert proceeds first by identifying the idiosyncratic risks associated with the venture and creating an appropriate model. Analyses usually model these types of risk as different scenarios, each with a specific probability of occurring. The expert computes the lost profits for each scenario, multiplies the lost profits by the probability of the event occurring, and then sums the weighted profits to arrive at expected lost profits. The result is a stream of future lost profits before adjustment for general economic variables such as inflation, stock market fluctuations, or wage growth. Because the expert has adjusted for idiosyncratic factors, the remaining risks of lost profits for a new business are the same as those for a similar, existing business. Then experts usually adjust lost profits for systematic risks using the CAPM (*see* Section VI.H) to estimate the cost of capital.

The actual calculation of expected damages is usually straightforward. Damages are the stream of expected lost profits discounted to present value. However, sometimes the alternatives and interactions between the possible outcomes become so complex that other methods are required. In such cases, experts often generate hundreds or thousands of possible outcomes using techniques such as Monte Carlo or bootstrap simulation. These techniques generate random values for the variables that change with different outcomes.[71] Expected damages are then the average of lost profits across all outcomes.

An alternative to calculating new business damages based on lost profits uses market valuations of the firm. For a publicly traded business, the valuation is implicit in the stock price—it is the market capitalization of the firm. For a new venture, the valuation is implicit in financing decisions. Startup firms are often financed by venture capitalists who invest funds in exchange for ownership in the venture. The valuation at the time of financing is the amount of financing divided by the ownership transferred. For example, if venture investors pay $4 million for 10% of the firm, the total value of the firm is $4 divided by 0.10, or $40 million.

3. Are the parties disputing the relevance and validity of the data on the value of a startup?

The expert seeking to establish economic loss on behalf of a new business will often face a lack of data and therefore will need to use additional resources for the

71. Monte Carlo relies on random draws from the hyposized distributions for the variables. Bootstrap takes the observed variables as the population of outcomes and relies on repeated random draws from this population.

analysis. Although the expert may have access to third-party data on factors such as overall success rates for comparable ventures, the expert will often need to rely on the plaintiff and other experts to refine the probability of success. If success reflects consumer preferences, then the expert may use market research techniques such as surveys. For example, a survey could evaluate the desirability of a new feature for a product and the premium that consumers will pay for it. Other sources of information include studies of success rates on behalf of venture capitalists. Such studies typically show the success rates for new businesses at different stages of investment and the actual returns that the venture capitalists have realized.

B. Issues Specific to Damages from Loss of Personal Income

As with all cases, many of the disputes that arise in estimating damages for lost personal income can be resolved by carefully applying the basic damages framework. Damages are the difference between the but-for and actual worlds, where the actual world reflects any mitigating factors. Estimating such damages also involves issues that are unique, such as calculating losses over a person's lifetime, valuing fringe benefits, estimating lost income in wrongful death cases, and calculating damages for economic losses other than lost wages. We discuss these issues below.

1. Calculating losses over a person's lifetime

In nearly all cases involving lost income, the effects continue past trial and sometimes until the plaintiff's death. Therefore, quantifying damages for loss of personal income necessarily involves projecting the plaintiff's work history and retirement. Conceptually, the estimate of income for each year, either but-for or actual, is the expected income multiplied by the probability that the person will be working for that year. The probability that the person will be working for that year is the product of the probability that the person will survive the year, the probability that the person will be in the labor force, the probability that the person will be employed, given that the person worked in the prior year.[72] We refer to this as the standard framework for calculating personal losses.

In many cases, such as those involving wrongful termination, the projection that the plaintiff was working is the same for both the but-for world and the actual world. However, in wrongful death cases and some personal injury cases, these projections may differ,[73] and the expert will need to compute separate projections for the but-for and actual worlds before taking the difference between the two.

72. Except for wrongful death cases, the probability that the persons will be working is usually 1 for both the but-for and actual cases. In wrongful death cases the probability is still usually 1 in the but-for case, but 0 in the actual.

73. This situation may arise in a personal injury case if, as a result of the accident, the injured person is less likely to be able to work. If so, then the person may have an increased likelihood of

These projections usually rely on data from the Bureau of Labor Statistics (BLS) and include tables on survival, labor force participation, and employment. The expert usually needs to manipulate this information in order to generate the conditional probabilities needed.

To simplify the calculations, sometimes the expert uses the person's expected lifespan and retirement age based on his or her age at trial using standard tables from the BLS. Then the expert need only sum the discounted losses for each year until the expected age at death. This method, often referred to as the life expectancy method, will considerably simplify the calculations associated with determining lost retirement benefits. However, the standard and the life expectancy methods will usually generate the same estimate of losses only if the expert is assuming that the expected discounted income in each year is the same—that is, that the expected increase in income is offset by the discount rate (*see* Section VI.D.3).

BLS data are generally only available by age and sex. Other data specify these statistics by race, location, or broad occupation categories but only by groups of ages. More specific tables are commercially available, but the reliability of these data may be disputed because of questions about the methodology used to generate the tables.

2. *Calculation of fringe benefits*

Fringe benefits are often a component of lost pay and may include medical insurance and retirement benefits such as social security. Although sick days and vacation are also fringe benefits, they are included already in lost-pay calculations. An exception occurs if these days are accrued but not taken, where, for example, the employee lost the cash payout that would have occurred at a normal termination or lost the benefit of future days off with pay.

a. Medical insurance benefits

In the following discussion, we assume that the plaintiff no longer has the benefit of the employer-provided insurance as a result of the actions of the defendant. Such situations typically arise in wrongful termination or wrongful death cases.

Calculating damages for lost medical insurance is straightforward if the plaintiff can purchase insurance under COBRA, or from his current employer, or on the open market. Then the value of the lost medical insurance is the employee's portion of the premium. If insurance coverage available to the plaintiff differs significantly between the but-for and the actual worlds, then the expert will need to project the impact of the difference in policy coverage.

leaving the labor force for each age compared with the likelihood prior to the incident. Similarly, the injured person may be more likely to retire at an earlier age.

If the plaintiff chooses not to purchase insurance even though the option is vailable, or the plaintiff is unable to purchase insurance,[74] then the plaintiff may argue that the value of insurance is the sum of his actual expenses less the premium in the but-for world. The defendant will likely respond that the plaintiff assumed the risk that the incurred medical expenses could exceed the plaintiff's portion of the premium, and therefore that the defendant's responsibility should be limited to the plaintiff's portion of the premium foregone. The plaintiff may counter that his pay was insufficient to afford the insurance.

b. Retirement benefits

For lost retirement benefits, the issues are similar to those involving lost medical benefits, but the calculations are more complex. There are basically two types of retirement plans: defined benefit plans and defined contribution plans. Defined benefit plans are those where the benefits paid out after retirement are guaranteed to be a definite amount upon retirement. In contrast, defined contribution plans are those where the employer makes a predefined contribution for the employee but the benefits paid out depend on the return earned on the money invested.

The expert can calculate both types of retirement plans on the basis of either the amounts paid in or the amounts paid out by the employer. If the expert uses the amount the employer paid for the benefits, which may be a function of the amount the plaintiff earned, then the calculation is analogous to computing the loss in plaintiff's earnings. The disadvantage of this approach is that the amounts paid in may not adequately predict the benefits paid out, particularly when the plan is a defined benefit plan. We discuss this topic below in connection with social security benefits, where the problem is particularly acute.

(1) Defined benefit plan

To determine the present value of the benefits received under a defined benefit plan, the calculation is simplified if the expert uses the life expectancy method to calculate the plaintiff's losses. In this situation, the expert must determine the number of years that the plaintiff would have worked at the firm upon retirement, his retirement age, his expected lifespan, and his salary at the firm over time. These factors must be consistent with the expert's belief about the projected trajectory of plaintiff's employment in both the but-for and actual worlds.

However, if the expert instead uses probability tables for each year, then the calculation is more complex. For each year in which the plaintiff may cease to be in the labor force for reasons other than death, the expert must determine the likelihood that the plaintiff would be receiving benefits from the plan because of

74. For example, a preexisting condition may make it impossible to purchase insurance on the open market or may limit the plaintiff's coverage to exclude a preexisting condition either permanently or for a period of time.

disability (if the plan permits) or retirement. Complicating this determination is that the payout from the plan may depend on the age of retirement. Thus, the calculation must incorporate the probability for each possible payout. Depending on the plan, defining possible outcomes can be extremely complex.

A special and common example of a defined benefit plan is Social Security.[75] Determining benefits from Social Security can be forbiddingly complex because the number of potential outcomes is so large. For example, a person can retire at almost any age, and disability payments are made if he is unable to work. In addition, calculating the benefit at any age depends on the person's average salary over the most recent 35 years. If social security benefits are critical to the magnitude of damages, the expert may choose to simplify the calculation by relying on the life expectancy method.

(2) Defined contribution plan

For a defined contribution plan, the expert's task is to project the employer's contribution, the number of years that the employee would have worked at the firm, as well as the employee's age at retirement. Generally, this determination is straightforward because it is based on the same factors the expert uses to project the employee's salary in the but-for and actual worlds. The present value of the employer's contributions will be the expected payouts from the plan.

3. Wrongful death

Traditionally, under common law, the right of recovery ended with a person's death and thus damages for wrongful death were not recoverable. Today, states have remedied this situation through the passage of wrongful death and survival statutes. A wrongful death action focuses on the impact of the decedent's death on persons other than the decedent. In contrast, a survival action continues the action the decedent could have maintained had he lived and compensates the decedent's estate for damages the decedent sustained. Some states have separate wrongful death and survival statutes; others have hybrid statutes that combine elements of both actions. Rules for recovery vary widely by state.

Generally, calculation of economic damages for wrongful death depends on whether the claimant is a relative of the decedent or is the estate. If the claimant is a relative of the decedent, economic damages are limited to the economic value that the relative would have received had the decedent lived. If the relative is a dependent, the recovery may be substantial, whereas if the decedent is a child or unmarried and childless and the only relatives are parents, the recovery may be small, because most parents receive little economic value from their children. In

75. Social Security is generally regarded as a defined benefit plan although it has some elements of a defined contribution plan.

contrast, if the beneficiary is the estate, the recovery may include all of the lost economic value.

Where the claimant is a relative, damages from lost wages may be reduced to reflect the decedent's own consumption spending had he continued living. Such expenses may be relatively small if the claimant is a spouse with children under the theory that much of the decedent's income would have been spent to support the dependents. If the decedent had no children or if there is another earner in the family, the offset for the spending of the decedent on himself may be higher.

4. *Shortened life expectancy*

An important issue is whether a plaintiff may recover compensation for shortened life expectancy caused by an injury. This issue may arise, for example, in medical malpractice cases in which a doctor fails to diagnose and treat a condition or where a surgeon fails to remove a medical device used during surgery. Some states allow such a recovery; others do not.[76]

A related issue is whether dependents in a wrongful death action may recover economic damages for support the decedent would have provided had the decedent lived—that is, whether such damages can be recovered over the remainder of the decedent's expected lifetime, had he lived. Again, rules regarding such recovery vary widely by state, but quantifying such damages requires a projection of the decedent's life expectancy using the methods discussed above.

5. *Damages other than lost income*

a. Loss of services

Economic damages may include loss of services in addition to lost wages. For example, in a case involving the death or disability of a housewife, the husband may seek recovery for money necessary to hire someone to take care of the children and the home.

b. Medical expenses

Damages for wrongful death or injury may also include past and future medical expenses. Recoverable medical expenses may include compensation for someone to provide medical assistance to the plaintiff, such as nursing care, or expenses for special equipment necessary for living. These expenses are usually calculated by an expert in this area. The role of the economic damages expert is usually confined to computing the present value of these expenses.

76. *See, e.g.*, Dillon v. Evanston Hospital, 771 N.E.2d 357 (Ill. 2002); Swain v. Curry, 595 So. 2d 168 (Fla. Dist. Ct. App. 1992).

c. Expenses not incurred

If the plaintiff is not employed, she may not be incurring certain expenses that she otherwise would have incurred had the wrongful termination or personal injury never taken place. Applying the but-for world analysis, the defendant may argue that these expenses should be an offset in the calculation of plaintiff's economic damages.[77] Examples of such expenses are union dues or transportation costs. *See* Section VIII.D for a general discussion. Legal standards vary by jurisdiction.

d. Other damages

Sometimes an expert may be asked to opine on damages for pain and suffering and other diminution in the quality of life. There has been some development in using hedonic models to estimate such losses, but the research is still preliminary. In general, the expert relies upon whoever is best positioned to place an economic value on the diminution of life suffered by the plaintiff. The expert may then be asked to calculate the present value of the estimate.

C. Damages with Multiple Challenged Acts: Disaggregation

Plaintiffs sometimes challenge a number of a defendant's acts and offer an estimate of the combined effect of those acts. If the court determines that only some of the challenged acts are illegal, the damages analysis needs to be adjusted to consider only those acts. This issue seems to arise most often in antitrust cases, but can arise in any type of case. Ideally the damages testimony would equip the factfinder to determine damages for any combination of the challenged acts, but that may be tedious. If there are, say, 10 challenged acts, it would take more than 1000 separate studies to determine damages for every possible combination of findings about the unlawfulness of the acts.

There have been several cases where the jury has found partially for the plaintiff, but the jury lacked assistance from the damages experts on how the damages should be calculated for the combination of acts the jury found to be unlawful. Although the jury has attempted to resolve the issue, appeals courts have sometimes rejected damages found by juries without supporting expert testimony.[78]

77. In wrongful death actions, these expenses may be included in the deduction for the amount the decedent would have spent on himself. *See supra* Section VIII.B.3.

78. *See e.g.,* Litton Sys., Inc. v. Honeywell, Inc., 1996 U.S. Dist. LEXIS 14662 (C.D. Cal. July 26, 1996) (granting new trial on damages only "[b]ecause there is no rational basis on which the jury could have reduced Litton's 'lump sum' damage estimate to account for Litton's losses attributable to conduct excluded from the jury's consideration, . . ."); Image Technical Servs., Inc. v. Eastman Kodak Co., 125 F.3d 1195, 1224 (9th Cir. 1997), *cert. denied,* 118 S. Ct. 1560 (1998) (plaintiffs "must segregate damages attributable to lawful competition from damages attributable to Kodak's monopolizing conduct").

One solution to this problem is to make the determination of the illegal acts before damages testimony is heard, termed "bifurcation" of liability and damages. The damages experts can adjust their testimony to consider only the acts found to be illegal.

In some situations, damages are the sum of separate damages for the various illegal acts. For example, there may be one injury in New York and another in Oregon. Then, the damages testimony may consider the acts separately, and disaggregation is not challenging.

When the challenged acts have effects that interact, it is not possible to consider damages separately and add up damages for each individual act. This is an area of great confusion. When the harmful acts substitute for each other, the sum of damages attributable to each separately is *less* than their combined effect. As an example, suppose that the defendant has used exclusionary contracts and anticompetitive acquisitions to ruin the plaintiff's business. However, the plaintiff's business could not survive if either the contracts or the acquisitions were found to be legal. Damages for the combination of acts are the value of the business, which would have thrived absent both the contracts and the acquisitions. Now consider damages if only the contracts but not the acquisitions are illegal. In the but-for analysis, the acquisitions are hypothesized to occur because they are not illegal, but not the contracts. But plaintiff's business cannot function in that but-for situation because the acquisitions alone were sufficient to ruin the business. Hence damages—the difference in value of the plaintiff's business in the but-for and actual situations—are zero. The same would be true for a separate damages measurement for the acquisitions, with the contracts taken to be legal but not the acquisitions. Thus, the sum of damages for the individual acts is zero, but the damages if both acts are illegal are the value of the business.

When the effects of the challenged conduct are complementary, the sum of damages for each type of conduct by itself will be *more* than damages for all types of conduct together. For example, suppose a party claims that a contract is exclusionary based on the combined effect of the contract's duration and its liquidated damages clause that includes an improper penalty provision. The actual amount of the penalty would cause little exclusion if the duration were brief, but substantial exclusion if the duration were long. Similarly, the actual duration of the contract would cause little exclusion if the penalty were small but substantial exclusion if the penalty were large. A damages analysis for the penalty provision in isolation compares but-for—without the penalty provision but with long duration—to actual, where both provisions are in effect. Damages are large. Similarly, a damages estimate for the duration in isolation gives large damages. The sum of the two estimates is nearly double the damages from the combined use of both provisions.

Thus, a request that the damages expert disaggregate damages for different combinations of challenged acts is far more than a request that the total damages estimate be broken down into components that add up to the damages attributable to the combination of all the challenged acts. In principle, a separate damages analysis—with its own carefully specified but-for scenario and analysis—needs to be done for every possible combination of illegal acts.

Example: Hospital challenges Glove Maker for illegally obtaining market power through the use of long-term contracts and the use of a discount program that gives discounts to consortiums of hospitals if they purchase exclusively from Glove Maker. The jury finds that Glove Maker has attempted to monopolize the market with its discount programs, but that the long-term contracts were legal because of efficiencies. Hospital states that its damages are the same as in the case in which both acts were unlawful because either act was sufficient to achieve the observed level of market power. Glove Maker argues that damages are zero because the lawful long-term contracts would have been enough to allow it to dominate the market.

Comment: The appropriate damages analysis is based on a careful new comparison of the market with and without the discount program. The but-for analysis should include the presence of the long-term contracts because they were found to be legal.

Apportionment, sometimes referred to as disaggregation, can arise in a different setting. A damages measure may be challenged as encompassing more than the harm caused by the defendant's harmful act. The expert may be asked to apportion his estimate of damages between the harm caused by the defendant and the harm caused by factors other than the defendant's misconduct. In this case, the expert is being asked to restate the improper actions, not to disaggregate the damages estimate for the improperly inclusive damages estimate. If the expert uses the standard format and thus properly isolates the effects of only the defendant's wrongful actions, no modification of the expert's estimate of damages is needed. In the standard format, the but-for analysis differs from the actual world only by hypothesizing the absence of the harmful act committed by the defendant. The comparison of the but-for world with the actual world automatically isolates the causal effects of the harmful act. No disaggregation of damages caused by the harmful act is needed once the standard format is applied.

D. Is There a Dispute About Whether the Plaintiff Is Entitled to All the Damages?

When the plaintiff is in some sense a conduit to other parties, the defendant may argue that the plaintiff is entitled to only those damages that it would have retained in the but-for scenario. In the following example, a regulated utility is arguably a conduit to the ratepayers:

Example: Generator Maker overcharges Utility. Generator Maker argues that the overcharge would have been part of Utility's rate base, and so Utility's regulator had set higher prices because of the overcharge. Utility, therefore, did not lose anything from the overcharge. Instead, the ratepayers paid the overcharge. Utility argues that it stands in for all the ratepayers and that the damages award will accrue to the ratepayers by the same principle—the regulator will set lower rates because the award will count as revenue for rate-making purposes.

Comment: In addition to the legal issue of whether Utility does stand in for ratepayers, there are two factual issues: Was the overcharge actually passed on to ratepayers? Will the award be passed on to ratepayers?

Similar issues can arise in the context of employment law.

Example: Plaintiff Sales Representative sues for wrongful denial of a commission. Sales Representative has subcontracted with another individual to do the actual selling and pays a portion of any commission to that individual as compensation. The subcontractor is not a party to the suit. Defendant Manufacturer argues that damages should be Sales Representative's lost profit measured as the commission less costs, including the payout to the subcontractor. Sales Representative argues that she is entitled to the entire commission.

Comment: Given that the subcontractor is not a plaintiff, and Sales Representative avoided the subcontractor's commission, the literal application of standard principles for damages measurement would appear to call for the lost-profit measure. The subcontractor, however, may be able to claim his share of the damages award. In that case, damages would equal the entire lost commission, so that, after paying off the subcontractor, Sales Representative receives exactly what she would have received absent the breach. Note that the second approach would place the subcontractor in exactly the same position as the Internal Revenue Service in our discussion of adjustments for taxes in Section VI.E.

The issue also arises acutely in the calculation of damages on behalf of a nonprofit corporation. When the corporation is entitled to damages for lost profits, the defendant may argue that the corporation intentionally operates its business without profit. The actual losers in such a case are the people who would have enjoyed the benefits from the nonprofit that would have been financed from the profits at issue.

E. Are the Defendants Disputing the Apportionment of Damages Among Themselves?

When the defendants are not jointly liable for the harmful acts, but rather each is responsible for its own harmful act, the damages expert needs to quantify damages separately for each defendant. The issues in apportionment among defendants are similar to those discussed above for disaggregation among the harmful acts.

1. Are the defendants disputing apportionment among themselves despite full information about their roles in the harmful event?

In the simplest case, there are no interactions among the harmful acts of different defendants, and the expert can proceed as if there were separate trials with separate damages analyses.

However if there are interactions among the harmful acts, then apportionment among defendants involves puzzles that cannot be resolved by economic principles. If either of the harmful acts of two defendants would have caused all the harm that occurred, then either defendant can argue for zero damages on the ground that the harm would have occurred anyway, because of the other defendant's act.

Example: Tire Maker supplies faulty tire and Landing Gear Maker supplies faulty landing gear. Either one would have resulted in the loss of the airplane upon landing. Airline measures damages as the value of the airplane and proposes that the two defendants split the amount equally. But Tire Maker asserts that the damages it owes the plaintiff are zero because the crash would have occurred anyway because of the Landing Gear Maker's faulty landing gear. Similarly, the Landing Gear Maker asserts the damages it owes the plaintiff to be zero because the crash would have occurred anyway because of the Tire Maker's faulty tire.

The issue also arises when the interaction is more complicated.

Example: Teenager drives through a red light and injures Driver. The injury is more serious than it would have been otherwise because Driver's airbag failed to deploy. Airbag Maker argues that it should pay nothing because there would have been no harm if Teenager had obeyed the red light. Teenager argues that Airbag Maker should pay the difference between the actual harm to Driver and the harm if the airbag had worked properly.

2. Are the defendants disputing the apportionment because the wrongdoer is unknown?

A second issue in apportioning damages arises when the harmful product is known, but more than one defendant made the product, and it is not known which made the product that caused the injury. One approach is to determine the probability that each defendant made the product that caused the plaintiff's loss. In some cases, a reasonable assumption may be that the probability that the defendant caused the plaintiff's losses may be determined from its market share. Thus, for example, a drug manufacturer's responsibility would be proportional to the likelihood that a plaintiff consumed one of its pills.

F. Is There Disagreement About the Role of Subsequent Unexpected Events?

Random events occurring after the harmful event can affect the plaintiff's actual loss. The effect might be either to amplify the economic loss from what might have been expected at the time of the harmful event or to reduce the loss.

Example: Housepainter uses faulty paint, which begins to peel a month after the paint job. Owner measures damages as the cost of repainting. Painter disputes on the ground that a hurricane that actually occurred 3 months after the paint job would have ruined a proper paint job anyway.

Comment: This dispute will need to be resolved on legal rather than economic grounds. Both sides can argue that their approach to damages will, on average over many applications, result in the right incentives for proper house painting.

The issue of subsequent random events should be distinguished from the legal principle of supervening events.[79] The subsequent events occur after the harmful act; there is no ambiguity about who caused the damage, only an issue of quantification of damages. Under the theory of a supervening event, there is precisely a dispute about who caused an injury. In the example above, there would be an

79. *See, e.g.*, Derdiarian v. Felix Contracting Corp., 414 N.E.2d 666 (N.Y. 1980) (interpreting state law to hold that a jury could find that the defendant is ultimately liable to plaintiff for negligence, even though a third person's negligence was a supervening event); Lavin v. Emery Air Freight Corp., 980 F. Supp. 93 (D. Conn. 1997) (holding that under Connecticut law, a party seeking to be excused from a promised performance as a result of a supervening event must show the performance was made impracticable, non-occurrence was an assumption at the time the contract was made, impracticability did not arise from the party's actions, and the party seeking to be excused did not assume a greater liability than the law imposed).

issue of the role of a supervening event if the paint had not begun to peel until after the hurricane.

Disagreements about the role of subsequent random events are particularly likely when the harmful event is fraud.

Example: Seller of property misstates condition of property. Buyer shows that he would not have purchased the property absent the misstatement. Property values in general decline sharply between the fraud and the trial. Buyer measures damages as the difference between the purchase price and the market value of the property at the time of trial. Seller measures damages as the difference between the purchase price and the market value at the time of purchase, assuming full disclosure.

Comment: Buyer may be able to argue that retaining the property was the reasonable course of action after uncovering the fraud; in other words, there may be no issue of mitigation here. In that sense, Seller's fraud caused not only an immediate loss, as measured by Seller's damages analysis, but also a subsequent loss. Seller, however, did not cause the decline in property values. The dispute needs to be resolved as a matter of law.

As a general matter, it is preferable to exclude the effects of random subsequent effects, especially if the effects are large in relation to the original loss.[80] The reason is that plaintiffs choose which cases to bring, which may influence the approach to damages. If random subsequent events are always included in damages, then plaintiffs will bring the cases that happen to have amplified damages and will not pursue those where the random later event makes damages negative. Such selection of cases will overcompensate plaintiffs. Similarly, if plaintiffs can choose whether to include the effects of random subsequent events, plaintiffs will choose to include those effects when they are positive and exclude them when they are negative. Again, the result will be to overcompensate plaintiffs.[81] If random subsequent events are always excluded, then the plaintiff is compensated for his loss, however temporary, and the defendant pays for the damages he actually caused.

80. *See* Franklin M. Fisher & R. Craig Romaine, *Janis Joplin's Yearbook and the Theory of Damages*, *in* Industrial Organization, Economics, and the Law 392, 399–402 (John Monz ed., 1991); Fishman v. Estate of Wirtz, 807 F.2d 520, 563 (7th Cir. 1986) (Easterbrook, J., dissenting in part).

81. *See* William B. Tye et al., *How to Value a Lost Opportunity: Defining and Measuring Damages from Market Foreclosure*, 17 Res. L. & Econ. 83 (1995). For a discussion of disclosure of expert reports under Federal Rule of Civil Procedure 26(a)(2), see Margaret A. Berger, The Admissibility of Expert Testimony, Section V.B.1, in this manual. For a discussion of disclosure of data supporting expert testimony, see Daniel L. Rubinfeld, Reference Guide on Multiple Regression, Section V, in this manual.

IX. Data Used to Measure Damages

A. Types of Data

1. Electronic data

Electronic data have four general formats: (1) proprietary, (2) electronic character, (3) scanned, and (4) survey.

Examples of proprietary formats are those used by the SAS statistical software, the Oracle database system, and Microsoft's Access and Excel. Although these formats are proprietary, the ones we have listed are de facto industry standards and are the most convenient ways to transmit data among experts and from parties to experts. All of these software systems can create data files in Excel format, which is the effective universal standard for sharing smaller bodies of data.

Electronic character representations are almost always in Adobe's Portable Document Format (PDF), a public domain standard. Essentially any computer software can produce a PDF document. The PDF format is convenient for the electronic sharing of documents formatted for visual presentation (as opposed to files formatted to be read by computers), but is not a useful way to move data, especially in large volumes. Reading data from a PDF document into analytical software requires endless human intervention.

Scanned documents are represented internally as pixels, not as characters. The automatic reading of scanned numerical documents into analytical software is close to impossible, because optical character recognition software is unreliable with numerical material and requires large amounts of human intervention, character by character.

Much confusion exists between electronic character documents and scanned documents, because both are part of the PDF standard. It is easy to tell them apart. With any amount of magnification, an electronic character document shows perfectly crisp characters, while a scanned document shows its granular pixels.

2. Paper data

Although the overwhelming majority of business records are kept in computer form today, historical data may be available only in paper form. The data usually reach the expert as scanned document images. Then, the expert needs to deal with the problems of accurately reading scanned data.

3. Sampling data

In some instances, the expert is faced with more information than is possible to process. This situation is most likely to arise if human review of each data record is part of the processing. Even if processing all of the data may be ultimately necessary, processing a sample of the data for preliminary analyses may be appropriate.

If the expert elects to study a sample of the data, the expert needs to have carefully considered how the information will be used to ensure that the data sample is large enough and contains sufficient information. Usually, this requires that the expert has constructed a model of damages and a related sampling plan that includes an estimate of the sampling error. Unless the expert is a trained statistician, the expert should seek outside help in designing the sampling plan.

4. *Survey data*

Another situation arises when the data can only be obtained by interviewing individuals. This often arises because damages hinge on consumer preference. For example, the issue may be how many bicycles of a certain brand would have been sold absent a misappropriated braking system. Another example might be the number of people who would have ordered their prescription contact lenses over the Internet if the wholesalers had not conspired to restrict sales to the Internet retailers.

A need for a survey can also arise when the plaintiffs comprise a class. In this situation, it may be prohibitive to interview every class member, and the damages expert will need to construct both a sampling plan and a survey instrument so that the results can be reliably used to estimate damages.

The principles in constructing a sampling to collect data for analysis from a dataset apply to constructing a sample of individuals to be surveyed: The expert needs to have carefully considered how the information will be used to ensure that the data sample is large enough and contains sufficient information. In addition, complexities arise because some respondents selected to be surveyed will not be reachable or will be unwilling to complete a survey. The expert must devise a plan to deal with such contingencies and be confident that such problems do not bias the results.

Care must also be taken in developing the survey instrument. Generally, it is advantageous to work with experts in survey to ensure that the responses can be reliably interpreted and are not biased.[82]

B. Are the Parties Disputing the Validity of the Data?

Validation of any dataset is critical. The expert needs to have a firm basis for relying on the chosen data. Opposing parties frequently try to impeach damages estimates by challenging the reliability of the data or an expert's validation of the data.

82. For a discussion of the issues in designing sample plans and survey instruments particularly for use in litigation, see Shari Diamond, Reference Guide on Survey Research, in this manual.

1. Criteria for determining validity of data

The validity of data is ultimately a matter of judgment. Experts often need to use data that are not mathematically precise, because the only relevant data may be known to contain some errors. Experts generally have an obligation to use data that are as accurate as possible, meaning that the expert has used every practical means to eliminate erroneous information. Experts should also perform cross checks with other data, to the extent possible, to demonstrate completeness and reliability. When data are inherently inaccurate because of random influences, validity requires absence of bias or adjustment for bias. Validation of data turns in part on commonsense indicators of accuracy and bias. The following is a list, in rough order of presumptive validity, of data sources often used in damages measurement:

- Official government publications and databases, such as from the Census Bureau, the BLS, and the Bureau of Economic Analysis;
- A company's audited financial statements and filings with the Securities and Exchange Commission;
- A company's accounting records maintained in the normal course of business;
- A company's operating reports prepared for management in the normal course of business;
- A survey designed by the damages expert with assistance from survey professionals, conforming to established standards of survey design and execution;
- A marketing research study conforming to established standards for these studies;
- Industry reports and other materials prepared by unaffiliated organizations and consultants;
- Newspaper articles;
- A company's study of damages from the harmful event, prepared in the normal course of business; and
- A company's study of damages, prepared for litigation.

Other factors can alter this presumptive order of validity. When audited financial statements are accused of being fraudulent, they lose their presumption of validity. The most fully researched articles in the best newspapers have a higher presumption of validity. Some private industry reports are highly reliable. Rules of evidence may also affect when and how these various data sources can be used in expert reports and at trial. However, when internal data are unavailable through no fault of the plaintiff, then courts often will make allowances for the lack availability if the expert has made every effort to demonstrate that the data relied upon are reasonable.

2. Quantitative methods for validation

One important aspect of validation is to verify that the data are complete. If a separate summary document is available that shows the number of records in the database together with summary statistics such as total amounts paid, completeness is easy to establish. Other methods for establishing completeness include examining serial numbers for records and finding other sources of information about transactions that should be in the data and verifying the presence of all or a sample of them.

Another validation method is to examine specific observations. For example, if the dataset consists of purchasing records, then the expert may examine all of the records for sampled customers, or the expert may examine the information for selected transactions. This type of validation is particularly useful if damages depend on certain types of transactions that are identified by additional data on the purchasing records but are not summarized in other records.

Another approach is to test the internal integrity of the data. For example, a company may keep separate records of sales of products at its stores and shipments to the stores. The expert can compare sales to shipments to establish data integrity.

Not surprisingly, validation of data usually reveals some inconsistent or missing data. Some ways to handle these issues are discussed in the next section.

C. Are the Parties Disputing the Handling of Missing Data?

In dealing with missing data, it is critical to ascertain why the data are missing and to attempt to isolate the extent of the missing data. If only a small fraction of data is missing and the pattern appears to be random, then potentially the issue of the missing data can be disregarded and inferences can be drawn using only the available data.

However, missing data are seldom random. For example, suppose that only 1% of transactions are missing, but the transactions that are missing are large ones accounting for a third of all volume. Validation by summarizing across different characteristics will usually identify missing data that are not random. Such errors might occur if all of the missing transactions were submitted in a different format that the program for reading the data does not handle. For example, a manufacturer might record sales to Wal-Mart separately from all other customers.

Identifying and adding the missing data to the database is the best correction for missing data, but this is often not possible. In this case, the expert needs to address the problem in another way. The simplest method is to "gross up" damages to reflect the missing data. For example, if 10% of the transactions are randomly missing, then the expert may correct for the missing transactions by dividing calculated damages by 0.9. This method implicitly assumes that the percent missing is known and that the missing and nonmissing transactions reflect the same damages.

A related approach is to rely on partial, detailed data to measure damages as a fraction of another variable, such as sales. With this approach, other reliable and complete company records such as audited financials may be used to identify the company's total revenues, and damages are then calculated as the fraction of total sales as calculated above. The expert may choose to patch together incomplete data from one source and infer complete, reliable data in other ways. Such ways can include using a survey of customers or workers to measure damages per dollar of sales or per dollar of earnings and then applying those ratios to reliable data on total sales or total earnings.

Example: Credit Card Issuer sells cardholders fraudulent overcharges for insurance against theft for computer purchases. But the insurance does not cover theft of additional purchases such as a printer. The overcharge is found to be 1% of the price of the computer. The transactions reflected in the only available data include computers bundled with printers. Defendant uses the assumption that all transactions over $800 include the purchase of a printer and deducts $150 as the average amount spent for the printer. Credit Card Issuer's damages estimate is 1% × (total purchases less $150 times the number of purchases for more than $800). The expert for the class of insurance purchasers surveys a sample of purchasers and finds that fewer printers were actually purchased in the sample than implied by Issuer's damages formula and thus calculates a higher total overcharge for the class.

Comment: The parties have competing approximations to solve the same problem in the data. The resolution depends on which one is more accurate. A proper survey would probably be the better answer unless the expert for the Issuer can offer additional evidence about the reliability of the approach used.

X. Standards for Disclosing Data to Opposing Parties

The usual procedure for disclosure of work performed by the damages expert in federal cases is to provide electronic data at the same time or soon after the delivery of the expert's Rule 26(a)(2) report.[83] The data enable the opposing expert to replicate and investigate the damages expert's work in preparation for the expert's deposition and the opposing expert's rebuttal analysis. Even the most complete Rule 26(a)(2) report falls far short of enabling replication of damages calculations

83. For a discussion of disclosure of expert reports under Federal Rule of Civil Procedure 26(a)(2), see Margaret A. Berger, The Admissibility of Expert Testimony, Section V.B.1, in this manual.

in all but the simplest damages case. The fundamental standard for data disclosure pursuant to Rule 26(a)(2) should be all the materials, starting from original data sources through all intermediate calculations up to the final computer output reflecting the results shown in the Rule 26(a)(2) report. This disclosure should also include all scripts (including programs or any instructions to any program used) as well as all data involved in any step of the computations in the format used by the corresponding software. In particular, Excel or other such programs should include the cell instructions in the worksheet. In addition to the backup for the calculations described in the expert's report, the disclosure should include the materials relating to any other opinion the expert has reached. If confidential data are involved, appropriate protective orders can be sought.

A. Use of Formats

Disclosure of data should be in standard formats. In general, the formats used by damages experts include Access, Oracle, and other relational databases; Excel, SAS, and Stata datasets; and flat files containing uniformly formatted data in character form. It is critical that the data be provided as actual data files on computer media such as DVDs or data disks, not paper or electronic printouts or reports formatted for visual presentation. As noted earlier, materials formatted for visual presentation are generally difficult to convert back to formats suitable for computer analysis.

B. Data Dictionaries

The disclosure should include data dictionaries when variable formats and descriptions are not obvious. Data dictionaries should state the format for each variable, the range of appropriate values, and how to interpret the data. This information is particularly important for historic data because specific data formatting may have been used to convey information. For example, positive income values might be used to indicate total household income but negative values might indicate the income of individuals in the household. This specific formatting may have evolved because the underlying data came from multiple sources or data storage was at a premium. Problems of this nature occur less often in more current data because the price of data storage has declined dramatically.

Often, the expert will receive multiple databases from an opposing party. In this situation, the expert needs the requisite information for linking the datasets together. Also, the expert needs to know how the data were compiled in order to understand how to interpret inconsistent information. For example, if the database consists of events corresponding to subscriptions to a newspaper, the expert may encounter overlapping dates for subscriptions to weekday-only service and subscriptions to both weekday and weekend service. Such issues become acute when the data have not been reviewed for errors or difficulties in interpretation.

In the example presented, the expert may be advised that the most reliable data are the starting dates and that the end dates should be ignored unless the entire subscription is terminated.

C. Resolution of Problems

Friction between parties over disclosure of electronic backup is unfortunately common. Some common accusations are

- Failing to disclose the intermediate steps that were performed to generate the data used in the final calculations from the source data;
- Disclosing data and other materials only on paper or as scanned images, not in electronically readable form;
- Disclosing data as reports formatted as tables (although the expert may have originally received the data in this format);
- Concealing the logic of an Excel spreadsheet by revealing only the cell values and not the formulas used to generate the values;
- Failing to provide data dictionaries explaining the meaning of the underlying data; and
- Omitting calculations related to opinions other than the actual damages calculation.

A judge, magistrate, or special master overseeing discovery should become familiar with these issues to resolve the disputes fairly and to ensure full disclosure of each expert's numerical work to the opposition.

A tool that may be effective, but that is rarely used in the United States, is to have the experts meet without attorneys and identify where they agree and where they disagree.[84] Such an arrangement would generally require the consent of the parties.

84. New Zealand allows such expert conferences in certain circumstances. For example, High Court Rule 9.44 provides:

> (1) The court may, on its own initiative or on the application of a party to a proceeding, direct expert witnesses to—(a) confer on specified matters; (b) confer in the absence of the legal advisers of the parties; (c) try to reach agreement on matters in issue in the proceeding; (d) prepare and sign a joint witness statement stating the matters on which the expert witnesses agree and the matters on which they do not agree, including the reasons for their disagreement; (e) prepare the joint witness statement without the assistance of the legal advisers of the parties. (2) The court must not give a direction under subclause (1)(b) or (e) unless the parties agree.

Judicature Act 1908 No. 89 (as at 24 May 2010), Schedule 2 High Court Rules, Part 9 Evidence, Subpart 5—Experts, Rule 9.44.

D. Special Masters and Neutral Experts

Court-appointed individuals with the appropriate backgrounds can be useful in cases with complex damages calculations. If such an individual is assisting the court, the parties are unlikely to fail to cooperate speedily in disclosing their underlying computer work, knowing that any failure of cooperation would be recognized immediately.

XI. Damages in Class Actions

A. Class Certification

Damages play a large and increasing role in the certification of a class. Courts are exhibiting an increasing tendency to deny certification unless the class has a well-developed method for measuring damages for individual class members. One aspect of this tightening of standards is the use of a damages model to limit the membership of the class to individuals who are known to have incurred losses from the harmful conduct. Whereas earlier standards for damages were mainly the assurance of a qualified expert that damages could be measured later in the proceeding, some courts now require the expert to present a more fully developed method for quantifying damages.[85] Disputes about the practicality of damages measurement are more and more likely in proposed class actions.[86]

A court operating under the rule that class certification requires a fully developed damages quantification will need to grant discovery prior to class certification to support the class's damages analysis and the defendant's opposition.

B. Classwide Damages

The class's damages expert normally measures and testifies to classwide damages using methods discussed elsewhere in this chapter. In many class actions, damages ultimately will be paid to class members who file claims in a phase that occurs after settlement or trial. In principle, the damages experts will need to forecast the number of claimants as well as the average amount of damages per claimant. The propensity of class members to file claims depends critically on the amount of their likely recovery. Asbestos victims have high claim rates, whereas individuals who overpaid their cell phone bills by a few dollars have low rates.

85. *See* David S. Evans, *The New Consensus on Class Certification: What It Means for the Use of Economic and Statistical Evidence in Meeting the Requirements of Rule 23* (Jan. 2009), *available at* http://ssrn.com/abstract=1330594.

86. *See, e.g., In Re* New Motor Vehicles Canadian Export Antitrust Litig., 522 F.3d 6 (1st Cir. 2008).

C. Damages of Individual Class Members

Damages experts may also have a role in the process of disbursing funds from verdict or settlement to individual class members. An expert can develop software that measures individual damages based on evidence supplied by individuals through a claims-processing facility. For example, in *Millsap v. McDonnell-Douglas,*[87] more than 1000 class members, victims of the defendant's challenged layoffs, completed sworn questionnaires describing their post-layoff experiences. McDonnell-Douglas supplied additional information from its employee records. The class's damages expert used standard methods for valuing claims for lost earnings to calculate estimates for each class member. Because the settlement compromised a number of disputes about the law and about the facts underlying the layoffs, the total cash from the settlement was less than the sum of these estimates, and class members received a fraction of the amount indicated by the damages model.

D. Have the Defendant and the Class's Counsel Proposed a Fair Settlement?

The classwide damages measure has a key role in resolving class-action cases, because courts refer to it in determining the fairness of proposed settlements. The court's careful review of the benefits proposed for the class is essential because the interests of the class's counsel are not aligned with those of the class members with respect to settlement.

Example: Lender required excessive escrow deposits for property taxes from a class of mortgage borrowers, although the excess was repaid at the end of each year. Under the terms of the proposed settlement negotiated between Lender's lawyers and those for the class, the excess is to be refunded to the class members immediately, with 30% of that amount paid to class counsel as fees.

Comment: This settlement is unreasonable and leaves the class worse off than they were under the excessive escrows. The loss to the class from placing funds in an escrow is the foregone interest on the amount in the escrow, which would likely be no more than 10% of the excess amount of the escrow. By granting 30% of the refund as fees to class counsel, the class members are at least 20% worse off than they would be if the excess were repaid with a delay.

87. No. 94-CV-633-H(M), 2003 WL 21277124 (N.D. Okla. May 28, 2003).

At one time, settlements granted coupons to class members rather than cash compensation, but this practice is now discouraged. The valuation of the coupons is controversial.[88]

Example: In a case brought by the Department of Justice, airlines were found culpable for price fixing. The settlement of the derivative consumer class action granted class members coupons for discounts on air travel. Alaska Airlines, not a defendant in the government case or earlier in the class action, petitioned the court to be added as a defendant so that it too could gain the marketing advantage of the coupons.[89]

Comment: Alaska's petition made it clear that the coupons were beneficial to the airlines, not costly, and so the corresponding value to the class was presumptively small.

XII. Illustrations of General Principles

In the sections below, we provide concrete examples of how damages may be calculated in two common situations: (1) lost personal earnings and (2) lost profits for a business. The discussions are intended to illustrate how to apply the general ideas presented in the previous sections.

A. Claim for Lost Personal Income

Claims for lost personal earnings generally arise from wrongful termination, discrimination, injury, or death. The earnings usually come from employment in a firm, but essentially the same issues arise if self-employment or partnership earnings are lost. Most damages studies for personal lost earnings closely fit the model of Figure 1. The but-for world is usually based on the projected employment trajectory absent the harmful act. Here we present an example of a moderately realistic lost personal income damages quantification.

A construction worker sues for lost personal income after he is severely injured when the defendant runs through a red light and hits him. He asserts that he is disabled and unable to work for the rest of his life. Moreover, his injuries

88. *See* Figueroa v. Sharper Image Corp., 519 F. Supp. 2d 1302–29 (S.D. Fla. 2007).

89. *In Re* Domestic Air Transp. Antitrust Litig., 148 F.R.D. 297 (N.D. Ga. 1993). According to a spokesman for Alaska Air, "The airlines using those coupons are going to see substantial additional ticket sales because of them. . . . We asked to be named in the case because, once we saw the settlement, we realized it was to our competitive disadvantage not to do so." Anthony Faiola, *In Settling with Airlines, There's No Free Ride; Coupons for Travelers, $16 Million for Lawyers*, Washington Post, Mar. 20, 1995, at A1. For a general discussion of coupon settlements, see Christopher R. Leslie, *A Market-Based Approach to Coupon Settlements in Antitrust and Consumer Class Action Litigation*, 49 UCLA L. Rev. 91 (2002).

are so severe that his lifespan has been shortened by 3.5 years. Although he was a construction worker at the time of the accident, he had been going to school to become a CPA. The plaintiff's damages study presumes that he will not be able to work at all in the future. The defendant argues that the plaintiff should have continued his education after the accident and worked as a CPA. The defendant also disputes the reliability of the reduced life expectancy calculation. The judge has ruled that a jury should decide if there is a sufficient basis to conclude that the calculation of lost personal earnings should reflect the plaintiff's reduced life expectancy.

1. Is there a dispute about projected earnings but for the harmful event?

A plaintiff who seeks compensation for lost earnings will normally estimate damages based on wages or salary; other cash compensation, such as commissions, overtime, and bonuses; and the value of fringe benefits. Employees in similar jobs whose earnings were not interrupted form a natural benchmark for earning growth between the harmful event and trial. The plaintiff may make the case that a promotion or job change would have occurred during that period. Disputes involving the more variable elements of cash compensation are likely to arise. The plaintiff may measure bonuses and overtime during a period when these parts of compensation were unusually high, while the defendant may choose a longer period, during which the average is lower.

In our example, the construction worker claims that he would have made $75,000 working for one more year in construction while completing his degree. After that, he would have worked as a CPA earning $100,000 a year until retirement at age 70 based on the average salary for all CPAs. As a result of his injury, he only receives $22,000 a year from disability payments. Table 4 shows these projections.

The defendant's damages study presumes that the plaintiff could have continued his education after the injury and begun working as a CPA a year later. However, he argues the plaintiff would have earned only $75,000 as a CPA because of the plaintiff's lackluster record as an undergraduate and his career as a construction worker, where his work resulted in a depreciation of the skills that a CPA needs. This salary is based on the median salary for all CPAs. Table 5 shows the defendant's projections.

2. Are the parties disputing the valuation of benefits?

Lost benefits are an important part of lost personal earnings damages. As discussed in Section VIII.B, strict adherence to the format of Figure 1 can help resolve these disputes.

In the example, plaintiff includes only disability payments because of the injury. Absent his injury, he would have received higher benefits than the disability payments after retirement at age 70. The defendant projects higher social

Table 4. Plaintiff's Estimate of Lost Personal Income

Age	Actual Earnings	Actual Social Sec Benefits	Total Actual Income	Probability of Surviving	Probability of Working	Expected Income	But-for Earnings	But-for Social Sec Benefits	But-for Total But-for Income	But-for Probability of Surviving	But-for Probability of Working	But-for Expected Income	Total Lost Income	Discount Rate	Discount Rate Index	Discounted Lost Income
56-57	0	22,008	22,008	1.00	0.00	22,008	75,000		75,000	1.00	0.90	67,500	45,492	0.01	1.00	45,492
57-58	0	22,008	21,727	0.99	0.00	21,727	87,083		87,083	0.99	1.00	86,341	64,615	0.01	0.99	63,975
58-59	0	22,008	21,428	0.97	0.00	21,428	100,000		100,000	0.98	1.00	98,238	76,810	0.01	0.98	75,297
59-60	0	22,008	21,107	0.96	0.00	21,107	100,000		100,000	0.97	1.00	97,258	76,151	0.01	0.97	73,911
60-61	0	22,008	20,761	0.94	0.00	20,761	100,000		100,000	0.96	1.00	96,195	75,434	0.01	0.96	72,491
61-62	0	22,008	20,387	0.93	0.00	20,387	100,000		100,000	0.95	1.00	95,039	74,653	0.01	0.95	71,029
62-63	0	22,008	19,984	0.91	0.00	19,984	100,000		100,000	0.94	1.00	93,788	73,804	0.01	0.94	69,527
63-64	0	22,008	19,556	0.89	0.00	19,556	100,000		100,000	0.92	1.00	92,448	72,892	0.01	0.93	67,988
64-65	0	22,008	19,104	0.87	0.00	19,104	100,000		100,000	0.91	1.00	91,024	71,920	0.01	0.92	66,417
65-66	0	22,008	18,629	0.85	0.00	18,629	100,000		100,000	0.90	1.00	89,514	70,885	0.01	0.91	64,813
66-67	0	22,008	18,130	0.82	0.00	18,130	100,000		100,000	0.88	1.00	87,916	69,786	0.01	0.91	63,177
67-68	0	22,008	17,605	0.80	0.00	17,605	100,000		100,000	0.86	1.00	86,220	68,615	0.01	0.90	61,501
68-69	0	22,008	17,052	0.77	0.00	17,052	100,000		100,000	0.84	1.00	84,414	67,362	0.01	0.89	59,780
69-70	0	22,008	16,467	0.75	0.00	16,467	100,000		100,000	0.82	1.00	82,483	66,016	0.01	0.88	58,006
70-71	0	22,008	15,849	0.72	0.00	15,849		31,152	31,152	0.80	0.00	25,053	9,203	0.01	0.87	8,007
71-72	0	22,008	15,197	0.69	0.00	15,197		31,152	31,152	0.78	0.00	24,365	9,168	0.01	0.86	7,897
72-73	0	22,008	14,506	0.66	0.00	14,506		31,152	31,152	0.76	0.00	23,626	9,121	0.01	0.85	7,778
73-74	0	22,008	13,775	0.63	0.00	13,775		31,152	31,152	0.73	0.00	22,832	9,058	0.01	0.84	7,648
74-75	0	22,008	13,005	0.59	0.00	13,005		31,152	31,152	0.71	0.00	21,982	8,977	0.01	0.84	7,505
75-76	0	22,008	12,200	0.55	0.00	12,200		31,152	31,152	0.68	0.00	21,074	8,875	0.01	0.83	7,346
76-77	0	22,008	11,364	0.52	0.00	11,364		31,152	31,152	0.65	0.00	20,113	8,748	0.01	0.82	7,170
77-78	0	22,008	10,505	0.48	0.00	10,505		31,152	31,152	0.61	0.00	19,098	8,594	0.01	0.81	6,973
78-79	0	22,008	9,628	0.44	0.00	9,628		31,152	31,152	0.58	0.00	18,035	8,408	0.01	0.80	6,755
79-80	0	22,008	8,740	0.40	0.00	8,740		31,152	31,152	0.54	0.00	16,927	8,187	0.01	0.80	6,512
80-81	0	22,008	7,852	0.36	0.00	7,852		31,152	31,152	0.51	0.00	15,780	7,928	0.01	0.79	6,244
81-82	0	22,008	6,973	0.32	0.00	6,973		31,152	31,152	0.47	0.00	14,602	7,629	0.01	0.78	5,949
82-83	0	22,008	6,112	0.28	0.00	6,112		31,152	31,152	0.43	0.00	13,401	7,289	0.01	0.77	5,627
83-84	0	22,008	5,283	0.24	0.00	5,283		31,152	31,152	0.39	0.00	12,188	6,906	0.01	0.76	5,279
84-85	0	22,008	4,494	0.20	0.00	4,494		31,152	31,152	0.35	0.00	10,976	6,481	0.01	0.76	4,905
85-86	0	22,008	3,758	0.17	0.00	3,758		31,152	31,152	0.31	0.00	9,777	6,019	0.01	0.75	4,510
86-87	0	22,008	3,082	0.14	0.00	3,082		31,152	31,152	0.28	0.00	8,605	5,523	0.01	0.74	4,097
87-88	0	22,008	2,475	0.11	0.00	2,475		31,152	31,152	0.24	0.00	7,474	5,000	0.01	0.73	3,673
88-89	0	22,008	1,941	0.09	0.00	1,941		31,152	31,152	0.21	0.00	6,400	4,459	0.01	0.73	3,243
89-90	0	22,008	1,484	0.07	0.00	1,484		31,152	31,152	0.17	0.00	5,394	3,911	0.01	0.72	2,816
90-91	0	22,008	1,102	0.05	0.00	1,102		31,152	31,152	0.14	0.00	4,469	3,367	0.01	0.71	2,401
91-92	0	22,008	793	0.04	0.00	793		31,152	31,152	0.12	0.00	3,634	2,841	0.01	0.71	2,005
92-93	0	22,008	551	0.03	0.00	551		31,152	31,152	0.09	0.00	2,895	2,344	0.01	0.70	1,638
93-94	0	22,008	369	0.02	0.00	369		31,152	31,152	0.07	0.00	2,256	1,887	0.01	0.69	1,306
94-95	0	22,008	236	0.01	0.00	236		31,152	31,152	0.06	0.00	1,716	1,480	0.01	0.69	1,014
95-96	0	22,008	145	0.01	0.00	145		31,152	31,152	0.04	0.00	1,272	1,127	0.01	0.68	765
96-97	0	22,008	84	0.00	0.00	84		31,152	31,152	0.03	0.00	916	832	0.01	0.67	559
97-98	0	22,008	46	0.00	0.00	46		31,152	31,152	0.02	0.00	640	594	0.01	0.67	395
98-99	0	22,008	24	0.00	0.00	24		31,152	31,152	0.01	0.00	433	410	0.01	0.66	270
99-100	0	22,008	11	0.00	0.00	11		31,152	31,152	0.01	0.00	283	272	0.01	0.65	177
100 and over	0	22,008	0	0.00	0.00	0		31,152	31,152	0.00	0.00	0	0	0.01	0.65	0
													Total Lost Personal Income			1,043,866

Table 5. Defendant's Estimate of Lost Personal Income

Age	Actual Earnings	Actual Social Sec Benefits	Total Actual Income	Probability of Surviving	Probability of Working	Expected Income	But-for Earnings	But-for Social Sec Benefits	But-for Total But-for Income	But-for Probability of Surviving	But-for Probability of Working	But-for Expected Income	Total Lost Income	Discount Rate	Discount Rate Index	Discounted Lost Income
56-57	0	0	0	1.00	0.00	0	75,000		75,000	1.00	0.80	60,000	60,000	0.05	1.00	60,000
57-58	75,000	0	74,361	0.99	1.00	74,361	75,000		75,000	0.99	1.00	74,361	0	0.05	0.95	0
58-59	75,000	0	73,678	0.98	1.00	73,678	75,000		75,000	0.98	1.00	73,678	0	0.05	0.91	0
59-60	75,000	0	72,944	0.97	1.00	72,944	75,000		75,000	0.97	1.00	72,944	0	0.05	0.86	0
60-61	75,000	0	72,146	0.96	1.00	72,146	75,000		75,000	0.96	1.00	72,146	0	0.05	0.82	0
61-62	75,000	0	71,280	0.95	1.00	71,280	75,000		75,000	0.95	1.00	71,280	0	0.05	0.78	0
62-63	75,000	0	70,341	0.94	1.00	70,341	75,000		75,000	0.94	1.00	70,341	0	0.05	0.75	0
63-64	75,000	0	69,336	0.92	1.00	69,336	75,000		75,000	0.92	1.00	69,336	0	0.05	0.71	0
64-65	75,000	0	68,268	0.91	1.00	68,268	75,000		75,000	0.91	1.00	68,268	0	0.05	0.68	0
65-66	75,000	0	67,135	0.90	1.00	67,135	75,000		75,000	0.90	1.00	67,135	0	0.05	0.64	0
66-67	75,000	0	65,937	0.88	1.00	65,937	75,000		75,000	0.88	1.00	65,937	0	0.05	0.61	0
67-68	75,000	0	64,665	0.86	1.00	64,665	75,000		75,000	0.86	1.00	64,665	0	0.05	0.58	0
68-69	75,000	0	63,310	0.84	1.00	63,310	75,000		75,000	0.84	1.00	63,310	0	0.05	0.56	0
69-70	75,000	0	61,863	0.82	1.00	61,863	75,000		75,000	0.82	1.00	61,863	0	0.05	0.53	0
70-71	0	30,864	24,821	0.80	0.00	24,821		31,152	31,152	0.80	0.00	25,053	232	0.05	0.51	117
71-72	0	30,864	24,140	0.78	0.00	24,140		31,152	31,152	0.78	0.00	24,365	225	0.05	0.48	108
72-73	0	30,864	23,408	0.76	0.00	23,408		31,152	31,152	0.76	0.00	23,626	218	0.05	0.46	100
73-74	0	30,864	22,621	0.73	0.00	22,621		31,152	31,152	0.73	0.00	22,832	211	0.05	0.44	92
74-75	0	30,864	21,779	0.71	0.00	21,779		31,152	31,152	0.71	0.00	21,982	203	0.05	0.42	84
75-76	0	30,864	20,879	0.68	0.00	20,879		31,152	31,152	0.68	0.00	21,074	195	0.05	0.40	77
76-77	0	30,864	19,927	0.65	0.00	19,927		31,152	31,152	0.65	0.00	20,113	186	0.05	0.38	70
77-78	0	30,864	18,922	0.61	0.00	18,922		31,152	31,152	0.61	0.00	19,098	177	0.05	0.36	63
78-79	0	30,864	17,868	0.58	0.00	17,868		31,152	31,152	0.58	0.00	18,035	167	0.05	0.34	57
79-80	0	30,864	16,771	0.54	0.00	16,771		31,152	31,152	0.54	0.00	16,927	156	0.05	0.33	51
80-81	0	30,864	15,634	0.51	0.00	15,634		31,152	31,152	0.51	0.00	15,780	146	0.05	0.31	45
81-82	0	30,864	14,467	0.47	0.00	14,467		31,152	31,152	0.47	0.00	14,602	135	0.05	0.30	40
82-83	0	30,864	13,277	0.43	0.00	13,277		31,152	31,152	0.43	0.00	13,401	124	0.05	0.28	35
83-84	0	30,864	12,076	0.39	0.00	12,076		31,152	31,152	0.39	0.00	12,188	113	0.05	0.27	30
84-85	0	30,864	10,874	0.35	0.00	10,874		31,152	31,152	0.35	0.00	10,976	101	0.05	0.26	26
85-86	0	30,864	9,686	0.31	0.00	9,686		31,152	31,152	0.31	0.00	9,777	90	0.05	0.24	22
86-87	0	30,864	8,525	0.28	0.00	8,525		31,152	31,152	0.28	0.00	8,605	80	0.05	0.23	18
87-88	0	30,864	7,405	0.24	0.00	7,405		31,152	31,152	0.24	0.00	7,474	69	0.05	0.22	15
88-89	0	30,864	6,341	0.21	0.00	6,341		31,152	31,152	0.21	0.00	6,400	59	0.05	0.21	12
89-90	0	30,864	5,344	0.17	0.00	5,344		31,152	31,152	0.17	0.00	5,394	50	0.05	0.20	10
90-91	0	30,864	4,428	0.14	0.00	4,428		31,152	31,152	0.14	0.00	4,469	41	0.05	0.19	8
91-92	0	30,864	3,600	0.12	0.00	3,600		31,152	31,152	0.12	0.00	3,634	34	0.05	0.18	6
92-93	0	30,864	2,868	0.09	0.00	2,868		31,152	31,152	0.09	0.00	2,895	27	0.05	0.17	5
93-94	0	30,864	2,235	0.07	0.00	2,235		31,152	31,152	0.07	0.00	2,256	21	0.05	0.16	3
94-95	0	30,864	1,700	0.06	0.00	1,700		31,152	31,152	0.06	0.00	1,716	16	0.05	0.16	2
95-96	0	30,864	1,260	0.04	0.00	1,260		31,152	31,152	0.04	0.00	1,272	12	0.05	0.15	2
96-97	0	30,864	908	0.03	0.00	908		31,152	31,152	0.03	0.00	916	8	0.05	0.14	1
97-98	0	30,864	634	0.02	0.00	634		31,152	31,152	0.02	0.00	640	6	0.05	0.14	1
98-99	0	30,864	429	0.01	0.00	429		31,152	31,152	0.01	0.00	433	4	0.05	0.13	1
99-100	0	30,864	280	0.01	0.00	280		31,152	31,152	0.01	0.00	283	3	0.05	0.12	0
100 +	0	30,864	0	0.00	0.00	0		31,152	31,152	0.00	0.00	0	0	0.05	0.12	0
													Total Lost Personal Income			61,104

security benefits based on a longer period and higher level of contributions to social security. The parties agree that the plaintiff would have retired at age 70 absent the accident.

3. Is there disagreement about how earnings should be discounted to present value?

Because personal lost earnings damages may accrue over the remainder of a plaintiff's working life, the issues of predicting future inflation and discounting earnings to present value are likely to generate quantitatively important disagreements. As we noted in Section VI.D, projections of future compensation can be calculated in constant dollars or escalated terms. In the first case, the interest rate used to discount future constant-dollar losses should be a real interest rate—the difference between the ordinary interest rate and the projected future rate of inflation. All else being the same, the two approaches will give identical calculations of damages.

In our example, both the plaintiff and defendant use constant dollars and use a real rate of interest for discounting. However, the plaintiff calculates the real rate of interest as 1%, relying on the implied rate from inflation-adjusted Treasury bonds. In contrast, the defendant uses a discount rate of 5% based on the historic real rate of return to investments in general.

4. Is there disagreement about subsequent unexpected events?

Disagreements about subsequent unexpected events are likely in cases involving personal earnings, as discussed in general in Section VIII.E. For example, the plaintiff may have suffered a debilitating illness that would have caused him to quit his job a year later even if the wrongful act had not occurred. Alternatively, the plaintiff may have been laid off as a result of employer hardship a year later notwithstanding the wrongful act. In these examples, the defendant may argue that damages should be limited to one year. The plaintiff might respond that subsequent events were unexpected at the time of the termination and therefore should be excluded from consideration in the calculation of damages. Thus, the plaintiff would argue that damages should be calculated without consideration of these events.

In our example, the defendant points out that the unemployment rate for construction workers was 50% beginning six months after the accident. The plaintiff argues that the unemployment rate for construction workers at the time of the accident was only 19% and therefore the revised unemployment rate after the accident is irrelevant.

5. Is there disagreement about retirement and mortality?

Closely related to the issue of unexpected events is how future damages should reflect the probability that the plaintiff will die or decide to retire. Sometimes an expert will assume a work-life expectancy and terminate damages at the end of that period. Tables of work-life expectancy incorporate the probability of both retirement and death. Another approach is to multiply each year's lost earnings by the probability that the plaintiff will be alive and working in that year. That probability declines gradually with age and can be inferred from data on labor force participation and mortality by age.

In our example, the plaintiff projects that his life expectancy was reduced by 3.5 years and uses revised survival rates as a result. The defendant disagrees, arguing that the survival tables relied upon by the plaintiff are unreliable. However, both agree that the plaintiff would have worked until age 70 absent the accident because the unemployment rate for CPAs is essentially zero in the area where the plaintiff lives.

6. Is there a dispute about mitigation?

Actual earnings before trial, although known, may be subject to dispute if the defendant argues that the plaintiff took too long to find a job or the job taken was not sufficiently remunerative. Even more problematic may be the situation in which the plaintiff continues to be unemployed. Parties disputing the length of job search frequently offer testimony from job placement experts. Testimony from a psychologist also may be offered if the plaintiff has suffered emotional trauma as a result of the defendant's actions. Recovery from temporarily disabling injuries may be the subject of testimony by experts in vocational rehabilitation.

In our example, the plaintiff argues that he is disabled and unable to work for the remainder of his life. The defendant argues that the plaintiff could have finished his education and could then have worked as a CPA. Both provide the testimony from experts in vocational rehabilitation to support their conclusions.

7. Is there disagreement about how the plaintiff's career path should be projected?

The issues that arise in projecting but-for and actual earnings after trial are similar to the issues that arise in measuring damages before trial. In addition, the parties are likely to disagree about the plaintiff's future increases in compensation. A damages analysis should be internally consistent. For example, the compensation paths for both but-for and actual earnings should be based on consistent assumptions about general economic conditions, about conditions in the local labor market for the plaintiff's type of work, the age-earnings profile for the career path, and particularly about the plaintiff's likely increased skills and earning capacity. The analysis probably should project a less successful career for mitigation if it is projecting a slow earnings growth absent the harm.

In our example, the plaintiff argues that he would have worked as a CPA but for the accident but that he is too injured to complete his education and work as a CPA. The defendant argues that working as a CPA is a viable option for the plaintiff. Although there is a disagreement about how much the plaintiff would have earned as a CPA, the plaintiff's argument that he is too disabled to work accounts for most of the damages. As shown in Tables 4 and 5, the plaintiff is seeking just over $1 million while the defendant calculates that damages are only $61,000. Differences of this magnitude between quantifications of lost personal earnings by plaintiffs and defendants are common. Our example illustrates some of the main reasons for the large differences.

B. Lost Profits for a Business

Claims for lost profits for a business generally arise from a lost stream of revenue. However, lost profits can also arise from increased costs. As an example, a breach of a supply contract may increase the victim firm's costs. Generally, an expert will likely be most involved in cases in which the plaintiff is seeking recovery for expectation, reliance, or restitution damages. Most damages studies will follow Figure 1 where earnings are the lost profits. For explication, the following is an example of a business lost profits case:

Plaintiff HSM makes cell phone handsets. Defendant TPC is a cell phone carrier. By denying HSM technical information and by informing HSM's potential customers that HSM's handsets are incompatible with TPC's network, TPC has imposed economic losses on HSM. TPC asserts that HSM has failed to mitigate its losses and overstates its lost revenues. Trial is set for the end of 2010. The respective damages analyses are shown in Tables 6 and Table 7 and discussed below.

Table 6. HSM's Damages Analysis (Dollars in Millions)

Year	(2) But-For Revenue	(3) But-For Costs	(4) But-For Earnings	(5) Actual Earnings	(6) Lost Earnings	(7) Discount Factor	(8) Damages
2008	$561	$374	$187	$34	$153	1.21	$185
2009	600	400	200	56	144	1.14	164
2010	639	426	213	45	168	1.07	180
2011	681	454	227	87	140	1.00	140
2012	726	484	242	96	147	0.96	141
2013	777	518	259	105	153	0.92	142
2014	828	552	276	116	160	0.89	142
2015	882	588	294	127	167	0.85	143
Total							$1236

Table 7. TPC's Damages Analysis (Dollars in Millions)

Year	(2) But-For Revenue	(3) But-For Costs	(4) But-For Earnings	(5) Mitigated Earnings	(6) Lost Earnings	(7) Discount Factor	(8) Damages
2008	$404	$303	$101	$79	$22	1.21	$27
2009	432	324	108	85	23	1.14	26
2010	460	345	115	81	34	1.07	36
2011	492	369	123	98	25	1.00	25
2012	524	393	131	108	23	0.87	20
2013	560	420	140	119	21	0.76	16
2014	596	447	149	130	19	0.66	12
2015	636	477	159	143	16	0.57	9
Total							$171

1. Is there a dispute about projected revenues?

Projecting lost revenues can be straightforward if the disrupted revenue stream occurs immediately following the bad act and the firm recovers relatively quickly. More complex cases can arise if the effect is delayed or the recovery is slow, intermittent, or nonexistent.

In the example above, the plaintiff's expert would argue that revenues would have been higher absent TPC's conduct and thus projects revenues based on the revenue growth prior to the bad act, which reflects increasing sales and increasing prices. The projected revenue for the plaintiff is shown in Table 6, column 2. The defendant's expert would argue that HSM's projections use a growth factor that improperly includes the period when HSM initially entered the market and, therefore, projects HSM's sales using the growth rate for the previous 2 years and assumes that prices would have remained unchanged. TPC's projection of HSM's revenue is shown in Table 7, column 2.

Some additional examples of complexities can found in antitrust cases. For example, assume a company is disadvantaged because a rival has constructed barriers to entry by entering into contracts that require customers to use its add-on products such as ink for a printer. In such cases, the plaintiff's expert may assert that the only suppliers in the but-for market for printer ink would consist of the defendant and the plaintiff, and that the profit would reflect pricing for a duopoly. The defendant may respond that there would be five firms in addition to the plaintiff who would have entered the market as suppliers, and that therefore the pricing would be close to that of a highly competitive market.

Other complexities may arise in intellectual property cases where the revenue stream is reduced because the intellectual property for a product has been misappropriated. In these cases, the expert may need to identify how much of the plaintiff's revenue stream should be attributed to the misappropriated intellectual property and how much should be attributed to other aspects of the product. For example, our printer manufacturer may believe that its printers are popular because of its proprietary method to increase the printing speed. However, the defendant may argue that the increase in printing speed has little to do with the popularity of the plaintiff's printer but rather the sharpness of the printing. Or the defendant may argue that at the time of the bad act the plaintiff's product was the fastest printer, but 2 years later, a noninfringing printer is faster and the plaintiff's sales therefore would have dropped to zero.

The projection of the revenue stream is likely to be the most controversial part of any damages estimate in a business case because it requires so many assumptions on the part of both experts with respect to the other players in the market and customer demand.

2. *Are the parties disputing the calculation of marginal costs?*

Another area of dispute that can arise is the measurement of marginal costs. Generally, if the business is an ongoing concern, then the costs can be determined from existing data. Often this is done either by directly modeling the costs needed for the additional revenues or using regression analysis that captures how costs have varied with revenues. The relevant concept is the measure of costs that would have been expended to generate the lost revenues.

In our example, plaintiff's expert would project that the additional costs would reflect the marginal cost ratio that was derived from a regression model of costs against revenues. The defendant's expert might use the average ratio of costs to revenues, arguing that this would be more appropriate because additional workers and equipment would have been needed to generate the increased revenues. The projected costs for both parties are shown in column 3 of Tables 6 and 7.

Costs are often expressed as a percentage of revenues, which simplifies the projection of costs. However, this approach can be problematic if there is reason to believe that the profit rate will change over time. The rate may change because the change in revenues will be so large as to require that an increasing percentage of fixed costs will need to be included, the mix of costs will change over time, or the components of cost will grow at disparate rates. If computing costs as a percentage of revenues is not viable, then the projected costs should reflect the same assumptions about growth and inflation that were used in the revenue projection.

3. *Is there a dispute about mitigation?*

Defendant's expert may argue that the plaintiff's *actual* profits are understated because the plaintiff failed to mitigate its losses. For example, the plaintiff's losses may have been minimized by closure of its business. Or the plaintiff perhaps should have invested in alternative facilities while its business was interrupted because it could not use its existing facilities.

In our example, the defendant's expert would argue that HSM could have mitigated its losses by obtaining the technical information it needed from other sources and could have counteracted TPC's disparagement with vigorous marketing. HSM's actual earnings are shown in column 5 of Table 6, and TPC's calculation of HSM's earnings with mitigation are shown in column 5 of Table 7.

4. Is there disagreement about how profits should be discounted to present value?

Generally, interest for lost earnings prior to trial is computed at a statutory rate, often not compounded. In our example, trial is at the end of year 2010 and the statutory rate is assumed to be 7% simple (i.e., without compounding). If the prejudgment rate is not set by law, economists favor the use of the cost of borrowing for the defendant, because damages are a forced loan to the defendant by the plaintiff.[90]

The rate used to discount future losses back to the time of the trial is not set by law and substantial disputes will arise about the discount rate. Generally, economists believe that the discount rate should equal the after-tax cost of capital for the plaintiff.

In our example, HSM argues that the proper discount rate should be based on a 4%, after-tax interest rate, obtained by applying HSM's corporate tax rate to TPC's medium-term borrowing rate. TPC, however, believes that the proper discount rate should be HSM's cost of capital, reflecting HSM's cost of equity and cost of debt. Column 7 of Tables 4 and 5 shows the respective discount rates after trial. The resulting damages are shown in column 8 of Tables 6 and 7.

5. Is there disagreement about subsequent unexpected events?

Disagreements about subsequent unexpected events are likely in cases involving lost profits. For example, the market for the plaintiff's goods may have suffered a substantial contraction a year after the bad act, with plaintiff likely to be forced into bankruptcy even if the wrongful act had not occurred. Or the costs of the plaintiff may have increased dramatically a year later because of shortages that would have necessitated that the plaintiff retool its business even if the wrongful act had not occurred. The plaintiff might respond that subsequent events were unexpected at the time of the bad act and so should be excluded from consideration in the calculation of damages. Plaintiff, therefore, would argue that damages should be calculated without consideration of these events. The defendant would respond that damages should be limited to 1 year because the unexpected events would have forced the closure of the plaintiff's business. This topic is discussed more fully in Section VIII.E.

90. *See* James M. Patell et al., *Accumulating Damages in Litigation: The Roles of Uncertainty and Interest Rates,* 11 J. Legal Stud. 341–64 (1982).

Glossary of Terms

appraisal. A method of determining the value of the plaintiff's claim on an earnings stream by reference to the market values of comparable earnings streams. For example, if the plaintiff has been deprived of the use of a piece of property, the appraised value of the property might be used to determine damages.

avoided cost. Cost that the plaintiff did not incur as a result of the harmful act. Usually it is the cost that a business would have incurred in order to make the higher level of sales the business would have enjoyed but for the harmful act.

but-for analysis. Restatement of the plaintiff's economic situation but for the defendant's harmful act. Damages are generally measured as but-for value less actual value received by the plaintiff.

capitalization factor. Factor used to convert a stream of revenue or profit into its capital or property value. A capitalization factor of 10 for profit means that a firm with $1 million in annual profit is worth $10 million.

compound interest. Interest calculation giving effect to interest earned on past interest. As a result of compound interest at rate r, it takes $(1 + r)(1 + r) = 1 + 2r + r^2$ dollars to make up for a lost dollar of earnings 2 years earlier.

constant dollars. Dollars adjusted for inflation. When calculations are done in constant 1999 dollars, it means that future dollar amounts are reduced in proportion to increases in the cost of living expected to occur after 1999.

discount rate. Rate of interest used to discount future losses.

discounting. Calculation of today's equivalent to a future dollar to reflect the time value of money. If the interest rate is r, the discount applicable to 1 year in the future is:

$$\text{discount rate} = 1/(1 + r).$$

The discount for 2 years is this amount squared; for 3 years, it is this amount to the third power, and so on for longer periods. The result of the calculation is to give effect to compound interest.

earnings. Economic value received by the plaintiff. Earnings could be salary and benefits from a job, profit from a business, royalties from licensing intellectual property, or the proceeds from a one-time or recurring sale of property. Earnings are measured net of costs. Thus, lost earnings are lost receipts less costs avoided.

escalation. Consideration of future inflation in projecting earnings or other dollar flows. The alternative is to make projections in constant dollars.

expectation damages. Damages measured on the principle that the plaintiff is entitled to the benefit of the bargain originally made with the defendant.

fixed cost. Cost that does not change with a change in the amount of products or services sold.

mitigation. Action taken by the plaintiff to minimize the economic effect of the harmful act. Also often refers to the actual level of earnings achieved by the plaintiff after the harmful act.

nominal interest rate. Interest rate quoted in ordinary dollars, without adjustment for inflation. Interest rates quoted in markets and reported in the financial press are always nominal interest rates.

prejudgment interest. Interest on losses occurring before trial.

present value. Value today of money due in the past (with interest) or in the future (with discounting).

price erosion. Effect of the harmful act on the price charged by the plaintiff. When the harmful act is wrongful competition, as in intellectual property infringement, price erosion is one of the ways that the plaintiff's earnings have been harmed.

real interest rate. Interest rate adjusted for inflation. The real interest rate is the nominal interest rate less the annual rate of inflation.

regression analysis. Statistical technique for inferring stable relationships among quantities. For example, regression analysis may be used to determine how costs typically vary when sales rise or fall.

reliance damages. Damages designed to reimburse a party for expenses incurred from reliance upon the promises of the other party.

restitution damages. Damages measured on the principle of restoring the economic equivalent of lost property or value.

variable cost. Component of a business's cost that would have been higher if the business had enjoyed higher sales. See also avoided cost.

Reference Guide on Exposure Science

JOSEPH V. RODRICKS

Joseph V. Rodricks, Ph.D., is Principal at Environ, Arlington, Virginia.

CONTENTS

I. Introduction

The sciences of epidemiology[1] and toxicology[2] are devoted to understanding the hazardous properties (the toxicity) of chemical substances. Moreover, epidemiological and toxicological studies provide information on how the seriousness and rate of occurrence of the hazard in a population (its risk) change as exposure to a particular chemical changes. To evaluate whether individuals or populations exposed to a chemical are at risk of harm,[3] or have actually been harmed, the information that arises from epidemiological and toxicological studies is needed, as is the information on the exposures incurred by those individuals or populations.

Epidemiologists and toxicologists can tell us, for example, how the magnitude of risk of benzene-induced leukemia changes as exposure to benzene changes. Thus, if there is a need to understand the magnitude of the leukemia risk in populations residing near a petroleum refinery, it becomes necessary to understand the magnitude of the exposure of those populations to benzene. Likewise, if an individual with leukemia claims that benzene exposure was the cause, it becomes necessary to evaluate the history of that individual's exposure to benzene.[4]

Understanding exposure is essential to understanding whether the toxic properties of chemicals have been or will be expressed. Thus, claims of toxic tort or product liability generally require expert testimony not only in medicine and in the sciences of epidemiology and toxicology, but also testimony concerning the nature and magnitude of the exposures incurred by those alleging harm. Similarly, litigation involving the regulation of chemicals said to pose excessive risks to health also requires litigants to present evidence regarding exposure. The need to understand exposure is a central topic in the reference guides in this publication on epidemiology and toxicology. This reference guide provides a view of how the magnitude of exposure comes to be understood.[5]

1. *See* Michael D. Green et al., Reference Guide on Epidemiology, in this manual.

2. *See* Bernard D. Goldstein & Mary Sue Henifin, Reference Guide on Toxicology, in this manual.

3. *See, e.g.,* Rhodes v. E.I. du Pont de Nemours & Co., 253 F.R.D. 365 (S.D. W. Va. 2008) (suit for medical monitoring costs because exposure to perfluoroctanoic acid in drinking water allegedly caused an increased risk of developing certain diseases in the future); *In re* Welding Fume Prods. Liab. Litig., 245 F.R.D. 279 (N.D. Ohio 2007) (exposure to manganese fumes allegedly increased the risk of later developing brain damage).

4. *See, e.g.,* Lambert v. B.P. Products North America, Inc., 2006 WL 924988 (S.D. Ill. 2006), 2006 U.S. Dist. LEXIS 16756 (plaintiff diagnosed with chronic lymphocytic leukemia was exposed to jet fuel allegedly containing excessive levels of benzene).

5. This chapter focuses on measuring exposure to toxic substances as a specific developing area of scientific investigation. This topic is distinct from the legal concept of "exposure," which is an element of a claim in toxic tort litigation. The legal concept of exposure relies on the evolving scientific understanding of the manner and extent to which individuals come into contact with toxic substances. However, the legal concept also reflects substantive legal principles and interpretations that vary across jurisdictions. *Compare* Parker v. Mobil Oil Corp., 793 N.Y.S.2d 434 (2005) (requiring findings of specific levels of exposure to benzene by plaintiff who claimed that his leukemia was the result of his

Not all questions concerning human exposures to potentially harmful substances require expert testimony. In those circumstances in which the magnitude of exposure is not relevant, or is clearly evident (e.g., because a plaintiff was observed to take the prescribed amount of a prescription medicine), expert testimony is not indicated. But if the magnitude of exposure is an important component of the needed evidence, and if that magnitude is not a simple question of fact, then expert testimony will be important.

II. Exposure Science

Exposure science is not yet a distinct academic discipline. Although some schools of public health may offer courses in exposure assessment, there are no academic degrees offered in exposure science. When regulatory and public health agencies began in the 1970s to examine toxicological risks in a quantitative way, it became apparent that quantitative exposure assessments would become necessary. Initially, exposure assessment was typically practiced by toxicologists and epidemiologists. As the breadth and complexity of the subject began to be recognized, it became apparent that scientists and engineers with a better grasp of the properties of chemicals (which affect how they behave and undergo change in different environments), and of the methods available to identify and measure chemicals in products and in the environment, would be necessary to provide scientifically defensible assessments. As the importance of exposure assessment grew and began to present significant scientific challenges, its practice drew increasing numbers of scientists and engineers, and some began to refer to their work as exposure science. Not surprisingly, most of the early expositions of exposure assessment came from government agencies that recognized the need to develop and refine the practice to meet their risk assessment needs. Indeed, various documents and reports used by the U.S. Environmental Protection Agency (EPA) remain essential sources for the practice of exposure assessment.[6] Academics and practitioners have written chapters on exposure science for major multiauthor reference works

17-year occupational exposure to gasoline containing benzene) *with* Westberry v. Gislaved Gummi AB, 178 F.3d 257 (4th Cir. 1999) (evidence of specific exposure level not required where evidence of talc in the workplace indicated that the worker was covered in talc and left footprints on the floor) *and* Allen v. Martin Surfacing, 263 F.R.D. 47 (D. Mass. 2009) (admissible expert testimony may be based on symptom accounts by those exposed rather than direct measurements of solvent concentrations). This chapter takes no position regarding exposure as a substantive legal concept.

6. U.S. Environmental Protection Agency, Exposure Assessment Tools and Models (2009), *available at* http://www.epa.gov/oppt/exposure/ (last visited June 6, 2011); National Exposure Research Laboratory, U.S. Environmental Protection Agency, Scientific and Ethical Approaches for Observational Exposure Studies, Doc. No. EPA 600/R-08/062 (2008), *available at* http://www.epa.gov/nerl/sots/index.html (last visited July 14, 2010); U.S. Environmental Protection Agency. Exposure Factors Handbook (1997).

on toxicology,[7] but most of the work in this area is still found in the primary reference works.

Although exposure science is not yet a distinct academic discipline, in this reference guide the phrase is retained and used to refer to the work of scientists and engineers ("exposure scientists") working in one or more aspects of exposure assessment.

A. What Do Exposure Scientists Do?

Human beings are exposed to natural and industrial chemicals from conception to death, and because almost all chemicals can become harmful if exposures exceed certain levels, understanding the magnitude and duration of exposures to chemicals is critical to understanding their health impacts. Exposure science is the study of how people can come into contact with (are exposed to)[8] chemicals that may be present in various environmental media (air, water, food, soil, consumer products of all types) and of the amounts of those chemicals that enter the body as a result of these contacts.[9] Exposure scientists also study whether and how those amounts change over time. The goal of exposure science is to quantify those amounts and time periods. The quantitative expression of those amounts is referred to as dose. Ultimately the dose incurred by populations or individuals is the measure needed by health experts to quantify risk of toxicity. Exposure science does not typically deal with the health consequences of those exposures.

The dose entering the body (through inhalation or ingestion, through the skin, and through other routes) is often referred to as the "exposure dose," to distinguish it from the dose that enters the bloodstream and reaches various organs of the body. The latter is typically only a fraction of the exposure dose and is identified through studies that can trace the fate of a chemical after it enters the body. The term "dose" as used in this reference guide is synonymous with "exposure dose," and doses reaching blood or various organs within the body are referred to as "target site doses" or "systemic doses,"

Exposure assessments can be directed at past, present, or even future exposures and can be narrowly focused (one chemical, one environmental medium, one population group) or very broad in scope (many chemicals, several environ-

7. P.J. Lioy, *Exposure Analysis and Its Assessment, in* Comprehensive Toxicology (I.G. Sipes et al. eds., 1997); D.J. Paustenbach & A. Madl, *The Practice of Exposure Assessment, in* Principles and Methods of Toxicology (Wallace Hayes ed., 5th ed. 2008).

8. *See, e.g.,* Kitzmiller v. Jefferson, 2006 WL 2473399, 2006 U.S. Dist. LEXIS 61109 (N.D. W. Va. 2006) (defendants offered expert's testimony that plaintiff's use of liquid cleaning agents containing benzalkonium chloride failed to show that she was exposed to benzalkonium chloride in the air); Hawkins v. Nicholson, 2006 WL 954654, 2006 U.S. App. Vet. Claims LEXIS 197, 21 Vet. App. 64 (Vet. App. 2006) (noting that "a veteran who served on active duty in Vietnam between January 9, 1962, and May 7, 1975, is entitled to a rebuttable presumption of exposure to Agent Orange").

9. The term "enter the body" also includes entering the external surface of the body.

mental media, several different population groups). This reference guide explores the various contexts in which exposure assessments are conducted and how their scope is determined.

B. *Who Qualifies as an Expert in Exposure Assessment?*

As noted, it is unlikely that any expert can present evidence of having an academic degree in exposure science. An expert's qualifications thus have to be tested by examining the expert's experience,[10] including his or her knowledge of and reliance on authoritative reference works.[11] Experts generally will have strong academic credentials in environmental science and engineering, chemistry, chemical engineering, statistics and mathematical model building, industrial hygiene, or other hard sciences related to the behavior of chemicals in the environment.

To the extent exposure assessments deal with the amounts and behaviors of chemicals in the body, individuals can qualify as experts if they can offer academic credentials or substantial experience in toxicology and in the measurement of chemicals in blood or in biological tissues. Certainly, toxicology, epidemiology, or medical credentials are needed if experts are to offer testimony on the health consequences associated with particular exposures.

Not all exposure assessments are complex; indeed, some, as will be seen, are relatively simple. Most toxicologists and epidemiologists have considerable training and experience assessing dose from medicines and other consumer products—and even from food. But if exposures result from chemicals moving from sources through one or more environmental media, it is unlikely that toxicologists or epidemiologists will be able to offer appropriate qualifications, because modeling or other forms of indirect measurement are needed to assess exposures. Further details on the qualifications of experts are offered in the closing sections of the reference guide.

C. *Organization of the Reference Guide*

The reference guide begins with a discussion of the various contexts in which exposure science is applied (Section III). Following that discussion is a section on chemicals and their various sources. Three broad categories of chemicals are discussed: (1) those that are produced for specific uses; (2) those that are byproducts of chemical production, use, and disposal and that enter the environment as contaminants; and (3) those that are created and released by the combustion of all types of organic substances (including tobacco) and of fuels used for energy

10. *See, e.g.*, Best v. Lowe's Home Ctrs, 2009 WL 3488367, 2009 U.S. Dist. LEXIS 97700 (E.D. Tenn. 2009) (a medical doctor with extensive industrial toxicology and product safety experience opined that the plaintiff could not have been exposed to the chemical at issue as alleged).

11. Most of the EPA's guidance documents on exposure assessment have been issued after extensive peer review and thus are considered authoritative.

production. Each of these categories can be thought of as a source for chemical exposure. Next, there is a discussion of the pathways chemicals follow from their sources to the environmental media to which humans are or could be in contact. Such contact is said to create an exposure. Chemicals can then move from these media of human contact and enter the body by different routes of exposure—by ingestion (in food or water, for example), by inhalation, or by direct skin contact (the dermal route). The section on exposure routes includes a discussion of how chemicals contact and enter the body and of how they behave within it. This last topic comprises the interface between exposure science and the sciences of epidemiology and toxicology. Traditionally, exposure scientists have described their work as ending with the description of dose to the body (exposure dose). As will be seen, some practitioners are focusing on the amounts of chemicals present in blood or various tissues of the body as a result of exposure. Unlike the toxicologist, the exposure scientist is not qualified to evaluate the health consequences of these so-called biomarkers of exposure.

This reference guide first presents all of the above material in nonquantitative terms—to describe and illustrate the various processes through which human exposures to chemicals are created (Sections III–V). The guide then focuses on the quantitative aspects (Sections VI and VII). Without some quantitative understanding of the magnitude of exposure, and of the duration of time over which exposure occurs, it becomes difficult to reach meaningful conclusions about health risks. Thus, the remaining sections are devoted to a critical quantitative concept in exposure science—that of dose—and are intended to integrate all of the earlier descriptive material. The reference guide ends with a review of the qualifications of exposure science experts and how they can be assessed.

III. Contexts for the Application of Exposure Science

There are perhaps four major contexts in which exposure science is applied: (1) consumer products, (2) contaminants in the environment and in consumer products, (3) chemicals in the workplace, and (4) disease causation.

A. Consumer Products

Many intentional uses of chemical substances lead to human exposures, and the health risks that are associated with those exposures need to be understood.[12] In some cases, laws and regulations require that health risks be understood in

12. *See, e.g.*, *In re* Stand 'n Seal, 623 F. Supp. 2d 1355 (N.D. Ga. 2009) (consumer use of spray-on product allegedly resulted in inhalation exposure to toxic substances, causing respiratory injuries).

advance of the marketing of such chemicals or products containing them. Thus, intentionally introduced food additives, pesticides, and certain industrial chemicals must have regulatory approvals before they are marketed, and manufacturers of such substances are required to demonstrate the absence of significant health risks (i.e., their safety) based on toxicology studies and careful assessments of expected exposures. Pharmaceuticals and other medical products must undergo similar premarket evaluations. The safety and efficacy of such products must be demonstrated through clinical studies (which are undertaken after animal toxicology studies have been done and have demonstrated the safety of such products for individuals who are involved in clinical trials). Human exposure assessments are central to the regulatory approval of these products.[13]

Many other consumer products require risk assessments, but premarket approvals are not generally required under our current laws. The list of such products is very long, and not all substances included in these products have been subjected to exposure and risk assessments, but regulatory initiatives in the United States and abroad are creating new requirements for more complete assessments of consumer safety.

B. Environmental and Product Contaminants

Byproducts of many industrial processes, including those created by combustion, have led to much environmental contamination (*see* Section IV for a discussion of the sources of such contamination).[14] Technically speaking, contamination refers to the presence of chemical substances in environmental media (including consumer products) in which such substances would not ordinarily be found. The term also may be used to refer to their presence in greater amounts than is usual.[15] The assessment of health risks from such contaminants depends upon an understanding of the magnitude and duration of exposure to them. Exposures may occur through the presence of contaminants in air, drinking water, foods, consumer products, or soils and dusts; in many cases, exposures may occur simultaneously through more than one of these media.

The results from exposure and risk assessments (which incorporate information regarding the toxic properties of the contaminants) are typically used by regulators and public health officials to determine whether exposed populations are at significant risk of harm. If regulators decide that the risks are excessive, they

13. B.D. Beck et al., *The Use of Toxicology in the Regulatory Process, in* Principles and Methods of Toxicology (A. Wallace Hayes ed., 5th ed. 2008).

14. *See, e.g.*, Orchard View Farms, Inc. v. Martin Marietta Aluminum, Inc., 500 F. Supp. 984, 1008 (D. Or. 1980) (failure to monitor fluoride emissions that harmed nearby orchards supported award of punitive damages).

15. For example, lead is naturally present in soils. It could be said that a sample of soil is contaminated with lead only if it were clear that the amounts present exceeded natural levels. The issue is complicated by the fact that natural levels are highly variable.

will take steps to reduce them, typically by using interventions that will reduce exposures (because the inherent toxic properties of the chemicals involved cannot be altered). Exposure scientists are called upon to assess the magnitude of exposure reduction (and therefore risk reduction) achieved through a given intervention.[16]

C. Chemicals in Workplace Environments

Workers in almost all industrial sectors are exposed to chemicals.[17] Exposures are created in industries involved in the extraction of the many raw materials used to manufacture chemical products (the mining, agricultural,[18] and petroleum industries). Raw materials are refined and otherwise processed in thousands of different ways and are eventually turned into manufactured chemical products that number in the tens of thousands. These products enter many channels of distribution and are incorporated into many other products (so-called downstream uses). Occupational exposures can occur at all of these various steps of manufacturing and use. Exposure also can occur from disposal of wastes. Exposure assessments in all of these various occupational settings are important to understand whether health risks are excessive and therefore require reduction.[19]

D. Claims of Disease Causation

In the above three situations, the exposures of interest are those that are currently occurring or that are likely to occur in the future. In those situations the exposure assessments are used to ascertain whether risks of harm are excessive (and thus require reduction) or to document safety (when risks are negligible). There are, however, many circumstances in which individuals claim they actually have been harmed by chemicals. Specifically, they allege that some existing medical condition has been caused by exposures occurring in the past, whether in the workplace, the environment, or through the use of various consumer products.[20]

16. National Research Council, Air Quality Management in the United States (2004).

17. *See, e.g.*, Kennecott Greens Creek Min. Co. v. Mine Safety & Health Admin., 476 F.3d 946 (D.C. Cir. 2007) (suit over regulations addressing miners' exposure to diesel particulate matter).

18. The term "agriculture" is applied here very broadly and includes the production of a wide variety of raw materials that have industrial and consumer product uses (including flavors, fragrances, fibers of many types, and some medicinal products). *See, e.g.*, Association of Irritated Residents v. Fred Schakel Dairy, 634 F. Supp. 2d 1081, 1083 (E.D. Cal. 2008) (methanol emissions from dairy allegedly resulted in exposure sufficient to create human health risks).

19. Office of Pesticide Programs, U.S. Environmental Protection Agency, General Principles for Performing Aggregate Exposure and Risk Assessments, *available at* http://www.epa.gov/pesticides/trac/science/aggregate.pdf (last visited July 14, 2010).

20. *See* Michael D. Green et al., *supra* note 1, in this manual, for a discussion on disease causation. Regulations and public health actions are usually driven by findings of excessive risk of harm (although sometimes evidence of actual harm).

Exposure science comes into play in these cases because the likelihood that any given disease or injury was induced because of exposure to one or more chemicals depends in large part on the size of that exposure.[21] Thus, with the advent of large numbers of so-called toxic tort claims has come the need to assess past exposures. Exposure scientists have responded to this need by adapting the methods of exposure assessment to reconstruct the past—that is, to produce a profile of individuals' past exposures.[22]

A plaintiff with a medical condition known from epidemiological studies to be caused by a specific chemical may not be able to substantiate his or her claim without evidence of exposure to that chemical of a sufficient magnitude.[23] Exposure experts are needed to quantify the exposures incurred; causation experts are then called upon to offer testimony on whether those exposures are of a magnitude sufficient to cause the plaintiff's condition. Chemicals known to cause diseases under certain exposure conditions will not do so under all exposure conditions.

Exposure reconstruction has a history of use by epidemiologists who are studying disease rates in populations that may be associated with past exposures.[24] Epidemiologists have paved the way for the use of exposure assessment methods to reconstruct the past. Although the methods for evaluating current and past exposures are essentially identical, the data needed to quantify past exposures are often more limited and yield less certain results than the data needed to evaluate current exposures. Assessment of past exposures is especially difficult when considering diseases with very long latency periods.[25] By the time disease occurs, documentary proof of exposure and magnitude may have disappeared. But courts regularly deal with evidence reconstructing the past, and assessment of toxic exposure is another application of this common practice.[26]

21. *See supra* notes 1 & 2. Causation may sometimes be established even if quantification of the exposure is not possible. *See, e.g.,* Best v. Lowe's Home Ctrs, Inc., 563 F.3d 171 (6th Cir. 2009) (doctor permitted to testify as to causation based on differential diagnosis).

22. Confounding factors must be carefully addressed. *See, e.g.,* Allgood v. General Motors Corp., 2006 WL 2669337, at *11 (S.D. Ind. 2006) (selection bias rendered expert testimony inadmissible); American Farm Bureau Fed'n v. EPA, 559 F.3d 512 (2009) (in setting particulate matter standards addressing visibility, the data relied on should avoid the confounding effects of humidity); Avila v. Willits Envtl. Remediation Trust, 2009 WL 1813125, 2009 U.S. Dist. LEXIS 67981 (N.D. Cal. 2009) (failure to rule out confounding factors of other sources of exposure or other causes of disease rendered expert's opinion inadmissible); Adams v. Cooper Indus. Inc., 2007 WL 2219212, 2007 U.S. Dist. LEXIS 55131 (E.D. Ky. 2007) (differential diagnosis includes ruling out confounding causes of plaintiffs' disease).

23. *See* Michael D. Green et al., Reference Guide on Epidemiology, in this manual.

24. *Id.*

25. W.T. Sanderson et al., *Estimating Historical Exposures of Workers in a Beryllium Manufacturing Plant,* 39 Am. J. Indus. Med. 145–57 (2001).

26. Courts have accepted indirect evidence of exposure. For example, differential diagnosis may support an expert's opinion that the exposure caused the harm. Best v. Lowe's Home Ctrs., Inc., 563 F.3d 171 (6th Cir. 2009). On occasion, qualitative evidence of exposure is admitted as evidence

IV. Chemicals

Before embarking on a description of the elements of exposure science, it is useful to provide a brief primer on some of the characteristics of chemicals that influence their behavior and that therefore affect the ways in which humans can be exposed to them. The primer also introduces some technical terms that frequently arise in exposure science.

A. Organic and Inorganic Chemicals

For both historical and scientific reasons, chemists divide the universe of chemicals into organic and inorganic compounds. The original basis for classifying chemicals as organic was the hypothesis, known since the mid-nineteenth century to be false, that organic chemicals could be produced only by living organisms. Modern scientists classify chemicals as organic if they contain the element carbon.[27] Carbon has the remarkable and nearly unique property that its atoms can combine with each other in many different ways, and, together with a few other elements—including hydrogen, oxygen, nitrogen, sulfur, chlorine, bromine—can create a huge number of different molecular arrangements. Each such arrangement is a unique chemical. Several million distinct organic chemicals are already known to chemists, and there are many more that will no doubt be found to occur naturally or that will be created by laboratory synthesis. All of life—at least on Earth—depends on carbon compounds and probably could not have evolved if carbon did not have its unique and extraordinary bonding properties.

All other chemicals are called inorganic. There are 90 elements in addition to carbon in nature (and several more that have been created in laboratories), and because these elements do not have the special properties of carbon, the number of different possible combinations of them is smaller than can occur with carbon.

Living organisms contain or produce organic chemicals by the millions. One of the most abundant organic chemicals on Earth is cellulose—a giant molecule containing thousands of atoms of carbon, hydrogen, and oxygen. Cellulose is produced by all plants and is their essential structural component. Chemically, cel-

that the magnitude was great enough to cause harm. *See, e.g.*, Westberry v. Gislaved Gummi AB, 178 F.3d 257 (4th Cir. 1999) (no quantitative measurement required where evidence showed plaintiff was covered in talc and left footprints); Allen v. Martin Surfacing, 263 F.R.D. 47 (D. Mass. 2009) (symptom accounts at the time of exposure formed the basis for expert's opinion that exposure was high enough to cause harm). And courts have accepted the government's reconstruction of exposure to radiation. Hayward v. U.S. Dep't of Labor, 536 F.3d 376 (5th Cir. 2008); Hannis v. Shinseki, 2009 WL 3157546 (Vet. App. 2009) (no direct measure of veteran's exposure to radiation was possible but VA's dose estimate was not clearly erroneous).

27. There are a few compounds of carbon that chemists still consider inorganic: These are typically simple molecules such as carbon monoxide (CO) and carbon dioxide (CO_2) and the mineral limestone, which is calcium carbonate ($CaCO_3$).

lulose is a carbohydrate (one that is not digested by humans), a group that together with proteins, fats, and nucleic acids are the primary components of life. But living organisms also produce huge numbers of other types of organic molecules. The colors of plants and animals and their odors and tastes are a result of the presence of organic chemicals. The numbers and structural varieties of naturally occurring chemicals are enormous.

Other important natural sources of organic chemicals are the so-called fossil fuels—natural gas, petroleum, and coal—all deposited in the Earth from the decay of plant and animal remains and containing thousands of degradation products. Most of these are simple compounds containing only carbon and hydrogen (technically known as hydrocarbons). The organic chemical industry depends upon these and just a few other natural products for everything it manufactures; the fraction of fossil fuels not used directly for energy generation is used as feedstock for the chemical industry. There are also inorganic chemicals—the minerals—present in living organisms, many essential to life. But the principal natural source of inorganic chemicals is the nonliving part of the Earth that humans have learned how to mine.

B. Industrial Chemistry

The modern chemical industry had its origins in the late nineteenth century when chemists, mostly European, discovered that it was possible to create in the laboratory chemicals that had previously been found only in nature. Most remarkably, scientists also discovered they could synthesize compounds not found in nature—substances never previously present on Earth. In other words, they found ways to alter through chemical reactions the bonds present in one compound so that a new compound was formed. The first compound synthesized in this way was a dye called aniline purple by the British chemist, William Henry Perkin, who discovered it.[28] The work of chemical synthesis grew out of the development of so-called structural theory in the nineteenth century and remains central to the science today. This theory explains that the number and type of chemical elements present, and the ways in which those elements are bonded to each other, are unique for each chemical compound and therefore distinguish one chemical from another.

In the late nineteenth century and up to World War II, coal was the major starting material for the organic chemical industry. When coal is heated in the absence of oxygen, coke and volatile byproducts called coal tars are created. All sorts of organic chemicals can be isolated from coal tar—benzene, toluene, xylenes, ethylbenzene, naphthalene, creosotes, and many others. The organic

28. This compound and others related to it became the bases for the first chemical industry, that devoted to dye production. Perkins' dye was later called "mauve" and its wide use led to what came to be called the Mauve Decade (1890s).

chemical industry also uses other natural products, such as animal fats, vegetable oils, and wood byproducts.

The move to petroleum as a raw materials source for the organic chemical industry began during the 1940s. Petrochemicals, as they are called, are now used to create thousands of useful industrial chemicals. The rate of commercial introduction of new chemicals shot up rapidly after World War II.

Among the thousands of products produced by the organic chemical industry and by related industries are medicines (most of which are organic chemicals of considerable complexity), dyes, agricultural chemicals, including substances used to eliminate pests (insecticides, fungicides, herbicides, rodenticides, and other "cides"), soaps and detergents, synthetic fibers and rubbers, paper chemicals, plastics and resins of great variety, adhesives, food additives, additives for drinking water, refrigerants, explosives, cleaning and polishing materials, cosmetics, and textile chemicals. Because of past disposal practices, chemicals primarily used as solvents (for many purposes) are among the most widespread environmental contaminants.

The history of human efforts to tap the inorganic earth for useful materials is complex and involves a blend of chemical, mining, and materials technologies. Included here is everything from the various silicaceous materials derived from stone (glasses, ceramics, clays, asbestos) to the vast number of metals derived from ores that have been mined and processed (iron, copper, nickel, cadmium, molybdenum, mercury, lead, silver, gold, platinum, tin, aluminum, uranium, cobalt, chromium, germanium, iridium, cerium, palladium, manganese, zinc, and many more). Other nonmetallic materials, such as chlorine and bromine, salt (sodium chloride), limestone (calcium carbonate), sulfuric acid, and phosphates, and various compounds of the metals, have hundreds of different uses, as strictly industrial chemicals and as consumer products. These inorganic substances reach, enter, and move about our environment, and we come into contact with them, sometimes intentionally, sometimes inadvertently. The number of organic and inorganic chemicals in commercial production exceeds 70,000, and the number of uses and products created from them far exceeds this number.

There are important health questions related to what is generally referred to as particulate matter (PM). Small particulates in the air usually arise from combustion of almost any organic material. The chemical composition of such particulates can vary depending upon source, but it is possible that their health effects depend more upon their physical size than their chemical composition. This issue is currently unresolved, but it is important to include PMs of all types as a class of chemical contaminants.

Finally, it is important to note that, in addition to PM, many chemicals are produced when fuels or other organic materials are burned. Organic chemicals take on oxygen atoms during combustion and yield large numbers of substances not present in the materials that are burned. Combustion also produces simple inorganic oxides of carbon, nitrogen, and sulfur, which are major air pollutants.

Burning tobacco introduces 4000 to 5000 chemicals into the lungs. Combustion products are another important source of environmental contamination.[29]

V. Human Exposures to Chemicals

As noted earlier, this section is entirely descriptive, rather than quantitative. It describes all the various physical processes that lead to human exposures to chemicals and introduces the terms that exposure scientists apply to those processes. Section VI illustrates how these various processes can be quantified and the types of data that are required to do so.

A. Exposure Sources—An Overview

Figure 1 provides a broad overview of most of the major sources of exposure. As shown, sources can be intended or unintended. Thus, many chemicals are intentionally used in ways that will lead to human exposures. Substances added to food and indeed food itself,[30] cosmetics, personal care products, fibers and the colorants added to them, and medical products of many types are included in this broad category. Direct ingestion of, or other types of direct contact with (on the skin or through inhalation), such products obviously creates exposures. Nicotine and tobacco combustion products might also be classified as intended exposures. Generally, these exposures are more readily quantifiable than those associated with unintended exposures.

Although the term is somewhat ambiguous, unintended exposures may be said to fall into two broad categories. There are deliberate uses of certain chemicals that, although not intended to lead to human exposures, will inevitably do so. Pesticides applied to food crops, some components of food packaging materials that may migrate into food, and many types of household products are not intended for direct human ingestion or contact, but exposures will nonetheless occur indirectly. Occupational exposures, although unintended, are similarly unavoidable. Also, many exposures to a very broad range of environmental contaminants are unintended (see Figure 1).

In all of these cases, such exposures are not described as intentional, in the sense that the term is applied to a pharmaceutical ingredient or a cosmetic, but most are not completely avoidable. Unintended exposures are generally more

29. National Research Council, Human Exposure Assessment for Airborne Pollutants: Advances and Opportunities (1991); J. Samet & S. Wang, *Environmental Tobacco Smoke, in* Environmental Toxicants (M. Lippmann ed., 2d ed. 2000).

30. The natural constituents of food include not only substances that have nutritional value, but also hundreds of thousands of other natural chemicals.

Figure 1. Opportunities for exposure: Sources of chemical releases.

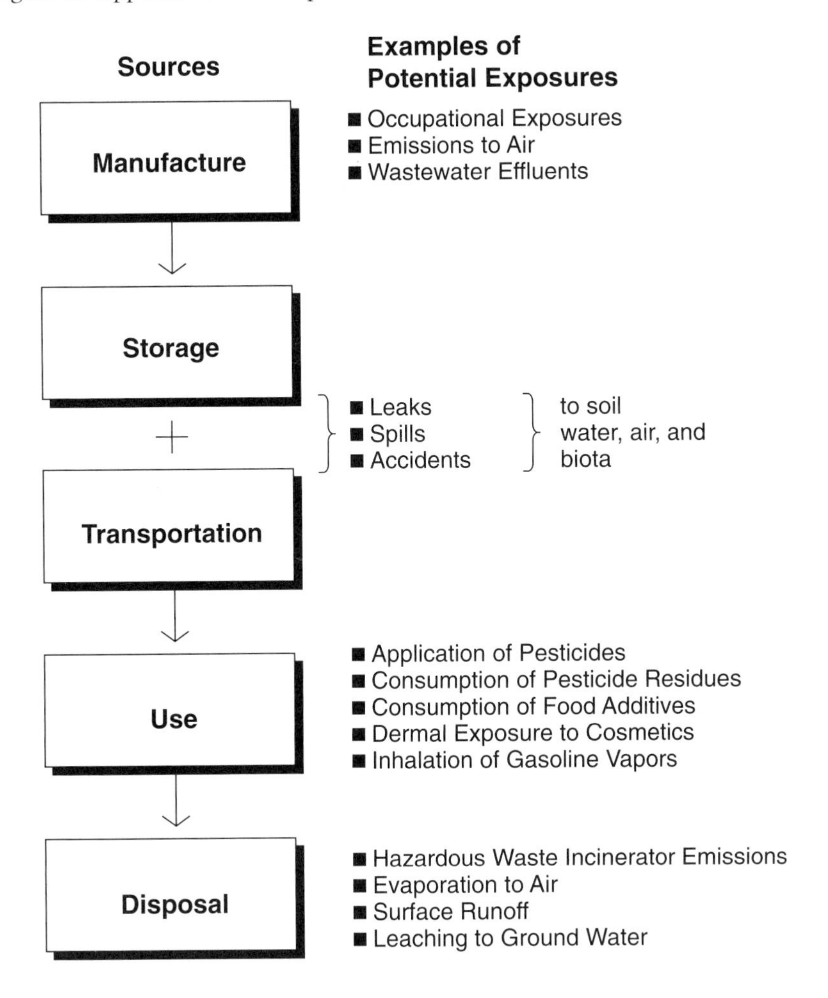

difficult to identify and quantify than are intended exposures.[31] In the case of the intended exposures, the pathway from source to humans is direct; in the case of unintended exposures, the pathway is indirect, sometimes highly so. Thus, the most important distinction for purposes of exposure assessment concerns the directness of the pathway from source to people.

31. There are significant differences in the laws regarding the regulation of substances that have been grouped as creating intended or unintended exposures.

B. The Goal of Exposure Assessment

Exposure assessment is generally intended to answer the following questions:

- Who has been, or could become exposed to a specific chemical(s) arising from one or more specific sources? Is it the entire general population, or is it a specific subpopulation (e.g., those residing near a certain manufacturing or hazardous waste facility, or infants and children), or is it workers?[32]
- What specific chemicals comprise the exposures?
- What are the pathways from the source of the chemical to the exposed population? Pathways include direct product use, or those (so-called indirect pathways) in which the chemical moves through one or more environmental media to reach the media to which people are exposed (air, water, foods, soils, and dusts). Understanding pathways is necessary to understanding exposure routes (below) and quantifying exposures.
- By what routes are people exposed? Routes include ingestion, inhalation, and dermal contact.[33] Identifying exposure routes is important because those routes affect the magnitude of ultimate exposures and because they often affect health outcomes.
- What is the magnitude and duration of exposure incurred by the population of interest? Dose is the technical term used for magnitude, and it is the amount of chemical entering the body or contacting the surface of the body, usually over some specified period of time (often over 1 day[34]). Duration refers to the number of days over which exposure occurs. Note that exposures can be intermittent or continuous and can be highly variable, especially for some air contaminants.

The ultimate goal of exposure assessment is to identify dose and duration. The concept of dose is further developed in Section VI. After a chemical enters or contacts the body, it can be absorbed (into the bloodstream), distributed to many organs of the body, metabolized (chemically altered by certain enzymes in cells of the liver and other organs), and then excreted. Understanding these processes is important to determining whether and how a chemical may cause adverse health effects. These processes mark the interface between exposure science and toxicology, epidemiology, and medicine. Understanding the dose is the necessary first step in understanding these processes; for purposes of this reference guide, the boundary of exposure science is set at understanding dose. However, some

32. *See, e.g.*, Hackensack Riverkeeper, Inc. v. Del. Ostego, 450 F. Supp. 2d 467 (D.N.J. 2006) (river and bay users alleged that hazardous waste runoff and emissions polluted the water).

33. Additional routes of exposure are relevant for some pharmaceuticals, diagnostics, and medical devices.

34. Shorter periods of time are used when the concern is very short-term exposures to chemicals that have extremely high toxicity—so-called acutely poisonous materials.

discussion of how it is possible to gain more direct measures of exposure (target site doses) by examining human blood and urine is included.

The completion of an exposure assessment provides the information needed (the dose and duration of exposure) by epidemiologists and toxicologists, who will have information on the adverse health effects of the chemicals involved and on the relationships between those effects and the dose and duration of exposure.[35] Recall that exposure assessments can be directed at exposures that occurred in the past, those that are currently occurring, or those that will occur in the future should certain actions be taken (e.g., the entry of a new product into the consumer market or the installation of new air pollution controls).

The discussion of each of these elements of exposure assessment is expanded in the following section, beginning with pathways.

C. Pathways

Assuming that the chemical of interest and its sources have been identified, exposure assessment focuses on the pathway the chemical follows to reach the population of interest.[36]

To ensure thoroughness in the assessment, all conceivable pathways should be explicitly identified, with the understanding that ultimately some pathways will be found to contribute negligibly to the overall exposure. Identifying pathways is also important to understanding exposure routes.

As noted earlier, the simplest pathways are those described as direct. Thus, a substance, such as a noncaloric sweetener or an emulsifier, once added to food, follows a simple and direct pathway to the people who ingest the food. The same can be said for pharmaceuticals, cosmetics, and other personal care products. Cal-

35. *See* reference guides on epidemiology and toxicology in this manual. *See also, e.g.*, White v. Dow Chem. Co., 321 Fed. App'x. 266, 2009 WL 931703 (4th Cir. 2009) (plaintiff must show more than possible exposure; must show concentration and duration); Anderson v. Dow Chem. Co., 255 Fed. Appx. 1, 2007 WL 1879170 (5th Cir. 2007) (lawsuit dismissed because uncontested data showed that magnitude and duration of exposure was insufficient to cause adverse health effects); Finestone v. Florida Power & Light Co., 272 Fed. App'x. 761, 2008 WL 931703 (4th Cir. 2009) (experts' testimony was properly excluded where their conclusions relied on unsupported assumptions).

36. SPPI-Somersville, Inc. v. TRC Cos., 2009 WL 2612227, at *16 (N.D. Cal. 2009) (groundwater contamination claim was dismissed because there was no current pathway to exposure); United States v. W.R. Grace Co., 504 F.3d 745 (9th Cir. 2007) (affirming exclusion of report, but not expert testimony based on the report, identifying which pathways of asbestos exposure were most associated with lung abnormalities); Grace Christian Fellowship v. KJG Investments Inc., 2009 WL 2460990, at *12 (E.D. Wis. 2009) (preliminary injunction was denied because the plaintiff did not establish that a complete pathway currently existed for toxins to enter the building); National Exposure Research Laboratory, U.S. Environmental Protection Agency, Scientific and Ethical Approaches for Observational Exposure Studies, Doc. No. EPA 600/R-08/062 (2008), *available at* http://www.epa.gov/nerl/sots/index.html (last visited July 14, 2010); U.S. Environmental Protection Agency. Exposure Factors Handbook (1997).

culating doses for such substances, as shown in Section VI, is generally a straightforward process. Even in such cases, however, complexities can arise. Thus, in the case of certain personal care products that are applied to the skin, there is a possibility of inhalation exposures to any substance in those products that can readily volatilize at room temperatures. One physical characteristic of chemicals that exposure scientists need to understand is their capacity to move from a liquid to a gaseous state (to volatilize). Not all chemicals are readily volatile (and almost all inorganic, metal-based substances are close to nonvolatile), but inhalation routes can be significant for those that are volatile, regardless of their sources.[37]

Indirect pathways of exposure can range from the relatively simple to the highly complex. Many packaging materials are polymeric chemicals—very large molecules synthesized by causing very small molecules to chemically bind to each other (or to other small molecules) to make very long chemical chains. These polymers (polyethylene, polyvinyl chloride, polycarbonates, and others) tend to be physically very stable and chemically quite inert (meaning they have very low toxicity potential). But it is generally not possible to synthesize polymers without very small amounts of the starting chemicals (those small molecules, usually called monomers) remaining in the polymers. The small molecules can often migrate from the polymer into materials with which the polymer comes into contact. If those materials are foods or consumer products, people consuming those foods or otherwise using those products will be exposed.

Some amount of the pesticides applied to food crops may remain behind in treated foods and be consumed by people.[38] This last pathway can become more complicated when treated crops are used as feed for animals that humans consume (meat and poultry and farm-raised fish) or from which humans obtain food (milk and eggs). Exposure scientists who study these subjects thus need to understand what paths pesticides follow when they are ingested by farm animals used as food. The same complex indirect pathways arise for some veterinary drugs used in animals from which humans obtain food.[39]

In the realm of environmental contamination, pathways can multiply and the problem of exposure assessment can become even more complex. Sources of environmental contamination include air emissions from manufacturing facilities and from numerous sources associated with the combustion of fuels and other

37. Inhalation exposures to nonvolatile chemicals can occur if they are caused to move into the air as dusts. *See* National Research Council, Human Exposure Assessment for Airborne Pollutants: Advances and Opportunities (1991).

38. Other pathways for pesticide exposure include spraying homes or fields. Kerner v. Terminix Int'l Co., 2008 WL 341363 (S.D. Ohio 2008) (pesticides allegedly misapplied inside home); Brittingham v. Collins, 2008 WL 678013 (D. Md. Feb. 26, 2008) (crop-dusting plane sprayed plaintiff's decedent); Haas v. Peake, 525 F.3d 1168 (Fed. Cir. 2008) (veteran claimed exposure to Agent Orange).

39. P. Frank & J.H. Schafer, *Animal Health Products, in* Regulatory Toxicology (S.C. Gad, ed., 2d ed. 2001).

organic materials.[40] Similar emissions to water supplies, including ground water used for drinking or for raising plants and animals, can result in human exposures through drinking water and food.[41] Contaminants of drinking water that are volatile can enter the air when water is used for bathing, showering, and cooking. A recent problem of much concern is the contamination of air in homes and other buildings because of the presence of volatile chemical contaminants in the water beneath those structures.[42]

Wastes from industrial processes and many kinds of consumer wastes can similarly result in releases to air and water.[43] In some cases, emissions to air can lead to the deposition of contaminants in soils and household dusts; this type of contamination is usually associated with nonvolatile substances. Some such substances may remain in soils for very long periods; others may migrate from their sites of deposition and contaminate ground water; whereas others may degrade relatively quickly.

All of these issues regarding the movement of chemicals from their sources and through the environment to reach human populations come under the heading of chemical fate and transport.[44] Transport concerns the processes that cause chemicals to follow certain pathways from their sources through the environment, and fate concerns their ultimate disposition—that is, the medium in which they finally reside and the length of time that they might reside there. Fate-and-transport scientists have models available to estimate the amount of chemical that will be present in that final environmental medium.[45] Some discussion of the nature of these models is offered in Section VI.

One final feature of pathways analysis that should be noted concerns the fact that some chemicals degrade rapidly when they enter the environment, others slowly, and some not at all or only exceedingly slowly. The study of environmental persistence of different chemicals is a significant feature of exposure science; its goal is to understand the chemical nature of the degradation products and the duration of time the chemical and its degradation products persist in any

40. *See, e.g.*, Natural Resources Defense Council, Inc. v. EPA, 489 F.3d 1250 (D.C. Cir. 2007) (vacating EPA rule for solid waste incinerators); Kurth v. ArcelorMittal USA, Inc., 2009 WL 3346588 (N.D. Ind. 2009) (defendant manufacturers allegedly emitted toxic chemicals, endangering schoolchildren); American Industrial Hygiene Association, Guideline on Occupational Exposure Reconstruction (S.M. Viet et al. eds., 2008).

41. United States v. Sensient Colors, Inc., 580 F. Supp. 2d 369, 373 (D.N.J. 2008) (leaching lead threatened to contaminate ground water used for drinking).

42. Interstate Technology & Regulatory Council (ITRC), Vapor Intrusion Pathway: A Practical Guideline. (Jan. 2007), *available at* http://www.itrcweb.org/Documents/VI-1.pdf.

43. American Farm Bureau Fed'n. v. EPA, 559 F.3d 512 (D.C. Cir. 2009) (EPA outdoor air pollution standards).

44. The common phrase used by exposure scientists is "fate and transport." In fact, transport takes place and has to be understood before fate is known.

45. In the context of exposure science, the term "final" refers to the medium through which people become exposed. A chemical may in fact continue to move to other media after that human exposure has occurred.

given environmental medium. Most inorganic chemicals are highly persistent; metals that become contaminants may change their chemical forms in small ways (lead sulfide may convert to lead oxide), but the metal persists forever (although it may migrate from one medium to another). Most organic chemicals degrade in the environment as a result of their exposure to light, to microorganisms present in soils and sediments, and to other environmental substances. But a few organic substances (e.g., polychlorinated biphenyls (PCBs) and the chlorinated dioxins, certain chlorinated pesticides such as DDT that were once widely used) are quite resistant to degradation and may persist for unexpectedly long periods (although even these ultimately degrade).[46]

Exposure scientists also need to be aware of the possibility that the degradation products of certain chemicals may be as or more toxic than the chemicals themselves. The once widely used solvents trichloroethylene and perchloroethylene (tetrachloroethylene) are commonly found in ground water. Under certain conditions, these compounds degrade by processes that lead to the replacement of some chlorine atoms by hydrogen atoms; one product of their degradation is the more dangerous chemical called vinyl chloride (monochloroethylene). The presence of such a degradation product in drinking water should not be ignored.

A description of pathways is the critical first step in exposure assessment and, especially for environmental contaminants, must be done with thoroughness. Are all conceivable pathways accounted for? Have some pathways been eliminated from consideration, and if so, why? Are any environmental degradation products of concern? Only with adequate description can adequate quantification (Section VI) be accomplished.

A graphical description of pathways is offered in Figure 2.

D. Exposure Routes

Pathways analysis leads to the identification of the environmental media in which the chemical of interest comes to be present and with which human contact can occur—the media of human exposure.

The inhalation of air containing the chemical of interest is one route of exposure.[47] The physical form of the chemical in air, which should be known from the pathways analysis, will influence what happens to the chemical during inhalation. Chemicals that are in the vapor phase will remain in that physical

46. K.W. Fried & K.K. Rozman, *Persistent Polyhalogenated Aromatic Hydrocarbons. in* Toxicology and Risk Assessment: A Comprehensive Introduction H. Greim & R. Snyder eds., 2009).

47. *See, e.g.*, Byers v. Lincoln Elec. Co., 607 F. Supp. 2d 840 (N.D. Ohio 2009) (welder inhaled toxic manganese fumes); O'Connor v. Boeing North American, Inc., 2005 WL 6035256 (C.D. Cal. 2005) (alleged failure to monitor ambient air emissions of radioactive particles); *In re* FEMA Trailer Formaldehyde Prod. Liab. Litig., 2009 WL 2382773 (E.D. La. 2009) (trailer residents exposed to formaldehyde).

Figure 2. Description of the many possible environmental pathways that chemicals may follow after releases from different sources.

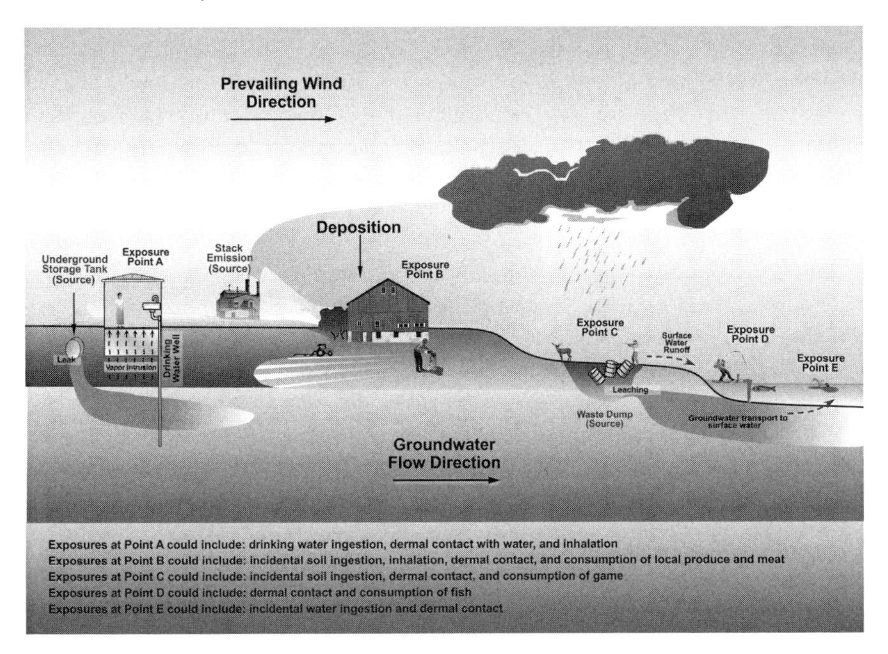

Source: Graphic created by Jason Miller.

state and will move to the lungs, where a certain fraction will pass through the lungs and enter the bloodstream. The extent to which different chemical substances pass through the lungs is dependent in large part upon their physical properties, particularly solubilities in both fatlike materials and water. Passage through cell membranes (of the cells lining the lungs) requires that substances have a degree of both fat solubility and water solubility. Predicting the extent of absorption through the lungs (or the gastrointestinal tract or skin, discussed below) cannot be accomplished with accuracy; knowledge in this area can be gathered only through measurement.

Certain fibrous materials (including but not limited to asbestos) and particulate matter and dusts may move through the airways and may reach the lungs, but some of these kinds of materials may be trapped in the nose and excreted. Generally, only very fine particles reach the lower lung area. Some particles may be deposited in the upper regions of the respiratory tract and then carried by certain physical processes to the pharynx and then be coughed up or swallowed. Thus, inhaled chemicals and particulates can enter the body through the gastrointestinal

(GI) tract or the respiratory tract.[48] Understanding risk requires information about these characteristics of the chemicals involved.

Ingestion is the second major route of exposure to substances in environmental media.[49] Chemicals that comprise or come to be present in foods, in drinking water, in soils and dusts,[50] and many of those that serve as medicines are all ingested. They are swallowed, enter the GI tract, and to greater or lesser degrees are absorbed into the bloodstream at various locations along that tract. This is often referred to as the oral route of exposure.

The largest organ of the body, the skin, is the third route of exposure for chemicals in products and the environment.[51] As with the GI tract and the lungs, chemicals are absorbed through the skin to greater or lesser degrees, depending on their physical and chemical characteristics. In some cases, toxic harm can occur directly within the respiratory or GI tracts or on the skin before absorption occurs.[52]

The pathways analysis allows the identification of all the routes by which chemicals from a given source may enter the body, because it identifies the media of human contact into which the chemicals migrate from their sources. Once the media of human contact are identified, the possible exposure routes are known.

E. Summary of the Descriptive Process

Once the exposure question to be examined has been defined, the exposure scientist sets out to identify all the relevant sources of exposure to the chemicals of interest. All the pathways the chemicals can follow from those sources to reach the population of interest are then described, with careful attention to the possibility that chemical degradation (to more or less toxic substances) can occur. The pathways analysis concludes with a description of what chemicals will be present in the various environmental media with which the exposed populations were, are, or could become exposed (air, water, foods, soils and dusts, consumer products). At this point, it becomes possible to identify the routes by which the chemicals can enter the body.

48. J.V. Rodricks, *From Exposure to Dose, in* Calculated Risks: The Toxicity and Human Health Risks of Chemicals in Our Environment (2d ed. 2007).

49. *See, e.g.*, Foster v. Legal Sea Foods, Inc., 2008 WL 2945561 (D. Md. 2008) (hepatitis A allegedly contracted from eating undercooked mussels); Winnicki v. Bennigan's, 2006 WL 319298 (D.N.J. 2006) (alleged foodborne illness contracted from defendant's restaurant led to renal failure and death); Palmer v. Asarco Inc., 2007 WL 2298422 (N.D. Okla. 2007) (children allegedly ingested dust and soil contaminated with lead).

50. Inadvertent exposures to these and other nonfood items are known to occur and can be especially common in children.

51. *See, e.g.*, United States v. Chamness, 435 F.3d 724 (7th Cir. 2006) (evidence that methamphetamine and the ingredients used in its manufacture are toxic to the eyes, mucous membranes, and skin supported sentencing enhancement for danger to human life).

52. J.V. Rodricks, *From Exposure to Dose, in* Calculated Risks: The Toxicity and Human Health Risks of Chemicals in Our Environment (2d ed. 2007).

Description by itself, however, often is inadequate. Attempts have to be made to quantify exposure, to arrive at estimates of the dose received by the exposed population, and to determine the duration of time over which that dose is received.

VI. Quantification of Exposure

A. Dose

The simplest dose calculations relate to situations in which direct exposures occur.[53] Thus, for example, consider the case of a substance directly added to food (and approved by the U.S. Food and Drug Administration (FDA) for such addition). Suppose the chemical is of well-established identity and is approved for use in nonalcoholic beverages at a concentration of 10 milligrams of additive for each liter of beverage (10 mg/L).[54] To understand the amount (weight) of the additive ingested each day, it is necessary to know how much of the beverage people consume each day. Data are available on rates of food consumption in the general population. Typically, those data reflect average consumption rates and also rates at the high end of consumption. To make sure that the additive is safe for use, FDA seeks to ensure the absence of risk for individuals who may consume at the high end, perhaps at the 95th percentile of consumption rates.[55] Surveys of intake levels for the beverage in our example reveal that the 95th percentile intake is 1.2 L per day for adults.

The weight of additive ingested by individuals at the 95th percentile of beverage consumption rate is thus obtained as follows:

$$10 \text{ mg/L} \times 1.2 \text{ L/day} = 12 \text{ mg/day}.$$

For a number of reasons, toxicologists express dose as weight of chemical per unit of body weight. For adults having a body weight (bw) of, on average, 70 kilograms (kg), the dose of additive is

$$12 \text{ mg/day} \div 70 \text{ kg bw} = 0.17 \text{ mg/kg bw per day.}$$[56]

53. *See, e.g.*, McLaughlin v. Sec'y of Dep't of Health & Human Servs., 2008 WL 4444142 (Fed. Cl. 2008) (plaintiff exposed to known dose of thimerosol in vaccine; study using four times that dose was not reliable evidence that exposure caused his autistic symptoms).

54. See Appendix A for a discussion of units used in exposure science.

55. J.V. Rodricks & V. Frankos, *Food Additives and Nutrition Supplements*, *in* Regulatory Toxicology 51–82 (C.P. Chengeliss et al. eds., 2d ed. 2001).

56. To gain approval for such an additive, FDA would require that no toxic effects are observable in long-term animal studies at doses of at least 17 mg/kg bw per day (100 times the high-end human intake).

Doses from other ingested products containing specified amounts of chemicals are calculated in much the same way. It generally would be assumed that the duration of exposure for a substance added to a food or beverage would be continuous and would cover a large fraction of a lifetime. For other products, particularly pharmaceuticals, exposure durations will vary widely; dose calculations would be the same, regardless of duration, but the potential for harm requires consideration of exposure duration.

It will be useful, before proceeding further, to illustrate dose calculations for exposures occurring by the inhalation and dermal routes.[57] Consider a hypothetical workplace setting in which a solvent is present in the air. Measurement by an industrial hygienist reveals its presence at a weight of 2 mg in each cubic meter (m^3) of air. Data on breathing rates reveal that a typical worker breathes in 10 m^3 of air each 8-hour workday.[58] Thus, the worker dose will be

$$2 \text{ mg/m}^3 \times 10 \text{ m}^3\text{/day} = 20 \text{ mg/day}$$
$$20 \text{ mg/day} \div 70 \text{ kg} = 0.28 \text{ mg/kg bw per day.}$$

As noted earlier, it is likely that only a fraction of this dose will reach and pass through the lungs and enter the bloodstream. As also noted earlier, if the chemical is a fiber or other particle, its dynamics in the respiratory tract will be different than that of a vapor, with a portion of the inhaled dose entering the GI tract.

Dose from skin exposure often is expressed as the weight of chemical per some unit of skin surface area (e.g., per m^2 of skin). The body surface area of an average (70 kg) adult is 1.8 m^2. Thus, consider a body lotion containing a chemical of interest. If the lotion is applied over the entire body, then it is necessary to know the total amount of lotion applied and then the total amount of chemical present in that amount of lotion. That last amount will then be divided by 1.8 to yield the skin dose in units of milligrams per square meter. If the chemical causes toxicity directly to the skin, that toxicity dose information also will be expressed in milligrams per square meter. Then risk is evaluated by examining the quantitative relationship between the toxic dose (milligrams per square meter) and the (presumably much lower) human dose expressed in the same units. If the chemical can penetrate the skin and produce toxicity within the body, then the dose determination must include an examination of the amount absorbed into the human body.[59]

57. *See, e.g.*, Henricksen v. ConocoPhillips Co., 605 F. Supp. 2d 1142, 1164 (E.D. Wash. 2009) (benzene exposure on skin and by inhalation); Bland v. Verizon Wireless (VAW) LLC, 2007 WL 5681791, at *9 (S.D. Iowa 2007) (inhalation exposure to Freon in "canned air" sprayed into water bottle). For a discussion of the importance of assessment of dose as a measure of exposure, see Bernard D. Goldstein & Mary Sue Henifin, Reference Guide on Toxicology, Section I.A.1.c, in this manual.

58. The 24-hour inhalation rate outside the workplace setting is ca. 20 m^3. The lack of direct proportion to time reflects the fact that breathing rates increase under exertion.

59. Rates of absorption of chemicals into the body, through the GI tract, the lungs, or the skin, usually must be obtained by measurement; they are not readily predicted.

One final matter concerning dose estimation concerns the importance of body size, in particular that of the infant and the growing child. In matters such as food and water intake, and breathing rates, small children are known to take in these media at higher rates per unit of their body weights than do adults.[60] Thus, when a small child is exposed to a food contaminant, that child will often receive a greater dose of the contaminant than will an adult consuming food with the same level of contaminant. Children also tend to ingest greater amounts of nonfood items, such as soils and dusts, than do adults. In some cases, nursing mothers excrete chemicals in their milk. The exposure scientist generally conducts separate assessments for children that take into account the possibility of periods of increased exposure during the developmental period.[61]

B. Doses from Indirect Exposure Pathways

Recall that the goal of exposure assessment is to identify the media through which people will be exposed to chemicals of interest that are emitted from sources of interest. As will be seen, the assessment, when completed, will reveal the amount of the chemical of interest in a certain weight or volume of each of the media with which people come into contact. Once this is known, dose calculations can proceed in the manner described in the preceding section.

In the preceding section, firm and readily available knowledge was available about the amount of chemical present in a given weight of food or consumer product (the body lotion example) or in a given volume (cubic meters) of air. These measures are called concentrations of the chemicals in the media of exposure (*see* Appendix A). When a chemical must move from one or more sources, and then through one or more environmental media, before it comes to be present in the media with which people have contact (the media of exposure), determining the concentrations of the chemical in the media of exposure becomes difficult.[62] Such a situation is clearly different from that in which a specific amount of an additive is directly added to a specific amount of food. The challenge faced by exposure scientists when the chemical comes to be present in the medium of human exposure not by direct and intentional addition, but by indirect means, through movement from source through the environment, is to find a reliable

60. *See, e.g.*, Northwest Coalition for Alternatives to Pesticides (NCAP) v. EPA, 544 F.3d 1043 (9th Cir. 2008) (dispute over how much lower allowable pesticide levels should be to account for children's greater susceptibility).

61. For some substances, susceptibility to toxicity is also enhanced during the same periods. See Section VII.B.

62. *See, e.g.*, Hannis v. Shinseki, 2009 WL 3157546 (Vet. App. 2009) (no direct measure of veteran's exposure to radiation was possible but VA's dose estimate was not clearly erroneous); Fisher v. Ciba Specialty Chem. Corp., 2007 WL 2302470 (S.D. Ala. 2007) (allowing expert's qualitative account of DDT and its metabolites spreading from defendant's plant to plaintiffs' property, because quantification would necessarily rely on speculative data).

way to estimate concentrations in the medium of human exposure.[63] Once concentrations are known, dose is readily calculated (as in Section VI.A), but reliably estimating concentrations can be difficult.

Two methods typically are used to estimate those concentrations. One involves direct measurement using the tools of analytical chemistry. The second involves the use of models that are intended to quantify the concentrations resulting from the movement of chemicals from the source to the media of human exposure.

C. Direct Measurement: Analytical Science

Once the media that could be subject to contamination have been identified through pathways analysis (Section V.C), one available choice for determining the concentrations of contaminants involves sampling those media and subjecting the samples taken to chemical analysis. The analysis will not only reveal the concentrations of chemicals in the media of concern, but should also confirm their identities. Environmental sampling and analysis is under way all over the world, at and near contaminated waste sites, in the vicinity of facilities emitting chemicals to air and water, and in many other circumstances.[64]

One purpose of such sampling and analysis is to determine whether products and environmental media contain substances at concentrations that meet existing regulatory requirements. In many circumstances, regulators have established limits on the concentrations of certain chemicals in foods, other products, water, air, and even soils. These limits generally are based on assessments of health risk and calculations of concentrations that are associated with what the regulators believe to be negligibly small risks. The calculations are made after first identifying the total dose of a chemical that is safe (poses a negligible risk) and then determining the concentration of that chemical in the medium of concern that should not be exceeded if exposed individuals (typically those at the high end of media contact) are not to incur a dose greater than the safe one. The most common concentration limits are regulatory tolerances for pesticide residues in food, Maximum Con-

63. *See, e.g.*, Knight v. Kirby Inland Marine Inc., 482 F.3d 347, 352–53 (5th Cir. 2007) (study of people with much longer exposure to organic solvents could not support conclusion that plaintiff's injuries were caused by such solvents); Kennecott Greens Creek Mining Co. v. Mine Safety & Health Admin., 476 F.3d 946, 950 (D.C. Cir. 2007) (because diesel particulate matter was difficult to monitor, MSHA's surrogate limits on total carbon and elemental carbon were reasonable).

64. *See, e.g.*, Genereux v. American Beryllia Corp., 577 F.3d 350, 366–67 (1st Cir. 2009) ("*all* beryllium operations should be periodically air-sampled, and a workspace may be dangerous to human health even though no dust is visible"); Allen v. Martin Surfacing, 2009 WL 3461145 (D. Mass. 2009) (where air sampling was not done, expert resorted to modeling plaintiff's decedent's exposure); Jowers v. BOC Group, Inc., 608 F. Supp. 2d 724, 738 (S.D. Miss. 2009) (OSHA measurements showed that 30% of welders experienced manganese fumes at higher than allowable concentrations); *In re* FEMA Trailer Formaldehyde Prod. Liab. Litig., 583 F. Supp. 2d at 776 (air sampling revealed formaldehyde levels higher than allowable).

taminant Levels (MCLs) for drinking water contaminants, National Ambient Air Quality Standards (NAAQS), and, for workplace exposure, Permissible Exposure Limits (PELs) or Threshold Limit Values (TLVs).[65] Much environmental sampling and analysis is done, by both government agencies and private organizations, for the purpose of ascertaining compliance with existing concentration limits (sometimes referred to as standards).

But sampling and analysis also are undertaken to investigate newly identified contamination or to ascertain exposures (and risks) in situations involving noncompliance with existing standards. As described earlier, information on concentrations in the media through which people are exposed is the necessary first step in estimating doses.

Although at first glance it might seem that direct measurements of concentrations would provide the most reliable data, there are limits to what can be gained through this approach.

- How can we be sure that the samples taken are actually representative of the media sampled?

  Standard methods are available to design sampling plans that have specified probabilities of being representative, but they can never provide complete assurance. Generally, when contamination is likely to be highly homogeneous, there is a greater chance of achieving a reasonably representative sample than is the case when it is highly heterogeneous. In the latter circumstance, obtaining a representative sample, even when very large numbers of samples are taken, may be unachievable.
- How can we be sure that the samples taken represent contamination over long periods?

  Sampling events may provide a good snapshot of current conditions, but in circumstances in which concentrations could be changing over time, and where the health concerns involve long-term exposures, snapshots could be highly misleading. This type of problem may be especially severe when attempts are being made to reconstruct past exposures, based on snapshots taken in the present.
- How can we be sure that the analytical work was done properly?

  Most major laboratories that routinely engage in this type of analysis have developed standard operating procedures and quality control proce-

65. PELs are official standards promulgated by the Occupational Safety and Health Administration. TLVs are guidance values offered by an organization called the American Conference of Governmental Industrial Hygienists. *See, e.g., In re* Howard, 570 F.3d 752, 754 (6th Cir. 2009) (challenging PELs for coal mine dust); Jowers v. BOC Group, Inc., 608 F. Supp. 2d 724, 735–36 (S.D. Miss. 2009) (PELs and TLVs for welders' manganese fume exposure); International Brominated Solvents Ass'n v. American Conf. of Gov. Indus. Hygienists, Inc., 625 F. Supp. 2d 1310 (M.D. Ga. 2008) (challenging TLVs for several chemicals); Miami-Dade County v. EPA, 529 F.3d 1049 (11th Cir. 2008) (MCLs for public drinking water).

dures. Laboratory certification programs of many types also exist to document performance. When analytical work is performed in certified, highly experienced laboratories, there is a reasonably high likelihood that the analytical results are reliable. But it is very difficult to confirm reliability when analytical work is done in laboratories or by individuals who cannot provide evidence of certification or of longstanding quality control procedures.

- How are data showing the absence of contamination to be interpreted?

  In most circumstances involving possible contamination of environmental media, the analysis of some (and sometimes many) of the samples will fail to find the contaminant. The analytical chemist will often report "ND" (for nondetect) for such samples. But an ND should never be considered evidence that the concentration of the contaminant is zero. In fact, most chemists will (and should) report that the contaminant is "BDL" (below detection limit). Every analytical method has a nonzero detection limit; the method is not sensitive to and cannot measure concentrations below that limit. Thus, for each sample reported as BDL, all that can be known is that the concentration of contaminant is somewhere below that limit. If there is clear evidence that the contaminant is present in some of the samples (its concentration exceeds the method's BDL), then it is usually assumed that all the samples of the same medium reported as BDL will actually contain some level of contaminant, often and for reliable reasons assumed to be one-half the BDL. Practices for dealing with BDL findings vary, but assuming that the BDL is actually zero is not one of the acceptable practices.

Sampling and measurement are no doubt useful, but are nonetheless limited in important ways. The alternative involves modeling. In fact, a combination of both approaches—one acting as a check on the other—is often the most useful and reliable.

D. Environmental Models

A model is an attempt to provide a mathematical description of how some feature of the physical world operates. In the matters at hand, a model refers to a mathematical description of the quantitative relationship between the amount of a chemical emitted from some source, usually over a specified period of time, to the concentrations of that chemical in the media of human exposure, again over some specified time period.[66]

66. *See, e.g.*, NCAP v. EPA, 544 F.3d 1043 (9th Cir. 2008) (EPA was permitted to rely on modeling in developing allowable pesticide residual levels); O'Neill v. Sherwin-Williams Co., 2009 WL 2997026, at *5 (C.D. Cal. 2009) (exposure model was inappropriate because it was based on a different type of paint than plaintiff was exposed to); Hayward v. U.S. Dep't of Labor, 536 F.3d 376

Models are idealized mathematical expressions of the relationship between two or more variables. They are usually derived from basic physical and chemical principles that are well established under idealized circumstances, but may not be validated under actual field conditions. Models thus cannot generate completely accurate predictions of chemical concentrations in the environment. In some cases, however, they are the only method available for estimating exposure—for example, in assessing the impacts of a facility before it is built or after it has ceased to operate. In such circumstances, they are necessary elements of exposure assessments and have been used extensively. Models are necessary if projections are to be made backward or forward in time or to other locations where no measurements have been made.

Typically, a model is developed by first constructing a flow diagram to illustrate the theoretical pathways of environmental contamination, as shown in Figure 2 and for a hazardous waste site in Appendix B. These models can be used to estimate concentrations in the relevant media based on several factors related to the nature of the site and the chemicals of interest. Model variables include the following:

1. The total amount of chemical present in or emitted from the media that are its sources;
2. The solubility of the chemical in water;
3. The chemical's vapor pressure (a measure of volatility);
4. The degree to which a chemical accumulates in fish, livestock, or crops (bioconcentration or bioaccumulation factor);
5. The nature of the soil present at the site; and
6. The volumes and movement of water around and beneath the site.

Some of this information derives from laboratory studies on the chemical (the first four points) and some from an investigation at the site (the remaining two points). The development of the data and modeling of the site often require the combined skills of chemists, environmental engineers, and hydrogeologists. In addition to the information listed above, time projection models also require information on the stability of the chemical of interest. As noted earlier, some chemicals degrade in the environment very quickly (in a matter of minutes), whereas others are exceedingly resistant to degradation. Quantitative information on rates of degradation is often available from laboratory and field studies.

Models that assess the exposures associated with air emissions consider the fact that the opportunity for people to be exposed to chemicals depends upon their activities and locations.[67] These models account for the activity patterns of

(5th Cir. 2008) (a model was used to reconstruct the dose of radiation that the employee was exposed to); Rodricks & Frankos, *supra* note 55.

67. *See, e.g.,* Palmer v. Asarco Inc., 2007 WL 2298422 (N.D. Okla. 2007) (children allegedly were exposed to lead by "hand-to-mouth activity ingestion of soil/house dust"); Henricksen

potentially exposed populations and provide estimates of the cumulative exposure over specified periods.

Perhaps the most widely used models are those that track the fate and transport pathways followed by substances emitted into the air. Knowledge of the amounts emitted per unit of time (usually obtainable by measurement) from a given location (a stack of a certain height, for example) provides the basic model input. Information on wind directions and velocities, the nature of the physical terrain surrounding the source, and other factors needs to be incorporated into the modeling. Some substances will remain in the vapor phase after emission, but chemical degradation (e.g., because of the action of sunlight) could affect media concentrations. Some models provide for estimating the distributions of soil concentrations for those substances (particulates of a certain size) that may fall during dispersion. Much effort has been put into developing and validating air dispersion models.[68] Similar models are available to track the movement of contaminants in both surface and ground waters.

The fate and transport modeling issue becomes more complex when attempts are made to follow a chemical's movement from air, water, and soils into the food chain and to estimate concentrations in the edible portions of plants and animals.[69] Most of the effort in this area involves the use of empirical data (e.g., What does the scientific literature tell us about the quantitative relationships between the concentration of cadmium in soil and its concentration in the edible portions of plants grown in that soil?). This type of empirical information, together with general data on chemical absorption into, distribution in, and excretion from living systems, is the usual approach to ascertain concentrations in these food media.[70]

Many models for environmental fate and transport analysis are available. It is not possible to specify easily which models have established validity and which have not; rather, some are preferred for some purposes and others are preferred for different purposes.

Perhaps the best that can be done to scrutinize the work of an expert in this area is to

- Require that the expert describe in full the basis for model selection;
- Ask the expert to describe the standing of the model with authoritative bodies such as EPA;
- Require the expert to state why other possible models are not suitable;

v. ConocoPhillips Co., 605 F. Supp. 2d 1142, 1164 (E.D. Wash. 2009) (expert calculated plaintiff's benzene exposure by adjusting study results to account for plaintiff's activities); Junk v. Terminix Int'l Co., 2008 WL 6808423 (S.D. Iowa 2008) (study measured chlorpyrifos exposure of inhabitants of houses sprayed indoors); *In re* W.R. Grace & Co., 355 B.R. 462 (Bankr. D. Del. 2006) (asbestos in attic insulation released by normal activity).

68. National Research Council, Models in Environmental Regulatory Decision Making (2007).

69. Ecologists also use modeling results to evaluate risks to wildlife, plants, and ecosystems.

70. National Research Council, *supra* note 68.

- Require that the expert describe the scientific basis and underlying assumptions of the model, and the ways in which the model has been verified;[71] and
- Require the expert to describe the likely size of error associated with model results.

Other issues pertaining to the sources and reliability of the data used in the application of a model can be similarly pursued.

Results from modeling are concentrations in media of concern over time. If sampling and analysis data are available for the same media, they can be compared with the modeling result, and efforts can be made to reconcile the two and arrive at the most likely values (or range of likely values).

E. Integrated Exposure/Dose Assessment

We have shown the various methods used to determine the concentrations of chemicals in products and in various environmental media and also the methods used to determine doses from each of the relevant media. Dose estimation as described in Section VI.A applies to each of the relevant routes of exposure.

In many cases, the dose issue concerns one chemical in one product and only one route of exposure. But numerous variations on this basic scenario are possible: one chemical in several products or environmental media, many chemicals in one product or environmental medium, or many chemicals in many environmental media. Even though some exposure situations can be complex and involve multiple chemicals through both direct and indirect pathways, the exposure assessment methods and principles described here can be applied. Exposures occurring by different routes can be added together, or they can be reported separately. The decisions on the final dose estimates and their form of presentation can be made only after discussions with the users of that information—typically the toxicologists and epidemiologists involved in the risk assessment.[72] The dose metrics emerging from the exposure assessment need to match the dose metrics that are used to describe toxicity risks.

One additional point should be highlighted. The principle that exposure to chemicals through foods and consumer products typically focuses on high-end consumers of those foods or products also applies in environmental settings. Thus,

71. This point is to ensure that the expert truly understands the model and its limits and that he or she is not simply using some "black box" computer software.

72. *See, e.g.*, American Farm Bureau Fed'n v. EPA, 559 F.3d 512 (D.C. Cir. 2009) (challenging EPA's risk assessment for fine PM); Miami-Dade County v. EPA, 529 F.3d 1049 (11th Cir. 2008) (assessment of risk of wastewater disposal methods to drinking water); Kennecott Greens Creek Min. Co. v. Mine Safety & Health Admin., 476 F.3d 946 (D.C. Cir. 2007) (risk assessment of diesel particulate matter to miners); Rowe v. E.I. du Pont de Nemours & Co., 2008 WL 5412912, 12 (D.N.J 2008) (risk assessment for proposed class).

for example, it is possible to assert with relatively high confidence that almost no one consumes more than 3.5 L of water a day and that almost everyone consumes less. If the dose calculation assumes a water consumption rate of 3.5 L/day, then the risk estimated for that dose is almost certainly an upper limit on the population risk, and regulatory actions based on that risk will almost certainly be highly protective. For regulatory and public health decisionmaking, such a precautionary approach has a great deal of precedent, although care must be taken to ensure adherence to scientific data and principles.[73]

This approach becomes problematic, however, if applied to assessments of exposures that may have been incurred in the past by individuals claiming to have been harmed by them. In such cases, it would seem that there is no basis for a precautionary approach; an approach based on attempts to accurately describe the individual's exposure would seem to be necessary. Whatever the case, the exposure scientist must be careful to ensure accurate description of the exposure concentration (and resulting dose), so that the users of the information can understand whether upper limits or more typical exposures and doses have been provided.

VII. Into the Body

A. Body Burdens

Section V described how chemicals in the environment contact the three major portals of entry into the body—the respiratory tract, the GI tract, and the skin. For some chemicals, the dose contacting one or more of those portals may be sufficient to cause harm before those chemicals are absorbed into the body; that is, they may cause one or more forms of toxicity to the respiratory system, to the GI tract, or to the skin. Although these forms of *contact* toxicity can be important, it is also important to consider the many forms of systemic toxicity. The latter refers to a large number of toxic manifestations that can affect any of the organs or organ systems of the body after a chemical is absorbed into the bloodstream and distributed within the body. Recall also that most chemicals are acted upon by certain large protein molecules, called enzymes, contained in cells, particularly those of the liver, the skin, and the lungs, and are converted to new compounds, called metabolites (the process leading to these changes is called metabolism). Metabolite formation

73. National Research Council, *Evolution and Use of Risk Assessment in the Environmental Protection Agency: Current Practice and Future Prospects, in* Science and Decisions: Advancing Risk Assessment (2008). Those who must comply with regulations that were developed based on a high degree of caution often protest that more accurate assessments should be used as their basis. For several reasons, truly accurate prediction of risk is difficult to achieve (*see* Bernard D. Goldstein & Mary Sue Henifin, Reference Guide on Toxicology, in this manual), while predicting an upper bound on the risk is not. At the same time, unless carefully done and described, upper-bound estimates may be so remote from reality that decisions based on them should be avoided.

is one of the body's mechanisms for creating compounds that are easily removed from the body by one or more excretion processes. Unfortunately, metabolism sometimes creates new compounds that are more toxic than the original (so-called parent) molecule, and, if the internal dose of toxic metabolite exceeds a so-called threshold, toxic harm may occur. Of course, not all toxicity is produced by metabolites; in some cases harm may be caused directly by the parent compound.[74]

As in the other areas of exposure science that have been discussed, it usually becomes important to move from description to quantification. Exposure scientists seek to understand the amount of chemical absorbed into the body after contact (i.e., the fraction of the dose that is absorbed), the amount of chemical reaching and distributed within the body (the blood concentration being the most easily measurable), and the rate of loss of the chemical from the body. The science devoted to understanding these important phenomena is called pharmacokinetics (drug rates). That name came to be used because most of the developmental work in this area related to the behavior of pharmaceuticals in the body, but the tools of pharmacokinetics have been extended to study all types of chemicals.

Pharmacokinetics is important because it reveals where in the body a chemical is most likely to cause harm (where the greatest concentrations, or target site doses, are reached for the longest period of time) and also the concentration—duration level necessary to cause harm. To understand these relationships, pharmacokinetic studies typically are carried out in conjunction with toxicity studies in animals, and their results are used to assess possible toxic risk in humans.[75]

Pharmacokineticists do not ordinarily characterize themselves as exposure scientists; more often they are toxicologists or pharmacologists. But they are in fact extending the usual work of exposure scientists into the body, and it is here that we see the interface between exposure science and toxicology and epidemiology.

B. Monitoring the Body (Biomonitoring)

As long as we live in a world of chemicals, we will be exposed to them. If analytical chemists developed sufficiently sensitive measuring techniques, it would not be far-fetched to say that we could find within the human body, at some level and for some period, virtually any of the tens of thousands of chemicals, natural and synthetic, with which it comes into contact. Some would be found only occasionally, some continuously; some would be found to persist for days, weeks,

74. J.V. Rodricks, *From Exposure to Dose, in* Calculated Risks: The Toxicity and Human Health Risks of Chemicals in Our Environment (2d ed. 2007)

75. See Bernard D. Goldstein & Mary Sue Henifin, Reference Guide on Toxicology, in this manual. *See also, e.g., In re* Fosamax Prod. Liab. Litig., 645 F. Supp. 2d 164, 186 (S.D.N.Y. 2009) (rat and dog studies showing a bisphosphonate caused jaw necrosis relevant to whether Fosamax, another type of bisphosphonate, could cause jaw necrosis in humans); Rose v. Matrixx Initiatives, Inc., 2009 WL 902311, at *14 (W.D. Tenn. 2009) (studies in animals of nasal spray effects could not be extrapolated to humans because olfactory physiology was too different).

or even longer, whereas others would persist for only minutes or hours. The concentrations in blood would likely vary over many orders of magnitude. Currently, we can measure only a few thousand chemicals in the body, a large share of them pharmaceuticals, nutrients, and substances of abuse. Some standards for occupational exposures are expressed as allowable blood or urine concentrations, and their measurement is a useful supplement to air monitoring.[76]

The environmental chemical that has perhaps received the most attention in this area of exposure science is lead (chemical symbol Pb). Indeed, lead may be the most studied of all environmental substances. After it was learned in the 1950s that the concentration of lead in blood could be easily measured, it became common to sample and test the blood level of lead (BPb) in individuals who had suffered one or more forms of this metal's toxicity. Some epidemiological studies of lead began to include BPb as the measure of exposure, and since the 1970s, hundreds of such studies involving lead have reported results using this measure.[77]

BPb is particularly useful for substances such as lead that have (or did have) a relatively large number of environmental sources.[78] The simple measure of BPb provides a single, integrated measure of exposures through multiple sources, pathways, and routes (although this measure reflects relatively recent and not long-term exposure).[79] This is perhaps the best example of the use of target site dose in risk assessment.

The Centers for Disease Control and Prevention (CDC) began, in the late 1970s, to take blood samples from a relatively large number of children as part of its National Health and Nutrition Examination Survey (NHANES). Children were selected because it was known that they take up more lead from their envi-

76. *See, e.g.*, Haas v. Peake, 525 F.3d 1168, 1177 (Fed. Cir. 2008) (presumption of dioxin exposure instituted because of the difficulty of measuring dioxin in the body); Young v. Burton, 567 F. Supp. 2d 121 (D.D.C. 2008) (hormone and enzyme levels allegedly altered by exposure to biotoxins in mold); Hazlehurst v. Sec'y of Dep't of Health & Human Servs., 2009 WL 332306, at *62 (Fed. Cl. 2009) (study measuring porphyrin in urine as a marker for mercury in the body); United States v. Bentham, 414 F. Supp. 2d 472 (S.D.N.Y. 2006) (cocaine use monitored by a "sweatpatch" on the skin).

77. National Center for Environmental Health, Centers for Disease Control and Prevention, Fourth National Report on Human Exposure to Environmental Chemicals (2009), *available at* http://www.cdc.gov/exposurereport/pdf/FourthReport.pdf (last visited July 1, 2010).

78. *See, e.g.*, Potter v. EnerSys, Inc., 2009 WL 3764031 (E.D. Ky. 2009) (alleged lead exposure from working on battery manufacturing site); City of North Chicago v. Hanovnikian, 2006 WL 1519578 (N.D. Ill. 2006) (alleged lead contamination of soil); Perry *ex rel.* Perry v. Frederick Inv. Corp., 509 F. Supp. 2d 11 (D.D.C. 2007) (residential lead paint exposure); Goodstein v. Continental Cas. Co., 509 F.3d 1042 (9th Cir. 2007) (environmental contamination from lead waste site); Evansville Greenway & Remediation Trust v. Southern Indiana Gas & Elec., 661 F. Supp. 2d 989 (S.D. Ind. 2009) (contamination of battery recycling site).

79. BPb usually is reported in units of micrograms (1 one-millionth of 1 gram) in each deciliter (one-tenth of a liter) of blood (μg/dL). More recently, noninvasive methods to measure lead levels in teeth and bones have become available; such measures reflect cumulative exposures over long periods, but their relationships to health are less clear than those based on BPb.

ronments (air,[80] water, food, paint, soils and dusts, emissions from lead and other metal smelters, consumer products, and more) than do adults; they are also, especially during early periods of development, more vulnerable to the adverse effects of lead than are adults. Nationwide, childhood BPb levels averaged 15–20 μg/dL during the 1970s, with substantial numbers of children having BPb levels well in excess of what was at the time thought to be the minimum BPb associated with adverse health effects (40 μg/dL). The most recent NHANES surveys reveal that average childhood levels are in the range of 2 μg/dL, although there remain substantial numbers of children with levels greater than the current CDC health guideline of 10 μg/dL.[81]

Lead is not the only chemical now being studied under the NHANES biomonitoring program. The most recent surveys involve nationwide sampling of blood and urine from close to 8000 children and adults for more than 100 different chemicals.[82] The program focuses on commonly used pesticides and consumer products and certain ubiquitous environmental contaminants, particularly those that persist in the body for long periods. Not surprisingly, most of these chemicals have been detected in some individuals. The NHANES program will continue, and similar programs are under way in government and research centers around the world.

The presence of a chemical in the body is not evidence that it is causing harm. And in some cases—those that involve chemicals, such as the metals and some organic compounds that occur naturally—the NHANES findings may simply reflect natural background levels.[83] In any case, data such as these provide far more direct measures of dose (often referred to as body burden), and in those cases (which are increasing in number) in which epidemiologists and toxicologists are able to relate disease rates to body burdens (instead of to external dose, as is the usual case), far more accurate measures of human risk should become available.

VIII. Evaluating the Scientific Quality of an Exposure Assessment

Exposure scientists may offer expert testimony regarding exposures to chemicals incurred by individuals or populations. Their assessments typically will include

80. At the time of the first NHANES lead survey, leaded gasoline, which emitted lead to air and to soil, was in wide use. That use, at least in the United States, came to an end in the 1980s. For a discussion of the routes of exposure to toxic substances, see Bernard D. Goldstein & Mary Sue Henifin, Reference Guide on Toxicology, Section III.A, in this manual.

81. There is developing evidence of IQ deficits in children at levels below 10 μg/dL.

82. National Center for Environmental Health, *supra* note 77.

83. Natural background levels of certain metals may, in some geological regions, be quite high and may even be associated with excess disease.

a description of how and when exposures have or could occur, the identities of the chemicals involved, the routes of exposure, the doses incurred, and the durations of exposure. In some cases, testimony will include a description and quantification of body burdens. If the exposure scientist is also an epidemiologist or toxicologist,[84] he or she may offer additional testimony on the health risks associated with those exposures or even regarding the question of whether such exposures have actually caused disease.

For purposes of this reference guide, it is assumed that questions regarding disease risk and causation are beyond the bounds of exposure science. Below is offered a set of questions that exposure scientists should be able to answer, with appropriate documentation and scientific reasoning, to support any given exposure assessment:

- Is the purpose of the assessment clear? Is the exposed population specified?
- What is the source(s) of exposure?
- When did the exposures occur: past? present? If they are occurring now, will they continue to occur?
- What is the assumed duration of exposure, and what is its basis?
- What are the pathways from the source to the exposed individuals? How has it been established that those pathways exist (past? present? future?).
- What is the concentration of the chemical in the media with which the exposed population comes into contact (past? present? future?). What is the basis for this answer: direct measurement? modeling?
- If the concentration is based on direct measurement, what procedures were followed in obtaining that measurement? Was media sampling sufficient to ensure that it was representative? If not, why is representativeness not important? Were validated analytical methods used by an accredited laboratory? If not, how can one be assured that the analytical results are reliable?
- If models were used, what is their reliability (see Section VI.D)? What is the variability over time in concentrations in the media of concern? How has the variability been determined?
- What is the variability among members of the population in their exposure to the chemical of concern? How is this known?
- What is known or assumed about the nature and extent of media contact by members of the exposed population? How has this been ascertained?
- What dose, over what period of time, by which routes, has been incurred? What calculations support this determination?

84. See Section IX, which deals with the question of the qualifications of exposure scientists. In many cases, the work of exposure experts is turned over to the health experts to incorporate into their evaluation of risk and disease causation. In some cases, usually the less complex ones, exposure assessments may be undertaken by the health experts.

- What is the likely error in the exposure estimates?
- What uncertainties are associated with the dose/duration findings? Is it a "most likely" estimate, or is it an "upper limit"? To what fraction of the population is the "upper limit" likely to apply?
- What has been omitted from the exposure assessment, and why?

These questions are perhaps the minimum that an expert should be able to address when offering testimony. Obviously, most such questions can be answered fully only if the expert can support the answers with documentation.

As noted in Section III.D, the evaluation of whether a current medical condition is causally related to exposures occurring in the past (prior to the onset or diagnosis of the medical condition) requires a retrospective examination of the conditions that led to those exposures. Thus, for example, a plaintiff suffering from leukemia and who alleges that benzene exposure in his or her workplace caused the disease may easily demonstrate the fact of benzene exposure. But ordinarily an estimation of the quantitative magnitude and duration of the incurred benzene exposure is necessary to evaluate the plausibility of the causation claim.[85] The methodological tools necessary to "reconstruct" the plaintiff's past exposure are identical to those used to estimate current exposures, but the availability of the data necessary to apply those methods may be limited or, in some cases, nonexistent.

Reconstruction of occupational exposures has been a relatively successful pursuit, because often historical industrial hygiene data are available involving the measurement of workplace air levels of chemicals. If it is possible, through the examination of employment records, to reconstruct an individual's job history, it may be possible to ascertain that individual's exposure history.[86] Guidelines for occupational exposure reconstruction have been published by the American Industrial Hygiene Association.[87] Clearly, experts presenting testimony regarding exposure reconstruction must be queried heavily on the sources of data used in their applications of exposure methods.

IX. Qualifications of Exposure Scientists

Exposure science is not yet a true academic discipline. Rather, scientists and engineers from diverse backgrounds have, over the past several decades, come together to give shape and substance and scientific rigor to what is clearly a criti-

85. *See* Michael D. Green et al., Reference Guide on Epidemiology, Section VII, in this manual.

86. T.W. Armstrong, *Exposure Reconstruction, in* Mathematical Models for Estimating Occupational Exposures to Chemicals (Charles B. Keil et al. eds., 2d ed. 2009).

87. American Industrial Hygiene Association, Guideline on Occupational Exposure Reconstruction (S.M. Viet et al. eds., 2008).

cal element in understanding toxicity risks and disease causation. Typically, those who have contributed to this developing field have come from backgrounds in industrial hygiene, environmental and analytical chemistry, chemical engineering, hydrogeology, and even behavioral sciences (pertaining to those aspects of human behavior that affect exposures).[88] Most toxicologists and epidemiologists have considerable experience in exposure science, as do pharmacologists who study drug kinetics and disposition. Many exposure assessments involve collaborative efforts among members of these various disciplines.

There are currently no certification programs available for exposure scientists, but increasingly exposure science research appears in publications such as *Environmental Health Perspectives*, *Risk Analysis*, and the *Journal of Exposure Science and Environmental Epidemiology*.

Certification programs do exist in occupational exposure science. Qualified industrial hygienists will almost always be certified (CIH). The *American Industrial Hygiene Association Journal* includes much scholarly work related to exposure science.

88. *See, e.g.*, Allen v. Martin Surfacing, 2009 WL 3461145, 2008 U.S. Dist. LEXIS 111658, 263 F.R.D. 47 (D. Mass. 2008) (industrial hygienist qualified to testify regarding concentration and duration of plaintiffs' decedent's exposure to toluene and other chemicals); Buzzerd v. Flagship Carwash of Port St. Lucie, Inc., 669 F. Supp. 2d 514 (M.D. Pa. 2009) (industrial hygienist qualified to opine on carbon monoxide exposure, but his conclusions were not based on reliable methodology).

Appendix A: Presentation of Data—Concentration Units

Choosing the proper units to express concentrations of chemicals in environmental media is crucial for precisely defining exposure. Chemical concentrations in environmental media usually are reported in one of two forms: as numeric ratios, such as parts per million or billion (ppm and ppb, respectively), or as unit weight of the chemical per weight or volume of environmental media, such as milligrams per kilogram (mg/kg) or milligrams per cubic meter (mg/m^3). Although concentrations expressed as parts per million or parts per billion are easier for some people to conceptualize, their use assumes that media are always sampled at standard temperature and pressure (25°C and 760 torr, respectively). Consequently, scientists prefer to express chemical concentrations as weight of chemical per unit weight or volume of media. This method also makes conversions to dose equivalents, usually expressed in terms of weight of chemical per unit body weight (mg/kg bw), more convenient.

To permit the presentation of results without excessive zeroes before or after the decimal point, appropriate units are needed. The choice of units depends on both the medium in which the chemical resides and the amount of chemical measured. For example, if 50 nanograms of chemical were found in 1 L of water, the appropriate units would be ng/L, rather than 0.00005 mg/L. If 50 grams were found instead, the appropriate units would be 50,000 mg/L, because milligrams are generally the largest units used to express the mass of a chemical in media (Table 1).

Table 1. Weight of Chemical per Unit Weight of Medium

Preferred Unit	Alternative Unit
mg/kg	ppm (parts per million)
μg/kg	ppb (parts per billion)
ng/kg	ppt (parts per trillion)
pg/kg	ppq (parts per quadrillion)

In water or food, concentration expressed by the preferred unit equals concentration expressed by alternative unit; thus, 2 mg/kg = 2 ppm. One mg (10^{-3} g) per kg (10^3 g) equals 1 part per million ($10^{-3}/10^3 = 10^{-6}$). Similarly, 1 μg (10^{-6} g) per kilogram (10^3 g) equals 1 part per billion ($10^{-6}/10^3 = 10^{-9}$), and so on (Table 2).

Note that in air, parts per million and parts per billion have different meanings than they do in water or food; to avoid confusion, it is always preferrable to express air concentrations in weight of chemical per unit volume (rather than weight) of air (usually cubic meters, m^3).

Table 2. Weight of Chemical per Unit Volume of Medium

Water	Air
mg/L = ppm	$mg/m^3 \neq ppm$
µg/L = ppb	$mg/m^3 \neq ppb$
ng/L = ppt	$ng/m^3 \neq ppt$

Appendix B: Hazardous Waste Site Exposure Assessment

Several principles of exposure assessment can be illustrated by examining the steps taken to evaluate a hazardous waste disposal site. From 1964 to 1972, more than 300,000 55-gallon drums of solid and liquid pesticide production wastes were buried in shallow trenches at a hazardous waste disposal site in Hardeman County, Tennessee. As early as 1965, county engineers had raised concerns that these operations might have affected the aquifer supplying drinking water to the City of Memphis, Tennessee. The State of Tennessee ordered the landfill to stop accepting hazardous waste in 1972; all operations were reported to have ceased by 1975. Testing in 1978 confirmed the presence of toxic chemicals in domestic wells, and by January 1979 all uses of the contaminated well water had been discontinued.

Among the chemicals of concern detected in the ground water were benzene, carbon tetrachloride, chlordane, chlorobenzene, chloroform, and several other pesticides or chemicals associated with pesticide production. As is often the case for ground water polluted by landfills, the observed concentrations fluctuated over a relatively wide range. For example, in a domestic well approximately 1500 feet north of the landfill, carbon tetrachloride concentrations ranged from 10 ppm to 20 ppm between November 1978 and November 1979; from May 1981 to June 1982, carbon tetrachloride levels varied from 18 ppm to 164 ppm.

The chemicals of greatest concern detected during ground-water monitoring near the Hardeman site included carbon tetrachloride, chloroform, and tetrachloroethylene. For each of these three chemicals, the concentrations detected in well water were significantly elevated over levels typically found in potable water. Health surveys conducted in 1978 and 1982 suggested that these chemicals might be causing a variety of health problems in nearby residents.

To confirm the cause-and-effect relationship suggested by the health surveys, an exposure assessment was conducted so that the findings of the health surveys could be compared to adverse health impacts predicted from exposure estimates and toxicological data from laboratory experiments. The exposure assessment for the Hardeman site focused on carbon tetrachloride, because of the high concentrations of this chemical found in the ground water and the severity of the potential health effects associated with exposure to it.

To estimate the range of possible exposures, the Hardeman site assessment considered exposures of both an adult and an infant. The exposure assessor then needed to identify the pathways of exposure that might be important. For the infant, the following exposure pathways were examined:

- Consumption of formula made using well water,
- Dermal absorption during bathing in contaminated water, and

- In-utero exposure of the fetus through exposure of the mother during pregnancy.

Adult exposures were evaluated for two pathways:

- Consumption of contaminated drinking water and
- Inhalation of carbon tetrachloride emanating from water during showers.

Because measurements of concentrations of carbon tetrachloride in the ground water were scant before 1978, estimates were modeled for these years; measured concentrations were used for 1978, the last year residents utilized ground water for drinking. Standard assumptions regarding the ingestion of water by adults (2 L/day) were used; water consumption by a child was assumed to be 0.5 L/day for 3 months following birth. Dermal absorption by infants was estimated by assuming that the child bathed in 30 L/day of well water, that 50% of this volume contacted the skin, and that 10% of the contaminant was absorbed through the skin. Three baths per week were assumed for the first 3 months after birth. In-utero exposure was estimated assuming equal concentrations of carbon tetrachloride in fetal and maternal blood. The concentration of carbon tetrachloride in air during showering was calculated assuming that it would quickly reach equilibrium with carbon tetrachloride in the shower water.

In Table 3, carbon tetrachloride exposure estimates for the infant and adult are compared with the minimum daily exposure producing liver damage in guinea pigs and the lifetime cumulative exposure producing liver cancer in mice. Daily exposure rates were based on a predicted yearly average exposure during the highest year of exposure. Monitoring data indicate that the concentration of carbon tetrachloride in the ground water may have varied by a factor of 10 around the mean. The maximum daily exposure rate may have been considerably higher than the estimates presented in the table, whereas the long-term averages may have been lower.

Table 3. Carbon Tetrachloride Exposure Estimates for Infants and Adults Compared with Minimum Daily Exposure Producing Liver Damage in Guinea Pigs and Lifetime Cumulative Exposure Producing Liver Cancer in Mice

	Daily Dose Rate (mg/kg/day)
Liver damage in guinea pigs	1.5
Estimated infant exposure	1.8
Estimated adult exposure	0.3
	Cumulative Dose (mg/kg)
40% Liver tumors in mice	1200
Estimated adult exposure	284

Glossary of Terms

absorbed dose. The amount of a substance that actually enters the body following absorption.

absorption. The penetration of a substance through a barrier (e.g., the skin, the gut, or the lungs).

acute exposure. An exposure of short duration and/or rapid onset. An acute toxic effect is one that develops during or shortly after an acute exposure to a toxic substance.

average daily dose (ADD). The average dose received on any given day during a period of exposure, expressed in mg/kg body weight per day. Ordinarily used in assessing noncancer risks.

bioavailability. The rate and extent to which a chemical or chemical breakdown product enters the general circulation, thereby permitting access to the site of toxic action.

body burden. The total amount of a chemical present or stored in the body. In humans, body burden is an important measure of exposure to chemicals that tend to accumulate in fat cells, such as DDT, PCBs, or dioxins.

chronic exposure. A persistent, recurring, or long-term exposure, as distinguished from an acute exposure. Chronic exposure may result in health effects (such as cancer) that are delayed in onset, occurring long after exposure has ceased.

direct exposure. Exposure of a subject who comes into contact with a chemical via the medium in which it was initially released to the environment. Examples include exposures mediated by cosmetics, other consumer products, some food and beverage additives, medical devices, over-the-counter drugs, and single-medium environmental exposures.

dose. The amount of a substance entering a person, usually expressed for chemicals in the form of weight of the substance (generally in milligrams (mg) or micrograms (μg)) per unit of body weight (generally in kilograms (kg)). It is necessary to specify whether the dose referred to is applied or absorbed. The time over which it is received must also be specified. The time of interest is typically 1 day. If the duration of exposure is specified, dose is actually a dose rate and is expressed as mg or μg/kg per day.

dose–response assessment. An analysis of the relationship between the dose administered to a group and the frequency or magnitude of the biological effect (response).

duration of exposure. Toxicologically, there are three categories describing duration of exposure: acute (one time), subchronic (repeated, for a fraction of a lifetime), and chronic (repeated, for nearly a lifetime).

environmental media. Air, water, soils, and food; consumer products may also be considered media. Chemicals may be directly and intentionally introduced into certain media. Others may move from their sources through one or more media before they reach the media with which people have contact.

exposure. The opportunity to receive a dose through direct contact with a chemical or medium containing a chemical. See also direct esposure; indirect exposure.

exposure assessment. The process of describing, for a population at risk, the amounts of chemicals to which individuals are exposed, or the distribution of exposures within a population, or the average exposure over an entire population.

frequency of exposure. The number of times an exposure occurs in a given period; exposure may be continuous, discontinuous but regular (e.g., once daily), or intermittent (e.g., less than daily, with no standard quantitative definition).

indirect exposure. Often defined as an exposure involving multimedia transport of chemicals from source to exposed individual. Examples include exposures to chemicals deposited onto soils from the air, chemicals released into the ground water beneath a hazardous waste site, or consumption of fruits or vegetables with pesticide residues.

intake. The amount of contact with a medium containing a chemical; used for estimating the dose received from a particular medium.

levels. An alternative term for expressing chemical concentration in environmental media. Usually expressed as mass per unit volume or unit weight in the medium of interest.

lifetime average daily dose (LADD). Total dose received over a lifetime multiplied by the fraction of lifetime during which exposure occurs, expressed in mg/kg body weight per day. Ordinarily used for assessing cancer risk.

models. Idealized mathematical expressions of the relationship between two or more factors (variables).

pathway. The connected media that transport a chemical from source to populations.

point-of-contact exposures. Exposure expressed as the product of the concentration of the chemical in the medium of exposure and the duration and surface area of contact with the body surface, for example, mg/cm^2-hours. Some chemicals do not need to be absorbed into the body but rather produce toxicity directly at the point of contact, for example, the skin, mouth, GI tract, nose, bronchial tubes, or lungs. In such cases, the absorbed dose is not the relevant measure of exposure; rather, it is the amount of toxic chemical coming directly into contact with the body surface.

population at risk. A group of subjects with the opportunity to be exposed to a chemical.

risk. The nature and probability of occurrence of an unwanted, adverse effect on human life or health or on the environment.

risk assessment. Characterization of the potential adverse effects on human life or health or on the environment. According to the National Research Council's Committee on the Institutional Means for Assessment of Health Risk, human health risk assessment includes the following: description of the potential adverse health effects based on an evaluation of results of epidemiologic, clinical, toxicological, and environmental research (hazard identification); extrapolation from those results to predict the type and estimate the extent of health effects in humans under given conditions of exposure (dose–response assessment); judgments regarding the number and characteristics of persons exposed at various intensities and durations (exposure assessment); summary judgments on the existence and overall magnitude of the public-health problem; and characterization of the uncertainties inherent in the process of inferring risk (risk characterization).

route of exposure. The way a chemical enters the body after exposure, that is, by ingestion, inhalation, or dermal absorption.

setting. The place or situation in which a person is exposed to the chemical. Setting is often modified by the activity a person is undertaking, for example, occupational or in-home exposures.

source. The activity or entity from which the chemical is released for potential human exposure.

subchronic exposure. An exposure of intermediate duration between acute and chronic.

subject. An exposed individual, whether a human or an exposed animal or organism in the environment. An exposed individual is sometimes also called a receptor.

systemic dose. A dose of a chemical within the body—that is, not localized at the point of contact. Thus, skin irritation caused by contact with a chemical is not a systemic effect, but liver damage due to absorption of the chemical through the skin is. Often referred to as target site dose.

total dose. The doses received by more than one route of exposure are added to yield the total dose.

References on Exposure

D.B. Barr, *Expanding the Role of Exposure Science in Environmental Health,* 16 J. Exposure Sci. Envtl. Epidemiol. 473 (2006).

Exposure Assessment in Occupational and Environmental Epidemiology (M.J. Nieuwenhuiysen ed., 2003).

S. Gad, Regulatory Toxicology (2d ed. 2001). Includes much discussion of pharmaceuticals, food ingredients, and other consumer products.

P. Lioy, *Exposure Science: A View of the Past and Milestones for the Future*, 118 Envtl. Health Persp. 1081–90 (2010).

U.S. Environmental Protection Agency, Guidelines for Exposure Assessment, Doc. No. EPA/600/Z-92/001 (1992), *available at* http://cfpub.epa.gov/ncea/cfm/recordisplay.cfm?deid=15263.

Reference Guide on Epidemiology

MICHAEL D. GREEN, D. MICHAL FREEDMAN, AND LEON GORDIS

Michael D. Green, J.D., is Bess & Walter Williams Chair in Law, Wake Forest University School of Law, Winston-Salem, North Carolina.

D. Michal Freedman, J.D., Ph.D., M.P.H., is Epidemiologist, Division of Cancer Epidemiology and Genetics, National Cancer Institute, Bethesda, Maryland.

Leon Gordis, M.D., M.P.H., Dr.P.H., is Professor Emeritus of Epidemiology, Johns Hopkins Bloomberg School of Public Health, and Professor Emeritus of Pediatrics, Johns Hopkins School of Medicine, Baltimore, Maryland.

CONTENTS

I. Introduction

Epidemiology is the field of public health and medicine that studies the incidence, distribution, and etiology of disease in human populations. The purpose of epidemiology is to better understand disease causation and to prevent disease in groups of individuals. Epidemiology assumes that disease is not distributed randomly in a group of individuals and that identifiable subgroups, including those exposed to certain agents, are at increased risk of contracting particular diseases.[1]

Judges and juries are regularly presented with epidemiologic evidence as the basis of an expert's opinion on causation.[2] In the courtroom, epidemiologic research findings are offered to establish or dispute whether exposure to an agent[3]

1. Although epidemiologists may conduct studies of beneficial agents that prevent or cure disease or other medical conditions, this reference guide refers exclusively to outcomes as diseases, because they are the relevant outcomes in most judicial proceedings in which epidemiology is involved.

2. Epidemiologic studies have been well received by courts deciding cases involving toxic substances. *See, e.g.*, Siharath v. Sandoz Pharms. Corp., 131 F. Supp. 2d 1347, 1356 (N.D. Ga. 2001) ("The existence of relevant epidemiologic studies can be a significant factor in proving general causation in toxic tort cases. Indeed, epidemiologic studies provide 'the primary generally accepted methodology for demonstrating a causal relation between a chemical compound and a set of symptoms or disease.'" (quoting Conde v. Velsicol Chem. Corp., 804 F. Supp. 972, 1025–26 (S.D. Ohio 1992))), *aff'd*, 295 F.3d 1194 (11th Cir. 2002); Berry v. CSX Transp., Inc., 709 So. 2d 552, 569 (Fla. Dist. Ct. App. 1998). Well-conducted studies are uniformly admitted. 3 Modern Scientific Evidence: The Law and Science of Expert Testimony § 23.1, at 187 (David L. Faigman et al. eds., 2007–08) [hereinafter Modern Scientific Evidence]. Since *Daubert v. Merrell Dow Pharmaceuticals*, 509 U.S. 579 (1993), the predominant use of epidemiologic studies is in connection with motions to exclude the testimony of expert witnesses. Cases deciding such motions routinely address epidemiology and its implications for the admissibility of expert testimony on causation. Often it is not the investigator who conducted the study who is serving as an expert witness in a case in which the study bears on causation. *See, e.g.*, Kennedy v. Collagen Corp., 161 F.3d 1226 (9th Cir. 1998) (physician is permitted to testify about causation); DeLuca v. Merrell Dow Pharms., Inc., 911 F.2d 941, 953 (3d Cir. 1990) (a pediatric pharmacologist expert's credentials are sufficient pursuant to Fed. R. Evid. 702 to interpret epidemiologic studies and render an opinion based thereon); Medalen v. Tiger Drylac U.S.A., Inc., 269 F. Supp. 2d 1118, 1129 (D. Minn. 2003) (holding toxicologist could testify to general causation but not specific causation); Burton v. R.J. Reynolds Tobacco Co., 181 F. Supp. 2d 1256, 1267 (D. Kan. 2002) (a vascular surgeon was permitted to testify to general causation); Landrigan v. Celotex Corp., 605 A.2d 1079, 1088 (N.J. 1992) (an epidemiologist was permitted to testify to both general causation and specific causation); Trach v. Fellin, 817 A.2d 1102, 1117–18 (Pa. Super. Ct. 2003) (an expert who was a toxicologist and pathologist was permitted to testify to general and specific causation).

3. We use the term "agent" to refer to any substance external to the human body that potentially causes disease or other health effects. Thus, drugs, devices, chemicals, radiation, and minerals (e.g., asbestos) are all agents whose toxicity an epidemiologist might explore. A single agent or a number of independent agents may cause disease, or the combined presence of two or more agents may be necessary for the development of the disease. Epidemiologists also conduct studies of individual characteristics, such as blood pressure and diet, which might pose risks, but those studies are rarely of interest in judicial proceedings. Epidemiologists also may conduct studies of drugs and other pharmaceutical products to assess their efficacy and safety.

caused a harmful effect or disease.[4] Epidemiologic evidence identifies agents that are associated with an increased risk of disease in groups of individuals, quantifies the amount of excess disease that is associated with an agent, and provides a profile of the type of individual who is likely to contract a disease after being exposed to an agent. Epidemiology focuses on the question of general causation (i.e., is the agent capable of causing disease?) rather than that of specific causation (i.e., did it cause disease in a particular individual?).[5] For example, in the 1950s, Doll and Hill and others published articles about the increased risk of lung cancer in cigarette smokers. Doll and Hill's studies showed that smokers who smoked 10 to 20 cigarettes a day had a lung cancer mortality rate that was about 10 times higher than that for nonsmokers.[6] These studies identified an association between smoking cigarettes and death from lung cancer that contributed to the determination that smoking causes lung cancer.

However, it should be emphasized that *an association is not equivalent to causation*.[7] An association identified in an epidemiologic study may or may not be

4. *E.g.*, Bonner v. ISP Techs., Inc., 259 F.3d 924 (8th Cir. 2001) (a worker exposed to organic solvents allegedly suffered organic brain dysfunction); Burton v. R.J. Reynolds Tobacco Co., 181 F. Supp. 2d 1256 (D. Kan. 2002) (cigarette smoking was alleged to have caused peripheral vascular disease); *In re* Bextra & Celebrex Mktg. Sales Practices & Prod. Liab. Litig., 524 F. Supp. 2d 1166 (N.D. Cal. 2007) (multidistrict litigation over drugs for arthritic pain that caused heart disease); Ruff v. Ensign-Bickford Indus., Inc., 168 F. Supp. 2d 1271 (D. Utah 2001) (chemicals that escaped from an explosives manufacturing site allegedly caused non-Hodgkin's lymphoma in nearby residents); Castillo v. E.I. du Pont De Nemours & Co., 854 So. 2d 1264 (Fla. 2003) (a child born with a birth defect allegedly resulting from mother's exposure to a fungicide).

5. This terminology and the distinction between general causation and specific causation are widely recognized in court opinions. *See, e.g.*, Norris v. Baxter Healthcare Corp., 397 F.3d 878 (10th Cir. 2005); *In re* Hanford Nuclear Reservation Litig., 292 F.3d 1124, 1129 (9th Cir. 2002) ("'Generic causation' has typically been understood to mean the capacity of a toxic agent . . . to cause the illnesses complained of by plaintiffs. If such capacity is established, 'individual causation' answers whether that toxic agent actually caused a particular plaintiff's illness."); *In re* Rezulin Prods. Liab. Litig., 369 F. Supp. 2d 398, 402 (S.D.N.Y. 2005); Soldo v. Sandoz Pharms. Corp., 244 F. Supp. 2d 434, 524–25 (W.D. Pa. 2003); Burton v. R.J. Reynolds Tobacco Co., 181 F. Supp. 2d 1256, 1266–67 (D. Kan. 2002). For a discussion of specific causation, see *infra* Section VII.

6. Richard Doll & A. Bradford Hill, *Lung Cancer and Other Causes of Death in Relation to Smoking: A Second Report on the Mortality of British Doctors,* 2 Brit. Med. J. 1071 (1956).

7. *See* Soldo v. Sandoz Pharms. Corp., 244 F. Supp. 2d 434, 461 (W.D. Pa. 2003) (Hill criteria [see *infra* Section V] developed to assess whether an association is causal); Miller v. Pfizer, Inc., 196 F. Supp. 2d 1062, 1079–80 (D. Kan. 2002); Magistrini v. One Hour Martinizing Dry Cleaning, 180 F. Supp. 2d 584, 591 (D.N.J. 2002) ("[A]n association is not equivalent to causation." (quoting the second edition of this reference guide)); Zandi v. Wyeth a/k/a Wyeth, Inc., No. 27-CV-06-6744, 2007 WL 3224242, at *11 (D. Minn. Oct. 15, 2007).

Association is more fully discussed *infra* Section III. The term is used to describe the relationship between two events (e.g., exposure to a chemical agent and development of disease) that occur more frequently together than one would expect by chance. Association does not necessarily imply a causal effect. Causation is used to describe the association between two events when one event is a necessary link in a chain of events that results in the effect. Of course, alternative causal chains may exist that do not include the agent but that result in the same effect. For general treatment of causation in tort law

causal.[8] Assessing whether an association is causal requires an understanding of the strengths and weaknesses of the study's design and implementation, as well as a judgment about how the study findings fit with other scientific knowledge. It is important to emphasize that all studies have "flaws" in the sense of limitations that add uncertainty about the proper interpretation of the results.[9] Some flaws are inevitable given the limits of technology, resources, the ability and willingness of persons to participate in a study, and ethical constraints. In evaluating epidemiologic evidence, the key questions, then, are the extent to which a study's limitations compromise its findings and permit inferences about causation.

A final caveat is that employing the results of group-based studies of risk to make a causal determination for an individual plaintiff is beyond the limits of epidemiology. Nevertheless, a substantial body of legal precedent has developed that addresses the use of epidemiologic evidence to prove causation for an individual litigant through probabilistic means, and the law developed in these cases is discussed later in this reference guide.[10]

The following sections of this reference guide address a number of critical issues that arise in considering the admissibility of, and weight to be accorded to, epidemiologic research findings. Over the past several decades, courts frequently have confronted the use of epidemiologic studies as evidence and have recognized their utility in proving causation. As the Third Circuit observed in *DeLuca v. Merrell Dow Pharmaceuticals, Inc.*: "The reliability of expert testimony founded on reasoning from epidemiologic data is generally a fit subject for judicial notice; epidemiology is a well-established branch of science and medicine, and epidemiologic evidence has been accepted in numerous cases."[11] Indeed,

and that for factual causation to exist an agent must be a necessary link in a causal chain sufficient for the outcome, see Restatement (Third) of Torts: Liability for Physical Harm § 26 (2010). Epidemiologic methods cannot deductively prove causation; indeed, all empirically based science cannot affirmatively prove a causal relation. *See, e.g.,* Stephan F. Lanes, *The Logic of Causal Inference in Medicine,* in Causal Inference 59 (Kenneth J. Rothman ed., 1988). However, epidemiologic evidence can justify an inference that an agent causes a disease. See *infra* Section V.

8. *See infra* Section IV.

9. *See In re* Phenylpropanolamine (PPA) Prods. Liab. Litig., 289 F. Supp. 2d 1230, 1240 (W.D. Wash. 2003) (quoting this reference guide and criticizing defendant's "ex post facto dissection" of a study); *In re* Orthopedic Bone Screw Prods. Liab. Litig., MDL No. 1014, 1997 U.S. Dist. LEXIS 6441, at *26–*27 (E.D. Pa. May 5, 1997) (holding that despite potential for several biases in a study that "may . . . render its conclusions inaccurate," the study was sufficiently reliable to be admissible); Joseph L. Gastwirth, *Reference Guide on Survey Research,* 36 Jurimetrics J. 181, 185 (1996) (review essay) ("One can always point to a potential flaw in a statistical analysis.").

10. *See infra* Section VII.

11. 911 F.2d 941, 954 (3d Cir. 1990); *see also* Norris v. Baxter Healthcare Corp., 397 F.3d 878, 882 (10th Cir. 2005) (an extensive body of exonerative epidemiologic evidence must be confronted and the plaintiff must provide scientifically reliable contrary evidence); *In re* Meridia Prods. Liab. Litig., 328 F. Supp. 2d 791, 800 (N.D. Ohio 2004) ("Epidemiologic studies are the primary generally accepted methodology for demonstrating a causal relation between the chemical compound and a set of symptoms or a disease. . . ." (quoting Conde v. Velsicol Chem. Corp., 804 F. Supp. 972,

much more difficult problems arise for courts when there is a paucity of epidemiologic evidence.[12]

Three basic issues arise when epidemiology is used in legal disputes, and the methodological soundness of a study and its implications for resolution of the question of causation must be assessed:

1. Do the results of an epidemiologic study or studies reveal an association between an agent and disease?
2. Could this association have resulted from limitations of the study (bias, confounding, or sampling error), and, if so, from which?
3. Based on the analysis of limitations in Item 2, above, and on other evidence, how plausible is a causal interpretation of the association?

Section II explains the different kinds of epidemiologic studies, and Section III addresses the meaning of their outcomes. Section IV examines concerns about the methodological validity of a study, including the problem of sampling error.[13] Section V discusses general causation, considering whether an agent is capable of causing disease. Section VI deals with methods for combining the results of multiple epidemiologic studies and the difficulties entailed in extracting a single global measure of risk from multiple studies. Additional legal questions that arise in most toxic substances cases are whether population-based epidemiologic evidence can be used to infer specific causation, and, if so, how. Section VII addresses specific causation—the matter of whether a specific agent caused the disease in a given plaintiff.

1025–26 (S.D. Ohio 1992))); Brasher v. Sandoz Pharms. Corp., 160 F. Supp. 2d 1291, 1296 (N.D. Ala. 2001) ("Unquestionably, epidemiologic studies provide the best proof of the general association of a particular substance with particular effects, but it is not the only scientific basis on which those effects can be predicted.").

12. *See infra* note 181.

13. For a more in-depth discussion of the statistical basis of epidemiology, see David H. Kaye & David A. Freedman, Reference Guide on Statistics, Section II.A, in this manual, and two case studies: Joseph Sanders, *The Bendectin Litigation: A Case Study in the Life Cycle of Mass Torts*, 43 Hastings L.J. 301 (1992); Devra L. Davis et al., *Assessing the Power and Quality of Epidemiologic Studies of Asbestos-Exposed Populations*, 1 Toxicological & Indus. Health 93 (1985). *See also* References on Epidemiology and References on Law and Epidemiology at the end of this reference guide.

II. What Different Kinds of Epidemiologic Studies Exist?

A. Experimental and Observational Studies of Suspected Toxic Agents

To determine whether an agent is related to the risk of developing a certain disease or an adverse health outcome, we might ideally want to conduct an experimental study in which the subjects would be randomly assigned to one of two groups: one group exposed to the agent of interest and the other not exposed. After a period of time, the study participants in both groups would be evaluated for the development of the disease. This type of study, called a randomized trial, clinical trial, or true experiment, is considered the gold standard for determining the relationship of an agent to a health outcome or adverse side effect. Such a study design is often used to evaluate new drugs or medical treatments and is the best way to ensure that any observed difference in outcome between the two groups is likely to be the result of exposure to the drug or medical treatment.

Randomization minimizes the likelihood that there are differences in relevant characteristics between those exposed to the agent and those not exposed. Researchers conducting clinical trials attempt to use study designs that are placebo controlled, which means that the group not receiving the active agent or treatment is given an inactive ingredient that appears similar to the active agent under study. They also use double blinding where possible, which means that neither the participants nor those conducting the study know which group is receiving the agent or treatment and which group is given the placebo. However, ethical and practical constraints limit the use of such experimental methodologies to assess the value of agents that are thought to be beneficial to human beings.[14]

When an agent's effects are suspected to be harmful, researchers cannot knowingly expose people to the agent.[15] Instead epidemiologic studies typically

14. Although experimental human studies cannot intentionally expose subjects to toxins, they can provide evidence that a new drug or other beneficial intervention also has adverse effects. *See In re* Bextra & Celebrex Mktg. Sales Practices & Prod. Liab. Litig., 524 F. Supp. 2d 1166, 1181 (N.D. Cal. 2007) (the court relied on a clinical study of Celebrex that revealed increased cardiovascular risk to conclude that the plaintiff's experts' testimony on causation was admissible); McDarby v. Merck & Co., 949 A.2d 223 (N.J. Super. Ct. App. Div. 2008) (explaining how clinical trials of Vioxx revealed an association with heart disease).

15. Experimental studies in which human beings are exposed to agents known or thought to be toxic are ethically proscribed. *See* Glastetter v. Novartis Pharms. Corp., 252 F.3d 986, 992 (8th Cir. 2001); Brasher v. Sandoz Pharms. Corp., 160 F. Supp. 2d 1291, 1297 (N.D. Ala. 2001). Experimental studies can be used where the agent under investigation is believed to be beneficial, as is the case in the development and testing of new pharmaceutical drugs. *See, e.g.*, McDarby v. Merck & Co., 949 A.2d 223, 270 (N.J. Super. Ct. App. Div. 2008) (an expert witness relied on a clinical trial of a new drug to find the adjusted risk for the plaintiff); *see also* Gordon H. Guyatt, *Using Randomized Trials in*

"observe"[16] a group of individuals who have been exposed to an agent of interest, such as cigarette smoke or an industrial chemical and compare them with another group of individuals who have not been exposed. Thus, the investigator identifies a group of subjects who have been exposed[17] and compares their rate of disease or death with that of an unexposed group. In contrast to clinical studies in which potential risk factors can be controlled, epidemiologic investigations generally focus on individuals living in the community, for whom characteristics other than the one of interest, such as diet, exercise, exposure to other environmental agents, and genetic background, may distort a study's results. Because these characteristics cannot be controlled directly by the investigator, the investigator addresses their possible role in the relationship being studied by considering them in the design of the study and in the analysis and interpretation of the study results (see *infra* Section IV).[18] We emphasize that the Achilles' heel of observational studies is the possibility of differences in the two populations being studied with regard to risk factors other than exposure to the agent.[19] By contrast, experimental studies, in which subjects are randomized, generally avoid this problem.

B. Types of Observational Study Design

Several different types of observational epidemiologic studies can be conducted.[20] Study designs may be chosen because of suitability for investigating the question of interest, timing constraints, resource limitations, or other considerations.

Most observational studies collect data about both exposure and health outcome in every individual in the study. The two main types of observational studies are cohort studies and case-control studies. A third type of observational study is a cross-sectional study, although cross-sectional studies are rarely useful in identifying toxic agents.[21] A final type of observational study, one in which data about

Pharmacoepidemiology, in Drug Epidemiology and Post-Marketing Surveillance 59 (Brian L. Strom & Giampaolo Velo eds., 1992). Experimental studies also may be conducted that entail the discontinuation of exposure to a harmful agent, such as studies in which smokers are randomly assigned to a variety of smoking cessation programs or have no cessation.

16. Classifying these studies as observational in contrast to randomized trials can be misleading to those who are unfamiliar with the area, because subjects in a randomized trial are observed as well. Nevertheless, the use of the term "observational studies" to distinguish them from experimental studies is widely employed.

17. The subjects may have voluntarily exposed themselves to the agent of interest, as is the case, for example, for those who smoke cigarettes, or subjects may have been exposed involuntarily or even without knowledge to an agent, such as in the case of employees who are exposed to chemical fumes at work.

18. *See* David A. Freedman, *Oasis or Mirage?* 21 Chance 59, 59–61 (Mar. 2008).

19. Both experimental and observational studies are subject to random error. See *infra* Section IV.A.

20. Other epidemiologic studies collect data about the group as a whole, rather than about each individual in the group. These group studies are discussed *infra* Section II.B.4.

21. See *infra* Section II.B.3.

individuals are not gathered, but rather population data about exposure and disease are used, is an ecological study.[22]

The difference between cohort studies and case-control studies is that cohort studies measure and compare the incidence of disease in the exposed and unexposed ("control") groups, while case-control studies measure and compare the frequency of exposure in the group with the disease (the "cases") and the group without the disease (the "controls"). In a case-control study, the rates of exposure in the cases and the rates in the controls are compared, and the odds of having the disease when exposed to a suspected agent can be compared with the odds when not exposed. The critical difference between cohort studies and case-control studies is that cohort studies begin with exposed people and unexposed people, while case-control studies begin with individuals who are selected based on whether they have the disease or do not have the disease and their exposure to the agent in question is measured. The goal of both types of studies is to determine if there is an association between exposure to an agent and a disease and the strength (magnitude) of that association.

1. Cohort studies

In cohort studies,[23] researchers define a study population without regard to the participants' disease status. The cohort may be defined in the present and followed forward into the future (prospectively) or it may be constructed retrospectively as of sometime in the past and followed over historical time toward the present. In either case, the researchers classify the study participants into groups based on whether they were exposed to the agent of interest (see Figure 1).[24] In a prospective study, the exposed and unexposed groups are followed for a specified length of time, and the proportions of individuals in each group who develop the disease of interest are compared. In a retrospective study, the researcher will determine the proportion of individuals in the exposed group who developed the disease from available records or evidence and compare that proportion with the proportion of another group that was not exposed.[25] Thus, as illustrated in Table 1,

22. For thumbnail sketches on all types of epidemiologic study designs, see Brian L. Strom, *Study Designs Available for Pharmacoepidemiology Studies, in* Pharmacoepidemiology 17, 21–26 (Brian L. Strom ed., 4th ed. 2005).

23. Cohort studies also are referred to as prospective studies and followup studies.

24. In some studies, there may be several groups, each with a different magnitude of exposure to the agent being studied. Thus, a study of cigarette smokers might include heavy smokers (>3 packs a day), moderate smokers (1 to 2 packs a day), and light smokers (<1 pack a day). *See, e.g.,* Robert A. Rinsky et al., *Benzene and Leukemia: An Epidemiologic Risk Assessment,* 316 New Eng. J. Med. 1044 (1987).

25. Sometimes in retrospective cohort studies the researcher gathers historical data about exposure and disease outcome of a cohort. Harold A. Kahn, An Introduction to Epidemiologic Methods 39–41 (1983). Irving Selikoff, in his seminal study of asbestotic disease in insulation workers, included several hundred workers who had died before he began the study. Selikoff was able to obtain information about exposure from union records and information about disease from hospital and autopsy

Figure 1. Design of a cohort study.

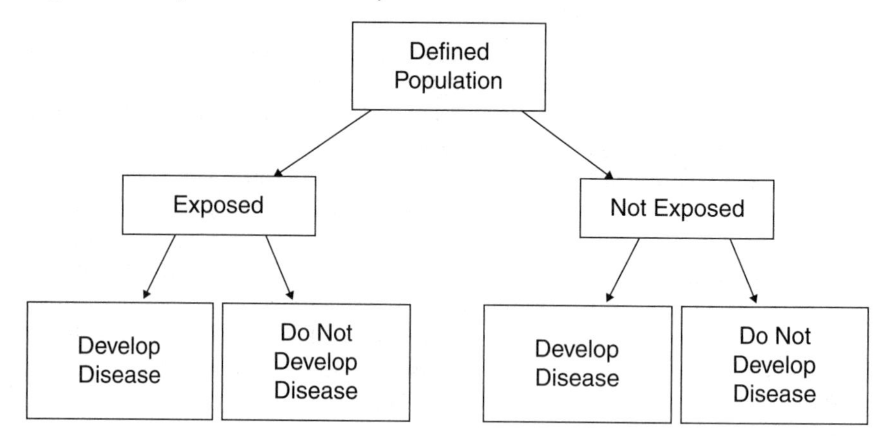

Table 1. Cross-Tabulation of Exposure by Disease Status

	No Disease	Disease	Totals	Incidence Rates of Disease
Not exposed	a	c	$a + c$	$c/(a + c)$
Exposed	b	d	$b + d$	$d/(b + d)$

a researcher would compare the proportion of unexposed individuals with the disease, $c/(a + c)$, with the proportion of exposed individuals with the disease, $d/(b + d)$. If the exposure causes the disease, the researcher would expect a greater proportion of the exposed individuals to develop the disease than the unexposed individuals.[26]

One advantage of the cohort study design is that the temporal relationship between exposure and disease can often be established more readily than in other study designs, especially a case-control design, discussed below. By tracking people who are initially not affected by the disease, the researcher can determine the time of disease onset and its relation to exposure. This temporal relationship is critical to the question of causation, because exposure must precede disease onset if exposure caused the disease.

As an example, in 1950 a cohort study was begun to determine whether uranium miners exposed to radon were at increased risk for lung cancer as com-

records. Irving J. Selikoff et al., *The Occurrence of Asbestosis Among Insulation Workers in the United States,* 132 Ann. N.Y. Acad. Sci. 139, 143 (1965).

26. Researchers often examine the rate of disease or death in the exposed and control groups. The rate of disease or death entails consideration of the number developing disease within a specified period. All smokers and nonsmokers will, if followed for 100 years, die. Smokers will die at a greater rate than nonsmokers in the earlier years.

pared with nonminers. The study group (also referred to as the exposed cohort) consisted of 3400 white, underground miners. The control group (which need not be the same size as the exposed cohort) comprised white nonminers from the same geographic area. Members of the exposed cohort were examined every 3 years, and the degree of this cohort's exposure to radon was measured from samples taken in the mines. Ongoing testing for radioactivity and periodic medical monitoring of lungs permitted the researchers to examine whether disease was linked to prior work exposure to radiation and allowed them to discern the relationship between exposure to radiation and disease. Exposure to radiation was associated with the development of lung cancer in uranium miners.[27]

The cohort design is used often in occupational studies such as the one just discussed. Because the design is not experimental, and the investigator has no control over what other exposures a subject in the study may have had, an increased risk of disease among the exposed group may be caused by agents other than the exposure of interest. A cohort study of workers in a certain industry that pays below-average wages might find a higher risk of cancer in those workers. This may be because they work in that industry, or, among other reasons, because low-wage groups are exposed to other harmful agents, such as environmental toxins present in higher concentrations in their neighborhoods. In the study design, the researcher must attempt to identify factors other than the exposure that may be responsible for the increased risk of disease. If data are gathered on other possible etiologic factors, the researcher generally uses statistical methods[28] to assess whether a true association exists between working in the industry and cancer. Evaluating whether the association is causal involves additional analysis, as discussed in Section V.

2. Case-control studies

In case-control studies,[29] the researcher begins with a group of individuals who have a disease (cases) and then selects a similar group of individuals who do not have the disease (controls). (Ideally, controls should come from the same source population as the cases.) The researcher then compares the groups in terms of past exposures. If a certain exposure is associated with or caused the disease, a higher proportion of past exposure among the cases than among the controls would be expected (see Figure 2).

27. This example is based on a study description in Abraham M. Lilienfeld & David E. Lilienfeld, Foundations of Epidemiology 237–39 (2d ed. 1980). The original study is Joseph K. Wagoner et al., *Radiation as the Cause of Lung Cancer Among Uranium Miners,* 273 New Eng. J. Med. 181 (1965).

28. *See* Daniel L. Rubinfeld, Reference Guide on Multiple Regression, Section II.B, in this manual; David H. Kaye & David A. Freedman, Reference Guide on Statistics, Section V.D, in this manual.

29. Case-control studies are also referred to as retrospective studies, because researchers gather historical information about rates of exposure to an agent in the case and control groups.

Figure 2. Design of a case-control study.

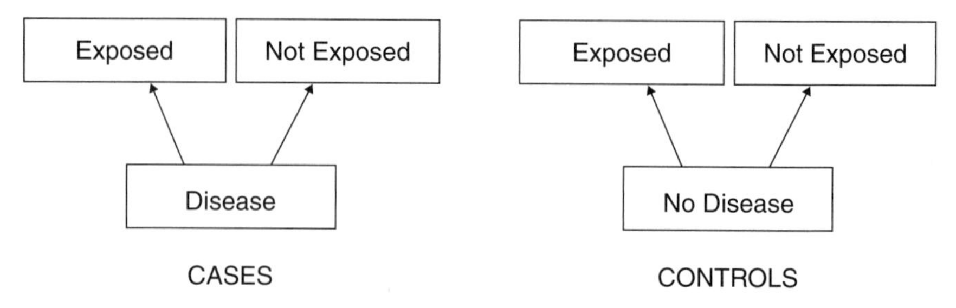

Thus, for example, in the late 1960s, doctors in Boston were confronted with an unusual number of young female patients with vaginal adenocarcinoma. Those patients became the "cases" in a case-control study (because they had the disease in question) and were matched with "controls," who did not have the disease. Controls were selected based on their being born in the same hospitals and at the same time as the cases. The cases and controls were compared for exposure to agents that might be responsible, and researchers found maternal ingestion of DES (diethylstilbestrol) in all but one of the cases but none of the controls.[30]

An advantage of the case-control study is that it usually can be completed in less time and with less expense than a cohort study. Case-control studies are also particularly useful in the study of rare diseases, because if a cohort study were conducted, an extremely large group would have to be studied in order to observe the development of a sufficient number of cases for analysis.[31] A number of potential problems with case-control studies are discussed in Section IV.B.

3. Cross-sectional studies

A third type of observational study is a cross-sectional study. In this type of study, individuals are interviewed or examined, and the presence of both the exposure of interest and the disease of interest is determined in each individual at a single point in time. Cross-sectional studies determine the presence (prevalence) of both exposure and disease in the subjects and do not determine the development of disease or risk of disease (incidence). Moreover, because both exposure and disease are determined in an individual at the same point in time, it is not possible to establish the temporal relation between exposure and disease—that is, that the

30. *See* Arthur L. Herbst et al., *Adenocarcinoma of the Vagina: Association of Maternal Stilbestrol Therapy with Tumor Appearance,* 284 New Eng. J. Med. 878 (1971).

31. Thus, for example, to detect a doubling of disease caused by exposure to an agent where the incidence of disease is 1 in 100 in the unexposed population would require sample sizes of 3100 for the exposed and nonexposed groups for a cohort study, but only 177 for the case and control groups in a case-control study. Harold A. Kahn & Christopher T. Sempos, Statistical Methods in Epidemiology 66 (1989).

exposure preceded the disease, which would be necessary for drawing any causal inference. Thus, a researcher may use a cross-sectional study to determine the connection between a personal characteristic that does not change over time, such as blood type, and existence of a disease, such as aplastic anemia, by examining individuals and determining their blood types and whether they suffer from aplastic anemia. Cross-sectional studies are infrequently used when the exposure of interest is an environmental toxic agent (current smoking status is a poor measure of an individual's history of smoking), but these studies can provide valuable leads to further directions for research.[32]

4. *Ecological studies*

Up to now, we have discussed studies in which data on both exposure and health outcome are obtained for each individual included in the study.[33] In contrast, studies that collect data only about the group as a whole are called ecological studies.[34] In ecological studies, information about individuals is generally not gathered; instead, overall rates of disease or death for different groups are obtained and compared. The objective is to identify some difference between the two groups, such as diet, genetic makeup, or alcohol consumption, that might explain differences in the risk of disease observed in the two groups.[35] Such studies may be useful for identifying associations, but they rarely provide definitive causal answers.[36] The difficulty is illustrated below with an ecological study of the relationship between dietary fat and cancer.

32. For more information (and references) about cross-sectional studies, see Leon Gordis, Epidemiology 195–98 (4th ed. 2009).

33. Some individual studies may be conducted in which all members of a group or community are treated as exposed to an agent of interest (e.g., a contaminated water system) and disease status is determined individually. These studies should be distinguished from ecological studies.

34. In *Cook v. Rockwell International Corp.*, 580 F. Supp. 2d 1071, 1095–96 (D. Colo. 2006), the plaintiffs' expert conducted an ecological study in which he compared the incidence of two cancers among those living in a specified area adjacent to the Rocky Flats Nuclear Weapons Plant with other areas more distant. (The likely explanation for relying on this type of study is the time and expense of a study that gathered information about each individual in the affected area.) The court recognized that ecological studies are less probative than studies in which data are based on individuals but nevertheless held that limitation went to the weight of the study. Plaintiff's expert was permitted to testify to causation, relying on the ecological study he performed.

In *Renaud v. Martin Marietta Corp.*, 749 F. Supp. 1545, 1551 (D. Colo. 1990), *aff'd*, 972 F.2d 304 (10th Cir. 1992), the plaintiffs attempted to rely on an excess incidence of cancers in their neighborhood to prove causation. Unfortunately, the court confused the role of epidemiology in proving causation with the issue of the plaintiffs' exposure to the alleged carcinogen and never addressed the evidentiary value of the plaintiffs' evidence of a disease cluster (i.e., an unusually high incidence of a particular disease in a neighborhood or community). *Id.* at 1554.

35. David E. Lilienfeld & Paul D. Stolley, Foundations of Epidemiology 12 (3d ed. 1994).

36. Thus, the emergence of a cluster of adverse events associated with use of heparin, a longtime and widely-prescribed anticoagulent, led to suspicions that some specific lot of heparin was responsible. These concerns led the Centers for Disease Control to conduct a case control study that concluded

If a researcher were interested in determining whether a high dietary fat intake is associated with breast cancer, he or she could compare different countries in terms of their average fat intakes and their average rates of breast cancer. If a country with a high average fat intake also tends to have a high rate of breast cancer, the finding would suggest an association between dietary fat and breast cancer. However, such a finding would be far from conclusive, because it lacks particularized information about an individual's exposure and disease status (i.e., whether an individual with high fat intake is more likely to have breast cancer).[37] In addition to the lack of information about an individual's intake of fat, the researcher does not know about the individual's exposures to other agents (or other factors, such as a mother's age at first birth) that may also be responsible for the increased risk of breast cancer. This lack of information about each individual's exposure to an agent and disease status detracts from the usefulness of the study and can lead to an erroneous inference about the relationship between fat intake and breast cancer, a problem known as an ecological fallacy. The fallacy is assuming that, on average, the individuals in the study who have suffered from breast cancer consumed more dietary fat than those who have not suffered from the disease. This assumption may not be true. Nevertheless, the study is useful in that it identifies an area for further research: the fat intake of individuals who have breast cancer as compared with the fat intake of those who do not. Researchers who identify a difference in disease or death in an ecological study may follow up with a study based on gathering data about individuals.

Another epidemiologic approach is to compare disease rates over time and focus on disease rates before and after a point in time when some event of interest took place.[38] For example, thalidomide's teratogenicity (capacity to cause birth defects) was discovered after Dr. Widukind Lenz found a dramatic increase in the incidence of limb reduction birth defects in Germany beginning in 1960. Yet, other than with such powerful agents as thalidomide, which increased the incidence of limb reduction defects by several orders of magnitude, these secular-trend studies (also known as time-line studies) are less reliable and less able to

that contaminated heparin manufactured by Baxter was responsible for the outbreak of adverse events. See David B. Blossom et al., *Outbreak of Adverse Event Reactions Associated with Contaminated Heparin,* 359 New Eng. J. Med. 2674 (2008); *In re* Heparin Prods. Liab. Litig. 2011 WL 2971918 (N.D. Ohio July 21, 2011).

37. For a discussion of the data on this question and what they might mean, see David Freedman et al., Statistics (4th ed. 2007).

38. In *Wilson v. Merrell Dow Pharmaceuticals, Inc.,* 893 F.2d 1149, 1152–53 (10th Cir. 1990), the defendant introduced evidence showing total sales of Bendectin and the incidence of birth defects during the 1970–1984 period. In 1983, Bendectin was removed from the market, but the rate of birth defects did not change. The Tenth Circuit affirmed the lower court's ruling that the time-line data were admissible and that the defendant's expert witnesses could rely on them in rendering their opinions. Similar evidence was relied on in cases involving cell phones and the drug Parlodel, which was alleged to cause postpartum strokes in women who took the drug to suppress lactation. *See* Newman v. Motorola, Inc., 218 F. Supp. 2d 769, 778 (D. Md. 2002); Siharath v. Sandoz Pharms. Corp., 131 F. Supp. 2d 1347, 1358 (N.D. Ga. 2001).

detect modest causal effects than the observational studies described above. Other factors that affect the measurement or existence of the disease, such as improved diagnostic techniques and changes in lifestyle or age demographics, may change over time. If those factors can be identified and measured, it may be possible to control for them with statistical methods. Of course, unknown factors cannot be controlled for in these or any other kind of epidemiologic studies.

C. Epidemiologic and Toxicologic Studies

In addition to observational epidemiology, toxicology models based on live animal studies (in vivo) may be used to determine toxicity in humans.[39] Animal studies have a number of advantages. They can be conducted as true experiments, and researchers control all aspects of the animals' lives. Thus, they can avoid the problem of confounding,[40] which epidemiology often confronts. Exposure can be carefully controlled and measured. Refusals to participate in a study are not an issue, and loss to followup very often is minimal. Ethical limitations are diminished, and animals can be sacrificed and their tissues examined, which may improve the accuracy of disease assessment. Animal studies often provide useful information about pathological mechanisms and play a complementary role to epidemiology by assisting researchers in framing hypotheses and in developing study designs for epidemiologic studies.

Animal studies have two significant disadvantages, however. First, animal study results must be extrapolated to another species—human beings—and differences in absorption, metabolism, and other factors may result in interspecies variation in responses. For example, one powerful human teratogen, thalidomide, does not cause birth defects in most rodent species.[41] Similarly, some known teratogens in animals are not believed to be human teratogens. In general, it is often difficult to confirm that an agent known to be toxic in animals is safe for human beings.[42] The second difficulty with inferring human causation from animal studies is that the high doses customarily used in animal studies require consideration of the dose–response relationship and whether a threshold no-effect dose exists.[43] Those matters are almost always fraught with considerable, and currently unresolvable, uncertainty.[44]

39. For an in-depth discussion of toxicology, see Bernard D. Goldstein & Mary Sue Henifin, Reference Guide on Toxicology, in this manual.

40. *See infra* Section IV.C.

41. Phillip Knightley et al., Suffer the Children: The Story of Thalidomide 271–72 (1979).

42. *See* Ian C.T. Nesbit & Nathan J. Karch, Chemical Hazards to Human Reproduction 98–106 (1983); Int'l Agency for Research on Cancer (IARC), Interpretation of Negative Epidemiologic Evidence for Carcinogenicity (N.J. Wald & Richard Doll eds., 1985) [hereafter IARC].

43. *See infra* Section V.C & note 119.

44. *See* Soldo v. Sandoz Pharms. Corp., 244 F. Supp. 2d 434, 466 (W.D. Pa. 2003) (quoting this reference guide in the first edition of the Reference Manual); *see also* General Elec. Co. v. Joiner, 522 U.S. 136, 143–45 (1997) (holding that the district court did not abuse its discretion in exclud-

Toxicologists also use in vitro methods, in which human or animal tissue or cells are grown in laboratories and are exposed to certain substances. The problem with this approach is also extrapolation—whether one can generalize the findings from the artificial setting of tissues in laboratories to whole human beings.[45]

Often toxicologic studies are the only or best available evidence of toxicity.[46] Epidemiologic studies are difficult, time-consuming, expensive, and sometimes, because of limited exposure or the infrequency of disease, virtually impossible to perform.[47] Consequently, they do not exist for a large array of environmental agents. Where both animal toxicologic and epidemiologic studies are available, no universal rules exist for how to interpret or reconcile them.[48] Careful assess-

ing expert testimony on causation based on expert's failure to explain how animal studies supported expert's opinion that agent caused disease in humans).

45. For a further discussion of these issues, see Bernard D. Goldstein & Mary Sue Henifin, Reference Guide on Toxicology, Section III.A, in this manual.

46. IARC, a well-regarded international public health agency, evaluates the human carcinogenicity of various agents. In doing so, IARC obtains all of the relevant evidence, including animal studies as well as any human studies. On the basis of a synthesis and evaluation of that evidence, IARC publishes a monograph containing that evidence and its analysis of the evidence and provides a categorical assessment of the likelihood the agent is carcinogenic. In a preamble to each of its monographs, IARC explains what each of the categorical assessments means. Solely on the basis of the strength of animal studies, IARC may classify a substance as "probably carcinogenic to humans." International Agency for Research on Cancer, *Human Papillomaviruses,* 90 Monographs on the Evaluation of Carcinogenic Risks to Humans 9–10 (2007), *available at* http://monographs.iarc.fr/ENG/Monographs/vol90/index.php; *see also* Magistrini v. One Hour Martinizing Dry Cleaning, 180 F. Supp. 2d 584, 600 n.18 (D.N.J. 2002). When IARC monographs are available, they are generally recognized as authoritative. Unfortunately, IARC has conducted evaluations of only a fraction of potentially carcinogenic agents, and many suspected toxic agents cause effects other than cancer.

47. Thus, in a series of cases involving Parlodel, a lactation suppressant for mothers of newborns, efforts to conduct an epidemiologic study of its effect on causing strokes were stymied by the infrequency of such strokes in women of child-bearing age. *See, e.g.,* Brasher v. Sandoz Pharms. Corp., 160 F. Supp. 2d 1291, 1297 (N.D. Ala. 2001). In other cases, a plaintiff's exposure to an overdose of a drug may be unique or nearly so. *See* Zuchowicz v. United States, 140 F.3d 381 (2d Cir. 1998).

48. *See* IARC, *supra* note 41 (identifying a number of substances and comparing animal toxicology evidence with epidemiologic evidence); Michele Carbone et al., *Modern Criteria to Establish Human Cancer Etiology,* 64 Cancer Res. 5518, 5522 (2004) (National Cancer Institute symposium concluding that "There should be no hierarchy [among different types of scientific methods to determine cancer causation]. Epidemiology, animal, tissue culture and molecular pathology should be seen as integrating evidences in the determination of human carcinogenicity.")

A number of courts have grappled with the role of animal studies in proving causation in a toxic substance case. One line of cases takes a very dim view of their probative value. For example, in *Brock v. Merrell Dow Pharmaceuticals, Inc.,* 874 F.2d 307, 313 (5th Cir. 1989), the court noted the "very limited usefulness of animal studies when confronted with questions of toxicity." A similar view is reflected in *Richardson v. Richardson-Merrell, Inc.,* 857 F.2d 823, 830 (D.C. Cir. 1988), Bell v. Swift Adhesives, Inc., 804 F. Supp. 1577, 1579–80 (S.D. Ga. 1992), and *Cadarian v. Merrell Dow Pharmaceuticals, Inc.,* 745 F. Supp. 409, 412 (E.D. Mich. 1989).

Other courts have been more amenable to the use of animal toxicology in proving causation. Thus, in *Marder v. G.D. Searle & Co.,* 630 F. Supp. 1087, 1094 (D. Md. 1986), *aff'd sub nom. Wheelahan v. G.D. Searle & Co.,* 814 F.2d 655 (4th Cir. 1987), the court observed: "There is a range of scientific

ment of the methodological validity and power[49] of the epidemiologic evidence must be undertaken, and the quality of the toxicologic studies and the questions of interspecies extrapolation and dose–response relationship must be considered.[50]

methods for investigating questions of causation—for example, toxicology and animal studies, clinical research, and epidemiology—which all have distinct advantages and disadvantages." In *Milward v. Acuity Specialty Products Group, Inc.,* 639 F.3d 11, 17-19 (1st Cir. 2011), the court endorsed an expert's use of a "weight-of-the-evidence" methodology, holding that the district court abused its discretion in ruling inadmissible an expert's testimony about causation based on that methodology. As a corollary to recognizing weight of the evidence as a valid scientific technique, the court also noted the role of judgment in making an appropriate inference from the evidence. While recognizing the legitimacy of the methodology, the court also acknowledged that, as with any scientific technique, it can be improperly applied. *See also* Metabolife Int'l, Inc. v. Wornick, 264 F.3d 832, 842 (9th Cir. 2001) (holding that the lower court erred in per se dismissing animal studies, which must be examined to determine whether they are appropriate as a basis for causation determination); *In re* Heparin Prods. Liab. Litig. 2011 WL 2971918 (N.D. Ohio July 21, 2011) (holding that animal toxicology in conjunction with other non-epidemiologic evidence can be sufficient to prove causation); Ruff v. Ensign-Bickford Indus., Inc., 168 F. Supp. 2d 1271, 1281 (D. Utah 2001) (affirming animal studies as sufficient basis for opinion on general causation.); *cf. In re* Paoli R.R. Yard PCB Litig., 916 F.2d 829, 853–54 (3d Cir. 1990) (questioning the exclusion of animal studies by the lower court). The Third Circuit in a subsequent opinion in *Paoli* observed:

> [I]n order for animal studies to be admissible to prove causation in humans, there must be good grounds to extrapolate from animals to humans, just as the methodology of the studies must constitute good grounds to reach conclusions about the animals themselves. Thus, the requirement of reliability, or "good grounds," extends to each step in an expert's analysis all the way through the step that connects the work of the expert to the particular case.

In re Paoli R.R. Yard PCB Litig., 35 F.3d 717, 743 (3d Cir. 1994); *see also* Cavallo v. Star Enter., 892 F. Supp. 756, 761–63 (E.D. Va. 1995) (courts must examine each of the steps that lead to an expert's opinion), *aff'd in part and rev'd in part,* 100 F.3d 1150 (4th Cir. 1996).

One explanation for these conflicting lines of cases may be that when there is a substantial body of epidemiologic evidence that addresses the causal issue, animal toxicology has much less probative value. That was the case, for example, in the Bendectin cases of *Richardson, Brock,* and *Cadarian.* Where epidemiologic evidence is not available, animal toxicology may be thought to play a more prominent role in resolving a causal dispute. *See* Michael D. Green, *Expert Witnesses and Sufficiency of Evidence in Toxic Substances Litigation: The Legacy of Agent Orange and Bendectin Litigation,* 86 Nw. U. L. Rev. 643, 680–82 (1992) (arguing that plaintiffs should be required to prove causation by a preponderance of the available evidence); Turpin v. Merrell Dow Pharms., Inc., 959 F.2d 1349, 1359 (6th Cir. 1992); *In re* Paoli R.R. Yard PCB Litig., No. 86-2229, 1992 U.S. Dist. LEXIS 16287, at *16 (E.D. Pa. 1992). For another explanation of these cases, see Gerald W. Boston, *A Mass-Exposure Model of Toxic Causation: The Control of Scientific Proof and the Regulatory Experience,* 18 Colum. J. Envtl. L. 181 (1993) (arguing that epidemiologic evidence should be required in mass-exposure cases but not in isolated-exposure cases); *see also* IARC, *supra* note 41; Bernard D. Goldstein & Mary Sue Henifin, Reference Guide on Toxicology, Section I.F, in this manual. The Supreme Court, in *General Electric Co. v. Joiner,* 522 U.S. 136, 144–45 (1997), suggested that there is no categorical rule for toxicologic studies, observing, "[W]hether animal studies can ever be a proper foundation for an expert's opinion [is] not the issue. . . . The [animal] studies were so dissimilar to the facts presented in this litigation that it was not an abuse of discretion for the District Court to have rejected the experts' reliance on them."

49. *See infra* Section IV.A.3.

50. *See* Ellen F. Heineman & Shelia Hoar Zahm, *The Role of Epidemiology in Hazard Evaluation,* 9 Toxic Substances J. 255, 258–62 (1989).

III. How Should Results of an Epidemiologic Study Be Interpreted?

Epidemiologists are ultimately interested in whether a causal relationship exists between an agent and a disease. However, the first question an epidemiologist addresses is whether an association exists between exposure to the agent and disease. An association between exposure to an agent and disease exists when they occur together more frequently than one would expect by chance.[51] Although a causal relationship is one possible explanation for an observed association between an exposure and a disease, an association does not necessarily mean that there is a cause–effect relationship. Interpreting the meaning of an observed association is discussed below.

This section begins by describing the ways of expressing the existence and strength of an association between exposure and disease. It reviews ways in which an incorrect result can be produced because of the sampling methods used in all observational epidemiologic studies and then examines statistical methods for evaluating whether an association is real or the result of a sampling error.

The strength of an association between exposure and disease can be stated in various ways,[52] including as a relative risk, an odds ratio, or an attributable risk.[53] Each of these measurements of association examines the degree to which the risk of disease increases when individuals are exposed to an agent.

A. Relative Risk

A commonly used approach for expressing the association between an agent and disease is relative risk ("RR"). It is defined as the ratio of the incidence rate (often referred to as incidence) of disease in exposed individuals to the incidence rate in unexposed individuals:

$$\text{RR} = \frac{\text{(Incidence rate in the exposed)}}{\text{(Incidence rate in the unexposed)}}$$

51. A negative association implies that the agent has a protective or curative effect. Because the concern in toxic substances litigation is whether an agent caused disease, this reference guide focuses on positive associations.

52. Another outcome measure is a risk difference. A risk difference is the difference between the proportion of disease in those exposed to the agent and the proportion of disease in those who were unexposed. Thus, in the example of relative risk in the text below discussing relative risk, the proportion of disease in those exposed is 40/100 and the proportion of disease in the unexposed is 20/100. The risk difference is 20/100.

53. Numerous courts have employed these measures of the strength of an association. *See, e.g., In re* Bextra & Celebrex Mktg. Sales Practices & Prod. Liab. Litig., 524 F. Supp. 2d 1166, 1172–74 (N.D. Cal. 2007); Cook v. Rockwell Int'l Corp., 580 F. Supp. 2d 1071, 1095 (D. Colo. 2006) (citing the second edition of this reference guide); *In re* W.R. Grace & Co., 355 B.R. 462, 482–83 (Bankr. D. Del. 2006).

The incidence rate of disease is defined as the number of cases of disease that develop during a specified period of time divided by the number of persons in the cohort under study.[54] Thus, the incidence rate expresses the risk that a member of the population will develop the disease within a specified period of time.

For example, a researcher studies 100 individuals who are exposed to an agent and 200 who are not exposed. After 1 year, 40 of the exposed individuals are diagnosed as having a disease, and 20 of the unexposed individuals also are diagnosed as having the disease. The relative risk of contracting the disease is calculated as follows:

- The incidence rate of disease in the exposed individuals is 40 cases per year per 100 persons (40/100), or 0.4.
- The incidence rate of disease in the unexposed individuals is 20 cases per year per 200 persons (20/200), or 0.1.
- The relative risk is calculated as the incidence rate in the exposed group (0.4) divided by the incidence rate in the unexposed group (0.1), or 4.0.

A relative risk of 4.0 indicates that the risk of disease in the exposed group is four times as high as the risk of disease in the unexposed group.[55]

In general, the relative risk can be interpreted as follows:

- If the relative risk equals 1.0, the risk in exposed individuals is the same as the risk in unexposed individuals.[56] There is no association between exposure to the agent and disease.
- If the relative risk is greater than 1.0, the risk in exposed individuals is greater than the risk in unexposed individuals. There is a positive association between exposure to the agent and the disease, which could be causal.
- If the relative risk is less than 1.0, the risk in exposed individuals is less than the risk in unexposed individuals. There is a negative association, which could reflect a protective or curative effect of the agent on risk of disease. For example, immunizations lower the risk of disease. The results suggest that immunization is associated with a decrease in disease and may have a protective effect on the risk of disease.

Although relative risk is a straightforward concept, care must be taken in interpreting it. Whenever an association is uncovered, further analysis should be

54. Epidemiologists also use the concept of prevalence, which measures the existence of disease in a population at a given point in time, regardless of when the disease developed. Prevalence is expressed as the proportion of the population with the disease at the chosen time. *See* Gordis, *supra* note 32, at 43–47.

55. *See* DeLuca v. Merrell Dow Pharms., Inc., 911 F.2d 941, 947 (3d Cir. 1990); Magistrini v. One Hour Martinizing Dry Cleaning, 180 F. Supp. 2d 584, 591 (D.N.J. 2002).

56. *See Magistrini*, 180 F. Supp. 2d at 591.

conducted to assess whether the association is real or a result of sampling error, confounding, or bias.[57] These same sources of error may mask a true association, resulting in a study that erroneously finds no association.

B. Odds Ratio

The odds ratio ("OR") is similar to a relative risk in that it expresses in quantitative terms the association between exposure to an agent and a disease.[58] It is a convenient way to estimate the relative risk in a case-control study when the disease under investigation is rare.[59] The odds ratio approximates the relative risk when the disease is rare.[60]

In a case-control study, the odds ratio is the ratio of the odds that a case (one with the disease) was exposed to the odds that a control (one without the disease) was exposed. In a cohort study, the odds ratio is the ratio of the odds of developing a disease when exposed to a suspected agent to the odds of developing the disease when not exposed.

Consider a case-control study, with results as shown schematically in a 2 × 2 table (Table 2):

Table 2. Cross-tabulation of cases and controls by exposure status

	Cases (with disease)	Controls (no disease)
Exposed	*a*	*b*
Not exposed	*c*	*d*

In a case-control study,

$$\text{OR} = \frac{\text{(Odds that a case was exposed)}}{\text{(Odds that a control was exposed).}}$$

57. *See infra* Sections IV.B–C.

58. A relative risk cannot be calculated for a case-control study, because a case-control study begins by examining a group of persons who already have the disease. That aspect of the study design prevents a researcher from determining the rate at which individuals develop the disease. Without a rate or incidence of disease, a researcher cannot calculate a relative risk.

59. If the disease is not rare, the odds ratio is still valid to determine whether an association exists, but interpretation of its magnitude is less intuitive.

60. *See* Marcello Pagano & Kimberlee Gauvreau, Principles of Biostatistics 354 (2d ed. 2000). For further detail about the odds ratio and its calculation, see Kahn & Sempos, *supra* note 31, at 47–56.

Looking at Table 2, this ratio can be calculated as

$$\frac{(a/c)}{(b/d)}.$$

This works out to *ad/bc*. Because we are multiplying two diagonal cells in the table and dividing by the product of the other two diagonal cells, the odds ratio is also called the cross-products ratio.

Consider the following hypothetical study: A researcher identifies 100 individuals with a disease who serve as "cases" and 100 people without the disease who serve as "controls" for her case-control study. Forty of the 100 cases were exposed to the agent and 60 were not. Among the control group, 20 people were exposed and 80 were not. The data can be presented in a 2 × 2 table (Table 3):

Table 3. Case-Control Study Outcome

	Cases (with disease)	Controls (no disease)
Exposed	40	20
Not exposed	60	80

The calculation of the odds ratio would be:

$$OR = \frac{(40/60)}{(20/80)} = 2.67.$$

If the disease is relatively rare in the general population (about 5% or less), the odds ratio is a good approximation of the relative risk, which means that there is almost a tripling of the disease in those exposed to the agent.[61]

61. The odds ratio is usually marginally greater than the relative risk. As the disease in question becomes more common, the difference between the odds ratio and the relative risk grows.

The reason why the odds ratio approximates the relative risk when the incidence of disease is small can be demonstrated by referring to Table 2. The odds ratio, as stated in the text, is *ad/bc*. The relative risk for such a study would compare the incidence of disease in the exposed group, or $a/(a + b)$, with the incidence of disease in the unexposed group or $c/(c + d)$. The relative risk would be:

$$\frac{a/(a+b)}{c/(c+d)} = \frac{a/(c+d)}{c/(a+b)}$$

When the incidence of disease is low, *a* and *c* will be small in relation to *b* and *d*, and the relative risk will then approximate the odds ratio of *ad/bc*. *See* Leon Gordis, Epidemiology 208–09 (4th ed. 2009).

C. *Attributable Risk*

A frequently used measurement of risk is the attributable risk ("AR"). The attributable risk represents the amount of disease among exposed individuals that can be attributed to the exposure. It also can be expressed as the proportion of the disease among exposed individuals that is associated with the exposure (also called the "attributable proportion of risk," the "etiologic fraction," or the "attributable risk percent"). The attributable risk reflects the maximum proportion of the disease that can be attributed to exposure to an agent and consequently the maximum proportion of disease that could be potentially prevented by blocking the effect of the exposure or by eliminating the exposure.[62] In other words, if the association is causal, the attributable risk is the proportion of disease in an exposed population that might be caused by the agent and that might be prevented by eliminating exposure to that agent (see Figure 3).[63]

Figure 3. Risks in exposed and unexposed groups.

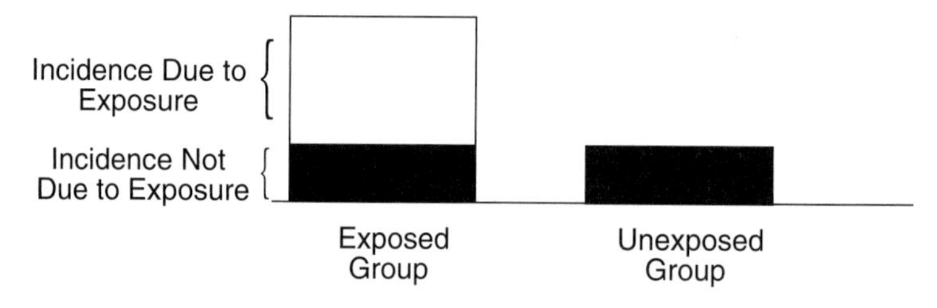

To determine the proportion of a disease that is attributable to an exposure, a researcher would need to know the incidence of the disease in the exposed group and the incidence of disease in the unexposed group. The attributable risk is

$$AR = \frac{(\text{incidence in the exposed}) - (\text{incidence in the unexposed})}{\text{incidence in the exposed}}$$

62. Kenneth J. Rothman et al., Modern Epidemiology 297 (3d ed. 2008); *see also* Landrigan v. Celotex Corp., 605 A.2d 1079, 1086 (N.J. 1992) (illustrating that a relative risk of 1.55 conforms to an attributable risk of 35%, that is, (1.55 − 1.0)/1.55 = .35, or 35%).

63. Risk is not zero for the control group (those not exposed) when there are other causal chains that cause the disease that do not require exposure to the agent. For example, some birth defects are the result of genetic sources, which do not require the presence of any environmental agent. Also, some degree of risk in the control group may be the result of background exposure to the agent being studied. For example, nonsmokers in a control group may have been exposed to passive cigarette smoke, which is responsible for some cases of lung cancer and other diseases. *See also* Ethyl Corp. v. EPA, 541 F.2d 1, 25 (D.C. Cir. 1976). There are some diseases that do not occur without exposure to an agent; these are known as signature diseases. *See infra* note 177.

The attributable risk can be calculated using the example described in Section III.A. Suppose a researcher studies 100 individuals who are exposed to a substance and 200 who are not exposed. After 1 year, 40 of the exposed individuals are diagnosed as having a disease, and 20 of the unexposed individuals are also diagnosed as having the disease.

- The incidence of disease in the exposed group is 40 persons out of 100 who contract the disease in a year.
- The incidence of disease in the unexposed group is 20 persons out of 200 (or 10 out of 100) who contract the disease in a year.
- The proportion of disease that is attributable to the exposure is 30 persons out of 40, or 75%.

This means that 75% of the disease in the exposed group is attributable to the exposure. We should emphasize here that "attributable" does not necessarily mean "caused by." Up to this point, we have only addressed associations. Inferring causation from an association is addressed in Section V.

D. Adjustment for Study Groups That Are Not Comparable

Populations often differ in characteristics that relate to disease risk, such as age, sex, and race. Those who live in Florida have a much higher death rate than those who live in Alaska.[64] Is sunshine dangerous? Perhaps, but the Florida population is much older than the Alaska population, and some adjustment must be made for the differences in age distribution in the two states in order to compare disease or death rates between populations. The technique used to accomplish this is called adjustment, and two types of adjustment are used—direct and indirect. In direct adjustment (e.g., when based on age), overall disease/death rates are calculated for each population as though each had the age distribution of another standard, or reference, population, using the age-specific disease/death rates for each study population. We can then compare these overall rates, called age-adjusted rates, knowing that any difference between these rates cannot be attributed to differences in age, since both age-adjusted rates were generated using the same standard population.

Indirect adjustment is used when the age-specific rates for a study population are not known. In that case, the overall disease/death rate for the standard/reference population is recalculated based on the age distribution of the population of interest using the age-specific rates of the standard population. Then, the actual number of disease cases/deaths in the population of interest can be compared with

64. *See* Lilienfeld & Stolley, *supra* note 35, at 68–70 (the mortality rate in Florida is approximately three times what it is in Alaska).

the number in the reference population that would be expected if the reference population had the age distribution of the population of interest.

This ratio is called the standardized mortality ratio (SMR). When the outcome of interest is disease rather than death, it is called the standardized morbidity ratio.[65] If the ratio equals 1.0, the observed number of deaths equals the expected number of deaths, and the mortality rate of the population of interest is no different from that of the reference population. If the SMR is greater than 1.0, the population of interest has a higher mortality risk than that of the reference population, and if the SMR is less than 1.0, the population of interest has a lower mortality rate than that of the reference population.

Thus, age adjustment provides a way to compare populations while in effect holding age constant. Adjustment is used not only for comparing mortality rates in different populations but also for comparing rates in different groups of subjects selected for study in epidemiologic investigations. Although this discussion has focused on adjusting for age, it is also possible to adjust for any number of other variables, such as gender, race, occupation, and socioeconomic status. It is also possible to adjust for several factors simultaneously.[66]

IV. What Sources of Error Might Have Produced a False Result?

Incorrect study results occur in a variety of ways. A study may find a positive association (relative risk greater than 1.0) when there is no true association. Or a study may erroneously result in finding that that there is no association when in reality there is. A study may also find an association when one truly exists, but the association found may be greater or less than the real association.

Three general categories of phenomena can result in an association found in a study to be erroneous: chance, bias, and confounding. Before any inferences about causation are drawn from a study, the possibility of these phenomena must be examined.[67]

65. *See* Taylor v. Airco, Inc., 494 F. Supp. 2d 21, 25 n.4 (D. Mass. 2007) (explaining SMR and its relationship with relative risk). For an example of adjustment used to calculate an SMR for workers exposed to benzene, see Robert A. Rinsky et al., *Benzene and Leukemia: An Epidemiologic Risk Assessment,* 316 New Eng. J. Med. 1044 (1987).

66. For further elaboration on adjustment, see Gordis, *supra* note 32, at 73–78; Philip Cole, *Causality in Epidemiology, Health Policy, and Law,* 27 Envtl. L. Rep. 10,279, 10,281 (1997).

67. *See* Cole, *supra* note 65, at 10,285. In *DeLuca v. Merrell Dow Pharmaceuticals, Inc.,* 911 F.2d 941, 955 (3d Cir. 1990), the court recognized and discussed random sampling error. It then went on to refer to other errors (e.g., systematic bias) that create as much or more error in the outcome of a study. For a similar description of error in study procedure and random sampling, see David H. Kaye & David A. Freedman, Reference Guide on Statistics, Section IV, in this manual.

The findings of a study may be the result of chance (or random error). In designing a study, the size of the sample can be increased to reduce (but not eliminate) the likelihood of random error. Once a study has been completed, statistical methods (discussed in Section IV.A) permit an assessment of the extent to which the results of a study may be due to random error.

The two main techniques for assessing random error are statistical significance and confidence intervals. A study that is statistically significant has results that are unlikely to be the result of random error, although any criterion for "significance" is somewhat arbitrary. A confidence interval provides both the relative risk (or other risk measure) found in the study and a range (interval) within which the risk likely would fall if the study were repeated numerous times. These two techniques (which are closely related) are explained in Section IV.A.

We should emphasize a matter that those unfamiliar with statistical methodology frequently find confusing: That a study's results are statistically significant says nothing about the importance of the magnitude of any association (i.e., the relative risk or odds ratio) found in a study or about the biological or clinical importance of the finding.[68] "Significant," as used with the adjective "statistically," does not mean important. A study may find a statistically significant relationship that is quite modest—perhaps it increases the risk only by 5%, which is equivalent to a relative risk of 1.05.[69] An association may be quite large—the exposed cohort might be 10 times more likely to develop disease than the control group—but the association is not statistically significant because of the potential for random error given a small sample size. In short, *statistical significance is not about the size of the risk found in a study*.

Bias (or systematic error) also can produce error in the outcome of a study. Epidemiologists attempt to minimize bias through their study design, including data collection protocols. Study designs are developed before they begin gathering data. However, even the best designed and conducted studies have biases, which may be subtle. Consequently, after data collection is completed, analytical tools are often used to evaluate potential sources of bias. Sometimes, after bias is identified, the epidemiologist can determine whether the bias would tend to inflate or dilute any association that may exist. Identification of the bias may permit the

68. *See* Modern Scientific Evidence, *supra* note 2, § 6.36 at 358 ("Statisticians distinguish between 'statistical' and 'practical' significance. . . ."); Cole, *supra* note 65, at 10,282. Understandably, some courts have been confused about the relationship between statistical significance and the magnitude of the association. *See* Hyman & Armstrong, P.S.C. v. Gunderson, 279 S.W.3d 93, 102 (Ky. 2008) (describing a small increased risk as being considered statistically insignificant and a somewhat larger risk as being considered statistically significant.); *In re* Pfizer Inc. Sec. Litig., 584 F. Supp. 2d 621, 634–35 (S.D.N.Y. 2008) (confusing the magnitude of the effect with whether the effect was statistically significant); *In re* Joint E. & S. Dist. Asbestos Litig., 827 F. Supp. 1014, 1041 (S.D.N.Y. 1993) (concluding that any relative risk less than 1.50 is statistically insignificant), *rev'd on other grounds*, 52 F.3d 1124 (2d Cir. 1995).

69. In general, small effects that are statistically significant require larger sample sizes. When effects are larger, generally fewer subjects are required to produce statistically significant findings.

epidemiologist to make an assessment of whether the study's conclusions are valid. Epidemiologists may reanalyze a study's data to correct for a bias identified in a completed study or to validate the analytical methods used.[70] Common biases and how they may produce invalid results are described in Section IV.B.

Finally, a study may reach incorrect conclusions about causation because, although the agent and disease are associated, the agent is not a true causal factor. Rather, the agent may be associated with another agent that is the true causal factor, and this latter factor confounds the relationship being examined in the study. Confounding is explained in Section IV.C.

A. What Statistical Methods Exist to Evaluate the Possibility of Sampling Error?[71]

Before detailing the statistical methods used to assess random error (which we use as synonymous with sampling error), two concepts are explained that are central to epidemiology and statistical analysis. Understanding these concepts should facilitate comprehension of the statistical methods.

Epidemiologists often refer to the true association (also called "real association"), which is the association that really exists between an agent and a disease and that might be found by a perfect (but nonexistent) study. The true association is a concept that is used in evaluating the results of a given study even though its value is unknown. By contrast, a study's outcome will produce an observed association, which is known.

Formal procedures for statistical testing begin with the null hypothesis, which posits that there is no true association (i.e., a relative risk of 1.0) between the agent and disease under study. Data are gathered and analyzed to see whether they disprove[72] the null hypothesis. The data are subjected to statistical testing to assess the plausibility that any association found is a result of random error or whether it supports rejection of the null hypothesis. The use of the null hypothesis for this testing should not be understood as the a priori belief of the investigator. When epidemiologists investigate an agent, it is usually because they hypothesize that the agent is a cause of some outcome. Nevertheless, epidemiologists prepare their

70. *E.g.*, Richard A. Kronmal et al., *The Intrauterine Device and Pelvic Inflammatory Disease: The Women's Health Study Reanalyzed*, 44 J. Clin. Epidemiol. 109 (1991) (a reanalysis of a study that found an association between the use of IUDs and pelvic inflammatory disease concluded that IUDs do not increase the risk of pelvic inflammatory disease).

71. For a bibliography on the role of statistical significance in legal proceedings, see Sanders, *supra* note 13, at 329 n.138.

72. *See, e.g.*, Daubert v. Merrell Dow Pharms., Inc., 509 U.S. 579, 593 (1993) (scientific methodology involves generating and testing hypotheses).

study designs and test the plausibility that any association found in a study was the result of random error by using the null hypothesis.[73]

1. *False positives and statistical significance*

When a study results in a positive association (i.e., a relative risk greater than 1.0), epidemiologists try to determine whether that outcome represents a true association or is the result of random error.[74] Random error is illustrated by a fair coin (i.e., not modified to produce more heads than tails [or vice versa]). On average, for example, we would expect that coin tosses would yield half heads and half tails. But sometimes, a set of coin tosses might yield an unusual result, for example, six heads out of six tosses,[75] an occurrence that would result, purely by chance, in less than 2% of a series of six tosses. In the world of epidemiology, sometimes the study findings, merely by chance, do not reflect the true relationships between an agent and outcome. Any single study—even a clinical trial—is in some ways analogous to a set of coin tosses, being subject to the play of chance. Thus, for example, even though the true relative risk (in the total population) is 1.0, an epidemiologic study of a particular study population may find a relative risk greater than (or less

73. *See* DeLuca v. Merrell Dow Pharms., Inc., 911 F.2d 941, 945 (3d Cir. 1990); United States v. Philip Morris USA, Inc., 449 F. Supp. 2d 1, 706 n.29 (D.D.C. 2006); Stephen E. Fienberg et al., *Understanding and Evaluating Statistical Evidence in Litigation,* 36 Jurimetrics J. 1, 21–24 (1995).

74. Hypothesis testing is one of the most counterintuitive techniques in statistics. Given a set of epidemiologic data, one wants to ask the straightforward, obvious question: What is the probability that the difference between two samples reflects a real difference between the populations from which they were taken? Unfortunately, there is no way to answer this question directly or to calculate the probability. Instead, statisticians—and epidemiologists—address a related but very different question: If there really is no difference between the populations, how probable is it that one would find a difference at least as large as the observed difference between the samples? *See* Modern Scientific Evidence, *supra* note 2, § 6:36, at 359 ("it is easy to mistake the *p*-value for the probability that there is no difference"); Expert Evidence: A Practitioner's Guide to Law, Science, and the FJC Manual 91 (Bert Black & Patrick W. Lee eds., 1997). Thus, the *p*-value for a given study does not provide a rate of error or even a probability of error for an epidemiologic study. In *Daubert v. Merrell Dow Pharmaceuticals, Inc.,* 509 U.S. 579, 593 (1993), the Court stated that "the known or potential rate of error" should ordinarily be considered in assessing scientific reliability. Epidemiology, however, unlike some other methodologies—fingerprint identification, for example—does not permit an assessment of its accuracy by testing with a known reference standard. A *p*-value provides information only about the plausibility of random error given the study result, but the true relationship between agent and outcome remains unknown. Moreover, a *p*-value provides no information about whether other sources of error—bias and confounding—exist and, if so, their magnitude. In short, for epidemiology, there is no way to determine a rate of error. *See* Kumho Tire Co. v. Carmichael, 526 U.S. 137, 151 (1999) (recognizing that for different scientific and technical inquiries, different considerations will be appropriate for assessing reliability); Cook v. Rockwell Int'l Corp., 580 F. Supp. 2d 1071, 1100 (D. Colo. 2006) ("Defendants have not argued or presented evidence that . . . a method by which an overall 'rate of error' can be calculated for an epidemiologic study.")

75. *DeLuca*, 911 F.2d at 946–47.

than) 1.0 because of random error or chance.[76] An erroneous conclusion that the null hypothesis is false (i.e., a conclusion that there is a difference in risk when no difference actually exists) owing to random error is called a false-positive error (also Type I error or alpha error).

Common sense leads one to believe that a large enough sample of individuals must be studied if the study is to identify a relationship between exposure to an agent and disease that truly exists. Common sense also suggests that by enlarging the sample size (the size of the study group), researchers can form a more accurate conclusion and reduce the chance of random error in their results. Both statements are correct and can be illustrated by a test to determine if a coin is fair. A test in which a fair coin is tossed 1000 times is more likely to produce close to 50% heads than a test in which the coin is tossed only 10 times. It is far more likely that a test of a fair coin with 10 tosses will come up, for example, with 80% heads than will a test with 1000 tosses. With large numbers, the outcome of the test is less likely to be influenced by random error, and the researcher would have greater confidence in the inferences drawn from the data.[77]

One means for evaluating the possibility that an observed association could have occurred as a result of random error is by calculating a *p*-value.[78] A *p*-value represents the probability that an observed positive association could result from random error even if no association were in fact present. Thus, a *p*-value of .1 means that there is a 10% chance that values at least as large as the observed relative risk could have occurred by random error, with no association actually present in the population.[79]

To minimize false positives, epidemiologists use a convention that the *p*-value must fall below some selected level known as alpha or significance level for the results of the study to be statistically significant.[80] Thus, an outcome is statistically significant when the observed *p*-value for the study falls below the preselected

76. *See* Magistrini v. One Hour Martinizing Dry Cleaning, 180 F. Supp. 2d 584, 592 (D.N.J. 2002) (citing the second edition of this reference guide).

77. This explanation of numerical stability was drawn from Brief for Professor Alvan R. Feinstein as Amicus Curiae Supporting Respondents at 12–13, Daubert v. Merrell Dow Pharms., Inc., 509 U.S. 579 (1993) (No. 92-102). *See also* Allen v. United States, 588 F. Supp. 247, 417–18 (D. Utah 1984), *rev'd on other grounds,* 816 F.2d 1417 (10th Cir. 1987). The *Allen* court observed that although "[s]mall communities or groups of people are deemed 'statistically unstable'" and "data from small populations must be handled with care [, it] does not mean that [the data] cannot provide substantial evidence in aid of our effort to describe and understand events."

78. *See also* David H. Kaye & David A. Freedman, Reference Guide on Statistics, Section IV.B, in this manual (the *p*-value reflects the implausibility of the null hypothesis).

79. Technically, a *p*-value of .1 means that if in fact there is no association, 10% of all similar studies would be expected to yield an association the same as, or greater than, the one found in the study due to random error.

80. Cook v. Rockwell Int'l Corp., 580 F. Supp. 2d 1071, 1100–01 (D. Colo. 2006) (discussing *p*-values and their relationship with statistical significance); *Allen*, 588 F. Supp. at 416–17 (discussing statistical significance and selection of a level of alpha); *see also* Sanders, *supra* note 13, at 343–44 (explaining alpha, beta, and their relationship to sample size); *Developments in the Law—Confronting*

significance level. The most common significance level, or alpha, used in science is .05.[81] A .05 value means that the probability is 5% of observing an association at least as large as that found in the study when in truth there is no association.[82] Although .05 is often the significance level selected, other levels can and have been used.[83] Thus, in its study of the effects of second-hand smoke, the U.S.

the New Challenges of Scientific Evidence, 108 Harv. L. Rev. 1481, 1535–36, 1540–46 (1995) [hereafter *Developments in the Law*].

81. A common error made by lawyers, judges, and academics is to equate the level of alpha with the legal burden of proof. Thus, one will often see a statement that using an alpha of .05 for statistical significance imposes a burden of proof on the plaintiff far higher than the civil burden of a preponderance of the evidence (i.e., greater than 50%). *See, e.g., In re* Ephedra Prods. Liab. Litig., 393 F. Supp. 2d 181, 193 (S.D.N.Y. 2005); Marmo v. IBP, Inc., 360 F. Supp. 2d 1019, 1021 n.2 (D. Neb. 2005) (an expert toxicologist who stated that science requires proof with 95% certainty while expressing his understanding that the legal standard merely required more probable than not). *But see* Giles v. Wyeth, Inc., 500 F. Supp. 2d 1048, 1056–57 (S.D. Ill. 2007) (quoting the second edition of this reference guide).

Comparing a selected *p*-value with the legal burden of proof is mistaken, although the reasons are a bit complex and a full explanation would require more space and detail than is feasible here. Nevertheless, we sketch out a brief explanation: First, alpha does not address the likelihood that a plaintiff's disease was caused by exposure to the agent; the magnitude of the association bears on that question. *See infra* Section VII. Second, significance testing only bears on whether the observed magnitude of association arose as a result of random chance, not on whether the null hypothesis is true. Third, using stringent significance testing to avoid false-positive error comes at a complementary cost of inducing false-negative error. Fourth, using an alpha of .5 would not be equivalent to saying that the probability the association found is real is 50%, and the probability that it is a result of random error is 50%. Statistical methodology does not permit assessments of those probabilities. *See* Green, *supra* note 47, at 686; Michael D. Green, *Science Is to Law as the Burden of Proof Is to Significance Testing,* 37 Jurimetrics J. 205 (1997) (book review); *see also* David H. Kaye, *Apples and Oranges: Confidence Coefficients and the Burden of Persuasion,* 73 Cornell L. Rev. 54, 66 (1987); David H. Kaye & David A. Freedman, Reference Guide on Statistics, Section IV.B.2, in this manual; Turpin v. Merrell Dow Pharms., Inc., 959 F.2d 1349, 1357 n.2 (6th Cir. 1992), *cert. denied,* 506 U.S. 826 (1992); *cf. DeLuca,* 911 F.2d at 959 n.24 ("The relationship between confidence levels and the more likely than not standard of proof is a very complex one . . . and in the absence of more education than can be found in this record, we decline to comment further on it.").

82. This means that if one conducted an examination of a large number of associations in which the true RR equals 1, on average 1 in 20 associations found to be statistically significant at a .05 level would be spurious. When researchers examine many possible associations that might exist in their data—known as data dredging—we should expect that even if there are no true causal relationships, those researchers will find statistically significant associations in 1 of every 20 associations examined. *See* Rachel Nowak, *Problems in Clinical Trials Go Far Beyond Misconduct,* 264 Sci. 1538, 1539 (1994).

83. A significance test can be either one-tailed or two-tailed, depending on the null hypothesis selected by the researcher. Because most investigators of toxic substances are only interested in whether the agent increases the incidence of disease (as distinguished from providing protection from the disease), a one-tailed test is often viewed as appropriate. *In re* Phenylpropanolamine (PPA) Prods. Liab. Litig., 289 F. Supp. 2d 1230, 1241 (W.D. Wash. 2003) (accepting the propriety of a one-tailed test for statistical significance in a toxic substance case); United States v. Philip Morris USA, Inc., 449 F. Supp. 2d 1, 701 (D.D.C. 2006) (explaining the basis for EPA's decision to use one-tailed test in assessing whether second-hand smoke was a carcinogen). *But see* Good v. Fluor Daniel Corp., 222 F. Supp. 2d 1236, 1243 (E.D. Wash. 2002). For an explanation of the difference

Environmental Protection Agency (EPA) used a .10 standard for significance testing.[84]

There is some controversy among epidemiologists and biostatisticians about the appropriate role of significance testing.[85] To the strictest significance testers,

between one-tailed and two-tailed tests, see David H. Kaye & David A. Freedman, Reference Guide on Statistics, Section IV.C.2, in this manual.

84. U.S. Environmental Protection Agency, Respiratory Health Effects of Passive Smoking: Lung Cancer and Other Disorders (1992); *see also Turpin,* 959 F.2d at 1353–54 n.1 (confidence level frequently set at 95%, although 90% (which corresponds to an alpha of .10) is also used; selection of the value is "somewhat arbitrary").

85. Similar controversy exists among the courts that have confronted the issue of whether statistically significant studies are required to satisfy the burden of production. The leading case advocating statistically significant studies is *Brock v. Merrell Dow Pharmaceuticals, Inc.,* 874 F.2d 307, 312 (5th Cir. 1989), *amended,* 884 F.2d 167 (5th Cir.), *cert. denied,* 494 U.S. 1046 (1990). Overturning a jury verdict for the plaintiff in a Bendectin case, the court observed that no statistically significant study had been published that found an increased relative risk for birth defects in children whose mothers had taken Bendectin. The court concluded: "[W]e do not wish this case to stand as a bar to future Bendectin cases in the event that new and statistically significant studies emerge which would give a jury a firmer basis on which to determine the issue of causation." *Brock,* 884 F.2d at 167.

A number of courts have followed the *Brock* decision or have indicated strong support for significance testing as a screening device. *See* Good v. Fluor Daniel Corp., 222 F. Supp. 2d 1236, 1243 (E.D. Wash. 2002) ("In the absence of a statistically significant difference upon which to opine, Dr. Au's opinion must be excluded under *Daubert.*"); Miller v. Pfizer, Inc., 196 F. Supp. 2d 1062, 1080 (D. Kan. 2002) (the expert must have statistically significant studies to serve as basis of opinion on causation); Kelley v. Am. Heyer-Schulte Corp., 957 F. Supp. 873, 878 (W.D. Tex. 1997) (the lower end of the confidence interval must be above 1.0—equivalent to requiring that a study be statistically significant—before a study may be relied upon by an expert), *appeal dismissed,* 139 F.3d 899 (5th Cir. 1998); Renaud v. Martin Marietta Corp., 749 F. Supp. 1545, 1555 (D. Colo. 1990) (quoting *Brock* approvingly), *aff'd,* 972 F.2d 304 (10th Cir. 1992).

By contrast, a number of courts are more cautious about or reject using significance testing as a necessary condition, instead recognizing that assessing the likelihood of random error is important in determining the probative value of a study. In *Allen v. United States,* 588 F. Supp. 247, 417 (D. Utah 1984), the court stated, "The cold statement that a given relationship is not 'statistically significant' cannot be read to mean there is no probability of a relationship." The Third Circuit described confidence intervals (i.e., the range of values that would be found in similar studies due to chance, with a specified level of confidence) and their use as an alternative to statistical significance in *DeLuca v. Merrell Dow Pharmaceuticals, Inc.,* 911 F.2d 941, 948–49 (3d Cir. 1990). *See also* Milward v. Acuity Specialty Products Group, Inc., 639 F.3d 11, 24-25 (1st Cir. 2011) (recognizing the difficulty of obtaining statistically significant results when the disease under investigation occurs rarely and concluding that district court erred in imposing a statistical significance threshold); Turpin v. Merrell Dow Pharms., Inc., 959 F.2d 1349, 1357 (6th Cir. 1992) ("The defendant's claim overstates the persuasive power of these statistical studies. An analysis of this evidence demonstrates that it is possible that Bendectin causes birth defects even though these studies do not detect a significant association."); *In re* Viagra Prods. Liab. Litig., 572 F. Supp. 2d 1071, 1090 (D. Minn. 2008) (holding that, for purposes of supporting an opinion on general causation, a study does not have to find results with statistical significance); United States v. Philip Morris USA, Inc., 449 F. Supp. 2d 1, 706 n.29 (D.D.C. 2006) (rejecting the position of an expert who denied that the causal connection between smoking and lung cancer had been established, in part, on the ground that any study that found an association that was not statistically significant must be excluded from consideration); Cook v. Rockwell Int'l Corp., 580 F. Supp.

any study whose *p*-value is not less than the level chosen for statistical significance should be rejected as inadequate to disprove the null hypothesis. Others are critical of using strict significance testing, which rejects all studies with an observed *p*-value below that specified level. Epidemiologists have become increasingly sophisticated in addressing the issue of random error and examining the data from a study to ascertain what information they may provide about the relationship between an agent and a disease, without the necessity of rejecting all studies that are not statistically significant.[86] Meta-analysis, as well, a method for pooling the results of multiple studies, sometimes can ameliorate concerns about random error.[87]

Calculation of a confidence interval permits a more refined assessment of appropriate inferences about the association found in an epidemiologic study.[88]

2d 1071, 1103 (D. Colo. 2006) ("The statistical significance or insignificance of Dr. Clapp's results may affect the weight given to his testimony, but does not determine its admissibility under Rule 702."); *In re* Ephedra Prods. Liab. Litig., 393 F. Supp. 2d 181, 186 (S.D.N.Y. 2005) ("[T]he absence of epidemiologic studies establishing an increased risk from ephedra of sufficient statistical significance to meet scientific standards of causality does not mean that the causality opinions of the PCC's experts must be excluded entirely.").

Although the trial court had relied in part on the absence of statistically significant epidemiologic studies, the Supreme Court in *Daubert v. Merrell Dow Pharmaceuticals, Inc.*, 509 U.S. 579 (1993), did not explicitly address the matter. The Court did, however, refer to "the known or potential rate of error" in identifying factors relevant to the scientific validity of an expert's methodology. *Id.* at 594. The Court did not address any specific rate of error, although two cases that it cited affirmed the admissibility of voice spectrograph results that the courts reported were subject to a 2%–6% chance of error owing to either false matches or false eliminations. One commentator has concluded, "*Daubert* did not set a threshold level of statistical significance either for admissibility or for sufficiency of scientific evidence." *Developments in the Law*, *supra* note 79, at 1535–36, 1540–46. The Supreme Court in *General Electric Co. v. Joiner*, 522 U.S. 136, 145–47 (1997), adverted to the lack of statistical significance in one study relied on by an expert as a ground for ruling that the district court had not abused its discretion in excluding the expert's testimony.

In *Matrixx Initiatives, Inc. v. Siracusano*, 131 S. Ct. 1309 (2011), the Supreme Court was confronted with a question somewhat different from the relationship between statistically significant study results and causation. *Matrixx* was a securities fraud case in which the defendant argued that unless adverse event reports from use of a drug are statistically significant, the information about them is not material, as a matter of law (materiality is required as an element of a fraud claim). Defendant's claim was premised on the idea that only statistically significant results can be a basis for an inference of causation. The Court, unanimously, rejected that claim, citing cases in which courts had permitted expert witnesses to testify to toxic causation in the absence of any statistically significant studies.

For a hypercritical assessment of statistical significance testing that nevertheless identifies much inappropriate overreliance on it, see Stephen T. Ziliak & Deidre N. McCloskey, The Cult of Statistical Significance (2008).

86. *See* Sanders, *supra* note 13, at 342 (describing the improved handling and reporting of statistical analysis in studies of Bendectin after 1980).

87. *See infra* Section VI.

88. Kenneth Rothman, Professor of Public Health at Boston University and Adjunct Professor of Epidemiology at the Harvard School of Public Health, is one of the leaders in advocating the use of confidence intervals and rejecting strict significance testing. In *DeLuca*, 911 F.2d at 947, the Third Circuit discussed Rothman's views on the appropriate level of alpha and the use of con-

A confidence interval is a range of possible values calculated from the results of a study. If a 95% confidence interval is specified, the range encompasses the results we would expect 95% of the time if samples for new studies were repeatedly drawn from the same population. Thus, the width of the interval reflects random error.

The narrower the confidence interval, the more statistically stable the results of the study. The advantage of a confidence interval is that it displays more information than significance testing. "Statistically significant" does not convey the magnitude of the association found in the study or indicate how statistically stable that association is. A confidence interval shows the boundaries of the relative risk based on selected levels of alpha or statistical significance. Just as the *p*-value does not provide the probability that the risk estimate found in a study is correct, the confidence interval does not provide the range within which the true risk must lie. Rather, the confidence interval reveals the likely range of risk estimates consistent with random error. An example of two confidence intervals that might be calculated for a given relative risk is displayed in Figure 4.

Figure 4. Confidence intervals.

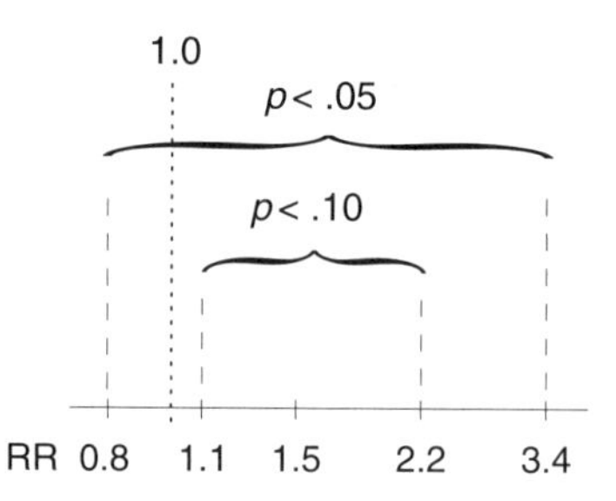

The confidence intervals shown in Figure 4 are for a study that found a relative risk of 1.5, with boundaries of 0.8 to 3.4 when the alpha is set to .05 (equivalently, a confidence level of .95), and with boundaries of 1.1 to 2.2 when alpha is set to .10 (equivalently, a confidence level of .90). The confidence interval for alpha equal to .10 is narrower because it encompasses only 90% of the expected test results. By contrast, the confidence interval for alpha equal to .05 includes the expected outcomes for 95% of the tests. To generalize this point, the lower the alpha chosen (and therefore the more stringent the exclusion of possible random error) the wider the confidence interval. At a given alpha, the width of the confidence interval is

fidence intervals. In *Turpin*, 959 F.2d at 1353–54 n.1, the court discussed the relationship among confidence intervals, alpha, and power. *See also* Cook v. Rockwell Int'l Corp., 580 F. Supp. 2d 1071, 1100–01 (D. Colo. 2006) (discussing confidence intervals, alpha, and significance testing). The use of confidence intervals in evaluating sampling error more generally than in the epidemiologic context is discussed in David H. Kaye & David A. Freedman, Reference Guide on Statistics, Section IV.A, in this manual.

determined by sample size. All other things being equal, the larger the sample size, the narrower the confidence boundaries (indicating greater numerical stability). For a given risk estimate, a narrower confidence interval reflects a decreased likelihood that the association found in the study would occur by chance if the true association is 1.0.[89]

For the example in Figure 4, the boundaries of the confidence interval with alpha set at .05 encompass a relative risk of 1.0, and the result would be said to be not statistically significant at the .05 level. Alternatively, if the confidence boundaries are defined as an alpha equal to .10, then the confidence interval no longer includes a relative risk of 1.0, and the result would be characterized as statistically significant at the .10 level.

2. *False negatives*

As Figure 4 illustrates, false positives can be reduced by adopting more stringent values for alpha. Using an alpha of .05 will result in fewer false positives than using an alpha of .10, and an alpha of .01 or .001 would produce even fewer false positives.[90] The tradeoff for reducing false positives is an increase in false-negative errors (also called beta errors or Type II errors). This concept reflects the possibility that a study will be interpreted as "negative" (not disproving the null

89. Where multiple epidemiologic studies are available, a technique known as meta-analysis (*see infra* Section VI) may be used to combine the results of the studies to reduce the numerical instability of all the studies. *See generally* Diana B. Petitti, Meta-analysis, Decision Analysis, and Cost-Effectiveness Analysis: Methods for Quantitative Synthesis in Medicine (2d ed. 2000). Meta-analysis is better suited to combining results from randomly controlled experimental studies, but if carefully performed it may also be helpful for observational studies, such as those in the epidemiologic field. *See* Zachary B. Gerbarg & Ralph I. Horwitz, *Resolving Conflicting Clinical Trials: Guidelines for Meta-Analysis,* 41 J. Clin. Epidemiol. 503 (1988). In *In re Bextra & Celebrex Marketing Sales Practices & Products Liability Litigation,* 524 F. Supp. 2d 1166 (N.D. Cal. 2007), the court relied on several meta-analyses of Celebrex at a 200-mg dose to conclude that the plaintiffs' experts who proposed to testify to toxicity at that dosage failed to meet the requirements of *Daubert.* The court criticized those experts for the wholesale rejection of meta-analyses of observational studies.

In *In re Paoli Railroad Yard PCB Litigation,* 916 F.2d 829, 856–57 (3d Cir. 1990), the court discussed the use and admissibility of meta-analysis as a scientific technique. Overturning the district court's exclusion of a report using meta-analysis, the Third Circuit observed that meta-analysis is a regularly used scientific technique. The court recognized that the technique might be poorly performed, and it required the district court to reconsider the validity of the expert's work in performing the meta-analysis. *See also* E.R. Squibb & Sons, Inc. v. Stuart Pharms., No. 90-1178, 1990 U.S. Dist. LEXIS 15788, at *41 (D.N.J. Oct. 16, 1990) (acknowledging the utility of meta-analysis but rejecting its use in that case because one of the two studies included was poorly performed); Tobin v. Astra Pharm. Prods., Inc., 993 F.2d 528, 538–39 (6th Cir. 1992) (identifying an error in the performance of a meta-analysis, in which the Food and Drug Administration pooled data from control groups in different studies in which some gave the controls a placebo and others gave the controls an alternative treatment).

90. It is not uncommon in genome-wide association studies to set the alpha at .00001 or even lower because of the large number of associations tested in such studies. Reducing alpha is designed to limit the number of false-positive findings.

hypothesis), when in fact there is a true association of a specified magnitude.[91] The beta for any study can be calculated only based on a specific alternative hypothesis about a given positive relative risk and a specific level of alpha selected.[92]

3. Power

When a study fails to find a statistically significant association, an important question is whether the result tends to exonerate the agent's toxicity or is essentially inconclusive with regard to toxicity.[93] The concept of power can be helpful in evaluating whether a study's outcome is exonerative or inconclusive.[94]

The power of a study is the probability of finding a statistically significant association of a given magnitude (if it exists) in light of the sample sizes used in the study. The power of a study depends on several factors: the sample size; the level of alpha (or statistical significance) specified; the background incidence of disease; and the specified relative risk that the researcher would like to detect.[95] Power curves can be constructed that show the likelihood of finding any given relative risk in light of these factors. Often, power curves are used in the design of a study to determine what size the study populations should be.[96]

The power of a study is the complement of beta ($1 - \beta$). Thus, a study with a likelihood of .25 of failing to detect a true relative risk of 2.0[97] or greater has a power of .75. This means the study has a 75% chance of detecting a true relative risk of 2.0. If the power of a negative study to find a relative risk of 2.0 or greater

91. *See also* DeLuca v. Merrell Dow Pharms., Inc., 911 F.2d 941, 947 (3d Cir. 1990).

92. *See* Green, *supra* note 47, at 684–89.

93. Even when a study or body of studies tends to exonerate an agent, that does not establish that the agent is absolutely safe. *See* Cooley v. Lincoln Elec. Co., 693 F. Supp. 2d 767 (N.D. Ohio 2010). Epidemiology is not able to provide such evidence.

94. *See* Fienberg et al., *supra* note 72, at 22–23. Thus, in *Smith v. Wyeth-Ayerst Labs. Co.,* 278 F. Supp. 2d 684, 693 (W.D.N.C. 2003) and *Cooley v. Lincoln Electric Co.,* 693 F. Supp. 2d 767, 773 (N.D. Ohio 2010), the courts recognized that the power of a study was critical to assessing whether the failure of the study to find a statistically significant association was exonerative of the agent or inconclusive. *See also* Procter & Gamble Pharms., Inc. v. Hoffmann-LaRoche Inc., No. 06 Civ. 0034(PAC), 2006 WL 2588002, at *32 n.16 (S.D.N.Y. Sept. 6, 2006) (discussing power curves and quoting the second edition of this reference guide); *In re* Phenylpropanolamine (PPA) Prods. Liab. Litig., 289 F. Supp. 2d 1230, 1243–44 (W.D. Wash. 2003) (explaining expert's testimony that "statistical reassurance as to lack of an effect would require an upper bound of a reasonable confidence interval close to the null value"); Ruff v. Ensign-Bickford Indus., Inc., 168 F. Supp. 2d 1271, 1281 (D. Utah 2001) (explaining why a study should be treated as inconclusive rather than exonerative based on small number of subjects in study).

95. *See* Malcolm Gladwell, *How Safe Are Your Breasts?* New Republic, Oct. 24, 1994, at 22, 26.

96. For examples of power curves, see Kenneth J. Rothman, Modern Epidemiology 80 (1986); Pagano & Gauvreau, *supra* note 59, at 245.

97. We use a relative risk of 2.0 for illustrative purposes because of the legal significance courts have attributed to this magnitude of association. *See infra* Section VII.

is low, it has substantially less probative value than a study with similar results but a higher power.[98]

B. *What Biases May Have Contributed to an Erroneous Association?*

The second major reason for an invalid outcome in epidemiologic studies is systematic error or bias. Bias may arise in the design or conduct of a study, data collection, or data analysis. The meaning of scientific bias differs from conventional (and legal) usage, in which bias refers to a partisan point of view.[99] When scientists use the term *bias,* they refer to anything that results in a systematic (nonrandom) error in a study result and thereby compromises its validity. Two important categories of bias are selection bias (inappropriate methodology for selection of study subjects) and information bias (a flaw in measuring exposure or disease in the study groups).

Most epidemiologic studies have some degree of bias that may affect the outcome. If major bias is present, it may invalidate the study results. Finding the bias, however, can be difficult, if not impossible. In reviewing the validity of an epidemiologic study, the epidemiologist must identify potential biases and analyze the amount or kind of error that might have been induced by the bias. Often, the direction of error can be determined; depending on the specific type of bias, it may exaggerate the real association, dilute it, or even completely mask it.

1. *Selection bias*

Selection bias refers to the error in an observed association that results from the method of selection of cases and controls (in a case-control study) or exposed and unexposed individuals (in a cohort study).[100] The selection of an appropriate

98. *See also* David H. Kaye & David A. Freedman, Reference Guide on Statistics, Section IV.C.1, in this manual.

99. A Dictionary of Epidemiology 15 (John M. Last ed., 3d ed. 1995); Edmond A. Murphy, The Logic of Medicine 239–62 (1976).

100. Selection bias is defined as "[e]rror due to systematic differences in characteristics between those who are selected for study and those who are not." A Dictionary of Epidemiology, *supra* note 98, at 153. In *In re "Agent Orange" Product Liability Litigation,* 597 F. Supp. 740, 783 (E.D.N.Y. 1985), *aff'd,* 818 F.2d 145 (2d Cir. 1987), the court expressed concern about selection bias. The exposed cohort consisted of young, healthy men who served in Vietnam. Comparing the mortality rate of the exposed cohort and that of a control group made up of civilians might have resulted in error that was a result of selection bias. Failing to account for health status as an independent variable tends to understate any association between exposure and disease in studies in which the exposed cohort is healthier. *See also In re* Baycol Prods. Litig., 532 F. Supp. 2d 1029, 1043 (D. Minn. 2007) (upholding admissibility of testimony by expert witness who criticized study based on selection bias).

control group has been described as the Achilles' heel of a case-control study.[101] Ideally, controls should be drawn from the same population that produced the cases. Selecting control participants becomes problematic if the control participants are selected for reasons that are related to their having the exposure being studied. For example, a study of the effect of smoking on heart disease will suffer selection bias if subjects of the study are volunteers and the decision to volunteer is affected by both being a smoker and having a family history of heart disease. The association will be biased upward because of the additional disease among the exposed smokers caused by genetics.

Hospital-based studies, which are relatively common among researchers located in medical centers, illustrate the problem. Suppose an association is found between coffee drinking and coronary heart disease in a study using hospital patients as controls. The problem is that the hospitalized control group may include individuals who had been advised against drinking coffee for medical reasons, such as to prevent the aggravation of a peptic ulcer. In other words, the controls may become eligible for the study because of their medical condition, which is in turn related to their exposure status—their likelihood of avoiding coffee. If this is true, the amount of coffee drinking in the control group would understate the extent of coffee drinking expected in people who do not have the disease, and thus bias upwardly (i.e., exaggerate) any odds ratio observed.[102] Bias in hospital studies may also understate the true odds ratio when the exposures at issue led to the cases' hospitalizations and also contributed to the controls' chances of hospitalization.

Just as cases and controls in case-control studies should be selected independently of their exposure status, so the exposed and unexposed participants in cohort studies should be selected independently of their disease risk.[103] For example, if women with hysterectomies are overrepresented among exposed women in a cohort study of cervical cancer, this could overstate the association between the exposure and the disease.

A further source of selection bias occurs when those selected to participate decline to participate or drop out before the study is completed. Many studies have shown that individuals who participate in studies differ significantly from those who do not. If a significant portion of either study group declines to participate, the researcher should investigate whether those who declined are different from those who agreed. The researcher can compare relevant characteristics of those who

101. William B. Kannel & Thomas R. Dawber, *Coffee and Coronary Disease,* 289 New Eng. J. Med. 100 (1973) (editorial).

102. Hershel Jick et al., *Coffee and Myocardial Infarction,* 289 New Eng. J. Med. 63 (1973).

103. When unexposed controls may differ from the exposed cohort because exposure is associated with other risk (or protective factors), investigators can attempt to measure and adjust for those differences, as explained in Section IV.C.3, *infra. See also* Martha J. Radford & JoAnne M. Foody, *How Do Observational Studies Expand the Evidence Base for Therapy?* 286 JAMA 1228 (2001) (discussing the use of propensity analysis to adjust for potential confounding and selection biases that may occur from nonrandomization).

participate with those who do not to show the extent to which the two groups are comparable. Similarly, if a significant number of subjects drop out of a study before completion, the remaining subjects may not be representative of the original study populations. The researcher should examine whether that is the case.

The fact that a study may suffer from selection bias does not necessarily invalidate its results. A number of factors may suggest that a bias, if present, had only limited effect. If the association is particularly strong, for example, bias is less likely to account for all of it. In addition, a consistent association across different control groups suggests that possible biases applicable to a particular control group are not invalidating. Similarly, a dose–response relationship (*see* Section V.C, *infra*) found among multiple groups exposed to different doses of the agent would provide additional evidence that biases applicable to the exposed group are not a major problem.

2. *Information bias*

Information bias is a result of inaccurate information about either the disease or the exposure status of the study participants or a result of confounding. In a case-control study, potential information bias is an important consideration because the researcher depends on information from the past to determine exposure and disease and their temporal relationship.[104] In some situations, the researcher is required to interview the subjects about past exposures, thus relying on the subjects' memories. Research has shown that individuals with disease (cases) tend to recall past exposures more readily than individuals with no disease (controls);[105] this creates a potential for bias called recall bias.

For example, consider a case-control study conducted to examine the cause of congenital malformations. The epidemiologist is interested in whether the malformations were caused by an infection during the mother's pregnancy.[106] A group of mothers of malformed infants (cases) and a group of mothers of infants with no

104. Information bias can be a problem in cohort studies as well. When exposure is determined retrospectively, there can be a variety of impediments to obtaining accurate information. Similarly, when disease status is determined retrospectively, bias is a concern. The determination that asbestos is a cause of mesothelioma was hampered by inaccurate death certificates that identified lung cancer rather than mesothelioma, a rare form of cancer, as the cause of death. *See* I.J. Selikoff et al., *Mortality Experience of Insulation Workers in the United States and Canada,* 220 Ann. N.Y. Acad. Sci. 91, 110–11 (1979).

105. Steven S. Coughlin, *Recall Bias in Epidemiological Studies,* 43 J. Clinical Epidemiology 87 (1990).

106. *See* Brock v. Merrell Dow Pharms., Inc., 874 F.2d 307, 311–12 (5th Cir. 1989) (discussion of recall bias among women who bear children with birth defects). We note that the court was mistaken in its assertion that a confidence interval could correct for recall bias, or for any bias for that matter. Confidence intervals are a statistical device for analyzing error that may result from random sampling. Systematic errors (bias) in the design or data collection are not addressed by statistical methods, such as confidence intervals or statistical significance. *See* Green, *supra* note 47, at 667–68; Vincent M. Brannigan et al., *Risk, Statistical Inference, and the Law of Evidence: The Use of Epidemiological Data in Toxic Tort Cases,* 12 Risk Analysis 343, 344–45 (1992).

malformation (controls) are interviewed regarding infections during pregnancy. Mothers of children with malformations may recall an inconsequential fever or runny nose during pregnancy that readily would be forgotten by a mother who had a normal infant. Even if in reality the infection rate in mothers of malformed children is no different from the rate in mothers of normal children, the result in this study would be an apparently higher rate of infection in the mothers of the children with the malformations solely on the basis of recall differences between the two groups.[107] The issue of recall bias can sometimes be evaluated by finding an alternative source of data to validate the subject's response (e.g., blood test results from prenatal visits or medical records that document symptoms of infection).[108] Alternatively, the mothers' responses to questions about other exposures may shed light on the presence of a bias affecting the recall of the relevant exposures. Thus, if mothers of cases do not recall greater exposure than controls' mothers to pesticides, children with German measles, and so forth, then one can have greater confidence in their recall of illnesses.

Bias may also result from reliance on interviews with surrogates who are individuals other than the study subjects. This is often necessary when, for example, a subject (in a case-control study) has died of the disease under investigation or may be too ill to be interviewed.

There are many sources of information bias that affect the measure of exposure, including its intensity and duration. Exposure to the agent can be measured directly or indirectly.[109] Sometimes researchers use a biological marker as a direct measure of exposure to an agent—an alteration in tissue or body fluids that occurs as a result of an exposure and that can be detected in the laboratory. Biological markers, however, are only available for a small number of toxins and usually only reveal whether a person was exposed.[110] Biological markers rarely help determine the intensity or duration of exposure.[111]

107. Thus, in *Newman v. Motorola, Inc.*, 218 F. Supp. 2d 769, 778 (D. Md. 2002), the court considered a study of the effect of cell phone use on brain cancer and concluded that there was good reason to suspect that recall bias affected the results of the study, which found an association between cell phone use and cancers on the side of the head where the cell phone was used but no association between cell phone use and overall brain tumors.

108. Two researchers who used a case-control study to examine the association between congenital heart disease and the mother's use of drugs during pregnancy corroborated interview data with the mother's medical records. *See* Sally Zierler & Kenneth J. Rothman, *Congenital Heart Disease in Relation to Maternal Use of Bendectin and Other Drugs in Early Pregnancy,* 313 New Eng. J. Med. 347, 347–48 (1985).

109. *See In re* Paoli R.R. Yard PCB Litig., No. 86-2229, 1992 U.S. Dist LEXIS 18430, at *9–*11 (E.D. Pa. Oct. 21, 1992) (discussing valid methods of determining exposure to chemicals).

110. *See* Gary E. Marchant, *Genetic Susceptibility and Biomarkers in Toxic Injury Litigation,* 41 Jurimetrics J. 67, 68, 73–74, 95–97 (2000) (explaining concept of biomarkers, how they might be used to provide evidence of exposure or dose, discussing cases in which biomarkers were invoked in an effort to prove exposure, and concluding, "biomarkers are likely to be increasingly relied on to demonstrate exposure").

111. There are different definitions of dose, but dose often refers to the intensity or magnitude of exposure multiplied by the time exposed. *See* Sparks v. Owens-Illinois, Inc., 38 Cal. Rptr. 2d 739,

Monitoring devices also can be used to measure exposure directly but often are not available for exposures that have occurred in the past. For past exposures, epidemiologists often use indirect measures of exposure, such as interviewing workers and reviewing employment records. Thus, all those employed to install asbestos insulation may be treated as having been exposed to asbestos during the period that they were employed. However, there may be a wide variation of exposure within any job, and these measures may have limited applicability to a given individual.[112] If the agent of interest is a drug, medical or hospital records can be used to determine past exposure. Thus, retrospective studies, which are often used for occupational or environmental investigations, entail measurements of exposure that are usually less accurate than prospective studies or followup studies, including ones in which a drug or medical intervention is the independent variable being measured.

742 (Ct. App. 1995). Other definitions of dose may be more appropriate in light of the biological mechanism of the disease.

For a discussion of the difficulties of determining dose from atomic fallout, see Allen v. United States, 588 F. Supp. 247, 425–26 (D. Utah 1984), *rev'd on other grounds*, 816 F.2d 1417 (10th Cir. 1987). The timing of exposure may also be critical, especially if the disease of interest is a birth defect. In *Smith v. Ortho Pharmaceutical Corp.*, 770 F. Supp. 1561, 1577 (N.D. Ga. 1991), the court criticized a study for its inadequate measure of exposure to spermicides. The researchers had defined exposure as receipt of a prescription for spermicide within 600 days of delivery, but this definition of exposure is too broad because environmental agents are likely to cause birth defects only during a narrow band of time.

A different, but related, problem often arises in court. Determining the plaintiff's exposure to the alleged toxic substance always involves a retrospective determination and may involve difficulties similar to those faced by an epidemiologist planning a study. Thus, in *John's Heating Service v. Lamb*, 46 P.3d 1024 (Alaska 2002), plaintiffs were exposed to carbon monoxide because of defendants' negligence with respect to a home furnace. The court observed: "[W]hile precise information concerning the exposure necessary to cause specific harm to humans and exact details pertaining to the plaintiff's exposure are beneficial, such evidence is not always available, or necessary, to demonstrate that a substance is toxic to humans given substantial exposure and need not invariably provide the basis for an expert's opinion on causation." *Id.* at 1035 (quoting Westberry v. Gislaved Gummi AB, 178 F.3d 257, 264 (4th Cir. 1999)); *see also* Alder v. Bayer Corp., AGFA Div., 61 P.3d 1068, 1086–88 (Utah 2002) (summarizing other decisions on the precision with which plaintiffs must establish the dosage to which they were exposed). *See generally* Restatement (Third) of Torts: Liability for Physical and Emotional Harm § 28 cmt. c(2) & rptrs. note (2010).

In asbestos litigation, a number of courts have adopted a requirement that the plaintiff demonstrate (1) regular use by an employer of the defendant's asbestos-containing product, (2) the plaintiff's proximity to that product, and (3) exposure over an extended period of time. *See, e.g.*, Lohrmann v. Pittsburgh Corning Corp., 782 F.2d 1156, 1162–64 (4th Cir. 1986); Gregg v. V-J Auto Parts, Inc., 943 A.2d 216, 226 (Pa. 2007).

112. Frequently, occupational epidemiologists employ study designs that consider all agents to which those who work in a particular occupation are exposed because they are trying to determine the hazards associated with that occupation. Isolating one of the agents for examination would be difficult if not impossible. These studies, then, present difficulties when employed in court in support of a claim by a plaintiff who was exposed to only one or fewer than all of the agents present at the worksite that was the subject of the study. *See, e.g.*, Knight v. Kirby Inland Marine Inc., 482 F.3d 347, 352–53 (5th Cir. 2007) (concluding that case-control studies of cancer that entailed exposure to a variety of organic solvents at job sites did not support claims of plaintiffs who claimed exposure to benzene caused their cancers).

The route (e.g., inhalation or absorption), duration, and intensity of exposure are important factors in assessing disease causation. Even with environmental monitoring, the dose measured in the environment generally is not the same as the dose that reaches internal target organs. If the researcher has calculated the internal dose of exposure, the scientific basis for this calculation should be examined for soundness.[113]

In assessing whether the data may reflect inaccurate information, one must assess whether the data were collected from objective and reliable sources. Medical records, government documents, employment records, death certificates, and interviews are examples of data sources that are used by epidemiologists to measure both exposure and disease status.[114] The accuracy of a particular source may affect the validity of a research finding. If different data sources are used to collect information about a study group, differences in the accuracy of those sources may affect the validity of the findings. For example, using employment records to gather information about exposure to narcotics probably would lead to inaccurate results, because employees tend to keep such information private. If the researcher uses an unreliable source of data, the study may not be useful.

The kinds of quality control procedures used may affect the accuracy of the data. For data collected by interview, quality control procedures should probe the reliability of the individual and whether the information is verified by other sources. For data collected and analyzed in the laboratory, quality control procedures should probe the validity and reliability of the laboratory test.

Information bias may also result from inaccurate measurement of disease status. The quality and sophistication of the diagnostic methods used to detect a disease should be assessed.[115] The proportion of subjects who were examined also should be questioned. If, for example, many of the subjects refused to be tested, the fact that the test used was of high quality would be of relatively little value.

113. *See also* Bernard D. Goldstein & Mary Sue Henifin, Reference Guide on Toxicology, Section I.D, in this manual.

114. Even these sources may produce unanticipated error. Identifying the causal connection between asbestos and mesothelioma, a rare form of cancer, was complicated and delayed because doctors who were unfamiliar with mesothelioma erroneously identified other causes of death in death certificates. *See* David E. Lilienfeld & Paul D. Gunderson, *The "Missing Cases" of Pleural Malignant Mesothelioma in Minnesota, 1979–81: Preliminary Report,* 101 Pub. Health Rep. 395, 397–98 (1986).

115. The hazards of adversarial review of epidemiologic studies to determine bias is highlighted by *O'Neill v. Novartis Consumer Health, Inc.,* 55 Cal. Rptr. 3d 551, 558–60 (Ct. App. 2007). Defendant's experts criticized a case-control study relied on by plaintiff on the ground that there was misclassification of exposure status among the cases. Plaintiff objected to this criticism because defendant's experts had only examined the cases for exposure misclassification, which would tend to exaggerate any association by providing an inaccurately inflated measure of exposure in the cases. The experts failed to examine whether there was misclassification in the controls, which, if it existed, would tend to incorrectly diminish any association.

The scientific validity of the research findings is influenced by the reliability of the diagnosis of disease or health status under study.[116] The disease must be one that is recognized and defined to enable accurate diagnoses.[117] Subjects' health status may be essential to the hypothesis under investigation. For example, a researcher interested in studying spontaneous abortion in the first trimester must determine that study subjects are pregnant. Diagnostic criteria that are accepted by the medical community should be used to make the diagnosis. If a diagnosis had been made at a time when home pregnancy kits were known to have a high rate of false-positive results (indicating pregnancy when the woman is not pregnant), the study will overestimate the number of spontaneous abortions.

Misclassification bias is a consequence of information bias in which, because of problems with the information available, individuals in the study may be misclassified with regard to exposure status or disease status. Bias due to exposure misclassification can be differential or nondifferential. In nondifferential misclassification, the inaccuracies in determining exposure are independent of disease status, or the inaccuracies in diagnoses are independent of exposure status—in other words, the data are crude, with a great deal of random error. This is a common problem. Generally, nondifferential misclassification bias leads to a shift in the odds ratio toward one, or, in other words, toward a finding of no effect. Thus, if the errors are nondifferential, it is generally misguided to criticize an apparent association between an exposure and disease on the ground that data were inaccurately classified. Instead, nondifferential misclassification generally underestimates the true size of the association.

Differential misclassification is systematic error in determining exposure in cases as compared with controls, or disease status in unexposed cohorts relative to exposed cohorts. In a case-control study this would occur, for example, if, in the

116. In *In re Swine Flu Immunization Products Liability Litigation,* 508 F. Supp. 897, 903 (D. Colo. 1981), *aff'd sub nom. Lima v. United States,* 708 F.2d 502 (10th Cir. 1983), the court critically evaluated a study relied on by an expert whose testimony was stricken. In that study, determination of whether a patient had Guillain-Barré syndrome was made by medical clerks, not physicians who were familiar with diagnostic criteria.

117. The difficulty of ill-defined diseases arose in some of the silicone gel breast implant cases. Thus, in *Grant v. Bristol-Myers Squibb,* 97 F. Supp. 2d 986 (D. Ariz. 2000), in the face of a substantial body of exonerative epidemiologic evidence, the female plaintiff alleged she suffered from an atypical systemic joint disease. The court concluded:

> As a whole, the Court finds that the evidence regarding systemic disease as proposed by Plaintiffs' experts is not scientifically valid and therefore will not assist the trier of fact. As for the atypical syndrome that is suggested, where experts propose that breast implants cause a disease but cannot specify the criteria for diagnosing the disease, it is incapable of epidemiologic testing. This renders the experts' methods insufficiently reliable to help the jury.

Id. at 992; *see also* Burton v. Wyeth-Ayerst Labs., 513 F. Supp. 2d 719, 722–24 (N.D. Tex. 2007) (parties disputed whether cardiology problem involved two separate diseases or only one; court concluded that all experts in the case reflected a view that there was but a single disease); *In re* Breast Implant Cases, 942 F. Supp. 958, 961 (E.D.N.Y. & S.D.N.Y. 1996).

process of anguishing over the possible causes of the disease, parents of ill children recalled more exposures to a particular agent than actually occurred, or if parents of the controls, for whom the issue was less emotionally charged, recalled fewer. This can also occur in a cohort study in which, for example, birth control users (the exposed cohort) are monitored more closely for potential side effects, leading to a higher rate of disease identification in that cohort than in the unexposed cohort. Depending on how the misclassification occurs, a differential bias can produce an error in either direction—the exaggeration or understatement of a true association.

3. *Other conceptual problems*

There are dozens of other potential biases that can occur in observational studies, which is an important reason why clinical studies (when ethical) are often preferable. Sometimes studies are limited by flawed definitions or premises. For example, if the researcher defines the disease of interest as all birth defects, rather than a specific birth defect, there should be a scientific basis to hypothesize that the effects of the agent being investigated could be so broad. If the effect is in fact more limited, the result of this conceptualization error could be to dilute or mask any real effect that the agent might have on a specific type of birth defect.[118]

Some biases go beyond errors in individual studies and affect the overall body of available evidence in a way that skews what appears to be the universe of evidence. Publication bias is the tendency for medical journals to prefer studies that find an effect.[119] If negative studies are never published, the published literature will be biased. Financial conflicts of interest by researchers and the source of funding of studies have been shown to have an effect on the outcomes of such studies.[120]

118. In *Brock v. Merrell Dow Pharmaceuticals, Inc.*, 874 F.2d 307, 312 (5th Cir. 1989), the court discussed a reanalysis of a study in which the effect was narrowed from all congenital malformations to limb reduction defects. The magnitude of the association changed by 50% when the effect was defined in this narrower fashion. *See* Rothman et al. *supra* note 61, at 144 ("Unwarranted assurances of a lack of any effect can easily emerge from studies in which a wide range of etiologically unrelated outcomes are grouped.").

119. Investigators may contribute to this effect by neglecting to submit negative studies for publication.

120. *See* Jerome P. Kassirer, On the Take: How Medicine's Complicity with Big Business Can Endanger Your Health 79–84 (2005); J.E. Bekelman et al., *Scope and Impact of Financial Conflicts of Interest in Biomedical Research: A Systematic Review*, 289 JAMA 454 (2003). Richard Smith, the editor in chief of the British Medical Journal, wrote on this subject:

> The major determinant of whether reviews of passive smoking concluded it was harmful was whether the authors had financial ties with tobacco manufacturers. In the disputed topic of whether third-generation contraceptive pills cause an increase in thromboembolic disease, studies funded by the pharmaceutical industry find that they don't and studies funded by public money find that they do.

Richard Smith, *Making Progress with Competing Interests*, 325 Brit. Med. J. 1375, 1376 (2002).

Examining a study for potential sources of bias is an important task that helps determine the accuracy of a study's conclusions. In addition, when a source of bias is identified, it may be possible to determine whether the error tended to exaggerate or understate the true association. Thus, bias may exist in a study that nevertheless has probative value.

Even if one concludes that the findings of a study are statistically stable and that biases have not created significant error, additional considerations remain. As repeatedly noted, an association does not necessarily mean a causal relationship exists. To make a judgment about causation, a knowledgeable expert[121] must consider the possibility of confounding factors. The expert must also evaluate several criteria to determine whether an inference of causation is appropriate.[122] These matters are discussed below.

C. Could a Confounding Factor Be Responsible for the Study Result?[123]

The third major reason for error in epidemiologic studies is confounding. Confounding occurs when another causal factor (the confounder) confuses the relationship between the agent of interest and outcome of interest.[124] (Confounding and selection bias (Section IV.B.1, *supra*) can, depending on terminology, overlap.) Thus, one instance of confounding is when a confounder is both a risk factor for the disease and a factor associated with the exposure of interest. For example, researchers may conduct a study that finds individuals with gray hair have a higher rate of death than those with hair of another color. Instead of hair color having an impact on death, the results might be explained by the confounding factor of age. If old age is associated differentially with the gray-haired group (those with gray hair tend to be older), old age may be responsible for the association found between hair color and death.[125] Researchers must separate the relationship between gray hair and risk of death from that of old age and risk of death. When researchers find an association between an agent and a disease, it is critical to determine whether the association is causal or the result of confounding.[126] Some

121. In a lawsuit, this would be done by an expert. In science, the effort is usually conducted by a panel of experts.

122. For an excellent example of the authors of a study analyzing whether an inference of causation is appropriate in a case-control study examining whether bromocriptine (Parlodel)—a lactation suppressant—causes seizures in postpartum women, see Kenneth J. Rothman et al., *Bromocriptine and Puerpal Seizures,* 1 Epidemiology 232, 236–38 (1990).

123. *See* Grassis v. Johns-Manville Corp., 591 A.2d 671, 675 (N.J. Super. Ct. App. Div. 1991) (discussing the possibility that confounders may lead to an erroneous inference of a causal relationship).

124. *See* Rothman et al., *supra* note 61, at 129.

125. This example is drawn from Kahn & Sempos, *supra* note 31, at 63.

126. Confounding can bias a study result by either exaggerating or diluting any true association. One example of a confounding factor that may result in a study's outcome understating an

epidemiologists classify confounding as a form of bias. However, confounding is a reality—that is, the observed association of a factor and a disease is actually the result of an association with a third, confounding factor.[127]

Confounding can be illustrated by a hypothetical prospective cohort study of the role of alcohol consumption and emphysema. The study is designed to investigate whether drinking alcohol is associated with emphysema. Participants are followed for a period of 20 years and the incidence of emphysema in the "exposed" (participants who consume more than 15 drinks per week) and the unexposed is compared. At the conclusion of the study, the relative risk of emphysema in the drinking group is found to be 2.0, an association that suggests a possible effect). But does this association reflect a true causal relationship or might it be the product of confounding?

One possibility for a confounding factor is smoking, a known causal risk factor for emphysema. If those who drink alcohol are more likely to be smokers than those who do not drink, then smoking may be responsible for some or all of the higher level of emphysema among those who do not drink.

A serious problem in observational studies such as this hypothetical study is that the individuals are not assigned randomly to the groups being compared.[128] As discussed above, randomization maximizes the possibility that exposures other than the one under study are evenly distributed between the exposed and the control cohorts.[129] In observational studies, by contrast, other forces, including self-selection, determine who is exposed to other (possibly causal) factors. The lack of randomization leads to the potential problem of confounding. Thus, for example, the exposed cohort might consist of those who are exposed at work to an agent suspected of being an industrial toxin. The members of this cohort may, however, differ from unexposed controls by residence, socioeconomic or health status, age, or other extraneous factors.[130] These other factors may be causing (or

association is vaccination. Thus, if a group exposed to an agent has a higher rate of vaccination for the disease under study than the unexposed group, the vaccination may reduce the rate of disease in the exposed group, thereby producing an association that is less than the true association without the confounding of vaccination.

127. Schwab v. Philip Morris USA, Inc., 449 F. Supp. 2d 992, 1199–1200 (E.D.N.Y. 2006), *rev'd on other grounds*, 522 F.3d 215 (2d Cir. 2008), describes confounding that led to premature conclusions that low-tar cigarettes were safer than regular cigarettes. Smokers who chose to switch to low-tar cigarettes were different from other smokers in that they were more health conscious in other aspects of their lifestyles. Failure to account for that confounding—and measuring a healthy lifestyle is difficult even if it is identified as a potential confounder—biased the results of those studies.

128. Randomization attempts to ensure that the presence of a characteristic, such as coffee drinking, is governed by chance, as opposed to being determined by the presence of an underlying medical condition.

129. *See* Rothman et al., *supra* note 61, at 129; *see also supra* Section II.A.

130. *See, e.g., In re* "Agent Orange" Prod. Liab. Litig., 597 F. Supp. 740, 783 (E.D.N.Y. 1984) (discussing the problem of confounding that might result in a study of the effect of exposure to Agent Orange on Vietnam servicemen), *aff'd*, 818 F.2d 145 (2d Cir. 1987).

protecting against) the disease, but because of potential confounding, an apparent (yet false) association of the disease with exposure to the agent may appear. Confounders, like smoking in the alcohol drinking study, do not reflect an error made by the investigators; rather, they reflect the inherently "uncontrolled" nature of exposure designations in observational studies. When they can be identified, confounders should be taken into account. Unanticipated confounding factors that are suspected after data collection can sometimes be controlled during data analysis, if data have been gathered about them.

To evaluate whether smoking is a confounding factor, the researcher would stratify each of the exposed and control groups into smoking and nonsmoking subgroups to examine whether subjects' smoking status affects the study results. If the relationship between alcohol drinking and emphysema in the smoking subgroups is the same as that in the all-subjects group, smoking is not a confounding factor. If the subjects' smoking status affects the relationship between drinking and emphysema, then smoking is a confounder, for which adjustment is required. If the association between drinking and emphysema completely disappears when the subjects' smoking status is considered, then smoking is a confounder that fully accounts for the association with drinking observed. Table 4 reveals our hypothetical study's results, with smoking being a confounding factor, which, when accounted for, eliminates the association. Thus, in the full cohort, drinkers have twice the risk of emphysema compared with nondrinkers. When the relationship between drinking and emphysema is examined separately in smokers and in nonsmokers, the risk of emphysema in drinkers compared with nondrinkers is not elevated in smokers or in nonsmokers. This is because smokers are disproportionately drinkers and have a higher rate of emphysema than nonsmokers. Thus, the relationship between drinking and emphysema in the full cohort is distorted by failing to take into account the relationship between being a drinker and a smoker.

Even after accounting for the effect of smoking, there is always a risk that an undiscovered or unrecognized confounding factor may contribute to a study's findings, by either magnifying or reducing the observed association.[131] It is, however, necessary to keep that risk in perspective. Often the mere possibility of uncontrolled confounding is used to call into question the results of a study. This was certainly the strategy of some seeking, or unwittingly helping, to undermine the implications of the studies persuasively linking cigarette smoking to lung cancer. The critical question is whether it is plausible that the findings of a given study could indeed be due to unrecognized confounders.

In designing a study, researchers sometimes make assumptions that cannot be validated or evaluated empirically. Thus, researchers may assume that a missing potential confounder is not needed for the analysis or that a variable used was adequately classified. Researchers employ a sensitivity analysis to assess the effect of those assumptions should they be incorrect. Conducting a sensitivity analysis

131. Rothman et al., *supra* note 61, at 129; *see also supra* Section II.A.

Table 4. Hypothetical Emphysema Study Data[a]

Drinking Status	Total Cohort				Smokers				Nonsmokers			
	Total	Cases	Incidence	RR	Total	Cases	Incidence	RR	Total	Cases	Incidence	RR
Nondrinkers	471	16	0.034	1.0[b]	111	9	0.081	1.0[b]	360	7	0.019	1.0[b]
Drinkers	739	41	0.069	2.0	592	48	0.081	1.0	147	3	0.020	1.0

[a] The incidence of disease is not normally presented in an epidemiologic study, but we include it here to aid in comprehension of the ideas discussed in the text.
[b] RR = relative risk. The relative risk for each of the cohorts is determined based on reference to the risk among nondrinkers; that is, the incidence of disease among drinkers is compared with nondrinkers for each of the three cohorts separately.

entails repeating the analysis using different assumptions (e.g., alternative corrections for missing data or for classifying data) to see if the results are sensitive to the varying assumptions. Such analyses can show that the assumptions are not likely to affect the findings or that alternative explanations cannot be ruled out.[132]

1. *What techniques can be used to prevent or limit confounding?*

Choices in the design of a research project (e.g., methods for selecting the subjects) can prevent or limit confounding. In designing a study, the researcher must determine other risk factors for the disease under study. When a factor or factors, such as age, sex, or even smoking status, are risk factors and potential confounders in a study, investigators can limit the differential distribution of these factors in the study groups by selecting controls to "match" cases (or the exposed group) in terms of these variables. If the two groups are matched, for example, by age, then any association observed in the study cannot be due to age, the matched variable.[133]

Restricting the persons who are permitted as subjects in a study is another method to control for confounders. If age or sex is suspected as a confounder, then the subjects enrolled in a study can be limited to those of one sex and those who are within a specified age range. When there is no variance among subjects in a study with regard to a potential confounder, confounding as a result of that variable is eliminated.

2. *What techniques can be used to identify confounding factors?*

Once the study data are ready to be analyzed, the researcher must assess a range of factors that could influence risk. In the hypothetical study, the researcher would evaluate whether smoking is a confounding factor by comparing the incidence of emphysema in smoking alcohol drinkers with the incidence in nonsmoking alcohol drinkers. If the incidence is substantially the same, smoking is not a confounding factor (e.g., smoking does not distort the relationship between alcohol drinking and the development of emphysema). If the incidence is substantially different, but still exists in the nonsmoking group, then smoking is a confounder, but does not wholly account for the association with alcohol drinking. If the association disappears, then smoking is a confounder that fully accounts for the association observed.

132. Kenneth Rothman & Sander Greenland, Modern Epidemiology (2d ed. 1998).

133. Selecting a control population based on matched variables necessarily affects the representativeness of the selected controls and may affect how generalizable the study results are to the population at large. However, for a study to have merit, it must first be internally valid; that is, it must not be subject to unreasonable sources of bias or confounding. Only after a study has been shown to meet this standard does its universal applicability or generalizability to the population at large become an issue. When a study population is not representative of the general or target population, existing scientific knowledge may permit reasonable inferences about the study's broader applicability, or additional confirmatory studies of other populations may be necessary.

3. What techniques can be used to control for confounding factors?

A good study design will consider potential confounders and obtain data about them if possible. If researchers have good data on potential confounders, they can control for those confounders in the data analysis. There are several analytic approaches to account for the distorting effects of a confounder, including stratification or multivariate analysis. Stratification permits an investigator to evaluate the effect of a suspected confounder by subdividing the study groups based on a confounding factor. Thus, in Table 4, drinkers have been stratified based on whether they smoke (the suspected confounder). To take another example that entails a continuous rather than dichotomous potential confounder, let us say we are interested in the relationship between smoking and lung cancer but suspect that air pollution or urbanization may confound the relationship. Thus, an observed relationship between smoking and lung cancer could theoretically be due in part to pollution, if smoking were more common in polluted areas. We could address this issue by stratifying our data by degree of urbanization and look at the relationship between smoking and lung cancer in each urbanization stratum. Figure 5 shows actual age-adjusted lung cancer mortality rates per 100,000 person-years by urban or rural classification and smoking category.[134]

Figure 5: Age-adjusted lung cancer mortality rates per 100,000 person-years by urban or rural classification and smoking category.

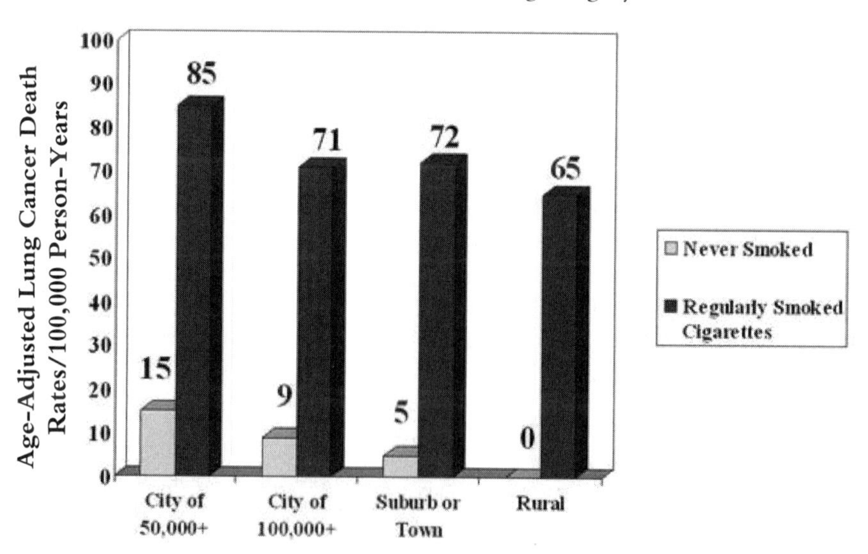

Source: Adapted from E. Cuyler Hammond & Daniel Horn, *Smoking and Death Rates—Report on Forty-Four Months of Follow-Up of 187,783 Men: II, Death Rates by Cause*, 166 JAMA 1294 (1958).

134. This example and Figure 4 are from Leon Gordis, Epidemiology 254 (4th ed. 2009).

For each degree of urbanization, lung cancer mortality rates in smokers are shown by the dark gray bars, and nonsmoker mortality rates are indicated by light gray bars. From these data we see that in every level (or stratum) of urbanization, lung cancer mortality is higher in smokers than in nonsmokers. Therefore, the observed association of smoking and lung cancer cannot be attributed to level of urbanization. By examining each stratum separately, we, in effect, hold urbanization constant, and still find much higher lung cancer mortality in smokers than in nonsmokers.

For each degree of urbanization, lung cancer mortality rates and smokers are shown by the dark-colored bars, and nonsmoker mortality rates are indicated by light-colored bars. For these data we see that in every level (or stratum) of urbanization, lung cancer mortality is higher in smokers than in nonsmokers. Therefore, the observed association of lung cancer cannot be attributed to level of urbanization. By examining each stratum separately, we are, in effect, holding urbanization constant, and we still find much higher lung cancer mortality in smokers than in nonsmokers.

Multivariate analysis controls for the confounding factor through mathematical modeling. Models are developed to describe the simultaneous effect of exposure and confounding factors on the increase in risk.[135]

Both of these methods allow for adjustment of the effect of confounders. They both modify an observed association to take into account the effect of risk factors that are not the subject of the study and that may distort the association between the exposure being studied and the disease outcomes. If the association between exposure and disease remains after the researcher completes the assessment and adjustment for confounding factors, the researcher must then assess whether an inference of causation is justified. This entails consideration of the Hill factors explained in Section V, *infra.*

V. General Causation: Is an Exposure a Cause of the Disease?

Once an association has been found between exposure to an agent and development of a disease, researchers consider whether the association reflects a true cause–effect relationship. When epidemiologists evaluate whether a cause–effect relationship exists between an agent and disease, they are using the term causation in a way similar to, but not identical to, the way that the familiar "but for," or sine qua non, test is used in law for cause in fact. "Conduct is a factual cause of

135. For a more complete discussion of multivariate analysis, see Daniel L. Rubinfeld, Reference Guide on Multiple Regression, in this manual.

[harm] when the harm would not have occurred absent the conduct."[136] This is equivalent to describing the conduct as a necessary link in a chain of events that results in the particular event.[137] Epidemiologists use causation to mean that an increase in the incidence of disease among the exposed subjects would not have occurred had they not been exposed to the agent.[138] Thus, exposure is a necessary condition for the increase in the incidence of disease among those exposed.[139] The relationship between the epidemiologic concept of cause and the legal question of whether exposure to an agent caused an individual's disease is addressed in Section VII.

As mentioned in Section I, epidemiology cannot prove causation; rather, causation is a judgment for epidemiologists and others interpreting the epidemiologic data.[140] Moreover, scientific determinations of causation are inherently tentative. The scientific enterprise must always remain open to reassessing the validity of past judgments as new evidence develops.

In assessing causation, researchers first look for alternative explanations for the association, such as bias or confounding factors, which are discussed in Section IV, *supra*. Once this process is completed, researchers consider how guidelines for inferring causation from an association apply to the available evidence. We emphasize that these guidelines are employed only *after* a study finds an association

136. Restatement (Third) of Torts: Liability for Physical and Emotional Harm § 26 (2010); *see also* Dan B. Dobbs, The Law of Torts § 168, at 409–11 (2000). When multiple causes are each operating and capable of causing an event, the but-for, or necessary-condition, concept for causation is problematic. This is the familiar "two-fires" scenario in which two independent fires simultaneously burn down a house and is sometimes referred to as overdetermined outcomes. Neither fire is a but-for, or necessary condition, for the destruction of the house, because either fire would have destroyed the house. *See* Restatement (Third) of Torts: Liability for Physical and Emotional Harm § 28 (2010). This two-fires situation is analogous to an individual being exposed to two agents, each of which is capable of causing the disease contracted by the individual. *See* Basko v. Sterling Drug, Inc., 416 F.2d 417 (2d Cir. 1969). A difference between the disease scenario and the fire scenario is that, in the former, one will have no more than a probabilistic assessment of whether each of the exposures would have caused the disease in the individual.

137. *See supra* note 7; *see also* Restatement (Third) of Torts: Liability for Physical and Emotional Harm § 26 cmt. c (2010) (employing a "causal set" model to explain multiple elements, each of which is required for an outcome).

138. "The imputed causal association is at the group level, and does not indicate the cause of disease in individual subjects." Bruce G. Charlton, *Attribution of Causation in Epidemiology: Chain or Mosaic?* 49 J. Clin. Epidemiology 105, 105 (1999).

139. *See* Rothman et al., *supra* note 61, at 8 ("We can define a cause of a specific disease event as an antecedent event, condition, or characteristic that was necessary for the occurrence of the disease at the moment it occurred, given that other conditions are fixed."); Allen v. United States, 588 F. Supp. 247, 405 (D. Utah 1984) (quoting a physician on the meaning of the statement that radiation causes cancer), *rev'd on other grounds*, 816 F.2d 1417 (10th Cir. 1987).

140. Restatement (Third) of Torts: Liability for Physical and Emotional Harm § 28 cmt. c (2010) ("[A]n evaluation of data and scientific evidence to determine whether an inference of causation is appropriate requires judgment and interpretation.").

to determine whether that association reflects a true causal relationship.[141] These guidelines consist of several key inquiries that assist researchers in making a judgment about causation.[142] Generally, researchers are conservative when it comes to assessing causal relationships, often calling for stronger evidence and more research before a conclusion of causation is drawn.[143]

The factors that guide epidemiologists in making judgments about causation (and there is no threshold number that must exist) are[144]

141. In a number of cases, experts attempted to use these guidelines to support the existence of causation in the absence of any epidemiologic studies finding an association. *See, e.g.*, Rains v. PPG Indus., Inc., 361 F. Supp. 2d 829, 836–37 (S.D. Ill. 2004) (explaining Hill criteria and proceeding to apply them even though there was no epidemiologic study that found an association); Soldo v. Sandoz Pharms. Corp., 244 F. Supp. 2d 434, 460–61 (W.D. Pa. 2003). There may be some logic to that effort, but it does not reflect accepted epidemiologic methodology. *See In re* Fosamax Prods. Liab. Litig., 645 F. Supp. 2d 164, 187–88 (S.D.N.Y. 2009); Dunn v. Sandoz Pharms. Corp., 275 F. Supp. 2d 672, 678–79 (M.D.N.C. 2003) ("The greater weight of authority supports Sandoz' assertion that [use of] the Bradford Hill criteria is a method for determining whether the results of an epidemiologic study can be said to demonstrate causation and not a method for testing an unproven hypothesis."); *Soldo*, 244 F. Supp. 2d at 514 (the Hill criteria "were developed as a mean[s] of interpreting an established association based on a body of epidemiologic research for the purpose of trying to judge whether the observed association reflects a causal relation between an exposure and disease." (quoting report of court-appointed expert)).

142. *See* Mervyn Susser, Causal Thinking in the Health Sciences: Concepts and Strategies in Epidemiology (1973); Gannon v. United States, 571 F. Supp. 2d 615, 624 (E.D. Pa. 2007) (quoting expert who testified that the Hill criteria are "'well-recognized' and widely used in the science community to assess general causation"); Chapin v. A & L Parts, Inc., 732 N.W.2d 578, 584 (Mich. Ct. App. 2007) (expert testified that Hill criteria are the most well-utilized method for determining if an association is causal).

143. Berry v. CSX Transp., Inc., 709 So. 2d 552, 568 n.12 (Fla. Dist. Ct. App. 1998) ("Almost all genres of research articles in the medical and behavioral sciences conclude their discussion with qualifying statements such as 'there is still much to be learned.' This is not, as might be assumed, an expression of ignorance, but rather an expression that all scientific fields are open-ended and can progress from their present state. . . ."); Hall v. Baxter Healthcare Corp., 947 F. Supp. 1387 app. B. at 1446–51 (D. Or. 1996) (report of Merwyn R. Greenlick, court-appointed epidemiologist). In *Cadarian v. Merrell Dow Pharmaceuticals, Inc.*, 745 F. Supp. 409 (E.D. Mich. 1989), the court refused to permit an expert to rely on a study that the authors had concluded should not be used to support an inference of causation in the absence of independent confirmatory studies. The court did not address the question whether the degree of certainty used by epidemiologists before making a conclusion of cause was consistent with the legal standard. *See* DeLuca v. Merrell Dow Pharms., Inc., 911 F.2d 941, 957 (3d Cir. 1990) (standard of proof for scientific community is not necessarily appropriate standard for expert opinion in civil litigation); Wells v. Ortho Pharm. Corp., 788 F.2d 741, 745 (11th Cir. 1986).

144. *See* Cook v. Rockwell Int'l Corp., 580 F. Supp. 2d 1071, 1098 (D. Colo. 2006) ("Defendants cite no authority, scientific or legal, that compliance with all, or even one, of these factors is required. . . . The scientific consensus is, in fact, to the contrary. It identifies Defendants' list of factors as some of the nine factors or lenses that guide epidemiologists in making judgments about causation. . . . These factors are not tests for determining the reliability of any study or the causal inferences drawn from it.").

1. Temporal relationship,
2. Strength of the association,
3. Dose–response relationship,
4. Replication of the findings,
5. Biological plausibility (coherence with existing knowledge),
6. Consideration of alternative explanations,
7. Cessation of exposure,
8. Specificity of the association, and
9. Consistency with other knowledge.

There is no formula or algorithm that can be used to assess whether a causal inference is appropriate based on these guidelines.[145] One or more factors may be absent even when a true causal relationship exists.[146] Similarly, the existence of some factors does not ensure that a causal relationship exists. Drawing causal inferences after finding an association and considering these factors requires judgment and searching analysis, based on biology, of why a factor or factors may be absent despite a causal relationship, and vice versa. Although the drawing of causal inferences is informed by scientific expertise, it is not a determination that is made by using an objective or algorithmic methodology.

These guidelines reflect criteria proposed by the U.S. Surgeon General in 1964[147] in assessing the relationship between smoking and lung cancer and expanded upon by Sir Austin Bradford Hill in 1965[148] and are often referred to as the Hill criteria or Hill factors.

145. *See* Douglas L. Weed, *Epidemiologic Evidence and Causal Inference,* 14 Hematology/Oncology Clinics N. Am. 797 (2000).

146. *See* Cook v. Rockwell Int'l Corp., 580 F. Supp. 2d 1071, 1098 (D. Colo. 2006) (rejecting argument that plaintiff failed to provide sufficient evidence of causation based on failing to meet four of the Hill factors).

147. Public Health Serv., U.S. Dep't of Health, Educ., & Welfare, Smoking and Health: Report of the Advisory Committee to the Surgeon General (1964); *see also* Centers for Disease Control and Prevention, U.S. Dep't of Health & Human Servs., The Health Consequences of Smoking: A Report of the Surgeon General (2004).

148. *See* Austin Bradford Hill, *The Environment and Disease: Association or Causation?* 58 Proc. Royal Soc'y Med. 295 (1965) (Hill acknowledged that his factors could only serve to assist in the inferential process: "None of my nine viewpoints can bring indisputable evidence for or against the cause-and-effect hypothesis and none can be required as a *sine qua non.*"). For discussion of these criteria and their respective strengths in informing a causal inference, see Gordis, *supra* note 32, at 236–39; David E. Lilienfeld & Paul D. Stolley, Foundations of Epidemiology 263–66 (3d ed. 1994); Weed, *supra* note 144.

A. Is There a Temporal Relationship?

A temporal, or chronological, relationship must exist for causation to exist. If an exposure causes disease, the exposure must occur before the disease develops.[149] If the exposure occurs after the disease develops, it cannot have caused the disease. Although temporal relationship is often listed as one of many factors in assessing whether an inference of causation is justified, this aspect of a temporal relationship is a necessary factor: Without exposure before the disease, causation cannot exist.[150]

With regard to specific causation, a subject dealt with in detail in Section VII, *infra*, there may be circumstances in which a temporal relationship supports the existence of a causal relationship. If the latency period between exposure and outcome is known,[151] then exposure consistent with that information may lend credence to a causal relationship. This is particularly true when the latency period is short and competing causes are known and can be ruled out. Thus, if an individual suffers an acute respiratory response shortly after exposure to a suspected agent and other causes of that respiratory problem are known and can be ruled out, the temporal relationship involved supports the conclusion that a causal relationship exists.[152] Similarly, exposure outside a known latency period constitutes evidence, perhaps conclusive evidence, against the existence of causation.[153] On the other hand, when latency periods are lengthy, variable, or not known and a

149. *See* Carroll v. Litton Sys., Inc., No. B-C-88-253, 1990 U.S. Dist. LEXIS 16833, at *29 (W.D.N.C. 1990) ("[I]t is essential for . . . [the plaintiffs' medical experts opining on causation] to know that exposure preceded plaintiffs' alleged symptoms in order for the exposure to be considered as a possible cause of those symptoms. . . .").

150. Exposure during the disease initiation process may cause the disease to be more severe than it otherwise would have been without the additional dose.

151. When the latency period is known—or is known to be limited to a specific range of time—as is the case with the adverse effects of some vaccines, the time frame from exposure to manifestation of disease can be critical to determining causation.

152. For courts that have relied on temporal relationships of the sort described, see Bonner v. ISP Technologies, Inc., 259 F.3d 924, 930–31 (8th Cir. 2001) (giving more credence to the expert's opinion on causation for acute response based on temporal relationship than for chronic disease that plaintiff also developed); Heller v. Shaw Industries, Inc. 167 F.3d 146 (3d Cir. 1999); Westberry v. Gislaved Gummi AB, 178 F.3d 257 (4th Cir. 1999); Zuchowicz v. United States, 140 F.3d 381 (2d Cir. 1998); Creanga v. Jardal, 886 A.2d 633, 641 (N.J. 2005); Alder v. Bayer Corp., AGFA Div., 61 P.3d 1068, 1090 (Utah 2002) ("If a bicyclist falls and breaks his arm, causation is assumed without argument because of the temporal relationship between the accident and the injury [and, the court might have added, the absence of any plausible competing causes that might instead be responsible for the broken arm].").

153. *See In re* Phenylpropanolamine (PPA) Prods. Liab. Litig., 289 F. Supp. 2d 1230, 1238 (W.D. Wash. 2003) (determining expert testimony on causation for plaintiffs whose exposure was beyond known latency period was inadmissible).

substantial proportion of the disease is due to unknown causes, temporal relationship provides little beyond satisfying the requirement that cause precede effect.[154]

B. How Strong Is the Association Between the Exposure and Disease?[155]

The relative risk is one of the cornerstones for causal inferences.[156] Relative risk measures the strength of the association. The higher the relative risk, the greater the likelihood that the relationship is causal.[157] For cigarette smoking, for example, the estimated relative risk for lung cancer is very high, about 10.[158] That is, the risk of lung cancer in smokers is approximately 10 times the risk in nonsmokers.

A relative risk of 10, as seen with smoking and lung cancer, is so high that it is extremely difficult to imagine any bias or confounding factor that might account for it. The higher the relative risk, the stronger the association and the lower the chance that the effect is spurious. Although lower relative risks can reflect causality, the epidemiologist will scrutinize such associations more closely because there is a greater chance that they are the result of uncontrolled confounding or biases.

154. These distinctions provide a framework for distinguishing between cases that are largely dismissive of temporal relationships as supporting causation and others that find it of significant persuasiveness. *Compare* cases cited in note 151, *supra*, *with* Moore v. Ashland Chem. Inc., 151 F.3d 269, 278 (5th Cir. 1998) (giving little weight to temporal relationship in a case in which there were several plausible competing causes that may have been responsible for the plaintiff's disease), *and* Glastetter v. Novartis Pharms. Corp., 252 F.3d 986, 990 (8th Cir. 2001) (giving little weight to temporal relationship in case studies involving drug and stroke).

155. Assuming that an association is determined to be causal, the strength of the association plays an important role legally in determining the specific causation question—whether the agent caused an individual plaintiff's injury. *See infra* Section VII.

156. *See supra* Section III.A.

157. *See* Miller v. Pfizer, Inc., 196 F. Supp. 2d 1062, 1079 (D. Kan. 2002) (citing this reference guide); Landrigan v. Celotex Corp., 605 A.2d 1079, 1085 (N.J. 1992). The use of the strength of the association as a factor does not reflect a belief that weaker effects occur less frequently than stronger effects. *See* Green, *supra* note 47, at 652–53 n.39. Indeed, the apparent strength of a given agent is dependent on the prevalence of the other necessary elements that must occur with the agent to produce the disease, rather than on some inherent characteristic of the agent itself. *See* Rothman et al., *supra* note 61, at 9–11.

158. *See* Doll & Hill, *supra* note 6. The relative risk of lung cancer from smoking is a function of intensity and duration of dose (and perhaps other factors). *See* Karen Leffondré et al., *Modeling Smoking History: A Comparison of Different Approaches,* 156 Am. J. Epidemiology 813 (2002). The relative risk provided in the text is based on a specified magnitude of cigarette exposure.

C. Is There a Dose–Response Relationship?

A dose–response relationship means that the greater the exposure, the greater the risk of disease. Generally, higher exposures should increase the incidence (or severity) of disease.[159] However, some causal agents do not exhibit a dose–response relationship when, for example, there is a threshold phenomenon (i.e., an exposure may not cause disease until the exposure exceeds a certain dose).[160] Thus, a dose–response relationship is strong, but not essential, evidence that the relationship between an agent and disease is causal.[161]

159. *See* Newman v. Motorola, Inc., 218 F. Supp. 2d 769, 778 (D. Md. 2002) (recognizing importance of dose–response relationship in assessing causation).

160. The question whether there is a no-effect threshold dose is a controversial one in a variety of toxic substances areas. *See, e.g.*, Irving J. Selikoff, Disability Compensation for Asbestos-Associated Disease in the United States: Report to the U.S. Department of Labor 181–220 (1981); Paul Kotin, *Dose–Response Relationships and Threshold Concepts*, 271 Ann. N.Y. Acad. Sci. 22 (1976); K. Robock, *Based on Available Data, Can We Project an Acceptable Standard for Industrial Use of Asbestos? Absolutely*, 330 Ann. N.Y. Acad. Sci. 205 (1979); Ferebee v. Chevron Chem. Co., 736 F.2d 1529, 1536 (D.C. Cir. 1984) (dose–response relationship for low doses is "one of the most sharply contested questions currently being debated in the medical community"); *In re* TMI Litig. Consol. Proc., 927 F. Supp. 834, 844–45 (M.D. Pa. 1996) (discussing low-dose extrapolation and no-dose effects for radiation exposure).

Moreover, good evidence to support or refute the threshold-dose hypothesis is exceedingly unlikely because of the inability of epidemiology or animal toxicology to ascertain very small effects. *Cf.* Arnold L. Brown, *The Meaning of Risk Assessment*, 37 Oncology 302, 303 (1980). Even the shape of the dose–response curve—whether linear or curvilinear, and if the latter, the shape of the curve—is a matter of hypothesis and speculation. *See* Allen v. United States, 588 F. Supp. 247, 419–24 (D. Utah 1984), *rev'd on other grounds*, 816 F.2d 1417 (10th Cir. 1987); *In re* Bextra & Celebrex Mktg. Sales Practices & Prod. Liab. Litig., 524 F. Supp. 2d 1166, 1180 (N.D. Cal. 2007) (criticizing expert for "primitive" extrapolation of risk based on assumption of linear relationship of risk to dose); Troyen A. Brennan & Robert F. Carter, *Legal and Scientific Probability of Causation for Cancer and Other Environmental Disease in Individuals*, 10 J. Health Pol'y & L. 33, 43–44 (1985).

The idea that the "dose makes the poison" is a central tenet of toxicology and attributed to Paracelsus, in the sixteenth century. *See* Bernard D. Goldstein & Mary Sue Henifin, Reference Guide on Toxicology, Section I.A, in this manual. It does not mean that any agent is capable of causing any disease if an individual is exposed to a sufficient dose. Agents tend to have specific effects, *see infra* Section V.H., and this dictum reflects only the idea that there is a safe dose below which an agent does not cause any toxic effect. *See* Michael A Gallo, *History and Scope of Toxicology*, in Casarett and Doull's Toxicology: The Basic Science of Poisons 1, 4–5 (Curtis D. Klaassen ed., 7th ed. 2008). For a case in which a party made such a mistaken interpretation of Paracelsus, see Alder v. Bayer Corp., AGFA Div., 61 P.3d 1068, 1088 (Utah 2002). Paracelsus was also responsible for the initial articulation of the specificity tenet. *See infra* Section V.H.

161. Evidence of a dose–response relationship as bearing on whether an inference of general causation is justified is analytically distinct from determining whether evidence of the dose to which a plaintiff was exposed is required in order to establish specific causation. On the latter matter, see *infra* Section VII; Restatement (Third) of Torts: Liability for Physical and Emotional Harm § 28 cmt. c(2) & rptrs. note (2010).

D. Have the Results Been Replicated?

Rarely, if ever, does a single study persuasively demonstrate a cause–effect relationship.[162] It is important that a study be replicated in different populations and by different investigators before a causal relationship is accepted by epidemiologists and other scientists.[163]

The need to replicate research findings permeates most fields of science. In epidemiology, research findings often are replicated in different populations.[164] Consistency in these findings is an important factor in making a judgment about causation. Different studies that examine the same exposure–disease relationship generally should yield similar results. Although inconsistent results do not necessarily rule out a causal nexus, any inconsistencies signal a need to explore whether different results can be reconciled with causality.

E. Is the Association Biologically Plausible (Consistent with Existing Knowledge)?[165]

Biological plausibility is not an easy criterion to use and depends upon existing knowledge about the mechanisms by which the disease develops. When biological plausibility exists, it lends credence to an inference of causality. For example, the conclusion that high cholesterol is a cause of coronary heart disease is plausible because cholesterol is found in atherosclerotic plaques. However, observations have been made in epidemiologic studies that were not biologically plausible at the time but subsequently were shown to be correct.[166] When an observation is inconsistent with current biological knowledge, it should not be discarded, but

162. In *Kehm v. Procter & Gamble Co.*, 580 F. Supp. 890, 901 (N.D. Iowa 1982), *aff'd*, 724 F.2d 613 (8th Cir. 1983), the court remarked on the persuasive power of multiple independent studies, each of which reached the same finding of an association between toxic shock syndrome and tampon use.

163. This may not be the legal standard, however. *Cf.* Smith v. Wyeth-Ayerst Labs. Co., 278 F. Supp. 2d 684, 710 n.55 (W.D.N.C. 2003) (observing that replication is difficult to establish when there is only one study that has been performed at the time of trial).

164. *See* Cadarian v. Merrell Dow Pharms., Inc., 745 F. Supp. 409, 412 (E.D. Mich. 1989) (holding a study on Bendectin insufficient to support an expert's opinion, because "the study's authors themselves concluded that the results could not be interpreted without independent confirmatory evidence").

165. A number of courts have adverted to this criterion in the course of their discussions of causation in toxic substances cases. *E.g., In re* Phenylpropanolamine (PPA) Prods. Liab. Litig., 289 F. Supp. 2d 1230, 1247–48 (W.D. Wash. 2003); Cook v. United States, 545 F. Supp. 306, 314–15 (N.D. Cal. 1982) (discussing biological implausibility of a two-peak increase of disease when plotted against time); Landrigan v. Celotex Corp., 605 A.2d 1079, 1085–86 (N.J. 1992) (discussing the existence vel non of biological plausibility); *see also* Bernard D. Goldstein & Mary Sue Henifin, Reference Guide on Toxicology, Section III.E, in this manual.

166. *See In re* Rezulin Prods. Liab. Litig., 369 F. Supp. 2d 398, 405 (S.D.N.Y. 2005); *In re* Phenylpropanolamine (PPA) Prods. Liab. Litig., 289 F. Supp. 2d 1230, 1247 (W.D. Wash. 2003).

the observation should be confirmed before significance is attached to it. The saliency of this factor varies depending on the extent of scientific knowledge about the cellular and subcellular mechanisms through which the disease process works. The mechanisms of some diseases are understood quite well based on the available evidence, including from toxicologic research, whereas other mechanism explanations are merely hypothesized—although hypotheses are sometimes accepted under this factor.[167]

F. Have Alternative Explanations Been Considered?

The importance of considering the possibility of bias and confounding and ruling out the possibilities is discussed above.[168]

G. What Is the Effect of Ceasing Exposure?

If an agent is a cause of a disease, then one would expect that cessation of exposure to that agent ordinarily would reduce the risk of the disease. This has been the case, for example, with cigarette smoking and lung cancer. In many situations, however, relevant data are simply not available regarding the possible effects of ending the exposure. But when such data are available and eliminating exposure reduces the incidence of disease, this factor strongly supports a causal relationship.

H. Does the Association Exhibit Specificity?

An association exhibits specificity if the exposure is associated only with a single disease or type of disease.[169] The vast majority of agents do not cause a wide vari-

167. *See* Douglas L. Weed & Stephen D. Hursting, *Biologic Plausibility in Causal Inference: Current Methods and Practice*, 147 Am. J. Epidemiology 415 (1998) (examining use of this criterion in contemporary epidemiologic research and distinguishing between alternative explanations of what constitutes biological plausibility, ranging from mere hypotheses to "sufficient evidence to show how the factor influences a known disease mechanism").

168. *See supra* Sections IV.B–C.

169. This criterion reflects the fact that although an agent causes one disease, it does not necessarily cause other diseases. *See, e.g.*, Nelson v. Am. Sterilizer Co., 566 N.W.2d 671, 676–77 (Mich. Ct. App. 1997) (affirming dismissal of plaintiff's claims that chemical exposure caused her liver disorder, but recognizing that evidence supported claims for neuropathy and other illnesses); Sanderson v. Int'l Flavors & Fragrances, Inc., 950 F. Supp. 981, 996–98 (C.D. Cal. 1996); *see also* Taylor v. Airco, Inc., 494 F. Supp. 2d 21, 27 (D. Mass. 2007) (holding that plaintiff's expert could testify to causal relationship between vinyl chloride and one type of liver cancer for which there was only modest support given strong causal evidence for vinyl chloride and another type of liver cancer).

When a party claims that evidence of a causal relationship between an agent and one disease is relevant to whether the agent caused another disease, courts have required the party to show that

ety of effects. For example, asbestos causes mesothelioma and lung cancer and may cause one or two other cancers, but there is no evidence that it causes any other types of cancers. Thus, a study that finds that an agent is associated with many different diseases should be examined skeptically. Nevertheless, there may be causal relationships in which this guideline is not satisfied. Cigarette manufacturers have long claimed that because cigarettes have been linked to lung cancer, emphysema, bladder cancer, heart disease, pancreatic cancer, and other conditions, there is no specificity and the relationships are not causal. There is, however, at least one good reason why inferences about the health consequences of tobacco do not require specificity: Because tobacco and cigarette smoke are not in fact single agents but consist of numerous harmful agents, smoking represents exposure to multiple agents, with multiple possible effects. Thus, whereas evidence of specificity may strengthen the case for causation, lack of specificity does not necessarily undermine it where there is a good biological explanation for its absence.

I. Are the Findings Consistent with Other Relevant Knowledge?

In addressing the causal relationship of lung cancer to cigarette smoking, researchers examined trends over time for lung cancer and for cigarette sales in the United States. A marked increase in lung cancer death rates in men was observed, which appeared to follow the increase in sales of cigarettes. Had the increase in lung cancer deaths followed a decrease in cigarette sales, it might have given researchers pause. It would not have precluded a causal inference, but the inconsistency of the trends in cigarette sales and lung cancer mortality would have had to be explained.

VI. What Methods Exist for Combining the Results of Multiple Studies?

Not infrequently, the scientific record may include a number of epidemiologic studies whose findings differ. These may be studies in which one shows an association and the other does not, or studies that report associations, but of different

the mechanisms involved in development of the disease are similar. Thus, in *Austin v. Kerr-McGee Refining Corp.*, 25 S.W.3d 280 (Tex. App. 2000), the plaintiff suffered from a specific form of chronic leukemia. Studies demonstrated a causal relationship between benzene and all leukemias, but there was a paucity of evidence on the relationship between benzene and the specific form of leukemia from which plaintiff suffered. The court required that plaintiff's expert demonstrate the similarity of the biological mechanism among leukemias as a condition for the admissibility of his causation testimony, a requirement the court concluded had not been satisfied. *Accord In re* Bextra & Celebrex Mktg. Sales Practices & Prod. Liab. Litig., 524 F. Supp. 2d 1166, 1183 (N.D. Cal. 2007); Magistrini v. One Hour Martinizing Dry Cleaning, 180 F. Supp. 2d 584, 603 (D.N.J. 2002).

magnitude.[170] In view of the fact that studies may disagree and that often many of the studies are small and lack the statistical power needed for definitive conclusions, the technique of meta-analysis was developed, initially for clinical trials.[171] Meta-analysis is a method of pooling study results to arrive at a single figure to represent the totality of the studies reviewed.[172] It is a way of systematizing the time-honored approach of reviewing the literature, which is characteristic of science, and placing it in a standardized framework with quantitative methods for estimating risk. In a meta-analysis, studies are given different weights in proportion to the sizes of their study populations and other characteristics.[173]

Meta-analysis is most appropriate when used in pooling randomized experimental trials, because the studies included in the meta-analysis share the most significant methodological characteristics, in particular, use of randomized assignment of subjects to different exposure groups. However, often one is confronted with nonrandomized observational studies of the effects of possible toxic substances or agents. A method for summarizing such studies is greatly needed, but when meta-analysis is applied to observational studies—either case-control or cohort—it becomes more controversial.[174] The reason for this is that often methodological differences among studies are much more pronounced than they are in randomized trials. Hence, the justification for pooling the results and deriving a single estimate of risk, for example, is problematic.[175]

170. *See, e.g.*, Zandi v. Wyeth a/k/a Wyeth, Inc., No. 27-CV-06-6744, 2007 WL 3224242 (Minn. Dist. Ct. Oct. 15, 2007) (plaintiff's expert cited 40 studies in support of a causal relationship between hormone therapy and breast cancer; many studies found different magnitudes of increased risk).

171. *See In re* Paoli R.R. Yard PCB Litig., 916 F.2d 829, 856 (3d Cir. 1990), *cert. denied,* 499 U.S. 961 (1991); Hines v. Consol. Rail Corp., 926 F.2d 262, 273 (3d Cir. 1991); Allen v. Int'l Bus. Mach. Corp., No. 94-264-LON, 1997 U.S. Dist. LEXIS 8016, at *71–*74 (meta-analysis of observational studies is a controversial subject among epidemiologists). Thus, contrary to the suggestion by at least one court, multiple studies with small numbers of subjects may be pooled to reduce the possibility of sampling error. *See In re* Joint E. & S. Dist. Asbestos Litig., 827 F. Supp. 1014, 1042 (S.D.N.Y. 1993) ("[N]o matter how many studies yield a positive but statistically insignificant SMR for colorectal cancer, the results remain statistically insignificant. Just as adding a series of zeros together yields yet another zero as the product, adding a series of positive but statistically insignificant SMRs together does not produce a statistically significant pattern."), rev'd, 52 F.3d 1124 (2d Cir. 1995); *see also supra* note 76.

172. For a nontechnical explanation of meta-analysis, along with case studies of a variety of scientific areas in which it has been employed, see Morton Hunt, How Science Takes Stock: The Story of Meta-Analysis (1997).

173. Petitti, *supra* note 88.

174. *See* Donna F. Stroup et al., *Meta-analysis of Observational Studies in Epidemiology: A Proposal for Reporting,* 283 JAMA 2008, 2009 (2000); Jesse A. Berlin & Carin J. Kim, *The Use of Meta-Analysis in Pharmacoepidemiology, in* Pharmacoepidemiology 681, 683–84 (Brian L. Strom ed., 4th ed. 2005).

175. On rare occasions, meta-analyses of both clinical and observational studies are available. *See, e.g., In re* Bextra & Celebrex Mktg. Sales Practices & Prod. Liab. Litig., 524 F. Supp. 2d 1166, 1175 (N.D. Cal. 2007) (referring to clinical and observational meta-analyses of low dose of a drug; both analyses failed to find any effect).

A number of problems and issues arise in meta-analysis. Should only published papers be included in the meta-analysis, or should any available studies be used, even if they have not been peer reviewed? Can the results of the meta-analysis itself be reproduced by other analysts? When there are several meta-analyses of a given relationship, why do the results of different meta-analyses often disagree? The appeal of a meta-analysis is that it generates a single estimate of risk (along with an associated confidence interval), but this strength can also be a weakness, and may lead to a false sense of security regarding the certainty of the estimate. A key issue is the matter of heterogeneity of results among the studies being summarized. If there is more variance among study results than one would expect by chance, this creates further uncertainty about the summary measure from the meta-analysis. Such differences can arise from variations in study quality, or in study populations or in study designs. Such differences in results make it harder to trust a single estimate of effect; the reasons for such differences need at least to be acknowledged and, if possible, explained.[176] People often tend to have an inordinate belief in the validity of the findings when a single number is attached to them, and many of the difficulties that may arise in conducting a meta-analysis, especially of observational studies such as epidemiologic ones, may consequently be overlooked.[177]

VII. What Role Does Epidemiology Play in Proving Specific Causation?

Epidemiology is concerned with the incidence of disease in populations, and epidemiologic studies do not address the question of the cause of an individual's disease.[178] This question, often referred to as specific causation, is beyond the

176. *See* Stroup et al., *supra* note 173 (recommending methodology for meta-analysis of observational studies).

177. Much has been written about meta-analysis recently and some experts consider the problems of meta-analysis to outweigh the benefits at the present time. For example, John Bailar has observed:

> [P]roblems have been so frequent and so deep, and overstatements of the strength of conclusions so extreme, that one might well conclude there is something seriously and fundamentally wrong with the method. For the present . . . I still prefer the thoughtful, old-fashioned review of the literature by a knowledgeable expert who explains and defends the judgments that are presented. We have not yet reached a stage where these judgments can be passed on, even in part, to a formalized process such as meta-analysis.

John C. Bailar III, *Assessing Assessments,* 277 Science 528, 529 (1997) (reviewing Morton Hunt, How Science Takes Stock (1997)); *see also Point/Counterpoint: Meta-analysis of Observational Studies,* 140 Am. J. Epidemiology 770 (1994).

178. *See* DeLuca v. Merrell Dow Pharms., Inc., 911 F.2d 941, 945 & n.6 (3d Cir. 1990) ("Epidemiological studies do not provide direct evidence that a particular plaintiff was injured by exposure to a substance."); *In re* Viagra Prods. Liab. Litig., 572 F. Supp. 2d 1071, 1078 (D. Minn. 2008) ("Epi-

domain of the science of epidemiology. Epidemiology has its limits at the point where an inference is made that the relationship between an agent and a disease is causal (general causation) and where the magnitude of excess risk attributed to the agent has been determined; that is, epidemiologists investigate whether an agent can cause a disease, not whether an agent did cause a specific plaintiff's disease.[179]

Nevertheless, the specific causation issue is a necessary legal element in a toxic substance case. The plaintiff must establish not only that the defendant's agent is capable of causing disease, but also that it did cause the plaintiff's disease. Thus, numerous cases have confronted the legal question of what is acceptable proof of specific causation and the role that epidemiologic evidence plays in answering that question.[180] This question is not a question that is addressed by epidemiology.[181] Rather, it is a legal question with which numerous courts

demiology focuses on the question of general causation (i.e., is the agent capable of causing disease?) rather than that of specific causation (i.e., did it cause a disease in a particular individual?)" (quoting the second edition of this reference guide)); *In re* Asbestos Litig,, 900 A.2d 120, 133 (Del. Super. Ct. 2006); Michael Dore, *A Commentary on the Use of Epidemiological Evidence in Demonstrating Cause-in-Fact,* 7 Harv. Envtl. L. Rev. 429, 436 (1983).

There are some diseases that do not occur without exposure to a given toxic agent. This is the same as saying that the toxic agent is a necessary cause for the disease, and the disease is sometimes referred to as a signature disease (also, the agent is pathognomonic), because the existence of the disease necessarily implies the causal role of the agent. *See* Kenneth S. Abraham & Richard A. Merrill, *Scientific Uncertainty in the Courts,* Issues Sci. & Tech. 93, 101 (1986). Asbestosis is a signature disease for asbestos, and vaginal adenocarcinoma (in young adult women) is a signature disease for in utero DES exposure.

179. *Cf. In re* "Agent Orange" Prod. Liab. Litig., 597 F. Supp. 740, 780 (E.D.N.Y. 1984) (Agent Orange allegedly caused a wide variety of diseases in Vietnam veterans and their offspring), *aff'd,* 818 F.2d 145 (2d Cir. 1987).

180. In many instances, causation can be established without epidemiologic evidence. When the mechanism of causation is well understood, the causal relationship is well established, or the timing between cause and effect is close, scientific evidence of causation may not be required. This is frequently the situation when the plaintiff suffers traumatic injury rather than disease. This section addresses only those situations in which causation is not evident, and scientific evidence is required.

181. Nevertheless, an epidemiologist may be helpful to the factfinder in answering this question. Some courts have permitted epidemiologists (or those who use epidemiologic methods) to testify about specific causation. *See* Ambrosini v. Labarraque, 101 F.3d 129, 137–41 (D.C. Cir. 1996); Zuchowicz v. United States, 870 F. Supp. 15 (D. Conn. 1994); Landrigan v. Celotex Corp., 605 A.2d 1079, 1088–89 (N.J. 1992). In general, courts seem more concerned with the basis of an expert's opinion than with whether the expert is an epidemiologist or clinical physician. *See* Porter v. Whitehall, 9 F.3d 607, 614 (7th Cir. 1992) ("curb side" opinion from clinician not admissible); Burton v. R.J. Reynolds Tobacco Co., 181 F. Supp. 2d 1256, 1266–67 (D. Kan. 2002) (vascular surgeon permitted to testify to general causation over objection based on fact he was not an epidemiologist); Wade-Greaux v. Whitehall Labs., 874 F. Supp. 1441, 1469–72 (D.V.I.) (clinician's multiple bases for opinion inadequate to support causation opinion), *aff'd,* 46 F.3d 1120 (3d Cir. 1994); *Landrigan,* 605 A.2d at 1083–89 (permitting both clinicians and epidemiologists to testify to specific causation provided the methodology used is sound); Trach v. Fellin, 817 A.2d 1102, 1118–19 (Pa. Super. Ct. 2003) (toxicologist and pathologist permitted to testify to specific causation).

have grappled.[182] The remainder of this section is predominantly an explanation of judicial opinions. It is, in addition, in its discussion of the reasoning behind applying the risk estimates of an epidemiologic body of evidence to an individual, informed by epidemiologic principles and methodological research.

Before proceeding, one more caveat is in order. This section assumes that epidemiologic evidence has been used as proof of causation for a given plaintiff. The discussion does not address whether a plaintiff must use epidemiologic evidence to prove causation.[183]

Two legal issues arise with regard to the role of epidemiology in proving individual causation: admissibility and sufficiency of evidence to meet the burden of production. The first issue tends to receive less attention by the courts but nevertheless deserves mention. An epidemiologic study that is sufficiently rigorous to justify a conclusion that it is scientifically valid should be admissible,[184] as it tends to make an issue in dispute more or less likely.[185]

182. *See* Restatement (Third) of Torts: Liability for Physical and Emotional Harm § 28 cmt. c(3) (2010) ("Scientists who conduct group studies do not examine specific causation in their research. No scientific methodology exists for assessing specific causation for an individual based on group studies. Nevertheless, courts have reasoned from the preponderance-of-the-evidence standard to determine the sufficiency of scientific evidence on specific causation when group-based studies are involved").

183. *See id.* § 28 cmt. c(3) & rptrs. note ("most courts have appropriately declined to impose a threshold requirement that a plaintiff always must prove causation with epidemiologic evidence"); *see also* Westberry v. Gislaved Gummi AB, 178 F.2d 257 (4th Cir. 1999) (acute response, differential diagnosis ruled out other known causes of disease, dechallenge, rechallenge tests by expert that were consistent with exposure to defendant's agent causing disease, and absence of epidemiologic or toxicologic studies; holding that expert's testimony on causation was properly admitted); Zuchowicz v. United States, 140 F.3d 381 (2d Cir. 1998); *In re* Heparin Prods. Liab. Litig. 2011 WL 2971918, at *7-10 (N.D. Ohio July 21, 2011).

184. *See* DeLuca v. Merrell Dow Pharms., Inc., 911 F.2d 941, 958 (3d Cir. 1990); *cf.* Kehm v. Procter & Gamble Co., 580 F. Supp. 890, 902 (N.D. Iowa 1982) ("These [epidemiologic] studies were highly probative on the issue of causation—they all concluded that an association between tampon use and menstrually related TSS [toxic shock syndrome] cases exists."), *aff'd,* 724 F.2d 613 (8th Cir. 1984).

Hearsay concerns may limit the independent admissibility of the study, but the study could be relied on by an expert in forming an opinion and may be admissible pursuant to Fed. R. Evid. 703 as part of the underlying facts or data relied on by the expert.

In *Ellis v. International Playtex, Inc.,* 745 F.2d 292, 303 (4th Cir. 1984), the court concluded that certain epidemiologic studies were admissible despite criticism of the methodology used in the studies. The court held that the claims of bias went to the studies' weight rather than their admissibility. *Cf.* Christophersen v. Allied-Signal Corp., 939 F.2d 1106, 1109 (5th Cir. 1991) ("As a general rule, questions relating to the bases and sources of an expert's opinion affect the weight to be assigned that opinion rather than its admissibility. . . . ").

185. Even if evidence is relevant, it may be excluded if its probative value is substantially outweighed by prejudice, confusion, or inefficiency. Fed. R. Evid. 403. However, exclusion of an otherwise relevant epidemiologic study on Rule 403 grounds is unlikely.

In *Daubert v. Merrell Dow Pharmaceuticals, Inc.,* 509 U.S. 579, 591 (1993), the Court invoked the concept of "fit," which addresses the relationship of an expert's scientific opinion to the facts of the case and the issues in dispute. In a toxic substance case in which cause in fact is disputed, an epi-

Far more courts have confronted the role that epidemiology plays with regard to the sufficiency of the evidence and the burden of production.[186] The civil burden of proof is described most often as requiring belief by the factfinder "that what is sought to be proved is more likely true than not true."[187] The relative risk from epidemiologic studies can be adapted to this 50%-plus standard to yield a probability or likelihood that an agent caused an individual's disease.[188] An important caveat is necessary, however. The discussion below speaks in terms of the magnitude of the relative risk or association found in a study. However, before an association or relative risk is used to make a statement about the probability of individual causation, the inferential judgment, described in Section V, that the association is truly causal rather than spurious, is required: "[A]n agent cannot be considered to cause the illness of a specific person unless it is recognized as a cause of that disease in general."[189] The following discussion should be read with this caveat in mind.[190]

demiologic study of the same agent to which the plaintiff was exposed that examined the association with the same disease from which the plaintiff suffers would undoubtedly have sufficient "fit" to be a part of the basis of an expert's opinion. The Court's concept of "fit," borrowed from *United States v. Downing,* 753 F.2d 1224, 1242 (3d Cir. 1985), appears equivalent to the more familiar evidentiary concept of probative value, albeit one requiring assessment of the scientific reasoning the expert used in drawing inferences from methodology or data to opinion.

186. We reiterate a point made at the outset of this section: This discussion of the use of a threshold relative risk for specific causation is not epidemiology or an inquiry an epidemiologist would undertake. This is an effort by courts and commentators to adapt the legal standard of proof to the available scientific evidence. See *supra* text accompanying notes 175–179. While strength of association is a guideline for drawing an inference of causation from an association, *see supra* Section V, there is no specified threshold required.

187. Kevin F. O'Malley et al., Federal Jury Practice and Instructions § 104.01 (5th ed. 2000); *see also* United States v. Fatico, 458 F. Supp. 388, 403 (E.D.N.Y. 1978) ("Quantified, the preponderance standard would be 50%+ probable."), *aff'd,* 603 F.2d 1053 (2d Cir. 1979).

188. An adherent of the frequentist school of statistics would resist this adaptation, which may explain why many epidemiologists and toxicologists also resist it. To take the step identified in the text of using an epidemiologic study outcome to determine the probability of specific causation requires a shift from a frequentist approach, which involves sampling or frequency data from an empirical test, to a subjective probability about a discrete event. Thus, a frequentist might assert, after conducting a sampling test, that 60% of the balls in an opaque container are blue. The same frequentist would resist the statement, "The probability that a single ball removed from the box and hidden behind a screen is blue is 60%." The ball is either blue or not, and no frequentist data would permit the latter statement. "[T]here is no logically rigorous definition of what a statement of probability means with reference to an individual instance. . . ." Lee Loevinger, *On Logic and Sociology,* 32 Jurimetrics J. 527, 530 (1992); *see also* Steve Gold, *Causation in Toxic Torts: Burdens of Proof, Standards of Persuasion and Statistical Evidence,* 96 Yale L.J. 376, 382–92 (1986). Subjective probabilities about unique events are employed by those using Bayesian methodology. *See* Kaye, *supra* note 80, at 54–62; David H. Kaye & David A. Freedman, Reference Guide on Statistics, Section IV.D, in this manual.

189. Cole, *supra* note 65, at 10,284.

190. We emphasize this caveat, both because it is not intuitive and because some courts have failed to appreciate the difference between an association and a causal relationship. *See, e.g.,* Forsyth v. Eli Lilly & Co., Civ. No. 95-00185 ACK, 1998 U.S. Dist. LEXIS 541, at *26–*31 (D. Haw. Jan. 5, 1998). *But see*

Some courts have reasoned that when epidemiologic studies find that exposure to the agent causes an incidence in the exposed group that is more than twice the incidence in the unexposed group (i.e., a relative risk greater than 2.0), the probability that exposure to the agent caused a similarly situated individual's disease is greater than 50%.[191] These courts, accordingly, hold that when there is group-based evidence finding that exposure to an agent causes an incidence of disease in the exposed group that is more than twice the incidence in the unexposed group, the evidence is sufficient to satisfy the plaintiff's burden of production and permit submission of specific causation to a jury. In such a case, the factfinder may find that it is more likely than not that the substance caused the particular plaintiff's disease. Courts, thus, have permitted expert witnesses to testify to specific causation based on the logic of the effect of a doubling of the risk.[192]

While this reasoning has a certain logic as far as it goes, there are a number of significant assumptions and important caveats that require explication:

1. *A valid study and risk estimate.* The propriety of this "doubling" reasoning depends on group studies identifying a genuine causal relationship and a reasonably reliable measure of the increased risk.[193] This requires attention

Berry v. CSX Transp., Inc., 709 So. 2d 552, 568 (Fla. Dist. Ct. App. 1998) ("From epidemiologic studies demonstrating an association, an epidemiologist may or may not infer that a causal relationship exists.").

191. An alternative, yet similar, means to address probabilities in individual cases is use of the attributable fraction parameter, also known as the attributable risk. *See supra* Section III.C. The attributable fraction is that portion of the excess risk that can be attributed to an agent, above and beyond the background risk that is due to other causes. Thus, when the relative risk is greater than 2.0, the attributable fraction exceeds 50%.

192. For a comprehensive list of cases that support proof of causation based on group studies, see Restatement (Third) of Torts: Liability for Physical and Emotional Harm § 28 cmt. c(4) rptrs. note (2010). The Restatement catalogues those courts that require a relative risk in excess of 2.0 as a threshold for sufficient proof of specific causation and those courts that recognize that a lower relative risk than 2.0 can support specific causation, as explained below. Despite considerable disagreement on whether a relative risk of 2.0 is required or merely a taking-off point for determining the sufficiency of the evidence on specific causation, two commentators who surveyed the cases observed that "[t]here were no clear differences in outcomes as between federal and state courts." Russellyn S. Carruth & Bernard D. Goldstein, *Relative Risk Greater than Two in Proof of Causation in Toxic Tort Litigation,* 41 Jurimetrics J. 195, 199 (2001).

193. Indeed, one commentator contends that, because epidemiology is sufficiently imprecise to accurately measure small increases in risk, in general, studies that find a relative risk less than 2.0 should not be sufficient for causation. The concern is not with specific causation but with general causation and the likelihood that an association less than 2.0 is noise rather than reflecting a true causal relationship. *See* Michael D. Green, *The Future of Proportional Liability, in* Exploring Tort Law (Stuart Madden ed., 2005); *see also* Samuel M. Lesko & Allen A. Mitchell, *The Use of Randomized Controlled Trials for Pharmacoepidemiology Studies, in* Pharmacoepidemiology 599, 601 (Brian L. Strom ed., 4th ed. 2005) ("it is advisable to use extreme caution in making causal inferences from small relative risks derived from observational studies"); Gary Taubes, *Epidemiology Faces Its Limits,* 269 Science 164 (1995) (explaining views of several epidemiologists about a threshold relative risk of 3.0 to seriously consider a causal relationship); N.E. Breslow & N.E. Day, *Statistical Methods in Cancer Research, in* The Analysis

to the possibility of random error, bias, or confounding being the source of the association rather than a true causal relationship as explained in Sections IV and V, *supra.*[194]

2. *Similarity among study subjects and plaintiff.* Only if the study subjects and the plaintiff are similar with respect to other risk factors will a risk estimate from a study or studies be valid when applied to an individual.[195] Thus, if those exposed in a study of the risk of lung cancer from smoking smoked half a pack of cigarettes a day for 20 years, the degree of increased incidence of lung cancer among them cannot be extrapolated to someone who smoked two packs of cigarettes for 30 years without strong (and questionable) assumptions about the dose–response relationship.[196] This is also applicable to risk factors for competing causes. Thus, if all of the subjects in a study are participating because they were identified as having a family history of heart disease, the magnitude of risk found in a study of smok-

of Case-Control Studies 36 (IARC Pub. No. 32, 1980) ("[r]elative risks of less than 2.0 may readily reflect some unperceived bias or confounding factor"); David A. Freedman & Philip B. Stark, *The Swine Flu Vaccine and Guillain-Barré Syndrome: A Case Study in Relative Risk and Specific Causation,* 64 Law & Contemp. Probs. 49, 61 (2001) ("If the relative risk is near 2.0, problems of bias and confounding in the underlying epidemiologic studies may be serious, perhaps intractable.").

194. An excellent explanation for why differential diagnoses generally are inadequate without further proof of general causation was provided in *Cavallo v. Star Enterprises,* 892 F. Supp. 756 (E.D. Va. 1995), *aff'd in relevant part,* 100 F.3d 1150 (4th Cir. 1996):

> The process of differential diagnosis is undoubtedly important to the question of "specific causation". If other possible causes of an injury cannot be ruled out, or at least the probability of their contribution to causation minimized, then the "more likely than not" threshold for proving causation may not be met. But, it is also important to recognize that a fundamental assumption underlying this method is that the final, suspected "cause" remaining after this process of elimination must actually be capable of causing the injury. That is, the expert must "rule in" the suspected cause as well as "rule out" other possible causes. And, of course, expert opinion on this issue of "general causation" must be derived from a scientifically valid methodology.

Id. at 771 (footnote omitted); *see also* Ruggiero v. Warner-Lambert Co., 424 F.3d 249, 254 (2d Cir. 2005); Norris v. Baxter Healthcare Corp., 397 F.3d 878, 885 (10th Cir. 2005); Meister v. Med. Eng'g Corp., 267 F.3d 1123, 1128–29 (D.C. Cir. 2001); Bickel v. Pfizer, Inc., 431 F. Supp. 2d 918, 923–24 (N.D. Ind. 2006); *In re* Rezulin Prods. Liab. Litig., 369 F. Supp. 2d 398, 436 (S.D.N.Y. 2005); Coastal Tankships, U.S.A., Inc. v. Anderson, 87 S.W.3d 591, 608–09 (Tex. Ct. App. 2002); *see generally* Joseph Sanders & Julie Machal-Fulks, *The Admissibility of Differential Diagnosis Testimony to Prove Causation in Toxic Tort Cases: The Interplay of Adjective and Substantive Law,* 64 Law & Contemp. Probs. 107, 122–25 (2001) (discussing cases rejecting differential diagnoses in the absence of other proof of general causation and contrary cases).

195 "The basic premise of probability of causation is that individual risk can be determined from epidemiologic data for a representative population; however the premise only holds if the individual is truly representative of the reference population." Council on Scientific Affairs, American Medical Association, *Radioepidemiological Tables* 257 JAMA 806 (1987).

196. Conversely, a risk estimate from a study that involved a greater exposure is not applicable to an individual exposed to a lower dose. *See, e.g., In re* Bextra & Celebrex Mktg. Sales Practices & Prod. Liab. Litig., 524 F. Supp. 2d 1166, 1175–76 (N.D. Cal. 2007) (relative risk found in studies of those who took twice the dose of others could not support expert's opinion of causation for latter group).

ing on the risk of heart disease cannot validly be applied to an individual without such a family history. Finally, if an individual has been differentially exposed to other risk factors from those in a study, the results of the study will not provide an accurate basis for the probability of causation for the individual.[197] Consider once again a study of the effect of smoking on lung cancer among subjects who have no asbestos exposure. The relative risk of smoking in that study would not be applicable to an asbestos insulation worker. More generally, if the study subjects are heterogeneous with regard to risk factors related to the outcome of interest, the relative risk found in a study represents *an average risk for the group* rather than a uniform increased risk applicable to each individual.[198]

3. *Nonacceleration of disease.* Another assumption embedded in using the risk findings of a group study to determine the probability of causation in an individual is that the disease is one that never would have been contracted absent exposure. Put another way, the assumption is that the agent did not merely accelerate occurrence of the disease without affecting the lifetime risk of contracting the disease. Birth defects are an example of an outcome that is not accelerated. However, for most of the chronic diseases of adulthood, it is not possible for epidemiologic studies to distinguish between acceleration of disease and causation of new disease. If, in fact, acceleration

197. *See* David H. Kaye & David A. Freedman, Reference Guide on Statistics, in this manual (explaining the problems of employing a study outcome to determine the probability of an individual's having contracted the disease from exposure to the agent because of variations in individuals that bear on the risk of a given individual contracting the disease); David A. Freedman & Philip Stark, *The Swine Flu Vaccine and Guillain-Barré Syndrome: A Case Study in Relative Risk and Specific Causation,* 23 Evaluation Rev. 619 (1999) (analyzing the role that individual variation plays in determining the probability of specific causation based on the relative risk found in a study and providing a mathematical model for calculating the effect of individual variation); Mark Parascandola, *What Is Wrong with the Probability of Causation?* 39 Jurimetrics J. 29 (1998).

198. The comment of two prominent epidemiologists on this subject is illuminating:

> We cannot measure the individual risk, and assigning the average value to everyone in the category reflects nothing more than our ignorance about the determinants of lung cancer that interact with cigarette smoke. It is apparent from epidemiological data that some people can engage in chain smoking for many decades without developing lung cancer. Others are or will become primed by unknown circumstances and need only to add cigarette smoke to the nearly sufficient constellation of causes to initiate lung cancer. In our ignorance of these hidden causal components, the best we can do in assessing risk is to classify people according to measured causal risk indicators and then assign the average observed within a class to persons within the class.

Rothman & Greenland, *supra* note 131, at 9; *see also* Ofer Shpilberg et al., *The Next Stage: Molecular Epidemiology,* 50 J. Clinical Epidemiology 633, 637 (1997) ("A 1.5-fold relative risk may be composed of a 5-fold risk in 10% of the population, and a 1.1-fold risk in the remaining 90%, or a 2-fold risk in 25% and a 1.1-fold for 75%, or a 1.5-fold risk for the entire population.").

is involved, the relative risk from a study will understate the probability that exposure accelerated the occurrence of the disease.[199]

4. *Agent operates independently.* Employing a risk estimate to determine the probability of causation is not valid if the agent interacts with another cause in a way that results in an increase in disease beyond merely the sum of the increased incidence due to each agent separately. For example, the relative risk of lung cancer due to smoking is around 10, while the relative risk for asbestos exposure is approximately 5. The relative risk for someone exposed to both is not the arithmetic sum of the two relative risks, that is, 15, but closer to the product (50- to 60-fold), reflecting an interaction between the two.[200] Neither of the individual agent's relative risks can be employed to estimate the probability of causation in someone exposed to both asbestos and cigarette smoke.[201]
5. *Other assumptions.* Additional assumptions include (a) the agent of interest is not responsible for fatal diseases other than the disease of interest[202] and (b) the agent does not provide a protective effect against the outcome of interest in a subpopulation of those being studied.[203]

Evidence in a given case may challenge one or more of these assumptions. Bias in a study may suggest that the study findings are inaccurate and should be estimated to be higher or lower or, even, that the findings are spurious, that is, do not reflect a true causal relationship. A plaintiff may have been exposed to a

199. *See* Sander Greenland & James M. Robins, *Epidemiology, Justice, and the Probability of Causation,* 40 Jurimetrics J. 321 (2000); Sander Greenland, *Relation of Probability of Causation to Relative Risk and Doubling Dose: A Methodologic Error That Has Become a Social Problem,* 89 Am. J. Pub. Health 1166 (1999). If acceleration occurs, then the appropriate characterization of the harm for purposes of determining damages would have to be addressed. A defendant who only accelerates the occurrence of harm, say, chronic back pain, that would have occurred independently in the plaintiff at a later time is not liable for the same amount of damages as a defendant who causes a lifetime of chronic back pain. *See* David A. Fischer, *Successive Causes and the Enigma of Duplicated Harm,* 66 Tenn. L. Rev. 1127, 1127 (1999); Michael D. Green, *The Intersection of Factual Causation and Damages,* 55 DePaul L. Rev. 671 (2006).

200. We use interaction to mean that the combined effect is other than the additive sum of each effect, which is what we would expect if the two agents operate independently. Statisticians employ the term interaction in a different manner to mean the outcome deviates from what was expected in the model specified in advance. *See* Jay S. Kaufman, *Interaction Reaction,* 20 Epidemiology 159 (2009); Sander Greenland & Kenneth J. Rothman, *Concepts of Interaction, in* Rothman & Greenland, *supra* note 131, at 329.

201. *See* Restatement (Third) of Torts: Liability for Physical and Emotional Harm § 28 cmt. c(5) (2010); Jan Beyea & Sander Greenland, *The Importance of Specifying the Underlying Biologic Model in Estimating the Probability of Causation,* 76 Health Physics 269 (1999).

202. This is because in the epidemiologic studies relied on, those deaths caused by the alternative disease process will mask the true magnitude of increased incidence of the studied disease when the study subjects die before developing the disease of interest.

203. *See* Greenland & Robins, *supra,* note 198, at 332–33.

dose of the agent in question that is greater or lower than that to which those in the study were exposed.[204] A plaintiff may have individual factors, such as higher age than those in the study, that make it less likely that exposure to the agent caused the plaintiff's disease. Similarly, an individual plaintiff may be able to rule out other known (background) causes of the disease, such as genetics, that increase the likelihood that the agent was responsible for that plaintiff's disease. Evidence of a pathological mechanism may be available for the plaintiff that is relevant to the cause of the plaintiff's disease.[205] Before any causal relative risk from an epidemiologic study can be used to estimate the probability that the agent in question caused an individual plaintiff's disease, consideration of these (and related) factors is required.[206]

Having additional evidence that bears on individual causation has led a few courts to conclude that a plaintiff may satisfy his or her burden of production even if a relative risk less than 2.0 emerges from the epidemiologic evidence.[207] For example, genetics might be known to be responsible for 50% of the incidence of a disease independent of exposure to the agent.[208] If genetics can be ruled out

204. *See supra* Section V.C; *see also* Ferebee v. Chevron Chem. Co., 736 F.2d 1529, 1536 (D.C. Cir. 1984) ("The dose–response relationship at low levels of exposure for admittedly toxic chemicals like paraquat is one of the most sharply contested questions currently being debated in the medical community."); *In re* Joint E. & S. Dist. Asbestos Litig., 774 F. Supp. 113, 115 (S.D.N.Y. 1991) (discussing different relative risks associated with different doses), *rev'd on other grounds,* 964 F.2d 92 (2d Cir. 1992).

205. *See* Tobin v. Astra Pharm. Prods., Inc., 993 F.2d 528 (6th Cir. 1993) (plaintiff's expert relied predominantly on pathogenic evidence).

206. *See* Merrell Dow Pharms., Inc. v. Havner, 953 S.W.2d 706, 720 (Tex. 1997); Smith v. Wyeth-Ayerst Labs. Co., 278 F. Supp. 2d 684, 708–09 (W.D.N.C. 2003) (describing expert's effort to refine relative risk applicable to plaintiff based on specific risk characteristics applicable to her, albeit in an ill-explained manner); McDarby v. Merck & Co., 949 A.2d 223 (N.J. Super. Ct. App. Div. 2008); Mary Carter Andrues, *Proof of Cancer Causation in Toxic Waste Litigation,* 61 S. Cal. L. Rev. 2075, 2100–04 (1988). An example of a judge sitting as factfinder and considering individual factors for a number of plaintiffs in deciding cause in fact is contained in *Allen v. United States,* 588 F. Supp. 247, 429–43 (D. Utah 1984), *rev'd on other grounds,* 816 F.2d 1417 (10th Cir. 1987), *cert. denied,* 484 U.S. 1004 (1988); *see also* Manko v. United States, 636 F. Supp. 1419, 1437 (W.D. Mo. 1986), *aff'd,* 830 F.2d 831 (8th Cir. 1987).

207. *In re* Hanford Nuclear Reservation Litig., 292 F.3d 1124, 1137 (9th Cir. 2002) (applying Washington law) (recognizing the role of individual factors that may modify the probability of causation based on the relative risk); Magistrini v. One Hour Martinizing Dry Cleaning, 180 F. Supp. 2d 584, 606 (D.N.J. 2002) ("[A] relative risk of 2.0 is not so much a password to a finding of causation as one piece of evidence, among others for the court to consider in determining whether an expert has employed a sound methodology in reaching his or her conclusion."); Miller v. Pfizer, Inc., 196 F. Supp. 2d 1062, 1079 (D. Kan. 2002) (rejecting a threshold of 2.0 for the relative risk and recognizing that even a relative risk greater than 2.0 may be insufficient); Pafford v. Sec'y, Dept. of Health & Human Servs., 64 Fed. Cl. 19 (2005) (acknowledging that epidemiologic studies finding a relative risk of less than 2.0 can provide supporting evidence of causation), *aff'd,* 451 F.3d 1352 (Fed. Cir. 2006).

208. *See generally* Steve C. Gold, *The More We Know, the Less Intelligent We Are? How Genomic Information Should, and Should Not, Change Toxic Tort Causation Doctrine,* 34 Harv. Envtl. L. Rev. 369 (2010); Jamie A. Grodsky, *Genomics and Toxic Torts: Dismantling the Risk-Injury Divide,* 59 Stan. L. Rev.

in an individual's case, then a relative risk greater than 1.5 might be sufficient to support an inference that the agent was more likely than not responsible for the plaintiff's disease.[209]

Indeed, this idea of eliminating a known and competing cause is central to the methodology popularly known in legal terminology as differential diagnosis[210] but is more accurately referred to as differential etiology.[211] Nevertheless, the logic is sound if the label is not: Eliminating other known and competing causes increases the probability that a given individual's disease was caused by exposure to the agent. In a differential etiology, an expert first determines other known causes of the disease in question and then attempts to ascertain whether those competing causes can be "ruled out" as a cause of plaintiff's disease[212] as in the

1671 (2007); Gary E. Marchant, *Genetic Data in Toxic Tort Litigation,* 14 J.L. & Pol'y 7 (2006); Gary E. Marchant, *Genetics and Toxic Torts,* 31 Seton Hall L. Rev. 949 (2001).

209. The use of probabilities in excess of .50 to support a verdict results in an all-or-nothing approach to damages that some commentators have criticized. The criticism reflects the fact that defendants responsible for toxic agents with a relative risk just above 2.0 may be required to pay damages not only for the disease that their agents caused, but also for all instances of the disease. Similarly, those defendants whose agents increase the risk of disease by less than a doubling may not be required to pay damages for any of the disease that their agents caused. *See, e.g.,* 2 American Law Inst., Reporter's Study on Enterprise Responsibility for Personal Injury: Approaches to Legal and Institutional Change 369–75 (1991). Judge Posner has been in the vanguard of those advocating that damages be awarded on a proportional basis that reflects the probability of causation or liability. *See, e.g.,* Doll v. Brown, 75 F.3d 1200, 1206–07 (7th Cir. 1996). To date, courts have not adopted a rule that would apportion damages based on the probability of cause in fact in toxic substances cases. *See* Green, *supra* note 192.

210. Physicians regularly employ differential diagnoses in treating their patients to identify the disease from which the patient is suffering. *See* Jennifer R. Jamison, Differential Diagnosis for Primary Practice (1999).

211. It is important to emphasize that the term "differential diagnosis" in a clinical context refers to identifying a set of diseases or illnesses responsible for the patient's symptoms, while "differential etiology" refers to identifying the causal factors involved in an individual's disease or illness. For many health conditions, the *cause* of the disease or illness has no relevance to its treatment, and physicians, therefore, do not employ this term or pursue that question. *See* Zandi v. Wyeth a/k/a Wyeth, Inc., No. 27-CV-06-6744, 2007 WL 3224242 (Minn. Dist. Ct. Oct. 15, 2007) (commenting that physicians do not attempt to determine the cause of breast cancer). Thus, the standard differential diagnosis performed by a physician is not to determine the cause of a patient's disease. *See* John B. Wong et al., Reference Guide on Medical Testimony, in this manual; Edward J. Imwinkelried, *The Admissibility and Legal Sufficiency of Testimony About Differential Diagnosis (Etiology): of Under — and Over — Estimations,* 56 Baylor L. Rev. 391, 402–03 (2004); *see also* Turner v. Iowa Fire Equip. Co., 229 F.3d 1202, 1208 (8th Cir. 2000) (distinguishing between differential diagnosis conducted for the purpose of identifying the disease from which the patient suffers and one attempting to determine the cause of the disease); Creanga v. Jardal, 886 A.2d 633, 639 (N.J. 2005) ("Whereas most physicians use the term to describe the process of determining which of several diseases is causing a patient's *symptoms,* courts have used the term in a more general sense to describe the process by which causes of the patient's *condition* are identified.").

212. Courts regularly affirm the legitimacy of employing differential diagnostic methodology. *See, e.g., In re* Ephedra Prods. Liab. Litig., 393 F. Supp. 2d 181, 187 (S.D.N.Y. 2005); Easum v. Miller, 92 P.3d 794, 802 (Wyo. 2004) ("Most circuits have held that a reliable differential diagnosis satisfies *Daubert* and provides a valid foundation for admitting an expert opinion. The circuits reason that a differential diagnosis is a tested methodology, has been subjected to peer review/publication, does not

genetics example in the preceding paragraph. Similarly, an expert attempting to determine whether an individual's emphysema was caused by occupational chemical exposure would inquire whether the individual was a smoker. By ruling out (or ruling in) the possibility of other causes, the probability that a given agent was the cause of an individual's disease can be refined. Differential etiologies are most critical when the agent at issue is relatively weak and is not responsible for a large proportion of the disease in question.

Although differential etiologies are a sound methodology in principle, this approach is only valid if general causation exists and a substantial proportion of competing causes are known.[213] Thus, for diseases for which the causes are largely unknown, such as most birth defects, a differential etiology is of little benefit.[214] And, like any scientific methodology, it can be performed in an unreliable manner.[215]

VIII. Acknowledgments

The authors are grateful for the able research assistance provided by Murphy Horne, Wake Forest Law School class of 2012, and Cory Randolph, Wake Forest Law School class of 2010.

frequently lead to incorrect results, and is generally accepted in the medical community." (quoting Turner v. Iowa Fire Equip. Co., 229 F.3d 1202, 1208 (8th Cir. 2000))); Alder v. Bayer Corp., AGFA Div., 61 P.3d 1068, 1084–85 (Utah 2002).

213. Courts have long recognized that to prove causation plaintiff need not eliminate *all* potential competing causes. *See* Stubbs v. City of Rochester, 134 N.E. 137, 140 (N.Y. 1919) (rejecting defendant's argument that plaintiff was required to eliminate all potential competing causes of typhoid); *see also* Easum v. Miller, 92 P.3d 794, 804 (Wyo. 2004). At the same time, before a competing cause should be considered relevant to a differential diagnosis, there must be adequate evidence that it *is* a cause of the disease. *See* Cooper v. Smith & Nephew, Inc., 259 F.3d 194, 202 (4th Cir. 2001); Ranes v. Adams Labs., Inc., 778 N.W.2d 677, 690 (Iowa 2010).

214. *See* Perry v. Novartis Pharms. Corp., 564 F. Supp. 2d 452, 469 (E.D. Pa. 2008) (finding experts' testimony inadmissible because of failure to account for idiopathic (unknown) causes in conducting differential diagnosis); Soldo v. Sandoz Pharms. Corp., 244 F. Supp. 2d 434, 480, 519 (W.D. Pa. 2003) (criticizing expert for failing to account for idiopathic causes); Magistrini v. One Hour Martinizing Dry Cleaning, 180 F. Supp. 2d 584, 609 (D.N.J. 2002) (observing that 90–95% of leukemias are of unknown causes, but proceeding incorrectly to assert that plaintiff was obliged to prove that her exposure to defendant's benzene was *the* cause of her leukemia rather than simply a cause of the disease that combined with other exposures to benzene). *But see* Ruff v. Ensign-Bickford Indus., Inc., 168 F. Supp. 2d 1271, 1286 (D. Utah 2001) (responding to defendant's evidence that most instances of disease are of unknown origin by stating that such matter went to the weight to be attributed to plaintiff's expert's testimony not its admissibility).

215. Numerous courts have concluded that, based on the manner in which a differential diagnosis was conducted, it was unreliable and the expert's testimony based on it is inadmissible. *See, e.g.,* Glastetter v. Novartis Pharms. Corp., 252 F.3d 986, 989 (8th Cir. 2001).

Glossary of Terms

The following terms and definitions were adapted from a variety of sources, including A Dictionary of Epidemiology (Miquel M. Porta et al. eds., 5th ed. 2008); 1 Joseph L. Gastwirth, Statistical Reasoning in Law and Public Policy (1988); James K. Brewer, Everything You Always Wanted to Know about Statistics, but Didn't Know How to Ask (1978); and R.A. Fisher, Statistical Methods for Research Workers (1973).

adjustment. Methods of modifying an observed association to take into account the effect of risk factors that are not the focus of the study and that distort the observed association between the exposure being studied and the disease outcome. See also direct age adjustment, indirect age adjustment.

agent. Also, risk factor. A factor, such as a drug, microorganism, chemical substance, or form of radiation, whose presence or absence can result in the occurrence of a disease. A disease may be caused by a single agent or a number of independent alternative agents, or the combined presence of a complex of two or more factors may be necessary for the development of the disease.

alpha. The level of statistical significance chosen by a researcher to determine if any association found in a study is sufficiently unlikely to have occurred by chance (as a result of random sampling error) if the null hypothesis (no association) is true. Researchers commonly adopt an alpha of .05, but the choice is arbitrary, and other values can be justified.

alpha error. Also called Type I error and false-positive error, alpha error occurs when a researcher rejects a null hypothesis when it is actually true (i.e., when there is no association). This can occur when an apparent difference is observed between the control group and the exposed group, but the difference is not real (i.e., it occurred by chance). A common error made by lawyers, judges, and academics is to equate the level of alpha with the legal burden of proof.

association. The degree of statistical relationship between two or more events or variables. Events are said to be associated when they occur more or less frequently together than one would expect by chance. Association does not necessarily imply a causal relationship. Events are said not to have an association when the agent (or independent variable) has no apparent effect on the incidence of a disease (the dependent variable). This corresponds to a relative risk of 1.0. A negative association means that the events occur less frequently together than one would expect by chance, thereby implying a preventive or protective role for the agent (e.g., a vaccine).

attributable fraction. Also, attributable risk. The proportion of disease in exposed individuals that can be attributed to exposure to an agent, as distinguished from the proportion of disease attributed to all other causes.

attributable proportion of risk (PAR). This term has been used to denote the fraction of risk that is attributable to exposure to a substance (e.g., *X* percent of lung cancer is attributable to cigarettes). Synonymous terms include attributable fraction, attributable risk, etiologic fraction, population attributable risk, and risk difference. See attributable risk.

background risk of disease. Also, background rate of disease. Rate of disease in a population that has no known exposures to an alleged risk factor for the disease. For example, the background risk for all birth defects is 3–5% of live births.

beta error. Also called Type II error and false-negative error. Occurs when a researcher fails to reject a null hypothesis when it is incorrect (i.e., when there is an association). This can occur when no statistically significant difference is detected between the control group and the exposed group, but a difference does exist.

bias. Any effect at any stage of investigation or inference tending to produce results that depart systematically from the true values. In epidemiology, the term bias does not necessarily carry an imputation of prejudice or other subjective factor, such as the experimenter's desire for a particular outcome. This differs from conventional usage, in which bias refers to a partisan point of view.

biological marker. A physiological change in tissue or body fluids that occurs as a result of an exposure to an agent and that can be detected in the laboratory. Biological markers are only available for a small number of chemicals.

biological plausibility. Consideration of existing knowledge about human biology and disease pathology to provide a judgment about the plausibility that an agent causes a disease.

case-comparison study. See case-control study.

case-control study. Also, case-comparison study, case history study, case referent study, retrospective study. A study that starts with the identification of persons with a disease (or other outcome variable) and a suitable control (comparison, reference) group of persons without the disease. Such a study is often referred to as retrospective because it starts after the onset of disease and looks back to the postulated causal factors.

case group. A group of individuals who have been exposed to the disease, intervention, procedure, or other variable whose influence is being studied.

causation. As used here, an event, condition, characteristic, or agent being a necessary element of a set of other events that can produce an outcome, such as a disease. Other sets of events may also cause the disease. For example, smoking is a necessary element of a set of events that result in lung cancer, yet there are other sets of events (without smoking) that cause lung cancer. Thus, a cause may be thought of as a necessary link in at least one causal chain that results in an outcome of interest. Epidemiologists generally speak of causation in a group context; hence, they will inquire whether an increased incidence of a disease in a cohort was "caused" by exposure to an agent.

clinical trial. An experimental study that is performed to assess the efficacy and safety of a drug or other beneficial treatment. Unlike observational studies, clinical trials can be conducted as experiments and use randomization, because the agent being studied is thought to be beneficial.

cohort. Any designated group of persons followed or traced over a period of time to examine health or mortality experience.

cohort study. The method of epidemiologic study in which groups of individuals can be identified who are, have been, or in the future may be differentially exposed to an agent or agents hypothesized to influence the incidence of occurrence of a disease or other outcome. The groups are observed to find out if the exposed group is more likely to develop disease. The alternative terms for a cohort study (concurrent study, followup study, incidence study, longitudinal study, prospective study) describe an essential feature of the method, which is observation of the population for a sufficient number of person-years to generate reliable incidence or mortality rates in the population subsets. This generally implies study of a large population, study for a prolonged period (years), or both.

confidence interval. A range of values calculated from the results of a study within which the true value is likely to fall; the width of the interval reflects random error. Thus, if a confidence level of .95 is selected for a study, 95% of similar studies would result in the true relative risk falling within the confidence interval. The width of the confidence interval provides an indication of the precision of the point estimate or relative risk found in the study; the narrower the confidence interval, the greater the confidence in the relative risk estimate found in the study. Where the confidence interval contains a relative risk of 1.0, the results of the study are not statistically significant.

confounding factor. Also, confounder. A factor that is both a risk factor for the disease and a factor associated with the exposure of interest. Confounding refers to a situation in which an association between an exposure and outcome is all or partly the result of a factor that affects the outcome but is unaffected by the exposure.

control group. A comparison group comprising individuals who have not been exposed to the disease, intervention, procedure, or other variable whose influence is being studied.

cross-sectional study. A study that examines the relationship between disease and variables of interest as they exist in a population at a given time. A cross-sectional study measures the presence or absence of disease and other variables in each member of the study population. The data are analyzed to determine if there is a relationship between the existence of the variables and disease. Because cross-sectional studies examine only a particular moment in time, they reflect the prevalence (existence) rather than the incidence (rate) of disease and can offer only a limited view of the causal association between the variables and disease. Because exposures to toxic agents often change over time, cross-sectional studies are rarely used to assess the toxicity of exogenous agents.

data dredging. Jargon that refers to results identified by researchers who, after completing a study, pore through their data seeking to find any associations that may exist. In general, good research practice is to identify the hypotheses to be investigated in advance of the study; hence, data dredging is generally frowned on. In some cases, however, researchers conduct exploratory studies designed to generate hypotheses for further study.

demographic study. See ecological study.

dependent variable. The outcome that is being assessed in a study based on the effect of another characteristic—the independent variable. Epidemiologic studies attempt to determine whether there is an association between the independent variable (exposure) and the dependent variable (incidence of disease).

differential misclassification. A form of bias that is due to the misclassification of individuals or a variable of interest when the misclassification varies among study groups. This type of bias occurs when, for example, it is incorrectly determined that individuals in a study are unexposed to the agent being studied when in fact they are exposed. See nondifferential misclassification.

direct adjustment. A technique used to eliminate any difference between two study populations based on age, sex, or some other parameter that might result in confounding. Direct adjustment entails comparison of the study group with a large reference population to determine the expected rates based on the characteristic, such as age, for which adjustment is being performed.

dose. Generally refers to the intensity or magnitude of exposure to an agent multiplied by the duration of exposure. Dose may be used to refer only to the intensity of exposure.

dose–response relationship. A relationship in which a change in amount, intensity, or duration of exposure to an agent is associated with a change—either an increase or a decrease—in risk of disease.

double blinding. A method used in experimental studies in which neither the individuals being studied nor the researchers know during the study whether any individual has been assigned to the exposed or control group. Double blinding is designed to prevent knowledge of the group to which the individual was assigned from biasing the outcome of the study.

ecological fallacy. Also, aggregation bias, ecological bias. An error that occurs from inferring that a relationship that exists for groups is also true for individuals. For example, if a country with a higher proportion of fishermen also has a higher rate of suicides, then inferring that fishermen must be more likely to commit suicide is an ecological fallacy.

ecological study. Also, demographic study. A study of the occurrence of disease based on data from populations, rather than from individuals. An ecological study searches for associations between the incidence of disease and suspected disease-causing agents in the studied populations. Researchers often conduct ecological studies by examining easily available health statistics, making these studies relatively inexpensive in comparison with studies that measure disease and exposure to agents on an individual basis.

epidemiology. The study of the distribution and determinants of disease or other health-related states and events in populations and the application of this study to control of health problems.

error. Random error (sampling error) is the error that is due to chance when the result obtained for a sample differs from the result that would be obtained if the entire population (universe) were studied.

etiologic factor. An agent that plays a role in causing a disease.

etiology. The cause of disease or other outcome of interest.

experimental study. A study in which the researcher directly controls the conditions. Experimental epidemiology studies (also clinical studies) entail random assignment of participants to the exposed and control groups (or some other method of assignment designed to minimize differences between the groups).

exposed, exposure. In epidemiology, the exposed group (or the exposed) is used to describe a group whose members have been exposed to an agent that may be a cause of a disease or health effect of interest, or possess a characteristic that is a determinant of a health outcome.

false-negative error. See beta error.

false-positive error. See alpha error.

followup study. See cohort study.

general causation. Issue of whether an agent increases the incidence of disease in a group and not whether the agent caused any given individual's disease. Because of individual variation, a toxic agent generally will not cause disease in every exposed individual.

generalizable. When the results of a study are applicable to populations other than the study population, such as the general population.

in vitro. Within an artificial environment, such as a test tube (e.g., the cultivation of tissue in vitro).

in vivo. Within a living organism (e.g., the cultivation of tissue in vivo).

incidence rate. The number of people in a specified population falling ill from a particular disease during a given period. More generally, the number of new events (e.g., new cases of a disease in a defined population) within a specified period of time.

incidence study. See cohort study.

independent variable. A characteristic that is measured in a study and that is suspected to have an effect on the outcome of interest (the dependent variable). Thus, exposure to an agent is measured in a cohort study to determine whether that independent variable has an effect on the incidence of disease, which is the dependent variable.

indirect adjustment. A technique employed to minimize error that might result when comparing two populations because of differences in age, sex, or another parameter that may independently affect the rate of disease in the populations. The incidence of disease in a large reference population, such as all residents of a country, is calculated for each subpopulation (based on the relevant parameter, such as age). Those incidence rates are then applied to the study population with its distribution of persons to determine the overall incidence rate for the study population, which provides a standardized mortality or morbidity ratio (often referred to as SMR).

inference. The intellectual process of making generalizations from observations. In statistics, the development of generalizations from sample data, usually with calculated degrees of uncertainty.

information bias. Also, observational bias. Systematic error in measuring data that results in differential accuracy of information (such as exposure status) for comparison groups.

interaction. When the magnitude or direction (positive or negative) of the effect of one risk factor differs depending on the presence or level of the other. In interaction, the effect of two risk factors together is different (greater or less) than the sum of their individual effects.

meta-analysis. A technique used to combine the results of several studies to enhance the precision of the estimate of the effect size and reduce the plausibility that the association found is due to random sampling error. Meta-analysis is best suited to pooling results from randomly controlled experimental studies, but if carefully performed, it also may be useful for observational studies.

misclassification bias. The erroneous classification of an individual in a study as exposed to the agent when the individual was not, or incorrectly classifying a study individual with regard to disease. Misclassification bias may exist in all study groups (nondifferential misclassification) or may vary among groups (differential misclassification).

morbidity rate. State of illness or disease. Morbidity rate may refer to either the incidence rate or prevalence rate of disease.

mortality rate. Proportion of a population that dies of a disease or of all causes. The numerator is the number of individuals dying; the denominator is the total population in which the deaths occurred. The unit of time is usually a calendar year.

model. A representation or simulation of an actual situation. This may be either (1) a mathematical representation of characteristics of a situation that can be manipulated to examine consequences of various actions; (2) a representation of a country's situation through an "average region" with characteristics resembling those of the whole country; or (3) the use of animals as a substitute for humans in an experimental system to ascertain an outcome of interest.

multivariate analysis. A set of techniques used when the variation in several variables has to be studied simultaneously. In statistics, any analytical method that allows the simultaneous study of two or more independent factors or variables.

nondifferential misclassification. Error due to misclassification of individuals or a variable of interest into the wrong category when the misclassification varies among study groups. The error may result from limitations in data collection, may result in bias, and will often produce an underestimate of the true association. See differential misclassification.

null hypothesis. A hypothesis that states that there is no true association between a variable and an outcome. At the outset of any observational or experimental study, the researcher must state a proposition that will be tested in the study. In epidemiology, this proposition typically addresses the existence of an association between an agent and a disease. Most often, the null hypothesis is a statement that exposure to Agent A does not increase the occurrence of Disease D. The results of the study may justify a conclusion that the null hypothesis (no association) has been disproved (e.g., a study that finds a strong association between smoking and lung cancer). A study may fail to disprove the null hypothesis, but that alone does not justify a conclusion that the null hypothesis has been proved.

observational study. An epidemiologic study in situations in which nature is allowed to take its course, without intervention from the investigator. For example, in an observational study the subjects of the study are permitted to determine their level of exposure to an agent.

odds ratio (OR). Also, cross-product ratio, relative odds. The ratio of the odds that a case (one with the disease) was exposed to the odds that a control (one without the disease) was exposed. For most purposes the odds ratio from a case-control study is quite similar to a risk ratio from a cohort study.

***p* (probability), *p*-value.** The *p*-value is the probability of getting a value of the test outcome equal to or more extreme than the result observed, given that the null hypothesis is true. The letter *p*, followed by the abbreviation "n.s." (not significant) means that $p > .05$ and that the association was not statistically significant at the .05 level of significance. The statement "$p < .05$" means that *p* is less than 5%, and, by convention, the result is deemed statistically significant. Other significance levels can be adopted, such as .01 or .1. The lower the *p*-value, the less likely that random error would have produced the observed relative risk if the true relative risk is 1.

pathognomonic. When an agent must be present for a disease to occur. Thus, asbestos is a pathognomonic agent for asbestosis. See signature disease.

placebo controlled. In an experimental study, providing an inert substance to the control group, so as to keep the control and exposed groups ignorant of their status.

power. The probability that a difference of a specified amount will be detected by the statistical hypothesis test, given that a difference exists. In less formal terms, power is like the strength of a magnifying lens in its capability to identify an association that truly exists. Power is equivalent to one minus Type II error. This is sometimes stated as Power $= 1 - \beta$.

prevalence. The percentage of persons with a disease in a population at a specific point in time.

prospective study. A study in which two groups of individuals are identified: (1) individuals who have been exposed to a risk factor and (2) individuals who have not been exposed. Both groups are followed for a specified length of time, and the proportion that develops disease in the first group is compared with the proportion that develops disease in the second group. See cohort study.

random. The term implies that an event is governed by chance. See randomization.

randomization. Assignment of individuals to groups (e.g., for experimental and control regimens) by chance. Within the limits of chance variation, randomization should make the control group and experimental group similar at the start of an investigation and ensure that personal judgment and prejudices of the investigator do not influence assignment. Randomization should not be confused with haphazard assignment. Random assignment follows a predetermined plan that usually is devised with the aid of a table of random numbers. Randomization cannot ethically be used where the exposure is known to cause harm (e.g., cigarette smoking).

randomized trial. See clinical trial.

recall bias. Systematic error resulting from differences between two groups in a study in accuracy of memory. For example, subjects who have a disease may recall exposure to an agent more frequently than subjects who do not have the disease.

relative risk (RR). The ratio of the risk of disease or death among people exposed to an agent to the risk among the unexposed. For instance, if 10% of all people exposed to a chemical develop a disease, compared with 5% of people who are not exposed, the disease occurs twice as frequently among the exposed people. The relative risk is 10%/5% = 2. A relative risk of 1 indicates no association between exposure and disease.

research design. The procedures and methods, predetermined by an investigator, to be adhered to in conducting a research project.

risk. A probability that an event will occur (e.g., that an individual will become ill or die within a stated period of time or by a certain age).

risk difference (RD). The difference between the proportion of disease in the exposed population and the proportion of disease in the unexposed population. $-1.0 \leq RD \geq 1.0$.

sample. A selected subset of a population. A sample may be random or nonrandom.

sample size. The number of subjects who participate in a study.

secular-trend study. Also, time-line study. A study that examines changes over a period of time, generally years or decades. Examples include the decline of tuberculosis mortality and the rise, followed by a decline, in coronary heart disease mortality in the United States in the past 50 years.

selection bias. Systematic error that results from individuals being selected for the different groups in an observational study who have differences other than the ones that are being examined in the study.

sensitivity. Measure of the accuracy of a diagnostic or screening test or device in identifying disease (or some other outcome) when it truly exists. For example, assume that we know that 20 women in a group of 1000 women have cervical cancer. If the entire group of 1000 women is tested for cervical cancer and the screening test only identifies 15 (of the known 20) cases of cervical cancer, the screening test has a sensitivity of 15/20, or 75%. Also see specificity.

signature disease. A disease that is associated uniquely with exposure to an agent (e.g., asbestosis and exposure to asbestos). See also pathognomonic.

significance level. A somewhat arbitrary level selected to minimize the risk that an erroneous positive study outcome that is due to random error will be accepted as a true association. The lower the significance level selected, the less likely that false-positive error will occur.

specific causation. Whether exposure to an agent was responsible for a given individual's disease.

specificity. Measure of the accuracy of a diagnostic or screening test in identifying those who are disease-free. Once again, assume that 980 women out of a group of 1000 women do not have cervical cancer. If the entire group of 1000 women is screened for cervical cancer and the screening test only identifies 900 women without cervical cancer, the screening test has a specificity of 900/980, or 92%.

standardized morbidity ratio (SMR). The ratio of the incidence of disease observed in the study population to the incidence of disease that would be expected if the study population had the same incidence of disease as some selected reference population.

standardized mortality ratio (SMR). The ratio of the incidence of death observed in the study population to the incidence of death that would be expected if the study population had the same incidence of death as some selected standard or known population.

statistical significance. A term used to describe a study result or difference that exceeds the Type I error rate (or *p*-value) that was selected by the researcher at the outset of the study. In formal significance testing, a statistically significant result is unlikely to be the result of random sampling error and justifies rejection of the null hypothesis. Some epidemiologists believe that formal significance testing is inferior to using a confidence interval to express the results of a study. Statistical significance, which addresses the role of random sampling error in producing the results found in the study, should not be confused with the importance (for public health or public policy) of a research finding.

stratification. Separating a group into subgroups based on specified criteria, such as age, gender, or socioeconomic status. Stratification is used both to control for the possibility of confounding (by separating the studied populations based on the suspected confounding factor) and when there are other known factors that affect the disease under study. Thus, the incidence of death increases with age, and a study of mortality might use stratification of the cohort and control groups based on age.

study design. See research design.

systematic error. See bias.

teratogen. An agent that produces abnormalities in the embryo or fetus by disturbing maternal health or by acting directly on the fetus in utero.

teratogenicity. The capacity for an agent to produce abnormalities in the embryo or fetus.

threshold phenomenon. A certain level of exposure to an agent below which disease does not occur and above which disease does occur.

time-line study. See secular-trend study.

toxicology. The science of the nature and effects of poisons. Toxicologists study adverse health effects of agents on biological organisms, such as live animals and cells. Studies of humans are performed by epidemiologists.

toxic substance. A substance that is poisonous.

true association. Also, real association. The association that really exists between exposure to an agent and a disease and that might be found by a perfect (but nonetheless nonexistent) study.

Type I error. Rejecting the null hypothesis when it is true. See alpha error.

Type II error. Failing to reject the null hypothesis when it is false. See beta error.

validity. The degree to which a measurement measures what it purports to measure; the accuracy of a measurement.

variable. Any attribute, condition, or other characteristic of subjects in a study that can have different numerical characteristics. In a study of the causes of heart disease, blood pressure and dietary fat intake are variables that might be measured.

References on Epidemiology

Causal Inferences (Kenneth J. Rothman ed., 1988).
William G. Cochran, Sampling Techniques (1977).
A Dictionary of Epidemiology (John M. Last et al. eds., 5th ed. 2008).
Anders Ahlbom & Steffan Norell, Introduction to Modern Epidemiology (2d ed. 1990).
Robert C. Elston & William D. Johnson, Basic Biostatistices for Geneticists and Epidemiologists (2008)
Encyclopedia of Epidemiology (Sarah E. Boslaugh ed., 2008).
Joseph L. Fleiss et al., Statistical Methods for Rates and Proportions (3d ed. 2003).
Leon Gordis, Epidemiology (4th ed. 2009).
Morton Hunt, How Science Takes Stock: The Story of Meta-Analysis (1997).
International Agency for Research on Cancer (IARC), Interpretation of Negative Epidemiologic Evidence for Carcinogenicity (N.J. Wald & R. Doll eds., 1985).
Harold A. Kahn & Christopher T. Sempos, Statistical Methods in Epidemiology (1989).
David E. Lilienfeld, *Overview of Epidemiology,* 3 Shepard's Expert & Sci. Evid. Q. 25 (1995).
David E. Lilienfeld & Paul D. Stolley, Foundations of Epidemiology (3d ed. 1994).
Marcello Pagano & Kimberlee Gauvreau, Principles of Biostatistics (2d ed. 2000).
Pharmacoepidemiology (Brian L. Strom ed., 4th ed. 2005).
Richard K. Riegelman & Robert A. Hirsch, Studying a Study and Testing a Test: How to Read the Health Science Literature (5th ed. 2005).
Bernard Rosner, Fundamentals of Biostatistics (6th ed. 2006).
Kenneth J. Rothman et al., Modern Epidemiology (3d ed. 2008).
David A. Savitz, Interpreting Epidemiologic Evidence: Strategies for Study Design and Analysis (2003).
James J. Schlesselman, Case-Control Studies: Design, Conduct, Analysis (1982).
Lisa M. Sullivan, Essentials of Biostatistics (2008).
Mervyn Susser, Epidemiology, Health and Society: Selected Papers (1987).

References on Law and Epidemiology

American Law Institute, Reporters' Study on Enterprise Responsibility for Personal Injury (1991).
Bert Black & David H. Hollander, Jr., *Unraveling Causation: Back to the Basics*, 3 U. Balt. J. Envtl. L. 1 (1993).
Bert Black & David Lilienfeld, *Epidemiologic Proof in Toxic Tort Litigation*, 52 Fordham L. Rev. 732 (1984).

Gerald Boston, *A Mass-Exposure Model of Toxic Causation: The Content of Scientific Proof and the Regulatory Experience,* 18 Colum. J. Envtl. L. 181 (1993).

Vincent M. Brannigan et al., *Risk, Statistical Inference, and the Law of Evidence: The Use of Epidemiological Data in Toxic Tort Cases,* 12 Risk Analysis 343 (1992).

Troyen Brennan, *Causal Chains and Statistical Links: The Role of Scientific Uncertainty in Hazardous-Substance Litigation,* 73 Cornell L. Rev. 469 (1988).

Troyen Brennan, *Helping Courts with Toxic Torts: Some Proposals Regarding Alternative Methods for Presenting and Assessing Scientific Evidence in Common Law Courts,* 51 U. Pitt. L. Rev. 1 (1989).

Philip Cole, *Causality in Epidemiology, Health Policy, and Law,* 27 Envtl. L. Rep. 10,279 (June 1997).

Comment, *Epidemiologic Proof of Probability: Implementing the Proportional Recovery Approach in Toxic Exposure Torts,* 89 Dick. L. Rev. 233 (1984).

George W. Conk, *Against the Odds: Proving Causation of Disease with Epidemiological Evidence,* 3 Shepard's Expert & Sci. Evid. Q. 85 (1995).

Carl F. Cranor, Toxic Torts: Science, Law, and the Possibility of Justice (2006).

Carl F. Cranor et al., *Judicial Boundary Drawing and the Need for Context-Sensitive Science in Toxic Torts After* Daubert v. Merrell Dow Pharmaceuticals, Inc., 16 Va. Envtl. L.J. 1 (1996).

Richard Delgado, *Beyond Sindell: Relaxation of Cause-in-Fact Rules for Indeterminate Plaintiffs,* 70 Cal. L. Rev. 881 (1982).

Michael Dore, *A Commentary on the Use of Epidemiological Evidence in Demonstrating Cause-in-Fact,* 7 Harv. Envtl. L. Rev. 429 (1983).

Jean Macchiaroli Eggen, *Toxic Torts, Causation, and Scientific Evidence After* Daubert, 55 U. Pitt. L. Rev. 889 (1994).

Daniel A. Farber, *Toxic Causation,* 71 Minn. L. Rev. 1219 (1987).

Heidi Li Feldman, Science and Uncertainty in Mass Exposure Litigation, 74 Tex. L. Rev. 1 (1995).

Stephen E. Fienberg et al., *Understanding and Evaluating Statistical Evidence in Litigation,* 36 Jurimetrics J. 1 (1995).

Joseph L. Gastwirth, Statistical Reasoning in Law and Public Policy (1988).

Herman J. Gibb, *Epidemiology and Cancer Risk Assessment, in* Fundamentals of Risk Analysis and Risk Management 23 (Vlasta Molak ed., 1997).

Steve Gold, Note, *Causation in Toxic Torts: Burdens of Proof, Standards of Persuasion and Statistical Evidence,* 96 Yale L.J. 376 (1986).

Leon Gordis, *Epidemiologic Approaches for Studying Human Disease in Relation to Hazardous Waste Disposal Sites,* 25 Hous. L. Rev. 837 (1988).

Michael D. Green, *Expert Witnesses and Sufficiency of Evidence in Toxic Substances Litigation: The Legacy of Agent Orange and Bendectin Litigation,* 86 Nw. U. L. Rev. 643 (1992).

Michael D. Green, *The Future of Proportional Liability, in* Exploring Tort Law (Stuart Madden ed., 2005).

Sander Greenland, *The Need for Critical Appraisal of Expert Witnesses in Epidemiology and Statistics*, 39 Wake Forest L. Rev. 291 (2004).

Khristine L. Hall & Ellen Silbergeld, *Reappraising Epidemiology: A Response to Mr. Dore*, 7 Harv. Envtl. L. Rev. 441 (1983).

Jay P. Kesan, *Drug Development: Who Knows Where the Time Goes?: A Critical Examination of the Post-*Daubert *Scientific Evidence Landscape,* 52 Food Drug Cosm. L.J. 225 (1997).

Jay P. Kesan, *An Autopsy of Scientific Evidence in a Post-*Daubert *World,* 84 Geo. L. Rev. 1985 (1996).

Constantine Kokkoris, Comment, DeLuca v. Merrell Dow Pharmaceuticals, Inc.: *Statistical Significance and the Novel Scientific Technique,* 58 Brook. L. Rev. 219 (1992).

James P. Leape, *Quantitative Risk Assessment in Regulation of Environmental Carcinogens,* 4 Harv. Envtl. L. Rev. 86 (1980).

David E. Lilienfeld, *Overview of Epidemiology*, 3 Shepard's Expert & Sci. Evid. Q. 23 (1995).

Junius McElveen, Jr., & Pamela Eddy, *Cancer and Toxic Substances: The Problem of Causation and the Use of Epidemiology,* 33 Clev. St. L. Rev. 29 (1984).

Modern Scientific Evidence: The Law and Science of Expert Testimony (David L. Faigman et al. eds., 2009–2010).

Note, *Development in the Law—Confronting the New Challenges of Scientific Evidence,* 108 Harv. L. Rev. 1481 (1995).

Susan R. Poulter, *Science and Toxic Torts: Is There a Rational Solution to the Problem of Causation?* 7 High Tech. L.J. 189 (1992).

Jon Todd Powell, Comment, *How to Tell the Truth with Statistics: A New Statistical Approach to Analyzing the Data in the Aftermath of* Daubert v. Merrell Dow Pharmaceuticals, 31 Hous. L. Rev. 1241 (1994).

Restatement (Third) of Torts: Liability for Physical and Emotional Harm § 28, cmt. c & rptrs. note (2010).

David Rosenberg, *The Causal Connection in Mass Exposure Cases: A Public Law Vision of the Tort System,* 97 Harv. L. Rev. 849 (1984).

Joseph Sanders, *The Bendectin Litigation: A Case Study in the Life-Cycle of Mass Torts,* 43 Hastings L.J. 301 (1992).

Joseph Sanders, *Scientific Validity, Admissibility, and Mass Torts After* Daubert, 78 Minn. L. Rev. 1387 (1994).

Joseph Sanders & Julie Machal-Fulks, *The Admissibility of Differential Diagnosis to Prove Causation in Toxic Tort Cases: The Interplay of Adjective and Substantive Law,* 64 L. & Contemp. Probs. 107 (2001).

Palma J. Strand, *The Inapplicability of Traditional Tort Analysis to Environmental Risks: The Example of Toxic Waste Pollution Victim Compensation,* 35 Stan. L. Rev. 575 (1983).

Richard W. Wright, *Causation in Tort Law*, 73 Cal. L. Rev. 1735 (1985).

Reference Guide on Toxicology

BERNARD D. GOLDSTEIN AND MARY SUE HENIFIN

Bernard D. Goldstein, M.D., is Professor of Environmental and Occupational Health and Former Dean, Graduate School of Public Health, University of Pittsburgh.

Mary Sue Henifin, J.D., M.P.H., is a Partner with Buchanan Ingersoll, P.C., Princeton, New Jersey.

CONTENTS

I. Introduction

The discipline of toxicology is primarily concerned with identifying and understanding the adverse effects of external chemical and physical agents on biological systems. The interface of the evidence from toxicological science with toxic torts can be complex, in part reflecting the inherent challenges of bringing science into a courtroom, but also because of issues particularly pertinent to toxicology. For the most part, toxicological study begins with a chemical or physical agent and asks what impact it will have, while toxic tort cases begin with an individual or a group that has suffered an adverse impact and makes claims about its cause. A particular challenge is that only rarely is the adverse impact highly specific to the toxic agent; for example, the relatively rare lung cancer known as mesothelioma is almost always caused by asbestos. The more common form of lung cancer, bronchial carcinoma, also can be caused by asbestos, but asbestos is a relatively uncommon cause compared with smoking, radon, and other known causes of lung cancer.[1] Lung cancer itself is unusual in that for the vast majority of cases, we can point to a known cause—smoking. However, for many diseases, there are few if any known causes, for example, pancreatic cancer. Even when there are known causes of a disease, most individual cases are often not ascribable to any of the known causes, such as with leukemia.

In general, there are only a limited number of ways that biological tissues can respond, and there are many causes for each response. Accordingly, the role of toxicology in toxic tort cases often is to provide information that helps evaluate the causal probability that an adverse event with potentially many causes is caused by a specific agent. Similarly, toxicology is commonly used as a basis for regulating chemicals, depending upon their potential for effect. Assertions related to the toxicological predictability of an adverse consequence in relation to the stringency of the regulatory law are not uncommon bases for legal actions against regulatory agencies.

Identifying cause-and-effect relationships in toxicology can be relatively straightforward; for example, when placed on the skin, concentrated sulfuric acid will cause massive tissue destruction, and carbon monoxide poisoning is identifiable by the extent to which carbon monoxide is attached to the oxygen-carrying portion of blood hemoglobin, thereby decreasing oxygen availability to the body. But even these two seemingly straightforward examples serve to illustrate the complexity of toxicology and particularly its emphasis on understanding dose–response relationships. The tissue damage caused by sulfuric acid is not specific to this chemical, and at lower doses, no effect will be seen. Carbon monoxide is not only an external poison but is a product of normal internal metabolism such

1. Contrast this issue with the relatively straightforward situation in infectious disease in which the disease name identifies the cause; for example, cholera is caused by *Vibrio cholerae*, tuberculosis by the *Mycobacterium tuberculosis*, HIV-AIDs by the HIV virus, and so on.

that about 1 out of 200 hemoglobin molecules will normally have carbon monoxide attached, and this can increase depending upon concomitant disease states. Furthermore, the complex temporal relation governing the uptake and release of carbon monoxide from hemoglobin also must be considered in assessing the extent to which an adverse impact may be ascribable to carbon monoxide exposure. Thus the diagnosis of carbon monoxide poisoning requires far more information than the simple presence of detectable carbon monoxide in the blood.

Complexity in toxicology is derived primarily from three factors. The first is that chemicals often change within the body as they go through various routes to eventual elimination.[2] Thus absorption, distribution, metabolism, and excretion are central to understanding the toxicology of an agent. The second is that human sensitivity to chemical and physical agents can vary greatly among individuals, often as a result of differences in absorption, distribution, metabolism, or excretion, as well as target organ sensitivity—all of which can be genetically determined. The third major source of complexity is the need for extrapolation, either across species, because much toxicological data are obtained from studies in laboratory animals, or across doses, because human toxicological and epidemiological data often are limited to specific dose ranges that differ from the dose suffered by a plaintiff alleging a toxic tort impact. All three of these factors are responsible for much of the complexity in utilizing toxicology for tort or regulatory judicial decisions and are described in more detail below.

Classically, toxicology is known as the science of poisons. It is the study of the adverse effects of chemical and physical agents on living organisms.[3] Although it is an age-old science, toxicology has only recently become a discipline distinct from pharmacology, biochemistry, cell biology, and related fields.

There are three central tenets of toxicology. First, "the dose makes the poison"; this implies that all chemical agents are intrinsically hazardous—whether they cause harm is only a question of dose.[4] Even water, if consumed in large quantities, can be toxic. Second, each chemical or physical agent tends to produce a specific pattern of biological effects that can be used to establish disease

2. Direct-acting toxic agents are those whose toxicity is due to the parent chemical entering the body. A change in chemical structure through metabolism usually results in detoxification. Indirect-acting chemicals are those that must first be metabolized to a harmful intermediate for toxicity to occur. For an overview of metabolism in toxicology, see R.A. Kemper et al., *Metabolism: A Determinant of Toxicity*, *in* Principles and Methods of Toxicology 103–178 (A. Wallace Hayes ed., 5th ed. 2008).

3. Casarett and Doull's Toxicology: The Basic Science of Poisons 13 (Curtis D. Klaassen ed., 7th ed. 2007).

4. A discussion of more modern formulations of this principle, which was articulated by Paracelsus in the sixteenth century, can be found in David L. Eaton, *Scientific Judgment and Toxic Torts—A Primer in Toxicology for Judges and Lawyers*, 12 J.L. & Pol'y 5, 15 (2003); Ellen K. Silbergeld, *The Role of Toxicology in Causation: A Scientific Perspective*, 1 Cts. Health Sci. & L. 374, 378 (1991). A short review of the field of toxicology can be found in Curtis D. Klaassen, *Principles of Toxicology and Treatment of Poisoning*, *in* Goodman and Gilman's The Pharmacological Basis of Therapeutics 1739 (11th ed. 2008).

causation.[5] Third, the toxic responses in laboratory animals are useful predictors of toxic responses in humans. Each of these tenets, and their exceptions, is discussed in greater detail in this reference guide.

The science of toxicology attempts to determine at what doses foreign agents produce their effects. The foreign agents classically of interest to toxicologists are all chemicals (including foods and drugs) and physical agents in the form of radiation, but not living organisms that cause infectious diseases.[6]

The discipline of toxicology provides scientific information relevant to the following questions:

1. What hazards does a chemical or physical agent present to human populations or the environment?
2. What degree of risk is associated with chemical exposure at any given dose?[7]

Toxicological studies, by themselves, rarely offer direct evidence that a disease in any one individual was caused by a chemical exposure.[8] However, toxicology can provide scientific information regarding the increased risk of contracting a disease at any given dose and help rule out other risk factors for the disease. Toxicological evidence also contributes to the weight of evidence supporting causal inferences by explaining how a chemical causes a specific disease through describing metabolic, cellular, and other physiological effects of exposure.

A. Toxicology and the Law

The growing concern about chemical causation of disease is reflected in the public attention devoted to lawsuits alleging toxic torts, as well as in litigation concerning the many federal and state regulations related to the release of potentially toxic compounds into the environment.

Toxicological evidence frequently is offered in two types of litigation: tort and regulatory. In tort litigation, toxicologists offer evidence that either supports

5. Some substances, such as central nervous system toxicants, can produce complex and nonspecific symptoms, such as headaches, nausea, and fatigue.

6. Forensic toxicology, a subset of toxicology generally concerned with criminal matters, is not addressed in this reference guide, because it is a highly specialized field with its own literature and methodologies that do not relate directly to toxic tort or regulatory issues.

7. In standard risk assessment terminology, hazard is an intrinsic property of a chemical or physical agent, while risk is dependent both upon hazard and on the extent of exposure. Note that this first "law" of toxicology is particularly pertinent to questions of specific causation, while the second "law" of toxicology, the specificity of effect, is pertinent to questions of general causation.

8. There are exceptions, for example, when measurements of levels in the blood or other body constituents of the potentially offending agent are at a high enough level to be consistent with reasonably specific health impacts, such as in carbon monoxide poisoning.

or refutes plaintiffs' claims that their diseases or injuries were caused by chemical exposures.[9] In regulatory litigation, toxicological evidence is used to either support or challenge government regulations concerning a chemical or a class of chemicals. In regulatory litigation, toxicological evidence addresses the issue of how exposure affects populations[10] rather than addressing specific causation, and agency determinations are usually subject to the court's deference.[11]

Dose is a central concept in the field of toxicology, and an expert toxicologist will consider the extent of a plaintiff's dose in making an opinion.[12] But dose has not been a central issue in many of the most important judicial decisions concerning the relation of toxicological evidence to toxic tort decisions. These have mostly been general causation issues: For example, is a silicon breast implant capable of causing rheumatoid arthritis, or is Bendectin capable of causing deformed babies.[13] However, in most specific causation issues involving exposure to a chemical known to be able to cause the observed effect, the primary issue will be whether there has been exposure to a sufficient dose to be a likely cause of this effect.

9. *See, e.g.*, Gen. Elec. Co. v. Joiner, 522 U.S. 136 (1997); Daubert v. Merrell Dow Pharms., Inc., 509 U.S. 579 (1993). Courts have held that toxicologists can testify as to disease causation related to chemical exposures. *See, e.g.*, Bonner v. ISP Techs, Inc., 259 F.3d 924, 928–31 (8th Cir. 2001); Paoli R.R. v. Monsanto Co., 915 F.2d 829 (3d Cir. 1990); Loudermill v. Dow Chem. Co., 863 F.2d 566, 569–70 (8th Cir. 1988).

10. Again, there are exceptions. For example, certain regulatory approaches, such as the control of hazardous air pollutants, are based on the potential impact to a putative maximally exposed individual rather than to the general population.

11. *See, e.g.*, Int'l Union, United Mine Workers of Am. v. U.S. Dep't of Labor, 358 F.3d 40, 43–44 (D.C. Cir. 2004) (determinations by Secretary of Labor are given deference by the court, but must be supported by some evidence, and cannot be capricious or arbitrary); N.M. Mining Ass'n v. N.M. Water Quality Control Comm., 150 P.3d 991, 995–96 (N.M. Ct. App. 2006) (action by a government agency is presumptively valid and will be given deference by the court. The court will only overturn a regulatory decision if it is capricious and arbitrary, or not supported by substantial evidence).

12. Dose is a function of both concentration and duration. Haber's rule is a century-old simplified expression of dose effects in which the effect of a concentration and duration of exposure is a constant (e.g., exposure to an agent at 10 parts per million for 1 hour has the same impact as exposure to 1 part per million for 10 hours). Exposure levels, which are concentrations, are often confused with dose. This can be particularly problematic when attempting to understand the implications of exposure to a level that exceeds a regulatory standard that is set for a different time frame. For example, assume a drinking water contaminant is a known cause of cancer. To avoid a 1 in 100,000 lifetime risk caused by this contaminant in drinking water, and assuming that the average person will drink approximately 2000 mL of water daily for a lifetime, the regulatory authority sets the allowable contaminant standard in drinking water at 10 μg/L. Drinking one glass of water containing 20 μg/L of this contaminant, although exceeding the standard, does not come close to achieving a "reasonably medically probable" cause of an individual case of cancer.

13. *See, e.g.*, *In re* Silicone Gel Breast Implants Prods. Liab. Litig., 318 F. Supp. 2d 879, 891 (C.D. Cal. 2004); Joseph Sanders, *From Science to Evidence: The Testimony on Causation in the Bendectin Cases*, 46 Stan. L. Rev. 1, 19 (1993).

B. Purpose of the Reference Guide on Toxicology

This reference guide focuses on the scientific issues that arise most frequently in toxic tort cases. Where it is appropriate, the guide explores the use of regulatory data and how the courts treat such data. It also provides an overview of the basic principles and methodologies of toxicology and offers a scientific context for proffered expert opinion based on toxicological data.[14] The reference guide describes research methods in toxicology and the relationship between toxicology and epidemiology, and it provides model questions for evaluating the admissibility and strength of an expert's opinion. Following each question is an explanation of the type of toxicological data or information that is offered in response to the question, as well as a discussion of its significance.

C. Toxicological Study Design

Toxicological studies usually involve exposing laboratory animals (in vivo research) or cells or tissues (in vitro research) to chemical or physical agents, monitoring the outcomes (such as cellular abnormalities, tissue damage, organ toxicity, or tumor formation), and comparing the outcomes with those for unexposed control groups. As explained below,[15] the extent to which animal and cell experiments accurately predict human responses to chemical exposures is subject to debate.[16] However, because it is often unethical to experiment on humans by exposing them to known doses of chemical agents, animal toxicological evidence often provides the best scientific information about the risk of disease from a chemical exposure.[17]

In contrast to their exposure to drugs, only rarely are humans exposed to environmental chemicals in a manner that permits a quantitative determination of adverse outcomes.[18] This area of toxicological study may consist of individual or multiple case reports, or even experimental studies in which individuals or groups of individuals have been exposed to a chemical under circumstances that permit analysis of dose–response relationships, mechanisms of action, or other aspects of

14. The use of toxicological evidence in regulatory decisionmaking is discussed in Casarett and Doull's Toxicology: The Basic Science of Poisons, *supra* note 3, at 13–14; Barbara D. Beck et al., *The Use of Toxicology in the Regulatory Process, in* Principles and Methods of Toxicology, *supra* note 2, at 45–102. For a more general discussion of issues that arise in considering expert testimony, *see* Margaret A. Berger, The Admissibility of Expert Testimony, Section IV, in this manual.

15. *See infra* Section I.D.

16. The controversy over the use of toxicological evidence in tort cases is described in Bernard D. Goldstein, *Toxic Torts: The Devil Is in the Dose*, 16 J.L. & Pol'y 551 (2008); Joseph V. Rodricks, *Evaluating Disease Causation in Humans Exposed to Toxic Substances*, 14 J.L. & Pol'y 39 (2006); Silbergeld, *supra* note 4, at 378.

17. *See, e.g.*, Office of Tech. Assessment, U.S. Congress, Reproductive Health Hazards in the Workplace 8 (1985).

18. However, it is from drug studies in which multiple animal species are compared directly with humans that many of the principles of toxicology have been developed.

toxicology. For example, individuals occupationally or environmentally exposed to polychlorinated biphenyls (PCBs) prior to prohibitions on their use have been studied to determine the routes of absorption, distribution, metabolism, and excretion for this chemical. Human exposure occurs most frequently in occupational settings where workers are exposed to industrial chemicals such as lead or asbestos; however, even under these circumstances, it is usually difficult, if not impossible, to quantify the amount of exposure. Moreover, human populations are exposed to many other chemicals and risk factors, making it difficult to isolate the increased risk of a disease that is the result of exposure to any one chemical.[19]

Toxicologists use a wide range of experimental techniques, depending in part on their area of specialization. Toxicological research may focus on classes of chemical compounds, such as solvents and metals; body system effects, such as neurotoxicology, reproductive toxicology, and immunotoxicology; and effects on physiological processes, including inhalation toxicology, dermatotoxicology, and molecular toxicology (the study of how chemicals interact with cell molecules). Each of these areas of research includes both in vivo and in vitro research.[20]

1. *In vivo research*

Animal research in toxicology generally falls under two headings: safety assessment and classic laboratory science, with a continuum between them. As explained in Section I.E, safety assessment is a relatively formal approach in which a chemical's potential for toxicity is tested in vivo or in vitro using standardized techniques often prescribed by regulatory agencies, such as the Environmental Protection Agency (EPA) and the Food and Drug Administration (FDA).[21]

The roots of toxicology in the science of pharmacology are reflected in an emphasis on understanding the absorption, distribution, metabolism, and excretion of chemicals. Basic toxicological laboratory research also focuses on the mechanisms of action of external chemical and physical agents. Such research is based on the standard elements of scientific studies, including appropriate experimental design using control groups and statistical evaluation. In general, toxicological research attempts to hold all variables constant except for that of the chemical exposure.[22] Any change in the experimental group not found in the control group is assumed to be perturbation caused by the chemical.

19. *See, e.g.*, Office of Tech. Assessment, U.S. Congress, *supra* note 17, at 8.

20. *See infra* Sections I.C.1, I.C.2.

21. W.J. White et al., *The Use of Laboratory Animals in Toxicology Research, in* Principles and Methods of Toxicology 1055–1102 (A. Wallace Hayes ed., 5th ed. 2008); M.A. Dorato et al., *The Toxicologic Assessment of Pharmaceutical and Biotechnology Products, in* Principles and Methods of Toxicology 325–68 (A. Wallace Hayes ed., 5th ed. 2008).

22. *See generally* Alan Poole & George B. Leslie, A Practical Approach to Toxicological Investigations (1989); Principles and Methods of Toxicology (A. Wallace Hayes ed., 2d ed. 1989); *see also* discussion on acute, short-term, and long-term toxicity studies and acquisition of data in Frank C. Lu, Basic Toxicology: Fundamentals, Target Organs, and Risk Assessment 77–92 (2d ed. 1991).

a. Dose–response relationships

An important component of toxicological research is dose–response relationships. Thus, most toxicological studies generally test a range of doses of the chemical. Animal experiments are conducted to determine the dose–response relationships of a compound by measuring how response varies with dose, including diligently searching for a dose that has no measurable physiological effect. This information is useful in understanding the mechanisms of toxicity and extrapolating data from animals to humans.[23]

b. Acute Toxicity Testing—Lethal Dose 50

To determine the dose–response relationship for a compound, a short-term lethal dose 50% (LD_{50}) may be derived experimentally. The LD_{50} is the dose at which a compound kills 50% of laboratory animals within a period of days to weeks. The use of this easily measured end point for acute toxicity to a large extent has been replaced, in part because recent advances in toxicology have provided other pertinent end points, and in part because of pressure from animal rights activists to reduce or replace the use of animals in laboratory research.[24]

c. No observable effect level

A dose–response study also permits the determination of another important characteristic of the biological action of a chemical—the no observable effect level (NOEL).[25] The NOEL sometimes is called a threshold, because it is the level above which observable effects in test animals are believed to occur and below which no toxicity is observed.[26] Of course, because the NOEL is dependent on the ability to

23. *See infra* Sections I.D, II.A.

24. Committee on Toxicity Testing and Assessment of Environmental Agents, National Research Council, Toxicity Testing in the 21st Century: A Vision and a Strategy (2007).

25. For example, undiluted acid on the skin can cause a horrible burn. As the acid is diluted to lower and lower concentrations, less and less of an effect occurs until there is a concentration sufficiently low (e.g., one drop in a bathtub of water, or a sample with less than the acidity of vinegar) that no effect occurs. This no observable effect concentration differs from person to person. For example, a baby's skin is more sensitive than that of an adult, and skin that is irritated or broken responds to the effects of an acid at a lower concentration. However, the key point is that there is some concentration that is completely harmless to the skin.

26. The significance of the NOEL was relied on by the court in *Graham v. Canadian National Railway Co.*, 749 F. Supp. 1300 (D. Vt. 1990), in granting judgment for the defendants. The court found the defendants' expert, a medical toxicologist, persuasive. The expert testified that the plaintiffs' injuries could not have been caused by herbicides, because their exposure was well below the reference dose, which he calculated by taking the NOEL and decreasing it by a safety factor to ensure no human effect. *Id.* at 1311–12 & n.11. *But see* Louderback v. Orkin Exterminating Co., 26 F. Supp. 2d 1298 (D. Kan. 1998) (failure to consider threshold levels of exposure does not necessarily render expert's opinion unreliable where temporal relationship, scientific literature establishing an association between exposure and various symptoms, plaintiffs' medical records and history of disease, and exposure to or

observe an effect, the level is sometimes lowered once more sophisticated methods of detection are developed.

d. Benchmark dose

For regulatory toxicology, the NOEL is being replaced by a more statistically robust approach known as the benchmark dose (BD). The BD is determined based on dose–response modeling and is defined as the exposure associated with a specified low incidence of risk, generally in the range of 1% to 10%, of a health effect, or the dose associated with a specified measure or change of a biological effect. To model the BD, sufficient data must exist, such as at least a statistically or biologically significant dose-related trend in the selected end point.[27]

e. No-threshold model and determination of cancer risk

Certain genetic mutations, such as those leading to cancer and some inherited disorders, are believed to occur without any threshold. In theory, the cancer-causing mutation to the genetic material of the cell can be produced by any one molecule of certain chemicals. The no-threshold model led to the development of the one-hit theory of cancer risk, in which each molecule of a cancer-causing chemical has some finite possibility of producing the mutation that leads to cancer. (See Figure 1 for an idealized comparison of a no-threshold and threshold dose–response.) This risk is very small, because it is unlikely that any one molecule of a potentially cancer-causing agent will reach that one particular spot in a specific cell and result in the change that then eludes the body's defenses and leads to a clinical case of cancer. However, the risk is not zero. The same model also can be used to predict the risk of inheritable mutational events.[28]

the presence of other disease-causing factors were all considered). *See also* DiPirro v. Bondo Corp., 62 Cal. Rptr. 3d 722, 750 (Cal. Ct. App. 2007) (judgment for the maker of auto touchup paint based on finding that there was substantial evidence in the record to show that the level of a particular toxin [toluene] present in the paint fell 1000 times below the NOEL of that toxin and therefore no warning label needed on paint can).

27. *See* S. Sand et al., *The Current State of Knowledge on the Use of the Benchmark Dose Concept in Risk Assessment*, 28 J. Appl. Toxicol. 405–21 (2008); W. Slob et al., *A Statistical Evaluation of Toxicity Study Designs for the Estimation of the Benchmark Dose in Continuous Endpoints*, 84 Toxicol. Sci. 167–85 (2005). Courts also recognize the benchmark dose. *See, e.g.*, Am. Forest & Paper Ass'n Inc. v. EPA, 294 F.3d 113, 121 (D.C. Cir. 2002) (EPA's use of benchmark dose takes into account comprehensive dose–response information unlike NOEL and thus its use was not arbitrary in determining that methanol should remain on the list of hazardous air pollutants); California v. Tri-Union Seafoods, LLC, 2006 WL 1544384 (Cal. Super. Ct. May 11, 2006) (benchmark dose should not be equated with LOEL (lowest observable effect level) and thus toxicologist's testimony regarding methylmercury in tuna was unreliable for purposes of California's Proposition 65).

28. For further discussion of the no-threshold model of carcinogenesis, see James E. Klaunig & Lisa M. Kamendulis, *Chemical Carcinogens, in* Casarett and Doull's Toxicology: The Basic Science of Poisons, *supra* note 3, at 329. *But see* V.P. Bond et al., *Current Misinterpretations of the Linear No-Threshold*

Figure 1. Idealized comparison of a no-threshold and threshold dose–response relationship.

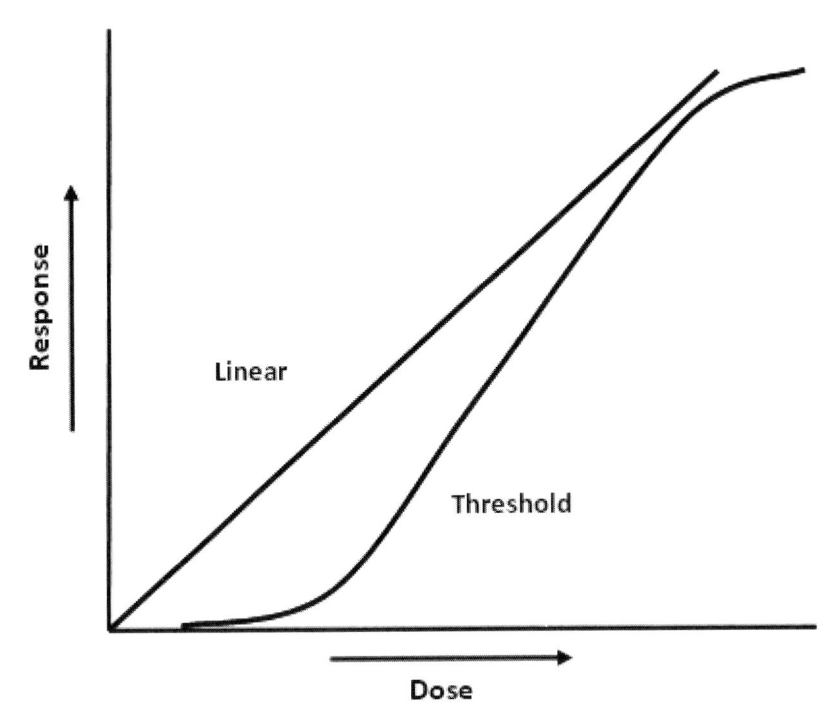

Hypothesis, 70 Health Physics 877 (1996); Marvin Goldman, *Cancer Risk of Low-Level Exposure*, 271 Science 1821 (1996).

Although the one-hit model explains the response to most carcinogens, there is accumulating evidence that for certain cancers there is in fact a multistage process and that some cancer-causing agents, so-called epigenetic or nongenotoxic agents, act through nonmutational processes, Committee on Risk Assessment Methodology, National Research Council, Issues in Risk Assessment 34–35, 187, 198–201 (1993). For example, the multistage cancer process may explain the carcinogenicity of benzo[a]pyrene (produced by the combustion of hydrocarbons such as oil) and chlordane (a termite pesticide). However, nonmutational responses to asbestos, dioxin, and estradiol cause their carcinogenic effects. The appropriate mathematical model to use to depict the dose–response relationship for such carcinogens is still a matter of debate. *Id.* at 197–201. Proposals have been made to merge cancer and noncancer risk assessment models. Committee on Improving Risk Analysis Approaches Used by the U.S. EPA, National Research Council, Toward a Unified Approach to Dose–Response Assessment 127–87 (2009).

Courts continue to grapple with the no-threshold model. *See, e.g., In re* W.R. Grace & Co. 355 B.R. 462, 476 (Bankr. D. Del. 2006) (the "no threshold model . . . flies in the face of the toxicological law of dose-response . . . doesn't satisfy *Daubert,* and doesn't stand up to scientific scrutiny"); Cano v. Everest Minerals Corp., 362 F. Supp. 2d 814, 853–54 (W.D. Tex. 2005) (even accepting the linear, no-threshold model for uranium mining and cancer, it is not enough to show exposure, you must show causation as well). Where administrative rulemaking is the issue, the no-threshold model has been accepted by some courts. *See, e.g.,* Coalition for Reasonable Regulation of Naturally

f. Maximum tolerated dose and chronic toxicity tests

Another type of study uses different doses of a chemical agent to establish over a 90-day period what is known as the maximum tolerated dose (MTD) (the highest dose that does not cause significant overt toxicity). The MTD is important because it enables researchers to calculate the dose of a chemical to which an animal can be exposed without reducing its lifespan, thus permitting the evaluation of the chronic effects of exposure.[29] These studies are designed to last the lifetime of the species.

Chronic toxicity tests evaluate carcinogenicity or other types of toxic effects. Federal regulatory agencies frequently require carcinogenicity studies on both sexes of two species, usually rats and mice. A pathological evaluation is done on the tissues of animals that died during the study and those that are sacrificed at the conclusion of the study.

The rationale for using the MTD in chronic toxicity tests, such as carcinogenicity bioassays, often is misunderstood. It is preferable to use realistic doses of carcinogens in all animal studies. However, this leads to a loss of statistical power, thereby limiting the ability of the test to detect carcinogens or other toxic compounds. Consider the situation in which a realistic dose of a chemical causes a tumor in 1 in 100 laboratory animals. If the lifetime background incidence of tumors in animals without exposure to the chemical is 6 in 100, a toxicological test involving 100 control animals and 100 exposed animals who were fed the realistic dose would be expected to reveal 6 control animals and 7 exposed animals with the cancer. This difference is too small to be recognized as statistically significant. However, if the study started with 10 times the realistic dose, the researcher would expect to get 10 additional cases for a total of 16 cases in the exposed group and 6 cases in the control group, a significant difference that is unlikely to be overlooked.

Unfortunately, even this example does not demonstrate the difficulties of determining risk. Regulators are responding to public concern about cancer by regulating risks often as low as 1 in 1,000,000—not 1 in 100, as in the example given above. To test risks of 1 in 1,000,000, a researcher would have to either increase the lifetime dose from 10 times to 100,000 times the realistic dose or

Occurring Substances v. Cal. Air Res. Bd., 19 Cal. Rptr. 3d 635, 641 (Cal. Ct. App. 2004) (use of the no-threshold model to establish no safe level of asbestos exposure by regulatory agency upheld).

29. Even the determination of the MTD can be fraught with controversy. *See, e.g.*, Simpson v. Young, 854 F.2d 1429, 1431 (D.C. Cir. 1988) (petitioners unsuccessfully argued that FDA improperly certified color additive Blue No. 2 dye as safe because researchers failed to administer the MTD to research animals, as required by FDA protocols); Valentine v. PPG Indus., Inc., 821 N.E.2d 580, 607–08 (Ohio Ct. App. 2004) (summary judgment for defendant upheld based in part on expert's observation that "there is no reliable or reproducible epidemiological evidence that shows that chemicals capable of causing brain tumors in animals at maximum tolerated doses over a lifetime can cause brain tumors in humans. The biological plausibility of those chemicals causing brain tumors in humans is lacking.").

See L.R. Rhomberg et al., *Issues in the Design and Interpretation of Chronic Toxicity and Carcinogenicity Studies in Rodents: Approaches to Dose Selection,* 37 Crit. Rev. Toxicol. 729–837 (2007).

expand the numbers of animals under study into the millions. However, increases of this magnitude are beyond the world's animal testing capabilities and are also prohibitively expensive. Inevitably, then, animal studies must trade statistical power for extrapolation from higher doses to lower doses.

Accordingly, proffered toxicological expert opinion on potentially cancer-causing chemicals almost always is based on a review of research studies that extrapolate from animal experiments involving doses significantly higher than that to which humans are exposed.[30] Such extrapolation is accepted in the regulatory arena. However, in toxic tort cases, experts often use additional background information[31] to offer opinions about disease causation and risk.[32]

2. *In vitro research*

In vitro research concerns the effects of a chemical on human or animal cells, bacteria, yeast, isolated tissues, or embryos. Thousands of in vitro toxicological tests have been described in the scientific literature. Many tests are for mutagenesis in bacterial or mammalian systems. There are short-term in vitro tests for just about every physiological response and every organ system, such as perfusion tests and DNA studies. Relatively few of these tests have been validated by replication in many different laboratories or by comparison with outcomes in animal studies to determine if they are predictive of whole animal or human toxicity.[33] However, these tests, and their validation, are becoming increasingly important.

30. *See, e.g.*, International Agency for Research on Cancer, World Health Organization, *Preamble, in* 63 IARC Monographs on the Evaluation of Carcinogenic Risks to Humans 9, 17 (1995); James Huff, *Chemicals and Cancer in Humans: First Evidence in Experimental Animals*, 100 Envtl. Health Persp. 201, 204 (1993); Joseph V. Rodricks, *Evaluating Disease Causation in Humans Exposed to Toxic Substances*, 14 J.L. & Pol'y 39 (2006).

31. Central to offering an expert opinion on specific causation is a comparison of the estimated risk with the likelihood of the adverse event if the individual had not suffered the alleged exposure. This will differ depending on factors specific to that individual, including age, gender, medical history, and competing exposures.

Researchers have developed numerous biomathematical formulas to provide statistical bases for extrapolation from animal data to human exposure. *See generally* S.C. Gad, Statistics and Experimental Design for Toxicologists (4th ed. 2005). *See also infra* Sections III, IV.

32. Policy arguments concerning extrapolation from high doses to low doses are explored in Troyen A. Brennan & Robert F. Carter, *Legal and Scientific Probability of Causation of Cancer and Other Environmental Disease in Individuals*, 10 J. Health Pol., Pol'y & L. 33 (1985). For a general discussion of dose issues in toxic torts, see also Bernard D. Goldstein, *Toxic Torts: The Devil Is in the Dose*, 16 J.L. & Pol'y 551–85 (2008).

33. *See* R. Julian Preston & George R. Hoffman, *Genetic Toxicology, in* Casarett and Doull's Toxicology: The Basic Science of Poisons, *supra* note 3, at 381, 391–404. Use of in vitro data for evaluating human mutagenicity and teratogenicity is described in John M. Rogers & Robert J. Kavlock, *Developmental Toxicology, in* Casarett and Doull's Toxicology: The Basic Science of Poisons, *supra* note 3, at 415, 436–40. For a critique of expert testimony using in vitro data, see Wade-Greaux v. Whitehall Laboratories, Inc., 874 F. Supp. 1441, 1480 (D.V.I. 1994), *aff'd*, 46 F.3d 1120 (3d Cir. 1994); *In re* Welding Fume Prods. Liab. Litig., 2006 WL 4507859, at *13 (N.D. Ohio Aug. 8, 2005)

The criteria of reliability for an in vitro test include the following: (1) whether the test has come through a published protocol in which many laboratories used the same in vitro method on a series of unknown compounds prepared by a reputable organization (such as the National Institutes of Health (NIH) or the International Agency for Research on Cancer (IARC)) to determine if the test consistently and accurately measures toxicity, (2) whether the test has been adopted by a U.S. or international regulatory body, and (3) whether the test is predictive of in vivo outcomes related to the same cell or target organ system.

D. Extrapolation from Animal and Cell Research to Humans

Two types of extrapolation must be considered: from animal data to humans and from higher doses to lower doses.[34] In qualitative extrapolation, one can usually rely on the fact that a compound causing an effect in one mammalian species will cause it in another species. This is a basic principle of toxicology and pharmacology. If a heavy metal, such as mercury, causes kidney toxicity in laboratory animals, it is highly likely to do so at some dose in humans. However, the dose at which mercury causes this effect in laboratory animals is modified by many internal factors, and the exact dose–response curve may be different from that for humans. Through the study of factors that modify the toxic effects of chemicals, including absorption, distribution, metabolism, and excretion, researchers can improve the ability to extrapolate from laboratory animals to humans and from higher to lower doses.[35] The mathematical depiction of the process by which an external dose moves through various compartments in the body until it reaches the target organ is often called physiologically based pharmacokinetics or toxicokinetics.[36]

Extrapolation from studies in nonmammalian species to humans is much more difficult but can be done if there is sufficient information on similarities in absorp-

(Toxicologist qualified to testify on relationship between welding fumes and Parkinson's disease including epidemiology and animal and in vitro toxicology studies).

34. *See* J.V. Rodricks et al., *Quantitative Extrapolations in Toxicology, in* Principles and Methods of Toxicology 365 (A. Wallace Hayes ed., 5th ed. 2008).

35. For example, benzene undergoes a complex metabolic sequence that results in toxicity to the bone marrow in all species, including humans. Robert Snyder, *Xenobiotic Metabolism and the Mechanism(s) of Benzene Toxicity*, 36 Drug Metab. Rev. 531, 547 (2004).

The exact metabolites responsible for this bone marrow toxicity are the subject of much interest but remain unknown. Mice are more susceptible to benzene than are rats. If researchers could determine the differences between mice and rats in their metabolism of benzene, they would have a useful clue about which portion of the metabolic scheme is responsible for benzene toxicity to the bone marrow. *See, e.g.*, Lois D. Lehman-McKeeman, *Absorption, Distribution, and Excretion of Toxicants, in* Casarett and Doull's Toxicology: The Basic Science of Poisons, *supra* note 3, at 131; Andrew Parkinson & Brian W. Ogilvie, *Biotransformation of Xenobiotics, in* Casarett and Doull's Toxicology: The Basic Science of Poisons, *supra* note 3, at 161.

36. For an analysis of methods used to extrapolate from animal toxicity data to human health effects, see references cited in notes 21 and 22, *supra*.

tion, distribution, metabolism, and excretion. Advances in computational toxicology have increased the ability of toxicologists to make such extrapolations.[37] Quantitative determinations of human toxicity based on in vitro studies usually are not considered appropriate. As discussed in Section I.F, in vitro or animal data for elucidating the mechanisms of toxicity are more persuasive when positive human epidemiological data or toxicological information also exists.[38]

E. Safety and Risk Assessment

Toxicological expert opinion also relies on formal safety and risk assessments. Safety assessment is the area of toxicology relating to the testing of chemicals and drugs for toxicity. It is a relatively formal approach in which the potential for toxicity of a chemical is tested in vivo or in vitro using standardized techniques. The protocols for such studies usually are developed through scientific consensus and are subject to oversight by governmental regulators or other watchdog groups.

After a number of bad experiences, including outright fraud, government agencies have imposed codes on laboratories involved in safety assessment, including industrial, contract, and in-house laboratories.[39] Known as good laboratory practices (GLPs), these codes govern many aspects of laboratory standards, including such details as the number of animals per cage, dose and chemical verification, and the handling of tissue specimens. GLPs are remarkably similar across agencies, but the tests called for differ depending on the mission. For example, there are major differences between FDA's and EPA's required procedures for testing drugs

37. *See* R.J. Kavlock et al., *Computational Toxicology: A State of the Science Mini Review*, 103 Toxicological Sci. 14–27 (2008). *See also* D. Malacarne et al., *Relationship Between Molecular Connectivity and Carcinogenic Activity: A Confirmation with a New Software Program Based on Graph Theory*, 101 Envtl. Health Persp. 331–42 (1993), for validation of the use of a computational structure-based approach to carcinogenicity originally proposed by H.S. Rosenkranz & G. Klopman, *Structural Basis of Carcinogenicity in Rodents of Genotoxicants and Non-genotoxicants*, 228 Mutat. Res. 105–24 (1990). Structure–activity relationships have also been used to extend the threshold concept in toxicology to look at low-dose exposures to agents present in foods or cosmetics. *See* R. Kroes et al., *Structure-Based Thresholds of Toxicological Concern (TTC): Guidance for Application to Substances Present at Low Levels in the Diet*, 42 Food Chem. Toxicol. 65–83 (2004).

38. An example of toxicological information in humans that is pertinent to extrapolation is the finding in human urine of a carcinogenic metabolite found in studies of the same compound in laboratory animals. *See, e.g.*, Goewey v. United States, 886 F. Supp. 1268, 1280–81 (D.S.C. 1995) (extrapolation of neurotoxic effects from chickens to humans unwarranted without human confirmation).

39. A dramatic case of fraud involving a toxicology laboratory that performed tests to assess the safety of consumer products is described in *United States v. Keplinger*, 776 F.2d 678 (7th Cir. 1985). Keplinger and the other defendants in this case were toxicologists who were convicted of falsifying data on product safety by underreporting animal morbidity and mortality and omitting negative data and conclusions from their reports. For further discussion of reviewing animal studies in light of the FDA's Good Laboratory Practice guidelines, see Eli Lilly & Co. v. Zenith Goldline Pharm., Inc. 364 F. Supp. 2d 820, 860 (S.D. Ind. 2005).

and environmental chemicals.[40] FDA requires and specifies both efficacy and safety testing of drugs in humans and animals. Carefully controlled clinical trials using doses within the expected therapeutic range are required for premarket testing of drugs because exposures to prescription drugs are carefully controlled and should not exceed specified ranges or uses. However, for environmental chemicals and agents, no premarket testing in humans is required by EPA. New European Union Regulation on Registration, Evaluation, Authorisation and Restriction of Chemicals (REACH) requires extensive testing of new chemicals and chemicals in commerce.[41] Moreover, because exposures are less predictable, doses usually are given in a wider range in animal tests for nonpharmaceutical agents.[42]

Because exposures to environmental chemicals may continue over a lifetime and affect both young and old, test designs called lifetime bioassays have been developed in which relatively high doses are given to experimental animals. The interpretation of results requires extrapolation from animals to humans, from high to low doses, and from short exposures to multiyear estimates. It must be emphasized that less than 1% of the 60,000 to 75,000 chemicals in commerce have been subjected to a full safety assessment, and there are significant toxicological data on

40. *See, e.g.*, 40 C.F.R. Parts 160, 792 (1993); Lu, *supra* note 22, at 89. There is a major difference between the information needed to establish a regulatory standard or tolerance, and that needed to establish causation for clinical or tort purposes.

41. For comparison of Toxic Substances Control Act (TSCA), 15 U.S.C. §§ 2601 et seq. (1978) and REACH, see E. Donald Elliott, Trying to Fix TSCA § 6: Lessons from REACH, Proposition 65, and the Clean Air Act, *available at* http://www.ucis.pitt.edu/euce/events/policyconf/07/PDFs/Elliott.pdf. For issues related to the intentional testing of environmental chemicals in humans, see Committee on the Use of Third Party Toxicity Research with Human Research Participation, National Research Council, Intentional Human Dosing Studies for EPA Regulatory Purposes: Scientific and Ethical Issues (2004).

42. It must be appreciated that the development of a new drug inherently requires searching for an agent that at useful doses has a biological effect (e.g., decreasing blood pressure), whereas those developing a new chemical for consumer use (e.g., a house paint) hope that at usual doses no biological effects will occur. There are other compounds, such as pesticides and antibacterial agents, for which a biological effect is desired, but it is intended that at usual doses humans will not be affected. These different expectations are part of the rationale for the differences in testing information available for assessing toxicological effects. Under FDA rules, approval of a new drug usually will require extensive animal and human testing, including a randomized double-blind clinical trial for efficacy and toxicity. In contrast, under TSCA, the only requirement before a new chemical can be marketed is that a premanufacturing notice be filed with EPA, including any toxicity data in the company's possession. EPA reviews this information, along with structure–activity relationship modeling, in order to determine whether any restrictions on release should be imposed. For existing chemicals, EPA may require companies to undertake animal and in vitro tests if the chemical may present an unreasonable risk to health. The lack of toxicity data for most chemicals in commerce has led EPA to propose methods of evaluation using in vitro toxicity pathway testing, followed by whole-animal testing where warranted. *See* Committee on Toxicity Testing and Assessment of Environmental Agents, National Research Council, Toxicity Testing in the 21st Century: A Vision and a Strategy (2007); U.S. Environmental Protection Agency, Strategic Plan for Evaluating the Toxicity of Chemicals (March 2009), *available at* http://www.epa.gov/spc/toxicitytesting.

only 10% to 20% of them. Under the current U.S. and international approaches to testing chemicals with high production volume, and with the advent of the REACH legislation, the extent of toxicological information is expanding rapidly.[43]

Risk assessment is an approach increasingly used by regulatory agencies to estimate and compare the risks of hazardous chemicals and to assign priority for avoiding their adverse effects.[44] The National Academy of Sciences defines four components of risk assessment: hazard identification, dose–response estimation, exposure assessment, and risk characterization.[45]

Risk assessment is not an exact science. It should be viewed as a useful framework to organize and synthesize information and to provide estimates on which policymaking can be based. In recent years, codification of the methodology used to assess risk has increased confidence that the process can be reasonably free of bias; however, significant controversy remains, particularly when actual data are limited and generally conservative default assumptions are used.[46]

Although risk assessment information about a chemical can be somewhat useful in a toxic tort case, at least in terms of setting reasonable boundaries regarding the likelihood of causation, the impetus for the development of risk assessment has been the regulatory process, which has different goals.[47] Because of their

43. *See* John S. Applegate, *The Perils of Unreasonable Risk: Information, Regulatory Policy, and Toxic Substances Control*, 261 Colum. L. Rev. 264–66 (1991) for a discussion of REACH and its potential impact on the availability of toxicological and risk information. *See* Sven O. Hanssen & Christina Ruden, *Priority Setting in the REACH System*, 90 Toxicological Sci. 304–08 (2005), for a discussion of the toxicological needs for REACH and its reliance on exposure.

44. The use of risk assessment by regulatory agencies was spurred by the Supreme Court's decision in *Industrial Union Dep't, AFL-CIO v. American Petroleum Institute,* 448 U.S. 607 (1980). A plurality of the court overturned the Occupational Safety and Health Administration's (OSHA) attempt to regulate benzene based on the intrinsic hazard of benzene being a human carcinogen. Instead, by requiring a risk assessment, the inclusion of exposure assessment and dose–response evaluation became a customary part of regulatory assessment. *See* John S. Applegate, *supra* note 43.

45. *See generally* National Research Council, Risk Assessment in the Federal Government: Managing the Process (1983); Bernard D. Goldstein, *Risk Assessment and the Interface Between Science and Law*, 14 Colum. J. Envtl. L. 343 (1989). Recently, a National Academy of Sciences panel has discussed potential approaches to updating the risk paradigm. *See* Committee on Improving Risk Analysis Approaches Used by the U.S. EPA, *supra* note 28.

46. An example of conservative default assumptions can be found in Superfund risk assessment. EPA has determined that Superfund sites should be cleaned up to reduce cancer risk from 1 in 10,000 to 1 in 1,000,000. A number of assumptions can go into this calculation, including conservative assumptions about intake, exposure frequency and duration, and cancer-potency factors for the chemicals at the site. *See, e.g.,* Robert H. Harris & David E. Burmaster, *Restoring Science to Superfund Risk Assessment*, 6 Toxics L. Rep. 1318 (1992).

47. *See* Committee on Improving Risk Analysis Approaches Used by the U.S. EPA, *supra* note 28. *See also* Rhodes v. E.I. du Pont de Nemours & Co., 253 F.R.D. 365, 377–78 (S.D. W. Va. 2008) (putative class-action plaintiffs alleging that contamination of their drinking water with industrial perfluorooctanoic acid entitled them to medical monitoring could not rely upon regulatory risk assessment that does not provide the requisite reasonable certainty required to show a medical monitoring injury). Risk assessment also has come under heavy criticism from those who prefer the precautionary

use of appropriately prudent assumptions in areas of uncertainty and their use of default assumptions when there are limited data, risk assessments often intentionally encompass the upper range of possible risks.[48] An additional issue, particularly related to cancer risk, is that standards based on risk assessment often are set to avoid the risk caused by lifetime exposure at this level. Exposure to levels exceeding this standard for a small fraction of a lifetime does not mean that the overall lifetime risk of regulatory concern has been exceeded.[49]

1. The use of toxicological information in risk assessment

Risk assessment as practiced by government agencies involved in regulating exposure to environmental chemicals is highly dependent upon the science of toxicology and on the information derived from toxicological studies. EPA, FDA, OSHA, the Consumer Product Safety Commission, and other international (e.g., the World Trade Organization), national, and state agencies use risk assessment as a means to protect workers or the public from adverse effects.[50] Acceptable risk levels, for example, 1 in 1000 to 1 in 1,000,000, are usually well below what

principle as an alternative. For advocacy of the precautionary principle, see Joel A. Tickner, *Precautionary Principle Encourages Policies That Protect Human Health and the Environment in the Face of Uncertain Risks,* 117 Pub. Health Rep. 493–97 (2002). Although variously defined, the precautionary principle in many ways is a hazard-based approach.

48. It is also claimed that standard risk assessment will underestimate true risks, particularly for sensitive populations exposed to multiple stressors, an issue of particular pertinence to discussions of environmental justice. Committee on Environmental Justice, Institute of Medicine, Toward Environmental Justice: Research, Education, and Health Policy Needs (1999). The EPA has been developing formal guidance for cumulative risk assessment, which has been defined as "the combined threats from exposure via all relevant routes to multiple stressors including biological, chemical, physical, and psychosocial entities." Michael A. Callahan & Ken Sexton, *If Cumulative Risk Assessment Is the Answer, What Is the Question?* Envtl. Health Persp. 799–806 (2007). *See also* International Life Sciences Institute, A Framework for Cumulative Risk Assessment Workshop Report (1999). A related issue is aggregate risk assessment, which focuses on exposure to a single agent through multiple routes. For example, swimming in water containing a volatile organic contaminant is likely to lead to exposure through the skin, through inhalation of the contaminant off-gassing just above the water surface, and through swallowing water. For a discussion of aggregate risk assessment, see International Life Science Institute, Aggregate Exposure Assessment Workshop Report (1998). For a study of a child's indoor exposure through different routes to a pesticide, see V.G. Zartarian et al., *A Modeling Framework for Estimating Children's Residential Exposure and Dose to Chlorpyrifos Via Dermal Residue Contact and Nondietary Ingestion,* 108 Envtl. Health Persp. 505–14 (2000).

49. A public health standard to protect against the lifetime risk of inhaling a known carcinogen will usually be based on lifetime exposure calculations of 24 hours a day, everyday for 70 years. This is more than 25,000 days and 600,000 hours. Exceeding this standard for a few hours would presumably have little impact on cancer risk. In contrast, for a short-term standard set to avoid a threshold-based risk, exceeding the standard for this short time may make a major difference, for example, an asthma attack caused by being outdoors on a day that the ozone standard is exceeded.

50. Pharmaceuticals intended for human use are an exception in that a tradeoff between desired and adverse effects may be acceptable, and human data are available prior to, and as a result of, the marketing of the agent.

can be measured through epidemiological study. Inevitably, this means that risk assessment is based solely on toxicological data—or, if epidemiological findings of an adverse effect are observed, then toxicological reasoning must be used to extrapolate to the appropriate lower dose standard aimed at protecting the public.

The four-part risk paradigm is heavily based on toxicological precepts. Hazard identification reflects the toxicological "law" of specificity of effects, and dose–response assessment is based upon "the dose makes the poison." The hazard identification process often uses "weight of evidence" approaches in which the toxicological, mechanistic, and epidemiological data are rigorously assessed to form a judgment regarding the likelihood that the agent produces a specific effect.[51] Establishing the appropriate dose–response curve, threshold, or "one-hit" is an exercise in toxicological reasoning. Even for those chemicals known to be carcinogens, a threshold model is appropriate if the toxicological mechanism of action can be demonstrated to depend upon a threshold. Exposure assessment requires knowledge of specific toxicological dynamics; for example, the impact on the lung of an air pollutant varies by factors such as inhalation rate per unit body mass, which is affected by exercise and by age; by the size of a particle or the solubility of a gas, both of which will affect the depth of penetrance into the more sensitive parts of the airways; by the competence of the usual airway defense mechanisms, such as mucus flow and macrophage function; and by the ability of the lung to metabolize the agent.[52]

F. Toxicological Processes and Target Organ Toxicity

The biological, chemical, and physical phenomena that are the basis of life are astounding in their complexity. As a result, human subcellular, cellular, and organ function are both delicately balanced and highly robust. Small changes caused by external chemical and physical agents can have major effects; yet, through the millennia, evolutionary pressures have led to the emergence of safety mechanisms that defend against adverse environmental stresses.

The specialization that is a hallmark of organ development in vertebrates inherently leads to diversity in the underlying processes that are the basis of organ function. Certain chemicals poison virtually all cells by affecting a basic biological process essential to life. For example, cyanide interferes with the conversion of oxygen to energy in a subcellular component known as mitochondria.[53] Other

51. *See* Section I.F for further discussion of weight-of-evidence approaches to potential human carcinogens.

52. Some toxic agents pass through the lung without producing any direct effects on this organ. For example, inhaled carbon monoxide produces its toxicity in essence by being treated by the body as if it is oxygen. Carbon monoxide readily combines with the oxygen combining site of hemoglobin, the molecule in red blood cells that is responsible for transporting oxygen from the lung to the tissues. By doing so, the effective transport and tissue utilization of oxygen is blocked.

53. Note that the diffuse toxicity of cyanide also reflects its ability to spread widely in the body. Certain mitochondrial poisons primarily affect the brain and active muscles, including the heart, which

chemical agents interfere selectively with an organ-specific process. For example, organophosphate pesticides, often known as nerve gases, specifically interact within the specialized intercellular nerve cell transmission of impulses—a process that is pertinent primarily to the nervous system. Table 1 provides arbitrarily selected examples of toxicological end points and agents of concern, which are not meant to be inclusive or exhaustive.

Despite this specialization, there are pathological processes common to diseases affecting many different organs. For example, chronic inflammation of the skin leads to fiber formation that is recognized as scarring. Similarly, cirrhosis of the liver can result from fibrogenic processes caused by repetitive inflammation of the liver, such as from the overuse of ethanol, and fibrosis of the lung is an important pathological process resulting from asbestos, silica, and other agents.[54] The potential for endocrine disruption by chemicals, particularly those that persist within the body, has become an increasing concern. Many of these persistent agents belong to families of chemically similar compounds, such as dioxins or PCBs, that may differ in their effect. Particularly challenging to standard toxicological approaches are agents that react with different receptors present on the surface or internal components of the cell. These receptors often belong to complex families of related cellular components that are continually interacting with the broad range of hormones produced by our bodies.[55] The intricate dynamic processes of normal endocrine activity include feedback loops that allow cyclic variation, such as in the menstrual cycle or in the variation of hormone and receptor levels that are linked to normal functions such as sleeping and sexual activity. These complex normal "up and down" variations produce conceptual difficulties when attempting to extrapolate the results from model systems to the functioning human.[56]

are particularly oxygen dependent. Others, unable to penetrate the blood-brain barrier, will primarily affect peripheral muscle including the heart.

54. Lung fibrosis is a key pathological finding in a group of diseases known as pneumoconiosis that includes coal miners' black lung disease, silicosis, asbestosis, and other conditions usually caused by occupational exposures.

55. As a simplification, agent–receptor interactions often are described as a key in a lock, with the key needing to be able to both fit into the lock and turn the mechanism. An example from the nervous system is the use in treating a heroin overdose of another opiate that has a much higher affinity for the receptor site but produces little effect once bound. When given to a normal person, this second opiate would have a mild depressant effect, but it can reverse a near fatal overdose of heroin by displacing the heroin from the receptor site. Thus the directionality of opiate effect depends upon the interaction of the components of the mixture. This interaction is even more complex when dealing with estrogenic agents that are naturally occurring as well as made within the body at different levels in response to different external and internal stimuli and at different time intervals.

56. The complexity of the interaction of a mixture of dioxins with receptors governing the endocrine system can be contrasted with that of the reaction of carbon monoxide with the hemoglobin oxygen receptor discussed in note 52. The latter is unidirectional in that any additional carbon monoxide will interfere with oxygen delivery, of which there cannot be too much under normal physiological conditions.

Table 1. Sample of Selected Toxicological End Points and Examples of Agents of Concern in Humans[a]

Organ System	Examples of End Points	Examples of Agents of Concern
Skin	allergic contact dermatitis	nickel, poison ivy, cutting oils
	chloracne	dioxins
	cancer	polycyclic aromatic hydrocarbons
Respiratory tract	nonspecific irritation (reactive airway disease)	formaldehyde, acrolein, ozone
	asthma	toluene diisocyanate
	chronic obstructive pulmonary disease	cigarette smoke
	fibrosis, pneumoconiosis	silica, mineral dusts, cotton dust
	cancer	cigarette smoke, arsenic, asbestos, nickel
Blood and the immune system	anemia	arsine, lead, methyldopa
	secondary polycythemia	cobalt
	methemoglobinemia	nitrites, aniline dyes, dapsone
	pancytopenia	benzene, radiation, chemotherapeutic agents
	secondary lupus erythematosus	hydralazine
	leukemia	benzene, radiation, chemotherapeutic agents
Liver and gastrointestinal tract	hepatic damage (hepatitis)	acetaminophen, ethanol, carbon tetrachloride, vitamin A
	cancer	aflatoxin, vinyl chloride
Urinary tract	kidney toxicity	ethylene and diethylene glycols, lead, melamine, aminoglycoside antibiotics
	bladder cancer	aromatic amines
Nervous system	nervous system toxicity	cholinesterase inhibitors, mercury, lead, *n*-hexane, bacterial toxins (botulinum, tetanus)
	Parkinson's disease	manganese
Reproductive and developmental toxicity	fetal malformations	thalidomide, ethanol

continued

Table 1. Continued

Organ System	Examples of End Points	Examples of Agents of Concern
Endocrine system	thyroid toxicity	radioactive iodine, perchlorate
Cardiovascular system	heart toxicity high blood pressure arrhythmias	anthracyclines, cobalt lead plant glycosides (e.g. digitalis)

[a]This table presents only examples of toxicological end points and examples of agents of concern in humans and is provided to help illustrate the variety of toxic agents and end points. It is not an exhaustive or inclusive list of organs, end points, or agents. Absence from this list does not indicate a relative lack of evidence for a causal relation as to any agent of concern.

The processes that result in the causation of cancer are also of particular interest to the public, to litigators, and to regulators. A common denominator for the various diseases that fall under the heading of cancer is uncontrolled cellular growth, usually reflecting the failure of the normal progression of precursor cells to maturation and cell death. Central to the mechanism of cancer causation is the production of a genetic change that leads a precursor cell to no longer conform to usual processes that control cell growth. In virtually all cancers, the overgrowth of cells can be traced to a single mutation, such that cancer cells are a clone of the one mutated precursor cell.[57] The understanding of the relationship between mutation and cancer led to some of the first toxicological tests to determine whether an external agent could cause cancer. Such tests have grown in sophistication because of the advances in molecular biology and computational toxicology that have occurred concomitantly with an increased understanding of the variety of potential pathways that lead to mutagenesis.[58]

Toxicological testing for chemical carcinogens ranges from relatively simple studies to determine whether the substance is capable of producing bacterial mutations to observation of cancer incidence as a result of long-term administration of the substance to laboratory animals. Between these two extremes are a multiplicity of tests that build upon the understanding of the mechanism of cancer causation. In vitro or in vivo tests may focus on the evidence of effects in DNA, such as the presence of adducts of the chemical or its metabolites bound to the DNA molecule or the cross-linking of the DNA molecule to protein. Researchers may look for changes in the nucleus of the cell suggestive of DNA damage that could

57. There may, in fact, be multiple mutations as the initial clone of cells undergoes further transformation before or after the cancer becomes clinically manifest.

58. Committee on Toxicity Testing and Assessment of Environmental Agents, National Research Council, Toxicity Testing in the 21st Century: A Vision and a Strategy (2007).

result in mutagenesis and carcinogenesis, for example, the micronucleus test or the comet assay. Certain mutagens cause an increase in the normal exchange of nuclear material among DNA components during normal cell division, which gives rise to a test known as the "sister chromatid exchange."[59] The direct observation of chromosomes to look for specific abnormalities, known as cytogenetic analysis, is providing more information about the pathways of carcinogenesis. For cancers such as acute myelogenous leukemia, it has long been recognized that those individuals who present with recognizable chromosomal abnormalities are more likely to have been exposed to a known human chemical leukemogen such as benzene.[60] But at this time there is no chromosomal abnormality that is unequivocally linked to a specific chemical or physical carcinogen.[61] These and other tests provide information that can be used in evaluating whether a chemical is a potential human carcinogen.

The many tests that are pertinent to estimating whether a chemical or physical agent produces human cancer require careful evaluation. The World Health Organization's (WHO's) IARC and the U.S. National Toxicology Program (NTP) have formal processes to evaluate the weight of evidence that a chemical causes cancer.[62] Each classifies chemicals on the basis of epidemiological evidence, toxicological findings in laboratory animals, and mechanistic considerations, and then assigns a specific category of carcinogenic potential to the individual chemical or exposure situation (e.g., employment as a painter).[63] Only a small percentage of

59. All of these tests require validation regarding their relevance to predicting human carcinogenesis, as well as to their technical reproducibility. *See* Raffaella Corvi et al., *ECVAM Retrospective Validation of In Vitro Micronucleus Test*, 23 Mutagenesis 271–83 (2008), for an example of an approach to validating a short-term assay for carcinogenesis.

60. F. Mitelman et al., *Chromosome Pattern, Occupation, and Clinical Features in Patients with Acute Nonlymphocytic Leukemia*, 4 Cancer Genet. & Cytogenet. 197, 214 (1981).

61. *See* Luoping Zhang et al., *The Nature of Chromosomal Aberrations Detected in Humans Exposed to Benzene*, 32 Crit. Rev. Toxicol. 1–42 (2002).

62. The U.S. National Toxicology Program issues a congressionally mandated Report on Carcinogens. The 12th report is available at http://ntp.niehs.nih.gov/ntp/roc/twelfth/roc12.pdf. IARC produces its reports through a monograph series that provides detailed description of the agents or processes under consideration as well as the findings of the IARC expert working group. See the IARC Web site for a list of these monographs (http://monographs.iarc.fr/).

63. IARC uses the following classifications:

> Group 1, The agent (mixture) is carcinogenic to humans;
> Group 2A, The agent (mixture) is probably carcinogenic to humans,
> Group 2B, The agent (mixture) is possibly carcinogenic to humans;
> Group 3, The agent (mixture) is not classifiable as to its carcinogenicity to humans; and
> Group 4, The agent (mixture) is probably not carcinogenic to humans.

Inherent in putting chemicals into distinct categories when there is a continuum for the strength of the evidence is that some chemicals will be very close to the dividing line between the discrete categories. Inevitably, small differences in the interpretation of the evidence for such chemicals will lead to disagreement regarding categorization.

the total chemicals in commerce are considered to be known human carcinogens. In the past, assignment to the highest category was dependent almost totally on epidemiological evidence, although animal data and mechanistic information were also considered. In recent years, with improved understanding of the mechanism of action of chemical carcinogens, there has been increased use of mechanistic data.[64] For example, higher credence is given to the likelihood that a chemical is a human carcinogen if the metabolite found to be responsible for carcinogenesis in a laboratory animal is also found in the blood or urine of humans exposed to this chemical, or if there is evidence of the same type of DNA damage in humans as there is in laboratory animals in which the agent does cause cancer.[65]

G. Toxicology and Exposure Assessment

In recent decades, exposure assessment has developed into a scientific field with the usual trappings of journals, learned societies, and research funding processes.

64. *See* Vincent James Cogliano et al., *Use of Mechanistic Data in IARC Evaluations,* 49 Envt. & Molecular Mutagenesis 100–09 (2008) for a discussion and for specific examples of the use of mechanistic data in evaluating carcinogens. The evolution in the approach to determining cancer causality is evident from reviewing the guidelines used to assemble the weight of evidence for causality by IARC and NTP, two of the organizations that have the lengthiest track record of responsibility for the hazard identification of carcinogens. Both have increased the weight given to mechanistic evidence in characterizing the overall strength of the total evidence used to classify the potential for a chemical or an exposure to be causal. IARC now permits classification in Group 1 when there is less than sufficient evidence in humans but sufficient evidence in animals and "strong evidence in exposed humans that the agent acts through a relevant mechanism of carcinogenicity" *Id.* at 103. The criteria used by NTP for listing a chemical as a known human carcinogen in its biannual Report on Carcinogens is "There is sufficient evidence of carcinogenicity from studies in humans,* which indicates a causal relationship between exposure to the agent, substance, or mixture, and human cancer." The asterisk is particularly notable in that it specifies that the evidence need not be solely epidemiological: "*This evidence can include traditional cancer epidemiology studies, data from clinical studies, and/or data derived from the study of tissues or cells from humans exposed to the substance in question that can be useful for evaluating whether a relevant cancer mechanism is operating in people." *See* National Toxicology Program, U.S. Dep't of Health and Human Servs., Report of Carcinogens (12th ed. 2011), at 4, *available at* http://ntp.niehs.nih.gov/ntp/roc/twelfth/roc12.pdf.

EPA also considers mechanism of action in its regulatory approaches and distinguishes further between mechanism of action and mode of action. *See* Katherine Z. Guyton et al., *Improving Prediction of Chemical Carcinogenicity by Considering Multiple Mechanisms and Applying Toxicogenomic Approaches,* 681 Mutation Res. 230, 240 (2009); Katherine Z. Guyton et al., *Mode of Action Frameworks: A Critical Analysis,* 11 J. Toxicol. & Envtl. Health Part B 16, 31 (2008).

65. A recent example is the IARC evaluation of formaldehyde that upgraded the categorization from 2A to 1 based upon epidemiological data that were strongly supported by the finding of nasal cancer in laboratory animals and by the presence of DNA-protein cross-links in the nasal tissue of the laboratory animals and of humans inhaling formaldehyde. However, epidemiological evidence associating formaldehyde with human acute myelogenous leukemia was questioned on the basis of the lack of mechanistic evidence, including questions about how such a highly reactive agent could reach the bone marrow following inhalation. *See Formaldehyde, 2-Butoxyethanol and 1-*tert*-Butoxypropan-2-ol, in* 88 IARC Monographs on the Evaluation of Carcinogenic Risks to Humans (2006).

Exposure assessment methodologies include mathematical models predicting exposure resulting from an emission source, which might be a long distance upwind; chemical or physical measurements of media such as air, food, and water; and biological monitoring within humans, including measurements of blood and urine specimens. An exposure assessment should also look for competing exposures. In this continuum of exposure metrics, the closer to the human body, the greater the overlap with toxicology.[66]

Exposure assessment is central to epidemiology as well. Many of the causal associations between chemicals and human disease have been developed from epidemiological studies relating a workplace chemical to an increased risk of the specific disease in cohorts of workers, often with only a qualitative assessment of exposure. An improved quantitative understanding of such exposures enhances the likelihood of observing causal relations.[67] It also can provide the information needed by the expert toxicologist to opine on the likelihood that a specific exposure was responsible for an adverse outcome.

H. Toxicology and Epidemiology

Epidemiology is the study of the incidence and distribution of disease in human populations. Clearly, both epidemiology and toxicology have much to offer in elucidating the causal relationship between chemical exposure and disease.[68] These

66. Toxicologists also have indirect means of approaching exposure through symptoms. For many agents, there is a known threshold for smell and a reasonable range of levels that might cause symptoms. For example, the use of toxicological expertise is appropriate in a situation in which chronic exposure to a volatile hydrocarbon is alleged to have occurred at levels at which acute exposure would be expected to render the individual unconscious. Toxicologists may also contribute knowledge of the extent of individual exposure based upon appropriate assumptions concerning inhalation rate or water use; for example, children inhale more per body mass than do adults, and outdoor workers in hot climates will drink more fluids.

67. In terms of general causation, accurate exposure assessment is important because a true effect can be missed because of the confounding caused by cohorts that often include workers with little exposure to the putative offending agent, thereby diluting the actual effect. *See* Peter F. Infante, *Benzene Exposure and Multiple Myeloma: A Detailed Meta-analysis of Benzene Cohort Studies,* 1976 Ann. N.Y. Acad. Sci. 90–109 (2006), for a discussion of this issue in relation to a meta-analysis of the potential causative role of benzene in multiple myeloma. On the other hand, an association between exposure and effect occurring solely by chance is more likely if the effect does not meet the expected standard of being more pronounced in those receiving the highest dose. *See* Bernard D. Goldstein, *Toxic Torts: The Devil Is in the Dose,* 16 J.L. & Pol'y 551–85 (2008). Setting regulatory standards based upon the observed effect in a cohort often requires a risk assessment, which in turn is dependent on understanding the extent of the exposure. This has led to extensive retrospective reconstruction of exposure in key cohorts.

68. *See* Michael D. Green et al., Reference Guide on Epidemiology, Section V, in this manual. For example, in *Norris v. Baxter Healthcare,* 397 F.3d 878, 882 (10th Cir. 2005), testimony was excluded as unreliable in which the expert ignored epidemiological studies that conflicted with the expert's opinion. However, epidemiological studies are not always necessary. Glastetter v. Novartis Pharms. Corp., 252 F.3d 986, 999 (8th Cir. 2001).

sciences often go hand in hand with assessments of the risks of chemical exposure, without artificial distinctions being drawn between them. However, although courts generally rule epidemiological expert opinion admissible, the admissibility of toxicological expert opinion has been more controversial because of uncertainties regarding extrapolation from animal and in vitro data to humans. This particularly has been true in cases in which relevant epidemiological research data exist. However, the methodological weaknesses of some epidemiological studies, including their inability to accurately measure exposure and their small numbers of subjects, render these studies difficult to interpret.[69] In contrast, because animal and cell studies permit researchers to isolate the effects of exposure to a single chemical or to known mixtures, toxicological findings offer unique information concerning dose–response relationships, mechanisms of action, specificity of response, and other information relevant to the assessment of causation.[70]

The gold standard in clinical epidemiology and in the testing of pharmaceutical agents is the randomized double-blind cohort study in which the control and intervention groups are perfectly matched. Although appropriate and very informative for the testing of pharmaceutical agents, it is generally unethical for chemicals used for other purposes. The randomized control design in essence is what is used in a classic toxicological study in laboratory animals, although matching is more readily achieved because the animals are genetically similar and have identical environmental histories.

Dose issues are at the interface between toxicology and epidemiology. Many epidemiological studies of the potential risk of chemicals do not have direct information about dose, although qualitative differences among subgroups or in comparison with other studies can be inferred. The epidemiology database includes many studies that are probing for the potential for an association between a cause and an effect. Thus a study asking all those suffering from a specific disease a multiplicity of questions related to potential exposures is bound to find some statistical association between the disease and one or more exposure conditions. Such studies generate hypotheses that can then be evaluated more thoroughly by subsequent studies that more narrowly focus on the potential cause-and-effect

69. *Id. See also* Michael D. Green et al., Reference Guide on Epidemiology, in this manual.

70. Both commonalities and differences between animal responses and human responses to chemical exposures were recognized by the court in *International Union, United Automobile, Aerospace and Agricultural Implement Workers of America, UAW v. Pendergrass,* 878 F.2d 389 (D.C. Cir. 1989). In reviewing the results of both epidemiological and animal studies on formaldehyde, the court stated: "Humans are not rats, and it is far from clear how readily one may generalize from one mammalian species to another. But in light of the epidemiological evidence [of carcinogenicity] that was not the main problem. Rather it was the absence of data at low levels." *Id.* at 394. The court remanded the matter to OSHA to reconsider its findings that formaldehyde presented no specific carcinogenic risk to workers at exposure levels of 1 part per million or less. *See also* Hopkins v. Dow Corning Corp., 33 F.3d 1116 (9th Cir. 1994); *In re* Accutane Prod. Liab., 511 F. Supp. 2d 1288, 1292 (M.D. Fla. 2007); United States v. Philip Morris USA, Inc., 449 F. Supp. 2d 1, 182 (D.D.C. 2006); Ambrosini v. Labarraque, 101 F.3d 129, 141 (D.C. Cir. 1996).

relation. One way to evaluate the strength of the association is to assess whether those epidemiological studies evaluating cohorts with relatively high exposure observe the association.[71]

The requirement in certain jurisdictions for epidemiological evidence of a relative risk greater than two (RR > 2) for general causation also has limited the utilization of toxicological evidence.[72] A firm requirement for such evidence means that if the epidemiological database showed statistically significant evidence that cohorts exposed to 10 parts per million of an agent for 20 years produced an 80% increase in risk, the court could not hear the case of a plaintiff alleging that exposure to 50 parts per million for 20 years of the same agent caused the adverse outcome. Yet to a toxicologist there would be little question that exposure to the fivefold higher dose would lead to more than a doubling of the risk, all other facets of the case being similar.

Even though there is little toxicological data on many of the 75,000 compounds in general commerce, there is far more information from toxicological studies than from epidemiological studies.[73] It is much easier, and more economical, to expose an animal to a chemical or to perform in vitro studies than it is to perform epidemiological studies. This difference in data availability is evident even for cancer causation, for which toxicological study is particularly expensive and time-consuming. Of the perhaps two dozen chemicals that reputable international authorities agree are known human carcinogens based on positive epidemiological studies, arsenic is the only one not known to be an animal carcinogen. Yet there are more than 100 known animal carcinogens for which there is no valid epidemiological database, and others for which the epidemiological database has been

71. For common chemicals, it is not unusual that a literature search reveals an association with virtually any disease. As an example of considering dose issues across epidemiological studies, see Luoping Zhang et al., *Formaldehyde Exposure and Leukemia: A New Meta-Analysis and Potential Mechanisms,* 681 Mutat. Res. 150–68 (2008). The subject of the strength of an epidemiological association and its relation to causality is considered in Michael D. Green et al., Reference Guide on Epidemiology, in this manual.

72. The basis for the use of RR > 2 is the translation of the preponderance of evidence, or "more likely than not," as a basis for tort law into at least a doubling of risk. An example is the Havner rule in Texas, which for general causation requires that there be at least two epidemiological studies with a statistically significant RR > 2 associating a putative cause with an effect (Merrell Dow Pharms. v. Havner, 953 S.W.2d 706, 716 (Tex. 1997)). For a discussion of the use by jurisdictions of relative risk > 2 for general and specific causation, see Russellyn S. Carruth & Bernard D. Goldstein, *Relative Risk GreaterThan Two in Proof of Causation in Toxic Tort Litigation,* 41 Jurimetrics 195 (2001); for the toxicological issues, see Bernard D. Goldstein, *Toxic Torts: The Devil Is in the Dose,* 16 J.L. & Pol'y 551–85 (2008).

73. *See generally* Committee on Toxicity Testing and Assessment of Environmental Agents, *supra* note 24. *See also* National Research Council, Toxicity Testing: Strategies to Determine Needs and Priorities (1984); Myra Karstadt & Renee Bobal, *Availability of Epidemiologic Data on Humans Exposed to Animal Carcinogens,* 2 Teratogenesis, Carcinogenesis & Mutagenesis 151 (1982); Lorenzo Tomatis et al., *Evaluation of the Carcinogenicity of Chemicals: A Review of the Monograph Program of the International Agency for Research on Cancer,* 38 Cancer Res. 877, 881 (1978).

equivocal.[74] To clarify any findings, regulators can require a repeat of an equivocal 2-year animal toxicological study or the performance of additional laboratory studies in which animals deliberately are exposed to the chemical. Such deliberate exposure is not possible in humans. As a general rule, unequivocally positive epidemiological studies reflect prior workplace practices that led to relatively high levels of chemical exposure for a limited number of individuals and that, fortunately, in most cases no longer occur now. Thus an additional prospective epidemiological study often is not possible, and even the ability to do retrospective studies is constrained by the passage of time.

In essence, epidemiological findings of an adverse effect in humans represent a failure of toxicology as a preventive science or of regulatory authorities or other responsible parties in controlling exposure to a hazardous chemical or physical agent. A corollary of the tenet that, depending upon dose, all chemical and physical agents are harmful, is that society depends upon toxicological science to discover these harmful effects and on regulators and responsible parties to prevent human exposure to a harmful level or to ensure that the agent is not produced. Epidemiology is a valuable backup approach that functions to detect failures of primary prevention. The two disciplines complement each other, particularly when the approaches are iterative.

II. Demonstrating an Association Between Exposure and Risk of Disease[75]

Once the expert has been qualified, he or she is expected to offer an opinion on whether the plaintiff's disease was caused by exposure to a chemical. To do so, the expert relies on the principles of toxicology to provide a scientifically valid

74. The absence of epidemiological data is due, in part, to the difficulties in conducting cancer epidemiology studies, including the lack of suitably large groups of individuals exposed for a sufficient period of time, long latency periods between exposure and manifestation of disease, the high variability in the background incidence of many cancers in the general population, and the inability to measure actual exposure levels. These same concerns have led some researchers to conclude that "many negative epidemiological studies must be considered inconclusive" for exposures to low doses or weak carcinogens. Henry C. Pitot III & Yvonne P. Dragan, *Chemical Carcinogenesis, in* Casarett and Doull's Toxicology: The Basic Science of Poisons 201, 240–41 (Curtis D. Klaassen ed., 5th ed. 1996).

75. Determinations about cause-and-effect relations by regulatory agencies often depend upon expert judgment exercised by assessing the weight of evidence. For a discussion of this process as used by the International Agency for Research on Cancer of the World Health Organization and the role of information about mechanisms of toxicity, see Vincent J. Cogliano et al., *Use of Mechanistic Data in IARC Evaluations*, 49 Envt'l & Molecular Mutagens 100 (2008). For the use of expert judgment in EPA's response to submission of information for premanufacture notification required under the Toxic Substances Control Act, 15 U.S.C. §§ 2604, 2605(e), 40 C.F.R. §§ 720 et seq., see Chemical Manufacturers Ass'n v. EPA, 859 F.2d 977 (D.C. Cir. 1988).

methodology for establishing causation and then applies the methodology to the facts of the case.

An opinion on causation should be premised on three preliminary assessments. First, the expert should analyze whether the disease can be related to chemical exposure by a biologically plausible theory. Second, the expert should examine whether the plaintiff was exposed to the chemical in a manner that can lead to absorption into the body. Third, the expert should offer an opinion about whether the dose to which the plaintiff was exposed is sufficient to cause the disease.

The following questions help evaluate the strengths and weaknesses of toxicological evidence.

A. On What Species of Animals Was the Compound Tested? What Is Known About the Biological Similarities and Differences Between the Test Animals and Humans? How Do These Similarities and Differences Affect the Extrapolation from Animal Data in Assessing the Risk to Humans?

All living organisms share a common biology that leads to marked similarities in the responsiveness of subcellular structures to toxic agents. Among mammals, more than sufficient common organ structure and function readily permit the extrapolation from one species to another in most instances. Comparative information concerning factors that modify the toxic effects of chemicals, including absorption, distribution, metabolism, and excretion, in the laboratory test animals and humans enhances the expert's ability to extrapolate from laboratory animals to humans.[76]

The expert should review similarities and differences between the animal species in which the compound has been tested and humans. This analysis should form the basis of the expert's opinion regarding whether extrapolation from animals to humans is warranted.[77]

76. *See generally supra* notes 35–36 and accompanying text.

77. The failure to review similarities and differences in metabolism in performing cross-species extrapolation has led to the exclusion of opinions based on animal data. *See In re* Silicone Gel Breast Implants Prods. Liab. Litig., 318 F. Supp. 2d 879, 891 (C.D. Cal. 2004); Fabrizi v. Rexall Sundown, Inc., 2004 WL 1202984, at *8 (W.D. Pa. June 4, 2004). Hall v. Baxter Healthcare Corp., 947 F. Supp. 1387, 1410 (D. Or. 1996); Nelson v. Am. Sterilizer Co., 566 N.W.2d 671 (Mich. Ct. App. 1997). *But see In re* Paoli R.R. Yard PCB Litig., 35 F.3d 717, 779–80 (3d Cir. 1994) (noting that humans and monkeys are likely to show similar sensitivity to PCBs), *cert. denied sub nom.* Gen. Elec. Co. v. Ingram, 513 U.S. 1190 (1995). As the Supreme Court noted in *General Electric Co. v. Joiner,* 522 U.S. 136, 144 (1997), the issue regarding admissibility is not whether animal studies are ever admissible to establish causation, but whether the particular studies relied upon by plaintiff's experts were sufficiently supported. *See* Carl F. Cranor et al., *Judicial Boundary Drawing and the Need for Context-Sensitive Science in Toxic Torts After* Daubert v. Merrell Dow Pharmaceuticals, Inc., 16 Va. Envtl. L.J. 1, 38 (1996).

In general, an overwhelming similarity is apparent in the biology of all living things, and there is a particularly strong similarity among mammals. Of course, laboratory animals differ from humans in many ways. For example, rats do not have gallbladders. Thus, rat data would not be pertinent to the possibility that a compound produces human gallbladder toxicity.[78] Note that many subjective symptoms are poorly modeled in animal studies. Thus, complaints that a chemical has caused nonspecific symptoms, such as nausea, headache, and weakness, for which there are no objective manifestations in humans, are difficult to test in laboratory animals.

B. Does Research Show That the Compound Affects a Specific Target Organ? Will Humans Be Affected Similarly?

Some toxic agents affect only specific organs and not others. This organ specificity may be due to particular patterns of absorption, distribution, metabolism, and excretion; the presence of specific receptors; or organ function. For example, organ specificity may reflect the presence in the organ of relatively high levels of an enzyme capable of metabolizing or changing a compound to a toxic form of the compound,[79] or it may reflect the relatively low level of an enzyme capable of detoxifying a compound. An example of the former is liver toxicity caused by inhaled carbon tetrachloride, which affects the liver but not the lungs because of extensive metabolism to a toxic metabolite within the liver but relatively little such metabolism in the lung.[80]

Some chemicals, however, may cause nonspecific effects or even multiple effects. Lead is an example of a toxic agent that affects many organ systems, including the blood, the central and peripheral nervous systems, the reproductive system, and the kidneys.

The basis of specificity often reflects the function of individual organs. For example, the thyroid is particularly susceptible to radioactive iodine in atomic fallout because thyroid hormone is unique within the body in that it requires iodine. Through evolution, a very efficient and specific mechanism has developed that

78. *See, e.g.*, Edward J. Calabrese, Multiple Chemical Interactions 583–89 tbl.14-1 (1991). Species differences that produce a qualitative difference in response to xenobiotics are well known. Sometimes understanding the mechanism underlying the species difference can allow one to predict whether the effect will occur in humans. Thus, carbaryl, an insecticide commonly used for gypsy moth control, among other things, produces fetal abnormalities in dogs but not in hamsters, mice, rats, and monkeys. Dogs lack the specific enzyme involved in metabolizing carbaryl; the other species tested all have this enzyme, as do humans. Therefore, it has been assumed that humans are not at risk for fetal malformations produced by carbaryl.

79. Certain chemicals act directly to produce toxicity, whereas others require the formation of a toxic metabolite.

80. Brian Jay Day et al., *Potentiation of Carbon Tetrachloride-Induced Hepatotoxicity and Pneumotoxicity by Pyridine,* 8 J. Biochem. Toxicol. 11 (1993).

concentrates any absorbed iodine preferentially within the thyroid, rendering the thyroid particularly at risk from radioactive iodine. In a test tube, the radiation from radioactive iodine can affect the genetic material obtained from any cell in the body, but in the intact laboratory animal or human, only the thyroid is at risk.

The unfolding of the human genome already is beginning to provide information pertinent to understanding the wide variation in human risk from environmental chemicals. The impact of this understanding on toxic tort causation issues remains to be explored.[81]

C. *What Is Known About the Chemical Structure of the Compound and Its Relationship to Toxicity?*

Understanding the structural aspects of chemical toxicology has led to the use of structure–activity relationships (SAR) as a formal method of predicting the potential toxicity of new chemicals. This technique compares the chemical structure of compounds with known toxicity and the chemical structure of compounds with unknown toxicity. Toxicity then is estimated based on the molecular similarities between the two compounds. Although SAR is used extensively by EPA in evaluating many new chemicals required to be tested under the registration requirements of TSCA, its reliability has a number of limitations.[82]

81. Committee on Applications of Toxicogenomic Technologies to Predictive Toxicology and Risk Assessment, National Research Council, Applications of Toxicogenomic Technologies to Predictive Toxicology and Risk Assessment (2007); Gary E. Marchant, *Toxicogenomics and Toxic Torts,* 20 Trends Biotech. 329 (2002). Genomics can also be misinterpreted. A recent example is the use of white blood cell gene expression to determine whether benzene was a cause of acute myelogenous leukemia (AML) in individual workers. M.T. Smith, 14 *Misuse of Genomics in Assigning Causation in Relation to Benzene Exposure,* Int'l J. Occup. Envtl. Health 144–46 (2008) describes why the failure to match a pattern of DNA expression in workers with AML who were previously exposed to benzene is not scientifically defensible as a means to establish the lack of causation, as said to have been done in workers' compensation cases in California. The wide range in the rate of metabolism of chemicals is at least partly under genetic control. A study of Chinese workers exposed to benzene found approximately a doubling of risk in people with high levels of either an enzyme that increased the rate of formation of a toxic metabolite or an enzyme that decreased the rate of detoxification of this metabolite. There was a sevenfold increase in risk for those who had both genetically determined variants. N. Rothman et al., *Benzene Poisoning, A Risk Factor for Hematological Malignancy, Is Associated with the NQO1 609C→T Mutation and Rapid Fractional Excretion of Chlorzoxazone,* 57 Cancer Res. 239–42 (1997). *See also* Frederica P. Perera, *Molecular Epidemiology: Insights into Cancer Susceptibility, Risk Assessment, and Prevention,* 88 J. Nat'l Cancer Inst. 496 (1996).

82. For example, benzene and the alkyl benzenes (which include toluene, xylene, and ethyl benzene) share a similar chemical structure. SAR works exceptionally well in predicting the acute central nervous system anesthetic-like effects of both benzene and the alkyl benzenes. Although there are slight differences in dose–response relationships, they are readily explained by the interrelated factors of chemical structure, vapor pressure, and lipid solubility (the brain is highly lipid). National Research Council, The Alkyl Benzenes (1981). However, only benzene produces damage to the bone marrow and leukemia; the alkyl benzenes do not have this effect. This difference is the result

D. Has the Compound Been the Subject of In Vitro Research, and if So, Can the Findings Be Related to What Occurs In Vivo?

Cellular and tissue culture research can be particularly helpful in identifying mechanisms of toxic action and potential target-organ toxicity. The major barrier to the use of in vitro results is the frequent inability to relate doses that cause cellular toxicity to doses that cause whole-animal toxicity. In many critical areas, knowledge that permits such quantitative extrapolation is lacking.[83] Nevertheless, the ability to quickly test new products through in vitro tests, using human cells, provides invaluable "early warning systems" for toxicity.[84]

E. Is the Association Between Exposure and Disease Biologically Plausible?

No matter how strong the temporal relationship between exposure and the development of disease, or the supporting epidemiological evidence, it is difficult to accept an association between a compound and a health effect when no

of specific toxic metabolic products of benzene in comparison with the alkyl benzenes. Thus SAR is predictive of neurotoxic effects but not bone marrow effects. *See* Preston & Hoffman, *supra* note 33, at 277. Advances in computational approaches show promise in improving SAR. *See* Committee on Toxicity Testing and Assessment of Environmental Agents, National Research Council, Toxicity Testing in the 21st Century: A Vision and a Strategy, ch. 4 (2007).

In *Daubert v. Merrell Dow Pharmaceuticals, Inc.*, 509 U.S. 579 (1993), the Court rejected a per se exclusion of SAR, animal data, and reanalysis of previously published epidemiological data where there were negative epidemiological data. However, as the court recognized in *Sorensen v. Shaklee Corp.*, 31 F.3d 638, 646 n.12 (8th Cir. 1994), the problem with SAR is that "'[m]olecules with minor structural differences can produce very different biological effects.'" (quoting Joseph Sanders, *From Science to Evidence: The Testimony on Causation in the Bendectin Cases*, 46 Stan. L. Rev. 1, 19 (1993)). *See also* Glastetter v. Novartis Pharms. Corp., 252 F.3d 986, 990 (8th Cir. 2001); Polski v. Quigley Corp., 2007 WL 2580550, at *6 (D. Minn. Sept. 5, 2007).

83. In Vitro Toxicity Testing: Applications to Safety Evaluation 8 (John M. Frazier ed., 1992). Despite its limitations, in vitro research can strengthen inferences drawn from whole-animal bioassays and can support opinions regarding whether the association between exposure and disease is biologically plausible. *See* Preston & Hoffman, *supra* note 33, at 278–93; Rogers & Kavlock, *supra* note 33, at 319–23.

84. Graham v. Playtex Prods., Inc., 993 F. Supp. 127, 131–32 (N.D.N.Y. 1998) (opinion based on in vitro experiments showing that rayon tampons were associated with higher risk of toxic shock syndrome was admissible in the absence of epidemiological evidence). *See also* Allgood v. General Motors Corp., 2006 WL 2669337, at *7 (S.D. Ind. Sept. 18, 2006); *In re* Ephedra Prods. Liab. Litig., 393 F. Supp. 2d 181, 194 (S.D.N.Y. 2005) (in vitro studies may be subject of proper inferences "although the gaps between such data and definitive evidence of causality are real and subject to challenge before the jury, they are not so great as to require the opinion to be excluded from evidence. Inconclusive science is not the same as junk science").

mechanism can be identified by which the chemical or physical exposure leads to the putative effect.[85]

III. Specific Causal Association Between an Individual's Exposure and the Onset of Disease

An expert who opines that exposure to a compound caused a person's disease engages in deductive clinical reasoning.[86] In most instances, cancers and other diseases do not wear labels documenting their causation. The opinion is based on an assessment of the individual's exposure, including the amount, the temporal relationship between the exposure and disease, and other disease-causing factors. This information is then compared with scientific data on the relationship between exposure and disease. The certainty of the expert's opinion depends on the strength of the research data demonstrating a relationship between exposure and the disease at the dose in question and the presence or absence of other disease-causing factors (also known as confounding factors).[87]

Particularly problematic are generalizations made in personal injury litigation from regulatory positions. Regulatory standards are set for purposes far different than determining the preponderance of evidence in a toxic tort case. For example, if regulatory standards are discussed in toxic tort cases to provide a reference point for assessing exposure levels, it must be recognized that there is a great deal of variability in the extent of evidence required to support different regulations.[88] The extent of evidence required to support regulations depends on

85. However, theories of bioplausibility, without additional data, have been found to be insufficient to support a finding of causation. *See, e.g.*, Golod v. Hoffman La Roche, 964 F. Supp. 841, 860–61 (S.D.N.Y. 1997); Hall v. Baxter Healthcare Corp., 947 F. Supp. 1387, 1414 (D. Or. 1996). *But see* Best v. Lowe's Home Centers, Inc., 2008 WL 2359986, at *8 (E.D. Tenn. June 5, 2008) (expert relied on temporal proximity in concluding that plaintiff lost his sense of smell due to chemical exposure).

86. For an example of deductive clinical reasoning based on known facts about the toxic effects of a chemical and the individual's pattern of exposure, see Bernard D. Goldstein, *Is Exposure to Benzene a Cause of Human Multiple Myeloma?* 609 Annals N.Y. Acad. Sci. 225 (1990).

87. Causation issues are discussed in Michael D. Green et al., Reference Guide on Epidemiology, Section V, and Wong et al., Reference Guide on Medical Testimony, Section IV, in this manual. *See also* David L. Bazelon, *Science and Uncertainty: A Jurist's View*, 5 Harv. Envtl. L. Rev. 209 (1981); Troyen A. Brennan, *Causal Chains and Statistical Links: The Role of Scientific Uncertainty in Hazardous-Substance Litigation*, 73 Cornell L. Rev. 469 (1988); Joseph Sanders, *Scientific Validity, Admissibility and Mass Torts After* Daubert, 78 Minn. L. Rev. 1387 (1994); Orrin E. Tilevitz, *Judicial Attitudes Towards Legal and Scientific Proof of Cancer Causation*, 3 Colum. J. Envtl. L. 344, 381 (1977).

88. *See, e.g.*, *In re* Paoli R.R. Yard PCB Litig., 35 F.3d 717, 781 (3d Cir. 1994) (district court abused its discretion in excluding animal studies relied upon by EPA), *cert. denied sub nom.* General

1. The law (e.g., the Clean Air Act National Ambient Air Quality Standard provisions have language focusing regulatory activity for primary pollutants on adverse health consequences to sensitive populations with an adequate margin of safety and with no consideration of economic consequences, while regulatory activity under TSCA clearly asks for some balance between the societal benefits and risks of new chemicals[89]);
2. The specific end point of concern (e.g., consider the concern caused by cancer and adverse reproductive outcomes versus almost anything else); and
3. The societal impact (e.g., the public's support for control of an industry that causes air pollution versus the public's relative lack of desire to alter personal automobile use patterns).

These three concerns, as well as others, including costs, politics, and the virtual certainty of litigation challenging the regulation, have an impact on the level of scientific proof required by the regulatory decisionmaker.[90]

In addition, regulatory standards traditionally include protective factors to reasonably ensure that susceptible individuals are not put at risk. Furthermore, standards often are based on the risk that results from lifetime exposure. Accordingly, the mere fact that an individual has been exposed to a level above a standard does not necessarily mean that an adverse effect has occurred.

A. *Was the Plaintiff Exposed to the Substance, and if So, Did the Exposure Occur in a Manner That Can Result in Absorption into the Body?*

Evidence of exposure is essential in determining the effects of harmful substances. Basically, potential human exposure is measured in one of three ways. First, when direct measurements cannot be made, exposure can be measured by mathematical modeling, in which one uses a variety of physical factors to estimate the transport of the pollutant from the source to the receptor. For example, mathematical models take into account such factors as wind variations to allow calculation of

Elec. Co. v. Ingram, 513 U.S. 1190 (1995); Molden v. Georgia Gulf Corp., 465 F. Supp. 2d 606, 613 (M.D. La. 2006) (Plaintiff failed to establish prima facie case due to failure to establish exposure at a level considered dangerous by regulatory agency); *In re* W.R. Grace & Co. 355 B.R. 462, 490 (Bankr. D. Del. 2006) (OSHA standards of exposure relevant to causation but not determinative for exposure occurring due to home attic insulation). *See also* John Endicott, *Interaction Between Regulatory Law and Tort Law in Controlling Toxic Chemical Exposure*, 47 SMU L. Rev. 501 (1994).

89. *See, e.g.*, Clean Air Act Amendments of 1990, 42 U.S.C. § 7412(f) (1994); Toxic Substances Control, Act, 15 U.S.C. § 2605 (1994).

90. These concerns are discussed in Stephen Breyer, Breaking the Vicious Circle: Toward Effective Risk Regulation (1993).

the transport of radioactive iodine from a federal atomic research facility to nearby residential areas. Second, exposure can be directly measured in the medium in question—air, water, food, or soil. When the medium of exposure is water, soil, or air, hydrologists or meteorologists may be called upon to contribute their expertise to measuring exposure. The third approach directly measures human receptors through some form of biological monitoring, such as blood tests to determine blood lead levels or urinalyses to check for a urinary metabolite indicative of pollutant exposure. Ideally, both environmental testing and biological monitoring are performed; however, this is not always possible, particularly in instances of past exposure.[91]

The toxicologist must go beyond understanding exposure to determine if the individual was exposed to the compound in a manner that can result in absorption into the body. The absorption of the compound is a function of its physiochemical properties, its concentration, and the presence of other agents or conditions that assist or interfere with its uptake. For example, inhaled lead is absorbed almost totally, whereas ingested lead is taken up only partially into the body. Iron deficiency and low nutritional calcium intake, both common conditions of inner-city children, increase the amount of ingested lead that is absorbed in the gastrointestinal tract and passes into the bloodstream.[92]

B. *Were Other Factors Present That Can Affect the Distribution of the Compound Within the Body?*

Once a compound is absorbed into the body through the skin, lungs, or gastrointestinal tract, it is distributed throughout the body through the bloodstream. Thus the rate of distribution depends on the rate of blood flow to various organs

91. *See, e.g.* Mitchell v. Gencorp Inc., 165 F.3d 778, 781 (10th Cir. 1999) ("[g]uesses, even if educated, are insufficient to prove the level of exposure in a toxic tort case"); Wright v. Willamette Indus., Inc., 91 F.3d 1105, 1107 (8th Cir. 1996); Ingram v. Solkatronic Chemical, Inc., WL 3544244, at *11–*18 (N.D. Okla. 2005) (no information on dose so causation cannot be evaluated); *In re* Three Mile Island Litig. Consol. Proceedings, 927 F. Supp. 834, 870 (M.D. Pa. 1996) (plaintiffs failed to present direct or indirect evidence of exposure to cancer-inducing levels of radiation); Valentine v. Pioneer Chlor Alkali Co., 921 F. Supp. 666, 678 (D. Nev. 1996). *But see* CSX Transp., Inc. v. Moody, 2007 WL 2011626, at *7 (Ky. Ct. App. July 13, 2007) (specific dose of solvent exposure not necessary as long as evidence of exposure that could cause plaintiff's toxic encephalopathy is presented including how often solvents were used, duration of exposure, and documentation of physical symptoms while plaintiff worked with solvents).

92. The term "bioavailability" is used to describe the extent to which a compound, such as lead, is taken up into the body. In essence, bioavailability is at the interface between exposure and absorption into the organism. For an example of the impact of bioavailability on a governmental decision, see Thomas H. Umbreit et al., *Bioavailability of Dioxin in Soil from a 2,4,5-T Manufacturing Site*, 232 Science 497–99 (1986), who found that the bioavailability of dioxins in the soil of Newark, New Jersey, was negligible compared with that of Times Beach, Missouri—the latter community having previously been evacuated because of dioxin soil contamination.

and tissues. Distribution and resulting toxicity also are influenced by other factors, including the dose, the route of entry, tissue solubility, lymphatic supplies to the organ, metabolism, and the presence of specific receptors or uptake mechanisms within body tissues.

C. What Is Known About How Metabolism in the Human Body Alters the Toxic Effects of the Compound?

Metabolism is the alteration of a chemical by bodily processes. It does not necessarily result in less toxic compounds being formed. In fact, many of the organic chemicals that are known human cancer-causing agents require metabolic transformation before they can cause cancer. A distinction often is made between direct-acting agents, which cause toxicity without any metabolic conversion, and indirect-acting agents, which require metabolic activation before they can produce adverse effects. Metabolism is complex, because a variety of pathways compete for the same agent; some produce harmless metabolites, and others produce toxic agents.[93]

D. What Excretory Route Does the Compound Take, and How Does This Affect Its Toxicity?

Excretory routes are urine, feces, sweat, saliva, expired air, and lactation. Many inhaled volatile agents are eliminated primarily by exhalation. Small water-soluble compounds are usually excreted through urine. Higher-molecular-weight compounds are often excreted through the biliary tract into the feces. Certain fat-soluble, poorly metabolized compounds, such as PCBs, may persist in the body for decades, although they can be excreted in the milk fat of lactating women.

E. Does the Temporal Relationship Between Exposure and the Onset of Disease Support or Contradict Causation?

In acute toxicity, there is usually a short time period between cause and effect. However, in some situations, the length of basic biological processes necessitates a longer period of time between initial exposure and the onset of observable disease. For example, in acute myelogenous leukemia, the adult form of acute leukemia, at least 1 to 2 years must elapse from initial exposure to radiation, benzene, or

93. Courts have explored the relationship between metabolic transformation and carcinogenesis. *See, e.g., In re* Methyl Tertiary Butyl Ether (MTBE) Prods. Liab. Litig., 2008 WL 2607852, at *2 (S.D.N.Y. July 1, 2008); Stites v. Sundstrand Heat Transfer, Inc., 660 F. Supp. 1516, 1519 (W.D. Mich. 1987).

cancer chemotherapy before the manifestation of a clinically recognizable case of leukemia, and the period of significantly higher risk from the last exposure usually persists for no more than about 15 years. A toxic tort claim alleging a shorter or longer time period between cause and effect is scientifically highly debatable. Much longer latency periods are necessary for the manifestation of solid tumors caused by agents such as asbestos and arsenic.[94]

F. If Exposure to the Substance Is Associated with the Disease, Is There a No Observable Effect, or Threshold, Level, and if So, Was the Individual Exposed Above the No Observable Effect Level?

For agents that produce effects other than through mutations, it is assumed that there is some level that is incapable of causing harm. If the level of exposure was below this no observable effect, or threshold, level, a relationship between the exposure and disease cannot be established.[95] When only laboratory animal

94. The temporal relationship between exposure and causation is discussed in *Rolen v. Hansen Beverage Co.*, 193 F. App'x 468, 473 (6th Cir. 2006) ("Expert opinions based upon nothing more than the logical fallacy of post hoc ergo propter hoc typically do not pass muster under *Daubert*."). *See also* Young v. Burton, 2008 WL 2810237, at *17 (D.D.C. July 22, 2008); Dellinger v. Pfizer, Inc., 2006 WL 2057654, at *10 (W.D.N.C. July 16, 2006) (temporal relationship between exposure and illness alone not sufficient for causation when exposure was over an 18-month period); Cavallo v. Star Enterprise, 892 F. Supp. 756, 769–74 (E.D. Va. 1995) (expert testimony based primarily on temporal connection between exposure to jet fuel and onset of symptoms, without other evidence of causation, ruled inadmissible). *But see In re* Stand 'N Seal, Prods. Liab. Litig., 623 F. Supp. 2d 1355, 1371–72 (N.D. Ga. 2009) (toxicologist's causation opinion that exposure to grout sealer caused chemical pneumonitis not subject to *Daubert* challenge based on a strong temporal relationship between exposure and acute onset of respiratory symptoms despite lack of dose response data); *In re* Ephedra Prods. Liab. Litig., 2007 WL 2947451, at *2 (S.D.N.Y. Oct. 9, 2007) (when exposure is known to produce quick biological effects, a temporal relationship between exposure and effect can be used to infer causation); Nat'l. Bank of Commerce v. Dow Chem. Co., 965 F. Supp. 1490, 1525 (E.D. Ark. 1996) ("[T]here may be instances where the temporal connection between exposure to a given chemical and subsequent injury is so compelling as to dispense with the need for reliance on standard methods of toxicology."). The issue of latency periods and the statute of limitations is considered in Carl F. Cranor, Toxic Torts: Science, Law and the Possibility of Justice 173 (2006).

95. *See, e.g.*, Allen v. Pennsylvania Eng'g Corp., 102 F.3d 194, 199 (5th Cir. 1996) ("Scientific knowledge of the harmful level of exposure to a chemical, plus knowledge that the plaintiff was exposed to such quantities, are minimal facts necessary to sustain the plaintiff's burden in a toxic tort case."); Redland Soccer Club, Inc. v. Dep't of the Army, 55 F.3d 827, 847 (3d Cir. 1995) (summary judgment for defendant precluded where exposure above cancer threshold level could be calculated from soil samples); Molden v. Georgia Gulf Corp., 465 F. Supp. 2d 606, 613 (M.D. La. 2006) (levels of phenol released into the air were not considered harmful by regulatory agencies); Adams v. Cooper Indus., Inc., 2007 WL 2219212, at *8 (E.D. Ky. July 30, 2007) (because plaintiffs' experts have not attempted to quantify or measure the amount or dosage of a substance to which a plaintiff was exposed, their opinions are unreliable as to specific causation). *But see* Byers v. Lincoln Elec. Co, 607 F. Supp.

data are available, the expert extrapolates the NOEL from animals to humans by calculating the animal NOEL based on experimental data and decreasing this level by one or more safety factors to ensure no human effect.[96] The NOEL can also be calculated from human toxicity data if they exist. This analysis, however, is not applied to substances that exert toxicity by causing mutations leading to cancer. Theoretically, any exposure at all to mutagens may increase the risk of cancer, although the risk may be very slight and not achieve medical probability.[97]

IV. Medical History

A. *Is the Medical History of the Individual Consistent with the Toxicologist's Expert Opinion Concerning the Injury?*

One of the basic and most useful tools in diagnosis and treatment of disease is the patient's medical history.[98] A thorough, standardized patient information ques-

2d 863, FN101 (N.D. Ohio 2009) (no welder could ever provide evidence of actual exposure levels after the fact "which is why the law does not require mathematical precision to show toxic exposure" to support claims that inhaled manganese in welding fumes caused neurological injury); Tamraz v. BOC Group, Inc., 2008 U.S. Dist. LEXIS 54932, at *9–*10 (N.D. Ohio July 18, 2008) (plaintiffs were able to provide substantial evidential to support estimates of actual workplace conditions and exposure for welder exposed to manganese).

96. *See, e.g., supra* note 26 & accompanying text; Robert G. Tardiff & Joseph V. Rodricks, Toxic Substances and Human Risk: Principles of Data Interpretation 391 (1988); Joseph V. Rodricks, Calculated Risks 230–39 (2006); Lu, *supra* note 22, at 84. For regulatory toxicology, NOEL is being replaced by a more statistically robust approach known as the benchmark dose. *See supra* note 27 & accompanying text. For example, EPA's use of the benchmark dose takes into account comprehensive dose–response information, unlike NOEL.

97. See sources cited *supra* note 28. *See also* Henricksen v. ConocoPhillips Co., 605 F. Supp. 2d 1142, 1164–65 (E.D. Wa. 2009) (toxicologists' opinion that exposure to gasoline containing benzene caused truck driver's acute mylogenous leukemia found unreliable where dose calculation was unreliable, and "no-threshold model" lacked scientific support). U.S. regulatory approaches aimed at protecting the general population tend to avoid setting a standard for a known human carcinogen, because any allowable level below the standard is at least theoretically capable of causing cancer. However, exposure to many chemical carcinogens, including benzene and arsenic, cannot be eliminated. Thus, agencies and Congress have developed a number of ingenious means to regulate carcinogens while not seeming to acquiesce in exposure of the general population to a carcinogen. These include FDA's approach to de minimis risk and EPA's setting of a zero maximum contaminant level goal for carcinogens in drinking water while setting a maximum contaminant level above zero that is "set as closely as possible to the MCLG, taking technology and cost data into account," http://safewater.custhelp.com/cgi-bin/safewater.cfg/php/enduser/std_adp.php?p_faqid=1319. In contrast, occupational standards, which also take into account feasibility, permit exposure to known human carcinogens. A generally outmoded approach for environmental or indoor air guidelines has been to divide the permissible OSHA standard by a factor accounting for the presumed lifetime exposure to the environmental chemical compared with 45 years at a 40-hour workweek.

98. For a thorough discussion of the methods of clinical diagnosis, see John B. Wong et al., Reference Guide on Medical Testimony, in this manual. *See also* Jerome P. Kassirer & Richard I.

tionnaire would be particularly useful for identifying the etiology, or causation, of illnesses related to toxic exposures; however, there is currently no validated or widely used questionnaire that gathers all pertinent information.[99] Nevertheless, it is widely recognized that a thorough medical history involves the questioning and examination of the patient as well as appropriate medical testing. The patient's written medical records also should be examined.

The following information is relevant to a patient's medical history: past and present occupational and environmental history and exposure to toxic agents; lifestyle characteristics (e.g., use of nicotine and alcohol); family medical history (i.e., medical conditions and diseases of relatives); and personal medical history (i.e., present symptoms and results of medical tests as well as past injuries, medical conditions, diseases, surgical procedures, and medical test results).

In some instances, the reporting of symptoms can be in itself diagnostic of exposure to a specific substance, particularly in evaluating acute effects.[100] For example, individuals acutely exposed to organophosphate pesticides report headaches, nausea, and dizziness accompanied by anxiety and restlessness. Other reported symptoms are muscle twitching, weakness, and hypersecretion with sweating, salivation, and tearing.[101]

B. Are the Complaints Specific or Nonspecific?

Acute exposure to many toxic agents produces a constellation of nonspecific symptoms, such as headaches, nausea, lightheadedness, and fatigue. These types of symptoms are part of human experience and can be triggered by a host of medical and psychological conditions. They are almost impossible to quantify or document beyond the patient's report. Thus, these symptoms can be attributed mistakenly to an exposure to a toxic agent or discounted as unimportant when in fact they reflect a significant exposure.[102]

Kopelman, Learning Clinical Reasoning (1991). A number of cases have considered the admissibility of the treating physician's opinion based, in part, on medical history, symptomatology, and laboratory and pathology studies.

99. Office of Tech. Assessment, U.S. Congress, *supra* note 17, at 365–89.

100. *But see* Moore v. Ashland Chem., Inc., 126 F.3d 679, 693 (5th Cir. 1997) (discussion of relevance of symptoms within 45 minutes of exposure); Armstrong v. Durango Georgia Paper Co. 2005 WL 2373443, at *5 (S.D. Ga. Sept. 27, 2005) (plaintiffs exhibited temporary symptoms widely recognized by the medical community as those associated with exposure to chlorine gas).

101. Environmental Protection Agency, Recognition and Management of Pesticide Poisonings (4th ed. 1989).

102. The issue of whether the development of nonspecific symptoms may be related to pesticide exposure was considered in *Kannankeril v. Terminix Int'l, Inc.*, 128 F.3d 802 (3d Cir. 1997). The court ruled that the trial court abused its discretion in excluding expert opinion that considered, and rejected, a negative laboratory test. *Id.* at 808–09. *See also* Kerner v. Terminix Int'l, Co., 2008 WL 341363, at *7 (S.D. Ohio Feb. 6, 2008) (expert testimony about causation admissible based on plaintiff's nonspecific symptoms because scientific literature has linked exposure to pyrethrins and pyrethroids to

In taking a careful medical history, the expert focuses on the time pattern of symptoms and disease manifestations in relation to any exposure and on the constellation of symptoms to determine causation. It is easier to establish causation when a symptom is unusual and rarely is caused by anything other than the suspect chemical (e.g., such rare cancers as hemangiosarcoma, associated with vinyl chloride exposure, and mesothelioma, associated with asbestos exposure). However, many cancers and other conditions are associated with several causative factors, complicating proof of causation.[103]

C. Do Laboratory Tests Indicate Exposure to the Compound?

Two types of laboratory tests can be considered: tests that are routinely used in medicine to detect changes in normal body status and specialized tests that are used to detect the presence of the chemical or physical agent.[104] For the most part, tests used to demonstrate the presence of a toxic agent are frequently unavailable from clinical laboratories. Even when available from a hospital or a clinical laboratory, a test such as that for carbon monoxide combined to hemoglobin is done so rarely that it may raise concerns regarding its accuracy. Other tests, such as the test for blood lead levels, are required for routine surveillance of potentially exposed workers. However, if a laboratory is certified for the testing of blood lead in workers, for which the OSHA action level is 40 micrograms per deciliter (μg/dl), it does not necessarily mean that it will give reliable data on blood lead levels at the much lower Centers for Disease Control and Prevention action level of 10 μg/dl.

D. What Other Causes Could Lead to the Given Complaint?

With few exceptions, acute and chronic diseases, including cancer, can be caused by either a single toxic agent or a combination of agents or conditions. In taking a careful medical history, the expert examines the possibility of competing causes, or confounding factors, for any disease, which leads to a differential diagnosis. In addition, ascribing causality to a specific source of a chemical requires that a history be taken concerning other sources of the same chemical. The failure of a physician to elicit such a history or of a toxicologist to pay attention to such a

numbness, tingling, burning sensations, and paresthesia); Wicker v. Consol. Rail Corp., 371 F. Supp. 2d 702, 732 (W.D. Pa. 2005).

103. Failure to rule out other potential causes of symptoms may lead to a ruling that the expert's report is inadmissible. *See, e.g.*, Perry v. Novartis Pharms. Corp., 564 F. Supp. 2d 452, 469 (E.D. Pa. 2008); Farris v. Intel Corp., 493 F. Supp. 2d 1174, 1185 (D.N.M. 2007); Hall v. Baxter Healthcare Corp., 947 F. Supp. 1387, 1413 (D. Or. 1996); Rutigliano v. Valley Bus. Forms, 929 F. Supp. 779, 786 (D.N.J. 1996).

104. *See, e.g.*, Kannankeril v. Terminix Int'l, Inc., 128 F.3d 802, 807 (3d Cir. 1997).

history raises questions about competence and leaves open the possibility of competing causes of the disease.[105]

E. Is There Evidence of Interaction with Other Chemicals?

An individual's simultaneous exposure to more than one chemical may result in a response that differs from that which would be expected from exposure to only one of the chemicals.[106] When the effect of multiple agents is that which would be predicted by the sum of the effects of individual agents, it is called an additive effect; when it is greater than this sum, it is known as a synergistic effect; when one agent causes a decrease in the effect produced by another, the result is termed antagonism; and when an agent that by itself produces no effect leads to an enhancement of the effect of another agent, the response is termed potentiation.[107]

Three types of toxicological approaches are pertinent to understanding the effects of mixtures of agents. One is based on the standard toxicological evaluation of common commercial mixtures, such as gasoline. The second approach is from studies in which the known toxicological effect of one agent is used to explore the mechanism of action of another agent, such as using a known specific inhibitor of a metabolic pathway to determine whether the toxicity of a second agent depends on this pathway. The third approach is based on an understanding of the basic mechanism of action of the individual components of the mixture, thereby allowing prediction of the combined effect, which can then be tested in an animal model.[108]

105. *See, e.g.*, Perry v. Novartis Pharms. Corp., 564 F. Supp. 2d 452, 471 (E.D. Pa. 2008) (plaintiff's experts failed to adequately account for the possibility that plaintiff's T-LBL was idiopathic, and thus their conclusion that exposure to Elidel was a substantial cause of plaintiff's cancer is unreliable and inadmissible); Bell v. Swift Adhesives, Inc., 804 F. Supp. 1577, 1580 (S.D. Ga. 1992) (expert's opinion that workplace exposure to methylene chloride caused plaintiff's liver cancer, without ruling out plaintiff's infection with hepatitis B virus, a known liver carcinogen, was insufficient to withstand motion for summary judgment for defendant).

106. *See generally* Edward J. Calabrese, Multiple Chemical Interactions 97–115, 220–221 (1991).

107. Courts have been called on to consider the issue of synergy. In *International Union, United Automobile, Aerospace & Agricultural Implement Workers of America v. Pendergrass*, 878 F.2d 389, 391 (D.C. Cir. 1989), the court found that OSHA failed to sufficiently explain its findings that formaldehyde presented no significant carcinogenic risk to workers at exposure levels of 1 part per million or less. The court particularly criticized OSHA's use of a linear low-dose risk curve rather than a risk-adverse model after the agency had described evidence of synergy between formaldehyde and other substances that workers would be exposed to, especially wood dust. *Id.* at 395.

108. *See generally* Calabrese, *supra* note 106. EPA has been addressing the issue of multiple exposures to different agents within a community under the heading of cumulative risk assessment. This approach is particularly of importance in dealing with environmental justice concerns. *See, e.g.*, Institute of Medicine, Toward Environmental Justice: Research, Education, and Health Policy Needs (1999); Michael A. Callahan & Ken Sexton, *If Cumulative Risk Assessment Is the Answer, What Is the Question?* 115 Envtl. Health Persp. 799–806 (2006).

F. Do Humans Differ in the Extent of Susceptibility to the Particular Compound in Question? Are These Differences Relevant in This Case?

Individuals who exercise inhale more than sedentary individuals and therefore are exposed to higher doses of airborne environmental toxins. Similarly, differences in metabolism, which are inherited or caused by external factors, such as the levels of carbohydrates in a person's diet, may result in differences in the delivery of a toxic product to the target organ.[109]

Moreover, for any given level of a toxic agent that reaches a target organ, damage may be greater because of a greater response of that organ. In addition, for any given level of target-organ damage, there may be a greater impact on particular individuals. For example, an elderly individual or someone with preexisting lung disease is less likely to tolerate a small decline in lung function caused by an air pollutant than is a healthy individual with normal lung function.

A person's level of physical activity, age, sex, and genetic makeup, as well as exposure to therapeutic agents (such as prescription or over-the-counter drugs), affect the metabolism of the compound and hence its toxicity.[110] Advances in human genetics research are providing information about susceptibility to environmental agents that may be relevant to determining the likelihood that a given exposure has a specific effect on an individual.

G. Has the Expert Considered Data That Contradict His or Her Opinion?

Multiple avenues of deductive reasoning based on scientific data lead to acceptance of causation in any field, particularly in toxicology. However, the basis for this deductive reasoning is also one of the most difficult aspects of causation to describe quantitatively. If animal studies, pharmacological research on mechanisms of toxicity, in vitro tissue studies, and epidemiological research all document toxic effects of exposure to a compound, an expert's opinion about causation in a particular case is much more likely to be true.[111]

109. *See generally* Calabrese, *supra* note 106.

110. The problem of differences in chemical sensitivity was addressed by the court in *Gulf South Insulation v. United States Consumer Product Safety Commission,* 701 F.2d 1137 (5th Cir. 1983). The court overturned the commission's ban on urea-formaldehyde foam insulation because the commission failed to document in sufficient detail the level at which segments of the population were affected and whether their responses were slight or severe: "Predicting how likely an injury is to occur, at least in general terms, is essential to a determination of whether the risk of that injury is unreasonable." *Id.* at 1148.

111. Consistency of research results was considered by the court in *Marsee v. United States Tobacco Co.,* 639 F. Supp. 466, 469–70 (W.D. Okla. 1986). The defendant, the manufacturer of snuff alleged to cause oral cancer, moved to exclude epidemiological studies conducted in Asia that demonstrate

The more difficult problem is how to evaluate conflicting research results. When different research studies reach different conclusions regarding toxicity, the expert must be asked to explain how those results have been taken into account in the formulation of the expert's opinion.

V. Expert Qualifications

The basis of the toxicologist's expert opinion in a specific case is a thorough review of the research literature and treatises concerning effects of exposure to the chemical at issue. To arrive at an opinion, the expert assesses the strengths and weaknesses of the research studies. The expert also bases an opinion on fundamental concepts of toxicology relevant to understanding the actions of chemicals in biological systems.

As the following series of questions indicates, no single academic degree, research specialty, or career path qualifies an individual as an expert in toxicology. Toxicology is a heterogeneous field. A number of indicia of expertise can be explored, however, that are relevant to both the admissibility and weight of the proffered expert opinion.

A. Does the Proposed Expert Have an Advanced Degree in Toxicology, Pharmacology, or a Related Field? If the Expert Is a Physician, Is He or She Board Certified in a Field Such as Occupational Medicine?

A graduate degree in toxicology demonstrates that the proposed expert has a substantial background in the basic issues and tenets of toxicology. Many universities have established graduate programs in toxicology. These programs are administered by the faculties of medicine, pharmacology, pharmacy, or public health.

Although most recent toxicology Ph.D. graduates have no other credentials, many highly qualified toxicologists are physicians or hold doctoral degrees

a link between smokeless tobacco and oral cancer. The defendant also moved to exclude evidence demonstrating that the nitrosamines and polonium-210 contained in the snuff are cancer-causing agents in some 40 different species of laboratory animals. The court denied both motions, finding:

> There was no dispute that both nitrosamines and polonium-210 are present in defendant's snuff products. Further, defendant conceded that animal studies have accurately and consistently demonstrated that these substances cause cancer in test animals. Finally, the Court found evidence based on experiments with animals particularly valuable and important in this litigation since such experiments with humans are impossible. Under all these circumstances, the Court found this evidence probative on the issue of causation.

Id. See also sources cited *supra* note 14.

in related disciplines (e.g., veterinary medicine, pharmacology, biochemistry, environmental health, or industrial hygiene). For a person with this type of background, a single course in toxicology is unlikely to provide sufficient background for developing expertise in the field.

A proposed expert should be able to demonstrate an understanding of the discipline of toxicology, including statistics, toxicological research methods, and disease processes. A physician without particular training or experience in toxicology is unlikely to have sufficient background to evaluate the strengths and weaknesses of toxicological research. Most practicing physicians have little knowledge of environmental and occupational medicine.[112] Generally, physicians are quite knowledgeable about the identification of effects and their treatment. The cause of these effects, particularly if they are unrelated to the treatment of the disease, is generally of little concern to the practicing physician. Subspecialty physicians may have particular knowledge of a cause-and-effect relationship (e.g., pulmonary physicians have knowledge of the relationship between asbestos exposure and asbestosis),[113] but most physicians have little training in chemical toxicology and lack an understanding of exposure assessment and dose–response relationships. An exception is a physician who is certified in medical toxicology as a subspeciality under the American Board of Medical Specialties' requirements, based on substantial training in toxicology and successful completion of rigorous examinations, including recertification exams.[114]

112. For recent documentation of how rarely an occupational history is obtained, see B.J. Politi et al., *Occupational Medical History Taking: How Are Today's Physicians Doing? A Cross-Sectional Investigation of the Frequency of Occupational History Taking by Physicians in a Major US Teaching Center*. 46 J. Occup. Envtl. Med. 550–55 (2004).

113. *See, e.g.*, Moore v. Ashland Chem., Inc., 126 F.3d 679, 701 (5th Cir. 1997) (treating physician's opinion admissible regarding causation of reactive airway disease); McCullock v. H.B. Fuller Co., 61 F.3d 1038, 1044 (2d Cir. 1995) (treating physician's opinion admissible regarding the effect of fumes from hot-melt glue on the throat, where physician was board certified in otolaryngology and based his opinion on medical history and treatment, pathological studies, differential etiology, and scientific literature); Benedi v. McNeil-P.P.C., Inc., 66 F.3d 1378, 1384 (4th Cir. 1995) (treating physician's opinion admissible regarding the causation of liver failure by mixture of alcohol and acetaminophen, based on medical history, physical examination, laboratory and pathology data, and scientific literature—the same methodologies used daily in the diagnosis of patients); *In re* Ephedra Prods. Liab. Litig., 478 F. Supp. 2d 624, 633 (S.D.N.Y. 2007) (opinion of treating physician will assist the trier of fact because a reasonable juror would want to know what inferences a treating physician would make); Morin v. United States, 534 F. Supp. 2d 1179, 1185 (D. Nev. 2005) (treating physician does not have sufficient expertise to offer opinion about whether exposure to jet fuel caused cancer in his patient).

Treating physicians also become involved in considering cause-and-effect relationships when they are asked whether a patient can return to a situation in which an exposure has occurred. The answer is obvious if the cause-and-effect relationship is clearly known. However, this relationship is often uncertain, and the physician must consider the appropriate advice. In such situations, the physician will tend to give advice as though the causality was established, both because it is appropriate caution and because of fears concerning medicolegal issues.

114. Before 1990, the American Board of Medical Toxicology certified physicians, but beginning in 1990, medical toxicology became a subspecialty board under the American Board of Emer-

Some physicians who are occupational health specialists also have training in toxicology. Knowledge of toxicology is particularly strong among those who work in the chemical, petrochemical, and pharmaceutical industries, in which the surveillance of workers exposed to chemicals is a major responsibility. Of the occupational physicians practicing today, only about 1000 have successfully completed the board examination in occupational medicine, which contains some questions about chemical toxicology.[115]

B. *Has the Proposed Expert Been Certified by the American Board of Toxicology, Inc., or Does He or She Belong to a Professional Organization, Such as the Academy of Toxicological Sciences or the Society of Toxicology?*

As of January 2008, more than 2000 individuals had received board certification from the American Board of Toxicology. To sit for the examination, the candidate must be involved full time in the practice of toxicology, including designing and managing toxicological experiments or interpreting results and translating them to identify and solve human and animal health problems. Diplomats must be recertified every 5 years. The Academy of Toxicological Sciences (ATS) was formed to provide credentials in toxicology through peer review only. It does not administer examinations for certification. Approximately 200 individuals are certified as Fellows of ATS.

gency Medicine, the American Board of Pediatrics, and the American Board of Preventive Medicine, as recognized by the American Board of Medical Specialties.

115. Clinical ecologists, another group of physicians, have offered opinions regarding multiple chemical hypersensitivity and immune system responses to chemical exposures. These physicians generally have a background in the field of allergy, not toxicology, and their theoretical approach is derived in part from classic concepts of allergic responses and immunology. This theoretical approach has often led clinical ecologists to find cause-and-effect relationships or low-dose effects that are not generally accepted by toxicologists. Clinical ecologists often belong to the American Academy of Environmental Medicine.

In 1991, the Council on Scientific Affairs of the American Medical Association concluded that until "accurate, reproducible, and well-controlled studies are available . . . multiple chemical sensitivity should not be considered a recognized clinical syndrome." Council on Scientific Affairs, American Med. Ass'n, Council Report on Clinical Ecology 6 (1991). In *Bradley v. Brown,* 42 F.3d 434, 438 (7th Cir. 1994), the court considered the admissibility of an expert opinion based on clinical ecology theories. The court ruled the opinion inadmissible, finding that it was "hypothetical" and based on anecdotal evidence as opposed to scientific research. *See also* Kropp v. Maine School Adm. Union No. 44, 471 F. Supp. 2d 175, 181–82 (D. Me. 2007) (expert physician does not rely upon scientifically valid methodologies or data in reaching the conclusion that plaintiff is hypersensitive to phenol vapors in indoor air); Coffin v. Orkin Exterminating Co., 20 F. Supp. 2d 107, 110 (D. Me. 1998); Frank v. New York, 972 F. Supp. 130, 132 n.2 (N.D.N.Y. 1997). *But see* Elam v. Alcolac, Inc., 765 S.W.2d 42, 86 (Mo. Ct. App. 1988) (expert opinion based on clinical ecology theories admissible).

The Society of Toxicology (SOT), the major professional organization for the field of toxicology, was founded in 1961 and has grown dramatically in recent years. It now has 6300 members.[116] Criteria for membership is based either on peer-reviewed publications or on the active practice of toxicology. Physician toxicologists can join the American College of Medical Toxicology and the American Academy of Clinical Toxicologists. There are also societies of forensic toxicology, such as the International Academy of Forensic Toxicology. Other organizations in the field are the American College of Toxicology, for which experience in the active practice of toxicology is the major membership criterion; the International Society of Regulatory Toxicology and Pharmacology; and the Society of Occupational and Environmental Health. For membership, the last two organizations require only the payment of dues.

C. What Other Criteria Does the Proposed Expert Meet?

The success of academic scientists in toxicology, as in other biomedical sciences, usually is measured by the following types of criteria: the quality and number of peer-reviewed publications, the ability to compete for research grants, service on scientific advisory panels, and university appointments.

Publication of articles in peer-reviewed journals indicates an expertise in toxicology. The number of articles, their topics, and whether the individual is the principal or senior author are important factors in determining the expertise of a toxicologist.[117]

Most research grants from government agencies and private foundations are highly competitive. Successful competition for funding and publication of the research findings indicate competence in an area.

Selection for local, national, and international regulatory advisory panels usually implies recognition in the field. Examples of such panels are the NIH Toxicology Study Section and panels convened by EPA, FDA, WHO, and IARC. Recognized industrial organizations, including the American Petroleum Institute and the Electric Power Research Institute, and public interest groups, such as the Environmental Defense Fund and the Natural Resources Defense Council,

116. There are currently 21 specialty sections of SOT that represent the different specialty areas involved in understanding the wide range of toxic effects associated with exposure to chemical and physical agents. These sections include mechanisms, molecular biology, inhalation toxicology, metals, neurotoxicology, carcinogenesis, risk assessment, and immunotoxicology.

117. Examples of reputable, peer-reviewed journals are the *Journal of Toxicology and Environmental Health; Toxicological Sciences; Toxicology and Applied Pharmacology; Science; British Journal of Industrial Medicine; Clinical Toxicology; Archives of Environmental Health; Journal of Occupational and Environmental Medicine; Annual Review of Pharmacology and Toxicology; Teratogenesis, Carcinogenesis and Mutagenesis; Fundamental and Applied Toxicology; Inhalation Toxicology; Biochemical Pharmacology; Toxicology Letters; Environmental Research; Environmental Health Perspectives; International Journal of Toxicology; Human and Experimental Toxicology;* and *American Journal of Industrial Medicine.*

employ toxicologists directly and as consultants and enlist academic toxicologists to serve on advisory panels. Because of a growing interest in environmental issues, the demand for scientific advice has outgrown the supply of available toxicologists. It is thus common for reputable toxicologists to serve on advisory panels.

Finally, a university appointment in toxicology, risk assessment, or a related field signifies an expertise in that area, particularly if the university has a graduate education program in that area.

VI. Acknowledgments

The authors greatly appreciate the excellent research assistance provided by Eric Topor and Cody S. Lonning.

Glossary of Terms

The following terms and definitions were adapted from a variety of sources, including Office of Technology Assessment, U.S. Congress, Reproductive Health Hazards in the Workplace (1985); Casarett and Doull's Toxicology: The Basic Science of Poisons (Curtis D. Klaassen ed., 7th ed. 2007); National Research Council, Biologic Markers in Reproductive Toxicology (1989); Committee on Risk Assessment Methodology, National Research Council, Issues in Risk Assessment (1993); M. Alice Ottoboni, The Dose Makes the Poison: A Plain-Language Guide to Toxicology (2d ed. 1991); and Environmental and Occupational Health Sciences Institute, Glossary of Environment Health Terms (1989) [update].

absorption. The taking up of a chemical into the body orally, through inhalation, or through skin exposure.

acute toxicity. An immediate toxic response following a single or short-term exposure to an agent or dosing.

additive effect. When exposure to more than one toxic agent results in the same effect as would be predicted by the sum of the effects of exposure to the individual agents.

antagonism. When exposure to one toxic agent causes a decrease in the effect produced by another toxic agent.

benchmark dose. The benchmark dose is determined on the basis of dose–response modeling and is defined as the exposure associated with a specified low incidence of risk, generally in the range of 1% to 10%, of a health effect, or the dose associated with a specified measure or change of a biological effect.

bioassay. A test for measuring the toxicity of an agent by exposing laboratory animals to the agent and observing the effects.

biological monitoring. Measurement of toxic agents or the results of their metabolism in biological materials, such as blood, urine, expired air, or biopsied tissue, to test for exposure to the toxic agents, or the detection of physiological changes that are due to exposure to toxic agents.

biologically plausible theory. A biological explanation for the relationship between exposure to an agent and adverse health outcomes.

carcinogen. A chemical substance or other agent that causes cancer.

carcinogenicity bioassay. Limited or long-term tests using laboratory animals to evaluate the potential carcinogenicity of an agent.

chronic toxicity. A toxic response to long-term exposure or dosing with an agent.

clinical ecologists. Physicians who believe that exposure to certain chemical agents can result in damage to the immune system, causing multiple-chemical hypersensitivity and a variety of other disorders. Clinical ecologists often have a background in the field of allergy, not toxicology, and their theoretical approach is derived in part from classic concepts of allergic responses and immunology. There has been much resistance in the medical community to accepting their claims.

clinical toxicology. The study and treatment of humans exposed to chemicals and the quantification of resulting adverse health effects. Clinical toxicology includes the application of pharmacological principles to the treatment of chemically exposed individuals and research on measures to enhance elimination of toxic agents.

compound. In chemistry, the combination of two or more different elements in definite proportions, which when combined acquire properties different from those of the original elements.

confounding factors. Variables that are related to both exposure to a toxic agent and the outcome of the exposure. A confounding factor can obscure the relationship between the toxic agent and the adverse health outcome associated with that agent.

differential diagnosis. A physician's consideration of alternative diagnoses that may explain a patient's condition.

direct-acting agents. Agents that cause toxic effects without metabolic activation or conversion.

distribution. Movement of a toxic agent throughout the organ systems of the body (e.g., the liver, kidney, bone, fat, and central nervous system). The rate of distribution is usually determined by the blood flow through the organ and the ability of the chemical to pass through the cell membranes of the various tissues.

dose, dosage. A product of both the concentration of a chemical or physical agent and the duration or frequency of exposure.

dose–response curve. A graphic representation of the relationship between the dose of a chemical administered and the effect produced.

dose–response relationships. The extent to which a living organism responds to specific doses of a toxic substance. The more time spent in contact with a toxic substance, or the higher the dose, the greater the organism's response. For example, a small dose of carbon monoxide will cause drowsiness; a large dose can be fatal.

epidemiology. The study of the occurrence and distribution of disease among people. Epidemiologists study groups of people to discover the cause of a disease, or where, when, and why disease occurs.

epigenetic. Pertaining to nongenetic mechanisms by which certain agents cause diseases, such as cancer.

etiology. A branch of medical science concerned with the causation of diseases.

excretion. The process by which toxicants are eliminated from the body, including through the kidney and urinary tract, the liver and biliary system, the fecal excretor, the lungs, sweat, saliva, and lactation.

exposure. The intake into the body of a hazardous material. The main routes of exposure to substances are through the skin, mouth, and lungs.

extrapolation. The process of estimating unknown values from known values.

good laboratory practice (GLP). Codes developed by the federal government in consultation with the laboratory testing industry that govern many aspects of laboratory standards.

hazard identification. In risk assessment, the qualitative analysis of all available experimental animal and human data to determine whether and at what dose an agent is likely to cause toxic effects.

hydrogeologists, hydrologists. Scientists who specialize in the movement of ground and surface waters and the distribution and movement of contaminants in those waters.

immunotoxicology. A branch of toxicology concerned with the effects of toxic agents on the immune system.

indirect-acting agents. Agents that require metabolic activation or conversion before they produce toxic effects in living organisms.

inhalation toxicology. The study of the effect of toxic agents that are absorbed into the body through inhalation, including their effects on the respiratory system.

in vitro. A research or testing methodology that uses living cells in an artificial or test tube system, or that is otherwise performed outside of a living organism.

in vivo. A research or testing methodology that uses living organisms.

lethal dose 50 (LD_{50}). The dose at which 50% of laboratory animals die within days to weeks.

lifetime bioassay. A bioassay in which doses of an agent are given to experimental animals throughout their lifetime. See bioassay.

maximum tolerated dose (MTD). The highest dose of an agent to which an organism can be exposed without it causing death or significant overt toxicity.

metabolism. The sum total of the biochemical reactions that a chemical produces in an organism.

molecular toxicology. The study of how toxic agents interact with cellular molecules, including DNA.

multiple-chemical hypersensitivity. A physical condition whereby individuals react to many different chemicals at extremely low exposure levels.

multistage events. A model for understanding certain diseases, including some cancers, based on the postulate that more than one event is necessary for the onset of disease.

mutagen. A substance that causes physical changes in chromosomes or biochemical changes in genes.

mutagenesis. The process by which agents cause changes in chromosomes and genes.

neurotoxicology. A branch of toxicology concerned with the effects of exposure to toxic agents on the central nervous system.

no observable effect level (NOEL). The highest level of exposure to an agent at which no effect is observed. It is the experimental equivalent of a threshold.

no-threshold model. A model for understanding disease causation that postulates that any exposure to a harmful chemical (such as a mutagen) may increase the risk of disease.

one-hit theory. A theory of cancer risk in which each molecule of a chemical mutagen has a possibility, no matter how tiny, of mutating a gene in a manner that may lead to tumor formation or cancer.

pharmacokinetics. A mathematical model that expresses the movement of a toxic agent through the organ systems of the body, including to the target organ and to its ultimate fate.

potentiation. The process by which the addition of one agent, which by itself has no toxic effect, increases the toxicity of another agent when exposure to both agents occurs simultaneously.

reproductive toxicology. The study of the effect of toxic agents on male and female reproductive systems, including sperm, ova, and offspring.

risk assessment. The use of scientific evidence to estimate the likelihood of adverse effects on the health of individuals or populations from exposure to hazardous materials and conditions.

risk characterization. The final step of risk assessment, which summarizes information about an agent and evaluates it in order to estimate the risks it poses.

safety assessment. Toxicological research that tests the toxic potential of a chemical in vivo or in vitro using standardized techniques required by governmental regulatory agencies or other organizations.

structure–activity relationships (SAR). A method used by toxicologists to predict the toxicity of new chemicals by comparing their chemical structures with those of compounds with known toxic effects.

synergistic effect. When two toxic agents acting together have an effect greater than that predicted by adding together their individual effects.

target organ. The organ system that is affected by a particular toxic agent.

target-organ dose. The dose to the organ that is affected by a particular toxic agent.

teratogen. An agent that changes eggs, sperm, or embryos, thereby increasing the risk of birth defects.

teratogenic. The ability to produce birth defects. (Teratogenic effects do not pass to future generations.) See teratogen.

threshold. The level above which effects will occur and below which no effects occur. See no observable effect level.

toxic. Of, relating to, or caused by a poison—or a poison itself.

toxic agent or toxicant. An agent or substance that causes disease or injury.

toxicology. The science of the nature and effects of poisons, their detection, and the treatment of their effects.

References on Toxicology

A Textbook of Modern Toxicology (Ernest Hodgson ed., 4th ed. 2010).
Casarett and Doull's Toxicology: The Basic Science of Poisons (Curtis D. Klaassen ed., 7th ed. 2007).
Committee on Toxicity Testing and Assessment of Environmental Agents, National Research Council, Toxicity Testing in the 21st Century: A Vision and a Strategy (2007).
Environmental Toxicants (Morton Lippmann ed., 3d ed. 2009).
Patricia Frank & M. Alice Ottoboni, The Dose Makes the Poison: A Plain-Language Guide to Toxicology (3d ed. 2011).
Genetic Toxicology of Complex Mixtures (Michael D. Waters et al. eds., 1990).
Human Risk Assessment: The Role of Animal Selection and Extrapolation (M. Val Roloff ed., 1987).
In Vitro Toxicity Testing: Applications to Safety Evaluation (John M. Frazier ed., 1992).
Michael A. Kamrin, Toxicology: A Primer on Toxicology Principles and Applications (1988).
Frank C. Lu, Basic Toxicology: Fundamentals, Target Organs, and Risk Assessment (4th ed. 2002).
National Research Council, Biologic Markers in Reproductive Toxicology (1989).
Alan Poole & George B. Leslie, A Practical Approach to Toxicological Investigations (1989).
Principles and Methods of Toxicology (A. Wallace Hayes ed., 5th ed. 2008).
Joseph V. Rodricks, Calculated Risks (2d ed. 2006).
Short-Term Toxicity Tests for Nongenotoxic Effects (Philippe Bourdeau et al. eds., 1990).
Toxic Interactions (Robin S. Goldstein et al. eds., 1990).
Toxic Substances and Human Risk: Principles of Data Interpretation (Robert G. Tardiff & Joseph V. Rodricks eds., 1987).
Toxicology (Hans Marquardt et al. eds., 1999).
Toxicology and Risk Assessment: Principles, Methods, and Applications (Anna M. Fan & Louis W. Chang eds., 1996).

Reference Guide on Medical Testimony

JOHN B. WONG, LAWRENCE O. GOSTIN, AND OSCAR A. CABRERA

John B. Wong, M.D., is Chief of the Division of Clinical Decision Making, Informatics, and Telemedicine at the Institute for Clinical Research and Health Policy Studies, Tufts Medical Center, and Professor of Medicine at Tufts University School of Medicine.

Lawrence O. Gostin, J.D., is Linda D. and Timothy J. O'Neill Professor of Global Health Law and Faculty Director of O'Neill Institute for National and Global Health Law, Georgetown University Law Center.

Oscar A. Cabrera, Abogado, LL.M., is Deputy Director of the O'Neill Institute for National and Global Health Law and Adjunct Professor of Law, Georgetown University Law Center.

CONTENTS

I. Introduction

Physicians are a common sight in today's courtroom. A survey of federal judges published in 2002 indicated that medical and mental health experts constituted more than 40% of the total number of testifying experts.[1] Medical evidence is a common element in product liability suits,[2] workers' compensation disputes,[3] medical malpractice suits,[4] and personal injury cases.[5] Medical testimony may also be critical in certain kinds of criminal cases.[6] The goal of this reference guide is to introduce the basic concepts of diagnostic reasoning and clinical decisionmaking, as well as the types of evidence that physicians use to make judgments as treating physicians or as experts retained by one of the parties in a case. Following this introduction (Section I), Section II identifies a few overarching theoretical issues that courts face in translating the methods and techniques customary in the medical profession in a manner that will serve the court's inquiry. Sections III and IV describe medical education and training, the organization of medical care, the elements of patient care, and the processes of diagnostic reasoning and medical judgment. When relevant, each subsection includes examples from case law illustrating how the topic relates to legal issues.

II. Medical Testimony Introduction

A. Medical Versus Legal Terminology

Because medical testimony is common in the courtroom generally and indispensable to certain kinds of cases, courts have employed some medical terms in ways

1. Joe S. Cecil, *Ten Years of Judicial Gatekeeping Under* Daubert, 95 Am. J. Pub. Health S74–S80 (2005).

2. *See, e.g.*, *In re* Bextra & Celebrex Mktg. Sales Practices and Prod. Liab., 524 F. Supp. 2d 1166 (N.D. Cal. 2007) (thoroughly reviewing the proffered testimony of plaintiff's expert cardiologist and neurologist in a products liability suit alleging that defendant's arthritis pain medication caused serious cardiovascular injury).

3. *See, e.g.*, AT&T Alascom v. Orchitt, 161 P.3d 1232 (Alaska 2007) (affirming the decision of the state workers' compensation board and rejecting appellant's challenges to worker's experts).

4. Schneider *ex rel.* Estate of Schneider v. Fried, 320 F.3d 396 (3d Cir. 2003) (allowing a physician to testify in a malpractice case regarding whether administering a particular drug during angioplasty was within the standard of care).

5. *See, e.g.*, Epp v. Lauby, 715 N.W.2d 501 (Neb. 2006) (detailing the opinions of two physicians regarding whether plaintiff's fibromyalgia resulted from an automobile accident with two defendants).

6. Medical evidence will be at issue in numerous kinds of criminal cases. *See* State v. Price, 171 P.3d 293 (Mont. 2007) (an assault case in which a physician testified regarding the potential for a stun gun to cause serious bodily harm); People v. Unger, 749 N.W.2d 272 (Mich. Ct. App. 2008) (a second-degree murder case involving testimony of a forensic pathologist and neuropathologist); State v. Greene, 951 So. 2d 1226 (La. Ct. App. 2007) (a child sexual battery and child rape case involving the testimony of a board-certified pediatrician).

that differ from their use by the medical profession. Differential diagnosis, for example, is an accepted method that a medical expert may employ to offer expert testimony that satisfies *Daubert*.[7] In the legal context, differential diagnosis refers to a technique "in which physician first rules in all scientifically plausible causes of plaintiff's injury, then rules out least plausible causes of injury until the most likely cause remains, thereby reaching conclusion as to whether defendant's product caused injury. . . ."[8] In the medical context, by contrast, differential diagnosis

7. *See, e.g.*, Feliciano-Hill v. Principi, 439 F.3d 18, 25 (1st Cir. 2006) ("[W]hen an examining physician calls upon training and experience to offer a differential diagnosis . . . most courts have found no *Daubert* problem."); Clausen v. M/V New Carissa, 339 F.3d 1049, 1058–59 (9th Cir. 2003) (recognizing differential diagnosis as a valid methodology); Mattis v. Carlon Elec. Prods., 295 F.3d 856, 861 (8th Cir. 2002) ("A medical opinion based upon a proper differential diagnosis is sufficiently reliable to satisfy [*Daubert*.]"); Westberry v. Gislaved Gummi AB, 178 F.3d 257, 262 (4th Cir. 1999) (recognizing differential diagnosis as a reliable technique).

8. Wilson v. Taser Int'l, Inc. 2008 WL 5215991, at *5 (11th Cir. Dec. 16, 2008) ("[N]onetheless, Dr. Meier did not perform a differential diagnosis or any tests on Wilson to rule out osteoporosis and these corresponding alternative mechanisms of injury. Although a medical expert need not rule out every possible alternative in order to form an opinion on causation, expert opinion testimony is properly excluded as unreliable if the doctor 'engaged in very few standard diagnostic techniques by which doctors normally rule out alternative causes and the doctor offered no good explanation as to why his or her conclusion remained reliable' or if 'the defendants pointed to some likely cause of the plaintiff's illness other than the defendants' action and [the doctor] offered no reasonable explanation as to why he or she still believed that the defendants' actions were a substantial factor in bringing about that illness.'"); Williams v. Allen, 542 F.3d 1326, 1333 (11th Cir. 2008) ("Williams also offered testimony from Dr. Eliot Gelwan, a psychiatrist specializing in psychopathology and differential diagnosis. Dr. Gelwan conducted a thorough investigation into Williams' background, relying on a wide range of data sources. He conducted extensive interviews with Williams and with fourteen other individuals who knew Williams at various points in his life.") (involving a capital murder defendant petitioning for habeus corpus offering supporting expert witness); Bland v. Verizon Wireless, L.L.C., 538 F.3d 893, 897 (8th Cir. 2008) ("Bland asserts Dr. Sprince conducted a differential diagnosis which supports Dr. Sprince's causation opinion. We have held, 'a medical opinion about causation, based upon a proper differential diagnosis is sufficiently reliable to satisfy *Daubert*.' A 'differential diagnosis [is] a technique that identifies the cause of a medical condition by eliminating the likely causes until the most probable cause is isolated.'") (stating expert's incomplete execution of differential diagnosis procedure rendered expert testimony unsatisfactory for *Daubert* standard) (citations omitted); Lash v. Hollis 525 F.3d 636, 640 (8th Cir. 2008) ("Further, even if the treating physician had specifically opined that the Taser discharges caused rhabdomyolysis in Lash Sr., the physician offered no explanation of a differential diagnosis or other scientific methodology tending to show that the Taser shocks were a more likely cause than the myriad other possible causes suggested by the evidence.") (finding lack of expert testimony with differential diagnosis enough to render evidence insufficient for jury to find causation in personal injury suit); Feit v. Great West Life & Annuity Ins. Co., 271 Fed. App'x. 246, 254 (3d Cir. 2008) ("However, although this Court generally recognizes differential diagnosis as a reliable methodology the differential diagnosis must be properly performed in order to be reliable. To properly perform a differential diagnosis, an expert must perform two steps: (1) 'Rule in' all possible causes of Dr. Feit's death and (2) 'Rule out' causes through a process of elimination whereby the last remaining potential cause is deemed the most likely cause of death.") (ruling that district court not in error for excluding expert medical testimony that relied on an improperly performed differential diagnosis) (citations omitted); Glastetter v. Novartis Pharms. Corp., 252 F.3d 986 (8th Cir. 2001).

refers to a set of diseases that physicians consider as possible causes for symptoms the patient is suffering or signs that the patient exhibits.[9] By identifying the likely potential causes of the patient's disease or condition and weighing the risks and benefits of additional testing or treatment, physicians then try to determine the most appropriate approach—testing, medication, or surgery, for example.[10]

Less commonly, courts often have used the term "differential etiology" interchangeably with differential diagnosis.[11] In medicine, etiology refers to the study of causation in disease,[12] but differential etiology is a legal invention not used by physicians. In general, both differential etiology and differential diagnosis are concerned with establishing or refuting causation between an external cause and a plaintiff's condition. Depending on the type of case and the legal standard, a medical expert may testify in regard to specific causation, general causation, or both. General causation refers to whether the plaintiff's injury could have been caused by the defendant, or a product produced by the defendant, while specific causation is established only when the defendant's action or product actually caused the harm.[13] An opinion by a testifying physician may be offered in support of both kinds of causation.[14]

Courts also refer to medical certainty or probability in ways that differ from their use in medicine. The standards "reasonable medical certainty" and "reasonable medical probability" are also terms of art in the law that have no analog for a practicing physician.[15] As is detailed in Section IV, diagnostic reasoning and medi-

9. Steadman's Medical Dictionary 531 (28th ed. 2006) (defining differential diagnosis as "the determination of which of two or more diseases with similar symptoms is the one from which the patient is suffering, by a systematic comparison and contrasting of the clinical findings.").

10. The Concise Dictionary of Medical-Legal Terms 36 (1998) (definition of differential diagnosis).

11. *See* Proctor v. Fluor Enters., Inc. 494 F.3d 1337 (11th Cir. 2007) (testifying medical expert employed differential etiology to reach a conclusion regarding the cause of plaintiff's stroke). *But see* McClain v. Metabolife Int'l, Inc., 401 F.3d 1233, 1252 (11th Cir. 2005) (distinguishing differential diagnosis from differential etiology, with the former closer to the medical definition and the latter employed as a technique to determine external causation).

12. Steadman's Medical Dictionary 675 (28th ed. 2006) (defining etiology as "the science and study of the causes of disease and their mode of operation. . . ."). For a discussion of the term "etiology" in epidemiology studies, see Michael D. Green et al., Reference Guide on Epidemiology, Section I, in this manual.

13. *See* Amorgianos v. Nat'l R.R. Passenger Corp., 303 F.3d 256, 268 (2d Cir. 2002).

14. *See, e.g.*, Ruggiero v. Warner-Lambert Co. 424 F.3d 249 (2d Cir. 2005) (excluding testifying expert's differential diagnosis in support of a theory of general causation because it was not supported by sufficient evidence).

15. *See, e.g.*, Dallas v. Burlington N., Inc., 689 P.2d 273, 277 (Mont. 1984) ("'[R]easonable medical certainty' standard; the term is not well understood by the medical profession. Little, if anything, is 'certain' in science. The term was adopted in law to assure that testimony received by the fact finder was not merely conjectural but rather was sufficiently probative to be reliable"). This reference guide will not probe substantive legal standards in any detail, but there are substantive differences in admissibility standards for medical evidence between federal and state courts. *See* Robin Dundis Craig, *When* Daubert *Gets* Erie*: Medical Certainty and Medical Expert Testimony in Federal Court*, 77 Denv. U. L. Rev. 69 (1999).

cal evidence are aimed at recommending the best therapeutic option for a patient. Although most courts have interpreted "reasonable medical certainty" to mean a preponderance of the evidence,[16] physicians often work with multiple hypotheses while diagnosing and treating a patient without any "standard of proof" to satisfy.

Statutes and administrative regulations may also contain terms that are borrowed, often imperfectly, from the medical profession. In these cases, the court may need to examine the intent of the legislature and the term's usage in the medical profession.[17] If no intent is apparent, the court may need to determine whether the medical definition is the most appropriate one to apply to the statutory language. Whether the language is a term of art or a question of law will often dictate the admissibility and weight of evidence.[18]

B. *Applicability of* Daubert v. Merrell Dow Pharmaceuticals, Inc.

The Supreme Court's decision in *Daubert v. Merrell Dow Pharmaceuticals, Inc.*,[19] changed the way that judges screen expert testimony. A 2002 study by the RAND Corporation indicated that after *Daubert*, judges began scrutinizing expert testimony much more closely and began more aggressively excluding evidence that does not meet its standards.[20] Despite the Court's subsequent decisions in *General Electric Co. v. Joiner*[21] and *Kumho Tire Co. v. Carmichael*[22] further defining the

16. *See, e.g.*, Sharpe v. United States, 230 F.R.D. 452, 460 (E.D. Va. 2005) ("It is not enough for the plaintiff's expert to testify that the defendant's negligence might or may have caused the injury on which the plaintiff bases her claim. The expert must establish that the defendant's negligence was 'more likely' or 'more probably' the cause of the plaintiff's injury . . . ").

17. *See, e.g.*, Feltner v. Lamar Adver., Inc., 83 F. App'x 101 (6th Cir. 2003) (holding that the statutory definition of "permanent total disability" under the Tennessee Workers Compensation Act was not the same as the medical definition); Endorf v. Bohlender, 995 P.2d 896 (Kan. Ct. App. 2000) (a medical malpractice case reversing a lower court's interpretation of the statutory phrase "clinical practice" because it did not comport with the legislature's intent that the statutory meaning reflect the medical definition).

18. *See, e.g.*, Coleman v. Workers' Comp. Appeal Bd. (Ind. Hosp.), 842 A.2d 349 (Pa. 2004) (holding that since the legislature did not define the medical term "physical examination," the common usage of the term is more appropriate than the strict medical definition).

19. 509 U.S. 579 (1993).

20. Lloyd Dixon & Brian Gill, Changes in the Standards for Admitting Expert Evidence in Federal Civil Cases Since the *Daubert* Decision (2002).

21. 522 U.S. 136 (1997) (holding that the trial court had properly excluded expert testimony extrapolated from animal studies and epidemiological studies).

22. 526 U.S. 137 (1999). In *Kumho*, the Court made clear that *Daubert* applies to all expert testimony and not just "scientific" testimony. Although the case involved a defect in tires, courts before *Kumho* were divided on whether expert medical opinion based on experience or clinical medical testimony were subject to *Daubert*. *See also* Joe S. Cecil, *Ten Years of Judicial Gatekeeping Under* Daubert, 95 Am. J. Pub. Health S74–S80 (2005). *See also* Lawrence O. Gostin, Public Health Law: Power, Duty, Restraint (2d ed. 2008).

Daubert standard, federal and state courts have sometimes employed conflicting interpretations of what *Daubert* requires from testifying physicians.

The standard of review is an important factor in understanding how *Daubert* has engendered seemingly inconsistent results. The Supreme Court adopted an abuse of discretion standard in *Joiner*[23] and affirmed it in *Kumho*.[24] Although in most product liability cases the courts reached the same conclusion, inconsistent determinations regarding the admissibility of similar evidence may not constitute an abuse of discretion under the federal standard of review or in states with a similar standard.[25]

C. Relationship of Medical Reasoning to Legal Reasoning

As Section II.A suggested, the goal that guides the physician—recommending the best therapeutic options for the patient—means that diagnostic reasoning and the process of ongoing patient care and treatment involve probabilistic judgments concerning several working hypotheses, often simultaneously. When a court requires a testifying physician to offer evidence "to a reasonable medical certainty" or "reasonable medical probability," it is supplying the expert with a legal rule to which his or her testimony must conform.[26] In other words, a lawyer often will

23. 522 U.S. at 143.

24. 526 U.S. at 142.

25. Hollander v. Sandoz Pharm. Corp., 289 F.3d 1193, 1207 (10th Cir. 2002); *see also* Brasher v. Sandoz Pharm. Corp., 160 F. Supp. 2d 1291, 1298 n.17 (N.D. Ala. 2001); Reichert v. Phipps, 84 P.3d 353, 358 (Wyo. 2004).

26. Courts have occasionally noted the tension between the medical reasoning and legal reasoning when applying the reasonable medical certainty or reasonable medical probability standards. *See* Clark v. Arizona, 548 U.S. 735, 777 (2006) ("When . . . 'ultimate issue' questions are formulated by the law and put to the expert witness who must then say 'yea' or 'nay,' then the expert witness is required to make a leap in logic. He no longer addresses himself to medical concepts but instead must infer or intuit what is in fact unspeakable, namely, the *probable relationship* between medical concepts and legal or moral constructs such as free will. These impermissible leaps in logic made by expert witnesses confuse the jury. . . ."); Rios v. City of San Jose, 2008 U.S. Dist. LEXIS 84923, at *4 (N.D. Cal. Oct. 9, 2008) ("In their fifth motion, plaintiffs seek to exclude the testimony of Dr. Brian Peterson who defendants designated to testify, among other subjects, about the 'proximate cause' of Rios' death. As the use of terms that also carry legal significance could confuse the jury, the motion is granted in part, and defendants are instructed to distinguish between medical and legal terms such as proximate cause to the extent possible. Where such terms must be used by the witness consistent with the language employed in his field of expertise, the parties shall craft a limiting instruction to advise the jury of the distinction between those terms and the issues they will be called upon to determine."); Norland v. Wash. Gen. Hosp., 461 F.2d 694, 697 (8th Cir. 1972) ("The use of the terms 'probable' and 'possible' as a basis for test of qualification or lack of qualification in respect to a medical opinion has frequently converted this aspect of a trial into a mere semantic ritual or hassle. The courts have come to recognize that the competency of a physician's testimony cannot soundly be permitted to turn on a mechanical rule of law as to which of the two terms he has employed. Regardless of which term he may have used, if his testimony is such in nature and basis of hypothesis as to judicially impress that the opinion expressed represents his professional judgment as to the most likely one among the

need to explain the legal standard to the physician, who will then shape the form and content of his or her testimony in a manner that serves the legal inquiry.[27]

Legal standards will shape how physicians testify in a number of other ways. Although treating physicians generally are concerned less about discovering the actual causes of the disease than treating the patient, the testifying medical expert will need to tailor his or her opinions in a way that conforms to the legal standard of causation. As Section IV will demonstrate, when analyzing the patient's symptoms and making a judgment based on the available medical evidence, a physician will not expressly identify a "proximate cause" or "substantial factor." For example, in order to recommend treatment, a physician does not necessarily need to determine whether a patient's lung ailment was more likely the result of a long history of tobacco use or prolonged exposure to asbestos if the optimal treatment is the same. In contrast, when testifying as an expert in a case in which an employee with a long history of tobacco use is suing his employer for possible injuries as a result of asbestos exposure in the workplace, physicians may need to make judgments regarding the likelihood that either tobacco or asbestos—or both—could have contributed to the injury.[28]

Physicians often will be asked to testify about patients from whom they have never taken a medical history or examined and make estimates about proximate cause, increased risk of injury, or likely future injuries.[29] The doctor may even need to make medical judgments about a deceased litigant.[30] Testifying in all such cases requires making judgments that physicians do not ordinarily make in their profession, making these judgments outside of physicians' customary patient encounters, and adapting the opinion in a way that fits the legal standard. The purpose of this guide is not to describe or recommend competing legal standards, whether it be the standard of proof, causation, admissibility, or the applicable standard of care in medical malpractice cases. Instead, it aims to introduce the practice of medicine to federal and state judges, emphasizing the tools and methods that

possible causes of the physical condition involved, the court is entitled to admit the opinion and leave its weight to the jury.").

27. There are several cases that demonstrate the difficulty that physicians sometimes have in adapting their testimony to the legal standard. *See* Schrantz v. Luancing, 527 A.2d 967 (N.J. Super. Ct. Law Div. 1986) (malpractice case in which the medical expert's opinion was inadequate because of her understanding of "reasonable medical certainty").

28. Physicians will testify as experts in cases in which the plaintiff's condition may be the result of multiple causes. In these cases, the divergence between medical reasoning and legal reasoning are very apparent. *See, e.g.*, Tompkin v. Philip Morris USA, Inc., 362 F.3d 882 (6th Cir. 2004) (affirming district court's conclusion that testimony offered by the defendant's expert regarding the decedent's work-related asbestos exposure was not prejudicial in a suit against a tobacco company on behalf of plaintiff's deceased husband); Mobil Oil Corp. v. Bailey, 187 S.W.3d 265 (Tex. Ct. App. 2006) (involving claims from a worker who had a long history of tobacco use that exposure to asbestos increased his risk of cancer).

29. *See, e.g.*, *Tompkin*, 362 F.3d 882.

30. *See, e.g.*, *id.*

doctors use to make decisions and highlighting the challenges in adapting them when testifying as medical experts.

Sections III and IV of this guide explain in great detail the practice of medicine, including medical education, the structure of health care, and, most importantly, the methods that physicians use to diagnose and treat their patients. Special attention is given to the physician–patient relationship and to the types of evidence that physicians use to make medical judgments. In an effort to make each issue more salient, examples from case law are offered when they are illustrative.

III. Medical Care

A. Medical Education and Training

1. Medical school

The Association of American Medical Colleges (AAMC) consists of 133 accredited U.S. medical schools and 17 Canadian medical schools.[31] The Liaison Committee on Medical Education performs the accreditation for AAMC and assesses the quality of postsecondary education by determining whether each institution or program meets established standards for function, structure, and performance. The goal of medical school is to prepare students in the art and science of medicine for graduate medical education.[32] Of the 4 years of medical school, the first 2 years are typically spent studying preclinical basic sciences involving the study of the normal structure and function of human systems (e.g., through anatomy, biochemistry, physiology, behavioral science, and neuroscience), followed by the study of abnormalities and therapeutic principles (e.g., through microbiology, immunology, pharmacology, and pathology). The final 2 years involve clinical experience, including rotations in patient care settings such as clinics or hospitals with required "core" clerkships in internal medicine, pediatrics, psychiatry, surgery, obstetrics/gynecology, and family medicine. All physicians who wish to be licensed must pass the United States Medical Licensing Examination Steps 1, 2, and 3.[33]

31. Association of American Medical Colleges, Membership, *available at* https://www.aamc.org/about/membership/ (last visited Feb. 12, 2011).

32. *See* Davis v. Houston Cnty., Ala. Bd. of Educ., 2008 WL 410619 (M.D. Ala. Feb. 13, 2008) (finding that an individual with no medical training was not qualified to give expert testimony).

33. Planned Parenthood Cincinnati Region v. Taft 444 F.3d 502, 515 (6th Cir. 2006), ("The State has not appealed the district court's order refusing to recognize Dr. Crockett as an expert in the critical review of medical literature. Although that order has not been placed before us, the only reason the district court gave for her ruling was that Dr. Crockett did not have any specific training in the critical review of medical literature beyond the training incorporated in her general medical school and residency training. This ruling ignored Dr. Crockett's testimony that her residency program at Georgetown University put particular emphasis on training residents in the critical review of medical literature, that she had taught classes on the subject, that she had done extensive reading and

In the United States, besides the more than 941,000 physicians, there are more than 61,000 doctors of osteopathy. The Commission on Osteopathic College Accreditation accredits 25 colleges of osteopathic medicine. Training is similar to that for medical physicians but with additional "special attention on the musculoskeletal system which reflects and influences the condition of all other body systems."[34] About 25% of current U.S. physicians are foreign medical graduates that include both U.S. citizens and foreign nationals.[35] Because educational standards and curricula outside the United States and Canada vary, the Education Commission for Foreign Medical Graduates has developed a certification exam to assess whether these graduates may enter Accreditation Council for Graduate Medical Education (ACGME) accredited residency and fellowship programs.[36]

self-education on the subject, and that she had critically reviewed medical literature for the FDA. If these qualifications are not sufficient to demonstrate expertise, this court is hard-pressed to imagine what qualifications would suffice."); Davis v. Houston Cnty., Ala. Bd. of Educ., 2008 WL 410619, at *4 (M.D. Ala. Feb. 13, 2008) ("The Board has moved to exclude all evidence of Freet's opinions and conclusions related to the cause of Joshua Davis's behavior at the football game contained in his deposition as well as Freet's letter to Malcolm Newman. The Board argues that Freet is not qualified to give expert testimony, and that Plaintiff failed to comply with Fed. R. Civ. P. 26(a)(2)(B) by not providing a report of Freet's testimony that includes all of the information required by Rule 26(a)(2)(B). . . . In order to consider Freet's expert opinions, this Court must find that Freet meets the requirements of Fed. R. Evid. 702. Rule 702 requires an expert to be qualified by 'knowledge, skill, experience, training, or education.' Freet is not a medical doctor and never attended medical school. The only evidence of Freet's qualifications are: approximately five years working for the Department of Veterans Affairs in the vocational rehabilitation program, followed by approximately seven years working in private practice as a 'licensed professional counselor.' There is no evidence in the record of Freet's educational background, or any details of the exact nature of Freet's work experience."); Therrien v. Town of Jay, 489 F. Supp. 2d 116, 117 (D. Me. 2007) ("Citing *Daubert v. Merrell Dow Pharmaceuticals, Inc.*, 509 U.S. 579, 113 S. Ct. 2786, 125 L. Ed. 2d 469 (1993) and Rule 702 of the Federal Rules of Evidence, Officer Gould's first objection is that Dr. Harding does not possess sufficient expertise to express expert opinions about 'the mechanism and timing of Plaintiff's injuries.' This objection is not well taken. Dr. Harding was graduated from Dartmouth College and Georgetown Medical School; he completed a residency in internal medicine, is board certified in internal medicine, and has been licensed to practice medicine in the state of Maine since 1978."). United States Medical Licensing Examination, Examinations, *available at* http://www.usmle.org/Examinations/index.html (last visited Aug. 9, 2011).

34. Association of American Medical Colleges, What is a DO? *available at* http://www.osteopathic.org/osteopathic-health/about-dos/what-is-a-do/Pages/default.aspx (last visited Feb. 12, 2011); Association of American Medical Colleges, About Osteopathic Medicine, *available at* http://www.osteopathic.org/osteopathic-health/about-dos/about-osteopathic-medicine/Pages/default.aspx (last visited Feb. 12, 2011).

35. American Medical Association, Physician Characteristics and Distribution in the U.S. (2009).

36. Commission for Foreign Medical Graduates, About ECFMG, *available at* http://www.ecfmg.org/about.html (last visited Feb. 12, 2011).

2. Postgraduate training

After graduating from medical school, most physicians undergo additional training in a residency program in a chosen specialty.[37] Residencies typically range from 3 to 7 years at teaching hospitals and academic medical centers where residents care for patients while being supervised by physician faculty and participating in educational and research activities.[38] After graduating from an accredited residency program, physicians become eligible to take their board certification examinations.[39] Physician licensure in many states requires the completion of a residency program accredited by the ACGME, the organization which is responsible for accrediting the more than 8700 residency programs in 26 specialties and 130 subspecialties.[40] Following residency, some physicians opt for additional subspecialty fellowship training. ACGME divides fellowship training[41] into (1) Dependent Subspecialty Programs in which the program functions in conjunction with an accredited specialty/core program and (2) Independent Subspecialty Programs in which the program does not depend on the accreditation status of a specialty program.[42] For osteopathic physicians, the American Osteopathic Association approves osteopathic postdoctoral

37. *See* Brown v. Harmot Med. Ctr., 2008 WL 55999 (W.D. Pa. Jan. 3, 2008). American Medical Association, Requirements for Becoming a Physician, *available at* http://www.ama-assn.org/ama/pub/education-careers/becoming-physician.page? (last visited Aug. 9, 2011).

38. *See* Planned Parenthood Cincinnati Region v. Taft, 444 F.3d 502, 515 (6th Cir. 2006). American Medical Association, Requirements for Becoming a Physician, *available at* http://www.ama-assn.org/ama/pub/education-careers/becoming-physician.page? (last visited Aug. 9, 2011).

39. *See* Therrien v. Town of Jay, 489 F. Supp. 2d 116, 117 (D. Me. 2007) (finding that a physician who completed a residency in internal medicine was qualified to give his opinion on trauma related to a § 1983 claim against a police department). American Medical Association, Requirements for Becoming a Physician, *available at* http://www.ama-assn.org/ama/pub/education-careers/becoming-physician.page? (last visited Aug. 9, 2011).

40. Accreditation Council for Graduate Medical Education, The ACMGE at a Glance, *available at* http://www.acgme.org/acWebsite/newsRoom/newsRm_acGlance.asp (last visited Feb. 12, 2011).

41. Accreditation Council for Graduate Medical Education, Specialty Programs with Dependent and Independent Subspecialties, *available at* http://www.acgme.org/acWebsite/RRC_sharedDocs/sh_progs_depIndSubs.asp (last visited Feb. 12, 2011).

42. John Doe 21 v. Sec'y of Health and Human Servs., 84 Fed. Cl. 19, 35–36 (Fed. Cl. 2008) ("The Government's expert, Dr. Wiznitzer, is a board-certified neurologist by the American Board of Psychiatry and Neurology, with a special qualification in Child Neurology. In addition, Dr. Wiznitzer is certified by the American Board of Pediatrics. Since 1986, Dr. Wiznitzer has been an Associate Pediatrician and an Associate Neurologist at University Hospital of Cleveland, Ohio. And, since 1992, Dr. Wiznitzer has been Director of the Autism Center at Rainbow Babies and Children's Hospital in Cleveland, Ohio. During the past 24 years, Dr. Wiznitzer also has been an Associate Professor of Pediatrics and Associate Professor of Neurology at Case Western Reserve University. Dr. Wiznitzer completed his residency in Pediatrics from Children's Hospital Medical Center in Cincinnati and served as a Fellow in Developmental Disorders, Pediatric Neurology, and Higher Cortical Functions. Dr. Wiznitzer also has received numerous awards and honors in the neurology field and his work has been widely published.") (citations omitted); Brown v. Hamot Med. Ctr., 2008 WL 55999, at *8–9

training programs.[43] The American Osteopathic Association established the Osteopathic Postdoctoral Training Institutions (OPTI), wherein each OPTI partners a community-based training consortium with one or more colleges of osteopathic medicine and one or more hospitals and possibly ambulatory care facilities.[44]

3. Licensure and credentialing

Medical Practice Acts defining the practice of medicine and delegating enforcement to state medical boards exist for each of the 50 states, the District of Columbia, and the U.S. territories. Besides awarding medical licenses, state medical boards also investigate complaints, discipline physicians who violate the law, and evaluate and rehabilitate physicians. The Federation of State Medical Boards represents the 70 medical boards of the United States and its territories, and its mission is "promoting excellence in medical practice, licensure, and regulation as the national resource and voice on behalf of state medical boards in their protection of the public."[45]

Credentialing typically involves verifying medical education, postgraduate training, board certification, professional experience, state licensure, prior credentialing outcomes, medical board actions, malpractice, and adverse clinical events. Credentialing or recredentialing by hospitals involves an assessment of a physician's professional or technical competence and performance by evaluating and monitoring the quality of patient care. This credentialing process defines physicians' scope of practice and hospital privileges, that is, the clinical services they may provide.

The American Board of Medical Specialties (ABMS) provides certification in 24 medical specialties (e.g., emergency medicine, internal medicine, obstetrics and gynecology, family medicine, pediatrics, surgery, and others) to provide[46] "assurance of a physician's expertise in a particular specialty and/or subspecialty

(W.D. Pa. Jan. 3, 2008) ("As the United States Court of Appeals for the Fifth Circuit has explained in another context, a medical residency is primarily an academic enterprise:

> [a] residency program is distinct from other types of employment in that the resident's "work" is what is academically supervised and evaluated. [T]he primary purpose of a residency program is not employment or a stipend, but the academic training and the academic certification for successful completion of the program. The certificate . . . tells the world that the resident has successfully completed a course of training and is qualified to pursue further specialized training or to practice in specified areas. . . . Successful completion of the residency program depends upon subjective evaluations by trained faculty members into areas of expertise that courts are poorly equipped to undertake in the first instance or to review. . . .").

43. American Osteopathic Association, Postdoctoral Training, *available at* http://www.osteopathic.org/inside-aoa/Education/postdoctoral-training/Pages/default.aspx (last visited Feb. 12, 2011).

44. *Id.*

45. Federation of State Medical Boards, FSMB Mission and Goals, *available at* http://www.fsmb.org/mission.html (last visited Feb. 12, 2011).

46. American Board of Medical Specialties, Who We Are and What We Do, *available at* http://www.abms.org/About_ABMS/who_we_are.aspx (last visited Feb. 12, 2011).

of medical practice."[47] Although the criteria vary depending on the field, board eligibility requires the completion of an appropriate residency, an institutional or valid license to practice medicine, and evaluation with written and—in some cases—oral examinations. Many boards also require an evaluation of practice performance for initial certification. Board certification documents the fulfillment of all criteria including passing the examinations. Originally, board certificates had no expiration, but a program of periodic recertification (every 6 to 10 years) was subsequently initiated to ensure that physicians remained current in their specialty. In 2006, the ABMS recertification process became the Maintenance of Certification to emphasize continuous professional development through a four-part process:

1. Licensure and professional standing;
2. Lifelong learning;
3. Cognitive expertise; and
4. Practice performance assessment in six core competencies
    a. patient care,
    b. medical knowledge,
    c. practice-based learning,
    d. interpersonal and communications skills,
    e. professionalism, and
    f. systems-based practice.[48]

In some cases, specialty organizations have opted to develop their own certification process outside of the ABMS (e.g., the American Board of Bariatric Medicine).[49]

The American Osteopathic Association (AOA) certifies osteopathic physicians in 18 osteopathic specialty boards (e.g., emergency medicine, internal medicine, obstetrics and gynecology, family medicine, pediatrics, surgery, and others).[50] The osteopathic continuous certification process involves (1) unrestricted licensure, (2) lifelong learning/continuing medical education, (3) cognitive assessment, (4) practice performance assessment and improvement, and (5) continuous AOA membership.[51]

47. Although specialization is a hallmark of modern medical practice, courts have not always required that medical testimony come from a specialist. *See* Gaydar v. Sociedad Instituto Gineco-Quirurgico y Planificacion Familiar, 245 F.3d 15, 24–25 (1st Cir. 2003) ("The proffered expert physician need not be a specialist in a particular medical discipline to render expert testimony relating to that discipline.").

48. American Board of Medical Specialties, ABMS Maintenance of Certification, *available at* http://www.abms.org/Maintenance_of_Certification/ABMS_MOC.aspx (last visited Feb. 12, 2011).

49. American Board of Bariatric Medicine, Certification, *available at* http://www.abbmcertification.org/ (last visited Feb. 12, 2011).

50. American Osteopathic Association, AOA Specialty Certifying Boards, *available at* http://www.osteopathic.org/inside-aoa/development/aoa-board-certification/Pages/aoa-specialty-boards.aspx (last visited Feb. 12, 2011).

51. *Id.*

4. Continuing medical education

For relicensure, state medical boards require continuing medical education so that physicians can acquire new knowledge and maintain clinical competence. The Accreditation Council for Continuing Medical Education (ACCME) identifies, develops, and promotes quality standards for continuing medical education for physicians. ACCME requires certain elements of structure, method, and organization in the development of continuing medical education materials to ensure uniformity across states and to help assure physicians, state medical boards, medical societies, state legislatures, continuing medical education providers, and the public that the education meets certain quality standards. For osteopathic physicians, the AOA Board of Trustees also oversees accreditation for osteopathic CME sponsors through the Council on Continuing Medical Education (CCME).[52] The AOA's Healthcare Facilities Accreditation Program (HFAP) reviews services delivered by medical facilities.[53]

B. Organization of Medical Care

The delivery of health care in the United States is highly decentralized and fragmented,[54] and is provided through clinics, hospitals, managed care organizations, medical groups, multispecialty clinics, integrated delivery systems, specialty standalone hospitals, imaging facilities, skilled nursing facilities, rehabilitation hospitals, emergency departments, and pharmacy-based and other walk-in clinics. When surveyed in 1996, patients viewed the health care system as a "nightmare to navigate."[55] Transitioning care from outpatient to inpatient hospitalization to recovery often involves multiple handoffs among different physicians and care providers with the need for accurate, timely, and complete transfer of information about the patient's acute and chronic medical conditions, medications, and treatments. Although hospitals increasingly belong to a network or system, most community physicians belong to practices involving 10 or fewer physicians.[56]

Concerns about the safety of the organization of medical care first arose from the Harvard Medical Practice Study which found that adverse events occurred in

52. American Osteopathic Association, Continuing Medical Education, *available at* http://www.osteopathic.org/inside-aoa/development/continuing-medical-education/Pages/default.aspx (last visited Feb. 12, 2011).

53. Healthcare Facilities Accreditation Program, About HFAP, *available at* http://www.hfap.org/about/overview.aspx (last visited Feb. 12, 2011).

54. Committee on Quality of Health Care in America, Institute of Medicine, Crossing the Quality Chasm: A New Health System for the 21st Century (2001) (hereinafter "2001 CQHCA Report").

55. *Id.* at 28.

56. *Id.* at 28.

3.7% of hospitalizations.[57] Following some highly publicized errors (fatal medication overdoses and amputation of the limb on the wrong side), the Institute of Medicine estimated that errors resulted in as many as 98,000 deaths in patients hospitalized during 1997.[58] The report highlights "The decentralized and fragmented nature of the health care delivery system (some would say 'nonsystem') also contributes to unsafe conditions for patients, and serves as an impediment to efforts to improve safety." While recognizing that "not all errors result in harm," the report defines safety as "freedom from accidental injury" and specifies two types of error: "the failure of a planned action to be completed as intended or the use of a wrong plan to achieve an aim."[59]

Subsequently, the Institute of Medicine recommended development of a learning health care delivery system "a system that both prevents errors and learns from them when they occur. The development of such a system requires, first, a commitment by all stakeholders to a culture of safety and, second, improved information systems."[60] Government and nongovernment institutions such as the Agency for Healthcare Research and Quality (designated as the federal lead for patient safety by the Healthcare Research and Quality Act of 1999 to "(1) identify the causes of preventable health care errors and patient injury in health care delivery; (2) develop, demonstrate, and evaluate strategies for reducing errors and improving patient safety; and (3) disseminate such effective strategies throughout the health care industry."),[61] the National Quality Forum (a nonprofit organization with multiple stakeholders developing and measuring performance standards), the Joint Commission (independent not-for-profit organization accrediting and certifying care quality and safety), Institute of Healthcare Improvement (independent not-for-profit organization fostering innovation that improves care), and the Leapfrog Group (a coalition of large employers rewarding performance) all have adopted as parts of their mission the assessment and promotion of safety at the healthcare system level. To deliver safe, effective, and efficient care, medical delivery systems having increasingly incorporated allied health professions, including nurses, nurse practitioners, physicians' assistants, pharmacists, and therapists into care delivery.

57. Troyen A. Brennan et al., *Incidence of Adverse Events and Negligence in Hospitalized Patients: Results of the Harvard Medical Practice Study I*, 324 New Eng. J. Med. 370–76 (1991); Lucian L. Leape et al., *The Nature of Adverse Events in Hospitalized Patients: Results of the Harvard Medical Practice Study II*, 324 New Eng. J. Med. 377–84 (1991).

58. Committee on Quality of Health Care in America, Institute of Medicine, To Err Is Human: Building a Safer Health System 26 (2000) (hereinafter "2000 CQHCA Report").

59. *Id* at 4, 54, 58.

60. Committee on Data Standards for Patient Safety, Institute of Medicine, Patient Safety: Achieving a New Standard for Care 1 (2005).

61. Agency for Healthcare Research and Quality, Advancing Patient Safety: A Decade of Evidence, Design and Implementation at 1, *available at* http://www.ahrq.gov/qual/advptsafety.htm (last visited Feb. 12, 2011.)

C. Patient Care

1. Goals

The Institute of Medicine (IOM) describes quality health care delivery as "[t]he degree to which health services for individuals and populations increase the likelihood of desired health outcomes and are consistent with current professional knowledge." The six specific aims for improving health care include

1. "Safe: avoiding injuries to patients from the care that is intended to help them;"
2. "Effective: providing services based on scientific knowledge to all who could benefit, and refraining from providing services to those not likely to benefit;"
3. "Patient-centered: providing care that is respectful of and responsive to individual patient preferences, needs, and values, and ensuring that patient values guide all clinical decisions;"
4. "Timely: reducing waits and sometimes harmful delays for both those who receive and those who give care;"
5. "Efficient: avoiding waste, including waste of equipment, supplies, ideas, and energy;" and
6. "Equitable: providing care that does not vary in quality because of personal characteristics such as gender, ethnicity, geographic location, and socioeconomic status."[62]

Health outcome goals include (1) improving longevity or life expectancy, (2) relieving symptoms (improving quality of life or reducing morbidity), and (3) preventing disease. These goals, however, may conflict with one another. For example, some patients may be willing to accept the chance of a reduced length of life to try to obtain a higher quality of life (e.g., if normal volunteers had a vocal cord cancer, about 20% of them would prefer radiation therapy instead of surgery to preserve their voice despite a reduction in survival[63]), whereas others may accept reduced quality of life to try to extend life (e.g., cancer chemotherapy). Some may accept a risk of dying from a procedure to prolong life or relieve symptoms (e.g., coronary revascularization), whereas others may prefer to avoid the near-term risk of the procedure or surgery despite future benefit (risk aversion). In *Crossing the Quality Chasm,* the IOM emphasized care delivery that should accommodate individual patient choices and preferences and be customized on the basis of patients needs and values.[64]

62. 2001 CQHCA Report, *supra* note 54, at 44, 5–6.

63. Barbara J. McNeil et al., *Speech and Survival: Tradeoffs Between Quality and Quantity of Life in Laryngeal Cancer,* 305 New Eng. J. Med. 982–87 (1981) (hereinafter "McNeil").

64. 2001 CQHCA Report, *supra* note 54, at 49.

The Charter on Medical Professionalism avers three fundamental principles: (1) patient welfare or serving the interest of the patient, (2) patient autonomy or empowering patients to make informed decisions, and (3) social justice or fair distribution of health care resources.[65] At times, the primacy of patient welfare places the physician in conflict with social justice—for example, a patient with an acute heart attack is in the emergency room with no coronary care unit (CCU) beds available, and the most stable patient in the CCU has a 2-day-old heart attack. Transferring the patient out of the CCU places him or her at a small risk for a complication, but the CCU bed is a limited societal resource that other patients should be able to access.[66] Similarly, patients may insist on an unneeded and costly test or treatment, and the first two principles would encourage physicians to acquiesce, yet these unnecessary tests or treatments expose patients to harm and expense and also diminish resources that would otherwise be available to others.[67]

2. *Patient-physician encounters*

A patient-physician encounter typically consists of four components: (1) patient history, (2) physical examination, (3) medical decisionmaking, and (4) counseling.[68] In many cases, patients seek medical attention because of a change in health that led to symptoms. During the patient history, physicians identify the chief complaint as the particular symptom that led the patient to seek medical evaluation. The history of the present illness includes the onset and progression of symptoms over time and may include eliciting pertinent symptoms that the patient does not exhibit. These "pertinent negatives" reduce the likelihood of certain competing diagnoses. A comprehensive encounter includes past medical history of prior illnesses, hospitalizations, surgeries, current medications, drug allergies, and lifestyle habits including smoking, alcohol use, illicit drug use, dietary habits, and exercise habits. Family history considers illnesses that have been diagnosed in related family members to identify potential genetic predispositions for disease. Social history usually includes education, employment, and social relationships and provides a socioeconomic context for developing or coping with illness and an employment context for exposure to environmental or toxin risks. Finally, the review of systems is a comprehensive checklist of symptoms that might or might not arise from the various organ systems and is an ancillary means to capture symp-

65. Medical Professionalism Project: ABIM Foundation, *Medical Professionalism in the New Millennium: A Physician Charter,* 136 Annals Internal Med. 243, 244 (2002).

66. Harold C. Sox et al., Medical Decision Making (2007).

67. Harold C. Sox, *Medical Professionalism and the Parable of the Craft Guilds*, 147 Annals Internal Med. 809–10 (2007).

68. *See generally* Davoll v. Webb, 194 F.3d 1116, 1138 (10th Cir. 1999) ("A treating physician is not considered an expert witness if he or she testifies about observations based on personal knowledge, including treatment of the party.").

toms that the patient may have unintentionally neglected to mention, but which may lead physicians to consider additional diagnostic possibilities.

Patients, particularly the elderly, also may seek care to monitor multiple chronic conditions. This places an emphasis on collaborative and continuous care that involves patients (and their families) and providers, long-term care goals and plans, and self-management training and support.[69] The organizational needs for condition management, however, differ substantially from those necessary to deliver health services for acute episodic complaints. Taking a patient history in this case involves determining the status of the multiple conditions and whether symptoms from those conditions have progressed, improved, or stabilized and of the ability of patients to manage their condition.

The physical examination may be directed or complete. Physical findings are referred to as signs (distinct from symptoms noted by the patient). Directed physical examination refers to the examination of the relevant organ systems that may cause the symptoms or that may have positive or negative findings related to suspected diseases. When the disease is a chronic condition, the examination may be used to monitor disease progression or resolution. The complete physical examination of all organ systems may be performed as part of any annual examination, for difficult diagnoses, or for diseases that affect multiple organ systems.

The medical decisionmaking step of the encounter involves performing an assessment and plan. After the history and physical examination—based on the diagnostic possibilities, their likelihood, and the risks and benefits of treatment for each—the physician decides whether to recommend diagnostic testing, empiric treatment or referral to specialty or subspecialty care for further diagnostic evaluation, or a therapeutic intervention. Particularly challenging diagnoses are those that present with atypical symptoms, occur rarely, mimic other diseases, or involve multiple organ systems. For example, symptoms may arise from different organ systems: Wheezing, which is consistent with asthma, could be caused by acid going up from the stomach into the esophagus and then into the lungs (gastroesophageal reflux), congestive heart failure, or vocal cord dysfunction, among other diagnostic possibilities. The final step in the encounter is counseling the patient regarding diagnoses, tests, and treatments including dietary and lifestyle changes, medications, medical devices, and procedural interventions.

IV. Medical Decisionmaking

A. Diagnostic Reasoning

Uncertainty in defining a disease makes diagnosis difficult: (1) the difference between normal and abnormal is not always well demarcated; (2) many diseases

69. 2001 CQHCA Report, *supra* note 54, at 27.

do not progress with certainty (e.g., progression of ductal carcinoma in situ of the breast to invasive breast cancer occurs less than 50% of the time) but rather increase the risk of a poor outcome (e.g., hypertension raises the risk of developing heart disease or stroke); and (3) symptoms, signs, and findings for one disease overlap with others.[70] Variation also exists in the ability of physicians to elicit particular symptoms (e.g., in a group of patients interviewed by many physicians, 23% to 40% of the physicians reported cough as being present), observe signs (e.g., only 53% of physicians detected cyanosis—a blue or purple discoloration of the skin resulting from lack of oxygen—when present), or interpret tests (e.g., only 51% of pathologists agreed with each other when examining PAP smear slides with cells taken from a woman's cervix to look for signs of cervical cancer).[71] Moreover, prognosis (response to disease or treatment) with alternative therapies is in many cases uncertain. In a report by the Royal College of Physicians:

> The practice of medicine is distinguished by the need for judgement in the face of uncertainty. Doctors take responsibility for these judgements and their consequences. A doctor's up-to-date knowledge and skill provide the explicit scientific and often tacit experiential basis for such judgements. But because so much of medicine's unpredictability calls for wisdom as well as technical ability, doctors are vulnerable to the charge that their decisions are neither transparent nor accountable.[72]

1. *Clinical reasoning process*

Studies of clinical problem solving suggest that physicians employ combinations of two diagnostic approaches ranging from hypothetico-deductive (deliberative and analytical) to pattern recognition (quick and intuitive).[73] In the hypothetico-deductive approach, based on partial information, such as patient age, gender, and chief complaint, physicians[74] begin to generate a limited list of potential diagnostic hypotheses (hypothesis generation). Over the past 50 years, cognitive scientists

70. David M. Eddy, *Variations in Physician Practice: The Role of Uncertainty*, 3 Health Affairs 74, 75–76 (1984).

71. *Id.* at 77–78.

72. Royal College of Physicians, RCP Bookshop. *Doctors in Society. Medical Professionalism in a Changing World technical supplement full text* at 11, *available at* http://bookshop.rcplondon.ac.uk/contents/pub75-411c044b-3eee-462d-936d-1dad7313e4a0.pdf (last visited Feb. 12, 2011).

73. Jerome P. Kassirer et al., Learning Clinical Reasoning (2d ed. 2009) (hereinafter "Kassirer et al."); Arthur S. Elstein & Alan Schwartz, *Clinical Problem Solving and Diagnostic Decision Making: Selective Review of the Cognitive Literature*, 324 BMJ 729–32 (2002) (hereinafter "Elstein"); Jerome P. Kassirer & G. Anthony Gorry, *Clinical Problem Solving: A Behavioral Analysis*, 89 Annals Internal Med. 245 (1978); Geoffrey Norman, *Research in Clinical Reasoning: Past History and Current Trends*, 39 Med Educ. 418–27 (2005).

74. Steven N. Goodman, *Toward Evidence-Based Medical Statistics, 1: The p Value Fallacy*, 130 Annals Internal Med. 995–1004 (1999) (hereinafter "Goodman").

have demonstrated that human short-term memory capacity is limited,[75] and so this initial list of possible diagnoses is a cognitive necessity and provides an initial context that physicians use to evaluate subsequent data. Based on their knowledge of the diagnoses on that list, physicians have expectations about what symptoms, risk factors, disease course, signs, or test results would be consistent with each diagnosis (deductive inference).

As physicians gather additional information, they evaluate those data for their consistency with the possibilities on their initial list and whether those data would increase or decrease the likelihood of each possibility (hypothesis refinement). If the data are inconsistent, additional diagnostic possibilities are considered (hypothesis modification). The information gathering continues as an iterative process at the same visit or over time during multiple visits with the same or other physicians. The final cognitive step (diagnostic verification) involves testing the validity of the diagnosis for its coherency (consistency with predisposing risk factors, physiological mechanisms, and resulting manifestations), its adequacy (the ability to account for all normal and abnormal findings and the disease time course), and its parsimony (the simplest single explanation as opposed to requiring the simultaneous occurrence of two or more diseases to explain the findings).[76]

At the other end of clinical reasoning are heuristics, quick automatic "rules of thumb" or cognitive shortcuts. In such cases, pattern recognition leads to rapid recognition and a quick diagnosis, improving cognitive efficiency.[77] For example, a black woman with large shadows of lymph nodes in her chest x ray would trigger a diagnosis of a disease known as sarcoidosis for many physicians. The simplifying assumptions involved in heuristics, however, are subject to cognitive biases. For example, episodic headache, sweating, and a rapid heartbeat form the classic triad seen in patients with a rare adrenal tumor known as a pheochromocytoma that also can cause hypertension. Physicians finding those three symptoms in a patient with hypertension may overestimate the patient's likelihood of having pheochromocytoma based on representativeness bias, overestimating the likelihood of a less common disease just because case findings resemble those found in that disease.[78] Other cognitive errors include availability (overestimating the

75. Elstein, *supra* note 73; George A. Miller, *The Magical Number Seven Plus or Minus Two: Some Limits on Our Capacity for Processing Information*, 63 Psychol. Rev. 81–97 (1956).

76. Kassirer et al., *supra* note 73, at 5-6.

77. Stephen G. Pauker & John B. Wong, *How (Should) Physicians Think? A Journey from Behavioral Economics to the Bedside,* 304 JAMA 1233–35 (2010).

78. For additional discussion and definition of terms, see Section IV.A.2. Applying Bayes' rule, about 100 in 100,000 patients with hypertension have pheochromocytoma; this symptom triad occurs in 91% of patients with pheochromocytoma (sensitivity) and does not occur in 94% of those without pheochromocytoma (specificity), and so 6% of those without pheochromocytoma would have this symptom triad. On the basis of Bayes' rule, 91 of the 100 individuals with pheochromocytoma (91% times 100) would have this triad, and 5994 without a pheochromocytoma (6% times 99,900) will have the triad. Thus, among the 100,000 hypertensive patients, 6085 will have the classic triad, suggesting the possibility of pheochromocytoma, but only 91 out of the 6085 or 1.5%, will indeed have pheochromcytoma.

likelihood of memorable diseases because of severity or media attention and underestimating common or routine diseases) and anchoring (insufficient adjustment of the initial likelihood of disease).[79]

Clinical intuition refers to rapid, unconscious processes that select the pertinent findings out of the multitude of available data.[80] Such expertise results from practice, is context sensitive, and cannot always be reduced to cause and effect.[81] Cognitive research into the development of expertise suggests two competing hypotheses. In instance- or exemplar-based memory, physicians store scripts or "stories" of prior recalled case examples, for example, visual information such as that in pathology, dermatology, or radiology, and match new cases to those stories. The alternative prototype memory hypothesis is based on a mental model of disease wherein experts store structured "facts" about the disease to create abstractions. These "prototypes" enable experts to link findings to one another, to connect findings to the possible diagnoses, and to predict additional findings necessary to confirm the diagnosis, even in the absence of prior experience with exactly such a case.[82]

Physicians typically apply hypothetico-deductive approaches when seeing patients with problems outside of their expertise or difficult problems with atypical issues within their expertise and apply intuitive pattern recognition for cases within their expertise or less challenging cases. However, diagnostic accuracy appears to depend more on mastery of domain knowledge than on the particular problem-solving method.[83]

2. *Probabilistic reasoning and Bayes' rule*

There is no correlation between physicians' ability to collect data thoroughly and their ability to interpret the data accurately.[84] Making quantitative predictions or interpretation of test results constitutes probabilistic reasoning and avoids the use of ambiguous qualitative terms such as "low" or "always" that may contribute to different management decisions.[85]

Over 200 years ago, the Reverend Bayes first wrote a paper published posthumously which now forms a critical concept in modern medicine. Ignored for

79. Kassirer et al., *supra* note 73; Elstein, *supra* note 73.

80. Trisha Greenhalgh, *Intuition and Evidence—Uneasy Bedfellows?* 52 Brit. J. Gen. Practice 395–400 (2002).

81. *Id.* at 396.

82. Kassirer et al., *supra* note 73; Elstein, *supra* note 73.

83. Elstein, *supra* note 73.

84. Arthur S. Elstein & Alan Schwartz, *Clinical Reasoning in Medicine, in* Clinical Reasoning in the Health Professions 223–34 (Joy Higgs et al. eds., 3d ed. 2008).

85. When physicians were asked to quantify "low probability," the estimates had a mean of ~37% with a range from 0% to ~80% and when asked to quantify "always," physicians had a mean of ~88% with a range from 70% to 100%. Geoffrey D. Bryant & Geoffrey R. Norman, *Expressions of Probability: Words and Numbers,* 302 New Eng. J. Med. 411 (1980).

nearly two centuries, his paper showed how to estimate the likelihood of disease following a test result using the likelihood of disease prior to testing and the specific test result obtained. Thus, Bayesian analysis refers to a method of combining existing evidence or a prior belief with additional evidence, for example, from test results. The additional evidence may be the presence or absence of a symptom, sign, test, or research study results.

The pretest suspicion of disease or, equivalently, the likelihood or prior probability of disease may be objective, that is, related to incidence (new cases over a specified period of time) or prevalence (existing cases at a particular point in time); based on clinical prediction rules (e.g., mathematical predictive models to estimate the likelihood of developing heart disease over the next 10 years using data from the Framingham Study); or subjective, that is, based on a clinician's estimated likelihood of disease prior to any testing.[86] Bayes' rule then combines that pretest suspicion with the observed test result. Those who have disease and a positive test are said to have true-positive test results. Those without disease who have a negative test are said to have true-negative test results. Tests, however, are almost always not perfectly accurate. That is, not everyone with disease has a positive test; these are called false-negative test results. Similarly, some individuals who are healthy may mistakenly have positive tests; these are called false-positive test results.

For example, consider screening mammography which is positive in 90% of women with breast cancer, and so the true-positive rate (or "sensitivity") of 90% is the likelihood of a positive test among those with disease. Mammography is negative in 93% of women without breast cancer, and so the true-negative rate (or "specificity") of 93% is the likelihood of a negative test among those who do not have disease (see Table 1).[87] Note that if the test is not negative, it must be positive, or vice versa, so that the sum of the columns in Table 1 must equal 100%.

Because a positive mammogram can occur among individuals with or without breast cancer, the interpretation of the likelihood of breast cancer with a positive mammogram can be problematic. Given that the prevalence of breast cancer among asymptomatic 40- to 50-year-old women is 8 in 1000, or 0.8%, Bayes' rule calculates the likelihood of breast cancer following a test result, for example, a positive mammogram (see Figures 1 and 2, Table 2).[88] This analysis helps explain in part why mammogram screening is controversial in women under age 50.

86. *See* Gonzalez v. Metro. Transp. Auth., 174 F.3d 1016, 1023 (9th Cir. 1999) (describing the implications of Bayes' rule for drug testing and noting that a test with the same false-positive rate will generate a higher proportion of false positives to true positives in a population with fewer drug users); *see generally* Michael O. Finkelstein & William B. Fairley, *A Bayesian Approach to Identification Evidence*, 83 Harv. L. Rev. 489 (1970). For a discussion of Baysian statistics, see David H. Kaye & David A. Freedman, Reference Guide on Statistics, Section IV.D, in this manual.

87. Gerd Gigerenzer, Calculated Risks: How to Know When Numbers Deceive You (2002) at 41 (hereinafter "Gigerenzer").

88. *Id.* at 45–48.

Table 1. 2 × 2 Test Characteristics of Screening Mammogram for Use in Bayes' Rule

	Breast Cancer	No Breast Cancer
Positive mammogram	90 true positives	7 false positives
Negative mammogram	10 false negatives	93 true negatives

Figure 1. Screening 1000 women for breast cancer.

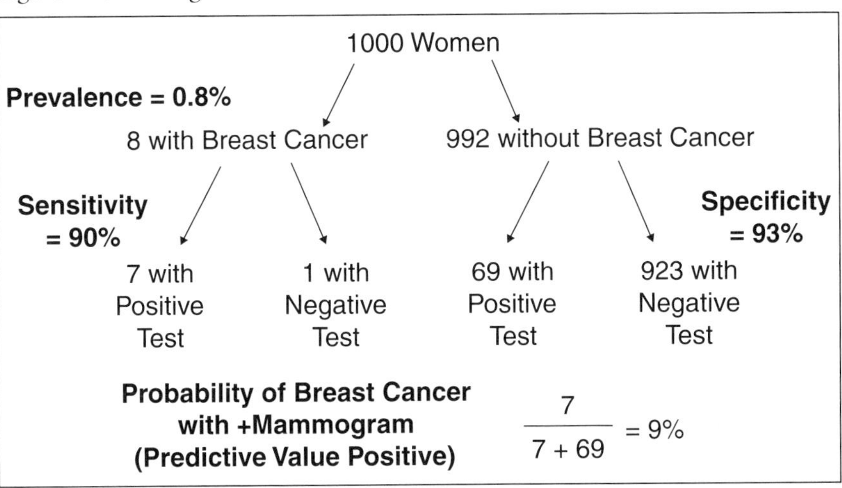

Figure 2. Likelihood of breast cancer after a positive or a negative mammogram.

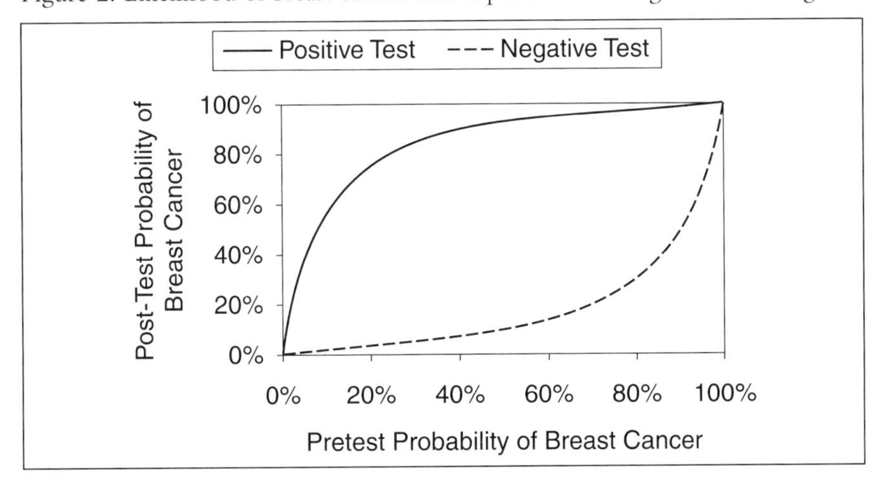

Table 2. Tabular and Formula Forms of Bayes' Rule

Tabular Form of Bayes' Rule

Condition	Pretest or Prior Probability (%)	Conditional Probability of Positive Test for the Condition (%)	Product of the Pretest and the Conditional Probabilities (%)	Posttest or Posterior Probability (%)
Breast cancer	0.8	90 sensitivity	0.72	9 = 0.72 ÷ 7.6
No breast cancer	99.2	7 1 − specificity	6.9	
			Sum = 7.6	

Formula Form of Bayes' Rule

$$\frac{pD+ \star\ pT+ | D+}{(pD+ \star pT+ | D+) + ((1-pD+) \star (1-pT- | D-))}$$

pD+ = prior probability of disease = 0.8%
pT+ | D+ = Sensitivity = True Positive Rate = 90%
pT− | D− = Specificity = True Negative Rate = 93%

$$\frac{0.008 \star 0.90}{(0.008 \star 0.90) + ((1-0.008) \star (1-0.93))}$$

$$= 9\%$$

Despite a test that has a 90% or higher rate on both sensitivity and specificity, a calculation using Bayes' theorem shows that having a low probability of breast cancer before testing means that even with a positive result on a screening mammogram, the likelihood that an average woman under age 50 has breast cancer is less than 10%.

The probability of breast cancer among those with a positive mammogram is termed the "predictive value positive." Similarly, if the test were negative, the likelihood of breast cancer in those with a negative mammogram ("false reassurance rate") would be 1 divided by 924 (1 woman with breast cancer and a negative test and 923 women without breast cancer who have negative tests in Figure 1), or about 0.1%. Interpreting a medical test result then depends on the pretest likelihood of disease and the test's sensitivity and specificity. Figure 2 illustrates the likelihood of breast cancer for differing pretest or prior probabilities of breast cancer.

The discriminating ability of a test can be succinctly summarized as a likelihood ratio. The likelihood ratio positive expresses how much more likely disease is to be present following a positive test result. It is the ratio of the true-positive rate to the false-positive rate (sensitivity divided by 1 minus the specificity), e.g., 12.5 (0.90 divided by 1 − 0.93) in the case of mammography. The likelihood ratio negative expresses how much less likely disease is to be present following a negative test result. It is 1 minus the ratio of the false-negative rate to the true-negative rate (1 minus the sensitivity divided by the specificity) or 0.11 (1 − 0.90 divided by 0.93) in the case of mammography. Likelihood ratios exceeding 10 or falling below 0.1 are believed to be strong discriminators causing "large" changes in the likelihood of disease; those between 5 and 10 or 0.1 and 0.2 cause "moderate" changes; and those between 2 and 5 or 0.2 and 0.5 cause "small" changes.[89] Note that even for a strongly discriminating test such as mammography, a positive or a negative test result does not change the likelihood of disease substantially for very low or very high probabilities of disease (see Figure 2), thereby highlighting the importance of the pretest likelihood of disease in interpreting test results.

Terms such as "sensitivity," "specificity," and "predictive value negative or positive" are called conditional probabilities because they express the likelihood of a particular result based on a particular condition (e.g., a positive test result among those with disease) or the likelihood of a particular condition among those with a particular result (e.g., disease among those with a positive test).[90] These kinds of expression, however, remove the base case probability (the pretest probability of disease, sometimes referred to as the prior probability of disease) as part of "normalization," so that Bayes' rule is required to interpret a test result. Moreover, confusion between sensitivity and predictive value positive may lead to errors in the interpretation of test results; for example, a 90% likelihood of having a positive mammogram in patients with breast cancer—the sensitivity—may be misinterpreted as the predictive value positive, implying that a woman with a positive mammogram has a 90% chance of having cancer. This misinterpretation ignores the role for pretest suspicion or likelihood of disease (or assumes that all

89. David A Grimes & Kenneth F Schulz, *Refining Clinical Diagnosis with Likelihood Ratios*, 365 Lancet 1500–05 (2005).

90. This terminology may be confusing. The predictive value negative (negative predictive value) is defined as the probability of no disease among those with a negative test. It also equals 1 minus the false reassurance rate. The false-alarm rate is defined as the probability of no disease among those with a positive test. It is also 1 minus the predictive value positive. The false reassurance rate may be confused with the false negative rate (among those with disease, the likelihood of a negative test) because both involve those with negative tests and those with disease but in one case the denominator is individuals with negative tests (false reassurance rate) and in the other case individuals with disease (false negative rate). Similarly, the false alarm rate may be confused with the false positive rate (among those with no disease, the likelihood of a positive test).

women undergoing the test have the disease). This confusion can be avoided by translating Bayes' rule into natural frequency expressions.[91] The natural frequency expression incorporates both the pretest likelihood and the conditional probabilities of the test results to yield the following statements (see Figure 1): Of 1000 women between 40 and 50 years old, 8 have breast cancer, and 7 of these will test positive. Of the remaining 992 who do not have breast cancer, about 69 will also test positive. When presented as a natural frequency (including the likelihood of disease), the likelihood of breast cancer becomes more transparent; thus 76 women will test positive, and 7 of the 76 will have breast cancer. When 48 physicians with an average of 14 years of professional experience were presented with the natural frequency version or the conditional probability version, 16 of 24 estimated the likelihood of breast cancer to exceed 50% with the conditional probability (sensitivity, specificity) version but only 5 of 24 did so with the natural frequency information.[92]

Just as mammography test results may be misinterpreted if Bayes' rule is not applied, the prosecutor's fallacy involves the misinterpretation of probabilistic information. For example, in *People v. Collins*, the prosecutor argued that 1 in 3 girls have blonde hair, 1 in 10 girls have a pony tail, 1 in 10 automobiles are partly yellow, 1 in 4 men have a mustache, 1 in 10 black men have a beard, and 1 in 1000 cars have an interracial couple in the car.[93] Multiplying these six probabilities together yields a 1 in 12 million joint probability of having all conditions present. Aside from being simply estimates and from assuming that the probabilities were independent of one another, the prosecutor made the statement that "The probability of the defendant matching on these six characteristics is 1 in 12 million," thereby assuming that someone other than the defendant being guilty is the same 1 in 12 million. However, if translated into natural frequency terms, 1 out of every 12 million couples would have these six characteristics, and so assuming that there are 24 million couples, there would be a 1 in 2 chance that the Collinses are innocent. The error results from confusing the probability of a positive test (having all six characteristics) among those with the disease (being guilty) and the probability of the disease (being guilty) among those with a positive test (having all six characteristics), that is, confusing the conditional probabilities—sensitivity and positive predictive value.

Bayes' rule becomes even more relevant in the genomic medicine era.[94] Suppose a genetic test has a sensitivity and specificity of 99.9%, and suppose the probability of disease is 1 in 1000 if a positive family history is present and 1 in 100,000 if no family history is present. Screening 1000 individuals with a positive family

91. Gigerenzer, *supra* note 87, at 42.

92. *Id.* at 43.

93. *Id.* at 152.

94. Isaac S. Kohane et al., *The Incidentalome. A Threat to Genomic Medicine*, 296 JAMA 212–15 (2006).

history for the gene results in 2 positive tests: 1 individual truly has disease, and in the other the test is a false positive. Screening 10 million individuals without a family history results in 10,100 positive tests in which 100 individuals have disease and 10,000 do not. Even with a specificity of 99.99%, if a test screens for 10,000 genes simultaneously, then 63% of individuals will have at least one false-positive test result. Based simply on the genetic test results alone, neither individuals nor physicians would be able to distinguish those with true-positive results from those with false-positive results, thereby potentially leading to inappropriate monitoring or treatment for all with positive test results.

Although a test is commonly thought of as a sample from a bodily fluid, tissue, or image, a test also could be the presence or absence of a symptom or physical sign. For example, both inhalation anthrax and influenza can cause symptoms of muscle aches, fever, and malaise. However, a critical symptom that helps distinguish one from the other is runny nose, which occurs in 14% of those with inhalation anthrax but in 78% to 89% of those with influenza or influenza-like illness. Thus, when faced with distinguishing between these diagnoses, patients with a runny nose given this symptom alone are about six times more likely to have influenza or a flu-like illness than to have anthrax.[95]

Sensitivity and specificity rely on setting a positivity criterion, the threshold level for determining normal above which tests are positive and below which the test is negative. If the criterion is made stricter (e.g., what is considered to be abnormal requires a higher test result), then sensitivity falls and specificity increases, and if the criterion is made laxer, then sensitivity rises and specificity falls. Depending on the context of the testing, it may be more appropriate to choose a laxer criterion (e.g., screening donated blood for HIV infection where the benefit is reducing transfusion-associated HIV transmission, and the risk is discarding some uninfected units of donated blood) or a stricter one (e.g., screening a low-prevalence population for HIV infection where the benefit is reducing false-positive diagnoses and the risk is missing some truly HIV-infected individuals).[96] Thus the benefits of finding and treating a person with disease versus the risk of treating a person without disease should help establish what is considered normal or abnormal.

The terms "sensitivity" and "specificity" apply to the simple situation in which disease is present or absent and a test can be positive or negative, but terminology and interpretation become more complicated when multiple diseases are under consideration and when multiple test results may occur.[97] For example, consider blood in the urine (hematuria), which could be caused by a urinary tract infection, a kidney stone, or a bladder cancer, among many other diseases. The

95. Nathaniel Hupert et al., *Accuracy of Screening for Inhalational Anthrax After a Bioterrorist Attack*, 139 Annals Internal Med. 337–45 (2003).

96. Klemens M. Meyer & Stephen G. Pauker, *Screening for HIV: Can We Afford the False Positive Rate?* 317 New Eng. J. Med. 238–41 (1987).

97. Kassirer et al., *supra* note 73, at 21–22.

terms "sensitivity" and "specificity" are no longer appropriate because disease is not simply present or absent. Instead, they are replaced by the term conditional probabilities, that is, sensitivity is replaced by the likelihood of blood in the urine with a urinary tract infection, or with a kidney stone, or with a bladder cancer. Similarly, a very positive test has a different interpretation than a weakly positive test, and Bayes' rule can quantify the difference. Results from multiple tests can be combined with Bayes' rule by applying Bayes' rule to the first test result and then reapplying Bayes' rule to subsequent test results. This approach assumes that the result of the first test does not affect the test characteristics (sensitivity or specificity) of the second test (i.e., that there is conditional independence of each test). When two tests are available, screening will usually occur first with the high-sensitivity test to detect a high proportion of those with disease (true positives), or "ruling in" disease. Those with a positive first test will then undergo a high-specificity test to reduce the number of individuals who do not have disease but a positive first test (false positive), or "ruling out" disease.

3. *Causal reasoning*

To select the most appropriate therapy, physicians seek to identify the cause of a patient's complaints and findings. While considering the presence or absence of risk factors (e.g., the presence of male gender, advanced age, high cholesterol, high blood pressure, diabetes mellitus, and smoking for the medical condition coronary heart disease), physicians will often use any type of evidence[98] that might support causation, for example, biological plausibility,[99] physiological drug effects, case reports, or temporal proximity[100] to an exposure.[101] Although physicians use epidemiological studies in their decisionmaking, "they are accustomed to using *any* reliable data to assess causality, no matter what their source" because they must make care decisions even in the face of uncertainty.[102] This is in contrast to the courts which require a higher standard than clinicians or regulators, and wherein causation cannot just be "possible" but where "a 'preponderance of evidence' establishes that an injury was caused by an alleged exposure."[103] For physicians, causal reasoning typically involves

98. Jerome P. Kassirer & Joe S. Cecil, *Inconsistency in Evidentiary Standards for Medical Testimony: Disorder in the Courts*, 288 JAMA 1382–87 (2002) (hereinafter "Kassirer & Cecil"); *see also* Section IV.C.2, for levels of evidence.

99. *See* Kennan v. Sec'y of Health & Human Servs., 2007 WL 1231592 (Ct. Fed. Cl. Apr. 5, 2007).

100. *But see* Wilson v. Taser Int'l, Inc., 303 F. App'x 708, 714 (11th Cir. 2008) ("[A]lthough a doctor usually may primarily base his opinion as to the cause of a plaintiff's injuries on this history where the patient 'has sustained a common injury in a way that it commonly occurs,' . . . Dr. Meier could not rely upon the temporal connection between the two events to support his causation opinion in this case.").

101. Kassirer & Cecil, *supra* note 98, at 1384.

102. *Id.* at 1394.

103. *Id.* at 1384.

understanding how abnormalities in physiology, anatomy, genetics, or biochemistry lead to the clinical manifestations of disease. Through such reasoning, physicians develop a "causal cascade" or "chain or web of causation" linking a sequence of plausible cause-and-effect mechanisms to arrive at the pathogenesis or pathophysiology of a disease. For example, kidney failure leads to poor drug excretion, resulting in symptoms or signs of drug toxicity.[104] Although probabilistic reasoning typically dominates initial hypothesis generation by physicians based on prevalence or incidence, pattern recognition of concomitant symptoms and signs could trigger a diagnosis. For example, cough, lung lesions, and enlarged breasts (gynecomastia) in a 37-year-old man could trigger the diagnosis of metastatic germ cell cancer.[105] More typically, physicians use causal reasoning in diagnostic refinement and verification to examine a diagnosis for its coherency, namely, asking whether its physiological mechanism would be expected to lead to the observed manifestations and whether it is adequate to account for all normal and abnormal findings and the disease time course. Once treatment has been implemented, physicians must make causal judgments in determining whether an alteration in patient status is the result of progression of disease or an adverse consequence of treatment, or whether the absence of improvement results from therapeutic ineffectiveness that should prompt a change in therapy or even reconsideration of the diagnosis.

Pathophysiological reasoning, however, also can lead to incorrect conclusions. In patients with heart failure with a weakened heart, a class of medications called beta blockers had been thought to be contraindicated because beta blockers would decrease the strength of the heart muscle contraction. Subsequent studies found that beta blockers in patients with heart failure usually had no ill effect and actually increased survival. Similarly, physicians once thought that atherosclerotic blockages in heart arteries slowly progressed to cause a heart attack, so that revascularizing those plaques through heart bypass surgery would prevent heart attacks.[106] Over the past 15 years, however, scientific evidence has emerged that small vulnerable atherosclerotic plaques (not amenable to revascularization because of their small size) can suddenly rupture and cause heart attacks. Not surprisingly, revascularization trials involving either bypass surgery or percutaneous interventions such as stenting or angioplasty do not diminish the risk of having a heart attack or improve survival for most patients.[107]

Although treating physicians[108] may testify with regard to both general and specific causation, as with use of evidence for causation, their standards for evi-

104. Kassirer et al., *supra* note 73, at 63–66.

105. *Id.* at 29.

106. David S. Jones, *Visions of a Cure: Visualization, Clinical Trials, and Controversies in Cardiac Therapeutics, 1968–1998*, 91 Isis 504–41 (2000).

107. Thomas A. Trikalinos et al., *Percutaneous Coronary Interventions for Non-acute Coronary Artery Disease: A Quantitative 20-Year Synopsis and a Network Meta-analysis*, 373 Lancet 911–18 (2009).

108. *See generally* Bland v. Verizon Wireless, LLC, 538 F.3d 893 (8th Cir. 2008) (upholding the district court's decision to reject a treating physician's evidence of causation under *Daubert*).

dence vary.[109] For example, some physicians may stop using a drug after the first reports of adverse effects, and others may continue to use a drug despite evidence of harm from randomized controlled trials. Determining whether an effect is a class effect or drug specific can be difficult. When considering beta blockers for patients with a weakened heart (heart failure), many studies have consistently demonstrated the benefit of beta blockers in reducing mortality in those with heart attacks often resulting in weakened heart function. However, in a randomized trial limited to patients with documented weakened heart, one particular beta blocker was found to not confer a survival benefit, and as a result the heart failure guidelines limited their beta blocker recommendation to just those three drugs with documented mortality benefit in trials.[110]

Although treating physicians may be aware of patient-specific risk factors such as smoking or family history, they may not routinely review specialized aspects of such data, for example, toxicology, industrial hygiene, environment, and some aspects of epidemiology. Additional experts may assist in distinguishing general from specific causation by using their specialized knowledge to weigh the relative contribution of each putative causative factor to determine "reasonable medical certainty" or "reasonable medical probability." The determination of general causation involves medical and scientific literature review and the evaluation of epidemiological data, toxicological data, and dose–response relationships. Consider for example, hormone replacement therapy for postmenopausal women. Multiple observational studies using methods such as case-control, cross-sectional, and cohort designs[111] suggested an association between hormone therapy and reduction in heart attack, but such designs are subject to confounding and bias and are particularly weak for causation because in case-control and cross-sectional studies, the sequence of the exposure and outcome is unknown. To resolve the question, the Women's Health Initiative (WHI) study randomized women to hormone replacement therapy or placebo and found a statistically significant increase in clot-related disorders—heart attack, stroke, and heart-related mortality over 5 years but most notable in the first year after initiation of hormone therapy.[112] Heart attacks are caused by blood clots and plaque rupture, and so the results were consistent with the known biological mechanism of estrogens in the clotting cascade. However, patients in the WHI were, on average, 63 years old and therefore not perimenopausal as analyzed in the observational studies. In a novel

109. Kassirer & Cecil, *supra* note 98, at 1384.

110. Mariell Jessup et al., *2009 Focused Update: ACCF/AHA Guidelines for the Diagnosis and Management of Heart Failure in Adults: A Report of the American College of Cardiology Foundation/American Heart Association Task Force on Practice Guidelines*, 119 Circulation 1977–2016 (2009).

111. *See* Michael D. Green et al., Reference Guide on Epidemiology, in this manual.

112. Jacques E. Rossouw et al., *Risks and Benefits of Estrogen Plus Progestin in Healthy Postmenopausal Women: Principal Results from the Women's Health Initiative Randomized Controlled Trial*, 288 JAMA 321–33 (2002); JoAnn E. Manson et al., *Estrogen Plus Progestin and the Risk of Coronary Heart Disease*, 349 New Eng. J. Med. 523–34 (2003).

approach, the observational Nurses' Health Study attempted to emulate the design and intention-to-treat (ITT) analysis aspect of the WHI randomized trial, and saw that the hormone replacement treatment effects were similar to those from the randomized trial, suggesting that "the discrepancies between the WHI and the Nurses' Health Study ITT estimates could be largely explained by differences in the distribution of time since menopause and length of followup."[113]

B. Testing

1. Screening

Screening on a population basis requires that (1) the condition be present in the population and affect quality and length of life; (2) the incidence or prevalence be sufficiently high to justify any risks associated with the test; (3) preventive or early treatment should be available; (4) an asymptomatic period for early detection must exist; (5) the screening test should be accurate, acceptable, and affordable; and (6) screening benefits should exceed harms. Screening for disease in asymptomatic, otherwise healthy patients has become widely accepted and promulgated.[114] Screening differs from diagnostic testing used to elucidate the cause of symptoms or loss of function because screening involves apparently healthy individuals.[115] Although screening may prevent the development of disease-related morbidity and mortality, positive test results (both false positive and true positive) may lead to interventions that could be unnecessary or even risky because of overdiagnosis and overtreatment.[116]

Normal ranges for biochemical tests are often based on the 95% confidence intervals in a normal healthy population—that is, although everyone is healthy, by convention, values outside the 2.5% lower and upper extremes are considered to be abnormal. Consequently, ordering six blood tests in a normal healthy individual yields only a 74% chance that all six tests will be normal; that is, there is a 26% chance that one or more may be abnormal. Similarly, when ordering 12 tests in a normal person, there is a 54% chance that all 12 will be normal and a 46% chance that 1 or more will be abnormal. So simply ordering tests in healthy individuals or in the absence of clinical suspicion of a disease may result in many

113. Miguel A. Hernán et al., *Observational Studies Analyzed Like Randomized Experiments: An Application to Postmenopausal Hormone Therapy and Coronary Heart Disease*, 19 Epidemiology 766–79 (2008).

114. Lisa M. Schwartz et al., *Enthusiasm for Cancer Screening in the United States*, 291 JAMA 71–78 (2004).

115. David A. Grimes & Kenneth F. Schulz, *Uses and Abuses of Screening Tests*, 359 Lancet 881–84 (2002) (hereinafter Grimes and Schulz); William C. Black, *Overdiagnosis: An Under Recognized Cause of Confusion and Harm in Cancer Screening*, 92 J. Nat'l Cancer Inst. 1280–82 (2000) (hereinafter "Black").

116. Grimes & Schulz, *supra* note 115, at 884; Black, *supra* note 115, at 1280.

false-positive test results that can lead to false alarms, anxiety, additional testing, and possible morbidity or mortality from subsequent testing or interventions.[117]

Even a valueless screening test may appear to be beneficial because of "lead-time bias." If screened or unscreened patients have the same prognosis from the time of onset of symptoms to death, then screened patients only appear to live longer because the time elapsed from diagnosis by screening to death exceeds that from diagnosis made at the time of symptom onset to death. A second bias, "length bias," also leads to overestimation of the benefit from screening.[118] Suppose that a randomized trial of screening or no screening is conducted over a limited length of time from study initiation to termination. The screening test detects patients with both aggressive and indolent forms of the disease. Among the unscreened patients, however, disease only becomes evident through the development of symptoms, which would be more likely in patients who have the aggressive form of the disease and a poorer prognosis. Thus screened patients with disease appear to have a better prognosis than unscreened patients with disease because a higher proportion of the screened patients have more indolent disease. Extending the concept of length bias further, screening can result in "pseudodisease" or "overdiagnosis," such as the identification of slow-growing cancers that even if untreated would never cause symptoms or reduce survival.[119] Although lung cancer is commonly thought to be one of the more aggressive cancers, an autopsy study found that one-third of lung cancers were unsuspected prior to autopsy, and nearly all of these patients with unsuspected lung cancer prior to autopsy died from other causes.[120] Lung cancer screening in these individuals would have resulted in pseudodisease or overdiagnosis because screening would have diagnosed their cancer but they would have died of something else (or from a severe adverse effect of the cancer treatment) before the cancer became evident.

To further illustrate bias in screening studies, the Mayo Lung Project was a randomized trial comparing screening for lung cancer with periodic chest X rays and sputum samples versus usual care. It found that screening did improve the likelihood of survival 5 years after diagnosis in those with lung cancer but surprisingly did not affect lung cancer deaths. Further analysis of the randomized trial found that the survival advantage of screening was attributable to the 46 extra

117. A radiologist described his own experience to illustrate the clinical aphorism that "the only 'normal' patient is one who has not yet undergone a complete work-up." He had a negative CT scan of the colon examination, but the CT scan also provided images outside the liver with radiologists identifying lesions in the kidneys, liver, and lungs. This resulted in additional CT scans, a liver biopsy, PET scan, video-aided thoracoscopy (a flexible scope inserted into the chest), and three wedge resections of the lung leading to multiple tubes, medications, and "excruciating pain" that required 5 weeks for recovery. William J. Casarella, *A Patient's Viewpoint on a Current Controversy*, 224 Radiology 927 (2002).

118. Grimes & Schulz, *supra* note 115, at 884.

119. Black, *supra* note 115, at 1280.

120. Charles K. Chan et al., *More Lung Cancer but Better Survival: Implications of Secular Trends in "Necropsy Surprise" Rates*, 96 Chest 291–96 (1989).

lung cancer cases detected by screening. These 46 cases had indolent (or, at worst, very slowly progressive) lung cancer; that is, these patients would have a normal life expectancy, and so, including their prognosis in those with screen-detected lung cancer inflates the apparent 5-year survival with screening because of length bias and overdiagnosis.[121] More recently, CT scan screening found lung cancer to be present in the same proportion of nonsmokers as smokers,[122] suggesting that many of the cancers detected in the nonsmokers were ones that would have never progressed. This overdiagnosis can lead to morbidity and mortality: CT scan screening for lung cancer results in a threefold increase in diagnosis and threefold increase in surgery with an average surgical mortality of 5% and serious complication rate exceeding 20%,[123] as well as potential risk from radiation exposure. A similar phenomenon occurs with breast cancer where screening increases surgeries by about one-third from overdiagnosis and with prostate cancer where the lifetime risk of dying from prostate cancer is about 3%, yet 60% of men in their sixties have prostate cancer, and so, screening and detecting all men with prostate cancer in their sixties would lead to treatment of many men who would not have died from prostate cancer.[124] In patients found to have cancer by screening, it is not possible to distinguish those whose cancers would have progressed from those in whom the cancer-appearing cells would not have progressed or spread.

2. *Diagnostic testing*

Based on the history and physical examination, physicians will establish diagnostic possibilities. They may then request additional tests to reduce uncertainty and to confirm the diagnosis, as part of diagnostic verification. Although, theoretically, all tests could be ordered, tests should be chosen on the basis of a clinical suspicion because of possible morbidity or even mortality from inappropriate testing. Normative prescriptive decision models for reasoning in the presence of uncertainty suggest that whether and which tests get ordered should depend on the sensitivity and specificity of the test as discussed in Section IV.A.2, *supra*, but also the risk of mortality or morbidity from the test, and the benefit and risk of treatment.[125] In general, for sufficiently low probabilities of disease, no tests should be ordered and no treatment given. For sufficiently high probabilities of disease,

121. Black, *supra* note 115.

122. William C. Black & John A. Baron, *CT Screening for Lung Cancer: Spiraling into Confusion?* 297 JAMA 995–97 (2007).

123. *Id.* at 996.

124. Karsten J. Jørgensen & Peter C. Gøtzsche, *Overdiagnosis in Publicly Organised Mammography Screening Programmes: Systematic Review of Incidence Trends*, 339 BMJ b2587 (2009); Michael J. Barry, *Prostate-Specific–Antigen Testing for Early Diagnosis of Prostate Cancer*, 344 New Eng. J. Med. 1373–77 (2001).

125. Stephen G. Pauker & Jerome P. Kassirer, *The Threshold Approach to Clinical Decision Making*, 302 New Eng. J. Med. 1109–17 (1980).

testing is unnecessary and treatment should be administered. For intermediate probabilities of disease, testing should be performed. When testing carries risks, the probabilities of disease for which testing should be done become narrower, and so physicians should be more likely to treat empirically or neither test nor treat. As sensitivity and specificity increase, the range of probabilities in which testing should be done expands.

Although an abnormal test result may be found, that abnormality may not be causing symptoms. For example, herniated lumbar discs are found in approximately 25% of healthy individuals without back pain; thus finding a herniated disc in patients with back pain may be an incidental finding. If signs such as a foot drop develop, additional muscle and nerve conduction studies might confirm evidence of nerve compromise from the herniated disc, but such tests are painful. Over time, sequential images show that the herniated disc has partial or complete resolution after 6 months without surgery. Therefore, a herniated disc may be seen with CT or MRI scanning in patients with or without symptoms, and so just having symptoms and evidence of a herniated disc would be an insufficient indication for back surgery.[126] In the absence of severe or progressive neurological deficits, elective disc surgery could be considered for patients with probable herniated discs who have persistent symptoms and findings consistent with sciatica (not just low back pain) for 4 to 6 weeks, but such "patients should be involved in decision making" (*see* Section IV.D.3, *infra*).[127]

Just as some therapies may eventually be found to be harmful or not beneficial, tests initially felt to be useful may be found to be less valuable.[128] Among other potential biases,[129] this may occur because of the choice of study population used to determine the test's sensitivity and specificity. For example, an FDA-approved rapid test for HIV infection has a reported specificity of 100%, implying that any positive tests must indicate truly infected individuals, yet one of the populations in which testing is recommended is women who have had prior children and are in labor but have not yet had an HIV test during the pregnancy.[130] In 15 multiparous women, this rapid HIV test resulted in one false-positive test result in the 15 women tested, yielding a specificity of 93%,[131] and so not all pregnant women with positive tests can be assumed to be truly infected.

126. Richard A. Deyo & James N. Weinstein, *Low Back Pain*, 344 New Eng. J. Med. 363–70 (2001); Richard A. Deyo et al., *Trends, Major Medical Complications, and Charges Associated with Surgery for Lumbar Spinal Stenosis in Older Adults*, 303 JAMA 1259–65 (2010).

127. Deyo & Weinstein, *supra* note 126, at 368.

128. David F. Ransohoff & Alvan R. Feinstein, *Problems of Spectrum and Bias in Evaluating the Efficacy of Diagnostic Tests*, 299 New Eng. J. Med. 926–30 (1978).

129. Penny Whiting et al., *Sources of Variation and Bias in Studies of Diagnostic Accuracy: A Systematic Review*, 140 Annals Internal Med. 189–202 (2004).

130. Food and Drug Administration, OraQuick® Rapid HIV-1 Antibody Test, *available at* http://www.fda.gov/downloads/BiologicsBloodVaccines/BloodBloodProducts/ApprovedProducts/PremarketApprovalsPMAs/ucm092001.pdf (last visited Mar. 2, 2011).

131. *Id.*

3. Prognostic testing

Once a diagnosis has been established, additional prognostic testing may be performed to establish the extent of disease (e.g., staging of a cancer) or to monitor response to therapy. Molecular profiling of disease may not only characterize prognosis but also treatment response. In women with breast cancer, for example, finding a genetic marker called the human epidermal growth factor receptor type 2 (HER2, also called HER2/neu) gene identified patients who responded poorly to any of the standard chemotherapeutic agents and hence had a poor prognosis. Illustrative of the emerging era of pharmacogenomics, adjuvant chemotherapy combined with a monoclonal antibody in HER2-positive breast cancer patients has been found to delay progression and prolong survival.[132]

C. Judgment and Uncertainty in Medicine

1. Variation in medical care

Studies over the past several decades show substantial geographic variation in the utilization rates for medical care within small areas or local regions (e.g., a three- to fourfold variation in the use of surgical procedures such as tonsillectomy when comparing children living in adjacent areas of similar demographics)[133] and between large areas or widespread regions (e.g., a 10-fold variation in the performance of other discretionary surgical procedures such as lower extremity revascularization, carotid endarterectomy, back surgery, and radical prostatectomy).[134] Even when limiting the analysis to 77 U.S. hospitals with reputations for high-quality care in managing chronic illness, the care that patients received in their last 6 months of life varied extensively, ranging from hospital stays of 9 to 27 days (threefold variation), intensive care unit stays of 2 to 10 days (fivefold variation); and physician visits of 18 to 76 (fourfold variation), depending on the hospital at which patients received their care.[135]

Four categories of variation are recognized: (1) underuse of effective care, (2) issues of patient safety, (3) concern for preference-sensitive care, and (4) notions of supply-sensitive services.[136] Effective care refers to treatments that are known to be beneficial and that nearly all patients should receive with little influence

132. Dennis J. Slamon et al., *Use of Chemotherapy Plus a Monoclonal Antibody Against HER2 for Metastatic Breast Cancer That Overexpresses* HER2, 344 New Eng. J. Med. 783–92 (2001).

133. John Wennberg & Alan Gittelsohn, *Small Area Variations in Health Care Delivery*, 182 Science 1102–08 (1973) (hereinafter "Wennberg & Gittelsohn").

134. John D. Birkmeyer et al., *Variation Profiles of Common Surgical Procedures*, 124 Surgery 917–23 (1998).

135. John E. Wennberg et al., *Use of Hospitals, Physician Visits, and Hospice Care During Last Six Months of Life Among Cohorts Loyal to Highly Respected Hospitals in the United States*, 328 BMJ 607 (2004).

136. John E. Wennberg, *Unwarranted Variations in Healthcare Delivery: Implications for Academic Medical Centres*, 325 BMJ 961–64 (2002) (hereinafter "Wennberg").

of patient preferences, for example, use of beta blockers following myocardial infarction. The *underuse of effective care* was illustrated by one prominent study that identified 439 high-quality process measures for 30 conditions and preventive care. In assessing the use of measures that were clearly recommended (i.e., clearly beneficial), they found that only about 50% of patients received these highly recommended care processes.[137] Issues of *patient safety* refer to the execution of care and the occurrence of iatrogenic complications (i.e., complications resulting from health care interventions). The IOM estimates that hospitalized patients risk one medication error for every day they are hospitalized, resulting in an estimated 7000 deaths annually (more than from workplace injuries) at an annual cost of $3.5 billion in 2006 dollars.[138] Concern for *preference-sensitive care* refers to treatment choices that should depend on patient health goals or preferences. Prostate surgery helps relieve symptoms of an enlarged prostate (such as frequent urination, waking up at night to urinate) but carries a risk of losing sexual function. Separate from the probability of losing sexual function, in preference-sensitive care, the decision to have prostate surgery depends on how much the enlarged prostate symptoms bother the patient and on how important sexual function is to them, that is, their preferences and values.[139] Finally, *supply-sensitive services* refer to care that depends not on evidence of effectiveness or patient preferences, but rather on the availability of services. Specifically, patients living in areas with more doctors or more hospitals experience more office visits, tests, and hospitalizations.[140]

2. Evidence-based medicine

The exceptional variation in the delivery of medical care was a major factor that led to a careful reexamination of physician diagnostic strategies, therapeutic decision making, and the use of medical evidence, but it was not the only one. Other circumstances that set the stage for an intense focus on medical evidence included (1) the development of medical research, including randomized controlled trials and other observational study designs; (2) the growth of diagnostic and therapeutic interventions;[141] (3) interest in understanding medical decisionmaking and how physicians reason;[142] and (4) the acceptance of meta-analysis as a method to com-

137. Elizabeth A. McGlynn et al., *The Quality of Health Care Delivered to Adults in the United States*, 348 New Eng. J. Med. 2635–45 (2003).

138. Committee on Identifying and Preventing Medication Errors, Institute of Medicine, Preventing Medication Errors (2006); 2000 CQHCA Report, *supra* note 58.

139. Michael J. Barry et al., *Patient Reactions to a Program Designed to Facilitate Patient Participation in Treatment Decisions for Benign Prostatic Hyperplasia*, 1995 Med. Care 771–82 (1995).

140. Wennberg, *supra* note 136, at 142.

141. Cynthia D. Mulrow & K.N. Lohr, *Proof and Policy from Medical Research Evidence*, 26 J. Health Pol., Pol'y & L. 249–66 (2001) (hereinafter "Mulrow & Lohr").

142. Robert S. Ledley & Lee B. Lusted, *Reasoning Foundations of Medical Diagnosis; Symbolic Logic, Probability, and Value Theory Aid Our Understanding of How Physicians Reason*, 130 Science 9–21 (1959).

bine data from multiple randomized trials.[143] In response to the above conditions, "evidence-based medicine" gained prominence in 1992.[144] It is aptly defined as "the conscientious, explicit and judicious use of current best evidence in making decisions about the care of the individual patient. It means integrating individual clinical expertise with the best available external clinical evidence from systematic research."[145]

Evidence-based medicine contrasts with the traditional informal method of practicing based on anecdotes, applying the most recently read articles, doing what a group of eminent experts recommend, or minimizing costs.[146] Rather, it is "the use of mathematical estimates of the risks of benefit and harm, derived from high-quality research on population samples, to inform clinical decision making in the diagnosis, investigation or management of individual patients."[147] In a paper from a joint workshop held by IOM and the Agency for Healthcare Research and Quality[148] that addressed what physicians consider to be sufficient evidence to justify their clinical practice and treatment decisions, Mulrow and Lohr wrote "evidence-based medicine stresses a structured critical examination of medical research literature: relatively speaking, it deemphasizes average practice as an adequate standard and personal heuristics."[149]

3. Hierarchy of medical evidence

With the explosion of available medical evidence, increased emphasis has been placed on assembling, evaluating, and interpreting medical research evidence. A fundamental principle of evidence-based medicine (*see also* Section IV.C.5, *infra*) is that the strength of medical evidence supporting a therapy or strategy is hierarchical. When ordered from strongest to weakest, systematic review of randomized trials (meta-analysis) is at the top, followed by single randomized trials, systematic reviews of observational studies, single observational studies,

143. *See* Michael D. Green et al., Reference Guide on Epidemiology, Section VI, in this manual; Video Software Dealers Ass'n v. Schwarzenegger, 556 F.3d 950, 963 (9th Cir. 2009) (analyzing a meta-analysis of studies on video games and adolescent behavior); Kennecott Greens Creek Min. Co. v. Mine Safety & Health Admin., 476 F.3d 946, 953 (D.C. Cir. 2007) (reviewing the Mine Safety and Health Administration's reliance on epidemiological studies and two meta-analyses).

144. Evidence-Based Medicine Working Group, *Evidence-Based Medicine. A New Approach to Teaching the Practice of Medicine,* 268 JAMA 2420–25 (1992).

145. David L. Sackett et al., *Evidence Based Medicine: What It Is and What It Isn't,* 312 BMJ 71–72, 71 (1996).

146. Trisha Greenhalgh, How to Read a Paper: The Basics of Evidence-Based Medicine (3d ed. 2006).

147. *Id.* at 1.

148. Clark C. Havighurst et al., *Evidence: Its Meanings in Health Care and in Law,* 26 J. Health Pol., Pol'y & L. 195–215 (2001).

149. Mulrow & Lohr, *supra* note 141, at 253.

physiological studies, and unsystematic clinical observations.[150] An analysis of the frequency with which various study designs are cited by others provides empirical evidence supporting the influence of meta-analysis followed by randomized controlled trials in the medical evidence hierarchy.[151] Although they are at the bottom of the evidence hierarchy, unsystematic clinical observations or case reports may be the first signals of adverse events or associations that are later confirmed with larger or controlled epidemiological studies (e.g., aplastic anemia caused by chloramphenicol,[152] or lung cancer caused by asbestos[153]). Nonetheless, subsequent studies may not confirm initial reports (e.g., the putative association between coffee consumption and pancreatic cancer).[154]

Just as in laboratory experiments, evidence about the benefits and risks of medical interventions arises through repetitive observations. A single randomized controlled trial relies on hypothesis testing, specifically assuming the null hypothesis that a new drug is equivalent to the comparator (e.g., placebo). As conceived nearly 100 years ago, interpreting the trial involved calculating the likelihood of the alpha error (*p*-value) wherein the study suggests that the drug or device is beneficial but the "truth" is that it is not, that is, a false-positive study result. Similarly, a beta error (1 minus power) is the likelihood of a study finding that the drug or device is not beneficial when the "truth" is that it is, that is, a false-negative study result (Table 3).

Table 3. Analogy Between Interpreting a Diagnostic Test and a Drug Study

	Truth	
	Drug +	Drug −
Study +	Power (true positive)	α Type I error (false positive)
Study −	β Type II error (false negative)	True negative

The choice of which specific error rates to use (e.g., false positive or *p*-value or alpha of 0.05) was suppose to depend on a judgment of the relative consequences of the two errors, missing an effective drug (Type II beta error) or

150. Gordon H. Guyatt et al., Users' Guides to the Medical Literature: A Manual for Evidence-Based Clinical Practice (2d ed. 2008) (hereinafter "Guyatt"); *see also* Michael D. Green et al., Reference Guide on Epidemiology, in this manual.

151. Nikolaos A. Patsopoulos et al., *Relative Citation Impact of Various Study Designs in the Health Sciences*, 293 JAMA 2362–66 (2005).

152. W.T.W. Clarke, *Fatal Aplastic Anemia and Chloramphenicol*, 97 Can. Med. Ass'n J. 815 (1967) (hereinafter "Clarke").

153. Michael Gochfeld, *Asbestos Exposure in Buildings*, Envtl. Med. 438, 440 (1995).

154. Brian MacMahon et al., *Coffee and Cancer of the Pancreas*, 304 New Eng. J. Med. 630–33 (1981) (hereinafter "MacMahon").

considering an ineffective drug to be effective (Type I alpha error).[155] The null hypothesis, however, assumes equivalence, and so it does not provide any measure of evidence outside of the particular study (e.g., prior studies or biological mechanism or plausibility). Thus, the null hypothesis assumption necessitates abandoning the ability to measure evidence or determine "truth" from a single experiment, so that hypothesis testing is thereby "equivalent to a system of justice that is not concerned with which individual defendant is found guilty or innocent (that is, 'whether each separate hypothesis is true or false') but tries instead to control the overall number of incorrect verdicts."[156] From a Bayesian perspective, the interpretation of a new study depends on whether prior studies showed benefit or harm and on the existence of a biological mechanism or plausibility (e.g., the association between coffee consumption and pancreatic cancer was a "false-positive" result because in further testing the initial finding was not validated and there was no known plausible biological mechanism).[157]

Cumulative meta-analysis of treatments enables the accumulation of randomized trial evidence to examine trends in efficacy or risks, overcoming issues of underpowered trials that have insufficient numbers of patients enrolled to reliably detect a benefit. For example, between 1959 and 1988, 33 randomized trials with streptokinase for acute myocardial infarction involving over 35,000 patients had been published. By combining the results of each trial as they occurred, a cumulative meta-analysis found "a consistent, statistically significant reduction in total mortality" with streptokinase use by 1973.[158] In contrast, for many years, physicians used a drug called lidocaine to prevent life-threatening heart rhythm disturbances, yet none of the randomized trials of lidocaine demonstrated any benefit, and finally cumulative meta-analysis found a trend toward harm. When the results of meta-analysis were compared with comments in textbooks and review articles,

> discrepancies were detected between the meta-analytic patterns of effectiveness in the randomized trials and the recommendations of reviewers [the review article author]. Review articles often failed to mention important advances or exhibited delays in recommending effective preventive measures. In some cases, treatments that have no effect on mortality or are potentially harmful continued to be recommended by several clinical experts.[159]

155. Goodman, *supra* note 74, at 998.

156. *Id.* at 998.

157. MacMahon, *supra* note 154, at 630.

158. Joseph Lau et al., *Cumulative Meta-Analysis of Therapeutic Trials for Myocardial Infarction*, 327 New Eng. J. Med. 248–54 (1992).

159. Elliott M. Antman et al., *A Comparison of Results of Meta-Analyses of Randomized Control Trials and Recommendations of Clinical Experts: Treatments for Myocardial Infarction*, 268 JAMA 240, 240 (1992).

4. Guidelines

Clinical practice guidelines are "systematically developed statements to assist practitioner and patient decisions about appropriate health care for specific clinical circumstances."[160] Such guidelines have been widely developed and issued by medical specialty associations, professional societies, government agencies, or health care organizations.[161] To avoid biases inherent in review articles (particularly single-authored ones) and to encourage transparency and acceptance, a standard method to develop clinical practice guidelines has emerged. It involves systematically searching for and reviewing the evidence (summarizing the evidence), grading the quality of evidence for each outcome (the certainty of the recommendation), and assessing the balance of benefits versus risks (the size of the treatment effect or the strength of the recommendation).[162] Additional considerations include values and preferences (patient health goals) and costs (resource allocation) where increasing variability or uncertainty in preferences or the presence of higher costs reduces the likelihood of making a strong recommendation.[163] The number, length, and diversity of guidelines developed by various professional organizations challenge practicing physicians. An attempt to quantify guideline development found exponential growth, with 8 guidelines published in 1990, 138 in 1996, and 855 by mid-1997, including 160 that were more than 10 pages long.[164]

With this proliferation, different professional organizations may issue guidelines on the same topic, but with competing recommendations. The composition of the panel and the processes for developing guideline recommendations may differ. For example, the U.S. Preventive Services Task Force (USPSTF) is "an independent panel of non-Federal experts in prevention and evidence-based medicine and is composed of primary care providers (such as internists, pediatricians, family physicians, gynecologists/obstetricians, nurses, and health behavior specialists)."[165] In their evaluation of mammography, the USPSTF "recommends against routine screening mammography in women aged 40 to 49 years" (*see*

160. Committee to Advise the Public Health Service on Clinical Practice Guidelines, Institute of Medicine, Clinical Practice Guidelines: Directions for a New Program 8 (Marilyn J. Field & Kathleen N. Lohr, eds. 1994).

161. *See generally* Sofamor Danek Group v. Gaus, 61 F.3d 929 (D.C. Cir. 1995) (reviewing guidelines issued by the Agency for Health Care Policy and Research in light of the Federal Advisory Committee Act); Levine v. Rosen, 616 A.2d 623 (Pa. 1992) (finding that differing guidance from two groups was evidence that reasonable physicians could follow either school of thought); Michelle M. Mello, *Of Swords and Shields: The Role of Clinical Practice Guidelines in Medical Malpractice Litigation*, 149 U. Pa. L. Rev. 645 (2001).

162. David Atkins et al., *Grading Quality of Evidence and Strength of Recommendations*, 328 BMJ 1490 (2004).

163. Gordon H. Guyatt et al., *Going from Evidence to Recommendations*, 336 BMJ 1049–51 (2008).

164. Arthur Hibble et al., *Guidelines in General Practice: The New Tower of Babel?* 317 BMJ 862–63 (1998).

165. U.S. Preventive Services Task Force (USPSTF), Agency for Healthcare Research and Quality, *available at* http://www.ahrq.gov/clinic/uspstfix.htm (last visited Mar. 2, 2011).

also Section IV.D.2).[166] In contrast, based on a writing group composed of its members who are "directly responsible for performing these screening tests," the Society of Breast Imaging and the American College of Radiology recommend "annual screening from age 40" with mammography for "women at average risk for breast cancer."[167] Similarly, for prostate cancer screening, the USPSTF update "concludes that the current evidence is insufficient to assess the balance of benefits and harms of prostate cancer screening in men younger than age 75 years."[168] In the American Urological Association update, a statement panel composed of urologists, oncologists, and other physicians made two recommendations: "The decision to use PSA for the early detection of prostate cancer should be individualized. Patients should be informed of the known risks and the potential benefits" and "Early detection and risk assessment of prostate cancer should be offered to asymptomatic men 40 years of age or older who wish to be screened with an estimated life expectancy of more than 10 years."[169]

Practice guidelines provide recommendations on how to evaluate and treat patients, but because they apply to the general case, their recommendations may not apply to a particular individual patient, or some extrapolation may be required, particularly when multiple diseases exist, as they frequently do in the elderly,[170] or when treatment entails competing risks. For example, anticoagulation is generally recommended for patients with atrial fibrillation (an abnormal heart rhythm disturbance) to prevent blood clots that could cause a stroke, yet anticoagulation can also lead to life-threatening bleeding; therefore, for individual patients, physicians must weigh the risk of developing clots versus the risk of bleeding. Consequently, guidelines typically include statements such as "clinical or policy decisions involve more considerations than this body of evidence alone. Clinicians and policymakers should understand the evidence but individualize decision making to the specific patient or situation."[171] Some physicians who rely on personal style, review articles, and colleagues to influence their clinical practice have been concerned with how guidelines affect clinical autonomy and health care costs.[172]

166. U.S. Preventive Services Task Force, *Screening for Breast Cancer: U.S. Preventive Services Task Force Recommendation Statement*, 151 Annals Internal Med. 716–26 (2009).

167. Carol H. Lee et al., *Breast Cancer Screening with Imaging: Recommendations from the Society of Breast Imaging and the ACR on the Use of Mammography, Breast MRI, Breast Ultrasound, and Other Technologies for the Detection of Clinically Occult Breast Cancer*, 7 J. Am. C. Radiology 18–27 (2010).

168. U.S. Preventive Services Task Force, *Screening for Prostate Cancer: U.S. Preventive Services Task Force Recommendation Statement*, 149 Annals Internal Med. 185–91 (2008).

169. American Urological Association, Prostate-Specific Antigen Best Practice Statement (rev. 2009), *available at* http://www.auanet.org/content/media/psa09.pdf (last visited Mar. 2, 2011).

170. Cynthia M. Boyd et al., *Clinical Practice Guidelines and Quality of Care for Older Patients with Multiple Comorbid Diseases: Implications for Pay for Performance*, 294 JAMA 716–24 (2005).

171. U.S. Preventive Services Task Force, *Screening for Carotid Artery Stenosis: U.S. Preventive Services Task Force Recommendation Statement*, 147 Annals Internal Med. 854–59 (2007).

172. Sean R. Tunis et al., *Internists' Attitudes About Clinical Practice Guidelines*, 120 Annals Internal Med. 956–63 (1994).

However, just as clinicians have been reluctant to apply guidelines in practice, courts have generally been slow to apply them in deciding cases.[173] There are political and legal issues that can arise with the development of guidelines.[174] Political sensitivities, conflicts of interest, and potential lawsuits often silence otherwise innovative and potentially useful guidelines. In 2006, the Connecticut Attorney General launched an antitrust suit against the Infectious Disease Society of America (IDSA) after IDSA promulgated guidelines recommending against the use of long-term antibiotics for the treatment of "chronic Lyme disease (CLD)."[175] Although the Centers for Disease Control and Prevention and the Food and Drug Administration (FDA) findings seemed to concur with IDSA's guidelines, a strong lobby representing patients afflicted with CLD and the physicians who treated them colored the Attorney General's decision to file suit.[176] Organizations can violate antitrust laws if their guideline-setting process is an unreasonable attempt to advance their members' economic interests by suppressing competition. IDSA settled without admitting guilt, but it is clear that organizations must be careful to maintain transparency in the guideline development process.[177]

Besides clinical practice guidelines, IOM defines other types of statements: (1) *medical review criteria* are systematically developed statements that can be used to assess the appropriateness of specific health care decisions, services, and outcomes; (2) *standards of quality* are authoritative statements of minimum levels of acceptable performance or results, excellent levels of performance or results, or the range of acceptable performance or results; and (3) *performance measures* are methods or instruments to estimate or monitor the extent to which the actions of a health care practitioner or provider conform to practice guidelines, medical review criteria, or standards of quality.

5. Vicissitudes of therapeutic decisionmaking

Medical decisionmaking often involves complexity, uncertainty, and tradeoffs[178] because of unique genetic factors, lifestyle habits, known conditions, medication histories, and ambiguity about possible diagnoses, test results, treatment benefits,

173. Arnold J. Rosoff, *Evidence-Based Medicine and the Law: The Courts Confront Clinical Practice Guidelines*, 26 J. Health Pol., Pol'y & L. 327–68 (2001).

174. One element in the near demise of the Agency for Health Care Policy and Research was a political audience receptive to complaints from an association of back surgeons who disagreed with the AHCPR practice guideline conclusions regarding low back pain. B.H. Gray et al., *AHCPR and the Changing Politics of Health Services Research*, Health Affairs, Suppl. Web Exclusives W3-283-307 (June 2003).

175. John D. Kraemer & Lawrence O. Gostin, *Science, Politics, and Values: The Politicization of Professional Practice Guidelines*, 301 JAMA 665–67 (2009).

176. *Id* at 666.

177. *Id* at 666.

178. John P.A. Ioannidis & Joseph Lau, *Systematic Review of Medical Evidence*, 12 J.L. & Pol'y 509–35 (2004).

and therapeutic harms. Given inherent diagnostic and therapeutic uncertainty, physicians often make treatment decisions in the face of uncertainty.

Donald Schön argued that regardless of the professional field, "An artful practice of the unique case appears anomalous when professional competence is modeled in terms of application of established techniques to recurrent events" and that specialization "fosters selective inattention to practical competence and professional artistry."[179] In the case of a patient with peanut allergies and heart disease, allergy guidelines recommend avoiding beta blockers, but heart disease guidelines recommend beta blockers because they have been shown to prolong life in patients with heart disease. An allergist would recommend against taking a beta blocker, yet a cardiologist would recommend taking it.[180]

Well-performed randomized trials provide the least biased estimates of treatment benefit and harm by creating groups with equivalent prognoses. Sticking strictly to the scientific evidence, some physicians may limit their use of medications to the specific drug at the specific doses found to be beneficial in such trials. Others may assume class effects until proven otherwise. Still others may consider additional factors such as out-of-pocket costs for patients or patient preferences. When physicians evaluate patients who might benefit from a treatment but who would have been excluded from the study in which the benefit was demonstrated, they must weigh the risks and benefits in the absence of definitive evidence of benefit or of harm. Indeed, because few medical recommendations are based on randomized trials (the least biased level of evidence) physicians frequently and necessarily face uncertainty in making testing and treatment decisions and tradeoffs: Very few treatments come without some risk, and in many disciplines, clear evidence of efficacy and risks of treatment are lacking. In cardiology (one of the better studied areas of medical care), nearly one-half of guideline recommendations are based on expert opinion, case studies, or standards of care.[181]

Applying well-designed studies to populations of patients represents another problem. The Randomized Aldactone Evaluation Study demonstrated that spironolactone reduced mortality and hospitalizations for heart failure and improved quality of life with minimal risk of seriously high levels of potassium (hyperkalemia).[182] Published in a prominent medical journal, prescriptions for spironolactone rose quickly because of familiarity with the medication and the poor prognosis of patients with heart failure. As opposed to the study population, however, community individuals were older, more frequently women, often

179. Donald A. Schön, The Reflective Practitioner: How Professionals Think in Action, at vii (1983).

180. John A. TenBrook et al., *Should Beta-Blockers Be Given to Patients with Heart Disease and Peanut-Induced Anaphylaxis? A Decision Analysis*, 113 J. Allergy & Clin. Immunol. 977–82 (2004).

181. Pierluigi Tricoci et al., *Scientific Evidence Underlying the ACC/AHA Clinical Practice Guidelines*, 301 JAMA 831–41 (2009).

182. Bertram Pitt et al., *The Effect of Spironolactone on Morbidity and Mortality in Patients with Severe Heart Failure. Randomized Aldactone Evaluation Study Investigators*, 341 New Eng. J. Med. 709–17 (1999).

had absolute or relative contraindications to treatment, and had not had tests of their heart function to establish the indication to treat or of their potassium level and kidney function to determine their risk for high potassium levels from treatment.[183] These factors increased the risk that spironolactone therapy in these patients might lead to high potassium levels that could be life-threatening. Indeed, hospitalizations per 1000 patients for high potassium rose from 2.4 in 1994 to 11.0 in 2001, resulting in an estimated 560 additional hospitalizations for high potassium and 73 additional hospital deaths in older patients with heart failure in Ontario.[184] Criteria for entry into randomized trials of drugs typically exclude individuals with concomitant medication use, medical comorbidities, and female gender, and they may limit participation by socioeconomic status or race and ethnicity, thereby limiting the ability to generalize the results of a trial to the clinical population being treated.[185] Physicians refer to randomized controlled studies as assessments of drug "efficacy" in restricted patient populations, whereas treatment in general clinical populations are often referred to as "effectiveness" studies.

To be sufficiently powered to demonstrate statistical significance,[186] randomized controlled trials usually require high event rates, prolonged followup, or large numbers of patients. Because of impracticality, expense, and the time period needed to obtain long-term outcomes, these trials may often choose a surrogate marker that is associated with a clinically important event or with survival. For example, statins were approved on the basis of their safety and efficacy in lowering cholesterol but were only demonstrated to improve survival in patients with known coronary heart disease years later.[187] Fast-track approval of new drugs for HIV infection was based on safety and efficacy in reducing viral levels (as a surrogate or substitute outcome measure felt to be related to survival) as opposed to demonstration of improved survival.

On the other hand, in the late 1970s, patients with frequent extra heartbeats (ventricular premature contractions) following a heart attack had an increased risk for sudden death. On that basis, those in the then-emerging field of cardiac electrophysiology believed that reducing ventricular premature beats (as a surrogate outcome measure) would decrease subsequent sudden cardiac death. In early randomized controlled trials, oral antiarrhythmic drugs such as encainide and flecainide were approved by FDA on the basis of their ability to suppress these extra heartbeats in patients who had had a myocardial infarction. Years after

183. Dennis T. Ko et al., *Appropriateness of Spironolactone Prescribing in Heart Failure Patients: A Population-Based Study*, 12 J. Cardiac Failure 205–10 (2006).

184. David N. Juurlink et al., *Rates of Hyperkalemia After Publication of the Randomized Aldactone Evaluation Study*. 351 New Eng. J. Med. 543–51 (2004).

185. Harriette G.C. Van Spall et al., *Eligibility Criteria of Randomized Controlled Trials Published in High-Impact General Medical Journals: A Systematic Sampling Review*, 297 JAMA 1233–40 (2007).

186. *See* Michael D. Green et al., Reference Guide on Epidemiology, in this manual.

187. *Randomised Trial of Cholesterol Lowering in 4444 Patients with Coronary Heart Disease: The Scandinavian Simvastatin Survival Study* (4S), 344 Lancet 1383–89 (1994).

approval of these drugs, however, a randomized controlled trial designed to demonstrate a survival benefit of these drugs was discontinued after only 10 months because of a statistically significant higher rate of mortality in patients receiving the drugs. Although these drugs effectively suppressed the extra heartbeats, the study found that they also increased the likelihood of fatal heart rhythm disturbances.[188]

Prior to approval by FDA, drugs and devices must undergo Phase 1, 2, and 3 clinical trials to demonstrate safety and efficacy. Following preliminary chemical discovery, toxicology, and animal studies, Phase 1 studies examine the safety of new drugs in healthy individuals. Phase 2 studies involve varying drug doses in individuals with the disease to explore efficacy and responses and adverse effects. Based on the dose or doses identified in Phase 2, a Phase 3 study examines drug response in a larger number of patients to again determine safety and efficacy in the hope of getting a new drug approved for sale by regulatory authorities. However, because fewer than 10,000 individuals have usually received the drug during all of these trials, uncommon adverse outcomes may not become apparent until usage is broadened and extended. For example, depending on dosage, between 1 in 24,200 and 1 in 40,500 patients who received the antibiotic chloramphenicol[189] developed fatal aplastic anemia (in which the bone marrow no longer produces any blood cells). This adverse effect was discovered only in the 1960s after chloramphenicol was initially considered safe and had been widely used during the 1950s.[190]

For all approved drug and therapeutic biological products, FDA has managed postmarketing safety surveillance since 1969 through the Adverse Event Reporting System. Health care professionals, including physicians, pharmacists, nurses, and others, and consumers, including patients, family members, lawyers, and others, are expected to report adverse events and medication errors. It is a voluntary system with the following limitations: (1) uncertainty that the drug caused the reported event, (2) no requirement for proof of a causal relationship between product and event, (3) insufficient detail to evaluate events, (4) incomplete reporting of all adverse events, and (5) inability to determine the incidence of an adverse events because the actual number of patients receiving a product and the duration of use of those products are unknown.

In 1999, rofecoxib (Vioxx), a Cox-2 selective nonsteroidal anti-inflammatory drug, was approved for pain relief in part on the basis of studies that suggested that it induced less gastrointestinal bleeding than other nonsteroidal anti-inflammatory drugs. In 2004, the manufacturer announced a voluntary worldwide withdrawal

188. *Preliminary Report: Effect of Encainide and Flecainide on Mortality in a Randomized Trial of Arrhythmia Suppression After Myocardial Infarction: The Cardiac Arrhythmia Suppression Trial (CAST) Investigators*, 321 New Eng. J. Med. 406–12 (1989).

189. Two pre-*Daubert* cases from the Fifth Circuit dealt with product liability suits against the manufacturer: Christophersen v. Allied-Signal Corp., 939 F.2d 1106 (5th Cir. 1991); Osburn v. Anchor Labs., 825 F.2d 908 (5th Cir. 1987). Clarke, *supra* note 152, at 515.

190. Clarke, *supra* note 152, at 815.

of rofecoxib when a prospective study confirmed that the drug increased the risk of myocardial infarctions (heart attacks) and stroke with chronic use.[191]

This section demonstrates some of the issues that physicians grapple with in treatment decisions. Some generally avoid using new drugs until sufficient experience with the medication provides an opportunity for unknown adverse effects to emerge following drug approval. Others may be quick to adopt new drugs, especially drugs perceived to have improved safety or efficacy such as through a novel mechanism of action. By withholding use of new drugs, more conservative physicians may avoid the occurrence of unforeseen adverse consequences, but they may also delay the use of new drugs that may benefit their patients. The converse may occur, of course, with physicians who are early adopters of new drugs, tests, or technologies.

Even in a randomized trial in which a drug is found to be beneficial, some patients who received the drug may have been harmed, emphasizing the need to individualize the balancing of risks and benefits and explaining in part why some physicians may not adhere to guideline recommendations. The fundamental dilemma articulated by Bernard in 1865 still haunts the clinician: The response of the "average" patient to therapy is not necessarily the response of the patient being treated.[192] Indeed, the average results of clinical trials do not apply to all patients in the trial. Even with well-defined inclusion and exclusion criteria, variation in outcome risk and, therefore, treatment benefit exists so that even "typical" patients included in the trial may not be likely to get the average benefits.

The Global Utilization of Streptokinase and tPA for Occluded Coronary Arteries Trial is a case in point. The trial suggested that accelerated tissue plasminogen accelerator (tPA) reduced mortality from acute myocardial infarction, with the tradeoff being an increased risk of bleeding from tPA.[193] In a reanalysis of this study, most (85%) of the survival benefit of tPA accrued to half of the patients (those at highest risk of dying from their heart attack). Some patients with very low risk of dying from their heart attack who received tPA likely were harmed because their risk of intracranial hemorrhage exceeded the benefit.[194] In practice then, even in a randomized controlled trial demonstrating survival benefit, *on average*, those benefits may not accrue to every patient in that trial that received treatment. Therefore, to optimize treatment decisions, physicians attempt to individualize treatment decisions based on their assessment of the patient's risk versus benefit. Even then, physicians may be reluctant to administer a medication such

191. *See generally In re* Vioxx Prods. Liab. Litig., 360 F. Supp. 2d 1352 (J.P.M.L. 2005).

192. Salim Yusuf et al., *Analysis and Interpretation of Treatment Effects in Subgroups of Patients in Randomized Clinical Trials*, 266 JAMA 93–98 (1991) (hereinafter "Yusuf").

193 *An International Randomized Trial Comparing Four Thrombolytic Strategies for Acute Myocardial Infarction. The GUSTO Investigators*, 329 New Eng. J. Med. 673–82 (1993).

194. David M Kent et al., *An Independently Derived and Validated Predictive Model for Selecting Patients with Myocardial Infarction Who Are Likely to Benefit from Tissue Plasminogen Activator Compared with Streptokinase*, 113 Am. J. Med. 104–11 (2002).

as tPA that can cause severe harm such as an intracranial hemorrhage. A single clinical experience with a patient who bled when given tPA might well color their judgment about the benefits of the treatment.

A fundamental principle of evidence-based medicine is that "Evidence alone is never sufficient to make a clinical decision."[195] Nearly all medical decisions involve some tradeoff between a benefit and a risk. Besides the options and the likelihood of the outcomes, patient preferences about the resulting outcomes should affect care choices, especially when there are tradeoffs such as a risk of complications or dying from a procedure or treatment versus some benefit such as living longer (provided the patient survives the short-term risk of the procedure) or improving their quality of life (relieving symptoms). Besides individualizing risk and benefit assessments, physicians may also deviate from guideline recommendations ("warranted variation") because of a particular patient's higher risk of adverse events or lower likelihood of benefit or because of patient preferences for the alternative outcomes, such as when risks occur at different times. For example, given a hypothetical choice between living 25 years for certain or a 50:50 chance of living 50 years or dying immediately, most individuals choose the 25 years for certain. Although both options yield, on average, 25 years, most individuals are risk averse and prefer to avoid the near-term risk of dying. When interviewed, some patients with "operable" lung cancer were quite averse to possible immediate death from surgery, and so, based on their preferences, these patients probably would opt for radiation therapy despite its poorer long-term survival.[196]

Besides risk aversion, some treatments may improve quality of life but place patients at risk for shortened life expectancy, and some patients may be willing to trade off quality of life for length of life. When presented with laryngeal cancer scenarios, some volunteer research subjects chose radiation therapy over surgery to preserve their voices despite a reduced likelihood of future survival. "These results suggest that treatment choices should be made on the basis of patients' attitudes toward the quality as well as the quantity of survival."[197]

To illustrate this principle, a National Institutes of Health Consensus Conference recommended breast-conserving surgery when possible for women with Stage I and II breast cancer[198] because well-designed studies with long-term followup on thousands of women demonstrated equivalence of lumpectomy and radiation therapy or mastectomy for survival and disease-free survival (being alive without breast cancer recurrence). In one study, lumpectomy and radiation appeared to have a lower risk of breast cancer recurrence with 5 women reported to have had breast cancer recurrences following lumpectomy and radiation versus

195. Guyatt, *supra* note 150, at 8; *see also supra* Section IV.C.3.

196. McNeil, *supra* note 63, at 986.

197. *Id.* at 982.

198. *NIH Consensus Conference: Treatment of Early-Stage Breast Cancer*, 265 JAMA 391–95 (1991).

10 women after mastectomy.[199] However, breast cancer that recurred in the breast that had been operated on was censored (i.e., deliberately not considered in the statistical analysis).[200] When including these censored cancer recurrences, 20 breast cancer recurrences occurred after lumpectomy versus 10 after mastectomy, and so lumpectomy actually had a higher overall risk of recurrence.[201] As expressed by one woman, "The decision about treatment for breast cancer remains an intensely personal one. The mastectomy I chose . . . felt a lot less invasive than the prospect of six weeks of daily radiation, not to mention the 14% risk of local recurrence."[202] In such a case, patient preferences[203] regarding tradeoffs involving breast preservation and increased risk of breast cancer recurrence or the need for radiation therapy associated with lumpectomy may play an important role in determining the optimal decision for any particular patient.[204]

D. Informed Consent

1. Principles and standards

Medical informed consent is an ethical, moral, and legal responsibility of physicians.[205] It is guided by four ethical principles: autonomy, beneficence, nonmalfeasance, and justice.[206] Autonomy refers to informed, rational decisionmaking after unbiased and thoughtful deliberation. Beneficence represents the moral obligation of physicians to act for the benefit of patients.[207] These two principles place physicians in conflict because they wish to provide the care they believe is best for the patient, but because that care usually involves some risk or cost, physicians also recognize that patient preferences may affect their recommendation. In a study examining the incidence of erectile dysfunction with use of a beta blocker medication known to be beneficial, heart disease patients were (1) blinded

199. Joan A. Jacobson et al., *Ten-Year Results of a Comparison of Conservation with Mastectomy in the Treatment of Stage I and II Breast Cancer,* 332 New Eng. J. Med. 907–11 (1995) (hereinafter "Jacobson").

200. Bernard Fisher et al., *Eight-Year Results of a Randomized Clinical Trial Comparing Total Mastectomy and Lumpectomy With or Without Irradiation in the Treatment of Breast Cancer,* 320 New Eng. J. Med. 822–28 (1989); Jacobson, *supra* note 199, at 998.

201. Jacobson, *supra* note 199, at 999.

202. Karen Sepucha et al., *Policy Support for Patient-Centered Care: The Need For Measurable Improvements In Decision Quality*, Health Affairs Supp. Web Exclusives VAR 54, VAR 62 (2004).

203. Proctor & Gamble Pharm., Inc. v. Hoffman-LaRoche, Inc., 2006 WL 2588002, at *10 (S.D.N.Y. 2006) (detailing the testimony of a physician stating that, in addition to efficacy, he considers patient preferences when determining treatment for osteoporosis).

204. Jerome P. Kassirer, *Adding Insult to Injury. Usurping Patients' Prerogatives*, 308 New Eng. J. Med. 898–901 (1983) (hereinafter "1983 Kassirer").

205. Timothy J. Paterick et al., *Medical Informed Consent: General Considerations for Physicians,* 83 Mayo Clinic Proc. 313–19 (2008) (hereinafter "Paterick").

206. Jaime S. King & Benjamin W. Moulton, *Rethinking Informed Consent: The Case for Shared Decision Making,* 32 Am. J.L. & Med. 429–501 (2006) (hereinafter "King & Moulton").

207. *Id.* at 435.

to the drug, (2) informed of the drug name only, or (3) informed about its erectile dysfunction adverse effect. Among those blinded, 3.1% developed erectile dysfunction compared with 15.6% of those given the drug name and 31.2% of those informed about adverse effects, showing that being informed increased the risk for adverse effects and might deprive patients of benefit from a drug because they stop taking it.[208] Physicians must balance the desire to provide beneficial care with the obligation to promote autonomous decisions by informing patients of potential adverse effects or tradeoffs.

State jurisdictions differ in their standards for disclosure, with half adopting the physician or professional standard (the information that other local physicians with similar skill levels would provide) and the other half adopting the patient or materiality standard (the information that a reasonable patient would deem important in decisionmaking).[209] The informed consent process involves the disclosure of alternative treatment options including no treatment and the risks and benefits associated with each alternative. Discussion should include severe risks and frequent risks, but the courts have not provided explicit guidance about what constitutes sufficient severity or frequency. Patients should be considered by the court to be competent and should have the capacity to make decisions (understanding choices, risks, and benefits). The decision should be voluntary—of free mind and free will, without coercion or manipulation. The language used should be understandable to the patient, and treatment should not proceed unless the physician believes the patient understands the options and their risks and benefits.

Patients may withdraw consent or refuse treatment. Such an action should engender additional discussion, and documentation may include the completion of a withdrawal-of-consent form. In certain situations, exceptions to medical consent may arise in emergencies, when the treatment is recognized by prudent physicians to involve no material risk to patients and when the procedure is unanticipated and not known to be necessary at the time of consent.[210]

The Merenstein case described an unpublished trial in which, during his residency, Dr. Merenstein examined a highly educated man. The examination included a discussion of the relevant risks and benefits regarding prostate cancer screening using the prostate-specific antigen (PSA) test based on recommendations from the U.S. Preventive Services Task Force, the American College of Physicians–American Society of Internal Medicine, the American Medical Association, the American Urological Association, the American Cancer Association, and the American Academy of Family Physicians. Dr. Merenstein testified that the patient declined the test because of the high false-positive rate, the risk of treatment-related adverse effects, and the low risk of dying from prostate cancer.

208. Antonello Silvestri et al., *Report of Erectile Dysfunction After Therapy with Beta-Blockers Is Related to Patient Knowledge of Side Effects and Is Reversed by Placebo*, 24 Eur. Heart J. 1928, 1928 (2003).

209. King & Moulton, *supra* note 206, at 430.

210. Paterick, *supra* note 205, at 315.

Another physician seeing the same patient subsequently ordered a PSA without any patient discussion. The PSA was high and the patient was diagnosed with incurable advanced prostate cancer. The plaintiff's attorney argued that despite the guidelines above, the standard of care in Virginia was to order the blood test without discussion, based on four physician witnesses. The jury ruled in favor of the plaintiff.[211]

To illustrate the importance of patient preferences, a woman with breast cancer described her experience: "But as the surgeon diagramed incision points on my chest with a felt-tip pen, my husband asked a question: Is it really necessary to transfer this back muscle? The doctor's answer shocked us. No, he said, he could simply operate on my chest. That would cut surgery and recovery time in half. He had planned the more complicated procedure because he thought it would have the best cosmetic result. 'I assumed that's what you wanted.'"[212] Instead the woman preferred the less invasive approach that shortened her recovery time.

In the research setting, a randomized trial with and without informed consent demonstrated that the process of getting informed consent altered the effect of a placebo when given to patients with insomnia. The first patient of each pair was randomized to no informed consent and the second to informed consent. Out of 56 patients randomized to informed consent, 26 declined to participate in the study (the patients without informed consent had no choice and were unaware of their participation in a study). The informed consent process created a "biased" group because the age and gender for those who declined participation differed significantly from those who did agree to be included in the study. The hypnotic activity of placebo was significantly higher without informed consent, and adverse events were found more commonly in the group receiving informed consent. The study suggests that the process of getting informed consent introduced biases in the patient population and affected the efficacy and adverse effects observed in this clinical trial, thereby potentially affecting the general applicability of any findings involving informed consent.[213]

Besides physicians, patients may get health information from the Internet, family, friends, and the media (newspapers, magazines, television). Among Internet users, 80% had searched for information on at least 1 of 15 major health topics but use varied from 62% to 89% by age, gender, education, or race/ethnicity.[214] Conducted between November 2006 and May 2007, a cross-sectional national survey of U.S. adults who had made a medical decision found that Internet use

211. King & Moulton, *supra* note 206, at 432–34; Daniel Merenstein, *A Piece of My Mind: Winners and Losers*, 291 JAMA 15–16 (2004).

212. Julie Halpert, *Health: What Do Patients Want?* Newsweek, Apr. 28, 2003, at 63–64.

213. R. Dahan et al., *Does Informed Consent Influence Therapeutic Outcome? A Clinical Trial of the Hypnotic Activity of Placebo in Patients Admitted to Hospital*, 293 Brit. Med. J. Clin. Res. Ed. 363–64 (1986).

214. Pew Internet, Health Topics, http://pewinternet.org/Reports/2011/HealthTopics.aspx (last visited Feb. 12, 2011).

averaged 28% but varied from 17% for breast cancer screening to 48% for hip/knee replacement among those 40 years of age and older.[215] However, even among Internet users, health care providers were felt to be the most influential source of information for medical decisions, followed by the Internet, family and friends, and then media.

2. *Risk communication*

Multiple health outcomes may result from alternative treatment choices, and how patients feel about the relative importance of those outcomes varies.[216] When patients with recently diagnosed curable prostate cancer were presented with 93 possible questions that might be important to patients like themselves, 91 of the questions were cited as relevant to at least one patient.[217] Communication skills should include patient problem assessment (appropriate questioning techniques, seeking patient's beliefs, checking patient's understanding of the problem); patient education and counseling (eliciting patient's perspective, providing clear instructions and explanations, assessing understanding); negotiation and shared decisionmaking (surveying problems and delineating options, arriving at mutually acceptable solutions); relationship development and maintenance (encouraging patient expression, communicating a supportive attitude, explaining any jargon, and using nonverbal behavior to enhance communication).[218]

Certain forms of risk communication, however, may be confusing and should be avoided: "single event probabilities, conditional probabilities (such as sensitivity and specificity), and relative risks."[219] An example of a single-event probability would be the statement that a particular medication results in a 30% to 50% chance of developing erectile dysfunction.[220] Although physicians are referring to patients, patients may misinterpret this as referring to their own sexual encounters and having an erectile dysfunction problem in 30% to 50% of their sexual encounters. The preferred natural frequency statement would be "out of 100 people like you taking this medication, 30 to 50 of them experience erectile dysfunction." The natural frequency statement specifies a reference class, thereby reducing misunderstanding.[221]

215. Mick P. Couper et al., *Use of the Internet and Ratings of Information Sources for Medical Decisions: Results from the DECISIONS Survey*, 30 Med. Decision Making 106S–14S (2010).

216. 1983 Kassirer, *supra* note 203, at 889.

217. Deb Feldman-Stewart et al., *What Questions Do Patients with Curable Prostate Cancer Want Answered?* 20 Med. Decision Making 7–19 (2000).

218. Michael J. Yedidia et al., *Effect of Communications Training on Medical Student Performance*, 290 JAMA 1157–65 (2003).

219. Gerd Gigerenzer & Adrian Edwards, *Simple Tools for Understanding Risks: From Innumeracy to Insight*, 327 BMJ 741–44 (2003).

220. Gigerenzer, *supra* note 87, at 4.

221. *Id.* at 4; *see also* Section IV.A.2.

Regarding relative risk, consider a statement that taking a cholesterol-lowering medication reduces the risk of dying by 22%.[222] This may be misinterpreted as saying that out of 1000 patients with high cholesterol, 220 of them can avoid dying by taking cholesterol-lowering medications. The actual data show that 32 deaths occur among 1000 patients taking the medication, and 41 deaths occur among 1000 patients taking the placebo. The relative risk reduction equals 9 divided by 41. A preferred way to express the benefit would be the absolute risk reduction (the difference between 41 and 32 deaths in 1000 patients), or to say that in 1000 people like you with high cholesterol, taking a cholesterol medication for 5 years helps 9 of them avoid dying.[223] Calculating an odds ratio, the cholesterol-lowering medication reduces the odds of dying by 23%; notice that neither the relative risk nor the odds ratio characterizes the number of events without treatment and that the odds ratio always magnifies the risk or benefit when compared with the relative risk. To illustrate further, a relative risk reduction of 20% has very different absolute risk reductions depending on the number of events without treatment. If 20 of 100 patients without treatment would die, then the absolute risk reduction is 4 of 100 or 4% (20% times 20), but if 20 of 100,000 patients without treatment would die, then the absolute reduction is 4 of 100,000 or 0.004%. The number needed to treat is an additional form of risk communication popularized as part of evidence-based medicine to account for the risk without treatment. It is the reciprocal of the absolute risk difference or 1 divided by the quantity 9 lives saved per 1000 ($1 \div (9/1000)$) treated with cholesterol medications in the above example. Therefore 111 patients need to be treated with a cholesterol medication for 5 years to save one of them, or in the illustrative example, with a relative risk reduction of 20%, either 25 or 25,000 would need to be treated to save 1 patient.

In the analysis of mammography for the U.S. Preventive Services Task Force, the number needed to be invited (NNI) for screening to avoid one breast cancer death was 1904 for 39- to 49-year-olds, 1339 for 50- to 59-year-olds, and 377 for 60- to 69-year-olds.[224] To account for possible harm, there is a corresponding determination of the number needed to harm (NNH) that is calculated in the same manner. Considering breast biopsy as a morbidity, 5 women need to undergo breast biopsy for every one woman diagnosed with breast cancer for 39- to 49-year-olds, and the corresponding numbers are 3 for women ages 50 to 59 and 2 for women ages 60 to 69 years old.[225] Estimates of overdiagnosis ranged mostly from 1% to 10%, and so, out of 100 women diagnosed with breast cancer from screening, 1 to 10 of them undergo treatment for a cancer that would never have caused any mortality.[226] Clearly no one can tell if any particular woman has

222. Gigerenzer, *supra* note 87, at 34.

223. *Id.* at 34–35.

224. Heidi D. Nelson et al., *Screening for Breast Cancer: An Update for the U.S. Preventive Services Task Force*, 151 Annals Internal Med. 727–37 (2009).

225. *Id.* at 732.

226. *Id.* at 731–732

been overdiagnosed because this is unobservable.[227] The estimated extent of overdiagnosis requires estimating mortality reductions in a screened population compared with an unscreened population over a long period. The difference between the two groups provides an estimate of the extent of overdiagnosis.

To summarize the evidence, "Mammography does save lives, more effectively among older women, but does cause some harm. Do the benefits justify the risks? The misplaced propaganda battle seems to now rest on the ratio of the risks of saving a life compared with the risk of overdiagnosis, two very low percentages that are imprecisely estimated and depend on age and length of followup."[228] In the USPSTF recommendations for mammography in 40- to 49-year-olds, the focus has been on the first part of their statement "The USPSTF recommends against routine screening mammography in women aged 40 to 49 years." Although screening has demonstrated benefits, in their view, the benefits of screening do not sufficiently and clearly outweigh the potential harms to make a recommendation that all women 40 to 49 years old have routine screening mammography from a public health or population perspective. Oft neglected, the USPSTF in their immediately subsequent sentence recognizes that individual preferences should affect the care that patients receive: "The decision to start regular, biennial screening mammography before the age of 50 years should be an individual one and take patient context into account, including the patient's values regarding specific benefits and harms."[229] The recommendation recognizes that depending on their experiences, values, and preferences, some women may seek the benefit in reducing breast cancer deaths and others may prefer to avoid possible morbidity (breast biopsy and worry) and potential overdiagnosis and overtreatment.

3. Shared Decisionmaking

The "professional values of competence, expertise, empathy, honesty, and commitment are all relevant to communicating risk: Getting the facts right and conveying them in an understandable way are not enough."[230] Shared and informed decisionmaking has emerged as one part of patient care. It distinguishes "problem solving" that identifies one "right" course that leaves little room for patient involvement from "decisionmaking" in which several courses of action may be reasonable and in which patient involvement should determine the optimal choice. In such cases, health care choices depend not only on the likelihood of alternative outcomes resulting from each strategy but also on the patient preferences for possible outcomes and their attitudes about risk taking to improve future survival or quality

227. Klim McPherson, *Screening for Breast Cancer—Balancing the Debate,* 341 BMJ 234–35 (2010).

228. *Id.* at 234.

229. U.S. Preventive Services Task Force, *Screening for Breast Cancer: U.S. Preventive Services Task Force Recommendation Statement*, 151 Annals Internal Med. 716, 716 (2009).

230. Adrian Edwards, *Communicating Risks*, 327 BMJ 691–92 (2003).

of life and the timing of that risk whether the risk occurs now or in the future.[231] Informed decisionmaking occurs

> when an individual understands the nature of the disease or condition being addressed; understands the clinical service and its likely consequences, including risks, limitations, benefits, alternatives, and uncertainties; has considered his or her preferences as appropriate; has participated in decision making at a personally desirable level; and either makes a decision consistent with his or her preferences and values or elects to defer a decision to a later time.[232]

Shared decisionmaking occurs "when a patient and his or her healthcare provider(s), in the clinical setting, both express preferences and participate in making treatment decisions."[233]

To assist with shared decisionmaking, health decision aids have been developed to help patients and their physicians choose among reasonable clinical options together by describing the "benefits, harms, probabilities, and scientific uncertainties."[234] In 2007, the legislature in the state of Washington became the first to establish and recognize in law a role for shared decisionmaking in informed consent.[235] The bill goes on to encourage the development, certification, use, and evaluation of decision aids. The consent form provides written documentation that the consent process occurred, but the crux of the medical consent process is the discussion that occurs between a physician and a patient. The physician shares his or her medical knowledge and expertise and the patient shares his or her values (health goals) and preferences. It is an opportunity to strengthen the patient–physician relationship through shared decisionmaking, respect, and trust.

V. Summary and Future Directions

Having sequenced the human genome, medical research is poised for exponential growth as the code for human biology (genomics) is translated into proteins (proteomics) and chemicals (metabolomics) to identify molecular pathways that lead to disease or that promote health. With advances in medical technologies in diagnosis and preventive and symptomatic treatment, the practice of medicine will be profoundly altered and redefined. For example, consider lymphoma, a blood cancer that used to be classified simply by appearance under the microscope as

231. Michael J. Barry, *Health Decision Aids to Facilitate Shared Decision Making in Office Practice*, 136 Annals Internal Med. 127–35 (2002).

232. Peter Briss et al., *Promoting Informed Decisions About Cancer Screening in Communities and Healthcare Systems*, 26 Am. J. Preventive Med. 67, 68 (2004).

233. *Id.* at 68.

234. Annette M. O'Connor et al., *Risk Communication in Practice: The Contribution of Decision Aids*, 327 BMJ 736, 736 (2003).

235. Bridget M. Kuehn, *States Explore Shared Decision Making*, 301 JAMA 2539–41 (2009).

either Hodgkin's or non-Hodgkin's lymphoma. As science has evolved, it is now further classified by cellular markers that identify the underlying cancer cells as one of two cells that help with immunity (protecting the body from infection and cancer): T cells or B cells. Current research is attempting to characterize those cells further by identifying underlying genetic and cellular markers and pathways that may distinguish these lymphomas and provide potential therapeutic targets. The growth in the research enterprise, both basic science and clinical translational (the translation of bench research to the bedside or basic science research into novel treatments or diagnostics), has greatly expanded research capacity to generate scientific research of all types.

With greatly expanded knowledge, research and specialization, judgments about admissibility and about what constitutes expertise become increasingly difficult and complex. The sifting of this research into sufficiently substantiated, competent, and reliable evidence, however, relies on the traditional scientific foundation: first, biological plausibility and prior evidence and, second, consistent repeated findings. The practice of medicine at its core will continue to be a physician and patient interaction with professional judgment and communication central elements of the relationship. Judgment is essential because of uncertainties in the underlying professional knowledge or because even if the evidence is credible and substantiated, there may be tradeoffs in risks and benefits for testing and for treatment. Communication is critical because most decisions involve tradeoffs, in which case individual patient preferences for the outcomes that may be unique to patients and that may affect decisionmaking should be considered.

In summary, medical terms shared in common by the legal and medical professions have differing meanings, for example, differential diagnosis, differential etiology, and general and specific causation. The basic concepts of diagnostic reasoning and clinical decisionmaking and the types of evidence used to make judgments as treating physicians or experts involve the same overarching theoretical issues: (1) alternative reasoning processes; (2) weighing risks, benefits, and evidence; and (3) communicating those risks.

Glossary of Terms

adequacy. In diagnostic verification, testing a particular diagnosis for its adequacy involves determining its ability to account for all normal and abnormal findings and the observed time course of the disease.

attending physician. The physician responsible for the patient's care at the hospital in which the patient is being treated.

Bayes' theorem (rule). A mathematical approach to integrating suspicion (pretest probability) with additional information such as from a test result (posttest probability) by using test characteristics (sensitivity and specificity) to demonstrate how well the test performs in individuals with and without the disease.

causal reasoning. For physicians, causal reasoning typically involves understanding how abnormalities in physiology, anatomy, genetics, or biochemistry lead to the clinical manifestations of disease. Through such reasoning, physicians develop a "causal cascade" or "chain or web of causation" linking a sequence of plausible cause-and-effect mechanisms to arrive at the pathogenesis or pathophysiology of a disease.

chief complaint. The primary or main symptom that caused the patient to seek medical attention.

coherency. In diagnostic verification, testing a particular diagnosis for its coherency involves determining the consistency of that particular diagnosis with predisposing risk factors, physiological mechanisms, and resulting manifestations.

conditional probability. The probability or likelihood of something given that something else has occurred or is present, for example, the likelihood of disease if a test is positive (posterior probability) or the likelihood of a positive test if disease is present (sensitivity). See Bayes' theorem or rule.

consulting physician. A physician, usually a specialist, asked by the patient's attending physician to provide an opinion regarding diagnosis, testing, or treatment or to perform a procedure or intervention, for example, surgery.

diagnostic test. A test ordered to confirm or exclude possible causes of a patient's symptoms or signs (distinct from screening test).

diagnostic verification. The last stage of narrowing the differential diagnosis to a final diagnosis by testing the validity of the diagnosis for its coherency, adequacy, and parsimony.

differential diagnosis. A set of diseases that physicians consider as possible causes for patients presenting with a chief complaint (hypothesis generation). As additional symptoms with further patient history, signs found on physical examination, test results, or specialty physician consultations become available, the likelihood of various diagnoses may change (hypothesis refinement) or new ones may be considered (hypothesis modification) until the diagnosis is nearly final (diagnostic verification).

differential etiology. Term used by the court or witnesses to establish or refute external causation for a plaintiff's condition. For physicians, etiology refers to cause.

external causation. External causation is established by demonstrating that the cause of harm or disease originates from outside the plaintiff's body, for example, defendant's action or product.

general causation. General causation is established by demonstrating, usually through scientific evidence, that a defendant's action or product causes (or is capable of causing) disease.

heuristics. Quick automatic "rules of thumb" or cognitive shortcuts often involving pattern recognition that facilitate rapid diagnostic and treatment decisionmaking. Although characteristic of experts, it may predispose to known cognitive errors. See Hypothetico-deductive.

hypothesis generation. A limited list of potential diagnostic hypotheses in response to symptoms, signs, and lab test results. See differential diagnosis.

hypothesis modification. A change in the list of diagnostic hypotheses (differential diagnosis) in response to additional information, e.g., symptoms, signs, and lab test results. See differential diagnosis.

hypothesis refinement. A change in the likelihood of the potential diagnostic hypotheses (differential diagnosis) in response to additional information, e.g., symptoms, signs, and lab test results. As additional information emerges, physicians evaluate those data for their consistency with the possibilities on the list and whether those data would increase or decrease the likelihood of each possibility. See differential diagnosis.

hypothetico-deductive. Deliberative and analytical reasoning involving hypothesis generation, hypothesis modification, hypothesis refinement, and diagnostic verification. Typically applied for problems outside an individual's expertise or difficult problems with atypical issues, it may avoid known cognitive errors. See Heuristics.

individual causation. See specific causation.

inductive reasoning. The process of arriving at a diagnosis based on symptoms, signs, and lab tests. See differential diagnosis.

inferential reasoning. See inductive reasoning.

overdiagnosis. Screening can lead to "pseudodisease" or "overdiagnosis," e.g., the identification of slow-growing cancers that even if untreated would never cause symptoms or reduce survival because the screening test cannot distinguish the abnormal-appearing cells that would become cancerous from those that would never do so. See overtreatment.

overtreatment. The treatment of patients with pseudodisease whose disease would never cause symptoms or reduce survival. The treatment may place patients at risk for treatment-related morbidity and possibly mortality. See overdiagnosis.

parsimony. In diagnostic verification, testing a particular diagnosis for its parsimony involves choosing the simplest single explanation as opposed to requiring the simultaneous occurrence of two diseases to explain the findings.

pathogenesis. See causal reasoning.

pathology test. Microscopic examination of body tissue typically obtained by a biopsy or during surgery to determine if the tissue appears to be abnormal (different than would be expected for the source of the tissue). The visual components of the abnormality are typically described (e.g., types of cells, appearance of cells, scarring, effect of stains or molecular markers that help facilitate identification of the components) and, on the basis of visual pattern, the abnormality may be classified, e.g., malignancy (cancer) or dysplasia (precancerous).

posttest probability. See predictive value.

predictive value or **posttest probability**. The suspicion or probability of a disease after additional information (such as from a test) has been obtained. The predictive value positive or positive predictive value is the probability of disease in those known to have a positive test result. The predictive value negative or negative predictive value is the probability of disease in those known to have a negative test result.

pretest probability. The suspicion or probability of a disease before additional information (such as from a test) is obtained.

prior probability. See pretest probability.

screening test. A test performed in the absence of symptoms or signs to detect disease earlier, e.g., cancer screening (distinct from diagnostic test).

sensitivity. Likelihood of a positive finding (usually referring to a test result but could also be a symptom or a sign) among individuals known to have a disease (distinct from specificity).

sign. An abnormal physical finding identified at the time of physical examination (distinct from symptoms).

specific causation or **individual causation.** Established by demonstrating that a defendant's action or product is the cause of a particular plaintiff's disease.

specificity. Likelihood of a negative finding (usually referring to a test result but could also be a symptom or a sign) among individuals who do not have a particular disease (distinct from sensitivity).

syndrome. A group of symptoms, signs, and/or test results that together characterize a specific disease.

symptom. The patient's description of a change in function, sensation, or appearance (distinct from sign).

References on Medical Testimony

Lynn Bickley et al., Bates' Guide to Physical Examination and History Taking (10th ed. 2008).
Gerd Gigerenzer. Calculated Risks. How to Know When Numbers Deceive You (2002).
Trisha Greenhalgh. How to Read a Paper: The Basics of Evidence-Based Medicine (4th ed. 2010).
Gordon Guyatt et al., Users' Guides to the Medical Literature: Essentials of Evidence-Based Clinical Practice (2d ed. 2009).
Jerome P. Kassirer et al., Learning Clinical Reasoning (2d ed. 2009).
Harold C. Sox et al., Medical Decision Making (2006).
Sharon E. Straus et al., Evidence-Based Medicine (4th ed. 2010).

Reference Guide on Neuroscience

HENRY T. GREELY AND ANTHONY D. WAGNER

Henry T. Greely, J.D., is Deane F. and Kate Edelman Johnson Professor of Law, Professor, by courtesy, of Genetics, and the Director of Center for Law and the Biosciences, Stanford University, Stanford, California.

Anthony D. Wagner, Ph.D., is Associate Professor of Psychology and Neuroscience, Stanford University, Stanford, California.

CONTENTS

I. Introduction

Science's understanding of the human brain is increasing exponentially. We know almost infinitely more than we did 30 years ago; however, we know almost nothing compared with what we are likely to know 30 years from now. The results of advances in understanding human brains—and of the minds they generate—are already beginning to appear in courtrooms. If, as neuroscience indicates, our mental states are produced by physical states of our brain, our increased ability to discern those physical states will have huge implications for the law. Lawyers already are introducing neuroimaging evidence as relevant to questions of individual responsibility, such as claims of insanity or diminished responsibility, either on issues of liability or of sentencing. In May 2010, parties in two cases sought to introduce neuroimaging in court as evidence of honesty; we are also beginning to see efforts to use it to prove that a person is in pain. These and other uses of neuroscience are almost certain to increase with our growing knowledge of the human brain as well as continued technological advances in accurately and precisely measuring the brain. This chapter strives to give judges some background knowledge about neuroscience and the strengths and weaknesses of its possible applications in litigation in order to help them become better prepared for these cases.[1]

The chapter begins with a brief overview of the structure and function of the human brain. It then describes some of the tools neuroscientists use to understand the brain—tools likely to produce findings that parties will seek to introduce in court. Next, it discusses a number of fundamental issues that must be considered when interpreting neuroscientific findings. Finally, after discussing, in general, the issues raised by neuroscience-based evidence, the chapter concludes by analyzing a few illustrative situations in which neuroscientific evidence is likely to appear in court in the future.

II. The Human Brain

This abbreviated and simplified discussion of the human brain describes the cellular basis of the nervous system, the structure of the brain, and finally our current understanding of how the brain works. More detailed, but still accessible, informa-

1. The Law and Neuroscience Project, funded by the John D. and Catherine T. MacArthur Foundation, is preparing a book about law and neuroscience for judges, which should be available by 2011. A Primer on Neuroscience (Stephen Morse & Adina Roskies eds., forthcoming 2011). The Project has already published a pamphlet written by neuroscientists for judges, with brief discussions of issues relevant to law and neuroscience. A Judge's Guide to Neuroscience: A Concise Introduction (M.S. Gazzaniga & J.S. Rakoff eds., 2010). One early book on a broad range of issues in law and neuroscience also deserves mention: Neuroscience and the Law: Brain, Mind, and the Scales of Justice (Brent Garland ed., 2004).

tion about the human brain can be found in academic textbooks and in popular books for general audiences.[2]

A. Cells

Like most of the human body the nervous system is made up of cells. Adult humans contain somewhere between 50 trillion and 100 trillion human cells. Each of those cells is both individually alive and part of a larger living organism.

Each cell in the body (with rare exceptions) contains each person's entire complement of human genes—his or her genome. The genes, found on very long molecules of deoxyribonucleic acid (DNA) that make up a human's 46 chromosomes, work by leading the cells to make other molecules, notably proteins and ribonucleic acid (RNA). We now believe that there are about 23,000 human genes. Cells are different from each other not because they contain different genes but because they turn on and off different sets of genes. All human cells seem to use the same group of several thousand "housekeeping" genes that run the cell's basic machinery, but skin cells, kidney cells, and brain cells differ in which other genes they use. Scientists count different numbers of "types" of human cells, with estimates ranging from a few hundred to a few thousand (depending largely on how narrowly or broadly one defines a cell type).

The most important cells in the nervous system are called neurons. Neurons pass messages from one neuron to another in a complex way that appears to be responsible for brain function, conscious or otherwise.

Neurons (Figure 1) come in many sizes, shapes, and subtypes (with their own names), but they generally have three features: a cell body (or "soma"), short extensions called dendrites, and a longer extension called an axon. The cell body contains the nucleus of the cell, which in turn contains the 46 chromosomes with the cell's DNA. The dendrites and axons both reach out to make connections with other neurons. The dendrites generally receive information from other neurons; the axons send information.

Communication between neurons occurs at areas called synapses (Figure 2), where two neurons almost meet. At a synapse, the two neurons will come within

2. The Society for Neuroscience, the very large scholarly society that covers a wide range of brain science, has published a brief and useful primer about the human brain called *Brain Facts*. The most recent edition, published in 2008, is available free at www.sfn.org/index.aspx?pagename=brainfacts.

Some particularly interesting books about various aspects of the brain written for a popular audience include Oliver W. Sacks, The Man Who Mistook His Wife for a Hat and Other Clinical Tales (1990); Antonio R. Damasio, Descartes' Error: Emotion, Reason, and the Human Brain (1994); Daniel L. Schacter, Searching for Memory: The Brain, the Mind, and the Past (1996); Joseph E. LeDoux, The Emotional Brain: The Mysterious Underpinnings of Emotional Life (1996); Christopher D. Frith, Making Up the Mind: How the Brain Creates Our Mental World (2007); and Sandra Aamodt & Sam Wang, Welcome to Your Brain: Why You Lose Your Car Keys But Never Forget How to Drive and Other Puzzles of Everyday Life (2008).

Figure 1. Schematic of the typical structure of a neuron.

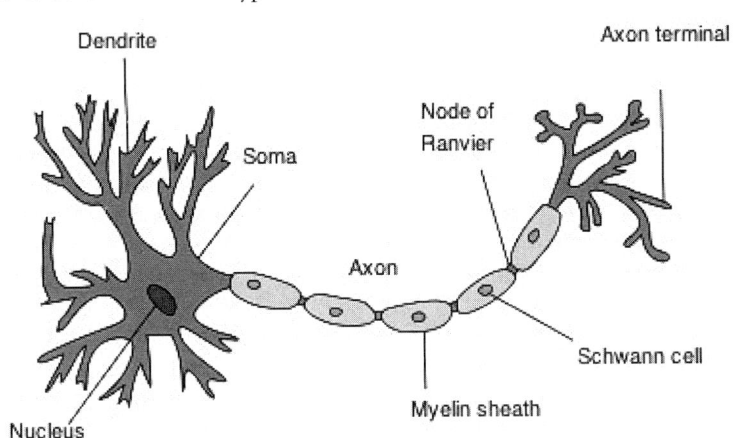

Source: Quasar Jarosz at en.wikipedia.

less than a micrometer (a millionth of a meter) of each other, with the presynaptic side, on the axon, separated from the postsynaptic side, on the dendrite, by a gap called the synaptic cleft. At synapses, when the axon (on the presynaptic side) "fires" (becomes active) it releases molecules, known as neurotransmitters, into the synaptic cleft. Some of those molecules are picked up by special receptors on the dendrite that is on the postsynaptic side of the cleft. More than 100 different neurotransmitters have been identified; among the best known are dopamine, serotonin, glutamate, and acetylcholine. Some of the neurotransmitters released into the synaptic cleft are picked up by special receptors on the postsynaptic side of the cleft by the dendrite.

At the postsynaptic side of the cleft, neurotransmitters binding to the receptors can have a wide range of effects. Sometimes they cause the receiving (postsynaptic) neuron to "fire," sometimes they suppress (inhibit) the postsynaptic neuron from firing, and sometimes they seem to do neither. The response of the receiving neuron is a complicated summation of the various messages it receives from multiple neurons that converge, through synapses, on its dendrites.

A neuron that does fire does so by generating an electrical current that flows down (away from the cell body) the length of its axon. We normally think of electrical current as flowing in things like copper wiring. In that case, free electrons move down the wire. The electrical currents of neurons are more complicated. Molecules with a positive or negative electrical charge (ions) move through the neuron's membrane and create differences in the electrical charge between the inside and outside of the neuron, with the current traveling along the axon, rather like a fire brigade passing buckets of water in only one direction

Figure 2. Synapse. Communication between neurons occurs at the synapse, where the sending (presynaptic) and receiving (postsynaptic) neurons meet. When the presynaptic neuron fires, it releases neurotransmitters into the synaptic cleft, which bind to receptors on the postsynaptic neuron.

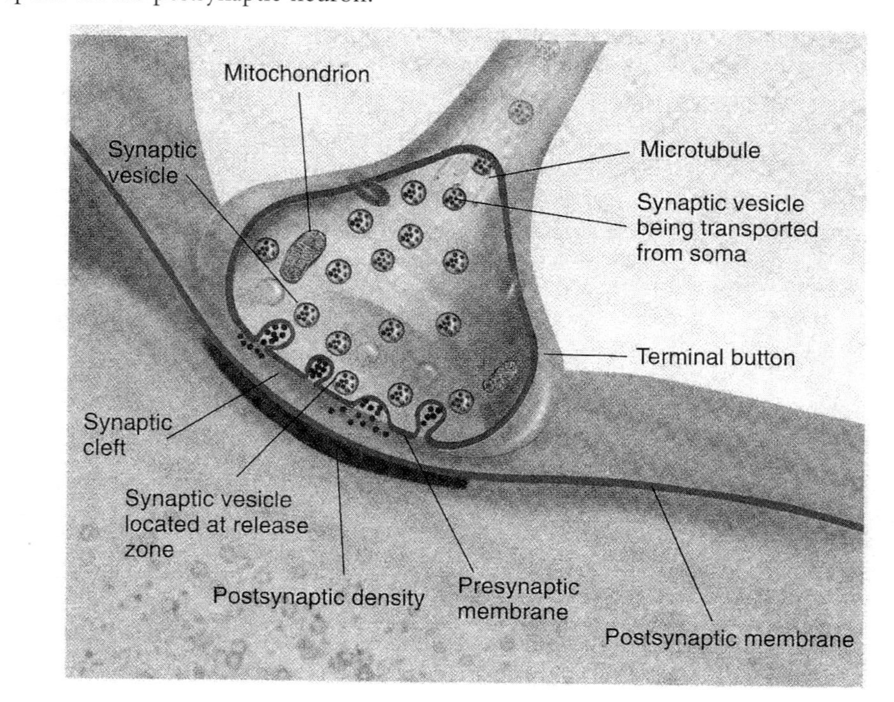

Source: From Carlson. Carlson, Neil R. Foundations of Physiological Psychology (with Neuroscience Animations and Student Study Guide CD-ROM), 6th. © 2005. Printed and electronically reproduced by permission of Pearson Education, Inc., Upper Saddle River, New Jersey.

down the line. Firing occurs in milliseconds. This process of moving ions in and out of the cell membrane requires that the cell use large amounts of energy. When the current reaches the end of the axon, it may or may not cause the axon to release neurotransmitters into the synaptic cleft. This complicated part-electrical, part-chemical system is how information passes from one neuron to another.

The axons of human neurons are all microscopically narrow, but they vary enormously in length. Some are micrometers long; others, such as neurons running from the base of the spinal cord to the toes, are several feet long. Longer axons tend to be coated with a fatty substance called myelin. Myelin helps insulate the axon and thus increases the strength and efficiency of the electrical signal, much like the insulation wrapped around a copper wire. (The destruction of this myelin sheathing is the cause of multiple sclerosis.) Axons coated with myelin appear white; thus areas of the nervous system that have many myelin-coated axons are referred to as "white matter." Cell bodies, by contrast, look gray, and so areas with many cell bodies and relatively few axons make up our "gray matter." White matter can roughly be thought of as the wiring that connects gray matter to the rest of the body or to other areas of gray matter.

What we call nerves are really bundles of neurons. For example, we all have nerves that run down our arms to our fingers. Some of those nerves consist of neurons that pass messages from the fingers, up the arm, to other neurons in the spinal cord that then pass the messages on to the brain, where they are analyzed and experienced. This is how we feel things with our fingers. Other nerves are bundles of neurons that pass messages from the brain through the spinal cord to nerves that run down the arms to the fingers, telling them when and how to move.

Neurons can connect with other neurons or with other kinds of cells. Neurons that control body movements ultimately connect to muscle cells—these are called motor neurons. Neurons that feed information into the brain start with specialized sensory cells (i.e., cells specialized for detecting different types of stimuli—light, touch, heat, pain, and more) that fire in response to the appropriate stimulus. Their firings ultimately lead, directly or through other neurons, into the brain. These are sensory neurons. These neurons send information only in one direction—motor neurons ultimately from the brain, sensory neurons to the brain. The paralysis caused by, for example, severe damage to the spinal cord both prevents the legs from receiving messages to move that would come from the brain through the motor neurons and keeps the brain from receiving messages from sensory neurons in the legs about what the legs are experiencing. The break in the spinal column prevents the messages from getting through, just as a break in a local telephone line will keep two parties from connecting. (There are, unfortunately, not yet any human equivalents to wireless service.)

Estimates of the number of cells in a human brain vary widely, from a few hundred billion to several trillion. These cells include those that make up blood vessels and various connective tissues in the brain, but most of them are specialized brain cells. About 80 billion to 100 billion of these brain cells are neurons; the other cells (and the source of most of the uncertainty about the number of cells) are principally another class of cells referred to generally as glial cells. Glial cells play many important roles in the brain, including, for example, producing and maintaining the myelin sheaths that insulate axons and serving as a special immune system for the brain. The full importance of glial cells is still being discovered; emerging data suggest that they may play a larger role in mental processes than as "support staff." At this point, however, we concentrate on neurons, the brain structures they form, and how those structures work.

B. Brain Structure

Anatomists refer to the brain, the spinal cord, and a few other nerves directly connecting to the brain as the central nervous system. All the other nerves are part of the peripheral nervous system. This reference guide does not focus on the peripheral nervous system, despite its importance in, for example, assessing some aspects of personal injuries. We also, less fairly, ignore the central nervous system other than the brain, even though the spinal cord, in particular, plays an important role in modulating messages going into and coming out of the brain.

The average adult human brain (Figure 3) weighs about 3 pounds and fills a volume of about 1300 cubic centimeters. If liquid, it would almost fill two standard wine bottles with a little space left over. Living brains have a consistency about like that of gelatin. Despite the softness of brains, they are made up of regular shapes and structures that are generally consistent from person to person. Just as every nondamaged or nondeformed human face has two eyes, two ears,

Figure 3. Lateral (left) and mid-sagittal (right) views of the human brain.

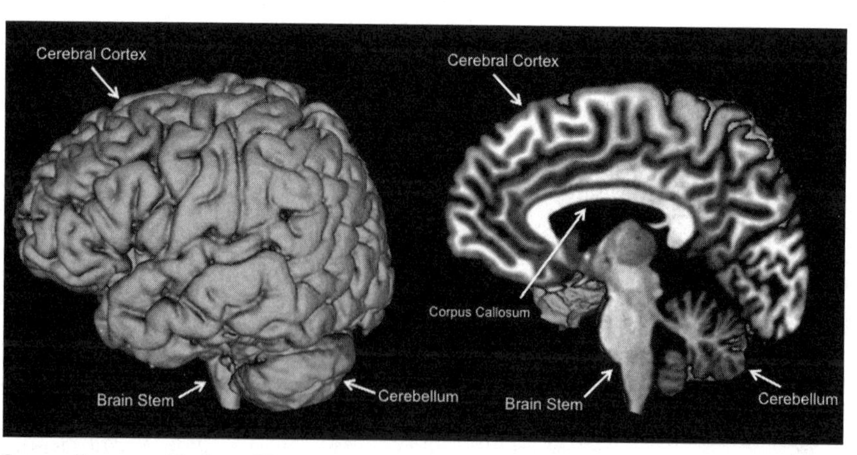

Source: Courtesy of Anthony Wagner.

one nose, and one mouth with standard numbers of various kinds of teeth, every normal brain has the same set of identifiable structures, both large and small.

Neuroscientists have long worked to describe and define particular regions of the brain. In some ways this is like describing parcels of land in property documents, and, like property descriptions, several different methods are used. At the largest scale, the brain is often divided into three parts: the brain stem, the cerebellum, and the cerebrum.[3]

The brain stem is found near the bottom of the brain and is, in some ways, effectively an extension of the spinal cord. Its various parts play crucial roles in controlling the body's autonomic functioning, such as heart rate and digestion. The brain stem also contains important regions that regulate processing in the cerebrum. For example, the substantia nigra and ventral tegmental area in the brain stem consist of critical neurons that generate the neurotransmitter dopamine. While the substantia nigra is crucial for motor control, the ventral tegmental area is important for learning about rewards. The loss of neurons in the substantia nigra is at the core of the movement problems of Parkinson's disease.

The cerebellum, which is about the size and shape of a squashed tennis ball, is tucked away in the back of the skull. It plays a major role in fine motor control and seems to keep a library of learned motor skills, such as riding a bicycle. It was long thought that damage to the cerebellum had little to no effect on a person's personality or cognitive abilities, but resulted primarily in unsteady gait, difficulty in making precise movements, and problems in learning movements. More recent studies of patients with cerebellar damage and functional brain imaging studies of healthy individuals indicate that the cerebellum also plays a role in more cognitive functions, including supporting aspects of working memory, attention, and language.

The cerebrum is the largest part of the human brain, making up about 85% of its volume. The cerebrum is found at the front, top, and much of the back of the human brain. The human brain differs from the brains of other mammals mainly because it has a vastly enlarged cerebrum.

There are several different ways to identify parts of, or locations in, the cerebrum. First, the cerebrum is divided into two hemispheres—the famous left and right brain. These two hemispheres are connected by tracts of white matter—of axons—most notably the large connection called the corpus callosum. Oddly, the right hemisphere of the brain generally receives messages from and controls the movements of the left side of the body, while the left hemisphere receives messages from and controls the movements of the right side of the body.

Each hemisphere of the cerebrum is divided into four lobes (Figure 4): The frontal lobe in the front of the cerebrum (behind the forehead), the parietal lobe at

3. The brain also is sometimes divided into the forebrain, midbrain, and hindbrain. This classification is useful for some purposes, particularly in describing the history and development of the vertebrate brain, but it does not entirely correspond to the categorization of cerebrum, brain stem, and cerebellum, and it is not used in this reference guide.

Figure 4. Lobes of a hemisphere. Each hemisphere of the brain consists of four lobes—the frontal, parietal, temporal, and occipital lobes.

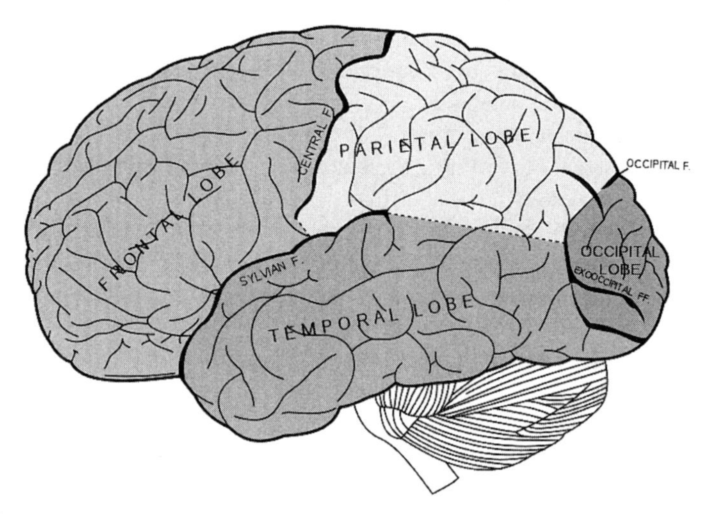

Source: http://commons.wikimedia.org/wiki/File:Gray728.svg. This image is in the public domain because its copyright has expired. This applies worldwide.

the top and toward the back, the temporal lobe on the side (just behind and above the ears), and the occipital lobe at the back. Thus, one could describe a particular region as lying in the left frontal lobe—the frontal lobe of the left hemisphere.

The surface of the cerebrum consists of the cortex, which is a sheet of gray matter a few millimeters thick. The cortex is not a smooth sheet in humans, but rather is heavily folded with valleys, called sulci ("sulcus" in the singular), and bulges, called gyri ("gyrus"). The sulci and gyri have their own names, and so a location can be described as in the inferior frontal gyrus in the left frontal lobe. These folds allow the surface area of the cortex, as well as the total volume of the cortex, to be much greater than in other mammals, while still allowing it to fit inside our skulls, similar to the way the many folds of a car's radiator give it a very large surface area (for radiating away heat) in a relatively small space.

The cerebral cortex is extraordinarily large in humans compared with other species and is clearly centrally involved in much of what makes our brains special, but the cerebrum contains many other important subcortical structures that we share with other vertebrates. Some of the more important areas include the thalamus, the hypothalamus, the basal ganglia, and the amygdala. These areas all connect widely, with the cortex, with each other, and with other parts of the brain to form complex networks.

The functions of all these areas are many, complex, and not fully understood, but some facts are known. The thalamus seems to act as a main relay that carries information to and from the cerebral cortex, particularly for vision, hearing, touch, and proprioception (one's sense of the position of the parts of one's body). It also is, importantly, involved in sleep, wakefulness, and consciousness. The hypothalamus has a wide range of functions, including the regulation of body temperature, hunger, thirst, and fatigue. The basal ganglia are a group of regions in the brain that are involved in motor control and learning, among other things. They seem to be strongly involved in selecting movements, as well as in learning through reinforcement (as a result of rewards). The amygdala appears to be important in emotional processing, including how we attach emotional significance to particular stimuli.

In addition, many other parts of the brain, in the cortex or elsewhere, have their own special names, usually with Latin or Greek roots that may or may not seem descriptive today. The hippocampus, for example, is named for the Greek word for seahorse. For most of us, these names will have no obvious rhyme or reason, but merely must be learned as particular structures in the brain—the superior colliculus, the tegmentum, the globus pallidus, the substantia nigra, the cingulate cortex, and more. All of these structures come in pairs, with one in the left hemisphere and one in the right hemisphere; only the pineal gland is unpaired. Brain atlases include scores of names for particular structures or regions in the brain and detailed information about the structures or regions.

Some of these smaller structures may have special importance to human behavior. The nucleus accumbens, for example, is a small subcortical region in each hemisphere of the cerebrum that appears important for reward processing and motivation. In experiments with rats that received stimulation of this region in return for pressing a lever, the rats would press the lever almost to the exclusion of any other behavior, including eating. The nucleus accumbens in humans appears linked to appetitive motivation, responding in anticipation of primary rewards (such as pleasure from food and sex) and secondary rewards (such as money). Through interactions with the orbital frontal cortex and dopamine-generating neurons in the midbrain (including the ventral tegmental area), the nucleus accumbens is considered part of a "reward network." With a hypothesized role in addictive behavior and in reward computations, more broadly, this putative reward network is a topic of considerable ongoing research.

All of these various locations, whether defined broadly by area or by the names of specific structures, can be further subdivided using directions: front and back, up and down, toward the middle, or toward the sides. Unfortunately, the directions often are not expressed in a straightforward manner, and several different terminological conventions exist. Locations toward the front or back of the brain can be referred to as either anterior or posterior or as rostral or caudal (literally, toward the nose, or beak, or the tail). Locations toward the bottom or top of the brain are termed inferior or superior or, alternatively, as ventral or dorsal (toward the stomach or toward the back). A location toward the middle of the brain is called medial; one toward the side is called lateral. Thus, different locations could be described, for example, as in the left anterior cingulate cortex, in the dorsal medial (or sometimes dorsomedial) prefrontal cortex, or in the posterior hypothalamus.

Finally, one other method often is used, a method created by Korbinian Brodmann in 1909. Brodmann, a neuroanatomist, divided the brain into about 50 different areas or regions (Figure 5). Each region was defined on the basis of the

Figure 5. Brodmann's areas. Brodmann divided the cortex into different areas based on the cell types and how they were organized.

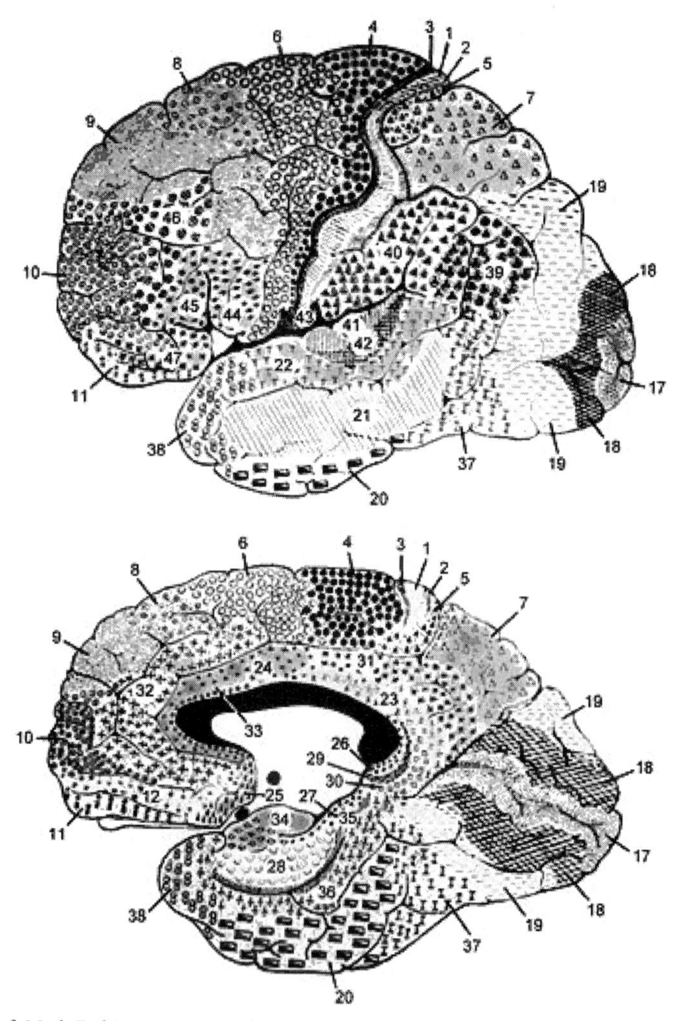

Source: Prof. Mark Dubin, University of Colorado.

kinds of neurons found there and how those neurons are organized. A location described by Brodmann area may or may not correspond closely with a structural location. Other organizational schemes exist, but Brodmann's remains the most widely used to describe the approximate locations of findings in modern human brain imaging studies.

C. Some Aspects of How the Brain Works

Most of neuroscience is dedicated to finding out how the brain works, but although much has been learned, considerably more remains unknown. We could use many different ways to describe what is known about how the brain works. This section discusses a few important aspects of brain function and makes several general points about the localization and distribution of functions, as well as brain plasticity, before commenting on the effects of hormones and other chemical influences on the brain.

Some brain functions are localized in, or especially dependent on, particular regions of the brain. This has been known for many years as a result of studies of people who, through traumatic injury, stroke, or cancer, have lost, or lost the use of, particular regions of their brains. For example, in the 1860s, French anatomist Paul Broca discovered through autopsies of patients that damage to a region in the left inferior frontal lobe (now known as Broca's area) caused an inability to speak. It is now known that some functions cannot normally be performed when particular brain areas are damaged or missing. The visual cortex, located at the back of the brain in the occipital lobes, is as necessary for vision as the eyes are; the hippocampus is necessary for the creation of many kinds of memory; and the motor cortex is necessary for voluntary movements. The motor cortex and the parallel somatosensory cortex, which is essential for processing sensory information such as the sense of touch from the body, are further subdivided, with particular regions necessary for causing motion or sensing feelings from the legs, arms, fingers, face, and so on. Other brain regions also will be involved in these actions or sensations, but these regions are necessary to them.

At the same time, the fact that a region is necessary to a particular class of sensations, behaviors, or cognition does not mean either that it is not involved in other brain functions or that other brain regions do not also contribute to these particular abilities. The amygdala, for example, is involved in our feelings of fear, but it is also involved broadly in emotional reactions, both positive and negative. It also modulates learning, memory, and even sensory perception. Although some functions are localized, others are widely distributed. For example, the visual cortex is essential to vision, but actual visual perception involves many parts of the brain in addition to the occipital lobes. Memories appear to be stored over much of the cortex. Networks of brain regions participate in many of these functions.

For example, if you touch something very hot with your left index finger, your spinal cord, through a reflex loop, will cause you to pull your finger back very quickly. Then the part of your right somatosensory cortex devoted to the index finger will be involved in receiving and initially interpreting the sensation. Other areas of your brain will recognize the stimulus as painful, your motor regions will be involved in waving your hand back and forth or bringing your finger to your mouth, widespread parts of your cortex may lead to your remembering other instances of burning yourself, and your hippocampus may play a role in making a new long-term memory of this incident. There is no brain region "for" burning your finger; many regions, both specific and general, contribute to the brain's response.

In addition, brains are at least somewhat "plastic" or changeable on both small and large scales. Anyone who can see has a working visual cortex, and it is always located in the back of the brain (in the occipital lobe), but its exact borders will vary slightly from person to person. In other cases, the brain may adjust and change in response to a person's behavior or changes in that person's anatomy. For example, a right-handed violinist may develop an enlarged brain region for controlling the fingers of the left hand, used in fingering the violin. If a person loses an arm to amputation, the parts of the motor and somatosensory cortices that had dealt with that arm may be "taken over" by other body parts. In some cases, this brain plasticity can be extreme. A young child who has lost an entire hemisphere of his or her brain may grow up to have normal or nearly normal functionality as the remaining hemisphere takes on the tasks of the missing hemisphere. Unfortunately, the possibilities of this kind of extreme plasticity do diminish with age, but rehabilitation after stroke in adults sometimes does show changes in the brain functions undertaken by particular brain regions.

The picture of the brain as a set of interconnected neurons that fire in networks or patterns in response to stimuli is useful but not complete. In addition to neuron firings, other factors affect how the brain works, particularly chemical factors.

Some of these are hormones, generated by the body either inside or outside the brain. They can affect how the brain functions, as well as how it develops. Sex hormones such as estrogen and testosterone can have both short-term and long-term effects on the brain. So can other hormones, such as cortisol, associated with stress, and oxytocin, associated with, among other things, trust and bonding. Endorphins, chemicals secreted by the pituitary gland in the brain, are associated with pain relief and a sense of well-being. Still other chemicals, brought in from outside the body, can have major effects on the brain, both in the short term and the long term. Examples include alcohol, caffeine, nicotine, morphine, and cocaine. These can trigger very specific brain reactions or can have broad effects.

III. Some Common Neuroscience Techniques

Neuroscientists use many techniques to study the brain. Some of them have been used for centuries, such as autopsies and the observation of patients with brain damage. Some, such as the intentional destruction of parts of the brain, can be used ethically only in research on nonhuman animals. Of course, research with nonhuman animals, although often helpful in understanding human brains, is of less value when examining behaviors that are uniquely developed among humans. The current revolution in neuroscience is largely the result of a revolution in the tools available to neuroscientists, as new methods have been developed to image and to intervene in living brains. These methods, particularly the imaging methods that allow more precise measurements of human brain structure and function in living people, are giving rise to increasing efforts to introduce neuroscientific evidence in court.

This section of this chapter focuses on several kinds of neuroimaging—computerized axial tomography (CAT) scans, positron emission tomography (PET) scans, single photon emission computed tomography (SPECT) scans, and magnetic resonance imaging (MRI), as well as an older method, electroencephalography (EEG), and its close relative, magnetoencephalography (MEG). Some of these methods show the structure of the brain, others show the brain's functioning, and some do both. These are not the only important neuroscience techniques; several others are discussed briefly at the end of this section. Genetic analysis provides yet another technique for increasing our understanding of human brains and behaviors, but this chapter does not deal with the possible applications of human genetics to understanding behavior.

A. Neuroimaging

Traditional imaging technologies have not been very helpful in studying the brain. X-ray images are the shadows cast by dense objects. Not only is the brain surrounded by our very dense skulls, but there are no dense objects inside the brain to cast these shadows. Although a few features of the brain or its blood vessels could be seen through methods that involved the injection of air into some of the spaces in the brain or of contrast media into the blood, these provided limited information. The opportunity to see inside a living brain itself only goes back to about the 1970s, with the development of CAT scans. This ability has since exploded with the development of several new techniques, three of which, with CAT, are discussed on the following pages.

1. CAT scans

The CAT scan is a multidimensional, computer-assisted X-ray machine. Instead of taking one X ray from a fixed location, in a CAT scan both the X-ray source and (180 degrees opposite the source) the X-ray detectors rotate around the person being scanned. Rather than exposing negatives to make "pictures" of dense objects, as in traditional X rays, the X-ray detectors produce data for computer analysis. A complete modern CAT scan includes data sufficient to reconstruct the scanned object in three dimensions. Computerized algorithms can then be used to produce an image of any particular slice through the object. The multiple angles and computer analysis make it possible to pick out the relatively small density differences within the brain that traditional X-ray technology could not distinguish and to use them to produce images of the soft tissue (Figure 6).

Figure 6. CAT scan depicting axial sections of the human brain. The ventral most (bottom) surface of the brain is at upper left and the dorsal most (top) surface is at the lower right.

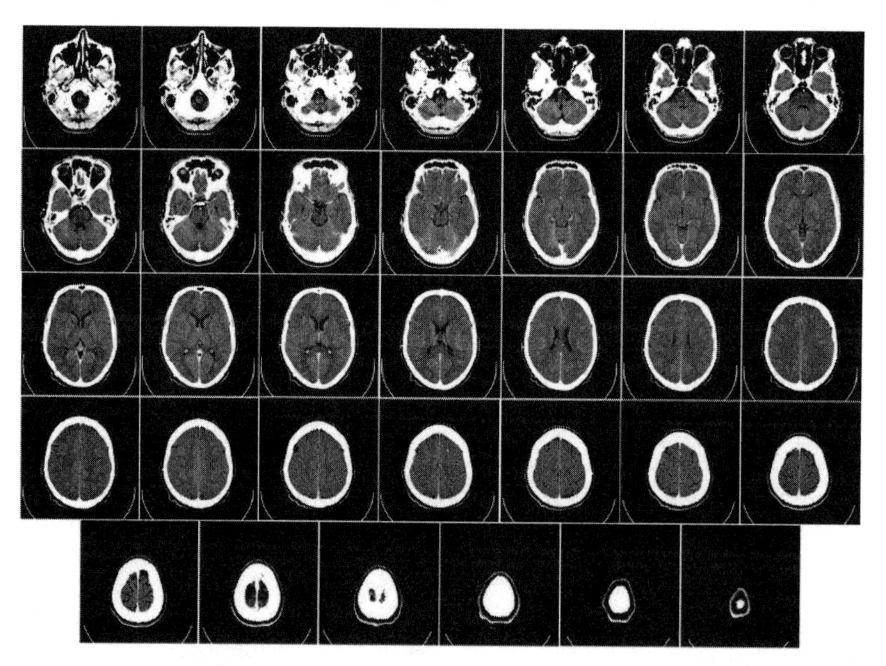

Source: http://en.wikipedia.org/wiki/File:CT_of_brain_of_Mikael_H%C3%A4ggstr%C3%B6m_large.png. Image in the public domain.

The CAT scan provides a structural image of the brain. It is useful for showing some kinds of structural abnormalities, but it provides no direct information about the brain's functioning. A CAT scan brain image is not as precise as the image produced from an MRI, but because the procedure is both quick and (relatively) inexpensive, CAT scanners are common in hospitals. Medically, brain CAT scans are used mainly to look for bleeding or swelling inside the brain, although they also will record sizeable tumors or other large structural abnormalities. For neuroscience, the great advantage of the CAT scan was its ability, for the first time, to reveal some details inside the skull, an ability that has been largely superseded for research by MRI. CAT scans have been used in courts to argue that structural changes in the brain, shown on the CAT scan, are evidence of insanity or other mental impairments. Perhaps their most notable use was in 1982 in the trial of John Hinckley for the attempted assassination of President Ronald Reagan. A CAT scan of Hinckley's brain that showed widened sulci (the "valleys" in the surface of the brain) was introduced into evidence to show that Hinckley suffered from organic brain damage in the form of shrinkage of his brain.[4]

2. *PET scans and SPECT scans*

Traditional X-ray machines and their more sophisticated descendant, the CAT scan, project X rays through the skull and create images based on how much of the X rays are blocked or absorbed. PET scans and SPECT scans operate very differently. In these methods, a substance that emits radiation is introduced into the body. That radiation then is detected from outside the body in a way that can determine the location of the radiation source. These scans generally are not used for determining the brain's structure, but for understanding how it is functioning. They are particularly good at measuring one aspect of brain structure—the density of particular receptors, such as those for dopamine, at synapses in some areas of the brain, such as the frontal lobes.

Radioactive decay of atoms can take several forms, producing alpha, beta, or gamma radiation. PET scanners take advantage of isotopes of atoms that decay by giving off positive beta radiation. Beta decay usually involves the emission of an electron; positive beta decay involves the emission of a positron, the positively charged antimatter equivalent of an electron. When positrons (antimatter) meet electrons (matter), the two particles are annihilated and converted into two photons of gamma radiation with a known energy (511,000 electron volts) that follow directly opposite paths from the site of the annihilation. Inside the body, the collision between the positron and electron and the consequent production of the gamma radiation photons takes place within a short distance (a millimeter or two) of the site of the initial radioactive decay that produced the positron.

4. The effects of this evidence on the verdict are unclear. *See* Lincoln Caplan, The Insanity Defense and the Trial of John W. Hinckley, Jr. (1984) for a discussion of the case and its consequences for the law.

PET scans, therefore, start with the introduction into a person's body of a radioactive tracer that decays by giving off a positron. One common tracer is fluorodeoxyglucose (FDG), a molecule that is almost identical to the simple sugar, glucose, except that one of the oxygen atoms in glucose is replaced by an atom of fluorine-18, an isotope of the element fluorine with nine protons and nine neutrons. Fluorine normally found in nature is fluorine-19, with 9 protons and 10 neutrons, and is stable. Fluorine-18 is very unstable and decays, through positive beta decay, quickly losing about half of its mass every 110 minutes (its half-life). The body treats FDG as though it were glucose, and so the FDG is concentrated where the body needs the energy supplied by glucose. A major clinical use of PET scans derives from the fact that tumor cells use energy, and hence glucose, at much higher rates than normal cells.

After giving the FDG time to become concentrated in the body, which usually takes about an hour, the person is put inside the scanner itself. There, the person is entirely surrounded by a very sensitive radiation detector, tuned to respond to gamma radiation of the energy produced by annihilated positrons. When two "hits" are detected by two sensors at about the same time, the source is known to be located on a line connecting the two. Very small differences in the timing of when the radiation is detected can help determine where along that the line the annihilation took place. In this way, as more gamma radiation from the decaying FDG is detected, the general location of the FDG within the body can be determined and, as a result, tissue that is using a lot of glucose, such as a tumor, can be located.

In neuroscience research, PET scans also can be taken using different molecules that bind more specifically to particular tissues or cells. Some of these more specific ligands use fluorine-18, but others use a different radioactive tracer that also decays by emitting a positron—oxygen-15. This can be used to determine what parts of the brain are using more or less oxygen. Oxygen-15, however, has a much shorter half-life (2 minutes) and so is more difficult and expensive to use than FDG. Similarly, carbon-11, with a half-life of 20 minutes, also can be used. Carbon-11 atoms can be introduced into various molecules that bind to important receptors in the brain, such as receptors for dopamine, serotonin, or opioids. This allows the study of the distribution and function of these receptors, both in healthy people and in people with various mental illnesses or neurological diseases.

The result of a PET scan is a record of the locations of positron decay events in the brain. Computer visualization tools can then create cross-sectional images of the brain, showing higher and lower rates of decay, with differences in magnitude typically depicted through the use of different colors (Figure 7).

PET scans are excellent for showing the location of various receptors in normal and abnormal brains. PET scans are also very good for showing areas of different glucose use and, hence, of different levels of metabolism. This can be very useful, for example, in detecting some kinds of brain damage, such as the damage that occurs with Alzheimer's disease, where certain regions of the brain become abnormally inactive, or in brain regions that have been damaged by a stroke.

Figure 7. PET scan depicting an axial section of the human brain.

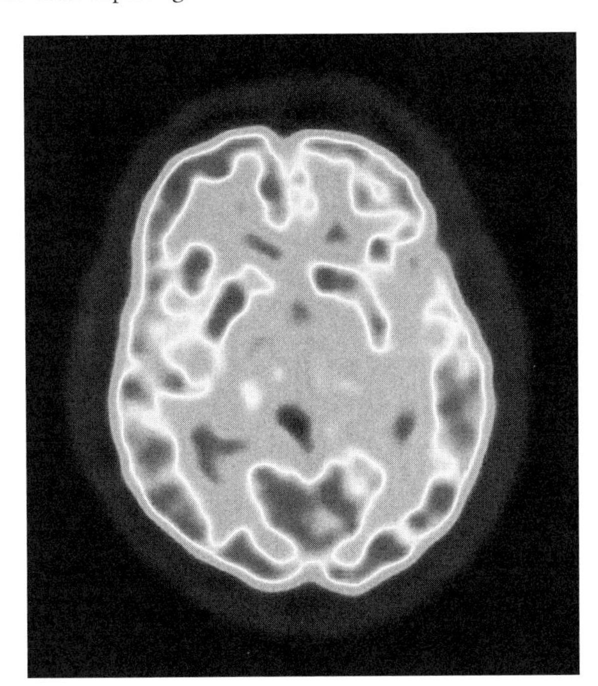

Source: http://en.wikipedia.org/wiki/Positron_emission_tomography. Image in the public domain.

In addition, the comparison (subtraction) of two PET scan measurements, one scan when a person is engaged in a task that is thought to require particular brain functions and a second control (or baseline) scan that is not thought to require these functions, allows researchers indirectly to measure brain function. PET scans were initially used in this way in research to show what areas of the brain were used when people experienced various stimuli or performed particular tasks. PET has been substantially superseded for this purpose by functional MRI, which is less expensive, does not involve radiation exposure, provides better spatial resolution, and allows a longer period of testing.

SPECT scans are similar to PET scans. Each can produce a three-dimensional model of the brain and display images of any cross section through the brain. Like PET scans, they require the injection of a radioactive tracer material; unlike PET scans, the radioactive tracer in SPECT directly emits gamma radiation rather than emitting positrons. These kinds of tracers are more stable, more accessible, and much cheaper than the positron-emitting tracers needed for PET scans. With a PET scan, the gamma detector entirely surrounds the person; with a SPECT scan, one to three gamma detectors are rotated around the body over about 15 to

20 minutes. As with PET scans, the SPECT tracers can be used to measure brain metabolism or to attach to specific molecular receptors in the brain. The spatial resolution of a SPECT scan, however, is poorer than with a PET scan, with an uncertainty of about 1 cm.

Both PET and SPECT scans are most useful if coupled with good structural images. Contemporary PET and SPECT scanners often include a simultaneous CAT scan; there is some experimental work aimed at providing simultaneous PET and MRI scans.

3. MRI—structural and functional

MRI was developed in the 1970s, first came into wide use in the 1980s, and is currently the dominant neuroimaging technology for producing detailed images of the brain's structure and for measuring aspects of brain function. MRI operates on completely different principles than either CAT scans or PET or SPECT scans; it does not rely on X rays passing through the brain or on the decay of radioactive tracer molecules inside the brain. Rather, MRI's workings involve more complicated physics. This section discusses the general characteristics of MRI and then focuses on structural MRI, diffusion tensor imaging, and finally, functional MRI.

The power of an MRI scanner is measured by the strength of its magnetic field, measured in units called tesla (T). The magnetic field of a small bar magnet is about 0.01 T. The strength of the Earth's magnetic field is about 0.00005 T. The MRI machines used for clinical purposes use magnetic fields of between 0.2 T and 3.0 T, with 1.5 T or 3.0 T being the systems most commonly used today. MRI machines for human research purposes have reached 9.4 T. In general, the stronger the magnetic field, the better the image, although higher fields also can create their own measurement difficulties, especially when imaging brain function. MRI machines achieve these high magnetic fields through using superconducting magnets, made by cooling the electromagnet with liquid helium at a temperature 4° (Celsius) above absolute zero. For this and other reasons, MRI systems are complicated, with higher initial and continuing maintenance costs compared with some other methods for functional imaging (e.g., electroencephalography; *see infra* Section III.B).

In most MRI systems (Figure 8), the subject, on an examination table, slides into a cylindrical opening in the machine so that the part of the body to be imaged is in the middle of the magnet. Depending on the kind of imaging performed, the examination or experiment can take from about 30 minutes to more than 2 hours; throughout the scanning process the subject needs to stay as motionless as possible to avoid corrupting the images. The main sensations for the subject are the loud thumping and buzzing noises made by the machine, as well as the machine's vibration.

MRI examinations appear to involve minimal risk. Unlike the other neuroimaging technologies discussed above, MRI does not involve any high-energy

Figure 8. MRI machine. Magnetic resonance imaging systems are used to acquire both structural and functional images of the brain.

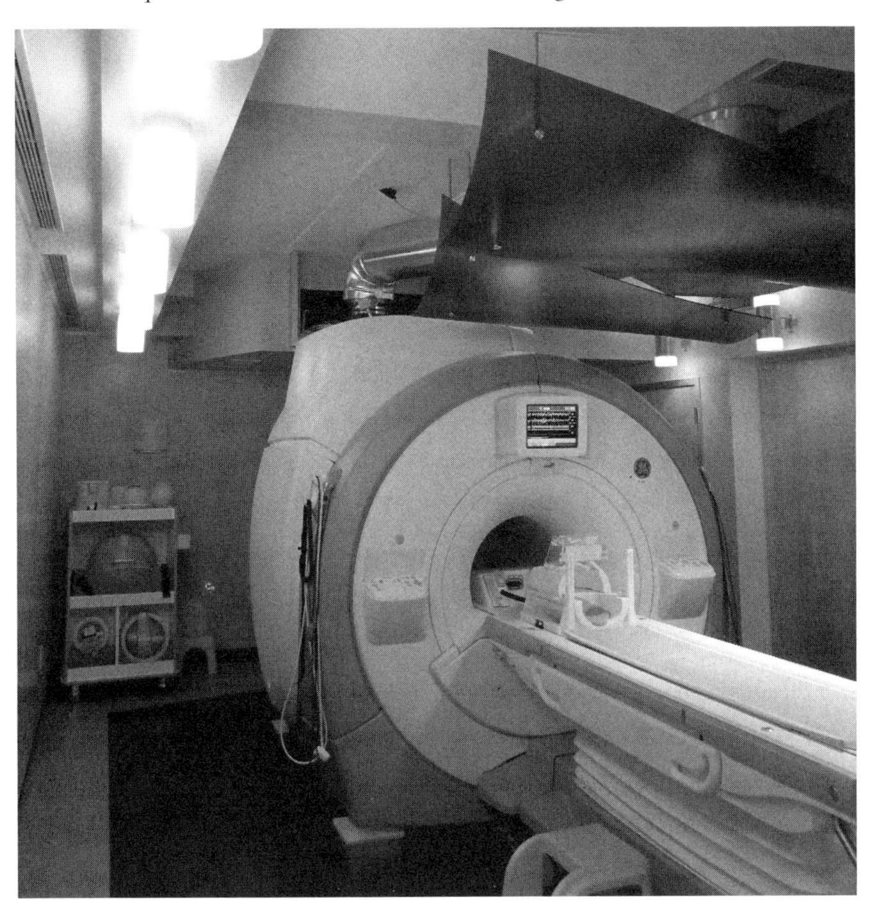

Source: Courtesy of Anthony Wagner.

radiation. The magnetic field seems to be harmless, at least as long as magnetizable objects are kept away from it. MRI subjects need to remove most metal objects; people with some kinds of implanted metallic devices, with tattoos with metal in their ink, or with fragments of ferrous metal anywhere in their bodies cannot be scanned because of the dangerous effects of the field on those bits of metal.

When the subject is positioned in the MRI scanner, the powerful field of the magnet causes the nuclei of atoms (usually the hydrogen nuclei of the body's water molecules) to align with the direction of the main magnetic field of the magnet. Using a brief electromagnetic pulse, these aligned atoms are then "flipped" out of alignment from the main magnetic field, and, after the pulse stops, the nuclei then rapidly realign with the strong main magnetic field. Because the nuclei spin (like a top), they create an oscillating magnetic field that is measured by a receiver coil. During structural imaging, the strength of the signal generated partially depends on the relative density of hydrogen nuclei, which varies from point to point in the body according to the density of water. In this manner, MRI scanners can generate images of the body's anatomy or of other scanned objects. Because an MRI scan can effectively distinguish between similar soft tissues, MRI can provide very-high-resolution images of the brain's anatomy, which is, after all, made up of soft tissue.

Structural MRI scans produce very detailed images of the brain (Figure 9). They can be used to spot abnormalities, large and small, as well as to see normal variation in the size and shape of brain features. Structural MRI can be used, for example, to see how brain features change as a person ages. Previously, getting that kind of detailed information about a brain required an autopsy or, at a minimum, extensive neurosurgery. This ability makes structural MRI both an important clinical tool and a very useful technique for research that tries to correlate human differences, normal and abnormal, with differences in brain structure, as well as for research that seeks to understand brain development.

Another structural imaging application of brain MRI has become increasingly prevalent over the past decade: diffusion tensor imaging (DTI). As noted above, neuronal tissue in the brain can be divided roughly into gray matter (the bodies of neurons) and white matter (neuronal axons that transmit signals over distance). DTI uses MRI to see what direction water diffuses through brain tissue. Tracts of white matter are made up of bundles of axons coated with fatty myelin. Water will diffuse through that white matter along the direction of the axons and not, generally, across them. This method can be used, therefore, to trace the location of these bundles of white matter and hence the long-distance connections between different parts of the brain. Abnormal patterns of these connections may be associated with various conditions, from Alzheimer's disease to dyslexia, some of which may have legal implications.

Functional MRI (fMRI) is perhaps the most exciting use of MRI in neuroscience for understanding brain function. This technique shows what regions of the brain are more or less active in response to the performance of particular tasks or the presentation of particular stimuli. It does not measure brain activity (the firing of neurons) directly but, instead, looks at how blood flow changes in response to brain activity and uses those changes, through the so-called BOLD response (the blood-oxygen-level dependent response), to allow the researcher to infer patterns of brain activity.

Structural MRI generally creates its images through detecting the density of hydrogen atoms in the subject and flipping them with radio pulses. For fMRI, the scanner detects changes in the ratio of oxygenated hemoglobin (oxyhemoglobin) and deoxygenated hemoglobin (deoxyhemoglobin) in particular locations in the brain. Hemoglobin is the protein in red blood cells that carries oxygen from the lungs to the body. On the basis of metabolic demands, hemoglobin molecules

Figure 9. Brain MRI scan depicting an axial (upper), coronal (lower left), and sagittal (lower right) image of the human brain.

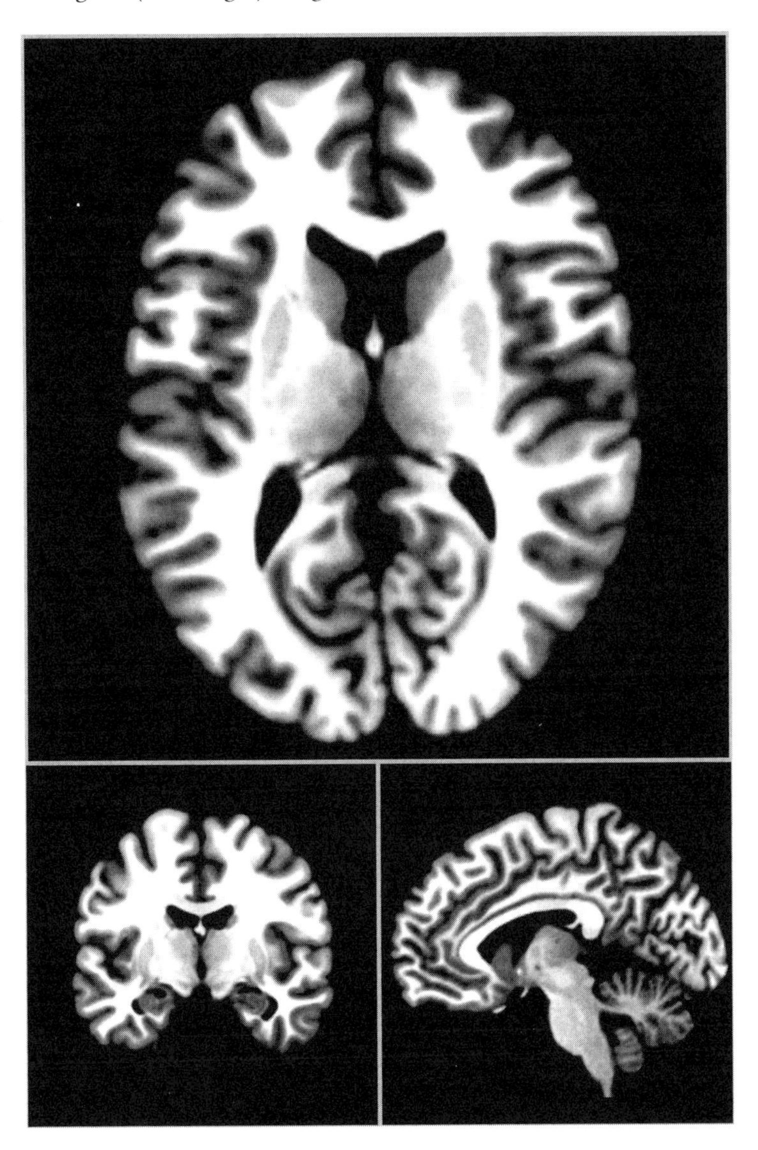

Source: Courtesy of Anthony Wagner.

supply oxygen for the body's needs. Accordingly, "fresher" blood will have a higher ratio of oxyhemoglobin to deoxyhemoglobin than more "used" blood. Importantly, because deoxyhemoglobin (which is found at a higher level in "used" blood) causes the fMRI signal to decay, a higher ratio of oxyhemoglobin to deoxyhemoglobin will produce a stronger fMRI signal.

Neural activity is energy intensive for neurons, and neurons do not contain any significant reserves of oxygen or glucose. Therefore, the brain's blood vessels respond quickly to increases in activity in any one region of the brain by sending more fresh blood to that area. This is the basis of the BOLD response, which measures changes in the ratio of oxyhemoglobin to deoxyhemoglobin in a brain region several seconds after activity in that region. In particular, when a brain region becomes more active, there is first, perhaps more intuitively, a *decline* in the ratio of oxyhemoglobin to deoxyhemoglobin immediately after activity in the region, apparently corresponding to the depletion of oxygen in the blood at the site of the activity. This decline, however, is very small and very hard to detect with fMRI. Immediately after this decrease, there is an infusion of fresh (oxyhemoglobin-rich) blood, which can take several seconds to reach maximum; it is this infusion that results in the *increase* in the oxy/deoxyhemoglobin ratio that is measured in BOLD fMRI studies. Because even this subsequent increase is relatively small and variable, fMRI experiments typically involve many trials of the same task or class of stimuli in order to be able to see the signal amidst the noise.

Thus, in a typical fMRI experiment the subject will be placed in the scanner and the researchers will measure differences in the BOLD response throughout his or her brain between different conditions. A subject might, for example, be told to look at a video screen on which images of places alternate with images of faces. For purposes of the experiment, the computer will impose a spatial map on the subject's brain, dividing it into thousands of little cubes, each a few cubic millimeters in size, referred to as "voxels." Either while the data are being collected (so-called "real-time fMRI"[5]) or after an entire dataset has been gathered, a computerized program will compare the BOLD signal for each voxel when the screen was showing *places* to that when the screen contained *faces*. Regions that showed a statistically significant increase in the BOLD response several seconds after the face was on the video screen compared with the effects several seconds after a screen showing a place appeared will be said to have been "activated" by seeing the face. The researchers will infer that those regions were, in some way, involved in how the brain processes images of faces. The results typically will be shown as a structural brain image on which areas of more or less activation, as shown by a statistical test, will be shown by different colors (Figure 10).[6]

5. Use of this "real-time" fMRI has been increasing, but it is not yet clear whether the claims for it will stand up.

6. This example is actually a simplified version of experiments performed by Professor Nancy Kanwisher at MIT in the early 2000s that explored a region of the brain called the fusiform face area

Figure 10. fMRI image. Functional MRI data reveal regions associated with cognition and behavior. Here, regions of the frontal and parietal lobes that are more active when remembering past events relative to detecting novel stimuli are depicted.

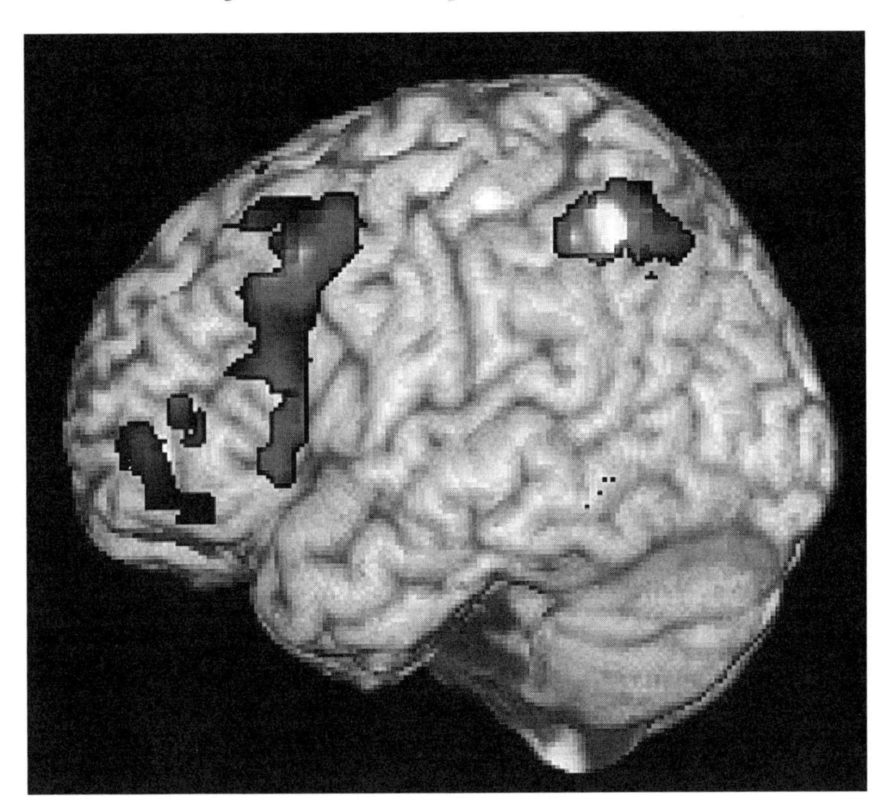

Source: Courtesy of Anthony Wagner.

Functional MRI was first proposed in 1990, and the first research results using BOLD-contrast fMRI in humans were published in 1992. The past decade has seen an explosive increase in the number of research articles based on fMRI, with nearly 2500 articles published in 2008—compared with about 450 in 1998.[7]

which is particularly involved in processing visions of faces. *See* Kathleen M. O'Craven & Nancy Kanwisher, *Mental Imagery of Faces and Places Activates Corresponding Stimulus-Specific Brain Regions*, 12 J. Cog. Neurosci. 1013 (2000).

7. *See* the census of fMRI articles from 1993 to 2008 in Carole A. Federico et al., *Intersecting Complexities in Neuroimaging and Neuroethics*, *in* Oxford Handbook of Neuroethics (J. Illes & B.J. Sahakian eds., 2011). This continued an earlier census from 1993 to 2001. Judy Illes et al., *From Neuroimaging to Neuroethics*, 5 Nature Neurosci. 205 (2003).

MRI (functional and structural) is quite safe, and MRI machines are widespread in developed countries, largely for clinical use but increasingly for research use as well. Although fMRI research is subject to many questions and controversies (discussed *infra* Section IV), this technique has been responsible for most of the recent interest in applying neuroscience to law, from criminal responsibility to lie detection.

B. EEG and MEG

EEG is the measurement of the brain's electrical activity as exhibited on the scalp; MEG is the measurement of the small magnetic fields generated by the brain's electrical activity. The roots of EEG go back into the nineteenth century, but its use increased dramatically in the 1930s and 1940s.

The process uses electrodes attached to the subject's head with an electrically conductive substance (a paste or a gel) to record electrical currents on the surface of the scalp. Multiple electrodes are used; for clinical purposes, 20 to 25 electrodes are commonly used, although arrays of more than 200 electrodes can be used. (In MEG, superconducting "squids"[8] are positioned over the scalp to detect the brain's tiny magnetic signals.) The electrical currents are generated by the neurons throughout the brain, although EEG is more sensitive to currents emerging from neurons closer to the skull. It is therefore more challenging to use EEG to reveal the functioning of structures deep in the brain.

Because EEG and MEG directly measure neural activity, in contrast to the measures of blood flow in fMRI, the timing of the neural activity can be measured with great precision (the temporal resolution), down to milliseconds. On the other hand, in comparison to fMRI, EEG and MEG are poor at determining the location of the sources of the currents (the spatial resolution). The EEG/MEG signal is a summation of the activity of thousands to millions of neurons at any one time. Any one pattern of EEG or MEG signal at the scalp has an infinite number of possible source patterns, making the problem of determining the brain source of measured EEG/MEG signal particularly challenging and the results less precise.

The results of clinical EEG and MEG tests can be very useful for detecting some kinds of brain conditions, notably epilepsy, and are also part of the process of diagnosing brain death. EEG and MEG are also used for research, particularly in the form of event-related potentials, which correlate the size or pattern of the EEG or MEG signal with the performance of particular tasks or the presentation of particular stimuli. Thus, as with the hypothetical fMRI experiment described above, one could look for any consistent changes in the EEG or MEG signal when a subject sees faces rather than a blank screen. Apart from the determina-

8. SQUID stands for superconducting quantum interference device (and has nothing to do with the marine animal). This device can measure extremely small magnetic fields, including those generated by various processes in living organisms, and so is useful in biological studies.

tion of brain death, where EEG is already used, the most discussed possible legally relevant uses of EEG have been lie detection and memory detection.

EEG is safe, cheap, quiet, and portable. MEG is safe and quiet, but the technology is considerably more expensive than EEG and is not easily portable. EEG methods can tolerate much more head movement by the subject than PET or MRI techniques, although movement is often a challenge for MEG. EEG and MEG have good temporal resolution, distinguishing between milliseconds, which makes them very attractive for research, but their spatial resolution is inadequate for many research questions. As a result, some researchers use a combination of methods, integrating MRI and EEG or MEG data (acquired simultaneously or at different times) using sophisticated data analysis techniques.

C. Other Techniques

Functional neuroimaging (especially fMRI) and EEG seem to be the techniques that are most likely to lead to efforts to introduce neuroscience-based evidence in court, but several other neuroscience techniques also might have legal applications. This section briefly describes four other methods that may be discussed in court: lesion studies, transcranial magnetic stimulation, deep brain stimulation, and implanted microelectrode arrays.

1. Lesion studies

One powerful way to test whether particular brain regions are associated with particular mental processes is to study mental processes after those brain regions have been destroyed or damaged. Observations of the consequences of such lesions, created by accidents or disease, were, in fact, the main way in which localization of brain function was originally understood.

For ethical reasons, the experimental destruction of brain tissue is limited to nonhuman animals. Nonetheless, in addition to accidental damage, on occasion human brains will need to be intentionally damaged for clinical purposes. Tumors may have to be removed or, in some cases, epilepsy may have to be treated by removing the region of the brain that is the focus for the seizures. Valuable knowledge may be gained from following these subjects.

Our understanding of the role of the hippocampus in creating memories, as one example, was greatly aided by study of a patient known as H.M.[9] When he was 27 years old, H.M. was treated for intractable epilepsy, undergoing an

9. H.M.'s name, not publicly released until his death, was Henry Gustav Molaison. Details of his life can be found in several obituaries, including Benedict Carey, *H.M., An Unforgettable Amnesiac, Dies at 82*, N.Y. Times, Dec. 4, 2008, at A1, and *H.M., A Man Without Memories*, The Economist, Dec. 20, 2008. The first scientific report of his case was W.B. Scoville & Brenda Milner, *Loss of Recent Memory After Bilateral Hippocampal Lesions*, 20 J. Neurol., Neurosurg. Psychiatry 11 (1957).

experimental procedure that surgically removed his left and right medial temporal lobes, including most of his two hippocampi. The surgery was successful, but from that time until his death in 2008, H.M. could not form new long-term memories, either of events or of facts. His short-term memory, also known as working memory, was intact, and he could learn new motor, perceptual, and (some) cognitive skills (his "procedural memory" still functioned). He also could remember his life's events from before his surgery, although his memories were weaker the closer the events were to the surgery. Those brain regions were clearly involved in *making* new long-term memories for facts or events, but not in storing old ones.

2. *Transcranial magnetic stimulation (TMS)*

TMS is a noninvasive method of creating a temporary, reversible functional brain "lesion." Using this technique, researchers disrupt the organized activity of the brain's neurons by applying an electrical current. The current is formed by a rapidly changing magnetic field that is generated by a coil held next to the subject's skull. The field penetrates the scalp and skull easily and causes a small current in a roughly conical portion of the brain below the coil. This current induces a change in the typical responses of the neurons, which can block the normal functioning of that part of the brain.

TMS can be done in a number of ways. In some approaches, TMS happens at the same time as the subject performs the task to be studied. These concurrent approaches include single pulses or paired pulses as well as rapid (more than once per second) repetitive TMS that is delivered during task performance. Another method uses TMS for an extended period, often several minutes, before the task is performed. This sequential TMS uses slow (less than once per second) repetitive TMS.

The effects of single-pulse/paired-pulse and concurrent repetitive TMS are present while the coil is generating the magnetic field, and can extend for a few tens of milliseconds after the stimulation is turned off. By contrast, the effects of pretask repetitive TMS are thought to last for a few minutes (about half as long as the actual stimulation). When TMS is repeated regularly in nonhumans, long-term effects have been observed. Therefore, guidelines regarding how much stimulation can be applied in humans have been established.

The Food and Drug Administration (FDA) has approved TMS as a treatment for otherwise untreatable depression. The neuroscience research value of TMS stems from its ability to alter brain function in a relatively small area (about 2 cm) in an otherwise healthy brain, thus allowing for targeted testing of the role of a particular brain region for a particular class of cognitive abilities. By blocking normal functioning of the affected neurons, this can be equivalent, in effect, to a temporary lesion of that area of the brain. TMS appears to have minimal risks, but its long-term effects are not known.

3. Deep brain stimulation (DBS)

DBS is an FDA-approved treatment for several neurological conditions affecting movement, notably Parkinson's disease, essential tremor, and dystonia. The device used in DBS includes a lead that is implanted into a specific brain region, a pulse generator (generally implanted under the shoulder or in the abdomen), and a wire connecting the two. The pulse generator sends an electric current to the electrodes in the lead, which in turn affect the functioning of neurons in an area around the electrodes.

The precise manner by which DBS affects brain function remains unclear. Even for Parkinson's disease, for which it is widely used, individual patients sometimes benefit in unpredictable ways from placement of the lead in different locations and from different frequency or power of the stimulation.

Researchers are continuing to experiment with DBS for other conditions, such as depression, minimally conscious state, chronic pain, and overeating that leads to morbid obesity. The results are sometimes surprising. In a Canadian trial of DBS for appetite control, the obese patient did not ultimately lose weight but did suddenly develop a remarkable memory. That research group is now starting a trial of DBS for dementia.[10] Other surprises have included some negative side effects from DBS, such as compulsive gambling, hypersexuality, and hallucinations. These kinds of unexpected consequences from DBS make it of continuing broader research interest.

4. Implanted microelectrode arrays

Ultimately, to understand the brain fully one would like to know what each of its 100 billion neurons is doing at any given time, analyzed in terms of their collective patterns of activity.[11] No current technology comes close to that kind of resolution. For example, although fMRI has a voxel size of a few cubic millimeters, it is looking at the blood flow responding to thousands or millions of neurons at each point in the brain. Conversely, while direct electrical recordings allow individual neurons to be examined, and manipulated, it is not easy to record from many neurons at once. While still on a relatively small scale, recent developments now offer one method for recording from multiple neurons simultaneously by using an implanted microelectrode array.

A chip containing many tiny electrodes can be implanted directly into brain tissue. Some of those electrodes will make useable connections with neurons and can then be used either to record the activity of that neuron (when it is firing or

10. *See* Clement Hamani et al., *Memory Enhancement Induced by Hypothalamic/Fornix Deep Brain Stimulation*, 63 Annals Neurol. 119 (2008).

11. See the discussion in Emily R. Murphy & Henry T. Greely, *What Will Be the Limits of Neuroscience-Based Mindreading in the Law? in* The Oxford Handbook of Neuroethics (J. Illes & B.J. Sahakian eds., 2011).

not) or to stimulate the neuron to fire. These kinds of implants have been used in research on motor function, both in monkeys and in occasional human patients. The research has aimed at understanding better what neuronal activity leads to motion and hence, in the long run, perhaps to a method of treating quadriplegia or other motion disorders.

These arrays have several disadvantages as research tools. Arrays require neurosurgery for their implantation, with all of its consequent risks of infection or damage. They also have a limited lifespan, because the brain's defenses eventually prevent the electrical connection between the electrode and the neuron, usually over the span of a few months. Finally, the arrays can only reach a tiny number of the billions of neurons in the brain; current arrays have about 100 microelectrodes.

IV. Issues in Interpreting Study Results

Lawyers trying to introduce neuroscience evidence will almost always be arguing that, when interpreted in the light of some preexisting research study, some kind of neuroscience-based test of the brain of a person in the case—usually a party, though sometimes a witness—is relevant to the case. It might be a claim that a PET scan shows that a criminal defendant was likely to have been legally insane at the time of the crime; it could be a claim that an fMRI of a witness demonstrates that she is lying. The judge will have to determine whether the scientific evidence is admissible at all under the Federal Rules of Evidence, and particularly under Rule 702. If the evidence is admissible, the finder of fact will need to consider the validity and strength of the underlying scientific finding, the accuracy of the particular test performed on the party or witness, and the application of the former to the latter.

Neuroscience-based evidence will commonly raise several scientific issues relevant to both the initial admissibility decision and the eventual determination of the weight to be given the evidence. This section of the reference guide examines seven of these issues: replication, experimental design, group averages, subject selection and number, technical accuracy, statistical issues, and countermeasures. The discussion focuses on fMRI-based evidence, because that seems likely to be the method that will be used most frequently in the coming years, but most of the seven issues apply more broadly.

One general point is absolutely crucial. The various techniques discussed in Section III, *supra*, are generally accepted scientific procedures, both for use in research and, in most cases, in clinical care. Each one is a good scientific tool in general. The crucial issue is not likely to be whether the techniques meet the requirements for admissibility when used for *some* purposes, but whether the techniques—*when used for the purpose for which they are offered*—meet those requirements. Sometimes proponents of fMRI-based lie detection, for example, have argued that the technique should be accepted because fMRI is the subject of more than 12,000 peer-reviewed publications. That is true, but irrelevant—the question is the application of fMRI to lie detection, which is the subject of far fewer, and much less definitive, publications.

A. Replication

A good general rule of thumb in science is never to rely on any experimental finding until it has been independently replicated. This may be particularly true with fMRI experiments, not because of fraud or negligence on the part of the experimenters, but because, for reasons discussed below, these experiments are very complicated. Replication builds confidence that those complications have not led to false results.

In many scientific fields, including much of fMRI research, replication is sometimes not as common as it should be. A scientist often is not rewarded for replicating (or failing to replicate) another's work. Grants, tenure, and awards tend to go to people doing original research. The rise of fMRI has meant that such original experiments are easy to conceive and to attempt—anyone with experimental expertise, access to research subjects (often undergraduates), and access to an MRI scanner (found at any major medical facility) can try his or her own experiments and, if the study design and logic are sound and the results are statistically significant, may well end up with published results. Experiments replicating, or failing to replicate, another's work are neither as exciting nor as publishable.

For example, as discussed in more detail below, more than 15 different laboratories have collectively published 20 to 30 peer-reviewed articles finding some statistically significant relationship between fMRI-measured brain activity and deception. None of the studies is an independent replication of another laboratory's work. Each laboratory used its own experimental design, its own scanner, and its own method of analysis. Interestingly, the published results implicate many different areas of the brain as being activated when a subject lies. A few of the brain regions are found to be important in most of the studies, but many of the other brain regions showing a correlation with deception differ from publication to publication. Only a few of the laboratories have published replications of their own work; some of those laboratories have actually published findings with different results from those in their earlier publications.

That a finding has been replicated does not mean it is correct; different laboratories can make the same mistakes. Neither does failure of replication mean that a result is wrong. Nonetheless, the existence of independent replication is important support for a finding.

B. Problems in Experimental Design

The most important part of an fMRI experiment is not the MRI scanner, but the design of the underlying experiment being examined in the scanner. A poorly designed experiment may yield no useful information, and even a well-designed experiment may lead to information of uncertain relevance.

A well-designed experiment must focus on the particular mental state or brain process of interest while minimizing any systematic biases. This can be especially difficult with fMRI studies. After all, these studies are measuring blood flow in the brain associated with neuronal responses in particular regions. If, for example, in an experiment trying to assess how the brain reacts to pain, the experimental subjects are consistently distracted at one point in the experiment by thinking about something else, the areas of brain activation will include the areas activated by the distraction. One of the earliest published lie detection experiments was designed so that the experimental subjects pushed a button for "yes" only when saying (honestly) that they held the card displayed; they pushed the "no" button both when they did not hold the card displayed and when they did hold it but were following instructions to lie. They were to say "yes" only 24 times out of 432 trials.[12] The resulting differences might have come from the differences in thinking about telling the truth or telling a lie—but they also may have come from the differences in thinking about pressing the "no" button (the most common action) and pressing the "yes" button (the less frequent response). The results themselves cannot distinguish between the two explanations.

Designing good experiments is difficult, but in some respects the better the experiment, the less relevant it may prove to a real situation. A laboratory experiment attempts to minimize distractions and differences among subjects, but such factors will be common in real-world settings. Perhaps more important, for some kinds of experiments it will be difficult, if not impossible, to reproduce in the laboratory the conditions of interest in the real world. As an extreme example, if one is interested in how a murderer's brain functions during a murder, one cannot conduct an experiment that involves having the subject commit a murder in the scanner. For ethical reasons, that condition of interest *cannot* be tested in the experiment.

The problem of trying to detect deception provides a different example. All published laboratory-based experiments involve people who know that they are taking part in a research project. Most of them are students and are being paid to participate in the project. They have received detailed information about the experiment and have signed a consent form. Typically, they are instructed to "lie" about a particular matter. Sometimes they are told what the lie should be (to deny that they see a particular playing card, such as the seven of clubs, on a screen in the scanner); sometimes they are told to make up a lie (about their most recent

12. Daniel D. Langleben et al., *Telling Truth from Lie in Individual Subjects with Fast Event-Related fMRI*, 26 Human Brain Mapping 262 (2005). See discussion in Nancy Kanwisher, *The Use of fMRI in Lie Detection: What Has Been Shown and What Has Not, in* Emilio Bizzi et al., Using Imaging to Identify Deceit: Scientific and Ethical Questions (2009), at 10, and in Anthony Wagner, *Can Neuroscience Identify Lies? in* A Judge's Guide to Neuroscience, *supra* note 1, at 30.

vacation, for example). In either case, they are following instructions—doing what they *should* be doing—when they tell the "lie."

This situation is different from the realistic use of lie detection, when a guilty person needs to tell a convincing story to avoid a high-stakes outcome such as arrest or conviction—and even an innocent person will be genuinely nervous about the possibility of an incorrect finding of deception. In an attempt to parallel these real-world characteristics, some laboratory-based studies have tried to give subjects some incentive to lie successfully; for example, the subjects may be told (falsely) that they will be paid more if they "fool" the experimenters. Although this may increase the perceived stakes, it seems unlikely that it creates a realistic level of stress. These differences between the laboratory and the real world do not mean that the experimental results of laboratory studies are unquestionably different from the results that would exist in a real-world situation, but they do raise serious questions about the extent to which the experimental data bear on detecting lies in the real world.

Few judges will be expert in the difficult task of designing valid experiments. Although judges may be able themselves to identify weaknesses in experimental design, more often they will need experts to address these questions. Judges will need to pay close attention to that expert testimony and the related argument, as "details" of experimental design may turn out to be absolutely crucial to the value of the experimental results.

C. The Number and Diversity of Subjects

Doing fMRI scans is expensive. The total cost of performing an hour-long research scan of a subject ranges from about $300 to $1000. Much fMRI research, particularly work without substantial medical implications, is not richly funded. As a result, studies tend to use only a small number of subjects—many fMRI studies use 10 to 20 subjects, and some use even fewer. In the lie detection literature, for example, the number of subjects used ranges from 4 to about 30.

It is unclear how representative such a small group would be of the general population. This is particularly true of the many studies that use university students as research subjects. Students typically are from a restricted age range, are likely to be of above-average intelligence and socioeconomic background, may not accurately reflect the country's ethnic diversity, and typically will underrepresent people with serious mental conditions. To limit possible confounding variables, it can make sense for a study design to select, for example, only healthy, right-handed, native-English-speaking male undergraduates who are not using drugs. But the very process of selecting such a restricted group raises questions about whether the findings will be relevant to other groups of people. They may be directly relevant, or they may not be. At the early stages of any fMRI research, it may not be clear what kinds of differences among subjects will or will not be important.

D. Applying Group Averages to Individuals

Most fMRI-based research looks for statistically significant associations between particular patterns of brain activation across a number of subjects. It is highly unlikely than any fMRI pattern will be found always to occur under certain circumstances in *every* person tested, or even that it will always occur under those circumstances in any one person. Human brains and their responses are too complicated for that. Research is highly unlikely to show that brain pattern "A" follows stimulus "B" each and every time and in every single person, although it may show that A follows B most of the time.

Consider an experiment with 10 subjects that examines how brain activation varies with the sensation of pain. A typical approach to analyzing the data is to take the average brain activation patterns of all 10 subjects combined, looking for the regions that, across the group, have the greatest changes—the most statistically significant changes—when the painful stimulus is applied compared with when it is absent. Importantly, though, the most significant region showing increased activation on average may not be the region with the greatest increase in activation in any particular one of the 10 subjects. It may not even be the area with the greatest activation in *any* of the 10 subjects, but it may be the region that was most consistently active across the brains of the 10 subjects, even if the response was small in each person.

Although group averages are appropriate for many scientific questions, the problem is that the law, for the most part, is not concerned with "average" people, but with individuals. If these "averaged" brains show a particular pattern of brain activation in fMRI studies and a defendant's brain does not, what, if anything, does that mean?

It may or may not mean anything—or, more accurately, the chances that it is meaningful will vary. The findings will need to be converted into an assessment of an individual's likelihood of having a particular pattern of brain activation in response to a stimulus, and that likelihood can be measured in various ways.

Consider the following simplified example. Assume that 1000 people have been tested to see how their brains respond to a particular painful stimulus. Each is scanned twice, once when touched by a painfully hot metal rod and once when the rod is room temperature. Assume that all of them feel pain from the heated rod and that no one feels pain from the room temperature rod. And, finally, assume that 900 of the 1000 show a particular pattern of brain activation when touched with the hot rod, but only 50 of the 1000 show the same pattern when touched with the room temperature rod.

For these 1000 people, using the fMRI activation pattern as a test for the perception of this pain would have a sensitivity of 90% (90% of the 1000 who felt the pain would be correctly identified and only 10% would be false negatives). Using the activation as a test for the *lack* of the pattern would have a specificity of 95% (95% of those who did not feel pain were correctly identified and only

5% were false positives). Now ask, of all those who showed a positive test result, how many were actually positive? This percentage, the positive predictive value, would be 94.7%—900 out of 950. Depending on the planned use of the test, one might care more about one of these measures than another and there are often tradeoffs between them. Making a test more sensitive (so that it misses fewer people with the sought characteristic) often means making it less specific (so that it picks up more people who do not have the characteristic in question). In any event, when more people are tested, these estimates of sensitivity, specificity, and positive predictive value become more accurate.

There are other ways of measuring the accuracy of a test of an individual, but the important point is that some such conversion is essential. A research paper that reveals that the average subject's brain (more accurately, the "averaged subjects' brain") showed a particular reaction to a stimulus does not, in itself, say anything useful about how likely any one person is to have the same reaction to that stimulus. Further analyses are required to provide that information. Researchers, who are often more interested in identifying possible mechanisms of brain action than in creating diagnostic tests, will not necessarily have analyzed their data in ways that make them useful for application to individuals—or even have obtained enough data for that to be possible. At least in the near future, this is likely to be a major issue for applying fMRI studies to individuals, in the courtroom, or elsewhere.

E. Technical Accuracy and Robustness of Imaging Results

MRI machines are variable, complicated, and finicky. The machines come in several different sizes, based on the strength of the magnet, with machines used for clinical purposes ranging from 0.2 T to 3.0 T and research scanners going as high as 9.4 T. Three companies dominate the market for MRI machines—General Electric, Siemens, and Philips—although several other companies also make the machines. Both the power and the manufacturer of an MRI system can make a substantial difference in the resulting data (and images). These variations can be more important with functional MRI (though they also apply to structural MRI) so that a result seen on a 1.5-T Siemens scanner might not appear on a 3.0-T General Electric machine. Similarly, results from one 3.0-T General Electric machine may be different from those on an identical model.

Even the exact same MRI machine may behave differently from day to day or month to month. The machines frequently need maintenance or adjustments and sometimes can be inoperable for days or even weeks at a time. Comparing results from even the same machine before and after maintenance—or a system upgrade—can be difficult. This can make it hard to compare results across different studies or between the group average of one study and results from an individual subject.

These issues concern not only the quality of the scans done in research, but, even more importantly, the credibility of the individual scan sought to be introduced at trial. If different machines were used, care must be taken to ensure that the results are comparable. The individual scans also can have other problems. Any one scan is subject not only to machine-derived artifacts and other problems noted above, but also to human-generated artifacts, such as those caused by the subject's movements during the scan.

Finally, another technical problem of a different kind comes from the nature of fMRI research itself. The scanner will record changes in the relative levels of oxyhemoglobin to deoxyhemoglobin for thousands of voxels throughout the brain. During data analysis, these signal changes will be tested to see if they show any change in the response between the experimental condition and the baseline or control condition. Importantly, with fMRI, there is no definitive way to quantify precisely how large a change there was in the neural response compared to baseline; hence, the researcher must set a somewhat arbitrary statistical cutoff value (a threshold) for saying that a voxel was activated or deactivated. A researcher who wants only to look at strong effects will require a large change from baseline; a researcher who wants to see a wide range of possible effects will allow smaller changes from baseline to count.

Neither way is "right"—we do not know whether there is some minimum change in the BOLD response that means an "important" amount of brain activation has taken place, and if such a true value exists, it is likely to differ across brain regions, across tasks, and across experimental contexts. What this means is that different choices of statistical cutoff values can produce enormous differences in the apparent results. And, of course, the cutoff values used in the studies and in the scan of the individual of interest must be consistent across repeated tests. This important fact often may not be known.

F. Statistical Issues

Interpreting fMRI results requires the application of complicated statistical methods.[13] These methods are particularly difficult, and sometimes controversial, for fMRI studies, partly because of the thousands of voxels being examined. Fundamentally, most fMRI experiments look at many thousands of voxels and try to determine whether any of them are, on average, activated or deactivated as a result of the task or stimulus being studied. A simple test for statistical significance asks whether a particular result might have arisen by chance more than 1 time in 20 (or 5%): Is it significant at the .05 level? If a researcher is looking at the results for thousands of different voxels, it is likely that a number of voxels will show an effect above the threshold *just* by chance. There are statistical ways to control the rate of these false positives, but they need to be applied carefully. At the same time, rigid control of false positives through statistical correction (or the use of

13. For a broad discussion of statistics, see David H. Kaye & David A. Freedman, Reference Guide on Statistics, in this manual.

a very conservative threshold) can create another problem—an increase in the false-negative rate, which results in failing to detect true brain responses that are present in the data but that fall below the statistical threshold. The community of fMRI researchers recognizes that these issues of statistical significance are difficult to resolve.

Over the past decade, other statistical techniques are increasingly being used in neuroimaging research, including techniques that do not look at the statistical significance of changes in the BOLD response in individual voxels, but that instead examine changes in the distributed patterns of activation across many voxels in a region of the brain or across the whole brain. These techniques include methods known as principal component analysis, multivariate analysis, and related machine learning algorithms. These methods, the details of which are not reviewed in this chapter, are being used increasingly in neuroimaging research and are producing some of the most interesting results in the field. The techniques are fairly complex, and determining how to interpret the results of these tests can be controversial. Thus, these methods alone may require substantial and potentially confusing expert testimony in addition to all the other expert testimony about the underlying neuroscience evidence.

G. Possible Countermeasures

When neuroimaging is being used to compare the brain of one individual—a defendant, plaintiff, or witness, for example—to others, the individual undergoing neuroimaging might be able to use countermeasures to make the results unusable or misleading. And at least some of those countermeasures may prove especially hard to detect.

Subjects can disrupt almost any kind of scanning, whether done for structural or functional purposes, by moving in the scanner. Unwilling subjects could ruin scans by moving their bodies, heads, or, possibly, even by moving their tongues. Blatant movements to disrupt the scan would be apparent, both from watching the subject in the scanner and from seeing the results, leading to a possible negative inference that the person was trying to interfere with the scan. Nonetheless, that scan itself would be useless.

More interesting are possible countermeasures for functional scans. Polygraphy may provide a useful comparison. Countermeasures have long been tried in polygraphy with some evidence of efficacy. Polygraphy typically looks at the differences in physiological measurements of the subject when asked anxiety-provoking questions or benign control questions. Subjects can use drugs or alcohol to try to dampen their body reactions when asked anxiety-provoking questions. They can try to use mental measures to control or affect their physiological reactions, calming themselves during anxiety-provoking questions and increasing their emotional reaction to control questions. And, when asked control questions, they can try to increase the physiological signs the polygraph measures through physical means.

For example, subjects might bite their tongues, step on tacks hidden in their shoes, or tighten various muscles to try to increase their blood pressure, galvanic skin response, and so on. The National Academy of Sciences report on polygraphs concluded that

> Basic science and polygraph research give reason for concern that polygraph test accuracy may be degraded by countermeasures, particularly when used by major security threats who have a strong incentive and sufficient resources to use them effectively. If these measures are effective, they could seriously undermine any value of polygraph security screening.[14]

Some of the countermeasures used by polygraph subjects can be detected by, for example, drug or alcohol tests or by carefully watching the subject's body. But purely mental actions cannot be detected. These kinds of countermeasures may be especially useful to subjects seeking to beat neuroscience-based lie detection. For example, some argue that deception produces different activation patterns than telling the truth because it is mentally harder to tell a lie—more of the brain needs to work to decide whether to lie and what lie to tell. If so, two mental countermeasures immediately suggest themselves: make the lie easier to tell (through, perhaps, memorization or practice) or make the brain work harder when telling the truth (through, perhaps, counting backward from 100 by sevens).

Countermeasures are not, of course, potentially useful only in the context of lie detection. A neuroimaging test to determine whether a person was having the subjective feeling of pain might be fooled by the subject remembering, in great detail, past experiences of pain. The possible uses of countermeasures in neuroimaging have yet to be extensively explored, but at this point they cast additional doubt on the reliability of neuroimaging in investigations or in litigation.

V. Questions About the Admissibility and the Creation of Neuroscience Evidence

The admissibility of neuroscience evidence will depend on many issues, some of them arising from the rules of evidence, some from the U.S. Constitution, and some from other legal provisions. Another often-overlooked reality is that judges may have to decide whether to order this kind of evidence to be created. Certainly, judges may be called to pass upon the requests for criminal defendants (or convicts seeking postconviction relief) to be able to use neuroimaging. They may also have to decide motions in civil or criminal cases to compel neuroimaging. One could even imagine requests for warrants to "search the brains" of possible

14. *See* National Research Council, The Polygraph and Lie Detection 5 (2003). This report is an invaluable resource for discussions of not just the scientific evidence about the reliability of the polygraph, but also for general background about the application of science to lie detection.

witnesses for evidence. This guide does not seek to resolve any of these questions, but points out some of the problems that are likely to be raised about admitting neuroscience evidence in court.

A. Evidentiary Rules

This discussion looks at the main evidentiary issues that are likely to be raised in cases involving neuroscience evidence. Note, though, that judges will not always be governed by the rules of evidence. In criminal sentencing or in probation hearings, among other things, the Federal Rules of Evidence do not apply,[15] and they apply with limitations in other contexts.[16] Nonetheless, even in those circumstances, many of the principles behind the Rules, discussed below, will be important.

1. Relevance

The starting point for all evidentiary questions must be relevance. If evidence is not relevant to the questions in hand, no other evidentiary concerns matter. This basic reminder may be particularly useful with respect to neuroscience evidence. Evidence admitted, for example, to demonstrate that a criminal defendant had suffered brain damage sometime before the alleged crime is not, in itself, relevant. The proffered fact of the defendant's brain damage must be relevant. It may be relevant, for example, to whether the defendant could have formed the necessary criminal intent, to whether the defendant should be found not guilty by reason of insanity, to whether the defendant is currently competent to stand trial, or to mitigation in sentencing. It must, however, be relevant to *something* in order to be admissible at all, and specifying its relevance will help focus the evidentiary inquiry. The question, for example, would not be whether PET scans meet the evidentiary requirements to be admitted to demonstrate brain damage, but whether they have "any tendency to make the existence of any fact that is of consequence to the determination of the action more probable or less probable than it would be without the evidence."[17] The brain damage may be relevant to a fact, but that fact must be "of consequence to the determination of the action."

2. Rule 702 and the admissibility of scientific evidence

Neuroscience evidence will almost always be "scientific . . . knowledge" governed by Rule 702 of the Federal Rules of Evidence, as interpreted in *Daubert v. Merrell Dow Pharmaceuticals*[18] and its progeny, both before and after the amendments to Rule 702 in 2000. Rule 702 allows the testimony of a qualified expert "if

15. Fed. R. Evid. 1101(d).
16. Fed. R. Evid. 1101(e).
17. Fed. R. Evid. 401.
18. Daubert v. Merrell Dow Pharms., Inc., 509 U.S. 579 (1993).

(1) the testimony is sufficiently based upon reliable facts or data, (2) the testimony is the product of reliable principles and methods, and (3) the witness has applied the principles and methods reliably to the facts of the case." In *Daubert*, the Supreme Court listed several, nonexclusive guidelines for trial court judges considering testimony under Rule 702. The Committee that proposed the 2000 Amendments to Rule 702 summarized these factors as follows:

> The specific factors explicated by the *Daubert* Court are (1) whether the expert's technique or theory can be or has been tested—that is, whether the expert's theory can be challenged in some objective sense, or whether it is instead simply a subjective, conclusory approach that cannot reasonably be assessed for reliability; (2) whether the technique or theory has been subject to peer review and publication; (3) the known or potential rate of error of the technique or theory when applied; (4) the existence and maintenance of standards and controls; and (5) whether the technique or theory has been generally accepted in the scientific community.[19]

The tests laid out in *Daubert* and in the evidentiary rules governing expert testimony have been the subjects of enormous discussion, both by commentators and by courts. And, to the extent some neuroscience evidence has been admitted in federal courts (and the courts of states that follow Rule 702 or *Daubert*), it has passed those tests. We do not have the knowledge needed to analyze them in detail, but we will merely point out a few aspects that seem especially relevant to neuroscience evidence.

Neuroscience evidence should often be subject to tests, as long as the point of the neuroscience evidence is kept in mind. An fMRI scan might provide evidence that someone was having auditory hallucinations, but it could not prove that someone was not guilty by reason of insanity. The latter is a legal conclusion, not a scientific finding. The evidence might be relevant to the question of insanity, but one cannot plausibly conduct a scientific test of whether a particular pattern of brain activation is always associated with legal insanity. One might offer neuroimaging evidence about whether a person is likely to have unusual difficulty controlling his or her impulses, but that is not, in itself, proof that the person acted recklessly. The idea of testing helps separate the conclusions that neuroscience might be able to reach from the legal conclusions that will be beyond it.

Daubert's stress on the presence of peer review and publication corresponds nicely to scientists' perceptions. If something is not published in a peer-reviewed journal, it scarcely counts. Scientists only begin to have confidence in findings after peers, both those involved in the editorial process and, more important, those who read the publication, have had a chance to dissect them and to search intensively for errors either in theory or in practice. It is crucial, however, to recognize that publication and peer review are not in themselves enough. The publications need to be compared carefully to the evidence that is proffered.

19. Fed. R. Evid. 702 advisory committee's note.

First, the published, peer-reviewed articles must establish the specific scientific fact being offered. An (accurate) assertion that fMRI has been the basis of more than 12,000 peer-reviewed publications will help establish that fMRI can be used in ways that the scientific community finds reliable. By themselves, however, those publications do not establish any particular use of fMRI. If fMRI is being offered as proof of deception, the 20 or 30 peer-reviewed articles concerning its ability to detect deception are most important, not the 11,980 articles involving fMRI for other purposes.

Second, the existence of several peer-reviewed publications on the same general method does not support the accuracy of any one approach if those publications are mutually inconsistent. There are now about 20 to 30 peer-reviewed publications that, using fMRI, find statistically significant differences in patterns of brain activation depending on whether the subjects were telling the truth or (typically) telling a lie when instructed to do so. Many of those publications find patterns that are different from, and often inconsistent with, the patterns described in the other publications. Multiple *inconsistent* publications do not add weight, and may indeed subtract it, from a scientific method or theory.

Third, the peer-reviewed publication needs to describe in detail the method about which the expert plans to testify. A commercial firm might, for example, claim that its method is "based on" some peer-reviewed publications, but unless the details of the firm's methods were included in the publication, those details were neither published nor peer reviewed. A proprietary algorithm used to generate a finding published in the peer-reviewed literature is not adequately supported by that literature.

The error rate is also crucial to most neuroscience evidence, in two different senses. One is the degree to which the machines used to produce the evidence make errors. Although these kinds of errors may balance out in a large sample used in published literature, any scan of any one individual may well be affected by errors in the scanning process. Second, and more important, neuroscience evidence will almost never give an absolute answer, but will give a probabilistic one. For example, a certain brain structure or activation pattern will be found in some percentage of people with a particular mental condition or state. These group averages will have error rates when they are applied to individuals. Those rates need to be known and presented.

The issue of standards and controls also is important in neuroscience. This area is new and has not undergone the kind of standardization seen, for example, in forensic DNA analysis. When trying to apply neuroscience findings to an individual, evidence from the individual needs to have been acquired in the same way, with the same standards and conditions, as the evidence from which the scientific conclusions were drawn—or, at least, in ways that can be made readily comparable. For example, there is no one standard in fMRI research for what statistical threshold should be used for a change in the BOLD signal to "count" as a meaningful activation or deactivation. An individual's scan would need to be analyzed under the same definition for activation as was used in the research supporting the method, and the effects of the chosen threshold on finding a false positive or false negative must be considered.

The final consideration, general acceptance in the scientific community, also needs to be applied carefully. There is clearly general acceptance in the scientific community that fMRI can provide scientifically and sometimes clinically useful information about the workings of human brains, but that does not mean there is general acceptance of any particular fMRI application. Similarly, there may be general acceptance that fMRI can provide some general information about the physical correlates of a particular mental state, but without general acceptance that it can do so reliably in an individual case.

3. Rule 403

Rule 702 is not the only test that neuroscience evidence will need to pass to be admitted in court. Even evidence admissible under that rule must still escape the exclusion provided by Rule 403:

> Although relevant, evidence may be excluded if its probative value is substantially outweighed by the danger of unfair prejudice, confusion of the issues, or misleading the jury, or by considerations of undue delay, waste of time, or needless presentation of cumulative evidence.

As discussed in detail in a recent article,[20] Rule 403 may be particularly important with some attempted applications of neuroscience evidence because of the balance it requires between the value of evidence to the decisionmaker and its costs.

The probative value of such evidence may often be questioned. Neuroscience evidence will rarely, if ever, be definitive. It is likely to have a range of uncertainties, from the effectiveness of the method in general, to questions of its proper application in this case, to whether any given individual's reactions are the same as those previously tested.

The other side of Rule 403, however, is even more troublesome. The time necessary to introduce such evidence, and to educate the jury (and judge) about it, will usually be extensive. The possibilities for confusion are likely to be great. And there is at least some evidence that jurors (or, to be precise, "mock jurors") are particularly likely to overestimate the power of neuroscience evidence.[21]

20. Teneille Brown & Emily Murphy, *Through a Scanner Darkly: Functional Neuroimaging as Evidence of a Criminal Defendant's Past Mental States*, 62 Stan. L. Rev. 1119 (2010).

21. *See* Deena Skolnick Weisberg et al., *The Seductive Allure of Neuroscience Explanations*, 20 J. Cog. Neurosci. 470 (2008); David P. McCabe & Alan D. Castel, *Seeing Is Believing: The Effect of Brain Images on Judgments of Scientific Reasoning*, 107 Cognition 343 (2008). These articles are discussed in Brown & Murphy, *supra* note 20, at 1199–1202. *But see* N.J. Schweitzer et al., *Neuroimages as Evidence in a Mens Rea Defense: No Impact*, Psychol. Pub. Pol'y L. (in press) providing the experimental results that seem to indicate that showing neuroimages to mock jurors does not affect their decisions.

A high-tech "picture" of a living brain, complete with brain regions shown in bright orange and deep purple (colors not seen in an actual brain), may have an unjustified appeal to a jury. In each case, judges will need to weigh possibilities of confusion or prejudice, along with the near certainty of lengthy testimony, against the claimed probative value of the evidence.

4. *Other potentially relevant evidentiary issues*

Neuroscience evidence will, of course, be subject in individual cases to all evidentiary rules, from the Federal Rules of Evidence or otherwise, and could be affected by many of them. Four examples follow where the application of several such rules to this kind of evidence may raise interesting issues; there are undoubtedly many others.

First, in June 2009 the U.S. Supreme Court decided *Melendez-Diaz v. Massachusetts*,[22] where the five-justice majority held that the Confrontation Clause required the prosecution to present the testimony at trial of state laboratory analysts who had identified a substance as cocaine. This would seem to apply to any use by the prosecution in criminal cases of neuroscience evidence about a scanned defendant or witness, although it is not clear who would have to testify. Would testimony be required from the person who observed the procedure, the person who analyzed the results of the procedure, or both? If the results were analyzed by a computerized algorithm, would the individual (or group) that wrote that algorithm have to testify? These questions, and others, are not unique to neuroscience evidence, of course, but will have to be sorted out generally after *Melendez-Diaz*.

Second, the Federal Rules of Evidence put special limits on the admissibility of evidence of character and, in some cases, of predisposition.[23] In some cases, neuroscience evidence offered for the purpose of establishing a regular behavior of the person might be viewed as evidence of character[24] or predisposition (or

22. 129 S. Ct. 2527 (2009).

23. Fed. R. Evid. 404, 405, 412–415, 608.

24. Evidence about lie detection has sometimes been viewed as "character evidence," introduced to bolster a witness's credibility. The Canadian Supreme Court has held that polygraph evidence is inadmissible in part because it violates the rule limiting character evidence.

> "What is the consequence of this rule in relation to polygraph evidence? Where such evidence is sought to be introduced, it is the operator who would be called as the witness, and it is clear, of course, that the purpose of his evidence would be to bolster the credibility of the accused and, in effect, to show him to be of good character by inviting the inference that he did not lie during the test. In other words, it is evidence not of general reputation but of a specific incident, and its admission would be precluded under the rule. It would follow, then, that the introduction of evidence of the polygraph test would violate the character evidence rule." R. v. Béland, 60 C.R. (3d) 1, ¶¶ 71–72 (1987).

The Canadian court also held that polygraph evidence violated another rule concerning character evidence, the rule against "oath-helping."

> "From the foregoing comments, it will be seen that the rule against oath-helping, that is, adducing evidence solely for the purpose of bolstering a witness's credibility, is well grounded in authority. It

lack of predisposition). Whether such evidence could be admitted might hinge on whether it was offered in a civil case or a criminal case, and, if in a criminal case, by the prosecution or the defendant.

Third, Federal Rule of Evidence 406 allows the admission of evidence about a habit or routine practice to prove that the relevant person's actions conformed to that habit or routine practice. It is conceivable that neuroscience evidence might be used to describe "habits of mind" and thus be offered under this rule.

The fourth example applies to neuroscience-based lie detection. Although New Mexico is the only U.S. jurisdiction that generally allows the introduction of polygraph evidence,[25] several jurisdictions allow polygraph evidence in two specific situations. First, polygraph evidence is sometimes allowed when both parties have stipulated to its admission in advance of the performance of the test. (This does lead one to wonder whether a court would allow evidence from a psychic or from a fortune-telling toy, like the Magic Eight Ball, if both parties stipulated to it.) Second, polygraph evidence is sometimes allowed to impeach or to corroborate a witness's testimony.[26] If a neuroscience-based lie detection technique were found to be as reliable as the polygraph, presumably those jurisdictions would have to consider whether to extend these exceptions to such neuroscience evidence.

B. Constitutional and Other Substantive Rules

In many contexts, courts will be asked to admit neuroscience evidence or to order, allow, or punish its creation. Such actions may implicate a surprisingly large number of constitutional rights, as well as other substantive legal provisions. Most of these would be rights against the creation or use of neuroscience evidence, although some would be possible rights to its use. And one constitutional provision, the Fourth Amendment, might cut both ways. Again, this section will not seek to discuss all possible such claims or to resolve any of them, but only to raise some of the most interesting issues.

1. Possible rights against neuroscience evidence

a. The Fifth Amendment privilege against self-incrimination

Could a person be forced to "give evidence" through a neuroscience technology, or would that violate his or her privilege against self-incrimination? This has

is apparent that, since the evidence of the polygraph examination has no other purpose, its admission would offend the well-established rule." R. v. Béland, 60 C.R. (3d) 1, ¶ 67(1) (The court also ruled against polygraph evidence as violating the rule against prior consistent statements and, because the jury needs no help in assessing credibility, the rule on the use of expert witnesses).

25. Lee v. Martinez, 96 P.3d 291 (N.M. 2004).

26. *See, e.g.*, United States v. Piccinonna, 885 F.2d 1529 (11th Cir. 1989) (en banc). *See also* United States v. Allard, 464 F.3d. 529 (5th Cir. 2006); Thornburg v. Mullin, 422 F.3d 1113 (10th Cir. 2005).

already begun to be discussed by legal scholars in the context of lie detection.[27] One issue is whether the neuroscience evidence is "testimonial evidence." If it were held to be "testimonial" it would be subject to the privilege, but if it were nontestimonial, it would, under current law, not be. Examples of nontestimonial evidence for purposes of the privilege against self-incrimination include incriminating information from a person's private diaries, a blood alcohol test, or medical X rays. An fMRI scan is nothing more than a computer record of radio waves emitted by molecules in the brain. It does not seem like "testimony." On the other hand, fMRI-based lie detection currently involves asking the subject questions to which he or she gives answers, either orally, by pressing buttons, or by some other form of communication. Perhaps those answers would make the resulting evidence "testimonial."

It is possible, however, that answers may not be necessary. Two EEG-based systems claim to be able to determine whether a person either recognizes or has "experiential knowledge" of an event (a memory derived from experience as opposed to being told about it).[28] Very substantial scientific questions exist about each system, but, assuming they were to be admitted as reliable, they would raise this question more starkly because they do not require the subject of the procedure to communicate. The subject is shown photographs of relevant locations or read a description of the events while hooked up to an EEG. The brain waves,

27. The law review literature, by both faculty and students, discussing the Fifth Amendment and neuroscience-based lie detection is already becoming voluminous. *See, e.g.*, Nita Farahany, *Incriminating Thoughts*, 64 Stan. L. Rev., Paper No. 11-17, *available at* SSRN: http://ssrn.com/abstract=1783101 (2011); Dov Fox, *The Right to Silence as Protecting Mental Control*, 42 Akron L. Rev. 763 (2009); Matthew Baptiste Holloway, *One Image, One Thousand Incriminating Words: Images of Brain Activity and the Privilege Against Self-Incrimination*, 27 Temp. J. Sci. Tech. & Envtl. L. 141 (2008); William Federspiel, *Neuroscience Evidence, Legal Culture, and Criminal Procedure*, 16 Wm. & Mary Bill Rts. J. 865 (2008); Sarah E. Stoller & Paul Root Wolpe, *Emerging Neurotechnologies for Lie Detection and the Fifth Amendment*, 33 Am. J.L. & Med. 359 (2007); Michael S. Pardo, *Neuroscience Evidence, Legal Culture, and Criminal Procedure*, 33 Am. J. Crim. L. 301 (2006); and Erich Taylor, *A New Wave of Police Interrogation? "Brain Fingerprinting," the Constitutional Privilege Against Self-Incrimination, and Hearsay Jurisprudence*, U. Ill. J.L. Tech. & Pol'y 287 (2006).

28. The first system is the so-called Brain Fingerprinting, developed by Dr. Larry Farwell. This method was introduced successfully in evidence at the trial court level in a postconviction relief case in Iowa; the use of the method in that case is discussed briefly in the Iowa Supreme Court's decision on appeal, Harrington v. Iowa, 659 N.W.2d 509, 516 n.6 (2003). (The Court expressed no view on whether that evidence was properly admitted. *See id.* at 516.) The method is discussed on the Web site of Farwell's company, Brain Fingerprinting Laboratories, www.brainwavescience.com. It is criticized from a scientific perspective in J. Peter Rosenfeld, *"Brain Fingerprinting": A Critical Analysis*, 4 Sci. Rev. Mental Health Practice 20 (2005). See also the brief discussion in Henry T. Greely & Judy Illes, *Neuroscience-Based Lie Detection: The Urgent Need for Regulation*, 33 Am. J.L. & Med. 377, 387–88 (2007).

The second system is called Brain Electrical Oscillation Signature (BEOS) and was developed in India, where it has been introduced in trials and has been important in securing criminal convictions. *See* Anand Giridharadas, *India's Novel Use of Brain Scans in Courts Is Debated*, N.Y. Times, Sept. 15, 2008, at A10.

it is asserted, demonstrate whether the subject recognizes the photographs or has "experiential knowledge" of the events—no volitional communication is necessary. It might be harder to classify these EEG records as "testimonial."

b. Other possible general constitutional protections against compulsory neuroscience procedures

Even if the privilege against self-incrimination applies to neuroscience methods of obtaining evidence, it only applies where someone invokes the privilege. The courts and other government bodies force people to answer questions all the time, often under penalty of criminal or civil sanctions or of the court's contempt power. For example, a plaintiff in a civil case alleging damage to his health can be compelled to undergo medical testing at a defendant's appropriate request. In that case, the plaintiff can refuse, but only at the risk of seeing his case dismissed. Presumably, a party could similarly demand that a party, or a witness, undergo a neuroimaging examination, looking for either structural or functional aspects of the person's brain relevant to the case. If the privilege against self-incrimination is not available, or is available but not attractive, could the person asked have any other protection?

The answer is not clear. One might try to argue, along the lines of *Rochin v. California*,[29] that such a procedure violates the Due Process Clause of the Fifth and Fourteenth Amendments because it intrudes on the person in a manner that "shocks the conscience." Alternatively, one might argue that a "freedom of the brain" is a part of the fundamental liberty or the right to privacy protected by the Due Process Clause.[30] Or one might try to use language in some U.S. Supreme Court First Amendment cases that talk about "freedom of thought" to argue that the First Amendment's freedoms of religion, speech, and the press encompass a broader protection of the contents of the mind. The Court never seems to have decided a case on that point. The closest case might be *Stanley v. Georgia*,[31] where the Court held that Georgia could not criminalize a man's private possession of pornography for his own use. None of these arguments is, in itself, strongly supported, but each draws some appeal from a belief that we should be able to keep our thoughts, and, by extension, the workings of our brain, to ourselves.

c. Other substantive rights against neuroscience evidence

At least one form of possible neuroscience evidence may already be covered by statutory provisions limiting its creation and use—lie detection. In 1988, Congress

29. 342 U.S. 165 (1952).

30. *See* Paul Root Wolpe, *Is My Mind Mine? Neuroethics and Brain Imaging*, *in* The Penn Center Guide to Bioethics (Arthur L. Caplan et al. eds., 2009).

31. 394 U.S. 557 (1969). In the context of finding that the First Amendment forbids criminalizing mere possession of pornography, in the home, for an adult's private use, the Court wrote "Our whole constitutional heritage rebels at the thought of giving government the power to control men's minds." The leap from that language, or that holding, to some kind of mental privacy, is not small.

passed the federal Employee Polygraph Protection Act (EPPA).[32] Under this Act, almost all employers are forbidden to "directly or indirectly, . . . require, request, suggest, or cause any employee or prospective employee to take or submit to any lie detector test" or to "use, accept, refer to, or inquire concerning the results of any lie detector test of any employee or prospective employee."[33] The Act defines a "lie detector" broadly, as "a polygraph, deceptograph, voice stress analyzer, psychological stress evaluator, or any other similar device (whether mechanical or electrical) that is used, or the results of which are used, for the purpose of rendering a diagnostic opinion regarding the honesty or dishonesty of an individual."[34] The Department of Labor can punish violators with civil fines, and those injured have a private right of action for damages.[35] The Act does provide narrow exceptions for polygraph tests in some circumstances.[36]

In addition to federal statutes, many states passed their own versions of the EPPA, either before or after the federal act. The laws passed after EPPA generally apply similar prohibitions to some employers not covered by the federal act (such as state and local governments), but with their own idiosyncratic set of exceptions. Many states have also passed laws regulating lie detection services. Most of these seem clearly aimed at polygraphy, but, in some states, the language used is quite broad and may well encompass neuroscience-based lie detection.[37]

States also may provide protection against neuroscience evidence that goes beyond lie detection and could prevent involuntary neuroscience procedures. Some states have constitutional or statutory rights of privacy that could be read to include a broad freedom for mental privacy. And in some states, such as California, such privacy rights apply not just to state action but to private actors as well.[38] Most employment cases would be covered by EPPA and its state equivalents, but such state privacy protections might be used to help decide whether courts could

32. Federal Employee Policy Protection Act of 1988, Pub. L. No. 100-347, § 2, 102 Stat. 646 (codified as 29 U.S.C. §§ 2001–2009 (2006)). See generally the discussion of federal and state laws in Greely & Illes, *supra* note 28, at 405–10, 421–31.

33. 29 U.S.C. § 2002 (1)–(2) (2006) (The section also prohibits employers from taking action against employees because of their refusal to take a test, because of the results of such a test, or for asserting their rights under the Act); and *id.* § 2001 (3)–(4) (2006).

34. *Id.* § 2001(3) (2006).

35. *Id.* § 2005 (2006).

36. *Id.* § 2001(6) (2006).

37. *See generally* Greely & Illes, *supra* note 28, at 409–10, 421–31 (for both state laws on employee protection and state laws more broadly regulating polygraphy).

38. "All people are by nature free and independent and have inalienable rights. Among these are enjoying and defending life and liberty, acquiring, possessing, and protecting property, and pursuing and obtaining safety, happiness, *and privacy*." (emphasis added). Calif. Const. art. I, § 1. The words, "and privacy" were added by constitutional amendment in 1972. The California Supreme Court has applied these privacy protections in suits against private actors: "In summary, the Privacy Initiative in article I, section 1 of the California Constitution creates a right of action against private as well as government entities." Hill v. Nat'l Collegiate Athletic Ass'n, 865 P.2d 633, 644 (Cal. 1994).

compel neuroimaging scans or whether they could be required in nonemployment relationships, such as school/student or parent/child.

d. Neuroscience evidence and the Sixth and Seventh Amendment rights to trial by jury

One might also argue that some kinds of neuroscience evidence could be excluded from evidence as a result of the federal constitutional rights to trial by jury in criminal and most civil cases. In *United States v. Scheffer,*[39] the Supreme Court upheld an express ban in the Military Rules of Evidence on the admission of any polygraph evidence against a criminal defendant's claimed Sixth Amendment right to introduce the evidence in his defense. Justice Thomas wrote the opinion of the Court holding that the ban was justified by the questionable reliability of the polygraph. Justice Thomas continued, however, in a portion of the opinion joined only by Chief Justice Rehnquist and Justices Scalia and Souter, to hold that the Rule could also be justified by an interest in the role of the jury:

> It is equally clear that Rule 707 serves a second legitimate governmental interest: Preserving the jury's core function of making credibility determinations in criminal trials. A fundamental premise of our criminal trial system is that "the *jury* is the lie detector." *United States v. Barnard,* 490 F.2d 907, 912 (CA9 1973) (emphasis added), *cert. denied,* 416 U.S. 959, 40 L. Ed. 2d 310, 94 S. Ct. 1976 (1974). Determining the weight and credibility of witness testimony, therefore, has long been held to be the "part of every case [that] belongs to the jury, who are presumed to be fitted for it by their natural intelligence and their practical knowledge of men and the ways of men." *Aetna Life Ins. Co. v. Ward,* 140 U.S. 76, 88, 35 L. Ed. 371, 11 S. Ct. 720 (1891).[40]

The other four justices in the majority, and Justice Stevens in dissent, disagreed that the role of the jury justified this rule, but the question remains open. Justice Thomas's opinion did not argue that exclusion was *required* as part of the rights to jury trials in criminal and civil cases under the Sixth and Seventh Amendments, respectively, but one might try to extend his statements of the importance of the jury as "the lie detector" to such an argument.[41]

39. 523 U.S. 303 (1998).

40. *Id.* at 312–13.

41. The Federal Rules of Criminal Procedure effectively give the prosecution a right to a jury trial, by allowing a criminal defendant to waive such a trial only with the permission of both the prosecution and the court. Fed. R. Crim. P. 23(a). Many states allow a criminal defendant to waive a jury trial unilaterally, thus depriving the prosecution of an effective "right" to a jury.

2. *Possible rights to the creation or use of neuroscience evidence*

a. The Eighth Amendment right to present evidence of mitigating circumstances in capital cases

In one of many ways in which "death is different," in *Lockett v. Ohio*,[42] the U.S. Supreme Court held that the Eighth Amendment guarantees a convicted defendant in a capital case a sentencing hearing in which the sentencing authority must be able to consider any mitigating factors. In *Rupe v. Wood*,[43] the Ninth Circuit, in an appeal from the defendant's successful habeas corpus proceeding, applied that holding to find that a capital defendant had a constitutional right to have polygraph evidence admitted as mitigating evidence in his sentencing hearing. The court agreed that totally unreliable evidence, such as astrology, would not be admissible, but that the district court had properly ruled that polygraph evidence was not that unreliable. (The Washington Supreme Court had previously decided that polygraph evidence should be admitted in the penalty phase of capital cases under some circumstances.[44]) Thus, capital defendants may argue that they have the right to present neuroscience evidence as mitigation even if it would not be admissible during the guilt phase.

b. The Sixth Amendment right to present a defense

The *Scheffer* case arose in the context of another right guaranteed by the Sixth Amendment, the right of a criminal defendant to present a defense. It seems likely that neuroscience evidence will first be offered by parties who have been its voluntary subjects and who will argue that it strengthens their cases. In fact, the main use of neuroimaging in the courts so far, at least in criminal cases, has been by defendants seeking to demonstrate through the scans some element of a defense or mitigation. If jurisdictions were to exclude such evidence categorically, they might face a similar Sixth Amendment challenge.

The Supreme Court has held that some prohibitions on evidence in criminal cases violate the right to present a defense. Thus, in *Rock v. Arkansas*,[45] the Court struck down a *per se* rule in Arkansas against the admission of hypnotically refreshed testimony, holding that it was "arbitrary or disproportionate to the purposes [it is] designed to serve." The *Scheffer* case probably provides the model for how arguments about exclusions of neuroscience evidence would play out. Eight of the Justices in *Scheffer* agreed that the reliability of polygraphy was sufficiently

42. 438 U.S. 586 (1978).

43. 93 F.3d 1434, 1439–41 (9th Cir. 1996). *But see* United States v. Fulks, 454 F.3d 410, 434 (4th Cir. 2006). *See generally* Christopher Domin, *Mitigating Evidence? The Admissibility of Polygraph Results in the Penalty Phase of Capital Trials*, 41 U.C. Davis L. Rev. 1461 (2010), which argues that the Supreme Court should resolve the resulting circuit split by adopting the Ninth Circuit's position.

44. State v. Bartholomew, 101 Wash. 2d 631, 636, 683 P.2d 1079 (1984).

45. 483 U.S. 44, 56, 97 L. Ed. 2d 37, 107 S. Ct. 2704 (1987).

questionable as to justify the *per se* ban on its use. Justice Stevens, however, dissented, finding polygraphy sufficiently reliable to invalidate its per se exclusion.

3. The Fourth Amendment

The Fourth Amendment raises some particularly interesting questions. It provides, of course, that,

> The right of the people to be secure in their persons, houses, papers, and effects, against unreasonable searches and seizures, shall not be violated, and no Warrants shall issue, but upon probable cause, supported by Oath or affirmation, and particularly describing the place to be searched, and the persons or things to be seized.

On the one hand, an involuntary neuroscience examination would seem to be a search or seizure, and thus "unreasonable" neuroscience examinations are prohibited. To that extent, the Fourth Amendment would appear to be a protection against compulsory neuroscience testing.

On the other hand, if, say, an fMRI scan or an EEG were viewed as a "search or seizure" for purposes of the Fourth Amendment, presumably courts could issue a warrant for such a search or seizure, given probable cause and the relevant procedural requirements. The use of such a warrant might (or might not) be limited by the privilege against self-incrimination or by some constitutional privacy right, but, if such rights did not apply, would such warrants allow our brains to be searched? This is, in a way, the ultimate result of the revolution in neuroscience, which identifies our incorporeal "mind" with our physical "brain" and allows us to begin to draw inferences from the brain to the mind. If the brain is a physical thing or a place, it could be searchable, even if the goal in searching it is to find out something about the mind, something that, as a practical matter, had never itself been directly searchable.

VI. Examples of the Possible Uses of Neuroscience in the Courts

Neuroscience may end up in court wherever someone's mental state or condition is relevant, which means it may be relevant to a vast array of cases. There are very few cases, civil or criminal, where the mental states of the parties are not at least theoretically relevant on issues of competency, intent, motive, recklessness, negligence, good or bad faith, or others. And even if the parties' own mental states were not relevant, the mental states of witnesses almost always will be potentially relevant—are they telling the truth? Are they biased against one party or another? The mental states of jurors and even of judges occasionally may be called into question.

There are some important limitations on the use of neuroscience in the courtroom. First, it is unlikely to be used that often, particularly if it remains expensive.

With the possible exception of lie detection or bias detection, most cases will not present a practical use for it. The garden variety breach of contract or assault and battery is not likely to provide a plausible context for convincing neuroscience evidence, especially if there is no evidence that the actor or the actions were odd or bizarre. And many cases will not provide, or justify, the resources necessary for a scan. Those costs could come down, but it seems unlikely that such evidence would commonly be admitted without expert testimony, and the costs of that seem likely to remain high.

Second, neuroscience evidence usually has a "time machine" problem. Neuroscience seems unlikely ever to be able to discern a person's state of mind in the past. Unless the legally relevant action took place inside an MRI scanner or other neuroscience tool, the best it may be able to do is to say that, based on your current mental condition or state, as shown by the current structure or functioning of your brain, you are more or less likely than average to have had a particular mental state or condition at the time of the relevant event. If the time of the relevant event is the time of trial (or shortly before trial)—as would be the case with the truthfulness of testimony, the existence of bias, or the existence of a particular memory—that would not be a problem, but otherwise it would be.

Nonetheless, neuroscience evidence seems likely to be offered into evidence for several issues, and in many of them, it already has been offered and even accepted. In some cases it will be, and has been, offered as evidence of "legislative facts," of realities relevant to a broader legal issue than the mental state of any particular party or witness. Thus, amicus briefs in two Supreme Court cases involving the punishment of juveniles—one about capital punishment and one about life imprisonment without possibility of parole—and to some extent the Court itself, have discussed neuroscience findings about adolescent brains.[46] Three of

46. In *Roper v. Simmons,* 543 U.S. 551 (2005), the Court held that the death penalty could not constitutionally be imposed for crimes committed while a defendant was a juvenile. Two amicus briefs argued that behavioral and neuroscience evidence supported this position. *See* Brief of Amicus Curiae American Medical Association et al., *Roper v. Simmons*; and Brief of Amicus Curiae American Psychological Association and the Missouri Psychological Association Supporting Respondent (No. 03-633).

In *Roper,* the Court itself did not substantially rely on the neuroscientific evidence and does not cite those amicus briefs. The Court's opinion noted the scientific evidence only in passing as one part of three relevant differences between adults and juveniles: "First, as any parent knows and as the scientific and sociological studies respondent and his *amici* cite tend to confirm, '[a] lack of maturity and an underdeveloped sense of responsibility are found in youth more often than in adults and are more understandable among the young. These qualities often result in impetuous and ill-considered actions and decisions.' [citation omitted]" 543 U.S. at 569. Justice Scalia, however, did take the majority to task for even this limited invocation of science and sociology. 543 U.S. at 616–18.

In *Graham v. Florida,* 2010 U.S. Lexis 3881, 130 S. Ct. 2011, 176 L. Ed. 2d 825 (2010), the Court held that defendants could not be sentenced to life without the possibility of parole for nonhomicide crimes committed while they were juveniles. Two amicus briefs similar to those discussed in *Roper* were filed. *See* Brief of Amicus Curiae American Medical Association (No. 08-7412) and Brief Amicus Curiae American Academy of Child & Adolescent Psychiatry (No. 08-7621) Supporting Neither

the handful of published cases in which fMRI evidence was offered in court concerned challenges to state laws requiring warning labels on violent videogames.[47] The states sought, without success, to use fMRI studies of the effects of violent videogames on the brains of children playing the games to support their statutes.[48]

These "wholesale" uses of neuroscience may (or may not) end up affecting the law, but the courts would be more affected if various "retail" uses of neuroscience become common, where a party or a witness is subjected to neuroscience procedures to determine something relevant only to that particular case. An incomplete list of some of the most plausible categories for such retail uses includes the following:

- Issues of responsibility, certainly criminal and likely also civil;
- Predicting future behavior for sentencing;
- Mitigating (or potentially aggravating) factors on sentencing;

Party; and Brief of Amicus Curiae American Psychological Association et al. Supporting Petitioners (Nos. 08-7412, 08-7621). The Court did refer more directly to the scientific findings in *Graham*, directly citing the amicus briefs:

> "No recent data provide reason to reconsider the Court's observations in *Roper* about the nature of juveniles. As petitioner's *amici* point out, developments in psychology and brain science continue to show fundamental differences between juvenile and adult minds. For example, parts of the brain involved in behavior control continue to mature through late adolescence." *See* Brief for American Medical Association et al. as Amici Curiae 16–24; Brief for American Psychological Association et al. as Amici Curiae 22–27.

Justice Thomas, in a dissent joined by Justice Scalia, reviewed some of the evidence from these amicus briefs:

> "In holding that the Constitution imposes such a ban, the Court cites 'developments in psychology and brain science' indicating that juvenile minds 'continue to mature through late adolescence,' *ante*, at 17 (citing Brief for American Medical Association et al. as *Amici Curiae* 16–24; Brief for American Psychological Association et al. as *Amici Curiae* 22–27 (hereinafter APA Brief)), and that juveniles are 'more likely [than adults] to engage in risky behaviors,'" *id*. at 7. But even if such generalizations from social science were relevant to constitutional rulemaking, the Court misstates the data on which it relies.

47. Entm't Software Ass'n v. Hatch, 443 F. Supp. 2d 1065 (D. Minn. 2006); Entm't Software Ass'n v. Blagojevich, 404 F. Supp. 2d 1051 (N.D. Ill. 2005); Entm't Software Ass'n v. Granholm, 404 F. Supp. 2d 978 (E.D. Mich. 2005). Each of the three courts held that the state statutes violated the First Amendment.

48. The courts, sitting in equity and so without juries, all considered the scientific evidence and concluded that it was insufficient to sustain the statutes' constitutionality. In *Blagojevich* the court heard testimony for the state directly from Dr. Kronenberger, the author of some of the fMRI-based articles on which the state relied, as well from Dr. Howard Nusbaum, for the plaintiffs, who attacked Dr. Kronenberger's study. After a substantial discussion of the scientific arguments, the district court judge, Judge Matthew Kennelly, found that "Dr. Kronenberger's studies cannot support the weight he attempts to put on them via his conclusions," and did not provide a basis for the statute. *Blagojevich*, 404 F. Supp. at 1063–67. Judge Kennelly's discussion of this point may be a good example of the kind of analysis neuroscience evidence may force upon judges.

- Competency, now or in the past, to take care of one's affairs, to enter into agreements or make wills, to stand trial, to represent oneself, and to be executed;
- Deception in current statements;
- Existence or nonexistence of a memory of some event and, possibly, some information about the status of that memory (true, false; new, old, etc.);
- Presence of the subjective sensation of pain;
- Presence of the subjective sensation of remorse; and
- Presence of bias against a party.

Many, but not all, of these issues have begun to be discussed in the literature. A few of them, such as criminal responsibility, mitigation, memory detection, and lie detection, are appearing in courtrooms; others, such as pain detection, have reached the edge of trial. This chapter does not discuss all of these topics and does not discuss any of them in great depth, but it will describe three of them—criminal responsibility, detection of pain, and lie detection—in order to provide a flavor of the possibilities.

A. Criminal Responsibility

Neuroscience may raise some deep questions about criminal responsibility. Assume we had excellent scientific evidence that a defendant could not help but commit the criminal acts because of a specific brain abnormality?[49] Should that affect the defendant's guilt and, if so, how? Should it affect his sentence or other subsequent treatment? The moral questions may prove daunting. Currently the law is not very interested in such deep questions of free will, but that may change.

Already, though, criminal law is concerned with the mental state of the defendant in many more specific contexts. A conviction generally requires both an *actus reus* and a *mens rea*—a "guilty act" and a "guilty mind." An unconscious person cannot "act," but even a conscious act is often not enough. Specific crimes often require specific intents, such as acting with a particular purpose or in a knowing or reckless fashion. Some crimes require even more defined mental states, such as a requirement for premeditation in some murder statutes. And almost all crimes can be excused by legal insanity. In these and other ways the mental state of the defendant may be relevant to a criminal case.

Neuroscience may provide evidence in some cases to support a defendant's claim of nonresponsibility. For example, a defendant who claims to have been insane at the time of the crime might try to support his or her claim by alleging that he or she is seeing and hearing hallucinations. Neuroimaging may be able

49. *See* Henry T. Greely, *Neuroscience and Criminal Responsibility Proving "Can't Help Himself" as a Narrow Bar to Criminal Liability*, *in* Law and Neuroscience, Current Legal Issues 13 (Michael Freeman ed. 2011).

to provide some evidence about whether the defendant is, in fact, hallucinating, at least at the time when he or she is in the scanner. Such imaging might show that the defendant had a stroke or tumor in a particular part of the brain, which then could be used to argue in some way against the defendant's criminal responsibility.[50]

Neuroimaging has been used more broadly in some criminal cases. For example, as noted above, in the trial of John Hinckley for the attempted assassination of President Reagan, the defense used CAT scans of Hinckley's brain to support the argument, based largely on his bizarre behavior, that he suffered from schizophrenia. The scientific basis for that conclusion, offered early in the history of brain CAT scans, was questionable at the time and has become even weaker since, but Hinckley was found not guilty by reason of insanity. More recently, in November 2009, testimony about an fMRI scan was introduced in the penalty phase of a capital case as mitigating evidence that the defendant suffered from psychopathy. The defendant was sentenced to death, but after longer jury deliberations than defense counsel expected.[51] (This appears to have been the first time fMRI results were introduced in a criminal case.[52])

Neuroscience evidence also may be relevant in wider arguments about criminal justice. Evidence about the development of adolescent brains has been referred to in appellate cases concerning the punishments appropriate for people who committed crimes while under age, including, as noted above, U.S. Supreme Court decisions. More broadly, some have urged that neuroscience will undercut much of the criminal justice system. The argument is that neuroscience ultimately will prove that no one—not even the sanest defendant—has free will and that this will fatally weaken the retributive aspect of criminal justice.[53]

50. In at least one fascinating case, a man who was convicted of sexual abuse of a child was found to have a large tumor pressing into his brain. When the tumor was removed, his criminal sexual impulses disappeared. When his impulses returned, so had his tumor. The tumor was removed a second time and, again, his impulses disappeared. J.M. Burns & R.H. Swerdlow, *Right Orbitofrontal Tumor with Pedophilia Symptom and Constructional Apraxia Sign*, 60 Arch. Neurology 437 (2003); *Doctors Say Pedophile Lost Urge After Tumor Removed*, USA Today, July 28, 2003. *See* Greely, *Neuroscience and Criminal Responsibility*, *supra* note 49 (offering a longer discussion of this case).

51. The defendant in this Illinois case, Brian Dugan, confessed to the murder but sought to avoid the death penalty. *See* Virginia Hughes, *Head Case*, 464 Nature 340 (2010) (providing an excellent discussion of this case).

52. Other forms of neuroimaging, particularly PET and structural MRI scans, have been more widely used in criminal cases. Dr. Amos Gur at the University of Pennsylvania estimates that he has used neuroimaging in testimony for criminal defendants about 30 times. *Id.*

53. *See, e.g.*, Robert M. Sapolsky, *The Frontal Cortex and the Criminal Justice System*, *in* Law and the Brain (Semir Zeki & Oliver Goodenough eds., 2006); Joshua Greene & Jonathan Cohen, *For the Law, Neuroscience Changes Nothing and Everything*, in Law and the Brain (Semir Zeki & Oliver Goodenough eds., 2006).

This argument has been forcefully attacked by Professor Stephen Morse. *See, e.g.*, Stephen J. Morse, *Determinism and the Death of Folk Psychology: Two Challenges to Responsibility from Neuroscience*, 9 Minn. J.L. Sci. & Tech. 1 (2008); Stephen J. Morse, *The Non-Problem of Free Will in Forensic Psychiatry*

The application of neuroscience evidence to individual claims of a lack of criminal responsibility should prove challenging.[54] Such claims will suffer from the time machine problem—the brain scan will almost always be from after, usually long after, the crime was committed and so cannot directly show the defendant's brain state (and hence, by inference, his or her mental state) at or before the time of the crime. Similarly, most of the neuroscience evidence will be from associations, not from experiments. It is hard to imagine an ethical experiment that would scan people when they are, or are not, committing particular crimes, leaving only indirect experiments. Evidence that, for example, more convicted rapists than nonrapists had particular patterns of brain activation when viewing sexual material might somehow be relevant to criminal responsibility, but it also might not.

Careful neuroscience studies, either structural or functional, of the brains of criminals are rare. It seems highly unlikely that a "responsibility region" will ever be found, one that is universally activated in law-abiding people and that is deactivated in criminals (or vice versa). At most, the evidence is likely to show that people with particular brain structures or patterns of brain functioning commit crimes more frequently than people without such structures or patterns. Applying this group evidence to individual cases will be difficult, if not impossible. All of the problems of technical and statistical analysis of neuroimaging data, discussed in Section IV, apply. And it is possible that the to-be-scanned defendants will be able to implement countermeasures to "fool" the expert analyzing the scan.

The use of neuroscience to undermine criminal responsibility faces another problem—identifying a specific legal argument. It is not generally a defense to a criminal charge to assert that one has a predisposition to commit a crime, or even a very high statistical likelihood, as a result of social and demographic variables, of committing a crime. It is not clear whether neuroscience would, in any more than a very few cases,[55] provide evidence that was not equivalent to predisposition evidence. (And, of course, prudent defense counsel might think twice before presenting evidence to the jury that his or her client was strongly predisposed to commit crimes.)

We are at an early stage in our understanding of the brain and of the brain states related to the mental states involved in criminal responsibility. At this point, about all that can be said is that at least some criminal defense counsel, seeking to represent their clients zealously, will watch neuroscience carefully for arguments they could use to relieve their clients from criminal responsibility.

and Psychology, 25 Behav. Sci. & L. 203 (2007); Stephen J. Morse, *Moral and Legal Responsibility and the New Neuroscience*, *in* Neuroethics: Defining the Issues in Theory, Practice, and Policy (Judy Illes ed., 2006); Stephen J. Morse, *Brain Overclaim Syndrome and Criminal Responsibility: A Diagnostic Note*, 3 Ohio St. J. Crim. L. 397 (2005).

54. A good short discussion of these challenges can be found in Helen Mayberg, *Does Neuroscience Give Us New Insights into Criminal Responsibility?* *in* A Judge's Guide to Neuroscience, *supra* note 1.

55. *See* Greely, *Neuroscience and Criminal Responsibility*, *supra* note 49 (arguing for a very narrow neuroscience-based defense).

B. Lie Detection

The use of neuroscience methods for lie detection probably has received more attention than any other issue raised in this chapter.[56] This is due in part to the cultural interest in lie detection, dating back in its technological phase nearly 90 years to the invention of the polygraph.[57] But it is also due to the fact that two commercial firms currently are offering fMRI-based lie detection services for sale in the United States: Cephos and No Lie MRI.[58] Currently, as far as we know,

56. For a technology whose results have yet to be admitted in court, the legal and ethical issues raised by fMRI-based lie detection have been discussed in an amazingly long list of scholarly publications from 2004 to the present. An undoubtedly incomplete list follows: Nita Farahany, *supra* note 27; Brown & Murphy, *supra* note 20; Anthony D. Wagner, *supra* note 12; Frederick Schauer, *Can Bad Science Be Good Evidence?: Neuroscience, Lie-Detection, and the Mistaken Conflation of Legal and Scientific Norms,* 95 Cornell L. Rev. 1191 (2010); Frederick Schauer, *Neuroscience, Lie-Detection, and the Law: A Contrarian View,* 14 Trends Cog. Sci. 101 (2010); Emilio Bizzi et al., Using Imaging to Identify Deceit: Scientific and Ethical Questions (2009); Joelle Anne Moreno, *The Future of Neuroimaged Lie Detection and the Law,* 42 Akron L. Rev. 717 (2009); Julie Seaman, *Black Boxes: fMRI Lie Detection and the Role of the Jury,* 42 Akron L. Rev. 931 (2009); Jane Campbell Moriarty, *Visions of Deception: Neuroimages and the Search for Truth,* 42 Akron L. Rev. 739 (2009); Dov Fox, *supra* note 27; Benjamin Holley, *It's All in Your Head: Neurotechnological Lie Detection and the Fourth and Fifth Amendments,* 28 Dev. Mental Health L. 1 (2009); Brian Reese, *Comment: Using fMRI as a Lie Detector—Are We Lying to Ourselves?* 19 Alb. L.J. Sci. & Tech. 205 (2009); Cooper Ellenberg, Student Article: *Lie Detection: A Changing of the Guard in the Quest for Truth in Court?* 33 Law & Psychol. Rev. 139 (2009); Julie Seaman, *Black Boxes,* 58 Emory L.J. 427 (2008); Matthew Baptiste Holloway, *supra* note 27; William Federspiel, *supra* note 27; Greely & Illes, *supra* note 28; Sarah E. Stoller & Paul R. Wolpe, *supra* note 27; Mark Pettit, *FMRI and BF Meet FRE: Braining Imaging and the Federal Rules of Evidence,* 33 Am. J.L. & Med. 319 (2007); Jonathan H. Marks, *Interrogational Neuroimaging in Counterterrorism: A "No-Brainer" or a Human Rights Hazard?* 33 Am. J.L. & Med. 483 (2007); Leo Kittay, *Admissibility of fMRI Lie Detection: The Cultural Bias Against "Mind Reading" Devices,* 72 Brook. L. Rev. 1351, 1355 (2007); Jeffrey Bellin, *The Significance (if Any) for the Federal Criminal Justice System of Advances in Lie Detector Technology,* Temp. L. Rev. 711 (2007); Henry T. Greely, *The Social Consequences of Advances in Neuroscience: Legal Problems; Legal Perspectives, in* Neuroethics: Defining the Issues in Theory, Practice and Policy 245 (Judy Illes ed., 2006); Charles N.W. Keckler, *Cross-Examining the Brain: A Legal Analysis of Neural Imaging for Credibility Impeachment,* 57 Hastings L.J. 509 (2006); Archie Alexander, *Functional Magnetic Resonance Imaging Lie Detection: Is a "Brainstorm" Heading Toward the "Gatekeeper"?* 7 Hous. J. Health L. & Pol'y (2006); Michael S. Pardo, *supra* note 27; Erich Taylor, *supra* note 27; Paul R. Wolpe et al., *Emerging Neurotechnologies for Lie-Detection: Promises and Perils,* 5 Am. J. Bioethics 38, 42 (2005); Henry T. Greely, *Premarket Approval Regulation for Lie Detection: An Idea Whose Time May Be Coming,* 5 Am. J. Bioethics 50–52 (2005); Sean Kevin Thompson, Note: *The Legality of the Use of Psychiatric Neuroimaging in Intelligence Interrogation,* 90 Cornell L. Rev. 1601 (2005); Henry T. Greely, *Prediction, Litigation, Privacy, and Property: Some Possible Legal and Social Implications of Advances in Neuroscience, in* Neuroscience and the Law: Brain, Mind, and the Scales of Justice 114–56 (Brent Garland ed., 2004); and Judy Illes, *A Fish Story? Brain Maps, Lie Detection, and Personhood,* 6 Cerebrum 73 (2004).

57. An interesting history of the polygraph can be found in Ken Alder, The Lie Detectors: The History of an American Obsession (2007). Perhaps the best overall discussion of the polygraph, including some discussion of its history, is found in the National Research Council report, *supra* note 14, commissioned in the wake of the Wen Ho Lee case, on the use of the technology for screening.

58. The Web sites for the two companies are at Cephos, www.cephoscorp.com (last visited July 3, 2010); and No Lie MRI, http://noliemri.com (last visited July 3, 2010).

evidence from fMRI-based lie detection has not been admitted into evidence in any court, but it was offered—and rejected—in two cases, *United States v. Semrau*[59] and *Wilson v. Corestaff Services, L.P.*,[60] in May 2010.[61] This section will begin by analyzing the issues raised for courts by this technology and then will discuss these two cases, before ending with a quick look at possible uses of this kind of technology outside the courtroom.

1. Issues involved in the use of fMRI-based lie detection in litigation

Published research on fMRI and detecting deception dates back to about 2001.[62] As noted above, to date between 20 and 30 peer-reviewed articles from about 15 laboratories have appeared claiming to find statistically significant correlations between patterns of brain activation and deception. Only a handful of the published studies have looked at the accuracy of determining deception in individual subjects as opposed to group averages. Those studies generally claim accuracy rates of between about 75% and 90%. No Lie MRI has licensed the methods used by one laboratory, that of Dr. Daniel Langleben at the University of Pennsylvania; Cephos has licensed the method used by another laboratory, that of Dr. Frank A. Kozel, first at the Medical University of South Carolina and then at the University of Texas Southwestern Medical Center. (The method used by a British researcher, Dr. Sean Spence, has been used on a British reality television show.)

All of these studies rely on research subjects, typically but not always college students, who are recruited for a study of deception. They are instructed to answer some questions truthfully in the scanner and to answer other questions inaccurately.[63] In the Langleben studies, for example, right-handed, healthy, male

59. No. 07-10074 M1/P, Report and Recommendation (W.D. Tenn. May 31, 2010).

60. 2010 NY slip op. 20176, 1 (N.Y. Super. Ct. 2010); 900 N.Y.S.2d 639; 2010 N.Y. Misc. LEXIS 1044 (2010).

61. In early 2009, a motion to admit fMRI-based lie detection evidence, provided by No Lie MRI, was made, and then withdrawn, in a child custody case in San Diego. The case is discussed in a prematurely entitled article, Alexis Madrigal, *MRI Lie Detection to Get First Day in Court*, WIRED SCI. (Mar. 16, 2009), *available at* http://blog.wired.com/wiredscience/2009/03/noliemri.html (last visited July 3, 2010). A somewhat similar method of using EEG to look for signs of "recognition" in the brain was admitted into one state court hearing for postconviction relief at the trial court level in Iowa in 2001, and both it and another EEG-based method have been used in India. As far as we know, evidence from the use of EEG for lie detection has not been admitted in any other U.S. cases. *See supra* note 28.

62. The most recent reviews of the scientific literature on this subject are Anthony D. Wagner, *supra* note 12; and S.E. Christ et al., *The Contributions of Prefrontal Cortex and Executive Control to Deception: Evidence from Activation Likelihood Estimate Meta-Analyses*, 19 Cerebral Cortex 2557 (2009). *See also* Greely & Illes, *supra* note 28 (for discussion of the articles through early 2007). The following discussion is based largely on those sources.

63. At least one fMRI study has attempted to investigate self-motivated lies, told by subjects who were *not* instructed to lie, but who chose to lie for personal gain. Joshua D. Greene & Joseph M. Paxton, *Patterns of Neural Activity Associated with Honest and Dishonest Moral Decisions*, 106 Proc. Nat'l Acad. Sci. 12,506 (2009). The experiment was designed to make it easy for subjects to realize they

University of Pennsylvania undergraduates were shown images of playing cards while in the scanner and asked to indicate whether they saw a particular card. They were instructed to answer truthfully except when they saw one particular card. Some of Kozel's studies used a different experimental paradigm, in which the subjects were put in a room and told to take either a watch or a ring. When asked in the scanner separately whether they had taken the watch and then whether they had taken the ring, they were to reply "no" in both cases—truthfully once and falsely the other time. When analyzed in various ways, the fMRI results showed statistically different patterns of brain activation (small changes in BOLD response) when the subjects were lying and when they were telling the truth.

In general, these studies are not guided by a consistent hypothesis about which brain regions should be activated or deactivated during truth or deception. The results are empirical; they see particular patterns that differ between the truth state and the lie state. Some have argued that the patterns show greater mental effort when deception is involved; others have argued that they show more impulse control when lying.

Are fMRI-based lie detection methods accurate? As a class of experiments, these studies are subject to all the general problems discussed in Section IV regarding fMRI scans that might lead to neuroscience evidence. So far there are only a few studies involving a limited number of subjects. (The method used by No Lie MRI seems ultimately to have been based on the responses of four right-handed, healthy, male University of Pennsylvania undergraduates.[64]) There have been, to date, no independent replications of any group's findings.

The experience of the research subjects in these fMRI studies of deception seems to be different from "lying" as the court system would perceive it. The subjects knew they were involved in research, they were following orders to lie, and they knew that the most harm that could come to them from being detected in a lie might be lesser payment for taking part in the experiment. This seems hard to compare to a defendant lying about participating in a murder. More fundamentally, it is not clear how one could conduct ethical but realistic experiments with lie detection. Research subjects cannot credibly be threatened with jail if they do not convince the researcher of the truth of their lies.

Only a handful of researchers have published studies showing reported accuracy rates with individual subjects and only with a small number of subjects.[65]

would be given more money if they lied about how many times they correctly predicted a coin flip. Investigators could not, however, determine if a subject lied in any particular trial.

64. Daniel D. Langleben et al., *Telling Truth from Lie in Individual Subjects with Fast Event-Related fMRI,* 26 Hum. Brain Mapping 262, 267 (2005).

65. See discussion in Anthony D. Wagner, *supra* note 12, at 29–35. Wagner analyzes 11 peer-reviewed, published papers. Seven come from Kozel's laboratory; three come from Langleben's. The only exception is a paper from John Cacciopo's group, which concludes " [A]lthough fMRI may permit investigation of the neural correlates of lying, at the moment it does not appear to provide a very accurate marker of lying that can be generalized across individuals or even perhaps across types

Some of the studies used complex and somewhat controversial statistical techniques. And although subjects in at least one experiment were invited to try to use countermeasures against being detected, no specific countermeasures were tested.

Beyond the scientific validity of these techniques lie a host of legal questions. How accurate is accurate enough for admissibility in court or for other legal system uses? What are the implications of admissible and accurate lie detection for the Fourth, Fifth, Sixth, and Seventh Amendments? Would jurors be allowed to consider the failure, or refusal, of a party to take a lie detector test? Would lie detection be available in discovery? Would each side get to do its own tests—and who would pay?

Accurate lie detection could make the justice system much more accurate. Incorrect convictions might become rare; so might incorrect acquittals. Accurate lie detection also could make the legal system much more efficient. It seems likely that far fewer cases would go to trial if the witnesses could expect to have their veracity accurately determined.

Inaccurate lie detection, on the other hand, holds the potential of ruining the innocent and immunizing the guilty. It is at least daunting to remember some of the failures of the polygraph, such as the case of Aldrich Ames, a Soviet (and then Russian) mole in the Central Intelligence Agency, who passed two Agency polygraph tests while serving as a paid spy.[66] The courts already have begun to decide whether and how to use these new methods of lie detection in the judicial process; the rest of society also will soon be forced to decide on their uses and limits.

2. *Two cases involving fMRI-based lie detection*

On May 31, 2010, U.S. Magistrate Judge Tu M. Pham of the Western District of Tennessee issued a 39-page report and recommendation on the prosecution's motion to exclude evidence from an fMRI-based lie detection report by Cephos in the case of *United States v. Semrau*.[67] The report came after a hearing on May 13–14 featuring testimony from Steve Laken, CEO of Cephos, for admission, and from two experts arguing against admission. (The district judge adopted the magistrate's report during the trial.)

The defendant in this case, a health professional accused of defrauding Medicare, offered as evidence a report from Cephos stating that he was being truthful

of lies by the same individuals." G. Monteleone et al., *Detection of Deception Using fMRI: Better Than Chance, But Well Below Perfection*, 4 Soc. Neurosci. 528 (2009). However, that study only looked at one brain region at a time, and it did not test combinations or patterns, which might have improved the predictive power.

66. *See* Senate Select Committee on Intelligence, Assessment of the Aldrich H. Ames Espionage Case and Its Implications for U.S. Intelligence (1994).

67. *See supra* note 59. The district court judge assigned to the case had a scheduling conflict on the date of the hearing on the prosecution's motion, and so the hearing was held before a magistrate judge from that district.

when he answered a set of questions about his actions and knowledge concerning the alleged crimes.

Judge Pham first analyzed the motion under Rule 702, using the *Daubert* criteria. He concluded that the technique was testable and had been the subject of peer-reviewed publications. On the other hand, he concluded that the error rates for its use in realistic situations were unknown. Furthermore, he found there were no standards for its appropriate use. To the extent that the publications relied on by Cephos to establish its reliability constituted such standards, those standards had not actually been followed in the tests of the defendant. Cephos actually scanned Dr. Semrau 2 times on 1 day, asking questions about one aspect of the criminal charges during the first scan and then about another aspect in the second scan. The company's subsequent analysis of those scans indicated that the defendant had been truthful in the first scan but deceptive in the second scan. Cephos then scanned him a third time, several days later, on the second subject but with revised questions, and concluded that he was telling the truth that time. Nothing in the publications relied upon by Cephos indicated that the third scan was appropriate. Finally, Judge Pham found that the method was not generally accepted in the relevant scientific community as sufficiently reliable for use in court, citing several publications, including some written by the authors whose methods Cephos used.

The magistrate judge then examined the motion under Rule 403 and found that the potential prejudicial effect of the evidence outweighed its probative value. He noted that the test had been conducted without the government's knowledge or participation, in a context where the defendant risked nothing by taking the test—a negative result would never be disclosed. He noted the jury's central role in determining credibility and considered the likelihood that the lie detection evidence would be a lengthy and complicated distraction from the jury's central mission. Finally, he noted that the probative value of the evidence was greatly reduced because the report only gave a result concerning the defendant's general truthfulness when responding to more than 10 questions about the events but did not even purport to say whether the defendant was telling the truth about any particular question.

Earlier that month, a state trial court judge in Brooklyn excluded another Cephos lie detection report in a civil case, *Wilson v. Corestaff Services, L.P.*[68] This case involved a claim by a former employee under state law that she had been subject to retaliation for reporting sexual harassment. The plaintiff offered evidence from a Cephos report finding that her main witness was truthful when he described how defendant's management said it would retaliate against the plaintiff.

That case did not involve an evidentiary hearing or, indeed, any expert testimony. The judge decided the lie detection evidence was not appropriate under New York's version of the *Frye* test, noting that, in New York, "courts have advised that the threshold question under *Frye* in passing on the admissibility

68. Wilson v. Corestaff Services, L.P., *supra* note 60.

of expert's testimony is whether the testimony is 'within the ken of the typical juror.'"[69] Because credibility is a matter for the jury, the judge concluded that this kind of evidence was categorically excluded under New York's version of *Frye*. He also noted that "even a cursory review of the scientific literature demonstrates that the plaintiff is unable to establish that the use of the fMRI test to determine truthfulness or deceit is accepted as reliable in the relevant scientific community."[70]

3. fMRI-based lie detection outside the courtroom

Lie detection might have applications to litigation without ever being introduced in trials. As is the case today with the polygraph, the fact that it is not generally admissible in court might not stop the police or the prosecutors from using it to investigate alleged crimes. Similarly, defense counsel might well use it to attempt to persuade the authorities that their clients should not be charged or should be charged with lesser offenses. One could imagine the same kinds of pretrial uses of lie detection in civil cases, as the parties seek to affect each other's perceptions of the merits of the case.

Such lie detection efforts could also affect society, and the law, outside of litigation. One could imagine prophylactic lie detection at the beginning of contractual relations, seeking to determine whether the other side honestly had the present intention of complying with the contract's terms. One can also imagine schools using lie detection as part of investigations of student misconduct or parents seeking to use lie detection on their children. The law more broadly may have to decide whether and how private actors can use lie detection, determining whether, for example, to extend to other contexts—or to weaken or repeal—the Employee Polygraph Protection Act.[71]

The current fMRI-based methods of lie detection provide one kind of protection for possible subjects—they are obvious. No one is going to be put into an MRI for an hour and asked to respond, repeatedly, to questions without realizing something important is going on. Should researchers develop less obtrusive or obvious methods of neuroscience-based lie detection, we will have to deal with the possibilities of involuntary and, indeed, surreptitious lie detection.

C. Detection of Pain

No matter where an injury occurs and no matter where it seems to hurt, pain is felt in the brain.[72] Without sensory nerves leading to the brain from a body

69. *Id.* at 6, *citing* People v. Cronin, 60 N.Y.2d 430, 458 N.E.2d 351, 470 N.Y.S.2d 110 (1983).

70. *See* Wilson, *supra* note 60, at 7.

71. *See supra* text accompanying note 32.

72. *See* Brain Facts, *supra* note 2, at 19–21, 49–50, which includes a useful brief description of the neuroscience of pain.

region, there is usually no experience of pain. Without the brain machinery and functioning to process the signal, no pain is perceived.

Pain turns out to be complicated—even the common pain that is experienced from an acute injury to, say, an arm. Neurons near the site of the injury called nociceptors transmit the pain signal to the spinal cord, which relays it to the brain. But other neurons near the site of the injury will, over time, adapt to affect the pain signal. Cells in the spinal cord can also modulate the pain signal that is sent to the brain, making it stronger or weaker. The brain, in turn, sends signals down to the spinal cord that cause or, at least, affect these modulations. And the actual sensation of pain—the "ouch"—takes place in the brain.

The immediate and localized sensation is processed in the somatosensory cortex, the brain region that takes sensory inputs from different body parts (with each body part getting its own portion of the somatosensory cortex) and processes them into a perceived sensation. The added knowledge that the sensation is painful seems to require the participation of other regions of the brain. Using fMRI and other techniques, some researchers have identified what they call the "pain matrix" in the brain, regions that are activated when experimental subjects, in scanners, are exposed to painful stimuli. The brain regions identified as part of the so-called pain matrix vary from researcher to researcher, but generally include the thalamus, the insula, parts of the anterior cingulate cortex, and parts of the cerebellum.[73]

Researchers have run experiments with subjects in the scanner receiving painful or not painful stimuli and have attempted to find activation patterns that appear when pain is perceived and that do not appear when pain is absent. (The subjects usually are given nonharmful painful stimuli such as having their skin touched with a hot metal rod or coated with a pepper-derived substance that causes a burning sensation.) Some have reported substantial success, detecting pain in more than 80% of the cases.[74] Other studies have found a positive correlation between the degree of activation in the pain matrix and the degree of subjective pain, both as reported by the subject and as possibly indicated by the heat of the rod or the amount of the painful substance—the higher the temperature or the concentration of the painful substance, the greater the average activation in the pain matrix.[75]

Other neuroscience studies of individual pain look not at brain function during painful episodes but at brain structure. Some researchers, for example, claim that different regions of the brain have different average size and neuron densities

73. A good review article on the uses of fMRI in studying pain is found in David Borsook & Lino R. Becerra, *Breaking Down the Barriers. fMRI Applications in Pain, Analgesia and Analgesics*, 2 Molecular Pain 30 (2006).

74. *See, e.g.*, Irene Tracey, *Imaging Pain*, 101 Brit J. Anaesth. 32 (2008).

75. *See, e.g.*, Robert C. Coghill et al., *Neural Correlates of Interindividual Differences in the Subjective Experience of Pain*, 100 Proc. Nat'l Acad. Sci. 8538 (2003).

in patients who have had long-term chronic pain than in those who have not had such pain.[76]

Pain is clearly complicated. Placebos, distractions, or great need can sometimes cause people not to sense, or perhaps not to notice, pain that could otherwise be overwhelming. Similarly, some people can become hypersensitive to pain, reporting severe pain when the stimulus normally would be benign. Amputees with phantom pain—the feeling of pain in a limb that has been gone for years—have been scanned while reporting this phantom pain. They show activation in the pain matrix. In some fMRI studies, people who have been hypnotized to feel pain, even when there is no painful stimulus, show activation in the pain matrix.[77] And in one fMRI study, subjects who reported feeling emotional distress, as a result of apparently being excluded from a "game" being played among research subjects, also showed, on average, statistically significant activation of the pain matrix.[78]

Pain also plays an enormous role in the legal system.[79] The existence and extent of pain is a matter for trial in hundreds of thousands of injury cases each year. Perhaps more importantly, pain figures into uncounted workers' compensation claims and Social Security disability claims. Pain is often difficult to prove, and the uncertainty of a jury's response to claimed pain probably keeps much litigation alive. We know that the tests for pain currently presented to jurors, judges, and other legal decisionmakers are not perfect. Anecdotes of and the assessments by pain experts both are convincing that some nontrivial percentage of successful claimants are malingering and only pretending to feel pain; a much greater percentage may be exaggerating their pain.

A good test for whether a person is feeling pain, and, even better, a "scientific" way to measure the amount of that pain—at least compared to other pains felt by that individual, if not to pain as perceived by third parties—could help resolve a huge number of claims each year. If such pain detection were reliable, it would make justice both more accurate and more certain, leading to faster, and

76. *See, e.g.*, Vania Apkarian et al., *Chronic Back Pain Is Associated with Decreased Prefrontal and Thalamic Gray Matter Density*, 24 J. Neurosci. 10,410 (2004); *see also* Arne May, *Chronic Pain May Change the Structure of the Brain*, 137 Pain 7 (2008); Karen D. Davis, *Recent Advances and Future Prospects in Neuroimaging of Acute and Chronic Pain*, 1 Future Neurology 203 (2006).

77. Stuart W. Derbyshire et al., *Cerebral Activation During Hypnotically Induced and Imagined Pain*, 23 NeuroImage 392 (2004).

78. Naomi I. Eisenberg, *Does Rejection Hurt? An fMRI Study of Social Exclusion*, 302 Science 290 (2003).

79. The only substantial analysis of the legal implications of using neuroimaging to detect pain is found in Adam J. Kolber, *Pain Detection and the Privacy of Subjective Experience*, 33 Am. J.L. & Med. 433 (2007). Kolber expands on that discussion in interesting ways in Adam J. Kolber, *The Experiential Future of the Law*, 60 Emory L.J. 585, 595–601 (2011). The possibility of such pain detection was briefly discussed earlier in two different 2006 publications: Henry T. Greely, *Prediction, Litigation, Privacy, and Property: Some Possible Legal and Social Implications of Advances in Neuroscience*, *supra* note 56, at 141–42; and Charles Keckler, *Cross-Examining the Brain*, *supra* note 56, at 544.

cheaper, resolution of many claims involving pain. The legal system, as well as the honest plaintiffs and defendants within it, would benefit.

A greater understanding of pain also might lead to broader changes in the legal system. For example, emotional distress often is treated less favorably than direct physical pain. If neuroscience were to show that, in the brain, emotional distress seemed to be the same as physical pain, the law might change. Perhaps more likely, if neuroscience could provide assurance that sincere emotional pain could be detected and faked emotional distress would not be rewarded, the law again might change. Others have argued that even our system of criminal punishment might change if we could measure, more accurately, how much pain different punishments caused defendants, allowing judges to let the punishment fit the criminal, if not the crime.[80] A "pain detector" might even change the practice of medicine in legally relevant ways, by giving physicians a more certain way to check whether their patients are seeking controlled substances to relieve their own pain or whether they are seeking them to abuse or to sell for someone else to abuse.

In at least one case, a researcher who studies the neuroscience of pain was retained as an expert witness to testify regarding whether neuroimaging could provide evidence that a claimant was, in fact, feeling pain. The case settled before the hearing.[81] In another case, a prominent neuroscientist was approached about being a witness against the admissibility of fMRI-based evidence of pain, but, before she had decided whether to take part, the party seeking to introduce the evidence changed its mind. This issue has not, as of the time of this writing, reached the courts yet, but lawyers clearly are thinking about these uses of neuroscience. (And note that in some administrative contexts, the evidentiary rules will not apply in their full rigor, possibly making the admission of such evidence more likely.)

Do either functional or structural methods of detecting pain work and, if so, how well? We do not know. These studies share many of the problems outlined in Section IV. The studies are few in number, with few subjects (and usually sets of subjects that are not very diverse). The experiments—usually involving giving college students a painful stimulus—are different from the experience of, for example, older people who claim to have low back pain. Independent replication is rare, if it exists at all. The experiments almost always report that, on average, the group shows a statistically significant pattern of activation that differs depending on whether they are receiving the painful stimulus, but the group average does not in itself tell us about the sensitivity or specificity of such a test when applied to individuals. And the statistical and technical issues are daunting.

In the area of pain, the issue of countermeasures may be the most interesting, particularly in light of the experiments conducted with hypnotized subjects. Does remembered pain look the same in an fMRI scan as currently experienced

80. Adam J. Kolber, *How to Improve Empirical Dessert*, 75 Brook. L. Rev. 429 (2009).

81. Greg Miller, *Brain Scans of Pain Raise Questions for the Law*, 323 Science 195 (2009).

pain? Does the detailed memory of a kidney stone pain look any different from the present sensation of low back pain? Can a subject effectively convince himself that he is feeling pain and so appear to the scanner to be experiencing pain? The answers to these questions are clear—we do not yet know.

Pain detection also would raise legal questions. Could a plaintiff be forced to undergo a "pain scan"? If a plaintiff offered a pain scan in evidence, could the defendant compel the plaintiff to undergo such a scan with the defendant's machine and expert? Would it matter if the scan were itself painful or even dangerous? Who would pay for these scans and for the experts to interpret them?

Detecting pain would be a form of neuroscience evidence with straightforward and far-reaching applications to the legal system. Whether it can be done, and, if so, how accurately it can be done, remain to be seen. So does the legal system's reaction to this possibility.

VII. Conclusion

Atomic physicist Niels Bohr is credited with having said "It is always hard to predict things, especially the future."[82] It seems highly likely that the massively increased understanding of the human brain that neuroscience is providing will have significant effects on the law and, more specifically, on the courts. Just what those effects will be cannot be accurately predicted, but we hope that this guide will provide some useful background to help judges cope with whatever neuroscience evidence comes their way.

82. This quotation has been attributed to many people, especially Yogi Berra, but Bohr seems to be the most likely candidate, even though it does not appear in anything he published. See discussion in Henry T. Greely, *Trusted Systems and Medical Records: Lowering Expectations*, 52 Stan. L. Rev. 1585, 1591–92 n.9 (2000). One of the authors, however, recently had a conversation with a scientist from Denmark, who knew the phrase (in Danish) as an old Danish saying and not something original with Bohr.

References on Neuroscience

Fundamental Neuroscience (Larry R. Squire et al. eds., 3d ed. 2008).
Eric R. Kandel et al., Principles of Neural Science (4th ed. 2000).
The Cognitive Neurosciences (Michael S. Gazzaniga ed., 4th ed. 2009).

Reference Guide on Mental Health Evidence

PAUL S. APPELBAUM

Paul S. Appelbaum, M.D., is the Elizabeth K. Dollard Professor of Psychiatry, Medicine, and Law, and Director, Division of Law, Ethics, and Psychiatry, Department of Psychiatry, Columbia University and New York State Psychiatric Institute.

CONTENTS

I. Overview of Mental Health Evidence

A. Range of Legal Cases in Which Mental Health Issues Arise

Evidence presented by mental health experts is common to a broad array of legal cases—criminal and civil. In the criminal realm, these include assessments of defendants' mental states at the time of their alleged offenses (e.g., criminal responsibility and diminished capacity[1]) and subsequent to the offenses, but prior to the initiation of the adjudicatory process (e.g., competence to consent to a search or waive *Miranda* rights[2]). As cases move toward adjudication, evaluation may be required of defendants' competence to stand trial or to represent themselves at trial.[3] Postconviction, mental health evidence may be introduced with regard to sentencing, including suitability for probation and conditions of probation.[4] Capital cases uniquely may raise questions regarding a condemned prisoner's competence to waive appeals or to be executed.[5] Postconfinement, mental health considerations may enter into parole determinations. Indeed, the development of

1. 18 U.S.C. § 17 (defining standard and burden of proof for insanity defense); Clark v. Arizona, 548 U.S. 735 (2006) (on the use of testimony for diminished capacity).

2. *See* Thomas Grisso, Evaluating Competencies: Forensic Assessments and Instruments (2002); Miranda v. Arizona, 384 U.S. 436 (1966) (holding confessions inadmissible unless suspect made aware of rights and waives them); Colorado v. Connelly, 479 U.S. 157 (1986) (holding that mental condition alone will not make a confession involuntary under the Fourth Amendment but may be used as a factor in assessing a defendant's voluntariness); United States v. Elrod, 441 F.2d 353 (5th Cir. 1971) (holding that a person of subnormal intelligence may be deemed incapable of giving consent). *See* Wayne R. LaFave, Search and Seizure 92–93 (2004); Wayne R. LaFave, Criminal Procedure 363–65 (2004); Brian S. Love, Comment: *Beyond Police Conduct: Analyzing Voluntary Consent to Warrantless Searches by the Mentally Ill and Disabled*, 48 St. Louis U. L.J. 1469 (2004).

3. Dusky v. United States, 362 U.S. 402 (1960) (establishing standard for competence to stand trial); Pate v. Robinson, 383 U.S. 375 (1966) (holding that the Due Process Clause of the Fourteenth Amendment does not allow a mentally incompetent criminal defendant to stand trial); Farretta v. California, 422 U.S. 806 (1975) (upholding defendant's right to refuse counsel and represent himself); Indiana v. Edwards, 554 U.S. 164 (2008) (finding that the standards for competency to stand trial and to represent oneself need not be the same).

4. Roger W. Haines, Jr., et al., Federal Sentencing Guidelines Handbook §§ 5B1.3(d)(5), 5D1.3(d)(5), 5H1.3 (2007–2008).

5. *See* Ford v. Wainwright, 477 U.S. 399 (1986) (upholding the common law bar against executing the insane and holding that a prisoner is entitled to a judicial hearing before he may be executed); Stewart v. Martinez-Villareal, 523 U.S. 637 (1998) (holding that death row prisoners are not barred from filing incompetence to be executed claims by dismissal of previous federal habeas petitions); Panetti v. Quarterman, 551 U.S. 930 (2007) (ruling that defendants sentenced to death must be competent at the time of their execution); Atkins v. Virginia, 536 U.S. 304 (2002) (finding that executing the mentally retarded constitutes cruel and unusual punishment under the Eighth Amendment); Rees v. Peyton, 384 U.S. 312 (1966) (formulating the test for competency to waive further proceedings as requiring that the petitioner "appreciate his position and make a rational choice with respect to continuing or abandoning further litigation or on the other hand whether he is suffering from a mental disease, disorder, or defect which may substantially affect his capacity in the premises.").

specialty services for probationers and parolees with mental disorders suggests that mental health professionals' input at this stage is likely to increase in the future.[6]

Mental health evidence in civil litigation is frequently introduced in personal injury cases, where emotional harms may be alleged with or without concomitant physical injury.[7] Issues of contract may turn on the competence of a party at the time that the contract was concluded or whether that person was subject to undue influence,[8] and similar questions may be at the heart of litigation over wills and gifts.[9] Broader questions of competence to conduct one's affairs are considered in guardianship cases,[10] and more esoteric ones may arise in litigation challenging a person's competence to enter into a marriage or to vote.[11] Suits alleging infringement of the statutory and constitutional rights of persons with mental disorders (e.g., under the Americans with Disabilities Act or the Civil Rights of Institutionalized Persons Act) often involve detailed consideration of psychiatric diagnosis and treatment and of institutional conditions.[12] Allegations of professional

6. Jennifer Skeem & Jennifer Eno Louden, *Toward Evidence-Based Practice for Probationers and Parolees Mandated to Mental Health Treatment*, 57 Psychiatric Servs. 333 (2006).

7. Dillon v. Legg, 441 P.2d 912 (Cal. 1968) (allowing recovery based on emotional distress not accompanied by physical injury); Molien v. Kaiser Foundation Hospitals, 616 P.2d 813 (Cal. 1980) (holding that plaintiff who is direct victim of negligent act need not be present when act occurs to recover for subsequent emotional distress); Rodriguez v. State, 472 P.2d 509 (Haw. 1970) (permitting recovery where a reasonable person would suffer serious mental distress as a result of defendant's behavior); Roes v. FHP, Inc., 985 P.2d 661 (Haw. 1999) (allowing assessment of damages for negligent infliction of emotional distress when plaintiff was in actual physical peril, even if no injury was suffered); Albright v. United States, 732 F.2d 181 (C.A.D.C. 1984) (holding that alleging mental distress is sufficient to confer standing); Cooper v. FAA, No. 07-1383 (N.D. Cal. Aug. 2008), *rev'd and remanded,* 596 F.3d 538 (9th Cir. 2010) (discussing mental distress as a result of disclosure of personal information); Sheely v. MRI Radiology Network, P.A., 505 F.3d 1173 (11th Cir. 2007) (holding damages available under § 504 of the Rehabilitation Act when emotional distress was foreseeable).

8. *See generally* E. Allan Farnsworth, Contracts 228–33 (2004); John Parry & Eric Y. Drogin, Mental Disability Law, Evidence, and Testimony 151–52, 185–86 (2007).

9. *See generally* William M. McGovern, Jr. & Sheldon F. Kurtz, Wills, Trusts and Estates Including Taxation and Future Interests 292–99 (2004); Parry & Drogin, *supra* note 8, at 149–51, 182–85.

10. Parry & Drogin, *supra* note 8, at 138–47, 177–81.

11. *Id.* at 54. Doe v. Rowe, 156 F. Supp. 2d 35 (D. Me. 2001) (finding a state law denying the vote to anyone under guardianship by reason of mental disability in violation of the Equal Protection Clause of the U.S. Constitution and Title II of the Americans with Disabilities Act (ADA)); Missouri Protection & Advocacy Servs. v. Carnahan, 499 F.3d 803 (8th Cir. 2007) (upholding a state law allowing disenfranchisement of persons under guardianship because it permits individualized determinations of capacity to vote).

12. Pennsylvania Dep't of Corrections v. Yeskey, 524 U.S. 206 (1998) (holding that ADA coverage extended to prisoners); Clark v. State of California, 123 F.3d 1267 (9th Cir. 1997) (finding state not immune on Eleventh Amendment grounds to suit alleging discrimination under ADA by developmentally disabled inmates); Gates v. Cook, 376 F.3d 323 (5th Cir. 2004) (upholding District Court's finding that prison conditions, including inadequate mental health provisions, violated the Eighth Amendment of the U.S. Constitution); Gaul v. AT&T, Inc., 955 F. Supp. 346 (D.N.J. 1997) (finding that depression and anxiety disorders may constitute a mental disability under the ADA); Anderson v. North Dakota State Hospital, 232 F.3d 634 (8th Cir. 2000) (finding that a plaintiff's fear

malpractice by mental health professionals, including failure to protect foreseeable victims of a patient's violence,[13] invariably call for mental health expert testimony, as do commitment proceedings for the hospitalization of persons with mental disorders[14] or who are alleged to be dangerous sexual offenders.[15]

1. Retrospective, contemporaneous, and prospective assessments

Depending on the questions at issue in a given proceeding, evaluators may be asked to assess the state of mind—including diagnosis and functional capacities—of a person at some point in the past, at present, or in the future.

Retrospective assessments are called for when criminal defendants assert insanity or diminished responsibility defenses, claiming that their state of mind at the time of the crime should excuse or mitigate the consequences of their behaviors, or when questions are raised about competence at some point in the past to waive legal rights (e.g., waiver of *Miranda* rights).[16] In civil contexts, challenges to the capacity of a now-deceased testator to write a will or of a party to enter into a contract, among other issues, will call for a similar look back at a person's functioning at some point in the past.[17] A variety of sources of information are available for such assessments. In some cases (e.g., in criminal proceedings), the defendant is likely to be available for clinical examination, whereas in other cases he or she will not be able to be assessed directly (e.g., challenges to a will). Although the person being evaluated will usually have an interest in portraying him- or herself in a particular light, a direct assessment can nonetheless be valuable in assessing the consistency of the reported symptoms with other aspects of the history and current status of the person. Whether or not the person can be assessed directly, information from persons who were in contact with the person before and during the time in question, including direct reports and contemporaneous

of snakes did not limit ability to work); Sinkler v. Midwest Prop. Mgmt., 209 F.3d 678 (7th Cir. 2000) (holding driving phobia did not substantially limit major life activity of working and hence was not an impairment under the ADA); McAlinden v. County of San Diego, 192 F.3d 1226 (9th Cir. 1999), *cert. denied*, 120 S. Ct. 2689 (2000) (reversing summary judgment against plaintiff who alleged that anxiety and somatoform disorders impaired major life activities of sexual relations and sleep); Steele v. Thiokol Corp., 241 F.3d 1248 (10th Cir. 2001) (finding major life activity under the ADA of interacting with others not substantially impaired by obsessive–compulsive disorder).

13. Tarasoff v. Regents of the Univ. of California, 551 P.2d 334 (Cal. 1976).

14. Addington v. Texas, 441 U.S. 418 (1979) (holding that standard of proof for involuntary commitment is clear and convincing evidence); O'Connor v. Donaldson, 422 U.S. 563 (1975) (holding unconstitutional the confinement of a nondangerous mentally ill person capable of surviving safely in freedom alone or with assistance).

15. Kansas v. Hendricks, 521 U.S. 346 (1997); Kansas v. Crane, 534 U.S. 407 (2002).

16. Predicting the Past: Retrospective Assessment of Mental States in Litigation (Robert I. Simon & Daniel W. Shuman eds., 2002); Bruce Frumkin & Alfredo Garcia, *Psychological Evaluations and Competency to Waive Miranda Rights*. 9 The Champion 12 (2003).

17. *See* Thomas G. Gutheil, *Common Pitfalls in the Evaluation of Testamentary Capacity*, 35 J. Am. Acad. Psychiatry & L. 514 (2007); Farnsworth, *supra* note 8, at 228–33.

records, is usually an essential part of the evaluation. Sometimes the available data from all of these sources are so limited or contradictory that they will not allow a judgment to be made of a person's state of mind at a point in the past. However, most experienced forensic evaluators appear to believe that conclusions regarding past mental state can often be reached with a reasonable degree of certainty if sufficient information is available.[18]

The most straightforward task for a mental health professional is to evaluate a person's current mental state. In criminal justice settings, concerns about a person's current competence to exercise or waive rights will call for such evaluations (e.g., competence to stand trial or to represent oneself at trial).[19] Civil issues calling for contemporaneous assessments include workers' compensation and other disability claims and litigation alleging emotional harms due to negligent or intentional torts, workplace discrimination, and other harm-inducing situations.[20] At the core of an assessment of current mental state is the diagnostic evaluation described below. As in all evaluations in legal contexts, careful consideration needs to be given to the possibility of secondary gain from manipulation of their presentation for persons being assessed.[21]

In contrast to contemporaneous assessments, the evaluation of a person's future mental state and consequent behaviors is fraught with particular difficulty, especially when the outcome being predicted occurs at a relatively low frequency.[22] Such predictive assessments may come into play in the criminal process when bail is set,[23] at sentencing,[24] and as part of probation and parole decisions.[25] They often involve

18. Robert I. Simon, *Retrospective Assessment of Mental States in Criminal and Civil Litigation: A Clinical Review* in Simon and Shuman, *supra* note 16 at 1, 8; McGregor v. Gibson, 248 F.3d 946, 962 (10th Cir. 2001) (stating that although disfavored, retrospective determinations of competence may be allowed in cases when a meaningful hearing can be conducted).

19. *See* Dusky v. United States, 362 U.S. 402 (1960) (holding that a criminal defendant must understand the charges and be able to participate in his defense); Godinez v. Moran, 509 U.S. 389 (1993) (holding that a defendant competent to stand trial is also sufficiently competent to plead guilty or waive the right to legal counsel).

20. *See, e.g.*, Kent v. Apfel, 75 F. Supp. 2d 1170 (D. Kan. 1999); Quigley v. Barnhart, 224 F. Supp. 2d 357 (D. Mass. 2002); Rivera v. City of New York, 392 F. Supp. 2d 644 (S.D.N.Y. 2005); Lahr v. Fulbright & Jaworski, L.L.P., 164 F.R.D. 204 (N.D. Tex. 1996).

21. *See* United States v. Binion, 132 F. App'x 89 (8th Cir. 2005) (upholding an obstruction of justice conviction and sentencing determination based on a finding that defendant had feigned mental illness). See discussion, *infra*, Section I.C.2.

22. Joseph M. Livermore et al., *On the Justifications for Civil Commitment*, 117 U. Pa. L. Rev. 75–96 (1968).

23. United States v. Salerno, 481 U.S. 739 (1987); United States v. Farris, 2008 WL 1944131 (W.D. Pa. May 1, 2008).

24. Tex. Code Crim. Proc. Ann. art. 37.071 (Vernon 1981); Barefoot v. Estelle, 463 U.S. 880 (1983).

25. *See* 28 C.F.R. § 2.19 (2008) for parole determination factors. For probation determination factors, see 18 U.S.C.A. § 356 (2008). *See generally* Neil Cohen, The Law of Probation and Parole §§ 2, 3 (2008).

estimates of the probable effectiveness of treatment, especially in the juvenile justice system, where the lack of amenability of juveniles to mental health treatment is frequently a key consideration in decisions regarding transfer to adult courts.[26] Predictions regarding behavior related to mental disorders are also seen in civil cases, for example, in the civil commitments of persons with mental disorders and in the newer statutes authorizing the commitment of dangerous sex offenders.[27] Damage assessments in civil cases alleging emotional harms will usually call for some estimate regarding the duration of symptoms and response to treatment.[28] The inescapable uncertainties of the course of mental disorders and their responsiveness to interventions create part of the difficulty in such assessments, but an equally important contribution is made by the unknowable contingencies of life. Will a person's spouse leave or will the person lose his job or his home? As a consequence, will the person return to drinking, stop taking medication, or reconnect with friends who have continued to engage in criminal behaviors? At best, predictive assessments can lead to general statements of probability of particular outcomes, with an acknowledgment of the uncertainties involved.[29]

2. *Diagnosis versus functional impairment*

A diagnosis of mental disorder per se will almost never settle the legal question in a case in which mental health evidence is presented. However, a diagnosis may play a role in determining whether a claim or proceeding can go forward. The clearest example in criminal law is embodied in the insanity defense, where the impairments of understanding, appreciation, and behavioral control that comprise the various standards must be based, in one popular formulation, on a "mental disease or defect."[30] In the absence of a diagnosis of mental disorder (including mental retardation and the consequences of injury to the brain), an affirmative

26. Michael G. Kalogerakis, Handbook of Psychiatric Practice in Juvenile Court 79–85 (1992).

27. *See* O'Connor v. Donaldson, 422 U.S. 563 (1975) (finding that a state may not confine a citizen who is nondangerous and capable of living by herself or with aid); for an example of a sex offender civil commitment statute, *see* Minn. Stat. § 253B.185 (2008). The constitutionality of civil commitment for dangerous sex offenders was upheld in *Kansas v. Hendricks,* 521 U.S. 346 (1997) (setting forth the procedures for the commitment of convicted sex offenders deemed dangerous due to a mental abnormality).

28. Gary B. Melton et al., Psychological Evaluations for the Courts: A Handbook for Mental Health Professionals and Lawyers 413–14 (2007).

29. For a more detailed discussion of predictive assessment regarding future dangerousness, see Section I.E.

30. The American Law Institute standard for the insanity defense reads, "a person is not responsible for criminal conduct if at the time of such conduct as a result of mental disease or defect he lacks substantial capacity either to appreciate the criminality of his conduct or to conform his conduct to the requirements of the law." Model Penal Code and Commentaries § 4.01(1) (Official Draft and Revised Comments 1985) (adopted by American Law Institute, May 24, 1962). The federal insanity defense was codified in the Insanity Defense Reform Act of 1984, *codified at* 18 U.S.C. § 17. *See also* Durham v. United States, 214 F.2d 862 (D.C. Cir. 1954) ("[A]n accused is not criminally responsible

defense of insanity will not prevail.[31] Comparable situations exist in civil commitment proceedings and work disability determinations.[32]

Even where the presence of a mental disorder is not an absolute prerequisite to claims involving mental state, it will often play a de facto threshold role. Thus, evidence in cases involving claims of incompetence (e.g., to engage in a contractual relationship) or emotional harms will often address the presence of a diagnosis, even though that may not strictly be required.[33] In these cases, failure to establish a diagnosis may be taken by a factfinder as an indicator of the probable lack of validity of the claim. That is, it may be assumed that unless an underlying disorder can be identified, the claimed impairments are bogus. Thus, conflicting testimony over the presence or absence of a diagnosis is common in cases in which mental health evidence is offered, even when not mandated by the operative legal standard.

Notwithstanding the threshold role played by a mental disorder diagnosis in many cases, the ultimate legal issue usually will turn on the impact of the mental disorder on the person's functional abilities.[34] Those abilities may relate to the person's cognitive capacities, including the capacity to make a legally relevant decision (e.g., granting consent for the police to conduct a warrantless search, altering a will) or the capacity to behave in a particular way (e.g., conforming one's conduct to the requirements of the law, cooperating with an attorney in one's own defense, resisting undue influence), or both (e.g., skill as a parent, competence to proceed with criminal adjudication). The former set of capacities can be denoted as *decisional capacities* and the latter set as *performative capacities*. Many of the legal questions to which mental health evidence may be relevant will involve a determination of the influence of a mental state or disorder on one or both of these sets of capacities. The mere presence of a mental disorder will almost always be insufficient for that purpose. Mental disorder in a criminal defendant, for example, if it does not interfere substantially with competence to stand trial, does not present a basis for postponing adjudication of the case.[35] Some degree of mental disorder, including dementia, without affecting relevant abilities, does not provide grounds for voiding a will.[36] The point can be generalized to all criminal and civil competency determinations, most assessments of emotional harms, and

if his unlawful act was the product of mental disease or defect."); *note* United States v. Brawner, 471 F.2d 969 (1972), which overturned the Durham Rule (or "product test").

31. Tennard v. Dretke, 542 U.S. 274 (2004); Bigby v. Dretke, 402 F.3d 551 (5th Cir. 2005).

32. Addington v. Texas, 441 U.S. 418 (1979) (setting the burden of proof required for involuntary civil commitment as requiring clear and convincing evidence); and Social Security Administration Listing of Impairments, *available at* http://www.ssa.gov/disability/professionals/bluebook/listing-impairments.htm.

33. Farnsworth, *supra* note 8, §§ 4.6–4.8, at 228–34.

34. Grisso, *supra* note 2.

35. United States v. Passman, 455 F. Supp. 794 (D.D.C. 1978); United States. v. Valierra, 467 F.2d 125 (9th Cir. 1972).

36. Rossi v. Fletcher, 418 F.2d 1169 (D.C. Cir. 1969); *In re* Estate of Buchanan, 245 A.D.2d 642 (3d Dept. 1997).

probably to the majority of cases in which mental health testimony is offered: Unless a mental disorder can be shown to have affected a person's functional capacity, decisional or performative, a diagnosis of mental disorder per se will not be determinative of the outcome.[37]

Despite its importance to the adjudicative process, mental health evidence is often introduced in the context of a serious stigma that attaches to mental disorders[38] and considerable confusion regarding their nature, consequences, and susceptibility to treatment.[39] Diagnoses of mental disorders often are perceived to be less reliable and more subjective than diagnoses of other medical conditions.[40] Symptoms of mental disorders may be seen as reflections of moral weakness or lack of will, and the impact of disorders on functional abilities may not be recognized, or occasionally may be exaggerated.[41] The potential impact and limits of current treatments are not widely understood. Indeed, even the various types of mental health professionals are frequently confused.[42] The remainder of Section I of this reference guide provides background to clarify these issues; Section II considers questions specifically related to the introduction of evidence by mental health experts.

B. Mental Health Experts

Evidence related to mental state and mental disorders may be presented by experts from a number of disciplines, but it is most commonly introduced by psychiatrists or psychologists.

1. Psychiatrists

Psychiatrists are physicians who specialize in the diagnosis and treatment of mental disorders.[43] After college, they complete 4 years of medical school, during

37. For a brief overview of competency evaluations, *see* Patricia A. Zapf & Ronald Roesch, *Mental Competency Evaluations: Guidelines for Judges and Attorneys*, 37 Ct. Rev. 28 (2000), *available at* http://aja.ncsc.dni.us/courtrv/cr37/cr37-2/CR37-2ZapfRoesch.pdf. For the underlying standard for competency to stand trial, see Dusky v. United States, 362 U.S. 402 (1960).

38. Bruce G. Link et al., *Measuring Mental Illness Stigma*, 30 Schizophrenia Bull. 511 (2004).

39. Bruce G. Link et al., *Stigma and Coercion in the Context of Outpatient Treatment for People with Mental Illnesses*, 67 Soc. Sci. & Med. 409 (2008).

40. Thomas A. Widiger, *Values, Politics, and Science in the Construction of the DSMs*, *in* Descriptions and Prescriptions: Values, Mental Disorders, and the DSMs 25 (John Z. Sadler ed., 2002).

41. Michael L. Perlin, *"Half-Wracked Prejudice Leaped Forth": Sanism, Pretextuality, and Why and How Mental Disability Law Developed as It Did*, 10 J. Contemp. Legal Issues 3 (1999); Michael L. Perlin, *"You Have Discussed Lepers and Crooks": Sanism in Clinical Teaching*, 9 Clinical L. Rev. 683 (2003); Michael L. Perlin, The Hidden Prejudice: Mental Disability on Trial (2000).

42. The degree of popular confusion is underscored by the results of a Web-based search for "psychiatrist vs. psychologist," which turns up a remarkably large number of Web sites attempting to explain the differences between the two professions.

43. Narriman C. Shahrokh & Robert E. Hales, American Psychiatric Glossary 157 (2003).

which they spend approximately 2 years in preclinical studies (e.g., physiology, pharmacology, genetics, pathophysiology), followed by 2 years of clinical rotations in hospital and clinic settings (e.g., medicine, surgery, pediatrics, obstetrics/gynecology, orthopedics, psychiatry).[44] Graduating medical students who elect to specialize in psychiatry enter residency programs of at least 4 years' duration.[45] Accredited residencies must currently offer at least 4 months in a primary care setting in internal medicine, family medicine, or pediatrics, and at least 2 months of training in neurology.[46] The remainder of a resident's time is spent learning psychiatry, including inpatient, outpatient, emergency, community, and consultation settings, and with exposure to the subspecialty areas of child and adolescent, geriatric, addiction, and forensic psychiatry. Residents will be taught how to use treatment techniques, among them medications and various forms of psychotherapy. Elective time is usually available to pursue particular interests in greater depth or to engage in research. Didactic seminars, including sessions on neuroscience, genetics, psychological theory, and treatment, and supervision sessions with experienced psychiatrists (and sometimes mental health professionals from other disciplines) complement the clinical experiences.[47]

After completion of 4 years of residency training, a psychiatrist is designated as "board eligible," that is, able to take the certification examination of the American Board of Psychiatry and Neurology in adult psychiatry.[48] Successful completion of this examination process results in the psychiatrist being designated "board certified." Psychiatrists who desire more intensive training in a subspecialty area of psychiatry—for example, child and adolescent or addiction psychiatry—can take a 1- or 2-year fellowship in that area. The psychiatrist who has completed an accred-

44. Medical schools in the United States are accredited by the Liaison Committee on Medical Education, which establishes general curricular and other standards that all schools must meet. Standards are available at http://www.lcme.org/standard.htm. Students can elect to extend their medical school training by taking additional time to conduct research or to obtain complementary training (e.g., in public health).

45. Residents who choose to combine adult and child psychiatry training can do so in a 5-year program, or can follow their 4 years of adult residency with 2 years of child training. Some residents will also extend their residency training by adding a year or more during which they conduct laboratory or clinical research.

46. Psychiatric residencies are accredited by the Accreditation Council on Graduate Medical Education. Program requirements are available at http://www.acgme.org/acwebsite/rrc_400/400_prindex.asp.

47. See descriptions of several leading psychiatry residency training programs on their Web sites: Columbia University (http://www.cumc.columbia.edu/dept/pi/residency/index.html); Johns Hopkins University (http://www.hopkinsmedicine.org/Psychiatry/for_med_students/residency_general/); Harvard/Longwood Psychiatry Residency (http://harvardlongwoodpsychiatry.org/).

48. Information regarding qualifications for board certification and the examination process is available from the American Board of Psychiatry and Neurology at http://www.abpn.com/Initial_Psych.htm.

ited fellowship[49] is eligible for additional board certification in that subspecialty.[50] Although fellowship training and board certification indicate expertise in a particular area of psychiatry, some psychiatrists are recognized by the courts as having developed equivalent levels of expertise by virtue of extensive clinical experience and self-designed instruction (e.g., continuing education courses, remaining current with the professional literature).[51]

Forensic psychiatry is the subspecialty that focuses on the interrelationships between psychiatry and the law.[52] Hence, forensic psychiatrists are particularly likely to offer evidence as part of court proceedings. Fellowship training in forensic psychiatry involves a 1-year program in which fellows are taught forensic evaluation for civil and criminal litigation and become involved in the treatment of persons with mental disorders in the correctional system.[53] They also learn about the rules and procedures for providing evidence in legal proceedings and for working with attorneys. However, training and/or board certification in forensic psychiatry are not necessarily the best qualification for expertise in a particular case. Although forensic psychiatrists are likely to have more expertise than general psychiatrists for certain kinds of evaluations that are the focus of forensic training (e.g., competence to stand trial, emotional harms), when issues are raised concerning other substantive areas of psychiatry (e.g., the effects of psychopharmacological agents on a civil defendant's ability to drive at the time of an accident that allegedly resulted in injury to the plaintiff), a psychiatrist who specializes in that area will often have greater expertise than someone with forensic training.

49. Accredited subspecialty training is currently available in addiction, child and adolescent, forensic, and geriatric psychiatry, and in psychosomatic medicine. Psychiatrists are also eligible for training in hospice and palliative medicine, pain medicine, and sleep medicine. *See* accreditation standards at http://www.acgme.org/acwebsite/rrc_400/400_prindex.asp. Fellowship programs also exist in some subspecialty areas for which accreditation and board certification are not available, e.g., research, psychopharmacology, and public and community psychiatry.

50. Typically, when new subspecialties are recognized and accreditation standards are developed, a certain period of time (e.g., 5 years) is allowed for psychiatrists who have gained expertise in that area by virtue of experience or alternative training to achieve board certification. Thus, many psychiatrists who are today board certified in a subspecialty have not completed a fellowship.

51. For a comparable determination involving a counselor, see Leblanc v. Coastal Mech. Servs., LLC, 2005 WL 5955027 (S.D. Fla. Sept. 7, 2005) (quoting Jenkins v. United States, 307 F.2d 637 (D.C. Cir. 1962) for the proposition that the determination of a psychologist's competence to render an expert opinion is a case-by-case matter based on knowledge, not claim to a professional title).

52. See the definition of forensic psychiatry offered by the American Academy of Psychiatry and the Law: "Forensic psychiatry is a medical subspecialty that includes research and clinical practice in the many areas in which psychiatry is applied to legal issues," *available at* http://www.aapl.org/org.htm. Psychiatrists who have been certified in adult or child psychiatry by the American Board of Psychiatry and Neurology, and who have completed a forensic psychiatry fellowship, can take the examination for subspecialty certification in forensic psychiatry. A description of the requirements for certification can be found at http://www.abpn.com/fp.htm. Board certification must be renewed by taking a recertification examination every 10 years.

53. See the accreditation standards in forensic psychiatry at http://www.acgme.org/acWebsite/downloads/RRC_progReq/406pr703_u105.pdf.

2. Psychologists

Psychologists have received graduate training in the study of mental processes and behavior.[54] Only a subset of psychologists evaluate and treat persons with psychological or behavioral problems; they may be termed clinical, counseling, health, neuro-, rehabilitation, or school psychologists. In contrast, many psychologists teach and/or pursue research in one of the academic aspects of the field (e.g., cognitive, developmental, or social psychology), or provide consultation of a nonclinical nature (e.g., organizational or industrial psychology).[55] Independent practice in psychology requires licensure from the appropriate state licensure board and generally requires a doctoral degree and postgraduate clinical experience. Although use of the term *psychologist* is restricted in many jurisdictions to licensed psychologists,[56] the term may be applied in some settings to persons with master's-level training in psychology.[57]

After college, students who enter graduate doctoral programs generally require 4 to 6 years to complete their training. Those who intend to pursue clinical work generally receive training in clinical, counseling, or school psychology.[58] Accredited programs in these areas are required to provide a minimum of three academic years of graduate study, and students are required in addition to take a year of clinical internship.[59] Course work must include study of biological aspects of behavior, cognitive and affective aspects of behavior, social aspects of behavior, history and systems of psychology, psychological measurement, research methodology, and techniques of data analysis. Students also must be taught about

54. The American Psychological Association defines the field of psychology in this way: "Psychology is the study of the mind and behavior. The discipline embraces all aspects of the human experience—from the functions of the brain to the actions of nations, from child development to care for the aged. In every conceivable setting from scientific research centers to mental health care services, 'the understanding of behavior' is the enterprise of psychologists." http://74.125.45.104/search?q=cache:JKti-_3SfkQJ:www.apa.org/about/+psychologist+definition&hl=en&ct=clnk&cd=9&gl=us.

55. *See id.* for the American Psychological Association's characterization of the subspecialties in psychology.

56. *See, e.g.*, Mass. Gen. Laws ch. 112, § 122; N.Y. Educ. Law § 7601.

57. Note that the American Psychological Association urges that the use of the term be restricted to persons with doctoral degrees in psychology: "Psychologists have a doctoral degree in psychology from an organized, sequential program in a regionally accredited university or professional school . . . it is [the] general pattern to refer to master's-level positions as counselors, specialists, clinicians, and so forth (rather than as 'psychologists')." http://74.125.45.104/search?q=cache:JKti-_3SfkQJ:www.apa.org/about/+psychologist+definition&hl=en&ct=clnk&cd=9&gl=us.

58. Other psychology programs offer training in experimental, social, and cognitive psychology, for example, with the intent of producing graduates who will pursue research or teaching careers, but will not engage in clinical work. United States v. Fishman, 743 F. Supp. 713, 723 (N.D. Cal. 1990) (excluding the expert testimony of a social psychologist holding a Ph.D. in sociology).

59. Accreditation of programs in clinical, counseling, and school psychology is undertaken by the Commission on Accreditation of the American Psychological Association. Accreditation standards are available at http://www.apa.org/ed/accreditation/.

individual differences in behavior, human development, dysfunctional behavior or psychopathology, and professional standards and ethics. A practicum experience, which usually involves placement in an agency or clinic that provides psychological services, is part of the training experience. Prior to receiving their degrees, students must also complete a 1-year clinical internship, which is often taken at a clinical facility that is separate from their graduate school.[60]

Psychology graduate programs award either the Ph.D. or Psy.D. ("professional psychology") degree.[61] Ph.D. programs generally place greater emphasis on research training, with students required to complete a research project and write a dissertation. Psy.D. programs ordinarily stress clinical issues and training and have less rigorous research requirements.[62] Supervised work may involve some combination of psychological treatment (e.g., individual or group psychotherapy) and the use of standardized testing techniques (i.e., "psychological tests"). Once licensed, psychologists can practice independently. At present, two states permit psychologists who complete additional training requirements to prescribe medications, although physicians' groups remain strongly opposed to the practice.[63]

Fellowships in subspecialty areas of psychology are becoming more common, although they are not always linked to subspecialty certification processes. Among the areas in which fellowships have been developed is forensic psychology, generally a 1-year program, with didactic and clinical training in forensic evaluation.[64] Certification in forensic psychology through an examination process is available for psychologists who have completed a fellowship in the field or who have at least 5 years of experience in forensic psychology.[65] As with psychiatry, whether the expertise of a forensic psychologist is relevant to a particular legal issue will vary and needs to be considered on a case-by-case basis.

60. *Id.*

61. See sample Ph.D. program curricula for programs at University of Illinois at Urbana-Champaign (http://www.psych.uiuc.edu/divisions/clinicalcommunity.php); Indiana University (http://bl-psy-appsrv.ads.iu.edu:8080/graduate/courses/clinical.asp); and University of California at Los Angeles (http://www.psych.ucla.edu/Grads/Areas/clinical.php). See sample Psy.D. curricula for programs at Massachusetts School of Professional Psychology (http://www.mspp.edu/academics/degree-programs/psyd/default.asp); and Wisconsin School of Professional Psychology (http://www.wspp.edu/courseswspp.html).

62. For a discussion of the so-called "Vail model" on which Psy.D. training is based, see John C. Norcross & Patricia H. Castle, *Appreciating the PsyD: The Facts*, 7 Eye on Psi Chi 22 (2002), *available at* http://www.psichi.org/pubs/articles/article_171.asp.

63. N.M. Stat. Ann. § 61-9 (2002); La. Rev. Stat. § 37:2371-78 (2004). Note that the New Mexico statute is set to expire in 2010 under a sunset provision. N.M. Stat. Ann. 61-9-19 (2002).

64. See, e.g., the description of the program at the University of Massachusetts Medical School at http://www.umassmed.edu/forensicpsychology/index.aspx.

65. Certification is provided by the American Board of Forensic Psychology. Requirements for candidates are available at: http://www.abfp.com/.

3. Other mental health professionals

Persons with a variety of other forms of training provide mental health services, including services that generally are referred to as psychotherapy or counseling, with individuals, couples, or groups. The best established of these professions is social work. Schools of social work offer 2-year programs that lead to a master's degree (MSW), and students can elect a track that is often referred to as psychiatric social work, which involves instruction and experience in psychotherapy.[66] Graduate social workers can obtain state licensure after the completion of a period of supervised practice and an examination, resulting in their designation as a "licensed independent clinical social worker (LICSW)," with variation in nomenclature across the states.[67] Social workers may offer psychotherapeutic or counseling services through social service agencies or in private practice. Recently, a subspecialty of forensic social work has begun to develop, involving social workers with experience in the criminal justice system.[68]

Another group that offers mental health services, which may include psychotherapy or counseling and medications, are master's- or doctoral-level nurses. A growing number of nursing schools are developing programs that are termed "psychiatric nursing."[69] Nursing practice is regulated by state law and hence varies across jurisdictions, but master's-level nurses (sometimes referred to as "nurse practitioners") can achieve a status that allows them to provide psychotherapy and to dispense medications, although they may need to have a supervisory arrangement with a physician for the latter.[70]

Other master's-level mental health professionals include persons who may be called psychologists, counselors, marital and family therapists, group therapists, and a variety of other terms.[71] Because state law generally does not regulate the

66. See, e.g., the curricula for social work training at Columbia (http://www.columbia.edu/cu/ssw/admissions/pages/programs_and_curriculum/index.html) and at Smith (http://www.smith.edu/ssw/geaa/academics_msw.php).

67. The Association of Social Work Boards provides an overview of state licensure requirements at http://www.datapathdesign.com/ASWB/Laws/prod/cgi-bin/LawWebRpts2DLL.dll/EXEC/0/0j6ws4m1dqx37r1ce43dq091bxya.

68. See the description of forensic social work offered by the National Association of Forensic Social Work at http://www.nofsw.org/html/forensic_social_work.html. Postgraduate certification programs for forensic social workers are also beginning to be developed, e.g., at the University of Nevada at Las Vegas (http://socialwork.unlv.edu/PGC_forensic_social_work.html).

69. See the listing of training programs in psychiatric nursing, with links to their curricula, provided by the American Psychiatric Nurses Association at http://www.apna.org/i4a/pages/index.cfm?pageid=3311.

70. Sharon Christian et al., Overview of Nurse Practitioner Scopes of Practice in the United States—Discussion, Center for Health Professions, University of California, San Francisco (2007), at http://www.acnpweb.org/i4a/pages/index.cfm?page id=3465.

71. See, e.g., the variety of mental health professionals listed by the National Alliance on Mental Illness at http://www.nami.org/Content/ContentGroups/Helpline1/Mental_Health_Professionals_Who_They_Are_and_How_to_Find_One.htm. United States v. Huber, 603 F.2d 387, 399 (2d Cir.

practice of psychotherapy—although the use of titles such as "psychologist" or "psychotherapist" may be restricted—there is no barrier to persons with variable levels of training in mental health opening independent practices.[72] This includes persons with degrees in educational psychology (M.Ed. or Ed.D.), clergypersons (who may have had some training in pastoral counseling in seminary), and members of disciplines unrelated to mental health. Because of the unregulated nature of their practices, they are largely beyond the reach of professional oversight and discipline.

Although psychiatrists and doctoral-level psychologists generally provide expert evidence related to mental health issues, courts will sometimes admit testimony from other mental health professionals.[73] Given that training and experience vary considerably, and titles may be used inconsistently, an individualized inquiry into the qualifications of the proposed expert is usually required.

1979) (affirming trial court's rejection of expert testimony on defendant's mental state from a professor of economics who was also a certified psychoanalyst).

72. The classic study, albeit now somewhat outdated, is Daniel Hogan, The Regulation of Psychotherapists (1979); *see also* Geoffrey Marczyk & Ellen Wertheimer, *The Bitter Pill of Empiricism: Health Maintenance Organizations, Informed Consent and the Reasonable Psychotherapist Standard of Care*, 46 Vill. L. Rev. 33 (2001).

73. Leblanc v. Coastal Mech. Servs., LLC, 2005 WL 5955027 (S.D. Fla. Sept. 7, 2005) (finding a marriage and family counselor holding a Ph.D. in family therapy, bachelor's and master's degrees in psychology, and a record of relevant publications may be qualified to offer helpful testimony about a plaintiff's alleged psychological condition); Jenkins v. United States, 307 F.2d 637, 646 (D.C. Cir. 1962) ("The critical factor in respect to admissibility is the actual experience of the witness and the probable probative value of his opinion. . . . The determination of a psychologist's competence to render an expert opinion based on his findings as to the presence or absence of mental disease or defect must depend upon the nature and extent of his knowledge. It does not depend upon his claim to the title 'psychologist.'"); United States v. Azure, 801 F.2d 336, 342 (8th Cir. 1986) ("The social worker was most likely qualified as an expert under Rule 702"); *see also* United States v. Raya, 45 M.J. 251 (1996) (finding that trial court's admission of expert testimony from a social worker on whether the victim suffered from PTSD was not an abuse of discretion) *and* United States v. Johnson, 35 M.J. 17 (1992) (holding social worker qualified to render opinion that child suffered trauma). Note, however, not all courts have been receptive to social worker testimony offered as expert opinion on the diagnosis of PTSD, *e.g.*, Neely v. Miller Brewing Co., 246 F. Supp. 2d 866 (S.D. Ohio 2003), Blackshear v. Werner Enters., Inc., 2005 WL 6011291 (E.D. Ky. May 19, 2005). For more restrictive approaches to testimony by non-Ph.D. psychologists, see also State v. Bricker, 321 Md. 86 (Md. Ct. App. 1990) (rejecting expert testimony from a nonpracticing psychologist who did not hold a doctorate and did not qualify for a reciprocal license under state law). People v. McDarrah, 175 Ill. App. 3d 284, 291 (1988) (affirming the trial court's rejection as an expert witness of a doctoral candidate who did not have the experience level required for state registration as a psychologist). Parker v. Barnhart, 67 F. App'x 495 (9th Cir. 2003) (finding error in an administrative law judge's failure to call a licensed psychologist, rather than another expert, as an expert witness for appropriate testimony). Earls v. Sexton, 2010 U.S. Dist. LEXIS 52980 (M.D. Pa. May 28, 2010) (allowing a nurse practitioner to testify in a negligence action concerning whether a motor vehicle accident caused psychiatric injuries).

C. Diagnosis of Mental Disorders

1. Nomenclature and typology—DSM-IV-TR and DSM-5

The standard nomenclature and diagnostic criteria for mental disorders in use in the United States are embodied in the *Diagnostic and Statistical Manual of Mental Disorders,* published by the American Psychiatric Association, and now in its fourth edition with revised text (DSM-IV-TR).[74] It is anticipated that the next edition of the manual (*DSM-5*) will appear in 2013.[75] According to the *DSM* framework, the presence of a mental disorder is typically diagnosed by a combination of the symptoms reported by the patient (e.g., sadness, difficulty falling asleep, anxiety) and signs observed by the clinician (e.g., attentional difficulties, sad affect, crying). To qualify for a *DSM* diagnosis, persons must meet a set of criteria that are characteristic of the disorder.[76] The presence of certain signs and symptoms may be

74. American Psychiatric Association, Diagnostic and Statistical Manual of Mental Disorders (4th ed. text rev. 2000) (hereinafter DSM-IV-TR).

75. An alternative nomenclature and set of criteria used internationally can be found in the *International Classification of Diseases,* now in its 10th edition (*ICD-10*), published by the World Health Organization. Although the DSM-IV-TR and ICD-10 nomenclature and criteria are generally similar, there are differences that can result in diagnostic variations in particular cases, depending on which criteria are applied.

76 E.g., A diagnosis of obsessive–compulsive disorder requires the following:

A. Either obsessions or compulsions:
   Obsessions as defined by (1), (2), (3), and (4):
   (1) recurrent and persistent thoughts, impulses, or images that are experienced, at some time during the disturbance, as intrusive and inappropriate and that cause marked anxiety or distress
   (2) the thoughts, impulses, or images are not simply excessive worries about real-life problems
   (3) the person attempts to ignore or suppress such thoughts, impulses, or images, or to neutralize them with some other thought or action
   (4) the person recognizes that the obsessional thoughts, impulses, or images are a product of his or her own mind (not imposed from without as in thought insertion)
   Compulsions as defined by (1) and (2):
   (1) repetitive behaviors (e.g., hand washing, ordering, checking) or mental acts (e.g., praying, counting, repeating words silently) that the person feels driven to perform in response to an obsession, or according to rules that must be applied rigidly
   (2) the behaviors or mental acts are aimed at preventing or reducing distress or preventing some dreaded event or situation; however, these behaviors or mental acts either are not connected in a realistic way with what they are designed to neutralize or prevent or are clearly excessive
B. At some point during the course of the disorder, the person has recognized that the obsessions or compulsions are excessive or unreasonable.
   NOTE: This does not apply to children.
C. The obsessions or compulsions cause marked distress, are time consuming (take more than 1 hour a day), or significantly interfere with the person's normal routine, occupational (or academic) functioning, or usual social activities or relationships.
D. If another Axis I disorder is present, the content of the obsessions or compulsions is not restricted to it (e.g., preoccupation with food in the presence of an Eating Disorder; hair pulling in the presence of Trichotillomania; concern with appearance in the presence of Body Dysmorphic Disorder; preoccupation with drugs in the presence of a Substance Use Disorder; preoccupation with having a serious illness in the presence of Hypochondriasis; preoccupation with sexual urges or fantasies in the presence of a Paraphilia; or guilty ruminations in the presence of Major Depressive Disorder).

mandatory for a diagnosis to be made, but in most cases—given the variable presentation of most mental disorders—only some proportion of signs and symptoms must be present (e.g., five out of nine).[77]

Since the influential third edition of *DSM* in 1980,[78] the manual has taken a "multiaxial" approach to diagnosis. That is, it recognizes that multiple aspects of a person's situation—not just the signs and symptoms of disorder—may be relevant to a full understanding of his or her situation. Currently, there are five *DSM* axes: Axis 1 is for the designation of most mental disorders, including substance abuse; Axis 2 covers disorders of personality and mental retardation, which may be present together with or independent of an Axis 1 disorder; Axis 3 addresses concurrent medical disorders; Axis 4 allows the designation of stressors confronting the person; and Axis 5 is a structured scale that speaks to the person's overall level of functioning.[79] A complete diagnosis in the *DSM* system requires some notation regarding all five axes, although clinicians commonly focus on Axes 1–3. More than one condition may be indicated on Axes 1–4; for example, major depressive disorder and alcohol abuse may coexist on Axis 1, and more than one personality disorder may be noted on Axis 2.

> E. The disturbance is not due to the direct physiological effects of a substance (e.g., a drug of abuse, a medication) or a general medical condition.

DSM-IV-TR at 462–463.

77. E.g., among the criteria required to be met for a diagnosis of Major Depressive Episode are:

> A. Five (or more) of the following symptoms have been present during the same 2-week period and represent a change from previous functioning; at least one of the symptoms is either (1) depressed mood or (2) loss of interest or pleasure.
> (1) depressed mood most of the day, nearly every day, as indicated by either subjective report (e.g., feels sad or empty) or observation made by others (e.g., appears tearful).
> **NOTE: In children and adolescents, can be irritable mood.**
> (2) markedly diminished interest or pleasure in all, or almost all, activities most of the day, nearly every day (as indicated by either subjective account or observation made by others)
> (3) significant weight loss when not dieting or weight gain (e.g., a change of more than 5% of body weight in a month), or decrease or increase in appetite nearly every day. Note: In children, consider failure to make expected weight gains.
> (4) insomnia or hypersomnia nearly every day
> (5) psychomotor agitation or retardation nearly every day (observable by others, not merely subjective feelings of restlessness or being slowed down)
> (6) fatigue or loss of energy nearly every day
> (7) feelings of worthlessness or excessive or inappropriate guilt (which may be delusional) nearly every day (not merely self-reproach or guilt about being sick)
> (8) diminished ability to think or concentrate, or indecisiveness, nearly every day (either by subjective account or as observed by others)
> (9) recurrent thoughts of death (not just fear of dying), recurrent suicidal ideation without a specific plan, or a suicide attempt or a specific plan for committing suicide.

DSM-IV-TR at 356.

78. American Psychiatric Association, Diagnostic and Statistical Manual of Mental Disorders (3d ed. 1980).

79. DSM-IV-TR at 27–37.

The *DSM* approach has been criticized on a number of grounds, at least one of which is relevant to the evidence likely to be presented in legal proceedings. By requiring that persons being evaluated meet a certain number of particular criteria (e.g., five out of nine signs and symptoms of major depressive disorder), the *DSM* all but guarantees that there will be people who fall just short of qualifying for a diagnosis but may nonetheless be experiencing significant symptoms and impairment.[80] When a mental disorder diagnosis is required as a threshold determination for legal purposes, this may preclude a claim or defense based on the presence of a disorder. In part, *DSM* compensates for this problem by allowing alternative "not otherwise specified" diagnoses to be assigned to persons who fail to meet the full criteria set (e.g., "depressive disorder, not otherwise specified" for persons who fall short of meeting criteria for major depressive disorder),[81] but the problem remains. Suggestions that a more dimensional approach to diagnosis be adopted, that is, one that recognizes a spectrum of extent and severity of symptoms along a continuum associated with a given disorder,[82] have so far been rejected in favor of continuing with the current categorical system.

The goal of the *DSM* is to provide a typology that is useful to clinicians and researchers and that reflects the latest psychiatric understanding of mental disorders.[83] Periodic revisions, such as the process now under way that will result in *DSM-5*, are accomplished by groups of experts, mostly psychiatrists, but include some experts from other disciplines and are ultimately subject to the review and approval of the Board of Trustees and Assembly of the American Psychiatric Association. Hence, the process is sometimes criticized as reflecting social or political biases, as opposed to science.[84] Although such effects cannot be ruled out, to the extent that they exist, they are likely to be associated with a small number of controversial categories and proposed categories (e.g., premenstrual dysphoric disorder,[85] paraphilic rapism[86]). In addition, the *DSM* itself recognizes—in a cautionary statement in the introduction to the text—that diagnostic criteria that are appropriate for clinical or research purposes may not map directly onto legally relevant categories.[87] Caution is therefore required in moving between clinical diagnoses and legal conclusions.

80. Harold A. Pincus et al., *Subthreshold Mental Disorders: Nosological and Research Recommendations*, *in* Advancing DSM: Dilemmas in Psychiatric Diagnosis 129 (Katharine A. Phillips et al. eds., 2002).

81. DSM-IV-TR at 381–82.

82. See the papers on dimensional approaches to psychiatric diagnosis published in the *International Journal of Methods in Psychiatric Research*, vol. 16, supplement.

83. Because of confusion regarding the connotations of the term "mental illness," the *DSM* eschews its use. All *DSM* conditions are referred to as "mental disorders." DSM-IV-TR at xxx–xxxi.

84. Widiger, *supra* note 40, at 25–41.

85. Anne E. Figert, Women and the Ownership of PMS: The Structuring of a Psychiatric Disorder (1996).

86. Herb Kutchins & Stuart A. Kirk, Making Us Crazy: DSM, the Psychiatric Bible and the Creation of Mental Disorders (2003).

87. The cautionary statement reads, in part: "The purpose of DSM-IV is to provide clear descriptions of diagnostic categories in order to enable clinicians and investigators to diagnose,

Given that the anticipated publication of *DSM-5* is not due until 2013,[88] it is not possible at this writing to specify the changes that will appear in the new edition. However, current indications are that the major categories of diagnoses described in the following section will be retained, although specific changes may be made to individual diagnostic criteria.[89] The task force directing the revision process is considering potential changes to the five-axis structure that has existed since 1980,[90] minor modifications to the core definition of a mental disorder,[91] and the introduction of structured assessments of dimensions that cut across diagnostic categories, such as depressed mood, anxiety, substance use, or sleep problems.[92] Proposed changes prior to publication can be tracked on a Web site established by the American Psychiatric Association, which also offers a time line of the steps in the process.[93]

2. *Major diagnostic categories*

Some hint of the number and diversity of mental disorders embodied in the current diagnostic typology is provided by the fact that *DSM-IV-TR* is approximately 900 pages long. However, the characteristics of the major categories of disorders that are likely to be relevant in legal proceedings can be summarized more concisely.[94]

communicate about, study, and treat people with various mental disorders. It is to be understood that inclusion here, for clinical and research purposes, of a diagnostic category such as Pathological Gambling or Pedophilia does not imply that the condition meets legal or other nonmedical criteria for what constitutes mental disease, mental disorder, or mental disability. The clinical and scientific considerations involved in categorization of these conditions as mental disorders may not be wholly relevant to legal judgments, for example, that take into account such issues as individual responsibility, disability determination, and competency." DSM-IV-TR at xxxvii.

88. American Psychiatric Association, DSM-5 Development, Timeline, *available at* http://www.dsm5.org/about/Pages/Timeline.aspx.

89. American Psychiatric Association, DSM-5 Development, Proposed Draft Revisions to DSM Disorders and Criteria, *available at* http://www.dsm5.org/ProposedRevisions/Pages/Default.aspx.

90. American Psychiatric Association, DSM-5 Development, Classification Issues Under Discussion, *available at* http://www.dsm5.org/ProposedRevisions/Pages/ClassificationIssuesUnderDiscussion.aspx.

91. American Psychiatric Association, DSM-5 Development, Definition of a Mental Disorder, *available at* http://www.dsm5.org/ProposedRevisions/Pages/proposedrevision.aspx?rid=465.

92. American Psychiatric Association, DSM-5 Development, Cross-Cutting Dimensional Assessment in DSM-5, *available at* http://www.dsm5.org/ProposedRevisions/Pages/Cross-CuttingDimensionalAssessmentinDSM-5.aspx.

93. American Psychiatric Association, DSM-5 Development, DSM-5: The Future of Psychiatric Diagnosis, *available at* http://www.dsm5.org.

94. These brief summaries of complex and variable conditions are meant to provide an orientation to the nature and course of major mental disorders. The current edition of the *DSM* itself or standard psychiatric textbooks should be consulted for more complete descriptions. Note that for a diagnosis of any disorder to be made per the *DSM*, the symptoms must be deemed to "cause clinically significant distress or impairment in social, occupational, or other important areas of functioning." DSM-IV-TR at 7.

- *Schizophrenia* is a complex psychotic[95] disorder, involving delusions, hallucinations, disorganization of thought, speech, and behavior, and social withdrawal. Social and occupational functioning are markedly impaired. The course is chronic, marked by periodic exacerbations, and often by slow deterioration over time.[96]
- *Bipolar disorder* (formerly called manic-depressive disorder) is a disturbance of mood marked by episodic occurrence of both mania and depression. During manic periods, persons experience elevated, expansive, or irritable mood, accompanied by such symptoms as grandiosity, racing thoughts and pressured speech, decreased sleep, and hypersexuality. The course is chronic, but intermittent, though some patients experience a downward trajectory.[97]
- *Major depressive disorder* involves one or more episodes of depression, typically involving depressed mood, loss of pleasure, weight loss, insomnia, feelings of worthlessness, diminished ability to think or concentrate, and thoughts of death. Episodes are often, but not always, recurrent.[98]
- *Substance disorders* include both substance abuse and substance dependence, the most common of which are alcohol abuse and dependence. Abuse consists of "a maladaptive pattern of substance use leading to clinically significant impairment or distress."[99] Dependence involves, in addition, signs of tolerance, withdrawal, and lack of success in restricting substance use. These are chronic, and often relapsing, disorders, though successful recovery, with or without treatment, is possible.[100]
- *Personality disorders* are inflexible, maladaptive, and enduring patterns of perceiving and relating to oneself, other people, and the external world that cause functional impairment and distress.[101]
- *Antisocial personality disorder* is often seen in criminal courts, because it is marked by a pervasive pattern of disregard for and violation of the rights of others. Personality disorders tend to be longstanding and difficult to treat.[102]
- *Dementia* is marked by progressive impairment of cognitive abilities, including memory, language, motor functions, recognition of objects, and executive functioning.[103] The most common form of dementia is Alzheimer's disease, the incidence of which increases with age and the cause of which remains unclear, although in many cases genetics seem

95. Psychotic conditions involve some degree of detachment from reality, characterized by delusional thinking and hallucinatory perceptions. *Id.* at 770.

96. *Id.* at 297–317.

97. *Id.* at 382–92.

98. *Id.* at 369–76.

99. *Id.* at 199.

100. *Id.* at 192–98.

101. *Id.* at 686.

102. *Id.* at 701–06.

103. *Id.* at 147–71.

to play a role.[104] Other causes of dementia include multiple small strokes ("multi-infarct dementia"), trauma, and infection with certain virus-like agents.

Additional disorders that may have special legal relevance include anxiety disorders (including post-traumatic stress disorder (PTSD)), dissociative disorders (such as dissociative identity disorder, formerly multiple personality disorder), impulse control disorders (such as kleptomania and pyromania), sexual disorders (especially the paraphilias, such as pedophilia), delirium, and mental retardation.[105]

The causes of mental disorders remain to be elucidated. However, as a general proposition, it appears that many mental disorders may derive from a genetic predisposition that is activated by particular environmental circumstances.[106] This hypothesis is supported by extensive studies of the genetics of mental disorders[107] and epidemiological studies showing a relationship between various environmental factors and occurrence of illness.[108] Only rarely at this point, however, have particular genes and given stressors been linked to a particular disorder. For example, a genetic variant in an enzyme that regulates neurotransmitter reuptake has been shown to predispose to depression, but only when the susceptible person has been exposed to stressful life events.[109]

104. Matthew B. McQueen & Deborah Blacker, *Genetics of Alzheimer's Disease*, *in* Psychiatric Genetics: Applications in Clinical Practice (Jordan W. Smoller et al. eds., 2008).

105. Rebrook v. Astrue, 2008 WL 822104 (N.D. W. Va. Mar. 26, 2008) (anxiety disorder); United States v. Holsey, 995 F.2d 960 (10th Cir. 1993) (dissociative disorder); Coe v. Bell, 89 F. Supp. 2d 922 (M.D. Tenn. 2000) (dissociative identity disorder); United States v. Miller, 146 F.3d 1281 (11th Cir. 1998) (impulse control disorder); United States v. McBroom, 991 F. Supp. 445 (D.N.J. 1998) (person receiving treatment for bipolar disorder and impulse control disorder sentenced for possession of child pornography); United States v. Silleg, 311 F.3d 557 (2d Cir. 2002) (pedophilia determination in a child pornography case); Fields v. Lyng, 705 F. Supp. 1134 (D. Md. 1988) (kleptomania); United States v. Warr, 530 F.3d 1152 (9th Cir. 2008) (sentencing of an arsonist diagnosed with pyromania upheld); Kansas v. Hendricks, 531 U.S. 346 (1997) (upholding commitment of man unable to control pedophilic impulses); United States v. Gigante, 996 F. Supp. 194 (E.D.N.Y. 1998) (dementia); Johnson v. City of Cincinnati, 39 F. Supp. 2d 1013 (S.D. Ohio 1999) (estate of man who died from police restraint during a seizure sued the city under 28 U.S.C. § 1983; Bertl v. City of Westland, 2007 WL 3333011 (E.D. Mich. Nov. 9, 2007) (finding that delirium tremens is an objectively serious medical need); Atkins v. Virginia, 536 U.S. 304 (2002) (banning the execution of the mentally retarded as a violation of the Eighth Amendment); *In re* Hearn, 418 F.3d 444 (5th Cir. 2005); Hamilton v. Southwestern Bell Tel. Co., 136 F.3d 1047, 1050 (5th Cir. 1998) (recognizing PTSD as a mental impairment for the purposes of the Americans with Disabilities Act).

106. Michael Rutter & Judy Silberg, *Gene-Environment Interplay in Relation to Emotional and Behavioral Disturbance*, 53 Ann. Rev. Psychol. 463 (2002).

107. Jordan W. Smoller et al., Psychiatric Genetics: Applications in Clinical Practice (2008).

108. Ezra Susser et al., Psychiatric Epidemiology: Searching for the Causes of Mental Disorders (2006).

109. Avshalom Caspi et al., *Role of Genotype in the Cycle of Violence in Maltreated Children*, 287 Science 851 (2002).

Active efforts are under way to explore this "diathesis/stress hypothesis" in other mental disorders as well.[110]

3. Approaches to diagnosis

The solicitation of symptoms and the observation of signs necessary for a mental disorder diagnosis can be accomplished with a variety of techniques.

a. Clinical examination

Direct clinical examination of the person whose condition is at issue is still the core of most mental health evaluations.[111] In contrast to general medicine, where examination involves the laying on of hands, evaluation of mental disorders is accomplished by careful elicitation of symptoms and observation of signs. A typical sequence of clinical examination involves exploring with the person being evaluated: the current presenting problem, including the specific symptoms experienced and the duration of such symptoms; past history of similar symptoms or other disorders and of treatment for those disorders; developmental history; social and occupational history; family history; medical history, including a review of current medical symptoms, medications taken, and substances used (e.g., alcohol, street drugs, cigarettes); and mental status examination.[112] The last category involves a structured assessment of the person's mental state, including motor function, speech, mood and affect, thought process and content, cognitive functioning, judgment, and insight, along with the presence of ideation or history of self-harm or harm toward others. Simultaneously, the clinician is observing the person's behavior and appearance to glean signs associated with mental disorders.[113] If indicated, a physical examination may be performed, if the evaluator is a psychiatrist who has maintained his or her general clinical skills, or requested.

The duration of a clinical examination sufficient to diagnose the person's condition will vary depending on the complexity of the case, the cooperativeness of the evaluee, and the questions being addressed. Examinations may take from one to several hours, sometimes spread over multiple sessions. When previous records of contact with mental health professionals are available, the clinician will ordinarily want to review them prior to the clinical examination, so that questions can be targeted more efficiently, and previous conclusions confirmed or

110. Margit Burmeister et al., *Psychiatric Genetics: Progress Amid Controversy*, 9 Nat. Rev. Genetics 527 (2008).

111. For an overview of the evaluation of mental health problems, see Linda B. Andrews, *The Psychiatric Interview and Mental Status Examination, in* The American Psychiatric Publishing Textbook of Clinical Psychiatry 3 (Robert E. Hales et al. eds., 2008).

112. American Psychiatric Association Work Group on Psychiatric Evaluation, *Practice Guideline for the Psychiatric Evaluation of Adults* (Supplement), 163 Am. J. Psychiatry 7 (2006) [hereinafter Psychiatric Evaluation of Adults].

113. Paula T. Trzepacz & Robert W. Baker, The Psychiatric Mental Status Examination (1993).

rejected.[114] Information from collateral sources (e.g., spouses, family members, friends, other health professionals) can be valuable in confirming the account given by an evaluee or in providing information not communicated by the evaluee, especially when an incentive may exist for the person being examined to exaggerate or downplay the nature and extent of symptoms.[115] In difficult cases, it may not be possible to distinguish with reasonable clinical certainty among two or more possible diagnoses; in such cases, clinicians may assign "rule out" diagnoses, indicating the range of possibilities and deferring a definitive diagnosis until more information is available.

b. Structured diagnostic interviews

When a diagnosis is based solely on a clinical examination, which is still most frequently the case, the clinician is being relied upon to conduct a complete evaluation and to apply the diagnostic criteria accurately. Studies that showed considerable variation in the results of clinical evaluations motivated the development, largely for research purposes, of structured diagnostic interviews.[116] Structured interviews provide a fixed set of questions—ensuring that important issues are not omitted from consideration—and a schema for applying the results to the diagnostic framework. Hence, they tend to show increased reliability over unassisted clinical evaluations. More complete diagnostic interviews may allow consideration of a large number of diagnostic categories;[117] focal interviews clarify whether a single disorder (e.g., obsessive–compulsive disorder[118]) or category of disorders is present (e.g., dissociative disorders[119]).

The disadvantages of structured diagnostic assessments include the time that may be required (i.e., more extensive instruments may take several hours to complete) and the fact that many persons respond negatively to an evaluation with a series of preset questions.[120] Many instruments require that the person conducting the interview be trained in their use; administration by untrained personnel may not achieve the level of reliability or validity demonstrated in research studies.[121]

114. Psychiatric Evaluation of Adults, *supra* note 112, at 16.

115. *Id.*

116. Robert Spitzer & Joseph Fleiss, *A Re-analysis of the Reliability of Psychiatric Diagnosis*, 125 Brit. J. Psychiatry 341 (1974).

117. See *generally* Michael B. First et al., User's Guide for the Structured Clinical Interview for DSM-IV Axis I Disorders: SCID-1 Clinician Version (1997).

118. *See, e.g.*, Wayne K. Goodman et al., The Yale Brown Obsessive Compulsive Scale (YBOCS), 1: Development, Use and Reliability, 46 Arch. Gen. Psychiatry 1006 (1989).

119. *See, e.g.*, Marlene Steinberg, Structured Clinical Interview for DSM-IV Dissociative Disorders (SCID-D) (1995).

120. Deborah Blacker, *Psychiatric Rating Scales*, *in* Comprehensive Textbook of Psychiatry 9th ed., 1032 (Benjamin J. Sadock, Virginia A. Sadock, & Pedro Ruiz, eds., 2009).

121. See, e.g., the recommended training requirements for the SCID, *available at* http://www.scid4.org/training/overview.html.

Although structured diagnostic interviews do not reflect the current standard of care for clinical purposes, there may be some value in their use for purposes of forensic evaluation in cases with particularly difficult diagnostic questions.

Diagnostic interviews should be distinguished from instruments that assess the nature and extent of psychiatric symptomatology.[122] The former yield a conclusion about the presence or absence of a psychiatric diagnosis; the latter allow estimates of the type and magnitude of symptoms experienced, regardless of diagnosis. As with diagnostic interviews, symptom measures may be broad in their scope or assess a single type of symptom.[123] Although they are not likely to be used for the purpose of diagnosis per se, the results of applying such instruments may be introduced in evidence to establish the severity of symptoms associated with a disorder.

c. Psychological and neuropsychological tests

Formal testing of psychological functions may be used to complement the clinical diagnostic process, but often it is not necessary for a diagnosis to be made.[124] Psychological tests such as the Minnesota Multiphasic Personality Inventory (MMPI) assess multiple dimensions of personality and mental state; research over many years of use has established correlations between patterns of performance on the MMPI and particular mental disorders, which may be helpful in establishing or confirming a diagnosis, particularly when the results of a clinical examination are inconclusive.[125] Tests of intelligence, such as the Wechsler Adult Intelligence Scale (WAIS-III), are important in establishing the presence of mental retardation and determining its severity.[126] Projective tests, such as the famed Rorschach ink-blot test or the Thematic Apperception Test, were once used more widely than they are today as a means of probing the nature and content of a person's thought processes; although results were said to be helpful for diagnostic purposes, questions about the reliability and validity of projective measures have limited their use.[127] Other tests target personality traits, such as psychopathy, or behavioral characteristics, such as impulsivity, and may be helpful but not determinative in making a diagnosis of mental disorder.[128]

122. *See, e.g.*, John E. Overall & Donald R. Gorham, *The Brief Psychiatric Rating Scale (BPRS): Recent Developments in Ascertainment and Scaling*, 24 Psychopharmacology Bull. 97 (1988).

123. Compare the BPRS, *supra* note 122, with the *Beck Depression Inventory*, *in* Aaron T. Beck et al., Manual for the Beck Depression Inventory-II (1996).

124. See discussion in John F. Clarkin et al., *The Role of Psychiatric Measures in Assessment and Treatment*, *in* Hales et al., *supra* note 111, at 73.

125. Starke R. Hathaway & John C. McKinley, Minnesota Multiphasic Personality Inventory-2 (1989).

126. David Wechsler, Wechsler Adult Intelligence Scale-III Administrative and Scoring Manual (1997).

127. Scott O. Lilienfeld et al., *The Scientific Status of Projective Techniques*, 1 Psychol. Sci. Publ. Int. 27 (2000).

128. *See, e.g.*, Robert Hare, The Psychopathy Checklist-Revised (1991); Ernest S. Barratt, *Impulsiveness and Aggression*, *in* Violence and Mental Disorder: Developments in Risk Assessment 61 (John Monahan & Henry J. Steadman eds., 1996).

No bright line distinguishes psychological from neuropsychological tests, but in general the latter are focused on assessing the integrity of the functioning of the brain itself.[129] Hence, they may be helpful—and sometimes even essential—in the diagnosis of states of impaired brain function, such as may occur in the wake of traumatic brain injury, infections such as meningitis, and learning disabilities. Neuropsychological testing usually involves the administration of a battery of measures, each targeting a relatively discrete area of function, such as attention, memory, verbal abilities, visual recognition, spatial perception, and the like. The tests selected by neuropsychologists as part of a battery will often vary on the basis of the person's history and suspected condition; thus, it is important before accepting a conclusion that "neuropsychological testing showed no signs of abnormality" to ascertain precisely which functions were specifically assessed.

Neuropsychological testing can be particularly helpful in the diagnosis of dementia, a condition that may lead to legal challenges to a person's decisional or performative capacities. Although the diagnosis may be suggested by elements of the person's history (e.g., forgetfulness, disorientation), serial testing of cognitive functions can provide strong evidence for a progressive disorder.[130] The most frequently used test is the Mini-Mental Status Examination (MMSE), a 20-question screening tool that can be applied by primary care and other clinicians in ordinary treatment settings. Structural and functional brain imaging can be helpful in ruling out other causes of the person's cognitive decline.[131]

d. Imaging studies

Progress has been made in recent years in the use of radiological techniques to assist in the diagnosis and evaluation of mental disorders. With the development of computer-assisted tomography (CAT or CT), a noninvasive technique became available for clinicians to visualize aspects of the gross structure of the brain.[132] CT scans, which use traditional X rays to provide computer-reconstructed pictures of "slices" through the brain, especially when combined with injection of radio-opaque dye into the bloodstream, permit the detection of intracranial masses (e.g., tumors), stroke, atrophy (e.g., associated with Alzheimer's disease and other dementias), and other deformations of brain structure. More recently, magnetic resonance imaging (MRI) has replaced CT scans in many of the situations in which they previously would have been used. MRI offers higher resolution of

129. Clarkin et al., *supra* note 124.

130. Diane B. Howieson & Muriel D. Lezak, *The Neuropsychological Examination*, *in* The American Psychiatric Publishing Textbook of Neuropsychiatry and Clinical Neurosciences 215–43 (Stuart C. Yudovsky & Robert E. Hales eds., 5th ed. 2008).

131. Marshall F. Folstein et al., *Mini-Mental State: A Practical Method for Grading the Cognitive State of Patients for the Clinician,* 12 J. Psychiatric Res. 189 (1975). *See* discussion *infra* Section I.C.3.d.

132. Robin A. Hurley et al., *Clinical and Functional Imaging in Neuropsychiatry*, *in* Yudofsky & Hales, *supra* note 130, 245.

brain structures without exposure to X-rays.[133] Regardless of whether CT or MRI is used, however, it is important to note that, despite evidence for the localization of some brain functions (e.g., speech, vision), the general tendency for the brain to function as an integrated network limits the conclusions that can be drawn about a person's functional abilities on the basis of structural studies alone.[134]

Functional imaging techniques have augmented the ability of clinicians to get inside the "black box" of the brain to more directly assess aspects of brain function. These include functional MRI (fMRI), single-photon emission computerized tomography (SPECT), and proton emission tomography (PET).[135] What they have in common is the capacity to detect changes in such characteristics of the brain as blood flow or oxygen saturation of the blood that presumably correlate with the activity of a given brain area. Thus, functional imaging can identify regions with aberrant patterns of activity that may be associated with impaired function in that area of the brain. Again, however, conclusions relevant to diagnosis or impairment of capacities are limited by the frequent absence of a tight correlation between functional imaging findings and actual functional impairment of a sort likely to have legal relevance.[136]

e. Laboratory tests

Use of standard laboratory tests may be helpful in ruling out general medical causes of abnormal mental states and behavior. For example, low levels of thyroid hormone may be associated with a state that resembles a major depressive episode, vitamin B-12 deficiency can lead to psychosis, and disturbance of the balance of electrolytes in the blood can cause states of delirium.[137] Each of these conditions is responsive to treatment of the underlying disorder, and all can lead to more severe and permanent impairments if untreated. Infectious diseases such as HIV, syphilis, and Lyme disease can present as mental disorders otherwise indistinguishable from depression, mania, and acute psychosis; all can be detected with appropriate blood tests.[138] Behavioral abnormalities that may be mistaken for mental disorders can be caused by several forms of epilepsy, which are usually detectable

133. *Id.*

134. William R. Uttal, The New Phrenology: The Limits of Localizing Cognitive Processes in the Brain (2003).

135. Yudofsky & Hales, Hurley et al., *supra* note 132 at 261–2.

136. Stephen J. Morse, *Moral and Legal Responsibility and the New Neuroscience*, *in* Neuroethics: Defining the Issues in Theory, Practice and Policy 33 (Judy Illes ed., 2006).

137. H. Florence Kim et al., *Laboratory Testing and Imaging Studies in Psychiatry*, *in* Hales et al., *supra* note 111, at 19–49.

138. Glenn J. Treisman et al., *Neuropsychiatric Aspects of HIV Infection and AIDS*, *in* Sadock, Sadock, & Ruiz, *supra* note 120, at 506–31; Brian A. Fallon, *Neuropsychiatric Aspects of Other Infectious Diseases (non-HIV)*, *in* Sadock, Sadock, & Ruiz, *supra* note 120, at 532–41.

by electroencephalogram (EEG).[139] When there is any reason to suspect, on the basis of a person's history or the findings of a clinical evaluation, that a general medical disorder may exist, laboratory testing is an essential aspect of a complete evaluation. On the other hand, despite many years of investigation of possible correlates of the major mental disorders in blood, urine, and other bodily fluids, there are no laboratory tests that can identify schizophrenia, bipolar disorder, major depression, or other mental disorders.[140]

f. Previous medical and mental health records

Among the most helpful adjunctive sources of information for a diagnostic assessment are the person's records of previous contact with the medical and mental health systems.[141] Past records can confirm a person's account or point to discrepancies that require further exploration (which is particularly important, as described in Section I.C.4, *infra*, when malingering is suspected). Such factors as age of onset, progression of illness, and variability of symptoms, all of which may affect diagnostic choices, can be determined from records of previous medical and mental health evaluations, as can susceptibility to treatment and need for other supportive interventions.

4. *Accuracy of diagnosis of mental disorders*

Diagnostic accuracy has two separate aspects: (1) reliability—the extent to which two or more examiners of the same person would derive the same diagnosis, and (2) validity—the extent to which the diagnosis corresponds to the person's actual mental state.[142] It is axiomatic that reliability is a necessary, but not sufficient, condition for validity. Prior to the introduction of *DSM-III* in 1980, several influential studies showed poor reliability of psychiatric diagnosis, even for major disorders such as schizophrenia.[143] Reliability improved with the new, criteria-based categories that were introduced at that point, but remains greater for broader categories of diagnosis, such as psychosis, than for finer distinctions, such as whether a person suffers from schizophrenia or the similar but not identical syndrome of schizoaffective disorder.[144] For most purposes, however, as discussed

139. H. Florence Kim, et al., *Neuropsychiatric Aspects of Seizure Disorders*, *in* Yudofsky & Hales, *supra* note 132, at 649–76.

140. Barry H. Guze & Martha J. Love, *Medical Assessment and Laboratory Testing in Psychiatry*, *in* Sadock, Sadock, & Ruiz, *supra* note 120, at 996.

141. Psychiatric Evaluation of Adults, *supra* note 112, at 16.

142. *See* Robert E. Kendell, *Five Criteria for an Improved Taxonomy of Mental Disorders*, *in* Defining Psychopathology in the 21st Century 3 (John E. Helzer & James J. Hudziak eds., 2002).

143. Spitzer and Fleiss, *supra* note 116.

144. Robert L. Spitzer et al., *DSM-III Field Trials: I. Initial Interrater Diagnostic Reliability*, 136 Am. J. Psychiatry 815 (1979); *see also* DSM-III at 467–72; Joseph D. Matarazzo, *The Reliability of Psychiatric and Psychological Diagnosis*, 3 Clinical Psychol. Rev. 103–45 (1983); Peter E. Nathan & James

below (*see* Section I.D), it is unlikely that differences within broader categories of diagnosis will have significance for the legal issue at stake.

The validity of a diagnosis of mental disorder depends on the underlying validity of the diagnostic criteria, that is, the extent to which they accurately characterize a particular psychiatric disorder; and on the validity of the judgment of the diagnosing clinician in a given case, that is, how well the clinician has applied the criteria. Diagnostic criteria can be judged on how well they identify a syndrome whose symptomatology, heritability, course, and treatment response, among other variables, differentiate it from similar disorders.[145] *DSM-IV-TR* diagnostic criteria vary along these dimensions, and it is impossible to make a general statement about the validity of the diagnostic framework as a whole.[146] Again, however, for most legal determinations it is the presence or absence of any mental disorder and associated levels of functional impairment that will be at issue, rather than distinctions among similar disorders. How well the criteria have been applied in a particular case can be determined more easily, whether by means of cross-examination or by virtue of conflicting expert testimony offered by the adverse party.

5. *Detection of malingering*

Because the diagnosis of mental disorders rests heavily on the elicitation of symptoms from the person being evaluated and observations of the person's behavior, the possibility of malingering—the deliberate simulation of symptoms of mental disorder—must always be considered.[147] Most commonly, the likelihood of malingering is assessed as part of a clinical evaluation. The pattern of symptoms reported by the person is compared with known syndromes, and the consistency of his or her behaviors is observed. Contrary to common belief, mental disorders are not easy to fake, especially when the deception must be sustained over a period of time.[148]

When deception is suspected, efforts to confirm it should begin during the clinical examination, as the person is offered the opportunity to endorse symptoms that are unlikely to occur naturally (e.g., "Do you ever feel as though the cars on the street are talking about you?") or do not fit the condition from which the

W. Langenbucher, *Psychopathology: Description and Classification*, 50 Ann. Rev. Psychol. 79 (1999). Reliability in actual clinical practice may well be less than has been demonstrated in research settings, especially when the latter make use of structured assessment instruments. *See, e.g.*, M. Katherine Shear et al., *Diagnosis of Nonpsychotic Patients in Community Clinics*, 157 Am. J. Psychiatry 581 (2000).

145. The Validity of Psychiatric Diagnosis (Lee N. Robins & James E. Barrett eds., 1989).

146. Kendell, *supra* note 142.

147. Phillip J. Resnick, *Malingering*, *in* Principles and Practice of Forensic Psychiatry 543 (Richard Rosner ed., 2003).

148. *Id.* at 544.

patient is claiming to suffer.[149] Psychological testing can be helpful in detecting deception; the MMPI-2, for example, has scales that correlate with persons who are both "faking bad" (i.e., fabricating symptoms) and "faking good" (i.e., hiding symptoms that actually exist).[150] Other instruments specifically for the assessment of malingering also have been developed, with varying degrees of validation.[151] Information from records of previous psychiatric or psychological evaluations can be helpful in determining the congruence of the person's current symptoms with past reports and behaviors. In addition, given the difficulty in maintaining a consistent pattern of deception over a sustained period, data provided by collateral sources (e.g., family members, roommates, prisoners in adjoining cells, correctional officers, nurses and other hospital staff, and others who have been in contact with the person) who have observed the person informally outside of the evaluator's presence can be crucial in distinguishing real from malingered disorders.[152]

The difficulty of simulating a mental disorder does not imply that it is impossible to do. Indeed, a skilled and determined person can sometimes fool even an experienced evaluator. Thus, the only honest response that a clinician can give in almost every circumstance to a question about the possibility of malingering is that it is always possible, but is more or less likely in this particular case, given the characteristics of the person being evaluated.[153]

D. Functional Impairment Due to Mental Disorders

1. Impact of mental disorders on functional capacities

Mental disorders can affect functional capacities in a variety of ways. Among these, *attention* and *concentration* may be impaired by the preoccupations that appear in anxiety and depressive disorders, or the grosser distractions (e.g., auditory hallucinations) of psychotic disorders.[154] *Perception* is often distorted in psychotic condi-

149. Paul S. Appelbaum & Thomas G. Gutheil, Clinical Handbook of Psychiatry and the Law 248–49 (2007).

150. Roger L. Greene, *Malingering and Defensiveness on the MMPI-II*, *in* Clinical Assessment of Malingering and Deception 159 (Richard Rogers ed., 2008). These scales, especially the most prominent of them, the "Fake Bad Scale (FBS)," are not without controversy that has sometimes led courts to rule them inadmissible. David Armstrong, *Malingerer Test Roils Personal-Injury Law: "Fake Bad Scale" Bars Real Victims, Its Critics Contend*, Wall Street J., Mar. 5, 2008, at A1. However, the bulk of the psychological literature appears to support the validity of the FBS and many of the other MMPI-based malingering scales. Nathaniel W. Nelson et al., *Meta-Analysis of the MMPI-2 Fake Bad Scale: Utility in Forensic Practice*, 20 Clinical Neuropsychologist 39 (2006).

151. Richard Rogers, *Structured Interviews and Dissimulation*, *in* Clinical Assessment of Malingering and Deception 301–22 (Richard Rogers ed., 2008).

152. Appelbaum & Gutheil, *supra* note 149.

153. Resnick, *supra* note 147.

154. Ronald A. Cohen et al., *Neuropsychiatric Aspects of Disorders of Attention*, *in* Yudofsky and Hales, *supra* note 132, at 405–44.

tions, as manifest by hallucinations of the auditory, visual, tactile, or other sensory systems.[155] *Cognition*, encompassing both the process and content of thought, is also often affected: Thought processes can be impeded by the slowing of thought in depression, its acceleration in mania, or the scrambling of thought experienced by persons with schizophrenia or other psychotic disorders; thought content may be altered by the odd reasoning to which persons with delusions appear to be prone.[156] *Motivation* to act, even in one's self-interest, is often globally reduced in states of intense depression and in schizophrenia.[157] *Judgment and insight* may be altered under the pressure of delusions.[158] *Control of behavior* can be weakened by the impulsivity seen in mania and psychosis, the drives of the impulse disorders, and the use of disinhibiting substances, especially alcohol.[159] Any of these impairments in principle could affect a person's relevant decisional and performative capacities.

This necessarily incomplete list of the ways in which mental disorders can affect functional capacities illustrates the vulnerability of almost every aspect of mental functioning to perturbation. Moreover, although it is common to divide mental functions into categories such as these for heuristic purposes, most neuroscientists recognize that the brain operates as a unified entity.[160] Thus, it is rare that impairments are limited to a single area of functioning. Impaired concentration, for example, inherently affects cognitive abilities, which in turn may alter judgment and therefore the person's choice of behaviors. Although focal deficits may occur, for example, the anxiety associated with exposure to a phobic stimulus such as a spider, more severe disorders will have a broader impact on a person's functional capacities as a whole.[161]

2. Assessment of functional impairment

Determining the nature and extent of past, present, or future functional impairment, therefore, is usually the most critical aspect of a mental health evaluation and subsequent presentation of mental health evidence.

155. Andre Aleman & Frank Laroi, Hallucinations: The Science of Idiosyncratic Perception (2008).

156. Ann A. Matorin & Pedro Ruiz, *Clinical Manifestations of Psychiatric Disorders*, *in* Sadock, Sadock, & Ruiz, *supra* note 120, at 1076–81.

157. *Id.* at 1092–93.

158. Phillipa A. Garety, *Insight and Delusions*, *in* Insight and Psychosis 66, 66–77 (Xavier F. Amador & Anthony S. David eds., 1998).

159. Eric Hollander et al., *Neuropsychiatric Aspects of Aggression and Impulse-Control Disorders*, *in* Yudofsky & Hales, *supra* note 132, at 535–66.

160. William R. Uttal, *supra* note 134.

161. The pervasive impact of schizophrenia on all aspects of personality and functioning is the most extreme example. *See* Michael J. Minzenberg et al., *Schizophrenia*, *in* Hales et al., *supra* note 111, at 407–56.

a. Clinical examination

As in establishing a diagnosis, the core of the assessment of functional impairment remains the clinical examination.[162] A diagnostic assessment may be integral to the functional assessment process, suggesting to the examiner areas of possible impairment to be explored in greater depth (e.g., attentional and concentration abilities in an anxiety disorder; impairments in motivation in a depressive disorder). Beginning in the 1970s, however, there was growing recognition among the mental health professions that merely establishing a diagnosis is insufficient to permit a conclusion to be drawn about a legally relevant capacity, because a broad range of functional impairments can be associated with almost any mental disorder.[163]

Thus, in addition to a diagnostic assessment, an adequate examination will explore the person's perspective on the alleged functional impairment and will probe for symptoms associated with such impairment. The process involves more than simply taking the person's word for the issue in question, for example, that she was not able to comprehend the details of the contract to which she is a party, or that he remains incapable of the careful calculations required in his job. Assessors compare the claimed impairments with the person's overall history and other areas of function, looking for congruence or incongruence. For example, the assertion by a plaintiff that because of being harassed on the job he has been unable to concentrate sufficiently to work will be more or less plausible depending on the consistency and extent of his symptoms and the degree to which the impairment may generalize to other areas of his life. Degrees of impairment that are out of scale with the extent of symptoms or the person's functional history are inherently suspect.[164]

In addition to questioning the evaluee directly, the use of collateral information can be essential to a valid assessment, particularly when the person has an incentive to malinger, which will often be the case in legal proceedings.[165] Family members, coworkers, and others who have had an opportunity to observe the person can provide invaluable information about the nature and extent of impairments, although one must always be alert to the possibility that informants will be motivated to assist the person by distorting or exaggerating their accounts. Records of performance, such as educational test results and work evaluations, especially if generated prior to the filing of the legal claim, may shed somewhat more objective light on the person's capacities.[166] To the extent that impairments

162. *See supra* Section I.C.2.a.

163. Michael Kindred, *Guardianship and Limitations upon Capacity*, *in* The Mentally Retarded and the Law 62 (The President's Committee on Mental Retardation, 1976); Laboratory of Community Psychiatry, Harvard Medical School, Competency to Stand Trial and Mental Illness (1973).

164. Richard Rogers, *Detection Strategies for Malingering and Defensiveness*, *in* Clinical Assessment of Malingering and Deception 14 (Richard Rogers ed., 2008).

165. Appelbaum & Gutheil, *supra* note 149.

166. Melton et al., *supra* note 28, at 53–55.

may be rooted in disruptions of brain functions per se, neuropsychological testing can also be helpful in documenting their nature and extent. Increasingly, however, it has been accepted that an unstructured clinical evaluation, even when supplemented by collateral information, is not necessarily the most accurate tool, standing on its own, for determining functional capacity.[167]

b. Structured assessment techniques

As with determination of diagnosis, the evaluation of the limitations of function due to mental disorders increasingly involves the use of structured assessment techniques.[168] Most commonly, these are standardized interviews or data-gathering protocols (e.g., based on a person's psychiatric record) designed to ensure that all relevant information is obtained. In addition, where research has established the validity of the instruments by demonstrating a correlation between the results and actual impairments, these techniques may allow a quantitative estimate to be made of the extent of actual functional deficiencies. A recent compendium of assessment instruments included structured evaluations that address criminal defendants' competence to stand trial, waiver of rights to silence and legal counsel, criminal responsibility and persons' parenting capacity, competence to manage one's affairs (i.e., need for a guardian or conservator), and competence to consent to medical treatment and research.[169] Given that this area is a rapidly developing focus of research, instruments to address other legally relevant functional capacities and states—propensity to commit violent or sexual offenses comes quickly to mind[170]—are continuously being tested and developed.

Although most assessment techniques rely on information gathered from the person being evaluated or from existing records, some approaches involve direct testing of the person's capacity to perform particular tasks. Examples include computerized assessment of driving capacities,[171] observation of tasks involving

167. Grisso, *supra* note 2. Surprisingly few studies exist of the reliability of clinical forensic evaluations. The only U.S. study of actual assessments showed good interrater reliability of evaluations of competence to stand trial, although many of the reports were deficient in other ways. Jennifer L. Skeem et al., *Logic and Reliability of Evaluations of Competence to Stand Trial,* 22 Law Hum. Behav. 519 (1998). A more recent Australian study found only fair to moderate reliability across assessments of competence to stand trial, but moderate to good reliability of criminal responsibility evaluations. Matthew Large et al., *Reliability of Psychiatric Evidence in Serious Criminal Matters: Fitness to Stand Trial and the Defence of Mental Illness*, 43 Austl. N.Z. J. Psychiatry 446 (2009).

168. *Id.*

169. *Id.*

170. *See, e.g.*, Christopher D. Webster et al., HCR-20 Assessing Risk for Violence Manual, Version 2 (1997); Vernon L. Quinsey et al., Violent Offenders: Appraising and Managing Risk (1998); John Monahan et al., COVR—Classification of Violence Risk, Professional Manual (2005).

171. Maria T. Schultheis et al., *The Neurocognitive Driving Test: Applying Technology to the Assessment of Driving Ability Following Brain Injury*, 48 Rehabilitation Psychol. 275 (2003).

the handling and management of money[172] and of parenting skills,[173] and direct measurement of such capacities as understanding and reasoning about medical information when a person's competence to decide about medical treatment is at issue.[174] In general, these approaches reduce the degree of inference required in drawing conclusions about a person's functioning because the person is observed performing something close to the precise tasks in question. Of course, such techniques may not be relevant when the legal issue relates to the impact of mental disorder on functional abilities at some time in the past or in the future, especially if the person's mental state at present may be different from what it was or will be. Nonetheless, these can be useful approaches to evaluation in appropriate legal contexts.

The advantages that attend the use of structured assessment instruments include the thoroughness of the evaluation, because the likelihood is reduced that variables that have been shown to be important to assessment will be omitted, and in many cases, a research base exists from which conclusions can be drawn regarding the degree of functional impairment of the person being assessed.[175] Indeed, in some jurisdictions, the use of structured assessments is required for particular purposes (e.g., evaluation of sexual offenders).[176] However, it remains true that the use of structured assessments performed for the purpose of being introduced in legal proceedings is variable and far from universal.[177] Grisso, the leading scholar in this area, suggests three reasons why this is still true: (1) it is easier and may be more lucrative (i.e., where a fixed rate is being paid per evaluation) for an examiner to avoid the frequently time-consuming use of a structured instrument; (2) many cases involve persons whose functional impairments—or lack of impairment—are obvious, and use of a structured assessment instrument would be "overkill"; and (3) perhaps paradoxically, the use of an assessment tool makes experts more vulnerable to attack on cross-examination.[178] To this list should be added the lack of knowledge of many expert witnesses regarding the existence of these instruments and a sense that their use denigrates the evaluator's expertise.

The vulnerability of testimony based on assessment instruments to cross-examination is worth special emphasis. Opinions offered on the basis of "clinical

172. Dan Marson et al., *Assessing Financial Capacity in Patients with Alzheimer's Disease: A Conceptual Model and Prototype Instrument*, 57 Archives Neurology 877 (2000).

173. Marc J. Ackerman & Kathleen Schoendorf, Ackerman-Schoendorf Scales for Parent Evaluation of Custody Manual (1992).

174. Thomas Grisso & Paul S. Appelbaum, MacArthur Competence Assessment Tool for Treatment (MacCAT-T) (1998).

175. Grisso, *supra* note 2, at 45–47.

176. *See, e.g.*, Va. Code Ann. § 37.2-903-C: "Each month the Director shall review the database and identify all such prisoners who are scheduled for release from prison within 10 months from the date of such review who receive a score of five or more on the Static-99 or a like score on a comparable, scientifically validated instrument designated by the Commissioner. . . ."

177. Grisso, *supra* note 2, at 481.

178. *Id.* at 481–82.

experience," which appears to be the norm, are difficult to challenge when expert witnesses in fact have appropriate training and a good deal of experience with the condition in question.[179] On the other hand, assessment instruments can be subjected to scrutiny with regard to the empirical database that supports their use, including their reliability and validity, their acceptance by the relevant professional community, and their probative value in a particular case. There may also be questions regarding the examiner's training and experience with the instrument and whether it was administered in the manner intended by its developers. All of these are legitimate questions, of course, and an argument can be made that the introduction of data from assessment instruments into evidence should be held to a more rigorous standard, because factfinders may give such data greater credence than unassisted clinical judgment.[180] But the undoubted consequence is that the arguably more reliable and perhaps more valid data from empirically derived assessment techniques are less likely to be introduced in evidence than evaluators' subjective judgments of unknown validity.[181]

E. Predictive Assessments

As noted above,[182] predictive assessments are the most challenging evaluations performed by mental health professionals.[183] The most common tasks involve the prediction of violence risk and of future functional impairment and responses to treatment.

1. Prediction of violence risk

The probability that a person may commit a violent act at some point in the future may come into play in the criminal process regarding determinations of suitability for diversion, bail, sentencing, probation, and parole, and in the civil process in hearings for civil commitment to psychiatric facilities and sexual offender treat-

179. *See* 4 Jack B. Weinstein, Weinstein's Federal Evidence § 702:02 n.1 (2d ed. 2008) on the liberal admissibility of expert testimony under Federal Rule of Evidence 702; § 702.02[4] nn.25–27 on the trial judge's broad discretion to admit or exclude expert testimony, to determine its helpfulness and relevancy, and the application of the "abuse of discretion" standard of review to determinations of whether a witness qualifies as an expert; § 702.04[1][c] on the typical "academic credentials plus experience" combination. Bryan v. City of Chicago, 200 F.3d 1092 (7th Cir. 2000) (an expert may qualify based on academic expertise and practical experience).

180. Christopher Slobogin, *Experts, Mental States, and Acts*, 38 Seton Hall L. Rev. 1009 (2008).

181. Grisso, *supra* note 2, at 482.

182. *See supra* Section I.A.1.

183. Yogi Berra, New York Yankees' Hall of Fame catcher and philosopher of everyday life, is purported to have said, "It's tough to make predictions, especially about the future." *See* http://www.famous-quotes-and-quotations.com/yogi-berra-quotes.html. For a discussion of the origin of this phrase, see Henry T. Greely & Anthony D. Wagner, Reference Guide on Neuroscience, Section VII, in this manual.

ment programs and when considering the imposition of liability on clinicians and facilities for failing to protect victims of patients' violence.[184] Although not all persons for whom such assessments must be made will have mental disorders, many will, and, in any event, psychiatrists and psychologists are seen by the courts as having expertise in this area and hence are almost invariably called upon for these evaluations.[185]

Persons with serious mental disorders, such as schizophrenia or bipolar disorder, are often considered by the general public to be at high risk for violence.[186] However, data on the relationship between serious mental disorders (schizophrenia is the disorder most frequently studied) and violence are variable. Although most studies suggest a moderately elevated risk, the proportion of violence accounted for by serious mental disorders is small, probably 3% to 5%, based on the best available U.S. estimates.[187] Data also suggest that the stereotype of violent mental patients who assault strangers in public places is inaccurate: Most violence by persons with serious mental disorders is directed at family members and friends and usually occurs in the living quarters of the perpetrator or the victim.[188] Much higher rates of violence are associated with substance use, especially alcohol use, and with traits such as psychopathy, often found in antisocial personality disorders.[189] Indeed, most of the strongest predictors of violence are common to both persons with serious mental disorders and those without, suggesting that the impact of the disorders per se is slight.[190]

a. Approaches to prediction of violence risk

Clinical evaluation of violence risk ordinarily focuses on those variables that have been shown in empirical research to have the strongest relationship to future violence,[191] whether the information is gleaned directly from the person or

184. *See, e.g.*, Kansas v. Hendricks, 521 U.S. 346 (1997); White v. Johnson, 153 F.3d 197 (5th Cir. 1998). A unanimous U.S. Supreme Court pointed to the importance of considering empirical data in identifying circumstances associated with increased risk of violence in *Chambers v. United States*, 130 S. Ct. 567, 691–93 (2009).

185. *See* Joanmarie Ilaria Davoli, *Psychiatric Evidence on Trial*, 56 SMU L. Rev. 2191 (2003).

186. Bernice Pescosolido et al., *The Public's View of the Competence, Dangerousness and Need for Legal Coercion Among Persons with Mental Illness*, 89 Am. J. Pub. Health 1339 (1999).

187. Jeffrey W. Swanson, *Mental Disorder, Substance Abuse, and Community Violence: An Epidemiologic Approach*, *in* Violence and Mental Disorder: Developments in Risk Assessment 101, 101–36 (John Monahan & Henry J. Steadman eds., 1994); Paul S. Appelbaum, *Violence and Mental Disorders: Data and Public Policy* (editorial), 163 Am. J. Psychiatry 1319 (2006).

188. Henry J. Steadman et al., *Violence by People Discharged from Acute Psychiatric Inpatient Facilities and by Others in the Same Neighborhoods*, 55 Archives Gen. Psychiatry 393 (1998).

189. John Monahan et al., Rethinking Risk Assessment: The MacArthur Study of Mental Disorder and Violence (2001).

190. *Id.* at 37–90; Simon Wessely, *The Epidemiology of Crime, Violence, and Schizophrenia*, 170 Brit. J Psychiatry 11 (1997).

191. Appelbaum & Gutheil, *supra* note 149, at 56.

derived from collateral informants or from a review of relevant records. These variables include a history of previous violence, age (violence risk peaks in the late teens and early twenties, declines slowly through the twenties and thirties, and drops off precipitously after age 40), male gender, lower socioeconomic status and employment instability, substance abuse, psychopathic personality traits, and childhood victimization.[192] The evaluation process is complicated by the fact that literally scores of variables show some significant correlation with future violence, but usually with little predictive power for each.[193] However, beginning with the variables noted above, the evaluator estimates the baseline risk of violence for the person and then adjusts that value by taking into account foreseeable perturbations to the current equilibrium. When previous violence has occurred, the risk estimate is adjusted to include those specific variables that have been associated with violence by this person in the past (e.g., being left by a girlfriend), including whether they are present at the time of evaluation or likely to recur in the future.[194]

The past two decades have seen the development of a growing number of structured assessment instruments specific to the prediction of future violence risk. Among the best known of these are the HCR-20,[195] the Violence Risk Assessment Guide (VRAG),[196] and the computerized Classification of Violence Risk (COVR).[197] A set of instruments also exists for the prediction of the risk of future sexual offenses.[198] Violence risk-assessment instruments have been developed in one of two ways: either by assembling known predictors from the research literature and combining them with variables drawn from clinical experience (e.g., HCR-20), or on the basis of statistical analysis of research data from large subject populations (e.g., VRAG and COVR). Attempts are then made to validate the instruments on populations similar to the ones with which it is anticipated they will be used. The more sophisticated measures yield estimates of the degree of risk, rather than dichotomous predictions that violence will or will not occur. In general, the most commonly used instruments have shown a correlation between the estimated degree of risk and future violence.[199]

192. *Id.*

193. Monahan et al., *supra* note 189, at 163–68.

194. Appelbaum & Gutheil, *supra* note 149.

195. Webster et al., *supra* note 170.

196. Quinsey et al., *supra* note 170.

197. Monahan et al., *supra* note 170.

198. Calvin M. Langton et al., *Actuarial Assessment of Risk for Reoffense Among Adult Sex Offender: Evaluating the Predictive Accuracy of the Static-2002 and Five Other Instruments*, 34 Crim. Just. & Behav. 37–59 (2007).

199. Kevin S. Douglas et al., *Assessing Risk for Violence Among Psychiatric Patients*, 67 J. Consulting Clinical Psychol. 917 (1999); Grant T. Harris et al., *Prospective Replication of the Violence Risk Appraisal Guide in Predicting Violent Recidivism Among Forensic Patients*, 26 Law & Hum. Behav. 377 (2002); John Monahan et al., *An Actuarial Model of Violence Risk Assessment for Persons with Mental Disorders*, 56 Psychiatric Servs. 810 (2005).

The literature on prediction is marked by strong and unresolved differences of opinion over the best basis for the ultimate risk estimate. Partisans of exclusive reliance on the quantitative predictions generated by structured assessment instruments, which is often referred to as "actuarial" prediction, argue that any attempts to modify the resulting risk estimates necessarily reduce accuracy.[200] Proponents of clinical evaluation note that exclusive reliance on instrumentation is unwise because of the inevitable questions about the applicability of the group data on which an instrument is based to the person being evaluated; the failure of a fixed set of questions ever to capture all the variables that may be relevant in a particular situation; and the potential uncooperativeness of evaluees with a structured process.[201] Compromise approaches include anchoring the estimate in the actuarial prediction, but allowing clinical judgment to modify the results on the basis of additional considerations, or using an instrument to structure the evaluation and ensure its completeness, but allowing the evaluator to reach a judgment on the basis of the totality of the information. This last approach has been termed "structured professional judgment,"[202] and at least one study has suggested that it is capable of yielding predictions with reasonable degrees of accuracy.[203] It is fair to say that the question of which approach is best remains unresolved.

b. Limitations of violence risk prediction

A voluminous research literature exists on violence risk prediction. Studies of predictions by psychiatrists and psychologists in the 1960s and 1970s showed poor accuracy in judging whether persons with mental disorders and sex offenders would be likely to be violent at some point after release.[204] Indeed, the most frequently cited conclusion was Monahan's statement that when mental health professionals predicted that a person would be violent, they were twice as likely to be wrong as right.[205] The cumulative impact of these findings stimulated a great deal of research to identify variables that predict violence and their incorporation into both clinical predictions and the structured assessment instruments described above.[206]

200. N. Zoe Hilton et al., *Sixty-Six Years of Research on the Clinical Versus Actuarial Prediction of Violence*, 34 Counseling Psychol. 400 (2006).

201. Thomas R. Litwak, *Actuarial Versus Clinical Assessments of Dangerousness*, 7 Psychol. Pub. Pol'y & L. 409 (2001); Andrew Carroll, *Are Violence Risk Assessment Tools Clinically Useful?* 41 Austrl. & N.Z. J. Psychiatry 301 (2007).

202. Kevin S. Douglas & P. Randall Kropp, *A Prevention-Based Paradigm for Violence Risk Assessment: Clinical and Research Applications*, 29 Crim. Just. & Behav. 617 (2002).

203. Kevin S. Douglas et al., *Evaluation of a Model of Violence Risk Assessment Among Forensic Psychiatric Patients*, 54 Psychiatric Servs. 1372 (2003).

204. John Monahan, The Clinical Prediction of Violent Behavior (1981).

205. *Id.* at 60.

206. John Monahan, *Clinical and Actuarial Predictions of Violence: II. Scientific Status, in* Modern Scientific Evidence: The Law and Science of Expert Testimony, vol. 1, at 122, 122–47 (David L. Faigman et al., 2007).

At this point, it is possible to identify several items of consensus from the research literature. Violence is not a unitary phenomenon; that is, it occurs for different reasons, related both to the motivations of the perpetrator and to the environmental context.[207] A bar-room brawl has different roots than a mugging; the precipitants of spouse abuse bear little similarity to the motivations underlying a killing that has been premeditated as an act of revenge. Thus, no single variable or set of variables can be relied upon in all cases to ascertain violence risk. Long-term prediction of violence is inherently inaccurate, due both to the intrinsic limitations in the prediction of low-frequency events[208] and to the difficulty that clinicians have in anticipating changes in the person and the environment over time and their effects on the person's behavior.[209] However, shorter-term prediction (i.e., days to weeks) holds greater potential for accuracy. Indeed, recent studies focused on shorter-term prediction, often from hospital emergency rooms, have found accuracies for predictions of violence in the range of 40% to 60%.[210] It is worth noting that even when the leading actuarial instruments are used to make dichotomous judgments of future violence—that is, a cutoff is set to simulate the clinical prediction process—their rates of accuracy are similar.[211] Mental health professionals, therefore, have been encouraged to move away from attempting to make dichotomous judgments of dangerousness and toward predictions couched in terms of the risk of future violence.[212] Even here, though, precision has not yet been attained—and may be unattainable. The state of the art probably allows well-trained clinicians, especially if they are using structured assessment instruments, to assign persons into high-, medium-, and low-risk groups with reasonable accuracy. At present, the hope of designating risk categories with greater precision than that for most categories of persons with mental disorders is likely illusory.[213] When quantitative data are available, however, precision in communication of risk

207. Paul S. Appelbaum, *Preface, in* Clinical Assessment of Dangerousness: Empirical Contributions ix–xiv (Georges-Franck Pinard & Linda Pagani eds., 2001).

208. Paul E. Meehl, Clinical Versus Statistical Prediction: A Theoretical Analysis and a Review of the Evidence (1954).

209. Jennifer L. Skeem et al., *Building Mental Health Professionals' Decisional Models into Tests of Predictive Validity: The Accuracy of Contextualized Predictions of Violence*, 24 Law & Hum. Behav. 607 (2000).

210. That is, 40% to 60% of those who have been predicted to be violent go on to commit violent acts. Note that because interventions to prevent the predicted violence (e.g., hospitalization) may be taken with many of these subjects, the figures probably underestimate the proportion of true positive predictions. In addition, when clinicians predicted that a person would not be violent, they almost always tended to be correct, with well over 90% of such predictions in most studies being accurate. *See, e.g.,* Charles W. Lidz et al., *The Accuracy of Predictions of Violence to Others*, 269 JAMA 1007 (1993); Dale E. McNiel & Renée L. Binder, *Clinical Assessment of the Risk of Violence Among Psychiatric Inpatients*, 148 Am. J. Psychiatry 1317 (1991).

211. See studies cited *supra* note 199.

212. Henry J. Steadman et al., *From Dangerousness to Risk Assessment: Implications for Appropriate Research Strategies, in* Mental Disorder and Crime (Sheilagh Hodgins ed., 1993).

213. Webster et al., *supra* note 170, at 10.

would undoubtedly be enhanced if they were utilized and if assessors specified their definitions of the categories being employed.[214]

The studies on the accuracy of prediction, whether clinical or actuarial, have typically involved the direct evaluation of the person about whom the prediction was being made. In many cases, considerable additional information about the person was available. Opinions about the risk of future violence by persons whom the evaluator has not examined have never been validated, and there are persuasive reasons to believe that such predictions are not likely to be highly accurate.[215] Such opinions have been introduced, for example, in death penalty cases in which the prosecution sought to prove that further violence was likely, but the defense denied the prosecution expert direct access to the defendant.[216] If such evidence is to be introduced, at a minimum, one would expect that the limitations on the assessor's knowledge of the evaluee and on the certainty with which conclusions can be reached would be noted.

2. *Predictions of future functional impairment*

Cases involving claims of emotional harms, along with disability and workers' compensation claims, often require that efforts be made to estimate the plaintiff's future functional impairment so that damages can be determined accordingly.[217] Techniques for the assessment of function were described above. See discussion *supra* Section I.D.2. However, these cases call for something more: predictions of the degree of change in functional impairment due to mental disorders that are likely to occur over time. In contrast to the structured assessment tools that assist in the prediction of future violence risk, no instruments have been developed

214. *See* Kelly M. Babchishin & R. Karl Hanson, *Improving Our Talk: Moving Beyond the "Low," "Moderate," and "High" Typology of Risk Communication*, 16 Crime Scene 11 (2009). Suggestions for improving the clarity of risk communications include distinguishing between the likelihood of future violence and the anticipated severity of the offense, specifying the period for which the prediction is being made (e.g., "over the next 6 months"), indicating the comparison population for the estimate (e.g., "risk is high compared with the general population" or "risk is high compared with the population of persons with similar histories of violence"), and providing both absolute and relative risks when quantitative data are available (e.g., "risk of future violence over the next year is between 8 and 12%, which is between 4 and 6 times greater than would be expected for the general population").

215. Brief of Amici Curiae American Psychiatric Association, Barefoot v. Estelle, 463 U.S. 880 (1983) (No. 82-6080).

216. *See* Barefoot v. Estelle, 463 U.S. 880 (1983); Ron Rosenbaum, *Travels with Doctor Death*, Vanity Fair, May 1990, at 141.

217. 20 C.F.R. § 404.1520a (2008). *See generally* Thomas P. Harding, *Psychiatric Disability and Clinical Decision Making: The Impact of Judgment Error and Bias*, 24 Clinical Psychol. Rev. 707 (2004); Harold A. Pincus et al., *Determining Disability Due to Mental Impairment: APA's Evaluation of Social Security Administration Guidelines*, 148 Am. J. Psychiatry 1037 (1991); Cille Kennedy, *SSA's Disability Determination of Mental Impairments: A Review Toward an Agenda for Research*, *in* The Dynamics of Disability: Measuring and Monitoring Disability for Social Security Programs 241 (Gooloo S. Wunderlich et al. eds., 2002); Dan B. Dobbs, The Law of Torts 1048–53, 1087–1110 (2000).

and validated to predict future functional status at this writing. Such predictions are complicated by the need for simultaneous estimates of several parameters that affect long-term functional outcome: variables intrinsic to the person (e.g., symptomatic fluctuation, changes in motivation to work), variables that relate to the environment (e.g., divorce, availability of new categories of jobs), and responses to treatment (see discussion *infra* Section I.F.6). Research aimed at identifying variables associated with some types of future functional impairment exists, but is largely focused on progressive disorders (e.g., Alzheimer's disease), and even here the accuracy of the predictions of forensic evaluators has not been determined.[218] Hence, acknowledgment of the uncertainties inherent in these predictions would appear to be unavoidable for experts undertaking this task.

F. Treatment of Mental Disorders

The nature of available treatments for mental disorders, the probability that they will be effective, the side effects that they may induce, and the existence of alternatives are likely to be material to a variety of legal cases. In criminal proceedings, for example, the continued confinement of a defendant in a psychiatric hospital on the basis of incompetence to stand trial will be based in part on the probability that treatment of the person will restore capacity;[219] involuntary treatment of the defendant will turn on a number of factors, including the likelihood of success and the side effects and their potential for impairing the defendant's defense.[220] Decisions about probation and parole of mentally disordered offenders may also relate to the likelihood that symptoms will remain in check, and courts may order ongoing treatment as a condition of release.[221] Among the civil cases for which treatment-related questions will be at issue are liability claims for malpractice and failure to protect third parties from patient violence, claims involving emotional harms (e.g., in calculating the cost of future care), and issues related to the deprivation of rights of prisoners in correctional facilities to have adequate mental health treatment.[222] Treatment of mental disorders today offers multiple options for most disorders, often with different levels of likely effectiveness and varying side-effect profiles. Planning treatment has become an increasingly complex task.

218. *See, e.g.*, Roy Martin et al., *Declining Financial Capacity in Patients with Mild Alzheimer Disease: A One-Year Longitudinal Study*, 16 Am. J. Geriatric Psychiatry 209 (2008).

219. *See* Jackson v. Indiana, 406 U.S. 715 (1972).

220. *See* Sell v. United States, 539 U.S. 166 (2003).

221. *See, e.g.*, United States v. Holman, 532 F. 2d 284 (4th Cir. 2008).

222. For an overview of the considerable body of case law on this issue, see Michael L. Perlin, 4 Mental Disability Law § 11-4.3 (2d ed. 1989).

1. Treatment with medication

The past 50 years have seen the ongoing introduction of new medications for the treatment of mental disorders. Currently, medications are a mainstay in the treatment of schizophrenia and bipolar disorder; indeed, it is a rare patient who can be treated successfully for these disorders without medication as part of the treatment plan.[223] Medications are also used commonly to treat and prevent the recurrence of depression, anxiety disorders, attention-deficit/hyperactivity disorder, and a large number of other conditions.[224] The field of psychopharmacology, as the treatment of mental disorders with medications is known, has become a complex and challenging part of psychiatric practice.

a. Targets of medication treatment

As a general rule, medications are targeted at the symptoms of mental disorders, which may occur in a large number of conditions, rather than being specific for the treatment of a given disorder. Psychotic phenomena such as delusions and hallucinations, for example, are generally responsive to antipsychotic medications, whether the underlying disorder is schizophrenia or bipolar disorder.[225] Antianxiety medications can be effective in primary anxiety disorders (e.g., agoraphobia) or in anxiety that develops secondary to another condition (e.g., depression).[226] Mood stabilizers, first introduced for bipolar disorder and its variants, can be helpful to some patients with personality disorders that are marked by fluctuations in mood.[227] Medications that aid patients in falling asleep work in many different disorders.[228]

Moreover, the same drug can have multiple effects. The best-known example is the selective serotonin reuptake inhibitors (SSRIs), the first and most famous of which is Prozac (the generic name is fluoxetine).[229] Originally introduced for the treatment of depression, for which they proved effective, SSRIs have since also proved helpful for anxiety, even in the absence of depression.[230] The newer antipsychotic medications, intended to target psychotic symptoms, can also be helpful for mania, even when psychosis per se is absent.[231] Indeed, one of the

223. Minzenberg et al., supra note 161; Steven L. Dubovsky et al., *Mood Disorders, in* Hales et al., *supra* note 111.

224. *See generally* Alan F. Schatzberg et al., Manual of Clinical Psychopharmacology (2003).

225. Stephen M. Stahl, Stahl's Essential Psychopharmacology: Neuroscientific Basis and Practical Applications 425 (2008).

226. *Id.* at 726.

227. C. Robert Cloninger & Dragan M. Svrakic, *Personality Disorders, in* Sadock, Sadock, & Ruiz, *supra* note 120, at 2236.

228. Stahl, *supra* note 225, at 831–39.

229. Prozac was the first SSRI to be introduced to the market, and its use was widely discussed in popular media, including Peter Kramer's bestseller, Listening to Prozac (1993).

230. Norman Sussman, *Selective Serotonin Reuptake Inhibitors, in* Sadock, Sadock, & Ruiz, *supra* note 120, at 3191.

231. Stahl, *supra* note 225, at 689–94.

antipsychotics is often prescribed to aid in sleep, as is one of the newer antidepressants.[232] These multiple effects of a single medication are probably due to their impact on more than one neurotransmitter system.

Another reality of contemporary psychopharmacology is that medications are often used for indications that have not been approved by the Food and Drug Administration (FDA).[233] FDA approval is required for a new medication to be marketed in the United States, and approval is granted only after evidence from clinical trials is presented to the agency demonstrating the efficacy of the drug for a particular purpose, within a given dosage range, and often with a particular population.[234] Once FDA has granted approval for a compound to be marketed, however, physicians are free to prescribe it for any purpose for which they believe it to be indicated, at a dosage of their choosing, and for whichever patients they believe will benefit—although pharmaceutical companies can advertise its use only for FDA-approved purposes. Because approval of a single indication for drug use makes the medication generally available for other purposes as well, and over time drugs lose patent protection, pharmaceutical companies often have little incentive to pursue FDA approval for additional indications.[235] Thus, many medications have long been used for purposes other than the one endorsed by FDA, often with impressive bodies of clinical experience supporting such use.[236]

As is true for many classes of medications, the precise mechanisms of action of most psychopharmacological compounds have not yet been established. Most appear to block or stimulate neuronal receptors in the brain, which trigger or inhibit the propagation of electrical impulses, and it has been assumed that this represents their primary mechanism of action.[237] Indeed, many compounds interact with multiple receptor systems, perhaps accounting for their efficacy against a variety of symptoms, as well as the diverse side effects they produce. But other

232. The antidepressant trazodone is a popular sleep-inducing medication, *id.* at 845; the antipsychotic quetiapine is also used for this purpose, *id.* at 848.

233. David C. Radley et al., *Off-Label Prescribing Among Office-Based Physicians*, 166 Archives Internal Med. 1021 (2006).

234. Celia J. Winchell, *Drug Development and Approval Process in the United States*, *in* Sadock. Sadock, & Ruiz, *supra* note 120, at 2988–96; FDA regulatory information on new drug approvals can be accessed at http://www.fda.gov/Drugs/DevelopmentApprovalProcess/default.htm.

235. Steven R. Salbu, *Off-Label Use, Prescription, and Marketing of FDA-Approved Drugs: An Assessment of Legislative and Regulatory Policy*, 51 Fla. L. Rev. 181 (1999); Rebecca Dresser, *The Curious Case of Off-Label Use*, 37 Hastings Center Rep. 9 (2007).

236. For example, the various formulations of valproic acid are among the most commonly used treatments for bipolar disorder, including maintenance treatment to prevent recurrence. Although an FDA indication was obtained for the treatment of acute mania, long-term maintenance use is "off-label." Schatzberg et al., *supra* note 224. *See also* Norman Sussman, *General Principles of Psychopharmacology*, *in* Sadock, Sadock, & Ruiz, *supra* note 120, at 2972–3.

237. Stahl, *supra* note 225, at 91–122.

mechanisms, such as initiating changes in DNA transcription, are also possible and remain to be fully explored.[238]

b. Categories of medications

Although a large number of medications are used to treat the symptoms of mental disorders, several major categories account for the largest number of prescriptions.

- Antipsychotic medications, first introduced in the 1950s, appear to have selective effects on psychotic symptoms such as delusions, hallucinations, and disordered thoughts.[239] The first generation of antipsychotics, marked by the introduction of chlorpromazine, often caused acute neuromuscular side effects, such as spasms of the muscles, along with a long-term risk of tardive dyskinesia, a condition characterized by involuntary movements of the muscles in the face, trunk, and extremities.[240] A second generation of these medications, introduced in the 1990s with great fanfare, presents lower risks of neuromuscular problems, but several of the most popular members of this group can cause weight gain, along with diabetes, hyperlipidemia, and increased cardiac risk.[241] There does not appear to be a difference in efficacy between the earlier and later medications.[242]
- Mood stabilizers were introduced for the treatment of bipolar disorder, which is characterized by episodic mood swings from mania to depression.[243] The first of these drugs was lithium, whose effect was discovered in the 1940s, but which was not widely adopted in the United States until the 1970s. Lithium can be very effective, but it often causes problematic side effects.[244] Subsequently, a number of medications that are also effective as treatment for seizure disorders were found to have mood stabilizing effects as well, and they are generally better tolerated.[245]

238. *Id.* at 41–89.

239. *Id.* at 425.

240. Irene Hurford & Daniel P. van Kammen, *First-Generation Antipsychotics, in* Sadock, Sadock, & Ruiz, *supra* note 120, at 3105–27.

241. Stephen R. Marder, Irene Hurford, & Daniel P. van Kammen, *Second-Generation Antipsychotics, in* Sadock, Sadock, & Ruiz, *supra* note 120, at 3206–40.

242. Jeffrey A. Lieberman et al., *Effectiveness of Antipsychotic Drugs in Patients with Chronic Schizophrenia*, 353 New Eng. J. Med. 1209 (2005); Peter B. Jones et al., *Randomized Controlled Trial of Effect on Quality of Life of Second- vs. First-Generation Antipsychotic Drugs in Schizophrenia: Cost Utility of the Latest Antipsychotic Drugs in Schizophrenia Study* (CUtLASS 1), 63 Archives Gen. Psychiatry 1079 (2006).

243. Stahl, *supra* note 225, at 667–719.

244. James W. Jefferson & John H. Greist, *Lithium, in* Sadock, Sadock, & Ruiz, *supra* note 120, at 3132–45.

245. Robert M. Post & Mark A. Frye, *Carbamazepine, in* Sadock, Sadock, & Ruiz, *supra* note 120, at 3073–89; Robert M. Post & Mark A. Frye, *Valpronte,* in Sadock, Sadock, & Ruiz, *supra* note 120 at 3278.

- Antidepressants include the older class of tricyclic compounds, which offered the first effective medication treatment for depression.[246] Again, a less-than-optimal side-effect profile led to efforts to discover alternatives. The SSRI medications turned out to have equal efficacy, but are generally better tolerated.[247] They too, though, have adverse side effects, including diminished sexual function, a numbing of emotional intensity, or increased anxiety.[248] Data suggesting that SSRIs may lead to suicidal ideation in some patients remain controversial, but have led to FDA-mandated "black-box" warnings for the drugs.[249] A group of non-SSRI, but chemically related, compounds has effects and side effects similar to those of the SSRIs, and the medications are often used interchangeably.[250]
- Antianxiety medications, which began with nonspecific sedatives, soon moved on to drugs with targeted effects on anxiety per se.[251] Benzodiazepines, including the well-known Valium and Librium, were used as mainstays of anxiety treatment for many years, but carry liabilities that include the potential for abuse and addiction. Today, the much safer SSRIs and related compounds are the drugs of choice for long-term treatment of anxiety, as they are for depression, with benzodiazepines often reserved for situations in which immediate effects are a priority.[252] Newer agents have been introduced from entirely different chemical classes specifically for anxiety.[253]

This is by no means a complete list of medications for the treatment of mental disorders, but represents a brief introduction to the major classes that are likely to be the focus of evidence presented in legal proceedings.

c. Polypharmacy

The use of more than one psychiatric medication for a patient—often called "polypharmacy"—is common for several reasons.[254] First, because medications

246. J. Craig Nelson, *Tricyclics and Tetracyclics*, *in* Sadock, Sadock, & Ruiz, *supra* note 120, at 3259.

247. Sussman, *supra* note 230.

248. *Id.*

249. *See* FDA guidance at http://www.fda.gov/cder/drug/antidepressants/default.htm.

250. These medications include drugs that selectively target the brain's norepinephrine transporters, the so-called selective norepinephrine reuptake inhibitors (SNRIs), along with medications that appear to act on both serotonin and norepinephrine systems. Michael E. Thase, *Selective Serotonin-Norepinephrine Reuptake Inhibitors*, *in* Sadock, Sadock, & Ruiz, *supra* note 120, at 3184–90.

251. Steven Dubovsky, *Benzodiazepine Receptor Agonists and Antagonists*, *in* Sadock, Sadock, & Ruiz, *supra* note 120, at 3044.

252. Stahl, *supra* note 225, at 765–71.

253. *See, e.g.*, Anthony J. Levitt, Ayal Schaffer, & Krista Lanctot, *Buspirone*, *in* Sadock, Sadock, & Ruiz, *supra* note 120, at 3060.

254. Sussman, *supra* note 236.

typically target symptoms rather than underlying disorders, and most disorders present with multiple symptoms, there may be an obvious rationale for the use of more than one agent (e.g., an antidepressant along with a sleep medication for a patient with depression who is experiencing insomnia). Second, some disorders that are imperfectly responsive to a single, initial medication may respond to an augmentation strategy involving the addition of a second medication, often from a different chemical class (e.g., an antidepressant medication can be augmented with lithium, thyroid hormone, or a second unrelated antidepressant).[255]

Although greater efficacy often can be obtained from combined treatment, there are risks as well. Multiplying medications increases the chance of adverse effects from both the individual medications and their interactions.[256] Hence, polypharmacy is best reserved for situations in which documented evidence of benefit exists or a compelling theoretical rationale is present. Failure to apply these principles accounts for the vaguely disreputable connotation that the term "polypharmacy" conveys.

d. Side effects

The specific side effects of several classes of medications have been referred to earlier. A general point to be noted, however, is that all medications have side effects, even commonly used drugs that are generally thought of as harmless, such as aspirin or acetaminophen.[257] Prescribers balance the positive effects of medication against the range of possible side effects in making recommendations for treatment to patients, who, of course, retain the right to decide that the adverse consequences do not warrant the possibility of therapeutic gains.[258] It is a reality, however, that the side effects of psychiatric medications limit the tolerability of many drugs, even among people who are benefiting from them. Moreover, some medications may have adverse effects with particular significance in legal settings.[259] These include sedation, which may be associated with antipsychotic or antianxiety medications and sometimes with other classes of drugs, and restricted expression of emotion, occasionally experienced with the first generation of antipsychotic medications. In the absence of previous exposure to a given medication, it is difficult to anticipate the side effects that may arise. Clinicians typically monitor those effects and adjust dosage or change medications accordingly.

255. Charles DeBattista & Alan F. Schatzberg, *Combination Pharmacotherapy*, in Sadock, Sadock, & Ruiz, *supra* note 120, at 3322.

256. Sussman, *supra* note 236.

257. *Id.* at 2684.

258. *See generally* Jessica W. Berg et al., Informed Consent: Legal Theory and Clinical Practice (2001).

259. Sell v. United States, 539 U.S. 166 (2003); Dora W. Klein, *Curiouser and Curiouser: Involuntary Medications and Incompetent Criminal Defendants After* Sell v. United States, 13 Wm. & Mary Bill Rts. J. 897 (2005); Debra A. Breneman, *Forcible Antipsychotic Medication and the Unfortunate Side Effects of* Sell v. United States, *539 U.S. 166, 123 S. Ct. 2174 (2004),* 27 Harv. J.L. & Pub. Pol'y 965 (2004).

e. Efficacy and effectiveness

Efficacy refers to a medication's ability to reduce or eliminate its target symptoms; effectiveness denotes the extent to which that effect can be achieved in ordinary clinical treatment.[260] An illustration of the difference is evident with antipsychotic medications, the efficacy of which in controlling the positive symptoms of psychosis has been demonstrated in numerous studies.[261] However, in real-world clinical settings, the effectiveness of these medications, particularly over the long term, is substantially limited by patients' reluctance to continue taking them, despite symptomatic relief.[262] This may be due in part to the nature of some mental disorders, especially schizophrenia, given that affected persons often deny their impairments.[263] But there is no question that the side effects of the medications lead many patients to stop them, because of their unwillingness to tolerate the weight gain, lethargy, sexual dysfunction, neuromuscular manifestations, or other side effects that often accompany the use of the drugs.[264] Because demonstrations of efficacy are required for FDA approval to market medications, it can be assumed that drugs for mental disorders are efficacious for their approved indications. However, their effectiveness may be more limited, and this can be an important consideration when predictions of long-term symptom control are called for in both criminal and civil contexts.

2. *Psychological treatments*

Although medications are a mainstay for treatment of serious mental disorders, a variety of psychological treatments may be important as either primary or adjunctive treatments.

a. Psychoanalysis and psychodynamic psychotherapy

Psychoanalysis was developed as a therapeutic technique by Sigmund Freud and is probably the form of psychotherapy that comes first to mind for most lay people.[265] It involves three to four sessions a week for many years, during which patients recline on a couch and free associate, with little direction from the analyst, whose job it is to analyze patients' developing unconscious attachment (or transference) to the analyst. Despite its ubiquity in *New Yorker* cartoons, psycho-

260. Gerard E. Hogarty et al., *Efficacy Versus Effectiveness*, 48 Psychiatric Servs. 1107 (1997).

261. Philip G. Janicak et al., Principles and Practice of Psychopharmacotherapy 118–27 (1993).

262. Lieberman et al., *supra* note 242.

263. Xavier F. Amador & Henry Kronengold, *The Description and Meaning of Insight in Psychosis*, *in* Insight and Psychosis 15, 15–32 (Xavier F. Amador & Anthony S. David eds., 1998).

264. Diana O. Perkins, *Predictors of Noncompliance in Patients with Schizophrenia*, 63 J. Clinical Psychiatry 1121 (2002).

265. T. Byram Karasu & Sylvia R. Karasu, *Psychoanalysis and Psychoanalytic Psychotherapy*, *in* Sadock, Sadock, & Ruiz, *supra* note 120, at 2746.

analysis is used with only a tiny percentage of patients, usually those who have less severe disorders, live in major urban centers, and can afford to pay for their own extended care. Efforts to demonstrate the efficacy of psychoanalysis have run into resistance from practitioners and considerable logistical problems; data supporting its use are therefore hard to find.[266] Thus, it is likely to have limited relevance when mental disorders are at issue in legal proceedings.

Psychodynamic psychotherapies, which are offshoots of psychoanalysis, are used more frequently and hence have more relevance to the law.[267] Based on similar notions of a dynamic unconscious, that is, processes out of the awareness of the person affect mood and behavior, psychodynamic therapies generally involve sessions once or twice a week, for periods ranging from months to several years, with patients sitting upright and greater activity on the part of the therapist in identifying conflicts and maladaptive behaviors. As in psychoanalysis, the underlying premise is that when unconscious motivations are made conscious, they become susceptible to control and alteration by the patient.

Psychodynamic therapies are easier to study and have a somewhat more robust set of data speaking to their efficacy—for example, in anxiety and depression.[268] It is often difficult, though, for patients with more severe disorders, such as schizophrenia and bipolar disorder, to tolerate the in-depth exploration and uncovering of intrapsychic conflicts that accompany the therapeutic process. But many patients with personality disorders, depression, and other conditions will attribute their stability to ongoing therapy.

b. Cognitive behavioral and related therapies

In contrast to the premises of psychodynamic therapies that mood and behavior are affected by unconscious conflicts, cognitive behavioral therapy (CBT) is based on the idea that conscious patterns of thought determine how one feels and behaves.[269] CBT is generally shorter term (weeks to months), highly structured, and focused on helping patients recognize and control maladaptive patterns of thinking. Patients are often given homework assignments to complete between sessions. A strong database supports its use in anxiety disorders, depression (where it can be as effective as medications and may be more likely to prevent relapse), and for control of some psychotic symptoms, and its use is steadily being extended to additional conditions.[270] Specialized forms of CBT have been developed for use with sex offenders, based on

266. Glen O. Gabbard et al., *The Place of Psychoanalytic Treatments Within Psychiatry*, 59 Archives Gen. Psychiatry 505 (2002).

267. Karasu, *supra* note 265.

268. Falk Leichsenring & Sven Rabung, *Effectiveness of Long-Term Psychodynamic Psychotherapy: A Meta-Analysis*, 300 JAMA 1551 (2008).

269. Cory F. Newman & Aaron T. Beck, *Cognitive Therapy*, *in* Sadock, Sadock, & Ruiz, *supra* note 120, at 2857–8.

270. Andrew C. Butler et al., *The Empirical Status of Cognitive-Behavioral Therapy: A Review of Meta-Analyses*, 26 Clinical Psychol. Rev. 17 (2006).

a model that is often termed "relapse prevention" that teaches patients to recognize situations that are likely to lead to recidivism and avoid them.[271] Dialectical behavior therapy is an offshoot of CBT that has shown success with patients with borderline personality disorders, an otherwise difficult disorder to treat.[272]

c. Other psychological therapies

Hundreds of forms of talking therapies have been catalogued, but it would be impossible to review them all here. Many have shown efficacy with particular disorders, and efforts have been made to identify common therapeutic elements, which may include the relationship with the therapist and the ability to instill hope for the future in the patient.[273] In addition to individual therapies, persons with mental disorders may benefit from group therapies of a variety of orientations, including psychodynamic and cognitive.[274] Group therapies can be especially helpful when socialization and relationships with other people are among the person's problems. Family and couples therapies generally target relationships within the family unit or marital dyad; because mental disorders are often disruptive to relationships, such approaches may be helpful adjuncts to treatments focused on the affected person's primary disorder.[275] Severely ill patients, including those with schizophrenia, may benefit from what is termed supportive therapy, which involves regular contacts aimed at identifying concrete problems in the person's life and helping to find solutions. It may also provide a nonthreatening outlet for social interaction when other relationships are limited.[276]

3. Treatment of functional impairments

Control of positive symptoms does not necessarily address deficits in function, particularly in the psychotic disorders. What may be required are techniques that focus on functional difficulties per se. Persons with schizophrenia, for example, given that the disorder often affects ability to function socially and occupationally, may need to be taught how to interact with other people, an approach known as social skills therapy.[277] Occupational therapy can provide them with a graded introduc-

271. *See, e.g.*, D. Richard Laws, Relapse Prevention with Sex Offenders (1989).

272. M. Zachary Rosenthal & Thomas R. Lynch, *Dialectical Behavior Therapy*, *in* Sadock, Sadock, & Ruiz, *supra* note 120, at 2884.

273. Jerome D. Frank & Julia B. Frank, Persuasion and Healing: A Comparative Study of Psychotherapy (1993).

274. Henry I. Spitz, *Group Psychotherapy*, *in* Sadock, Sadock, & Ruiz, *supra* note 120, at 2832.

275. Henry I. Spitz & Susan Spitz, *Family and Couple Therapy*, *in* Sadock, Sadock, & Ruiz,*supra* note 120, at 2584.

276. Peter J. Buckley, *Applications of Individual Supportive Psychotherapy to Psychiatric Disorders: Efficacy and Indications*, *in* Textbook of Psychotherapeutic Treatments (Glen O. Gabbard ed., 2009).

277. Melinda Stanley & Deborah C. Beidel, *Behavior Therapy*, *in* Sadock, Sadock, & Ruiz, *supra* note 120, at 2795–6.

tion (or reintroduction) to the workplace, with patients taught how to maintain focus and deal with the demands of the work setting.[278] More focal impairments can be addressed, as well. Thus, defendants found incompetent to stand trial can be taught about the nature of the courtroom and the expectations they must meet to be found competent to proceed. Studies of such programs have shown higher rates of restoration of competence than occurs with treatment of the primary disorder alone.[279] Comparable programs are available for anger management,[280] control of spousal abuse,[281] and training in parenting skills,[282] among other areas of function that are often the target of legal proceedings.

4. Electroconvulsive and other brain stimulation therapies

The therapeutic effect of seizure induction by electrical or chemical means on psychosis and depression was first demonstrated in the 1930s.[283] Electroconvulsive therapy (ECT) became the most popular of these approaches in the era before efficacious medications existed for mental disorders. The early techniques for ECT involved application of an electrical current to the brain of patients while they were awake. Not only was this often terrifying for the patients, but the resulting violent seizures could cause bone fractures and other complications. Contemporary use of ECT is quite different, with patients anesthetized prior to the procedure and paralyzing agents used to prevent muscular contractions.[284] Although temporary confusion and memory loss often occur, long-term adverse effects are uncommon, making ECT a safe procedure—indeed, for elderly patients with complex medical problems, it may be preferable to the use of medications. Unfortunately, the graphic images associated with early ECT use, embodied in novels and films, dominate the popular mind and often lead to a distorted perception of the treatment.[285]

278. *See generally* Jennifer Creek, Occupational Therapy and Mental Health: Principles, Skills and Practice (2002).

279. *See, e.g.*, Alex M. Siegel & Amiram Elwork, *Treating Incompetence to Stand Trial*, 14 L. & Hum. Behav. (1990); Barry W. Wall et al., *Restoration of Competency to Stand Trial: A Training Program for Persons with Mental Retardation*, 31 J. Am. Acad. Psychiatry L. 189 (2003).

280. Raymond DiGiuseppe & Raymond C. Tafrate, *Anger Treatment for Adults: A Meta-Analytic Review*, 10 Clinical Psychol.: Sci. & Prac. 70 (2006).

281. Julia C. Babcock et al., *Does Batterers' Treatment Work? A Meta-Analytic Review of Domestic Violence Treatment*, 23 Clin. Psychol. Rev. 1023 (2004). Note that in contrast to anger management and parenting training, the data on the efficacy of treatment for batterers indicates that effects are limited at best.

282. Kathryn M. Bigelow & John R. Lutzker, *Training Parents Reported for or at Risk for Child Abuse and Neglect to Identify and Treat Their Children's Illnesses*, 15 J. Fam. Violence 311 (2000).

283. Joan Prudic, *Electroconvulsive Therapy*, *in* Sadock, Sadock, & Ruiz, *supra* note 120, at 3285–3301.

284. *Id.*

285. Ken Kesey, One Flew Over the Cuckoo's Nest (1962); the popular movie version appeared in 1976 (see description at http://www.imdb.com/title/tt0073486/); Garry Walter & Andrew McDonald, *About to Have ECT? Fine, but Don't Watch It in the Movies: The Sorry Portrayal of ECT in Film*, 21 Psychiatric Times 65 (2004), *available at* http://www.psychiatrictimes.com/display/

ECT is used today primarily for the acute treatment of depression, for which it has been demonstrated to be effective.[286] Although it can also have a therapeutic effect on psychotic symptoms, it is not commonly used for that purpose. An exception involves states of catatonic stupor or excitement, both of which can be life threatening and for which ECT can provide immediate relief.[287] For patients responsive to ECT but not to medications, maintenance ECT (i.e., periodic, perhaps monthly, treatments) can be used.[288] In most cases, though, ECT is reserved for patients who have not responded to one or more medications or whose conditions are sufficiently severe (e.g., acute suicidal urges) that a more rapidly acting intervention than medication—which can take up to 6 to 8 weeks before an effect is seen—is indicated. ECT's history continues to haunt its current use, with many states imposing statutory or regulatory restrictions.[289] However, it can be a safe and effective treatment—and in some cases a life-saving one. The mechanism of effect for ECT remains unclear.

Given that brain function is integrally linked to electrical transmission of impulses between nerve cells, it is not surprising that other efforts have been made to use electrical stimulation for therapeutic purposes. Electrical stimulation of the vagus nerve has been approved by FDA for the treatment of depression, although the supporting data are generally thought to be weak.[290] The therapeutic use of transcranial magnetic stimulation, in which a strong magnetic field is applied externally, is being explored, including for depression, autism, and other disorders.[291] Successful use of implanted devices for deep brain stimulation (DBS) for Parkinson's disease have led to trials of DBS for obsessive–compulsive disorder and depression;[292] further experimentation in other disorders seems likely.

article/10168/48111; C. Lauber et al., *Can a Seizure Help? The Public's Attitude Toward Electroconvusive Therapy*, 134 Psychiatry Res. 205 (2005); Balkrishna Kalayam & Melvin J. Steinhart, *A Survey of Attitudes on the Use of Electroconvulsive Therapy*, 32 Hosp. Community Psychiatry 185 (1981); Richard Abrams, Electroconvulsive Therapy (1997).

286. Daniel Pagnin et al., *Efficacy of ECT in Depression: A Meta-Analytic Review*, 20 J. Electroconvulsive Therapy 13 (2004).

287. Barbara M. Rohland et al., *ECT in the Treatment of the Catatonic Syndrome*, 29 J. Affective Disorders 255 (1993).

288. Prudic, *supra* note 283, at 3297.

289. For a review, though now somewhat out of date, see William J. Winslade et al., *Medical, Judicial, and Statutory Regulation of ECT in the United States,* 141 Am. J. Psychiatry 1349 (1984). Restrictive regulations appear to reduce the incidence of ECT use in the Unied States; Richard C. Hermann et al., *Variation in ECT Use in the United States*, 152 Am. J. Psychiatry 869 (1995).

290. Although approved by the FDA for use in depression, concern over the weak database for TMS led the Centers for Medicare & Medicaid Services to withhold approval for payment for the procedure. Miriam Shuchman, *Approving the Vagus-Nerve Stimulator for Depression*, 356 New Eng. J. Med. 1604 (2007).

291. Philip B. Mitchell & Colleen K. Loo, *Transcranial Magnetic Stimulation for Depression*, 40 Austrl. N.Z. J. Psychiatry 406 (2006).

292. Helen S. Mayberg et al., *Deep Brain Stimulation for Treatment-Resistant Depression*, 45 Neuron 651 (2005); Benjamin D. Greenberg et al., *Three-Year Outcomes in Deep Brain Stimulation for Highly*

5. *Psychosurgery*

Direct surgical intervention to alter brain function in mental disorders has an unfortunate history.[293] Prefrontal leucotomy or lobotomy was developed in the 1930s as a treatment for intractable disorders such as schizophrenia, and became popular in the United States after World War II. Although there was never persuasive evidence of its efficacy, lobotomies were performed in many facilities, often in primitive conditions, on thousands of patients. Consequences frequently included a dulling of sensation and emotion. Interest in lobotomies faded in the late 1950s, because it became clear that the procedures were not having a positive effect, and they are not used today. Surgical interventions are used only rarely for psychiatric disorders, and only then for otherwise untreatable conditions. The most common procedures today involve parallel focal lesions in each of the two halves of the brain, which seems to help intractable and disabling obsessive–compulsive disorder and depression.[294] But psychosurgery for the treatment of psychiatric disorders is, in any form, extremely uncommon.

6. *Prediction of responses to treatment*

In a number of legal contexts, experts are called on to anticipate the responses of persons with mental disorders to treatment. For example, likely effectiveness must be considered before a court orders treatment over objections for a defendant who is incompetent to stand trial,[295] and the probable impact of future treatment may need to be estimated in determining damages in emotional harm cases.[296] The difficulty with these projections relates to several parameters that are inherently challenging to predict:

- *Effectiveness of treatment.* Even highly effective treatments for mental disorders do not work in all cases, and when they do work, they may provide varying levels of relief.[297]

Resistant Obsessive-Compulsive Disorder, 31 Neuropsychopharmacology 2384 (2006).

293. Elliot S. Valenstein, Great and Desperate Cures: The Rise and Decline of Psychosurgery and Other Radical Treatments for Mental Illness (1987).

294. Scott L. Rauch et al., *Neurosurgical Treatments and Deep Brain Stimulation, in* Sadock, Sadock, & Ruiz, *supra* note 120, at 2983–90.

295. Sell v. United States, 539 U.S. 166 (2003).

296. Melton et al., *supra* note 28.

297. For example, only 45% to 60% of patients receiving antidepressant medication for uncomplicated major depression show clinically significant responses to the first medication they receive, and of those who fail to respond, a similar percentage will respond positively to a second medication. A. John Rush & Andrew A. Nierenberg, *Mood Disorders: Treatment of Depression, in* Sadock, Sadock, & Ruiz, *supra* note 120, at 1734–9. Rates of response in unselected populations of patients with depression are lower. Madhukar H. Trivedi et al., *Evaluation of Outcomes with Citalopram for Depression Using Measurement-Based Care in STAR*D: Implications for Clinical Practice,* 163 Am. J. Psychiatry 28 (2006).

- *Adherence.* Treatment has no chance of being effective if a person declines to pursue or to continue it, a particular issue in cases where the court lacks control over the person's future behavior.[298] Tolerability of side effects may play an important role in these decisions.
- *Fluctuations in the course and responsiveness of the disorder.* Many mental disorders are chronic, and tend to wax and wane in intensity. Although adjustments in treatment can sometimes bring more severe symptoms under good control, that is not always possible. Moreover, for reasons that are not understood, previously responsive disorders may become resistant to the therapeutic effects of medication.[299]
- *Environmental conditions.* Unpredictable stresses in a person's life may exacerbate symptoms, reduce the effectiveness of treatment, or lead to diminished adherence.

However, given that estimates sometimes must be made of probable treatment effects, there are several indicators to which clinicians can turn.[300] Previous treatment response is the best predictor of future response; it is likely, for example, that someone whose previous delusions have rapidly resolved with antipsychotic medication will have a similar response in the future. In the absence of a documented history of successful treatment, estimates should be based on evidence indicating base rates of response for the person's disorder, along with any specific prognostic factors present in the person's case (e.g., a schizophrenic disorder that develops slowly over many years and that is associated with gradual functional decline generally has a poorer prognosis than one with rapid onset and good premorbid functioning). To a greater or lesser extent, however, it needs to be acknowledged that there is always uncertainty associated with these predictions.

298. Rates of nonadherence to medications among patients with psychiatric disorders are in the range of 50% or more. Although these figures are perhaps somewhat higher than those seen in other chronic conditions, long-term treatment with medication in general is marked by high rates of noncompliance with prescribed medications. Lars Osterberg & Terrence Blaschke, *Adherence to Medication*, 353 New Eng. J. Med. 487 (2005).

299. So-called "poop-out" during treatment of depression is a commonly encountered example. *See, e.g.*, Sarah E. Byrne & Anthony J. Rothschild, *Loss of Antidepressant Efficacy During Maintenance Therapy: Possible Mechanisms and Treatments*, 59 J. Clin. Psychiatry 279 (1998).

300. *See, e.g.*, for predictors of response to treatment for depression, Stuart M. Sotsky et al., *Patient Predictors of Response to Psychotherapy and Pharmacotherapy: Findings in the NIMH Treatment of Depression Collaborative Research Program*, 148 Am. J. Psychiatry 997 (1991); for predictors of response to treatment for schizophrenia, Delbert G. Robinson et al., *Predictors of Treatment Response from a First Episode of Schizophrenia or Schizoaffective Disorder*, 156 Am. J. Psychiatry 544 (1999).

G. Limitations of Mental Health Evidence

Certain limitations exist where mental health evidence is concerned that may not come into play with other types of scientific evidence. Both retrospective assessments of past mental states and prospective estimates of future behavior depend on estimates of variables that are inherently difficult to know with a high degree of certainty. Even contemporaneous assessments of functional abilities depend, in part, on the evaluee's self-report of such difficult-to-measure attributes as distress, motivation, and judgment. Where empirically validated assessment tools are used, the usual concerns about measurement error are present. Two additional problematic areas involve the use of psychodynamic theory and testimony that speaks to the ultimate legal issue.

1. Limits of psychodynamic theory

Psychoanalysis developed a complex theory of the mind that included both functional elements (i.e., ego, superego, and id) and processes by which unconscious motivations are brought to bear on conscious thought and behavior (e.g., displacement, projection, reaction formation), largely in the service of protecting the conscious mind from unbearable conflict.[301] Freud's basic schemata, which underwent evolution even during his lifetime, subsequently have been subject to permutation and elaboration by a large number of theorists. These schemata form the theoretical basis for the dynamic psychotherapies and have been incorporated into popular culture, as reflected in the work of historians, literary theorists, novelists, and cartoonists, among others.[302] However, although these concepts have proven useful in a variety of fields, many of them have been resistant to empirical testing. Even when ample evidence exists to support a psychodynamic construct—e.g., recovery of unconscious, nontraumatic memories[303] or repression[304]—it has been difficult to prove the postulated functional role for the process. Nonetheless, psychodynamic concepts—and the use of psychodynamic therapies—remain mainstays in many psychiatry and psychology training programs. Testimony based on these concepts is often introduced, for example, in discussions of a defendant's mental state at the time of the crime, in relation to defenses of insanity, dimin-

301. William W. Meissner, *Classical Psychoanalysis*, *in* Sadock & Sadock, *supra* note 120, at 701–46.

302. *See, e.g.*, Psychoanalytic Literary Criticism (Maud Ellmann ed., 1994); Peter Loewenberg, *Psychoanalytic Models of History: Freud and After*, *in* Psychology and Historical Interpretation (William M. Runyan ed., 1980).

303. Matthew H. Erdelyi, The Recovery of Unconscious Memories: Hypermnesia and Reminiscence (1996).

304. David S. Holmes, *The Evidence for Repression: An Examination of Sixty Years of Research*, *in* Repression and Dissociation: Implications for Personality Theory, Psychopathology, and Health 85–102 (Jerome L. Singer ed., 1990).

ished capacity, self-defense, provocation, duress, and entrapment.[305] It may also play a role in civil cases, regarding questions as disparate as a parent's capacity to raise a child and whether a testator was subject to undue influence.[306] Because these concepts were generally accepted in the relevant fields, although there have always been skeptics, the test of admissibility under *Frye v. United States* and similar state rules was usually met.[307] The reinvigorated admissibility requirements promulgated under *Daubert v. Merrell Dow Pharmaceuticals, Inc.* and *Kumho Tire Co. v. Carmichael*, with their emphasis on empirical verification of the bases for the expert's testimony, have called the future of testimony based on most psychodynamic concepts into question.[308]

Questions about testimony based on psychodynamic theory can be raised with regard both to the legitimacy of the underlying constructs (e.g., displacement of affect) and to the techniques by which the examiner can know that such a mechanism came into play in a particular case (e.g., the displacement of the defendant's unconscious rage at his mother led to a loss of behavioral control that resulted in an assault on another woman). Slobogin has argued, with regard to criminal defendants, that frankly speculative testimony about psychodynamic influences on the crime should be held to a lesser standard of admissibility than required under *Daubert*.[309] In part, he suggests that the very concepts on which the law relies—such as extreme emotional stress and reasonable apprehension of harm—are themselves not easily susceptible to determinations that would meet *Daubert's* reliability considerations. Thus, if defendants are to be able to introduce evidence that would overcome the presumptions against them, testimony that relies on accepted but inherently unprovable constructs is essential. Moreover,

305. Christopher Slobogin, Proving the Unprovable: The Role of Law, Science, and Speculation in Adjudicating Culpability and Dangerousness (2007).

306. Robertson v. McCloskey, 676 F. Supp. 351 (D.D.C. 1988) (declining to admit psychodynamic testimony under the *Frye* standard); United States v. Libby, 461 F. Supp. 2d 3 n.6 (D.D.C. 2006) (noting that although psychodynamic testimony was not admissible under the *Frye* standard, that does not necessarily hold under *Daubert*, and that "there can be little doubt that today . . . the science of memory is well established and accepted in the scientific community . . . has been well tested and subjected to peer review"); United States v. Fishman, 743 F. Supp. 713 (N.D. Cal. 1990) (excluding testimony on "thought reform" theory from a qualified mental health professional).

307. Frye v. United States, 293 F. 1013 (1923).

308. Daubert v. Merrell Dow Pharms., Inc., 509 U.S. 579 (1993); Kumho Tire Co. v. Carmichael, 526 U.S. 137 (1999).

309. Slobogin, *supra* note 305, at 39–57. Note that Slobogin's argument is not limited to testimony rooted in psychodynamic concepts, but extends to other mental health evidence that intends to speak to aspects of a person's mental state at some point in the past, knowledge of which is unlikely ever to meet scientific standards of proof. Under the *Daubert* standard, the judge serves as the gatekeeper for scientific testimony. The admissibility of evidence is determined on the basis of relevance and reliability. Reliability factors offered as examples include falsifiability, peer review, the known or potential rate of error, and general acceptance by the relevant scientific community. *Daubert,* 509 U.S. at 593–95.

Slobogin claims that this is, in fact, why trial courts usually resist efforts to exclude mental health testimony.[310]

Granting the legitimacy of Slobogin's analysis, there is still reason for caution in a wholesale embrace of psychodynamic theories. Because persuasive empirical demonstrations of either the concepts themselves or their application in particular cases is unlikely, their speculative—even if plausible—nature should be recognized. Moreover, to say that such testimony should not be held to reliability-based standards of admissibility is not to say that no relevant standards exist. Idiosyncratic concepts and conclusions that would not be generally accepted by clinicians with appropriate training might well run afoul of prevailing rules for admissibility, because they lack support even under the older standard of acceptance in the relevant professional community,[311] and to the extent that techniques are available for generating testable data, they would appear to be preferable. This appears, in fact, to be the way in which courts generally approach such evidence.[312]

2. Ultimate issue testimony

Whether mental health experts should testify—or be permitted to testify—to the ultimate legal issue in a case has been the subject of longstanding controversy.[313] The question arises, for example, in criminal cases where experts often have commented directly on whether a defendant is competent to stand trial or whether the legal standard for insanity has been met.[314] Similar issues can arise in civil settings, in which experts may be asked to testify directly about a person's capacity to manage affairs or to serve as a custodial parent, or regarding whether a person was competent to sign a contract at an earlier point in time.[315] Some mental health experts find themselves encouraged or pressured by attorneys to draw conclusions about the ultimate issue, and judges have been known to exclude testimony in which experts are unwilling to take that step on the grounds that the evidence that they would otherwise provide lacks probative value.[316] Concerns arise over the fact that conclusions about the ultimate issue in a case are matters to be decided by the factfinder, on whose legitimate territory an expert who speaks to the issue may be encroaching and whose deliberations may be preempted.[317]

310. For a response to Slobogin's argument, see Edward J. Imwinkelried, *The Case Against Abandoning the Search for Substantive Accuracy*, 38 Seton Hall L. Rev. 1031 (2008).

311. Frye v. United States, 293 F. 1013 (D.C. Cir. 1923).

312. Slobogin, *supra* note 305, at 21–29.

313. *See* Fed. R. Evid. 704. *See* Anne Lawson Braswell, *Resurrection of the Ultimate Issue Rule: Federal Rule of Evidence 704(b) and the Insanity Defense*, 72 Cornell L. Rev. 620 (1987).

314. But see discussion below regarding the current prohibition on this practice in federal courts.

315. *See* Restatement (Second) of Contracts § 15.

316. Appelbaum & Gutheil, *supra* note 149, at 221.

317. Insanity Defense Workgroup, *American Psychiatric Association Position on the Insanity Defense*, 140 Am. J. Psychiatry 681, 686 (1983); American Bar Association, ABA Criminal Justice Standards:

Proponents of ultimate issue testimony often include attorneys and judges, who may be concerned that an expert who provides a clinical formulation without tying it directly to the ultimate legal issue will leave a group of confused jurors unable to discern the connection on their own.[318] Many experts themselves share similar concerns or worry that mental health issues will simply be ignored if their relevance to the legal question at hand is not made clear; moreover, they note that courts have applied such rules erratically.[319] They counter concerns about such testimony having an undue impact on jurors' deliberations by noting that members of juries appear to be little influenced by whether or not ultimate issue testimony is offered by an expert.[320]

Moreover, efforts to restrict testimony on the ultimate issue often quickly run into line-drawing problems. As an example, after a jury found John W. Hinckley, Jr. not guilty by reason of insanity of the attempted assassination of President Reagan, the verdict led to wholesale revision of laws governing the insanity defense at the federal and state levels.[321] Among the changes wrought by the Federal Insanity Defense Reform Act of 1984 was a prohibition on experts directly addressing the question of insanity.[322] The Federal Rules of Evidence were amended to effect this change: "No expert witness testifying with respect to the mental state or condition of a defendant in a criminal case may state an opinion or inference as to whether the defendant did or did not have the mental state or condition constituting an element of the crime or of a defense thereto."[323] Although it seems clear that, according to the terms of the rule, the expert is precluded from opining directly that a defendant lacked criminal responsibility, it is less clear whether the expert could say that the defendant could not "appreciate the wrongfulness of his acts," the language used in the statute to define the relevant standard.[324] And if that, too, were prohibited, could the expert say that the defendant "could not grasp how wrong his behavior was," and if so, would that language be likely to have any different impact on a jury than simply speaking in the words of the statute? Empirical data exist to suggest that the answer to that question is no.[325]

Still, a large number of mental health and legal scholars oppose experts addressing the ultimate legal question, and during the high-pitched debate follow-

Mental Health, Standard 7-6.6 (1984). Note that the APA position was recently withdrawn as outdated and replaced by a briefer statement that does not address the question of ultimate issue testimony.

318. Ralph Slovenko, *Commentary: Deceptions to the Rule on Ultimate Issue Testimony*, 34 J. Am. Acad. Psychiatry & L. 22 (2006).

319. Alec Buchanan, *Psychiatric Evidence on the Ultimate Issue,* 34 J. Am. Acad. Psychiatry & L. 14 (2006).

320. Solomon M. Fulero & Norman J. Finkel, *Barring Ultimate Issue Testimony: An "Insane" Rule?* 15 L. & Hum. Behav. 495 (1991).

321. Henry J. Steadman, Before and After Hinckley: Evaluating Insanity Defense Reform (1993).

322. 18 U.S.C. § 17.

323. Fed. R. Evid. 704(b).

324. *Id.*

325. Fulero & Finkel, *supra* note 320.

ing the Hinckley trial, both the American Psychiatric Association and the American Bar Association adopted positions against ultimate issue testimony.[326] In addition to the argument that such testimony trenches on the function of the jury, opponents often point to the legal and moral nature of the question whether someone is criminally responsible.[327] Although mental health expertise may be helpful in determining the person's mental state at the relevant time, determining whether the resulting impairment was sufficient to negate responsibility requires the application of the relevant legal standard and a moral judgment of the fairness or unfairness of punishing the person for his or her behavior. Psychiatrists and psychologists have no particular expertise on legal or moral issues; hence, opponents of ultimate issue testimony urge that they should not be permitted to speak to those issues. Such preclusion may also reduce the much bemoaned "battle of the experts," because a good deal of disagreement may derive from views of how data from the evaluation should be applied to the ultimate legal question, rather than from differences regarding the person's mental state. Although testimony on the ultimate legal issue is now barred in federal courts in insanity defense cases (18 U.S.C. § 17), it remains common in many states, and even in federal jurisdictions it may be offered in other sorts of cases.[328]

II. Evaluating Evidence from Mental Health Experts

To this point, we have considered the kind of evidence that is likely to be offered by mental health experts and some of the challenges that such testimony presents. The remainder of the chapter addresses those factors that should enter into consideration of the value and impact of such testimony.

A. What Are the Qualifications of the Expert?

The appropriate qualifications of a mental health professional whose testimony is proffered will depend on the nature of the evidence that will be presented. However, a number of relevant parameters can be identified.

326. Insanity Defense Workgroup, *supra* note 317; American Bar Association, *supra* note 317. *See also* Grisso, *supra* note 2, at 208; Fulero & Finkel, *supra* note 320, at 496.

327. Mark S. Brodin, *Behavioral Science Evidence in the Age of* Daubert*: Reflections of a Skeptic*, 73 U. Cin. L. Rev. 867 (2005); Michele Cotton, *A Foolish Consistency: Keeping Determinism Out of the Criminal Law*, 18 B.U. Pub. Int. L.J. 1, 21–23 (2005); Ric Simmons, *Conquering the Province of the Jury: Expert Testimony & the Professionalization of Fact-Finding*, 74 U. Cin. L. Rev. 1013 (2006).

328. Fed. R. Evid. 704. Pennsylvania's law represents a typical formulation: "Testimony in the form of an opinion or inference otherwise admissible is not objectionable because it embraces an ultimate issue to be decided by the trier of fact." Pa. R. Evid. 704.

1. Training

Most mental health expert testimony is given by psychiatrists or doctoral-level clinical psychologists. Given the differences in the education and training of each profession, their testimony is not necessarily interchangeable. As a rule, psychiatrists are prepared by their training to speak to the diagnosis of mental disorders, including medical issues that may play a role in a particular case, and to treatment approaches, including psychopharmacological treatment.[329] They should be capable of testifying, within the limits of existing knowledge and the information available to them, regarding the impact of a disorder on a person's behavior and functional abilities. Psychologists' training, in contrast, may provide deeper knowledge of the theoretical and experimental bases for understanding the function of the mind, both normal and abnormal.[330] As a general matter, doctoral-level clinical psychologists will be prepared by their training to provide evidence regarding diagnosis and psychotherapeutic treatment of mental disorders, the results of psychological and neuropsychological testing, and the roots of normal and abnormal behavior.

However, although the core elements of training in psychiatry and psychology may be similar across training programs, the variability is substantial.[331] Moreover, variation in subspecialty (in psychiatry) or specialty (in psychology) training—for example, in geriatric psychiatry or neuropsychology—contributes to further differentiation among experts. Thus, inquiries regarding the specific training afforded an expert may be necessary. This is particularly true when an expert is testifying about topics that would ordinarily fall outside disciplinary boundaries, for example, a psychiatrist discussing the results of psychological testing or a psychologist offering evidence regarding the effect of medication on a person's behavior. The same is true for experts who are testifying beyond the range of their specialty or subspecialty training. In addition, in recent years, expert testimony on mental health issues has been admitted at times from nonpsychiatric physicians and mental health professionals of other disciplines.[332] These include

329. See discussion of psychiatrists' training in Section I.B.1, *supra*.

330. See discussion of psychologists' training in Section I.B.2, *supra*.

331. *See, e.g.*, Khurshid A. Khurshid et al., *Residency Programs and Psychotherapy Competencies: A Survey of Chief Residents*, 29 Academic Psychiatry 452 (2005); Committee on Incorporating Research into Psychiatry Residency Training, Institute of Medicine, Research Training in Psychiatric Residency: Strategies for Reform 91–132 (Michael T. Abrams et al. eds., 2003); Charles J. Gelso, *On the Making of a Scientist-Practitioner: A Theory of Research Training in Professional Psychology*, S(1) Training and Education in Professional Psychology 3–16 (2006); Brendan A. Maher, *Changing Trends in Doctoral Training Programs in Psychology: A Comparative Analysis of Research-Oriented Versus Professional-Applied Programs*, 10 Psychol. Sci. 475 (1999).

332. Campbell v. Metropolitan Prop. & Cas. Ins. Co., 239 F.3d 179 (2d Cir. 2001) (professor of pediatrics with substantial relevant publications found qualified to testify on neurological injuries resulting from lead paint exposure); Carroll v. Otis Elevator Co., 896 F.2d 210 (7th Cir. 1990) (experimental psychologist found qualified to give expert testimony on likelihood that product design

social work and nursing, and in the future arguably could include master's-level psychologists, marriage and family therapists, physician assistants, and additional disciplines as well. Specific inquiry into relevant training will probably be needed at least until testimony from such disciplines becomes more widely accepted and their specific qualifications more generally known.

2. Experience

Experience is relevant to the qualifications of mental health experts in at least two ways. First, as the Federal Rules of Evidence recognize, experience may substitute for training as a basis for concluding that a witness has special expertise.[333] Many experts in forensic psychiatry and forensic psychology, for example, lack formal training in conducting evaluations of the sort provided in forensic fellowships, because such training programs have become widely available only fairly recently. In addition, formal training is simply unavailable (or at least difficult to acquire) in a number of substantive areas of clinical psychiatry and psychology. For example, most professionals who acquire special knowledge about particular mental disorders will do so by pursuing their interest through reading and following the literature and by means of clinical contact with patients with the disorders, as opposed to formal training. Thus, experience must often be relied upon as a stand-in for more conventional credentials.

The second way in which experience can be material to expert qualifications relates to the attrition of skills and knowledge over time. Mental health professionals often complete their training within several years of their 30th birthdays and may engage in practice, including the provision of expert testimony, over the subsequent four or five decades. Brief exposure to information about a particular disorder[334] or some experience in evaluating and treating the condition may fade from memory several decades later unless reinforced in a direct way. Just as problematic is the possibility that additional knowledge about the condition has

would cause children to press escalator's emergency stop button); United States v. Withorn, 204 F.3d 790 (8th Cir. 2000) (trial court properly admitted testimony from midwife on alleged sexual assault on basis of bachelor's degree, some postgraduate work, and clinical experience). *But see* United States v. Moses, 137 F.3d 894 (8th Cir. 1998) (social worker lacked expertise to opine that victim of alleged child abuse would suffer trauma from facing the accused abuser in the courtroom).

333. "If scientific, technical, or other specialized knowledge will assist the trier of fact to understand the evidence or to determine a fact in issue, a witness qualified as an expert by knowledge, skill, experience, training, or education, may testify thereto in the form of an opinion or otherwise, if (1) the testimony is based upon sufficient facts or data, (2) the testimony is the product of reliable principles and methods, and (3) the witness has applied the principles and methods reliably to the facts of the case." Fed. R. Evid. 702 (2000).

334. Although this discussion in framed in terms of a particular disorder, the condition in issue may not constitute a disorder in a formal sense. Rather it may involve a symptom (e.g., auditory hallucinations), a mental state not linked to a specific disorder (e.g., dissociation), or a behavioral propensity (e.g., violent behavior). The argument in this section is generally applicable to all these categories of phenomena.

been gained in the interim, familiarity with which might alter an expert's evaluation or opinion. Training regarding a mental disorder or treatment, therefore, may be a necessary but not sufficient aspect of an expert's qualifications, in the absence of ongoing experience. Indicia of such experience may include evaluating or treating patients with the disorder, teaching trainees how to assess or treat the disorder, systematically reviewing the literature on the disorder, attending continuing education sessions concerning the disorder, and conducting research on the disorder.

Although experience, including ongoing experience, with the condition at issue is important in establishing expertise for the purpose of providing evidence in a case, there is a danger that experience can be overemphasized as a criterion of expertise as well. Assuming a baseline degree of adequate training and some ongoing experience in a field or with a condition, it is not clear that additional experience necessarily enhances an expert's authoritativeness. Experts will sometimes boast of the number of evaluations they have performed of a particular type of evaluee (e.g., alleged or convicted murderers) or of a given kind (e.g., assessments of competence to stand trial). However, if evaluations are performed inadequately or used as the basis for invalid conclusions, especially if there is no feedback loop to correct the expert's errors, mere experience may only have the effect of reinforcing bad clinical habits. Indeed, studies of diagnostic performance by mental health professionals divided into groups by the duration of their clinical experience have shown no consistent correlation between years of experience and reliability.[335] An explanation for the failure to find a consistent effect of expertise may be that, despite less clinical experience, recently trained clinicians are more familiar with the contemporary diagnostic framework and are less tempted to use their clinical experience as a substitute for generally accepted criteria (e.g., "I know schizophrenia when I see it, regardless of what the criteria say"). It is of interest that few studies have compared the performance of experienced forensic psychiatrists and psychologists to their nonforensic colleagues.[336] Although it might be expected that experts with forensic training would be more sensitive to the unique aspects of forensic examinations discussed above, for example, the importance of maintaining a level of suspicion regarding secondary gain and of confirming the evaluee's account, when possible, with collateral information, that hypothesis remains to be tested. One small study has shown that forensic psychiatrists may be less susceptible to some kinds of hindsight bias than their clinical

335. Here reliability is being used in its technical sense of agreement across more than one rater. For an example of the failure to find a consistent effect of previous experience, *see, e.g.*, Sean H. Yutzy et al., *DSM-IV Field Trial: Testing a New Proposal for Somatization Disorder*, 152 Am. J. Psychiatry 97 (1995).

336. There are, however, data to suggest, as might be expected, that clinicians with forensic training have higher levels of knowledge regarding relevant legal issues, *e.g.*, Gary B. Melton et al., Community Mental Health Centers and the Courts: An Evaluation of Community-Based Forensic Services 43–55 (1985).

colleagues,[337] but additional research would be helpful before firm conclusions are drawn.

3. Licensure and board certification

a. Licensure

Possession of a valid professional license is usually considered a threshold requirement in the qualification of an expert in legal proceedings. Licensure of physicians (including psychiatrists) is governed by a licensure board in each state.[338] Although criteria may differ somewhat, generally a physician who has graduated from an accredited American medical school, passed a sequence of tests designed to ensure adequate levels of knowledge and clinical judgment,[339] and completed 1 or 2 years of residency training is eligible for full licensure.[340] Prior to that point, a temporary license, allowing practice under supervision, is usually issued. Graduates of medical schools that are not in the United States are usually subject to a different set of requirements, often requiring longer periods of residency training and individual review of qualifications. Once licensure is attained in a state, should a physician desire to acquire a license in another state, the process is variable. Some states will grant such a license fairly easily; others, such as California, will require that the physician take and pass a test of general medical knowledge if a certain period of time (e.g., 10 years in California) has passed since the original sequence of testing was completed.[341]

For clinical psychologists, standards for licensure differ somewhat by state, but generally after completion of an accredited Ph.D. program in the United States (including a 1-year internship), they are required to complete 2 years of clinical work under the supervision of a licensed psychologist and to pass a national licensure examination.[342] Because the states do not restrict the practice of psychotherapy per se, but regulate the use of professional titles instead, an unlicensed psychologist can engage in many aspects of the clinical practice of psychology, including all forms of psychotherapy, but will not be able to use the title of psychologist. For psychologists who are seeking licensure in another jurisdiction, many states will grant reciprocity—that is, they will not engage in an independent

337. Herbert W. LeBourgeois et al., *Hindsight Bias Among Psychiatrists*, 35 J. Am. Acad. Psychiatry & L. 67 (2007).

338. A summary of the requirements for medical licensure in each jurisdiction is available from the Federation of State Medical Boards at http://www.fsmb.org/usmle_eliinitial.html.

339. See a description of the tests and the examination process at http://www.usmle.org/General_Information/general_information_about.html.

340. Federation of State Medical Boards, *supra* note 338.

341. Cal Bus. & Prof. Code §§ 2080–99, 2184.

342. Details of requirements in each state can be found at the Web site of the Association of State and Provincial Psychology Boards at http://www.asppb.net.

process of reviewing the applicant's credentials, relying instead on the review conducted by the initial licensure board.

b. Board certification

Board certification represents a level of qualifications beyond those required for licensure in either medicine or psychology. Although well-trained, competent psychiatrists may have reasons for not attaining board certification (e.g., examination anxiety that interferes with performance, a career centered on nonclinical research for which clinical board certification is thought to be unnecessary), the tests are designed to be passed by a competent psychiatrist and do not require exceptional levels of clinical skill. Thus, in most cases board certification can be viewed as reflecting attainment of an adequate level of clinical competence to engage in independent psychiatric practice. Whether a court chooses to admit testimony from a psychiatrist who has not been board certified may depend on the reasons why certification has not been achieved and on the specific question(s) that will be addressed in the psychiatrist's testimony.[343]

Professional psychology also has a board certification process, administered by the American Board of Professional Psychology.[344] Certification is only offered in psychology specialties, but these include such general clinical fields as clinical psychology, counseling psychology, and group psychology. As in subspecialty certification in psychiatry, candidates are expected to exhibit advanced competence in the specialty area, defined specifically for each specialty. Board certification is less common among psychologists than among psychiatrists, in part perhaps because the process is more recent.[345] Given this, it is less likely that certification will be applied as a minimum standard for expert testimony in psychology than in psychiatry or other areas of medicine.

343. For examples of the scope of judicial discretion on this issue, see, e.g., Hall v. Quarterman, 534 F.3d 365 (5th Cir. 2008) (finding that a state requirement that only a licensed expert may testify in a civil commitment hearing as to mental retardation did not extend to expert testimony on the same topic); Oberlander v. Oberlander, 460 N.W.2d 400 (1990) (reversing as abuse of discretion the trial court's exclusion of expert testimony from a psychologist who was licensed in the neighboring state); Williams v. Brown, 244 F. Supp. 2d 965 (N.D. Ill. 2003) (finding that psychiatrists who were not board-certified child psychiatrists may nonetheless testify about the condition of juvenile plaintiffs).

344. A description of the process and eligibility requirements for the examination process can be found at http://www.abpp.org/abpp_certification_specialties.htm.

345. A recent study suggests that approximately 85% of psychiatrists become board certified in the 8 years following completion of residency training. Dorthea Juul et al., *Achieving Board Certification in Psychiatry: A Cohort Study,* 160 Am. J. Psychiatry 563 (2003). In contrast, it was estimated that in 2000 only 3.5% of psychologists had achieved board certification. Frank M. Dattilio, *Board Certification in Psychology: Is It Really Necessary?* 33 Prof. Psychol.: Res. & Prac. 54 (2002).

4. Prior relationship with the subject of the evaluation

A presumption may exist among some attorneys, judges, and jurors that a mental health professional who has had a treatment relationship with the person whose mental state is in question is better qualified to testify about aspects of that mental state than an evaluator who is meeting the person for the first time. The logic seems strong: A professional who has known the person for some period of time, perhaps a substantial one, should be better able to offer conclusions about the person's diagnosis, treatment requirements, and the impact of the person's mental state on the person's function and behavior. Thus, it may seem surprising that the ethics guidelines produced by both the American Academy of Psychiatry and the Law, the leading organization of forensic psychiatrists, and the American Psychological Association's division of forensic psychologists point to problems inherent in such situations.[346] Although neither set of guidelines construes testimony involving current or former patients as unethical, they both have words of caution to offer and discourage clinicians from playing both clinical and expert roles.[347]

The professional literature on this issue, and the ethics guidelines themselves, cite several reasons why having a treating professional perform the evaluation for

346. American Academy of Psychiatry and the Law: Ethics Guidelines for the Practice of Forensic Psychiatry, May 2005, https://www.aapl.org/ethics.htm; Committee on Ethical Guidelines for Forensic Psychologists (Division 41 of the American Psychological Association and the American Academy of Forensic Psychology), *Specialty Guidelines for Forensic Psychologists*, 15 L. & Hum. Behav. 655 (1991).

347. The forensic psychiatry guidelines are explicitly discouraging of this practice:

> Psychiatrists who take on a forensic role for patients they are treating may adversely affect the therapeutic relationship with them. Forensic evaluations usually require interviewing corroborative sources, exposing information to public scrutiny, or subjecting evaluees and the treatment itself to potentially damaging cross-examination. The forensic evaluation and the credibility of the practitioner may also be undermined by conflicts inherent in the differing clinical and forensic roles. Treating psychiatrists should therefore generally avoid acting as an expert witness for their patients or performing evaluations of their patients for legal purposes.

American Academy of Psychiatry and the Law: Ethics Guidelines for the Practice of Forensic Psychiatry, Sec. IV (May 2005), *available at* https://www.aapl.org/ethics.htm. In contrast, the forensic psychology guidelines could be seen as being somewhat more permissive:

> "D. Forensic psychologists recognize potential conflicts of interest in dual relationships with parties to a legal proceeding, and they seek to minimize their effects.
>
> 1. Forensic psychologists avoid providing professional services to parties in a legal proceeding with whom they have personal or professional relationships that are inconsistent with the anticipated relationship.
>
> 2. When it is necessary to provide both evaluation and treatment services to a party in a legal proceeding (as may be the case in small forensic hospital settings or small communities), the forensic psychologist takes reasonable steps to minimize the potential negative effects of these circumstances on the rights of the party, confidentiality, and the process of treatment and evaluation."

Committee on Ethical Guidelines for Forensic Psychologists (Division 41 of the American Psychological Association and the American Academy of Forensic Psychology): *Specialty Guidelines for Forensic Psychologists*, 15 Law & Hum. Behav. 655 (1991).

legal purposes may not be prudent.[348] First, offering testimony, even if it is supportive of the patient's legal claim, may interfere with the therapeutic relationship. Not only will it often come as a shock to a patient to hear herself described in diagnostic terms, but details of the treating clinician's view of the patient revealed under both direct and cross-examination may alienate the person from the clinician. The treating clinician, in fact, may be aware of more information that is not relevant to the legal question than an evaluator called in specifically for purposes of providing evidence, and hence may be even more likely to reveal it during testimony. At best, when this happens it impedes the therapeutic process and takes time away from the primary therapeutic goals; at worst, it may lead the person to abandon treatment. This effect is likely to be exacerbated if the testimony is adverse to the patient's legal position.

Second, the underlying assumption regarding the desirability of having the clinician testify may be flawed. That is, although the clinician may have known the person for a long time as a patient, the clinical process may never have required the clinician to collect the type of information that would be relevant to the legal question. Even if that information was discussed, the treating clinician is less likely to have approached it with the degree of caution that a forensic evaluator would be likely to employ or to have attempted to verify the information through collateral sources. Indeed, even after agreeing to participate as an expert witness, a clinician may be unaware of the importance of assessing the veracity of the person's claim or afraid that doing so may lead to strains in the therapeutic relationship.

A third problem is that the clinician, having formed an alliance with the person as a patient, perhaps over a considerable period of time, may feel a natural allegiance to the person and a desire, even if not a conscious one, to support the person's contentions in the case. Thus, presentation of evidence may undergo subtle distortion, or may be subject to conscious manipulation by a clinician who sees his or her role as being the patient's advocate. Fourth, there is an ethical problem when the clinician is subpoenaed to testify over the patient's objection. The preexisting therapeutic relationship was premised on the information that the patient revealed being used for treatment purposes. It places the clinician whose testimony cannot support the person's legal claim in an extremely awkward position to be compelled now to use that information to the patient's detriment.[349]

348. Larry H. Strasburger et al., *On Wearing Two Hats: Role Conflict in Serving as Both Psychotherapist and Expert Witness*, 154 Am. J. Psychiatry 448 (1997); Ronald Schouten, *Pitfalls of Clinical Practice: The Treating Clinician as Expert* Witness, 1 Harv. Rev. Psychiatry 405 (1993); Stuart Greenberg & Daniel Shuman, *Irreconciliable Conflict Between Therapeutic and Forensic Roles*, 28 Prof. Psychol.: Res. & Prac. 50 (1997); Appelbaum & Gutheil, *supra* note 149, at 236–39.

349. Although all states have psychotherapist–patient and/or physician–patient testimonial privilege statutes that limit testimony by treating psychiatrists and psychologists (and often other mental health professionals) without the patient's consent, the exceptions in many of these statutes—including the so-called patient-litigant exception that is invoked when patients place their mental state at issue

Thus, in contrast to what might seem the logical assumption—that the treating clinician is the best qualified person to testify regarding the patient—there are multiple reasons to avoid relying on the treater, and in fact to discourage that person from serving as an expert witness in the case.

B. How Was the Assessment Conducted?

The reliability and validity of an expert opinion related to mental health issues depends heavily on the manner in which the assessment that forms the basis for the conclusions was conducted.

1. Was the evaluee examined in person?

Given the range of cases in which mental health experts provide testimony and the various questions to which they are asked to respond, situations arise in which the experts are providing evidence without having examined the person about whom they are testifying.[350] Such circumstances may arise when direct evaluation is impossible, for example, in contests over testamentary capacity, where often only after the testator is deceased will a claim regarding the person's capacity be litigated. Other civil litigation in which there may be issues regarding the state of mind of a deceased person include contractual capacity, wrongful death, and medical malpractice claims.[351] Testimony regarding a person who cannot be evaluated directly is less likely to occur in criminal cases, but a highly contentious example occurs in death penalty cases in Texas; defendants have the right to decline evaluation by prosecution experts,[352] but such experts frequently testify on the basis of a hypothetical question that reflects some of the facts regarding the defendants' history and behavior.[353]

in a case—are sufficiently numerous that this situation cannot be ruled out. Jaffee v. Redmond, 518 U.S. 1 n.13 (1996); Bruce J. Winick, *The Psychotherapist-Patient Privilege: A Therapeutic Jurisprudence View*, 50 U. Miami L. Rev. 249 (1996).

350. In addition, on some occasions, testimony will provide contextual information for the decisionmaker, for example, how a person in a given situation or with a given disorder would usually respond, without being applied directly to a specific person. John Monahan & Laurens Walker, *Social Authority: Obtaining, Evaluating, & Establishing Social Science in Law*, 134 U. Pa. L. Rev. 477 (1986); John Monahan & Laurens Walker, *Social Science Research in Law: A New Paradigm*, 43 Am. Psychol. 465 (1988).

351. Farnsworth, *supra* note 8, § 3:11. For a case study of the use of postmortem analysis in the USS *Iowa* explosion investigation, see Charles Patrick Ewing & Joseph T. McCann, Minds on Trial: Great Cases in Law and Psychology 129–39 (2006); *see also* Norman Poythress et al., *APA's Expert Panel in the Congressional Review of the USS* Iowa *Incident*, 48 Am. Psychol. 8 (1993). *See* Moon v. United States, 512 F. Supp. 140 (D. Nev. 1981) (finding that hospital psychiatrists were negligent in diagnosing as schizophrenic a patient who later committed suicide); Urbach v. United States, 869 F.2d 829 (5th Cir. 1989) (finding no medical malpractice where a mental patient on furlough from a VA hospital was arrested and beaten to death in a Mexican prison).

352. Estelle v. Smith, 451 U.S. 454 (1981).

353. Barefoot v. Estelle, 463 U.S. 880 (1983); Satterwhite v. Texas, 486 U.S. 249 (1988).

Conclusions about persons who have not been directly examined may be drawn on the basis of available records, including medical, mental health, police, educational, armed services, and other records; information from informants who have been or are in contact with the person, which may derive from interviews by the expert, prior testimony, depositions, police reports, and other sources; and on some occasions observations by the expert of the person's behavior, for example, in a prison or courtroom setting.[354] Although it may be possible to draw valid conclusions on the basis of such data, conclusions generally are more limited and have a lesser degree of certainty than when a direct evaluation has taken place. The ethics statements of the major forensic psychiatry and forensic psychology organizations offer words of caution about such testimony.[355] There are several reasons why caution is warranted.

Expert knowledge in mental health can be viewed as comprising two components: the knowledge of how to conduct an evaluation to obtain relevant data and the knowledge of how to weigh those data to reach a conclusion.[356] When a direct examination of the person cannot be carried out, the expert must rely on information accumulated by others, sometimes for other purposes. The likelihood

354. Kirk Heilbrun et al., *Third Party Information in Forensic Assessment, in* Handbook of Psychology, Vol. 11: Forensic Psychology 69 (Alan M. Goldstein ed., 2003). Testimony offered in capital sentencing contexts without examination of the defendant has been particularly controversial, *see, e.g.,* Bennett v. State, 766 S.W.2d 227, 232 (Tex. Crim. App. 1989) (Teague, J., dissenting) ("[W]hen Dr. Grigson testifies at the punishment stage of a capital murder trial he appears to the average lay juror . . . to be the second coming of the Almighty. . . . Dr. Grigson is extremely good at persuading jurors to vote to answer the [future dangerousness] issue in the affirmative."); "*They Call Him Dr. Death,*" Time. June 1, 1981; Rosenbaum, *supra* note 216.

355. The Ethics Guidelines for the Practice of Forensic Psychiatry of the American Academy of Psychiatry and Law (available at https://www.aapl.org/ethics.htm) note:

> For certain evaluations (such as record reviews for malpractice cases), a personal examination is not required. In all other forensic evaluations, if, after appropriate effort, it is not feasible to conduct a personal examination, an opinion may nonetheless be rendered on the basis of other information. Under these circumstances, it is the responsibility of psychiatrists to make earnest efforts to ensure that their statements, opinions and any reports or testimony based on those opinions, clearly state that there was no personal examination and note any resulting limitations to their opinions.

The comparable guidelines for forensic psychology state:

> Forensic psychologists avoid giving written or oral evidence about the psychological characteristics of particular individuals when they have not had an opportunity to conduct an examination of the individual adequate to the scope of the statements, opinions, or conclusions to be issued. Forensic psychologists make every reasonable effort to conduct such examinations. When it is not possible or feasible to do so, they make clear the impact of such limitations on the reliability and validity of their professional products, evidence, or testimony.

Committee on Ethical Guidelines for Forensic Psychologists (Division 41 of the American Psychological Association and the American Academy of Forensic Psychology), *Specialty Guidelines for Forensic Psychologists*, 15 Law & Hum. Behav. 655 (1991).

356. Paul S. Appelbaum, *Hypotheticals, Psychiatric Testimony, and the Death Sentence*, 12 Bull. Am. Acad. Psychiatry & L. 169 (1984); *see also* American Psychiatric Ass'n amicus brief in *Barefoot, supra* note 215.

that all the data that the expert would have wanted to obtain will be available in such circumstances is remote. This is true even when the data have been generated by another mental health professional, for example, in medical or mental health records, both because that person may not have asked all the questions that the testifying expert would have asked and because all of the person's responses may not have been fully recorded. The intangible aspects of an evaluation, including the person's relatedness, affect, and degree of cooperation, may be especially difficult to convey. Because many of the diagnostic categories require that other possibilities have been excluded first,[357] the absence of pertinent negative information (e.g., the person does not abuse substances) can restrict the ability to make definitive diagnoses. Moreover, to the extent that the data available to the expert have been shaped by someone with an interest in the outcome of the case, as when an expert testifies in sole reliance on information in a hypothetical question that is designed to mirror the defendant's or plaintiff's situation, these problems are compounded.

Thus, the major professional organizations in forensic mental health agree that evidence based on sources other than a direct evaluation of the person should be framed with due regard for its limitations and that those limitations should be made clear in reports or testimony by the expert. Failure to do so may represent unethical behavior on the part of the expert witness[358] and should probably cast doubt on the credibility of the evidence presented.

2. Did the evaluee cooperate with the assessment?

Even when a direct evaluation has taken place, the degree of cooperativeness of the person may affect the validity of the data obtained.[359] Civil plaintiffs and criminal defendants have obvious reasons to distrust experts who are examining them on behalf of adverse parties, and may be less than forthcoming in such interactions. However, even when an evaluation is being conducted by an expert hired by the person's own attorney, his or her cooperativeness may be limited by the symptoms of the disorder. For example, the person who is experiencing paranoid delusions may be suspicious and fearful even of an expert with whom his or her attorney encourages cooperation (indeed, even of the attorney). As a consequence, it is important for the expert to clarify, in the presentation of the evidence and

357. For example, DSM-IV-TR criteria for Major Depressive Episode require both that the symptoms on which a diagnosis is based not be due to the direct physiological effects of a drug (licit or illicit) that has been ingested or to a general medical condition; and that they not be better accounted for by a diagnosis of Bereavement after the death of a loved one. DSM-IV-TR at 356. Other major diagnostic categories carry similar requirements to rule out the possibility that the person's presentation is due to other causes before making the diagnosis in question.

358. For one highly publicized case of a psychiatric expert witness who was expelled from the American Psychiatric Association on these grounds, see Ron Rosenbaum, *supra* note 216; Estelle v. Smith, 451 U.S. 454 (1981).

359. Melton et al., *supra* note 28, at 46.

conclusions based on the evaluation, the extent to which the evaluee cooperated with the examination process.

3. *Was the evaluation conducted in adequate circumstances?*

Mental health evaluations often involve discussions of sensitive material, including histories of abuse, use of illegal substances, sexual practices, intimate fears and fantasies, and potentially embarrassing symptoms. Although some persons may be reluctant to speak freely about these issues with an evaluator whom they barely know—and who may reveal this information in the courtroom—the reassurance that they are talking with a mental health professional often substantially mitigates those concerns.[360] However, when the evaluation takes place in a setting that is less than private, the likelihood of such disclosures is reduced.[361] This is often a problem in correctional institutions, where interviews may take place where guards or other inmates can overhear them. Medical hospitals are another location where privacy may be compromised, with nursing staff or other patients nearby. Even if no one is within earshot, interview sites that are noisy or subject to other distractions may interfere with the evaluee's ability to attend to the questions and respond accurately; this can be a particular problem for people with mental disorders that may impair concentration and attention. Whenever possible, a competent evaluator tries to obtain a venue that is free of these intrusions, and when it is not possible, the situation should be noted as a limitation on the completeness of the evaluation in the report or testimony.

Attorneys sometimes ask to sit in on the evaluation. Their presence can raise similar concerns, even when they are representing the person being evaluated, because the type of information discussed in a mental health evaluation may be quite different from what a client usually discloses to an attorney.[362] Particularly when the examination is being conducted by an expert for an adverse party, attorneys may be tempted to object to questions or to signal the person regarding their answers. Thus, if an attorney is present, as will sometimes be unavoidable, the ground rules should include having the attorney sit out of the line of sight of the evaluee and not interrupt the examination. An alternative is to have the evaluation audiotaped or videotaped, a technique that some experts now use routinely. Empirical data on

360. Indeed, a considerable literature exists on the question of whether evaluees may too easily be induced to speak frankly with someone who is introduced as a mental health professional, but whose role is very different than would obtain in treatment settings and who may reach opinions adverse to the person's interests. See, e.g., Daniel Shuman, *The Use of Empathy in Forensic Evaluations*, 3 Ethics & Behav. 289 (1993); Strasburger et al., *supra* note 348; Greenberg & Shuman, *supra* note 348.

361. Melton et al., *supra* note 28, at 47. Distraction can be a particular problem when formal psychological tests are used; *see, e.g.*, Kirk Heilbrun, *The Role of Psychological Testing in Forensic Assessment*, 16 Law & Hum. Behav. 257 (1992).

362. Robert I. Simon, *"Three's a Crowd": The Presence of Third Parties During the Forensic Psychiatric Examination,* 27 J. Psychiatry & L. 3 (1999); Robert L. Goldstein, *Consequences of Surveillance of the Forensic Psychiatric Examination: An Overview*, 145 Am. J. Psychiatry 1234 (1988).

the impact of taping on evaluees' willingness to be forthcoming are lacking, but experienced forensic examiners have expressed the view that evaluees rapidly adjust to the recording equipment, with little impact on the evaluation.[363]

A final consideration is the time available for the examination.[364] Time constraints may result from correctional rules (e.g., prisoners are only available during given periods of time), medical illnesses or mental disorders (e.g., the evaluee has limited strength or attention), or limitations on resources (e.g., the party employing the expert only has funds for a certain number of hours of work). Appropriate duration of a direct examination is difficult to specify for all situations. It is likely to depend on the question being asked, the complexity of the person's history and presentation, and the person's degree of cooperation with the evaluation. Needless to say, the duration of an examination, standing alone, is not a good indicator either of its quality or of the validity of the conclusions that were drawn. However, an expert should be able to assess the time necessary to perform an adequate evaluation and, if sufficient time is not available, should indicate the limitations on the resulting opinions that are offered.

4. Were the appropriate records reviewed?

The importance for the evaluator of having access to the person's records will vary somewhat depending on the legal question being addressed, but can often be critical to the validity of the evaluation.[365] When retrospective assessments are being conducted—for example, an evaluation of a defendant's state of mind at the time of a crime that occurred months to years before the examination, or an assessment of a person's capacity to enter into a contract at some distant prior date—reviewing contemporary or nearly contemporary records can provide crucial insights into the person's symptoms and functioning at that time. However, even when contemporaneous function or future behavior is being assessed, having access to available records may still be of great importance. Because distinctions between mental disorders can depend in part on the pattern of symptoms over time, accurate diagnosis often is dependent on having a view of the person's prior psychiatric history.[366] In addition, when malingering is a consideration, as it will frequently be, the consistency of the person's presentation over time can be an important datum in the assessment.[367] And given that past behavior is generally the

363. AAPL Task Force, *Videotaping of Forensic Psychiatric Evaluations*, 27 J. Am. Acad. Psychiatry & L. 345 (1999).

364. Melton et al., *supra* note 28, at 47.

365. Kirk Heilbrun et al., *supra* note 354; see also discussion in Section I.C.3.f, *supra*.

366. Diagnosis and subcategorization of bipolar disorder, for example, is dependent not only on assessing the person's current symptoms—whether manic or depressed—but also on ascertaining whether mania or depression was present in the past if it is not apparent at present. *See* DSM-IV-TR at 388–89.

367. *See generally* Section I.C.5, *supra*.

best predictor of future behavior, especially where violence is concerned, knowledge of a person's previous history can be essential for predictions of reasonable accuracy.[368] Thus, regardless of the focus of the evaluation, an effort should be made to obtain all relevant available records.

Which records are relevant will depend somewhat on the nature of the legal question being asked.[369] Whenever possible, records of past mental health evaluations or treatment should be obtained. Medical records often contain information about patients' psychiatric symptoms, alcohol and drug use, and functional levels, and thus can be useful as well. Light can be shed on both patterns of symptoms and functional impairment by educational, work, and military records. Educational records may be especially helpful where disorders of early onset are suspected, and work and military records are often illuminating when occupational disability is at issue. In criminal cases, particularly those involving assessments of the defendant's state of mind at the time of the crime, police records can often be valuable, including interviews with witnesses or the defendant, and the results of physical evaluations—including pictures—of the crime scene. It can be helpful to compare the data obtained by these means with the defendant's own accounts of the episode that led to the arrest. Diaries or other accounts written by the person whose mental state is at issue are sometimes available and, to the extent that they were generated prior to the initiation of legal proceedings, can be enlightening regarding the person's state of mind and motivation, the influence of third parties, and the like. When there has been previous litigation involving the person being evaluated, depositions or transcripts of testimony can be helpful for information about state of mind and factual data.

5. Was information gathered from collateral informants?

In addition to reviewing records, interviewing informants with relevant data can provide important perspectives on the person being evaluated.[370] Family members and friends, including coworkers, often can report on patterns of behavior indicative of symptoms of mental disorder or of functional impairment. They may know about prior treatment for mental disorders, including hospitalization, or histories of involvement with the criminal justice system. Current or former therapists can share useful impressions of diagnosis and comment on levels of function, although to the extent that their interactions with the person are subsumed under a psychotherapist–patient or physician–patient privilege, and do not fall under one of the exceptions in that jurisdiction, it may not be possible to contact them without the person's consent. Witnesses to an alleged crime or workplace

368. *See generally* Section I.E.1.a, *supra*.

369. *See, e.g.*, Deborah Giori-Guarnieri et al., *AAPL Practice Guideline for Forensic Psychiatric Evaluation of Defendants Raising the Insanity Defense*, 30 J. Am. Acad. Psychiatry & L. 22 (Supplement) (2002).

370. Heilbrun et al., *supra* note 354.

harassment can similarly round out a picture of the person and help to confirm or disconfirm the evaluator's impressions. Access to collateral informants may be complicated by legal restrictions or, if they are close to the person being evaluated, by their reluctance to speak to an expert working for an adverse party. When contact does occur, the assessor needs to take into account possible distortions by the informant in the service of helping, or sometimes of harming, the interests of the person who is the subject of the evaluation.

6. *Were medical diagnostic tests performed?*

Dualistic views of human behavior, in which mind and body are seen as distinctly separate entities, have been rejected by scientists who study thought and behavior, and clinicians who treat mental disorders.[371] The relevant fields, including cognitive science, neuroscience, psychology, psychiatry, and philosophy, now acknowledge the brain as the seat of mentation and behavior, and recognize that all mental phenomena, including abnormal mental states, result from perturbations in the function of the brain. At some level, there must be a physical concomitant of every mental phenomenon, and sometimes the physical influences on abnormal behavior are gross enough to be detected by existing techniques, which may reveal potentially treatable conditions. Thus, identification of the causes of abnormal thought or behavior and formulation of a diagnosis may require an evaluation of the person's physical state, along with the mental state.[372] If there is any reason to suspect that an identifiable general medical disorder lies at the root of the person's condition (e.g., a sudden and unprecedented appearance of symptoms, disproportionate impairment of aspects of cognitive function), medical testing, including EEGs and imaging studies, may be indicated.[373]

371. *See, e.g.*, DSM-IV-TR, *supra*, at xxx, "the term *mental disorder* unfortunately implies a distinction between 'mental' and 'physical' disorders that is a reductionistic anachronism of mind-body dualism." *See also* Kenneth S. Kendler, *Toward a Philosophical Structure for Psychiatry*, 162 Am. J. Psychiatry 433 (2005).

372. *See generally* Section I.C.3.e, *supra*.

373. Identification of structural or electrical abnormalities, however, does not necessarily imply that they impaired the person's functioning or were responsible for the person's behavior. For discussion of a well-known case in which this issue was raised, see Stephen Morse, *Brain and Blame*. 84 Geo. L.J. 527 (1996). For a more general discussion of the introduction of findings of abnormalities demonstrated on brain imaging in court, see Dean Mobbs et al., *Law, Responsibility and the Brain*, 5 PLoS Biology 693 (2007). Moreover, as with structural findings, the mere presence of a functional abnormality is not sufficient to establish a causal link to the person's mentation or behavior. Growing legal and neuroscience literatures are being generated on the use of functional imaging data in court. *See, e.g.*, Neal Feigenson, *Brain Imaging and Courtroom Evidence: On the Admissibility and Persuasiveness of fMRI*, 2 Int'l J.L. Context 233 (2006); Hal S. Wortzel et al., *Forensic Applications of Cerebral Single Photon Emission Computed Tomography in Mild Traumatic Brain Injury*, 36 J. Am. Acad. Psychiatry & L. 310 (2008).

7. *Was the evaluee's functional impairment assessed directly?*

As previously discussed, mental health evidence will often focus on the extent to which a person is capable of performing a particular task or set of tasks, that is, testimony will relate to a person's impairment on one or more functional abilities.[374] Sometimes an evaluator will be able to infer from an examination of the person's mental state and information from other sources whether the person is or was capable of performing the task at hand (e.g., standing trial, returning to work, managing property). However, another option for evaluation exists, namely direct assessment of the relevant function.[375] Where a functional ability that relates to a discrete task or set of tasks is at issue, a competent evaluator should have considered direct assessment of performance on those tasks and be able to explain a decision not use such a technique. It should be noted, though, that conclusions drawn even from direct assessments of function involve a degree of inference. A person claiming occupational impairment as a result of anxiety induced by longstanding harassment on the job, for example, might respond very differently to the demands of a work-related task in the actual workplace compared with the safe confines of a mental health professional's office. Therefore, when actual observation of functional capacity is employed, the evaluator should be prepared to comment on the ecological validity of the test, that is, the degree to which the environment in which the test took place resembled the real-world environment in the person's life.[376] Although observations in very different settings may have some value as part of the broader dataset available in an evaluation, they do not carry the same weight as conclusions reached in environments similar to those at issue in the case.

8. *Was the possibility of malingering considered?*

In almost every mental health evaluation for legal purposes, the person being evaluated has an incentive to exaggerate or confabulate symptoms or to distort the impact of actual symptoms on his or her functional abilities.[377] Thus, the possibility of malingering should be considered by the evaluator in every assessment. Techniques for detecting malingering are described above.[378] Although such

374. *See generally* Section I.D, *supra*.

375. *See* Section I.D.2.b, *supra*.

376. Additional issues related to the use of functional tests are discussed in Section II.C, *infra*.

377. There are situations in which the incentive runs in the opposite direction. For example, a defendant facing relatively minor charges for whom an evaluation of competence to stand trial was ordered may have every reason to minimize his or her level of symptoms, preferring to go to trial rapidly rather than spend an extended period of time in a psychiatric facility being treated to restore competence. A second example is a defendant whose risk for violence is being evaluated prior to a bail hearing, who also has a powerful incentive to downplay the presence of risk factors associated with violence and to minimize a past history of violence.

378. *See* Section I.C.5, *supra*.

techniques are not foolproof, and well-prepared evaluees can sometimes mislead mental health professionals regarding the existence or severity of disorders, successful malingering over time is a difficult task. However, uncovering distortions of the degree of actual symptoms or exaggerations of their impact is usually more challenging than detecting wholesale invention of disorders that are not present. Competent evaluators should be able to explain how they took into account the possibility of malingering and why they believe that their conclusions are valid and to acknowledge that their degree of certainty can never be absolute.

C. Was a Structured Diagnostic or Functional Assessment Instrument or Test Used?

Notwithstanding the advantages of structured assessment techniques, they raise a set of concerns that must be addressed to determine their relevance to the question at issue and the weight that should be given to their results.

1. Has the reliability and validity of the instrument or test been established?

Reliability and validity are key concepts in test development.[379] Each contains several subcategories. Reliability refers to the reproducibility of results obtained with a particular test. That is, it is an estimate of the precision of an assessment technique. *Interrater reliability* is a measure of whether different examiners using the same test or instrument with the same subject come out with similar results, an important characteristic for an assessment approach that will be used by many raters. *Test-retest reliability* assesses the stability of results from an instrument or test over time; poor correspondence of results between time periods may indicate either an unreliable technique or a condition subject to periodic changes in status. It is an axiom of test and instrument development that good reliability is a prerequisite for having a valid assessment technique, but does not in itself guarantee validity.

Validity connotes the degree to which an instrument or test yields results that accurately reflect reality. *Construct validity* refers to the extent that an instrument or test reflects the theoretical construct that it purports to measure (e.g., anxiety or depression). Elements of construct validity include *discriminant validity*, which is the degree to which the test distinguishes between related conditions or states, and *convergent validity*, the extent to which the results of this test resemble results of other instruments that assess the same or a similar construct. *Content validity* describes the adequacy or thoroughness with which a test has sampled the variables associated with a given domain (e.g., does a measure of ability to work assess all relevant aspects of a given occupation?). Finally, *predictive validity* denotes

379. For the discussion in the following two paragraphs, see generally American Psychological Association, Standards for Educational and Psychological Testing (1999).

the ability of an instrument or test to foretell a person's condition or behavior at some point in the future.

When the results of an evaluation using an instrument or test are offered in evidence, clarification of the extent to which reliability and validity have been demonstrated is an essential aspect of determining admissibility and weight. Indeed, based on its discussion in *Daubert,* when the U.S. Supreme Court referred to the "reliability" of a scientific technique, it was encompassing both reliability and validity as usually understood in the social sciences.[380] Which aspects of reliability and validity are relevant to a particular case will depend on the purpose for which the data from the test are being introduced. For example, if the evidence is addressing change in a person's test results over time, a measure's test-retest reliability becomes crucial. If more than one evaluator was involved, interrater reliability may be key. Discriminant validity will be relevant when two states or conditions must be distinguished from each other and predictive validity when forecasts of future mental state or behavior are being made. Careful evaluators will only use instruments or tests that have had the relevant types of reliability and validity confirmed in peer-reviewed publications and will be prepared to cite such data should questions be raised. Of course, some tests are so widely used over a sustained period that their reliability and validity are generally accepted (e.g., the MMPI-2) and do not ordinarily need to be demonstrated again prior to introducing data based on an evaluation in which they were employed. However, the reliability and validity of some longstanding tests (e.g., the Rorschach ink-blot test) remain controversial,[381] and data even from established tests can be used to reach conclusions of uncertain validity. Thus, novel uses of instruments or tests may also require that their psychometric characteristics for that purpose be demonstrated.

2. Does the person being evaluated resemble the population for which the instrument or test was developed?

Reliability and validity once established are not necessarily universally applicable. If an assessment technique is being used on someone drawn from a different population than the one for which the instrument or test was developed, and the new group is likely to differ in some material way, reliability and/or validity may need to be reestablished. An example with regard to reliability might be the use with a child of an instrument that was developed to measure symptoms of mental disorders in adults.[382] Either the nature of the symptoms that adults experience or

380. Daubert v. Merrell Dow Pharms., Inc., 509 U.S. 579, 589 (1993).

381. Lilienfeld et al., *supra* note 127.

382. The frequently differing presentations of mental disorders in children have led to the development of instruments intended specifically for use in that population. *See, e.g.,* David Shaffer et al., *NIMH Diagnostic Interview Schedule for Children, Version IV (NIMH DISC-IV): Description, Differences from Previous Versions, and Reliability of Some Common Diagnoses,* 39 J. Am. Acad. Child & Adolescent Psychiatry 28 (2000).

the ability of adults to describe their symptoms could be substantially different with children, leading to greater difficulty in applying the instrument or test. Thus, it might be prudent for an evaluator to ascertain that data exist showing good reliability in this new population before using this assessment approach. An example involving validity is the use of predictive scales, such as instruments to assess risk of future violence, with a different group than the one from which the predictive algorithm was derived.[383] Concretely, if a predictive test is based on a criminal, but nonmentally disordered sample, applying it to persons with mental disorders—for whom very different variables may affect their behavior—is dubious in the absence of data demonstrating that it is valid in the latter group and vice versa.

It should be emphasized, however, that reestablishing reliability and validity is only necessary when the original group and the new population are likely to differ in some relevant way. Why an instrument developed in California, for example, would not be as reliable and valid when used in Texas is not at all clear. Moreover, the nature of the instrument or test will play a role. Diagnostic tests are likely to differ in their characteristics across populations only if the disorders or the ways in which they manifest themselves are different, which will not usually be the case. Predictive tests, however, may be more sensitive to cultural, socioeconomic, geographic, and other considerations that could introduce new predictors of future conditions or behaviors into the mix. In addition, tests that involve comparisons with broader populations are said to be "normed" against those groups,[384] and the comparative data (e.g., the evaluee is in the lowest quartile of performance) may be invalid unless the test is renormed for the group of which the person being evaluated is a member. Thus, whether additional reliability and validity testing is required for a new use, or whether a test must be renormed before being used in this way, is necessarily a fact-specific determination.

3. Was the instrument or test used as intended by its developers?

Established reliability and validity are necessary but not sufficient to determine whether an instrument or test has yielded reliable and valid results. Unless the assessment approach was applied in the manner intended by the developers, the data on reliability and validity may simply not be applicable to a particular use. Three possible areas of deviation relate to training in, administration of, and scoring of the assessment tool.

a. Training

Some instruments and tests are so straightforward in their use that little or no training is required. Reading the instructions accompanying the assessment tool might be

383. *See, e.g.,* John Monahan et al., *The Classification of Violence Risk*, 24 Behav. Sci. & L. 721 (2006).

384. For a good discussion of norming in the forensic context, *see* Grisso, *supra* note 2, at 56–59.

sufficient. In some cases, though, training may be required to ask the questions properly, especially when followup probing of responses is necessary or when evaluees are asked to perform tasks that must be conducted in a particular way. Diagnostic instruments, in particular, may have complex "skip-out" rules, that is, procedures for determining when to include or omit certain questions based on the person's responses to previous questions.[385] When information is acquired at least in part from existing records, rather than from the evaluee directly, rules may exist for how the information should be identified and abstracted. All of these characteristics of an assessment approach may require elaborate training for proper implementation.[386] Sometimes the training can be acquired from test manuals, but for more complex instruments or tests, face-to-face training with an opportunity to practice administration is necessary. Developers of such instruments or tests may offer such training in 1-day or multiday seminars that professionals can arrange to take.[387] Thus, a key question in assessing data based on an instrument or test is whether proper use requires special training, and if so, whether the assessor was trained in the technique.

b. Administration

Even if training was obtained, the reliability and validity of an instrument or test will depend on whether the assessor administered the test in the proper way. Many assessment tools require that questions be asked in a given sequence and that they be phrased in a particular way. After an incorrect response, it may be permissible to ask the question again, but only a certain number of times. Probing of responses may be needed, but only certain probes may be permitted. Some tests are timed, with a given period allotted for the completion of a particular task. Deviations from any of these requirements could make the published data on the psychometric characteristics of the tool inapplicable to its use in a particular instance. Thus, a second crucial question is whether the instrument or test was administered in the same way as it was when its reliability and validity were established.

c. Scoring

Assessment tools generally require that evaluees' responses be scored in some way. For some instruments and tests, the scoring is simple and self-evident, for example, the number of positive responses is totaled to yield the score for the test, or evaluees themselves are asked to indicate the severity of their symptoms on a

385. The Diagnostic Interview Schedule, which is widely used in epidemiological studies of mental disorders in the United States, is an example. See a description of the latest version of the instrument at http://epi.wustl.edu/CDISIV/dishome.aspx.

386. Indeed, some psychological and neuropsychological tests should be administered only by psychologists trained in their use.

387. The creator of the popular Psychopathy Check List (PCL-R), for example, offers an extensive training program for clinicians and researchers desiring to learn proper administration of the instrument. See the Web site at http://www.hare.org/training/.

1-to-7 scale. Or the results could be calculated by a computer program that automatically applies the relevant algorithm, generates statistical data, and even draws comparisons with broader groups, such as the general population or persons with a particular disorder. Often, however, particularly when evaluees' verbal or narrative responses are elicited, more complex scoring rules exist. An instrument assessing the severity of symptoms, for example, may require the person administering it to categorize responses along a numerical scale,[388] and specific capacity assessment tools frequently require similar judgments to be applied.[389] Published data on the reliability of scoring may indicate that it is possible for an instrument to be scored in the same way by many different raters, but unless the person administering the instrument in this particular circumstance adheres to the usual rules, the results of the evaluation may not be comparable to those that would be obtained by another rater and may be invalid as well. Hence, a third important question when such evidence is introduced deals with whether the rules for scoring responses were properly applied.

D. How Was the Expert's Judgment Reached Regarding the Legally Relevant Question?

In evaluating testimony from mental health experts, as noted in the preceding sections, their training and the manner in which they conduct their assessments is vital information. However, the value of an expert's opinion also depends on the process by which the data were assessed and a conclusion was reached.

1. Were the findings of the assessment applied appropriately to the question?

a. Were diagnostic and functional issues distinguished?

Mental health professionals without experience in performing particular forensic evaluations may fail to recognize that the legal question being asked deals with a person's functional capacity, not with some aspect of their clinical state per se.[390] As a result, they may mistakenly base their opinions on the presence of a particular diagnosis or symptom cluster rather than on the person's capacity to perform in the legally relevant manner. Studies over many years indicate that this has occurred frequently in testimony regarding defendants' competence to stand trial, in which experts often conflated the presence of psychosis with incompetence, and concluded that any psychotic defendant was ipso facto incapable of proceed-

388. *E.g.*, the Brief Psychiatric Rating Scale. *See* Overall & Gorham, *supra* note 122.

389. *E.g.*, the MacArthur Competence Assessment Tool for Treatment; Thomas Grisso & Paul S. Appelbaum, MacArthur Competence Assessment Tool for Treatment (MacCAT-T) (1998).

390. *See* Dusky v. United States, 362 U.S. 402 (1960); Thomas Grisso, Competency to Stand Trial Evaluations: A Manual for Practice 1–23 (1988).

ing to trial.[391] Similar problems may occur in hearings on guardianship or contests regarding testimonial capacity, where the person's ability to manage or dispose of assets might be thought incorrectly to turn solely on the clinical question of whether dementia is present, as opposed to the legal issue of whether the person retains the necessary capacities despite his or her condition.[392] This problem may be more likely to occur—and to go undetected—when experts are allowed or encouraged to address the ultimate legal issue in their testimony.[393] When experts are permitted to testify to the ultimate question, the importance of probing their reasoning is magnified.[394] Experts can be asked to identify the relevant functional capacities and to speak directly to the impact of the person's mental state on those capacities.[395] That allows their reasoning processes and the correctness of their assumptions about the relevant functional standard to be tested.

b. Were the limitations of the assessment and the conclusions acknowledged?

Most assessments are imperfect. Evaluees are less than cooperative. Records are unavailable. Evidence from witnesses is conflicting. Inadequate time is available. Or the evaluator may simply have forgotten to ask about some piece of information that would have been helpful. Experts should be able to identify the limitations of their evaluations, and the possible impact of those less-than-optimal aspects of the assessments. It is unlikely that an expert would be prepared to offer testimony if he or she believed that the limitations rendered the opinions invalid. But competent experts should be able to explain why, despite the limitations (which can occur even in the best evaluations by the most experienced experts), their evaluations were adequate to allow them to draw the conclusions that they intend to present.

A comparable set of limitations can occur when conclusions are drawn and opinions formulated. Just as all assessment tools have error rates, so do expert witnesses, although their rates are difficult to subject to statistical analysis. Errors may be introduced by inadequacies in the data available or the uncertainties inherent in particular determinations, especially predictions of future mental states and behaviors. As noted above, it is often impossible to specify the contingencies that may arise in a person's life that could influence their mental states and actions. Thus, any prediction, no matter how firmly grounded in available data, has a

391. *See, e.g.*, A. Louis McGarry, *Competence for Trial and Due Process Via the State Hospital*, 122 Am. J. Psychiatry 623 (1965). More recent studies suggest that this is now a less common problem, as educational efforts among mental health professionals who do such work have had a positive impact. Robert A. Nicholson & Karen E. Kugler, *Competent and Incompetent Criminal Defendants: A Quantitative Review of Comparative Research*, 109 Psychol. Bull. 355 (1991).

392. *See* Parry & Drogin, *supra* note 8, at 149–51.

393. *See* Section I.G.2, *supra*.

394. *See* Parry & Drogin, *supra* note 8, at 429–31.

395. Buchanan, *supra* note 319.

degree of uncertainty attached to it that a competent expert should be expected to acknowledge.

c. Are opinions based on valid empirical data rather than theoretical formulations?

From the development of Freud's theories in the late nineteenth and early twentieth centuries until the present, many mental health professionals have based their clinical approaches on psychoanalytically inspired concepts. Some of these concepts have been confirmed scientifically (e.g., the existence of unconscious mental states), whereas others have not (e.g., dreams always represent the fantasied fulfillment of wishes). Although psychoanalytical theories and the psychodynamic psychotherapies that derive from them have declined in popularity in recent decades, many mental health professionals have received psychodynamic training and use the concepts they have learned to assess and treat their patients. Regardless of the possible utility of these theories from a clinical perspective, which is controversial and may depend on the condition being treated, they are arguably more problematic when they serve as the basis for conclusions offered as part of legal proceedings. Nor are psychoanalytical theories the only ones that mental health professionals use; alternative approaches may be based on theories that have a greater or lesser degree of empirical support.

To the extent that expert opinions are introduced to inform the judgments of legal factfinders, it is important for them to be based, insofar as possible, on empirically validated conclusions rather than on untested or untestable theories. That appears to be the import of the U.S. Supreme Court's decision in *Kumho Tire*.[396] As Slobogin plausibly maintains, some legal questions (such as those concerning past mental states) may not easily lend themselves to approaches based on scientific methods, but expert opinions may nonetheless be of assistance to the finders of fact.[397] At a minimum, it would seem fair for an expert to indicate when that is the case, so that the factfinder can make an informed judgment about the appropriate degree of reliance to be had on that opinion. And when empirically tested approaches are available, it would appear to be incumbent on an expert to use them or to be prepared to explain why they were not employed.

396. Kumho Tire Co. v. Carmichael, 526 U.S. 137 (1999) (holding that the *Daubert* standard for admitting expert testimony also applies to nonscientists).

397. Christopher Slobogin, *supra* note 305.

III. Case Example

A. Facts of the Case

John, a 25-year-old Army veteran who saw combat in Iraq, had begun to have anomalous experiences in the 4 years since his discharge from active duty. At first, he believed that people were staring at him, though he was not sure why. Later, he came to the conclusion that they thought he was a drug addict or a criminal, ideas confirmed when he heard voices coming through the walls of his apartment, which he attributed to the neighbors, saying, "He's using drugs" and "He steals things." To avoid people's stares, John left his apartment less often, spending most of his time listening to loud music, which helped to drown out the voices. He also found that alcohol made it easier to ignore the voices, and began to drink up to a gallon of wine each day.

One evening when the voices were particularly loud and insistent, he began banging on the walls of his apartment and yelling that he would kill the neighbors if they did not stop talking about him. Thirty minutes later, the police arrived to take him to the local Department of Veterans Affairs (VA) hospital, where he was admitted to the psychiatric unit. Over the course of his hospitalization, he received antipsychotic medication and participated in group therapy. By the end of his hospital stay, although he still wondered whether people were staring at him oddly, he no longer heard people's voices making derogatory statements about him. He denied having thoughts of hurting himself and other people. When asked whether he would continue taking his medication and would attend outpatient sessions, he said he would. Fourteen days after admission, John was discharged to outpatient care.

Immediately after discharge, John stopped his medication, and he never saw his outpatient therapist. As he became more suspicious of his neighbors, he again began to hear them talking about him, and he resumed drinking several bottles of wine each day to deal with the situation. Three weeks after discharge, while he was on his way to the grocery store to pick up more wine, a passerby accidentally bumped into John. Reacting with fury, John pummeled the older man with his fists, then began beating him with a broomstick that he found on the sidewalk nearby. It took four people who lived in nearby buildings to pull John off his victim.

In the wake of the assault, the victim brought suit against the VA for negligence in John's treatment. The suit alleged that VA mental health staff should have known that John was dangerous as a result of his mental disorder and not fit for discharge. Damages were claimed as a result of physical injuries and the development of PTSD.

B. Testimony of the Plaintiff's Expert on Negligence

At trial, the plaintiff introduced testimony from a board-certified forensic psychiatrist, Dr. A, who was 20 years out of residency training and had not directly treated patients for the past 13 years. Dr. A had reviewed the medical records of John's treatment and the police records of the assault, but he had not examined John directly. On direct examination, he testified that John had a diagnosis of schizophrenia, with a number of risk factors for violence, including having killed enemy combatants in Iraq, excessive alcohol consumption, and delusions of persecution. It was Dr. A's opinion that the VA treatment team had failed to abide by the standard of care because they had not used a structured violence risk-assessment instrument to determine John's dangerousness. Moreover, although they had obtained a CT brain scan that had shown frontal lobe injury from an old automobile accident, the team had failed to recognize that this constituted an additional risk factor for violence. However, Dr. A believed that, even on the basis of the available information, at the time of hospital discharge it was reasonably foreseeable that John would be violent, and thus he should not have been allowed to leave the hospital.

C. Questions for Consideration

1. Given that Dr. A had devoted himself entirely to forensic evaluations and had not actually treated a patient for 13 years, should he have been considered qualified to offer opinions about whether John's evaluation and treatment had conformed to the standard of care?
2. How reliable were Dr. A's conclusions regarding John's diagnosis and likelihood of committing an act of violence, given that he did not examine John or speak directly to anyone who had been in contact with him, but relied solely on hospital and police records?
3. What information would be needed to determine whether the failure to use a structured violence risk-assessment tool should be considered evidence of negligence? What information would be needed to determine whether the alleged failure to recognize the relationship between CT evidence of frontal brain damage and the risk of violence should be considered evidence of negligence?
4. Is the assertion that John's violence was reasonably foreseeable sufficient to establish a prima facie case for the plaintiff? If not, what type of data should Dr. A have presented to support his testimony?

D. Testimony of the Plaintiff's Expert on Damages

A second expert, Dr. B, a clinical psychologist in general clinical practice, offered testimony on the mental health consequences of the assault. Dr. B had been treating the victim prior to the assault and had been seeing him weekly for cognitive behavioral therapy since the assault. She testified that the patient described having intrusive thoughts about the attack, nightmares, difficulty concentrating, and startle responses when people came near him without his having noticed them. He also felt overwhelming anxiety walking down the street where the attack had occurred. Dr. B diagnosed the victim as suffering from PTSD and had used a structured assessment tool to help make the diagnosis. On cross-examination, she admitted that she had only seen three or four cases of PTSD in her 5 years of practice and that the diagnosis was based entirely on the victim's report of his symptoms. Although she had not considered the possibility that the victim was malingering, she considered it very unlikely. Because of his symptoms, she concluded to a reasonable degree of psychological certainty that he was disabled from working in his job as a middle manager for a utility company. On cross-examination, she admitted that she did not know exactly what his job entailed and had not determined how each of his symptoms might interfere with his work—but she nonetheless believed that normal work performance was not possible given his condition.

E. *Questions for Consideration*

1. Should Dr. B be qualified as an expert with regard to the damages suffered by the plaintiff?
2. To what extent should the following considerations affect the weight given to Dr. B's testimony:
   a. Dr. B had been treating the plaintiff prior to the attack, and continued to treat him afterward.
   b. Dr. B has seen only three or four cases of PTSD in her practice.
   c. Dr. B's diagnosis was made on the basis of the patient's self-report, without corroboration from collateral informants, and she had not considered the possibility that he might be malingering.
3. What information regarding the structured assessment tool that was used in making the diagnosis of PTSD would be needed to determine whether the results of the assessment should be admissible?
4. Was an appropriate evaluation done with regard to the extent of the victim's work disability? If not, what additional information should have been obtained and by what means? Should the testimony as offered have been admissible?

References on Mental Health Diagnosis and Treatment

American Psychiatric Association, American Psychiatric Association Practice Guidelines for the Treatment of Psychiatric Disorders: Compendium 2006 (2006).

American Psychiatric Association, Diagnostic and Statistical Manual of Mental Disorders DSM-IV-TR (4th ed. Text Rev. 2000).

American Psychiatric Publishing Textbook of Clinical Psychiatry (Robert E. Hales et al. eds., 5th ed. 2008).

Kaplan and Sadock's Comprehensive Textbook of Psychiatry (Benjamin J. Sadock et al. eds, 9th ed. 2009).

Alan F. Schatzberg et al., Manual of Clinical Psychopharmacology (6th ed. 2007).

Stephen M. Stahl, Essential Psychopharmacology: The Prescriber's Guide (3d ed. 2009).

References on Mental Health and Law

Paul S. Appelbaum, *A Theory of Ethics for Forensic Psychiatry*, 25 J. Am. Acad. Psychiatry L. 233 (1997).

Paul S. Appelbaum & Thomas G. Gutheil, Clinical Handbook of Psychiatry and the Law (4th ed. 2007).

Deborah Giorgi-Guarnieri et al., *American Academy of Psychiatry and the Law Practice Guideline for Forensic Psychiatric Evaluation of Defendants Raising the Insanity Defense,* 30 J. Am. Acad. Psychiatry L. S1 (2002).

Thomas Grisso, Evaluating Competencies: Forensic Assessments and Instruments (2d ed. 2002).

Gisli H. Gudjonsson, The Psychology of Interrogation and Confessions (2003).

Glenn J. Larrabee, Forensic Neuropsychology: A Scientific Approach (2005).

Gary B. Melton et al., Psychological Evaluations for the Courts: A Handbook for Mental Health Professionals and Lawyers (3d ed. 2007).

Douglas Mossman et al., *American Academy of Psychiatry and the Law Practice Guideline for the Forensic Psychiatric Evaluation of Competence to Stand Trial,* 35 J. Am. Acad. Psychiatry L. S3 (2007).

Mental Disorder, Work Disability, and the Law (Richard J. Bonnie & John Monahan eds., 1997).

John Monahan, *The Scientific Status of Research on Clinical and Actuarial Predictions of Violence*, *in* Modern Scientific Evidence: The Law and Science of Expert Testimony (David L. Faigman et al. eds., 2007).

Michael L. Perlin, Mental Disability Law, Civil and Criminal (2d ed. 2002).

Retrospective Assessment of Mental States in Litigation: Predicting the Past (Robert I. Simon & Daniel W. Shuman eds., 2002).
Richard Rogers, Clinical Assessment of Malingering and Deception (3d ed. 2008).
Christopher Slobogin, Proving the Unprovable: The Role of Law, Science, and Speculation in Adjudicating Culpability and Dangerousness (2006).
Robert M. Wettstein, Treatment of Offenders with Mental Disorders (1998).

Reference Guide on Engineering

CHANNING R. ROBERTSON, JOHN E. MOALLI, AND DAVID L. BLACK

Channing R. Robertson, Ph.D., is Ruth G. and William K. Bowes Professor, School of Engineering, and Professor, Department of Chemical Engineering, Stanford University, Stanford, California.

John E. Moalli, Sc.D., is Group Vice President & Principal, Exponent, Menlo Park, California.

David L. Black, J.D., is Partner, Perkins Coie, Denver, Colorado.

CONTENTS

"Scientists investigate that which already is; Engineers create that which has never been."
Albert Einstein

I. What Is Engineering?

A. Thinking About Engineering and Science

Although this is a reference manual on *scientific* evidence, the Supreme Court in *Kumho Tire Co., Ltd. v. Carmichael*[1] extended the *Daubert v. Merrell Dow Pharmaceuticals, Inc.*[2] decision on admissibility of scientific evidence to encompass nonscientific expert testimony as well.[3] Put another way, experts not proffered as "scientists" also are held to the *Daubert* standard.[4] So then we might ask, who are these nonscience experts and where do they come from? Many emerge from the realm of engineering and hence the relevance of "engineering" or "technical" expert testimony to this manual.

The Court's distinction between these two kinds of expert testimony might suggest that there is a bright line dividing science and engineering. Indeed, a great deal has been written and discussed about this matter and arguments made for why science and engineering are either similar or different. It is a conversation that resonates among philosophers, historians, "scientists," "engineers," politicians, and lawyers. Apparently even Albert Einstein had a point of view on this issue as attested to by the above quotation. Perhaps this deceptively attractive dichotomy is best resolved by recognizing that at the end of the day engineering and science can be as different as they are alike.

There is no shortage of "sound bites" that attempt to categorize science from engineering and vice versa. Consider, for instance, the notion that engineering is nothing more than "applied science." This is a too often recited, simple and uninformed view and one that has long been discredited.[5] Indeed, it is not the case that science is only about knowing and experimentation, and that engineering is only about doing, designing, and building. These are false asymmetries that defy reality. The reality is that who is in science or who is in engineering or who is doing science or who is doing engineering are questions to be answered based on the merit of accomplishments and not on pedigree alone.

1. 526 U.S. (1999).
2. 509 U.S. 579 (1993).
3. *See* Margaret A. Berger, The Admissibility of Expert Testimony, in this manual.
4. *See* David Goodstein, How Science Works, in this manual, for a discussion of science and scientists.
5. Walter G. Vincenti, What Engineers Know and How They Know It (1990).

B. Engineering Disciplines and Fields of Practice

One can think of engineering in terms of its various disciplines as they relate to the academic enterprise and the names of departments or degrees with which they are associated, for instance electrical engineering or chemical engineering. One also can consider the technological context in which engineering is practiced as in the case of nanotechnology, aerospace engineering, biotechnology, green buildings, or clean energy.

In the same sense that some struggle trying to identify the differences and likenesses between science and engineering, others pursue a different kind of identity crisis by staking out their turf through title assignment. It is pointless to list titles of engineering disciplines because such a list would be incomplete and not stand the test of time as disciplines come and go, merge, diverge, and evolve. Bioengineering, biochemical engineering, molecular engineering, nanoengineering, and biomedical engineering are relative newcomers and have emerged in response to discoveries in the sciences that underlie biological and physiological processes. Software engineering and financial engineering are two other examples of disciplines that have developed in recent years.

In the end, it is not the names of disciplines that are critical, they being no more than labels. Names of disciplines are at best imprecise descriptors of the activities taking place within those disciplines and ought not to be relied on for accurate characterizations of pursuits that may or may not be occurring within them.

C. Cross-Disciplinary Domains

Whereas engineering disciplines are often associated with their scientific roots (i.e., mechanical engineering and physics, electrical engineering and physics, chemical engineering and chemistry, bioengineering and biology, biomedical engineering and physiology) some lack this kind of direct association (i.e., aerospace engineering, materials engineering, civil engineering, polymer engineering, marine engineering). Indeed, there are software engineers, hardware engineers, financial engineers, and management engineers. There is no shortage of adjectives here.

Nonetheless, these and many other such discipline titles have meant or mean something to someone, and new ones are emerging all the time as the historical barriers that once separated and defined the "classic" engineering disciplines continue to disintegrate and become a thing of the past. No longer can we rely on discipline names to inform us of specific enterprises and activities. There is, after all, nothing wrong with this as long as it is recognized that they ought not be used as reliable descriptors to subsume all possible activities that might be occurring within a domain. One must reach into a domain and investigate what kind of engineering is being conducted and resist the temptation to draw conclusions based on name only. Doing otherwise could easily lead to an unreliable and inaccurate characterization.

To provide a tangible example, consider cases involving personal injury in which central questions often revolve around the specifics of how a particular trauma occurred. In situations where proximate cause is an issue, the trier of fact can benefit from a thorough understanding of the mechanics that created an injury. The engineering and scientific communities are increasingly called on to provide expert testimony that can assist courts and juries in coming to this type of understanding. What qualifies an individual to offer expert opinions in this area is often a matter of dispute. As gatekeepers of admission of scientific evidence, courts are required to evaluate the qualifications of experts offering opinions regarding the physical mechanics of a particular injury. As pointed out earlier, however, this gatekeeping function should not rise and fall on whether a person is referred to or refers to himself or herself as a scientist or engineer.

Specifically, one cross-disciplinary domain deals with the study of injury mechanics, which spans the interface between mechanics and biology. The traditional role of the physician is the diagnosis (identification) of injuries and their treatment, not necessarily a detailed assessment of the physical forces and motions that created injuries during a specific event. The field of biomechanics (alternatively called biomechanical engineering) involves the application of mechanical principles to biological systems, and is well suited to answering questions pertaining to injury mechanics. Biomechanical engineers are trained in principles of mechanics (the branch of physics concerned with how physical bodies respond to forces and motion), and also have varying degrees of training or experience in the biological sciences relevant to their particular interest or expertise. This training or experience can take a variety of forms, including medical or biological coursework, clinical experience, study of real-world injury data, mechanical testing of human or animal tissue in the laboratory, studies of human volunteers in non-injurious environments, or computational modeling of injury-producing events.

Biomechanics by its very nature is diverse and multidisciplinary; therefore courts may encounter individuals being offered as biomechanical experts with seemingly disparate degrees or credentials. For example, qualified experts may have one or more advanced degrees in mechanical engineering, bioengineering, or related engineering fields, the basic sciences or even may have a medical degree. The court's role as gatekeeper requires an evaluation of an individual's specific training and experience that goes beyond academic degrees. In addition to academic degrees, practitioners in biomechanics may be further qualified by virtue of laboratory research experience in the testing of biological tissues or human surrogates (including anthropomorphic test devices, or "crash-test dummies"), experience in the reconstruction of real-world injury events, or experience in computer modeling of human motion or tissue mechanics. A record of technical publications in the peer-reviewed biomechanical literature will often support these experiences. Such an expert would rely on medical records to obtain information regarding clinical diagnoses, and would rely on engineering and physics training to understand the mechanics of the specific event that created the injuries. A practitioner whose experience spans the interface between mechanics (i.e., engineering) and biology (i.e., science), considered in the context of the facts of a particular case, can be of significant assistance in answering questions pertaining to injury mechanism and causation.

This example illustrates the futility of trying to untangle engineering from science and vice versa and to the inappropriateness of using semantics, dictionary definitions, or labels (i.e., degree names) to parse, dissect, or portray the intellectual activities of an expert witness. In the end, it is their background and experience that are the dominant defining factors—not whether they are a scientist and/or an engineer and not by the titles they hold.

II. How Do Engineers Think?

A. Problem Identification

Although a somewhat overworked part of our lexicon, it is indeed the case that "necessity is the mother of invention." Engineering breeds a culture of technological responsiveness. All the "science" explaining a solution to a problem need not be known before an engineer can solve a problem.

Take steam engines, for example. Their history goes back several thousand years and their utility forged the beginning of the industrial revolution late in the seventeenth century. It was not until the middle of the nineteenth century that the science of thermodynamics began to gain a firm ground and offer explanations for the how and why of steam power.[6] In this instance, technology came first—science second. This, of course, is not always the case, but demonstrates that one does not necessarily precede the other and notions otherwise ought to be discarded. So here the problem was one of wanting to produce mechanical motions from a heat source, and engineers designed and built systems that did this even though the science base was essentially nonexistent.

To reinforce the point that technology can precede science, consider the design of the shape of aircraft wings. This, of course, was driven by the desire of humans to fly, a problem already solved in nature since the time of the dinosaurs but one that had eluded humankind for tens of thousands of years. Practical solutions to this problem began to emerge with the Wright brothers' first motive-powered flight and continued into the twentieth century before the "science" of fluid flow over wing structures had been fully elucidated. Once that happened, wings could be designed to reduce drag and increase lift using a set of "first principles" rather than relying solely on the results of empirical testing in wind tunnels and prototype aircraft.[7]

6. Pierre Perrot, A to Z of Thermodynamics (1998).

7. The pioneering aerodynamicist Walter Vincenti provides a detailed and fascinating account of this. *See* Vincenti, *supra* note 5, ch. 2; *see also* John D. Anderson, *Ludwig Prandtl's Boundary Layer*, Physics Today, December 2005, at 42–48.

So, in short, engineers create, design, and construct because interesting and challenging problems arise in the course of human events and emergent societal needs. Whether a science base exists or only partially exists is just one of a myriad of constraints that shapes the process. Other constraints might include, but are not limited to, the availability of materials; device shape, size, and/or weight; cost; demand; efficiency; safety; robustness; and utility. It has been said, and possibly overstated, but it does make the point, that if engineers waited until scientists completed their work, they might well still be starting fires with flint stones.

B. Solution Paradigms

So when faced with a vexing and challenging problem, along with its particular or peculiar constraints, an engineer seeks a path to follow that has a reasonable chance of leading to a solution. In so doing an engineer must contend with uncertainty and be comfortable with it. In very few instances will everything be known that is required to proceed with a project. Assumptions need to be made and here it is critical that the engineer understand the difference between what is incidental and what is essential. There are excellent assumptions, good assumptions, fair assumptions, poor assumptions, and very bad assumptions. Along this spectrum the engineer must carefully pick and choose to make those assumptions that ensure the robustness, safety, and utility of a design without undue compromise. This is the sort of wisdom that comes from experience and is not often well honed in the novice engineer.

This impreciseness that accompanies uncertainty can be used as a perceived disadvantage for the engineer in the role of expert witness. Yet it is this very uncertainty that lies at the heart of technological innovation and is not to be viewed as so much a weakness as it is a strength. To overcome uncertainty in design under the burden of constraints is the hallmark of great design, and although subtle and not always well understood by those who seek precision (i.e., why can't you define your error rate?), this is the way the world works and one must accept it for what it is. Assumptions and approximations are key elements of the engineering enterprise and must be regarded as such. And as with all things, hindsight might suggest that a particular assumption or approximation was not appropriate. Even so, given what was known, it may well have been the right thing to do at the time it was made.

In addition to evolving business opportunities and changing financial markets, technological innovation results from the continuing and many times unexpected advances in science and technology that occur as time passes. Buildings constructed in Los Angeles in the 1940s would never be built there in the same way now. We have a much better understanding of earthquakes and the forces they exert on structures now than then. Airbags were not placed in automobiles until recently because we did not have cost-effective systems and materials in place to accurately measure deceleration and acceleration forces, trigger explosives, contain the explosion, and do this on a timescale that was effective without harming an occupant more-so than the impending collision. It is unavoidable that as we learn from new discoveries about the natural world and accumulate more experience with our designed systems, products, and infrastructure, engineers will be in an increasingly better place to move forward with improved and new designs. It is both an evolutionary and a revolutionary process, one that produces both failures and successes.

III. How Do Engineers Make Things?

A. The Design Process—How Engineers Use This Guiding Principle

The genesis of nearly every object, thing, or environment conceived by engineers is the design process. Surprisingly, although products designed using it can be incredibly complex, the general tenets of the design process are relatively simple, and are illustrated in Figure 1.

The progression is iterative from two perspectives: (1) Changes in the design resulting from testing and validation lead to new formulations that are retested. (2) After the design is complete, performance data from the field can also lead to design changes.

As a first step, engineers begin with a concept—an idea that addresses a need, concern, or function desired by society. The concept is refined through research, appropriate goals and constraints are identified, and one or more prototypes are constructed. Although confined to a sentence here, this stage can take a significant amount of time to complete.

In the next phase of the design process, the prototypes are tested and evaluated against the design requirements, and refinements, perhaps even significant changes, are made. The process is iterative, as faults identified during the testing phase manifest themselves as changes in the concept, and the testing and evaluation process is restarted after having been reset to a higher point on the learning curve. As knowledge is gained with each iteration, the design progresses and is eventually validated, although as alternative solutions are considered, it is possible that certain undesirable characteristics in the design cannot be completely mitigated through changes in design and should be guarded against to minimize their impact on safety or other constraints. A classic example of this step in the design process is the installation of a protective shield over the blade in a table saw; although the saw may have the unwanted characteristic of cutting fingers or arms, the blade clearly cannot be eliminated (designed out) in a functioning product. As a last resort, anomalies that cannot be designed out or guarded against can be addressed through warnings. Not every design is amenable to guarding or warning, but instead the iterative process of testing and prototype revision is relied

Figure 1. Schematic of the engineering design process.

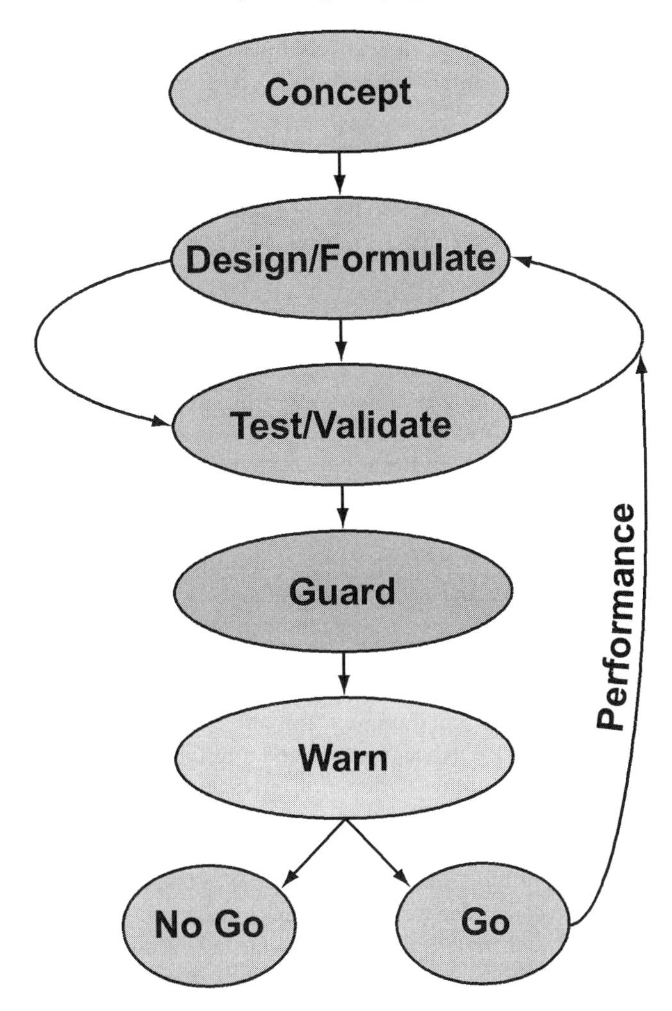

upon to perfect designs. Indeed, in some instances, an acceptable design solution cannot be found and the work is abandoned.

The testing process itself can be complex, ranging from simple evaluations to examine a certain characteristic to multifaceted procedures that evaluate the prototype in conditions it is anticipated to see in the real world. The latter type of evaluation is often denoted as *end-use testing*, and is very effective in identifying faults in the prototype. Because many designs cannot be evaluated over their anticipated life cycle because of time constraints (a product expected to last for 20 years cannot be tested for 20 years in the development process), the testing cycle is often accelerated. For example, if it is known that a pressure vessel will see 50,000 cycles over a 10-year lifetime, those cycles can be performed in several months and the resultant effects on vessel performance established. Another method of accelerating the evaluation cycle involves testing at an elevated temperature and using scientific theory and principles to equate the temperature increase to a timescale reduction. The efficacy of this approach is highly dependent on correct execution, but done properly and with appropriate care, it allows product development to go forward rather than having good or even great designs languish on the drawing boards because there is no feasible way to validate them under the exact end-use environment.

Regulations, standards, and guidelines also play an important role in testing of products during the design process. Federal requirements are imposed on design and testing of aircraft, medical devices, and motor vehicles, for example, and mostly govern how those products are evaluated by engineers. Standards organizations such as the American Society for Testing and Materials (ASTM), the American National Standards Institute (ANSI), and the European Committee for Standardization (CEN) promulgate test methods and associated performance requirements for a large number of objects and materials, and are relied on by engineers as they evaluate their designs. It is critical to understand, however, that ASTM, ANSI, CEN, and other such national and international standards organizations describe testing methods that engineers use to obtain reliable data about either the products they are evaluating (or components thereof), but most often they do not in and of themselves provide a means to evaluate a finished product in its actual end-use environment. It is also important to understand the difference between a performance standard and a testing standard—the former actually specifies values (strength, ductility, environmental resistance) that a product must achieve, whereas the latter simply describes how a test to measure a parameter should be conducted. It is the engineer's job to use the correct testing procedures from those that have been approved and on which he or she can rely. Or, alternatively, if no approved test exists, the engineer must create one that is reproducible, repeatable, reliable, and efficacious. Furthermore, it is the engineer's job to ensure the relevance of such testing to the overall and final product performance in its end-use environment. No testing or standards organization can foresee, nor do they claim to do so, all possible combinations of product components, design choices, and functional end-use requirements. Therefore, testing of a design in accordance with a testing standard does not necessarily validate the design, nor does it necessarily mean that the design will function in its end-use environment.

After testing and validation are complete, and the product is introduced to the market, the design process is still not finished. As field experience is gained, and products are used by consumers and sometimes returned to the manufacturer, engineers often fine-tune and perfect designs based on newly acquired data. In this part of the design process, engineers will analyze failures and performance problems in products returned from the field, and adjust product parameters appropriately.[8] The process of continual product improvement, illustrated by an arrow from the "Go" stage to the "Design/Formulate" and "Test/Validate" stages in Figure 1, is taught to engineers as a method to effectively optimize designs. Such refinements of product design are often the topic of inquiry in depositions of engineers and others involved in product design, and frequently misunderstood as an indication that the initial design was defective.[9] The engineering design process anticipates review and ongoing refinement of product design as a means of developing better and safer products. In fact, retrospective product modification is mandated as company practice in some industries, and regulated or suggested by the government in others. For example, examination of FDA guidelines for medical device design will show a process that mirrors the one described above.

Another important component of the design process relates to changes in technology that render a design, design feature, or even tools used by an engineer obsolete. Engineers consider obsolescence to be a consequence of advancement, and readily adjust designs, or create new designs, as new technology becomes available. This concept is apparent in the automotive industry, where tremendous advances in restraint systems and impact protection have greatly reduced the risk of fatal injuries from driving (see discussion below). Although vehicles with lap belts as the sole means of occupant protection would today be considered unacceptable, they were by no means deficient when introduced in the 1950s. From the engineer's perspective, errors and omissions in the design process can render a design defective; however, changes in technology can render a design obsolete, not retrospectively defective.

Of course even well-designed products can fail, especially if they are not manufactured or used in the manner intended by the design engineer. For example, a steel structure may be adequately designed, but if the welds holding it together are not properly made, the structure can fail. Similarly, a well-designed plastic component manufactured in such a way as to overheat and degrade its constituents may also be prone to premature failure. In terms of misuse of a product, most engineers are trained to consider foreseeable misuse as part of the design process, and one can generally expect to encounter a debate over what is reasonably foreseeable and what is not.

8. Although feedback on product performance and failure analysis on returned products is most often used to perfect designs, the iterative nature of the process can also cause the design to progress toward failure when cost becomes the driving factor.

9. Although the reasons for subsequent refinements in product design may be explored in depositions, Federal Rule of Evidence 407 bars the introduction of evidence of such improvements at trial as evidence of a defect in a product or a product's design.

B. The Design Process—How Engineers Think About Safety and Risk in Design

Almost everything that an engineer designs involves some aspect of safety, and the elegance and efficiency of designs are often forced to balance safety with competing parameters such as cost and physical constraints. The legal dilemmas that often arise from this balance are a direct result of the way an engineer must deal with safety in the reality of the engineering world (i.e., assertions that safety must be considered over everything else or that a particular design should or could be safer). Therefore, a discussion of how safety factors into design, and "how safe is safe enough" is prudent for an understanding of engineering and engineering design. It is critical that the reader note that in the framework of this discussion, risk is something engineers constantly face, and while we discuss what levels of risk are acceptable, the context is clearly engineering, and no legal construct is intended.

There is practically no product that cannot be made safer by reducing the product benefits (making it more inconvenient) or increasing the product cost, or both. In product design, safety is just one of the many variables factored into the design, as also is cost, and often safety and cost trade off directly on the product price point. In product design there are rarely instances where small cost changes render a substantial improvement in the risks. Safety always has a cost; the question is whether the consumer will find it reasonable in the face of what else the design has to offer. Conversely, the claim that the product is as safe as possible is almost never true either.

The simple and completely correct answer to the question "How safe is safe enough?" is "It depends." Exactly what safety is, and what conditions determine its adequacy, that is, what adequate safety depends on, are the topics briefly discussed in the following sections.

1. What is meant by "safe"?

Few words are used more often in the context of a product liability tort than the words "safe" and "unsafe," and their close cousin, "defective." Because the word "safe" is commonly used in so many different contexts, it is seldom, if ever, used with precision. Indeed, its common use has given it a number of meanings, some of which are in conflict.

Intuitively we understand the word and have a grasp of what a speaker probably means when declaring a product or environment "safe." We have to say "probably" because some would mean by a "safe" product one that presents no risk to the user under normal circumstances, and others would mean no risk to the user under any circumstances. Still others who ask the question "how safe is safe enough?" clearly evidence an understanding that safety is a continuum and not an absolute. Although "safe" is a simple word, it is used in so many ways that rigorous definition presents much of the complexity of other deceptively simple but widely used four letter words, for example, "good."

Fortunately, there is a whole field of scholarship, science, and technology related to the study of "safety." The field was spawned during the industrial revolution, when it came to be recognized that preventable industrial accidents were simply economically, if not morally, unacceptable.[10] For the remainder of this discussion, we examine the concept of safety as it relates to the possibility of physical harm to persons.

Safety is technically defined, and empirically measured, by the concept of "risk." And often a speaker who declares a product or environment "safe" does indeed mean to say that the product or environment is risk-free. However, as we will discuss in more detail, there is no product or product environment that attains the ideal status of "risk-free."[11] Every product manufactured by man, with his imperfections, and every environment, no matter how carefully constructed, presents some risk in its use, even if this risk is extremely small. This fact of life is easily illustrated.

For example, the U.S. Consumer Product Safety Commission (CPSC) estimates that nationally in the year 2007 alone, there were approximately 42,000 injuries serious enough to require treatment at a hospital emergency room associated with the use, and more often the misuse, of first-aid equipment. Thousands of these injuries were associated with the use of first-aid kits. The CPSC maintains the National Electronic Injury Surveillance System (NEISS), which monitors a statistically selected sample of all the emergency rooms in the United States, so that data collected on each consumer injury associated with the categories of consumer products that fall under the jurisdiction of the CPSC can be extrapolated to a national estimate.

It is not immediately obvious how so many injuries could be associated with first-aid equipment. And this reaction is an excellent lesson about the reliability of intuition for determining risk.[12] One soon learns that the cotton swabs in a first-aid kit can puncture eardrums; the ointments, pills, and antibiotic creams can be ingested by infants; the ice packs can cause thermal burns to the skin; and the cotton can become lodged in all sorts of unintended places.

With the understanding that there are no risk-free products, then we have no choice but to define safe in terms of the amount of risk. Of course with "risk" defining "safe," the task of defining "safety," or "safe enough," has been replaced

10. The rigorously scientific portion of this field is a product of the past 50 years. Although it has no single father, the seminal contributions of Dr. Chauncey Starr, ultimately recognized through his receipt in 1990 of the National Medal of Technology from President George H.W. Bush, deserve mention. http://www.rpi.edu/about/hof/starr.html.

11. S.C. Black & F. Niehaus, *How Safe Is "Too" Safe*, 22 IAEA Bull. 1 (1980); Water Quality: Guidelines, Standards and Health 207 (Lorna Fewtrell & Jamie Bartram eds., 2001).

12. The emergency room treatment records of CPSC's NEISS can be retrieved by anyone online at the CPSC Web site, www.cpsc.gov.

with the task of defining "risk" or "acceptable risk." This then is the definition of safe; something is "safe" when it presents "acceptable risk"[13] (the reader is again reminded that we are discussing an engineering, not legal, construct).

2. What is meant by "risk"?

"Risk" is another of those deceptively simple four-letter words that society uses in a wide variety of ways. Perhaps a few decades ago, when the field was developing its rigorous intellectual underpinnings, a term other than "risk" could have been chosen. But now there is a Harvard Center for Risk Analysis,[14] a Food Safety Risk Analysis Clearinghouse at the University of Maryland,[15] and many other academically oriented risk analysis organizations too numerous to name. There is a convenient Web site, Risk World, that lists many of the other Web sites that reference risk analysis.[16] Risk as the technical term for safety is now too institutionalized to be changed, and for the remainder of this discussion we are concerned with safety as the risk of physical harm, that is, health or safety risk.

The concept of risk is slightly more complicated and significantly more rigorous than the concept of safety. Again we have an intuitive understanding of the concept of "risk," and that it involves some concept of probability, more specifically the probability of some "bad" thing. In the case of safety the "bad" thing is injury or physical harm.

Risk is often empirically measured and expressed quantitatively, and a "risk" number always contains units of frequency (or probability) and severity. This is a substantial advantage over the concept of "safe." It would make no sense to say, "this product was found to be 2.73 safe." Risk on the other hand is the measure of safety. For example, the fatal risk of driving in the United States in 2007 was 1.36 fatalities for every hundred million vehicle-miles traveled (note: 100 million = 100,000,000 = 10^8).[17] This is a risk number because it contains a severity, "fatal," and the frequency, per every 10^8 miles. This fatality risk is not a complete measure of the safety or risk of U.S. vehicular travel in 2007 because the same 10^8 vehicle miles traveled that produced the 1.36 fatalities also produced 82 injuries; "injuries" as defined by the National Highway Traffic Safety Administration (NHTSA).[18]

13. International Organization for Standardization & International Electrotechnical Commission, *ISO/IEC Guide 51: Safety Aspects—Guidelines for Their Inclusion in Standards* (2d ed. 1999); William W. Lowrance, Of Acceptable Risk: Science and the Determination of Safety 8 (1976); Fred A. Manuele, On the Practice of Safety 58 (3d ed. 2003); National Safety Council, Accident Prevention Manual—Engineering & Technology 6 (Philip Hagan et al. eds., 12th ed. 2000).

14. http://www.hcra.harvard.edu/.

15. http://www.foodriskclearinghouse.umd.edu/.

16. http://www.riskworld.com/websites/webfiles/ws5aa013.htm.

17. National Highway Traffic Safety Administration, Motor Vehicle Traffic Crash Fatality Counts and Estimates of People Injured for 2007, DOT HS 811 034 (Sept. 2008, updated Feb. 2009) (hereinafter "NHTSA, Motor Vehicle Traffic Crash Fatality Counts").

18. *Id.*, slide 9.

These two different risk metrics, one for injuries and one for fatalities, naturally invite the question of a single metric that characterizes the risk, and therefore safety, of highway travel in the United States.

Sadly, the answer is that the single metric does not exist. For decades the risk analysis community has worked on developing some calculus through which injuries of differing severity could be rigorously combined and expressed as a defensible "average" severity. Some safety data are collected in a form that naturally lends itself to this exercise. When occupational injury severity is characterized by a "lost workday" metric, that is, the more severe injuries obviously result in more lost days of work, then the average number of lost work days is a defensible average severity with which one can characterize a population of occupational injuries. But this exercise quickly breaks down in the face of permanently disabling occupational injuries and deaths. Obviously, one could impute an entire career's worth of lost workdays in the case of fatal injury or permanent injury, but then these injuries would completely overwhelm all other types of occupational injury. And the issue of whether a permanently disabling injury is really of the same severity as a fatal injury remains unresolved.

Similarly the CPSC attempted soon after its creation in the early 1970s, to develop a geometric sliding scale to numerically categorize the differing consumer product–associated injury severities being treated in the hospital emergency rooms that the agency monitored. The CPSC scale had six to eight severity categories over the years, to which numerical weights were applied, ranging from 10 for severity category 1, mild injuries and sprains, to 34,721 for severity category 8, all deaths, in its original configuration. The weighting for deaths has changed and has been as low as 2516. An amputation was accorded a weight of 340, and fell into category 6, unless it resulted in hospitalization, at which time it became a category 7 with a weight of 2516. In the end, this scheme has proved generally unsatisfactory, but it still appears in the occasional CPSC document, and is used to generate a "mean severity" for emergency room–treated injuries.[19] Even if somehow a calculus for comparing and combining various injury severities could be developed, the challenge of how to compare the risk of differing injury frequencies at different severity levels would remain. There is practically no chance that the relationship would be linear, and the nonlinear characteristics would be highly subjective.

Instead of trying to develop a calculus to combine severities with differing frequency, it has become the custom and practice in the risk analysis community to express risk frequency or probability by stratified severity. That is, if a level of severity is specified, then the risk likelihood is stated. There is no agreement on the proper stratification, but rather a de facto consensus that fatal injuries are the most severe, and fatality risk is commonly measured. In addition, calculations of accident risk with no injury, injury risk, and hospitalized injury risk are often seen

19. U.S. Consumer Product Safety Commission, 1995 Annual Report to Congress A-5 (1995).

in the risk literature. Rather than being combined into a single metric, these risks are expressed as independent risk frequencies or probabilities. Specialized average severities, such as average number of lost workdays as a risk metric for the severity of average occupational injury, are occasionally used.

The upshot of our inability to develop a severity calculus means that risk metrics cease to be parameters with units of both frequency and severity, and become merely frequencies or likelihood of an injury of a given severity. Frequencies are easier to compute and merely require what is called "numerator" data (i.e., the actual number of adverse events for which the risk is being calculated) and "denominator" data (i.e., some measure of the opportunity to have the adverse event). In the previously cited 1.36 fatal injuries per 10^8 miles of vehicle travel, the numerator datum was the 41,059 deaths in 2007 traffic accidents and the denominator datum was the 3,029,822 million vehicles miles traveled by all vehicles in 2007.[20] The division of these two numbers gives 1.36 deaths per 10^8 miles. Vehicle-miles traveled (VMT) is one obvious measure of the opportunity to have a vehicular accident. However, it is not the only measure. If the data are available, vehicle hours can be substituted for vehicle miles, and then the fatal risk can be expressed as a frequency per vehicle hour. Measures such as miles and hours are often called "exposure" data, and must be some empirical measure of the opportunity to encounter the hazard (the adverse event itself) for which the risk is being calculated. The "correct" exposure measure is usually determined by the analysis being performed. Miles is appropriate for on-road vehicles, because travel is what the automotive products are intended to produce. For off-road recreational vehicles, where recreation as opposed to travel is the purpose of the product, hours of use would probably be a more appropriate exposure measure.

3. Risk metric calculation assumptions

Having determined that the fatality risk of driving in the United States is 1.36 deaths per 10^8 VMT, does that mean one's risk of dying in a traffic accident is 1.36×10^{-8} every time a mile is driven? No. With the danger of presenting too much detail, we can use this one risk parameter as a tool to briefly illustrate that many assumptions are inherent in any risk calculation, and that questions should arise in the court's mind when encountering any number that purports to represent the "risk" and therefore "safety" of any product or activity.

First, the number 1.36 is the gross fatality risk for vehicular travel in the United States. It is the risk we as a society de facto accept for the benefits of vehicular travel. Some of those deaths are pedestrians, motorcyclists, passengers, and bicyclists, and their deaths are part of the risk society must accept to have motorized vehicular travel. But, the fatal risk to you as a driver, by your "exposure" driving a mile, clearly does not involve any pedestrian risk or bicycle risk

20. NHTSA, Motor Vehicle Traffic Crash Fatality Counts, *supra* note 18, slide 40.

or motorcycle risk or vehicle passenger risk. Thus, fatally injured pedestrians (4654),[21] pedal cyclists (698),[22] motorcyclists (5154),[23] vehicle passengers (8657),[24] and others (147 skate boarders, etc.)[25] have to be subtracted from the 41,059 traffic deaths in 2007 numerator datum, to compute a "fatal risk of you being a driver" number, because none of them was driving a car. That leaves us with 21,647[26] fatally injured vehicle drivers in 2007. Because all of the vehicles had to have a driver to go even a mile, it might be tempting to just use the 3,029,822 million vehicle miles number in 2007 as the denominator without adjustment. But, to be accurate, the motorcycle operators were "vehicle" drivers, and so we cannot remove their 5154 deaths from the numerator without removing the approximately 13,610 million[27] vehicle miles those motorcycles were driven from the denominator datum. Because the motorcycle operator fatal injury risk per mile is 37.86 per 10^8 VMT,[28] more than 52 times that of an automobile driver, removing the motorcycle data entirely when trying to compute an automotive risk number is sound. If we do the appropriate adjustments, then we compute a fatality risk for a nonmotorcycle vehicle driver of 0.718 deaths per 10^8 VMT.

Now, we can again ask the question, "Is 0.718×10^{-8} one's risk of being killed every time a mile is driven?" The answer is now "Possibly, but unlikely." This risk number is the composite risk for all drivers in society for 2007. And, because of lifestyle choices, this number might serendipitously be accurate for some but not for everyone. Every driver has significant control over the majority of his or her risk of being killed on the road. For example, 33.6% of the fatally injured drivers, 7283, almost exactly one-third, had blood alcohol levels at or above 0.08 g/dL.[29] Exactly how much a blood alcohol level of 0.08 increases one's risk of dying per mile driven is a topic of some debate, but the consensus would fall somewhere between 3 and 5 times. You are much more likely to be killed if you drive on the weekends during the early morning hours. Even restricting ourselves to passenger vehicle fatalities in the daytime, when 82% of vehicle occupants wear their seat belts, 45% of the drivers killed in the daytime were unrestrained by their seat belts.[30] Numerous other decisions that we make concerning our driving or circumstances that affect us, such as the size of the car we drive, cell phone use,

21. *Id.*

22. National Highway Traffic Safety Administration, Traffic Safety Facts 2007 Data: Pedestrians, DOT HS 810 994, at 3 (hereinafter "NHTSA, Pedestrians").

23. National Highway Traffic Safety Administration, Traffic Safety Facts 2007 Data: Motorcycles, DOT HS 810 990, at 1 (hereinafter "NHTSA, Motorcycles").

24. NHTSA, Motor Vehicle Traffic Crash Fatality Counts, *supra* note 18, slides 52, 74, 85.

25. *Id.*, slide 9.

26. *Id.*, slide 40.

27. *Id.*

28. NHTSA, Motorcycles, *supra* note 23, at 1.

29. National Highway Traffic Safety Administration, Traffic Safety Facts: 2007 Traffic Safety Annual Assessment—Alcohol-Impaired Driving Fatalities, DOT HS 811 016, at 2 (Aug. 2008).

30. NHTSA, Pedestrians, *supra* note 22, at 3.

regard for yellow lights, aggressiveness, medication, vision correction, etc., may contribute in some way to the likelihood that we will be fatally injured driving the next mile, but are beyond the scope of this brief discussion.

4. Risk metric evaluation

With some understanding of what comprises the calculation of a risk metric, we can now turn to the more important questions related to its meaning. A fair question about the vehicular risk we just examined might be: "Is a fatal motor vehicle risk of 1.36 × 10^8 VMT good or bad?" Should society be ashamed or proud? This question for vehicle safety, and every other arena of risk analysis, can only be answered comparatively. The only absolute risk standard is "zero," but this ideal can never be achieved. So, to answer the question of how "good" the 1.36 number is, we can look to several comparisons. A logical starting point might be previous years; are we getting better or worse? Fortunately, with a few singular exceptions (such as motorcycles), everything is getting safer, and has been for the past 100 years. Although 1.36 people dead for every 10^8 VMT is surely not desirable, in 1966, that same number was over 5.[31]

The data in Table 1 are for the previous decade:[32]

Table 1. Fatalities per 100 Million Vehicle Miles Traveled

Year	Fatalities per 10^8 VMT
1996	1.69
1997	1.64
1998	1.58
1999	1.55
2000	1.53
2001	1.51
2002	1.51
2003	1.48
2004	1.44
2005	1.45
2006	1.42

These data illustrate the fact that a risk number such as 1.36 deaths per 10^8 VMT in isolation is practically meaningless. But when put in a historical context, or in the context of other products or activities, a perspective is gained to evaluate the magnitude of the risk. As can be seen in Table 1, we as a society are making steady progress on reducing the fatal risk of driving, and our current risk number

31. Matthew L. Wald, *Deaths on Motorcycles Rise Again*, N.Y. Times, Aug. 15, 2008, at A11.

32. NHTSA, Motor Vehicle Traffic Crash Fatality Counts, *supra* note 18, slides 52, 74, 85.

does not look too bad. Similarly, if our risk number were presented in the context of the fatal highway risk of other industrialized nations, it would compare very favorably as well.

5. What is meant by "acceptable risk"?[33]

With an understanding of how risk is calculated, and that risk must necessarily be viewed in a comparative context, we now turn our discussion back to our original question "how safe is safe enough?" which in light of what we have learned must be rephrased "how much risk is acceptable?"

How much risk is acceptable is not a simple question; books are written with "acceptable risk" in the title. However, a simple answer to this question is typically another question: "acceptable to whom?" As individuals we exhibit radically different de facto risk acceptance, and the same individual will exhibit significantly different risk acceptance throughout his or her lifetime. Certainly, as compared with a stuntman, the average person would have widely variant views on what is an acceptable risk. And, neither could nor probably should make this decision for society as a whole. There is no absolute standard of how much risk is too much or too little, but innumerable federal, state, and voluntary standards prescribe maximum risk levels, and we touch on them briefly.

Risk acceptance has been studied extensively, and there are more than a dozen factors that influence how much risk is acceptable either to an individual, or to society as a whole, in a given situation. And they are not always the same factors. Examining and discussing all these factors is beyond the scope of this guide, but a few of the most important are illustrated.

Probably the single most important factor for determining how much risk is "acceptable" is how much "benefit" we gain from accepting the risk. We are willing to accept substantial risk for substantial benefit. Motorized vehicular transportation confers tremendous benefits in our society and almost the entire population participates, and by our participation we indisputably evidence our de facto "acceptance" of the known risks for the known benefits, even if we do not find the risks of driving intellectually "acceptable." That does not mean we have to like the level of risk, or that we "accept" the current level of risk in the sense that we do not need to do anything about it. Indeed we spend billions and billions of dollars trying to reduce the level of risk associated with motorized vehicular travel. That being said, the overwhelming majority of the current population finds the current level of motorized vehicular travel risks low enough, given the benefits, to participate. This would not be too surprising if the level of risk associated with motorized transportation were low, because the benefits of motorized transport are clearly high. However, the fatal risk associated with motorized vehicle travel is not low.

33. Although acceptable risk is also a legal concept, we are merely using engineering vernacular in this chapter, and no legal construct is intended.

Returning to our fatality risk for a nonmotorcycle vehicle driver of 0.718 deaths per 10^8 VMT, this translates into a fatal risk of 0.718×10^{-8} for each mile. That 10^{-8} term makes this number quite small, and the fatal risk per mile low. However, very few people drive just a mile in a week or year. In fact, according to the Federal Highway Administration, the average U.S. driver logs about 13,500 miles behind the wheel every year.[34] That means for the year, the average U.S. driver faces $0.718 \times 1.35 \times 10^4 \times 10^{-8} = 0.97 \times 10^{-4}$ risk of a fatal accident every year, or about 1/10,000. But, very few people drive for a year. It is not uncommon to drive 60 years. Certainly the mileage we drive when young and old is less each year, and when middle-aged, more, but for the purpose of calculation let's assume the average value for 60 years. Then the risk of driving for one's adult lifetime, on average, is $0.97 \times 6 \times 10 \times 10^{-4} = 5.82 \times 10^{-3}$ Stated another way, if we drive for a lifetime, even at the low fatal risk of 2007, the average driver runs a risk of 0.00582 of being killed in his lifetime in a vehicular accident, a little more than a chance of 1/200. So, one out of every 200 drivers will die in his or her lifetime from the activity of driving a vehicle.

Needless to say, this number does not look so small any more. This brings us to the most commonly advanced argument against de facto "risk acceptance" being the measure of "acceptable risk" or "safe enough." Critics argue that no one can be said to "accept" a risk if they do not know what the risk is. Logically this is true, but it ignores the fact that even if we cannot cite a specific risk parameter, that does not mean we do not have an intuitive grasp of the risk. For example, in the case of motorized vehicle travel above, relatively few people can go through the calculation above and derive the number 1/200. But, we all have personally known in our lifetime more than one person (not just luminaries such as James Dean, Jayne Mansfield, Princess Grace Kelly, General George Patton, and Princess Diana, and even Barack Obama Sr., father of our current President) who has died in a vehicular accident. For the 1/200 number to be true, since we all know a few hundred people, we must know at least a couple who have died in vehicular accidents. Therefore, even though we may not be able to calculate the number, society has an excellent grasp of the risk associated with vehicular travel.

This 1/200 risk of fatal vehicular injury also illustrates the important difference between a "unit of participation risk" and a "lifetime risk." Because, fortunately, average lifetime is so long, when a risk to which we are constantly exposed is summed over a lifetime, the resulting fraction can become uncomfortably large. For example, the lifetime risk of developing cancer from merely exposure to the background levels of environmental chemicals has been estimated to between 1/1000 to 1/100.[35]

Indeed, people who study common perceptions of risk have found that people do a fair job of estimating the national death toll from a great many com-

34. http://www.fhwa.dot.gov/ohim/onh00/bar8.htm.

35. C.C. Travis & S.T. Hester, *Global Chemical Pollution*, 25 Envtl. Sci. & Tech. 814–19 (1991).

mon risks, such as vehicular travel, but typically overestimate risks, such as airplane crashes, that have significant publicity associated with them.[36] We all know there is a small, but highly controllable, risk of drowning when we go swimming. Yet most of the U.S. population participates in this activity on some basis.

After the benefits gained from assuming the risk, probably the second most important factor determining the acceptability of a given risk level is "control." Is the risk under our control or in the hands of fate? We are willing to voluntarily assume up to 1000 times more risk if we perceive we are assuming the risk voluntarily and it is under our control. This is certainly a substantial factor in the acceptance of the risk of motorized vehicular travel. It is also observed very commonly in sports recreation activities. We perceive that the overwhelming majority of this risk is under our direct control, so we are almost universally willing to accept it for the perceived benefits on a societal basis. On the other hand, if we perceive the risk is imposed on us involuntarily, and it is out of our control, such as a nuclear power plant being built in our city, then the amount of risk we are willing to "accept" being imposed on us is dramatically less.

Another important factor in determining if a particular risk is "acceptable" is the cost of reducing or eliminating the risk. This issue is commonly encountered in product-related injury tort litigation, and it is often not a simple one. As mentioned above, there is practically no product that cannot be made safer by reducing the product benefits or increasing the product cost, or both.

Unfortunately, plaintiffs and defendants often muddy the intellectual landscape related to safety in products litigation. Plaintiffs will often assert that the product should be completely risk free, an impossible ideal to achieve, even if the product is being misused. Defendants will often assert that safety is the "highest" priority in their product's design. However, this cannot be true either. If safety were the highest priority in any product's design, the cost would be uneconomical, because at no point in the design, no matter how low the risk, would the level of risk be as low as could be achieved with more cost. Everyone knows that a big car is safer than a small car. And this is demonstrably true. It is particularly true when the big car hits the small car, and death risk in the small car is commonly 8 to 10 times higher than that of occupants of the big car in such collisions. Big cars also present less risk to their occupants, even hitting stationary obstacles such as trees. But, big cars cost more than small cars. If safety were the highest priority in vehicle design, we would all have to pay for vehicles with the weight, complexity of design, and handwork found in nameplates such as Mercedes. In the real world we can choose among more than 300 car models. Some of the very smallest and lightest mass-produced models are very inexpensive relative to a Mercedes, but they also do not remotely protect their occupants to the degree of a Mercedes. All cars must provide their occupants a minimum level of protection by compliance with the Federal Motor

36. Baruch Fischhoff et al., Acceptable Risk (1981).

Vehicle Safety Standards (FMVSS). But even with the FMVSS, there is demonstrably more risk associated with driving a small car.

How much risk is "acceptable" is further complicated by the fact that risk cannot be spread uniformly in society. The fatal risk of motorized vehicle travel is borne by those relatively few who die, and by the rest of us only by taxes and insurance premiums. Unfortunately, some purchasers are willing to assume the additional risk of a small car in the showroom for the very substantial cost savings, but change their minds after a collision demonstrates the complete cost of the tradeoff. Our economic system permits purchasers to trade cost for safety in innumerable other products, from helmets, to tools, to furniture and houses. In reality, consumers and manufacturers must engage in consideration of cost versus safety virtually every day, because there are few products where a safer and more expensive model is not available, and no products exist that cannot be made safer by being made less convenient and/or more expensive. Denying or obfuscating this process does not advance safety, science, engineering, or justice.

In light of all the preceding considerations, we last examine the question of whether there is any absolute level of risk low enough that it almost always is regarded as "acceptable" and therefore "safe." Unfortunately, there are a multitude of such levels from a myriad of sources. In the United States, Chauncey Starr in 1969 quantified the risk of disease to the general population as one fatality per million hours of exposure, and after studying risk acceptance and participation of society in many activities concluded that "the statistical risk set by disease appears to be a psychological yardstick for establishing the level of acceptability of other risks."[37] Starr observed the de facto level of risk people accept, not necessarily that which they would say is "acceptable," was about one in a million chance of fatality per hour, or unit of, exposure. If an activity presents this level of fatal risk, and a person wants the benefit of that activity, he or she will almost always accept this level of risk for the perceived benefit. As a consequence of this initial observation, "one in a million risk" calculations are now commonplace in the risk literature.[38]

As the risk level rises above this threshold, a decreasing fraction of the population will find the risk worth the benefits. This is why very high risk sports, such as skydiving, have many fewer participants. Let us return one more time to our driver fatality risk of 0.718×10^{-8} for each mile. This can be conveniently converted into a risk per hour by recognizing that the average driving speed in the United States is about 30 miles per hour.[39] That means in an hour, the fatal risk to the average driver is $30 \times 0.718 \times 10^{-8} = 0.214 \times 10^{-6}$ or about 0.2 per million hours or 2 in 10 million hours. It is perhaps more appropriate to return

37. C. Starr, *An Overview of the Problems of Public Safety*, in *Proceedings of Symposium on Public Safety* 18 (1969).

38. R. Wilson & C. Crouch, Risk-Benefit Analysis 208–09 (2d ed. 2001).

39. See, for example, government calculations at http://www.epa.gov/OMS/models/ap42/apdx-g.pdf or http://nhts.ornl.gov/briefs/Is%20Congestion%20Slowing%20us%20Down.pdf.

to the 1.36×10^{-8} per mile traveled for the overall risk to society of motorized vehicular travel, not just the driver risk, to compute the risk level that society de facto accepts for the benefits of motorized vehicular transport. Then the risk per hour becomes $30 \times 1.36 \times 10^{-8} = 0.408 \times 10^{-6}$ or a little less than half a fatality per million hours of exposure. This is well below the "one in a million" threshold, and thus 98%+ of the society will participate in this activity. As a final sanity check on our work, let's return to the 37.86 risk of fatal injury per 10^8 VMT for motorcycles. This translates into a risk per hour of $30 \times 37.86 \times 10^{-8} = 1.136 \times 10^{-5}$ or more than 11 fatal injuries per million exposure hours. This is above the one-in-a-million threshold, and, understandably, motorcycle riding is regarded as an unacceptable risk by a large fraction of the population.

This threshold of "one in a million" as the "acceptable" risk level has many variants. In the United Kingdom, for example, the Health and Safety Executive[40] adopted the following levels of risk, in terms of the probability of an individual dying in any one year:

- 1 in 1000 as the "'just about tolerable risk'" for any substantial category of workers for any large part of a working life;
- 1 in 10,000 as the "'maximum tolerable risk'" for members of the public from any single nonnuclear plant;
- 1 in 100,000 as the "'maximum tolerable risk'" for members of the public from any new nuclear power station;
- 1 in 1,000,000 as the level of "'acceptable risk'" at which no further improvements in safety need to be made.

There are essentially innumerable regulations promulgated by different agencies within states and the federal government that are beyond the scope of this guide, but which mandate expenditures to maintain certain maximum risk levels either implicitly or explicitly. These regulations cover everything from food additives to acceptable levels of remediation at toxic Superfund sites. Regrettably, there is little or no coordination among regulating agencies, and no standardized procedures for addressing risk within the federal government, or within the states. As a result, the amount spent to "save a life," which should be termed "forestall a fatality" (because everyone eventually dies) varies by six orders of magnitude. Table 2 lists a number of regulations, the year that they were mandated, their issuing agency, and the cost they effectively mandate be expended "per life saved." Needless to say, these are estimates, and the data are somewhat dated, but the relative costs will be approximately the same. Executive orders from recent Presidents starting with Reagan have attempted to introduce "cost-effectiveness" in one form or another into the regulatory process, but with little observable effect at this writing.

40. Water Quality: Guidelines, Standards and Health 208–09 (L. Fewtrell & J. Bartram eds., 2001).

Table 2. Relative Cost of Selected Regulations as a Function of Lives Saved

Regulation	Year	Agency	Cost per Life Saved (Millions of Dollars in 1990)
Unvented space heater ban	1980	CPSC	0.1
Aircraft cabin fire protection	1985	FAA	0.1
Aircraft seat cushion flammability	1984	FAA	0.5
Trenching and excavation standards	1989	OSHA	1.8
Rear lap/shoulder belts for cars	1989	NHTSA	3.8
Asbestos occupational exposure limit	1972	OSHA	9.9
Ethylene oxide occupational exposure limit	1984	EPA	24.4
Acrylonitrile occupational exposure limit	1978	OSHA	61.3

Note: CPSC = Consumer Product Safety Commission; EPA = Environmental Protection Agency; FAA = Federal Aviation Administration; NHTSA = National Highway Traffic Safety Administration; OSHA = Occupational Safety and health Administration.

Source: W. Kip Viscusi & Ted Gayer, *Safety at Any Price?* Regulation 54, 58 (Fall 2002).

Finally, although we acknowledge that this section on safety is quite extensive, we also believe it is extremely important for the court to recognize how engineers think about safety. Engineers are dedicated to making safe products. At the same time, they recognize that every increment in safety has an expense associated with it. Just as there is no product or environment that is risk-free, there is no bright-line threshold that universally divides safe and unsafe products; safety is not binary. For each properly designed product, there is a unique set of constraints (including cost), and a safe-enough level exists that balances constraints with acceptable risk.

C. The Design Process—Examples in Which This Guiding Principle Was Not Followed

To illustrate ways in which flawed design processes lead to adverse outcomes, a number of examples are selected covering a range of incidents that occurred during the past century. In each instance, the link in the design process that was either missing or corrupted is highlighted and discussed. The reader may wish to refer to Figure 1 when considering these examples.

1. Inadequate response to postmarket problems: Intrauterine devices (IUDs)

Insertion of objects into a woman's uterus has long been a means of contraception. In the twentieth century, IUDs were designed, manufactured, and mass marketed around the world. Many of them were associated with adverse health consequences, in particular, pelvic inflammatory disease, which led to long-term disabilities and even death in substantial numbers of women. An example was the Dalkon Shield, marketed and sold by A.H. Robbins. The health problems of wearers of this device put its manufacturer in bankruptcy and led Congress to pass legislation to enhance medical device regulation generally, including most IUDs. Thus, those authorities corrected the flawed design process employed by A.H Robins which had led it to conclude that the product could be initially marketed and even continued to be marketed in the face of reports of serious health problems and death.

The Copper-7 IUD, marketed and sold by G.D. Searle, represented a somewhat different situation. That device received FDA approval as a drug. After it reached the market, Searle received reports of health problems. In litigation brought by women who used the product, some courts concluded that the risk associated with its use was "unacceptable."[41]

With all IUDs, the inserted device has a "string" attached to it that passes from the uterus through the cervix and into the vagina. The "string" is used for the purposes of removal as well as to provide certainty to the woman that the IUD remains in place and has not been expelled. But, to provide these functions, it compromises a biological firewall that ensures sterility of the uterus—the cervix. Therefore, in choosing the string material and fabrication method, designers had to assess choices that if properly made, reduced, if not eliminated, the potential for bacteria to migrate from the vagina into the uterus. With both the Dalkon Shield and the Copper-7, the designers set aside this consideration and traded it for the ability to enhance manufacturability and appearance by using strings that resulted in the unacceptable transmission of infectious agents into the uterus. These design choices were made for the purpose of reducing expense and gaining a competitive marketing edge, not to enhance consumer safety, and therefore led to unacceptable risk. They turned out to be lethal choices, two more examples of failures to adhere to the well-established and time-honored design process.

2. Initial design concept: Toxic waste site

For 17 years, over 35 million gallons of industrial waste were deposited in pits dug into the ground in what had been presumably certified to be a granite-lined impermeable geological formation that would not leak. These were known as the Stringfellow Acid Pits located near the Riverside suburb of Glen Avon, California, some 50 miles east of Los Angeles. History proved otherwise and millions of gallons of toxic materials escaped containment and contaminated groundwater supplies and exposed local inhabitants to chemical vapors.[42]

41. *See* Robinson v. G.D. Searle & Co., 286 F. Supp. 2d 1216 (N.D. Cal. 2003); Kociemba v. G.D. Searle & Co., 683 F. Supp. 1577 (D. Minn. 1988).

42. *See* State v. Underwriters at Lloyd's London, 54 Cal. Rptr. 3d 343 (Cal. Ct. App. 2006), *pet. for review granted,* 156 P.3d 1014 (Cal. 2007) for general overview and United States v. Stringfellow, No.

In this instance, the design process was flawed from the very beginning (i.e., the incomplete and incorrect geological analysis) and led to an "engineered" site for the containment of toxic wastes, which had no chance of performing properly. This is an excellent example showing that once the design process is corrupted, everything that follows in the design cascade, although perhaps done correctly, will most likely not lead to a successful design outcome.

3. Forseeable safety hazards: Air coolers

In low-humidity locales, it is possible to "air condition" a structure using evaporative cooling of water. This is done in devices known as "swamp coolers." They either sit beside or in most instances on the roofs of the structures being cooled. The operation is simple. They consist of an enclosure or a box in which a small pump is used to saturate porous panels through which air is drawn thereby evaporating the water and cooling the air that is directed into the interior spaces. The pumps are electrically powered and are known to short-circuit and fail, thus becoming a potential source of ignition and fire. A simple design solution is to make the box inflammable. This, of course, is the case when the box is metal, but then one has to be concerned with corrosion and subsequent maintenance. To obviate the corrosion issue, the box can be made of plastic. Plastic does not corrode but it is potentially flammable unless flame retardants are added as part of the materials formulation. Foreseeing this occurrence and making the conscious choice not to add flame retardants is an abdication of the design process and with that comes tragic consequences.

This scenario was played out in *Vanasen v. Tradewinds*,[43] where a 5-year-old girl was killed as the result of a foreseeable pump failure, subsequent electrical short circuit, and ignition of a non-fire retardant plastic swamp cooler attached to the roof of her home. Again, failure to adhere to the straightforward tenets of the design process (i.e., designing out the known tendency of many plastics to burn) is tantamount to "rolling the dice" and hoping for the best. Experience teaches us time and again that taking design "shortcuts" seldom translates into an acceptable design outcome.

4. Failure to validate a design: Rubber hose for radiant heating

Radiant heating has been in use since Roman times, and a common variant of this heating method involves placement of tubes that circulate heated fluids beneath floors, thus warming the floors that then in turn heat the surrounding structure. Even though metallic tubes once frequented this application, their cumbersome

CV 83-2501 JMI, 1993 WL 565393 (C.D. Cal. Nov. 30, 1993) for discussion of specific findings of fact by the special master; P. Kemezis, *Stringfellow Cleanup Settlement: Companies Agree to Pay $150 Million*, Chemical Week, Aug. 12, 1992, at 11; http://www.dtsc.ca.gov/PressRoom/upload/t-01-99.pdf.

43. Tulare County, CA Sup. Ct., No. 93-161828.

installation and susceptibility to corrosion led to the development of plastic tubes. One manufacturer recognized that rubber hose would be even easier to install than the somewhat rigid plastic conduits, and engaged a major rubber company to design a hose for the radiant heating market. The rubber company supplied a hose formulation that was designed for and used in automotive cooling applications, which made some sense given that similar fluids at similar temperatures are circulated in both cases. The rubber company failed to test the newly developed hose under end-use conditions, and thereby neglected to detect a failure mode caused by hose hardening and embrittlement. Engineering experts for the plaintiffs conducted a simple end-use test that verified that the hose would degrade under foreseeable conditions, thus completing the step in the design process that was not performed by the rubber company.[44]

5. Proper design—improper assembly: Kansas City Hyatt Regency Hotel

On July 17, 1981, during a tea dance in the vast atrium at the Hyatt Regency Hotel in Kansas City, two elevated walkways collapsed onto the people celebrating in the lobby, killing 114 of them and injuring more than 200.

The determination of what happened focused on the design and construction of the walkways. The 40-story complex featured a unique main lobby design consisting of a 117-foot by 145-foot atrium that rose to a height of 50 feet. Three walkways spanned the atrium at the second, third, and fourth floors. The second-floor walkway was directly below the fourth, and the third was offset to the side of the other two walkways. The third- and fourth-floor walkways were suspended directly from the atrium roof trusses, while the second-floor walkway was suspended from the fourth-floor walkway. During construction, the design, fabrication, and installation of the walkway hanger system were changed from that originally intended by the design engineer. Instead of one hanger rod connecting the second- and fourth-floor walkways to the roof trusses, two rods were used—one to connect the second- to the fourth-floor walkway, and another to connect the fourth-floor walkway to the roof, thus doubling the stresses in the ill-conceived connection.

Just prior to the collapse, about 2000 people had gathered in the atrium to participate in and watch a dance contest, including dozens who filled the walkways. At 7 p.m., the walkways on the second, third, and fourth floor were packed with visitors as they looked down to the lobby, also full of people. It was the second- and fourth-floor walkways—the ones that experienced the design changes—that collapsed. Clearly then, in the iterative cycle of the design process, modifications to the original design need to be validated, and failure to do so can

44. http://www.entraniisettlement.com/PDFs/PreliminaryApprovalAmended.pdf; J. Moalli et al., *Failure Analysis of Nitrile Radiant Tubing*, ANTEC 2006 Plastics: Annual Technical Conference, Society of Plastics Engineers, May 7–11, 2006, Charlotte, NC (2006).

have severe consequences. Further details of this event can be found in the second edition of this manual.[45]

6. Failure to validate a design: Tacoma Narrows Bridge

Spanning a strait, the third longest suspension bridge of its time, the Tacoma Narrows Bridge opened on July 1, 1940. In November of that same year, it collapsed into Puget Sound. During the design process, engineers failed to adequately account for the effects of aerodynamic flutter on the structure, a phenomenon in which forces exerted by winds couple with the natural mode of vibration of the structure to establish rapid and growing oscillations. In essence, the bridge self-destructed.

It is fair to say, however, that aerodynamic flutter was not well understood at the time this bridge was constructed. Indeed, the term was not coined until the late 1940s, years after the bridge collapsed. The root cause of this unfortunate circumstance was a desire to build a bridge with enhanced visual elegance (i.e., long and narrow) and to use an untested girder system that offered significant cost savings. This should have led to a thorough testing and validation program to ensure that venturing into uncharted waters in bridge design would not result in unintended or unanticipated consequences. Indeed, after the bridge was constructed and put into use on July 1, 1940, it gained a reputation for its unusual oscillations and was known as "Galloping Gertie." It was only then that engineers built a scale model of the bridge and began testing its behavior in a wind tunnel. Those studies were completed and remedies proposed in November 1940, just days before the bridge fell into the Tacoma Narrows channel.

A substantial departure from the norm of appropriate testing and validation is an unacceptable application of the design process, and the collapse of this bridge is an all too sobering reminder of this. Stated in another way, end-use testing should not be done by the "consumer" and in cases where this occurs, a clear violation of the design process tenets has taken place.[46]

7. Failure to conform to standards and validate a design: Automotive lift

Automotive lifts are often used in dealerships and service stations to raise vehicles and provide access to components on the bottom of the vehicle for service. To reduce the propensity for injury, ANSI and the American Lift Institute (ALI) promulgate standards that specify, among other things, the minimum resistance on the horizontal swing-arm restraints. The lift in question had a label on the lift support structure that indicated it was in compliance with these specifications, so

45. Henry Petroski, *Reference Guide on Engineering Practice and Methods*, *in* Reference Manual on Scientific Evidence 577, 601–02 (2d ed. 2000).

46. This example is also further discussed in the second edition of this manual.

when a Jeep Wrangler fell from the lift and injured the owner of a service station, verification of conformity to the standards was assessed by the plaintiff.

Testing by the plaintiff's expert revealed that the swing-arm lift restraints resisted only 30% of the criteria specified in the standard, and that simple reconfiguration of the restraint components could create a conforming lift. Furthermore, the plaintiff's expert calculated that for the vehicle-lift configuration in question, the amount of force required to provide positive restraint was less than that required by the standards, and therefore the accident would have been prevented had the standards been met. Finally, the plaintiff's expert opined that the label on the lift that claimed compliance with the standard would tend to convey to the end-user of the product that the presence of the swing-arm restraint added a layer of insurance for the operator in the event that there was an imperfect placement of the vehicle over the lifting pads.

In response, the lift manufacturer claimed that the intended swing-arm restraining forces arose from the friction created when the lifting pad contacted the vehicle undercarriage, and further argued that the swing-arm restraint was nothing but "fluffery" forced upon lift manufacturers to remain competitive in the marketplace. The jury found for the plaintiff, implicitly recognizing the tenant of the design process that calls for testing and validation of design claims and features.

8. Lack of sufficient information and collective expertise to consummate a design: Dam collapse

After 2 years of construction the St. Francis dam in southern California was completed in 1926 and the reservoir behind it began to fill. As the reservoir reached near capacity behind the 195-foot-high concrete arch dam, the eastern abutment gave way shortly before midnight on March 12, 1928, unleashing a wall of water over 100 feet high that eventually dissipated into the Pacific Ocean some 50 miles downstream. The flood killed more than 600 people and most likely more. The collapse of the St. Francis dam is one of the worst American civil engineering failures of the twentieth century.[47]

The dam was designed and certified by a single individual, William Mulholland, chief engineer and general manager of the Los Angeles Department of Water & Power (at the time known as the Bureau of Water Works & Supply). Mulholland had no formal education and was a self-taught individual. While the ultimate physical cause of the failure was the proximity of a paleomegalandslide to the eastern dam abutment, a geological anomaly that geologists argue today as to whether such a feature could have been detected in the 1920s, the inquest that followed the disaster determined that improper engineering, design, and governmental inspection was where the responsibility for this tragedy resided.

47. St. Francis Dam Disaster Revisited (Doyce B. Nunis. Jr., ed., 2002).

Indeed, we now know that the design of this structure failed to meet accepted design principles already in place in the 1920s. The dam height was increased by 10 feet at the start of construction, and another 10 feet midway through construction, bringing the final capacity to 38,000 acre feet. No modifications were made to the base to accommodate this additional capacity, and there were a number of weaknesses in the design of the base. It is estimated that the factor of safety, which was meant to be above 4 in the initial design, may have been as low as 0.77 on the dam that was actually constructed.

Geoforensics expert J. David Rogers enumerated many other design deficiencies associated with the St. Francis Dam, among them the lack of hydraulic uplift theory being incorporated into the dam's design; lack of uplift relief wells on the sloping abutment sections of the dam; failure to batter the upstream face of the dam to reduce tensile forces via cantilever action; failure to analyze arch stresses of the main dam; failure to remove high-water-content cement paste (laitance layer) between concrete lifts; failure to account for the mass concrete heat-of-hydration; failure to recognize the tendency of the Vasquez formation to slake upon submersion and failure to provide the dam with grouted contraction joints; failure to recognize that the dam concrete would eventually become saturated; and failure to wash concrete aggregate before incorporation in the dam's concrete.[48]

In this instance there simply was no credible design process from concept, through design, execution, and postconstruction surveillance. As a result, a massive failure ensued.[49]

9. Operation outside of design intent and specifications: Space shuttle Challenger

On January 28, 1986 the space shuttle *Challenger* and its accompanying liquid hydrogen and oxygen external tank (ET) disintegrated over the Atlantic Ocean after only about 70 seconds of flight. The two attached solid rocket boosters (SRB) separated from the shuttle and ET and were remotely destructed by the range safety officer. All seven of the NASA crewmembers were killed.

We now know the physical reason for this catastrophe. Two rubber "O"-rings placed at the aft joint where two sections of the right SRB came together had failed to "extrude" themselves as the SRB metal shell deformed during the early moments of ignition. Because of this, hot gases escaped through the breach created by the ineffective seal at the O-ring joint and led to the separation of the aft strut that attached the right SRB to the ET. This was followed by failure of the

48. J. D. Rogers, *The St. Francis Dam Disaster Revisited*, 77 Southern California Q. (1-2) (2003); J. D. Rogers, *The St. Francis Dam Disaster Revisited*, 40 Ventura County Q. (3-4) (2003).

49. Donald C. Jackson & Norris Hundley, *Privilege and Responsibility: William Mulholland and the St. Francis Dam Disaster*, California History (Fall 2004).

aft dome of the liquid hydrogen tank. The massively uneven thrust created by the escaping hydrogen gas altered the trajectory of the shuttle and aerodynamic forces destroyed it. The failure of the "O"-rings to alter their conformations with SRB shell deformation was attributed to the low ambient temperature at the time of launch. The O-rings had "hardened" and as a result lost their required flexibility.

Two investigations into the circumstances surrounding this disaster took place. Reports and findings were issued by the Presidential Rogers Commission[50] and the U.S. House Committee on Science and Technology.[51] While both reports agreed on the technical causes of the catastrophe (failure of the "O"-rings to perform as intended), their conclusions as to the root cause were stated somewhat differently but in the end pointed to the same basic issue. The Rogers Commission concluded that the National Aeronautics and Space Administration (NASA) and the O-ring manufacturer, Morton Thiokol, failed to respond adequately to a known design flaw in the O-ring system and communicated poorly in reaching the decision to launch the shuttle under extremely low ambient temperature conditions. The House Committee concluded that there was a history of poor decisionmaking over a period of several years by NASA and Morton Thiokol in that they failed to act decisively to solve the increasingly serious anomalies in the SRB joints.

Another way of stating what both reports essentially say is that the design process resulting in the double O-ring (now a triple O-ring system) was flawed. Moreover, NASA managers knew of this problem as early as 1977. Warnings by engineers not to launch that cold morning were disregarded. Each SRB consisted of six pieces, three welded together in the factory and the remaining three fastened together at the launch facility in Florida using the double O-ring seal system. Thiokol engineers lacked sufficient data to guarantee seal performance of the O-rings below 53 degrees Fahrenheit (°F). Temperature at launch hovered at 31°F. When originally designed, the O-rings were intended to remain in circumferential grooves. After several shuttle launches, it became evident that the SRB shell was deforming and that hot gases could escape but that the O-rings were "extruding" to seal these temporary breaches. As a result, the design specifications were changed to accommodate this process. The design itself, however, remained unchanged.

If one considers that the original design concept was to ensure a seal between the SRB field-joined sections using two O-rings, the question on the table is whether the actual design and subsequent execution were consistent with the design process. Clearly this was not the case. First, the system performed differently than expected (i.e., extrusion occurred). Validation and testing to ensure that

50. Rogers Commission, Report of the Presidential Commission on the Space Shuttle Challenger Accident (1986).

51. Committee on Science and Technology, *Investigation of the Challenger Accident*, H.R. Rep. No. 99-1016, (Oct. 29, 1986).

this aberrant behavior of the original seal system actually was acceptable never was done other than to monitor shuttle launches and hope for the best. Second, the O-rings were known to have insufficient resiliency at temperatures substantially higher than those encountered on the day of the *Challenger* launch; therefore, launching at such a low ambient temperature equated to misuse of the system. The unfortunate truth of all this is that an unsound design process most certainly will produce a flawed product.

10. Foreseeable failure and lack of design change in light of field experience: Air France 4590

On July 25, 2000, Air France flight 4590, a Concorde supersonic passenger jet departed Charles de Gaulle Airport and crashed into a nearby hotel killing 100 passengers, 9 crew, and 4 others on the ground. The physical cause was readily determined. The Concorde was designed to take off without flaps or leading-edge slats as a weight-saving measure. Because of this, it required a very high takeoff roll speed to become airborne. This placed unusually high stresses on the tires. A piece of titanium metal approximately 1 × 16 inches was lying on the departure runway. It had fallen from a thrust reverser assembly on a Continental Airlines DC-10 that had departed minutes earlier. During its takeoff roll, the Concorde struck the metal debris and this punctured and subsequently shredded one of its tires. The tire remnants broke an electrical cable and created a shock wave that fractured a fuel tank. The fuel ignited and an engine caught fire. The plane had reached a ground speed such that the pilot elected that it was prudent to continue the takeoff rather than abort. The crew shut down the burning engine. Unable to retract the landing gear, and now experiencing problems with the remaining engines, the crew was unable to climb and the aircraft rolled substantially to the left and contacted the ground.[52]

In this instance, a design decision was made to save weight by not having retractable flaps and slats. This led to higher than normal landing and takeoff speeds. This in turn placed additional demands on the tires. They would be rotating at higher speeds and contain much increased kinetic energy. This meant that when one or more failed, the rubber shrapnel would be released with additional force. This led to a greater risk of puncture of the aircraft structure and therefore special consideration to ensure that the aircraft skin could maintain integrity in the foreseeable event of a tire rupture. Making the skin more resilient to puncture implied additional weight and this would work against the primary reasoning for not having the slats and flaps. And there we have the design conundrum.

Having made what was initially regarded as a reasonable compromise in the aircraft design, the manufacturer subsequently gained experience with the Concorde, learning that tire failures could be potentially catastrophic (the type

52. http://www.bea-fr.org/docspa/2000/f-sc000725a/htm/f-sc000725a.html.

of experience illustrated by the "Performance" arrow from the "Go" stage to the "Design/Formulate" and "Test/Validate" stages in Figure 1). Between July 1979 and February 1981, there were four documented tire ruptures on takeoff. In two of these instances, substantial damage was done to the aircraft structure, but the planes were able to land without incident. Despite having these critical data related to the initial design assumptions and associated compromises in hand, no remedial changes were made to either the tire or aircraft design. After the 2000 crash, design changes were made to the electrical cables, the fuel tanks were lined with Kevlar, and specially designed burst-resistant tires were put into use. The Concorde fleet was retired from service in 2003, with declining passenger revenues cited as the major cause.

In the case of the Concorde, the record appears to indicate that designers chose not to alter the design, even in the face of significant data, until a fatal accident occurred. Although these actions may be consistent with the above discussion on risk, and how it is perceived, the crash is illustrative of how the fundamentally simple design process works, and that departures from it can have serious consequences.

IV. Who Is an Engineer?

A. Academic Education and Training

Having earned a bachelor's degree in an engineering curriculum is generally sufficient to enter the professional workplace and begin to immediately solve a wide variety of problems. It is less so the case for students who graduate with degrees in the basic sciences such as physics, chemistry, or biology or in mathematics. Typically, but not always, these basic science students will go on to earn graduate degrees.

It is also the case that some students who have earned an engineering degree will continue to the master's or even doctorate level of study. In 2004, U.S. colleges and universities awarded approximately 75,000 bachelor's degrees, 36,000 master's degrees, and 6000 doctoral degrees in all areas of engineering.[53]

One can think of the educational process as providing engineering students with a toolkit from which they select "tools" to enable them to either individually or in teams participate in scientific and technological innovation. Because these students are educated, as opposed to having been trained, one can never be quite sure how they will choose to use their tools, or add to the kit, or delete from the kit. Although carpenters share a common toolkit, we know the structures they build can be appreciably different in size, shape, and scope. So it is with engineers.

53. *Report 1004D: Total Numbers of Bachelor's, Master's and Doctoral Degrees Awarded per Million Population Since AY1945-46—Including Data for Degrees Awarded to US Citizens Since AY1970-71*, Engineering Trends Quarterly Newsletter, Oct. 2004, at 1.

One example that scientists and engineers can be one and the same is epitomized by Renaissance humanism during a period almost five centuries past. There, Leonardo da Vinci, with a minimalist toolkit by today's standards, lived a life equally as an engineer and a scientist, and indeed an artist. Four centuries later, Buckminster Fuller seamlessly combined elements of geometry (aka "science"), structures (aka "engineering"), and architecture (aka "art") to conceive and develop an entirely new approach to architectural design. Architects Norman Foster and Frank Ghery seized on recent advances in computer science and engineering to provide innovative platforms for architectural design that paved the way for radical changes in structural and visual renderings. Striking examples include the Guggenheim Museum Bilbao, the Walt Disney Concert Hall Los Angeles, the Experience Music Project Seattle, the City Hall London, the Beijing Airport, and the Reichstag Berlin.

Searching for ways to create or define the "bright line" that classifies da Vinci, Fuller, Foster, or Gehry as engineers, scientists, architects, or artists is as empty an exercise today as it would have been five centuries ago. This, of course, does not preclude one from considering himself or herself as an "engineer" or a "scientist"; however, the subtler point is that one can also be both or either at different points in time or at the same time. This can be overlooked or ignored in the quest for limiting or excluding expert testimony.

B. Experience

Without knowing how an engineer or scientist will use his or her toolkit and to what extent it will be replenished or modified as time goes on, it is not possible to begin to even second-guess what any particular individual may do to shape his or her career as time passes. There is a great deal of truth to the notion of "learning on the job." Indeed, as one's career unfolds, the number of opportunities expands and with that comes additional skills and an ever-increasing ability to make wise and informed choices and decisions. Being an engineer affords one the opportunity to continually remodel oneself as new and unexpected problems and challenges become evident.

And so it is with the passage of time that the "title" of one's degree becomes an increasingly murky description of who one is and what one does. This is why it is so critical when evaluating whether an "engineer" is testifying within his or her realm of expertise that titles do not overshadow the actual context of a degree (i.e., the name may not reflect the knowledge attributes accurately) and the experience base at hand. Even though it is an all too common tactic to attempt to confine expert witness testimony to the asserted domain of his or her named academic credentials, it is one that may necessarily lead to less-informed testimony than otherwise would be the case. This is a high price to pay when the desired outcome is finding the right path to both truth and justice.

C. Licensing, Registration, Certification, and Accreditation

Licenses are required for engineering professionals in all 50 states and the District of Columbia, *if their services are offered directly to the public and they would affect public health and safety.* Licensed engineers are called professional engineers (PEs). In general, to become a PE, a person must have a degree from an ABET[54]-accredited engineering college or university, have a specified time of practical and pertinent work experience, and pass two examinations. The first examination—Fundamentals of Engineering (FE)[55]—can be taken after 3 years of university-level education, or can be waived in lieu of pertinent experience. The FE examination is a measure of minimum competency to enter the profession. Many colleges and universities encourage students to take the FE exam as an outcome assessment tool following the completion of the education coursework. Students who pass this examination are called engineering interns (EIs) or engineers in training (EIT) and take the second examination after some work experience. This is the Principles and Practice of Engineering examination. The earmark that distinguishes a licensed/registered PE is the authority to sign and seal or "stamp" documents (reports, drawings, and calculations) for a study, estimate, design, or analysis, thus taking legal responsibility for it.

Many engineering professionals do not seek a PE license because *their services are not offered directly to the public* or they have no need to sign, seal, or "stamp" engineering documents. Whether an individual is licensed as a PE is neither sufficient nor necessary to establish his or her competency as an engineer. Furthermore, the two examinations test only for knowledge gained and assimilated at the undergraduate level. It is therefore common for professors of engineering in colleges and universities not to have PE licensure—indeed, they are the ones who teach and prepare those who do take these examinations. Despite this, a common litigation practice is to attempt to preclude "engineering" testimony offered by professionals who have had no need to obtain PE licensure as if this was intended to be some sort of requirement for practicing in the profession or for testifying in court. Such an approach is unwarranted and inconsistent with the way in which engineers behave and think about the work they do.

54. Founded in 1932 as the Engineer's Council for Professional Development (ECPD), it was later renamed ABET (Accreditation Board for Engineering and Technology). In the United States, accreditation is a nongovernmental, peer-review process that ensures the quality of the postsecondary education that students receive. Educational institutions or programs volunteer to undergo this review periodically to determine if certain criteria are being met. ABET accreditation is assurance that a college or university program meets the quality standards established by the profession for which it prepares its students. The quality standards that programs must meet to be ABET-accredited are set by the ABET professions themselves. This is made possible by the collaborative efforts of many different professional and technical societies. These societies and their members work together through ABET to develop the standards, and they provide the professionals who evaluate the programs to make sure that they meet those standards.

55. In the past, this examination was known as the Engineer in Training (EIT) exam.

PE licensure is quite different from board certification for a physician or bar certification for a lawyer. Physicians and lawyers may not practice their professions without having such board certification. Such is not always the case for engineers and therefore it is not appropriate or correct to construe this to be so. The title "engineer" is legally protected in many states, meaning that it is unlawful to use it to offer engineering services *to the public* unless permission is specifically granted by that state, through a professional engineering license, an "industrial exemption," or certain other nonengineering titles such as "operating engineer." Employees of state or federal agencies may also call themselves engineers if that term appears in their official job title. In some states, businesses generally cannot offer engineering services *to the public* or have a name that implies that it does so unless it employs at least one PE. For example, New York requires that the owners of a company offering engineering services be PEs. In summary, licensing procedures and requirements are state specific, but such licensure is not a requirement to testify in federal court.

As a postscript to this discussion, civil engineers often seek PE registration because of their association with public works projects. This can be traced directly back to the failure and subsequent legacy of the St. Francis dam collapse in southern California in the late 1920s. More about this disaster is discussed in Section III.C.8.

V. Evaluating an Engineer's Qualifications and Opinions

A. *Qualification Issues and the Application of* Daubert *Standards*[56]

Engineers are treated like other witnesses when it comes to determining whether they can testify as factual or expert witnesses. Thus, if they have information regarding facts in dispute, an engineer can be a fact witness describing that information. In the context of the design of a product or the conception of an allegedly protectable method or device, that may take the form of describing what the engineer did to create the product or construct at issue, how he or she conceived of the subject of that product or construct, and how the product or allegedly

56. *Daubert* standards were established in the trilogy of cases, *Kumho Tire Co. v. Carmichael,* 526 U.S. 137 (1999), *General Elec. Co. v. Joiner,* 522 U.S. 136 (1997), and *Daubert v. Merrell Dow Pharms., Inc.,* 509 U.S. 579 (1993), and refer to factors to be considered when assessing the admissibility of expert testimony. *See generally* Margaret A. Berger, The Admissibility of Expert Testimony, in this manual.

protectable property compares to other designs or intellectual property that are claimed to relate to the subject of the dispute.

As discussed above, engineers can also be expert witnesses. Like any other proffered expert, an engineer's training, background, and experience play a role in qualifying him or her to provide expert opinions. Education, licensing, professional activities, patents, professional society involvement, committee work, standards development, professional consulting experience, and involvement in a business based on similar technology or engineering principles can all help to fortify an engineering expert's qualifications. The work that the engineer did to acquire facts about the matter at issue (described below) and to test the engineer's hypothesis as to how the incident in question occurred and what caused it provide still stronger bases for allowing the engineer to testify as an expert witness.

Ultimately, the court's application of *Daubert* standards to the qualifications asserted by the engineer and the opinions that the engineer seeks to give determines whether the engineer may testify. In the role of gatekeeper of scientific or technical testimony, the trial judge determines whether the engineering testimony is both "relevant" and "reliable." The relevance and reliability of engineering testimony is judged in the context of the design process and the way that engineers approach a problem as described above. And as the court clarified in *Kumho Tire*, *Daubert* extends to all expert testimony, including testimony based on experience alone.[57]

B. *Information That Engineers Use to Form and Express Their Opinions*

Under Federal Rule of Civil Procedure 26(a)(2)(B)(i), the expert report must contain the basis and reasons for all opinions expressed, and certainly the expectation is that oral testimony will do the same. Apart from opinions based purely on knowledge, skill, experience, training, or education, nearly all expert opinion is based on observations, calculations, experimentation, or some combination thereof.

1. *Observations*

a. Inspections

When called as an expert in a products liability case, engineers will often complete a physical inspection of a failed product or accident scene. ASTM and the National Fire Protection Association (NFPA) have published several standard practices that

57. *See* Margaret A. Berger, The Admissibility of Expert Testimony, in this manual.

offer guidance on inspections and related issues.[58] Although it is not required that engineers adopt and follow these standards, if the court has questions as to whether the techniques or procedures used by an engineer are reasonable, reference to the standards can certainly be helpful.

As a first step in the inspection process, engineers will typically document evidence or the accident scene using photography and videography. It may be worth noting that just as 2009 represents the first year that the official presidential portrait is digital, most engineers will record photos and video digitally. Other measurements and readings can also be made at the initial inspection, as engineers establish the state of the evidence and attempt to determine if it has been altered subsequent to the incident.

One important issue that often arises during an inspection is the destruction of evidence, and engineers sometimes argue as to whether testing is truly destructive. ASTM E 860 provides some guidance that could be useful to the court in terms of providing a reference to engineers:

> Destructive testing—testing, examination, re-examination, disassembly, or other actions likely to alter the original, as-found nature, state or condition of items of evidence, so as to preclude or adversely affect additional examination or testing.

In terms of inspections, destruction of evidence typically relates to disassembly or displacement of parts, and disputes can usually be resolved by establishing an agreed-on protocol between parties. If items that have physically broken or separated are at issue, it should be remembered that two fracture surfaces are created, each a mirror image of the other, and one can be preserved while the other is evaluated. Microscopic examination of failure surfaces, also known as fractography, is commonly used by engineers to determine the cause of failure. Fractography can be used to establish such things as how the product failed (overload versus a fatigue or time-dependent failure) and whether manufacturing defects (poor welds, voids, inclusions) exist.

b. Experiments and testing

After performing inspections of the evidence, engineers develop hypotheses as to the cause of what they are investigating and evaluate these hypotheses. One common method of testing a hypothesis is experimentation, and engineers are educated and trained to conduct experiments, often to the displeasure of their

58. Although not intended to be an exhaustive list, these standards include:

- ASTM E 860—*Standard Practice for Examining and Preparing Items That Are or May Become Involved in Criminal or Civil Litigation*,
- ASTM E 2332—*Standard Practice for Investigation and Analysis of Physical Component Failures*,
- ASTM E 1188—*Standard Practice for Collection and Preservation of Information and Physical Items by A Technical Investigator*, and
- NFPA 921—*Guide for Fire and Explosion Investigations*.

client-attorneys who would rather not perform any test for which the outcome is uncertain. Engineers can design tests to study kinematics (motions) and kinetics (forces) and to recreate accidents; to evaluate physical, mechanical, and chemical properties of materials; or to assess specific characteristics against claims in a patent. Because the circumstances surrounding accident and product failure investigation can be quite complex, and often novel as well, engineers sometimes must design experiments that have never before been performed. This notion, experiments conducted for the first time for purposes of litigation, has been the topic of much debate.

Although it is typically suggested that such work is biased and therefore ought to be excluded, an experiment that is designed and executed for the purposes of litigation is not inherently suspect. If the experiment has a well-defined protocol that can be interpreted and duplicated by others, articulates underlying assumptions, uses instrumentation and equipment that is properly calibrated, and is demonstrated to be reliable and reproducible, it should not be summarily discarded simply because it is new. It is often the case that the precise matter in dispute has not been the subject of engineering or scientific studies, because in the normal course of events, the problem at hand was never addressed in a public forum and no peer-reviewed literature spoke directly to it. In typical engineering problems, because a multitude of factors can vary, it is often difficult to find suitable pre-existing information, and the question at hand may not have been asked in such a way as is before the court.

The fact that problem identification occurs within the course of a legal dispute does not mean that the problem cannot then be explored directly using either the scientific method or the engineering design process or both to ascertain and understand the physical or chemical behavior of the issue at hand. In point of fact, an experiment that is designed for litigation will better fit the issues standing before the court, and either the plaintiff or the defendant is free to pursue this and to subsequently criticize the results. Not only will experiments designed to specifically address the matter at issue be more directly relevant to questions at hand, they will also provide data the court can use in thoughtful deliberation.

Indeed our personal experience has found this not only to be helpful in adjudicating complex issues for which no directly relevant prior work had been done, but in the end, after the litigation had been completed, peer-reviewed articles were written about the work that was done for the purposes of studying an issue for litigation.[59]

59. Richard D. Hurt & Channing R. Robertson, *Prying Open the Door of the Tobacco Industry's Secrets About Nicotine: The Minnesota Tobacco Trial,* 280 JAMA 1173 (1998); John Moalli et al., *supra* note 44; Monique E. Muggli et al., *Waking a Sleeping Gaint: The Tobacco Industry's Response to the Polonium-210 Issue*, 98 Am. J. Pub. Health 1643 (2008); M.S. Warner et al., *Performance of Polypropylene as the Copper-7 Intrauterine Device Tailstring*, 2 J. Applied Biomaterials 73 (1991); Richard Hurt et al., *Open Doorway to Truth: Legacy of the Minnesota Tobacco Trial*, 84 Mayo Clinic Proc. 444 (2009).

Of course not all situations require novel techniques to be developed, and in those instances an abundance of standards for testing materials and products exist. Typically promulgated by organizations such as ASTM, ANSI, CEN, and others, these standards envelop everything from sample preparation, to sampling procedures, to test equipment operation and calibration, to analysis of data acquired during testing. Although such a broad array of standards and guidelines exist, it is possible that some portion of even the more novel test may not be covered. It is also common for engineers to follow a standard to the maximum extent allowed by the circumstances and state of the evidence, and to note deviations from that standard in their protocols and reports.

2. *Calculations*

A substantial portion of an engineer's education is spent learning how to calculate things, so it should come as no surprise that when litigation is involved, engineers would be making calculations as well. As part of this education, engineers learn how to derive equations based on scientific and mathematical principles, and consequently become aware of the limitations of a particular equation or expression. Although it would be convenient if a single equation could be used to solve every engineering problem, this is clearly not the case, and so engineers must learn what principles to apply, and when to apply them.

The difference between a good calculation and a marginal one is related to how applicable the equations used in the calculation are to the situation at hand, and how valid the underlying assumptions are. As mentioned above, it is the rare case in which an engineering analysis contains no assumptions. For example, there are well-known equations that relate the pressure inside a cylindrical vessel to the stresses in the wall of that vessel. These equations assume, however, that the wall thickness of the pressure vessel is small compared with the inner diameter, and if this is not the case, significant error may result. If an engineer uses the more simplified approach, he should assess whether his analysis is conservative (i.e., how the assumptions affect the overall calculated result).

In the modern age, it is simple to download programs from the Internet that will make calculations based on input variables. These programs can save engineers considerable time, because they can reduce hours of "paper" calculation to minutes. Used blindly, though, without proper understanding of core assumptions or approximations, these programs can be precarious. Computer programs should always be validated, and the simplest way to accomplish that task is to have the program calculate a range of solutions for which the result is already known. The program is then validated within that range.

3. *Modeling—mathematical and computational*

When hand calculations become overly tedious, or are too simplified to handle a highly complex problem, engineers will often use computer models to examine systems, processes, or phenomena. Quite distinct from the simple programs mentioned above used to solve an equation or two, these computer models employ enormous bodies of code that can solve thousands of equations. One of the most common techniques employed by these programs is the finite element method (FEM), which can be used to solve problems in stress analysis, heat transfer, and fluid flow behavior.

FEM is dependent on the computational power of computers, and basically divides the system or component into small units, or elements, of uniform geometry. This mesh, as it is called, reflects the geometry of the actual system or component as closely as possible. Boundary conditions are established on the basis of known applied loads, and the fundamental equations of Newtonian mechanics are solved by iterative calculations for each individual cell. The resulting loads and displacements (or stresses and strains) in each cell are then summed at each increment of time to give an overall picture of the load/displacement (or stress/strain) history of the system or component. The literally millions of calculations required for each time step can only be handled by a computer. These data can then be used to determine the loads and displacements at the time of failure, information that otherwise could not be obtained from hand (or "back of the envelope") calculations.

In its early stages, FEM code could only be found in universities and corporate and governmental laboratories, and was executed by doctoral-level engineers who used separate programs to postprocess results into usable graphical output. Today, commercial FEM programs are widely available, and are capable of generating eye-catching graphics that appeal to juries. Other software programs are available that create similar graphics for car-crash or mechanical simulations. This tool is as much an accepted part of the engineering design community as the slide rule was in the 1960s. In addition, engineers involved in determining the cause of failure of mechanical systems have been using FEM since the 1980s to determine the loads and strains at critical points in complex geometries as part of root-cause analysis efforts. This is often a principal means to determine what actually caused something to break, and ultimately to determine whether a design or manufacturing defect or overload or abuse was ultimately at fault. FEM can, in certain circumstances, be a valuable tool to assess the cause of a design failure.

To be sure, FEM, like any scientific tool, must be properly applied and interpreted within its limitations. It can be abused and misused, and because the output from these models can be made to appear extremely realistic, especially when coupled with computer graphics, their use needs to be carefully considered. To summarily reject FEM as a simulation, though, would be to deprive a modern-day engineer of a tool that is regularly used. There is an old adage in the modeling world, called "garbage in—garbage out," or GIGO, that gets to the heart of the issue. No matter how sophisticated the software, or how realistic the output seems, if the data fed to the program are inaccurate, the results will be poor, and thus can be misleading. The proper way to evaluate the efficacy of the model or simulation is to validate it, and this is usually done by processing known scenarios or input conditions, and making certain the results are representative of the known output within the validated range. Regardless of the qualifications of the engineer, if any mathematical model has not been validated within the boundaries at issue, its use in the courtroom should be carefully considered. Additionally, once the model is used in litigation, engineers should be prepared to provide a fully executable copy of the model if requested during discovery.

4. Literature

Engineers are trained to rely on literature as part of their work, and the literature they employ is nearly as varied as engineers themselves. Structural and mechanical engineers use codes and regulations when they design everything from buildings to bridges, and pressure vessels to heating systems (an extended discussion on the use and misuse of codes is provided below). Engineers rely on published standard methods when they conduct run-of-the-mill tests, scientific literature to test the efficacy of complex calculations and experiments, and textbooks to validate techniques and methods from their educational training.

It is common for engineers to gather literature that addresses an issue about which they are testifying. Industrial engineers may gather literature related to warnings, materials engineers may collect literature related to development and processing of a compound, and mechanical engineers may assemble literature related to stress analysis. Inevitably, literature exists that is not concurrent with the engineer's perspective, and a proper analysis of the available literature should include this as well, with the engineer addressing discrepancies directly.

Engineers may also rely on scientific and technical literature to assess the state of knowledge at a given period in time. This is especially useful in matters involving intellectual property (discussions related to prior art, best mode, and the like) or product design (state-of-the-art analysis). The appropriateness of reliance on this type of literature should not only be weighed by its applicability to the case in discussion, but also by the engineer's mastery and frequency of use of the particular subject. The topic of peer review is often raised concerning scientific and technical literature, and although the peer review process aids in the promotion of sound science and engineering, its presence does not ensure accuracy or validity, and its absence does not imply that a reference is scientifically unsound.

5. Internal documents

Engineers called as experts by either party in a products or personal injury case will likely review documents produced during discovery that relate to the design process of the product in question. From these documents, engineers can often assess whether appropriate actions were taken during the product design process, including product development, product testing and validation, warning and risk communication, and safety and risk assessment. Because the specific constraints imposed on a design are not always apparent from internal engineering documents, and understanding constraints can be critical in terms of effective critical review, engineers called as experts may need to review deposition testimony relating to the design to supplement what they learn from the documents themselves.

VI. What Are the Types of Issues on Which Engineers May Testify?

Because engineers are problem solvers, their work frequently becomes the subject of disputes, which eventually involve lawyers and courtrooms. Many times these disputes involve the sort of "scientific, technical or other specialized knowledge" that may be best understood with the help of one or more engineers.[60] Stated differently, these issues may be difficult for a jury of laypersons, or even judges, to understand and resolve without the assistance of an engineer who was not directly involved in the facts of the case. As a result, disputes involving engineering concepts and principles may be properly the subject of expert testimony from one or more witnesses qualified in the field of engineering.

Just as there are a multitude of disciplines within engineering, there are a multitude of issues upon which engineers may be called upon to testify. Some examples follow.

A. Product Liability

Generally speaking, a product may be defective if it contains a design defect, a manufacturing defect, or inadequate warnings or instructions. Therefore, disputes regarding the efficacy or safety of products typically involve questions regarding whether the product was properly designed, tested, manufactured, sold, or marketed. These issues are examined from the perspective of what was known at the time of first sale and also what was done after information became available about the product's performance.

1. Design

The conception and design of a product is often a focus of dispute in a product liability case. An understanding of the way that engineers think and the engineering design process described above is essential to determine the nature of and extent to which engineering testimony should be admitted. For example, in medical device litigation, it may be significant to know the purpose for which the medical device was designed and the process by which the design at issue was

60. Fed. R. Evid. 702.

achieved. To gain that understanding, testimony from the product designer as well as testimony by engineers with experience in design may be helpful.[61]

The adequacy of testing done on a product is closely related to the issue of design defect. This is true whether the testing in question occurred before the product was first sold ("premarket") or after the product had been on the market for a time and information regarding its performance became available ("postmarket").[62] Engineering testimony may be helpful to the court and to the trier of fact in these circumstances as well.

An engineer's examination of products that have failed in use may result in valuable evidence for a court and trier of fact to consider. For example, an engineer skilled in fractography can testify regarding how and why a product failed.[63] Such testimony may prove helpful to the court and a trier of fact on such issues as whether the subject product was defective as originally designed, whether an alternative design could have been used, what the cost of such response would be and whether the manufacturer's response to such incidents of product failure was reasonable.

61. Russell v. Howmedica Osteonics Corp., No. C06-4078-MWB, 2008 WL 913320 (N.D. Iowa 2008) (biomechanical expert allowed to testify that medical device's inability to handle weight loads was a design defect and hence caused the plaintiff's injuries, and that the defendant's failure to warn surgeons of this fact that caused the failure of the device); Poust v. Huntleigh Healthcare, 998 F. Supp. 478 (D.N.J. 1998) (engineer with expertise in medical device use, safety, and design allowed to testify about defects concerning lack of instructions, the alarm, lack of fail-safe mechanism, and lack of pressure gauge in pneumatic compression device); *see also* Dunton v. Arctic Cat, Inc., 518 F. Supp. 2d 296 (D. Me. 2007) (admitting expert testimony of mechanical engineer and product designer regarding, among other things, purpose and design of certain components of allegedly defectively designed snowmobile); Floyd v. Pride Mobility Prods. Corp., No. 1:05-CV-00389, 2007 WL 4404049 (S.D. Ohio 2007) (three engineering experts, including mechanical engineer with expertise as product designer, allowed to testify about defects in design of scooter); Tunnell v. Ford Motor Co., 330 F. Supp. 2d 731 (W.D. Va. 2004) (engineer allowed to testify about feasibility of proposed safer auto design).

62. *See, e.g.,* Smith v. Ingersoll-Rand Co., 214 F.3d 1235 (10th Cir. 2000) (human factors engineering expert allowed to testify that defendant's failure to conduct human factors analysis of milling machine was inadequate); Montgomery v. Mitsubishi Motors Corp., 448 F. Supp. 2d 619 (E.D. Penn. 2006) (engineer allowed to testify improper or deficient testing rendered vehicle design defective and unsafe, based in part on his review of test results of another engineer); *accord* Phelan v. Synthes (U.S.A.), 35 F. App'x 102 (4th Cir. 2002) (biomedical engineer not allowed to testify about inadequacy of premarket testing of medical device when underlying opinion that device was unreasonably dangerous was not supported by reliable methodology).

63. *See* Parkinson v. Guidant Corp., 315 F. Supp. 2d 754 (W.D. Pa. 2004) (metallurgist who reviewed fractographs was allowed to testify in product liability action that manufacturing flaws caused the premature fracture of guidewire used in angioplasty); Hickman v. Exide, Inc., 679 So. 2d 527 (La. Ct. App. 1996) (expert in, among other things, fracture analysis was allowed to testify about cause of explosion of car battery in product liability action); Reif v. G & J Pepis-Cola Bottlers, Inc., No. CA87-05-041, 1988 WL 14052 (Ohio Ct. App. Feb. 15, 1988) (fractography expert was allowed to testify about cause of break in broken glass bottle).

2. *Manufacturing*

The manufacture of a product and the quality process through which uniformity of ingredients, processes, and the final product are ensured may properly be the subject of product safety litigation. Testimony of engineers with experience in designing and implementing manufacturing systems to ensure product quality may be critical in resolving product disputes and helpful to the court and trier of fact.[64]

3. *Warnings*

Warnings issues in product liability cases are at the intersection of factual evidence and legal standards and thus are particularly difficult for the court and/or other trier of fact to resolve. Many product disputes involve claims concerning the adequacy of warnings that accompanied the product when it was first sold. In these cases, the focus may be on what was known through the conception and design phases of the design process and the necessity for and adequacy of warnings that accompanied the product in view of that knowledge. Other disputes center upon the warnings that were added or could have been added after the product had been used and the company received feedback from users of the product. The reasonableness of the company's response to these reports may be an issue. Thus, the case may be decided on the basis of whether the company conducted, or failed to conduct, design and testing activities in view of that information or whether the company modified the product or communicated to users of the product what it knew.

But not all warnings issues are properly the subject of expert testimony, particularly with respect to products that are regulated by federal law.[65] Properly qualified engineers may be able to provide opinions that could help the court and the trier of fact to understand such issues with respect to such products, but

64. *See, e.g.*, Galloway v. Big G Express, Inc., 590 F. Supp. 2d 989 (E.D. Tenn. 2008) (defendant's expert with significant experience in engineering fields, including product design, allowed to testify about manufacturing process used by the defendant); Schmude v. Tricam Indus., Inc., 550 F. Supp. 2d 846 (E.D. Wis. 2008) (discussing generally the propriety of admitting testimony of expert who studied mechanical engineering and had degree in product design regarding manufacturing process for rivets used in ladder that collapsed); Yanovich v. Sulzer Orthopedics, Inc., No. 1:05 CV 2691, 2006 WL 3716812 (N.D. Ohio 2006) (discussing testimony of engineering experts regarding manufacture of medical device). *See also* Pineda v. Ford Motor Co., 520 F.3d 237 (3d Cir. 2008) (metallurgical engineer allowed to testify about explicit procedure for replacing allegedly defective product in order to reduce likelihood of product failure).

65. The FDA's drug approval process may preempt state law product liability claims based on a failure to warn. *See, e.g.*, Riegel v. Medtronic, Inc., 552 U.S. 312, 128 S. Ct. 999 (2008). *See also* Bates v. Dow Agrosciences LLC, 332 F.3d 323 (2005) (discussing preemption of state law product liability claims by the Federal Insecticide, Fungicide, and Rodenticide Act). *But see* Wyeth v. Levine, 555 U.S. 555, 129 S. Ct. 1187 (2009) (holding that FDA approval of a drug did not preempt state law tort claim based on inadequate drug warnings).

nonetheless may not be allowed to testify based on the substantive law applicable to such products.[66]

For example, industrial engineers, or engineers educated in human factors, may have training that allows them not only to testify when warnings are necessary from an engineering perspective (recall the discussion above about the design process), but also about the efficacy of warnings, and development of risk communications including text, pictures, auditory, or visual signals.

4. Other issues

Issues regarding the sale and marketing of products often concern promises made regarding the expected performance of the product, including both the positive results that a product is able to achieve and, especially, what possible harm that a product may cause. The efficacy of a product may be proved by a straightforward comparison between premarket data on product performance and sales and marketing claims, and engineers may provide helpful testimony regarding the interpretation of such data. There may be a dispute about whether the claims made about the product's safety exceeded the testing results that had been obtained for the product or led to a hazardous situation because the product was not properly tested. These may also be the subject of appropriately qualified engineering testimony.

Common personal injury cases may also present issues on which engineering testimony may be helpful. Such disputes often turn on testimony as to how a particular trauma occurred. Our discussion of biomechanical engineering highlights some of these issues.[67] In a car accident case, properly qualified engineers may provide opinion testimony regarding how an accident occurred, including reconstructing the conduct of each of the parties and how that conduct affected the accident. In a slip-and-fall case, engineering testimony can concern such basic issues as why the injured person slipped and what could have been done to prevent it.

In addition to the above, engineers may also testify about various aspects of a party's damages and give an opinion about whether those alleged damages were caused by the conduct in question. Testimony about causation in a products dispute often involves both factual and legal questions. Through experience, training, and activities in the case, engineers may have the ability to understand the interrelationship between events and thus can provide helpful testimony on whether the asserted damages had a relationship to the asserted misconduct so as to have

66. *See* Pineda v. Ford Motor Co., 520 F.3d 237 (3d Cir. 2008) (metallurgical engineer permitted to testify that safety manual should have contained warning about glass failure in SUV); Michaels v. Mr. Heater, Inc., 411 F. Supp. 2d 992 (W.D. Wis. 2006) (human factors engineering expert allowed to testify about the adequacy of product warning); Nesbitt v. Sears, Roebuck & Co., 415 F. Supp. 2d 530 (E.D. Pa. 2005) (expert with practical experience as engineer allowed to testify that the plaintiff would have responded to an additional warning); Santoro v. Donnelly, 340 F. Supp. 2d 464 (S.D.N.Y. 2004) (mechanical engineer allowed to testify about adequacy of warnings for fireplace heater).

67. *Supra* Section I.C.

been "caused" by it.[68] But issues regarding the standard to apply for the sufficiency of causal proof may be both scientific and legal issues. Thus, the adequacy or admissibility of an engineer's opinion on causation will be evaluated in light of the law, as well as the adequacy of the science that forms the basis for the opinion.[69]

Situations where property damages are asserted may pose special problems on which engineering testimony may be appropriate. For example, determining whether a product problem is an isolated occurrence or whether it is part of a widespread product problem may be difficult to resolve in the absence of engineering testing and analysis, which aims at determining a product defect and a product breakdown process.

B. Special Issues Regarding Proof of Product Defect

Although the definition of what is defective may be the subject of a jury instruction at trial,[70] proof of a product defect may involve identifying key facts that

68. *See, e.g.*, Nemir v. Mitsubishi Motors Corp., 381 F.3d 540 (6th Cir. 2004) (automotive safety engineer allowed to testify that defective seatbelt latching mechanism caused plaintiff's injuries); Babcock v. General Motors, 299 F.3d 60 (1st Cir. 2002) (structural and mechanical engineer allowed to give testimony about impact speed, cause of injuries, how the product allegedly ultimately failed, and testing procedures for the product); McCullock v. H.B. Fuller Co., 61 F.3d 1038 (2d Cir. 1995) (engineer allowed to testify regarding whether plaintiff was within "breathing zone" for hot-melt glue in workplace); Perez v. Townsend Eng'g Co., 545 F. Supp. 2d 461 (M.D. Penn. 2008) (engineer allowed to testify that product was defective, that defect caused plaintiff's injury, and that alternative design would have prevented injury); Farmland Mut. Ins. Co. v. AGCO Corp., 531 F. Supp. 2d 1301 (D. Kan. 2008) (electrical engineer allowed to testify about cause of farm equipment fire); Phillips v. Raymond Corp., 364 F. Supp. 2d 730 (N.D. Ill. 2005) (biomechanical engineer testified as to the mechanics of plaintiff's injury resulting from allegedly defective forklift); Tunnell v. Ford Motor Co., 330 F. Supp. 2d 731 (W.D. Va. 2004) (engineer allowed to testify there was an absence of evidence that the accident was caused by electrical arcing); Figueroa v. Boston Scientific Corp., 254 F. Supp. 2d 361 (S.D.N.Y. 2003) (expert with substantial experience, education, and knowledge in engineering field allowed to testify about cause of damage to plaintiff); Yarchak v. Trek Bicycle Corp., 208 F. Supp. 2d 470 (D.N.J. 2002) (expert in forensic and safety engineering, among other subjects, allowed to testify that bicycle seat caused the plaintiff's erectile dysfunction); Traharne v. Wayne Scott Fetzer Co., 156 F. Supp. 2d 690 (N.E. Ill. 2001) (electrical engineer allowed to testify about cause of deceased's electrocution); Bowersfield v. Suzuki Motor Corp., 151 F. Supp. 2d 625 (E.D. Pa. 2001) (engineer allowed to testify about causation of automobile passenger's injuries).

69. *See generally* Margaret A. Berger, The Admissibility of Expert Testimony, in this manual; *see also* Michael D. Green et al., Reference Guide on Epidemiology, Section V, in this manual.

70. Restatement (Third) of Torts § 2 (1998) provides that the general definition of a product defect is as follows:

> A product is defective when, at the time of sale or distribution, it contains a manufacturing defect, is defective in design, or is defective because of inadequate instructions or warnings. A product:
>
> (a) contains a manufacturing defect when the product departs from its intended design even though all possible care was exercised in the preparation and marketing of the product;
>
> (b) is defective in design when the foreseeable risks of harm posed by the product could have been reduced or avoided by the adoption of a reasonable alternative design by the seller or other distributor,

relate to that definition. These issues may be the subjects of the testimony of an engineer. The issue of whether a product is unreasonably dangerous may involve proof of available alternative designs, the existence of modifications to the product that would make it safer (which directs us back to the discussion on risk, above; all products can be made safer at the expense of cost and/or convenience) and what consumers expect the product to do or not to do, to identify a few such issues.[71] Engineers may be asked to engage in testing to determine a cause and/or mechanism of failure and as a basis for an opinion regarding product defect. Such testing may include accelerated testing and end-use testing to replicate the conditions that the products see in use.

To understand the product and its expected or anticipated uses, engineers may review documents regarding the product at issue and published literature about like products or product elements. Visits to sites where the product is or was in use may provide information to engineers about recurring characteristics of product performance and aspects of the environment of use, which bear on that performance. Visual examination, measurements made at the site and experiments conducted at the site and in the laboratory may provide valuable information regarding the characteristics of the product that affect product performance or nonperformance.

In sum, the engineer's problem-solving approach using the design process as described above can provide valuable information about the nature and cause of product problems and the limitations of the design of the product at issue, including the characteristics of the environment of use and the choice of materials for the subject product. Armed with this and other information, properly qualified engineers can provide valuable opinions on issues going to the heart of the question of whether the product at issue is defective and caused the claimed damages.[72]

> or a predecessor in the commercial chain of distribution, and the omission of the alternative design renders the product not reasonably safe;
>
> (c) is defective because of inadequate instructions or warnings when the foreseeable risks of harm posed by the product could have been reduced or avoided by the provision of reasonable instructions or warnings by the seller or other distributor, or a predecessor in the commercial chain of distribution, and the omission of the instructions or warnings renders the product not reasonably safe.

See also Restatement (Second) of Torts § 402A (1965), which defines a defect as one that makes a product "unreasonably dangerous."

71. Martinez v. Triad Controls, Inc., 593 F. Supp. 2d 741 (E.D. Pa. 2009) (engineer allowed to testify about design defects and warnings); Page v. Admiral Craft Equip. Corp., No. 9:02-CV-15, 2003 WL 25685212 (E.D. Tex. 2003) (mechanical engineer allowed to testify about defect in design of bucket and safer alternative).

72. "While an expert's legal conclusions are not admissible, an opinion as to the ultimate issue of fact is admissible, so long as it is based upon a valid scientific methodology and supported by facts. *See* Fed. R. Evid. 704. The 'ultimate issue of fact,' as used in Rule 704, means that the expert furnishes an opinion about inferences that should be drawn from the facts and the trier's decision on such issue necessarily determines the outcome of the case." Strickland v. Royal Lubricant Co., Inc., 911 F. Supp. 1460, 1469 (M.D. Ala. 1995).

C. Intellectual Property and Trade Secrets

Engineering testimony may be helpful in disputes regarding patents and other forms of intellectual property. Knowledge has become a key source to wealth in our economy,[73] and we increasingly depend on innovation and the protection of innovation.[74] The federal government's power to protect patents and copyrights is one of only a handful of enumerated powers in the U.S. Constitution.[75] Engineers are at the very heart of technology and innovation and therefore often become natural contributors to the resolutions of disputes involving these subjects.

The issues for factual or expert engineering testimony in this area are closely allied to those highlighted in the above description of the product design process. Key issues concern conception and development of the invention or protected trade secret, commercialization and sales/marketing of the protected concept, infringement or theft of the protected concept, and damages, including proof of willfulness or bad intent. There are a myriad of situations in which engineering testimony may be received. We will highlight a few of them.

The patentability of an idea is measured by its advance over prior art in the relevant field. Almost all new inventions are combinations or uses of known elements. What constitutes prior art and what is the relevant field for such art are thus questions that relate to the conception stage of the design process. Who is a qualified engineer to testify about these issues is answered under the *Daubert* standard. Thus, engineering testimony may be helpful on such issues as whether the invention is new or novel and whether it is non-obvious to one who has ordinary skill in the art. And engineers can help to define the description of the person with ordinary skill and interpret what such a person would learn from the art in question. Prior art is meant to include all prior work in the field. It sometimes connotes "public" prior art, not hidden or unknown art. A properly qualified engineer witness can provide relevant and reliable testimony regarding these and other prior art–related questions.

The rules for using engineering experts in patent infringement proceedings in federal courts are reasonably well defined. For example, under the U.S. Supreme Court's decision in *KSR International Co. v. Teleflex, Inc.*,[76] non-obviousness is

73. *See, e.g.*, Thomas A. Stewart, Intellectual Capital (1997).

74. "There is established within the [National Institute of Standards and Technology] a program linked to the purpose and functions of the Institute, to be known as the 'Technology Innovation Program' for the purpose of assisting United States businesses and institutions of higher education or other organizations, such as national laboratories and nonprofit research institutions, to support, promote, and accelerate innovation in the United States through high-risk, high-reward research in areas of critical national need." 15 U.S.C. § 278n. *See also* Prioritizing Resources and Organization for Intellectual Property (PRO-IP) Act of 2008, Pub. L. No. 110-403, 122 Stat. 4256 (2008); America Creating Opportunities to Meaningfully Promote Excellence in Technology, Education, and Science Act, Pub. L. No. 110-69 (2007).

75. U.S. Const. art. I, § 8.

76. 550 U.S. 398 (2007). *See also* Dennison Mfg. Co. v. Panduit Corp., 475 U.S. 809 (1986).

ultimately a question of law for the court to decide. However, the underlying factual determinations, including the secondary factors involved in determining patent validity, remain jury questions concerning which expert engineering testimony may be admitted.[77] Questions regarding the scope and teachings of prior art also may invite engineering testimony and interpretation.[78]

A similar analysis applies to trade secret matters. The nature and scope of the claimed trade secret, including the existence of steps taken to protect the trade secret are all issues where the engineer as a witness may be involved. This is true of protected processes and methods as well as devices or other products of the subject trade secret.[79]

When an assertion is made that intellectual property or protected trade secrets have been infringed, engineering testimony may be necessary on a number of issues as courts attempt to resolve issues regarding identifying features of the challenged device, methods or processes that infringe the protected property.[80] Additional issues may relate to knowledge regarding the protected property and conception and design of the subject of the alleged infringement.[81]

Proof of damages may involve a number of issues that relate to or derive from an engineer's analysis of the scope of claims or protected methods and processes, commercial viability of the subject intellectual property, and scope of protection afforded by the subject patent. Qualification of engineers as witnesses to provide testimony in these areas may present its own challenges under *Daubert* as to both reliability and relevance.

D. Other Cases

There are many other areas where engineering testimony may be helpful to the court and trier of fact. Because the range of such possible situations is virtually limitless, we will list only a few examples.

77. *See* Finisar Corp. v. DirecTV Group, Inc., 523 F.3d 1323, 1338 (Fed. Cir. 2008).

78. *See, e.g.*, Rosco, Inc. v. Mirror Lite Co., 506 F. Supp. 2d 137 (E.D.N.Y. 2007) (mechanical engineer allowed to present testimony of his review of patent for teaching or suggestion as to meaning of the claims).

79. Am. Heavy Moving & Rigging Co. v. Robb Technologies, LLC, No. 2:04-CV-00933-JCM (GWF), 2006 WL 2085407 (D. Nev. 2006) (in case involving misuse of trade secrets, engineer appointed by the court to assist in making discovery rulings).

80. *See, e.g.*, The Post Office v. Portec, Inc., 913 F.2d 802 (10th Cir. 1990).

81. *See* State Contracting & Eng'g Corp. v. Condotte Am., Inc., 346 F.3d 1057 (Fed. Cir. 2003) (expert in civil and structural engineering testified about whether different pieces of prior art are in the same field of endeavor as patents at issue); Philips Indus., Inc. v. State Stove & Mfg. Co., 522 F.2d 1137 (6th Cir. 1975) (use of engineering expert to establish the presence of design concept in prior art); Mayview Corp. v. Rodstein, 385 F. Supp. 1122 (C.D. Cal. 1974) (tool engineer testifying about concept of balance in hand-tool design in prior art).

- Claims of personal injury or property damage resulting from the spread of a toxic substance may involve a number of issues where engineering testimony may be both reliable and relevant.[82]
- Environmental disputes regarding the necessity for and nature of an environmental problem and the responsibility for and cost of its cleanup involve numerous issues concerning which properly qualified engineers may provide reliable and relevant evidence.[83]
- The testimony of an engineer can be helpful in determining causation in both product liability cases and nonproducts liability cases as well. Recent cases in the electrical engineering area demonstrate the range of possible situations where such issues may arise and engineering testimony may be admitted.[84] For example, an electrical engineer was allowed to testify about lightning in *Walker v. Soo Line Railroad Co.*[85] The plaintiff in that case filed suit under the Federal Employees Liability Act. Claiming that he had been injured by lightning while working in a railroad tower, the plaintiff sought to introduce the testimony of the chairman of the electrical engineering department at the University of Florida to the effect that lightning could have struck a number of places in the yard and penetrated the tower without a direct hit. The district court excluded the evidence and the Seventh Circuit reversed, finding that the jury would have been helped by hearing the engineer's testimony about the ways in which lightning could have struck the tower, even if he could not testify which of the locations was struck or if any of them were struck at all.
- In a slightly different context than might be expected, an electricity transmission line planning engineer testified as an expert at a administrative hearing in *California Public Utilities Commission v. California Energy Resources Conservation & Development Commission.*[86] The dispute in that case was the extent of the CERCDC's jurisdiction over transmission line siting, and more specifically the interpretation of a section in California's

82. *See, e.g.*, Jaasma v. Shell Oil Co., 412 F.3d 501 (3d Cir. 2005) (civil and environmental engineer permitted to testify about environmental status of real property, which was relevant to damages and efforts to mitigate); *In re* Train Derailment Near Amite La., No. Civ. A. MDL. 1531, 2006 WL 1561470 (E.D. La. 2006) (court relied on declaration of environmental engineer regarding exposure to airborne contaminants in concluding that claims of potential class were not based on actual physical harm).

83. Olin Corp. v. Lloyd's London, 468 F.3d 120 (2d Cir. 2006) (admission of environmental civil engineer testimony on issue of property damage in pollution liability insurance coverage case not an abuse of discretion).

84. *See, e.g.*, Newman v. State Farm Fire & Cas. Co., 290 F. App'x. 106 (10th Cir. 2008) (electrical engineer was allowed to testify about the origin of fire that destroyed insureds' house); McCoy v. Whirlpool Corp., 287 F. App'x 669 (10th Cir. 2008) (electrical engineer was allowed to testify in product liability case that manufacturing defect in dishwasher caused fire).

85. 208 F.3d 581 (7th Cir. 2000).

86. 50 Cal. App. 3d 437 (Cal. Ct. App. 1984).

Public Resources Code which defined "electric transmission line" as "any electric power line carrying electric power from a thermal power plant located within the state to a point of junction with any interconnected transmission system."[87] The engineer, employed by Pacific Gas & Electric, testified at the hearing before the CERCD about the use of certain terms in the industry related to that definition and about electricity transmission principles. The court subsequently relied on that testimony in part in determining that "electric transmission line" had a plain meaning, and that the plain meaning cut off the CERCDC's jurisdiction at the first point at which a power line emanating from a thermal power plant joined to the interconnected transmission grid.

- Civil engineers have been allowed to testify in a broad range of circumstances also, including those involving an improper application of the design process in the building of a bridge. Numerous examples of situations involving roles of engineers involved in building bridges, which ultimately failed, have been described in an earlier version of this guide.[88] In each of those situations, engineering testimony from engineers who were involved in the design of the bridge or who had experience in designing other bridges or who had experience with design generally could be qualified to testify in an inquiry or lawsuit about the causes and financial implications of the failures.

VII. What Are Frequent Recurring Issues in Engineering Testimony?

A. Issues Commonly in Dispute

Following are several issues with which engineers frequently are confronted in the course of attempting to give testimony as experts. Each of them is controlled by the specific fact pattern that gives rise to the case and the way in which the case is presented. In this section we describe the *issues as they are perceived by the engineer in the courtroom.* Because this is not a treatise on the procedural or substantive law at issue, we do not summarize the state of the law on each issue. We assume that the court and other readers of this guide are familiar with the applicable law on these issues.

87. *Id.* at 440.

88. Petroski, *supra* note 45, at 593–94, 597–600, 604–06, 608–09, 612–13.

1. *Qualifications*

In an earlier section of this guide (Section III.C.2), we referred to the *Stringfellow* case.[89] An important aspect of this litigation revolved around estimates of the mass of toxic materials released over a long time period some 20 to 30 years after the fact. To reconstruct events long past, a chemical engineer used aerial photographs of the site taken during its period of operation to estimate the surface areas of the toxic waste ponds. His qualifications were challenged under *Frye v. United States*[90] because he was not a photogrammist. Despite having a background and credentials to support the work he did using the aerial photographs, but not the pedigree, the court found that a photogrammist would need to confirm his findings. In the end a photogrammist corroborated the engineer's work. This is but one example of an all too common situation where an engineering expert's qualifications have been challenged based on "name" rather than on relevant and documented experience. Under *Daubert,* there may be even more pressure on the court to assess who can or cannot testify as an expert. But this example illustrates that a court should be cautious about drawing conclusions about an expert's qualifications based solely on titles, licenses, registration, and other such documentation.

2. *Standard of care*

Another common issue for engineers to confront in their testimony is the standard of care. Engineers do not think of the concepts of standard of care and duty of care as they relate to tort law, particularly negligence. Instead, for many engineers, "standard of care" means "how we do it in my office" or some variation thereof.

Following the Oklahoma City bombing, a structural engineering expert prepared a report regarding blast damage and progressive collapse for the U.S. Attorney prosecuting McVeigh. In an attempt to block this testimony, McVeigh's defense team obtained an affidavit from an engineer with a well-known structural engineering firm to the effect that the prosecution expert's report did not meet minimal standards for a building condition report because it did not include detailed architectural and structural drawings, measurements, and specifications, all of which were irrelevant to the issues at bar. The defense expert engineer argued essentially that he and his firm were leaders in the field of building assessment reports and therefore what they did set the standard. He was wrong on two counts: (1) the practice of any single firm or office does not establish the standard of care, and (2) the standard of care for one technical purpose (condition assessment of commercial buildings with leaky curtain walls) cannot be applied to another technical purpose (determination of number of bombs employed to destroy a building) just because both involve buildings and engineers.

89. United States v. Stringfellow, No. CV 83-2501 JMI, 1993 WL 565393 (C.D. Cal. Nov. 30, 1993).

90. 293 F. 1013 (D.C. Cir. 1923).

The phrase "standard of care" has various meanings and connotations to engineers that are somewhat discipline specific. Standard of care in the medical sciences may be different than standard of care in some other context. In engineering, it can be said that the standard of care is met whenever the design process was properly employed at the point in time that the event or incident happened. Although the design process itself is "fixed," when properly applied to a problem in the 1940s and again to the same problem in 2009, the design outcome can be quite different and indeed might be expected to be so. Even so, the standard of care may be met each time.

3. *State of the art*

"State of the art" has a specific meaning in the law and may be the subject of a particular statute in many jurisdictions. In addition, state of the art can be a distinct defense in many states.[91] To engineers, however, its meaning may be slightly different.

Simply put, this phrase refers to the current stage of development of a particular technology or technological application. It does not imply that it is the best one can ever hope for but is merely a statement that at whatever point in time referenced, technology was in a certain condition or form. For instance, the Intel 4004 4-bit microprocessor was state of the art in 1971 whereas the Intel 64-bit microprocessor was the state of the art in 2006. Of course, there is the question as to whether in either of these cases those microprocessors were state of the art for just Intel, for all American semiconductor companies, or for all semiconductor companies in the world. The question of the context in which this phrase is used often lies at the heart of disputes. Because appropriate context may be difficult to pin down, experts are often challenged with defining the "state of the art" in relation to a particular technology or application. The answer from an engineering perspective is often an assumption, nothing more, nothing less. As such, from an engineering perspective, it is best to accept this phrase as a general colloquialism that is difficult to define even though it is simple to state.

4. *Best practice*

Although this term is used colloquially and oftentimes in "business" activities, to engineers it is not a phrase that is easily quantifiable and suffers from meaning different things to different people. Despite this, it generally refers to the notion that at any point in time there exists a method, technique, or process that is preferred over any other to deliver a particular outcome. That being said, there is great latitude in how one goes about determining that preference and associating it with the desired outcome. So, although it sounds good, this phrase is fraught

91. *See, e.g.*, Ariz. Rev. Stat. § 12-683(1) (2009); Colo. Rev. Stat. § 13-21-403(1)(a) (2009); Ind. Code § 34-20-5-1(1) (2009).

with ambiguity. In the end, the more important issue is whether there was adherence to the design process.

5. Regulations, standards, and codes

An issue that often arises in matters involving buildings and structures is the distinction between design codes and physics (political laws vs. physical laws) in the context of failure analysis. Design codes and standards are very conservative political documents. They are conservative because they are intended to address the worst-case scenario with a comfortable margin of safety. But buildings do not fail because of code violations—they fail according to the laws of physics. They do not fail when the code-prescribed loads exceed the code-prescribed strength of the materials—they fail when the actual imposed loads exceed the physical strength of the components. Buildings fail not when the laws of man are ignored but when the laws of physics are violated. Examples of this are most common in the context of earthquake-damaged structures. Buildings are not designed to resist 100% of expected earthquake forces. Rather, they are designed to resist only a fraction of the expected load (typically about one-eighth) without permanently deforming. The code implicitly recognizes that buildings are much stronger than assumed in design and also have considerable ability to absorb overloads without failure or collapse. Yet following an earthquake, engineers may inappropriately compare the ground accelerations recorded by the U.S. Geological Survey with design values in the code.

In the Northridge, California earthquake, recorded acceleration values were 2–3 times greater than the design code values. Many engineers concluded that the buildings had been "overstressed" by 200–300% and were thus extensively damaged, even if that damage was not visually apparent. In a line of reasoning remarkably similar to that of the plaintiff's expert in *Kumho*,[92] the damage was "proved" analytically, even though it could not be physically seen (or disproved) in the building itself. (If the same logic were applied to cars, every car that sustained an impact greater than the design capacity of the bumper would be a total loss.) If this approach was accepted, the determination of damage could only be done by a few wizards with supposedly sophisticated, yet often unproven, analytical tools. The technical issues in the Northridge situation were thus removed from the realm of observation and common sense (where a jury has a chance of understanding the issues) to the realm of arcane analysis where the experts have the final say.

This is not to say that standards and codes do not have their place in the courtroom. We described above how standards are often used by engineers to conduct tests, and cases that involve malpractice or standard-of-care may often critically examine if a particular code was followed in the course of a design. On

92. 526 U.S. 137 (1999). In *Kuhmo*, the expert inferred the defect from an alleged set of conditions, even though the alleged defect was not observed.

the other hand, failure to use a code, or comparison of code values to actual values does not guarantee that a disaster will occur. Common sense is often the best judge in these situations—if a code value is exceeded, yet no damage is observed, it is likely that the conservative nature of the code met its objectives and protected its subject.

6. Other similar incidents

From an evidentiary perspective, evidence of similar or like circumstances has a number of evidentiary hurdles to overcome before it can be admitted into evidence.[93] To an engineer, however, the concept of similarity or "other similar incidents" (OSIs) has a somewhat different meeting and describes the types of circumstances and documentation of such circumstances that an engineer can rely on as a basis for his or her opinions. Although this section focuses primarily on product design issues, the underlying theme is nonetheless broadly applicable across the domain of engineering forensics.

Sometimes these other events are recorded in documentary form and relate to events regarding product performance characteristics, product failures, product anomalies, product performance anomalies, operational problems associated with product use, product malfunctions, or other types of product failures. These events are sometimes alleged by a party to a dispute to be substantially similar in kind to an event or circumstance that had precipitated the subject case. Alleged OSIs can be documented in multiple forms: (1) written narratives from various sources (consumers, employees of the manufacturer, bystanders to a reported event, insurers' representatives, investigators, law enforcement personnel, owners of a location involved in the dispute at bar, etc.) who might prepare and submit a record of observation to a legal entity who retains those records of submission; (2) telephonic reports of the same character and source as written reports, but documented through telephone reports made to a recording representative or office staff responsible for collecting event reports of interest to a legal entity; (3) electronic submissions of the same character and source as written narratives; (4) reports in a standardized format that are intended to record and document events of interest (the forms may be in written or electronic media; (5) images of events in film or electronic media that may or may not also have been recorded and submitted in alternative formats. As a result, each may have its evidentiary hurdles to overcome before it is admitted into evidence.

Similarly, each OSI may have legal issues regarding authentication, which may be overcome by the repository where the underlying documentation is

93. For evidence of other similar incidents (OSIs) to be admissible, the proponent must show that the OSIs are (1) relevant, *see* Fed. R. Evid. 401; (2) "substantially similar" to the defect alleged in the case at bar; and (3) the probative value of the evidence outweighs its prejudicial effect, *see* Fed. R. Evid. 403. Some courts merge the first two requirements; to be relevant, the OSIs must be substantially similar to the incident at issue.

found. The repositories of documents and reports that may be alleged to be OSIs to an issue at bar can have many original purposes, and a collection of such documents may serve multiple purposes for the owner institution. Such document collections may be used by the owner of the repository for various administrative purposes, accounting, claims management and resolution, an archive of information and/or data, database management, institutional knowledge building, warranty management, in-service technology performance assessment and discovery, service records, customer interactions, and satisfaction of regulatory specifications or requirements, to name a few. Discovery requests may call for the owner of the materials to search and retrieve records, documents, and reports from such repositories even if the collections and repositories themselves may not have been constructed for the purposes of document search and retrieval. Sometimes engineers can be of use in searching and retrieving potentially relevant materials.

OSIs are discovered and may be offered into evidence to (1) demonstrate prior knowledge on the part of the record owner regarding an alleged defect or danger manifest to the consuming public that is causally related to the issue at bar; (2) demonstrate by the number, volume, or rate of reports that a defect exists; and/or (3) demonstrate careless disregard for the safety of others.[94] To be admitted or relied upon by an engineering expert, the proponent must demonstrate that the event recorded and reported is "substantially similar" to the issue at bar.[95] Testifying engineers can be useful in identifying and describing the specific characteristics that must be known and shown to make an assessment of similarity, including specifying objective parameters for determinations of the degree of similarity or dissimilarity and detailing the objective parameters and physical measurements necessary and sufficient to determine substantial similarity. The conditions that are necessary and sufficient to demonstrate substantial similarity include the following: (1) the product or circumstance in the alleged OSI must be of like design to the product or condition at issue in the instant case; (2) the product or circumstance in the alleged OSI must be of like function to the product or condition at issue in the instant case; (3) the application to which the product had been subjected must be like the application to which the product at issue in the instant case was subjected; and (4) the condition of the product, its state of repair, and/or its relevant state of wear must be like the state of repair and the relevant state of wear of the product that had been involved in the instant case.[96] Engineers can contribute to a technical understanding of each of these dimensions and, in some cases, they may be able

94. *See, e.g.*, Sparks v. Mena, No. E2006-02473-COA-R3-CV, 2008 WL 341441, at *2 (Tenn. Ct. App. Feb. 6, 2008); Francis H. Hare, Jr. & Mitchell K. Shelly, *The Admissibility of Other Similar Incident Evidence: A Three-Step Approach*, 15 Am. J. Trial Advoc. 541, 544–45 (1992).

95. *See, e.g.*, Bitler v. A.O. Smith Corp., 391 F.3d 1114, 1126 (10th Cir. 2004); Whaley v. CSX Transp. Inc., 609 S.E.2d 286, 300 (S.C. 2005); Cottrell, Inc. v. Williams, 596 S.E.2d 789, 793–94 (Ga. Ct. App. 2004).

96. *See, e.g.*, Brazos River Auth. v. GE Ionics, Inc., 469 F.3d 416, 427 (5th Cir. 2006); Steele v. Evenflo Co., 147 S.W.3d 781, 793 (Mo. Ct. App. 2004).

to apply objective measures to questions of substantial similarity and thus quantify the level of similarity between an event proffered as an OSI and the instant case.

The reverse is also true. Failure to establish likeness in any of these dimensions is failure to demonstrate substantial similarity to the circumstances of the subject case.[97] If one or more of the necessary and sufficient conditions are unknown or unknowable, the test of substantial similarity also fails; the lack of demonstrable similarity is a lack of substantial similarity.

To demonstrate like design, a product or condition need not be identical in all aspects of form.[98] It must simply be similar in form to the product or condition at issue in the instant case.[99] Consider a machine control design with a feature alleged to have been the proximate cause of an injury-producing event that gave rise to a product liability lawsuit. Events proffered as OSIs that involve products having an identical control design meet the test of "likeness" in design. In addition, other control designs that differ in aspects not related to the feature that is alleged to have served as the proximate cause for the instant injury event may also be considered to be "like" if the relevant design elements on the two products cannot be differentiated. Engineers can assess the design elements of the control, determine which features may be relevant to questions of design likeness, and provide testimony to answer such questions.

Like function can be demonstrated if the operational purpose of the product or condition defined in the alleged OSI is similar to the function of the product or condition in the instant case. In the control design hypothesized above, a control that is applied to command the dichotomous functional states to start and stop (either "on" or "off") a crane winch might serve the same operational purpose to start and stop another type of equipment or winch. In such a case, the functions and purpose of the control design may be alike. If however, that same control design is applied to a machine in which the operational purpose is not simply to command a dichotomous "on" or "off" signal, but rather its purpose is to provide a modulated signal to which the machine response is a continuously variable function of control placement, the control design function is unlike the purpose of dichotomous positioning. Engineers can provide assessments and analyses of the functions embedded in a specific design and assist in the determination of likeness or lack of likeness between an instant condition and one proffered as an OSI.

Like application can be demonstrated if it can be shown that the operational conditions to which the product is subject are alike in the proffered OSI and in the instant case.[100] The environmental exposure to which a product is subjected must be of like condition. A control design function can vary with temperature,

97. *See, e.g.*, Peters v. Nissan Forklift Corp. N. Am., No. 06-2880, 2008 WL 262552, at *2 (E.D. La. Feb. 1, 2008); Whaley v. CSX Transp., Inc., 609 S.E.2d 286, 300 (S.C. 2005).

98. *See, e.g.*, Bitler v. A.O. Smith, 391 F.3d 1114, 1126 (10th Cir. 2004).

99. *Id.*

100. *See, e.g.*, Steele v. Evenflo Co., 147 S.W.3d 781, 793 (Mo. Ct. App. 2004).

air or water exposure, reactions to corrosive elements, reactions to acid or base contaminants, and in potential interactions with surrounding materials and components that can be of differing electrochemical potential. Engineers with the appropriate technical background can evaluate operating conditional applications and determine if the conditions that obtain for a proffered OSI are similar to those that had obtained in an instant case, thereby assisting the determination of substantial similarity.

Differing environmental exposures resulting from differing applications may render an event proffered as an OSI unlike and not substantially similar. Further, like applications must comprehend that the load and stress conditions to which a product or condition is placed is substantially similar to the circumstances that obtained in the instant case to which the OSIs are being proffered for comparison. In our control design identified above, the control device may be manually actuated through a lever. Levers of differing length will apply differing forces to the control device and produce differing operational stresses upon the control device itself. The durability and performance of the control design itself can be affected by these differing operating applications, and anomalies or failures under one application may not be at all similar to those that obtain under differing circumstances in which the operating loads and applied stresses are different. Engineers are well qualified to assess conditions of comparative loading and applied stresses.

A like state of repair can be demonstrated if there is reasonable evidence that products involved in the proffered OSI are (1) in a specific working order, (2) in a condition of adjustment (if possible to adjust), (3) in a state of wear, and (4) within an expected range of tolerance that would not differentiate the product or condition from that which obtained in the product or condition involved in the instant case. Additionally, the products or conditions reported in the proffered OSIs must be shown to be free of modification from an original design state, or must be shown to be in a state of modification that is reflective of the product or condition involved in the instant case.[101] An absence of evidence to demonstrate a state of likeness in application, operating environment, state of repair and wear, or state of modification is not sufficient to show similarity. Engineers with appropriate background can review data and information about modifications and service conditions related to wear and wear rates, as well as assess information related to the state of repair or disrepair, and thereby contribute to understanding of the level of similarity or dissimilarity among specific events and operational conditions.

For evidentiary reasons, OSIs generally are not admissible to demonstrate the truth of the matter recorded therein.[102] Event records are necessarily reports of noteworthy events made after the fact by parties who may or may not have an interest in establishing a specific fact pattern, may or may not be qualified to

101. *See, e.g.*, Cottrell, Inc. v. Williams, 596 S.E.2d 789, 791, 794 (Ga. Ct. App. 2004).
102. Fed. R. Evid. 801 & 802.

make the observations and assertions included in such reports, and may or may not have any specialized training necessary to evaluate proper system function or state of repair. The persons who report events collected and offered as OSIs may not be fully informed of the set of circumstantial conditions that are necessary and sufficient to determine causation of the reported event. Thus, often-reported events have incomplete or insufficient data and information to determine substantial similarity. Even if informed, persons reporting events may not have the correct observational powers, tools, and insights necessary for accurate evaluation and reporting. The individuals who make reports regarding recorded events may be unable to factually assess and accurately report all of the conditions relevant to determination of event causation and resolution of questions regarding substantial similarity. Reports of events made by parties who may have an interest in economic recovery or other compensation may not always accurately disclose known or knowable facts that could bear on determinations of causation and substantial similarity. Furthermore, some parties may have an economic or other interest in the outcome of a report or claim. Therefore, such reports, if offered to prove the truth of the other incidents, are typically excluded as hearsay (unless the business records exception applies).[103]

B. Demonstratives and Simulations

Computer animations, simulation models, and digital displays have become more common in television and movies, especially in entertainment media concerning forensic investigation, law enforcement, and legal drama. The result is an increased expectation among the court and juries that visual graphics and displays will be used by engineering experts and other expert witnesses to explain and illustrate their testimony. Additionally, boxed presentation software such as PowerPoint, is often a technology used. Attorneys and their clients typically expect their experts will use computer animations, simulations, and/or exhibits to educate the jury and demonstrate the bases for their opinions. When used correctly, these tools can make the expert's testimony understandable and can leave a lasting impression with the trier of fact of that party's theory of the case. For that very reason, the role of the court as the gatekeeper for use of these demonstratives has become increasingly critical. As the technology underlying these tools rapidly advances, the court's task likewise becomes more difficult. In assessing the validity of these tools, the court is often forced to decide whether the visual display accurately represents the evidence and/or is supported by the

103. *See* Willis v. Kia Motors Corp., No. 2:07CV062-PA, 2009 WL 2351766 (N.D. Miss. July 29, 2009) (finding customer complaints of similar accidents were not hearsay because they were offered to notice, not the truth of the matter asserted, and even if they were hearsay, they fell under the business records exception of Fed. R. Evid. 803(6)).

expert's opinions and qualifications.[104] To assist the court in this difficult task, we present some guidance regarding the types of technology presently in use and the strengths and weaknesses of each.

A primary basis for misunderstanding and uncertainty is the difference between a computer animation and a computer simulation. An animation is a sequence of still images graphically illustrated (two dimensions) or modeled (three dimensions), and are often textured and rendered to create the illusion of motion. A cartoon is a simple example. There are no constraints inherent in an animation, and the laws of physics, or any other science, do not necessarily apply (a black mouse can be dressed in red shorts with yellow shoes and be made to dance, sing, and fly). The lack of imposed restriction does not make the animation deficient a priori; if the still images that comprise the animation are accurate in their representation of individual snapshots of time, then the animation itself can be proven precise. The converse, of course, is also true.

Animations contain key frames that define the starting and ending points of actions, with sequences of intermediate frames defining how movements are depicted. For example, a series of still photographs can depict the path of a vehicle vaulting off an embankment, with a single image at the takeoff, mid-flight, and landing positions each correct in its representation. However, when an animation of the event is created, the intermediate frames fill in the missing areas, and if so desired, contrary to known physical phenomena, the animation could show the vault trajectory of the vehicle to remain flat and then suddenly drop, similar to the inaccurate representation of motion experienced by a cartoon coyote momentarily contemplating his fate after chasing a bird off a cliff. Thus, in an animation, some of the inputs (stills) may represent reality, but the sum of the parts (intermediate frames) may not.

Unlike an animation, a simulation is a computer program that relies on source data and algorithms to mathematically model a particular system (see, e.g., the discussion on finite element modeling, above), and allows the user to rapidly and inexpensively gain insight into the operation and sensitivity of that system to certain constraints. Perhaps the most common example of a simulation can be found daily as a computer-generated image showing the predicted growth of a storm system.

On the surface, a simulation would seem to provide more accuracy than an animation. However, this is not necessarily the case. The simulation model is only as accurate as its input data and/or constraining variables and the equations that form its calculation stream. Simulation models also require a sensitivity analysis—just because a model produces an answer does not mean that it is the best model or

104. *See* Lorraine v. Market Am. Ins. Co., 241 F.R.D. 534 (D. Md. 2007) (distinguishing between demonstrative computer animations and scientific computer simulations and discussing the evidentiary requirements, including authentication, for each); People v. Cauley, 32 P.3d 602 (Colo. Ct. App. 2001) (same).

the most correct answer. For example, a computer model depicting the motions of a vehicle prior to and after an impact with a pole may be correct if it matches the known physical evidence (e.g., tire marks and vehicle damage). However, whether the model is accurate depends on the accuracy of the inputs for tire friction, vehicle stiffness, vehicle weight, location of the vehicle's center of gravity, etc. Even if the inputs are accurate, once a solution is found, other solutions may exist that also match the evidence. Assessing the accuracy of each solution requires an iterative process of making changes to those variables believed to have the greatest effect on the output. Simply put, the difference between a vehicle accident simulation model that predicts 10 inches of crush deformation and two complete revolutions post impact versus 14 inches of crush and three complete revolutions may depend on just a few selected vehicle characteristics. Thus, compared to an animation, in a simulation model, the sum of the constraining variables and equations may represent reality, but some of the user-selected inputs may not.

The difficulty for the court is the need to decide whether some or all of the computer animation or simulation accurately represents the facts and/or opinions of the expert.[105] This is not an easy endeavor, but can usually be executed in a reasonable fashion for simulations by evaluating whether the simulation has been validated. If the underlying program predicts the behavior of vehicles in a crash, it can be validated by crashing vehicles under controlled conditions, and comparing the actual results to those predicted by the simulation. If the software in question predicts the response of a complex object to applied forces, it can be validated by modeling a simple object, the response of which can be calculated by hand, and comparing the simulation to those known results.[106]

Similarly, for animations, engineers need to establish authenticity, relevance, and accuracy in representing the evidence using visual means.[107] They may rely on blueprint drawings, CAD (computer-aided design) drawings, U.S. Geological Survey data, photogrammetry, geometric databases (vehicles, aircraft, etc.), eyewitness statements, and field measurements to establish accuracy of an animation.

VIII. Epilogue

Most engineers are not educated in the law and to them the setting of a deposition or a courtroom is peculiar and often uncomfortable. The rules are different

105. *See id.*

106. *See* Lorraine v. Markel Am. Ins. Co., 241 F.R.D. 534 (D. Md. 2007); Livingston v. Isuzu Motors, Ltd., 910 F. Supp. 1473 (D. Mont. 1995) (finding computer simulation of rollover accident by expert to be reliable and admissible under *Daubert* whether computer program was made up of various physical laws and equations commonly understood in science, program included case-specific data, and expert's computer simulation methodology had been peer reviewed).

107. *See, e.g.*, Friend v. Time Mfg. Co., No. 03-343-TUC-CKL, 2006, WL 2135807 (D. Ariz. July 28, 2006); People v. Cauley, 32 P.3d 602 (Colo. Ct. App. 2001).

from those to which they are accustomed. The conversations are somewhat alien. Treading in this unfamiliar territory is a challenge. And so, although it is important for the engineer to "fit" into this environment, it is equally important for the triers of fact and the court to understand the engineer's world. We hope this chapter has provided a glimpse into that world, and by considering it, the reader will have some insight as to why engineers respond to questions as they do. The foundation that underlies and supports essentially all that has been done and all that will be done by engineers is the design process. It is the roadmap for innovation, invention, and reduction to practice that characterizes those who do engineering and who call themselves "engineers." It is the key metric against which products and processes can be and should be evaluated.

IX. Acknowledgments

The authors would like to thank the following for their significant contributions: Dr. Roger McCarthy, Robert Lange, Dr. Catherine Corrigan, Dr. John Osteraas, Michael Kuzel, Dr. Shukri Souri, Dr. Stephen Werner, Dr. Robert Caligiuri, Jeffrey Croteau, Kerri Atencio, and Jess Dance.

Appendix A
Biographical Information of Committee and Staff

Jerome P. Kassirer (Co-Chair) served as the Editor-in-Chief of the *New England Journal of Medicine* (1991–1999). He is currently Distinguished Professor of Medicine at Tufts University School of Medicine where he has also served as vice chairman of the Department of Medicine. Dr. Kassirer has served on the American College of Physician's Board of Governors and Board of Regents, chaired the National Library of Medicine's Board of Scientific Counselors, and is past chairman of the American Board of Internal Medicine. He is a member of the Association of American Physicians, the Institute of Medicine of the National Academies, and the American Academy of Arts and Sciences. Dr. Kassirer's current interests are reliable approaches to the assessment of the quality of health care, professionalism, ethical scientific conduct, and patient involvement in decisionmaking. He has been highly critical of for-profit medicine, abuses of managed care, and political intrusion into medical decisionmaking. Dr. Kassirer received his M.D. from the University of Buffalo School of Medicine and trained in internal medicine at Buffalo General Hospital. He trained in nephrology at the New England Medical Center. His latest book, on financial conflicts of interest in medicine, entitled *On the Take: How Medicine's Complicity with Big Business Can Endanger Your Health*, was published by Oxford University in 2004. He has also published extensively on nephrology, medical decisionmaking, and the diagnostic process.

Gladys Kessler (Co-Chair) was appointed to the United States District Court for the District of Columbia in July 1994. She received a B.A. from Cornell University and her LL.B. from Harvard Law School. Following graduation, Judge Kessler was employed by the National Labor Relations Board, served as legislative assistant to a U.S. senator and a U.S. congressman, worked for the New York City Board of Education, and then opened a public interest law firm. In June 1977, she was appointed Associate Judge of the Superior Court of the District of Columbia. From 1981 to 1985, Judge Kessler served as Presiding Judge of the Family Division and was a major architect of one of the nation's first Multi-Door Courthouses. She was president of the National Association of Women Judges from 1983 to 1984, served on the Executive Committee of the ABA's Conference of Federal Trial Judges and the U.S. Judicial Conference's Committee on Court Administration and Management. She is a board member and has been chair of the board of directors of Our Place, D.C., an organization devoted to serving the needs of incarcerated women returning to the community. She now chairs the District of Columbia Commission on Disabilities and Tenure.

Ming W. Chin was appointed to the California Supreme Court in March 1996. Before being named to the high court, Justice Chin served from 1990 to 1996 on the First District Court of Appeal, Division Three, San Francisco. Prior to his appointment to the Court of Appeal, Justice Chin served on the bench of the Alameda County Superior Court. Previously, Justice Chin was a partner in an Oakland law firm specializing in business and commercial litigation. He also served as a prosecutor in the Alameda County District Attorney's office. Justice Chin earned his bachelor's degree in political science and law degree from the University of San Francisco. After his graduation from law school, Justice Chin served 2 years as a Captain in the U.S. Army, including a year in Vietnam, where he was awarded the Army Commendation Medal and the Bronze Star. Justice Chin chairs the Judicial Council of California's Court Technology Advisory Committee, as well as the California Commission for Impartial Courts. He frequently lectures on DNA, genetics, and the courts. Justice Chin served as chair of the Judicial Council's Science and the Law Steering Committee. In 2009 the Judicial Council named him California Jurist of the Year. He is an author of *California Practice Guide: Employment Litigation* (The Rutter Group 2011). He is also an author of *California Practice Guide: Forensic DNA* (The Rutter Group, to be published in 2012).

Pauline Newman is a Judge on the United States Court of Appeals for the Federal Circuit. She received a B.A. degree from Vassar College in 1947, M.A. in pure science from Columbia University in 1948, Ph.D. in chemistry from Yale University in 1952, and LL.B. from New York University School of Law in 1958. She was admitted to the New York bar in 1958 and to the Pennsylvania bar in 1979. Judge Newman worked as research scientist for the American Cyanamid Company from 1951 to 1954; as patent attorney and house counsel for the FMC Corp. from 1954 to 1984; and, since 1969, as director of the FMC Patent, Trademark, and Licensing Department. On leave from FMC Corp. in 1961–62, she worked for the United Nations Educational, Scientific and Cultural Organization as a science policy specialist in the Department of Natural Sciences. Offices in scientific and professional organizations include member of Council of the Patent, Trademark and Copyright Section of the American Bar Association, 1982–84; board of directors of the American Patent Law Association, 1981–84; vice president of the United States Trademark Association, 1978–79, and member of its board of directors, 1975–76, 1977–79; member of board of governors of the New York Patent Law Association, 1970–74; president of the Pacific Industrial Property Association, 1978–80; member of executive committee of the International Patent and Trademark Association, 1982–84; member of board of directors of the American Chemical Society, 1973–75, 1976–78, 1979–81; member of board of directors of the American

Institute of Chemists, 1960–66, 1970–76; member of the board of trustees of Philadelphia College of Pharmacy and Science, 1983–84; member of patent policy board of State University of New York, 1983–84; member of national board of Medical College of Pennsylvania, 1975–84; and member of board of directors of Research Corp., 1982–84. Service on government committees included State Department Advisory Committee on International Intellectual Property, 1974–84; Advisory Committee to the Domestic Policy Review of Industrial Innovation, 1978–79; Special Advisory Committee on Patent Office Procedure and Practice, 1972–74; and member of the U.S. Delegation to the Diplomatic Conference on the Revision of the Paris Convention for the Protection of Industrial Property, 1982–84. Judge Newman received the Wilbur Cross Medal of Yale University Graduate School, 1989; the Jefferson Medal of the New Jersey Patent Law Association, 1988; the Award for Outstanding Contributions in the Intellectual Property Field of the Pacific Industrial Property Association, 1987; Vanderbilt Medal of New York University School of Law, 1995; and Vassar College Distinguished Achievement Award, 2002. She was Distinguished Professor of Law, George Mason University (adjunct faculty), served on the Council on Foreign Relations, and was appointed judge of the U.S. Court of Appeals for the Federal Circuit by President Reagan and entered upon duties of that office on May 7, 1984.

Kathleen McDonald O'Malley was appointed to the United States Court of Appeals for the Federal Circuit by President Barack H. Obama on December 27, 2010. Prior to joining the Federal Circuit, Judge O'Malley was a District Judge on the United States District Court for the Northern District of Ohio, a position to which she was appointed by President William J. Clinton on October 12, 1994. Prior to her appointment to the bench, Judge O'Malley served as First Assistant Attorney General and Chief of Staff in the Office of the Attorney General for the State of Ohio from 1992 to 1994, and Chief Counsel in that office from 1991 to 1992. From 1983 to 1991, Judge O'Malley was in private practice, where she focused on complex corporate and intellectual property litigation; she was with Porter, Wright, Morris & Arthur from 1985 to 1991 and with Jones Day from 1983 to 1985. As an educator, Judge O'Malley has taught patent litigation at Case Western Reserve University School of Law and is a regular lecturer on issues arising in complex litigation, including intellectual property matters. Judge O'Malley began her legal career as a law clerk to the Honorable Nathaniel R. Jones, United States Court of Appeals for the Sixth Circuit, from 1982 to 1983. She received her J.D. degree from Case Western Reserve University School of Law, Order of the Coif, in 1982, and her undergraduate degree from Kenyon College in Gambier, magna cum laude and Phi Beta Kappa, in 1979. She received an honorary Doctor of Laws degree from Kenyon College in 1995.

Jed S. Rakoff has been a United States District Judge for the Southern District of New York since 1996. Prior to his appointment, he was a partner at Fried,

Frank, Harris, Shriver & Jacobson LLP. From 1980 to 1990, he was a partner at Mudge, Rose, Guthrie, Alexander & Ferdon LLP. Judge Rakoff was an Assistant U.S. Attorney for the Southern District of New York from 1973 to 1980 and chief of the Business and Securities Fraud Prosecutions Unit from 1978 to 1980. Before joining the U.S. Attorney's Office, Judge Rakoff spent 2 years in private practice as an associate attorney at Debevoise & Plimpton LLP. He served as a law clerk to the Honorable Abraham L. Freedman, U.S. Court of Appeals for the Third Circuit, in 1969–70. Judge Rakoff is coauthor of five books and author of more than 110 published articles, more than 375 speeches, and more than 900 judicial opinions. He has been a lecturer in law at Columbia Law School since 1988. He was a member of the Board of Managers, Swarthmore College, from 2004 to 2008. Judge Rakoff currently serves as a Trustee for the William Nelson Cromwell Foundation and from 2007–10 served as a member of the Governance Board for the MacArthur Foundation Initiative on Law and Neuroscience. From 1998–2011, he was chair of the Criminal Justice Advisory Board, Southern District of New York; from 2003–11 he was chair of the Second Circuit Bankruptcy Committee; and from 2006–09 he was chair of the Honors Committee of the New York City Bar Association. Since 2001 he has served as chair of the Grievance Committee of the Southern District of New York. He is a Judicial Fellow at the American College of Trial Lawyers and was chair of the Downstate New York Chapter in 1993–94. Judge Rakoff is the former director of the New York Council of Defense Lawyers and former chair of the Criminal Law Committee, New York City Bar Association. He has been a Judicial Fellow at the American Board of Criminal Lawyers since 1995. Judge Rakoff received a B.A. from Swarthmore College in 1964, an M.Phil. from Oxford University in 1966, and a J.D. from Harvard Law School in 1969. He was awarded honorary LL.D.s from Swarthmore College in 2003 and St. Francis University in 2005.

Channing R. Robertson is Ruth G. and William K. Bowes Professor and former Dean of Faculty and Academic Affairs, School of Engineering, and Professor, Department of Chemical Engineering, Stanford University. He was named a Bass University Fellow in Undergraduate Education in 2010. Dr. Robertson received his B.S. Chemical Engineering, from the University of California, Berkeley; M.S. in Chemical Engineering, from Stanford University; and Ph.D. in Chemical Engineering—emphasis on fluid mechanics and transport phenomena, from Stanford University. Professor Robertson began his career at the Denver Research Center of the Marathon Oil Company and worked in the areas of enhanced oil recovery, geophysical chemistry, and polyurethane chemistry. Since 1970, he has been on the faculty of Stanford's Department of Chemical Engineering. He has educated and trained over 40 Ph.D. students, holds seven patents, and has published over 140 articles. He is past director of the Stanford-NIH Graduate Training Program in Biotechnology. He was co-director of the Stanford initiative in biotechnology known as BioX, which in part includes the Clark Center for Biomedical Engineering and Sciences. He directed the summer Stanford Engineering Executive Program. He received the 1990 Stanford Associates Award for service to the University, the Stanford Associates Centennial Medallion Award in 1991, the 1991 Richard W. Lyman Award, the Society of Women Engineers Award for Teacher of the Year 2000 at Stanford, the Stanford Society of Chicano/Latino Engineers & Scientists Faculty of the Year Award in 2004, and the Lloyd W. Dinkelspiel Award for Distinctive Contributions to Undergraduate Education in 2009. He is a Founding Fellow of the American Institute of Medical and Biological Engineering. Professor Robertson serves on the Scientific Advisory Committee on Tobacco Product Regulation of the World Health Organization and served on the Panel on Science, Technology, and Law, National Research Council, National Academy of Sciences, 1999–2006. Because of his interests in biotechnology, he has consulted widely in the design of biomedical diagnostic devices. He has also served as an expert witness in several trials, including the Copper-7 intrauterine contraceptive cases (United States and Australia), the Stringfellow Superfund case, and most recently the Minnesota tobacco trial. He has cofounded 2 and consulted with over 30 Silcion Valley startups during the past three decades.

Joseph V. Rodricks is an internationally recognized expert in the field of toxicology and risk analysis, and their uses in the regulation and evaluation of toxic tort and product liability cases. Since 1980, he has consulted for hundreds of manufacturers, for government agencies, and the World Health Organization, and he has served on 30 boards and committees of the National Academy of Sciences and the Institute of Medicine. He has more than 120 publications on toxicology and risk analysis, and has lectured nationally and internationally on these topics. Dr. Rodricks was formerly Deputy Associate Commissioner, Health Affairs, and Toxicologist, U.S. Food and Drug Administration (1965–80), and is a visiting professor at The Johns Hopkins University School of Public Health. He has been certified as a Diplomate, American Board of Toxicology, since 1982. Dr. Rodricks' experience includes chemical products and contaminants in foods, food ingredients, air, water, hazardous wastes, the workplace, consumer products, and medical devices and pharmaceutical products. He is the author of *Calculated Risks* (Cambridge University Press), a nontechnical introduction to toxicology and risk analysis that is now available in a fully revised and updated second edition, for which he won an award from the American Medical Writers Association.

Allen Wilcox is Senior Investigator, Epidemiology Branch at the National Institute of Environmental Health Sciences, NIH, and Editor-in-Chief of *Epidemiology*. His research is primarily on human reproduction, with research topics ranging from fertility and early pregnancy loss to fetal growth and birth defects. Dr. Wilcox earned his undergraduate and medical degrees at the University of Michigan, Ann Arbor, and his M.P.H. (maternal and child health) and Ph.D. (epidemiology) at the University of North Carolina School of Public Health at

Chapel Hill, where he is an adjunct professor in the Department of Epidemiology. The school recognized him with its Distinguished Alumni Award in 2006. Other distinctions include the Distinguished Service Medal (highest award of the U.S. Public Health Service); election as a Fellow of the American College of Epidemiology; and election as president of the Society of Epidemiologic Research, the American Epidemiologic Society, and the Society for Pediatric Epidemiologic Research. In 2008, he received the National Maternal and Child Health Epidemiology Award. He holds an honorary doctorate from the University of Bergen (Norway). He is the author of *Fertility and Pregnancy: An Epidemiologic Perspective* (Oxford University Press 2010).

Sandy L. Zabell is Professor of Mathematics and Statistics at Northwestern University. He received his A.B. from Columbia College in 1968, his A.M. in biochemistry and molecular biology from Harvard University in 1971, and his Ph.D. in mathematics from Harvard University in 1974. He was Assistant Professor of Statistics at the University of Chicago from 1974 to 1979, and joined Northwestern University as Associate Professor of Mathematics in 1980. He is a Fellow of the American Statistical Association and the Institute of Mathematical Statistics. In the past he has served as an associate editor of the *American Mathematical Monthly* and the *Journal of Mathematical Analysis and Applications,* and book review editor of the *Annals of Probability*. His principal research interests revolve around mathematical probability (in particular, large deviation theory) and Bayesian statistics (in particular, the study of exchangeability). He has also written extensively on the history and philosophical foundations of probability and statistics, is an affiliated faculty member of the Northwestern Philosophy Department, and the author of *Symmetry and its Discontents* (Cambridge University Press, 2006). Professor Zabell has had a longstanding involvement in the legal applications of statistics, including serving on two panels of the National Research Council, and teaching courses on statistics at both the University of Chicago and Northwestern Law Schools. One of his primary interests at present is forensic science, in particular, the statistical issues arising from the use of DNA in human identification. He has spoken numerous times at forensic science conferences, and lectured on forensic DNA identification in courses at Northwestern. He is also interested in the statistical proof of employment discrimination and the legal uses of sampling. In addition to his scholarly interests, he has assisted legal counsel over the years in more than 200 cases, both civil and criminal.

Staff

Joe S. Cecil is a Senior Research Associate and Project Director in the Division of Research at the Federal Judicial Center. Currently, he is directing the Center's Program on Scientific and Technical Evidence. As part of this program, he served as principal editor of the first and second editions of the Center's Reference Manual on Scientific Evidence. He has published several articles on the use of court-appointed experts and is currently examining changes in dispositive motion practice in federal district courts over the past 30 years. Dr. Cecil received his J.D. and a Ph.D. in psychology from Northwestern University. He serves on the editorial boards of social science and legal journals. He has served as a member of several panels of NAS, and currently is serving as a member of the National Academies Committee on Science, Technology, and Law. Other areas of research interest include federal civil and appellate procedure, jury competence in complex civil litigation, claim construction in patent litigation, and judicial governance.

Anne-Marie Mazza is the Director of the Committee on Science, Technology, and Law. Dr. Mazza joined the National Research Council in 1995. She has served as Senior Program Officer with both the Committee on Science, Engineering, and Public Policy and the Government-University-Industry Research Roundtable. In 1999, she was named the first director of the Committee on Science, Technology, and Law, a newly created activity designed to foster communication and analysis among scientists, engineers, and members of the legal community. Dr. Mazza has been the study director on numerous Academy reports including *Review of the Scientific Approaches Used During the FBI's Investigation of the 2001 Anthrax Mailings* (2011), *Managing University Intellectual Property in the Public Interest* (2010); *Strengthening Forensic Science in the United States: A Path Forward* (2009); *Science and Security in a Post-9/11 World* (2007); Daubert *Standards: Summary of Meetings* (2006); *Reaping the Benefits of Genomic and Proteomic Research: Intellectual Property Rights, Innovation, and Public Health* (2005); *Intentional Human Dosing Studies for EPA Regulatory Purposes: Scientific and Ethical Issues* (2004); *Ensuring the Quality of Data Disseminated by the Federal Government* (2003). Dr. Mazza received an NRC distinguished service award in 2008. In 1999–2000, Dr. Mazza divided her time between the National Academies and the White House Office of Science and Technology Policy (OSTP), where she served as a Senior Policy Analyst responsible for issues associated with a Presidential Review Directive on the government-university research partnership. Before joining the Academy, Dr. Mazza was a Senior Consultant with Resource Planning Corporation. Dr. Mazza received a B.A., an M.A., and a Ph.D. from the George Washington University.

Steven Kendall is Associate Program Officer for the Committee on Science, Technology, and Law. Mr. Kendall has contributed to numerous Academy reports including *Review of the Scientific Approaches Used During the FBI's Investigation of the 2001 Anthrax Mailings* (2011), *Managing University Intellectual Property in the Public Interest* (2010); and *Strengthening Forensic Science in the United States: A Path Forward* (2009). He is currently a Ph.D. candidate in the Department of the History of Art and Architecture at the University of California, Santa Barbara, where he is completing a dissertation on nineteenth-century British painting. Mr. Kendall received his M.A. in Victorian Art and Architecture at the University of London. Prior to joining the NRC in 2007, he worked at the Smithsonian American Art Museum and the Huntington in San Marino, California.

Guruprasad Madhavan is a Program Officer with the Board on Population Health and Public Health Practice, and the Committee on Science, Engineering, and Public Policy at the National Academies. Previously, he served as a Program Officer for the Committee on Science, Technology, and Law and as a Christine Mirzayan Science and Technology Policy Fellow with the Board on Science, Technology and Economic Policy. He has worked on such National Academies' publications as *Direct-to-Consumer Genetic Testing: Summary of a Workshop* (2010); *Managing University Intellectual Property in the Public Interest* (2010); and *Rising Above the Gathering Storm, Revisited* (2010). Dr. Madhavan completed his Ph.D. in biomedical engineering at the State University of New York (SUNY) at Binghamton where his research was directed toward developing noninvasive, nonpharmacological, neuromuscular stimulation approaches for enhancing circulation. He received his B.E. (honors with distinction) in instrumentation and control engineering from the University of Madras, and M.S. in biomedical engineering from SUNY Stony Brook. Following his medical device industry experience as a research scientist at AFx, Inc. and Guidant Corporation in California, Dr. Madhavan completed his M.B.A. in leadership and healthcare management from SUNY Binghamton. Among other honors, he was selected as an outstanding young scientist to attend the 2008 World Economic Forum Annual Meeting of the New Champions, and 1 among 14 people as the "New Faces of Engineering" of 2009 in *USA Today*. He is co-editor of *Career Development in Bioengineering and Biotechnology* (Springer 2008) and *Pathological Altruism* (Oxford University Press 2011).

Index

A

B

C

D

E

F

G

H

I

J

K

L

M

N

O

P

Q

R

S

T

Z